Applied Mathematical Sciences
Volume 158

For further volumes:
http://www.springer.com/series/34

Herbert Oertel
Editor

Prandtl–Essentials of Fluid Mechanics

Third Edition

With Contributions by
P. Erhard, D. Etling, U. Müller, U. Riedel,
K.R. Sreenivasan, J. Warnatz

Translated by Katherine Asfaw

With 536 Illustrations

Herbert Oertel
University of Karlsruhe
Institute for Fluid Mechanics
Kaiserstr. 12
D-76131 Karlsruhe
Germany

Ludwig Prandtl
University of Göttingen,
Dir. MPI für Strömungsforschung, † 1953

Editors:

ISSN 0066-5452
ISBN 978-1-4614-2487-1 e-ISBN 978-1-4419-1564-1
DOI 10.1007/978-1-4419-1564-1
Springer New York Dordrecht Heidelberg London

Originally published in the German language by Vieweg+Teubner, 65189 Wiesbaden, Germany as **"Oertel: Prandtl – Führer durch die Strömungslehre. 12th revised and enlarged edition"**

Softcover reprint of the hardcover 3rd edition 2010

Printed on acid-free paper

Springer is part of Springer Science+Business Media (www.springer.com)

Preface

Ludwig Prandtl, with his fundamental contributions to hydrodynamics, aerodynamics, and gas dynamics, greatly influenced the development of fluid mechanics as a whole, and it was his pioneering research in the first half of the last century that founded modern fluid mechanics. His book *Führer durch die Strömungslehre*, which appeared in 1942, originated from previous publications in 1913, *Lehre von der Flüssigkeit und Gasbewegung*, and 1931, *Abriß der Strömungslehre*. The title *Führer durch die Strömungslehre*, or *Essentials of Fluid Mechanics*, is an indication of Prandtl's intentions to guide the reader on a carefully thought-out path through the different areas of fluid mechanics. On his way, the author advances intuitively to the core of the physical problem, without extensive mathematical derivations. The description of the fundamental physical phenomena and concepts of fluid mechanics that are needed to derive the simplified models has priority over a formal treatment of the methods. This is in keeping with the spirit of Prandtl's research work.

The first edition of Prandtl's *Führer durch die Strömungslehre* was the only book on fluid mechanics of its time and, even today, counts as one of the most important books in this area. After Prandtl's death, his students Klaus Oswatitsch and Karl Wieghardt undertook to continue his work, and to add new findings in fluid mechanics in the same clear manner of presentation.

When the ninth edition went out of print and a new edition was desired by the publishers, we were glad to take on the task. The first four chapters of this book keep to the path marked out by Prandtl in the first edition, in 1942. The original historical text has been linguistically revised, and leads, after the *Introduction*, to chapters on *Properties of Liquids and Gases*, *Kinematics of Flow*, and *Dynamics of Fluid Flow*. These chapters are taught to science and engineering students in introductory courses on fluid mechanics even today. We have retained much of Prandtl's original material in these chapters, but added a section on the *Topology of a Flow* in Chapter 3 on *Flows of Non-Newtonian Media and Aerodynamics* in Chapter 4. Chapter 5 on *Fundamental Equations of Fluid Mechanics* enlarges the material in the original, and forms the basis for the treatment of different branches of fluid mechanics that appear in subsequent chapters.

The major difference from previous editions lies in the treatment of additional topics of fluid mechanics. The field of fluid mechanics is continuously growing, and has now become so extensive that a selection had to be made. I am greatly indebted to my colleagues K.R. Sreenivasan, U. Müller, J. Warnatz, U. Riedel, D. Etling, and P. Erhard, who revised individual chapters in their own research areas, keeping Prandtl's purpose in mind and presenting the latest developments of the last seventy years in Chapters 6 to 12. Some of these chapters can be found in some form in Prandtl's book, but have undergone substantial revisions; others are entirely new. The original chapters on *Wing Aerodynamics*, *Heat Transfer*, *Stratified Flows*, *Turbulent Flows*, *Multiphase Flows*, *Flows in the Atmosphere and the Ocean*, and *Turbomachinery* have been revised, while the chapters on *Instabilities and Turbulent Flows*, *Flows with Chemical Reactions*, *Microflows* and *Biofluid Mechanics* are new. References to the literature in the individual chapters have intentionally been kept to those few necessary for comprehension and completion. The extensive historical citations may be found by referring to previous editions.

Essentials of Fluid Mechanics is targeted to science and engineering students who, having had some basic exposure to fluid mechanics, wish to attain an overview of the different branches of fluid mechanics. The presentation postpones the use of vectors and eschews the use integral theorems in order to preserve the accessibility to this audience. For more general and compact mathematical derivations we refer to the references. In order to give students the possibility of checking their learning of the subject matter, Chapters 2 to 5 are supplemented with problems. The book will also give the expert in research or industry valuable stimulation in the treatment and solution of fluid-mechanical problems.

We hope that we have been able, with the treatment of the different branches of fluid mechanics, to carry on the work of Ludwig Prandtl as he would have wished. Chapters 1–5, 7, and 12 were written by H. Oertel, Chapter 6 by K.R. Sreenivasan and H. Oertel, Chapter 8 by U. Müller, Chapter 9 by J. Warnatz and U. Riedel, Chapter 10 by D. Etling, and Chapter 11 by P. Erhard. Thanks are due to those colleagues whose numerous suggestions have been included in the text.

I thank Katherine Aswaf for the translation and typesetting of the English manuscript and K. Fritsch-Kirchner for the completion of the text files. The extremely fruitful collaboration with Springer-Verlag also merits particular praise.

Karlsruhe, July 2009 *Herbert Oertel*

Contents

1. Introduction

The development of modern fluid mechanics is closely connected to the name of its founder, *Ludwig Prandtl*. In 1904 it was his famous article on fluid motion with very small friction that introduced *boundary-layer theory*. His article on *airfoil theory*, published the following decade, formed the basis for the calculation of friction drag, heat transfer, and flow separation. He introduced fundamental ideas on the modeling of turbulent flows with the *Prandtl mixing length* for turbulent momentum exchange. His work on gas dynamics, such as the Prandtl–Glauert correction for compressible flows, the theory of shock waves and expansion waves, as well as the first photographs of supersonic flows in nozzles, reshaped this research area. He applied the methods of fluid mechanics to meteorology, and was also pioneering in his contributions to problems of elasticity, plasticity, and rheology.

Prandtl was particularly successful in bringing together theory and experiment, with the experiments serving to verify his theoretical ideas. It was this that gave Prandtl's experiments their importance and precision. His famous experiment with the tripwire, through which he discovered the turbulent boundary layer and the effect of turbulence on flow separation, is one example. The tripwire was not merely inspiration, but rather was the result of consideration of discrepancies in *Eiffel's* drag measurements on spheres. Two experiments with different tripwire positions were enough to establish the generation of turbulence and its effect on the flow separation. For his experiments Prandtl developed wind tunnels and measuring apparatus, such as the Göttingen wind tunnel and the Prandtl stagnation tube. His scientific results often seem to be intuitive, with the mathematical derivation present only to serve the physical understanding, although it then does indeed deliver the decisive result and the simplified physical model. According to Werner Heisenberg, Prandtl was able to "see" the solutions of differential equations without calculating them.

Selected *individual examples* aim to introduce the reader to the path to understanding of fluid mechanics prepared by Prandtl and to the contents and modeling in each chapter. As an example of the *dynamics of flows* (Chapter 4), the different regimes in the *flow past a vehicle*, an incompressible flow, and in the *flow past an automobile*, a compressible flow, are described.

H. Oertel (ed.), *Prandtl-Essentials of Fluid Mechanics*,
Applied Mathematical Sciences 158, DOI 10.1007/978-1-4419-1564-1_1,

In *flow past a vehicle*, we differentiate between the free flow over the surface and the flow between the vehicle moving with velocity U_∞ and the street which is at rest. At the stagnation point, where the pressure is at its maximum, the flow divides, and is accelerated along the hood and past the spoiler along the base of the vehicle. This leads to a pressure drop and to a negative downward pressure on the street, as shown in Figure 1.1. The flow again slows down at the windshield, and is decelerated downstream along the roof and the trunk. This leads to a pressure increase with a positive lift, while the negative downward pressure on the street along the lower side of the vehicle remains.

Viscous flow (Section 4.2) on the upper and lower sides of the vehicle is restricted to the *boundary-layer flow*, which becomes the viscous wake at the back edge of the vehicle. In the wind tunnel experiment the flow is made visible with smoke, and this shows that downstream from the back of the automobile, a backflow region forms. This is seen in the figure as the black region. Outside the boundary layer and the wake, the flow is essentially *inviscid* (Section 4.1).

In order to be able to understand the different flow regimes, and therefore to establish a basis for the aerodynamic design of a motor vehicle, Prandtl worked out the carefully prepared path (Chapters 2 to 4) from the properties of liquids and gases, to kinematics, and to the dynamics of inviscid and viscous flows. By following this path, too, the reader will successively gain physical understanding of this first flow example.

The second flow example considers compressible *flow past a wing* with a *shock wave* (Sections 4.3 and 4.4.5). The free flow toward the wing has the

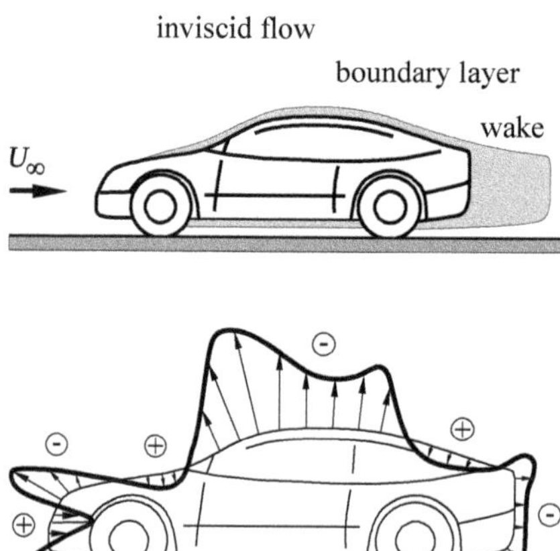

wake flow visualization

Fig. 1.1. Flow past a vehicle

velocity of a civil aircraft U_∞, a large subsonic velocity. Figure 1.2 shows the flow regimes on a cross-section of the wing and the negative pressure distribution, with the flow again made visible with small particles. From the stagnation point, the stagnation line bifurcates to follow the suction side (upper side) and the pressure side (lower side) of the wing. On the upper side, the flow is accelerated up to supersonic velocities, an effect that is connected with a large pressure drop. Further downstream, the flow is again decelerated to the subsonic regime via a compression shock wave. This shock wave interacts with the boundary layer and causes it to thicken, leading to increased drag.

On the lower side the flow is also accelerated from the stagnation point. However, in the nose region the acceleration is not as great as on the suction side, and so no supersonic velocities occur along the pressure side. From about the middle of the wing onwards, the flow is again decelerated. The pressures above and below then approach one another, leading to the wake region downstream of the trailing edge.

A thin *boundary layer* is formed on the suction and pressure sides of the wing. The suction and pressure side boundary layers meet at the trailing edge and form the wake flow downstream. As in the example of the flow past a motor vehicle, both the flow in the boundary layers and the flow in the wake are *viscous*. Outside these regions the flow is essentially *inviscid*.

The pressure distribution in Figure 1.2 results in a lift, which, for the wing of the civil aircraft, has to be adapted to the number of passengers to be transported. In designing the wing, the design engineer has to keep the

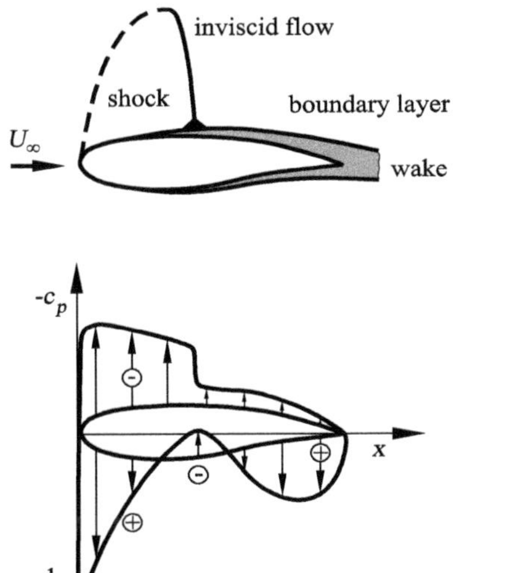

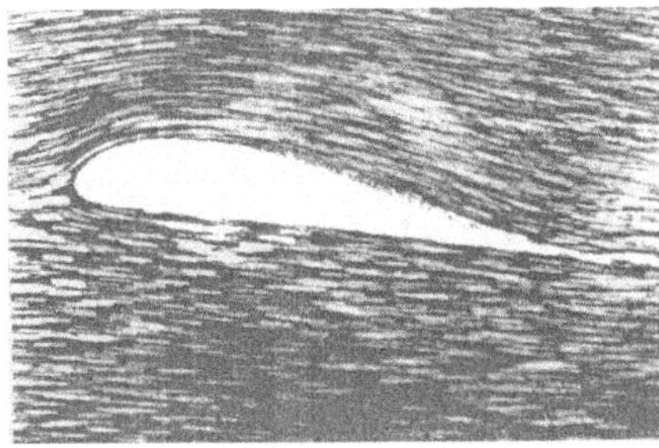

Fig. 1.2. Flow past a wing

drag of the wing as small as possible to save fuel. This is done by shaping the wing appropriately.

Different equations for computing each flow result from the different properties of each flow regime. To good approximation, the boundary-layer equations hold in the boundary-layer regime. In contrast, computing the wake flow and the flow close to the trailing edge is more difficult. In these regimes, the Navier–Stokes equations have to be solved. The inviscid flow in the region in front of the shock can be treated using the potential equation, a comparatively simple task. The inviscid flow behind the shock outside the boundary layer has to be computed with the Euler equations, since the flow there is rotational. In the shock boundary-layer interaction region, again the Navier–Stokes equations have to be solved.

In contrast to Prandtl's day, numerical software is now available for solving the different partial differential equations. Because of this, in Chapter 5 we present the *fundamental equations* of laminar and turbulent flows as a basis for the following chapters dealing with the different branches of fluid mechanics. Following the same procedure as Prandtl, the mathematical solution algorithms and methods may be found by referral to the texts and literature cited.

As will be shown in Chapters 6 to 12, no withstanding of numerically computed flow fields, it is necessary to consider the physical modeling in the different regimes. There are still no closed theories of turbulent flows, of multiphase flows, or of the coupling of flows with chemical reactions out of thermal or chemical equilibrium. For this reason, Prandtl's method of intuitive connection of theory and experiment to physical modeling is still very much up-to-date.

The fascinating complexity of turbulence has attracted the attention of scientists for centuries (Chapter 6). For example, the swirling motion of fluids that occurs irregularly in space and time is called turbulence. However, this randomness, is not without some order, as is apparent from casual observation. Turbulent flows are a paradigm for spatially extended nonlinear dissipative systems in which many length scales are excited simultaneously and coupled strongly. The phenomenon has been studied extensively in engineering and in such diverse fields as astrophysics, oceanography, and meteorology.

Figure 1.3 shows a turbulent jet of water emerging from a circular orifice into a tank of still water. The fluid from the orifice is made visible by mixing small amounts of a fluorescing dye and illuminating it with a thin light sheet. The picture illustrates swirling structures of various sizes amidst an avalanche of complexity. The boundary between the turbulent flow and the ambient is usually rather sharp and convoluted on many scales. The object of study is often an ensemble average of many such realizations. Such averages obliterate most of the interesting aspects seen here, and produce a smooth object that grows linearly with distance downstream. Even in such smooth objects, the averages vary along the length and width of the flow, these variations being a measure of the spatial inhomogeneity of the turbulence. The inhomogeneity is typically stronger along the smaller dimension of the flow. The fluid velocity measured at any point in the flow is an irregular function of time. The degree of order is not as apparent in time traces as in spatial cuts, and a range of intermediate scales behaves like fractional Brownian motion.

In contrast, Figure 1.4 shows homogeneous and isotropic turbulence produced by sweeping a grid of bars at a uniform speed through a tank of still water. Unlike the jet turbulence of Figure 1.3, turbulence here does not have a preferred direction or orientation. On average, it does not possess significant spatial inhomogeneities or anisotropies. The strength of the structures, such as they are, is weak in comparison with such structures in Figure 1.3. Homogeneous and isotropic turbulence offers considerable theoretical simplifications, and is the object of many studies.

In many fluid-mechanical problems, the onset of turbulent flow is due to *instabilities*. An example of this is *thermal cellular convection* in a horizontal fluid layer heated from below and under the effect of gravity. The base beneath the fluid has a higher temperature than the free surface. If a critical temperature difference between the free surface and the base is exceeded, the fluid is suddenly set into motion and, as in Figure 1.5, it forms hexagonal cell structures in the center of which fluid rises and on whose edges the fluid sinks. The phenomenon is known as thermal cellular convection. If the fluid

Fig. 1.3. Turbulent jet of water

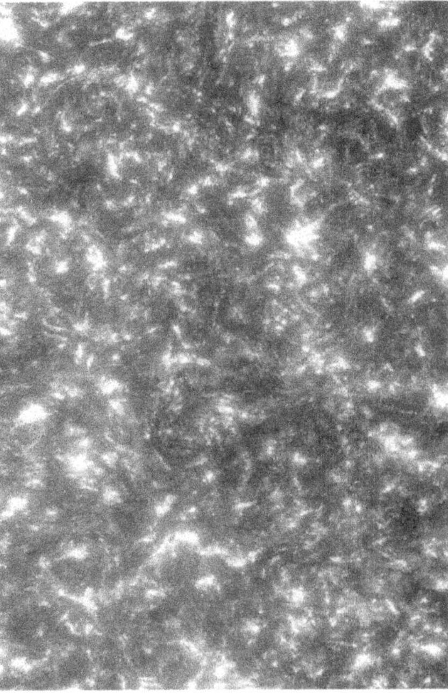

Fig. 1.4. Homogeneous and isotropic turbulent flow

is covered by a plate, instead of hexagonal cells periodically spaced rolling structures are formed without surface tension. The reason for the instabilities is the same in both cases. Cold, denser fluid is layered above warmer fluid, and this tends to flow toward lower layers. The smallest perturbation to this layering leads to the onset of the equalizing motion, providing critical temperature difference is exceeded.

The transition to turbulent convection flow takes place with increasing temperature difference via several time-dependent intermediate states. The size of the hexagonal structures or the long convection rolls changes, but the original cellular structure of the instability can still be seen in the turbulent convection flow.

Convection flows with *heat and mass transport* are treated in Chapter 7. These occur frequently in nature and technology, and it is via such flows that heat exchange in the atmosphere determines the weather. The example of a tropical cyclone is shown in Figure 1.10. The extensive heat adjustment between the equator and the North Pole leads to convection flows in the oceans, such as the Gulf Stream (Figure 1.11). Convection flows in the center of the Earth are also the cause of continental drift and are responsible for the Earth's magnetic field. In energy technology and environmental technology flows are connected with heat and mass transport, and with phase transitions, as in steam generators and condensers. Convection flows are used in cooling

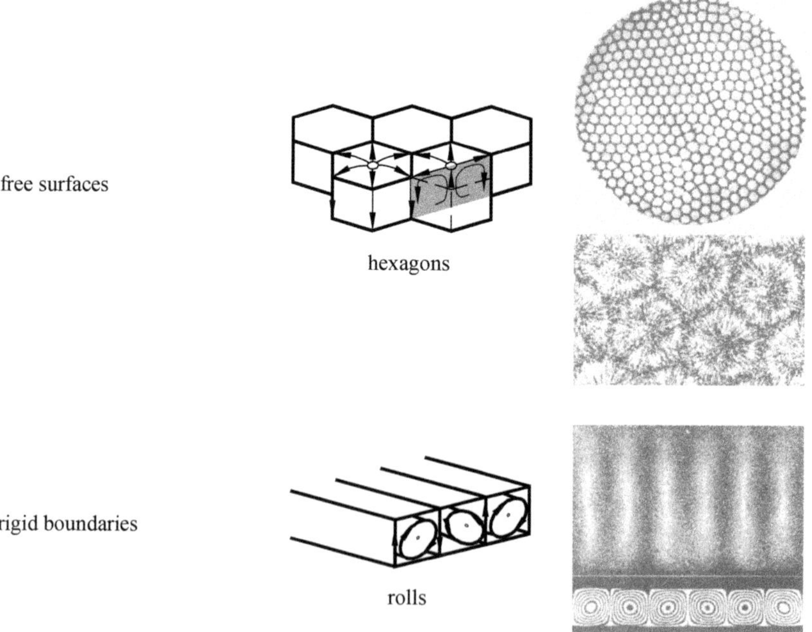

Fig. 1.5. Thermal cellular convection

towers to transport the waste heat from power stations. Other examples of convection flows are the propagation of waste air and gas in the atmosphere and of cooling and waste water in lakes, rivers, and oceans, heating systems and air-conditioning in buildings, circulation of fluids in solar collectors and heat accumulators.

Figure 1.6 shows experimental results on thermal convection flows. In contrast to *forced convection flows*, these are *free convection flows*, where the flow is due to only lift forces. These may be caused by temperature or concentration gradients in the gravitational field. A heated horizontal circular cylinder initially generates a rising laminar convection flow in the surrounding medium, which is at rest, until the transition to turbulent convection flow is caused by thermal instabilities. Similar thermal convection flows occur at vertical and horizontal heated plates.

Multiphase flow (Chapter 8) is the flow form that appears most frequently in nature and technology. Here the word *phase* is meant in the thermodynamic sense and implies either the solid, liquid, or gaseous state, any of which can occur simultaneously in a one-component or multicomponent system of substances. Impressive examples of multiphase flows in nature are storm clouds containing raindrops and hailstones, and snow dust in an avalanche or a cloud of volcano ash.

In power station engineering and chemical process engineering, multiphase flows are an important means of transporting heat and material. Two-phase, or binary, flows determine the processes in the steam generators, condensers, and cooling towers of steam power stations. The rain from the cooling water of a wet cooling tower is shown in Figure 1.7. The water drops lose their heat by evaporation to the warmed rising air. Multiphase, multicomponent flows are used in the extraction, transportation, and processing of oil and natural gas. Such flow forms are also very much involved in distillation and rectification

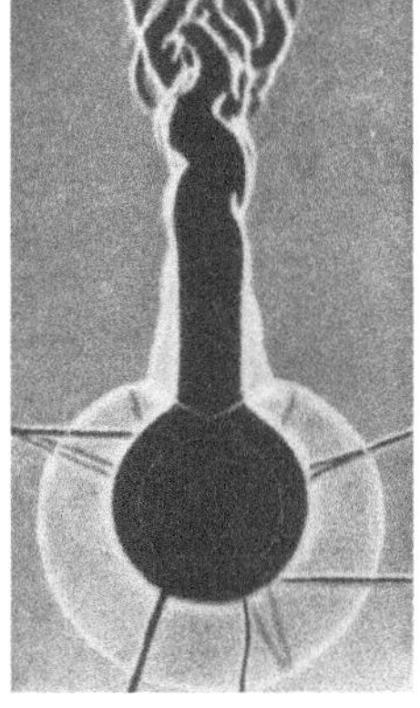

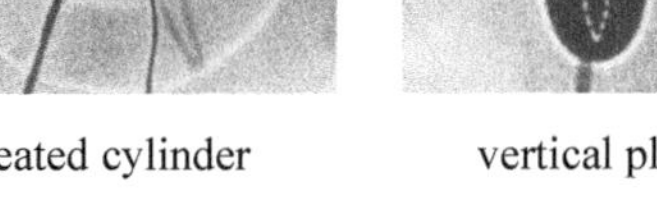

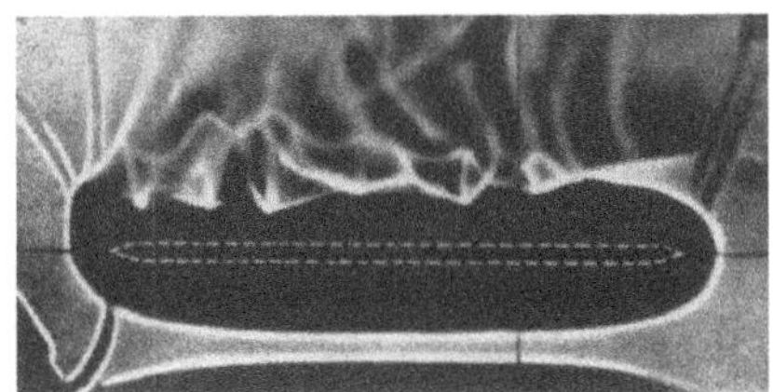

heated cylinder vertical plate horizontal plate

Fig. 1.6. Thermal convection flows

Fig. 1.7. Wet cooling tower

processes in the chemical industry. They also appear as cavitation effects on underwater wing surfaces in fast flows. The example in Figure 1.8 shows a cavitating underwater foil. Phenomena of this kind are highly undesirable in flow machinery since they can lead to serious material damage.

Turbulent reactive flows (Chapter 9) are very important for a great number of applications in energy, chemical, and combustion technology. The optimization of these processes places great demands on the accuracy of the numerical simulation of turbulent flows. Because of the complexity of the interaction between turbulent flow, molecular diffusion, and chemical reaction kinetics, there is a great need for improved models to describe these processes.

Turbulent flames are characterized by a wide spectrum of time and length scales. The typical length scales of the turbulence extend from the dimensions of the combustion chamber right down to the smallest vortex in which turbulent kinetic energy is dissipated. The chemical reactions that cause the combustion have a wide spectrum of time scales. Depending on the overlapping of the turbulent time scales with the chemical time scales, there are regimes with a strong or weak interaction between chemistry and turbulence. Because of this, a joint description of turbulent diffusion flames generally always requires an understanding of turbulent mixing and combustion.

Fig. 1.8. Cavitation at an underwater foil

A complete description of turbulent flames therefore has to resolve all scales from the smallest to the largest, which is why a numerical simulation of technical combustion systems is not possible on today's computers and why averaging techniques in the form of turbulence models have to be used. However, if turbulence models are to describe such aspects of technical application as mixing, combustion, and formation of emissions realistically, it is necessary to be able to better determine the parameters of such models from detailed investigations.

One promising approach is the use of *direct numerical simulation*, the generation of artificial laminar and turbulent flames with the computer. For a small spatial area, the conservation equations for reactive flows are solved, taking all turbulent fluctuations into account, and thus describing a small but realistic section of a flame. This can then be used to describe real flames.

The formation of closed regions of fresh gas that penetrate into the exhaust are an interesting phenomenon of turbulent premixed flames. The time resolution of this transient process can be investigated by means of direct numerical simulation and is important in determining the region of validity of current models and in the development of new models to describe turbulent combustion. Figure 1.9 shows the concentration of OH and CO radicals, as well as the vortex strength in a turbulent methane premixed flame.

Many different flows in nature (Chapter 10) can be seen on Earth and in space. The *flow processes in the atmosphere* range from small winds to the tropospherical jet stream of strong winds surrounding the globe. One particularly impressive atmospheric phenomenon is the tropical cyclone, known in the Caribbean and the United States under the name hurricane. Hurricanes form in the summer months above the warm waters off the African coast close to the equator and move with a southeasterly flow first toward the Caribbean and then northeastwards along the east coast of the United States. Wind speeds of up to 300 km/h can occur in these tropical wind storms, with much resulting damage on land. An example of a cyclone is shown in Figure 1.10. This figure shows the path and a satellite image of Hurricanes Ivan and

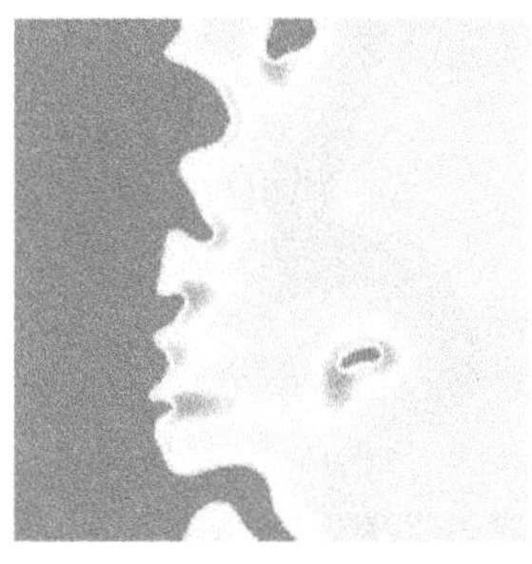

OH concentration

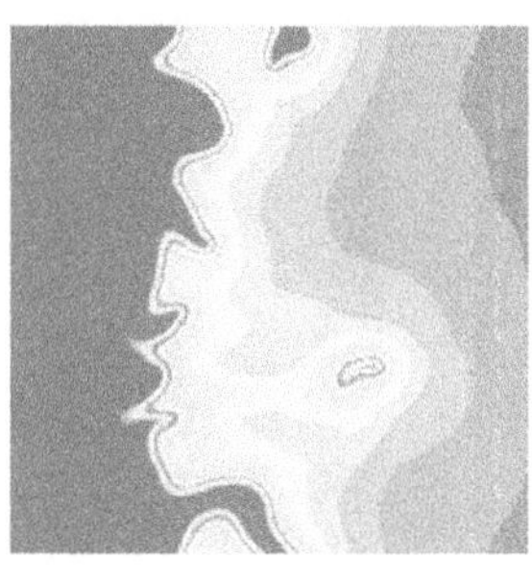

CO concentration

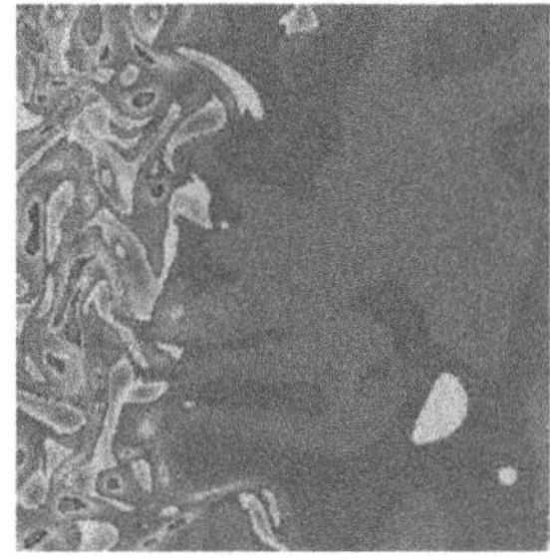

vorticity

Fig. 1.9. Turbulent premixed methane flame

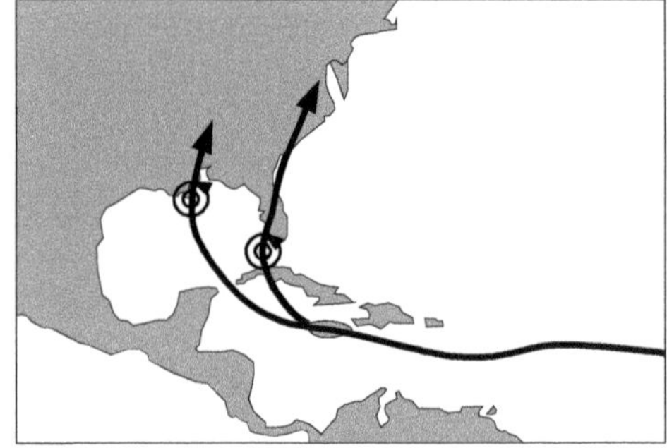

Fig. 1.10. Path of Hurricanes Ivan and Charley 2004

Charley which passed over the Caribbean islands and the southeast coast of the United States in 2004, and continued their path as a low-pressure region across the Atlantic as far as Europe.

The *flow processes in the ocean* extend from small phenomena such as water waves to large sea currents. An example of the latter is the Gulf Stream, which as a warm surface current can be tracked practically from the African coast, past the Caribbean to western and northern Europe. Thanks to its relatively high water temperature, it ensures a mild climate along of the coast of Britain and Norway. In order to compensate the warm surface current directed towards the pole, a cold deep current forms, and this flows from the north Atlantic along the east coast of North and South America, toward the south. Both of these large flow systems are shown in Figure 1.11.

Microflows, a new area of fluid mechanics, are discussed in Chapter 11. Through advances in manufacturing technology, the flow processes and transport processes through microchannels and past micro-objects are becoming relevant for technical applications. Modern manufacturing methods permit very small structures considerably less than one millimeter in size to be made

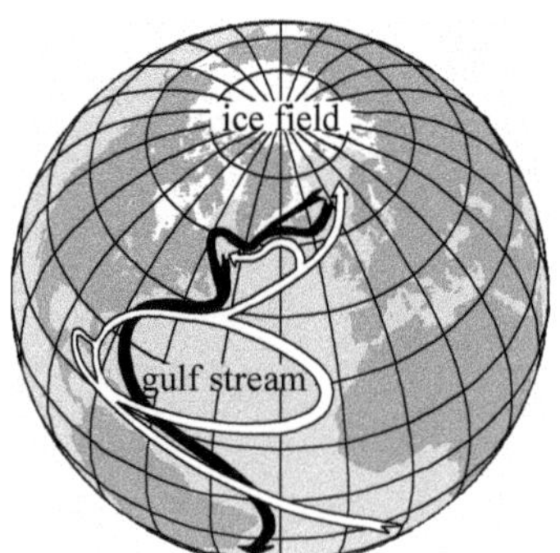

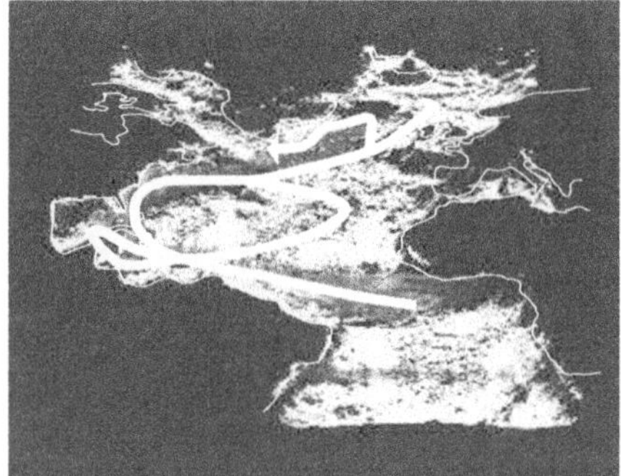

Fig. 1.11. Large ocean currents in the Atlantic

from various materials such as silicon, glass, metal or plastic. Complex fluidic functions then take place in tiny spaces.

An inkjet printer head is an example of a microfluidic system. The ink is ejected through a matrix of apertures about 45 μm diameter and generates points of color on the paper. Figure 1.12 shows the ejection of a single droplet of ink from the printer head. The pressure is built up in the cavity by piezo crystals or through application of heat and evaporation. Similar systems are used for highly precise dosage in process engineering.

In a second example, the favorable surface to volume ratio in microchannels is used to construct a compact micro heat exchanger. Figure 1.12 shows a crossflow heat exchanger made of a pile of metal sheets with microchannels of cross-section 100 x 200μm etched onto it. In a cube of side 14mm at temperature differences of up to 80K, heat can be transferred at rates of up to 14kW. The large transfer surface is advantageous not only for heat transfer, but can also be used in catalytic coating to improve material transfer in chemical reactions. Similar heat exchangers can be used as microreactors, where the temperature of the chemical reaction in a passage can be controlled very precisely by a heat carrier in a second passage. Chemical reactions that otherwise would be quite impossible can thus be made possible or optimized.

Depending on the fluid, flows through and past very small geometries cannot be treated using continuum mechanics. Corrections to the continuum mechanical equations or even molecular methods are necessary to represent correctly the physics of flows at these small length scales.

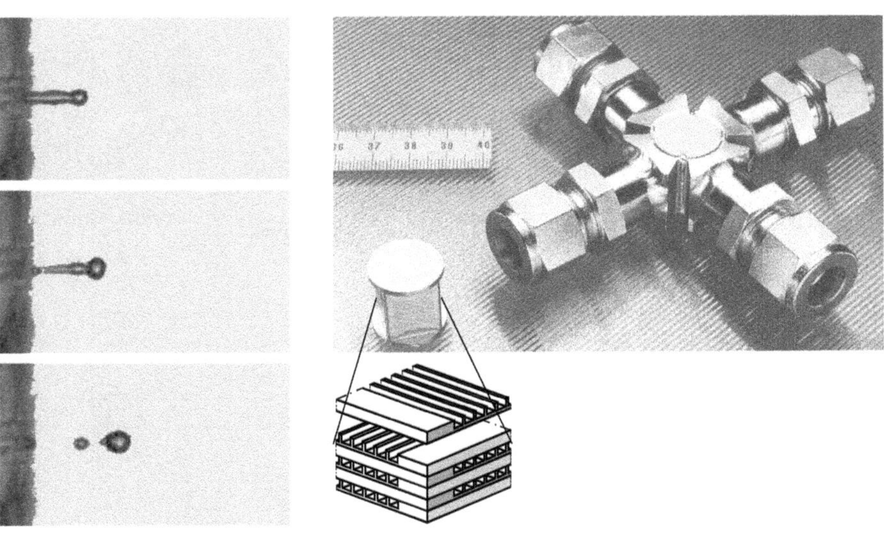

C. Maier 2004 *K. Schubert et al.* 2001

Fig. 1.12. Examples of microfluidic components

In contrast to the previous examples of flows, *biofluid mechanics* in Chapter 12 deals with flows that are characterized by flexible biological surfaces. One distinguishes between flows past animals in the air or in water, such as a bird in flight or a fish swimming, and internal flows, such as the closed human blood circulation.
The human heart consists of two separate pump chambers, the left and right ventricles. The right ventricle is filled with blood low in oxygen from the circulation around the body, and on contraction it is emptied into the lung circulatory system. The reoxygenated blood in the lung is passed into the circulation around the body via the left ventricle. A simple representation of the flow throughout one cardiac cycle is shown in Figure 1.13. The atria and ventricles of the heart are separated by the atrioventricular valves, which regulate the flow into the ventricles. They prevent backward flow of the blood during contraction of the ventricles. During relaxation of the ventricles, the pulmonary valves prevent backward flow of the blood out of the lung arteries, while the aortic valves prevent backward flow out of the aorta into the left ventricle.
During the cardiac cycles, the ventricles undergo periodic contraction and relaxation, ensuring the pulsing blood flow in the circulatory system around the body. This pump cycle is associated with changes in pressure in the ven-

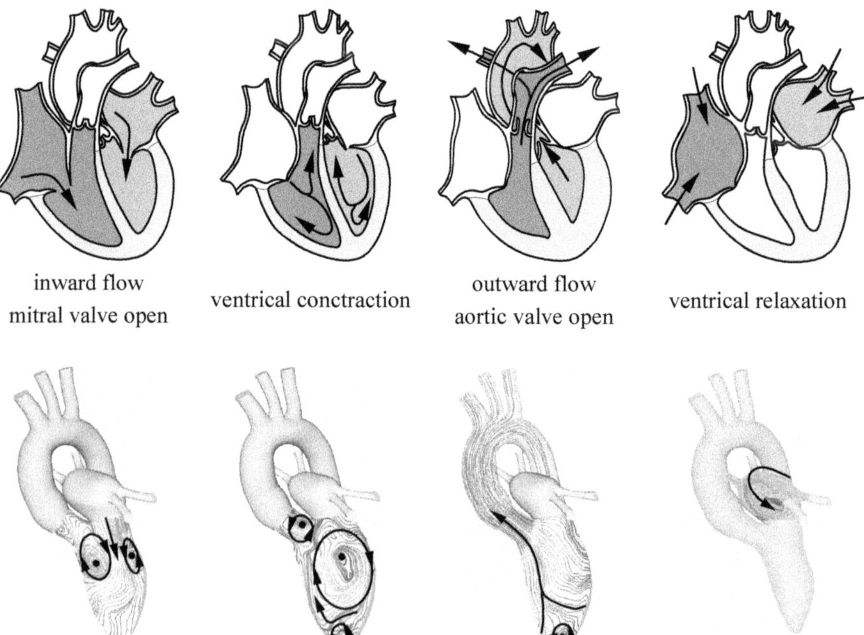

Fig. 1.13. Flow of the human heart

tricles and arteries. The pressure differences control the opening and closing of the cardiac valves. In a healthy heart, the pulsing flow is laminar and does not separate. Defects in the pumping behavior of the heart and heart failure lead to turbulent flow regimes and backflow in the ventricles, increasing flow losses in the heart.

The flow simulation of Figure 1.13 shows the streamlines of the inward flow in the left ventricle accompanied by a ring vortex. The mitral valve is open and the aortic valve closed. Large inward flow velocities directed downward and with a maximal velocity of about 0.5 m/s can be seen. After a quarter of the cardiac cycle the ring vortex branches, and the blood begins to flow through the top of the ventricle. When the ventricle contracts, the aortic and mitral valves are closed. The left ventricle is completely filled with blood, and the flow velocities calculated are very small. As the blood flows out of the ventricle, the mitral valve is closed and the aortic valve open. The streamlines show the blood flow jet into the aorta. As the ventricle relaxes, both cardiac valves are closed. The flow into the left atrium can be seen.

2. Properties of Liquids and Gases

2.1 Properties of Liquids

Liquids are distinguished from solids by the fact that their particles are readily displaced. Whereas forces of finite magnitude are required to deform a solid, no force at all is required to alter the shape of a liquid, provided only that sufficient time is allowed for the change of shape to take place. When the shape is altered quickly, liquids do display a resistance, but this vanishes very quickly after the motion is finished. This ability of liquids to oppose a change in shape is called *viscosity*. We will discuss viscosity in depth in Section 4.2. As well as the usual liquids that are easy to move, there are also very viscous liquids whose resistance to change of shape is considerable, but which vanishes again at rest. Starting out from the viscous state, all phase transitions to (amorphous) solid bodies are possible. Heated glass, for example, passes through all possible transitions; in asphalt and similar substances these transitions occur at normal temperatures. For example, depending on the temperature, if a barrel of asphalt is tipped over, the asphalt will flow out within a few days or weeks. The mass that flows out forms a flat cake. Although it continually flows, one can walk on it without making footprints. Footprints will be left, however, if one stands still for a longer time on the asphalt. Hammering on the asphalt causes the mass to shatter like glass.

In the study of the *equilibrium* of liquids, we consider states of rest or sufficiently slow motion. The resistance to change of shape may then be set to zero, and we obtain *a definition of the liquid state: In a liquid in equilibrium, all resistance to change of shape is equal to zero.*

According to the kinetic theory of material, atoms or molecules are in constant motion. The kinetic energy of this motion is observed as heat. From this point of view, liquids differ from solids in that the particles do not oscillate about fixed positions, but rather more or less frequently swap places with neighboring particles. If the liquid is in a state of stress, such exchanges of place are favored. They cause the material to yield in the direction of the stress differences. In the state of rest this yielding causes the stress differences to vanish. During the change of shape, stresses arise that are larger the faster the change of shape takes place.

The gradual softening of amorphous bodies with increasing temperature may be explained as follows: If the body is heated, i.e. the energy of the

H. Oertel (ed.), *Prandtl-Essentials of Fluid Mechanics*,
Applied Mathematical Sciences 158, DOI 10.1007/978-1-4419-1564-1_2,

molecular motion is increased, initially some particles situated where the oscillation amplitudes just happen to be particularly large change place. On further heating, the exchange of place becomes more and more frequent, until eventually it occurs everywhere. For crystalline solid bodies the transition from a solid to a liquid state takes place discontinuously by melting, i.e. by the disintegration of the regular crystal structure.

A further property of liquids is their great resistance to change in volume. It is not possible to force 1 liter of water into a container half a liter in size. If the same amount of water is placed in a container 2 liters in size, only half of the container is filled. However, water is not fully incompressible. At high pressures it can be pressed together by noticeable amounts (4% reduction in volume at a pressure of about 100 bar). Other liquids behave in a similar way.

2.2 State of Stress

We now consider more closely the state of stress of a liquid in equilibrium.

We first note that forces are always interactions between masses. For example, if one mass m_1 attracts another mass m_2 with a force $\boldsymbol{F}$, this force $\boldsymbol{F}$ also acts on m_1 as the effect of m_2, as an attraction in the direction of m_2. The two forces act in opposite directions (Newton's principle of action and reaction). For a *system of masses* separated from other masses, we distinguish between two types of force: the *internal forces*, which act between two masses belonging to the system, and which therefore always act *opposite in pairs*, and the *external forces*, which act between each system mass and a mass situated outside the system, and which therefore occur only once in the system. If we sum over all the forces acting on the masses in the system, the internal forces always cancel each other out in twos, so that only the external forces remain.

For the equilibrium of the system it is necessary that the sum of all the forces acting on each individual mass *vanish* (vector sum). If we sum this over all masses of the system, only the sum of all the *external forces* remains. Because each individual sum vanishes because of the equilibrium, the *sum* of the *external forces on the system* also *vanishes*. This law, which assumes no more about the mass system than that it is in equilibrium, is highly useful in many different applications. We obtain three statements:

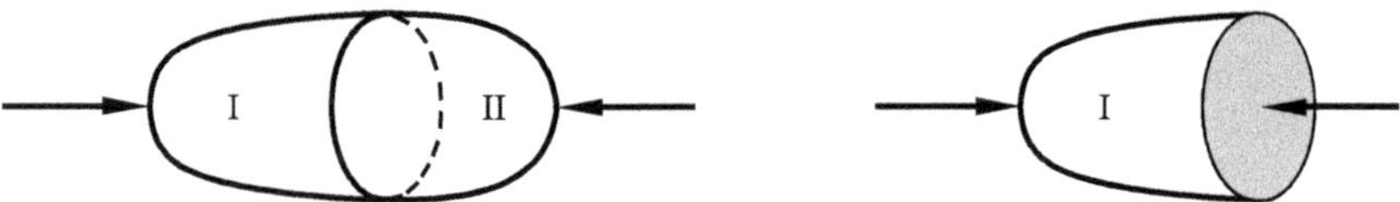

Fig. 2.1. Forces on a mass system

$$\sum F_x = 0, \qquad \sum F_y = 0, \qquad \sum F_z = 0,$$

with the components F_x, F_y, F_z of the external forces in the x, y, and z directions.

As well as the above law, there is an analogous law for the torques of the external forces. Their sum also must vanish in equilibrium.

For both elastic solid bodies and liquid bodies we are interested in the *state of stress inside the body.* This arises via the internal forces that act between the smallest particles of the body. In general, we are content with knowing the average state in a region that already contains a large number of particles. Imagine the body cut and one of the two pieces (labeled I in Figure 2.1) to be part of a mass system. Then all forces that came from a particle in region II and acted on a particle in region I, and which were previously internal forces, have now become external forces. If the whole body was in an external state of stress (indicated in Figure 2.1 by two arrows), internal stresses also occur. Imagining the cut carried out, forces act through the interface from the particles to the right of the cut on particles to the left of the cut. We add all these forces together to a resultant force, which then exactly maintains the equilibrium of the forces acting on part I. This gives us a clear statement on the resultant of the forces in the section. This approach could equally well have been applied to part II. We would have obtained an equally large resultant force pointing in the opposite direction (precisely the force acting from part I on part II).

By *stresses* we mean *forces per unit area in a section.* In the above example, we obtain the mean stress in the section when we divide the resultant force in the section due to equilibrium by the surface area of the section. We see that the *stress in a surface* is a vector, just as the force is.

The *method of sections*, i.e. the manner of transforming internal forces to external forces by imagining a cut, has further applications. With a number of planes of section through a body whose state of stress is to be investigated,

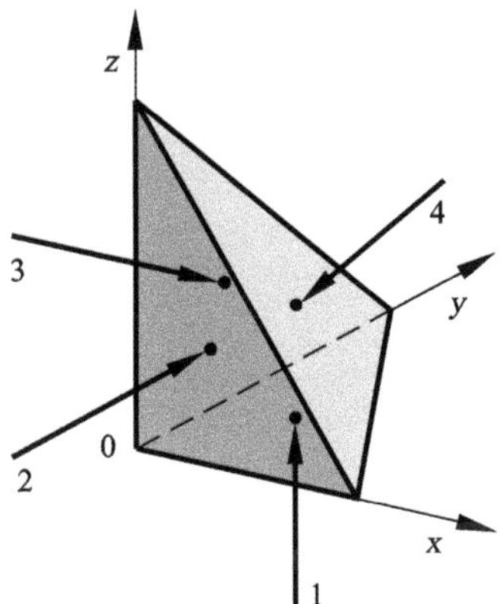

Fig. 2.2. Stress forces on a tetrahedron

we can select a small body (parallelepiped, prism, tetrahedron, etc.) and investigate its equilibrium. In the simplest case, all forces that hold a body in equilibrium are stress forces. From the equilibrium of such a body, we can derive several important laws; one is proved here as an example.

If we know the stress vectors for three planes of section that together form a corner of a body, then the stress vectors for all other planes of section are also known.

As proof, we cut the corner with a fourth plane, whose stress is to be determined. This gives rise to the tetrahedron shown in Figure 2.2. The forces 1, 2, and 3 are then obtained by multiplication of the given stress vectors by the surface areas of the associated triangles. There is only one direction and magnitude of the force 4 that maintains equilibrium with the sum of forces 1, 2, and 3. This force divided by the associated triangular surface area is the desired stress. For the calculation it is useful to select the surfaces 1, 2, and 3 as the coordinate planes.

We menely note that the *state of stress*, which represents the whole of the stress vectors in all possible cut directions through a point, can be related to an ellipsoid, and is therefore a tensor. According to the derived law, the state of stress in a point (and also its ellipsoid) is given if the stress vectors in three planes of section are known. Corresponding to the three principal axes of every ellipsoid, three orthogonal planes of section can be given for every state of stress to which the associated stress vectors are normal. The three stresses distinguished in this manner are called *principal stresses*.

2.3 Liquid Pressure

The state of stress of a liquid in equilibrium is particularly simple. A resistance to change of shape, thus against displacement of the particles against each other, can be compared to the friction of solid bodies. If there is no friction between two bodies that are in contact, the force must always be perpendicular to the contact surface between both bodies, so that no work

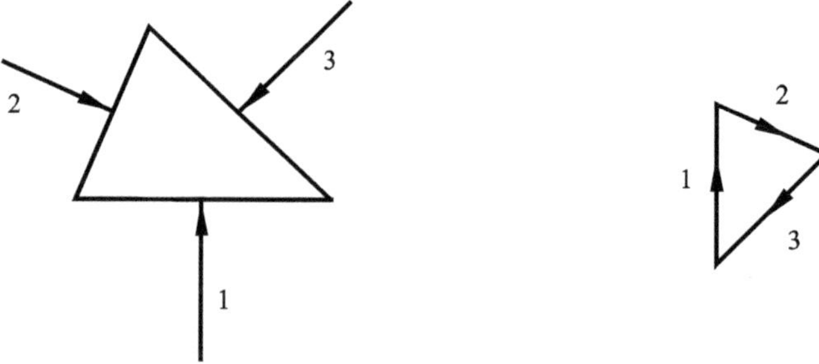

Fig. 2.3. Forces on the front side of a prism and force equilibrium

is done by a sliding motion along this surface. Similarly, the absence of a resistance to change of shape is distinguished by the fact that the *stress*, here called the *pressure*, is *always perpendicular to every plane of section*. This property, that the pressure is perpendicular to the associated surface, can be taken as a *definition of the liquid state*. It is completely equivalent to the definition given in Section 2.1.

By a simple equilibrium approach, a further property of the liquid pressure may immediately be derived. We cut a small three-sided prism out of the liquid. The faces of the prism are perpendicular to the edges of the prism. We consider the equilibrium of the forces that act on the prism from the rest of the liquid. The pressure forces on the faces are equal and directed opposite to each other. They therefore maintain equilibrium and do not have to be considered further. The forces on the side surfaces are perpendicular to the surfaces, and are therefore in a plane perpendicular to the prism's edges. Figure 2.3 shows a front view of the prism with the forces, as well as the triangle that the forces must form so that they are in equilibrium. Since the sides of the force triangle are perpendicular to the sides of the prism, both triangles have the same angles and are therefore similar. This means that the three pressure forces behave like the associated prism sides. In order to determine the pressures relative to the unit surface area, the pressure forces have to be divided by the respective prism surface areas. The prism surface areas all have the same height and are therefore in the same ratio to each other as their base lines and as the associated forces. Therefore, the *pressure per unit area* is equally large on all three prism surfaces. Since the prism was arbitrarily chosen, we can conclude that *the pressure at one point in the liquid is equally large in all directions*. The stress ellipsoid is a sphere in this case. In order to describe a state of stress of this kind, also called the *hydrostatic state of stress*, we need only the numerical value of the pressure p. The pressure p means the force acting on a unit surface area.

Pressure Distribution of a Liquid without Gravity

Every liquid is subject to the force of gravity. In many cases, in particular at high pressures, the effect of gravity can be neglected, thereby simplifying matters greatly. Again we set up the force equilibrium on a prism, this time with a longitudinal shape. We consider the change in equilibrium on displacement along the prism axis. The pressure varies with position. The cross-section of the prism is the same as its front surface, here again assumed perpendicular to the axis of the prism, and is denoted by A (see Figure 2.4). This cross-section is assumed to be so small that the change in pressure within A can be

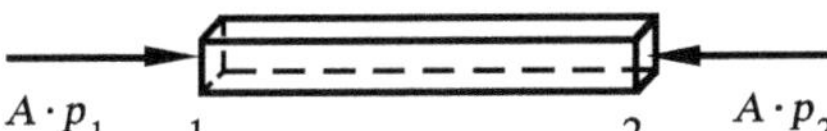

Fig. 2.4. Pressure forces on a longitudinal prism

neglected. If the pressure at one end of the prism is p_1 and at the other p_2, the forces $A \cdot p_1$ and $A \cdot p_2$ act in opposite directions parallel to the axis of the prism. All pressure forces on the side faces of the prism are assumed to be perpendicular to these faces and are therefore also perpendicular to the prism axis. They do not contribute to the force component parallel to the prism axis, irrespective of how the pressure is distributed along it. Equilibrium demands that the forces $A \cdot p_1$ and $A \cdot p_2$ in the direction under consideration must balance each other. We must have

$$A \cdot p_1 = A \cdot p_2 \qquad \text{or} \qquad p_1 = p_2.$$

Since the position of the prism was chosen arbitrarily, *in the absence of gravity* (and other external forces) *the pressure at all positions in the liquid is equal.*

If the liquid fills narrow, curved spaces, so that it is not possible to place a prism between two arbitrary points in the liquid, the above procedure can be repeated as often as necessary. We start out from point 1 to point 2, from this point in another direction to point 3, etc., until the required endpoint n is reached. From $p_1 = p_2$, $p_2 = p_3$, etc., we then obtain $p_1 = p_n$.

In extremely narrow spaces, after a change in the liquid pressure, e.g. following an external stress, considerable time may pass until equilibrium is reached. For plastic potter's clay (consisting of very fine solid particles, with the spaces between filled with water), this time may be days, or, in the case of layers of clay in the earth, even years. During this time the water flows from positions of higher to those of lower pressure (see Section 4.2.8), while the solid frame yields elastically.

We summarize as follows: *The pressure in a liquid in equilibrium is everywhere perpendicular to the surface on which it acts and* in the absence of gravity and other mass forces *is everywhere and in all directions equal.*

Whatever holds for the pressure inside the liquid is also true for the pressure on the walls of the vessel containing the liquid. To clarify this, we imagine a cut through the liquid very close to the wall and at some distance from it, and connect these two faces with a cylindrical surface perpendicular to the cut (see Figure 2.5). The equilibrium of the body of water enclosed in this manner yields the force component $\boldsymbol{F}$ that the section of wall perpendicular to the plane of section experiences, that is, the force $A \cdot p$. This approach has

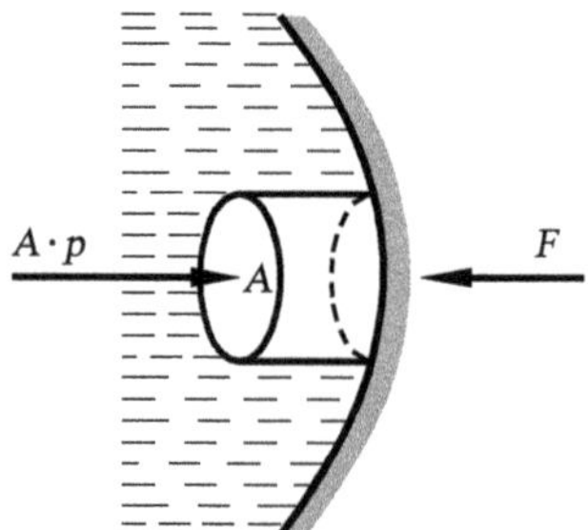

Fig. 2.5. Pressure force on the wall of a vessel

the advantage that we immediately see that uneven parts of the wall do not change the result. Figure 2.5 shows the force $\boldsymbol{F}$ acting from the wall onto the body of liquid under consideration. The pressure force of the liquid on the wall has the opposite direction.

Pressure Distribution with Gravity

The effect of gravity on a given mass m is caused by a force of attraction to the center of the Earth of magnitude $m \cdot g$, where g, the gravitational acceleration, is equal to 9.81 m/s^2 at a latitude of 50° N. This value is not exact as the rotation of the Earth has been neglected. In fact, the force of gravity is due to the force of attraction and the centrifugal force. In the northern hemisphere, the direction of a plumb line intersects the axis of the Earth somewhat south of the center of the Earth.

The force $m \cdot g$ is called the weight of the mass m. Because the amount of a liquid is frequently measured according to its volume, the term *density* ρ is introduced for the *mass of a unit volume*. An amount of a liquid of volume V and density ρ therefore has a mass of $\rho \cdot V$ and a *weight* of $g \cdot \rho \cdot V$. The product $g \cdot \rho$ is therefore the weight of a volume unit and is called the specific weight γ. Because the gravitational acceleration g is not the same at all positions on the Earth, the magnitude of the specific weight also varies from place to place. On the other hand, the density is independent of the strength of the gravitational force.

The *basic task of hydrostatics*, i.e. the study of the equilibrium of liquids, is to *determine the pressure distribution of a homogeneous liquid.*

We again consider the equilibrium of a bounded prism in a liquid to displacement in the axial direction and initially use the prism of Figure 2.4. Its axis is horizontal and is therefore at right angles to gravity. Therefore, the weight of the prism has no component in the axial direction, and so all the arguments from Section 2.3 may be repeated. Here again we obtain $p_1 = p_2$. By repeating this procedure for many prisms lined up with horizontal axes, we find that in all points in a horizontal plane the pressure must have the same value.

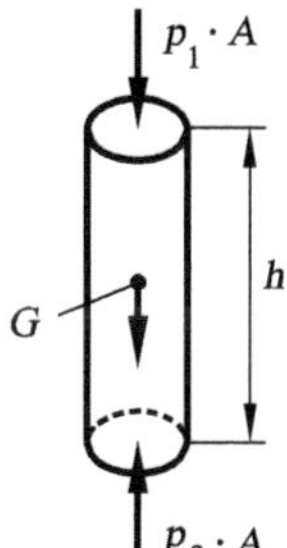

Fig. 2.6. Balance of forces on a vertical cylinder element

A relation between different horizontal planes is obtained by considering the equilibrium of a prism or cylinder with vertical axis in respect of displacement in the vertical direction. In this case the weight of the prism has to be taken into account in the equilibrium of the forces. Corresponding to Figure 2.6, the pressure force $p_1 \cdot A$ on the upper face and the weight $G = \gamma \cdot V = \gamma \cdot A \cdot h$ are directed downward. The pressure force $p_2 \cdot A$ acts upward on the lower face. Equilibrium requires that

$$\gamma \cdot A \cdot h + p_1 \cdot A = p_2 \cdot A.$$

Therefore,

$$p_2 - p_1 = \gamma \cdot h. \tag{2.1}$$

The pressure difference between the positions 1 and 2 is equal to the weight of the vertical column of liquid of cross-section 1 between them. Repeated application of this procedure leads to the following result: *The pressure increases in the direction of the force of gravity by the amount γ for each unit of length. It is constant in every horizontal plane.*

If we introduce an x, y, z coordinate system whose z axis points vertically upward in the opposite direction to gravity, and if p_0 is the pressure in the horizontal plane $z = 0$, the pressure p at an arbitrary position is given by

$$p = p_0 - \gamma \cdot z. \tag{2.2}$$

This relation holds in large spaces filled with liquid, in communicating vessels, in arbitrary pipe systems, in the gaps in gravel or sand, etc. The only assumption is a *homogeneous*, connected liquid at rest.

We determine the force that a body submerged in a liquid experiences due to liquid pressures as follows. We first imagine the body replaced by liquid. The new section of liquid has the same shape as the body and has the same specific weight as the remaining liquid. It is kept in equilibrium by the pressure forces on its surface. The resultant of the pressure forces must point vertically upward, through the center of gravity of the new part of liquid. The size of this resultant force, called the lift, is equal to the product of the displaced volume V and the specific weight γ of the liquid. This law was discovered by *Archimedes* and reads thus: *The loss of weight of a body submerged in a liquid is equal to the weight of the fluid it displaces.* If a body is weighed in a submerged state and in air, where it also experiences a small lift, there is a reduction in weight of $G_{\text{liq}} - G_{\text{air}} = V \cdot (\gamma_{\text{liq}} - \gamma_{\text{air}})$. This can

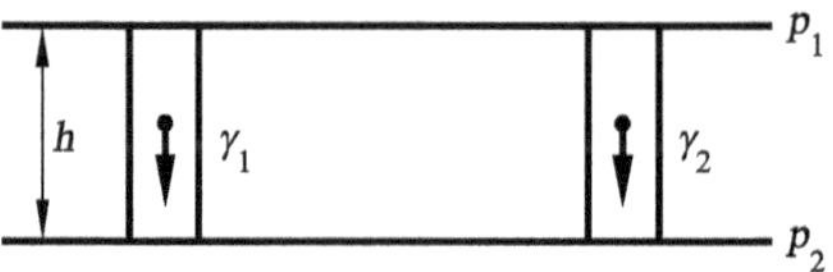

Fig. 2.7. Balance of forces on two horizontally displaced cylinder elements

be determined for a known specific weight γ_{liq} or a known volume V. The quantity γ_{air} can be computed using the concepts introduced in Section 2.5.

If the liquid is *inhomogeneous* (e.g. at different positions in a liquid with a nonuniform temperature distribution, salt solution with different salt content at different positions), the procedure with the prism with the horizontal axis can be applied without any change. Here, too, the pressure is the same in all horizontal planes. Two such horizontal planes a (not too large) distance h a part are selected (see Figure 2.7), with the upper plane at pressure p_1 and the lower at pressure p_2. We consider two vertical prisms with height h and mean specific weights of γ_1 and γ_2 for the left and right prisms, respectively.

The balance of forces requires that on the left $p_2 - p_1 = \gamma_1 \cdot h$ and on the right $p_2 - p_1 = \gamma_2 \cdot h$. This is possible only if $\gamma_1 = \gamma_2$. Otherwise, there would be no equilibrium, and the liquid would be set in motion. We can refine this approach by assuming the height h to be very small and carrying out the procedure for arbitrarily many pairs of neighboring horizontal planes. We obtain the result that *in an inhomogeneous liquid, equilibrium is possible only if the density is constant in every horizontal layer.* This result already contains the answer to the question of the equilibrium of two liquids of different densities that are layered above one another and do not mix. Their equilibrium requires that the interface must be a horizontal surface. We can directly apply the approach of Figure 2.7 to two homogeneous liquids layered above one another, whose interface is between the two horizontal planes and is initially unknown, and again we arrive at the same result.

Considering the *stability* of such a layering of liquids, we note that the liquid with the lower density always must be situated above the denser liquid. The reverse stratification is unstable. The smallest disturbance will put it into motion.

The proof of this can again be drawn from Figure 2.7. We assume a disturbed, slightly inclined interface between the two horizontal planes and determine the pressure differences in the interface. In the stable case, this inclination of the interface tends to decrease, whereas in the unstable case it tends to increase.

Similar statements hold for densities that vary continuously. The system is stable if the density everywhere decreases as we move upward. In contrast to the stable, layered inhomogeneous liquid, the homogeneous liquid is a case of neutral equilibrium. Any parts of the liquid may be arbitrarily displaced without generating any forces that would disturb the equilibrium.

For the pressure distribution in the inhomogeneous liquid, for every layer in which the density is sufficiently inhomogeneous, (2.1) in differential form holds:

$$\mathrm{d}p = -\gamma \cdot \mathrm{d}z. \tag{2.3}$$

If γ is given as a function of the height z, integration leads to the relation

$$p = p_0 - \int_0^z \gamma \cdot \mathrm{d}z. \tag{2.4}$$

2.4 Properties of Gases

Gases differ from liquids in that at large pressures they can be pressed together into a very small space. If more space become available than in the initial state, the gas always fills it uniformly, with a corresponding drop in pressure. Apart from this, their behavior is very similar to that of liquids. For gases at rest, all resistance to change of shape also vanishes, and they also have a certain viscosity to internal displacement. As long as there is no change in volume, the behavior of a gas is qualitatively no different from that of a liquid that fills the same space without having a free surface.

The most important gas is the air in our atmosphere. Other gases have essentially the same behavior. As we will discuss in more detail in what follows, the air on the surface of the Earth is under approximately constant pressure of around 1 bar or 10^5 N/m^2. At higher altitudes the air pressure is lower (cf. Section 2.5).

There are several devices available to measure air pressure (gas pressure). Devices that show pressure differences are called *manometers*. If they show absolute pressures of the surrounding gas, they are called *barometers*. Liquid columns can be used for both sorts of measurement (see Section 2.6). Devices in which the pressure to be measured acts on a spring are also frequently used. In order to measure the absolute pressure of the air, one can, for example, connect a metal can that has been pumped empty of air to a flexible lid with a strong spring, so that the tension of the spring just prevents the lid from being pushed in by the external air pressure. If this device is brought to a position with a different air pressure, the pressure change can be read from the deflection of the pointer (aneroid barometer).

The law according to which the pressure of the gas changes for a given change in volume was first discovered by *R. Boyle* in 1662 and then independently by *Mariotte* in 1679, and is therefore called the *Boyle-Mariotte law.* According to this law, at constant temperature the pressure is inversely proportional to the volume. Therefore, if a fixed amount of gas is pressed together to half of its volume, its pressure doubles. If the volume is doubled, the pressure sinks by half. This law is expressed by the equation

$$p \cdot V = p_1 \cdot V_1, \tag{2.5}$$

where p_1 is the initial pressure, V_1 the initial volume, and p and V the values of these quantities for the gas in some given state.

The volume of a gas also changes greatly with temperature. *Gay-Lussac* found in 1816 that the expansion of a gas for a change in temperature of 1 °C at constant pressure is always 1/273.2 of its volume at 0 °C. This is valid to good approximation for all gases and temperatures. This behavior is described by the equation

$$V = V_0 \cdot (1 + \alpha \cdot \vartheta)\,, \tag{2.6}$$

where V_0 is the volume at 0 °C, ϑ the temperature in °C and $\alpha = 1/273.2$ °C the *coefficient of expansion*. At moderate pressures, this value of α is valid not only for air, but also to good approximation for other gases, such as steam and helium.

Since (2.6) is independent of the pressure, it may be combined with (2.5). We therefore obtain an equation applicable at all pressures and temperatures:

$$p \cdot V = p_0 \cdot V_0 \cdot (1 + \alpha \cdot \vartheta)\,. \tag{2.7}$$

Here p_0 is an arbitrary but fixed initial pressure and V_0 the volume at the initial pressure p_0 and at 0 °C. Equation (2.7) is frequently called the *Mariotte–Gay-Lussac* law. It is also called the *equation of state*, since it connects the three state variables pressure, volume, and temperature. It is called the *equation of state of the ideal gas*, since the behavior of real gases deviates somewhat from this equation. For gases at normal densities these deviations may be neglected, but they are very important if the gas is greatly compressed, and particularly if the temperature is reduced so far that the gas begins to condense.

These deviations are treated in detail in thermodynamics. Here we mention only one of the deviations. According to (2.5), at very high pressures the gas volume is very small. Equation (2.7) can be used to calculate at which pressure the density of water, or that of gold, is reached. However, in reality this is impossible. There is a limiting volume below which the gas cannot be compressed, however large the pressure, i.e. a volume at which the molecules have attained their densest possible structuring. This fact can be included in equation (2.7), by writing

$$p \cdot (V - V') = p_0 \cdot (V_0 - V') \cdot (1 + \alpha \cdot \vartheta)\,,$$

with the small limiting volume V'. For every finite p, V is somewhat larger than V'. For volumes V that are large compared to V', the results of this equation are essentially no different from those of (2.5) or (2.7).

As a gas is compressed, heat is generated. The *Boyle–Mariotte* law, which is valid only for constant temperatures, can be observed only if the gas has enough time during compression to release the heat generated and to assume the surrounding temperature. The same is true for the cooling associated with expansion. If the gas is not given enough time to equalize its temperature differences, the ratio of the pressure to the initial pressure increases more strongly than the ratio of the volumes decreases. Thermodynamics states that in the case in which there is no exchange of the heat generated, i.e. when the compression or expansion takes place quickly, instead of (2.5) we have the equation

$$p \cdot V^{\kappa} = p_1 \cdot V_1^{\kappa}, \tag{2.8}$$

where $\kappa = c_p/c_v$ is the ratio of the specific heat at constant pressure to the specific heat at constant volume. For dry air, $\kappa = 1.4$. Whereas a compression or expansion that obeys (2.5) is an *isothermal change of state*, a change according to (2.8) is called *adiabatic* compression or expansion. Heating is associated with adiabatic compression, and this can be calculated from (2.7) and (2.8), while cooling is associated with adiabatic expansion.

The behavior of gases discussed in this section can be explained by the assumption of gas kinetics that the molecules of the gas move at large velocities, colliding with each other and with their surrounding walls. The pressure is the summation of these collisions, and the temperature is the same as the kinetic energy of the particles. The temperature increases on compression, as the velocity of the particles is increased due to elastic reflection as the walls move together.

2.5 Gas Pressure

The condition for the equilibrium of a gas is the same as that for the equilibrium of a liquid. The relations of Section 2.3 can therefore be carried over. In many cases, e.g. for moderate vertical extensions of a gas, the specific weight of the gas can be assumed to be spatially constant. Equations (2.1) and (2.2) of the previous section can be applied; i.e. the gas may be considered to be a homogeneous liquid. For greater vertical extensions (to the order of kilometers) this is no longer permissible. The pressure differences are so great here that, because of the compressibility of the gas, the densities above and below are different. Temperature differences are also frequently important. Here the equation for inhomogeneous liquids must be used. Equation (2.3) is divided by γ and integrated. We obtain

$$\int_p^{p_0} \frac{\mathrm{d}p}{\gamma} = z. \tag{2.9}$$

Depending on how the temperature depends on the height, this integral yields different results. The most important case is that of *constant temperature*. According to the *Boyle–Mariotte* law ($p \cdot V = \text{const}$), the specific weight γ is directly proportional to the pressure:

$$\gamma = \gamma_0 \cdot \frac{p}{p_0}. \tag{2.10}$$

Therefore,

$$\int_p^{p_0} \frac{\mathrm{d}p}{\gamma} = \frac{p_0}{\gamma_0} \cdot \int_p^{p_0} \frac{\mathrm{d}p}{p} = \frac{p_0}{\gamma_0} \cdot \ln\left(\frac{p_0}{p}\right), \tag{2.11}$$

As can be seen from (2.1), p_0/γ_0 is the height of a column of liquid with constant specific weight γ_0, and with pressure p_0 at the lower end and a pressure of zero at the upper end. This height is called the *height of the uniform atmosphere.* With regard to the real atmosphere, it is nothing more than a computational quantity.

As an example we determine its numerical value. We therefore require the value of γ_0. In order to determine γ_0 we proceed as follows: We weigh a container with a faucet out of which the air has been pumped. We then open the faucet and wait for the temperature to equalize, as the air in the container is initially heated by the work done by the external atmosphere as it flows into the container. We then weigh the container a second time. Since it was empty before and is now filled with air, its weight has increased by the weight G of the air inside it. We then determine the volume V of the container, by, for example, pumping the air out of the container again, opening the faucet under water and again weighing the container filled with water. The measured quantities give us the value $\gamma_0 = G/V$ associated with the pressure p_0 on the ground. For every other ground pressure p_0, γ_0 can be calculated similarly. Assuming that p_0 is equal to 1 bar, for moderately damp air of temperature ϑ, the Gay-Lussac law yields

$$\gamma_0 = \frac{12.45}{1 + \alpha \cdot \vartheta} \ \mathrm{N/m^3}. \tag{2.12}$$

In dynamics, the density $\rho = \gamma/g$ is used as a measure of the mass inertia. At room temperature, we can choose a mean value of 11.8 $\mathrm{N/m^3}$ for γ. With $g = 9.81\ \mathrm{m/s^2}$ we then obtain a mean value for ρ of 1.20 $\mathrm{Ns^2/m^4}$.

In order to compute p_0/γ_0 in (2.11), p_0 has to be expressed in the same mass system as γ_0. With 1 bar $= 10^5\ \mathrm{N/m^2}$, we obtain

$$\frac{p_0}{\gamma_0} = \frac{100000}{12.45} \cdot (1 + \alpha \cdot \vartheta) = 8030 \cdot (1 + \alpha \cdot \vartheta)\,.$$

The unit of p_0/γ_0 is m. The height of the uniform atmosphere for moderately damp air is (independent of the pressure but dependent on the temperature) $8030 \cdot (1 + \alpha \cdot \vartheta)$ m. We set this equal to H_0. Equation (2.9) applied to two different heights yields

$$z_1 = H_0 \cdot \ln\left(\frac{p_0}{p_1}\right), \qquad z_2 = H_0 \cdot \ln\left(\frac{p_0}{p_2}\right).$$

Therefore,

$$z_1 - z_2 = H_0 \cdot \ln\left(\frac{p_2}{p_1}\right). \tag{2.13}$$

This is the so-called *barometric height formula.* By inverting (2.13), we obtain the dependence of the pressure on the height:

$$p = p_1 \cdot e^{-\frac{z - z_1}{H_0}}\,. \tag{2.14}$$

Considering the balance of the forces, in analogy to Figure 2.6, we see that the weight of a column of air with base area A that extends from position z

upward to the edge of the atmosphere is equal to $A \cdot p$. Therefore, p is directly equal to the weight of the column of air with cross-section 1 situated above position z. Figure 2.8 shows (2.14) graphically. The pressure decreases continuously but ever more weakly with increasing height. For large heights the pressure is equal to zero. The pressure decrease with height can be measured in the free atmosphere with a pressure-measuring device (barometer) on a tower or mountain. It can even be measured in a multistoried house. If the air temperatures are also measured, the observed pressure differences can be used to determine the difference in height. This method is used in aircraft to determine the altitude. If this height difference is known, this method can also be used to determine the mean specific weight of the air layer situated between two positions. If the temperature of the mass of air is not constant, the height equation can still be applied to height sections in which the temperature differences are not very large. The height H_0 associated with each height section is then determined for the mean value of the temperature in this section.

Finally, we turn to the question of when the equilibrium of a layered mass of gas is *stable* and when it is *unstable*. The condition that the specific weight of the upper layers must be smaller than that of the lower layers is not sufficient, because as a mass of gas moves upward or downward the pressure and thus the density of the mass of gas changes. The correct answer to the question is the following: The system is stable if a part of the gas at a greater height and at the new pressure is denser than its new surroundings, or if a part of the gas at a lower height and at the new pressure has a lower density than its new surroundings. In these cases the part of the gas will tend to return to its original position. There is a stratification (temperature distribution) in a mass of gas that corresponds to a homogeneous liquid, which therefore implies neutral equilibrium for the mass of gas. In order for this to hold, each part of the gas taken from an arbitrary position must have the same density as its surroundings after displacement, as if it had always belonged there. A part of a gas behaves adiabatically under a change of pressure as long as it has

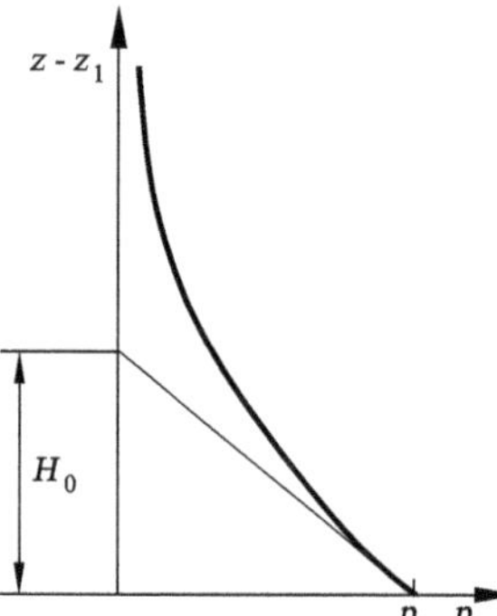

Fig. 2.8. Pressure distribution in an atmosphere of constant temperature

no possibility to exchange heat. If the stratification is such that pressure and density satisfy the equation of state (2.8) at all heights (i.e. p is proportional to γ^{κ}), every raised or lowered gas part always reaches a neighborhood with the temperature that it has itself due to its own adiabatic change of state. Therefore, it has no possibility to exchange heat with its surroundings. It can be shown that this *adiabatic stratification* has the following in common with a homogeneous liquid: It can be made by strong mixing of an originally different type of stratification, such as an inhomogeneously layered salt solution.

In the air of the atmosphere, adiabatic stratification is characterized by the fact that the temperature decreases by 1 °C with an increase of height of 100 m. A lesser temperature decrease already indicates stability, while a temperature increase with height indicates even stronger stability. A larger temperature decrease than 1 °C per 100 m generally does not occur in the free atmosphere, since it would correspond to an unstable state. However, it is found close to the surface of the earth if the ground is hotter than the air. The air is then not in equilibrium, but rather is in motion with vertical upward and downward streams.

The pressure distribution in the adiabatically layered atmosphere can also be computed with (2.9), by setting $\gamma = \gamma_0 \cdot (p/p_0)^{1/\kappa}$. Integration yields

$$z = \frac{\kappa \cdot H_0}{\kappa - 1} \cdot \left(1 - \left(\frac{p}{p_0}\right)^{\frac{\kappa-1}{\kappa}}\right) \quad \text{or} \quad p = p_0 \cdot \left(1 - \frac{\kappa - 1}{\kappa} \cdot \frac{z}{H_0}\right)^{\frac{\kappa}{\kappa-1}} .$$

The equation of state $p/\rho = R \cdot T$, with the density $\rho = \gamma/g$, the absolute temperature $T = (273.2 + \vartheta/1\ ^\circ\mathrm{C})$ K and the gas constant R, with $p_0/\gamma_0 = H_0$, yields

$$\frac{R \cdot T}{g} = \frac{p}{\gamma} = H_0 - \frac{\kappa - 1}{\kappa} \cdot z, \quad \text{and so} \quad \frac{\mathrm{d}z}{\mathrm{d}T} = -H_0 \cdot \frac{\kappa}{\kappa - 1} \cdot \frac{R}{g} .$$

For moderately damp air, $R/g = 29.4$ m/K and $\mathrm{d}z/\mathrm{d}T = -102$ m/K.

If we replace κ in the above equations by a different number n, we obtain an interpolation formula that describes states of layering that actually occur in the atmosphere. These states of layering are called *polytropic*. For stable stratification, $n < \kappa$.

2.6 Interaction Between Gas Pressure and Liquid Pressure

As long as the pressure difference between the air in a container and the external air in the atmosphere is not too large, it can be measured with a U-tube manometer (cf. Figure 2.9). Neglecting the weight of the air, we obtain the following relations. At position A, the liquid pressure is equal to the air pressure p_1 in the container. In the other limb of the U-tube, the pressure at the same height B is the same (communicating containers). Say the free liquid surface in this limb is at C. There the liquid pressure is equal to the pressure p_0 of the atmosphere. According to the relations in Section 2.3,

$$p_1 = p_0 + \gamma \cdot h$$

if the height $\overline{\mathrm{BC}}$ is set equal to h. A U-tube filled with liquid is therefore suitable for measuring such pressure differences. It is used in various different forms. In order not to have to read the liquid heights at two positions (A and C in Figure 2.9), one of the limbs is frequently reshaped as a large pot in which the movement of the surface becomes very small (see Figure 2.10). To zero the device, both openings have to be connected to the atmosphere. For very small pressure differences the reading of the heights is refined using, for example, a moveable microscope, or with a magnifying projection of a scale swimming on the surface of the liquid, according to *A. Betz.*

The use of the liquid manometer has led to a particular type of pressure units, where the pressure is expressed by the height of a liquid column. For example, 1 mm WC (water column, or WG water gauge) is equal to 1 kp/m^2 = 9.81 Pa.

Water is not very suitable as a measurement liquid, since it wets the walls of the glass pipe very irregularly. All fat-soluble liquids (alcohol, toluol, xylol, etc.) are much more suitable. For larger pressure differences mercury is recommended, as in its pure state it permits very precise adjustment in a glass tube that is not too narrow. Because of specific weight of 133.370 N/m^3 at 0 °C, 1 mm Hg (mercury) is equal to 13.6 kp/m^2 = 133.4 Pa. The pressure unit 1 mm Hg is also called 1 torr, in honor of *Torricelli.* In recent times, membrane pressure gauges with digital data memory and piezopressure gauges that exploit the piezoelectric effect have been used.

If we pump some air out of the container in Figure 2.9, so that the pressure there becomes lower than the atmospheric pressure, the liquid in limb A of the U-tube will be higher than the liquid in limb B. Figure 2.11 shows a somewhat altered arrangement for the same experiment. The setup in Figure 2.9 is called an *overpressure manometer*, while that in Figure 2.11 is called a *vacuum manometer*. The pressure is measured from the height h.

Here we mention something about the history of pressure measurement: The question arose of how high a liquid can be sucked. In the middle ages,

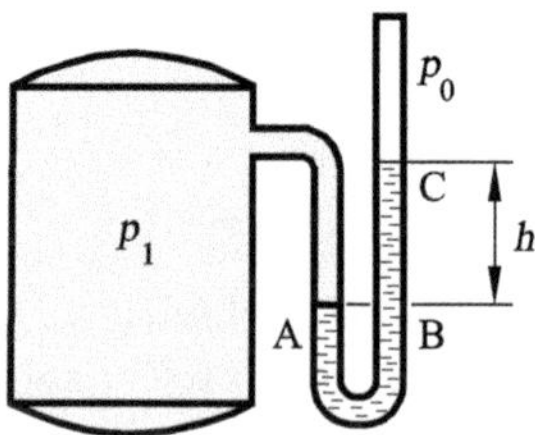

Fig. 2.9. Hydrostatic pressure measurement (U-tube manometer)

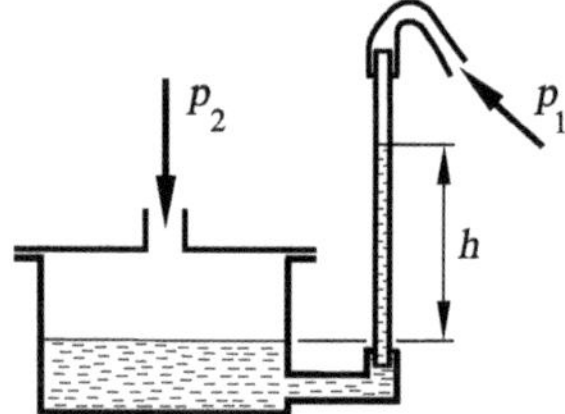

Fig. 2.10. Liquid manometer

the rising of a column of liquid due to suction was explained by the idea of *horror vacui*, that "nature abhors a vacuum." No investigations had been carried out into whether the *horror vacui* was arbitrarily strong, or whether it had a limit. It was the misfortune of Florentine pump makers, who built a water pump with the suction valve more than 10 m above the water surface and were unable to pump water as high as they wanted, that encouraged *Galileo* to look into the problem. Meanwhile, it was his pupil *Torricelli* who first recognized the facts, and this because of an experiment with mercury that he prompted his friend *Viviani* to perform in 1643. From our point of view, the answer to the question above is not difficult. Suction is merely compressing more weakly than the atmosphere compresses. The pressure in the container in Figure 2.11 is at its lowest when all the air has been pumped out of the container. Then it is equal to zero. The column of liquid can rise only so high that its height h corresponds to the air pressure p_0 ($h = p_0/\gamma$). *Viviani's* experiment was as follows: He took a glass tube two ells (120 cm) in length with a glass bubble blown on one end, and filled it completely with mercury from the other open end, closing this end with his finger. He then turned the tube upside down and placed the closed end in a flat container filled with mercury, and removed his finger. The column of mercury sank to a height of 1 1/4 ell (75 cm) above the surface of the mercury in the container and left an empty space behind. *Torricelli* correctly recognized that the mercury column retained the equilibrium with the outer air pressure. He observed that the mercury column did not always have the same height and concluded that the air pressure undergoes certain fluctuations. This fact is today of great importance for meteorology. *Torricelli* already concluded that the air pressure on a mountain must be higher than that in the valley, and that therefore the height of the mercury column is lower on the summit than down below. This was demonstrated several years later by *Perrier*, on the encouragement of *Pascal*, whereby he measured the height of the mercury column on the Puy de Dome and at the foot of this 975 m high mountain and noted a difference of 3 inches. The name *barometer* for this pressure

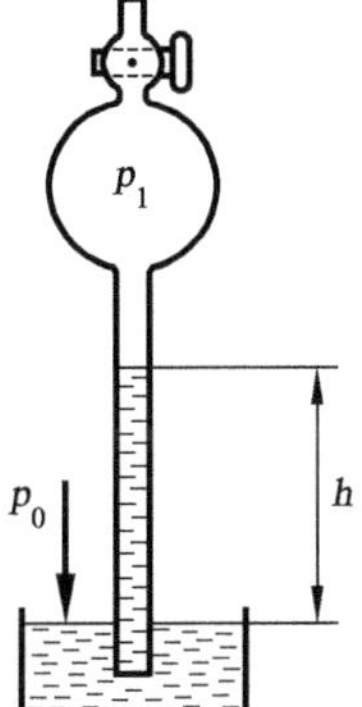

Fig. 2.11. Barometer

gauge comes from *Pascal.* The word (derived from the Greek barys, meaning heavy) indicates that the weight of the air column above the liquid is what is measured.

At this point we mention another unit of pressure based on the barometer, the *physical atmosphere.* The mean level of a barometer at sea level is about 760 mm Hg. It has been agreed to define this barometer level at 0 °C mercury temperature as the normal state of the atmosphere and to call the associated air pressure one "physical atmosphere". The qualifier "physical" is used because the technical atmosphere used by engineers is equal to 1 kp/cm^2. Since the specific weight of mercury at 0 °C is equal to 13.595 p/cm^3 and 1 cm^3 therefore weighs 13.595 p, a mercury column of 76 cm therefore corresponds to a pressure of

$$76\ \text{cm} \cdot 13.595\ \text{p/cm}^3 = 1033.2\ \text{p/cm}^2 = 1.0132 \cdot 10^5\ \text{Pa}.$$

This pressure also corresponds to a water column of height 10.332 m (water barometer). The suction height of pumps must therefore be lower than this value.

Since the force of gravity plays a role in the definition of the physical atmosphere, and this does not have the same value at all positions on Earth, for greater precision in the definition of pressure units a particular value of the acceleration due to gravity g has been chosen. The value 980.665 cm/s^2 has been determined as the normal value of gravitational acceleration at the 45th degree of latitude at sea level. For a different acceleration due to gravity g, the pressure of the normal atmosphere is $(1.0332 \cdot 980.665)/g$ local kiloponds per square centimeter. To get away from this somewhat arbitrary setting, a pressure unit was introduced to the CGS system: one million times the pressure unit 1 dyn/cm^2 is called the bar. At the normal value of gravitational acceleration, one bar corresponds to a mercury column of height 750.06 mm.

2.7 Equilibrium in Other Force Fields

In Sections 2.3 to 2.6, a homogeneous gravitational field was used; i.e. the acceleration due to gravity was assumed to be everywhere equally strong and orientated in the same direction. This assumption suffices for most applications. However, if we consider regions of Earth that are no longer small compared to Earth's radius, the variations of the acceleration due to gravity in its magnitude and direction have to be taken into account. For a liquid at rest relative to a uniformly rotating container, in addition to the acceleration due to gravity, the centrifugal acceleration also has to be considered. In what follows we consider the quite general question of the equilibrium of a homogeneous or inhomogeneous liquid in a general force field, whose force per unit mass (i.e. acceleration) varies in strength and direction from place to place.

The considerations for a general force field lead directly from the ideas in Section 2.3. It follows from this section that the pressure cannot change in every direction perpendicular to the force field at hand (equilibrium of a small prism according to Figure 2.4 with the axis perpendicular to the direction of the force). Condensing all directions perpendicular to the force direction to one point, the pressure on the surface element perpendicular to the force direction must be constant. For the case in which the adjoining surface elements can be integrated into one finite surface, i.e. when the force field has *normal surfaces*, the pressure is constant along all such normal surfaces. If a force field has no normal surface, then equilibrium is not possible in a liquid in this force field.

In contrast to the previous sections, where g denoted the strength of the gravitational field of the Earth, g will now denote the strength of a general force field. From the equilibrium at a small prism as in Figure 2.6 with height $\mathrm{d}h$ parallel to the force direction and pressure increase $\mathrm{d}p$, we find that the pressure in the direction of the force increases according to the equation

$$\mathrm{d}p = g \cdot \rho \cdot \mathrm{d}h. \tag{2.15}$$

In the discussion below, we assume that the force field has normal surfaces. We consider two such normal surfaces with pressures p and $p + \mathrm{d}p$. At two positions 1 and 2 in Figure 2.12, according to (2.15) we have on the one hand $\mathrm{d}p = g_1 \cdot \rho_1 \cdot \mathrm{d}h_1$, and on the other hand $\mathrm{d}p = g_2 \cdot \rho_2 \cdot \mathrm{d}h_2$. If ρ is either constant or a function of p (homogeneous liquid or homogeneous gas, cf. Sections 2.3 and 2.5), then $p_1 = p_2$ and $\rho_1 = \rho_2$. This yields $g_1 \cdot \mathrm{d}h_1 = g_2 \cdot \mathrm{d}h_2$, where $g \cdot \mathrm{d}h$ is the work done by the force in the transition from one normal surface to the other. This work has the same value at all positions between the normal surfaces. The force field has a potential. The normal surfaces are therefore surfaces of constant potential. Introducing the potential U at a point with the equation

$$\mathrm{d}U = -g \cdot \mathrm{d}h \tag{2.16}$$

(the minus sign because in (2.15) $\mathrm{d}h$ in the direction of g is assumed positive), we obtain

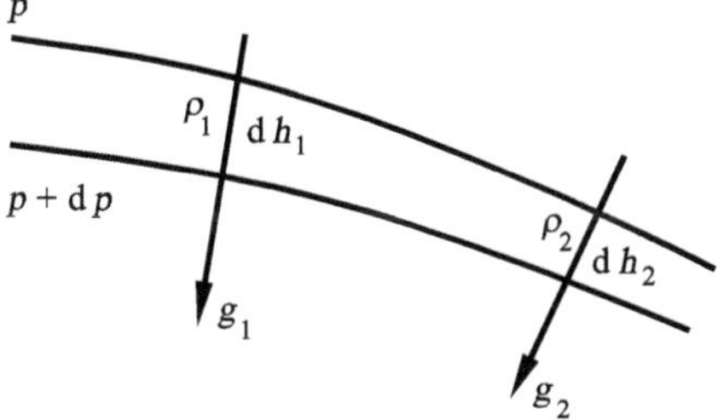

Fig. 2.12. Normal surfaces of a force field

$$\mathrm{d}p = -\rho \cdot \mathrm{d}U, \qquad \text{or} \qquad \mathrm{d}U = -\frac{\mathrm{d}p}{\rho}. \tag{2.17}$$

This yields the potential difference between two points A and B:

$$U_\mathrm{A} - U_\mathrm{B} = \int_\mathrm{A}^\mathrm{B} \frac{\mathrm{d}p}{\rho}. \tag{2.18}$$

In the case of a homogeneous liquid or a homogeneous gas assumed here, the right-hand side can be computed, and we obtain the pressure directly as a function of the potential. These results may be summarized as follows:

In the case of a homogeneous liquid or a homogeneous gas, equilibrium is possible only if the force field has a potential. The surfaces of constant potential that lie perpendicular to the force are simultaneously surfaces of constant pressure. The pressure increases in the direction of the force. We have $\mathrm{d}p = -\rho \cdot \mathrm{d}U$.

For an inhomogeneous liquid, it may happen that although $g_1 \cdot \mathrm{d}h_1$ is not equal to $g_2 \cdot \mathrm{d}h_2$, by suitable distribution of the density, we still have

$$\rho_1 \cdot g_1 \cdot \mathrm{d}h_1 = \rho_2 \cdot g_2 \cdot \mathrm{d}h_2.$$

It is seen that the equilibrium is unstable, as if the liquid were displaced along the normal surface, an action requiring no work, the distribution of the density would be changed and the equilibrium disturbed. Therefore, if we want to restrict ourselves to stable states, we may consider only force fields that have a potential. However if $g_1 \cdot \mathrm{d}h_1$ is equal to $g_2 \cdot \mathrm{d}h_2$, for equilibrium to exist we must have $\rho_1 = \rho_2$. Therefore, we can make the following assertion:

A stable state of an inhomogeneous liquid is possible only if the force field has a potential. The surfaces of constant potential are simultaneously surfaces of constant pressure and constant density.

Equations (2.17) and (2.18) may therefore be applied here too. The conditions for stability of the stratification are the same as those discussed for the homogeneous gravitational field in Sections 2.3 and 2.5.

Apart from magnetic force fields, the force fields that occur in physics almost always have a potential. However, the demand that the density ρ be constant on all surfaces of constant potential is of importance. This condition can be violated if the liquid or gas is locally heated, with a reduction in density at that region. In this case equilibrium is no longer possible, and the heated fluid and its surroundings are set into motion. This process comes to rest only if the warmer parts lie above the colder layers, and so the condition of constant density on surfaces of constant potential is again satisfied.

The free surface of a liquid or the interface between two immiscible liquids of different densities always follows a surface of constant potential. For this reason, surfaces of constant potential (*equipotential surfaces*) are also called *level surfaces* (free surface or level of an imaginary liquid). In surveying, the surface of the sea forms the level surface to which all heights are referred.

The discussions above will now be clarified in a simple example. Inside a container rotating uniformly about a vertical axis is a homogeneous liquid that is at rest relative to the rotating motion. We consider the equilibrium of this liquid. We first determine an expression for the potential, which is additively made up of a part due to gravity and a part due to the centrifugal force. Using cylindrical coordinates r and z (see Figure 2.13), we see that the contribution to the potential from gravity is $U_1 = U_0 + g \cdot z$, where g is the acceleration due to gravity and U_0 an arbitrarily chosen starting potential. In order to determine the contribution to the potential from the centrifugal force, we note that the strength of the centrifugal force field is $\omega^2 \cdot r$, where ω is the angular velocity with which the container and the liquid both rotate. Integrating in the direction of the centrifugal acceleration, i.e. in the direction of r, we obtain the second contribution to the potential:

$$U_2 = -\frac{\omega^2 \cdot r^2}{2}.$$

This yields the potential at a point of the liquid:

$$U = U_1 + U_2 = U_0 + g \cdot z - \frac{\omega^2 \cdot r^2}{2}.$$

The equipotential surfaces are found with the condition $U = \text{const}$:

$$z = \text{const} + \frac{\omega^2 \cdot r^2}{2 \cdot g}.$$

The free surfaces and all surfaces of equal pressure are paraboloids with the same parameter g/ω^2. Integration of (2.17) leads to the relation $p = p_0 - \rho \cdot U$ for the pressure. With $\rho \cdot g = \gamma$ we obtain

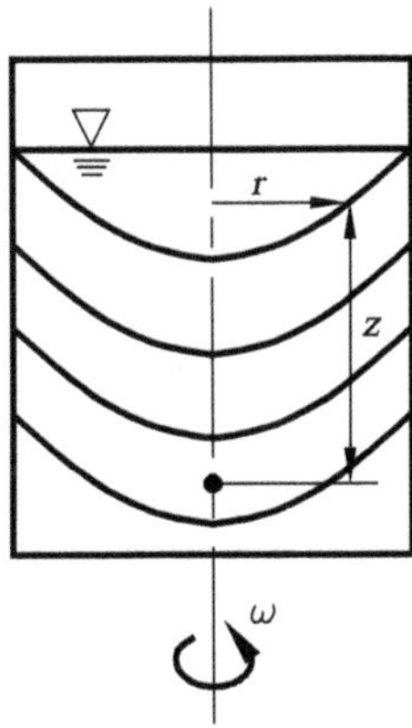

Fig. 2.13. Liquid in a rotating container

$$p = \text{const} + \gamma \cdot \left(-z + \frac{\omega^2 \cdot r^2}{2 \cdot g} \right).$$

2.8 Surface Stress (Capillarity)

Free surfaces of liquids tend to shrink and form *minimal surfaces*. This behavior can be explained with a stress state in the surface that would be taken on by a uniformly stretched thin skin. The origin of this tendency to shrink is as follows: Each liquid molecule close to the surface is pulled into the interior of the liquid by the attraction of the neighboring molecules (intermolecular forces). Because of this, only as many molecules as are absolutely necessary to form the surface remain on the surface. The same behavior is also found on interfaces between two liquids that do not mix. The stress that keeps the surface in equilibrium is called *surface stress*. On flat interfaces the surface stress causes no pressure differences, since the resulting surface stress force is equal to zero. At curved surfaces pressure differences are necessary to establish equilibrium. We consider a small rectangle of a curved surface with sides of length $\mathrm{d}s_1$ and $\mathrm{d}s_2$ (see Figure 2.14). The pressure difference $p_1 - p_2$ on the surface $\mathrm{d}s_1 \cdot \mathrm{d}s_2$ leads to a force $(p_1 - p_2) \cdot \mathrm{d}s_1 \cdot \mathrm{d}s_2$. The surface stress is the force per unit length that keeps the surface in equilibrium. It has the magnitude C (C = capillary constant). Therefore, on the four edges of the rectangle we obtain two forces $C \cdot \mathrm{d}s_1$ on the sides $\mathrm{d}s_1$ and two forces $C \cdot \mathrm{d}s_2$ on the sides $\mathrm{d}s_2$. The two forces on the sides $\mathrm{d}s_2$ are at an angle $\mathrm{d}\alpha = \mathrm{d}s_1/R_1$ to each other. This leads to a resultant $C \cdot \mathrm{d}s_2 \cdot \mathrm{d}\alpha = C \cdot \mathrm{d}s_2 \cdot \mathrm{d}s_1/R_1$. The two other forces, which form the angle $\mathrm{d}\beta = \mathrm{d}s_2/R_2$, yield a resultant $C \cdot \mathrm{d}s_1 \cdot \mathrm{d}s_2/R_2$.

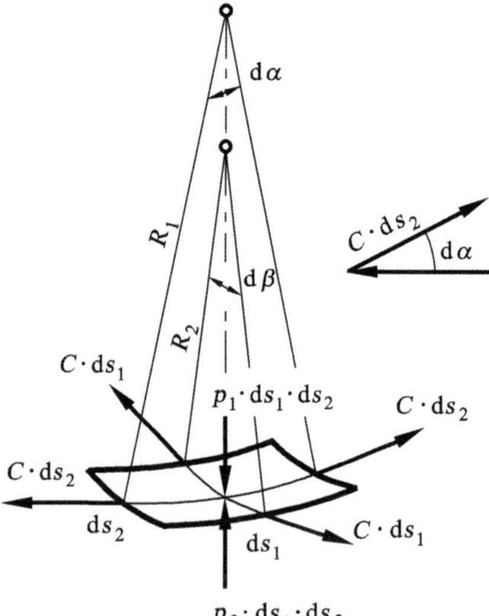

Fig. 2.14. Surface stress and pressure on a curved liquid surface

From the equilibrium of the three forces we obtain

$$p_1 - p_2 = C \cdot \left(\frac{1}{R_1} + \frac{1}{R_2} \right) . \tag{2.19}$$

As seen in Figure 2.14, R_1 and R_2 are the radii of curvature of the curves of section of the surface with two orthogonal planes perpendicular to the tangential plane. Equation (2.19) leads to the geometric relation that the sum $1/R_1 + 1/R_2$ is independent of the direction, since the pressure difference $p_1 - p_2$ does not depend on the direction.

In *liquids* that are in equilibrium, the pressure dependent on the specific weight varies with height, according to the law $p = p_0 - \gamma \cdot z$. Therefore, at the interface of two liquids with specific weights γ_1 and γ_2, we find that the associated pressures are $p_1 = p_0 - \gamma_1 \cdot z$ and $p_2 = p_0 - \gamma_2 \cdot z$. With (2.19) we then obtain the relation between the curvature and the height at the interface:

$$\frac{1}{R_1} + \frac{1}{R_2} = \frac{\gamma_2 - \gamma_1}{C} \cdot z . \tag{2.20}$$

Figure 2.15 shows two examples of such surfaces. The capillary constant C can be determined by measurement of the geometries occurring.

It can be seen from (2.20) that for very small differences in the specific weights, we find an n-fold geometrically similar increase in the different surface forms (R_1, R_2 and z are n times as large) if the term $(\gamma_2 - \gamma_1)/C$ is reduced by the factor $1/n^2$. For $\gamma_2 = \gamma_1$ the effect of gravity vanishes. These surfaces are the so-called minimal surfaces. If for $\gamma_2 - \gamma_1 \to 0$ we simultaneously set the plane $z = 0$ at infinity, we find from (2.20) that $1/R_1 + 1/R_2$ is constant. This result yields minimal surfaces with a given volume content, the simplest example of which is the sphere. These minimal surfaces may be obtained experimentally using soap films. In the interior of spherically shaped soap bubbles is an overpressure of magnitude $p_1 - p_2 = 4 \cdot C/R$ (There are two surfaces of the soap solution in air to be

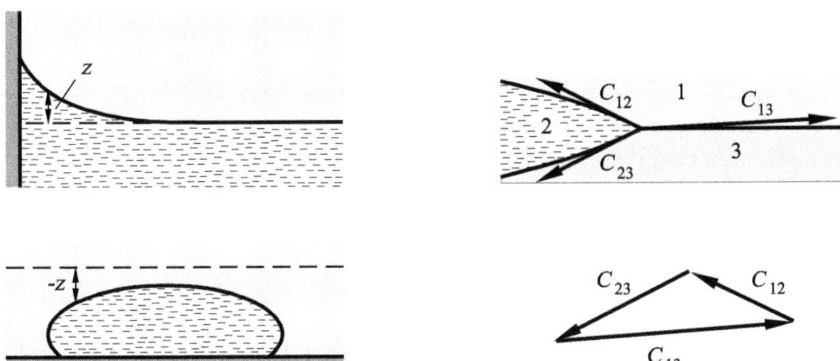

Fig. 2.15. Capillary surfaces of a liquid

Fig. 2.16. Equilibrium of three surface stresses

taken into account, which is why the factor $2 \cdot C$ instead of C is used in (2.19).)

If three liquids meet along an edge, the balance of forces of the three surface stresses C_{12}, C_{13}, and C_{23} yields certain angles at which the three interfaces join (see Figure 2.16). It may happen that C_{13} is larger than the sum of C_{12} and C_{23}. In this case no equilibrium is possible. For example, this happens when air, mineral oil, and water meet. The mineral oil then coats the entire surface, possibly with a very thin layer. This behavior is observed in the spreading of drops of motor oil on wet roads. If the oil is replaced by melted fat, this assumes the shape of flat lenses between the water and the air (globules of fat in soup). Figure 2.16 shows this case. If one of the three materials is solid, the balance of forces of the three surface stresses can be set up only with the components in the possible direction of displacement, parallel to the solid surface. Using the *wetting angle* α (see Figure 2.17), we obtain $C_{12} \cdot \cos(\alpha) + C_{23} = C_{13}$, i.e.

$$\cos(\alpha) = \frac{C_{13} - C_{23}}{C_{12}}. \tag{2.21}$$

If C_{12} (surface stress at the interface of the two liquids 1 and 2) is already known and α is measured, we can obtain the difference $C_{13} - C_{23}$. However, C_{13} and C_{23} cannot be individually determined. If the difference is negative, the angle α is greater than $\pi/2$ as with, for example, air, mercury, and glass. The lower picture in Figure 2.15 shows such a drop of mercury. The case $C_{13} - C_{23} > C_{12}$ may also occur. Then the entire solid body is coated by liquid 2. This occurs in the case of petroleum. Liquids are observed to rise considerably in narrow tubes. If r is the inner radius of the tube, then, simplifying the liquid surface as a spherical shell (r small compared to h), we see from Figure 2.18 that the spherical radius is $R = r/\cos(\alpha)$, with the wetting angle α. Therefore, according to (2.20), we obtain

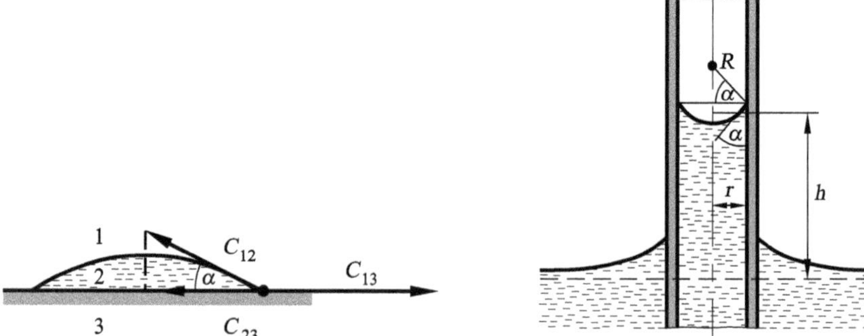

Fig. 2.17. Wetting angle on a solid surface

Fig. 2.18. Capillary rise in a tube

$$h = \frac{2 \cdot C_{12}}{\gamma_2 - \gamma_1} \cdot \frac{\cos(\alpha)}{r}. \tag{2.22}$$

The height h can become very large if r is very small (suction effect of blotting paper, fine clay, etc.).

In (2.22) we can eliminate $\cos(\alpha)$ using (2.21) and multiply both sides by $\pi \cdot r^2 \cdot (\gamma_2 - \gamma_1)$. This yields the equation

$$(\gamma_2 - \gamma_1) \cdot \pi \cdot r^2 \cdot h = (C_{13} - C_{23}) \cdot 2 \cdot \pi \cdot r.$$

The weight of the column of liquid, reduced by its lift, is equal to the resulting tensile force on the tube wall. If the *tensile force* is negative, i.e., $\alpha > \pi/2$ as in the case of mercury, h becomes negative (Figure 2.18 reflected in the horizontal plane). For wetted surfaces $C_{13} - C_{23}$ may be replaced by C_{12}. Then $\cos(\alpha) = 1$; i.e., $\alpha = 0$. This yields the maximum value of h. On measurement of h and r we obtain

$$C_{12} = \frac{1}{2} \cdot (\gamma_2 - \gamma_1) \cdot h \cdot r.$$

Another method of determining C_{12} is the measurement of capillary waves, to be discussed in Section 4.1.8.

Values of C_{12} at 20°C:	water to air	0.073 N/m,
	oil to air	0.025 to 0.030 N/m,
	mercury to air	0.472 N/m.

2.9 Problems

2.1

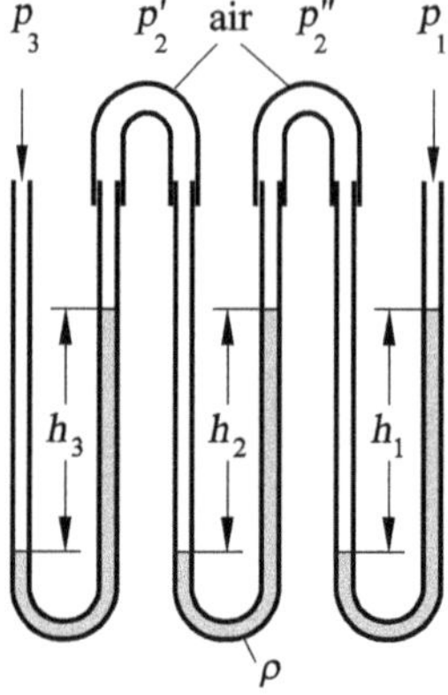

Three identical U-tubes are connected in a row. In each U-tube is a liquid with density ρ. The levels of the liquids show the height differences h_1, h_2, and h_3. The effect of gravity on the air may be neglected. How great is the pressure difference $\Delta p = p_3 - p_1$ between the free ends of the first and third tubes?

$$\Delta p = p_3 - p_1 = \rho \cdot g \cdot (h_1 + h_2 + h_3) .$$

2.2

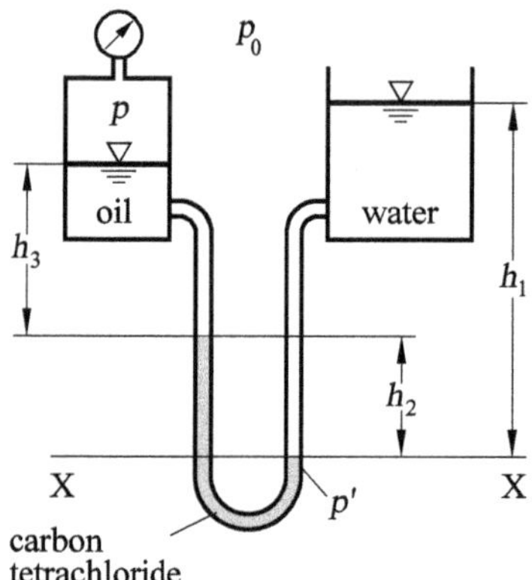

An open water container and a container that is closed to the atmosphere by a manometer are connected by a U-tube, whose lower part is filled with carbon tetrachloride (tet) (CCl_4). The height of the water column (density of water $\rho_w = 1000$ km/m^3) is $h_1 = 0.4$ m, the column of oil (density of oil $\rho_{\text{oil}} = 950$ kg/m^3) has the height $h_3 = 0.13$ m, and the height h_2 of the CCl_4 column is $h_2 = 0.1$ m.

What is the density ρ_{tet} of the CCl_4 filling if an excess pressure compared to the atmospheric pressure of 1200 N/m^2 is read from the manometer?

$$\rho_{\text{tet}} = 1541.76 \text{ kg/m}^3 .$$

2.3

The pressure p_0 and the temperature T_0 are known for the atmosphere at sea level $z = 0$ (specific gas constant of air $R = 287\ \mathrm{m^2/(s^2 \cdot K)}$, $p_0 = 101300\ \mathrm{N/m^2}$, $T_0 = 283$ K).

(a) Assuming that the state of the gas in the atmosphere changes isothermally, determine the dependence of the pressure and the density of the atmosphere on the height z.

$$p = p_0 \cdot e^{-\frac{z}{H_0}}, \quad \rho = \rho_0 \cdot e^{-\frac{z}{H_0}}, \quad H_0 = \frac{R \cdot T_0}{g} \quad .$$

(b) Assuming that the state of the gas in the atmosphere changes polytropically, determine the dependence of the pressure and the density of the atmosphere on the height z:

$$\frac{p}{p_0} = \left(\frac{\rho}{\rho_0}\right)^n, \quad \frac{p}{p_0} = \left(1 - \frac{n-1}{n} \cdot \frac{z}{H_0}\right)^{\frac{n}{n-1}}, \quad \frac{\rho}{\rho_0} = \left(1 - \frac{n-1}{n} \cdot \frac{z}{H_0}\right)^{\frac{1}{n-1}} .$$

2.4

A balloon is suspended in an isothermal atmosphere (air pressure on the ground $p_0 = 1.013$ bar, air density on the ground $\rho_0 = 1.225\ \mathrm{kg/m^3}$) at a height $z_0 = 500$ m. How far will the balloon sink if a change in the weather causes the air density on the ground to change to $\rho_0' = 1.0\ \mathrm{kg/m^3}$ while the air pressure remains the same? The volume V of the balloon is not to change as the height varies.

$$z_x = H_0' \cdot \left[\ln\left(\frac{\rho_0'}{\rho_0}\right) + \frac{z_0}{H_0}\right], \quad \Delta z = 272.41\ \mathrm{m}.$$

2.5

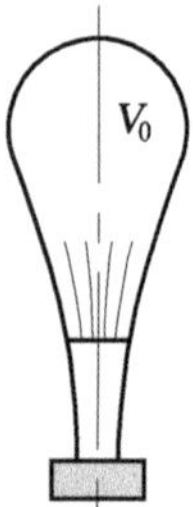

on the ground

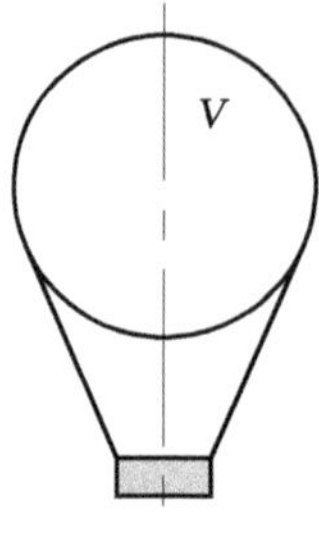

at a height
above the ground

A stratospheric balloon is partially filled with the buoyant gas hydrogen H_2 on the ground. As the balloon rises, it inflates with an increase in volume of the filling. This leads to an additional lift. On the ground, the balloon has volume $V_0 = 450\ \text{m}^3$, while its maximum volume is $V_1 = 1400\ \text{m}^3$.

(a) What is the greatest possible weight of the load G_{max} that can be lifted (the balloon itself is part of the weight, but the buoyant gas is not) if the stratospheric balloon is to reach a maximum height of $z_{max} = 1.2$ km in a polytropic atmosphere? On the ground, the air pressure is $p_0 = 1.013$ bar and the air density is $\rho_0 = 1.234\ \text{kg/m}^3$. The density of hydrogen in the balloon has the value $\rho_{H_2,0} = 0.087\ \text{kg/m}^3$ on the ground. The temperature $T_{1km} = 280$ K at an altitude of 1 km, and the specific gas constant of the air $R = 287\ \text{m}^2/(\text{s}^2 \cdot \text{K})$ are also known.

$$G_{max} = 3955.8\ \text{N}.$$

(b) At what height z_1 does the balloon reach its largest volume $V_1 = 1400\ \text{m}^3$? Until the maximum volume is reached, the hydrogen in the balloon is to have the same temperature and pressure as the atmosphere at all heights.

$$z_1 = H_0 \cdot \frac{n}{n-1} \cdot \left[1 - \left(\frac{V_0}{V_1}\right)^{n-1}\right], \quad z_1 = 10224.1\ \text{m}.$$

2.6

A number of small solids are moving on the surface of a liquid. Show that the surface stress causes the solids to move toward each other, whether they are wetted by the liquid or not. They move away from each other if one solid is wetted and the other is not wetted by the liquid.

2.7

How much work W must be done to atomize a volume V of liquid into spherically shaped droplets of radius R? The surface energy of the volume V before the atomization is assumed to be negligible.

$$W = \frac{3C}{R} \cdot V.$$

3. Kinematics of Fluid Flow

The flows of liquids and gases have so much in common that it is practical to treat them together. In contrast to liquids, gases are compressible. However, whether the compressibility is important depends on the *flow process under consideration*. At small velocities and for moderate height dimensions of the gas, the pressure changes remain small compared to the mean pressure. The volume changes are then so small that they can be neglected. Gas flows are then no different from flows of incompressible liquids. If we neglect volume changes of 1%, we can apply the equations for incompressible flows to flows in the atmosphere at mean temperatures. This remains the case for velocities of up to 50 m/s and for height dimensions of up to 100 m (cf. Sections 2.5 and 4.1.2). At flow velocities of 150 m/s, the volume changes are about 10%. If the flow velocities reach the magnitude of the velocity of sound (about 340 m/s), the volume changes become so large that the flow is greatly affected by them. At flow velocities that are greater than the velocity of sound, the flow has a completely different character from that of an incompressible liquid.

In this chapter we mainly consider incompressible flows. In order not to have to speak of *liquids* and *gases*, we use the word *fluid* as a collective term for liquids and gases. For the purposes of this usage, gases are referred to as compressible fluids (Section 4.3).

The *kinematics* of a *flow* describes the motion of the fluid without taking into account the forces that cause this motion. The goal of kinematics is to describe the dependence of the motion of the fluid elements on time for a given velocity field.

3.1 Methods of Representation

The flow of a fluid can be described by determining the position of every fluid particle at every point in time. A particle's change of position in time then yields its velocity and acceleration. To distinguish between the different particles, we mathematically introduce a particular coordinate system, fixed to the fluid particles but moving in space. We first consider a family of surfaces with $a =$ const, where a is given as some initial position as a function of the spatial coordinates x, y, and z. We select two further families of surfaces $b =$ const and $c =$ const such that a surface with $a =$ const, a surface with

H. Oertel (ed.), *Prandtl-Essentials of Fluid Mechanics*,
Applied Mathematical Sciences 158, DOI 10.1007/978-1-4419-1564-1_3,

$b =$ const, and a surface with $c =$ const meet only at a single point. A fluid particle at this point of intersection is then fully defined by the values of a, b, and c at a fixed but arbitrary time. A fluid particle retains these *fluid coordinates* a, b, and c as its initial or rest position throughout its motion. This means that each surface $a =$ const, $b =$ const, or $c =$ const as an initial position is always made up of the same fluid particles. The original choice of the fluid coordinates is arbitrary and is determined only by practical considerations. For example, Cartesian coordinates may be chosen in some initial or rest position as the fluid coordinates. The paths of the fluid particles in the flow are called *particle paths.*

Another manner of describing flows is by means of *streaklines.* These are the lines connecting all positions reached by the particle paths of all particles that passed through a single point in the flow field at a given point in time. In an experiment, a certain point in the flow field can be defined by color or smoke. Snapshots of the color or smoke filaments are then streaklines.

In order to determine the motion, i.e. the change in position of all fluid particles, the values of the current position coordinates x, y, z of the particles have to be stated as functions of time and of the fluid coordinates a, b, c of the initial position of the particle. We obtain

$$x = \mathrm{F}_1(a,b,c,t), \qquad y = \mathrm{F}_2(a,b,c,t), \qquad z = \mathrm{F}_3(a,b,c,t). \tag{3.1}$$

To fully describe the state of the flowing fluid, we need to know the pressure p and, in the case of a compressible flow, the density ρ. In general, we use a simpler representation that describes the flow state at every position and time more closely, without having to consider each individual particle. If the flow is *steady*, it is sufficient to state the magnitude and direction of the velocity at each position in the space through which the fluid flows, and to make corresponding statements about the pressure and, if necessary, the density. However, if the flow changes in time, this information is necessary for the unsteady flow at all times. Mathematically, we state the three orthogonal velocity components u, v, w (and the pressure p and the density ρ, if necessary) as functions of the spatial coordinates x, y, z and the time t. For u, v, w we obtain the equations

$$u = \mathrm{f}_1(x,y,z,t), \qquad v = \mathrm{f}_2(x,y,z,t), \qquad w = \mathrm{f}_3(x,y,z,t). \tag{3.2}$$

The system of equations (3.1) is named for *Lagrange* (fluid particle reference frame), and the system (3.2) for *Euler* (spatially fixed reference frame), although both systems were known to *Euler*. The systems of equations (3.1) and (3.2) are called the *fundamental equations of kinematics.* For the calculation of a fluid particle path, the three equations

$$\mathrm{d}x = u \cdot \mathrm{d}t, \qquad \mathrm{d}y = v \cdot \mathrm{d}t, \qquad \mathrm{d}z = w \cdot \mathrm{d}t \tag{3.3}$$

have to be integrated, using the system of equations (3.2). Since the three constants of integration may be directly interpreted as the fluid coordinates a, b, c, we again obtain the system of equations (3.1).

For *another representation* of the *instantaneous state of the flow* of a fluid, *streamlines* are used. These run in the direction of the flow at all points; i.e. their tangents everywhere have the direction of the velocity vector.

The differential equations of the *streamlines* read

$$\frac{\mathrm{d}z}{\mathrm{d}y} = \frac{w}{v}, \qquad \frac{\mathrm{d}z}{\mathrm{d}x} = \frac{w}{u}, \qquad \frac{\mathrm{d}y}{\mathrm{d}x} = \frac{v}{u}. \tag{3.4}$$

In a steady flow, the streamlines are the same as the paths of the fluid particles. This is not the case in an unsteady flow, since the streamlines provide an illustration of the *instantaneous* velocity directions, while the particle paths illustrate the velocity directions held *by one particle over time.* Streamlines of a single flow, just like *pathlines*, look completely different if the reference frame is changed. For example, if the observer of the motion of a body through a fluid is at rest relative to the undisturbed fluid, or if the observer moves with the body such that the body is at rest and the fluid flows toward it, then two quite different streamline portraits will be seen.

Streamlines can be made visible by sprinkling small particles onto the surface of the fluid or mixing them in with the fluid. These particles then follow the motion of the fluid. In snapshot with a short exposure time, each particle generates a short dash on the film. If the sprinkled particles are dense enough, these dashes provide a streamline portrait. A picture of the pathlines is found if a long exposure time is used and the number of sprinkled particles is small. Figures 3.1 and 3.2 show simultaneous shots of the motion of a plate through a fluid at rest in two different reference frames. Figure 3.1 was taken

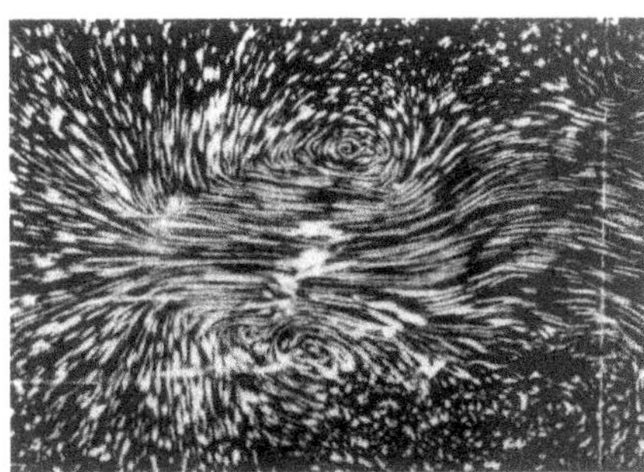

Fig. 3.1. Flow past a moving plate, camera at rest. The path of the plate can be seen from the tracks of the side walls, *F. Ahlborn* 1909

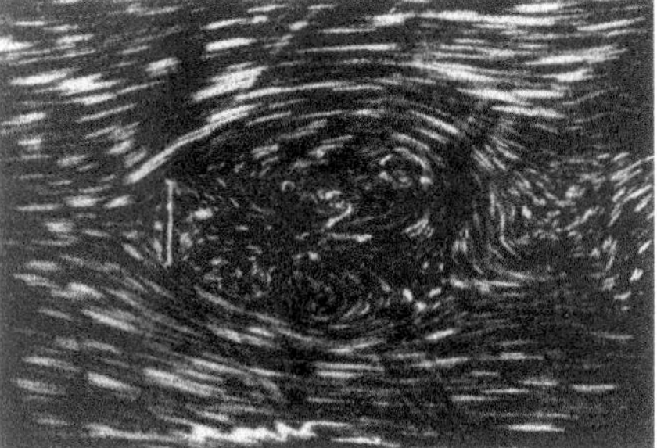

Fig. 3.2. Flow past a moving plate, camera moving with the plate, *F. Ahlborn* 1909

by a camera at rest, while Figure 3.2 was taken by a camera moving with the plate. The shots are by *F. Ahlborn*, 1909. Club moss was used to make the flow visible.

A further example of an unsteady flow is shown in Figure 3.3. This shows the streaklines, particle paths, and streamlines of the periodic vortex separation of a cylinder moving with constant velocity U_∞ through a fluid at rest. The first three flow portraits of the so-called Kármán vortex street (see also Figure 4.91) are shown for an observer at rest. The periodically separating vortices move past the observer with velocity c. An observer moving with the vortices in the very same flow sees a completely different portrait, observing streamlines similar to cat's eyes.

If the velocity field is continuous everywhere, on taking a streamline through all points on a small closed curve, we can form a tube. This has the particular property that by definition, at the time under consideration the fluid inside it flows parallel to the streamlines, as in a solid tube. If the fluid were to flow through the wall of the tube, this would assume that a velocity component is perpendicular to the wall, i.e. perpendicular to the streamlines, thereby contradicting their definition. Such tubes are called *stream tubes*, and their contents are called *stream filaments*. In steady flows, the stream tubes

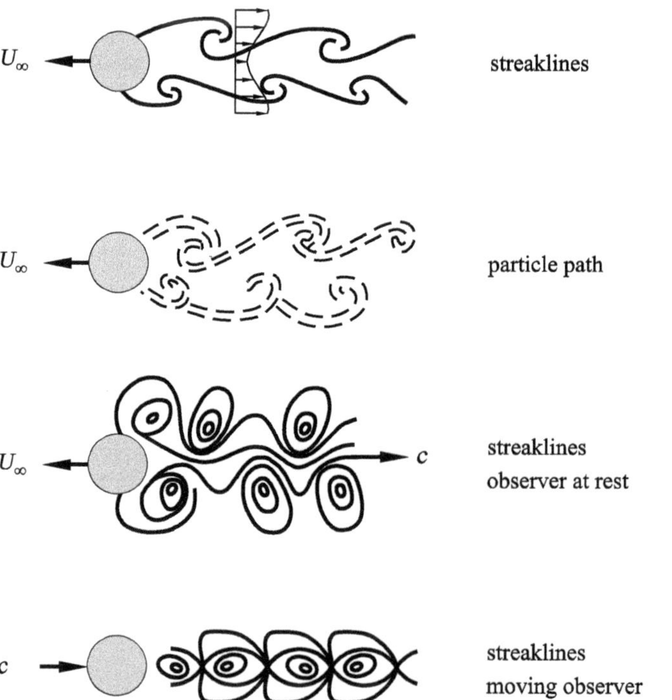

Fig. 3.3. Kármán vortex street, observer at rest and moving observer

do not change, and the fluid particles inside them flow as if in a solid tube. In contrast, in general, in unsteady flows at a later instant in time different particles from those earlier are joined together in stream tubes. We can imagine the entire space filled with the fluid divided up into such stream tubes, and hence obtain a vivid picture of the fluid flow.

In many simpler cases of flows, particularly flows through pipes and channels, it is permissible to consider the entire space filled with the fluid as a single stream filament. It is then not the different velocities in a cross-section that are of interest, but only the mean velocity, which we can calculate. This idea is used by engineers in practical calculations (see Section 4.1). The representation of the change in flow quantities along a stream filament permits the development of the one-dimensional theory of flows.

3.2 Acceleration of a Flow

In the last section we saw that the flow picture is dependent on the frame of reference. We now consider the two different ways of treating a flow mathematically. In the *Euler picture* we assume a *fixed* observer. This manner of description corresponds to using a measuring apparatus that is fixed in position to measure local flow quantities, and will be used exclusively in deriving the fluid-mechanical fundamental equations in the following chapters.

The *Lagrange picture* assumes a frame of reference moving with a particle or fluid element. The mathematical relationship between the two pictures is, for example for the acceleration of the flow $\boldsymbol{a} = \mathrm{d}\boldsymbol{v}/\mathrm{d}t = \mathrm{d}^2\boldsymbol{x}/\mathrm{d}t^2$, the *total differential* of the given velocity vector $\boldsymbol{v} = (u, v, w)$. For the u component $u(x, y, z, t)$ of the velocity vector we have

$$\mathrm{d}u = \frac{\partial u}{\partial t} \cdot \mathrm{d}t + \frac{\partial u}{\partial x} \cdot \mathrm{d}x + \frac{\partial u}{\partial y} \cdot \mathrm{d}y + \frac{\partial u}{\partial z} \cdot \mathrm{d}z.$$

So the total time derivative of u is

$$\frac{\mathrm{d}u}{\mathrm{d}t} = \frac{\partial u}{\partial t} + \frac{\partial u}{\partial x} \cdot \frac{\mathrm{d}x}{\mathrm{d}t} + \frac{\partial u}{\partial y} \cdot \frac{\mathrm{d}y}{\mathrm{d}t} + \frac{\partial u}{\partial z} \cdot \frac{\mathrm{d}z}{\mathrm{d}t},$$

with

$$\frac{\mathrm{d}x}{\mathrm{d}t} = u, \qquad \frac{\mathrm{d}y}{\mathrm{d}t} = v, \qquad \frac{\mathrm{d}z}{\mathrm{d}t} = w,$$

from which we obtain

$$\underbrace{\frac{\mathrm{d}u}{\mathrm{d}t}}_{\mathrm{S}} = \underbrace{\frac{\partial u}{\partial t}}_{\mathrm{L}} + \underbrace{u \cdot \frac{\partial u}{\partial x} + v \cdot \frac{\partial u}{\partial y} + w \cdot \frac{\partial u}{\partial z}}_{\mathrm{C}}, \tag{3.5}$$

where

S is the substantial rate of change (Lagrange picture),
L is the local rate of change at a fixed position (Euler picture),
and C is the convective spatial changes due to convection from place to place (effect of the velocity field $\boldsymbol{v} = (u, v, w)$).

For the acceleration $\boldsymbol{a}$ of the flow field, which we will need in the following chapters, we obtain

$$\boldsymbol{a} = \frac{\mathrm{d}\boldsymbol{v}}{\mathrm{d}t} = \frac{\partial \boldsymbol{v}}{\partial t} + u \cdot \frac{\partial \boldsymbol{v}}{\partial x} + v \cdot \frac{\partial \boldsymbol{v}}{\partial y} + w \cdot \frac{\partial \boldsymbol{v}}{\partial z} = \frac{\partial \boldsymbol{v}}{\partial t} + (\boldsymbol{v} \cdot \nabla)\boldsymbol{v}, \tag{3.6}$$

with the nabla operator $\nabla = (\partial/\partial x, \partial/\partial y, \partial/\partial z)$ and $(\boldsymbol{v} \cdot \nabla)$ the scalar product of the velocity vector $\boldsymbol{v}$ and the nabla operator ∇.

For Cartesian coordinates this yields

$$\boldsymbol{a} = \begin{pmatrix} a_x \\ a_y \\ a_z \end{pmatrix} = \begin{pmatrix} \frac{\mathrm{d}u}{\mathrm{d}t} \\ \frac{\mathrm{d}v}{\mathrm{d}t} \\ \frac{\mathrm{d}w}{\mathrm{d}t} \end{pmatrix} = \begin{pmatrix} \frac{\partial u}{\partial t} + u \cdot \frac{\partial u}{\partial x} + v \cdot \frac{\partial u}{\partial y} + w \cdot \frac{\partial u}{\partial z} \\ \frac{\partial v}{\partial t} + u \cdot \frac{\partial v}{\partial x} + v \cdot \frac{\partial v}{\partial y} + w \cdot \frac{\partial v}{\partial z} \\ \frac{\partial w}{\partial t} + u \cdot \frac{\partial w}{\partial x} + v \cdot \frac{\partial w}{\partial y} + w \cdot \frac{\partial w}{\partial z} \end{pmatrix},$$

and for $(\boldsymbol{v} \cdot \nabla)\boldsymbol{v}$,

$$\boldsymbol{v} \cdot \nabla = \begin{pmatrix} u \\ v \\ w \end{pmatrix} \cdot \begin{pmatrix} \frac{\partial}{\partial x} \\ \frac{\partial}{\partial y} \\ \frac{\partial}{\partial z} \end{pmatrix} = u \cdot \frac{\partial}{\partial x} + v \cdot \frac{\partial}{\partial y} + w \cdot \frac{\partial}{\partial z},$$

$$(\boldsymbol{v} \cdot \nabla)\boldsymbol{v} = \left(u \cdot \frac{\partial}{\partial x} + v \cdot \frac{\partial}{\partial y} + w \cdot \frac{\partial}{\partial z} \right) \begin{pmatrix} u \\ v \\ w \end{pmatrix}$$

$$= \begin{pmatrix} u \cdot \frac{\partial u}{\partial x} + v \cdot \frac{\partial u}{\partial y} + w \cdot \frac{\partial u}{\partial z} \\ u \cdot \frac{\partial v}{\partial x} + v \cdot \frac{\partial v}{\partial y} + w \cdot \frac{\partial v}{\partial z} \\ u \cdot \frac{\partial w}{\partial x} + v \cdot \frac{\partial w}{\partial y} + w \cdot \frac{\partial w}{\partial z} \end{pmatrix}.$$

In the case of a *steady flow*, all partial derivatives with respect to time vanish, so $\partial/\partial t = 0$, while the substantial derivative with respect to time $\mathrm{d}/\mathrm{d}t$ can indeed be nonzero when convective changes occur. In unsteady flows both $\partial/\partial t \neq 0$ and $\mathrm{d}/\mathrm{d}t \neq 0$ occur.

3.3 Topology of a Flow

The original text by Prandtl will now be supplemented by some conclusions drawn from the kinematic fundamental equations (3.2), which, besides the streamlines, particle paths, and streaklines, permit in particular an improved description of three-dimensional flows. Analysis of the topology of a flow serves to provide an understanding of the *critical points* (singularities) that are produced by the velocity vector field and their relations to each other. A critical point is characterized by the fact that the direction of the velocity vector is undetermined at that point. For the flow portrait in Figure 3.2 we use the terminology of critical points and obtain the description of the structure of the flow field (Figure 3.4) with two half-saddle points S′, the stagnation points of the flow, and a saddle point S that divides the backflow region of periodically separating vortices from the wake flow. In what follows, the vortices themselves will be called foci F. Following the description of Figure 3.2, as a moving observer we see a snapshot of foci (vortices) periodically swimming downstream from a plate in a perpendicular flow. Thus the unsteady wake flow is uniquely described in the moving reference frame.

The theory of *critical points* (x_0, y_0, z_0) of a steady flow takes the three-dimensional vector field $\boldsymbol{v}(x, y, z) = (u, v, w)$ as its starting point. We assume that this is continuous and twice differentiable.

At a critical point, the direction field of the vector quantity under consideration is undetermined. As we consider the velocity vector $\boldsymbol{v}$ in what follows, we mean that at a critical point the magnitude of the velocity vanishes and that in these points no direction is associated with the streamlines according to (3.4). Closer investigation of the space directly surrounding a critical point is possible, however, if the vector field can be approximated by a series expansion (3.7) about the singular point (x_0, y_0, z_0). Without loss of generality, we now assume that $(x_0, y_0, z_0) = (0, 0, 0)$. In critical points, the components of the velocity vector $\boldsymbol{v}$ are analytic functions of the spatial coordinates:

$$\begin{aligned} \dot{x} = u &= \sum_{\mathrm{i}=0}^{\mathrm{N}} \sum_{\mathrm{j}=0}^{\mathrm{N-i}} \sum_{\mathrm{k}=0}^{\mathrm{N-i-j}} U_{\mathrm{i,\,j,\,k}} \cdot x^{\mathrm{i}} \cdot y^{\mathrm{j}} \cdot z^{\mathrm{k}} + \mathrm{O}_1(\mathrm{N}+1), \\ \dot{y} = v &= \sum_{\mathrm{i}=0}^{\mathrm{N}} \sum_{\mathrm{j}=0}^{\mathrm{N-i}} \sum_{\mathrm{k}=0}^{\mathrm{N-i-j}} V_{\mathrm{i,\,j,\,k}} \cdot x^{\mathrm{i}} \cdot y^{\mathrm{j}} \cdot z^{\mathrm{k}} + \mathrm{O}_2(\mathrm{N}+1), \\ \dot{z} = w &= \sum_{\mathrm{i}=0}^{\mathrm{N}} \sum_{\mathrm{j}=0}^{\mathrm{N-i}} \sum_{\mathrm{k}=0}^{\mathrm{N-i-j}} W_{\mathrm{i,\,j,\,k}} \cdot x^{\mathrm{i}} \cdot y^{\mathrm{j}} \cdot z^{\mathrm{k}} + \mathrm{O}_3(\mathrm{N}+1), \end{aligned} \tag{3.7}$$

with

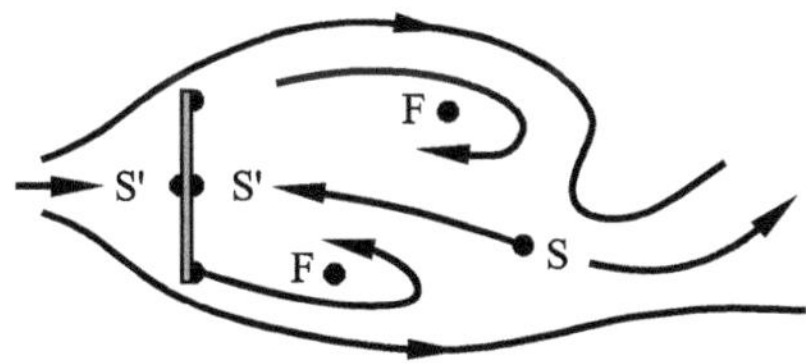

Fig. 3.4. Structure of the flow past a moving plate, snapshot in the moving reference frame

$$U_{\mathrm{i,\,j,\,k}} = \frac{1}{(\mathrm{i}+\mathrm{j}+\mathrm{k})!} \cdot \frac{\partial^{\mathrm{i}+\mathrm{j}+\mathrm{k}} u}{\partial x^{\mathrm{i}} \cdot \partial y^{\mathrm{j}} \cdot \partial z^{\mathrm{k}}},$$
$$V_{\mathrm{i,\,j,\,k}} = \frac{1}{(\mathrm{i}+\mathrm{j}+\mathrm{k})!} \cdot \frac{\partial^{\mathrm{i}+\mathrm{j}+\mathrm{k}} v}{\partial x^{\mathrm{i}} \cdot \partial y^{\mathrm{j}} \cdot \partial z^{\mathrm{k}}},$$
$$W_{\mathrm{i,\,j,\,k}} = \frac{1}{(\mathrm{i}+\mathrm{j}+\mathrm{k})!} \cdot \frac{\partial^{\mathrm{i}+\mathrm{j}+\mathrm{k}} w}{\partial x^{\mathrm{i}} \cdot \partial y^{\mathrm{j}} \cdot \partial z^{\mathrm{k}}},$$

where the $\mathrm{O_i}$ are error functions that are determined by terms of order $\mathrm{N}+1$.

We first consider the case of a critical point in the *free flow.* It suffices to carry out a series expansion from (3.7) up to order $\mathrm{N} = 1$. This leads to a system of first-order differential equations:

$$\dot{\boldsymbol{x}} = \mathbf{A} \cdot \boldsymbol{x}, \qquad \boldsymbol{x} = (x, y, z), \qquad \dot{\boldsymbol{x}} = \frac{\mathrm{d}\boldsymbol{x}}{\mathrm{d}t},$$

$$\begin{pmatrix} \dot{x} \\ \dot{y} \\ \dot{z} \end{pmatrix} = \begin{pmatrix} a_{11} & a_{12} & a_{13} \\ a_{21} & a_{22} & a_{23} \\ a_{31} & a_{32} & a_{33} \end{pmatrix} \cdot \begin{pmatrix} x \\ y \\ z \end{pmatrix}. \tag{3.8}$$

The coefficients a_{ij} are the components of the gradients of the velocity vector. In the general case, the trajectories of the system of equations (3.8) are the pathlines of the flow field, which are identical to the streamlines in the steady case.

To consider critical points on *solid walls* we now assume that the velocity $\boldsymbol{v}$ is given in coordinates normal to the wall, where z is the direction normal to the wall. In contrast to points in the free flow, the condition $\boldsymbol{v} = 0$ on a solid wall is no longer a sufficient criterion for the existence of a critical point, since the no-slip condition means that $\boldsymbol{v} = 0$ is identically satisfied there anyway. However, in identifying a critical point, lack of knowledge of the direction of the integral curves of the vector field is decisive. As the direction field of the velocity passes over to the direction field of the wall shear stress vector $\boldsymbol{\tau}_w$ in the limiting case of vanished distance z from the wall, $\boldsymbol{\tau}_w$ is now the relevant quantity. Therefore, critical points on the wall require the vanishing of the wall shear stress $\boldsymbol{\tau}_w$.

It follows from the no-slip condition that the quantity $\boldsymbol{v}/z$ tends toward a constant value for $z \to 0$ and that the vector field of this quantity has the same integral curves as the field of the wall shear stress.

It is therefore practical to avoid considering the critical character of the surface $z = 0$ and instead to consider the Taylor expansion of the quantity $\boldsymbol{v}/z$.

With $\boldsymbol{x}' = \dot{\boldsymbol{x}}/z$, (3.7) with $\mathrm{N} = 2$ leads to the following series expansion:

$$x' = \frac{u}{z} = U_{1,0,1} \cdot x + U_{0,1,1} \cdot y + U_{0,0,2} \cdot z + \mathrm{O}_1(3),$$
$$y' = \frac{v}{z} = V_{1,0,1} \cdot x + V_{0,1,1} \cdot y + V_{0,0,2} \cdot z + \mathrm{O}_2(3),$$
$$z' = \frac{w}{z} = W_{0,0,2} \cdot z + \mathrm{O}_3(3).$$

Because of the relation $U_{\mathrm{i,\,j,\,0}} = V_{\mathrm{i,\,j,\,0}} = W_{\mathrm{i,\,j,\,0}} = 0$, this expansion also takes the no-slip condition into account.

In contrast to (3.8), second-order derivatives of the velocity field now appear. If we restrict ourselves to the linear terms in the spatial directions x, y, and z, we obtain, in complete analogy to the free flow, again a system of first-order differential equations with a different matrix of coefficients $\mathbf{A}$:

$$\boldsymbol{x}' = \mathbf{A} \cdot \boldsymbol{x}, \qquad \begin{pmatrix} x' \\ y' \\ z' \end{pmatrix} = \begin{pmatrix} \frac{\dot{x}}{z} \\ \frac{\dot{y}}{z} \\ \frac{\dot{z}}{z} \end{pmatrix} = \begin{pmatrix} a_{11} & a_{12} & a_{13} \\ a_{21} & a_{22} & a_{23} \\ a_{31} & a_{32} & a_{33} \end{pmatrix} \cdot \begin{pmatrix} x \\ y \\ z \end{pmatrix}. \tag{3.9}$$

Classification of critical points in the given flow field has therefore been reduced to investigation of the singular points of ordinary differential equations with constant coefficients, whose mathematical theory is well understood. The difference between critical points in the free flow and those on solid walls is merely in the different matrices of coefficients $\mathbf{A}$ ((3.8) or (3.9)).

Calculating the eigenvalues of this matrix according to $\det[\mathbf{A} - \lambda \cdot \mathbf{I}] = 0$ leads to the characteristic polynomial

$$\lambda^3 + \mathrm{P} \cdot \lambda^2 + \mathrm{Q} \cdot \lambda + \mathrm{R} = 0, \tag{3.10}$$

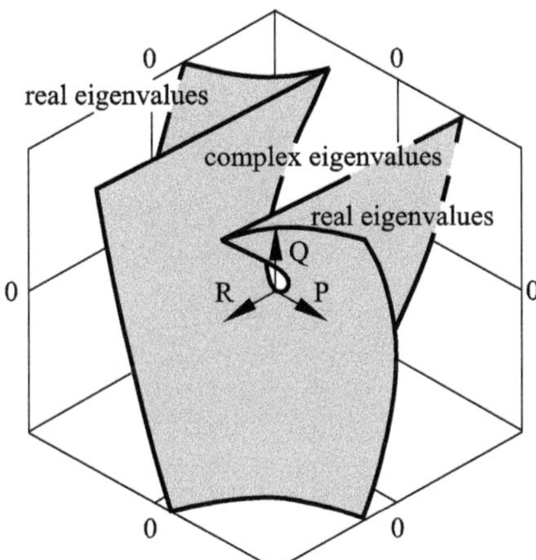

Fig. 3.5. Real and complex eigenvalues of the characteristic polynomial (3.10)

with the three real-valued matrix invariants

$$\begin{aligned}
\mathrm{P} &= -\mathrm{tr}(\mathbf{A}) = -(\lambda_1 + \lambda_2 + \lambda_3),\\
\mathrm{Q} &= \frac{1}{2} \cdot \left[\mathrm{P}^2 - \mathrm{tr}(\mathbf{A}^2)\right] = \lambda_1 \cdot \lambda_2 + \lambda_2 \cdot \lambda_3 + \lambda_3 \cdot \lambda_1,\\
\mathrm{R} &= -\det(\mathbf{A}) = -\lambda_1 \cdot \lambda_2 \cdot \lambda_3.
\end{aligned}$$

The solutions of the cubic equation (3.10) may initially be classified according to the value of the discriminant D, with

$$\mathrm{D} = 27 \cdot \mathrm{R}^2 + (4 \cdot \mathrm{P}^2 - 18 \cdot \mathrm{Q}) \cdot \mathrm{P} \cdot \mathrm{R} + (4 \cdot \mathrm{Q} - \mathrm{P}^2) \cdot \mathrm{Q}^2. \tag{3.11}$$

For $\mathrm{D} > 0$ we obtain one real eigenvalue and a pair of complex conjugate eigenvalues, while for $\mathrm{D} < 0$ we have three real eigenvalues. This is shown in Figure 3.5. The surface defined by the condition $\mathrm{D} = 0$ divides the space spanned by the three invariants P, Q, and R into two half-spaces.

A first overview of the flow behavior close to the critical points is obtained by considering the eigenvectors for the two-dimensional flow with $\mathrm{R} = 0$. The associated characteristic equation $\lambda^2 + \mathrm{P} \cdot \lambda + \mathrm{Q} = 0$ leads to the simplified discriminant $\Delta = 4 \cdot \mathrm{Q} - \mathrm{P}^2$. This divides the P-Q plane into a region of real eigenvalues and a region of complex eigenvalues in the shape of a parabola. Figure 3.6 shows the eigenvectors associated with the critical points in the P-Q plane.

The eigenvectors associated with each eigenvalue determine the direction of the tangent to the streamline running into or out of the critical point. If the real eigenvalue or the real part of the complex eigenvalue is negative, the trajectories move toward the critical point, while positive real values mean that the trajectory runs away from the critical point.

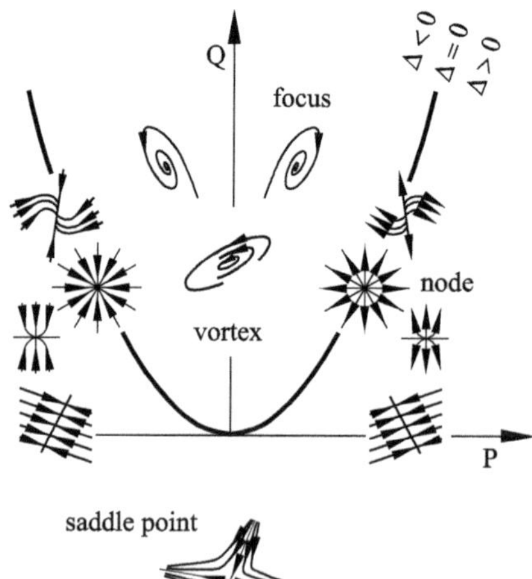

Fig. 3.6. Eigenvectors of the critical points for $\mathrm{R} = 0$, two-dimensional flow

If there are two real eigenvalues with different signs ($Q < 0$), two tangents of the eigenvectors lead into the critical point and two lead out of it. The critical point is therefore a *saddle point*. When Q is positive, for $\Delta > 0$ we obtain a *node* with two real eigenvalues with the same sign. For $\Delta < 0$ we obtain a *focus* with two complex conjugate eigenvalues.

On the boundary lines between the various regions, i.e. the axes $P = 0$ or $Q = 0$ or the parabola $P^2 = 4 \cdot Q$, degenerate cases are found, such as *vortices*, *sinks*, and *sources* (degenerate nodes). For example, for $P = 0$ only saddle points ($Q < 0$) or vortex points ($Q > 0$) are kinematically possible. For $P = 0$ and $Q = 0$ the critical point is degenerate, so that further terms in the expansion (3.7) are required for its description.

For *three-dimensional flows*, flow states are also associated with the eigenvalues in Figure 3.5. Figure 3.7 shows some selected examples. For example, the node focus structure is found in whirlwinds; saddle foci and unstable vortices occur in the vortex formation in the atmosphere; while nodes and node saddle points appear in numerous technical problems involving separation of bodies in a flow, as well as in the human heart of Figure 1.13.

As a supplement to Figure 3.4, the *flow past an automobile* is shown in Figure 3.8. In the vertical plane A1, in the wake we can identify three half-saddle points S′ (stagnation and separation points) on the rear and a saddle point S in the flow field. The backflow regime is characterized by two foci F. If we place the plane of section A2 into the wake of the automobile, we see a focus, a saddle point, and a node. The superposition of the flow structure of

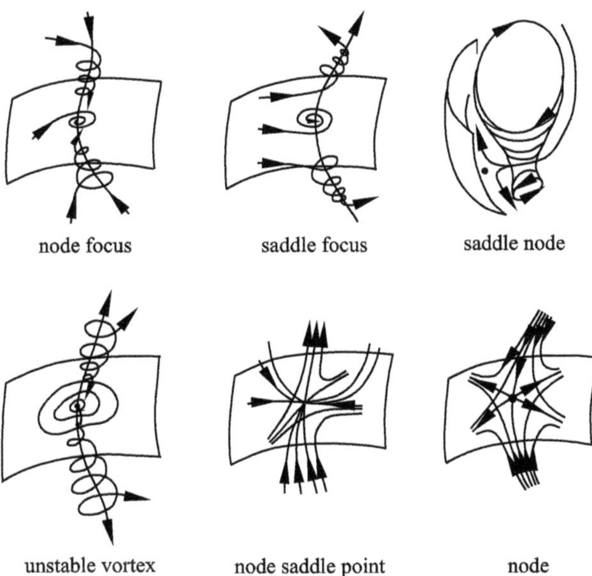

Fig. 3.7. Examples of the structure of three-dimensional flows

both planes initially looks confusing. With some visual imagination, however, it is possible to construct the three-dimensional structure of the automobile flow out of the planes shown. On the lid of the trunk a horseshoe vortex forms, which then passes into the wake flow. The shear layer between the street and the underbody of the automobile forms the backflow regime of wind tunnel experiments, bounded downstream by the saddle point in the plane A1.

Another example describes the *flow structure at a delta wing at an angle of attack*, found on supersonic aircraft (see Section 4.4.8). The aerodynamic lift is essentially generated by the underpressure inside the separated vortex at the leading edge of the wing. Figure 3.9 shows the primary vortex separation (foci) as well as the reattachment lines on the wing, made visible by the convergence of the wall streamlines. Downstream from the primary leading edge separation, the three-dimensional transverse flow on the wing causes secondary separation to occur. This leads to two further foci F and one saddle S on each half of the wing. Therefore, the structure of the flow indicates a total of three foci, one saddle, and the half-saddle of the separation and reattachment lines on the upper side of each half-wing. However, the vortex strength of the secondary separation is small compared to that of the pri-

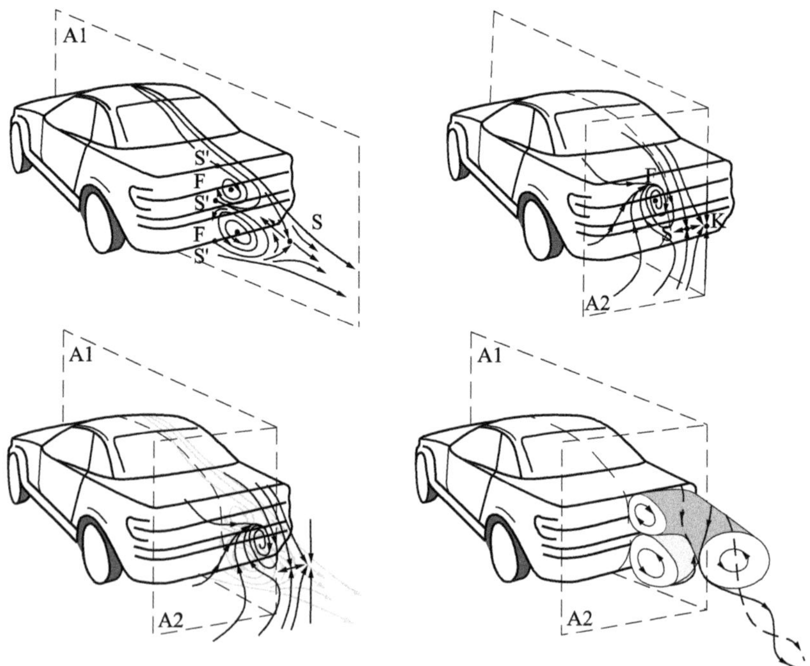

Fig. 3.8. Structure of the wake flow of an automobile

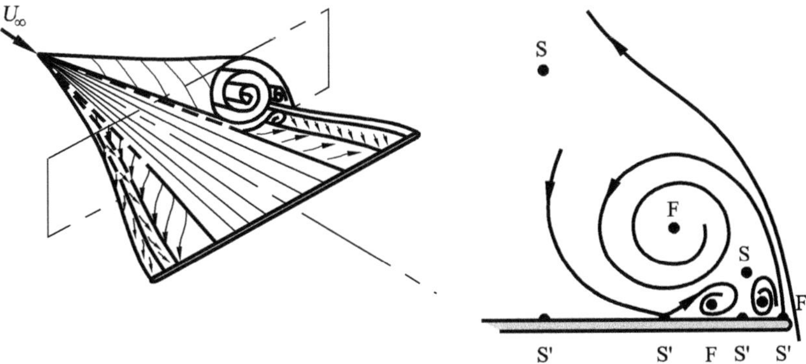

Fig. 3.9. Wall streamlines and structure of the flow past a delta wing at an angle of attack

mary vortices, so that it is these that essentially determine the aerodynamic properties of the delta wing.

These very complex examples of separated flows show how useful it can be in describing these flows to analyze the topology of the critical points solely on the basis of the kinematic fundamental equations (3.2). This is not only a description of the flow field, but a well-defined classification of the description.

3.4 Problems

3.1

A two-dimensional flow field is described by the velocity components $u = a \cdot x$ and $v = -a \cdot y$ (a is a positive constant).

(a) Compute the streamlines of the flow field

$$y = \frac{C}{x}, \qquad C = \text{constant of integration.}$$

(b) What is the rotation ω of the flow field?

$$\omega = 0 \qquad \text{for all} \quad (x, y).$$

(c) A particle of dust is placed at time $t_0 = 0$ on the point (x_0, y_0) on an arbitrary streamline. At what time t_e does the dust particle reach the point (x_1, y_1) of the streamline? It is assumed that the dust particle has a very small mass, so that no slippage occurs between it and the flow.

$$t_e = \frac{1}{a} \cdot \ln \left(\frac{x_1}{x_0} \right) .$$

3.2

The unsteady two-dimensional flow of an incompressible fluid in the x, y plane for $x > 0$ and $y > 0$ is given by the velocity components

$$u(x,t) = -[A + B \cdot \sin(\omega \cdot t)] \cdot x, \qquad v(y,t) = -[A + B \cdot \sin(\omega \cdot t)] \cdot y,$$

with the constants $A > B > 0$.
(a) Determine the component $y(t)$ of the trajectory vector for the fluid particle that at time $t = 0$ is situated at the point $\mathrm{P}(x_\mathrm{P}, y_\mathrm{P})$.

$$y(t) = y_\mathrm{P} \cdot \exp\left(A \cdot t + \frac{B}{\omega} \cdot [1 - \cos(\omega \cdot t)]\right).$$

(b) Determine the equation of the streamline that passes through the point P.

$$y(x) = \frac{x_\mathrm{P} \cdot y_\mathrm{P}}{x}.$$

(c) Determine an implicit equation for the time difference Δt that elapses as a fluid particle passes from point $\mathrm{P}(x_\mathrm{P}, y_\mathrm{P})$ to point $\mathrm{Q}(x_\mathrm{Q}, y_\mathrm{Q})$ with $y_\mathrm{Q} = 3 \cdot y_\mathrm{P}$.

$$A \cdot \Delta t + \frac{B}{\omega} \cdot [1 - \cos(\omega \cdot \Delta t)] = \ln(3).$$

(d) Determine the dependence on space and time of the x and y components a_x and a_y of the substantial acceleration in the flow field.

$$a_x = -B \cdot \omega \cdot \cos(\omega \cdot t) \cdot x + [A + B \cdot \sin(\omega \cdot t)]^2 \cdot x,$$
$$a_y = B \cdot \omega \cdot \cos(\omega \cdot t) \cdot y + [A + B \cdot \sin(\omega \cdot t)]^2 \cdot y.$$

3.3

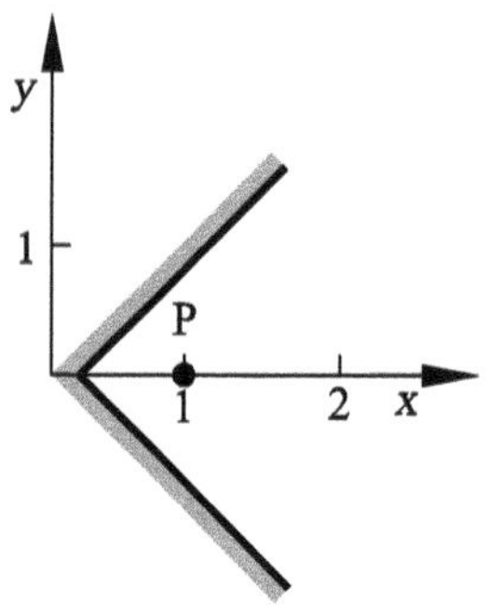

The steady irrotational two-dimensional flow of an incompressible fluid along an inside corner has the velocity components

$$u = \alpha \cdot y, \qquad v = \alpha \cdot x,$$

with $\alpha > 0$. The boundary of the semi-infinite flow field is given by the two straight lines $y = +x$ and $y = -x$, for $x \geq 0$.

(a) How many stagnation points exist in the flow field? State their coordinates.

One stagnation point in the flow field with $x_s = 0, y_s = 0$.

(b) Determine the equation $y = f(x)$ of the streamline that passes through the point $P_1(x_1 = 1, y_1 = 0)$.

$$y = \pm\sqrt{x^2 - 1}.$$

(c) Consider another point P_2 with x-coordinate $x_2 = 2$ on the same streamline as passes through P_1. How much time Δt elapses as a fluid element moves along this streamline from point P_1 to P_2?

$$\Delta t = \frac{1}{\alpha} \cdot \ln(2 + \sqrt{3}).$$

3.4

The velocity components of a steady three-dimensional incompressible flow field with dimensionless velocity vector $\boldsymbol{v} = (u, v, w)$ are given in a Cartesian (x, y, z) coordinate system as $u = x^2 + 2 \cdot z^2$ and $w = y^2 - 2 \cdot y \cdot z$.
(a) For the case in which the velocity field $\boldsymbol{v} = (u, v, w)$ satisfies the continuity equation, calculate the general form of the component v of the velocity field in the y-direction.

$$v(x, y, z) = -2 \cdot x \cdot y + y^2 + C(x, z), \qquad C(x, z) \text{ an arbitrary function.}$$

(b) Investigate whether the flow in question is irrotational for all (x, y, z).
(c) Calculate the acceleration $a_x(x, y, z)$ of the flow in question in the x-direction.

$$a_x = 2 \cdot x^3 + 4 \cdot x \cdot z^2 + 4 \cdot y^2 \cdot z - 8 \cdot y \cdot z^2.$$

3.5

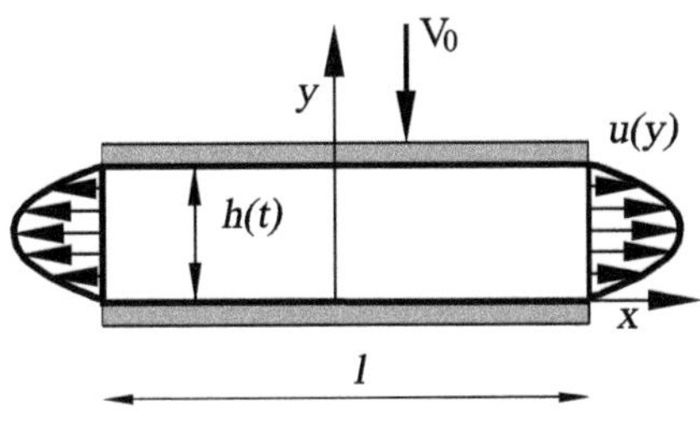

A gap of length l and time-dependent height $h(t)$ is filled with an incompressible fluid. The upper boundary moves downward with constant velocity V_0. The velocity distribution at the outlet is

$$u(y) = 4 \cdot U_0 \cdot \left(\frac{y}{h(t)} - \left(\frac{y}{h(t)} \right)^2 \right).$$

(a) Determine the function of the gap height for $h(t = 0) = h_0$.

$$h(t) = -V_0 \cdot t + h_0.$$

(b) Using the continuity equation, calculate the maximum velocity U_0 at the outlet of the gap.

$$U_0 = \frac{3}{4} \cdot \frac{l}{h(t)} \cdot V_0 .$$

3.6

A two-dimensional flow field satisfies the following differential equation

$$\frac{\mathrm{d}y}{\mathrm{d}x} = \frac{x+y}{x} .$$

(a) Determine the characteristic polynomial of the equation.

$$\lambda^2 - 2 \cdot \lambda + 1 = 0 .$$

(b) What type of singularity is at hand?

A node.

(c) What is the equation for the family of integral curves?

$$y = x \cdot \ln |x| + C \cdot x .$$

4. Dynamics of Fluid Flow

4.1 Dynamics of Inviscid Liquids

4.1.1 Continuity and the Bernoulli Equation

In flows material does not vanish, nor does new material appear. The velocity fields therefore have to satisfy the law of conservation of mass. This law is easiest to formulate for steady flows if the shape of the streamlines is already known. We consider a stream filament through every cross-section of which the same amount of mass flows per unit time. If this mass were not the same in two cross-sections, the mass content of the stream filament between two cross-sections would have to decrease or increase, contradicting the idea of a steady state. If A is the cross-section of the stream filament at a certain position, w the mean velocity in this cross-section, and ρ the associated density, then per unit time, the fluid volume $A \cdot w$ flows through the cross-section. The fluid mass flowing through the cross-section per unit time is $\rho \cdot A \cdot w$. *Continuity* requires that $\rho \cdot A \cdot w$ must have the same value in all cross-sections of a stream filament. This implies that a stream filament of a steady flow cannot terminate in the interior of the fluid. It may extend from one boundary of the fluid space under consideration to the other boundary of the space, or it can turn back on itself.

If we consider an incompressible flow, the relations for the mass flowing through a cross-section also hold for the volume. Since more volume cannot pass through one cross-section of a stream filament than through another cross-section at any time, the restriction to steady flows may be dropped. In general, for incompressible flows we have

$$A \cdot w = \text{const.} \tag{4.1}$$

i.e. the velocity is inversely proportional to the cross-section of the stream filament. If we divide the space through which the fluid flows into a large number of stream tubes through which the same amount of fluid passes per unit time, at large velocities many stream filaments will crowd together, and in places where the velocity is small, the stream filaments will expand out further. The number of stream filaments that pass through a unit of area is proportional to the velocity at this position. Therefore, in incompressible

H. Oertel (ed.), *Prandtl-Essentials of Fluid Mechanics*,
Applied Mathematical Sciences 158, DOI 10.1007/978-1-4419-1564-1_4,

flows it is not only the direction of the stream tubes but also their density that serve to illustrate the flow.

The relations that are discussed here are particularly useful if the entire flow may be treated as a single stream filament. The prescribed cross-sections correspond to the stream filament cross-section. From the relation

$$A \cdot w = \dot{V}$$

we can determine the mean velocity at every position of such an incompressible flow. Here $\dot{V}$ indicates the volume transported per unit time.

Similarly, for compressible flows we have

$$\rho \cdot A \cdot w = \dot{M},$$

with the mass transported per unit time $\dot{M}$. Since in this case the density ρ generally can be determined only in connection with the pressure, the velocity cannot be determined from the continuity alone (cf. Section 4.3).

In dealing with steady, incompressible flows, this representation leaves us with only one independent variable, namely the distance of the relevant cross-section along the central line of the tube from some given starting point. The treatment of the flow is then *one-dimensional*, in contrast to the three-dimensional treatment where the spatial variation of the velocity and the other quantities is taken into account. For water, all one-dimensional flows come under the collective name *hydraulics*. In contrast, three-dimensional flows are grouped under the name *hydrodynamics*. For flows that occur in air travel and in other areas of application of air flows, the term *aerodynamics* is used.

If the fluid is bounded at one position by a solid body or by another fluid, continuity requires that no hole may form at this position, nor may the two fluids seep into each other. In order to avoid both of these situations, the velocity components perpendicular to the bounding surface must be identical on both sides of this bounding surface. If we consider a body at rest in a moving fluid, or fixed solid walls, the velocity components of the fluid perpendicular to the surface of the body or to the wall must vanish at the boundary. According to continuity alone, the velocity components parallel to the wall may take on any value.

We now consider the forces acting in a flowing fluid. We have learned that there are two forces acting on a fluid at rest: *gravity* (and other mass forces) and *pressure force*. These two forces are also to be found in a moving liquid. Whereas the two forces are in equilibrium in a fluid at rest, in a moving liquid this is not the case. In addition, the *liquid friction*, to be regarded as a resistance to change of shape, appears. This will be discussed in depth in Section 4.2 but will be neglected in this section. The fluids that are technically most important (water, air, etc.) have a very small viscosity and in some cases demonstrate only very small friction drag, so that it is justified to neglect this. For this reason we first develop the fundamental flow laws for *inviscid fluids* and will only later consider the alterations in these laws when friction is present. For this reason the following discussions are based on *inviscid liquids*. Initially, we consider an incompressible flow.

In order to develop the dynamic relation between the pressure and the mass force on the one hand, and the state of motion on the other hand, we go back to *Newton's equation* force = mass · acceleration, the basis of dynamics. We compute the simultaneous states along a stream filament. To do this we need the component of acceleration in the direction of motion, as presented in Section 3.2 for three-dimensional flow. In the case of one-dimensional flow, we denote the arc length along the streamline by s, the time by t, and the velocity by w. The change in velocity as s changes by $\mathrm{d}s$ and t changes by $\mathrm{d}t$ is then

$$\mathrm{d}w = \frac{\partial w}{\partial s} \cdot \mathrm{d}s + \frac{\partial w}{\partial t} \cdot \mathrm{d}t.$$

Here $\partial w/\partial t$ is the partial derivative (at fixed s), and $\mathrm{d}w/\mathrm{d}t$ the total derivative (for a fixed fluid element).

This yields the acceleration

$$\frac{\mathrm{d}w}{\mathrm{d}t} = w \cdot \frac{\partial w}{\partial s} + \frac{\partial w}{\partial t}. \tag{4.2}$$

The term $w \cdot (\partial w/\partial s)$ is the part of the acceleration that arises from the fact that the particle moves to positions with different velocities, and $\partial w/\partial t$ is the part due to the change in time of the flow state at one position. In steady flows the second term is equal to zero. The first term can also be written in the form $\partial(w^2/2)/\partial s$.

In order to apply the equation force = mass · acceleration, a cylindrical element with cross-section $\mathrm{d}A$ and length $\mathrm{d}s$ is again selected from the flowing liquid. The discussion of equilibrium in Section 2.3 was performed on a similar cylindrical element. The axis of the cylindrical element is in the direction of flow (Figure 4.1). The mass of the cylindrical element is $\rho \cdot \mathrm{d}A \cdot \mathrm{d}s$.

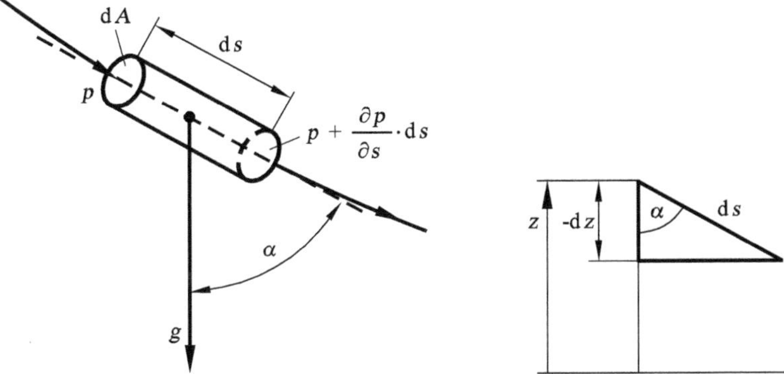

Fig. 4.1. Balance of forces on a cylinder element

If the motion is inviscid, the following forces act on the cylinder element: a force of pressure, due to the pressure difference, and a mass force. Let the pressure on the end of the cylinder element lying upstream have the value p. It then acts on that end surface area $\mathrm{d}A$ with a force $p \cdot \mathrm{d}A$. On the downstream end the pressure has the value $p + (\partial p/\partial s) \cdot \mathrm{d}s$, so that the resultant of the two pressure forces is $p \cdot \mathrm{d}A - (p + (\partial p/\partial s) \cdot \mathrm{d}s) \cdot \mathrm{d}A = -(\partial p/\partial s) \cdot \mathrm{d}s \cdot \mathrm{d}A$. In addition, a mass force also acts on the liquid, with an effect on a unit of mass equal to g (e.g. the force of gravity). If the direction of the mass force and the direction of flow form an angle α, the mass $\rho \cdot \mathrm{d}A \cdot \mathrm{d}s$ experiences the following force component in the direction of flow:

$$\rho \cdot \mathrm{d}A \cdot \mathrm{d}s \cdot g \cdot \cos(\alpha).$$

In the equation force = mass · acceleration, every term now has the factor $\mathrm{d}A \cdot \mathrm{d}s$, which can thus be canceled (i.e. the volume of the arbitrarily chosen cylinder element has no effect on the result). Dividing by ρ, we obtain

$$-\frac{1}{\rho} \cdot \frac{\partial p}{\partial s} + g \cdot \cos(\alpha) = \frac{\partial}{\partial s}\left(\frac{w^2}{2}\right) + \frac{\partial w}{\partial t}. \tag{4.3}$$

Usually, the only mass force is gravity. Then g's magnitude and direction are constant, and for $\cos(\alpha)$ we can write $-\partial z/\partial s$ using the vertical coordinate z (Figure 4.1).

If the flow is steady ($\partial w/\partial t = 0$) and the density ρ is assumed to be constant, then all terms are derivatives with respect to s. Equation (4.3) can then be integrated *along the stream filament.* From

$$\frac{1}{\rho} \cdot \frac{\partial p}{\partial s} + g \cdot \frac{\partial z}{\partial s} + \frac{\partial}{\partial s}\left(\frac{w^2}{2}\right) = 0$$

we obtain

$$\frac{p}{\rho} + g \cdot z + \frac{w^2}{2} = \text{const.} \tag{4.4}$$

This equation, which is known as the *Bernoulli equation,* is the fundamental equation for the one-dimensional treatment of inviscid flows. If we divide all terms of (4.4) by g, the terms have the dimensions of a length and can be interpreted as heights. Introducing the weight of the unit volume $\rho \cdot g = \gamma$, as in the previous chapter, we obtain the Bernoulli equation in the form

$$\frac{p}{\gamma} + z + \frac{w^2}{2 \cdot g} = \text{const.} \tag{4.5}$$

According to Section 2.3, p/γ is the height of the liquid column that generates the pressure p by its weight, and is therefore called the *pressure height.* Here z is the height of the position under consideration above an arbitrarily fixed horizontal plane and is called the *position height,* and $w^2/(2 \cdot g)$ is the height that a body would have to fall to achieve the velocity w by free-fall, and is therefore called the *velocity height.* According to the Bernoulli equation, the sum of the pressure height, the position height, and the velocity height is constant along a streamline. The value of the constant may change from streamline to streamline. This occurs in particular when the streamlines originate in different places. However, if all streamlines come from one region where static conditions reign (i.e. rest or uniform, rectilinear motion), the constant is the same for all streamlines. The Bernoulli equation is then also valid perpendicular to the streamlines in the entire space. According to Section 2.3, in a fluid at rest $p/\gamma + z = \text{const}$. This is in agreement with the Bernoulli equation for $w = 0$ or $w = \text{const}$. The special flow state described here is identical to the steady potential motion to be described later.

Integration can also be carried out for other mass forces if they have a potential U, since $g \cdot \cos(\alpha)$ can then be set equal to $-\partial U/\partial s$. If the flow is compressible, integration is also possible as long as the flow is homogeneous, i.e. if the density depends only on the pressure. Then $\int(\mathrm{d}p/\rho) = \mathrm{F}(p)$ is a function of the pressure, and we have $(1/\rho) \cdot (\partial p/\partial s) = \partial \mathrm{F}/\partial s$. Integration with respect to s yields the general form of the Bernoulli equation for steady motion:

$$\mathrm{F} + U + \frac{w^2}{2} = \text{const.} \tag{4.6}$$

4.1.2 Consequences of the Bernoulli Equation

The Bernoulli equation can solve a great number of applications in a very simple manner. We present some important examples below.

Discharge from a Vessel Under the Effect of Gravity

Following the streamlines in the vessel in Figure 4.2 from the flow outlet B, we see that they lead to the surface of the water A, whose level sinks as the water flows out of the vessel. The water particles at A, like the particles in the free jet at B, are under atmospheric pressure p_0. The weight of the air has been neglected. This is possible if it suffices to state the pressure to the second decimal place. If the surface area of the water is large compared to the outlet at B, the velocity at A is so small that its square may be neglected

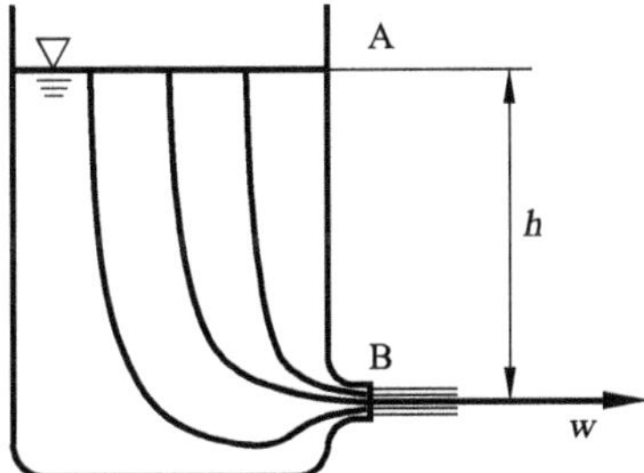

Fig. 4.2. Discharge from a vessel

compared to that of the velocity at B. With z_{A} and z_{B} the position heights of A and B, the Bernoulli equation states that

$$\frac{p_0}{\rho} + g \cdot z_{\mathrm{B}} + \frac{w_{\mathrm{B}}^2}{2} = \frac{p_0}{\rho} + g \cdot z_{\mathrm{A}} + 0.$$

Therefore, with $z_{\mathrm{A}} - z_{\mathrm{B}} = h$ we have

$$\frac{w_{\mathrm{B}}^2}{2 \cdot g} = z_{\mathrm{A}} - z_{\mathrm{B}} = h,$$

or

$$w_{\mathrm{B}} = \sqrt{2 \cdot g \cdot h}. \tag{4.7}$$

The velocity at B is thus as large as if the water particles had free-fallen from the height h. The relation given in (4.7) is called *Torricelli's discharge formula.*

The cross-section of the jet is generally not the same as that of the outlet. For a jet that exits from a circular opening in a thin wall, the jet cross-section is about 0.61 to 0.64 times the outlet cross-section. This behavior, also called *contraction*, is due to the fact that the liquid inside the vessel flows radially toward the outlet and at the edge of the outlet cannot be suddenly deviated from the radial direction to the direction of the jet axis. Such flows are shown in the upper illustrations in Figure 4.3. In the case of a rounded opening, the deviation of the stream filament can take place within the outlet, and the contraction is approximately equal to 1. The discharge $\dot{V}$ (volume per second) through an opening of cross-section A is

$$\dot{V} = \alpha \cdot A \cdot \sqrt{2 \cdot g \cdot h},$$

with the contraction α. If the opening in a thin wall is not circular, α deviates only slightly from the value of the circular outlet, but the jets that form have in general a much more complicated form. For example, a jet that comes out of a square outlet forms a thin cross-shaped cross-section. A jet that comes out of a rectangular outlet forms a band perpendicular to the long side of the rectangle.

Discharge from a Vessel Under the Effect of Internal Overpressure

The vessel in the lower picture in Figure 4.3 is under pressure p_1. In the outer region, the pressure is atmospheric pressure p_0. For a streamline that runs horizontally, we have $z_{\mathrm{A}} = z_{\mathrm{B}}$. If again the velocity at A can be taken to be negligibly small, the Bernoulli equation yields

$$\frac{p_0}{\rho} + \frac{w^2}{2} = \frac{p_1}{\rho} + 0,$$

i.e.

$$w = \sqrt{\frac{2 \cdot (p_1 - p_0)}{\rho}} = \sqrt{\frac{2 \cdot g \cdot (p_1 - p_0)}{\gamma}}. \tag{4.8}$$

We denote the height $(p_1 - p_0)/\gamma$ by h. This is the height of a liquid column with the specific weight γ between whose upper and lower ends the pressure difference is $p_1 - p_0$. Then (4.8) again yields $w = \sqrt{2 \cdot g \cdot h}$.

Equation (4.8) permits us to estimate the magnitude of the velocity up to which it is possible to treat a gas as an incompressible liquid. The limiting velocity w_1 depends of the size of the density fluctuations that can be permitted. Because of $p \cdot V^\kappa = \text{const}$ or $p = \text{const} \cdot \rho^\kappa$, we have $\Delta p/p \approx \kappa \cdot \Delta\rho/\rho$. Therefore, $\Delta p \approx \kappa \cdot p_0 \cdot \Delta\rho/\rho$. If we select the admissible density change to be $\Delta\rho/\rho = 0.01$, for air at a normal pressure of $p_0 = 1$ bar $= 10^5$ N/m^2 we obtain a pressure difference of $\Delta p = 1.405 \cdot 10^5 \cdot 0.01$ N/m$^2 = 1405$ N/m^2. With a mean value of $\rho = 1.21$ Ns^2/m^4 we obtain the following limiting velocity:

$$w_1 = \sqrt{\frac{2 \cdot \Delta p}{\rho}} = \sqrt{2322 \text{ m}^2/\text{s}^2} \approx 48 \text{ m/s} \quad .$$

If we permit density variations of 10%, we obtain a velocity $\sqrt{10}$ times larger, i.e. about 150 m/s. The density variations have two effects. Kinematically, the stream filament cross-sections change, and dynamically, the magnitude of the pressure change associated with an acceleration is affected.

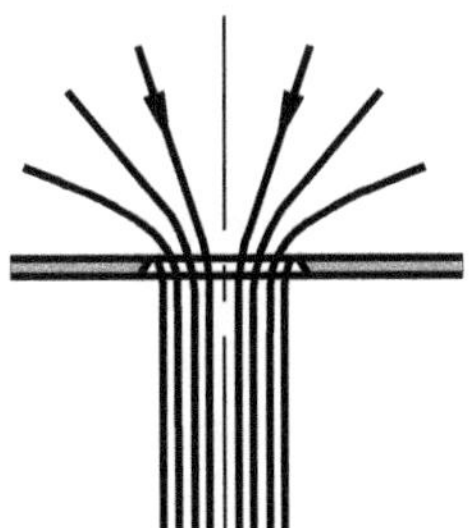

outflow from an opening in a flat wall

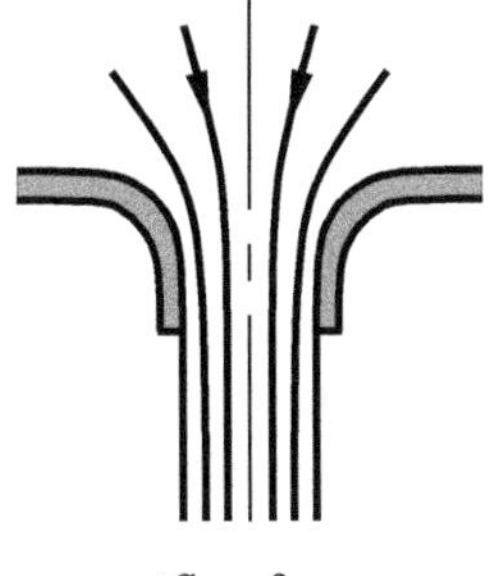

outflow from a rounded opening

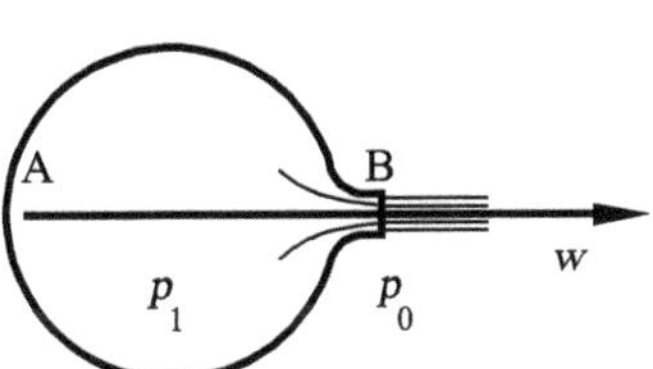

outflow from a vessel

Fig. 4.3. Different discharges

Stagnation Point Flow

If an obstacle is situated in a uniform liquid flow of velocity w_0, the flow dams up directly in front of the obstacle and branches out to all sides in order to pass the obstacle (Figure 4.4). In the central point of the stagnation region, the *stagnation point*, the flow comes completely to rest. For the streamline that passes through the stagnation point, with the pressure p_S at the stagnation point and the unperturbed pressure p_∞ in the free flow at the same height, the Bernoulli equation therefore yields

$$\frac{p_S}{\rho} + 0 = \frac{p_\infty}{\rho} + \frac{w_\infty^2}{2}, \qquad \text{and so} \qquad p_S = p_\infty + \rho \cdot \frac{w_\infty^2}{2}.$$

The pressure increase $p_S - p_\infty = \rho \cdot w_\infty^2/2$ is known as the *stagnation pressure* or *dynamic pressure.* Measurement of this pressure increase is a method of determining flow velocities. If a body with velocity U_∞ is moved through air (or liquid) at rest, the above flow is observed in the reference frame moving with the body. The velocity w_∞ is directed in the opposite direction to U_∞, and its magnitude is equal to U_∞. In this case, a pressure increase of $\rho \cdot U_\infty^2/2$ is also observed. If the obstacle at the stagnation point has a bore hole, the pressure p_S passes through this into the interior and can be led to a measuring device. In order to measure the pressure $p_S = p + \rho \cdot w^2/2$ in a flow, we need only a simple bent tube as an obstacle (Figure 4.5). This is called the *Pitot tube* after its inventor.

To every point in the flowing liquid, as well as the pressure at hand p (which a pressure gauge moving with the liquid would measure) we can also assign the pressure p_S that a Pitot tube would measure. The pressure p is called the *static pressure*, the pressure p_S the *total pressure.* Therefore, we have *total pressure* = *static pressure* + *dynamic pressure.* From the Bernoulli equation

$$\frac{p}{\rho} + g \cdot z + \frac{w^2}{2} = \text{const}$$

we can introduce the total pressure $p_S = p + \rho \cdot w^2/2$, and so obtain

$$\frac{p_S}{\rho} + g \cdot z = \text{const}; \qquad \text{or} \qquad p_S + \gamma \cdot z = \text{const}.$$

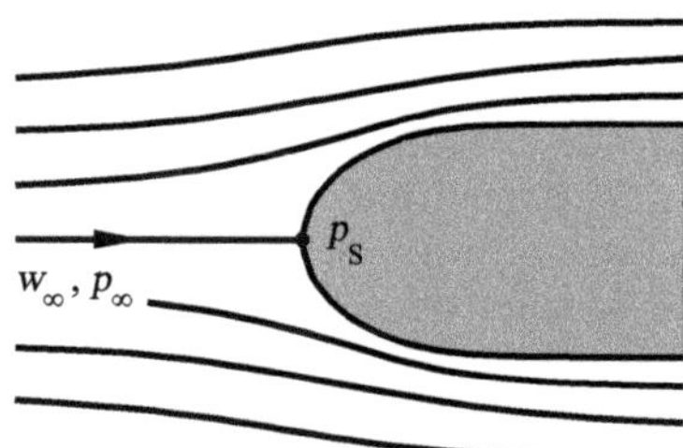

Fig. 4.4. Stagnation point flow

i.e. p_S is distributed according to static laws. This means that p_S is constant in all horizontal planes if all streamlines have the same constant.

In order to use the derived relations to determine flow velocities, as well as knowing p_S, we also have to measure the static pressure p. This is much more difficult than determining p_S, since the static pressure is disturbed by introducing a probe at the place where it is to be measured. For details on carrying out such pressure measurements, see Section 4.1.3.

The following investigations are not restricted to inviscid liquids but rather (if need be with small alterations) also hold for moderately strongly viscous liquids. However, our first investigation assumes an incompressible fluid of constant density.

The pressure in such a liquid can be decomposed into two parts, one of which represents the pressure that would arise if the liquid were at rest. This equilibrium pressure is denoted by p', and $p' = \text{const} - \gamma \cdot z$. If we set the pressure that actually acts in the flowing liquid to $p = p' + p^*$, then p^* represents the difference in the pressure in the case of motion compared to the case at rest. If the Bernoulli equation may be applied, i.e. if $p + \gamma \cdot z + \rho \cdot w^2/2 = \text{const}$, and if we take the value of p' into account, it follows that $p^* + \rho \cdot w^2/2 = \text{const}$. Therefore, p^* is distributed as in the case of a weightless liquid with an inert mass. The position height z has no effect on p^*. Every particle of a liquid under the effect of gravity experiences just enough lift from its neighboring particles to be suspended. This result can be carried over to viscous flows. In the following approach we will therefore not take the effect of gravity into motions in water or in air. This means that instead of the pressure p, the pressure difference p^* is always taken into account. For simplicity we will again write p instead of p^*.

If the pressure of an air or water flow is determined by external pressure gauges at rest, to which tubes from a moving pressure sampling point (probe) lead, the weight of the liquid in the tubes acts just so that the indicated pressure is independent of the height of the pressure sampling position. Therefore, the device indicates a pressure of type p^*. If the probe is a Pitot tube directed against the flow, the device at rest indicates constant pressure on a streamline. If all streamlines have the same constant, the pressure reading is the same for the entire region.

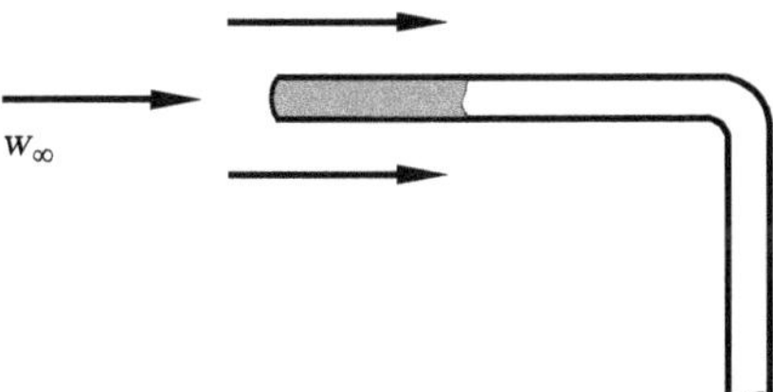

Fig. 4.5. Pitot tube

Spiral Flow

The Bernoulli equation treats pressures along a streamline. We can also obtain statements about the pressure differences in the *direction perpendicular* to the flow if we consider the transverse acceleration instead of the longitudinal acceleration. This has the direction of the principal normal vector to the trajectory and the magnitude w^2/r. Here r is the radius of curvature of the trajectory. If we consider the force on a prism element whose axis lies in the direction of the principal normal vector, the component in the direction of the radius r yields

$$\frac{w^2}{r} = \frac{1}{\rho} \cdot \frac{\partial p}{\partial s'}. \tag{4.9}$$

Here $\mathrm{d}s'$ is an element of arc in the direction of the principal normal vector, and p is to be interpreted as p^*. Equation (4.9) expresses the effect of the centrifugal force in a curvilinear flow. The pressure increases in the radial direction by $\rho \cdot w^2/r$ per unit length. This relation connects neighboring stream filaments. It is important to note that for a *rectilinear flow* ($r = \infty$) *there is no pressure difference perpendicular to the direction of flow.* In the above special case where the constants of the Bernoulli equation have the same value for all streamlines, a particularly simple result is found for a curvilinear flow. From $\int (\mathrm{d}p/\rho) + w^2/2 = \text{const.}$ (4.4), on differentiation with respect to s' we can obtain a second expression for $(1/\rho) \cdot \partial p/\partial s'$, namely, $(1/\rho) \cdot \partial p/\partial s' = -w \cdot \partial w/\partial s'$. Inserting this into (4.9), we obtain

$$\frac{\partial w}{\partial s'} + \frac{w}{r} = 0. \tag{4.10}$$

As will also be shown later, in Section 4.1.5, it follows from this that the individual liquid elements experience no rotation in a curvilinear flow. The circulation along a rectangle formed by two radial sections of length $\mathrm{d}s'$ and two streamline arcs vanishes if (4.10) is satisfied.

An example is the *flow in a spiral casing* (see Figure 4.6). All streamlines start in the parallel flow at A. The velocity is to be equal on all stream filaments, so that if the pressure is the same in the parallel flow, the Bernoulli constant is identical on all streamlines. The radii of curvature of the individual streamlines can be approximately set to the radius r from the midpoint O,

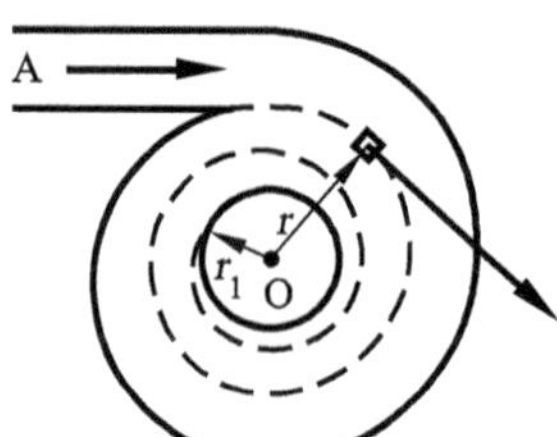

Fig. 4.6. Spiral casing

and the element of arc ds' can be set equal to dr. Then $dw/dr + w/r = 0$, or $dw/w = -dr/r$. Integration yields $\ln(w) = \ln(C) - \ln(r)$, i.e. $w = C/r$, with the constant of integration C. The velocity increases toward the midpoint. If the height of the spiral casing is constant, continuity implies that the radial component of the velocity is also proportional to $1/r$. Therefore, the angle between the streamlines and the radii is everywhere the same, and the streamlines are logarithmic spirals. The pressure is obtained from the Bernoulli equation as $p = \text{const} - \rho \cdot C^2/(2 \cdot r^2)$. If the liquid exits into the surroundings at the internal radius r_1 of the casing with pressure p_0, the pressure at another part of the spiral casing can be calculated with

$$p = p_0 + \rho \cdot \frac{C^2}{2} \cdot \left(\frac{1}{r_1^2} - \frac{1}{r^2} \right) .$$

Very large overpressures can occur at A if the radius of the outlet is small.

Unsteady Flow

For *unsteady flows*, a change in the flow state results in an additional pressure term to the previous pressures. This investigation is restricted to longitudinal acceleration, where, according to (4.2), the term $\partial w/\partial t$ (rate of change of velocity at a fixed place) also appears. Using the ideas that led to the Bernoulli equation, starting from the full equation (4.3), we see that the term $\int_0^s (\partial w/\partial t) \cdot ds$ is added to the left-hand side of (4.4). If the flow is in a pipe with constant cross-section in which the velocity is the same at every cross-section (the velocity across the cross-sections is also assumed constant, since we assumed an inviscid flow), $\partial w/\partial t$ is independent of the position. The integral can be set equal to $(dw/dt) \cdot s$.

An example is the start of *discharge through a faucet pipe* of length l (Figure 4.7). Along the pipe axis, assumed to be horizontal, we have

$$\frac{p}{\rho} + \frac{w^2}{2} + \frac{dw}{dt} \cdot s = \text{const} = \frac{p_\infty}{\rho} + g \cdot h .$$

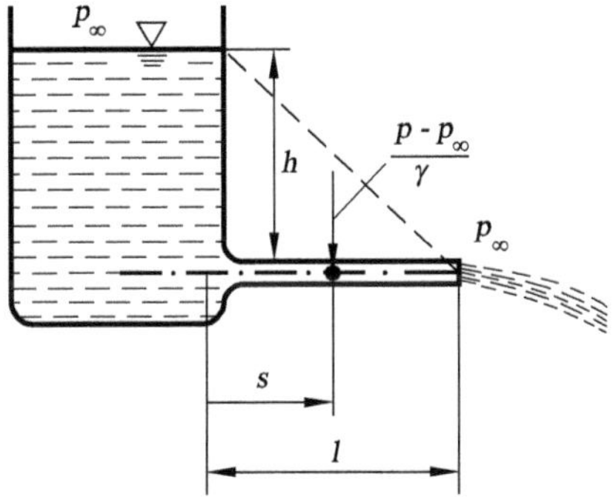

Fig. 4.7. Start of discharge

As long as $\mathrm{d}w/\mathrm{d}t$ is nonzero, the pressure p sinks along the pipe in proportion to s. The pressure at the end of the pipe ($s = l$) is equal to the ambient pressure p_∞. We obtain:

$$\frac{p_\infty}{\rho} + \frac{w^2}{2} + \frac{\mathrm{d}w}{\mathrm{d}t} \cdot l = \frac{p_\infty}{\rho} + g \cdot h,$$

i.e.

$$\frac{\mathrm{d}w}{\mathrm{d}t} = \frac{1}{l} \cdot \left(g \cdot h - \frac{w^2}{2} \right). \tag{4.11}$$

At the start of the discharge, we have the simple relation $\mathrm{d}w/\mathrm{d}t = g \cdot h/l$ at $w = 0$. As w increases, $\mathrm{d}w/\mathrm{d}t$ decreases more and more and tends to zero for large values of t; i.e. the flow becomes steady and w becomes equal to $\sqrt{2 \cdot g \cdot h}$. The precise rate of increase of w is obtained by integrating (4.11), although this will not be considered here. An estimation of the time T that approximately elapses until the steady state is reached is obtained as follows: We assume a constant acceleration $\mathrm{d}w/\mathrm{d}t$ until w reaches the value $w_1 = \sqrt{2 \cdot g \cdot h}$. Therefore, w_1/T can be introduced into (4.11) instead of $\mathrm{d}w/\mathrm{d}t$. At time $t = 0$ we obtain

$$T = \frac{w_1 \cdot l}{g \cdot h} = \frac{2 \cdot l}{w_1}.$$

Another example of unsteady flow of a liquid is the *oscillation of a column of liquid* in a bent pipe open at both ends under the effect of the Earth's gravitational field (Figure 4.8). The pipe has a constant cross-section. The length of the liquid column measured along the axis of the pipe is l. The deflection at some point in time in the direction of the pipe axis is x. Because of continuity, the deflection at both ends and at any point in the middle is the same. The velocity is the same everywhere, namely, $w = \mathrm{d}x/\mathrm{d}t$; i.e. $w \cdot \partial w/\partial s = 0$. Therefore, the acceleration is $\mathrm{d}^2x/\mathrm{d}t^2$. The right end is raised $h_1 = x \cdot \sin(\alpha)$ above the zero level, and the other end is lowered by $h_2 = x \cdot \sin(\beta)$. The height difference between the levels of the liquid at the ends is $h_1 + h_2 = x \cdot (\sin(\alpha) + \sin(\beta))$. The pressure at both ends is the ambient pressure p_∞. The extended Bernoulli equation applied to both ends yields

$$g \cdot x \cdot (\sin(\alpha) + \sin(\beta)) + l \cdot \frac{\mathrm{d}^2x}{\mathrm{d}t^2} = 0.$$

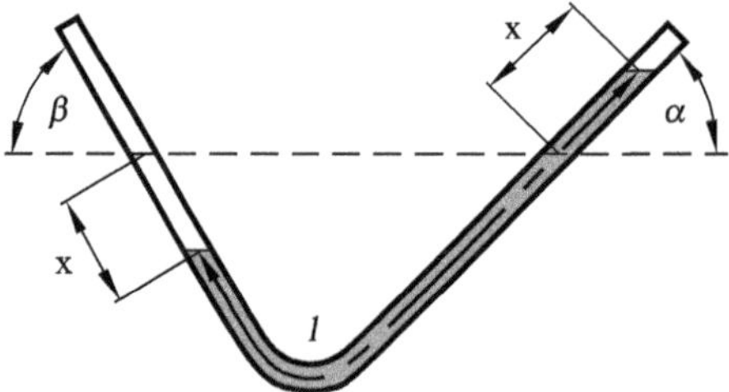

Fig. 4.8. Oscillation of a water column

The solution of this differential equation, in agreement with the result for elastic oscillations, is $x = A \cdot \cos(\omega \cdot t + \vartheta)$, with $\omega = \sqrt{g \cdot (\sin(\alpha) + \sin(\beta))/l}$. This yields a period of oscillation of

$$T = \frac{2 \cdot \pi}{\omega} = 2 \cdot \pi \cdot \sqrt{\frac{l}{g \cdot (\sin(\alpha) + \sin(\beta))}}.$$

For a vertical U-tube ($\sin(\alpha) = \sin(\beta) = 1$) we obtain $T = 2 \cdot \pi \cdot \sqrt{l/(2 \cdot g)}$. This corresponds to the period of a pendulum of half the length of the liquid column.

4.1.3 Pressure Measurement

The slot in a flow shown in Figure 4.9 is of interest for pressure measurement. At the start of motion, a flow occurs in the slot (Figure 4.9 left). Vortices and interfaces initially form at the edges. After the vortices have floated off, assuming that the distance between both edges is small enough, a flow corresponding to the right-hand side of Figure 4.9 forms. Inside the slot, the fluid is essentially at rest. The pressure in the slot is the same as the pressure in the flowing fluid, since it is constant in the part at rest and must pass continuously over to that of the flowing fluid at the interface. If the interior of the slot is connected to a pressure gauge via a pipe, it is possible to measure the pressure in the flowing liquid. Instead of a slot, any shape of hole, such as one with a circular cross-section, can be used. The edges of the hole or the slot must be smooth. No sharp edge may stand in the way of the flow, since the pressure in the arched interface that would then occur would be considerably different from the pressure in the neighboring parts of the fluid. A slight rounding of the edges of the hole is permissible.

The left picture of Figure 4.10 shows a practical arrangement for a pressure measurement setup at the wall of a pipe. In order to measure the pressure in the interior of the fluid, a very thin disk (Ser disk, Figure 4.10) with a hole through the middle can be applied to the end of a thin pipe. Although this measurement uses the same ideas, it is very sensitive to a change in direction of the air stream to the plane of the disk. A manometric capsule is less sensitive. It correctly measures the pressure up to an angular deviation of about 5°. At larger angles it indicates a pressure that is too low.

By relating such a pressure measurement to the measurement of the total pressure in Figure 4.5, we can measure the velocity pressure (dynamic pres-

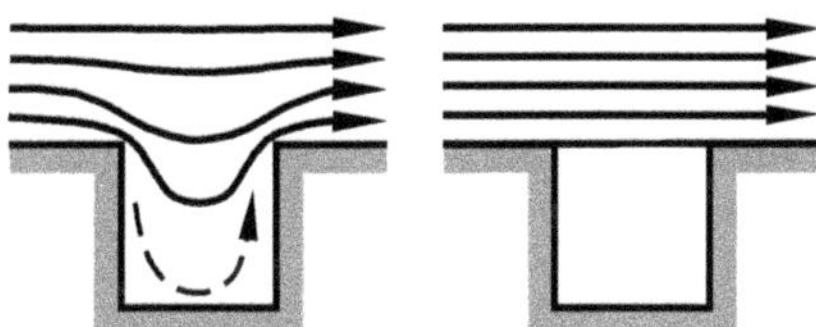

Fig. 4.9. Flow at a slot

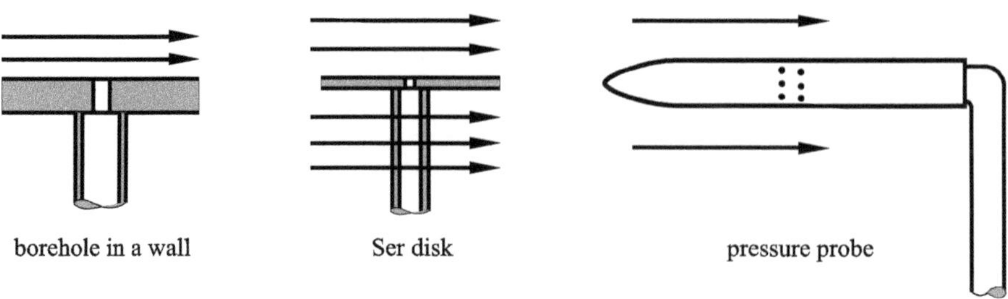

Fig. 4.10. Pressure measurement

sure or stagnation pressure) as the pressure difference $p_{\mathrm{d}} = \rho \cdot w^2/2$. When the density ρ is known, we can then compute the velocity w. In the atmosphere at normal pressure with a density of $\rho = 1.21\ \mathrm{Ns^2/m^4}$, the stagnation pressure at $w = 10$ m/s is $p_{\mathrm{d}} = 60.5\ \mathrm{N/m^2}$. In water at the same velocity with $\rho = 1\,050\ \mathrm{Ns^2/m^4}$ the stagnation pressure is considerably larger, namely, $p_{\mathrm{d}} = 50\,000\ \mathrm{N/m^2}$.

The manometric capsule in Figure 4.10 can be combined with the Pitot tube in Figure 4.5 in one device. This is the *Prandtl stagnation tube* for velocity measurement (Figure 4.11). It is relatively insensitive to deviations of its axis to the direction of flow.

Pressure measurement via bore holes is used in many flows. The pressure difference on the surface of a body in a flow (e.g. the wing of an airplane) is measured through a series of boreholes like those in Figure 4.10, each connected to a pressure gauge.

Figure 4.12 shows a very famous early attempt to demonstrate the pressure difference in a pipe that first contracts and then expands. This experiment illustrates the Bernoulli equation. The pressure can be adjusted by the faucet at the end of the pipe. If the faucet is opened, an underpressure occurs at b. The pressure recovery in the pipe behind the narrowest cross-section is somewhat smaller when friction is taken into account than for the inviscid theory.

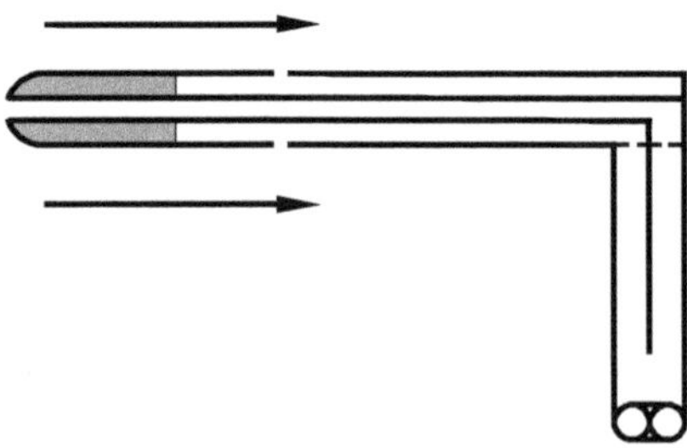

Fig. 4.11. Prandtl stagnation tube

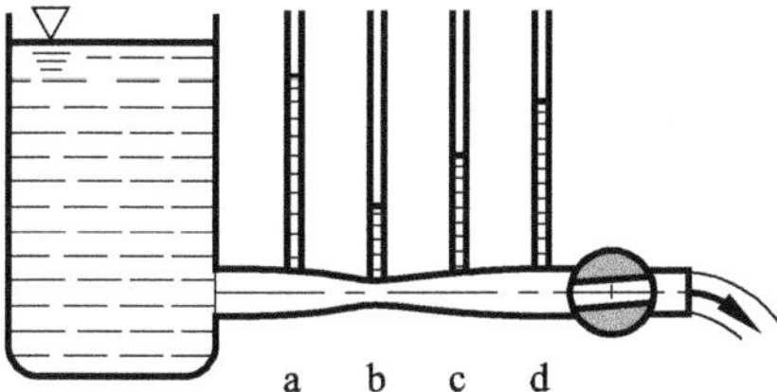

Fig. 4.12. Pressure decrease in a contraction

4.1.4 Interfaces and Formation of Vortices

If two liquids meet behind an edge (Figure 4.13), in general the constant of the Bernoulli equation is not the same for both flows. Since the pressure is the same along the surface that divides the two flows (interface), the magnitude of the velocity is different in both flows. Even if the Bernoulli constant for both flows is the same, the direction of the flows can be different on both sides. The velocity changes discontinuously across the interface. In the first case, the jump in velocity is longitudinal, and in the second case, transversal. Such interfaces are frequently observed. However, they are unstable and therefore do not remain in their original form for very long. Small perturbations can amplify quickly, so that the velocity differences increase in some positions and decrease in others. This causes the interface to decay into a great number of vortices. This is an important process in understanding the motion of fluids, and will be investigated more closely.

Fluctuations in the free stream cause the interface in Figure 4.13 to acquire a slightly wavy shape, sketched in Figure 4.14. The waves move forwards with the average velocity of the two streams, indicated in Figure 4.13 by the dashed line. In Figure 4.14 a reference frame has been chosen that moves with this average velocity, and so the crests and troughs of the waves are fixed in space. In this reference frame, the upper liquid flows to the right, and the lower to the left. If we analyze the pressure ratios in this flow, both the Bernoulli equation and (4.9) state that the transversal pressure increase is such that there is overpressure in the wave crests and underpressure in the wave troughs, assuming steady flow (indicated in Figure 4.14 by + and −). This pressure distribution shows that the flow cannot be steady. The

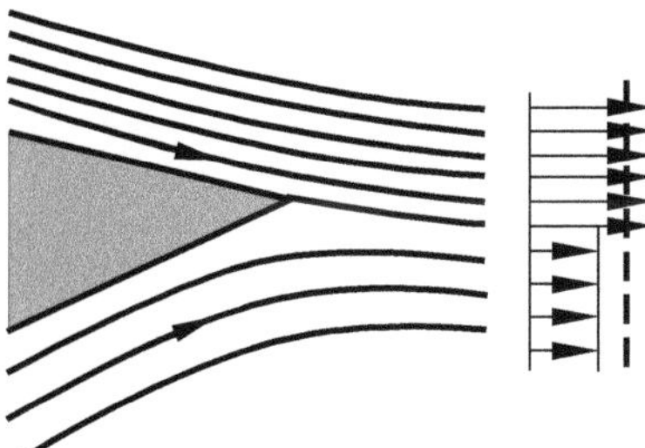

Fig. 4.13. Confluence of two liquids

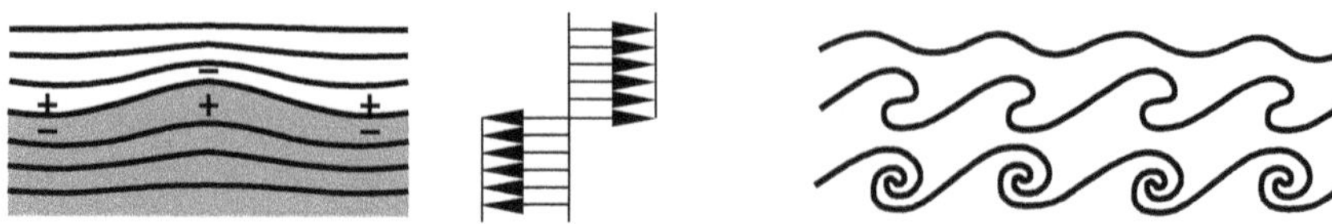

Fig. 4.14. Development of vortices from waves

liquid in the overpressure regions is set into motion, and it flows toward the neighboring underpressure region. This causes the waves to become stronger which leads to an instability. The subsequent behavior of such an interface is shown in Figure 4.14. It ends in decay into individual vortices.

The flapping of a flag in the wind has a similar origin. The pressure distribution in Figure 4.14 does not change if the direction of the lower flow is opposite, i.e. if it has the same direction as the upper flow. A slight bulge in the flag tends to strengthen (since the bulges move slightly with the wind, the process is actually somewhat more complicated).

At this point we consider yet another type of interface whose formation coincides with the formation of a vortex. If a fluid flows past an edge, at the start there is a flow around the edge, as shown in the left sketch in Figure 4.15. The velocity at the edge is very large. According to the theory for inviscid liquids, it would be infinitely large. It is observed that the velocity at the edge decreases with the formation of a vortex. This behavior can be considered as a particular principle, that the flow attempts to avoid infinite velocities and instead forms interfaces. In Section 4.2.6 we will show that it is the friction in the fluid, affecting the flow close to solid walls, that is behind this principle. If we assume a vortex behind the edge, so that fluid passes around the edge from behind, the conditions for merging of two flows at the edge are satisfied, and an interface forms (Figure 4.15 right). The interface is rolled up by the vortex, and fluid is supplied to it so that it can grow. In fact, both the vortex and the interface form a single unit that starts off very small (Figure 4.16). As it grows, the vortex moves away from the edge, and the interface decays into individual vortices, as described above, while new pieces of the interface continue to form at the edge.

Analogous processes occur at the edges of a round hole in a flat wall. The front edge of the interface rolls up and forms a vortex ring, which moves downstream forming a bounding liquid jet (Figure 4.17). Vortex rings can be produced by taking a box with a flexible rear wall and with a circular hole in the front wall, filling it with smoke and then hitting the rear wall.

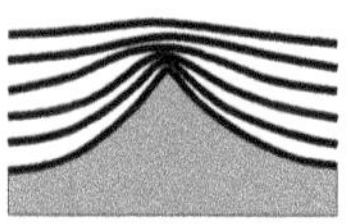

Fig. 4.15. Flow past an edge

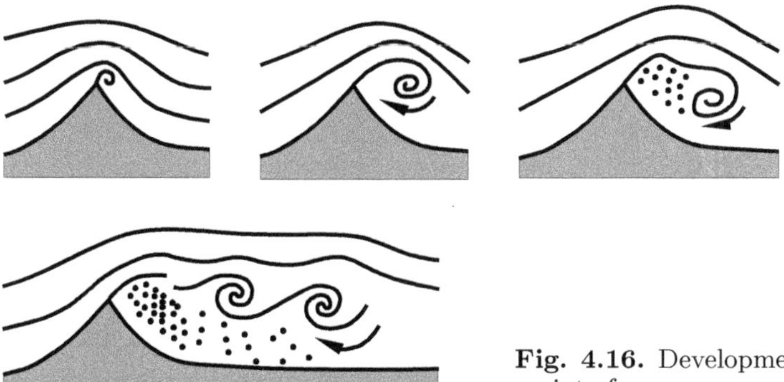

Fig. 4.16. Development and decay of an interface

A momentary flow out of the hole is generated, and so a jet is not formed, but rather only a vortex ring, which moves onwards and is seen as a smoke ring. Such vortex rings are very stable structures and decay only when their energy is almost completely dissipated by friction.

Transversal jumps in the velocity occur as a flow merges behind a finite plate that is tilted at a small angle to the direction of motion. On the pressure side, the streamlines move apart to the left and right under the effect of the overpressure that forms. On the suction side, the streamlines then bend inwards, due to the underpressure. Viewed from the middle of the plate, at the trailing edge perpendicular to the flow direction the flow on the pressure side has a velocity component toward the side edges, while on the suction side the flow is directed toward the middle. In the steady case the requirement that the pressure be continuous and the fact that all streamlines have the same origin means that the magnitude of the velocity is the same on both sides of the interface. The velocity jump is therefore wholly transversal. We know from experience that such interfaces roll inward from the ends of the plate and that two vortices arise that extend along the entire length passed by the plate. Figure 4.18 indicates this process. It shows the shape of the interface at different sections behind the plate. These processes are very important

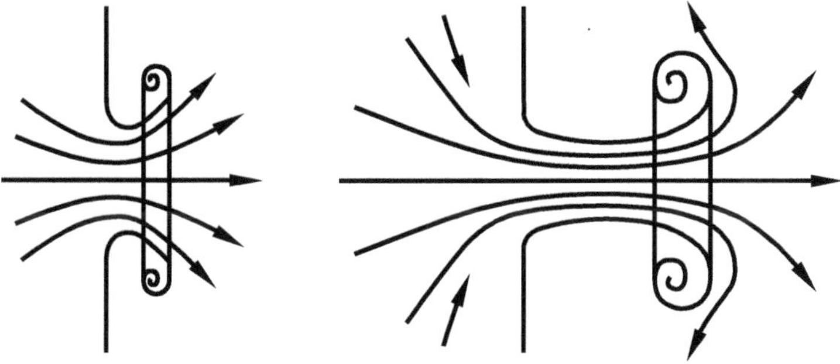

Fig. 4.17. Jet formation

Fig. 4.18. Interface behind a tilted plate

in understanding the flow past the wing of an airplane (Section 4.4). The vortices can be made visible by producing balls of smoke from cigar smoke in air at rest, taking a ruler tilted at a small angle and moving its free end quickly through the balls of smoke.

4.1.5 Potential Flow

In the previous sections, essentially only the average values of the flow variables were determined. However, the aim of hydrodynamics is to be able to determine the velocity at every point in space of the homogeneous inviscid flow. More mathematics than is assumed here is required to understand the relevant methods, and so in what follows we present only some more general explanations of the properties of inviscid flows and some simple examples. First of all, some concepts have to be explained.

By liquid lines and liquid surfaces we mean those lines and surfaces that are continually formed by the same liquid particles.

A line integral along a given line between points A and B is the integral over the product of the velocity components in the direction of ds with the line element $\mathbf{ds}$; i.e.

$$\Lambda = \int_{\mathrm{A}}^{\mathrm{B}} w \cdot \mathrm{d}s \cdot \cos(\alpha) = \int_{\mathrm{A}}^{\mathrm{B}} \boldsymbol{w} \cdot \mathbf{d}\boldsymbol{s}$$

(α is the angle between $\boldsymbol{w}$ and $\mathbf{d}\boldsymbol{s}$, while $\boldsymbol{w} \cdot \mathbf{d}\boldsymbol{s}$ is the scalar product of $\boldsymbol{w}$ and $\mathbf{d}\boldsymbol{s}$). For an unsteady flow, these line integrals are to be formed for an *instantaneous state* of the velocity distribution.

The magnitude of the line integral of a *closed* line is called the circulation Γ; i.e. with the sign $\oint$ for an integral along a closed line we have

$$\Gamma = \oint \boldsymbol{w} \cdot \mathbf{d}\boldsymbol{s}. \tag{4.12}$$

Thomson's law reads: *In an inviscid homogeneous liquid, the circulation along a closed liquid line remains constant in time.*

From this law we can draw the following important consequences:

If the motion of a liquid starts from rest, before the start of motion the circulation for every closed liquid line is equal to zero. Therefore, it remains zero at all times for this line. If the line integral along every closed line in some region is equal to zero, then the line integral from a point A to a point B is independent of the path, whatever path is chosen inside this region. We can move back along the previous integration path from B to A (the magnitude of the line integral from A to B is then canceled out, since the direction of $\mathbf{d}\boldsymbol{s}$ is opposite), and take another path to B. We obtain $\int_{\mathrm{A}}^{\mathrm{B}}$ plus an integral along a closed line, which is equal to zero. This again yields the integral $\int_{\mathrm{A}}^{\mathrm{B}}$, as was to be proved. If the point A is fixed, the line integral $\int_{\mathrm{A}}^{\mathrm{B}} \boldsymbol{w} \cdot \mathbf{d}\boldsymbol{s}$ assigns a numerical value to every point B. This value is denoted by Φ and is called the potential at point B. Moving from B to a point C a distance ds away, to form the integral $\int_{\mathrm{A}}^{\mathrm{C}}$ we can select the path via B. This yields

$$\int_{\mathrm{A}}^{\mathrm{C}} = \int_{\mathrm{A}}^{\mathrm{B}} + \boldsymbol{w} \cdot \mathbf{d}\boldsymbol{s} \quad \text{or} \quad \Phi_{\mathrm{C}} = \Phi_{\mathrm{B}} + w \cdot \mathrm{d}s \cdot \cos(\alpha) = \Phi_{\mathrm{B}} + w \cdot \mathrm{d}h \tag{4.13}$$

if dh is the projection of $\mathbf{d}\boldsymbol{s}$ onto the direction of $\boldsymbol{w}$. For $\alpha = 90°$ we have $\cos(\alpha) = 0$ and furthermore $\Phi_{\mathrm{C}} = \Phi_{\mathrm{B}}$. The segment $\mathrm{d}s = \overline{\mathrm{BC}}$ is therefore always perpendicular to the direction of $\boldsymbol{w}$ if $\Phi_{\mathrm{C}} = \Phi_{\mathrm{B}}$. All points for which $\Phi = \Phi_{\mathrm{B}}$ form a surface that passes through the point B. This surface divides the region where $\Phi > \Phi_{\mathrm{B}}$ from the region where $\Phi < \Phi_{\mathrm{B}}$. The tangential plane to this surface at the point B is perpendicular to the velocity vector $\boldsymbol{w}$ at point B. Therefore, in general, the *streamlines*, which always have the direction of the velocity vector, are *everywhere perpendicular to the surfaces* $\Phi = \text{const}$.

For arbitrary values of α, (4.13), with $\Phi_{\mathrm{C}} - \Phi_{\mathrm{B}} = \mathrm{d}\Phi$, yields

$$\frac{\partial \Phi}{\partial s} = w \cdot \cos(\alpha), \tag{4.14}$$

or

$$\frac{\mathrm{d}\Phi}{\mathrm{d}h} = w, \tag{4.15}$$

where dh is perpendicular to the surface $\Phi = \text{const}$. In vector notation this is written as

$$\boldsymbol{w} = \operatorname{grad} \Phi. \tag{4.16}$$

This combines (4.15) with the statement that $\boldsymbol{w}$ is perpendicular to the surfaces $\Phi = \text{const}$. The magnitude and direction of the velocity are equal to the greatest ascent of Φ, i.e. to its gradient.

These geometric interpretations of the potential and the gradient match those of the force potential U in physics, and it is from here that the name potential has been taken. Now, the gradient of the force potential is a field strength, while the gradient of the potential defined here is a velocity. This potential is therefore also called the *velocity potential.* Another difference is that the field strength is $\boldsymbol{g} = -\operatorname{grad} U$, while we set $\boldsymbol{w} = +\operatorname{grad} \Phi$. From the above discussions and using the potential and the circulation it follows that *every motion of a homogeneous inviscid fluid from rest has a potential.* Such motion is called a *potential flow.* It is characterized by the fact that the fluid particles experience no rotation. The circulation along a small closed curve is a measure of the rotation, and according to Thomson's law, this is equal to zero.

In a contrasting example, we consider a liquid that rotates like a rigid body with an angular velocity ω. For a circle with radius r centered at the origin of the reference frame, the velocity is equal to $\omega \cdot r$. Any motion of translation does not contribute to the circulation, and so does not have to be taken into account in computing the circulation. The direction of the velocity is tangential to the circumference of the circle. The line integral along the circumference is $\Gamma = 2 \cdot \pi \cdot r \cdot \omega \cdot r = 2 \cdot \pi \cdot r^2 \cdot \omega$. Dividing this equation by the area of the circle $A = \pi \cdot r^2$, we obtain $\Gamma/A = 2 \cdot \omega$. Thus Γ/A is a suitable measure for rotation. If the surface A is arbitrarily placed in space and forms an angle α with the axis of rotation, we obtain $\Gamma/A = 2 \cdot \omega \cdot \sin(\alpha)$, which is a maximum if the axis of rotation is perpendicular to A.

In potential flow the circulation for lines in the interior of the flow field is equal to zero. The flow in the interior is *irrotational.* In spite of this, vortices can occur in the motion of a homogeneous inviscid liquid from rest. If we consider the processes in the formation of an interface (Section 4.1.4), we see that all lines drawn at rest in the interior of the liquid move and deform to avoid the interface. None of the lines intersects the interface. *Thomson's law makes no statement about the relations of the regions on either side of an interface to each other.* Therefore, the fact that interfaces and vortices can arise in an inviscid liquid is not a contradiction of Thomson's law.

In real liquids, which have some friction, a shear layer forms instead of an interface. However, this is frequently very thin. The particles in the shear layer always come from the immediate neighborhood of the surface of the solid body, where, even for small viscosity, the friction may not be neglected. Exact analysis of the processes in shear layers must therefore take the friction into account. However, in general, considering an interface instead of a shear layer suffices in investigating the external processes. The effects of friction are explained in Section 4.2.

In Section 4.1.2 we derived (4.10) from the pressure drop perpendicular to the streamline for flows where the constant of the Bernoulli equation has the same value for all streamlines in a region. If the radius of curvature of the streamline is r, we can obtain the circulation about a small quadrilateral element formed from two streamlines and two normals (Figure 4.19):

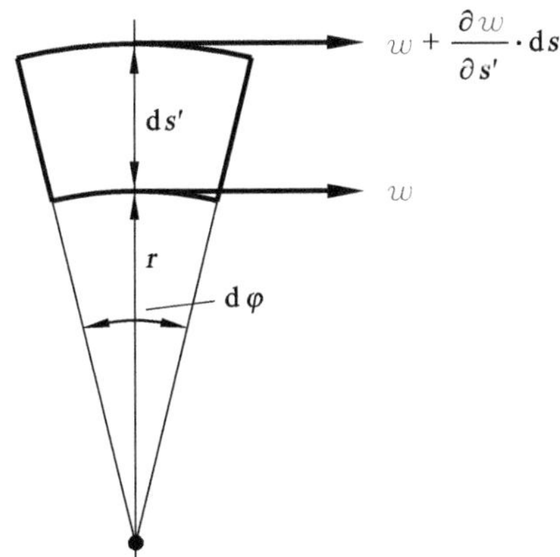

Fig. 4.19. Circulation around an infinitesimal quadrilateral

$$w \cdot r \cdot \mathrm{d}\varphi - \left(w + \frac{\partial w}{\partial s'} \cdot \mathrm{d}s' \right) \cdot (r + \mathrm{d}s') \cdot \mathrm{d}\varphi$$
$$= -\,\mathrm{d}s' \cdot \mathrm{d}\varphi \cdot \left(r \cdot \frac{\partial w}{\partial s'} + w + \frac{\partial w}{\partial s'} \cdot \mathrm{d}s' \right).$$

The normals do not contribute to the circulation. The last term in parentheses is of higher order and so can be discarded. The remainder of the parenthetical expression on the right-hand side is equal to zero, according to (4.10). This means that the above flows, for which the Bernoulli constant has the same value on all streamlines in a region, are motions with circulation equal to zero for every small element; i.e. they are potential flows. Conversely, the Bernoulli equation also holds perpendicular to the streamlines in every steady potential flow.

Potential Equation

The derivation of the potential equation of a general three-dimensional flow is carried out using the angular velocity. The angular velocity ω has three components (rotations about the coordinate axes):

$$\omega_x = \frac{1}{2} \cdot \left(\frac{\partial w}{\partial y} - \frac{\partial v}{\partial z} \right),$$
$$\omega_y = \frac{1}{2} \cdot \left(\frac{\partial u}{\partial z} - \frac{\partial w}{\partial x} \right), \qquad (4.17)$$
$$\omega_z = \frac{1}{2} \cdot \left(\frac{\partial v}{\partial x} - \frac{\partial u}{\partial y} \right).$$

If all three of these contributions to the rotation are to be zero, we must have $\partial v/\partial x = \partial u/\partial y$, etc. If a velocity potential Φ is introduced, i.e. if we set $u = \partial\Phi/\partial x$, $v = \partial\Phi/\partial y$, and $w = \partial\Phi/\partial z$, these conditions are identically satisfied. We have $\partial(\partial\Phi/\partial y)/\partial x = \partial(\partial\Phi/\partial x)/\partial y$, etc. This is always satisfied for regular multivariable functions. With $\partial v/\partial x = \partial u/\partial y$ and $\partial w/\partial x = \partial u/\partial z$, (3.5) leads to

$$\frac{\mathrm{d}u}{\mathrm{d}t} = \frac{\partial u}{\partial t} + u \cdot \frac{\partial u}{\partial x} + v \cdot \frac{\partial u}{\partial y} + w \cdot \frac{\partial u}{\partial z} = \frac{\partial u}{\partial t} + u \cdot \frac{\partial u}{\partial x} + v \cdot \frac{\partial v}{\partial x} + w \cdot \frac{\partial w}{\partial x}$$
$$= \frac{\partial u}{\partial t} + \frac{\partial}{\partial x}\left(\frac{u^2 + v^2 + w^2}{2}\right).$$

We obtain similar equations for $\mathrm{d}v/\mathrm{d}t$ and $\mathrm{d}w/\mathrm{d}t$. Inserting these expressions into the three Euler equations (5.76), multiplying each by $\mathrm{d}x$, $\mathrm{d}y$, and $\mathrm{d}z$, respectively and adding them, all terms may be integrated without any restriction of the path of integration. With $\int(\mathrm{d}p/\rho) = \mathrm{F}(p)$ we have

$$\frac{\partial \Phi}{\partial t} + \frac{u^2 + v^2 + w^2}{2} + \mathrm{F} + U = \text{const.} \tag{4.18}$$

The constant on the right-hand side still depends on the time, since the integration was carried out at a fixed time (e.g. the pressure may change in time due to external effects). It is therefore better to replace const with $\mathrm{f}(t)$.

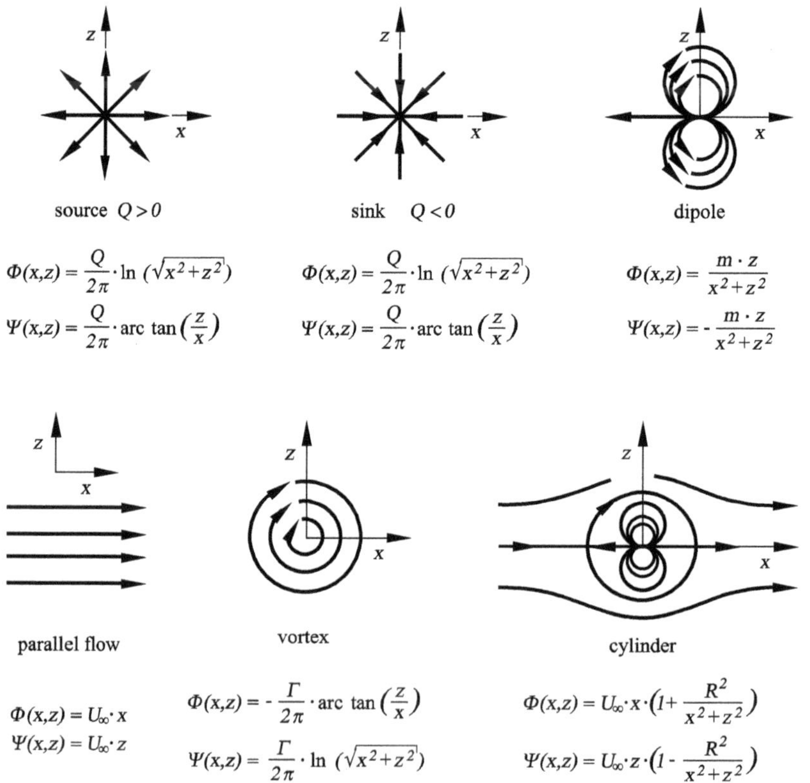

Fig. 4.20. Elementary solutions of potential flows

The expression $\partial\Phi/\partial t$ is obtained from $\Phi = \int(u \cdot \mathrm{d}x + v \cdot \mathrm{d}y + w \cdot \mathrm{d}z)$ and $\int(\partial u/\partial t) \cdot \mathrm{d}x = \partial(\int u \cdot \mathrm{d}x)/\partial t$, etc. For steady flows, (4.18) becomes the ordinary Bernoulli equation (4.4).

The relation of the velocity components u, v, and w to the potential Φ arises from (4.14); $\mathrm{d}s$ is replaced by $\mathrm{d}x$, $\mathrm{d}y$, and $\mathrm{d}z$ in turn; and we obtain

$$u = \frac{\partial\Phi}{\partial x}, \qquad v = \frac{\partial\Phi}{\partial y}, \qquad w = \frac{\partial\Phi}{\partial z}. \tag{4.19}$$

Using (4.19), the continuity equation for incompressible flows $\partial u/\partial x + \partial v/\partial y + \partial w/\partial z = 0$ (4.48) yields

$$\frac{\partial^2\Phi}{\partial x^2} + \frac{\partial^2\Phi}{\partial y^2} + \frac{\partial^2\Phi}{\partial z^2} = 0. \tag{4.20}$$

This equation is called the *Laplace equation.* This is a linear second-order partial differential equation, and so solutions may be represented as linear superpositions of elementary solutions. Figure 4.20 shows a summary of those elementary solutions applied in the following flow examples. The Laplace equation also appears in connection with electrostatic potentials and is valid in the parts of the field that have no charge and for which the dielectric constant is constant. Solutions of (4.20) known from electrostatics can also be applied here, such as the solution for a point charge or a dipole.

Stagnation Point Flow

One of the simplest forms of a potential is $\Phi = 0.5 \cdot (\mathrm{a} \cdot x^2 + \mathrm{b} \cdot y^2 + \mathrm{c} \cdot z^2)$. It follows from (4.20) that $\mathrm{a} + \mathrm{b} + \mathrm{c} = 0$ must hold. If the system is rotationally symmetric with respect to the z axis, we can set $\mathrm{b} = \mathrm{a}$. Then (4.20) says that $\mathrm{c} = -2 \cdot \mathrm{a}$, and so the potential is

$$\Phi = \frac{\mathrm{a}}{2} \cdot (x^2 + y^2 - 2 \cdot z^2),$$

with $u = \mathrm{a} \cdot x$, $v = \mathrm{a} \cdot y$, and $w = -2 \cdot \mathrm{a} \cdot z$. The streamlines in the y-z plane ($x = 0$) are given by the differential equation

$$\frac{\mathrm{d}z}{\mathrm{d}y} = \frac{w}{v} = -\frac{2 \cdot z}{y},$$

which when integrated yields

$$\ln(z) = \mathrm{const} - 2 \cdot \ln(y), \qquad \text{or} \qquad y^2 \cdot z = \mathrm{const}$$

(cubic hyperbolas, Figure 4.21).

If the motion is steady, i.e. if a is constant in time, then the pressure is

$$p = \mathrm{const} - \frac{\rho}{2} \cdot (u^2 + v^2 + w^2) = \mathrm{const} - \frac{\rho \cdot \mathrm{a}^2}{2} \cdot (x^2 + y^2 + 4 \cdot z^2).$$

The pressure is a maximum for $x = y = z = 0$. The surfaces of equal pressure are ellipsoids with axial ratio $1 : 1 : 0.5$ (Figure 4.21).

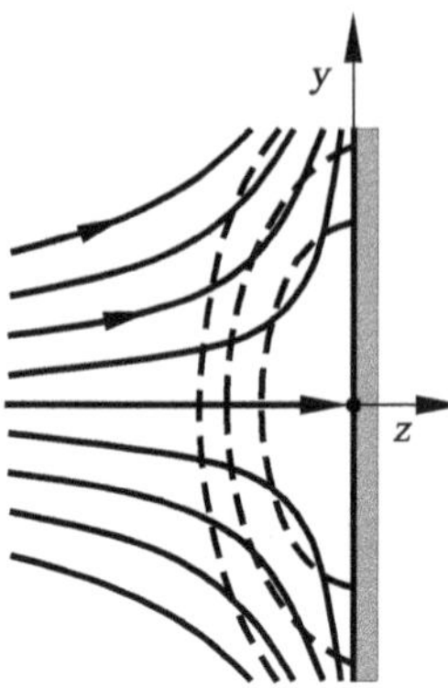

Fig. 4.21. Stagnation point flow (streamlines and isobars (dashed))

Sources and Sinks

According to the note following (4.20), known solutions of electrostatic potentials are also solutions for possible potential flows, as long as the boundary conditions can be satisfied. In fact, the electrostatic field of a point charge leads to an important flow, the *source* or *sink* flow. The potential reads $\Phi = \pm\ \mathrm{C}/r$, where r is the distance from a point O, and C is a constant. The potential is therefore constant on spheres with center 0. The velocity is always in the radial direction, since it is perpendicular to surfaces of constant potential. It has magnitude $|\mathrm{C}|/r^2$. The amount of fluid that flows through a sphere of radius r (surface $4{\cdot}\pi{\cdot}r^2$) per unit time is $Q = 4{\cdot}\pi{\cdot}r^2{\cdot}\mathrm{C}/r^2 = 4{\cdot}\pi{\cdot}\mathrm{C}$. For a source at point 0, this amount appears per second, while for a sink, this amount vanishes per unit time. This case is physically impossible. However, a thin tube, for example, can be used to suck fluid at the point O, and a flow approximating that described then occurs close to the suction site (only approximately, since the finite volume of the pipe affects the flow).

A further very useful application of the source and sink flow is the following: If a rod-shaped body moves forward in the direction of the rod axis with velocity U_∞, fluid is constantly being displaced at its front end, while at its trailing end, fluid flows together in the space that has become free (Figure 4.22). The flow in the neighborhood of the front part therefore behaves as if a source were placed there. From the flow close to the back part of the rod

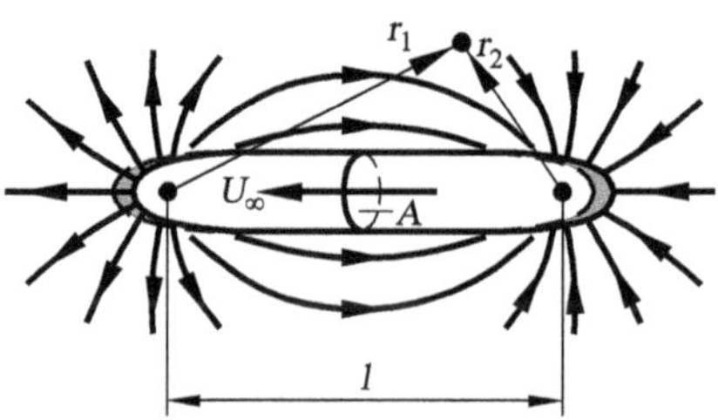

Fig. 4.22. Potential flow past a moving body, reference frame at rest

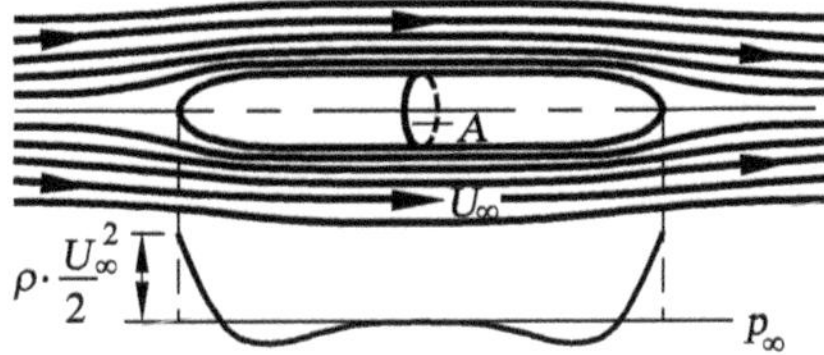

Fig. 4.23. Potential flow and pressure distribution past a moving body, reference frame moving with the body

it appears as if there is a sink. In fact, the flow is described by the equation

$$\Phi = \mathrm{C} \cdot \left(\frac{1}{r_2} - \frac{1}{r_1} \right) .$$

In order that this equation delivers the exact solution of the flow, the ends of the rod have to have a certain rounded form. However, even if the ends have another form, this equation is still a useful equation. The strength Q of the source and the sink is equal to $A \cdot U_\infty$. Here A is the cross-section of the rod, i.e. $\mathrm{C} = A \cdot U_\infty / (4 \cdot \pi)$. The flow is unsteady due to the forward motion of the rod and the velocity distribution around the rod. However, if we consider the flow from a reference frame that moves with the body, it is steady.

For this flow, the body is at rest and the fluid moves past the body. Mathematically, this flow is described by the potential $\Phi' = \Phi + U_\infty \cdot x$. Its streamlines are shown in Figure 4.23. The pressure distribution along the surface of the body is qualitatively sketched below, as obtained from the Bernoulli equation.

The flow past other slender rotationally symmetric bodies can be described by continuous distributions of sources along the axis. If the distance between the source and the sink is reduced and the source strength is increased to the same degree as its distance is decreased, we obtain a dipole as the limiting case. The flow from Figure 4.23 then becomes the flow past a sphere (Figure 4.24). With the radius of the sphere R, the associated poten-

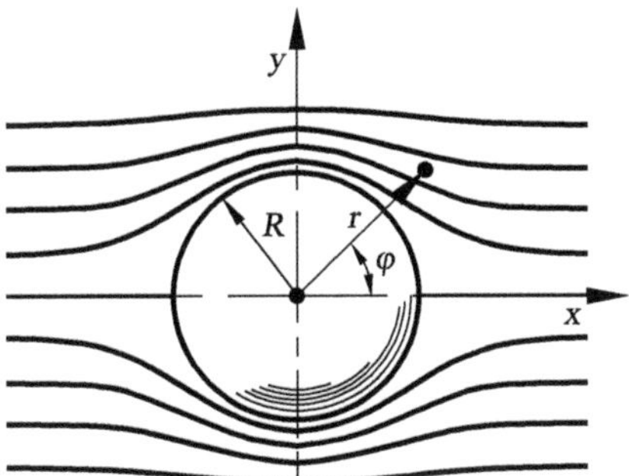

Fig. 4.24. Potential flow past a sphere

tial is $\Phi = U_\infty \cdot x \cdot (1 + R^3/(2 \cdot r^3))$. In a real flow past a sphere, the effects of friction mean that the wake looks different (see Section 4.2.6).

Two-Dimensional Motion

If all streamlines are two-dimensional curves in parallel planes and if the flow state is the same on a straight line perpendicular to the family of planes, the flow is called a planar flow. If one of these planes is chosen to be the x-y plane, the velocity component $w = 0$ and the velocity components u and v are functions of x and y only. It can be shown that both the real and imaginary parts of every analytic function of the complex variable $x + i \cdot y$ is a potential that satisfies (4.20). Let the complex variable be called $z = x + i \cdot y$, and the function $\mathrm{F}(z)$, with real part Φ and imaginary part Ψ. We have

$$\frac{\partial \mathrm{F}}{\partial x} = \frac{\mathrm{dF}}{\mathrm{d}z} \cdot \frac{\partial z}{\partial x} \qquad \text{and} \qquad \frac{\partial \mathrm{F}}{\partial y} = \frac{\mathrm{dF}}{\mathrm{d}z} \cdot \frac{\partial z}{\partial y}.$$

Because of

$$\frac{\partial z}{\partial x} = 1 \qquad \text{and} \qquad \frac{\partial z}{\partial y} = i,$$

we also have

$$\frac{\mathrm{dF}}{\mathrm{d}z} = \frac{\partial \mathrm{F}}{\partial x} = \frac{1}{i} \cdot \frac{\partial \mathrm{F}}{\partial y}.$$

With $\mathrm{F} = \Phi + i \cdot \Psi$ this yields

$$\frac{\partial \Phi}{\partial x} + i \cdot \frac{\partial \Psi}{\partial x} = \frac{1}{i} \cdot \frac{\partial \Phi}{\partial y} + \frac{\partial \Psi}{\partial y}.$$

Both the real and imaginary parts of this equation must hold. With $1/i = -i$, it follows that

$$\frac{\partial \Phi}{\partial x} = \frac{\partial \Psi}{\partial y} = u \qquad \text{and} \qquad \frac{\partial \Phi}{\partial y} = -\frac{\partial \Psi}{\partial x} = v, \tag{4.21}$$

and we obtain

$$\frac{\partial^2 \Phi}{\partial x^2} + \frac{\partial^2 \Phi}{\partial y^2} = \frac{\partial^2 \Psi}{\partial y \partial x} - \frac{\partial^2 \Psi}{\partial x \partial y} = 0;$$

i.e. the Laplace equation (4.20) is identically satisfied. The function Ψ also satisfies $\partial^2\Psi/\partial x^2 + \partial^2\Psi/\partial y^2 = 0$, and so Ψ is also a flow potential. It follows from (4.21) that the flows associated with the potentials Φ and Ψ are orthogonal to each other at all positions and their velocities have the same magnitude. The two gradient directions α and β are given by $\tan(\alpha) = (\partial\Phi/\partial y)/(\partial\Phi/\partial x) = v/u$ and $\tan(\beta) = (\partial\Psi/\partial y)/(\partial\Psi/\partial x) = u/(-v)$, i.e. $\tan(\beta) = -1/\tan(\alpha)$. The magnitude of the gradient in both cases is equal to $\sqrt{u^2 + v^2}$. The lines of constant potential of one flow are therefore streamlines of the other flow. The velocity is always perpendicular to the potential surface. The function that is constant on streamlines is called the stream function. If Φ is the *potential*, then Ψ is the *stream function*. The stream function has another graphical meaning: The difference in the function value of two points represents the amount of volume flowing between both points per unit time in a layer of thickness 1.

The properties of the lines of equal potential and equal stream function present us with a graphical method of determining both systems of lines for given boundary conditions. We begin with a rough draft of the streamlines, draw an orthogonal system onto this draft, and improve the sketch until the mesh is everywhere sufficiently square. Characteristic of this are equal lengths of the midlines in the squares and the orthogonality of the two families of diagonal curves, satisfying the equations $\Phi + \Psi =$ const. and $\Psi - \Phi =$ const., drawn through the corners of the squares. Figures 4.23, 4.26, 4.27 and 4.30 were all sketched in this manner. Figure 4.25 shows an example of a graphically constructed solution.

We now present simple examples of two-dimensional flows. The *plane stagnation point* flow is given by the function $\mathrm{F} = (\mathrm{a}/2) \cdot z^2$:

$$\Phi + i \cdot \Psi = \frac{\mathrm{a}}{2} \cdot (x^2 + 2 \cdot i \cdot x \cdot y - y^2),$$

i.e.

$$\Phi = \frac{\mathrm{a}}{2} \cdot (x^2 - y^2) \quad \text{and} \quad \Psi = \mathrm{a} \cdot x \cdot y.$$

The streamlines $\Psi =$ const are equal-sided hyperbolas. The velocity components u and v satisfy the equations

$$u = \frac{\partial \Phi}{\partial x} = \mathrm{a} \cdot x, \qquad v = \frac{\partial \Phi}{\partial y} = -\mathrm{a} \cdot y.$$

The two-dimensional source flow is obtained from $\mathrm{F} = \mathrm{b} \cdot \ln(z)$, with $\ln(z) = \ln(r) + i \cdot \varphi$, with radius r and central angle φ in polar coordinates (i.e. $\Phi =$ const. on circles $r =$ const., and $\Psi =$ const. on radial straight lines $\varphi =$ const.).

Another example is the *flow at two walls* that form an angle α with each other. If the point of intersection is at the origin and the first wall is along the x axis, the function reads $\mathrm{F} = (\mathrm{a}/n) \cdot z^n$, with $n = \pi/\alpha$. Introducing polar coordinates, we have $z = x + i \cdot y = r \cdot (\cos(\varphi) + i \cdot \sin(\varphi))$ and

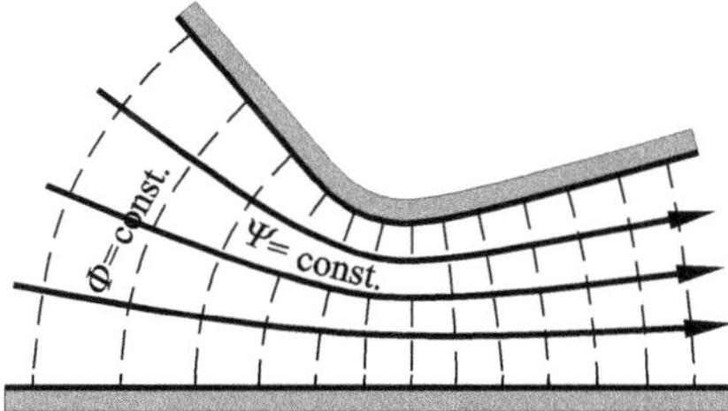

Fig. 4.25. Graphical construction of Φ and Ψ

$$z^n = r^n \cdot (\cos(n \cdot \varphi) + i \cdot \sin(n \cdot \varphi)).$$

This yields a stream function $\Psi = (\mathrm{a}/n) \cdot r^n \cdot \sin(n \cdot \varphi)$. For $\varphi = 0, \pi/n, 2 \cdot \pi/n, \ldots$, i.e. for $\varphi = 0, \alpha, 2 \cdot \alpha, \ldots$ we have $\Psi = 0$. The shape of the streamlines for different values of α is seen in Figure 4.26. For $\alpha < \pi$ the velocity at the origin is 0, while for $\alpha > \pi$, it is ∞. Taking the limit to $\alpha = 0$ leads to the function

$$\mathrm{F} = \mathrm{a}' \cdot e^{\mu \cdot z} = \mathrm{a}' \cdot e^{\mu \cdot x} \cdot (\cos(\mu \cdot y) + i \cdot \sin(\mu \cdot y)).$$

The distance between the two walls is $h = \pi/\mu$. The flow deviated about a right angle $\mathrm{F}' = \mathrm{a}' \cdot e^{-i \cdot \mu \cdot z} = \mathrm{a}' \cdot e^{\mu \cdot y} \cdot (\cos(\mu \cdot x) - i \cdot \sin(\mu \cdot x))$ can be used to describe wave processes (Figure 4.40).

The flow past a *circular cylinder* of radius R is given by $\mathrm{F} = U \cdot (z + R^2/z)$. The stream function is then found to be $\Psi = U \cdot \sin(\varphi) \cdot (r - R^2/r)$. For the x axis on which $\sin(\varphi) = 0$ and for the circle of radius R for which $r - R^2/r = 0$, the value of the stream function is $\Psi = 0$. The streamline portrait and the potential line portrait of this flow are very similar to those in Figure 4.25.

There are a great many further examples of potential flows, and a great number of different methods can be used to find suitable solutions. For example, the complex relation $z = \mathrm{f}(\zeta)$, where $\zeta = \xi + i \cdot \eta$ is another complex number, assigns to every ξ, η a pair of values x, y. To each point in the ξ-η plane there is an associated point in the x-y plane. This is called a mapping. One line corresponds to one line; the point of intersection of two lines corresponds to the point of intersection of the associated lines. Specifically, the relations analogous to (4.21) hold. A right-angular mesh is again mapped onto a right-angular (but in general curvilinear) mesh. The scale of the mapping is the same in both directions, so that infinitely small scales are mapped in a geometrically similar manner. This type of mapping is therefore also called a conformal mapping. The previous examples of two-dimensional flows are also conformal mappings if Φ and Ψ are replaced by ξ and η. The last example (a flow past a circular cylinder) shows how half of the Φ-Ψ plane is mapped onto a region bounded by two pieces of the x axis with a semicircle of radius R between them.

If F is an analytic function of z, and z is an analytic function of ζ, then F is also an analytic function of ζ; i.e. $\mathrm{F} = \Phi + i \cdot \Psi$ also yields a possible

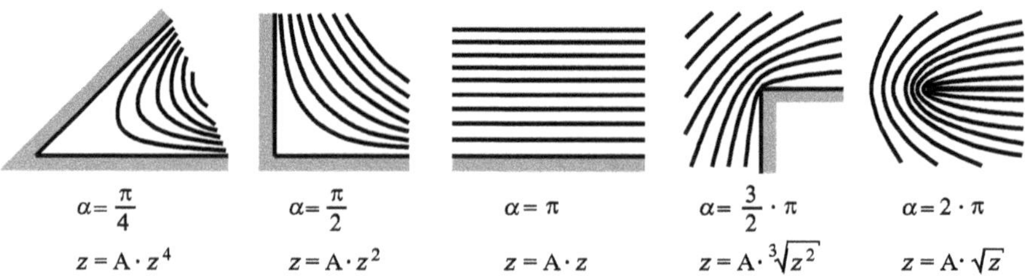

Fig. 4.26. Flows $F = \mathrm{A} \cdot z^n$

flow in the ζ plane. Any flow in the x-y plane can be mapped by any mapping of the x-y plane to the ξ-η plane to a new flow in the ξ-η plane. The process can be repeated as often as required. This is an important concept for hydrodynamics.

There are different methods of mapping the outer region of a airfoil-like contour to the outer region of a circle. Therefore, the flow past the circle can also yield a flow past a wing, etc.

The differential quotient $\mathrm{dF}/\mathrm{d}z$ is equal to $u - i \cdot v$ (the conjugate value of the complex velocity $u + i \cdot v$). Calling this quantity w, then $w = \mathrm{dF}/\mathrm{d}z$ is also an analytic function of z or of F. The relation of the Φ-Ψ plane to the u-v plane is also a conformal mapping. There are cases in which statements can be made about the velocities that suffice to completely determine the region in the w plane. If a liquid jet exits through a *gap* between two walls (Figure 4.27), the direction is given for the limiting streamline as long as it flows along a flat wall. The direction is not known for the boundaries of the free jet, but the magnitude of the velocity is known. Because of the Bernoulli equation this must be constant if the pressure is constant. This yields a boundary to the region (Figure 4.27, right). It now remains only to correctly describe the singularities that occur in order to obtain F as a function of w. We determine the inverse function $w = w(\mathrm{F})$. From $\mathrm{dF}/\mathrm{d}z = w(\mathrm{F})$ we obtain $z = \int(\mathrm{dF}/w(\mathrm{F}))$. Separating the real and imaginary parts, we finally determine the x and y values to each value of Φ and Ψ and thus obtain the streamline portrait.

This brief overview gives an idea of the complex methods used in determining potential flows.

Although the circulation vanishes in all small regions of all potential flows, there are flows in which a circulation occurs in the entire flow field. The condition for this is that the region in which the flow occurs is multiply connected. This multiple connection is characterized by the fact that there are curves that cannot be pulled together to zero by continuous changes without leaving the region. Examples of multiply connected spaces are a room with a column in the middle, or the space surrounding a ring. If the circulation along such a curve is equal to Γ, the circulation along every other curve that

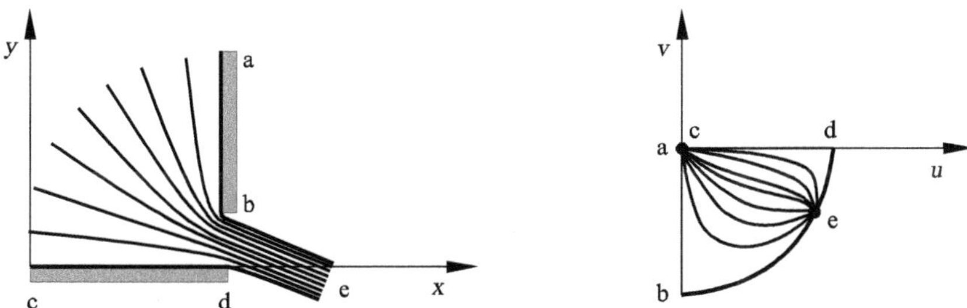

Fig. 4.27. Flow and velocity field in the discharge from a gap

arises from continuous change of this curve without leaving the region is also equal to Γ if the flow is otherwise irrotational (i.e. the circulation in every simply connected region vanishes). The potential found from the line integral between a fixed point and each point in space is ambiguous in such flows. For every turn it increases by the amount Γ.

The simplest case of a two-dimensional flow of this type is described by the potential $\Phi = \mathrm{C} \cdot \varphi$. Here φ is a central angle (Figure 4.28). This potential, which also satisfies (4.20) in complex notation $\mathrm{F} = -i \cdot \mathrm{C} \cdot \ln(z)$, increases by $2 \cdot \pi \cdot \mathrm{C}$ for each turn ($\varphi_2 = \varphi_1 + 2 \cdot \pi$). The value by which the potential increases is the circulation Γ. The surfaces of constant potential in this case are planes through the axis, and the streamlines are therefore circles. The velocity $w = \mathrm{d}\Phi/\mathrm{d}s$ is found with $\mathrm{d}s = r \cdot \mathrm{d}\varphi$ to be $w = \mathrm{C}/r$. The flow therefore corresponds to the flow in the example in Figure 4.6. For $r = 0$ we would obtain $w = \infty$, and so the flow has physical meaning only outside a core of finite diameter (Figure 4.28). The core can either be formed by a solid body or can consist of rotating liquid (in which there is no potential). It can also consist of another (lighter) nonrotating liquid, such as air, if water forms the surrounding liquid (hollow vortex). The effect of Earth's gravity causes the surface of such a hollow vortex to assume a shape as in Figure 4.29. Its form is found from the Bernoulli equation to be

$$z = z_0 - \frac{w^2}{2 \cdot g} = z_0 - \frac{\mathrm{C}^2}{2 \cdot g \cdot r^2}.$$

Such funnels can be observed in flowing bodies of water, or on emptying a bathtub. In these cases the flows already had circulation from other causes.

4.1.6 Wing Lift and the Magnus Effect

A further application of potential flows with circulation is in determining the lift of wings (Section 4.4.3). The flow past a wing in Figure 4.30 (top picture) can be generated by superposition of an ordinary potential flow (without circulation) and a flow with circulation around the wing. The flow past the

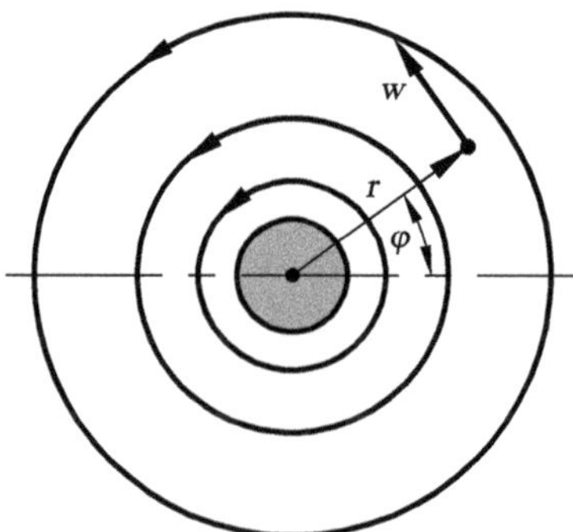

Fig. 4.28. Potential flow with circulation

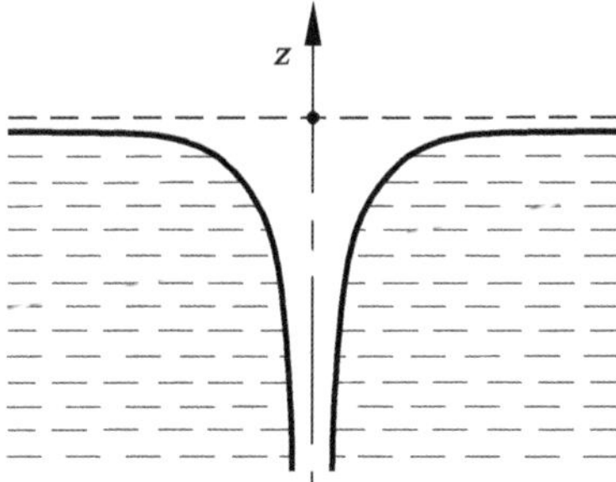

Fig. 4.29. Hollow vortex

wing itself therefore has a circulation. Even without any calculation we see that the flow with circulation on the upper side of the wing strengthens the potential flow, and that on the lower side acts against the potential flow. According to the Bernoulli equation, this implies a pressure decrease on the wing and a pressure increase on the lower side of the wing; i.e. a lift occurs. *M. W. Kutta* and *N. Y. Joukowski* independently discovered that this force is proportional to Γ. Its size per unit length is equal to $\rho \cdot \Gamma \cdot U_\infty$, with the free-stream velocity U_∞ of the wing. This law will be proved in Section 4.1.7.

According to Thomson's law, no circulation can occur in a motion from rest even in multiply connected spaces, since at rest the circulation on every line is equal to zero. Therefore, even in motion the circulation remains zero. In fact, the circulation generally occurs over an interface. For example, in the

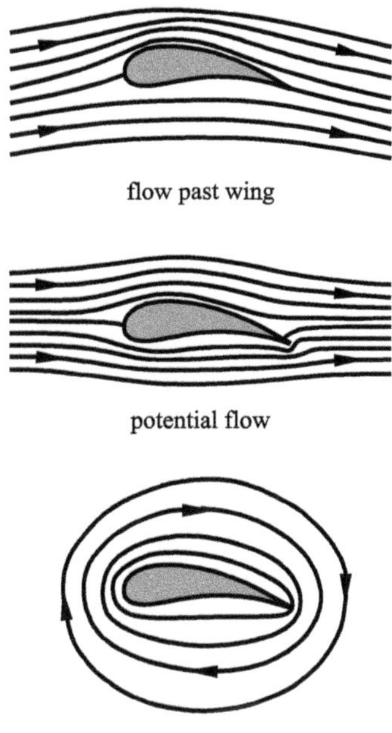

Fig. 4.30. Flow past a wing

spiral casing in Figure 4.6, a vortex forms on the sharp edge at the start of motion, as shown in Figure 4.15. The vortex later floats away at O, and only its circulation remains for the duration of the flow.

In the flow past a wing the solution is similar. At the start of motion an interface forms at the trailing edge, as shown in Figure 4.31. Later, the vortex arising from the interface moves downstream. A circulation remains at the wing, and this is equal to but in the opposite direction of that of the vortex. The lines that contain both the wing and the vortex still retain the circulation zero, as Thomson's law requires.

In order that the wing generates a doubly connected flow region, the wing has to be bounded on the sides by two parallel walls, or it must be assumed that it is infinitely extended out to both sides. In real wings neither one nor the other is true. The circulation about the wing, which is present here, too, and is necessary to bring about the lift, is generated by an interface with a transversal velocity jump.

A circulation like that at the wing also occurs at a rotating circular cylinder in a flow parallel to its axis. This time, the occurrence is due to friction. It produces a force per unit length perpendicular to the flow that is equal to $\rho \cdot \Gamma \cdot U_\infty$ and is called the transverse drive. In cases of triangular and quadrilateral prisms that rotate about their longitudinal axes, and of spheres, etc., this force occurs. The action of the force always takes place from the side where rotation and flow are orientated opposite to each other, to the side where they have the same direction. This effect is named after its discoverer *H. G. Magnus* (1852) as the *Magnus effect*.

Spherically shaped bullets often acquire an unintentional rotation about transverse axes, and their flight paths deviate to one side. This behavior was the origin of the investigations into the Magnus effect. Such sideward deviations can be seen in the flights of sliced tennis and golf balls in the air. *A. Flettner* (1926) exploited the effect in his rotor ship to drive ships by the wind. Instead of a sail, a perpendicular, rapidly rotating cylinder was used. Disks are applied at the ends (Figure 4.32, left), since otherwise, the air that does not pass around at the ends of the cylinder would penetrate the underpressure region on the suction side and partially destroy the flow there. The experiments with such ships were successful. However, the regular motor ship was economically superior, so that the Flettner drive did not catch on.

The effect of the Flettner rotor can be understood with a simple experiment. A rotating cylinder by a small electric motor driven is found on a cart running on tracks. If air is blown toward the cylinder from a small ventilator perpendicular to the tracks, the cart moves forward on the tracks. If the ventilator is turned so that the wind forms a different angle with the tracks,

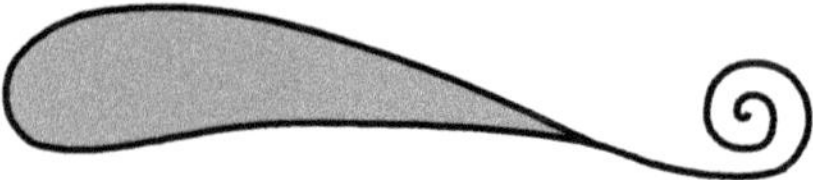

Fig. 4.31. Start-up vortex of a wing

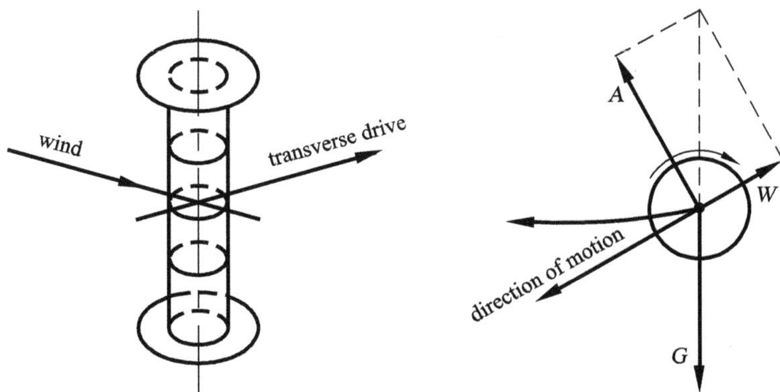

Fig. 4.32. Rotating cylinder

the behavior of the cylinder sail can be investigated under different angles of attack. It is possible to make the cart move at an acute angle to the wind. If the direction of rotation of the cylinder is reversed, the cart travels in the opposite direction.

A light cylinder rotating rapidly about a horizontal axis does not fall down vertically if it is allowed to drop, but rather its flight path becomes a flat gliding flight. Apart from the lift A perpendicular to its path, it experiences a drag W in the direction of the flight path, which in the most favorable case (longitudinal cylinder with end disks) is considerably smaller than the lift. The resultant of these two forces keeps the weight G of the cylinder in equilibrium (Figure 4.32, right) and prevents it from falling vertically.

4.1.7 Balance of Momentum for Steady Flows

The balances of momentum of general mechanics, known as the law of center of mass and the law of areas, are also applied to steady and unsteady flows of liquids whose time averages can be considered to be steady motion. The value of these balances of momentum lies in the fact that they contain statements only about the states on the boundaries of a region, and so processes can be predicted without the necessity of fully understanding their details.

The momentum of a mass is the product of mass and velocity. The momentum is a vector and, like the velocity, has three components. The *rate of change of the momentum is equal to the resultant force acting on the mass.* In Section 2.2 we saw that in summing over all masses of a mechanical system, all internal forces cancel out according to the principle of action and reaction, and only the external forces, acting from masses outside the system, remain.

In a *steadily* flowing liquid mass with arbitrary boundaries, the momentum changes only when the boundaries of the liquid mass shift due to the flow. Inside the liquid mass, each particle has been replaced by another, which has taken on its velocity. What happens on the boundaries can be shown by considering a stream filament. The balance of momentum states that every partial mass that belongs to the system remains in the system, and no new partial masses are added to the system. The bounding surfaces selected for the application of the balance of momentum therefore move with the flow. For the stream filament in Figure 4.33, at 1 the mass $dm_1 = \rho \cdot A_1 \cdot w_1 \cdot dt$ vanishes in time dt. At 2 the mass $dm_2 = \rho \cdot A_2 \cdot w_2 \cdot dt$ appears. Because of continuity we must have $dm_1 = dm_2 = dm$. In the time dt the stream filament at 2 therefore contributed the positive amount $dm \cdot w_2$ to the total change of momentum, thus per unit time $(dm/dt) \cdot w_2 = \rho \cdot A_2 \cdot w_2^2$ (in the direction of w_2). Similarly, at 1 the negative amount $-(dm/dt) \cdot w_1 = -\rho \cdot A_1 \cdot w_1^2$ is contributed (in the opposite direction to w_1). The vector sum of these changes in momentum per unit time is equal to the resultant of all the external forces acting on the stream filament. Instead of the changes in momentum, we can also look at their reactions, i.e. the forces of the same magnitude but in the opposite direction. The vector sum of these reaction forces is in equilibrium with the forces acting on the stream filament. This procedure is the same as that used in the introduction of the inertial forces in d'Alembert's principle of the mechanics of rigid bodies. The liquid flow in Figure 4.33 at 1 corresponds to a reaction force $\rho \cdot A_1 \cdot w_1^2$ in the direction of the incoming flow, and at 2 it corresponds to a reaction force $\rho \cdot A_2 \cdot w_2^2$ in the opposite direction to the outgoing flow. This formulation completes the transition to a surface fixed in space. The changes in momentum (or their reaction forces) and the pressure forces are carried over to the boundary surface fixed in space. In order to apply the balances of momentum correctly, it is practical to surround the liquid mass with a closed surface, the *control surface.* This is shown in bold in some of the following figures. For all incoming and outgoing stream filaments, the reaction forces must form an equilibrium system with all the external forces that act on the liquid inside the control surface, according to the laws of statics. This means that both the sum of the forces and the sum of the moments of the forces must be equal to zero for all coordinate axes. In practice, it is often the forces that the liquid causes to act on the walls of its container rather than the forces acting on the liquid that are of interest. Very frequently, only the equation of one component is needed to solve the particular problem.

In the case of *unsteady* flows there is an additional term in the balance of forces. This is due to the change of momentum inside the liquid. If the unsteady flow has a constant average value of momentum, as is often the case

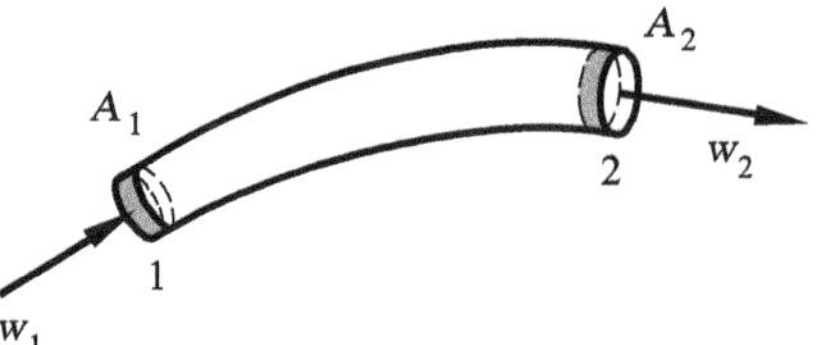

Fig. 4.33. Change of momentum in a stream filament

in turbulent flows, these contributions inside the liquid cancel each other out on average, and so the balances of momentum can be applied as for steady flows.

Reaction Forces in Curved Channels

A liquid flows with a velocity w_1 and a pressure p_1 into the curved channel (Figure 4.34). The transport of momentum through the surface A_1 is equal to $\rho \cdot A_1 \cdot w_1^2$. This is the same as a force acting from the inflowing liquid in the direction of flow. A pressure force $p_1 \cdot A_1$ in the same direction also has to be taken into account. A corresponding reaction force $A_2 \cdot (\rho \cdot w_2^2 + p_2)$ acts as the liquid flows out of the channel. It is directed opposite to the velocity (therefore always toward the interior of the control surface). The resultant of the two forces is the actual force acting from the liquid flow onto the channel through the pressure forces at the wall.

Reaction Forces in Jets

A jet that exits a region with pressure p_1 through an opening into a region with pressure p_2 has a momentum of magnitude $J = \rho \cdot A \cdot w^2$ per unit time, where A is the cross-section of the jet. With $w = \sqrt{2 \cdot (p_1 - p_2)/\rho}$ (Section 4.1.2) we obtain $J = 2 \cdot A \cdot (p_1 - p_2)$. This corresponds to twice the force that would act on a piston of the size of the jet cross-section from the pressure difference $p_1 - p_2$. This momentum must have an equivalent in the pressure distribution. It follows that a loss in the wall pressure arises, compared to that of the closed vessel, by the vanishing of the overpressure p_1 at the opening and the pressure reduction close to the opening due to the outgoing flow. This loss corresponds to the pressure on twice the jet cross-section. The vanishing of the pressure is expressed as a reaction force of the exiting jet. This reaction force can be detected by placing a vessel with a side opening on a moveable cart. The cart with the vessel moves in the opposite direction to the exiting jet.

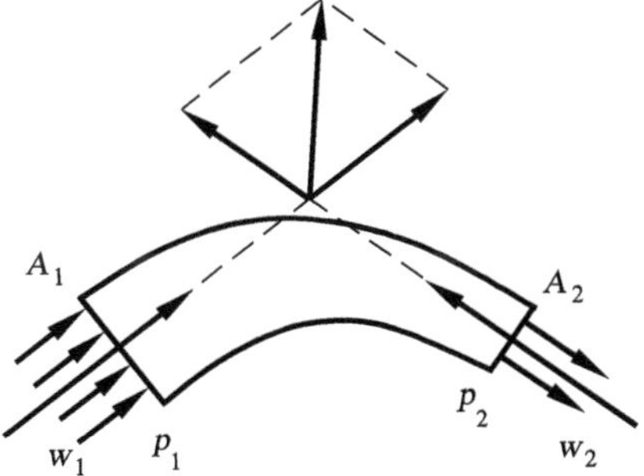

Fig. 4.34. Reaction forces at a curved pipe

A similar experiment can be performed using the *Segner waterwheel* (Figure 4.35). A weight can be lifted by the exiting water, or it can carry out other work.

In the case of the *Borda outlet* (Figure 4.36), the so-called contraction coefficient, the ratio of the jet cross-section to the hole cross-section, can be determined from the magnitude of the momentum. Since the entire overpressure p_1 acts on all wall surfaces whose pressure forces have components in the jet direction, the vanishing of the overpressure in the outlet cross-section A must be equal to the jet momentum. We have $A \cdot (p_1 - p_2) = 2 \cdot A_S \cdot (p_1 - p_2)$ or $A_S = (1/2) \cdot A$.

Sudden Expansion

If a liquid flow with velocity w_1 exits a cylindrical pipe section into a larger cylindrical pipe, the jet mixes with the surrounding liquid. After the mixing, it flows downstream almost uniformly with a mean velocity w_2. We can use the balance of momentum to compute the pressure increase $p_1 - p_2$ associated with the mixing, without having to know the details of the mixing process. In the liquid at rest in the larger pipe that surrounds the incoming jet, the same pressure p_1 is at hand as in the jet (cf. Section 4.3.5, free jet). For the control surface sketched in Figure 4.37, of which only the forces on the two facing surfaces contribute to the force balance, we have

$$\frac{\mathrm{d}m}{\mathrm{d}t} \cdot (w_1 - w_2) = A_2 \cdot (p_2 - p_1).$$

With $\mathrm{d}m/\mathrm{d}t = \rho \cdot A_2 \cdot w_2$ we obtain

$$p_2 - p_1 = \rho \cdot w_2 \cdot (w_1 - w_2).$$

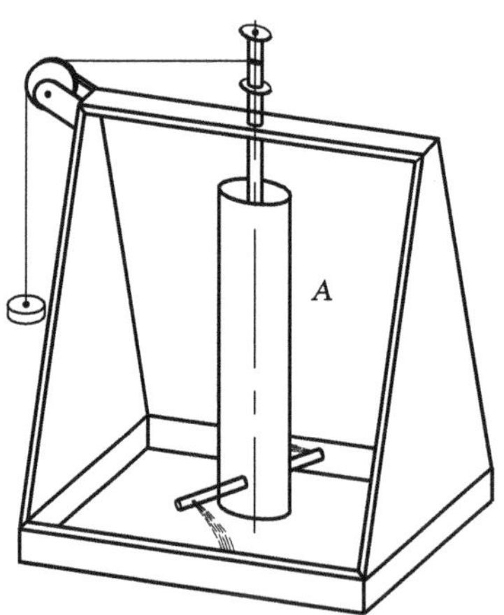

Fig. 4.35. Segner waterwheel

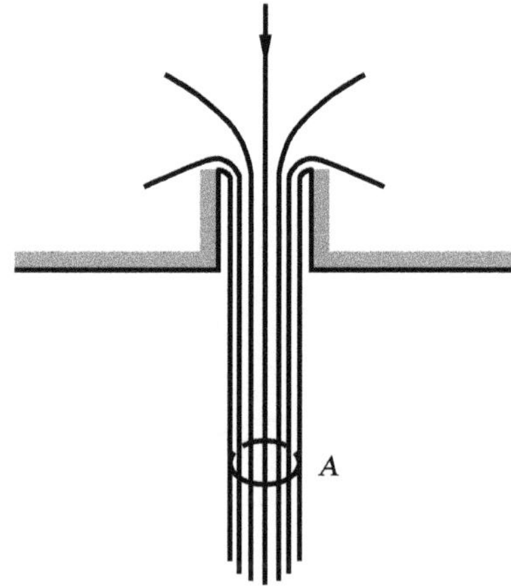

Fig. 4.36. Borda outlet

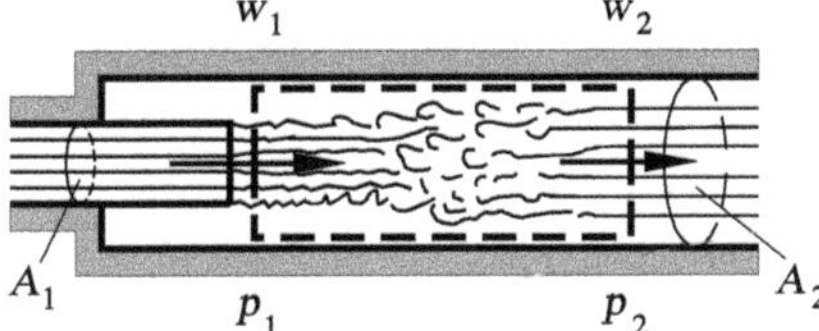

Fig. 4.37. Sudden expansion (diffusor)

In a pipe that expands gradually, the Bernoulli equation would yield $p_2 - p_1 = 0.5 \cdot \rho \cdot (w_1^2 - w_2^2)$ for the pressure difference. However, for sudden expansion, the pressure loss is $p_2' - p_2 = 0.5 \cdot \rho \cdot (w_1 - w_2)^2$. This equation is the same as the equation for the loss of kinetic energy in an inelastic collision between solid bodies, and for this reason the loss on sudden expansion is often called the *impact loss* even though no impact takes place. It is only the mixing of the velocities that collision and sudden expansion have in common.

Suspension of Heavy Bodies in Air

In order to keep a load suspended in air at rest, it is necessary to keep accelerating new masses of air downwards. Let w be the final velocity with which the air moves downwards, for simplicity assumed to be uniform. Here $\mathrm{d}m/\mathrm{d}t = \dot{m}$ is the mass of air set into motion per unit time. If there are no great pressure differences in the mass moving downward, the resulting force is equal to the momentum $J = \dot{m} \cdot w$. To good approximation, this approach can be carried out for a freely suspended helicopter propeller at a sufficient distance from the ground. An air jet with momentum $J = \dot{m} \cdot w$ directed vertically downward then forms. If the helicopter is far enough above the ground, the air jet mixes with the surrounding air at rest and is slowed down. The momentum is unchanged as the moving mass increases correspondingly. As the jet hits the ground, it transfers the weight of the propeller to the ground as a pressure force, thereby losing its momentum.

In the case of an airplane, the mass of air moving downward is formed by the vortex system remaining in the air. However, in this case, the pressure field is also important. Whether the equivalent of the lift is found as a momentum force or as a pressure force is dependent on the shape of the control surface. A pressure increase occurs on the ground below the airplane, namely, the transfer of the weight of the airplane to the ground.

Cascade, Kutta–Joukowski Theorem

In order to investigate the interaction between the blades of a turbine or of a propeller with fluid flowing past, we first consider the simpler case of a two-dimensional *cascade*. The two-dimensional cascade consists of many equally large infinitely long blades oriented parallel to each other. The balances of momentum for the force components parallel and perpendicular to the plane of the cascade, together with the Bernoulli equation and the continuity equation, yield information on the relation of the blade forces to the flow velocity. Figure 4.38 shows a cascade with a flow as seen by an observer at rest with respect to the blades. The row of blades shown is that of a propeller. The camber of the turbine blades is inverted, and the force components point in the opposite direction. However, the following discussion holds for both blade forms. The velocity components parallel and perpendicular to the cascade are u and v, with the corresponding forces per unit length of a blade F_x and F_y (positive in the directions shown in Figure 4.38). Index 1 refers to the incoming flow, and index 2 to the outgoing flow.

It is assumed that there are no losses in the flow. It is then a potential flow with circulation about the blades. In the balance of momentum we make use of the fact that the velocities at some distance behind and in front of the cascade are almost constant. The flow between the blades does not need to be known in more detail. We need only to ensure that no separation occurs, as can happen on ineffectively shaped blades.

With the distance between the blades a, continuity requires that

$$Q = v_1 \cdot a = v_2 \cdot a.$$

Here Q is the amount of liquid that passes between two blades per unit time in a layer of depth a parallel to the blade axis. Thus $v_1 = v_2$, and so in

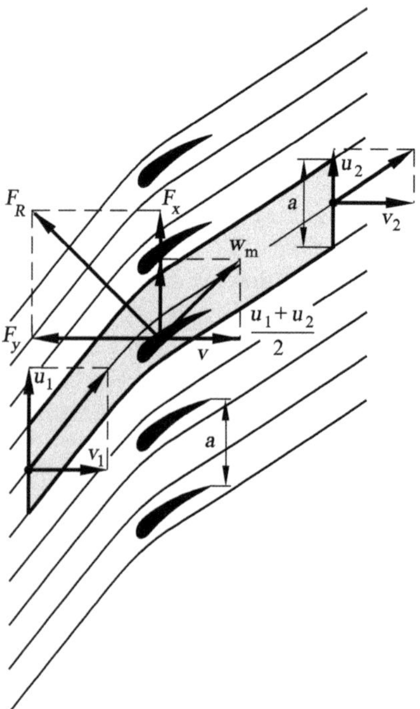

Fig. 4.38. Cascade

what follows we denote this velocity component only by $v = v_1 = v_2$. With $w^2 = u^2 + v^2$ (resultant velocity w), the Bernoulli equation yields

$$p_1 + \frac{\rho}{2} \cdot (u_1^2 + v^2) = p_2 + \frac{\rho}{2} \cdot (u_2^2 + v^2),$$

or

$$p_2 - p_1 = \frac{\rho}{2} \cdot (u_1^2 - u_2^2). \tag{4.22}$$

For the balance of momentum, we select a control surface whose boundaries in the cascade consist of two identical streamlines lying a distance a apart. The rest of the boundary consists of two straight lines of length a parallel to the plane of the cascade. The control surface is shown in bold in Figure 4.38. Let the depth of all surfaces be 1. Nothing flows through the two streamline surfaces. Because of their equal length with respect to the cascade, all quantities on these streamlines are the same, and so they also have the same pressure distribution. For this reason they contribute neither to the momentum nor to the resultant of the pressure forces. Only the contributions from the surfaces parallel to the cascade plane have to be computed for the force balance. The mass flowing through the cascade per unit time is $\rho \cdot Q = \rho \cdot a \cdot v$. We obtain

$$F_x = 0 + \rho \cdot a \cdot v \cdot (u_1 - u_2) = \rho \cdot a \cdot v \cdot (u_1 - u_2), \tag{4.23}$$

$$F_y = a \cdot (p_2 - p_1) + 0 = a \cdot (p_2 - p_1). \tag{4.24}$$

It makes sense to introduce the circulation about a blade into these equations. We again use the bold line to compute this. The two streamlines run in opposite directions and yield two equally large but opposite contributions. However, the two straight pieces yield $a \cdot u_1$ and $-a \cdot u_2$, and so the circulation becomes

$$\Gamma = a \cdot (u_1 - u_2). \tag{4.25}$$

Using (4.22) and the relation

$$u_1^2 - u_2^2 = (u_1 - u_2) \cdot (u_1 + u_2),$$

(4.23) and (4.24) yield

$$F_x = \rho \cdot \Gamma \cdot v, \tag{4.26}$$

$$F_y = \rho \cdot \Gamma \cdot \frac{u_1 + u_2}{2}. \tag{4.27}$$

The ratio $F_y/F_x = ((u_1 + u_2)/2)/v$ shows that the resultant of F_x and F_y is perpendicular to the velocity resulting from $(u_1 + u_2)/2$ and v. This can easily be seen by considering the similar triangles in Figure 4.38. Calling the resultant force F_{R} and the resultant mean velocity w_{m}, we also obtain

$$F_{\mathrm{R}} = \rho \cdot \Gamma \cdot w_{\mathrm{m}}. \tag{4.28}$$

This is the theorem of *Kutta and Joukowski*. It can also be proved otherwise. Joukowski derived it by using a control surface for the balance of momentum

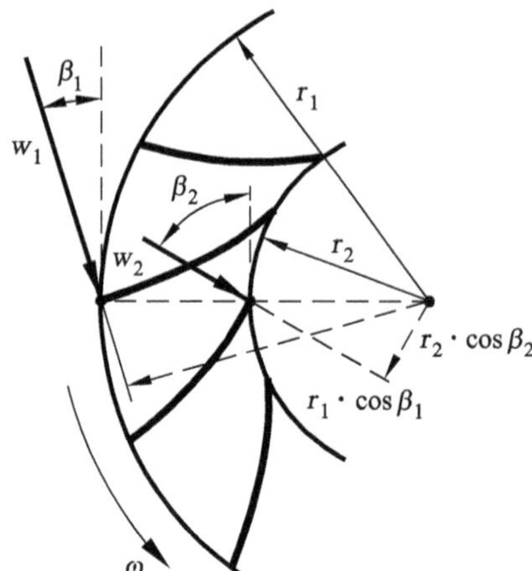

Fig. 4.39. Water turbine

that consisted of a circular cylinder with a very large radius. The axis of the airfoil is the axis of the cylinder. Half of F_R is then obtained as the momentum force and the other half as the resultant of the pressure forces. This theorem is important because it allows us to determine the circulation associated with a given lift through which the vortices behind the airfoil are determined.

Moments of Momentum, Euler's Turbine Equation

As well as moments of forces in statics, we can also form moments of momentum forces. A principle analogous to the center-of-gravity principle holds here: *The rate of change of the moment of momentum is equal to the resulting moment of the forces.* This principle is also called the *conservation of angular momentum.* As for the conservation of momentum, in the case of steady liquid flows it passes over to the principle of equilibrium of the moments of the external forces and the moments of the reaction forces of the liquid.

As an example we derive *Euler's turbine equation.* An amount of water $\dot{m}$ flows through a turbine per unit time (Figure 4.39). The absolute entry velocity is w_1, its angle with the direction of motion of the turbine is β_1, and the entry radius is r_1. The water flows through the rotating turbine in a direction given approximately by the blades. The relative exit velocity at radius r_2 together with the circumferential velocity of the turbine at that point gives the absolute exit velocity w_2 in the direction β_2. The torque acting from the water on the turbine is therefore equal to

$$\dot{m} \cdot (w_1 \cdot r_1 \cdot \cos(\beta_1) - w_2 \cdot r_2 \cdot \cos(\beta_2)). \tag{4.29}$$

Instead of considering the moment of momentum of the unit mass to be a product of the velocity w with $r \cdot \cos(\beta)$, the moment of momentum can also be seen as a product of the circumferential component $w \cdot \cos(\beta)$ with the radius r.

The best working conditions of the turbine occur when the flow exits in the radial direction, i.e. when $\cos(\beta_2) = 0$. Then the lost kinetic energy of the exiting water is smallest. The work output for this case is found from the product of the torque with the angular velocity w of the turbine:

$$L = \dot{m} \cdot r_1 \cdot \omega \cdot w_1 \cdot \cos(\beta_1). \tag{4.30}$$

If we apply the same principle to a circular fluid motion in which there is no turbine wheel, the torque must be equal to 0. We obtain

$$w_1 \cdot r_1 \cdot \cos(\beta_1) = w_2 \cdot r_2 \cdot \cos(\beta_2). \tag{4.31}$$

If the angles β are small enough, we can set $\cos(\beta) = 1$ and we have $w \cdot r =$ const. This result was already obtained in another manner in Section 4.1.2.

4.1.8 Waves on a Free Liquid Surface

Plane Suspension Waves

In most cases in dealing with a free liquid surface it is permissible to neglect the mass of the air particles set into motion by the liquid compared to the mass of the liquid. In order to do this, the pressure of the free surface must be equal to the air pressure p_∞. Observations have shown that in the simplest form of wave motion the individual particles of the water surface describe paths that are approximately circles. In a reference frame that moves with the translational velocity of the wave crests and troughs, the flow is a steady flow to which the Bernoulli equation can be applied (Figure 4.40). The radius of the circular path of a particle lying on the surface is r, and the period of revolution is T. Therefore, the velocity on the circle is $2 \cdot \pi \cdot r/T$. If the translational velocity of the wave is equal to c, the flow velocity on the crest of the wave in the above reference frame is $w_1 = c - 2 \cdot \pi \cdot r/T$ and that in the trough is $w_2 = c + 2 \cdot \pi \cdot r/T$. The difference in height is $h = 2 \cdot r$. Because the pressures are equal, we have

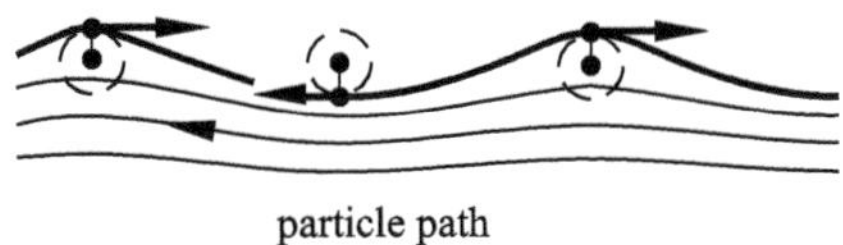

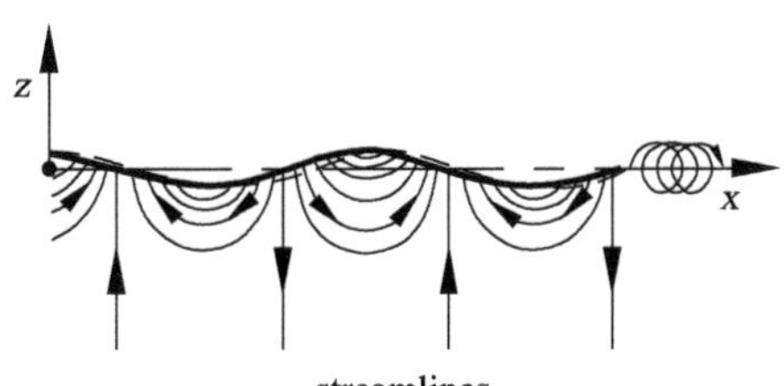

Fig. 4.40. Wave motion

$$w_2^2 - w_1^2 = 2 \cdot g \cdot h = 4 \cdot g \cdot r.$$

The left-hand side yields $8 \cdot \pi \cdot c \cdot r/T$, and so we obtain

$$c = g \cdot \frac{T}{2 \cdot \pi}. \tag{4.32}$$

The radius r cancels out; i.e. the wave velocity does not depend on the height of the crest of the wave. If the wavelength λ is given instead of the period T, we also need a relation between the translation of the crests and troughs with the velocity c and the period of oscillation. This is

$$\lambda = c \cdot T. \tag{4.33}$$

Eliminating T from (4.32) using (4.33), we obtain

$$c = \sqrt{g \cdot \frac{\lambda}{2 \cdot \pi}}. \tag{4.34}$$

In contrast to sound waves, in water waves the wave velocity depends greatly on the wavelength. Longer waves move faster than short waves. The waves can interfere with each other without being essentially perturbed. When short and long waves are superimposed, the short waves remain behind the long waves. The streamlines of the wave motion in a reference frame at rest relative to the unperturbed water are shown in the lower picture in Figure 4.40. The streamlines show that the motion of water decreases greatly with $\exp(-2 \cdot \pi \cdot (z_1 - z)/\lambda)$ with increasing depth below the surface. At a depth of one wavelength, the motion is only about 1/500 of that at the surface.

The surface waves are potential motions, according to the ideas presented in Section 4.1.5. For waves with small amplitudes, the potential is $\Phi = \mathrm{a}_1 \cdot e^{\mu \cdot z} \cdot \cos(\mu \cdot (x - c \cdot t))$, with $\mu = 2 \cdot \pi/\lambda$. For finite amplitudes, a Fourier series appears in place of the cosine. The amplitudes of each term follow from the condition that the pressure must be constant at all points on the surface. A more precise theoretical approach shows that (4.34) holds only for shallow waves and that the translational velocity is independent of the wave height. For high waves the wave velocity becomes somewhat larger. In this case, the paths of the water particles are no longer closed, and the particles move further forwards in the crest of the wave than they do backwards in the trough (Figure 4.40, lower right). Water transport occurs in the wave. According to *G. G. Stokes* (1847), the highest possible steady form of a wave has a summit with an angle of 120°. When more energy is supplied to the wave, the crest begins to foam.

For short wavelengths, the surface stress acts in addition to gravity. As this smooths the wavy surface, it leads to an increase of the translational velocity. For the capillary constant C (tensile stress in the surface) we have

$$c = \sqrt{\frac{g \cdot \lambda}{2 \cdot \pi} + \frac{2 \cdot \pi \cdot C}{\rho \cdot \lambda}}. \tag{4.35}$$

In the case of long waves, only the first term is important. If the wavelength is very short, the second term dominates. For the wavelength $\lambda_1 = 2 \cdot \pi \cdot \sqrt{C/(g \cdot \rho)}$, c has a minimum $c_1 = \sqrt[4]{4 \cdot g \cdot C/\rho}$. For water with $\rho = 1000\ \mathrm{Ns^2/m^4}$ and $C = 0.073\ \mathrm{N/m}$ we can determine that $\lambda_1 = 1.71$ cm and $c_1 = 23.1$ cm/s (simultaneously group velocity). Waves with a wavelength larger than λ_1 are called *gravity waves*, while those whose wavelength is shorter than λ_1 are called *capillary waves.*

Wave Groups

We distinguish between the velocity with which the wave fronts progress, the so-called phase velocity c, and the translational velocity of a wave group, the so-called *group velocity* c^*. To derive the group velocity, we consider the superposition of two waves with the same amplitude but slightly different wavelengths. This holds not only for water waves, but quite generally for waves whose phase velocity depends on the wavelength, i.e. for which there is *dispersion.* Consider a simple sine wave:

$$y = A \cdot \sin(\mu \cdot x - \nu \cdot t).$$

If x is increased by $2 \cdot \pi/\mu$ or t by $2 \cdot \pi/\nu$, the sine function has the same value as before. Therefore, $\lambda = 2 \cdot \pi/\mu$ is the wavelength and $T = 2 \cdot \pi/\nu$ is the period of oscillation. For $\mu \cdot x - \nu \cdot t = \mathrm{const}$, i.e. $x = \mathrm{const} + (\nu/\mu) \cdot t$, the argument of the sine function is constant in time. Therefore, y is also constant in time. This means that the entire wave form moves with velocity $c = \nu/\mu$. We now superimpose a second wave y' on this wave. It has the same amplitude, but slightly different values of μ and ν, denoted by μ' and ν'. Therefore, $y' = A \cdot \sin(\mu' \cdot x - \nu' \cdot t)$ and

$$y + y' = A \cdot [\sin(\mu \cdot x - \nu \cdot t) + \sin(\mu' \cdot x - \nu' \cdot t)]$$

is the result of the superposition. At positions where the two oscillations act in the same direction, the amplitude is equal to $2 \cdot A$. At positions where the oscillations are in opposite directions, the total amplitude is equal to 0. This process is called *a beat.* By applying the equation

$$\sin(\alpha) + \sin(\beta) = 2 \cdot \sin\left(\frac{\alpha + \beta}{2}\right) \cdot \cos\left(\frac{\alpha - \beta}{2}\right),$$

we obtain

$$y + y' = 2 \cdot A \cdot \sin\left(\frac{\mu + \mu'}{2} \cdot x - \frac{\nu + \nu'}{2} \cdot t\right) \cdot \cos\left(\frac{\mu - \mu'}{2} \cdot x - \frac{\nu - \nu'}{2} \cdot t\right).$$

Fig. 4.41. Beats

In this expression the factor sin(...) represents a wave with the average values of μ and μ', and ν and ν'. The factor $2 \cdot A \cdot \cos(\dots)$, which changes only slowly for small $\mu - \mu'$ and $\nu - \nu'$, can be considered as a varying amplitude (cf. Fig. 4.41). The *wave group* comes to an end when the cosine is equal to 0. The translational velocity at this position, the group velocity c^*, is therefore equal to $(\nu - \nu')/(\mu - \mu')$. For long groups (slow beats), we have

$$c^* = \mathrm{d}\nu/\mathrm{d}\mu. \tag{4.36}$$

Since no energy transport can take place through the nodes of the beats, the translational velocity of the wave energy is identical to the group velocity. This can be strictly proven for simple wave trains.

For water waves determined by gravity, (4.32) yields

$$\nu = \frac{2 \cdot \pi}{T} = \frac{g}{c}.$$

According to (4.34),

$$c = \sqrt{\frac{g \cdot \lambda}{2 \cdot \pi}} = \sqrt{\frac{g}{\mu}}.$$

This leads to the relation

$$\nu = \sqrt{g \cdot \mu}.$$

Therefore, with (4.36), we have the group velocity

$$c^* = \frac{\mathrm{d}\nu}{\mathrm{d}\mu} = \frac{1}{2} \cdot \sqrt{\frac{g}{\mu}} = \frac{1}{2} \cdot c. \tag{4.37}$$

The wave group progresses with velocity $0.5 \cdot c$, or in other words, the wave fronts move at twice the speed of the progression of the wave group. At the back end of the group new waves keep forming, to vanish again at the front end. This can be seen well for the waves that occur when a stone is thrown into water at rest.

Ship Waves

Ship waves belong to another type of wave group. We can produce a figure very similar to the waves of a ship by considering the waves generated when a

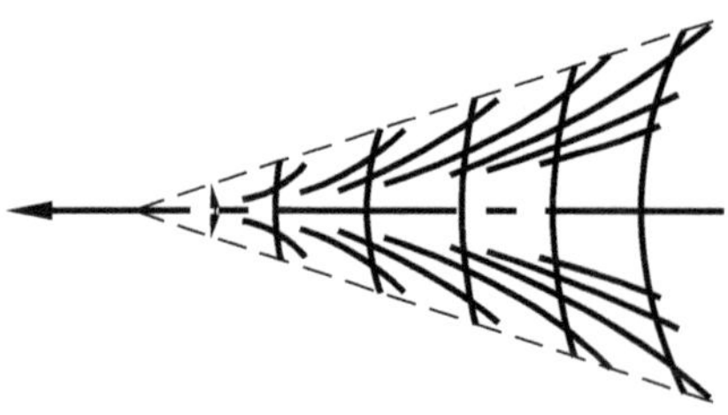

Fig. 4.42. Wave system of a pressure perturbation moving uniformly across the surface of water

point pressure perturbation moves with constant velocity over the surface of a still, deep body of water. According to the calculations of *Lord Kelvin, V. W. Ekman* (1905), and others, a wave system such as that shown in Figure 4.42 is obtained. The lines extending outward in this figure represent wave crests. This wave system moves with the pressure perturbation. According to (4.35), the wavelength of the transverse waves is $\lambda = 2 \cdot \pi \cdot c^2/g$. Here c is the translational velocity of the pressure perturbation. The length of the wave group is equal to half the distance covered by the pressure perturbation.

When a ship moves through water, one such wave system is generated at its bow and another at its stern, and these systems interfere with each other.

The group velocity of capillary waves is analogous to the group velocity of gravity waves. It is larger than the phase velocity, about 1.5 times in the case of very small waves. For a pressure perturbation moving with constant velocity, the wave group leads the position of generation. In fact, a fishing line or some other obstacle at rest in water flowing at more than 23.3 cm/s generates capillary waves upstream and gravity waves downstream. The gravity waves have approximately the shape shown in Figure 4.42. The capillary waves fill the space in front in an arc-like manner. At velocities below 23.3 cm/s no waves occur.

Interfaces Between Two Liquids

If two liquids with different specific weights are layered on top of each other, the interface can carry out wave motion. For two liquids at rest with densities ρ_1 and ρ_2 layered on top of each other, the theory yields a phase velocity

$$c = \sqrt{\frac{g \cdot \lambda}{2 \cdot \pi} \cdot \frac{\rho_1 - \rho_2}{\rho_1 + \rho_2} + \frac{2 \cdot \pi \cdot C}{\lambda \cdot (\rho_1 + \rho_2)}}.$$

If the upper liquid flows with a velocity w_1 above the lower liquid, according to the theory only the longer waves are stable. The shorter waves are unstable, as shown in Section 4.1.4 for the motion of two streams of liquid along an interface. This can lead to a mixing of the two liquids in an intermediate zone, whereby the flow becomes stable again. With increasing velocity w_1, the boundary between instability and stability shifts toward longer wavelengths. Such waves can occur between two layers of air of different densities, as can occur in the atmosphere. This is sometimes made visible by cloud formation (Helmholtz waves).

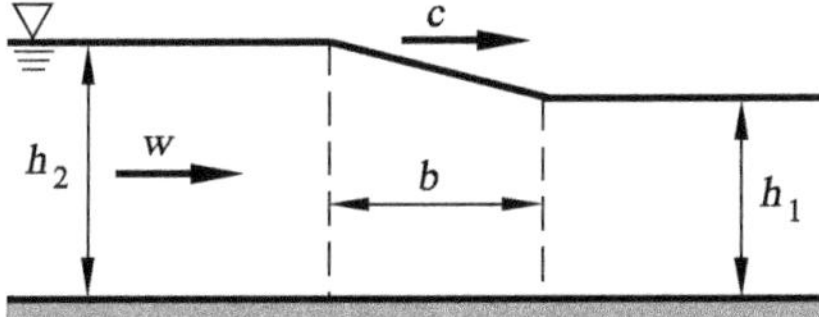

Fig. 4.43. Surge on the surface of water

Surge

The equations given in this section are valid for waves in deep water. The relations are altered when the depth of the water becomes small compared to the wavelength. For water depths of more than half a wavelength, the previous equations are precise enough. For shallower water, the water particles move on elliptical paths. The relations between the wavelength and the translational velocity become more complicated. For very shallow water, or for very large wavelengths, the water particles on the surface essentially move back and forth horizontally, and rise and fall only very slightly in comparison. In this case we can obtain new simple relations. Periodic waves with an approximately sinusoidal form are again considered. For the very flat elliptical paths of the particles, the effect of the vertical accelerations on the pressure distribution can be neglected. The pressure changes only statically in every vertical line, and the differences in level of the water cause only horizontal accelerations.

We now carry out an even simpler approach. We consider a low *surge* (Figure 4.43). This approach is closely related to the treatment of pressure expansion in a compressible medium (Section 4.3.1). We assume that a surge, in which the height of the water over the flat ground increases from h_1 to h_2, expands to the right with a velocity c. Before the arrival of the surge, the water is at rest, and after the water level has risen, it has the velocity w to the right.

This velocity is necessary to increase the water level from h_1 to h_2 by sideward compression of the water mass in the transition region of width b. For simplicity, we assume that the water level in the transition region has a constant slope $(h_2 - h_1)/b$. If the velocity w is small compared to the expansion velocity c, the water level increases with a velocity $v = c \cdot (h_2 - h_1)/b$.

On setting the depth perpendicular to the plane of Figure 4.43 to 1, continuity requires that $h_2 \cdot w = b \cdot v$, or

$$h_2 \cdot w = c \cdot (h_2 - h_1). \tag{4.38}$$

The width of the surge b has canceled out of this equation, and it does not depend on this quantity. Equation (4.38) is also correct if the profile of the surge is not linear. The surge can then be decomposed into a number of surges with linear profiles. Adding the continuity equations of each surge, on the right-hand side of the equation we again obtain $h_2 - h_1$, and on the left-hand side the individual velocity differences again sum to w. However, this is true only when the differences of each h_2 can be neglected. It also follows from (4.38) that for a very small velocity w, $h_2 - h_1$ also must be small. This equation is therefore valid only for low surges, where the previous neglecting is indeed permissible.

As well as the kinematic relation (4.38) we also need a dynamic relation, obtained in the following discussion. The water mass of width b is in accelerated motion, since its particles have velocity 0 on the right boundary and velocity w on the left boundary. The time in which the surge moves past a particle is $\tau = b/c$. Thus the acceleration of a particle is $w/\tau = w \cdot c/b$. The water mass of width b and depth 1 perpendicular to the plane of the figure is $\rho \cdot b \cdot h_m$ (h_m is the mean water level). The pressure at the same height on both sides of the surge differs by the amount $\gamma \cdot (h_2 - h_1)$. The horizontal total force on the water mass below the surge is (neglecting small quantities) equal to $h_m \cdot \gamma \cdot (h_2 - h_1)$. With the equation force = mass · acceleration, and $\gamma = \rho \cdot g$, we obtain

$$w \cdot c = g \cdot (h_2 - h_1). \tag{4.39}$$

Here, too, the width of the surge b has canceled out. It can again be shown that (4.39) is also valid for a surge with another profile, if $h_2 - h_1$ is small compared to h_1 and h_2.

For simplicity we now replace h_2 in (4.38) by h_m. This is also permissible within the range of neglect already carried out. Then dividing (4.39) by (4.38), we obtain

$$c^2 = g \cdot h_m. \tag{4.40}$$

Positive and negative surges following one another form waves. The translational velocity of such waves is independent of the form of the wave. It is obtained from (4.40). As with sound waves, there is no dispersion, and thus $c^* = c$. Long waves in shallow water progress with velocity $c = \sqrt{g \cdot h}$ (fundamental wave velocity).

When several low surges follow one another, with each one leading to a further increase in the water level, because of the greater water depth, the velocity $\sqrt{g \cdot h}$ of the subsequent surge is larger than that of the previous surge. What is of more importance is that the subsequent surge moves in a water mass that is already in motion with velocity w. Therefore the subsequent surge overtakes the previous one, and a surge with a large amplitude occurs. This approach can also be applied to the form of a single surge. For example, the surge with the form shown in Figure 4.43 can be taken as a series of very many small surges that fill up the interval b. From the above consideration it follows that the interval b becomes smaller and smaller until a steep step occurs. This can also be seen in nature: For waves in shallow water, the crests of the waves move faster than the troughs of the waves, and they collapse on top of each other (breakers).

Surges of finite height can also be treated in a similar manner using the conservation of momentum, as in the example of the flow with sudden expansion in Section 4.1.7. The flow is then considered from a reference frame moving with the surge, so that the process is steady. The velocity of the finitely high surge is larger than that of the fundamental wave. Here, too, there is a loss in kinetic energy, equivalent to the foaming of a collapsing water mass.

Open Channel

When water flows in a river, the velocity of the surge and fundamental waves are apparent in a similar manner to the velocity of sound in gas flows (compare Sections 4.3.1 and 4.3.3). If the flow velocity is smaller than the velocity of surge, banking up the water in the river (e.g. by means of a weir) leads to an increase in the water level upstream. If the flow velocity is greater than the fundamental wave velocity, a finitely high steady surge occurs in front of the weir or at the weir, a so-called water jump. Upstream from this surge, the water flow is completely unaffected by the banking. Unevenness at the sides of the channel generates small oblique waves that are very similar to the oblique sound waves discussed in Section 4.3.3. The two types of motion in a water channel with flow velocities larger or smaller than the fundamental wave velocity are called *streaming* and *shooting.*

For a given volume flux $\dot{V}$ per unit width, we compute the water depth in Figure 4.44, and obtain the drop in water level from the level at rest from the Bernoulli equation as $h = w^2/(2 \cdot g)$. The local water depth necessary for a volume flux $\dot{V}$ per unit width follows from continuity as $a = \dot{V}/w$. The distance to the associated channel point below the water level at rest is

$$z = h + a = \frac{w^2}{2 \cdot g} + \frac{\dot{V}}{w}.$$

For a certain value of the velocity w, z is a minimum. A similar result is found for the stream filament cross-section of a gas flow (cf. Section 4.3.1). This minimum is found by differentiating the equation with respect to the velocity:

$$\frac{w_1}{g} - \frac{\dot{V}}{w_1^2} = 0, \qquad \text{i.e.} \qquad w_1 = \sqrt[3]{\dot{V} \cdot g}.$$

We obtain

$$h_1 = \frac{1}{2} \cdot \sqrt[3]{\frac{\dot{V}^2}{g}} \qquad \text{and} \qquad a_1 = \sqrt[3]{\frac{\dot{V}^2}{g}} = 2 \cdot h_1.$$

Therefore, $w_1 = \sqrt{g \cdot a_1}$; i.e. w_1 is equal to the surge velocity at the water depth a_1. If water flows over a flat weir crest, the water depth a_1 above the highest point of the crest of the weir is equal to 2/3 the depth z_1 of the weir crest below the surface of the water. The velocity there is $\sqrt{2 \cdot g \cdot z_1/3}$. The volume flux is found to be

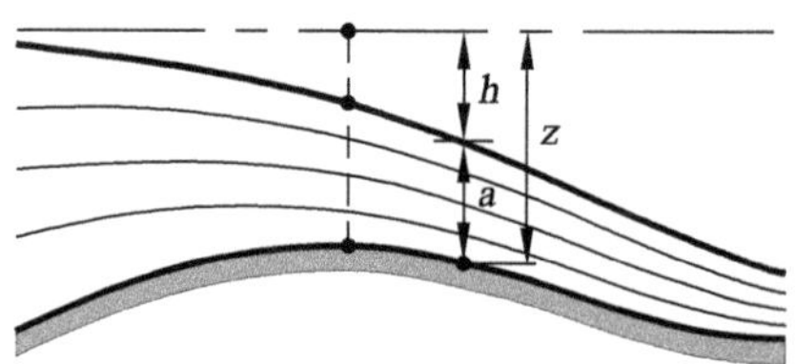

Fig. 4.44. Flow over the crest of a weir

$$\dot{V} = a_1 \cdot w_1 = \frac{2}{3} \cdot z_1 \cdot \sqrt{\frac{2}{3} \cdot g \cdot z_1}. \tag{4.41}$$

The water shoots downstream from the crest of the weir. This generally passes over to streaming motion again via a water jump.

For strongly curved weirs we can no longer assume that the flow velocity is the same in the entire cross-section. However, the qualitative relations remain valid.

These equations derived for open channels can also be used in a much more extensive application. For a slightly tilted but otherwise arbitrary channel floor (Figure 4.45), and taking a family of heights of the water surface (dashed-dotted lines), we can sketch the water depths a (two each for each position and each water level) associated with a fixed value of the volume flux $\dot{V}$. This yields the given forms of the water surface. Only the line passing through the double point from I to IV, which corresponds to the lowest possible water level at rest, yields the flow shown in Figure 4.44. The lines associated with the higher water levels, those passing from I to II and from III to IV, also occur in practice. The dashed curves shown in Figure 4.45, associated with lower water levels at rest, can occur behind a water jump as they move upward. This is associated with a loss in energy.

In the left-hand picture in Figure 4.45, the velocity is smaller than the expansion velocity of the fundamental wave. At the peak of the rise in the ground, there is a drop in the water level. In the middle picture the velocity is greater than the fundamental wave velocity. The water surface then rises more than the rise in the ground. In the case of a water jump (right-hand picture), the flow velocity from the crest of the weir to the water jump is larger and then subsequently smaller than that of the associated fundamental wave. As changes in the flow state can progress only with the fundamental wave

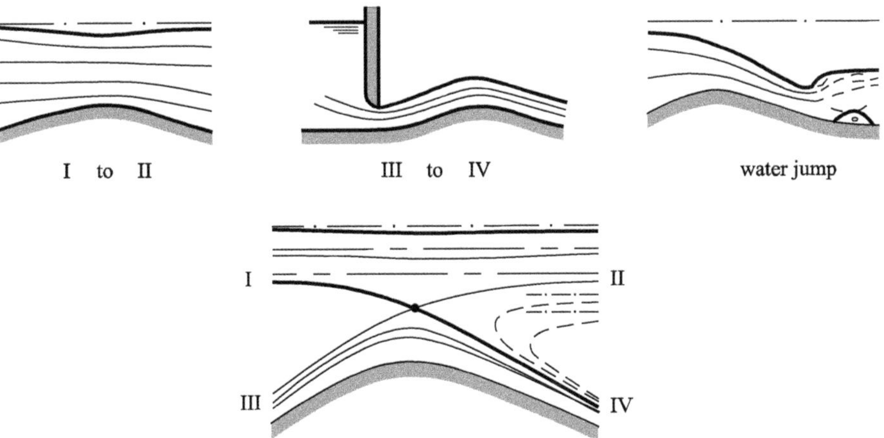

Fig. 4.45. Further examples of the flow over the crest of a weir

velocity, the shooting flow between the weir crest and the water jump cannot be changed by a rise in the water level, and a sudden transition occurs.

In the above discussions, the effect of vertical acceleration was ignored. In shooting flow, taking the vertical acceleration into account leads only to slight quantitative corrections. However, in the streaming motion the character changes as standing waves occur downstream from the perturbation position. The wavelength satisfies (4.34), where the local flow velocity replaces the translational velocity c.

4.1.9 Problems

4.1

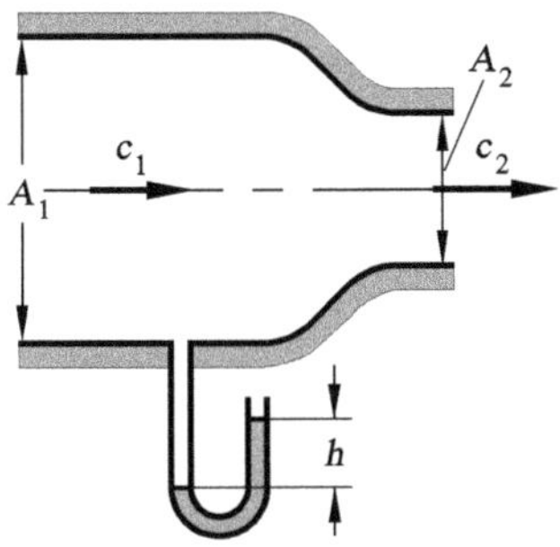

A U-tube manometer filled with water is connected to a wind tunnel nozzle with contraction ratio $A_1/A_2 = 4$ in front of the contraction. When in operation, the manometer indicates a height difference of $h = 94$ mm. What is the exit velocity w_2 at the cross-section A_2 if the density of water in the U-tube is $\rho_w = 1000\ \mathrm{kg/m^3}$ and the density of air is $\rho_a = 1.226\ \mathrm{kg/m^3}$?

$$w_2 = \sqrt{2 \cdot \frac{\rho_w}{\rho_a} \cdot \frac{g \cdot h}{\left(1 - \left(\frac{A_2}{A_1}\right)^2\right)}} = 40\ \mathrm{m/s}.$$

4.2

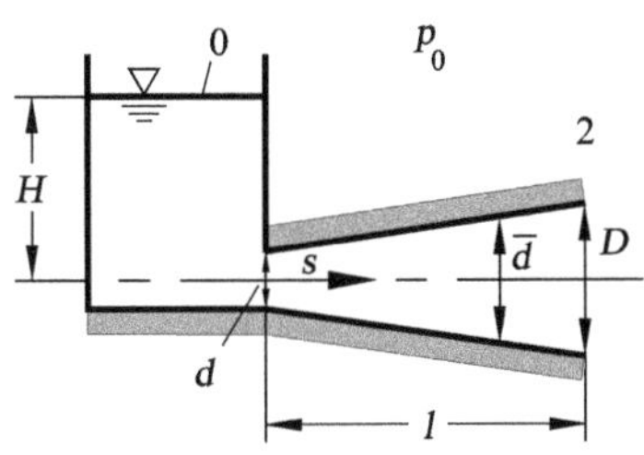

A large container is filled to height H with water. A long diffusor of length l is attached to the container. The diameter of the diffusor entry cross-section is d, and that of its exit cross-section is D. At time $t = 0$ the diffusor is closed at the exit point 2. For $t > 0$ the diffusor is suddenly opened at position 2 so that the water can flow out.

(a) Compute the steady exit velocity $w_{2,e}$ at position 2, which is $w_2(t)$ for $t \to \infty$.

$$w_{2,e} = \sqrt{2 \cdot g \cdot H}.$$

(b) Calculate the exit velocity $w_2(t)$ for $t > 0$.

$$\frac{w_2(t)}{w_{2,e}} = \tanh\left(\frac{d}{D} \cdot \frac{t}{\tau}\right), \qquad \tau = \frac{2 \cdot l}{w_{2,e}}.$$

4.3

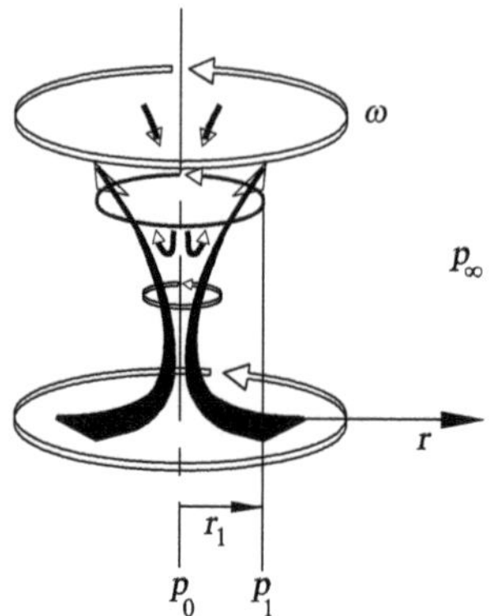

Pressure measurements are carried out in a tornado. In the center of the tornado at position r_0 a pressure $p_0 = 0.8$ bar is measured. At another point 1 at a distance $r_1 = 50$ m from the center, a pressure $p_1 = 0.85$ bar is measured. The pressure a large distance away from the tornado is $p_\infty = 1$ bar.

(a) What is the maximum circumferential velocity $w_{\max}$ occurring in the tornado and how large is the pressure p_m at this point? The flow is incompressible, the streamlines are concentric circles, and the external flow in the tornado is inviscid. At the center of the vortex there is a viscous rigid body rotation with constant angular velocity. Gravity is not to be taken into account.

$$w_{\max} = \sqrt{\frac{p_\infty - p_0}{\rho}} = 127.7 \text{ m/s}, \qquad p_m = \frac{p_\infty + p_0}{2} = 0.9 \text{ bar}.$$

(b) At what distance r_m from the center of the tornado does the maximum circumferential velocity $w_{\max}$ occur?

$$r_m = r_1 \cdot \sqrt{\frac{p_\infty - p_0}{2 \cdot (p_1 - p_0)}} = 70.7 \text{ m}.$$

4.4

A two-dimensional flow field is described by the velocity components $u = U \cdot \frac{y}{L}$, $v = U \cdot \frac{x}{L}$, where U and L are constants, U having the dimensions of velocity and L the dimensions of length.

Investigate whether the given flow field is a potential flow and determine the associated potential function Φ. What is the stream function Ψ for the given flow field?

$$\Phi(x, y) = U \cdot \frac{x \cdot y}{L}, \qquad \Psi(x, y) = \frac{U}{2 \cdot L} \cdot (y^2 - x^2).$$

4.5

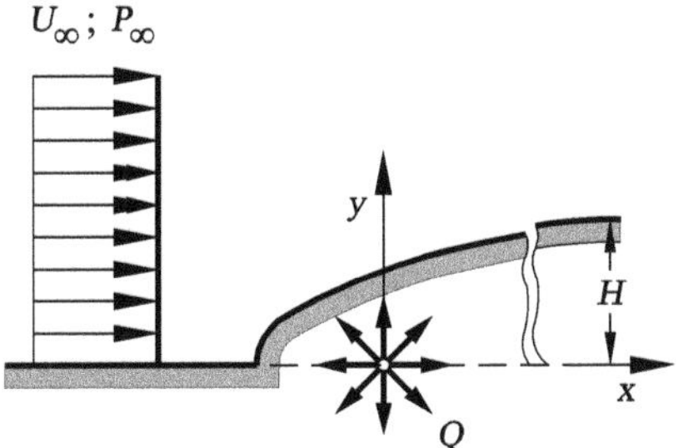

A section through a ridge of height H, whose extension perpendicular to the plane of the figure can be taken to be infinite, has the form of a two-dimensional semibody in the x, y plane. The flow past the ridge is a potential flow and has velocity U_∞.

(a) What source strength Q must be chosen to mathematically reproduce the flow?

$$Q = 2 \cdot H \cdot U_\infty .$$

(b) In what region of the x, y plane must a glider with vertical descent velocity v_s (relative to the air) remain so that it does not drop in altitude?

$$x_s^2 + \left(y_s - \frac{Q}{4 \cdot \pi \cdot v_s} \right)^2 = \left(\frac{Q}{4 \cdot \pi \cdot v_s} \right)^2 .$$

(c) What is the highest position $(x_{\max}, y_{\max})$ at which the glider can use the up-current of air without losing altitude?

$$x_{\max} = 0, \qquad y_{\max} = \frac{Q}{2 \cdot \pi \cdot v_s} .$$

4.6

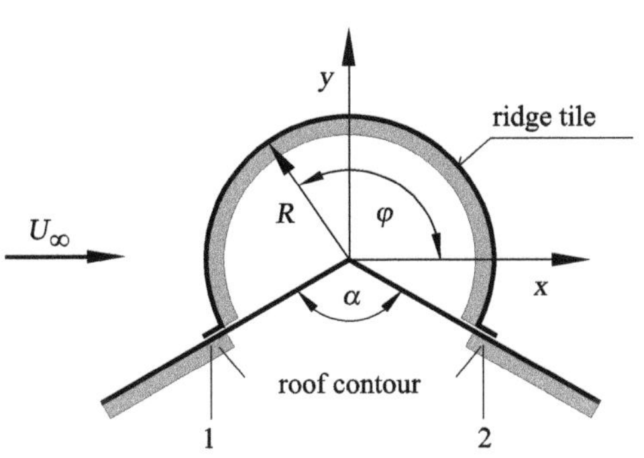

A model to describe the inviscid flow past the ridge of a roof is obtained by superimposing a flow with velocity U_∞ (U_∞ = 120 km/s) past a circular cylinder of radius R on the flow of a potential vortex. The radius R of the roof ridge is $R = 7.5$ cm; the ridge angle α is $\alpha = 120°$.

(a) What circulation Γ of the potential vortex must be chosen to correctly model the inviscid flow past the ridge?

$$\Gamma = 2 \cdot \pi \cdot U_\infty \cdot R.$$

(b) What is the force F_A acting on the ridge if the pressure of the flow below the ridge is p_∞ and the ridge has length $b = 1$ m (b perpendicular to the plane of the figure)? The density ρ of the flow is $\rho = 1.226$ kg/m^3.

$$F_A = \left(\sqrt{3} + \frac{4}{3} \cdot \pi\right) \cdot \rho \cdot U_\infty^2 \cdot R \cdot b = 604.9 \text{ N}.$$

4.7

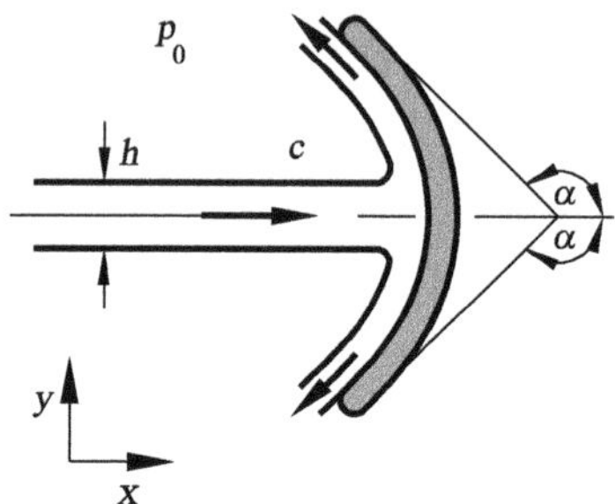

A two-dimensional water jet with density $\rho = 1000$ kg/m^3 exits with velocity $w = 20$ m/s from a rectangular nozzle of height $H = 25$ mm and width $b = 20$ mm. It is deviated by a guide plate by $\alpha = 135°$. What is the force F that the water jet applies to the guide plate?

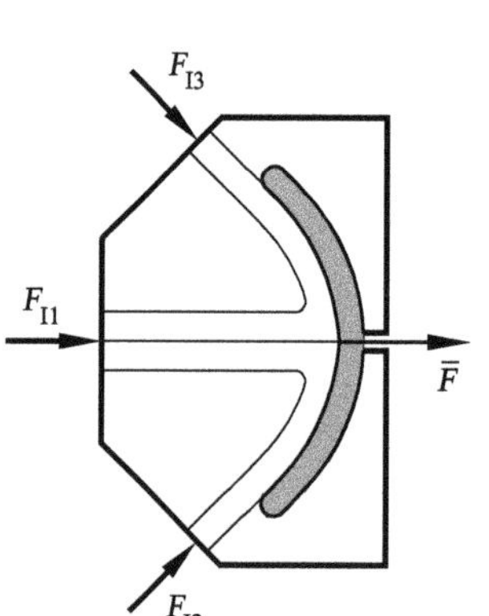

$$F = -\left(1 + \frac{\sqrt{2}}{2}\right) \cdot \rho \cdot w^2 \cdot h \cdot b$$
$$= 341.42 \text{ N}.$$

4.8

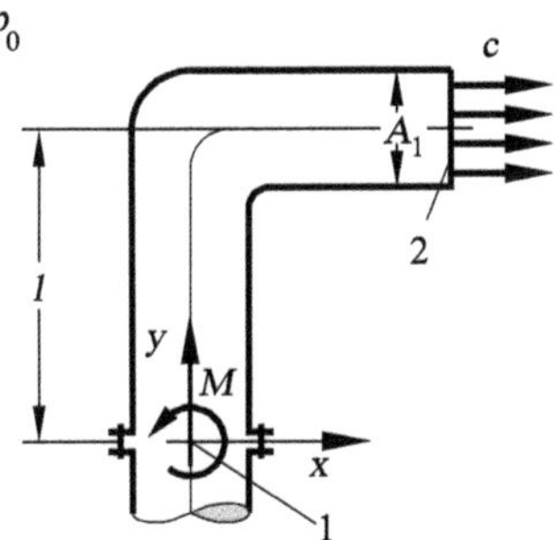

The sketch shows an elbow bend with constant cross-section A_1 fitted to a pipe at position 1 by means of a flanged joint. At position 2, water with density ρ exits the pipe with velocity w into the open air. What is the moment M acting on the flanged joint?

$$M = \rho \cdot l \cdot w^2 \cdot A_1.$$

4.9

A characteristic of the linear theory of wave motion is that waves with the same amplitude but opposite directions of propagation produce a standing wave. Determine the velocity potential of the standing surface wave in a deep body of water, where the vertical extension of the free surface can be written as

$$h(x,t) = A \cdot e^{i \cdot \omega \cdot t} \cdot \cos(a \cdot x).$$

Show that the streamlines of the motion assume the following form:

$$e^{a \cdot z} \cdot \sin(a \cdot x) = \text{const}.$$

4.2 Dynamics of Viscous Liquids

4.2.1 Viscosity (Inner Friction), the Navier–Stokes Equation

All liquids and gases have a *viscosity.* This is expressed as an internal friction when the shape of the fluid is changed. Particularly viscous liquids are honey, glycerin, and thick oils. In order to understand viscosity, we consider the flow between two parallel plates, where the upper plate moves with velocity U while the lower plate remains at rest (Figure 4.46). Because of the friction, the liquid at the plates has the same velocity as the plates themselves (*no-slip condition*). The layers between the plates glide over each other with velocities $u(y)$ that are proportional to the distance y from the lower plate:

$$u = U \cdot \frac{y}{a}.$$

The liquid friction is expressed as a force that causes a resistance to the motion of the upper plate and has magnitude $\tau = \mu \cdot U/a$ per unit surface area. In general, for the shear stress we have

$$\tau = \mu \cdot \frac{\mathrm{d}u}{\mathrm{d}y}, \tag{4.42}$$

where μ indicates the *dynamic viscosity.* The ansatz (4.42) is valid for *Newtonian media.*

With this knowledge we can already treat some examples of laminar flows. One of these is the flow of a viscous liquid in a straight pipe with a circular cross-section. The pressure difference $p_1 - p_2$ causes the force $(p_1 - p_2) \cdot \pi \cdot r^2$ on a cylindrical liquid element of radius r (Figure 4.47). The countervailing force is produced by the friction on the surface shell $2 \cdot \pi \cdot r \cdot l$. This is τ per unit area and yields in total $2 \cdot \pi \cdot r \cdot l \cdot \tau$. Setting both forces equal, we obtain

$$-\tau = \frac{p_1 - p_2}{l} \cdot \frac{r}{2}, \tag{4.43}$$

where τ has been given a minus sign because the frictional force acts to oppose the flow. Equation (4.42) yields $\mathrm{d}u/\mathrm{d}r = \tau/\mu$. Integrating this and using the no-slip condition, we obtain

$$u(r) = \frac{p_1 - p_2}{4 \cdot \mu \cdot l} \cdot (R^2 - r^2), \tag{4.44}$$

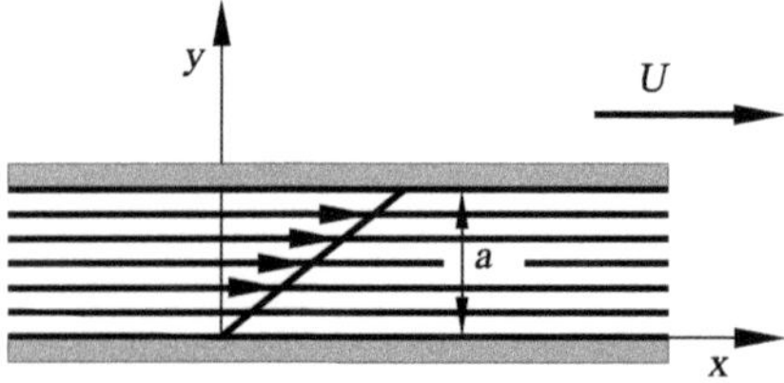

Fig. 4.46. Shear flow between parallel plates

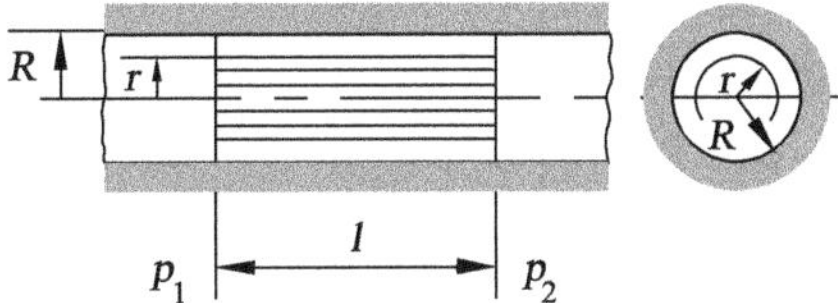

Fig. 4.47. Laminar pipe flow

with the radius of the pipe R. The velocity distribution is a paraboloid of revolution (cf. Figure 4.53). The volume per unit time flowing through the pipe is found to be

$$Q = \int_0^R u \cdot 2 \cdot \pi \cdot r \cdot \mathrm{d}r = \frac{\pi \cdot R^4}{8 \cdot \mu} \cdot \frac{p_1 - p_2}{l}. \tag{4.45}$$

If the amount flowing through the pipe is measured, this equation allows us to determine the dynamic viscosity μ precisely. The flux is proportional to the pressure drop per unit length and the fourth power of the pipe's radius. *G. H. L. Hagen* (1839) and *J. L. M. Poiseuille* (1840) both obtained (4.45) by experiment independently of each other, and for this reason it is called the *Hagen–Poiseuille law*. We note here that the *Hagen–Poiseuille law* is valid only for laminar pipe flow. The law for turbulent pipe flow will be found in Section 4.2.5.

Navier–Stokes Equation

The *general theory of liquid friction* tells us that when the shape of a single liquid element is changed, stresses arise that are similar to those of elastic bodies. The difference lies in the fact that these stresses are not proportional to the changes of shape, but rather to the velocities of the changes of shape. The equations for the nine stress components (three each on the three surfaces perpendicular to the coordinate axes (Figure 4.48)) therefore read

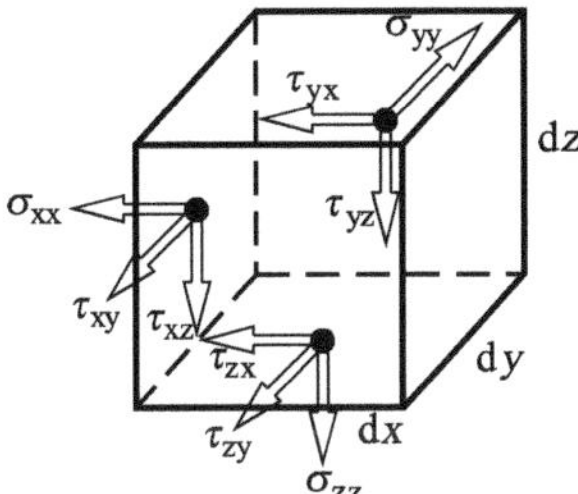

Fig. 4.48. Normal and shear stress at volume element $\mathrm{d}V = \mathrm{d}x \cdot \mathrm{d}y \cdot \mathrm{d}z$

$$\begin{aligned}
\sigma_{xx} &= 2 \cdot \mu \cdot \frac{\partial u}{\partial x}, &\quad \tau_{xy} = \tau_{yx} &= \mu \cdot \left(\frac{\partial u}{\partial y} + \frac{\partial v}{\partial x}\right), \\
\sigma_{yy} &= 2 \cdot \mu \cdot \frac{\partial v}{\partial y}, &\quad \tau_{yz} = \tau_{zy} &= \mu \cdot \left(\frac{\partial v}{\partial z} + \frac{\partial w}{\partial y}\right), \\
\sigma_{zz} &= 2 \cdot \mu \cdot \frac{\partial w}{\partial z}, &\quad \tau_{zx} = \tau_{xz} &= \mu \cdot \left(\frac{\partial w}{\partial x} + \frac{\partial u}{\partial z}\right).
\end{aligned} \tag{4.46}$$

Forces per volume arise with the components f'_x, f'_y, f'_z. The component f'_x satisfies

$$f'_x = \frac{\partial \sigma_{xx}}{\partial x} + \frac{\partial \tau_{xy}}{\partial y} + \frac{\partial \tau_{xz}}{\partial z}. \tag{4.47}$$

Similar equations are obtained for f'_y and f'_z. With (4.46), for Newtonian media and constant values μ and ρ, and using the continuity equation for incompressible flow (Section 5.1)

$$\frac{\partial u}{\partial x} + \frac{\partial u}{\partial y} + \frac{\partial u}{\partial z} = 0, \tag{4.48}$$

(4.47) yields

$$f'_x = \mu \cdot \left(\frac{\partial^2 u}{\partial x^2} + \frac{\partial^2 u}{\partial y^2} + \frac{\partial^2 u}{\partial z^2}\right).$$

Analogous expressions hold for f'_y and f'_z.

In viscous liquids, the frictional forces $\boldsymbol{f}'$ per volume occur in addition to the pressure forces of inviscid flow discussed in the previous section, and in addition to any mass forces $\boldsymbol{f}$ present. These forces determine the acceleration of the liquid particle. By taking into account the frictional forces on the right-hand side of the Euler equations, we obtain the *Navier–Stokes equations* of viscous flow. Employing the Δ-operator $\partial^2/\partial x^2 + \partial^2/\partial y^2 + \partial^2/\partial z^2$, the Navier–Stokes equations for incompressible flow read (Section 5.2.1)

$$\begin{aligned}
\rho \cdot \frac{\mathrm{d}u}{\mathrm{d}t} &= f_x - \frac{\partial p}{\partial x} + \mu \cdot \Delta u \quad , \\
\rho \cdot \frac{\mathrm{d}v}{\mathrm{d}t} &= f_y - \frac{\partial p}{\partial y} + \mu \cdot \Delta v \quad , \\
\rho \cdot \frac{\mathrm{d}w}{\mathrm{d}t} &= f_z - \frac{\partial p}{\partial z} + \mu \cdot \Delta w \quad .
\end{aligned} \tag{4.49}$$

Here $\mathrm{d}u/\mathrm{d}t$, for example, means

$$\frac{\partial u}{\partial t} + u \cdot \frac{\partial u}{\partial x} + v \cdot \frac{\partial u}{\partial y} + w \cdot \frac{\partial u}{\partial z}.$$

For a flow in which the u component predominates and that changes most in $\mu \cdot (\partial^2 u/\partial y^2)$ in the y direction, $\tau_{xy} = \mu \cdot \partial u/\partial y$ is the dominant stress. Therefore, the term $\mu \cdot (\partial^2 u/\partial y^2)$ of the frictional force f'_x will predominate. This, then, interacts with the pressure gradient $-\partial p/\partial x$, the inertial force $-\rho \cdot (\partial u/\partial t)$, and, if present, the mass force $\boldsymbol{f}$ per volume.

Non-Newtonian Fluids

The Navier–Stokes equations derived above are valid for *Newtonian fluids*. These differ from *non-Newtonian fluids* such as liquid tar, magma, plastic melts, polymer solutions, and suspensions like blood. In non-Newtonian fluids the frictional stresses acting on the fluid element can be dependent on both the instantaneous state of motion and the motion of the fluid in the past. The fluid can therefore have a memory.

In order to characterize the flowing properties of the fluid, e.g. for the shear flow of Figure 4.46, the shear stress τ_{xy} is plotted as a function of the shear velocity $\mathrm{d}u/\mathrm{d}y$. Some examples of Newtonian and non-Newtonian fluids are shown in Figure 4.49. To contrast with Newtonian fluids, we speak of non-Newtonian fluids when the functional relation (4.42) is nonlinear. The curves for fluids that cannot resist a shear rate pass through the origin. For so-called yielding fluids, a finite shear stress occurs even when the velocity gradient vanishes. These fluids behave partly as solid bodies and partly as fluids. The curve for pseudoplastic fluids such as melts or high polymers have a reduction in the slope as the shear stress grows. In contrast, dilatant fluids such as suspensions indicate an increase in the slope with increasing shear stress. The behavior of an idealized Bingham medium occurs for toothpaste or mortar. The finite value of τ_{xy} at $\mathrm{d}u/\mathrm{d}y = 0$ follows the linear course of a Newtonian fluid. In addition, some non-Newtonian fluids show a time dependence of the shear stress. Even if the shear rate is kept constant, the shear stress changes. A frequently used ansatz for non-Newtonian media is

$$\tau_{xy} = K \cdot \left| \frac{\mathrm{d}u}{\mathrm{d}y} \right|^n , \tag{4.50}$$

where K and n are material constants. For $n < 1$ we have the pseudoplastic fluid, $n = 1$ with $K = \mu$ is the Newtonian fluid, and $n > 1$ is the dilatant fluid. Note that ansatz (4.50) yields unrealistic values for the origin of Figure 4.49.

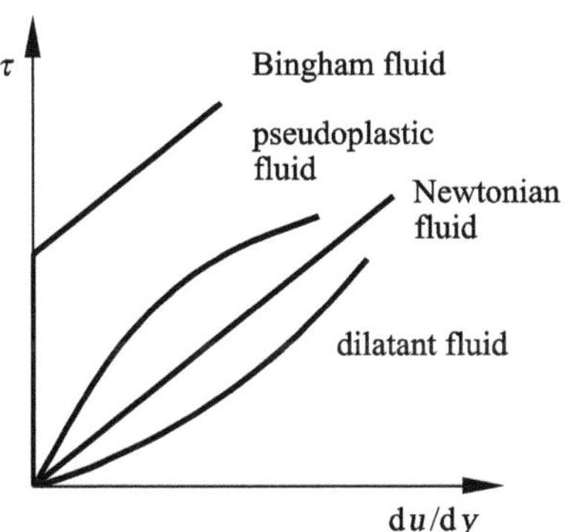

Fig. 4.49. Shear stress τ for Newtonian and non-Newtonian fluids

Numerous other laws have been derived for non-Newtonian fluids, mostly from experimental results. Selected flow examples are presented in Section 4.2.11. In what now follows we assume that the fluid at hand is Newtonian.

4.2.2 Mechanical Similarity, Reynolds Number

The question arises of when flows will be geometrically similar for similar geometries (geometrically similar channels, geometrically similar bodies in a flow). This means that when the mass force is neglected, the ratios between the pressure force, the frictional force, and the inertial force are the same in both flows. Because the forces are in equilibrium, it suffices to consider one ratio. We select the ratio of inertial force to frictional force. The different geometrically similar flows are to be characterized by characteristic lengths l_1, l_2 (e.g. diameter or length of a body, diameter of a pipe) and by characteristic velocities u_1, u_2 (e.g. velocity of a body or mean velocity in a certain pipe cross-section). The different densities and viscosities are denoted by ρ_1 and ρ_2, and μ_1 and μ_2, respectively. The x component of the inertial force reads

$$-\rho \cdot \frac{\mathrm{d}u}{\mathrm{d}t} = -\rho \cdot \left(u \cdot \frac{\partial u}{\partial x} + \cdots \right).$$

For similar flows this behaves according to $\rho_1 \cdot u_1^2/l_1$ to $\rho_2 \cdot u_2^2/l_2$. At corresponding positions, the u values behave like the characteristic velocities u_1, u_2. The lengths x and y behave like the characteristic lengths l_1 and l_2. The frictional forces, on the other hand, behave according to the expression $\mu \cdot (\partial^2 u/\partial y^2)$ like $\mu \cdot u/l^2$, where $\partial^2 u$ indicates a small velocity difference of second order. It behaves like the velocity u. The quantity ∂y^2 is the square of a small length difference and behaves like l^2.

The demand for mechanical similarity requires that $\rho \cdot u^2/l$ and $\mu \cdot u\ l^2$ be in a fixed ratio to one another:

$$\frac{\rho \cdot u^2}{l} / \frac{\mu \cdot u}{l^2} = \frac{\rho \cdot u \cdot l}{\mu}.$$

Therefore, mechanical similarity of the two systems 1 and 2 is expected when

$$\frac{\rho_1 \cdot u_1 \cdot l_1}{\mu_1} = \frac{\rho_2 \cdot u_2 \cdot l_2}{\mu_2} \tag{4.51}$$

holds. The ratio of the inertial forces to the viscous forces is called the *Reynolds number*. The ratio μ/ρ is called the *kinematic viscosity* and is denoted by ν.

The flow drag of a viscous liquid can be characterized by the value of its Reynolds number $\mathrm{Re} = \rho \cdot u \cdot l/\mu = u \cdot l/\nu$. Small Reynolds numbers indicate predominating frictional forces, and large Reynolds numbers predominating inertial forces.

In the limit of very small Reynolds numbers the flow is called *creeping flow*. For this case, an analytical solution of the Navier–Stokes equations

(4.49) is known for the flow past a sphere. These flows are characterized by the fact that the acceleration terms drop away, and only the pressure and frictional forces are in equilibrium with each other, as in the case of very viscous motor oils or in very small geometric dimensions.

The frictional forces at a volume element are proportional to $\mu \cdot u/l^2$. Because they are in equilibrium, the pressure forces obey the same relation, so that here geometrical similarity always implies mechanical similarity. Comparable volumes behave like l^3, so that the total drag forces must be proportional to $\mu \cdot u \cdot l$. The drag of a sphere flow is computed according to the *Stokes solution* of the Navier–Stokes equations as

$$W = 6 \cdot \pi \cdot \mu \cdot u \cdot R. \tag{4.52}$$

In the case of small drops falling to earth, the drag is to be set equal to the difference between weight and lift. Therefore, for drops with radius R and density ρ_t in a surrounding fluid of density ρ, we have

$$6 \cdot \pi \cdot \mu \cdot u \cdot R = \frac{4 \cdot \pi}{3} \cdot (\rho_t - \rho) \cdot g \cdot R^3.$$

This corresponds to a rate of fall of

$$u = \frac{2}{9} \cdot \frac{(\rho_t - \rho)}{\mu} \cdot g \cdot R^2. \tag{4.53}$$

This equation is valid for Reynolds numbers smaller than 1. For water droplets in air we obtain $u = 1.2 \cdot 10^8 \cdot R^2$, which is valid for droplets whose radius is smaller than 10^{-2} mm, i.e. for fine mist.

4.2.3 Laminar Boundary Layers

In the limit of very large Reynolds numbers, the inertial force dominates. A thin boundary layer forms on the surface of a body, and in this boundary layer the velocity of the inviscid outer flow is decelerated to the value zero at the wall (no-slip condition). The boundary layer is thinner, the smaller the viscosity. In the boundary layer the frictional forces are of the same order of magnitude as the inertial forces.

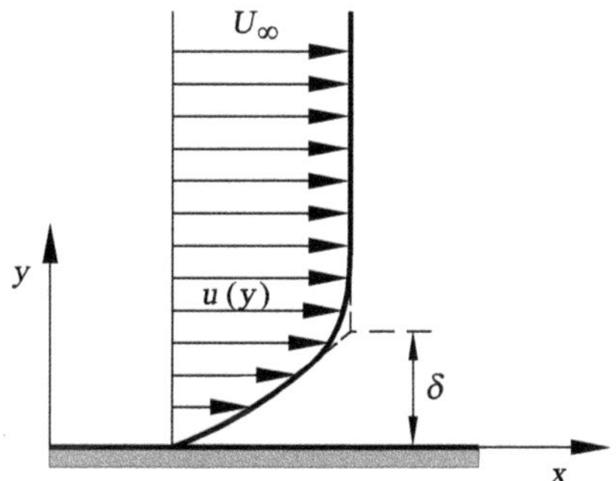

Fig. 4.50. Velocity distribution close to the wall

Figure 4.50 shows the velocity distribution in a boundary layer. If the dimension of the body in the flow direction is of order of magnitude l and the thickness of the boundary layer of order of magnitude δ, then the frictional force on the volume element $\mu \cdot (\partial^2 u/\partial y^2)$ is of order of magnitude $\mu \cdot U_\infty/\delta^2$. The inertial force is of order of magnitude $\rho \cdot U_\infty^2/l$. The order of magnitude of both expressions is the same if

$$\delta \propto \sqrt{\frac{\mu \cdot l}{\rho \cdot U_\infty}}. \tag{4.54}$$

An estimate leading to the same result is also obtained by considering the momentum for the boundary-layer flow along a flat plate. Let the length of the plate be l, the width b, the velocity of the outer flow U_∞, the thickness of the boundary layer δ (Figure 4.51). The mass transported per second in the boundary layer is proportional to $\rho \cdot b \cdot \delta \cdot U_\infty$. In the free stream this mass has velocity U_∞, while in the boundary layer it loses a certain amount of this velocity. The corresponding momentum loss is computed from mass times velocity loss and is proportional to $\rho \cdot b \cdot \delta \cdot U_\infty^2$. The momentum loss must be equal to the frictional force acting from the wall on the liquid. According to (4.42), this frictional force is proportional to $l \cdot b \cdot \mu \cdot U_\infty/\delta$. The proportionality of these two expressions leads to

$$\delta \propto \sqrt{\frac{\mu \cdot l}{\rho \cdot U_\infty}} = \sqrt{\frac{\nu \cdot l}{U_\infty}}.$$

The ratio δ/l is therefore proportional to $\sqrt{\nu/(U_\infty \cdot l)}$. With $U_\infty \cdot l/\nu = \mathrm{Re}_l$ and $U_\infty \cdot \delta/\nu = \mathrm{Re}_\delta$ we obtain $\delta/l \propto 1/\sqrt{\mathrm{Re}_l}$ and $\mathrm{Re}_\delta \propto \sqrt{\mathrm{Re}_l}$.

We can also introduce the time during which the individual liquid element flows along the body. For elements that are not too close to the surface, this time is of order of magnitude $t \propto l/U_\infty$, so that

$$\delta \propto \sqrt{\nu \cdot t}. \tag{4.55}$$

This equation can also be applied to unsteady boundary-layer flows of bodies suddenly set into motion. It shows that the boundary-layer thickness at the start of motion increases in proportion to the square root of the time.

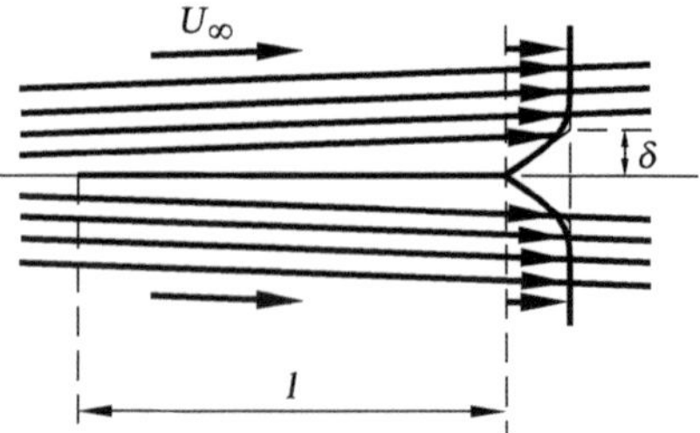

Fig. 4.51. Flow along a plate

Local shear stresses at the wall cause a friction drag corresponding to the flow in Figure 4.50. For the plate (Figure 4.51), the shear stress $\tau_w = \mu \cdot (\partial u/\partial y)_{y=0}$ is of order of magnitude

$$\tau_w \propto \mu \cdot \frac{U_\infty}{\delta} \propto \sqrt{\frac{\mu \cdot \rho \cdot U_\infty^3}{l}}.$$

If b is the width of the plate, then the total surface area is given by $A = 2 \cdot b \cdot l$. Therefore, the drag is

$$W \propto 2 \cdot b \cdot l \cdot \tau_w = \mathrm{K} \cdot b \cdot \sqrt{\mu \cdot \rho \cdot l \cdot U_\infty^3}, \tag{4.56}$$

with the constant K.

Boundary-layer theory can be traced back to *L. Prandtl* (1904). In his famous paper on the motion of liquids with very small friction, he presented the mathematical basis of flows for very large Reynolds numbers. His student *H. Schlichting* (1950) set out its application to almost all areas of fluid mechanics in his book *Boundary-Layer Theory.*

In a boundary-layer flow, the pressure gradient perpendicular to the wall may be neglected. Similarly, the velocity gradient along the wall is neglected compared to the velocity gradient perpendicular to the wall. Of the terms on the right-hand side of (4.49) only the term $\mu \cdot (\partial^2 u/\partial y^2)$ remains. This is of the same order of magnitude as $\rho \cdot u \cdot \partial u/\partial x$.

In a two-dimensional flow, small curvature in the boundary layer may also be neglected. The x coordinate is set equal to the arc length of the streamline along the wall. We obtain the *Prandtl boundary-layer equation* for the velocity component in the x direction:

$$\frac{\partial u}{\partial t} + u \cdot \frac{\partial u}{\partial x} + v \cdot \frac{\partial u}{\partial y} = -\frac{1}{\rho} \cdot \frac{\mathrm{d}p}{\mathrm{d}x} + \nu \cdot \frac{\partial^2 u}{\partial y^2}, \tag{4.57}$$

$$\frac{\partial u}{\partial x} + \frac{\partial v}{\partial y} = 0. \tag{4.58}$$

Since the vertical pressure gradient may be neglected, the pressure p of the outer flow is impressed on the boundary layer. This follows from (4.49) for the velocity component in the y direction. At the wall with $u = 0$ and $v = 0$, the left side of (4.57) vanishes. Therefore, we have

$$\left.\frac{\partial^2 u}{\partial y^2}\right|_{y=0} = \frac{1}{\mu} \cdot \frac{\partial p}{\partial x}. \tag{4.59}$$

If there is a pressure drop in the flow direction ($\partial p/\partial x$ negative), the velocity profile is convexly curved. On the other hand, if there is a pressure rise ($\partial p/\partial x$ positive) the velocity profile close to the wall is concave and thus has a turning point. If the pressure increase is too large, a backflow can occur close to the wall, and the boundary-layer flow separates. The *separation point* where the boundary-layer flow leaves the wall is given by the condition $(\partial u/\partial y)_{y=0} = 0$. Since the profile must have concave curvature for flow separation to occur, the separation point lies in the pressure increase region.

The *boundary-layer thickness* δ is introduced as the distance from the wall where $u = 0.99 \cdot U_\infty$. The *displacement thickness* δ_1 of a boundary layer is calculated from

$$\delta_1 = \int_0^\infty \left(1 - \frac{u}{U_\infty}\right) \cdot dy. \tag{4.60}$$

This is the distance the external inviscid flow is displaced from the wall of the body by the presence of the boundary layer. The *momentum thickness*

$$\delta_2 = \int_0^\infty \frac{u}{U_\infty} \cdot (1 - \frac{u}{U_\infty}) \cdot dy \tag{4.61}$$

is a measure of the relative momentum loss of the fluid compared to the inviscid flow.

4.2.4 Onset of Turbulence

Pipe Flow

In the flow of viscous liquids through long straight pipes, at higher velocities and thus at larger Reynolds numbers the Hagen–Poiseuille law given in (4.45) is replaced by another law. The pressure drop becomes considerably larger and is approximately proportional to the second power of the flux. Simultaneously, velocity fluctuations are superimposed on the flow. In a laminar

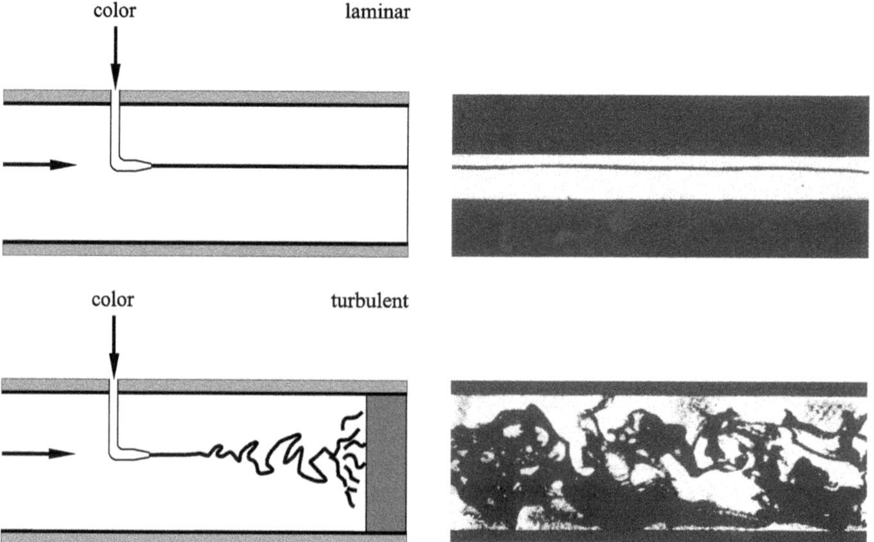

Fig. 4.52. Laminar and turbulent pipe flow, *O. Reynolds* 1883

flow a colored filament forms a straight line. At larger Reynolds numbers the colored filament is torn apart, and downstream the color is spread uniformly throughout the liquid. The linear motion is called *laminar*, and the swirled motion is called *turbulent*.

This experiment was first performed by *O. Reynolds* (1883). Figure 4.52 shows the colored filament in laminar and turbulent pipe flow. The main motion of the flow takes place in the direction of the axis of the pipe. Because of the flow fluctuations, a great amount of mixing occurs in the turbulent flow, leading to a transverse motion perpendicular to the main motion. This transverse motion causes an exchange of momentum in the transverse direction. For this reason, the velocity distribution across the diameter of the pipe is much more uniform and full for turbulent flow than for laminar pipe flow (see Figure 4.53).

In his experiments, *O. Reynolds* (1883) discovered that the transition from laminar to turbulent flow always takes place at almost exactly the same Reynolds number $\mathrm{Re}_d = u_\mathrm{m} \cdot d/\nu$, where $u_\mathrm{m} = \dot{V}/A$ is the mean flow velocity (d pipe diameter, $\dot{V}$ volume flux, A pipe cross-sectional area). The numerical value of the *critical Reynolds number* at which the transition occurs is

$$\mathrm{Re}_\mathrm{crit} = \left(\frac{u_\mathrm{m} \cdot d}{\nu}\right)_\mathrm{crit} = 2300. \tag{4.62}$$

Therefore, pipe flows whose Reynolds number is $\mathrm{Re} < \mathrm{Re}_\mathrm{crit}$ are laminar, and those for which $\mathrm{Re} > \mathrm{Re}_\mathrm{crit}$ are turbulent. The numerical value of the critical Reynolds number depends greatly on the pipe intake and the incoming flow. *O. Reynolds* (1883) already suspected that the critical Reynolds number will be larger if the disturbances in the incoming flow are smaller. This was confirmed experimentally. Values of $\mathrm{Re}_\mathrm{crit}$ up to 40 000 were able to be measured. On the other hand, a lower limit to the critical Reynolds number of about 2000 was measured. Below this Reynolds number the flow remains laminar, even for very strong disturbances. We now know from results from stability theory that the laminar–turbulent transition is caused by three-dimensional disturbances. Pipe flow is stable with respect to two-dimensional disturbances.

Associated with the laminar–turbulent transition is also a change in the pipe drag law. Whereas for laminar flow the pressure drop is proportional to the first power of the mean flux velocity u_m, for turbulent flows this pressure

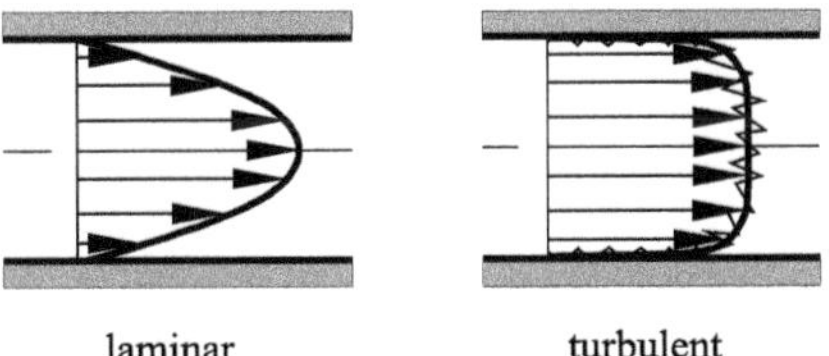

Fig. 4.53. Velocity distributions of laminar and turbulent pipe flow

drop is almost proportional to the square of the mean flux velocity. This large flux drag is due to the turbulent mixing motion.

The laminar–turbulent *transition* is a *stability problem.* The laminar flow is influenced by small perturbations, which, in the case of pipe flow, can be caused by the intake. At small Reynolds numbers, i.e. at large enough values of ν, the damping action of the viscosity is large enough to permit these perturbations to die away again. It is only at large enough Reynolds numbers that the damping does not suffice, so that the perturbations are amplified and finally the transition to the turbulent flow form is started. In the next section we will see that, first of all, two-dimensional perturbations occur in two-dimensional boundary layers, to be followed by three-dimensional perturbations later on in the transition.

As already mentioned, stability theory investigations of the parabolic velocity profile of the pipe flow show that this is stable with respect to two-dimensional perturbations. In contrast to the boundary-layer flows treated in the following section, the laminar–turbulent transition in pipe flows begins with three-dimensional perturbations.

Reynolds Ansatz

The *mathematical description* of turbulent flows can be derived from the experimental results in Figure 4.52. The flow quantities, such as the u component of the velocity, can be written down as a superposition of the time-averaged velocities $\overline{u}(x, y, z)$ and the additional fluctuations $u'(x, y, z, t)$.

From Figure 4.54, the *Reynolds ansatz* for turbulent flows can be written as

$$\boldsymbol{u}(x, y, z, t) = \overline{\boldsymbol{u}}(x, y, z) + \boldsymbol{u}'(x, y, z, t). \tag{4.63}$$

With the velocity component u taken as an example, the definition of the time average at a fixed position reads

$$\overline{u} = \frac{1}{T} \cdot \int_0^T u(x, y, z, t) \cdot \mathrm{d}t. \tag{4.64}$$

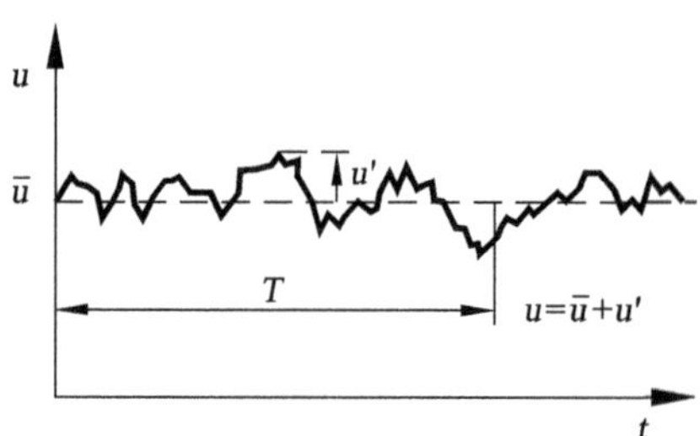

Fig. 4.54. Reynolds ansatz for the x component of the velocity u

The time T is a suitably large time interval with the condition that an increase in T leads to no further change in the time-averaged value $\overline{u}$. From the definition of the time average we can determine that the time-averaged values of the fluctuation quantities must vanish:

$$\overline{u'} = 0, \qquad \overline{v'} = 0, \qquad \overline{w'} = 0. \tag{4.65}$$

Boundary-Layer Flow

The appearance of turbulence is not restricted to flows in pipes and channels. It is also seen in *boundary layers.* The Reynolds number $U_\infty \cdot \delta/\nu$ is now formed with the boundary-layer thickness δ and the velocity U_∞ outside the boundary layer. For bodies in a flow, the boundary-layer thickness close to the stagnation line is very thin. The flow is initially laminar and becomes turbulent downstream, as a critical Reynolds number is exceeded. The thickness of the laminar boundary layer on the plate increases with $\sqrt{x}$, where x is the distance from the leading edge. The critical Reynolds number of the plate boundary layer is

$$\mathrm{Re}_{\mathrm{crit}} = \left(\frac{U_\infty \cdot x}{\nu}\right)_{\mathrm{crit}} = 5 \cdot 10^5. \tag{4.66}$$

In the case of a plate in a longitudinal flow, as in the case of pipe flow, the critical Reynolds number can also be raised when the incoming flow is less perturbed (lower intensity of turbulence).

The experimental results of the investigations into the laminar–turbulent transition in the boundary layer are summarized in Figure 4.55. The laminar boundary-layer flow is superimposed with two-dimensional perturbing waves at a critical Reynolds number $\mathrm{Re}_{\mathrm{crit}}$. These waves are called *Tollmien–Schlichting waves.* Further downstream, three-dimensional perturbations are imposed on the flow. This leads to characteristic Λ-vortex formation with

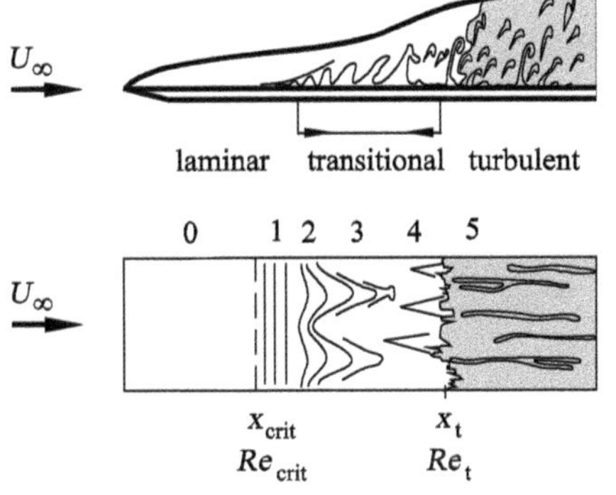

0 stable, laminar flow
1 unstable Tollmien-Schlichting waves
2 three-dimensional waves, Λ-vortex
3 vortex decay
4 formation of turbulent spots
5 turbulent flow

Fig. 4.55. Sketch of the laminar–turbulent transition in the boundary layer of a flat plate in a longitudinal flow

local shear layers in the boundary layer. The decay of the Λ-vortices causes *turbulent spots*, which begin the transition to a turbulent boundary-layer flow. At Re_t the transition process is completed, and downstream from this point the boundary layer is turbulent.

As can be seen in Figure 4.55, the boundary-layer thickness grows greatly at the laminar–turbulent transition.

Stability Theory

The onset of the laminar–turbulent transition can be treated with *stability theory*. The attempts to do this began in the nineteenth century and eventually led to success in 1930. The theoretical investigations are based on the idea that small perturbations are present in the laminar flow. In the case of a pipe flow this can be due to the intake, whereas for boundary layers of bodies in a flow they can be caused by roughness of the wall or by perturbations in the outer flow. The theory follows the behavior in time of such perturbations superimposed on the laminar flow, whose shapes in each case must still be determined more precisely. The decisive question is whether the perturbation motion dies away or is amplified. If the perturbations die away in time, the basic flow is said to be *stable*. If the perturbations grow in time, the basic flow is *unstable*; i.e. the transition to turbulent flow is possible.

In this manner a *stability theory* of laminar flow can be developed, whose aim is to calculate theoretically the *critical Reynolds number* for a given laminar flow (see Section 6.2). In this stability investigation, the motion is decomposed into the basic flow whose stability is to be investigated and the superimposed perturbation motion. The basic flow, which can be taken to be steady, will now be denoted by the velocity components U_0, V_0, W_0 and the pressure P_0. This basic flow is a solution of the Navier–Stokes equations (4.49). The time-varying perturbation motion has associated quantities u', v', w', and p'. The resulting flow is obtained with the perturbation ansatz

$$u = U_0 + u', \quad v = V_0 + v', \quad w = W_0 + w', \quad p = P_0 + p'. \tag{4.67}$$

In most cases it is assumed that the perturbation quantities are small compared to the values of the basic flow.

For a two-dimensional incompressible basic flow (U_0 and V_0) and a two-dimensional perturbation (u' and v'), the flow resulting from (4.67) satisfies the two-dimensional Navier–Stokes equations. The basic flow $U_0(y)$ is chosen to be particularly simple, so that U_0 depends only on y. The velocity component V_0 vanishes. The boundary-layer flow approximately satisfies this condition, since the dependence of the basic flow U_0 on the longitudinal coordinate x is much smaller than on the transverse coordinate y. This is called the *parallel flow assumption*. For the pressure of the basic flow $P_0(x, y)$, the dependence on x also has to be taken into account, since it is the pressure drop $\partial P_0/\partial x$ that produces the flow. Therefore, the basic flow at hand has the form

$$U_0(y), \qquad V_0 = 0, \qquad P_0(x, y). \tag{4.68}$$

Onto this basic flow is superimposed a two-dimensional perturbation motion (Figure 4.56), which is also dependent on time. The associated velocity components and the pressure are

$$u'(x, y, t), \qquad v'(x, y, t), \qquad p'(x, y, t). \tag{4.69}$$

From (4.67) we obtain the resulting flow as

$$u = U_0 + u', \qquad v = v', \qquad p = P_0 + p'. \tag{4.70}$$

The basic flow (4.68) is a solution of the Navier–Stokes equations by assumption. The resulting flow (4.70) also has to satisfy the Navier–Stokes equations. The superimposed perturbation motion (4.69) is assumed to be *small*; i.e. all square terms of the perturbation motion are neglected compared to the linear terms. The aim of the stability investigation is to determine whether the perturbation motion dies away or is amplified for a given basic flow, in which case the basic flow is called stable or unstable.

Inserting (4.70) into the Navier–Stokes equations, and neglecting square terms in the perturbation velocities, we obtain

$$\frac{\partial u'}{\partial t} + U_0 \cdot \frac{\partial u'}{\partial x} + v' \cdot \frac{\mathrm{d}U_0}{\mathrm{d}y} + \frac{1}{\rho} \cdot \frac{\partial P_0}{\partial x} + \frac{1}{\rho} \cdot \frac{\partial p'}{\partial x} = \nu \cdot \left(\frac{\mathrm{d}^2 U_0}{\mathrm{d}y^2} + \Delta u' \right),$$

$$\frac{\partial v'}{\partial t} + U_0 \cdot \frac{\partial v'}{\partial x} + \frac{1}{\rho} \cdot \frac{\partial P_0}{\partial y} + \frac{1}{\rho} \cdot \frac{\partial p'}{\partial y} = \nu \cdot \Delta v',$$

$$\frac{\partial u'}{\partial x} + \frac{\partial v'}{\partial y} = 0.$$

Here Δ is the Laplace operator $\partial^2/\partial x^2 + \partial^2/\partial y^2$.

Note that the basic flow satisfies the Navier–Stokes equations (approximately, in the case of the boundary layer), and so this equation simplifies to

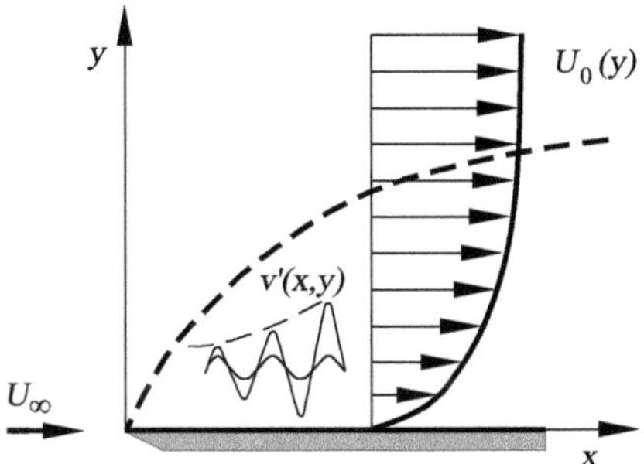

Fig. 4.56. Basic flow $U_0(y)$ and perturbation wave $v'(x, y)$ of the plate boundary layer

$$\frac{\partial u'}{\partial t} + U_0 \cdot \frac{\partial u'}{\partial x} + v' \cdot \frac{\mathrm{d}U_0}{\mathrm{d}y} + \frac{1}{\rho} \cdot \frac{\partial p'}{\partial x} = \nu \cdot \Delta u',$$
$$\frac{\partial v'}{\partial t} + U_0 \cdot \frac{\partial v'}{\partial x} + \frac{1}{\rho} \cdot \frac{\partial p'}{\partial y} = \nu \cdot \Delta v', \tag{4.71}$$
$$\frac{\partial u'}{\partial x} + \frac{\partial v'}{\partial y} = 0.$$

These are three equations for u', v', and p'. The associated boundary conditions require that the perturbation velocities u' and v' vanish at the bounding walls (no-slip condition) and at infinity. The pressure p' may be eliminated from (4.71), so that together with the continuity equation, we obtain two equations for u' and v'.

In order to describe the components of the perturbation velocities for the *Tollmien–Schlichting waves* we use the wave ansatz

$$u' = \hat{u}(y) \cdot \exp(\mathrm{i} \cdot a \cdot x - \mathrm{i} \cdot \omega \cdot t), \quad v' = \hat{v}(y) \cdot \exp(\mathrm{i} \cdot a \cdot x - \mathrm{i} \cdot \omega \cdot t), \tag{4.72}$$

with the wave number a, the angular frequency ω, and the amplitude functions $\hat{u}, \hat{v}$ of the perturbation waves. For the time-amplified stability problem at hand, ω is complex:

$$\omega = \omega_\mathrm{r} + \mathrm{i} \cdot \omega_\mathrm{i},$$

with the real part of the angular frequency ω_r and the rate of amplification in time ω_i. If $\omega_\mathrm{i} < 0$, the perturbation wave is damped, and the laminar boundary-layer flow is *stable.* With $\omega_\mathrm{i} > 0$ the boundary layer is *unstable*, and the Tollmien–Schlichting waves are amplified in time. It is useful, in addition to a and ω, to introduce the phase velocity of the perturbation wave:

$$c = \frac{\omega}{a} = c_\mathrm{r} + \mathrm{i} \cdot c_\mathrm{i}.$$

Inserting the wave ansatz (4.72) into the perturbation differential equation for u' and v', we obtain, for example, the *Orr–Sommerfeld equation* for the amplitude function $\hat{v}(y)$:

$$(a \cdot U_0 - \omega) \cdot \frac{\mathrm{d}^2 \hat{v}}{\mathrm{d}y^2} + a \cdot \left(a \cdot \omega - a^2 \cdot U_0 - \frac{\mathrm{d}^2 U_0}{\mathrm{d}y^2} \right) \cdot \hat{v}$$
$$+ \mathrm{i} \cdot \frac{1}{\mathrm{Re}_d} \cdot \left(\frac{\mathrm{d}^4 \hat{v}}{\mathrm{d}y^4} - 2 \cdot a^2 \cdot \frac{\mathrm{d}^2 \hat{v}}{\mathrm{d}y^2} + a^4 \cdot \hat{v} \right) = 0. \tag{4.73}$$

Quantities made dimensionless with the characteristic velocity at the edge of the boundary layer U_δ, the characteristic length $d = \sqrt{\nu \cdot x / U_\delta}$ and the characteristic time d/U_δ are introduced. The Orr–Sommerfeld equation is a fourth-order ordinary differential equation that with the boundary conditions at the wall and in the unperturbed free stream

$$y = y_{\mathrm{w}}: \qquad \hat{v} = 0, \qquad \frac{\mathrm{d}\hat{v}}{\mathrm{d}y} = 0,$$

$$y \rightarrow \infty: \qquad \hat{v} = 0, \qquad \frac{\mathrm{d}\hat{v}}{\mathrm{d}y} = 0, \tag{4.74}$$

is an *eigenvalue problem* with the Reynolds number Re_d as parameter. This is generally solved numerically with spectral methods. The solutions of the eigenvalue problem are presented in the form of stability diagrams (Figure 4.2.4). The stability diagram is produced by plotting the wave number a against the Reynolds number Re_d. The pairs of values (Re_d, a) associated with the roots of the imaginary part of the complex eigenvalue ω are drawn in the diagram. This neutral curve divides the stable perturbations from the unstable perturbations. It is also called the *indifference curve*, since in the case $\omega_{\mathrm{i}} = 0$, the perturbation amplitudes retain their original value. In the region inside the indifference curve $\omega_{\mathrm{i}} > 0$ holds; i.e. the flow is unstable. In the region outside the indifference curve ω_{i} assumes negative values, and the basic flow under investigation is stable at the Reynolds number at hand to perturbations with the associated wave number a.

Thus a critical Reynolds number $\mathrm{Re}_{\mathrm{crit}}$ can be determined, above which a given laminar flow becomes unstable. A tangent to the indifference curve is drawn in Figure 4.57 parallel to the a axis. The point of intersection of this tangent with the abscissa yields the value of the *critical Reynolds number* $\mathrm{Re}_{\mathrm{crit}}$. For a Blasius boundary layer, the critical Reynolds number formed with the length along the plate is

$$\mathrm{Re}_{\mathrm{crit}} = \left(\frac{U_\delta \cdot x}{\nu} \right)_{\mathrm{crit}} = 5 \cdot 10^5. \tag{4.75}$$

With the critical Reynolds number $\mathrm{Re}_{\mathrm{crit}} = 5 \cdot 10^5$, Figure 4.57 yields the critical wave number $a_{\mathrm{crit}} = 2 \cdot \pi / \lambda_{\mathrm{crit}}$, with which the critical wavelength λ_{crit} of the perturbations at hand can be calculated. This means physically that the laminar basic flow is stable with respect to perturbations of any wavelength for Reynolds numbers smaller than $\mathrm{Re}_{\mathrm{crit}}$, since in this Reynolds number regime $\omega_{\mathrm{i}} < 0$ holds for all wave numbers. Forming the critical Reynolds

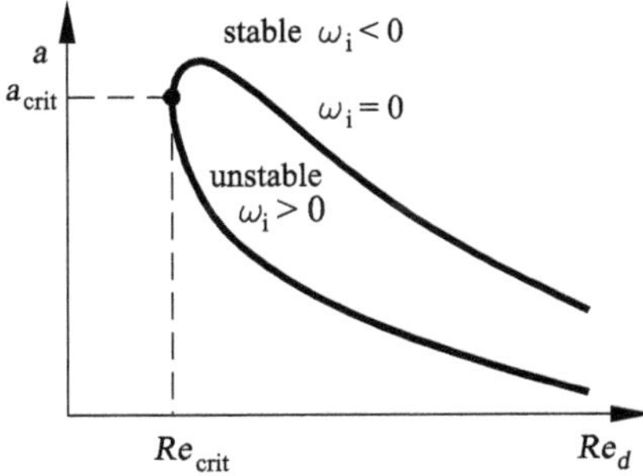

Fig. 4.57. Stability diagram of the plate boundary layer

number with the characteristic length $d = \sqrt{\nu \cdot x / U_\delta}$, we obtain the value

$$\mathrm{Re}_{\mathrm{crit}} = \frac{U_\delta \cdot d}{\nu} = 302. \tag{4.76}$$

This value also makes sense for comparisons with the instability of compressible boundary layers. The onset of Tollmien–Schlichting waves in a compressible boundary-layer flow at an adiabatic wall also yields $\mathrm{Re}_{\mathrm{crit}} = 302$. It is only for isothermal boundaries that differences occur.

Laminar–Turbulent Transition Control

H. Schlichting (1968) presented a summary of how to influence the laminar–turbulent transition in two-dimensional boundary-layer flow. Certain measures can be taken to shift the transition downstream. This leads to a reduction in the drag. The laminar–turbulent transition can be influenced using moving surfaces, acceleration of the boundary layer by blowing or by a pressure gradient, suction of the boundary layer, and cooling of the surface. In what follows we investigate the effect of the pressure gradient that occurs on a wing profile on acceleration.

Whereas in the flow past a plate, similar velocity profiles form at different distances from the leading edge of the plate, the pressure gradient $\partial p/\partial x$ along the wing profile causes different laminar boundary-layer profiles. In the region where the pressure decreases, $\partial p/\partial x < 0$, the velocity profiles have no turning point, and in the region of increasing pressure, $\partial p/\partial x > 0$, velocity profiles with a turning point are found. Whereas all velocity profiles on a plate in a longitudinal flow have the same critical Reynolds number $\mathrm{Re}_{\mathrm{crit}} = 302$, the limit of stability for each boundary-layer profile on a wing is different. In the region of decreasing pressure the critical Reynolds numbers $\mathrm{Re}_{\mathrm{crit}}$ are

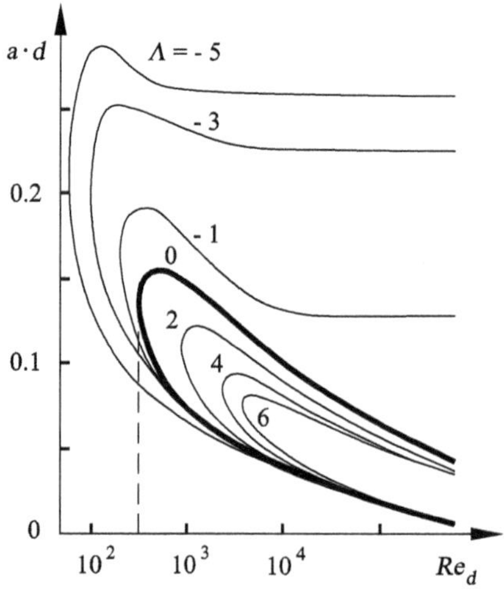

Fig. 4.58. Stability diagram for laminar boundary-layer profiles for pressure decrease $\Lambda > 0$ and pressure increase $\Lambda < 0$

larger than those of the flow past a plate, while they are smaller in the region of increasing pressure.

The pressure gradient on the wing profile can be described using the form parameter Λ:

$$\Lambda = -\frac{\delta^2}{\mu \cdot U_\delta} \cdot \frac{\partial p_\delta}{\partial x} = \frac{\delta^2}{\nu} \cdot \frac{\partial U_\delta}{\partial x},$$

with the boundary-layer thickness δ and the velocity at the edge of the boundary layer U_δ. The form parameter Λ takes on values between $\Lambda = +12$ and $\Lambda = -12$, where the laminar boundary-layer flow separates for $\Lambda = -12$.

At the front stagnation point of the profile $\Lambda = 7.05$ and at the pressure minimum $\Lambda = 0$, $\Lambda > 0$ means a decrease in the pressure, and $\Lambda < 0$ a pressure increase. The velocity profiles for $\Lambda < 0$ have a turning point. Figure 4.58 shows the stability diagram of laminar boundary-layer profiles with pressure decrease and increase. For the velocity profile in the region of decreasing pressure $\Lambda > 0$, both branches of the indifference curve tend to zero for $\mathrm{Re}_d \to \infty$, as well as for the plate boundary layer with $\Lambda = 0$. On the other hand, for the velocity profile in the region of increasing pressure, $\Lambda < 0$, the upper branch of the indifference curve has a nonzero asymptote, so that a finite wavelength regime of amplified perturbations is present even for $\mathrm{Re}_d \to \infty$. It can be seen that for boundary layers in the pressure increase region the unstable region of perturbation wavelengths enclosed by the indifference curve is much larger than in the region of decreasing pressure.

Fig. 4.59. Propagation of a turbulent perturbation

Propagation of Turbulent Perturbations

Up until now we have treated the onset of the laminar–turbulent transition using stability theory. In what follows, perturbations in the transition region will be considered, as have already been introduced in Figure 4.55 as *turbulent spots*.

Figure 4.59 shows the propagation of local turbulent perturbations in the transition regime of the laminar–turbulent transition on the plate boundary layer. The time sequence of the turbulent perturbation propagation shows that turbulence generated from a perturbation propagates further downstream of its own accord. The perturbation was introduced into the boundary layer by momentarily sucking some fluid out of the boundary layer. The camera traveled with the perturbation, so that the same group of vortices could be observed at all times. Spatially, new vortices keep forming downstream, until eventually the turbulent boundary-layer flow is fully developed. There is as yet no theory of the propagation process of turbulent perturbations, just as there is as yet no exact theory of the fully developed turbulent state (see Chapter 6).

4.2.5 Fully Developed Turbulence

Many technical flows are turbulent. According to the Reynolds ansatz (4.63), this means that the time-average primary motion is overlaid with turbulent fluctuations. By way of illustration, Figure 4.60 shows some shots of turbulent flow in a water channel.

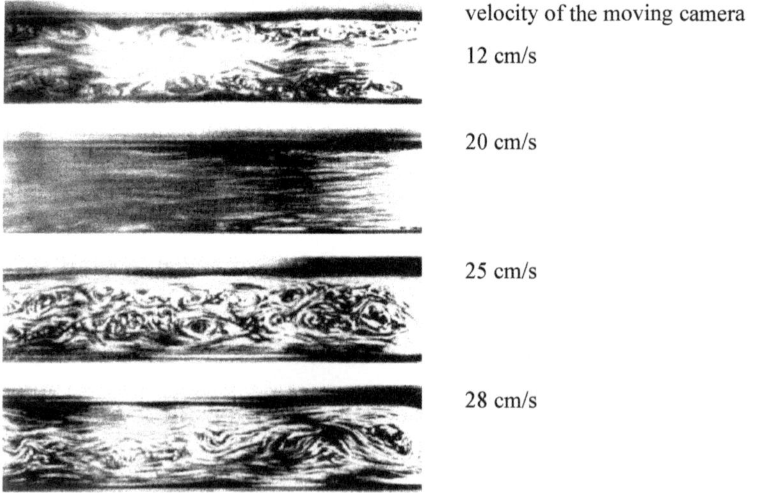

Fig. 4.60. Turbulent flow in a water channel, moving camera. Shots taken by *J. Nikuradse* (1929), published by *W. Tollmien* (1931)

One and the same flow portrait was taken at the same flux velocity and at different camera velocities. In the four pictures it can be seen whether the longitudinal velocity of the flow is larger or smaller than that of the camera. When the camera velocity is smaller, the turbulence structure at the wall can be seen. At larger camera velocities the turbulence structures within the flow become visible. We can see the positions at which the longitudinal velocity of the flow instantaneously coincides with the camera motion.

The longitudinal and transverse momentum exchange of turbulent flow shown in the figure causes a turbulent mixing motion that is essentially responsible for the larger drag of the turbulent flow.

Prandtl Mixing Length

The velocity fluctuations cause apparent stresses, e.g. the turbulent shear stress $\tau' = -\rho \cdot \overline{u' \cdot v'}$. This must be related to the distribution of the mean velocities, and to do this, the so-called *Prandtl mixing length* is essential. This is the path on which a fluid element loses its individuality by turbulent mixing with the surrounding liquid.

In Figure 4.61 a liquid element in the boundary layer under consideration is displaced from position y with mean velocity $\overline{u}(y)$ by distance l. The velocity difference from the velocity at the new position is $\overline{u}(y + l) - \overline{u}(y)$. To first approximation, this can be written as $l \cdot (\partial \overline{u}/\partial y)$. This value is of the order of magnitude of the fluctuation u'. The size of v' is obtained using the assumption that two fluid elements that enter the layer under consideration from different sides approach each other or move away from each other with relative velocity $2 \cdot l \cdot (\partial \overline{u}/\partial y)$. For reasons of continuity, the transverse velocity is of the same order of magnitude. Therefore, v' also has order of magnitude $l \cdot (\partial \overline{u}/\partial y)$. In forming the average $\overline{u' \cdot v'}$ we must pay attention to the signs of the u and v components. Negative u' are associated with positive v', and positive u' with negative v'. The product $u' \cdot v'$ is therefore always negative. The apparent shear stress becomes positive and is of order of magnitude $\rho \cdot (l \cdot (\partial \overline{u}/\partial y))^2$.

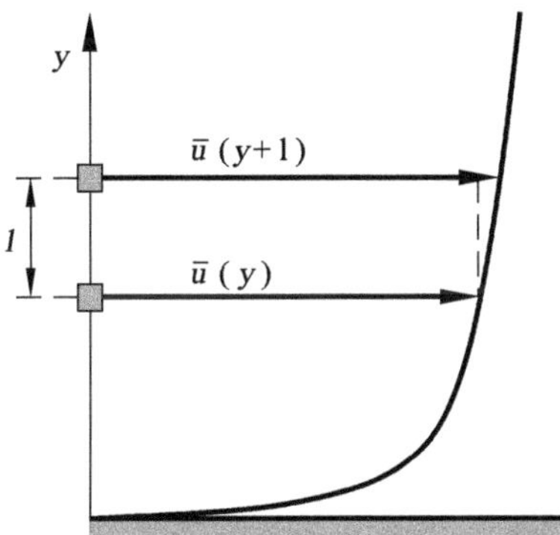

Fig. 4.61. Prandtl mixing length

For the turbulent shear stress τ' we obtain

$$\tau' = \rho \cdot l^2 \cdot \left|\frac{\partial \overline{u}}{\partial y}\right| \cdot \frac{\partial \overline{u}}{\partial y}. \tag{4.77}$$

It follows from (4.77) that the apparent stresses of the turbulent mixing motion vary in proportion to the square of the velocity. In fact, all hydraulic drags demonstrate essentially this behavior. The length l, called the *Prandtl mixing length*, has a certain similarity to the mean free path of kinetic gas theory. There the momentum transport due to molecular motion is considered in a similar manner to the momentum transport of fluid elements in the case of turbulent flow. The mixing length l of the turbulent motion is in general dependent on the position. There is as yet no general theory to predict its size in each case. However, suitable assumptions can be found for a number of individual cases, which then lead to well-confirmed results (see Section 6.3).

Free Jet

In the case of a free jet with sufficiently large Reynolds number (see Figure 1.3), it is advisable to set the mixing length l in each cross-section proportional to the jet width at that point, i.e. $l = \alpha \cdot b$. Here b is the half-diameter of a parabolic velocity distribution whose maximum velocity and volume flux are those of the actual flow, and α is a constant of proportionality with $\alpha \approx 0.125$. Such a manner of determining a velocity profile is necessary because the viscous flow passes over diffusely into the external liquid. The velocity of the round free jet decreases with increasing distance from the orifice, with the velocity distribution bell-shaped in all cross-sections (see Figure 4.62). Since the pressure in the jet is approximately that of the surroundings, it is mainly the apparent shear stresses that reduce the velocity with the distance and simultaneously sweep along more and more liquid at rest with the jet. The apparent shear stress τ' increases radially from zero at the center of the free jet to a maximum value, and then decreases to zero again.

Because of the approximately constant pressure, it makes sense to assume that the momentum of the jet $J = \rho \cdot \int u^2 \cdot \mathrm{d}A$ is equally large for all x values. Then J is proportional to $\rho \cdot u_1^2 \cdot \pi \cdot b^2$, where u_1 is the maximum velocity in the cross-section A of the free jet. Since J is constant, it follows that u_1 is proportional to $1/b$ and therefore also proportional to $1/x$. The flow is shown in Figure 4.62. If b is the half-value diameter, for which $u/u_1 = 0.5$, then for $x/d > 10$ (d is the jet diameter at $x = 0$) we have $b/x = 0.0848$ and

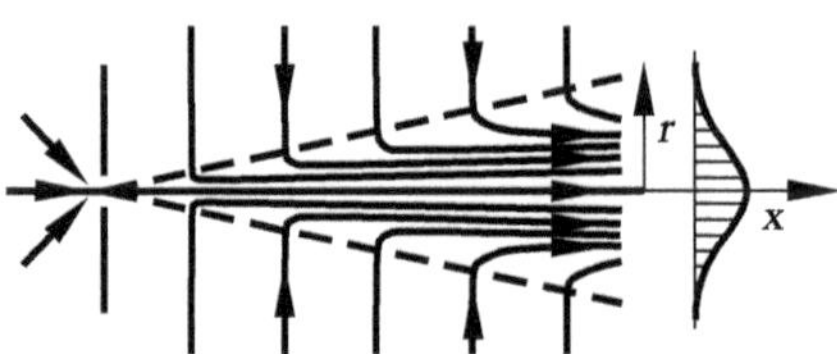

Fig. 4.62. Streamlines of an expanding free jet

further $u_1(x)/u_1(x=0) = 6.57 \cdot d/x$. The amount of liquid flowing in the axial direction $\int u \cdot \mathrm{d}A$ is proportional to $u_1 \cdot \pi \cdot b^2$ and increases linearly with the distance x. The liquid in the space therefore flows with radial velocity $v \propto \sqrt{J/\rho}/r$ toward the jet.

With $l = \alpha \cdot b$ the shear stress τ' from (4.77) has mean value τ'_{m} over the cross-section when $\partial\overline{u}/\partial y$ is approximated by $-2 \cdot u_1/b$. We obtain

$$\tau'_{\mathrm{m}} = -4 \cdot \rho \cdot l^2 \cdot \left(\frac{u_1^2}{b}\right) = -4 \cdot \alpha^2 \cdot \rho \cdot u_1^2.$$

Shear Layer

Another case of turbulent expansion is the dispersion of the edge of the jet flow past a corner (Figure 4.63). Here u_1 is constant. With $l = \alpha \cdot b$, τ'_{m} is proportional to $\alpha^2 \cdot \rho \cdot u_1^2$ and thus also constant. In what follows we set the width of the jet perpendicular to the plane of the figure equal to 1, and so the momentum loss of the approaching flow is proportional to $\rho \cdot u_1^2 \cdot b$. The associated drag is proportional to $\tau'_{\mathrm{m}} \cdot x$; i.e. $b \propto \alpha^2 \cdot x$, as in the case of the free jet. The liquid at rest in the surroundings experiences an increase in momentum of equal size.

Wall Turbulence

In flows along a wall, the mixing length must tend to zero as we come closer to the wall. This implies that $\partial\overline{u}/\partial y$ becomes very small inside the flow, but takes on large values close to the wall. The no-slip condition holds at the wall with $y = 0$. Because of this, a thin friction layer (*viscous sublayer*) forms directly at the wall, in which $\partial\overline{u}/\partial y = \tau_{\mathrm{w}}/\mu$ holds approximately where τ_{w} is the wall shear stress.

To treat this theoretically, we assume that the wall is smooth and the shear stress constant $\overline{\tau} = \tau_{\mathrm{w}}$. For simplicity, the wall is assumed to extend to infinity in the x and z directions. We then have

$$\overline{\tau} = \tau_{\mathrm{w}} = \mu \cdot \frac{\mathrm{d}\overline{u}}{\mathrm{d}y} - \rho \cdot \overline{u' \cdot v'}. \tag{4.78}$$

The mean velocity is dependent only on y and is completely determined by τ_{w}, ρ, and ν. This relation can be stated in dimensionless form. We introduce the shear stress velocity $u_\tau = \sqrt{\tau_{\mathrm{w}}/\rho}$. The ratio ν/u_τ is a characteristic length.

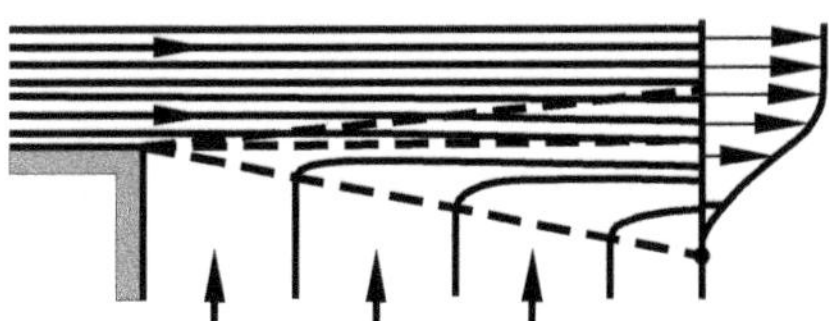

Fig. 4.63. Streamlines of corner flow

The total shear stress in the layer close to the wall consists of the mean value of the friction stresses and the apparent stresses of the turbulence, and for positive gradient $\mathrm{d}\overline{u}/\mathrm{d}y$ it is

$$\tau_\mathrm{w} = \mu \cdot \frac{\mathrm{d}\overline{u}}{\mathrm{d}y} + \rho \cdot l^2 \cdot \left(\frac{\mathrm{d}\overline{u}}{\mathrm{d}y}\right)^2 . \tag{4.79}$$

The first term of (4.79) holds in the viscous sublayer, while the second term is valid in the layer above this close to the wall. The velocity distribution can be written in the form

$$\frac{\overline{u}}{u_\tau} = \mathrm{f}\left(\frac{y \cdot u_\tau}{\nu}\right), \tag{4.80}$$

where f is a function of $y \cdot u_\tau/\nu$. Within the viscous sublayer $y \cdot u_\tau/\nu \leq 1$ we have $\mathrm{f}(y \cdot u_\tau/\nu) = y \cdot u_\tau/\nu$. At large distances from the wall, $y \cdot u_\tau/\nu > 50$, $\mu \cdot (\mathrm{d}\overline{u}/\mathrm{d}y)$ tends to zero and $-\overline{u' \cdot v'}$ becomes approximately u_τ^2. The flow is determined only by the quantities u_τ and y. Assuming that $l = \kappa \cdot y$, we obtain

$$\frac{\mathrm{d}\overline{u}}{\mathrm{d}y} = \frac{1}{\kappa} \cdot \frac{u_\tau}{y}, \tag{4.81}$$

where κ is the *Kármán constant.* The experimental value of κ is approximately 0.4. Integrating (4.81) yields

$$\overline{u} = u_\tau \cdot \left(\frac{1}{\kappa} \cdot \ln(y) + \mathrm{C}\right), \tag{4.82}$$

or, using (4.80),

$$\frac{\overline{u}}{u_\tau} = \mathrm{f}\left(\frac{y \cdot u_\tau}{\nu}\right) = \frac{1}{\kappa} \cdot \ln\left(\frac{y \cdot u_\tau}{\nu}\right) + \mathrm{C}_1. \tag{4.83}$$

Equation (4.83) is known as the *logarithmic wall law.* According to measurements by *J. Nikuradse* (1932), for smooth pipes the value $\kappa = 0.4$ with a constant of integration $\mathrm{C}_1 = 5.5$ is found.

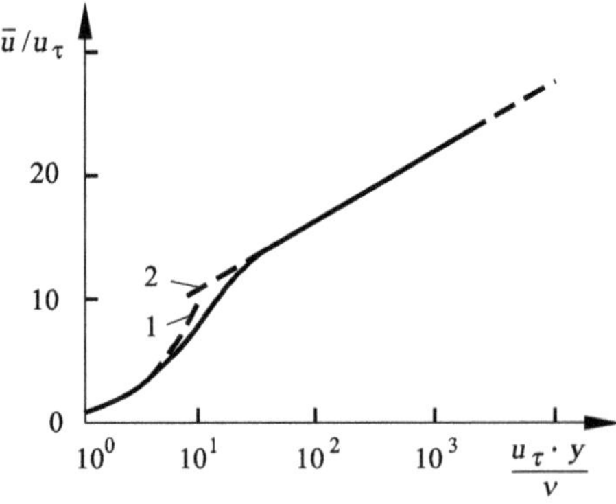

Fig. 4.64. Logarithmic wall law and velocity distribution in the viscous sublayer

Figure 4.64 shows experimentally determined velocity distributions. The logarithmic wall law can be seen for values greater than $y \cdot u_\tau/\nu = 50$ (curve 2). Curve 1 shows the velocity distribution $u/u_\tau = y \cdot u_\tau/\nu$ of the viscous sublayer.

In the turbulent flow past a *rough wall*, as well as the viscous shear stress $\mu \cdot (\mathrm{d}\overline{u}/\mathrm{d}y)$, there are also additional forces on the wall due to the roughness. These are summarized as a resultant frictional force whose mean value is now represented as a wall shear stress τ_w. A direct effect of the wall roughness on the viscous sublayer can be seen if its thickness is of the order of magnitude of the height of the roughness. There is an alteration in the value of the constant of integration C_1. The spatially averaged roughness height k introduces a further characteristic length. Of importance here is the Reynolds number $\mathrm{Re}_k = k \cdot u_\tau/\nu$ formed with the roughness. If Re_k is large, ν/u_τ may be neglected compared to k. Equation (4.82), together with $\mathrm{C} = \mathrm{C}_2 - (1/\kappa) \cdot \ln(k)$, yields

$$\frac{\overline{u}}{u_\tau} = \frac{1}{\kappa} \cdot \ln\left(\frac{y}{k}\right) + \mathrm{C}_2. \tag{4.84}$$

For small values of Re_k, instead of C_2 we obtain a function of $k \cdot u_\tau/\nu$, which for very small values of Re_k has the form $\mathrm{C}_1 + (1/\kappa) \cdot \ln(k \cdot u_\tau/\nu)$. Equation (4.84) then becomes (4.83). A wall with little roughness is called *hydraulically smooth*.

Pipe Flows

For turbulent flow through pipes with constant cross-section, the shear stress velocity u_τ is also the characteristic velocity:

$$u_\tau = \sqrt{\frac{\tau_w}{\rho}} = \sqrt{\frac{p_1 - p_2}{2 \cdot \rho} \cdot \frac{R}{l}}. \tag{4.85}$$

In the interior of the pipe flow the viscosity is of no importance, and so the radius of the pipe R is the only characteristic length. We obtain

$$\overline{u}_\mathrm{max} - \overline{u}(y) = u_\tau \cdot \mathrm{F}\left(\frac{y}{R}\right), \tag{4.86}$$

with the universal function F, maximum velocity $\overline{u}_\mathrm{max}$ in the middle of the pipe, and distance from the wall $y = R - r$. This law is true for both smooth and rough pipes at very large Reynolds numbers. The function F has to be determined experimentally. For the mean flux velocity w, we can derive the following relation from (4.86):

$$w = \overline{u}_\mathrm{max} - 2 \cdot u_\tau \cdot \int_0^1 \left(1 - \frac{y}{R}\right) \cdot F\left(\frac{y}{R}\right) \cdot \mathrm{d}\left(\frac{y}{R}\right). \tag{4.87}$$

As we approach the wall, (4.82) is again valid outside the viscous sublayer. We set $\mathrm{C} = (\overline{u}_\mathrm{max}/u_\tau) - (1/\kappa) \cdot \ln(R) + A$. The value A is a further characteristic number of the turbulent pipe flow. With $A = -0.6$, and for very small values y/R, we have

$$\overline{u}_{\max} - \overline{u} = u_\tau \cdot \left(0.6 - 2.5 \cdot \ln\left(\frac{y}{R}\right)\right). \qquad (4.88)$$

Equations (4.85) and (4.88) are sufficient to calculate the velocity distribution and the pressure drop in smooth and rough pipes with (4.83) and (4.84) for the wall law.

Boundary-Layer Flows

Turbulent boundary layers are bounded on one side by a fixed wall and on the other side by the inviscid outer flow. As the thickness of the boundary layer increases in the direction of flow, liquid keeps entering the boundary layer from the outer flow and free turbulence forms at the edge of the boundary layer. Depending on the surface (smooth or rough), the wall flow treated above forms close to the wall.

In the plate boundary layer, the wall law (4.83) is valid only in the layer close to the wall. In the outer part of the plate boundary layer, the deviations from the wall law are always greater than those in pipe flows. An *outer law* has therefore been formulated for the plate boundary layer in the form

$$\frac{U_\infty - \overline{u}}{u_\tau} = \mathrm{G}\left(\frac{y}{\delta}\right), \qquad (4.89)$$

with the function G, the boundary-layer thickness δ, and the velocity U_∞ in the outer flow. For turbulent boundary layers we have

$$\frac{U_\infty - \overline{u}}{u_\tau} = -\frac{1}{\kappa} \cdot \ln\left(\frac{y}{\delta}\right) + \frac{\pi(x)}{\kappa} \cdot \left(2 - \mathrm{w}\left(\frac{y}{\delta}\right)\right).$$

This equation is also valid in the *wake flow*. The wake function $\mathrm{w}(y/\delta)$ and the parameter $\pi(x)$ have to be determined empirically, with the parameter $\pi(x)$ dependent only on $p(x)$ and possibly the turbulence of the outer flow.

Instead of using the boundary-layer thickness, we prefer to use the reference length $\delta_1 \cdot U_\infty / u_\tau$ formed with the displacement thickness δ_1. Equation (4.89) then becomes

$$U_\infty - \overline{u} = u_\tau \cdot \mathrm{F}\left(\frac{y \cdot u_\tau}{\delta_1 \cdot U_\infty}\right),$$

where F is a dimensionless function, which, because of the definition of δ_1 (4.60), satisfies the condition

$$\int_0^\infty \mathrm{F}\left(\frac{y \cdot u_\tau}{\delta_1 \cdot U_\infty}\right) \cdot \mathrm{d}\left(\frac{y \cdot u_\tau}{\delta_1 \cdot U_\infty}\right) = 1.$$

Figure 4.65 shows the experimentally determined outer law of the plate boundary-layer flow. The validity of this outer law is not quite as natural as the validity of the same law (4.86) for the pipe flow, since in this case the shear stress distribution is dependent on the velocity distribution. For this reason F in the plate boundary layer depends on the local friction coefficient expression place $c_f = 2 \cdot (u_\tau / U_\infty)^2$. The velocity distribution is dependent on the turbulence of the outer flow. As we approach the wall, the velocity distribution becomes that of the logarithmic wall law (4.82). After determining the constant of integration C, (4.89) takes on the form

$$U_\infty - \overline{u} = u_\tau \cdot \left(-\frac{1}{\kappa} \cdot \ln\left(\frac{y \cdot u_\tau}{\delta_1 \cdot U_\infty} \right) + \mathrm{K} \right). \qquad (4.90)$$

The constant K has a value of about -1.5. If we join (4.90) to the logarithmic wall law (4.83), we obtain an equation for the local friction coefficient c_f as a function of the Reynolds number $\mathrm{Re}_1 = U_\infty \cdot \delta_1 / \nu$:

$$\frac{1}{\sqrt{\frac{c_f}{2}}} = \frac{1}{\kappa} \cdot \ln\left(\frac{U_\infty \cdot \delta_1}{\nu} \right) + \mathrm{C}_1 + \mathrm{K}. \qquad (4.91)$$

After inserting numerical values from experiment, we obtain for smooth plates

$$\frac{1}{\sqrt{\frac{c_f}{2}}} = 2.5 \cdot \ln\left(\frac{U_\infty \cdot \delta_1}{\nu} \right) + 3.7. \qquad (4.92)$$

In the same manner, the friction coefficient for rough surfaces can also be computed. We introduce the quantity

$$I = \int_0^\infty \mathrm{F}^2 \cdot \mathrm{d}\left(\frac{y \cdot u_\tau}{\delta \cdot U_\infty} \right).$$

From the function in Figure 4.65, we obtain the value $I = 6.2$. We thus derive the relation

$$\delta_2 = \delta_1 \cdot \left(1 - \sqrt{\frac{c_f}{2}} \cdot I \right) \qquad (4.93)$$

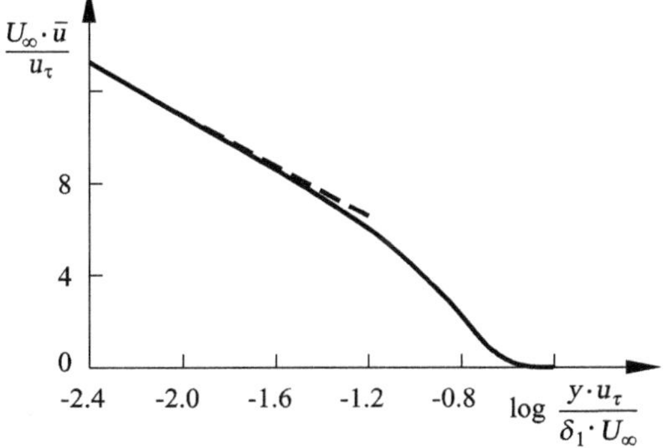

Fig. 4.65. Outer law of the turbulent plate boundary layer

between the momentum thickness δ_2 (4.61) and the displacement thickness δ_1. With (4.92) and (4.93) we can integrate the momentum equation of the laminar boundary layer and compute the friction drag of plates in turbulent flow.

The outer law can also be applied to boundary layers with *variable pressure*. It has been observed that the velocity profiles at different pressure distributions behave approximately like a single-parameter family of curves corresponding to (4.90) for small y values. Only the constant K is different. Thus we have a fixed relation between K and the integral I.

Since the wall law (4.83) can also be applied at different pressure distributions, (4.91) and (4.93) are valid with the correct numerical values for K and U_∞ even when the pressure along the wall varies. The friction coefficient decreases in boundary layers as the pressure rises. *Ludwieg* and *Tillmann* (1949) derived the equation

$$c_f = 0.246 \cdot 10^{(-0.678 \cdot H)} \cdot \mathrm{Re}_2^{-0.268} \tag{4.94}$$

from their measurements, with $H = \delta_1/\delta_2$ and $\mathrm{Re}_2 = U_\infty \cdot \delta_2/\nu$.

The velocity profiles of turbulent boundary layers where the pressure varies can be approximately characterized by the shape parameter $H = \delta_1/\delta_2$. However, a further relation between the pressure distribution and the shape parameter is required. We obtain a second differential equation for the change of H with the local pressure gradient:

$$\delta_2 \cdot \frac{\mathrm{d}H}{\mathrm{d}x} = -\mathrm{M} \cdot \frac{\delta_2}{U_\infty} \cdot \frac{\mathrm{d}U_\infty}{\mathrm{d}x} - \mathrm{N}. \tag{4.95}$$

Here M and N are functions of H and Re_2 (for rough surfaces they are also functions of k/δ_2), which have to be determined experimentally.

4.2.6 Flow Separation and Vortex Formation

The decelerated friction layers on the surfaces of bodies can form free interfaces or vortices (cf. Section 4.1.4). If the outer flow is accelerated by a pressure drop in the direction of flow, the liquid particles in the friction layer also experience an acceleration in the direction of motion. The flow will therefore retain its direction along the surface of the body in the entire boundary layer. On the other hand, if the pressure decreases in the direction opposite to the flow direction, the outer flow is decelerated. Slower fluid elements of the friction layer are then slowed down even more. If the deceleration is large enough, the flow separates from the wall, and a backflow region appears. Figure 4.66 illustrates the steady separation process for a given pressure distribution p. The *interface* that occurs due to the separation rolls up into one or more vortices. Because of the backflow close to the wall, the streamline portrait of the boundary-layer flow close to the separation position A indicates a great thickening of the boundary layer. Related to this is the transport of fluid mass out of the boundary layer into the outer flow. At the separation

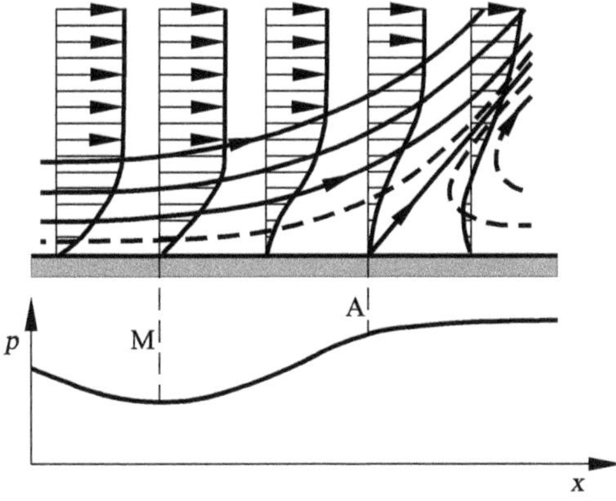

Fig. 4.66. Separation process (velocity maximum M, separation point A)

point the wall streamline departs the wall at a certain angle. The position of the point of separation is that point on the wall where the velocity gradient perpendicular to the wall vanishes, i.e. the point where the wall shear stress τ_{w} becomes zero:

$$\tau_{\mathrm{w}} = \mu \cdot \left.\frac{\partial u}{\partial y}\right|_{\mathrm{w}} = 0 \qquad \text{(separation).} \tag{4.96}$$

Figure 4.67 shows a sequence of photos depicting the onset of flow separation at a circular cylinder set into motion in a liquid. At the start of motion, a potential flow arises. At a later point in time, the flow separates from the

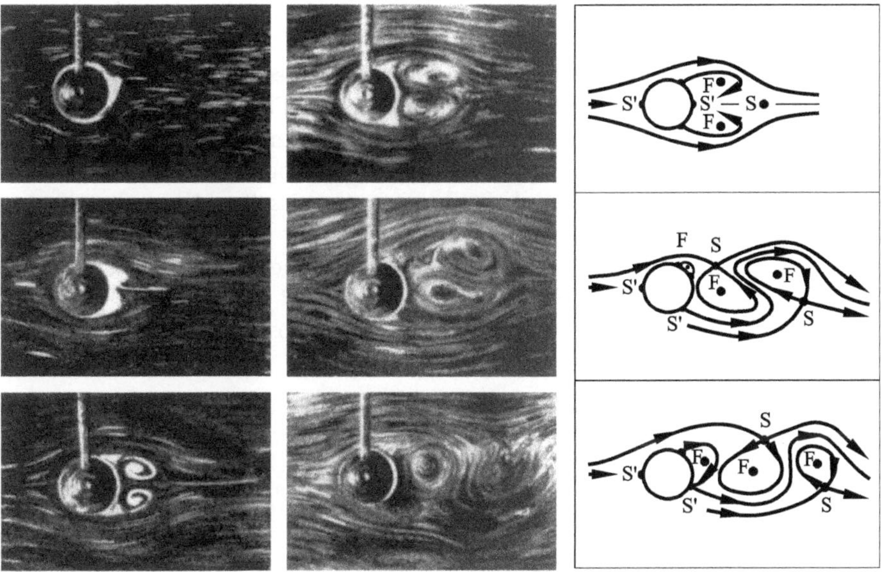

Fig. 4.67. Development of the vortex system behind a nonrotating cylinder

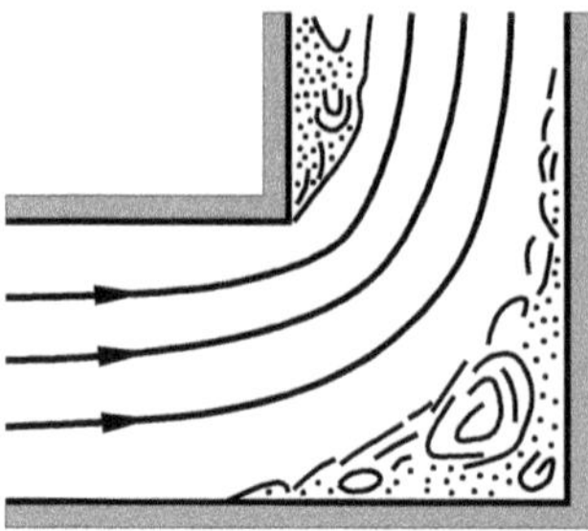

Fig. 4.68. Flow past a corner

cylinder. A backflow region with pronounced vortices forms in the wake flow. The interface in the liquid can be seen clearly where the aluminum specks gather. If we analyze the structure of the snapshots of the flow past a cylinder introduced in Section 3.3, we see the four half-saddles S′, the stagnation points, and the separation points on the cylinder, as well as the saddle point S and the two foci F of the wake flow. The sequence of photos shows that the vortices of the backflow region grow as time passes and eventually become unstable. After a critical time, a *Kármán vortex street* with periodically departing vortices forms. The structure of the vortex street is characterized by a succession of foci F and saddles S. The same separation process also occurs in the flow in a channel that expands in the direction of flow (diffusor, see Figure 4.72). In front of the narrowest cross-section the pressure in the flow direction decreases. Here the flow is attached to the walls. After the narrowest cross-section the channel widens, and the pressure in the flow direction increases. This causes the boundary layer to separate from the two walls and a backflow region to form. The actual flow then takes place only in the core region of the channel cross-section.

If liquid flows past a turning in a channel, a pressure drop perpendicular to the direction of flow occurs in the curved part. The velocity on the outer wall then decreases, and the flow separates, as shown in Figure 4.68. Further downstream, the pressure drop caused by the turning dies away, the velocity on the outer wall increases, and the flow attaches itself to the wall again.

Similar flow separations also form in the intake into an elbow bend, as well as in front of a sudden contraction in a channel. In both the case of a house in a wind flow (see Figure 4.69) and that of a pillar in a river, flow separation occurs on the ground upstream of the obstacle and in the wake.

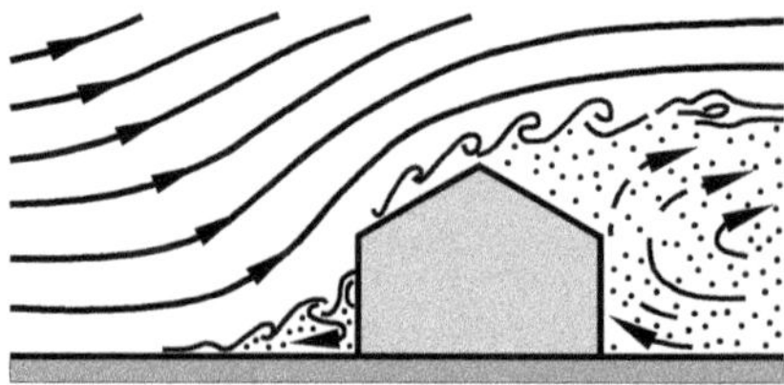

Fig. 4.69. Flow past a house

In industry one attempts to avoid a separation of the flow in spite of the pressure rise, in order to keep flow losses small. This is achieved by permitting channels to expand only gradually, or by designing the shape of bodies sufficiently narrow so that the acceleration of the outer flow prevails over the pressure rise. This is generally successful when the boundary layer in the decelerated part is turbulent.

In a flow with pressure increase, the flow past a body can remain laminar up to the point of separation if the surface is very smooth and the approach flow free of turbulence. Just in front of the separation point, the boundary-layer profile has a turning point. This is a sufficient criterion for the onset of the instability in the boundary layer. The laminar–turbulent transition begins, leading to a reattachment of the turbulent boundary-layer flow downstream if the Reynolds number is large enough. The reattachment of the turbulent boundary-layer flow depends both on the Reynolds number formed with the radius of curvature and on the change in the surface curvature of the wall. Laminar flow separation with turbulent reattachment frequently occurs on thin wing profiles with sharp nose curvature and sufficiently large angles of attack. Figure 4.70 shows the transition from the separating boundary-layer flow at low Reynolds numbers to the attached flow at larger Reynolds numbers. The photos correspond to values $2 \cdot 10^4$, $5 \cdot 10^4$, and $6 \cdot 10^4$ for the Reynolds number $U \cdot r/\nu$ formed with the radius of curvature r.

In a turbulent flow, the turbulent mixing causes the separation point of a body in a flow to be displaced downstream. Thus the backflow region in the wake of the body becomes considerably smaller. Related to this is a considerable reduction in the pressure drag, seen as a jump in the drag coefficient c_w = f(Re). *L. Prandtl* (1914) was able to show this in his famous experi-

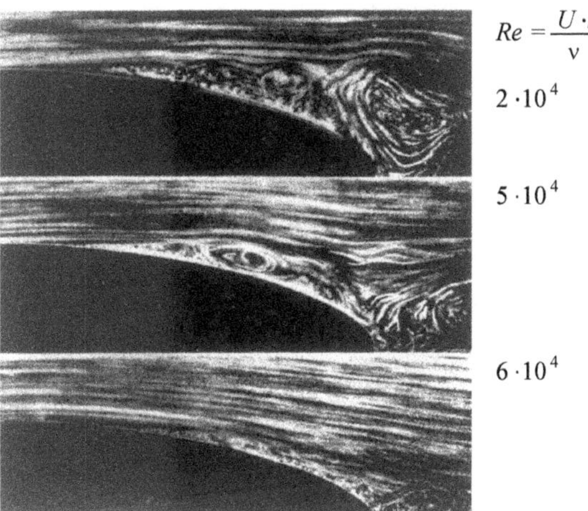

Fig. 4.70. Laminar separation and turbulent reattachment with increasing Reynolds number

ment in which he laid a thin tripwire on a sphere and artificially caused the laminar boundary layer to become turbulent at a smaller Reynolds number. He achieved a reduction in the drag that without the tripwire would have occurred only at a larger Reynolds number.

Influencing the Flow Separation

Rotation

Flow separation is generally undesirable, since it causes losses. There are many different ways of artificially influencing the boundary layer so that separation is prevented. For example, by causing a cylinder in a transverse flow to *rotate*, so that the circumferential velocity is equal to or larger than the maximum flow velocity at the circumference of the cylinder, an acceleration of the boundary layer occurs on the side in which the liquid and the wall move in the same direction. No separation then occurs on this side. On the other side, the wall moves against the liquid and decelerates the boundary layer, so that first backflow and then separation of a vortex is observed. A vortex with opposite circulation remains at the cylinder. The vortex formation at the start of formation can be seen in Figure 4.71. The flow structure is sketched for the last three snapshots of the vortex separation.

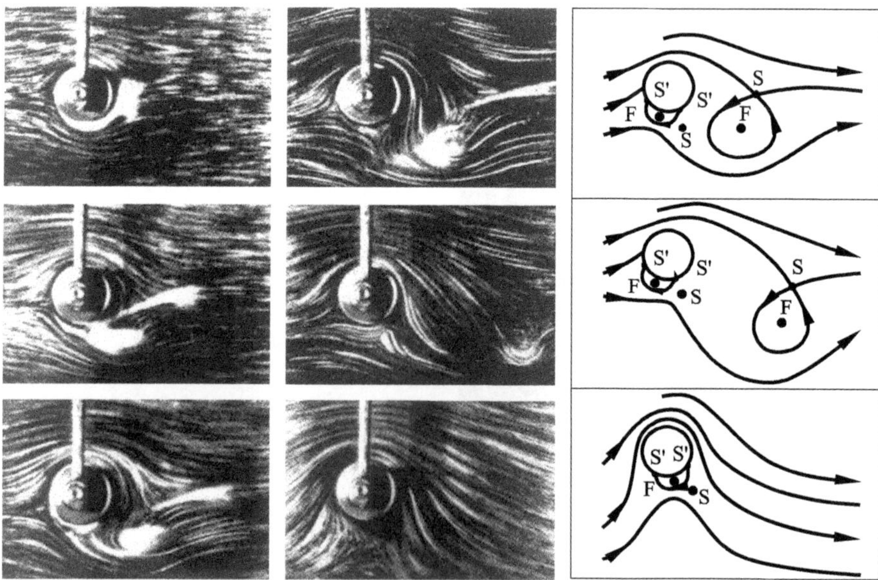

Fig. 4.71. Development of the flow past a rotating cylinder

Suction

Another very effective method of avoiding boundary-layer separation is *suction*. In this case, the fluid in the boundary layer is sucked into the interior of the body through small slits or pores in the wall of the body in the backflow region. If the suction is sufficiently strong, the accumulation of decelerated fluid is avoided, and the boundary-layer separation can be prevented. An example of the effect of boundary-layer suction is shown in Figure 4.72. The flow in a strongly divergent channel is observed. Without suction, separation occurs. If the backflow region is sucked away on both sides of the diffusor, the flow fills the entire channel cross-section, and the flow separation is avoided.

Tangential Blowing

The separation of the boundary layer can also be prevented by *tangential blowing* into the boundary layer. A wall jet blown into the boundary layer through a slit in the contour parallel to the main flow direction can supply enough kinetic energy to the boundary layer to prevent separation. According to this principle, for example, the maximum lift of a wing can be greatly increased, although at the expense of a large drag.

The arrangement of the *flap* on the wing in Figure 4.73 can also prevent separation. In this case, the pressure increase to be overcome by the boundary layer of the wing is smaller than without the flap, and separation is prevented up to considerably larger angles of attack.

This arrangement is somewhat related to the application of auxiliary wings to improve flows in pipe bends. This is exemplified by the usual deviation blades in wind tunnels. Auxiliary wings are also used in other flows

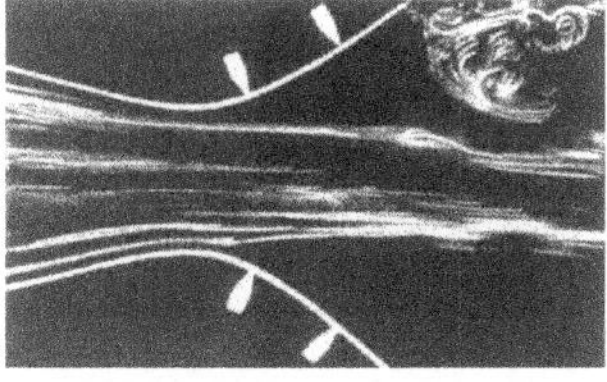

without suction

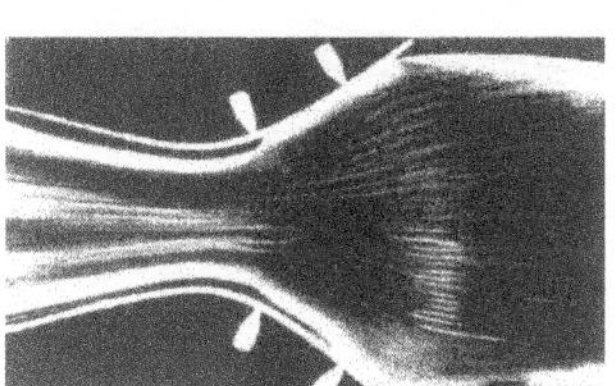

with suction at the wall

The white marks indicate the position of the invisible suction slits

Fig. 4.72. Flow in a strongly divergent channel

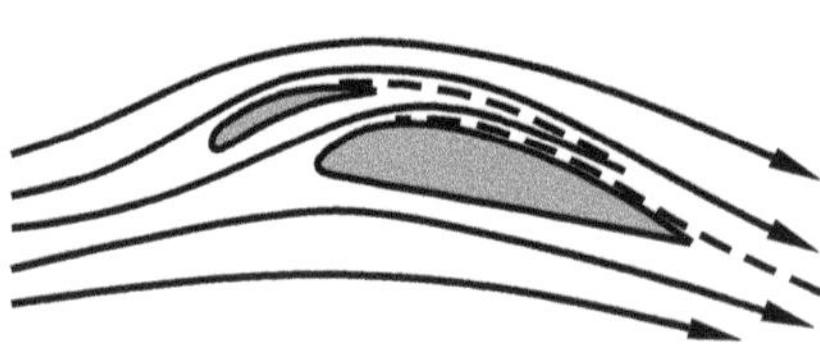

Fig. 4.73. Slotted wing

Fig. 4.74. Deviation by auxiliary wings

to achieve sharp bends without large losses (Figure 4.74). No separation occurs, because the pressure distribution along the auxiliary wing causes the pressure on the wall, which the pressure sides of the auxiliary wings face, to be larger than in the flow without auxiliary wings. The pressure rise that the boundary layer has to overcome is therefore smaller.

Paint Visualization

The streamlines of separated flows directly at the wall can be made visible using *paint visualization.* In water flows, oil-bound paint is used, while in air flows a mixture of dyes and petroleum is applied. If the flow is permitted to act on the paint of the wall for a characteristic time (for water about 5 minutes), a pattern forms in the direction of the mean velocity of the viscous layer close to the wall. This allows conclusions to be drawn about the course of the flow, in particular, separation points. Such paint visualization indicates only the streamlines in the layers close to the wall and not those in the core flow.

Figures 4.75 and 4.76 show two portraits of water flows taken by *A. Hinderks.* Figure 4.75 shows the flow at the bottom of a channel that contains a flat plate placed perpendicular to the flow. The wide white stripe that is drawn around the plate indicates a *horseshoe vortex* that evades the overpressure region in front of the plate. The two foci of the vortex indicate a spiral flow directed inward behind the plate that displays two vortices extending into the core flow.

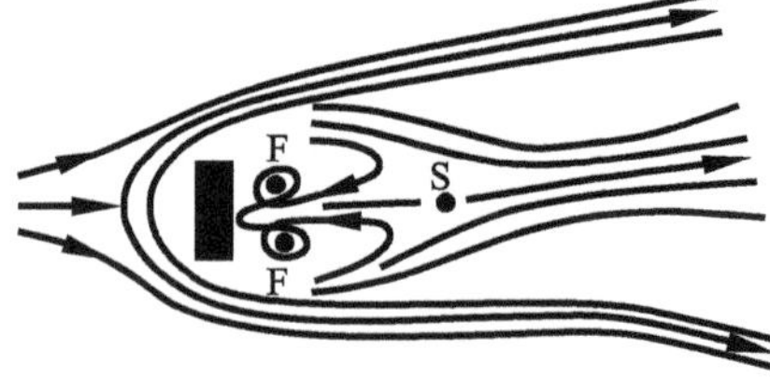

Fig. 4.75. Paint visualization and structure of a wall flow perturbed by a vertical plate (horseshoe vortex) *A. Hinderks*

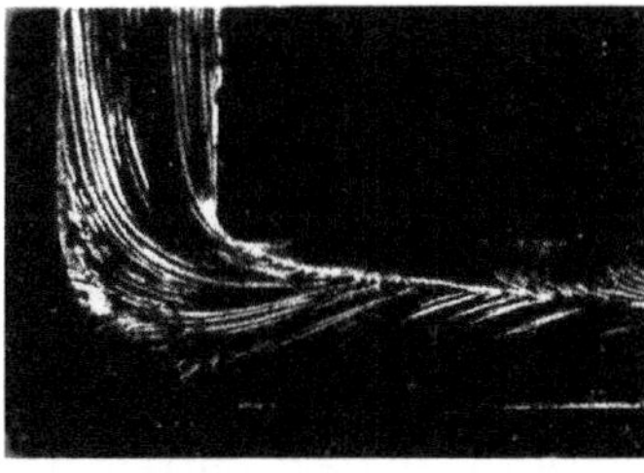

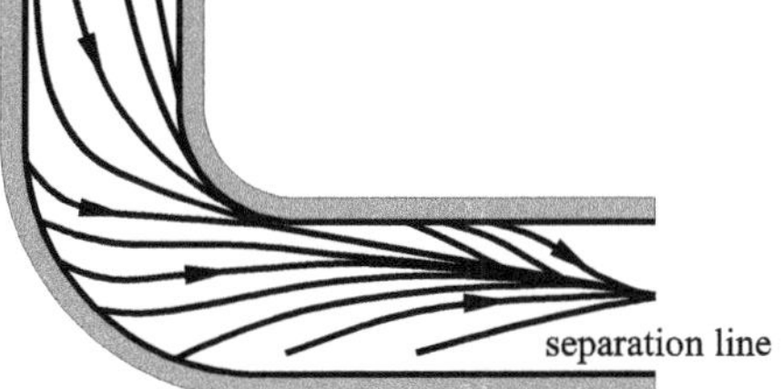

Fig. 4.76. Paint visualization and structure of the flow through an elbow bend *A. Hinderks*

Figure 4.76 shows the wall flow in a curved rectangular channel. The deviation of the wall layer to the inner side of the curve can be seen clearly. The convergence of the wall streamlines downstream from the bend indicate the separation at the inner side due to the pressure increase.

4.2.7 Secondary Flows

Elbow Bends

We consider the flow of a fluid along a plane wall. It is deviated by a sideward pressure gradient parallel to the wall. The layers close to the wall are deviated more strongly than the outer flow because of their lower velocity. This leads to a *secondary flow*, superimposed onto the main flow in the pipe.

For inviscid flow, (4.9) yields the ratio of the radii of curvature $r_1/r_0 = w_1^2/w_0^2$. In fact, the flow is viscous. The friction at the wall, in combination with the sideward pressure gradient, causes a deviation of the boundary layer in the direction of the lower pressure. The deviation in the laminar case has maximum 45° and in the turbulent case maximum 25° to 30°. As liquid flows through a curved pipe, because of its greater velocity the core flow attempts to flow as straight as possible. In contrast, the slower edge layers are greatly deviated and tend toward the inner side of the arc of the bend. The main

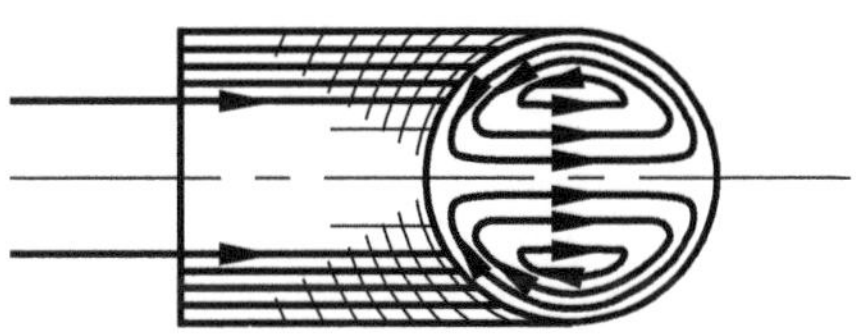

Fig. 4.77. Secondary flows

flow (parallel to the center line of the pipe) in the curved stretch of the pipe is therefore superimposed with a perpendicular secondary flow. This flows inward in the edge layers and outward in the core. The left illustration of Figure 4.77 shows the secondary flow in an elbow bend. It causes the position of maximum velocity to be displaced in the direction of the external arc of the bend.

In the natural course of rivers, the secondary flow in curves also has the effect that sediments (sand, pebbles) that move with the bottom current are transported away from the outer side of the curve and accumulate on the inner side. The outer bed of the river is deepened, and the inner bed made shallower. The larger flow velocity at the outer bank causes the curvature of the river to keep increasing, and so natural rivers tend whenever possible to have a very sinuous course (*meander formation*).

Rotating Vessel

Another example of a secondary flow is the flow that occurs at the bottom of a round rotating vessel (Figure 4.77, right). Because of the lower velocity in the ground layer, the *centrifugal force* there is less than that in the middle of the vessel. This causes the bottom current to be directed inward. Everyday observation shows that small particles at the base of the vessel move toward the middle of the base and accumulate there. This can be explained with the bottom current.

Channels with Rectangular and Triangular Cross-Sections

The flow through straight channels of noncircular cross-section also causes secondary flows. These cause transverse flow in the corners of the channels, as shown in Figure 4.78. The occurrence of the secondary flows can be explained by the fact that liquid is conveyed into the interior of the channel from points of larger shear stress and therefore at positions of smaller shear stress, e.g. in the corners, liquid flows from the interior to the wall. Thus at large wall shear stress positions the velocity is decreased, and at positions of lower wall shear stress the velocity is increased. This leads to a leveling out of the wall shear stress.

Oscillating Bodies

Secondary flows also occur at oscillating bodies. If $U(x) \cdot \cos(\omega \cdot t)$ is the velocity outside the boundary layer, according to *H. Schlichting* (1932) there is an additional velocity u' with the following magnitude close to the wall outside the boundary layer:

$$u' = \frac{3}{4} \cdot \frac{U}{\omega} \cdot \frac{\partial U}{\partial x}.$$

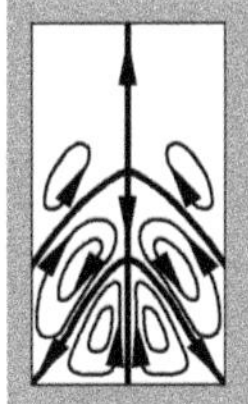

Fig. 4.78. Secondary flows in channels with triangular and rectangular cross-sections

It is directed from positions of larger velocity toward those of smaller velocity. Figure 4.79 shows a snapshot of the water flow around a circular cylinder that oscillates back and forth. The camera is moving with the cylinder. The metal particles that make the flow visible generate wide bands over a long exposure time. The flow approaches the cylinder from above and below and moves away in the direction of oscillation at both sides. The asymmetry of this picture is due to a weak eigenmotion of the water in the experimental vessel.

4.2.8 Flows with Prevailing Viscosity

As was also discussed in Section 4.2.2, at large viscosity and small Reynolds numbers, the inertial forces may be neglected compared to the frictional forces. These *creeping flows* have a flow drag proportional to the first power of the velocity. The groundwater flow and the bearing lubrication will be discussed in more detail in this section.

Groundwater Flow

An example of a flow with prevailing viscosity is the groundwater flow in soil. The flow between the individual grains of sand is, in analogy to the Hagen–Poiseuille law for pipe flows, a creeping flow proportional to the pressure drop

Fig. 4.79. Secondary flows at an oscillating body, after *H. Schlichting* 1932

and inversely proportional to the dynamic viscosity μ:

$$u = -\frac{k}{\mu} \cdot \frac{\partial p}{\partial x}, \qquad v = -\frac{k}{\mu} \cdot \frac{\partial p}{\partial y}, \qquad w = -\frac{k}{\mu} \cdot \frac{\partial p}{\partial z}. \tag{4.97}$$

The permeability k has the dimension of a surface and depends only on the porous medium. With the continuity equation

$$\frac{\partial u}{\partial x} + \frac{\partial v}{\partial y} + \frac{\partial w}{\partial z} = 0$$

we obtain

$$\frac{\partial^2 p}{\partial x^2} + \frac{\partial^2 p}{\partial y^2} + \frac{\partial^2 p}{\partial z^2} = 0. \tag{4.98}$$

The same relation holds for the pressure p as for the velocity potential Φ of the inviscid flow. Groundwater flows are therefore potential flows, as discussed in Section 4.1.5. The essential difference is that the pressure p must be physically single-valued and continuous, whereas Φ can be discontinuous at interfaces, and in flows with circulation is multivalued.

Equations (4.97) and (4.98) can be used to treat the groundwater flow close to a well, for example. The removal of water is taken into account; i.e. as well as the velocity distribution, the drop in the water table close to the well is also considered.

The proportionality assumed between the velocity and the pressure drop is true only as long as the Reynolds number formed with the diameter of the grains d remains sufficiently small. The limit is at $\mathrm{Re}_d = u \cdot d/\nu \approx 10$.

Bearing Lubrication

A further example of flows with prevailing viscosity is the flow in lubricated bearings and guides in machines. In between machine parts that move against one another (journals and bearings, or sliding blocks and guides) are gap flows of thin layers of oil. These protect the solid bodies from touching each other. The ability of a journal bearing and a guide shoe to take on large loads at low friction is the result of the flow process in the oil layer.

A first example is the sliding block on a flat guide. For simplicity we assume that the sliding surfaces are infinitely extended perpendicular to the direction of motion. This is the assumption of a plate flow. We select a reference frame at rest with respect to the sliding block. The guide of the sliding block moves to the right with velocity v, and so we assume a steady flow.

We first consider the flow through a gap of height h with an upper wall at rest (sliding block) and a parallel lower wall moving with velocity v (guide). The x axis points in the direction of motion, while the y axis is perpendicular to the walls. The pressure increase $\mathrm{d}p/\mathrm{d}x$ is abbreviated by p'. Because the layer is so thin, p' is independent of h. The flow velocity in the x direction is u. According to the remarks in Section 4.2.1, the inertial force can be neglected, as can $\partial^2 u/\partial x^2$ compared to $\partial^2 u/\partial y^2$:

$$\mu \cdot \frac{\partial^2 u}{\partial y^2} = p'. \tag{4.99}$$

Integration yields

$$\mu \cdot \frac{\partial u}{\partial y} = p' \cdot y + \mathrm{C}_1,$$

$$\mu \cdot u = p' \cdot \frac{y^2}{2} + \mathrm{C}_1 \cdot y + \mathrm{C}_2. \quad . \tag{4.100}$$

The no-slip condition for $y = 0$, that u is equal to the relative velocity U of the guide compared to the sliding block, is satisfied by $\mathrm{C}_2 = \mu \cdot U$. For $y = h$ we must have $u = 0$. Thus C_1 satisfies

$$\mathrm{C}_1 = -\left(\frac{\mu \cdot U}{h} + \frac{p' \cdot h}{2}\right).$$

This yields the velocity distribution in the gap:

$$u = \frac{p'}{2 \cdot \mu} \cdot (y^2 - h \cdot y) + \frac{U}{h} \cdot (h - y). \tag{4.101}$$

The positive frictional force per unit area at the lower wall is

$$\tau_0 = -\mu \cdot \left.\frac{\partial u}{\partial y}\right|_{y=0} = -\mathrm{C}_1 = \frac{\mu \cdot U}{h} + \frac{p' \cdot h}{2}, \tag{4.102}$$

and at the upper wall is

$$\tau_h = -\mu \cdot \left.\frac{\partial u}{\partial y}\right|_{y=h} = \frac{\mu \cdot U}{h} - \frac{p' \cdot h}{2}. \tag{4.103}$$

In discussing these results we note that a pressure increase in the direction of the positive x axis corresponds to a positive p'. A negative p' means a pressure drop.

The amount of liquid per unit depth of the gap flow can be calculated using

$$Q = \int_0^h U \cdot \mathrm{d}y.$$

This yields

$$Q = \frac{u \cdot h}{2} - \frac{p' \cdot h^3}{12 \cdot \mu}. \tag{4.104}$$

We now calculate the load-bearing sliding block with varying pressure gradient p' in the x direction (see Figure 4.80). Since v is the constant velocity of the sliding block, continuity ($Q = \text{const}$) requires that the gap height must change with x. If h varies in the x direction, (4.104) gives us

$$p' = \frac{\mathrm{d}p}{\mathrm{d}x} = 12 \cdot \mu \cdot \left(\frac{U}{2 \cdot h^2} - \frac{Q}{h^3}\right). \tag{4.105}$$

Then $p(x)$ is obtained by integrating this equation. At the beginning and the end of the sliding block the pressure p is set equal to the surrounding pressure p_0. This yields the unknown value for Q, and so $p(x)$ is determined. If l is the length of the gap, further integration allows us to calculate the resultant pressure force of the flow in the sliding block with $\int_0^l p \cdot \mathrm{d}x$, as well as the moment $\int_0^l p \cdot x \cdot \mathrm{d}x$. The ratio of the moment to the force determines the distance of the working point of the force from the position $x = 0$. The viscous force is calculated using (4.102) with $\int_0^l \tau_0 \cdot \mathrm{d}x$, and so we can determine the magnitude, direction, and position of the resultant force on the sliding block for any given function h of the gap height. Frequently it is the resultant pressure force that is given, and the gap height is to be calculated.

The viscous force can also be calculated using τ_h. Here we must recall that the pressure p on the surface inclined to the direction of motion by $\tan\delta = \mathrm{d}h/\mathrm{d}x$ generates a force component in the direction of motion. Since the pressure at the end of the sliding block is p_0, this force component is equal to $-\int_0^l (p - p_0) \cdot (\mathrm{d}h/\mathrm{d}x) \cdot \mathrm{d}x$. Partial integration with $p = p_0$ for $x = 0$ and $x = l$ yields the force component $+\int_0^l p' \cdot h \cdot \mathrm{d}x$. Taking (4.102) and (4.103) into account, this is in agreement with the viscous force calculated from τ_0.

The simplest case of a varying gap height occurs when the sliding block and the guiding surface are flat but inclined at a small angle δ to each other. The sliding block extends from $x = 0$ to $x = l$. The two planes meet at a distance a from the leading edge of the sliding block at $x = 0$ (Figure 4.80). The height of the gap is

$$h = (a - x) \cdot \delta.$$

Integrating (4.105) yields the two integrals

$$\int_0^x \frac{\mathrm{d}x}{h^3} = \frac{1}{2 \cdot \delta^3} \cdot \left(\frac{1}{(a-x)^2} - \frac{1}{a^2} \right) = \frac{2 \cdot a \cdot x - x^2}{2 \cdot \delta^3 \cdot a^2 \cdot (a-x)^2}$$

and

$$\int_0^x \frac{\mathrm{d}x}{h^2} = \frac{1}{\delta^2} \cdot \left(\frac{1}{a-x} - \frac{1}{a} \right) = \frac{x}{\delta^2 \cdot a \cdot (a-x)}.$$

Therefore, the pressure distribution is

$$p = p_0 + \frac{6 \cdot \mu \cdot x}{\delta^2 \cdot a \cdot (a-x)} \cdot \left(v - \frac{Q \cdot (2 \cdot a - x)}{\delta \cdot a \cdot (a-x)} \right). \tag{4.106}$$

According to (4.106), $p = p_0$ at the position $x = 0$. Since $p = p_0$ at $x = l$, too, the expression in parentheses in (4.106) must vanish:

$$Q = \frac{U \cdot \delta \cdot a \cdot (a - l)}{2 \cdot a - l}. \tag{4.107}$$

Again replacing $\delta \cdot (a - x)$ by h, we obtain

$$p = p_0 + \frac{6 \cdot \mu \cdot U \cdot x \cdot (l - x)}{h^2 \cdot (2 \cdot a - l)}. \tag{4.108}$$

To estimate the mean pressure, the pressure p_1 is assumed in the center of the sliding block ($x = l/2$). This pressure is not the pressure maximum, since h varies with x. However, if the variation in the x direction is not too large, it is of the correct order of magnitude of the maximum. According to (4.108) we use $h = \delta \cdot (a - l/2) = h_m$ to obtain

$$p_1 - p_0 = \frac{3}{2} \cdot \frac{\mu \cdot U \cdot l^2}{h_m^2 \cdot (2 \cdot a - l)}.$$

If the pressure distribution is approximated by a parabola, the mean overpressure p_m is approximately $2 \cdot (p_1 - p_0)/3$, i.e.

$$p_m = \frac{\mu \cdot U \cdot l^2}{h_m^2 \cdot (2 \cdot a - l)}. \tag{4.109}$$

This equation shows that even at relatively small μ, very small mean gap thicknesses h_m can generate very large pressures. According to (4.108), the reduction of h in the direction of flow means that the pressure maximum lies behind the center. Therefore, the point of application of the resultant force is also behind the center. Figure 4.80 shows such a distribution according to (4.108). Below this pressure distribution is a sketch of the associated velocity distribution in the gap, the different curvature of which makes the pressure difference clearly visible.

The pressure distribution and the position of the pressure force depend on the ratio l/a. Therefore, *A. G. M. Mitchell* (1905) had the idea of applying a flexible attachment to the sliding block somewhat behind the middle of the guide surface (Figure 4.81). This causes a certain inclined position to occur automatically (or more precisely, a certain a). For a larger inclination, the

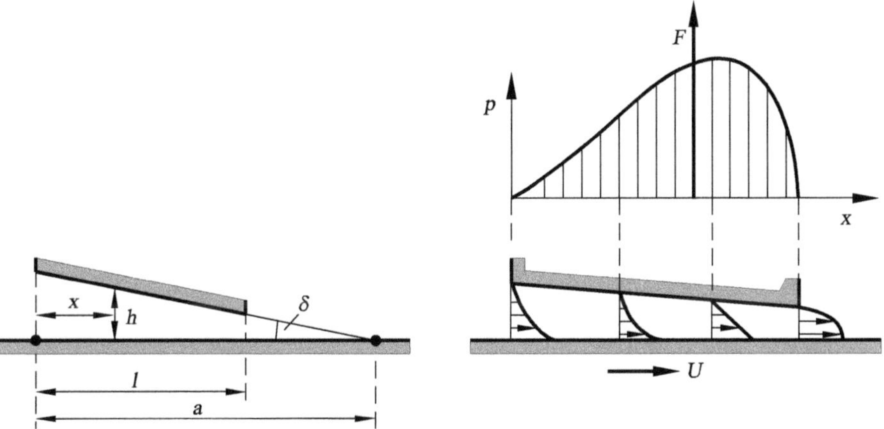

Fig. 4.80. Flow in the gap between sliding block and guide

pressure middle point is further back, and at a weaker inclination it is further forward, and so the correct position is particularly stable. In this manner *A. G. M. Mitchell* was able to achieve a sliding block that worked equally well under all loads. In fact, a certain amount of the oil that passes the leading edge of such sliding blocks flows out through the side edges. This causes a reduction in the pressure in the interior. Qualitatively, however, the process may still be described as above.

The shear stresses at the sliding block are, because of the pressure distribution, smaller at the entrance and larger at the exit than in simple gap friction. The shear stresses on the sliding track are opposite. The corresponding values can be determined from (4.102), (4.103), (4.105), and (4.107).

We will now estimate the viscous force. This estimation is more precise the larger the ratio a/l is chosen to be. The distribution of the shear stress is assumed to be approximately trapezoidal. The mean viscous force per unit area can therefore be set equal to the viscous force in the middle. There the magnitude of p' is very small, and (4.102) yields

$$\tau_m \approx \frac{\mu \cdot U}{h_m}.$$

Equation (4.109) is used to eliminate the lubrication layer thickness h_m:

$$h_m = \sqrt{\frac{\mu \cdot U \cdot l^2}{p_m \cdot (2 \cdot a - l)}}. \tag{4.110}$$

This leads to

$$\tau_m = \sqrt{\frac{\mu \cdot U \cdot p_m}{l}} \cdot \sqrt{\frac{2 \cdot a - l}{l}}. \tag{4.111}$$

The expression $\mu \cdot v/l$ is the very small shear stress that occurs in a layer of oil of thickness l. According to the order of magnitude, the actual shear stress is the geometric mean of this small shear stress and the mean load of the sliding block. This slippage resistance varies for fixed values of l and a in proportion to the square root of the viscosity, the load, and the velocity. This law does not only hold for the mean values considered, but is also obtained with a more precise calculation.

The *friction coefficient* is given by

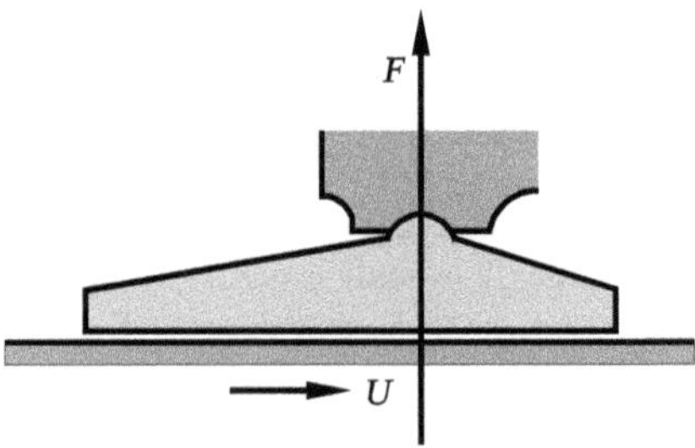

Fig. 4.81. Sliding block, *A. G. M. Mitchell* (1905)

$$C_f, g = \frac{\tau_m}{p_m}.$$

For fixed values of l and a, i.e. when the dimensions of the sliding block are given according to Figure 4.80, it is proportional to $\sqrt{\mu \cdot U/(p_m \cdot l)}$. The relations are more complicated for a journal in a bearing, where bearing play occurs. Two further unknowns appear due to the displacement of the center of the bearing in the horizontal and vertical directions. Essentially, a wedge-like layer of oil is formed here too, through which the oil from the rotating journal is transported from the wide side to the narrow side (Figure 4.82). The calculation is simplified by assuming that the eccentricity of the journal e is small compared to the bearing play s. This is valid for fast-running and moderately loaded journals in completely closed bearings. In this case,

$$h = s + e \cdot \cos(\varphi + \alpha),$$

with the central angle φ and the angle α between the force direction and the direction of the connecting line between the center of the bearing and the center of the journal. The angle α is about 90°. The point of the smallest distance between the journal and the bearing is in front of this, in the direction of rotation opposite the direction of the journal pressure.

An analogous calculation to that of the sliding block leads to the result that e/s is proportional to the dimensionless size $L = (p_m \cdot s^2/(\mu \cdot v \cdot r)$. Here p_m is the mean bearing pressure, r is the radius of the journal, and v the circumferential velocity. The bearing coefficient L can be derived from (4.109) for the sliding block:

$$\frac{l}{2 \cdot a - l} = \frac{p_m \cdot h_m^2}{\mu \cdot U \cdot l}.$$

The left-hand side of this equation corresponds to e/s. On the right, h_m appears in place of s, and l in place of r.

The effects of varying bearing load, different bearing play, different oil viscosity, and circumferential velocity are taken into account in the bearing

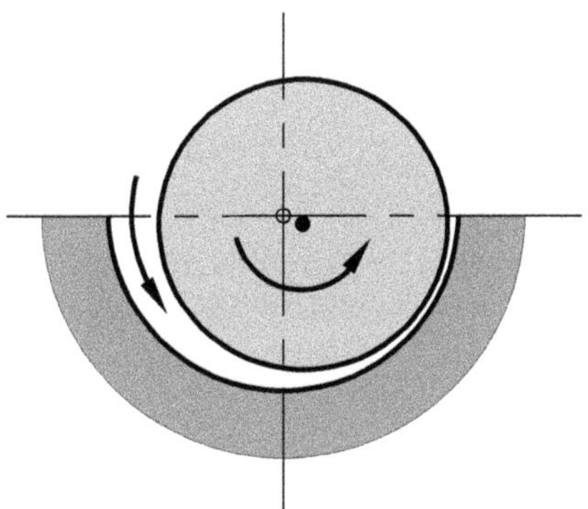

Fig. 4.82. Journal in a bearing

coefficient. The friction coefficient C_f, g of a bearing (circumferential force to bearing load) can be expressed analogously to that for the sliding block. We obtain $C_f, g \sim \sqrt{\mu \cdot U/(p_m \cdot r)}$. *O. Walger* (1932) experimentally found the value 2.4.

Until now we have assumed that in the bearing, the oil film wets the journal completely, preventing any metallic contact. Because of the production tolerances with which bearing and journal, or sliding block and guide, can be manufactured, metallic contact does occur if the gap width h is too small. Similarly, in applying the derived equations we must rule out the possibility that negative pressures occur in the oil film. In this case, the oil film will separate. Separation of the oil film generally occurs in bearings under a great load. Similar conditions occur as for a journal only partially surrounded by a bearing. We will not, however, further investigate the extended theory of such bearings.

At heavy loads, heating of the oil leads to considerable deviations compared to the derived equations. *G. Vogelpohl* (1938) showed that oils whose viscosity decreases to a lesser degree with increasing temperature are more suitable for heavily loaded bearings. He also indicated that a great part of the bearing load is hydrodynamically carried by so-called mixed friction, by the oil contained between the two-sided surface roughness. Only a very small part of the load is carried by the peaks of the roughness in mechanical contact.

4.2.9 Flows Through Pipes and Channels

The mean value of the wall shear stress τ_{w} for the turbulent channel flow can be calculated from $\lambda' \cdot \rho \cdot w^2/2$. Here λ' is a number and w the mean velocity. The pressure drop in a pipe or channel of length l must keep the shear stresses in the wall in equilibrium. With the cross-sectional area A and the wetted cross-sectional circumference U we have

$$(p_1 - p_2) \cdot A = \tau_{\mathrm{w}} \cdot l \cdot U = \lambda' \cdot \rho \cdot \frac{w^2}{2} \cdot l \cdot U, \tag{4.112}$$

i.e.

$$\frac{p_1 - p_2}{l} = \lambda' \cdot \frac{U}{A} \cdot \frac{\rho \cdot w^2}{2}. \tag{4.113}$$

In an open channel or river, the free surface is not part of the wetted circumference. The quotient A/U is called the hydraulic radius r_h. For a body of water flowing under the effect of gravity, such as a river, a drop in the water level $i = (z_1 - z_2)/l$ is given (Figure 4.83). This is dependent on the pressure drop along a horizontal line via the relation $p_1 - p_2 = g \cdot \rho \cdot (z_1 - z_2) = g \cdot \rho \cdot l \cdot i$. Therefore, (4.112) yields

$$\tau_w = g \cdot \rho \cdot r_h \cdot i, \tag{4.114}$$

and from (4.113),

$$i = \frac{1}{g \cdot \rho} \cdot \frac{p_1 - p_2}{l} = \frac{\lambda'}{r_h} \cdot \frac{w^2}{2 \cdot g}. \tag{4.115}$$

This leads to

$$w = \sqrt{\frac{2 \cdot g}{\lambda'} \cdot r_h \cdot i}.$$

For rivers and channels, this equation is written in the form

$$w = \mathrm{C} \cdot \sqrt{r_h \cdot i} \tag{4.116}$$

and is known as the *Chézy equation.* The value of C, which is a function of the hydraulic radius and the wall roughness, varies at water depths of 0.5 m to 3 m from 80 $\mathrm{m}^{(1/2)} \cdot \mathrm{s}^{-1}$ for channels of smooth wood or smoothly plastered masonry to 30 to 50 $\mathrm{m}^{(1/2)} \cdot \mathrm{s}^{-1}$ for walls of earth, to 24 to 49 $\mathrm{m}^{(1/2)} \cdot \mathrm{s}^{-1}$ for shingle.

Pipes with Circular Cross-Section

For pipes with radius R, the hydraulic radius r_h is

$$r_h = \frac{A}{U} = \frac{\pi \cdot R^2}{2 \cdot \pi \cdot R} = \frac{R}{2} = \frac{d}{4}. \tag{4.117}$$

Inserting $4/d$ for U/A and λ for $4 \cdot \lambda'$ in (4.113), we obtain

$$\frac{p_1 - p_2}{l} = \frac{\lambda}{d} \cdot \frac{\rho \cdot w^2}{2}, \tag{4.118}$$

where λ is called the loss coefficient. The loss coefficient for laminar and turbulent pipe flows is shown in Figure 4.84 as a function of the Reynolds number Re_d. *Laminar pipe flow* satisfies the Hagen–Poiseuille law (4.45). With the flux Q, the mean velocity is $w = Q/(\pi \cdot R^2)$. This leads to a pressure loss in the pipe of

$$\frac{p_1 - p_2}{l} = \frac{8 \cdot \mu \cdot w}{R^2} = 32 \cdot \mu \cdot \frac{w}{d^2}. \tag{4.119}$$

Comparison with (4.118) leads to an expression for the loss coefficient λ:

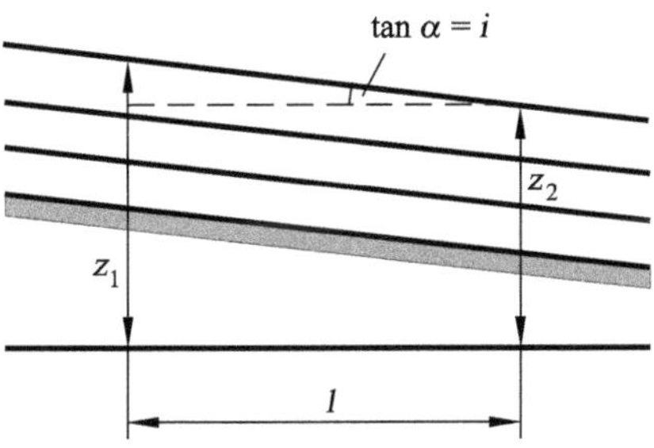

Fig. 4.83. Flow in a channel

$$\lambda = \frac{64 \cdot \mu}{\rho \cdot w \cdot d} = \frac{64}{\mathrm{Re}_d}. \tag{4.120}$$

There are many experimental results on the behavior of *turbulent flows.* Up to a Reynolds number of about 80 000, the *Blasius law* is valid:

$$\lambda = \frac{0.3164}{\mathrm{Re}^{\frac{1}{4}}}. \tag{4.121}$$

Stability theory for the Hagen–Poiseuille pipe flow (see Section 4.2.4) shows that the laminar–turbulent transition occurs at the critical Reynolds number $\mathrm{Re}_{\mathrm{crit}} = 2300$, so that in Figure 4.84, (4.120) passes over to (4.121) in a transition region.

L. Prandtl (1932) stated an implicit equation for the loss coefficient of smooth pipes for Reynolds numbers smaller than 10^6:

$$\frac{1}{\sqrt{\lambda}} = 2 \cdot \lg(\mathrm{Re}_d \cdot \sqrt{\lambda}) - 0.8. \tag{4.122}$$

To obtain this equation the equations in Section 4.2.5 are used, taking into account the logarithmic wall law (4.83).

Using (4.84), the evaluation of experimental results for *rough pipes* with fully developed flow yields the following extension to (4.122):

$$\frac{1}{\sqrt{\lambda}} = 1.74 - 2 \cdot \lg\left(\frac{18.7}{\mathrm{Re}_d \cdot \sqrt{\lambda}} + \frac{2 \cdot k}{d}\right). \tag{4.123}$$

Here the roughness k is the spatial average of the surface roughness of the pipe walls. For very large Reynolds numbers, the loss coefficient becomes independent of the Reynolds number. The viscous sublayer of the turbulent pipe boundary layer then covers the roughness of the pipe surface.

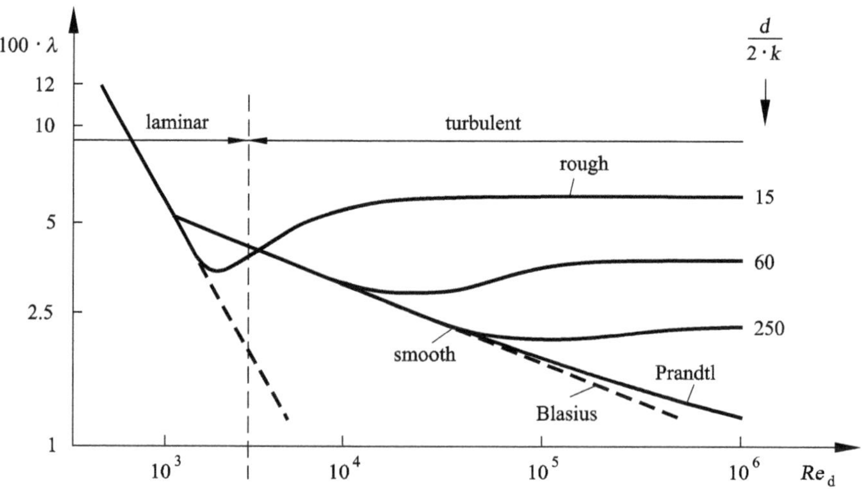

Fig. 4.84. *Nikuradse diagram*: loss coefficient λ for smooth and rough pipes

The first measurements of losses in rough pipes were carried out by *J. Nikuradse* (1933). Filtered sand of different grain sizes was stuck onto the inside of pipes. These experiments by *J. Nikuradse* (1933) gave the diagram of Figure 4.84 its name.

Intake Flow

Equations (4.119) to (4.123), as well as Figure 4.84, are valid for *fully developed pipe flow*. This is approximately the case from a distance of about 60 pipe diameters d from the *intake* of a pipe. In the intake cross-section of the pipe, the velocity is almost uniformly distributed. The deceleration caused by the friction begins at the wall of the pipe. In the flow, which is initially laminar, a growing layer of decelerated liquid forms downstream (Figure 4.85). The velocity then has to increase in the core flow, so that the same mass flows through every cross-section. This acceleration of the core flow in the intake stretch of the pipe is associated with a pressure decrease along the pipe axis that can be calculated with the Bernoulli equation. Further downstream, the friction zone encompasses the entire pipe cross-section, and the well-known Hagen–Poiseuille flow occurs. According to observations by *L. Schiller* (1922), this occurs after a distance of $l = 0.03 \cdot d \cdot \mathrm{Re}_d$. When the critical Reynolds number $\mathrm{Re}_{\mathrm{crit}} = 2300$ is exceeded, the laminar–turbulent transition occurs, and a turbulent fully developed pipe flow forms.

If the flow at the intake cross-section of the pipe is already turbulent, the distance l until the onset of the fully developed pipe flow is considerably shorter.

Pipe Flow with Cross-Sectional Variation

When a pipe suddenly contracts (Figure 4.86), as well as the inviscid pressure drop there are also viscous pressure losses. A sharp-edged contraction in the pipe or an orifice causes a contraction of the flow. According to *J. L. Weisbach* (1845), the contraction coefficient can be calculated using $\alpha = 0.63 + 0.37 \cdot (A_1/A_0)^3$. If the contraction is followed by a sudden expansion (orifice), the associated pressure loss is

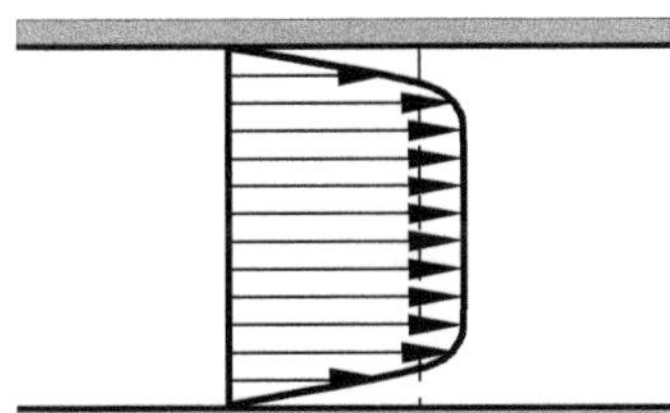

Fig. 4.85. Velocity distribution of the intake flow

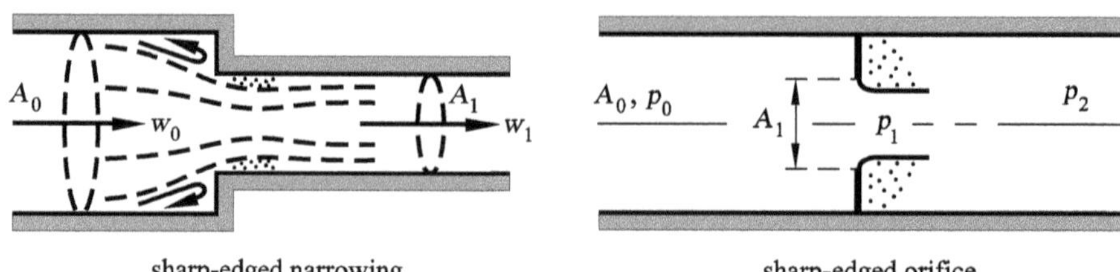

Fig. 4.86. Contraction in a pipe

$$p_0 - p_2 = \frac{\rho \cdot w_0^2}{2} \cdot \left(\frac{A_0}{\alpha \cdot A_1} - 1 \right)^2 .$$

Orifices as in Figure 4.86 or *Venturi nozzles* as in Figure 4.87 are used to measure volume fluxes. For the orifice, the inviscid pressure loss calculated with the Bernoulli equation is

$$p_0 - p_1 = \frac{\rho \cdot w_0^2}{2} \cdot \left[\left(\frac{A_0}{\alpha \cdot A_1} \right)^2 - 1 \right] .$$

If the pressure difference $p_0 - p_1$ is measured with boreholes in front of and behind the contraction, a known contraction coefficient α allows w_0 and thus the volume flux $A_0 \cdot w_0$ to be computed. Experimentally, the following equation is obtained for $A_1/A_0 < 0.7$:

$$\alpha = 0.598 + 0.4 \cdot \left(\frac{A_1}{A_0} \right)^2 .$$

For the gradual expansion of the Venturi nozzle in Figure 4.87, the pressure recovery is considerably larger than for the sudden expansion of the aperture. The pressure loss in the nozzle can be described with

$$p_0 - p_2 = \xi \cdot \frac{\rho}{2} \cdot (w_1^2 - w_2^2),$$

where ξ is an empirical drag coefficient to be determined for every nozzle. The values for Venturi nozzles lie between 0.15 and 0.2. The contraction coefficient α can be set equal to 1 if flow separation is avoided.

Cross-sectional expansion in *diffusors* leads to pressure recovery. Assuming that the flow is inviscid, the velocity is constant at all cross-sections.

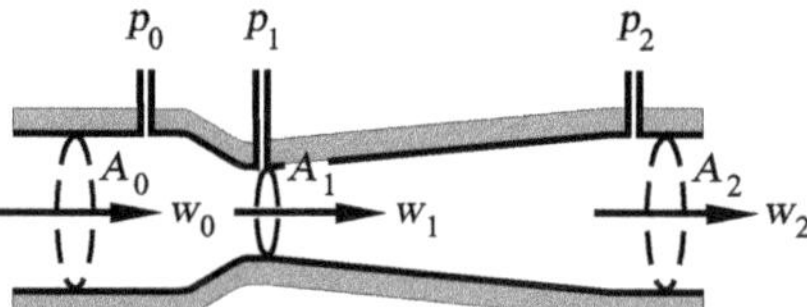

Fig. 4.87. Venturi nozzle

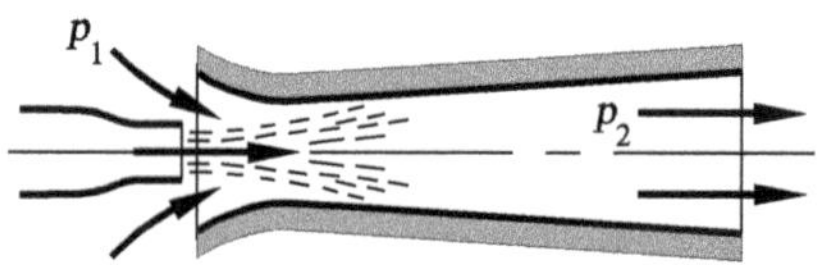

Fig. 4.88. Jet pump

Under the effect of friction, the flow close to the wall is decelerated. If the opening angle of the diffusor is too large, flow separation will occur.

The pressure increase $p_2 - p_1$ in a sudden or gradually expanding pipe is used to draw off liquid in *jet pumps*, as shown in Figure 4.88. In order to attain a pressure difference of 1 bar for a water jet air pump, the jet velocity w_1 must be about 20 m/s. Another example is the *Bunsen burner*, where a gas jet exiting from a nozzle draws in air and mixes with it.

4.2.10 Drag of Bodies in Liquids

Newton Drag Law

I. Newton concluded that the drag of a body moving in a liquid must be proportional to the surface area A of the body, the density ρ of the liquid, and the square of the velocity v. This result may be understood in the following simple approach. The body must displace a fluid mass $M = \rho \cdot A \cdot v$ per second. Here each mass element obtains a velocity that is set proportional to the velocity of the body. The drag is therefore proportional to the momentum imparted per second

$$M \cdot v = \rho \cdot A \cdot v^2 .$$

Newton's theory assumes that the drag of a body in a liquid can be treated using the collision laws of solid bodies. Newton considered the medium to be made up of mass particles at rest that are pushed away by the moving body. However, the resulting drag does not take into account the hydrodynamic flow past the body and the wake flow of the body.

This will be explained using this example of the *flow past a dihedron* (Figure 4.89). The flow past a dihedron must differ from the flow past two plates that are far apart and inclined in the same manner as the plates of the dihedron. In this latter case, the flow can pass between the two plates, while it cannot in the flow past a dihedron. The drag of a dihedron in a flow is about 60% of the drag of two isolated plates, according to experiments by *G. Eiffel* (1907). However, according to the Newtonian theory, both objects ought to have the same drag.

Another example is the flow past a circular disk and past two circular cylinders of the length of one diameter and of twice the diameter. Drag coefficients of 1.12, 0.91, and 0.85 respectively were measured. The reason that the longer cylinder has a smaller drag than the shorter is due to the fact that the flow along the surface of the cylinder reattaches itself and the wake becomes smaller. The suction effect of the wake flow on the rear end surface is smaller than in the other two cases.

Pressure Drag and Friction Drag

Hydrodynamic drag is made up of a pressure part and a friction part. The associated drag coefficients are therefore

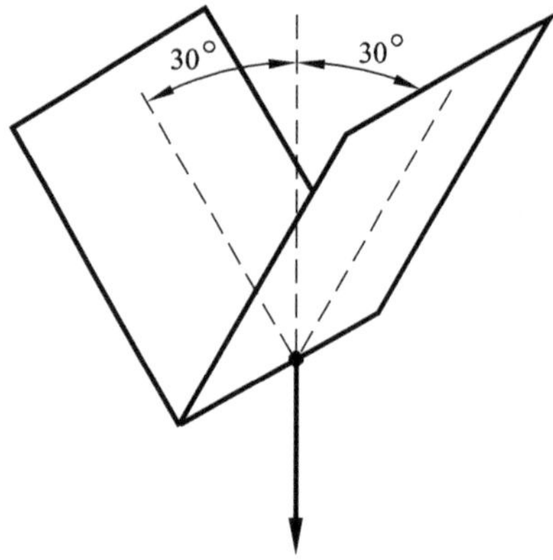

Fig. 4.89. Flow past a dihedron

$$c_w = c_d + c_f. \tag{4.124}$$

The total drag coefficient c_w is defined by

$$c_w = \frac{W}{\frac{\rho}{2} \cdot v^2 \cdot A},$$

with the drag force W, the dynamic pressure $(\rho/2)\cdot v^2$, and the cross-sectional area A. The pressure drag coefficient c_d and the global friction drag coefficient c_f are

$$c_d = \frac{W_d}{\frac{\rho}{2} \cdot v^2 \cdot A}, \qquad c_f = \frac{W_f}{\frac{\rho}{2} \cdot v^2 \cdot A},$$

where W_d is the pressure force and W_f the force due to friction,

The drag coefficient c_w is in general a function of the Reynolds number $\mathrm{Re}_l = v \cdot l/\nu$:

$$c_w = \mathrm{f}(\mathrm{Re}_l). \tag{4.125}$$

If the friction may be neglected, as in the example of a plate in a transverse flow, there is for Reynolds number $Re_D > 10^3$ no dependence on the Reynolds number, and the c_w value is constant. For a circular plate, the c_w value is 1.12. For a plate in a longitudinal flow, on the other hand, the friction drag coefficient c_f dominates, and the pressure drag coefficient c_d is small enough to be neglected.

The total drag can always be decomposed into pressure and friction parts. Assuming that the pressure drag depends greatly on the shape of the body, while the friction drag depends essentially on the size of the surface of the body and not on the shape of the surface, the drag can also be decomposed into a shape drag and a surface drag. Strictly speaking, the friction drag also depends on the shape of the surface, so that this decomposition is only approximately valid.

For bodies that move on the free surface of a liquid, there is an additional particular kind of pressure drag, the *wave drag*. This is caused by the wave system generated by the body. Since the wave motion is under the effect of gravity (the surface forces are not taken into account), the dimensionless characteristic number is the *Froude number.* It is formed with the velocity v, the length l, and the gravitational acceleration g:

$$\mathrm{Fr} = \frac{v}{\sqrt{g \cdot l}}. \tag{4.126}$$

The expected wave system will be geometrically similar for, for example, two different-sized versions of a ship (e.g. model and ship) if the Froude number has the same value.

The wave drag varies with small changes in the shape of a ship and in the velocity. If the body of the ship is made longer, the wave drag may increase or decrease, depending on how the stern and bow waves interfere with each other. The drag becomes larger if the stern lies in a trough of the bow wave system, and smaller if it is at a crest of the bow system.

Potential Flow

A *potential flow of an inviscid liquid* causes no drag in the direction of motion and no lift in the perpendicular direction. This can be proved using the balance of momentum if the control volume surrounds the body in the flow at some distance from it. The perturbation velocities caused by the displacement effect of the body die away quickly to all sides of the body. If the control volume is allowed to grow to infinity, the contributions to the momentum tend to zero. Since the balance of momentum must have the same result for all control volumes, the drag is therefore zero.

Of the different attempts to treat the drag within the framework of the theory of inviscid liquids, we consider the *Kirchhoff flow past a plate* and the *Kármán vortex street.*

Kirchhoff Flow past a Plate

In inviscid flow past a flat plate (Figure 4.90), the flow divides at the stagnation point and forms the discontinuity surfaces introduced in Section 4.1.4. In the wake of the plate, the liquid is at rest and forms the so-called dead water. In this region the pressure is constant. Therefore, we have the condition that the pressure on the interface must also be constant. According to the Bernoulli equation, the velocity on the interface is therefore also constant. If this condition is met, the inviscid theory leads only to those solutions in which the interfaces extend to infinity and the velocity on the interface is equal to the velocity of the unperturbed flow at infinity. The pressure distribution has a maximum at the stagnation point and tapers off at the edges to the pressure of the unperturbed flow. In the wake the pressure is the constant pressure of the unperturbed flow. The *pressure drag coefficient* c_d is proportional to the surface area of the plate and to the stagnation pressure.

G. R. Kirchhoff (1869) calculated the constant value $c_d = 2 \cdot \pi/(4+\pi) = 0.88$ for an infinitely long plate.

In reality, the interfaces are unstable, and they decay, forming vortices (see Section 4.1.4). In the wake of the plate, a time-averaged backflow region forms with a considerably lower pressure than the unperturbed pressure. This leads to a suction effect in the wake, generating a considerably larger drag than the inviscid Kirchhoff calculation. For the infinitely long plate, the total drag coefficient is $c_w = 1.98$. For a square plate, liquid flows over the side edges into the wake and thus greatly reduces the underpressure. The resulting total drag coefficient is $c_w = 1.17$.

Therefore, the Kirchhoff drag calculation does not agree sufficiently with reality. A better agreement with the calculation is obtained for the case in which, because of the low pressure and at a sufficiently high velocity, cavitation causes the wake to fill with liquid vapor. For this case the interfaces are stable, and the conditions of the inviscid theory are approximately satisfied.

Kármán Vortex Street

In certain circumstances, in the longitudinal flow past a plate, periodic separation of vortices takes place at the trailing edge (Figure 4.91). This observation prompted *T. von Kármán* (1912) to investigate the stability of parallel vortex filaments. Stability was obtained for a ratio of the distance h between the two vortex rows to the separation l of $h/l = 0.281$. The vortex rows actually observed come very close to this given ratio h/l of the inviscid stability theory. Figure 4.91 shows that the friction causes the vortices to move further apart downstream.

The periodic separation of vortices generates a drag that was also calculated by *T. von Kármán* (1912). It is a success for the inviscid theory that from a photographic measurement of the vortex system and a measurement of the vortex velocity the drag coefficient of the vortex-generating body can be determined.

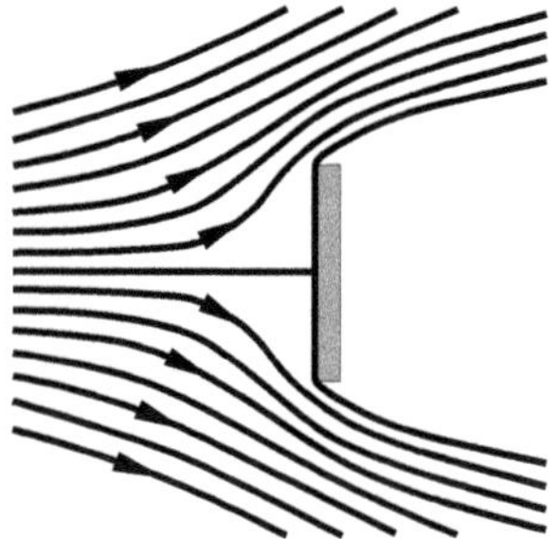

Fig. 4.90. Kirchhoff flow at a flat plate

Friction Drag of the Flow past a Plate

The friction drag of the plate is referred to the total surface area of the body A. The drag force is

$$W_f = \int_A \tau_w \cdot \sin(x, \boldsymbol{n}) \cdot \mathrm{d}A = c_f \cdot A \cdot \frac{\rho \cdot v^2}{2}. \tag{4.127}$$

Here x is the direction of the free stream, $\boldsymbol{n}$ is the local normal vector to the surface, and c_f is the friction drag coefficient. For a rectangular plate of width b and length l in a longitudinal flow, $A = 2 \cdot b \cdot l$.

The friction drag for the laminar *plate boundary-layer flow* is proportional to $\sqrt{l}$. For turbulent flow and smooth surfaces, and sufficiently large Reynolds numbers, it is proportional to about $l^{0.8}$ to $l^{0.85}$, while for rough surfaces it is proportional to $l^{0.65}$ to $l^{0.75}$. Introducing the Reynolds number formed with l, $\mathrm{Re}_l = v \cdot l / \nu$, we obtain the curves shown in Figure 4.92, in which c_f and Re_l are plotted logarithmically. The unbroken and dashed lines indicate different equations for the calculation of the friction drag coefficient. For laminar flow, curve 1 is valid:

$$c_f = \frac{1.33}{\sqrt{\mathrm{Re}_l}}. \tag{4.128}$$

If the plate boundary layer is turbulent from the start, it is curve 2 that holds:

$$c_f = \frac{0.074}{\mathrm{Re}_l^{0.2}}. \tag{4.129}$$

If the boundary-layer flow starts off laminar and becomes turbulent at the critical Reynolds number $5 \cdot 10^5$, curve 3 holds:

$$c_f = \frac{0.074}{\mathrm{Re}_l^{0.2}} - \frac{1700}{\mathrm{Re}_l}. \tag{4.130}$$

This equation can be applied for Reynolds numbers up to $5 \cdot 10^6$. For Reynolds numbers up to $5 \cdot 10^8$, *H. Schlichting* (1934) presented the following interpolation formula (curve 4):

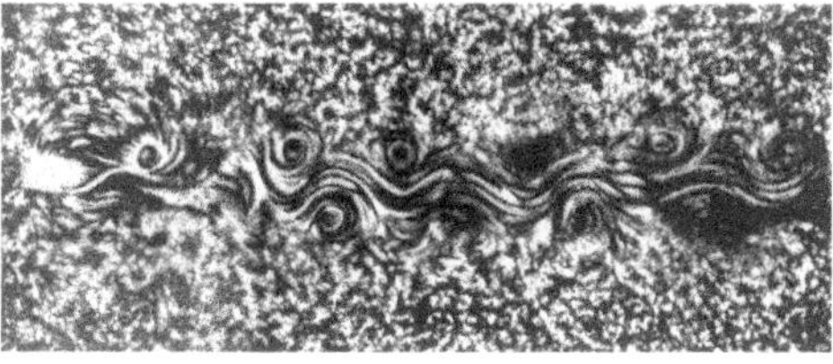

vortex street behind a plate

computed streamlines
T. von Kármán 1912

Fig. 4.91. Kármán vortex street

$$c_f = \frac{0.455}{(\log(\mathrm{Re}_l))^{2.58}}. \tag{4.131}$$

Curve 5 is the interpolation equation adapted to experiments by *T. von Kármán and K. Schönherr* (1932):

$$\sqrt{c_f} = \frac{0.242}{\log(\mathrm{Re}_l \cdot c_f)}. \tag{4.132}$$

The behavior of turbulent flows on rough surfaces discussed in Section 4.2.9 also permits calculation of the friction drag of *rough plates*. It is to be expected that the drag for a given length and a given roughness height k for fully developed flow is proportional to the square of the velocity. The friction drag coefficient c_f is larger the larger the ratio k/l. As for a fixed k, this ratio sinks with increasing length, c_f decreases for increasing Reynolds numbers at constant velocity.

The calculation of the drag of rough plates was initially carried out by *L. Prandtl and H. Schlichting* (1934) based on measurements of *J. Nikuradse* (1922) on rough pipes. The results are shown in Figure 4.93 for smooth and rough surfaces.

Relation of the Drag to the Situation in the Wake

Figure 4.94 shows the time-averaged wake profile for a body moving with velocity U_∞. The frame of reference is at rest. The wake flow contains the liquid set into motion by the drag of the body. The liquid flows past the front of the body to all sides as in a source flow (Section 4.1.5). The source strength Q is the same as the strength of the wake and is closely related to the drag. With the wake velocity w, relative to the liquid at rest, we obtain the source strength at a sufficiently large distance from the body:

$$Q = \int_N w \cdot \mathrm{d}A. \tag{4.133}$$

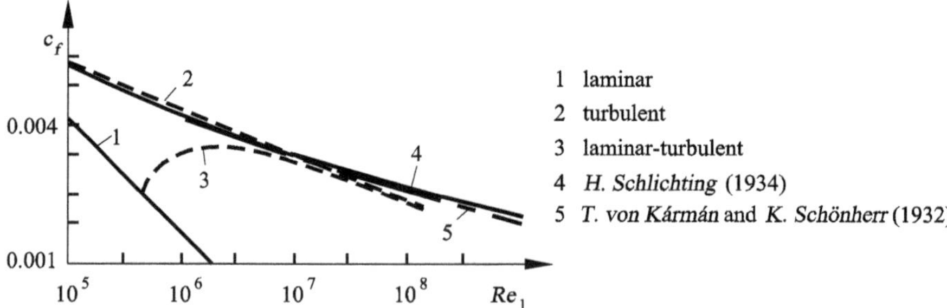

Fig. 4.92. Dependence of the friction drag c_f of smooth plates on the Reynolds number Re_l

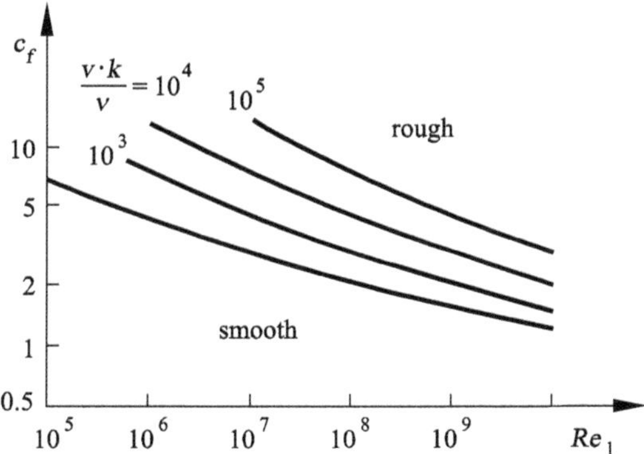

Fig. 4.93. Dependence of the friction drag c_f of smooth and rough plates on the Reynolds number Re_l

The integration takes place over the wave surface N. Applying the balance of momentum to the source and wake flows, we obtain

$$W = \rho \cdot Q \cdot U_\infty. \tag{4.134}$$

It can be seen from (4.133) and (4.134) that the drag can be determined by measuring the wake. *W. Betz* (1925) was the first to indicate this possibility to measure the drag.

The velocity relative to the body is $U_\infty - w$ in the wake. With a Pitot tube (see Section 4.1.3) at rest relative to the body, the total pressure $p_g = p + (\rho/2) \cdot (U_\infty - w)^2$ is measured. If p_{g_0} is the unperturbed total pressure $p_0 + (\rho/2) \cdot U_\infty^2$, then the drag at a sufficiently large distance behind the body is calculated from (4.133) and (4.134) as

$$W = \int\limits_N (p_{g_0} - p_g) \cdot \mathrm{d}A. \tag{4.135}$$

The term $(\rho/2) \cdot w^2$ is neglected.

The inviscid consideration of the flow past a body with drag also permits us to draw an important conclusion about the pressure field. This is generated by the source. The radial velocity is $w_r = Q/(4 \cdot \pi \cdot r^2)$ for the point source or $Q_1/(2 \cdot \pi \cdot r)$ for the line source of the plane flow with source strength Q_1

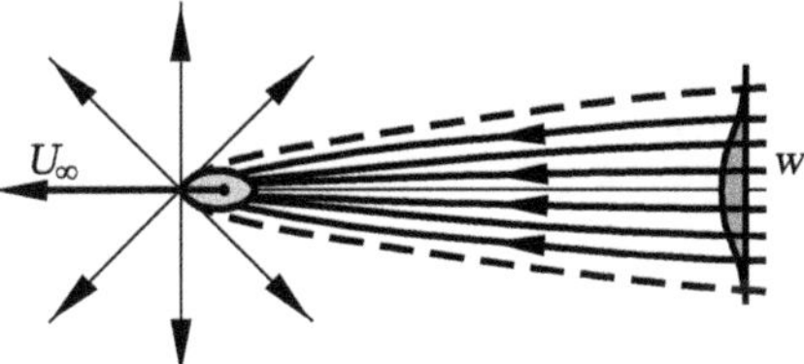

Fig. 4.94. Wake flow of a moving body, reference frame at rest

per unit length. Restricting ourselves to first-order precision, in forming the square of the resultant velocity at a large distance from the source, we only need consider the x component $u = w_r \cdot \cos(\varphi)$. Neglecting the second-order term, the expression $(\rho/2) \cdot (U_\infty + u)^2 - U_\infty^2 = (\rho/2) \cdot (2 \cdot U_\infty \cdot u + u^2)$ in the *Bernoulli equation* yields

$$p - p_0 = -\rho \cdot U \cdot u = -\rho \cdot \frac{Q \cdot U_\infty}{4 \cdot \pi \cdot r^2} \cdot \cos(\varphi) \quad \text{or} \quad -\rho \cdot \frac{Q_1 \cdot U_\infty}{2 \cdot \pi \cdot r} \cdot \cos(\varphi).$$

Using (4.134), it then follows that

$$p - p_0 = -\frac{W \cdot \cos(\varphi)}{4 \cdot \pi \cdot r^2} \quad \text{or} \quad -\frac{W_1 \cdot \cos(\varphi)}{2 \cdot \pi \cdot r}.$$

Here W_1 is the drag per unit length for the line source of the plane flow. The contributions are still considerable at large distances, particularly for the line source. This must be taken into account in measurement if, for example, a fixing device for the measuring apparatus perturbs the flow transverse to its direction. The wake flow, in which, since it is a viscous flow, the *Bernoulli equation* does not hold, delivers a lower-order contribution to the pressure field.

We note the following with respect to the viscous wake flow. For Reynolds numbers $\text{Re}_d < 1$ there exist analytic solutions by *C. W. Oseen* (1910) for the sphere and by *H. Lamb* (1911) for the cylinder. These solutions are in good agreement with the measurements shown in Figure 4.96. With increasing Reynolds numbers, a steady backflow region initially forms behind the cylinder (Figure 4.95), then becoming the laminar Kármán vortex street. The statements about the drag are then valid for the time-averaged velocity profile in the wake. The drag coefficients c_w as functions of the the Reynolds

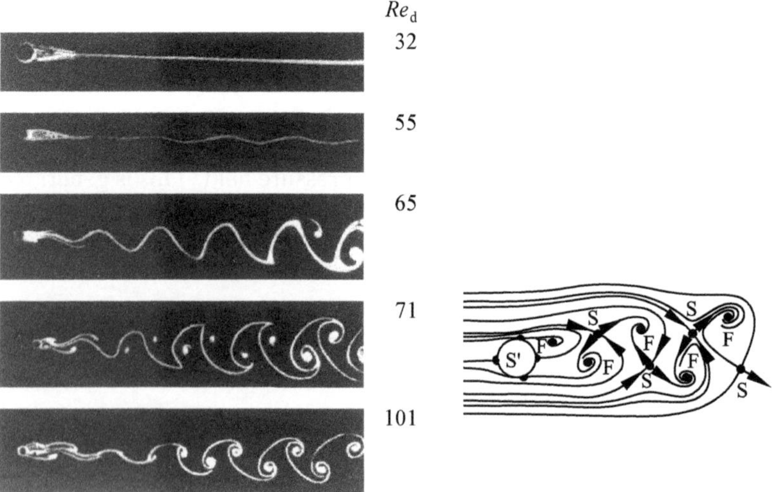

Fig. 4.95. Kármán vortex street behind a circular cylinder, *F. Homann* (1936)

number Re_d formed with the diameter of the body d are shown in Figure 4.96 for a sphere, cylinder, and disk. The drag coefficient is determined by the position of the separation point on the body. Whether the boundary-layer flow on the body is laminar or turbulent is of importance. For the turbulent boundary layer, the separation point is displaced downstream, causing the drag to be greatly decreased (see Section 4.2.6).

This behavior was first determined in the investigation of the drag of *spheres*. This drops at the Reynolds number $4 \cdot 10^5$ to values $c_w = 0.12$. With increasing Reynolds number, the c_w value increases again to about 0.18. *L. Prandtl* in his famous tripwire experiment showed that it is indeed the transition to a turbulent boundary layer that is responsible for the reduction in drag. If a thin wire (thickness the order of magnitude of the viscous sublayer) is placed around a sphere somewhat upstream of the point where separation would occur in laminar flow, the lower drag is observed even below the Reynolds number $4 \cdot 10^5$. Because of the forced turbulent boundary-layer flow, the separation point is displaced by the wire from about 80° to between 111° and 120°.

For a creeping flow $\mathrm{Re}_d < 1$, the *Stokes* law $c_w = 24/\mathrm{Re}_d$ is valid for the flow past a sphere.

For a circular cylinder, the transition from large drag values to small drag values is at about $\mathrm{Re}_d = 5 \cdot 10^5$. The drag drops from $c_w = 1.2$ to $c_w = 0.3$. For a creeping flow, instead of the *Stokes* solution, it is the *Lamb* solution that is valid:

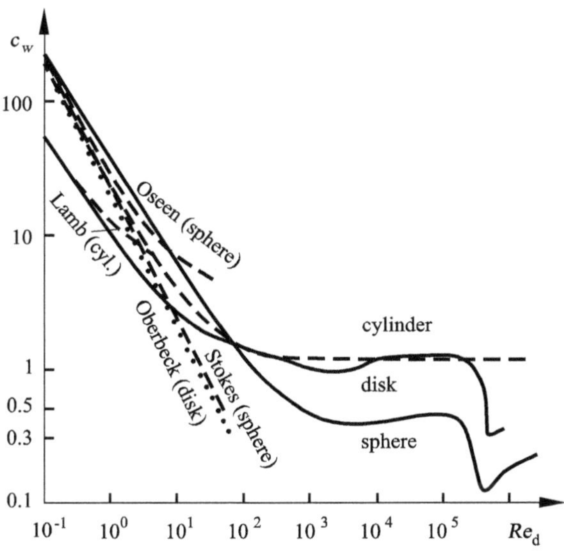

Fig. 4.96. Dependence of the drag coefficient c_w of sphere, cylinder, and disk on the Reynolds number Re_d

$$c_w = \frac{8 \cdot \pi}{\mathrm{Re}_d \cdot (2 - \ln(\mathrm{Re}_d))}.$$

In the case of a circular disk, the separation point is fixed, so that the laminar–turbulent transition plays no role in the boundary layer of the body. For this reason, the drag coefficient remains at a value of $c_w = 1.18$.

Low Drag Airships

In aircraft technology, bodies with small air drag are of particular importance. This has led to the design of body shapes where flow separation is avoided, to so-called streamline bodies. For streamline bodies, the pressure distribution calculated with the potential equation is in very good agreement with the measured pressure distribution (see Figure 4.97). There have to be deviations at the trailing edge. Here the boundary layer of the body becomes the shear layer of the wake flow, and so the measured pressure distribution lacks the inviscid pressure increase to the stagnation pressure. The experimentally determined drag coefficient is $c_w = 0.04$. This is 1/28 of the drag of a circular disk with the same diameter.

As well as avoiding flow separation, attempts are also made to keep the friction drag small. This is possible if the flow remains laminar on a large part of the surface. It is useful to note that an accelerated flow can be kept laminar more easily than a decelerated flow. The acceleration on the body must take place in such a way that the velocity maximum is as far as possible downstream. This is achieved by placing the widest point of the profile as far downstream as possible. However, the surface must be completely free from roughness, since otherwise the laminar–turbulent transition would be caused too soon.

4.2.11 Flows in Non-Newtonian Media

In Section 4.2.1 we treated nonlinear flow properties of non-Newtonian fluids. As an example of a non-Newtonian flow, we now consider a fully developed *circular pipe flow* whose shear strength obeys the power law (4.50).

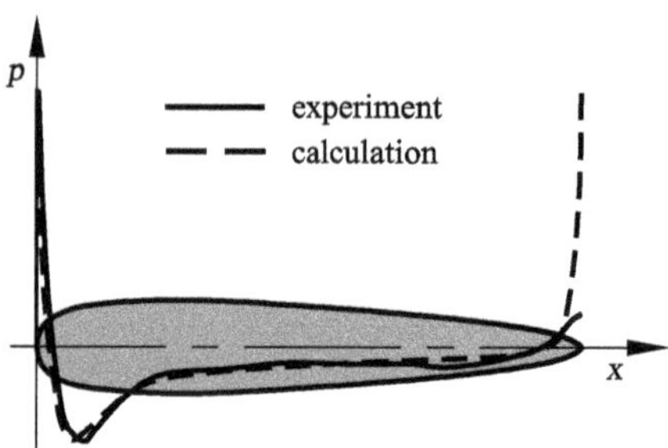

Fig. 4.97. Pressure distribution on an airship model, *G. Fuhrmann* (1910)

Pipe Flow

The driving force of the fully developed pipe flow is the constant pressure difference Δp. As in the flow of a Newtonian liquid, the pressure gradient along the pipe is constant, $\mathrm{d}p/\mathrm{d}z = -\Delta p/l$. To determine the solution we use the continuity equation for incompressible liquids (Section 5.1)

$$\nabla \cdot \boldsymbol{v} = 0 \tag{4.136}$$

and the Navier–Stokes equation for steady flows with a gravitational field (Section 5.2.1)

$$\rho \cdot (\boldsymbol{v} \cdot \nabla) \cdot \boldsymbol{v} = -\nabla \boldsymbol{p} + \nabla \cdot \boldsymbol{\tau}. \tag{4.137}$$

Here $\boldsymbol{\tau}$ is the tensor of the normal and shear stresses. The ansatz

$$v_r = 0, \qquad v_\varphi = 0, \qquad v_z = u(r), \qquad p = p(z) \tag{4.138}$$

satisfies the continuity equation, and the left-hand side of (4.137) is equal to zero, and $\boldsymbol{\tau}$ has only two nonvanishing components. For $\tau_{rz} = \tau_{zr}$ and using (4.50), we have

$$\tau_{zr} = \tau_{rz} = K \cdot \left| \frac{\mathrm{d}u}{\mathrm{d}r} \right|^{n-1} \cdot \frac{\mathrm{d}u}{\mathrm{d}r}. \tag{4.139}$$

With this, the z component of (4.137) alone yields a contribution:

$$0 = -\frac{\mathrm{d}p}{\mathrm{d}z} + \frac{1}{r} \cdot \frac{\mathrm{d}}{\mathrm{d}r}(r \cdot \tau_{rz}). \tag{4.140}$$

The r and φ components of (4.137) are identically satisfied. On integrating (4.140) we obtain

$$\tau_{rz} = \frac{\mathrm{d}p}{\mathrm{d}z} \cdot \frac{r}{2} + \frac{\mathrm{C}_1}{r}.$$

The shear stress τ_{rz} has a finite value for $r = 0$. This implies that the constant of integration C_1 must be equal to zero. With the ansatz (4.139) we obtain

$$K \cdot \left| \frac{\mathrm{d}u}{\mathrm{d}r} \right|^{n-1} \cdot \frac{\mathrm{d}u}{\mathrm{d}r} = \frac{\mathrm{d}p}{\mathrm{d}z} \cdot \frac{r}{2}.$$

As the pressure decreases in the direction of the z axis, $\mathrm{d}p/\mathrm{d}z = -\Delta p/l$ is negative, and so $\mathrm{d}u/\mathrm{d}r$ must also be negative:

$$\frac{\mathrm{d}u}{\mathrm{d}r} = -\left(\frac{\Delta p}{2 \cdot K \cdot l} \right)^{\frac{1}{n}} \cdot r^{\frac{1}{n}}.$$

Integrating this we obtain

$$u(r) = -\frac{n}{n+1} \cdot \left(\frac{\Delta p}{2 \cdot K \cdot l} \right)^{\frac{1}{n}} \cdot r^{\frac{n+1}{n}} + \mathrm{C}_2,$$

where C_2 is determined from the no-slip condition at the wall $u(R) = 0$, with the radius of the pipe R. This yields

$$u(r) = -\frac{n}{n+1} \cdot \left[\frac{R^{n+1}}{2 \cdot K} \cdot \frac{\Delta p}{l}\right]^{\frac{1}{n}} \cdot \left[1 - \left(\frac{r}{R}\right)^{\frac{n+1}{n}}\right]. \tag{4.141}$$

For $n = 1$, (4.141) is identical to the velocity profile of a Newtonian liquid. For $n < 1$, there is a steeper velocity gradient, as shown in Figure 4.98. The volume flux Q is calculated from (4.141) as

$$Q = \int_0^{2\cdot\pi} \int_0^R u(r) \cdot r \cdot \mathrm{d}r \cdot \mathrm{d}\varphi = \frac{n}{3 \cdot n + 1} \cdot \pi \cdot R^3 \cdot \left(\frac{R}{2 \cdot K} \cdot \frac{\Delta p}{l}\right)^{\frac{1}{n}}. \tag{4.142}$$

This yields the mean velocity u_m:

$$u_m = \frac{Q}{\pi \cdot R^2} = \frac{n}{3 \cdot n + 1} \cdot R \cdot \left(\frac{R}{2 \cdot K} \cdot \frac{\Delta p}{l}\right)^{\frac{1}{n}}.$$

For $n = 1$ and $K = \mu$ we again obtain the *Hagen–Poiseuille* law for the pipe law of a Newtonian liquid.

Weissenberg Effect

Shear flows of liquids with high molecular weights have non-Newtonian effects that can be associated with the normal stresses. As an example we consider the *Weissenberg effect*. A non-Newtonian fluid moves between two concentric cylinders with radii R_1 and R_2 (Figure 4.99), of which the inner cylinder rotates with constant angular velocity ω. The liquid has a free surface on which the surrounding pressure acts. The height of the liquid surface is so large that the flow on the bottom of the cylinder has no effect on the form of the free surface.

In cylindrical coordinates, only the φ component of the velocity $v_\varphi(r)$ is nonzero. There is therefore a shear flow between the two cylinders. The

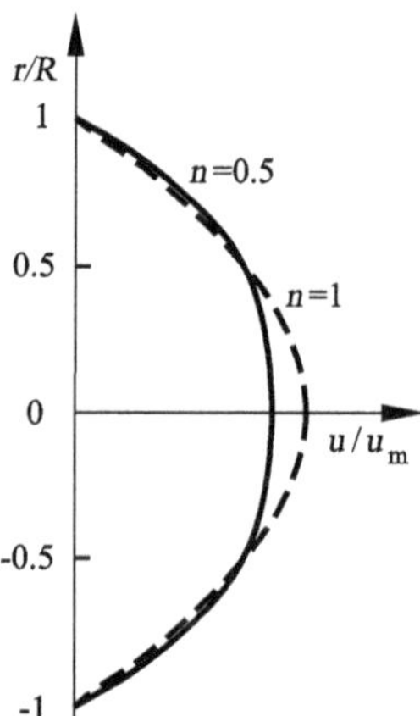

Fig. 4.98. Velocity distribution of a non-Newtonian liquid in a circular pipe

pressure is dependent only on r. The stress tensor of the non-Newtonian fluid is to have the following form:

$$\boldsymbol{\tau} = \begin{pmatrix} 0 & \tau_{r\varphi} & 0 \\ \tau_{\varphi r} & \sigma_{\varphi\varphi} & 0 \\ 0 & 0 & 0 \end{pmatrix}, \tag{4.143}$$

where $\sigma_{\varphi\varphi}$ and $\tau_{r\varphi}$ are dependent only on r. The r and φ components of the Navier–Stokes equation for steady flows (4.137) are

$$-\rho \cdot \frac{v_\varphi^2}{r} = -\frac{\mathrm{d}p}{\mathrm{d}r} - \frac{\sigma_{\varphi\varphi}}{r}, \tag{4.144}$$

$$0 = \frac{1}{r} \cdot \frac{\mathrm{d}}{\mathrm{d}r}(r \cdot \tau_{r\varphi}) + \frac{\tau_{r\varphi}}{r} = \frac{1}{r^2} \cdot \frac{\mathrm{d}}{\mathrm{d}r}(r^2 \cdot \tau_{r\varphi}). \tag{4.145}$$

The z component of (4.137) is satisfied identically. Using the Newtonian ansatz in cylindrical coordinates for the shear stress $\tau_{r\varphi} = \mu \cdot (\mathrm{d}v_\varphi/\mathrm{d}r - v_\varphi/r)$, (4.145) yields

$$0 = \mu \cdot \frac{\mathrm{d}}{\mathrm{d}r}\left(\frac{1}{r} \cdot \frac{\mathrm{d}}{\mathrm{d}r}(r \cdot v_\varphi)\right). \tag{4.146}$$

Integration of this expression allows us to determine the velocity distribution. This is identical to the corresponding velocity distribution of a Newtonian liquid:

$$v_\varphi(r) = \mathrm{A} \cdot r + \mathrm{B} \cdot \frac{1}{r}. \tag{4.147}$$

With the boundary conditions $v_\varphi(r = R_1) = \omega \cdot R_1$ and $v_\varphi(r = R_2) = 0$ we obtain the constants

$$\mathrm{A} = -\frac{\omega \cdot R_1^2}{R_2^2 - R_1^2} \quad \text{and} \quad \mathrm{B} = \frac{\omega \cdot R_1^2 \cdot R_2^2}{R_2^2 - R_1^2}.$$

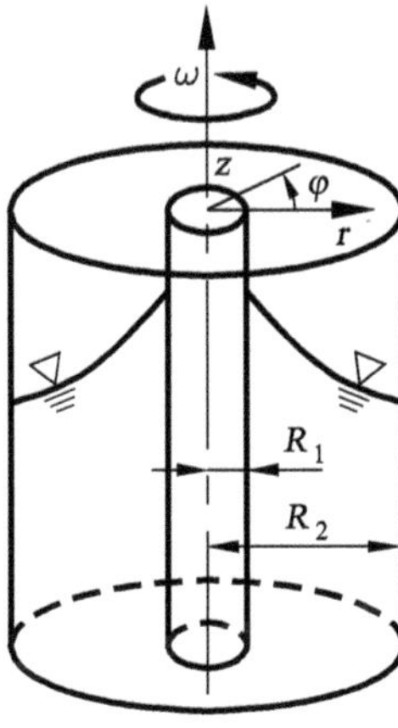

Fig. 4.99. Flow between two concentric cylinders, rotating inner cylinder

Equation (4.144) yields an equation for the pressure:

$$\frac{\mathrm{d}p}{\mathrm{d}r} = \frac{\mathrm{d}(\ln(r))}{\mathrm{d}r} \cdot \frac{\mathrm{d}p}{\mathrm{d}(\ln(r))} = -\frac{\sigma_{\varphi\varphi}}{r} + \rho \cdot \frac{v_\varphi^2}{r},$$

or

$$\frac{\mathrm{d}p}{\mathrm{d}(\ln(r))} = -\sigma_{\varphi\varphi} + \rho \cdot v_\varphi^2. \tag{4.148}$$

Formally, we can replace $\sigma_{\varphi\varphi}$ by the normal stress difference $\sigma_{\varphi\varphi} - \sigma_{rr}$. By assumption, the constant external pressure acts on the free surface. Therefore, the change in liquid height h is proportional to the pressure gradient:

$$\frac{\mathrm{d}h}{\mathrm{d}r} = \frac{1}{\rho \cdot g} \cdot \frac{\mathrm{d}p}{\mathrm{d}r}. \tag{4.149}$$

For liquids with high molecular weights, $\sigma_{\varphi\varphi} - \sigma_{rr} > 0$. For sufficiently large values of the difference in the normal stresses, (4.148) and (4.149) declare that the surface level of the liquid h at the rotating inner cylinder is greater than that at the outer cylinder at rest. This rise of the liquid at the rotating inner cylinder was described by *K. Weissenberg* (1947) and can be observed in many viscoelastic liquids.

Jet Expansion

Another normal stress effect occurs when a viscoelastic liquid exits as a free jet from a nozzle or an opening in a cylindrical pipe. A non-Newtonian liquid jet exiting downward from a vertical pipe (Figure 4.100) first expands before gravity causes it to contract again. Assuming that the flow at the cross-section of the opening is a fully developed Hagen–Poiseuille flow, the Navier–Stokes equation in the radial direction reduces to

$$\frac{\mathrm{d}(p - \sigma_{rr})}{\mathrm{d}r} = -\frac{1}{r} \cdot (\sigma_{\varphi\varphi} - \sigma_{rr}). \tag{4.150}$$

With (4.150), together with a balance of momentum around the opening and the normal stress functions, the jet expansion can be related to the normal stresses of the non-Newtonian fluid, as in the case of the Weissenberg effect. The expansion of the jet is larger, the smaller the pipe radius. This corresponds to the aspect of the Weissenberg effect in which the rise of the liquid at the rotating cylinder is greater, the smaller the diameter of the inner cylinder.

Fig. 4.100. Expansion of a liquid jet

4.2.12 Problems

4.10

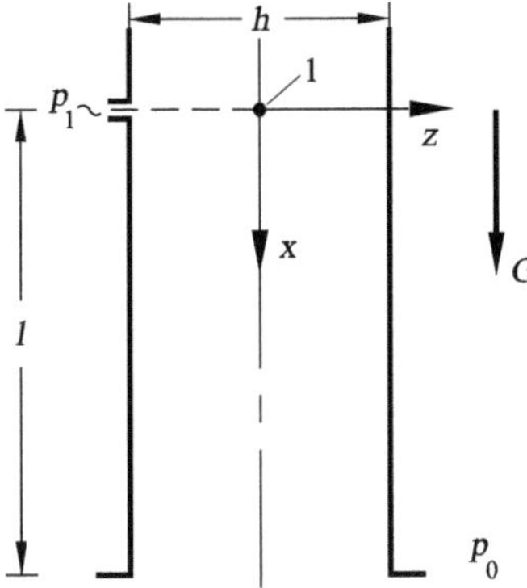

In a vertical channel a fluid with constant density ρ and dynamic viscosity μ flows under the effect of gravity g. The channel has width h, and its depth b perpendicular to the plane of the figure is much larger than h (two-dimensional flow). At position 1 ($x = 0$) there is a pressure borehole at which the static pressure p_1 of the flow can be measured. The distance between the pressure borehole and the exit cross-section is l. At the exit cross-section, the pressure is the surrounding pressure p_0.

It is assumed that the channel flow is a fully developed, steady, laminar flow with a pressure gradient. The following are to be determined:

(a) The dependence of the velocity profile $u(x, y)$ on the pressure gradient $\partial p/\partial x$.

$$u(y) = \frac{h^2}{8 \cdot \mu} \cdot \left(\rho \cdot g - \frac{\mathrm{d}p}{\mathrm{d}x}\right) \cdot \left(1 - 4 \cdot \left(\frac{y}{h}\right)^2\right) .$$

(b) The pressure $p = f(x, y)$.

$$p(x) = \frac{p_0 - p_1}{l} \cdot x + p_1 .$$

(c) The pressure $p_{1,\dot{m}}$ at position 1 that is necessary to move a given mass flux $\dot{m}$.

$$p_{1,\dot{m}} = p_0 + l \cdot \left(\frac{12 \cdot \mu \cdot \dot{m}}{\rho \cdot h^2 \cdot b} - \rho \cdot g\right) .$$

4.11

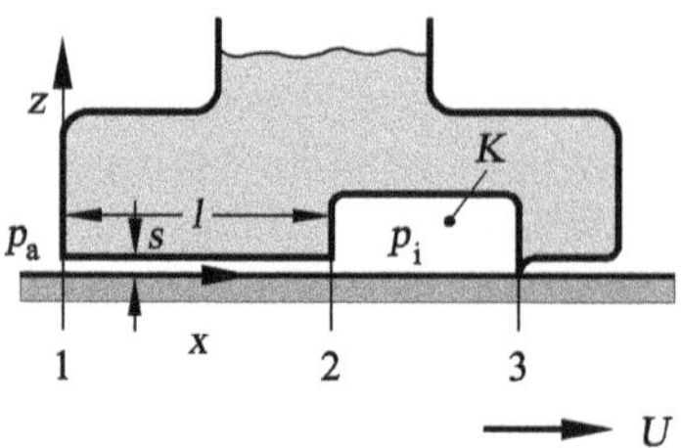

A machine part is set at rest above a horizontal plane wall that moves with the constant velocity U in such a manner that the left part of the lower side and the moving wall form a two-dimensional gap with length l, height s, and depth b (perpendicular to the plane of the figure).

The gap and the attached chamber K contain oil (Newtonian medium with constant dynamic viscosity μ). The moving wall drags the oil in the lower part of the gap into the chamber K, which then flows back into the upper part of the gap out of the chamber.

No oil can exit at the sealing lip (position 3). The pressure at the left end of the gap at position 1 is p_a, and that at the right end at position 2 is the chamber pressure p_i. The flow over the entire length l is fully developed and laminar.

(a) What is the differential equation for the velocity $u(x, y)$ and the relation for the dependence of the pressure p on p_a and p_i?

$$\frac{\mathrm{d}^2 u}{\mathrm{d}y^2} = \frac{1}{\mu} \cdot \frac{\mathrm{d}p}{\mathrm{d}x}, \qquad p(x) = \frac{p_i - p_a}{l} \cdot x + p_a.$$

(b) Determine the velocity profile $u(y)$ and the pressure p_i.

$$u(y) = \frac{1}{2 \cdot \mu} \cdot \frac{p_a - p_i}{l} \cdot s^2 \cdot \left[\frac{y}{s} - \left(\frac{y}{s}\right)^2\right] + U \cdot \left[1 - \frac{y}{s}\right],$$

$$p_i = \frac{6 \cdot \mu \cdot l}{s^2} \cdot U + p_a.$$

4.12

In order to determine whether a given steady, incompressible fundamental velocity profile $U_0(z)$ is stable or unstable, we need the perturbation differential equations. These can be derived from the Navier–Stokes equations using the following perturbation ansatz:

$$u = U_0(z) + u', \qquad w = w', \qquad p = p_0 + p'.$$

Inserting this ansatz into the Navier–Stokes equations and then linearizing, we obtain the linearized perturbation differential equations to determine the flow quantities u', w', and p'. They read

$$\frac{\partial u'}{\partial x} + \frac{\partial w'}{\partial z} = 0,$$

$$\frac{\partial u'}{\partial t} + U_0 \cdot \frac{\partial u'}{\partial x} + w' \cdot \frac{\mathrm{d}U_0}{\mathrm{d}z} = -\frac{1}{\rho} \cdot \frac{\partial p'}{\partial x} + \nu \cdot \left(\frac{\partial^2 u'}{\partial x^2} + \frac{\partial^2 u'}{\partial z^2} \right),$$

$$\frac{\partial w'}{\partial t} + U_0 \cdot \frac{\partial w'}{\partial x} = -\frac{1}{\rho} \cdot \frac{\partial p'}{\partial z} + \nu \cdot \left(\frac{\partial^2 w'}{\partial x^2} + \frac{\partial^2 w'}{\partial z^2} \right),$$

The perturbation quantities u', w', and p' are modeled using the wave trial solution:

$$u'(x,z,t) = \hat{u}(z) \cdot e^{i \cdot (a \cdot x - \omega \cdot t)}, \qquad w'(x,z,t) = \hat{w}(z) \cdot e^{i \cdot (a \cdot x - \omega \cdot t)},$$
$$p'(x,z,t) = \hat{p}(z) \cdot e^{i \cdot (a \cdot x - \omega \cdot t)},$$

with a the complex wave number and ω the complex angular frequency.
(a) Insert the wave trial solution into the perturbation differential equations and determine a system of differential equations for the unknowns $\hat{u}$, $\hat{w}$, and $\hat{p}$.

$$a \cdot \hat{u} + \frac{\mathrm{d}\hat{\omega}}{\mathrm{d}z} = 0,$$

$$(a \cdot U_0 - \omega) \cdot \hat{u} - i \cdot \frac{\mathrm{d}U_0}{\mathrm{d}z} \cdot \hat{w} = -\frac{1}{\rho} \cdot a \cdot \hat{p} + i \cdot \nu \cdot \left(a^2 \cdot \hat{u} - \frac{\mathrm{d}^2 \hat{u}}{\mathrm{d}z^2} \right),$$

$$(a \cdot U_0 - \omega) \cdot \hat{w} = i \cdot \frac{1}{\rho} \frac{\mathrm{d}\hat{p}}{\mathrm{d}z} + i \cdot \nu \cdot \left(a^2 \cdot \hat{u} - \frac{\mathrm{d}^2 \hat{w}}{\mathrm{d}z^2} \right).$$

(b) Transform the differential equations obtained to express the unknown wave amplitudes $\hat{u}$, $\hat{w}$, and $\hat{p}$ in a single equation to determine $\hat{w}$.

$$\left[(a \cdot U_0 - \omega) \cdot \left(\frac{\mathrm{d}^2}{\mathrm{d}z^2} - a^2 \right) - a \cdot \frac{\mathrm{d}^2 U_0}{\mathrm{d}z^2} + i \cdot \nu \cdot \left(\frac{\mathrm{d}^2}{\mathrm{d}z^2} - a^2 \right)^2 \right] \hat{w} = 0 \quad .$$

4.13

Air (kinematic viscosity ν, density ρ) flows with velocity U_∞ past a thin plate of length l and width B. The flow is two-dimensional and incompressible.

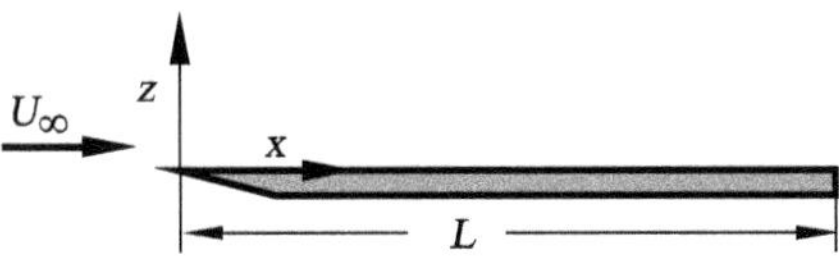

A laminar boundary layer forms on the forward part of the plate, while downstream, after the critical Reynolds number $\mathrm{Re}_{l_{\text{crit}}}$ is exceeded, a turbulent boundary layer develops.

(a) How is the total drag of the boundary-layer flow made up and what is the relative contribution of each individual drag?

pressure drag $W_d (\approx 0\%)$ and friction drag $W_f (\approx 100\%)$.

(b) Calculate the position x_{crit} of the laminar–turbulent transition if the critical Reynolds number is $\text{Re}_{l_{\text{crit}}} = 5 \cdot 10^5$. ($U = 10$ m/s, $\rho = 1.2$ kg/m^3, $\nu = 1.511 \cdot 10^{-5}$ m^2/s, $l = 2$ m, $B = 2$ m.)

$$x_{\text{crit}} = \frac{\text{Re}_{l_{\text{crit}}} \cdot \nu}{U} = 0.76 \text{ m}.$$

(c) Calculate the total drag W on the upper side of the plate if the drag coefficient can be approximated by $c_{f_l} = 0.664/\sqrt{\text{Re}_x}$ up to the position x_{crit} and by $c_{f_t} = 0.0609 \cdot (\text{Re}_x)^{-1/5}$ after the position x_{crit}.

$$W = \left(\int_0^{x_{\text{crit}}} \frac{0.644}{\sqrt{\frac{u \cdot x}{\nu}}} \cdot dx + \int_{x_{\text{crit}}}^{L} \frac{0.0609}{\left(\frac{u \cdot x}{\nu}\right)^{\frac{1}{5}}} \cdot dx \right) \cdot \frac{1}{2} \cdot \rho_\infty \cdot c_\infty^2 \cdot B,$$

$$W = 0.379 \text{ N}.$$

4.14

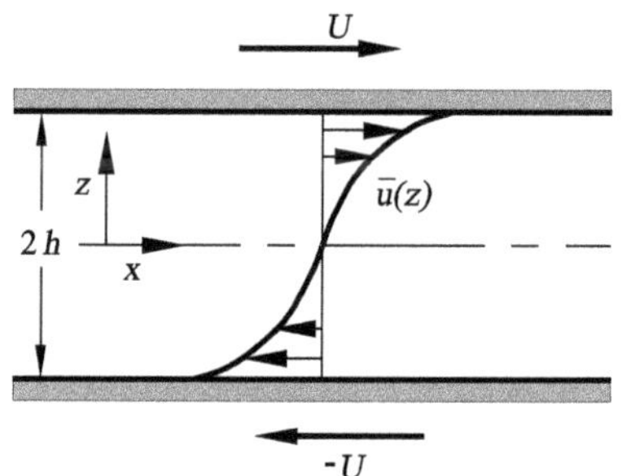

The turbulent Couette flow of constant density ρ between two infinitely extended plates moving with velocity U in opposite directions has a time-averaged velocity profile $\overline{u}(y)$. The turbulent Reynolds shear stresses are calculated using the Prandtl mixing length:

$$l(y) = K(h^2 - y^2).$$

(a) Determine the constant K such that the condition

$$-\left.\frac{dl}{dy}\right|_{y=\pm h} = \pm\kappa$$

is satisfied.

$$K = \frac{\kappa}{2h}.$$

(b) Determine the equation of the turbulent shear stresses τ_t for the given distribution of the Prandtl mixing length.

$$\tau_t = -\rho\overline{u'v'} = \rho \left[\frac{\kappa}{2h}(h^2 - y^2)\right]^2 \left(\frac{d\overline{u}}{dy}\right)^2.$$

(c) For the Couette flow $\overline{p} = \text{const}$, this means that $\mu(d\overline{u}/dy) - \rho\overline{u'v'}$ is also constant. Outside the viscous sublayer, the viscous shear stress $\mu(d\overline{u}/dy)$ may be neglected compared to the turbulent shear stress. Calculate the velocity profile $\overline{u}(y)$ at the upper wall $y' = y + h$.

$$\frac{\overline{u}(y')}{u^*} = \frac{1}{\kappa} \ln \left(\frac{y'/h}{2 - (y'/h)} \right), \qquad \text{with} \quad u^* = \frac{\kappa}{2h}(h^2 - y^2)\frac{\mathrm{d}\overline{u}}{\mathrm{d}y}.$$

4.15

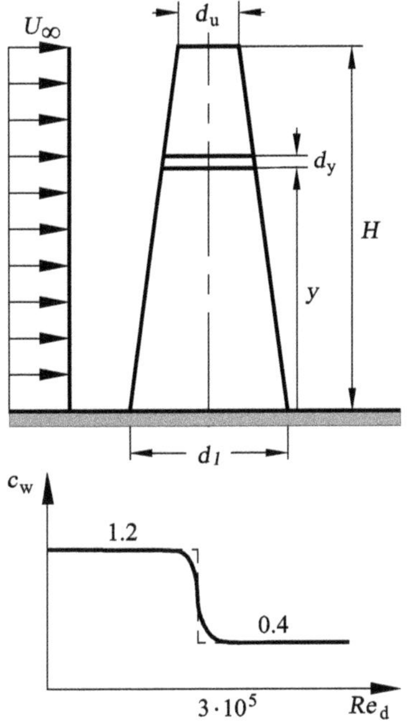

A stream of air of velocity $U = 1.62$ m/s flows past the entire length of a factory smokestack of height $H = 100$ m whose diameter decreases linearly from bottom $d_l = 6$ m to top $d_u = 0.5$ m, where the index l stands for lower and u for upper (kinematic viscosity of air $\nu = 15 \cdot 10^{-6}$ m^2/s, density of air $\rho = 1.234$ kg/m^3). To determine the wind load on the smokestack, the drag coefficient c_w of a segment of height dy is assumed to depend on Re$_d$ as for the circular cylinder $c_w = f(\text{Re}_\alpha)$.
With the idealized assumption that the drag coefficient has the constant numerical value $c_{w,l} = 1.2$ in the subcritical regime (Re$_d < 3.5 \cdot 10^6$) and jumps discontinuously to the constant numerical value $c_{w,u} = 0.4$ in the supercritical regime (Re$_d > 3.5 \cdot 10^6$), determine the wind load W on the smokestack.

$$W = \frac{\rho}{2} \cdot U_\infty^2 \cdot \left[c_{w,u} \cdot \left(\frac{d_u - d_l}{2 \cdot H} \cdot y_{\text{crit}}^2 + d_u \cdot y_{\text{crit}} \right) + c_{w,l} \cdot \frac{d_u - d_l}{2 \cdot H} \cdot (H^2 - y_{\text{crit}}^2) + c_{w,l} \cdot d_l \cdot (H - y_{\text{crit}}) \right] \quad ,$$

$W = 331.2\ N.$

4.16

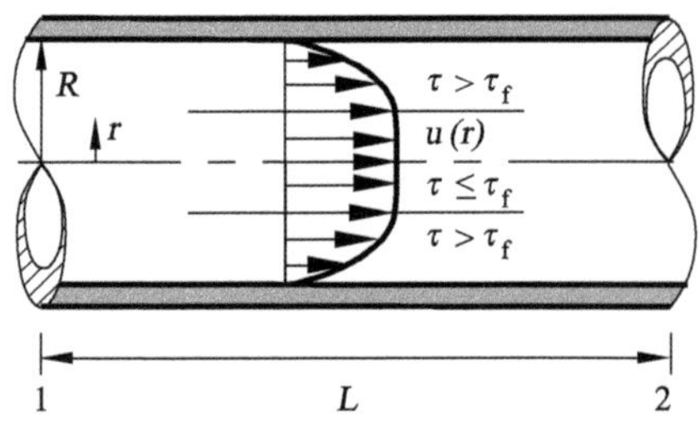

A non-Newtonian Bingham material flows through a pipe of length l. The flow function $\mathrm{d}u/\mathrm{d}r = f(\tau)$ of the Bingham medium is written as

$$f(\tau) = 0 \quad \text{for} \quad 0 \leq \frac{\tau}{\tau_f} \leq 1,$$

$$\frac{f(\tau)}{\mu} \left(\frac{\tau}{\tau_f} - 1 \right) \quad \text{for} \quad \frac{\tau}{\tau_f} \geq 1.$$

Below the flow stress τ_f the Bingham material behaves like a solid elastic body, and above τ_f it behaves like a Newtonian medium. Two zones form in the pipe. In the edge zone the Newtonian medium flows with a parabolic velocity profile. The core zone behaves like a solid body.

(a) Calculate the dependence of the volume flux Q on the general flow function $f(\tau)$.

$$Q = \frac{\pi R^3}{\tau_w^3} \cdot \int_0^{\tau_w} r^2 f(\tau) \mathrm{d}\tau, \qquad \text{with} \quad \tau_w = \tau(R) = \frac{R}{2} \frac{\mathrm{d}p}{\mathrm{d}\tau}.$$

(b) Insert the flow function of the Bingham medium for $f(\tau)$ and calculate the volume flux Q.

$$Q = \frac{\pi R^4 (p_1 - p_2)}{8 \, \mu \, l} \cdot f(\xi), \qquad \text{with} \quad \xi = \frac{(p_1 - p_2) R}{\mu \, l},$$

$$f(\xi) = 1 - \frac{4}{3} \left(\frac{2}{\xi} \right) + \frac{1}{3} \left(\frac{2}{\xi} \right)^4 .$$

4.3 Dynamics of Gases

Considerable density or volume changes occur in flows of gases and vapors where large pressure differences appear. Volume changes and the pressure differences necessary to cause them occur basically in the following cases:

Large height extensions of gas masses under the effect of gravity

Such flows occur in the free atmosphere. They will be treated in Section 10.2.

Large velocities in a gas flow

These occur in pressure compensation between two containers of different pressures, or when a body moves with a very large velocity in a gas. In practice, these flows occur in vapor and gas turbines and similar flow machinery. On the other hand, they are also found in the motion of rockets and airplanes, as well as in airplane propellers and jet engines. The fluid mechanics of compressible media is also called *gas dynamics.*

Large acceleration

This occurs in gases that are at rest or in motion if parts of the wall or body carry out greatly accelerated motion. Examples are the consequences of the sudden opening and closing of flaps and valves, and the expansion of explosions.

Large temperature differences

These can occur when heat is transferred, even at small flow velocities. Such flows with heat transfer will be treated in Chapter 7.

4.3.1 Pressure Propagation, Velocity of Sound

We consider a gas at rest in a pipe. A piston is moved and causes a pressure increase that propagates in the gas at rest as shown in Figure 4.101. We as-

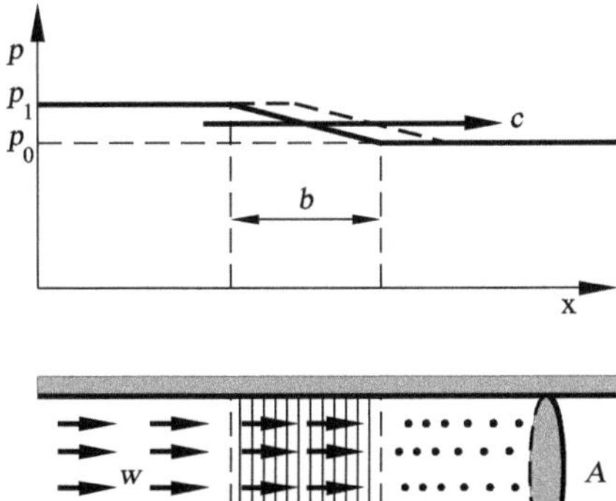

Fig. 4.101. Pressure wave in a pipe

sume that the pressure distribution and the entire flow state move to the right with velocity c without any change in their form. As the gas is compressed, it has the flow velocity w behind the pressure increase. We assume that the pressure increase $p_1 - p_0$ is small compared to the pressure p_0. Similarly, we assume that the density change $\rho_1 - \rho_0$ and w are small. The increase of mass per unit time in the pipe is $A \cdot (\rho_1 - \rho_0) \cdot c$, and the mass flowing into the pipe per unit time is $A \cdot \rho_1 \cdot w$. From *continuity* it follows that

$$\rho_1 \cdot w = (\rho_1 - \rho_0) \cdot c. \tag{4.151}$$

With the approaching momentum flux per unit time $A \cdot w \cdot \rho_1 \cdot w$, the increase in momentum per unit time $A \cdot w \cdot \rho_1 \cdot c$ and the resulting force $A \cdot (p_1 - p_0)$, the *equation of motion* leads to

$$p_1 - p_0 + \rho_1 \cdot w^2 = \rho_1 \cdot w \cdot c. \tag{4.152}$$

The assumptions made above mean that the square term $\rho_1 \cdot w^2$ can be neglected. Using (4.151), (4.152) yields

$$c^2 = \frac{p_1 - p_0}{\rho_1 - \rho_0}.$$

The expression on the right-hand side depends only on the compression law of the fluid. Assuming that the disturbances are small, it can be replaced by the differential quotient $\mathrm{d}p/\mathrm{d}\rho$:

$$c^2 = \frac{\mathrm{d}p}{\mathrm{d}\rho}. \tag{4.153}$$

The propagation velocity c of small pressure perturbations is therefore independent of the size of the pressure change and of the width of the transition region. It is dependent only on the compression law of the fluid. The quantity c is called the *velocity of sound* of small pressure perturbations (sound waves).

According to the isentropic law $p = \text{const} \cdot \rho^\kappa$, gases satisfy

$$c^2 = \frac{\mathrm{d}p}{\mathrm{d}\rho} = \kappa \cdot \text{const} \cdot \rho^{(\kappa-1)} = \kappa \cdot \frac{p}{\rho}. \tag{4.154}$$

With the equation of state of ideal gases $p = R \cdot \rho \cdot T$ (R the material-specific gas constant) we have

$$c = \sqrt{\kappa \cdot \frac{p}{\rho}} = \sqrt{\kappa \cdot R \cdot T}.$$

Therefore, the velocity of sound in a gas is dependent only on the temperature. For air at 0° C, i.e. $T = 273$ K, we obtain

$$c = \sqrt{\kappa \cdot \frac{p_0}{\rho_0}} = 331\ \frac{\mathrm{m}}{\mathrm{s}}.$$

Expansion of Pressure Waves

In a reference frame moving with the flowing gas, the pressure perturbation expands with the velocity of sound c relative to the gas. Relative to the flow velocity w, the pressure perturbation moves downstream with velocity $c + w$ and upstream with velocity $c - w$. If w is larger than c, the pressure perturbation will not propagate upstream.

If the flow velocity w is smaller than the velocity of sound c, the perturbations expand in the form of a spherical wave in all directions. If the flow velocity is greater than the velocity of sound, all spherical waves move within a cone downstream of the position A where the perturbation first appeared (Figure 4.102). If a sound source A moves with velocity $w > c$ through a gas at rest, the situation is similar. The perturbations expand inside a cone downstream of the sound source. The apex angle of this so-called *Mach cone* can be determined as follows. Within the time interval τ, a point-shaped perturbation will develop to a sphere of radius $c \cdot \tau$, whose midpoint has moved a distance $w \cdot \tau$ away. The cone touches the spheres tangentially, so that

$$\sin(\alpha) = \frac{c \cdot \tau}{w \cdot \tau} = \frac{c}{w} = \frac{1}{M}, \tag{4.155}$$

where α is called the *Mach angle* and M the *Mach number*. For $M < 1$ the flow is said to be subsonic, for $M \approx 1$ transonic, and for $M > 1$ supersonic.

The same relations can also be applied to the motion of bodies in air at rest. If the body moves with supersonic velocity, the perturbations caused by the body expand within a Mach cone. Figure 4.103 shows an example of the head wave of a bullet flying with supersonic velocity. The pressure differences are so large that the approximation of small perturbations is no longer valid, and the head wave propagates with supersonic velocity. The angle of the head wave is therefore larger than the Mach angle α.

The continuity equation (4.151) and the equation of motion (4.152) for the propagation velocity of a wave front are based on the assumption of

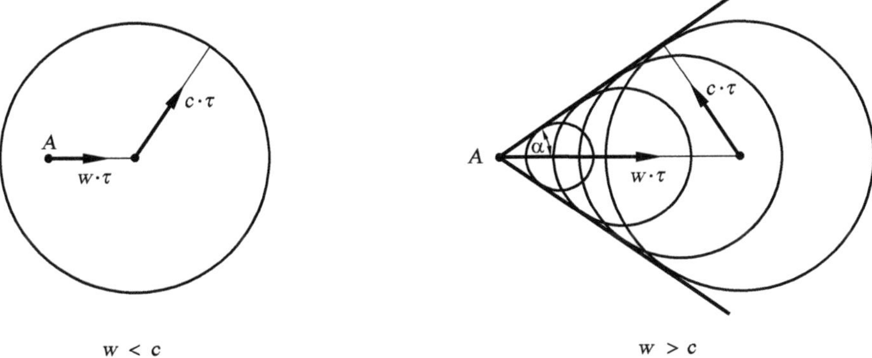

Fig. 4.102. Expansion of a pressure wave

unchanging wave shape. This is satisfied for small perturbations of the free-stream state, or in the case of the shock waves treated in Section 4.3.4. Finite, continuous pressure changes, on the other hand, alter their wave form as they propagate. This can be explained by considering the finite pressure change to be a series of many small changes. Each perturbation then moves in the state altered by the previous wave. If w_0 is the flow velocity in front of the wave, the change in the flow velocity is computed by (4.151) to be

$$w_1 - w_0 = \frac{c \cdot (\rho_1 - \rho_0)}{\rho_1}. \tag{4.156}$$

The change in density $\mathrm{d}\rho = \rho_1 - \rho_0$ is related to the change in pressure $\mathrm{d}p$ and the change in the velocity of sound $\mathrm{d}c$. Equation (4.154) leads to an expression for $\mathrm{d}c$:

$$\begin{aligned} 2 \cdot c \cdot \mathrm{d}c = 2 \cdot c \cdot (c_1 - c_0) &= \kappa \cdot \frac{\mathrm{d}p}{\rho} - \kappa \cdot \frac{p}{\rho^2} \cdot \mathrm{d}\rho \\ &= \frac{\mathrm{d}p}{\mathrm{d}\rho} \cdot (\kappa - 1) \cdot \frac{\mathrm{d}\rho}{\rho} = \frac{c^2 \cdot (\kappa - 1) \cdot (\rho_1 - \rho_0)}{\rho_1}. \end{aligned}$$

The density change $\rho_1 - \rho_0$ is thus related to the change in the velocity of sound $c_1 - c_0$, and (4.156) leads us to

$$w_1 - w_0 = \frac{2}{\kappa - 1} \cdot (c_1 - c_0). \tag{4.157}$$

In a two-dimensional sound wave, the flow velocity changes $2/(\kappa - 1)$ times (for air five times) as much as the magnitude of the change of the velocity of sound. This result is valid also for strong perturbations.

For the compression wave shown in Figure 4.104, the velocity of sound in the wave is larger than the velocity of sound in front of the wave. According to (4.157), the flow velocity is therefore also larger. The propagation velocity of each part of the wave is equal to the sum of the local velocity of sound and the local flow velocity $c + w$. Therefore, the perturbation moves ever faster as the depth of the wave increases. The wave becomes steeper with time and forms a vertical jump, the *shock wave*, to be treated in Section 4.3.4.

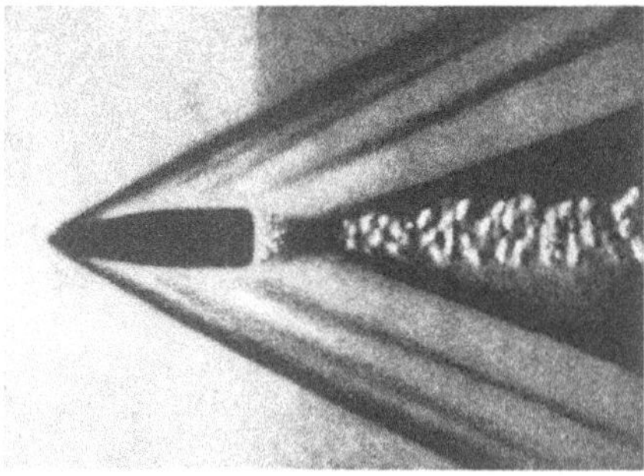

Fig. 4.103. Schlieren photograph of a bullet, *C. Cranz* (1926)

On the other hand, if, as in Figure 4.104, an expansion wave moves to the right into a medium at rest, $w_0 = 0$, the gas in the wave flows to the left. Following (4.157), because $c_1 < c_0$, w_1 becomes negative. The perturbations behind the wave front move slower, the smaller the pressure becomes. Such an expansion wave becomes flatter with time.

4.3.2 Steady Compressible Flows

In a compressible, inviscid flow, the generalized Bernoulli equation (4.4) is valid for a stream filament. Neglecting the effect of gravity, this reads

$$\mathrm{f} + \frac{w^2}{2} = \mathrm{f}_0 = \text{const}, \tag{4.158}$$

with the pressure function $\mathrm{f}(p) = \int (\mathrm{d}p/\rho)$. For isentropic changes of state

$$\rho = \rho_0 \cdot \left(\frac{p}{p_0}\right)^{\frac{1}{\kappa}},$$

the evaluation of the integral yields

$$\mathrm{f} = \frac{\kappa}{\kappa - 1} \cdot \frac{p_0}{\rho_0} \cdot \left(\frac{p}{p_0}\right)^{\frac{\kappa-1}{\kappa}}. \tag{4.159}$$

If p_0 is the reservoir pressure at $w_0 = 0$, e.g. the reservoir pressure in a vessel where exit processes are to take place, then

$$w = \sqrt{2 \cdot (\mathrm{f}_0 - \mathrm{f})} = \sqrt{\frac{2 \cdot \kappa}{\kappa - 1} \cdot \frac{p_0}{\rho_0} \cdot \left(1 - \left(\frac{p}{p_0}\right)^{\frac{\kappa-1}{\kappa}}\right)}. \tag{4.160}$$

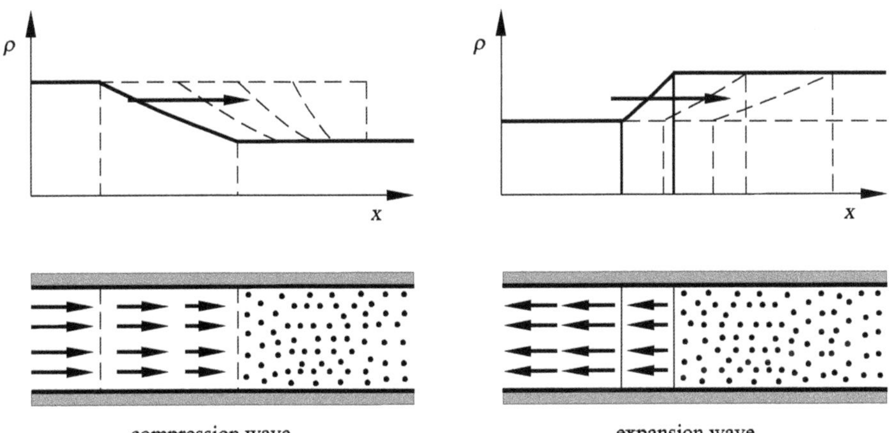

Fig. 4.104. Expansion of compression and expansion waves

If the gas is expanded as far as a vacuum ($p = 0$), (4.160) yields the maximum velocity as

$$w_{\max} = \sqrt{\frac{2 \cdot \kappa}{\kappa - 1} \cdot \frac{p_0}{\rho_0}} = \sqrt{\frac{2}{\kappa - 1}} \cdot c_0. \tag{4.161}$$

For air at $0°C$ the expansion has a maximum velocity of

$$w_{\max} = 740 \ \frac{\mathrm{m}}{\mathrm{s}}.$$

This is a hypothetical limiting value. Because a temperature and a condensation of the gas at absolute zero cannot be attained, this value cannot be reached. In hypersonic wind tunnels driven with air, a limiting value is obtained that is about 10% smaller than the theoretical value (4.161).

The relationship between w and p is shown in Figure 4.105. The figure also contains the dependence of the specific volume $v = 1/\rho$ on the pressure as follows from the isentropic equation. The shaded area $\int_p^{p_0} v \cdot \mathrm{d}p$ indicates the difference $\mathrm{F}_0 - \mathrm{F}$. For steady, compressible flow (see Section 4.1.1), the continuity equation states that the same mass flows through all cross-sections of a stream filament per unit time. Along the stream filament we have

$$A \cdot \rho \cdot w = \mathrm{const.} \tag{4.162}$$

The dependence of the stream filament cross-section A on the pressure p is given by the dependence of the function $1/(\rho {\cdot} w) = v/w$. This can be explained as follows, using (4.160) and (4.162). At $p = p_0$, $w = 0$ and therefore $A = \infty$. If p is reduced, w increases gradually, initially without much of a change in ρ. If p is very small and is reduced even further, w approaches the value $w_{\max}$ and then changes only slightly. However, ρ also decreases without limit as p decreases without limit; i.e. A must increase and tend toward ∞.

Between the regime where A decreases and that where the stream filament cross-section increases, there must clearly exist a minimum of A. This is found at the point where the relative increase of the velocity $\mathrm{d}w/w$ is just as large as the relative decrease of the density $-\mathrm{d}\rho/\rho$. This is the case at the point

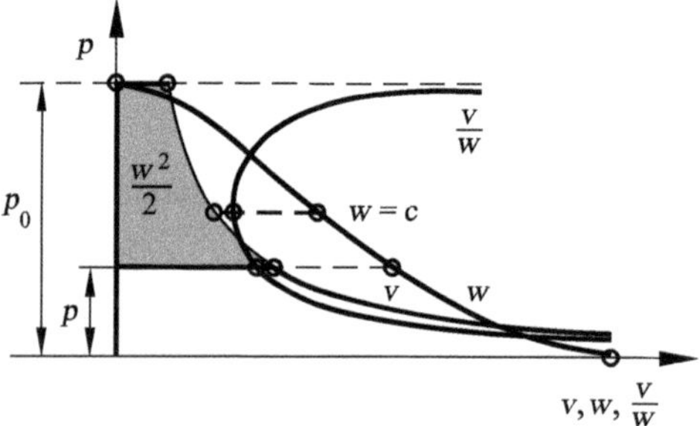

Fig. 4.105. Dependence of the specific volume v, velocity w, and v/w on the pressure p

where the flow velocity is equal to the velocity of sound. Because of the isentropic cooling that has taken place, this velocity of sound is not that of the initial state. It is smaller, corresponding to the reduced temperature (about 315 m/s in air at rest at 20°C). After the minimum is exceeded, the flow velocity is larger than the velocity of sound. In such a flow, a reduction of the pressure (increase in velocity) causes an increase in the cross-section. If the pressure is raised (decrease in velocity), the cross-section decreases. A continuous acceleration of the gas from the subsonic state to the supersonic state initially requires a contraction and, after the velocity of sound has been passed, an expansion of the stream tube. Such an arrangement is called a *Laval nozzle.*

In the case of a simple opening without expansion, as soon as the back pressure is small enough, the fluid in the opening flows with the velocity of sound. In air, the critical pressure ratio of back pressure to reservoir pressure is about 0.53 of the reservoir pressure. In general, the critical pressure ratio of an ideal gas is

$$\frac{p'}{p_0} = \left(\frac{2}{\kappa+1}\right)^{\frac{\kappa}{\kappa-1}} .$$

The associated velocity is

$$w' = c' = \sqrt{\frac{2 \cdot \kappa}{\kappa+1} \cdot \frac{p_0}{\rho_0}} .$$

The discharge amount is therefore independent of the back pressure. Outside the outlet, the cross-section of the gas jet expands due to the inertia of the gas flow to such a large degree that an underpressure occurs within it. This underpressure causes the flow to become convergent, and it compresses again to a pressure that is about that of the pressure in the outlet. This process repeats itself periodically (Figure 4.106).

The outlet pressure p_m can be measured with a borehole in the nozzle close to the outlet (cf. Figure 4.107). For external pressures p_2 that are smaller than the *critical pressure* p' it is constant and is equal to the critical pressure. For higher back pressures p_2, p_m is the same as p_2. If p_2 is gradually reduced from the value p_0, the discharge amount

$$Q = A \cdot \rho_m \cdot w_m = A \cdot \left(\frac{p_2}{p_0}\right)^{\frac{1}{\kappa}} \cdot \sqrt{\frac{2 \cdot \kappa}{\kappa-1} \cdot p_0 \cdot \rho_0 \cdot \left(1 - \left(\frac{p_2}{p_0}\right)^{\frac{\kappa-1}{\kappa}}\right)} \qquad (4.163)$$

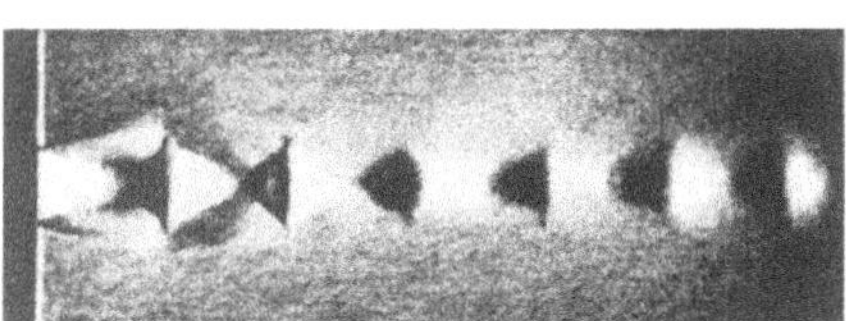

Fig. 4.106. Supersonic jet, *L. Mach* (1897)

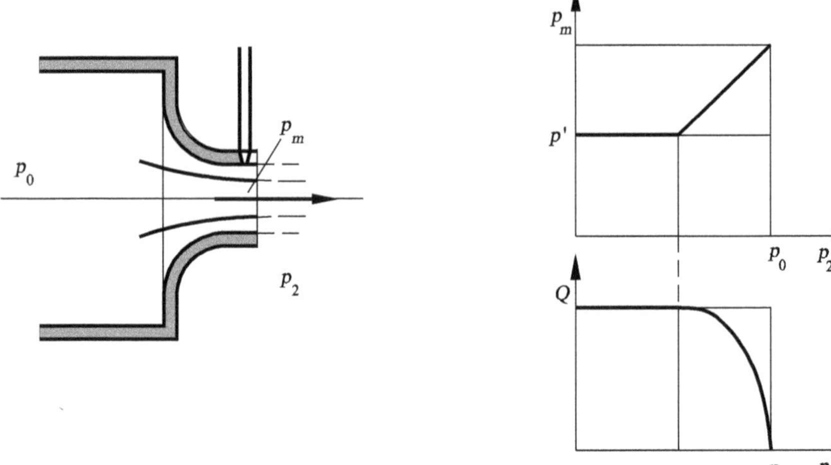

Fig. 4.107. Dependence of the discharge and the outlet pressure on p_2, measurement of the outlet pressure

increases gradually to a maximum value at the critical pressure

$$Q_{\max} = \left(\frac{2}{\kappa+1}\right)^{\frac{1}{\kappa-1}} \cdot A \cdot \sqrt{\frac{2 \cdot \kappa}{\kappa+1} \cdot p_0 \cdot \rho_0} . \tag{4.164}$$

For further reduction of p_2, $Q = Q_{\max}$ remains constant. The dependence of p_m and Q on p_2 are shown in Figure 4.107. This behavior can be understood with the pressure expansion discussed in Section 4.3.1. A chamber is connected to the end of the outlet of the nozzle in which the pressure can be regulated by means of a throttle (Figure 4.108). Let the pressure in the chamber p_2 be larger than the critical pressure p'. If p_2 is lowered by further opening of the throttle, an expansion wave moves into the nozzle and causes a new flow state. For further reduction of p_2 the velocity of sound is eventually reached in the outlet. If the pressure p_2 is further reduced, the perturbations

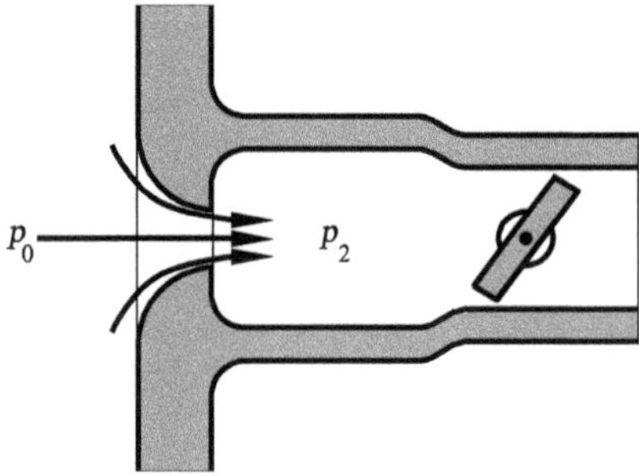

Fig. 4.108. Throttle

that expand with the velocity of sound can no longer propagate upstream into the outlet. The state there remains constant.

Laval Nozzle Flow

In order to attain regulated expansion at supercritical pressure ratios, the Swedish engineer *G. de Laval* (1883) applied the form of delivery nozzle shown in Figure 4.109 in the construction of his steam turbine. If the pressure in front of the nozzle p_0 is given, associated values of w and v/w can be determined for every lower pressure p corresponding to Figure 4.105 for the inviscid flow. With the relation $Q = A \cdot \rho \cdot w = A \cdot w/v$ for the flow, for every given value of Q we can determine the value of v/w associated with every cross-section A. Figure 4.105 can be used to determine the associated pressure. In the flow through the Laval nozzle, the minimum of the stream filament cross-section is at the same point as the minimum of the cross-section of the nozzle. At this point the delivery has a maximum and can be calculated as in the case of a simple outlet from (4.164). The pressure in the nozzle is shown as the heavy line in Figure 4.109 that leads to the lower final pressure p_{u}. Since two pressures are always associated with one value of v/w from Figure 4.105, the pressure may take a second path at the narrowest point, leading to the upper final pressure p_{o}, the outer pressure p_2.

If we determine the pressure course associated with smaller delivery amounts, we obtain the lines above p_{o}. The dependence of the delivery Q on the pressure p_2 at the end of the nozzle is shown in the right-hand diagram in Figure 4.109. The delivery grows from zero to Q_{max}. Moving down from the pressure p_{o}, the velocity of sound is reached in the narrowest cross-section, and the delivery remains constant for further reduction in the pressure.

It turns out that the flow for outer pressures between p_{o} and p_u is not loss-free. Observations by *A. Stodola* showed that in this regime discontinuous compression (compression shock waves) occur, to be treated in Section 4.3.4.

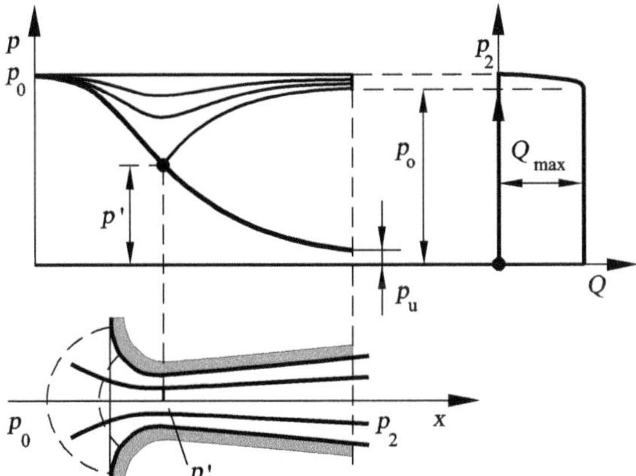

Fig. 4.109. Flow through a Laval nozzle

To do this, as well as the continuity and Bernoulli equations, we also need the *conservation of energy.*

4.3.3 Conservation of Energy

There are many different ways in which flows can be associated with losses in mechanical energy. The losses can be caused by friction, turbulence, or discontinuous processes such as shock waves. The mechanical energy that is destroyed is converted to heat energy. In the case of gases this heat energy can be of use in further expansion.

The conservation of energy needed to describe the losses will be derived in analogy to the derivation of the conservation of momentum in Section 4.1.7 for one-dimensional inviscid flows.

We consider the change in energy of a bounded part of a steady gas flow. Here we consider a part of a stream filament (Figure 4.33). The change in the bounded gas volume in the time $\mathrm{d}t$ consists of the vanishing of the mass particle $\mathrm{d}m = \rho_1 \cdot A_1 \cdot w_1 \cdot \mathrm{d}t$ at A_1 and the addition of a mass particle $\mathrm{d}m' = \rho_2 \cdot A_2 \cdot w_2 \cdot \mathrm{d}t$ at A_2. From continuity it follows that $\mathrm{d}m = \mathrm{d}m'$. As the gas mass is shifted, there is a change in the energy content that must be equal to the energy supplied from outside in the time interval $\mathrm{d}t$. The energy content of a mass particle consists of its *kinetic energy*, its *potential energy* and its *heat energy* e. If the potential energy is due only to gravity, the energy content of the mass $\mathrm{d}m$ is equal to $\mathrm{d}m \cdot (w^2/2 + g \cdot z + e)$. The energy transfer to the mass contained in the stream filament consists of the pressure work on the end surfaces and the heat transfer through the side surface. The friction work is neglected. The pressure work on the surface A_1 is $A_1 \cdot p_1 \cdot w_1 \cdot \mathrm{d}t$. With the specific volume v_1 and $\mathrm{d}m = \rho_1 \cdot A_1 \cdot w_1 \cdot \mathrm{d}t = A_1 \cdot w_1 \cdot \mathrm{d}t / v_1$ this yields $\mathrm{d}m \cdot p_1 \cdot v_1$, and for the pressure work on the surface A_2, similarly $\mathrm{d}m \cdot p_2 \cdot v_2$. The heat transfer between A_1 and A_2 is denoted by $q_{1,2} \cdot \mathrm{d}m$. The change in the energy content is therefore

$$\mathrm{d}m \cdot \left(\frac{w_2^2}{2} + g \cdot z_2 + e_2\right) - \mathrm{d}m \cdot \left(\frac{w_1^2}{2} + g \cdot z_1 + e_1\right) = \mathrm{d}m \cdot (p_1 \cdot v_1 - p_2 \cdot v_2 + q_{1,2}).$$

This yields

$$\frac{w_2^2}{2} + g \cdot z_2 + e_2 + p_2 \cdot v_2 = \frac{w_1^2}{2} + g \cdot z_1 + e_1 + p_1 \cdot v_1 + q_{1,2}$$

or at an arbitrary position of the end cross-section

$$\frac{w^2}{2} + g \cdot z + e + p \cdot v = \mathrm{const} + q. \tag{4.165}$$

In differential form we obtain the equation

$$w \cdot \mathrm{d}w + g \cdot \mathrm{d}z + \mathrm{d}e + \mathrm{d}(p \cdot v) = \mathrm{d}q, \tag{4.166}$$

where $e + p \cdot v$ is the *enthalpy* h. For ideal gases with constant specific heat we have

$$e = \frac{1}{\kappa - 1} \cdot p \cdot v = c_v \cdot T,$$

$$h = e + p \cdot v = \frac{\kappa}{\kappa - 1} \cdot p \cdot v = c_p \cdot T,$$

where c_v and c_p are the specific heats at constant volume and pressure, respectively. For a steady flow without heat transfer the total energy remains constant, because the friction energy present is wholly changed into heat. Gravity may be neglected for stratification flows, so that the energy equation takes on the following form:

$$h + \frac{w^2}{2} = \text{const.} \tag{4.167}$$

According to the first law of thermodynamics, for every mass element of the gas, the heat supplied through heat conduction and the friction work transformed into heat are used to raise the internal energy and to carry out expansion work. The friction work $\mathrm{d}W_R$ done on a mass element satisfies

$$\mathrm{d}q + \mathrm{d}W_R = \mathrm{d}e + p \cdot \mathrm{d}v. \tag{4.168}$$

Adding (4.168) and (4.166), and using $\mathrm{d}(p \cdot v) = p \cdot \mathrm{d}v + v \cdot \mathrm{d}p$, we obtain

$$w \cdot \mathrm{d}w + g \cdot \mathrm{d}z + v \cdot \mathrm{d}p + \mathrm{d}W_R = 0. \tag{4.169}$$

After integration we obtain the Bernoulli equation extended by the friction term W_R:

$$\frac{w^2}{2} + g \cdot z + \int v \cdot \mathrm{d}p + W_R = \text{const.} \tag{4.170}$$

4.3.4 Theory of Normal Shock Waves

In a parallel flow of velocity w_1 and pressure p_1, the specific volume v_1 is compressed discontinuously to the smaller specific volume v_2 by means of a steady *normal shock wave* in the plane AA (Figure 4.110), with reduction of the velocity to w_2 and increase of the pressure to p_2. The following equations hold for the change of the state quantities and the velocity across the normal shock wave:

Continuity equation:

$$m = \frac{w_1}{v_1} = \frac{w_2}{v_2}, \tag{4.171}$$

Momentum equation:

$$m \cdot (w_1 - w_2) = p_2 - p_1, \tag{4.172}$$

Energy equation (without heat transfer q):

$$\frac{w_1^2}{2} + h_1 = \frac{w_2^2}{2} + h_2, \tag{4.173}$$

where m is the mass flux per unit area. The enthalpy h is a function of p and v. With (4.171), w_1 and w_2 can be eliminated in (4.172). This yields $p_2 - p_1 = (v_1 - v_2) \cdot m^2$. With the energy equation (4.173) we obtain

$$(p_2 - p_1) \cdot \frac{v_1 + v_2}{2} = h_2 - h_1.$$

This dependence of p_2 on v_2 for given p_1 and v_1 is called the *Hugoniot curve.*

If three state quantities, e.g. p_1, v_1, and p_2, are given, we can then determine the fourth, v_2. Thus we obtain m and also the velocities w_1 and w_2. The normal shock wave satisfies

$$w_1 \cdot w_2 = c'^2,$$

with the critical velocity of sound c' of the approaching flow. One of the velocities w_1 and w_2 is larger than the velocity of sound c', while the other is smaller. Theoretically, (compression) shock waves and discontinuous expansion processes are both possible. However, it is only the compression shock wave in which the entropy increases, and so, according to the second law of thermodynamics, only this is physically possible.

Equations (4.171) to (4.173) for the steady shock wave can also be applied to an unsteady compression wave by changing the frame of reference. If we superimpose the velocity w_1 onto the flowing fluid in Figure 4.110, the velocity of the shock plane becomes zero. The shock moves with velocity $U = w_1$ to the left, and the gas behind the shock follows with velocity $w = w_1 - w_2$.

The momentum equation for unsteady shock motion yields $p_2 - p_1 = \rho \cdot U \cdot w$. The velocity of propagation U of the shock is always greater than the velocity of sound and can become arbitrarily large for arbitrarily large pressure differences. Such large propagation velocities can be observed in explosions.

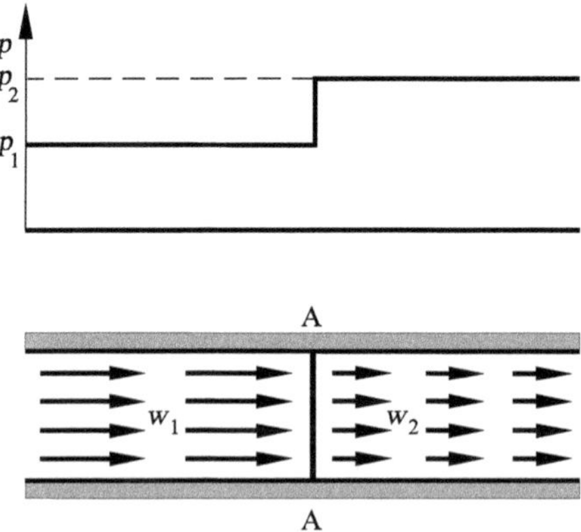

Fig. 4.110. Normal shock wave

In shock waves, the quantity $(w_1^2 - w_2^2)/2$ contained in $h_1 - h_2$ leads to increases in the heat content. For a curved shock wave, such as the head waves in Figures 4.103 or 4.111, the different stream filaments experience different heating, so that the gas mass behind the flow is no longer homogeneous and therefore no longer irrotational.

Shock in Front of Blunt Bodies

In the supersonic flow past blunt bodies a steady shock appears in front of the body (Figure 4.111). It can be calculated close to the stagnation streamline using the equations of the normal shock wave. The pressure jump across the shock propagates sideways as an oblique shock wave. With increasing distance from the body the pressure increase in the shock wave becomes less, and the oblique shock wave passes over to a normal conical wave. At large velocities the shock lies close to the body, while at lower free-stream velocities the distance to the shock becomes larger.

The flow portrait looks similar for a body moving with supersonic velocity. The shock wave is heard as the sonic boom of supersonic aircraft or of the bullet in Figure 4.103. The pressure increase at the stagnation point S is proportional to the square of the velocity for both large and small velocities:

$$p_s - p_\infty = \frac{\rho_\infty \cdot w_\infty^2}{2} \cdot c_p .$$

The pressure coefficient c_p is a function of the Mach number. The pressure increase consists of two parts, a continuous part behind the shock and a discontinuous part across the shock (shock part). For comparison, we consider the pressure coefficient c_{p_0} of an imaginary isentropic (loss free) deceleration of the flow up to the stagnation point. The dependence of the values of c_p of the shock part and of c_{p_0} on the Mach number can be read from the following table:

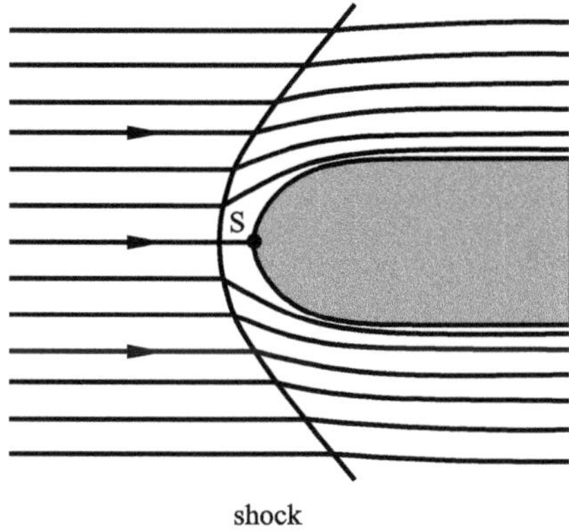

Fig. 4.111. Shock

$M = w/c$	0	0.5	1.0	1.5	2	3	∞
c_p	1	1.065	1.275	1.53	1.655	1.75	1.85
shock part	-	-	0	0.92	1.25	1.48	1.65
c_{p0}	1	1.065	1.275	1.69	2.48	4.85	∞

We draw an analogy to the behavior of the stagnation pressure and note that the drag at even very large velocities is again proportional to the square of the velocity.

Normal Shock in the Laval Nozzle

If the outer pressure p_2 at the end of a Laval nozzle (Figure 4.109) is between p_o and p_u, a normal shock occurs in the nozzle. This leads from supersonic flow to subsonic flow. The pressure distributions in Figure 4.109 with the same mass fluxes and the same total energy can be extended, and they are shown in Figure 4.112. The transition of the normal pressure distribution from p_1 to p_u to the pressure distribution depending on the outer pressure p_2 is caused by the shock. The position of the shock is uniquely determined by the momentum equation. In fact, the processes inside the Laval nozzle are more complicated. Instead of the normal shock, shock branching with oblique shocks can also occur. The sudden pressure increase that occurs when the shocks interact with the wall boundary layer can lead to flow separation.

Figure 4.113 shows schlieren photographs by *L. Prandtl* (1907) for different pressure ratios at the end of the nozzle. The first pictures shows the unperturbed acceleration of compressed air from an initial pressure $p_0 = 7$ bar to atmospheric pressure. The nozzle walls have been roughened so that the crossed perturbations (characteristic lines) of steady sound waves are made visible in the supersonic part of the nozzle. The second picture shows the density through the nozzle when the velocity of sound is not attained. The

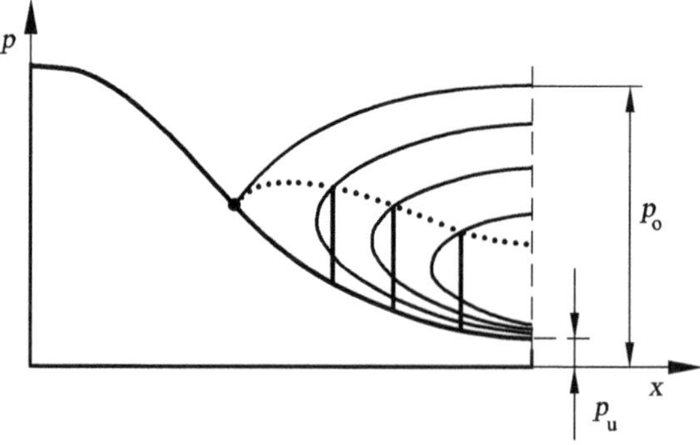

Fig. 4.112. Pressure distribution in a Laval nozzle with shock waves

density drops up to the narrowest cross-section, and then increases again. Even for this roughened wall, there are no perturbations seen in the flow field. The third picture shows a shock wave downstream of the narrowest cross-section of the nozzle. The steady sound waves in the supersonic part in front of the shock and the continued increase in density of the decelerated subsonic flow are seen. If the outer pressure p_2 is further reduced, the shock wave moves toward the end of the nozzle. Because of the interaction with the wall boundary layer, shock branching and separation of the boundary layer occur, as seen in the fourth picture of Figure 4.113.

4.3.5 Flows past Corners, Free Jets

Supersonic Flow past a Corner

We first consider a supersonic flow in which a small pressure drop occurs discontinuously at a point A of the wall (Figure 4.114). This pressure drop propagates with the Mach angle α and leads to an acceleration of the flow in the direction normal to the pressure jump. This causes an increase in the flow velocity and a simultaneous deviation of the flow. If a further continuous pressure drop occurs at point A, this propagates in the altered flow with a

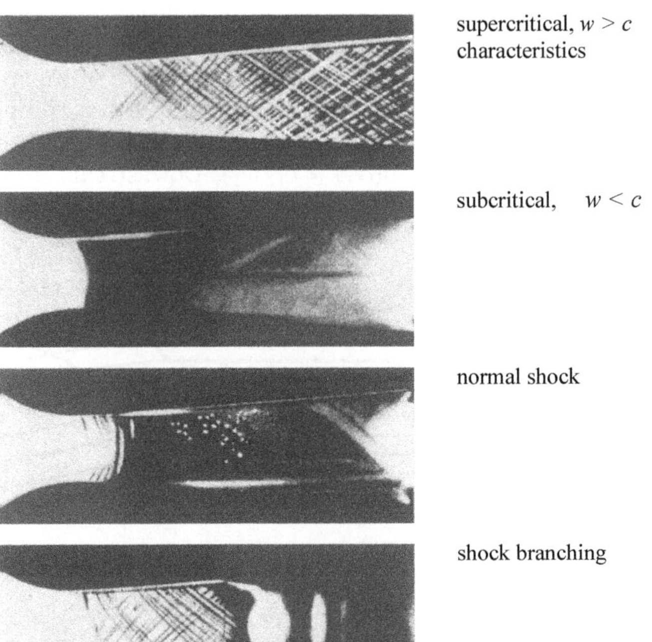

Fig. 4.113. Schlieren photograph of Laval nozzle flows, *L. Prandtl* (1907)

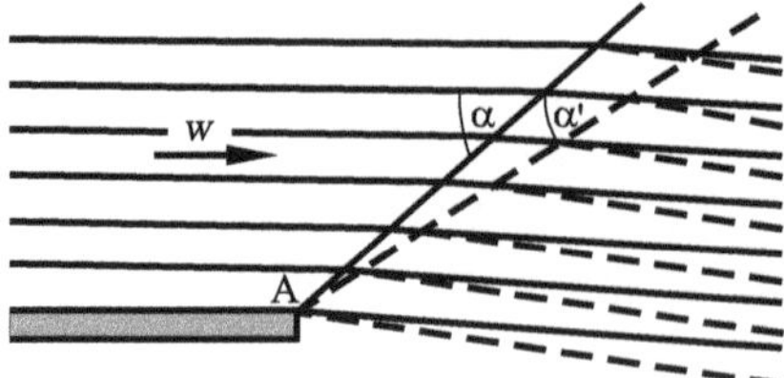

Fig. 4.114. Supersonic flow with pressure drop

different Mach angle $\alpha' < \alpha$ and causes a further increase and deviation of the flow velocity.

This *Prandtl–Meyer expansion*, which in reality takes place continuously, can be theoretically treated as a potential flow. Along any ray (characteristic) originating from point A, the pressure as well as the magnitude and the direction of the velocity are constant. Each characteristic forms the Mach angle with the flow direction. The velocity component perpendicular to the characteristic is equal to the velocity of sound associated with the flow state at hand.

The course of the expansion flow from the velocity of sound to the maximum velocity (expansion into the vacuum) is shown in Figure 4.115. The deviation of the ray is 129°. Since the characteristics are rays along which the pressure and velocity are constant, sections of the flow enclosed by two characteristic free jets can be combined with straight-lined flows. If, for example, a supersonic flow moves parallel to a wall with velocity w_1, and the pressure p_2 after the end of the wall (A in the right-hand picture of Figure 4.115) is smaller than the pressure p_1 in the parallel flow, the flow propagates in an unchanged fashion as far as characteristic 1, which forms the Mach angle α_1 with the flow direction ($\sin(\alpha_1) = c_1/w_1$). Downstream from this characteristic, an expansion between characteristics 1 and 2 leads from

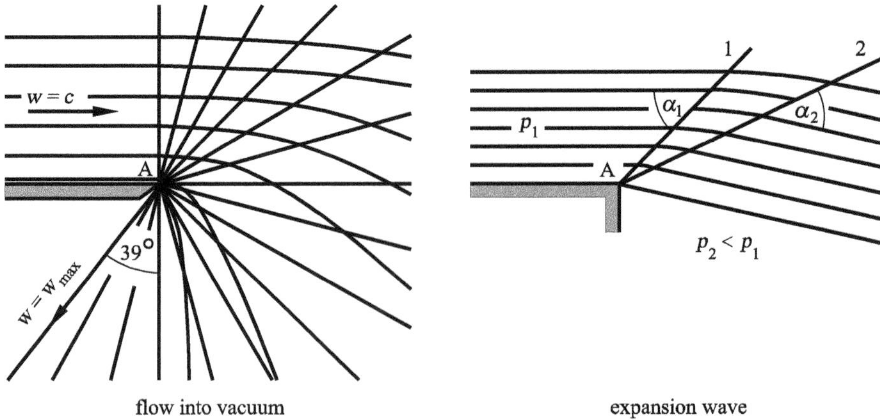

Fig. 4.115. Prandtl–Meyer corner flow

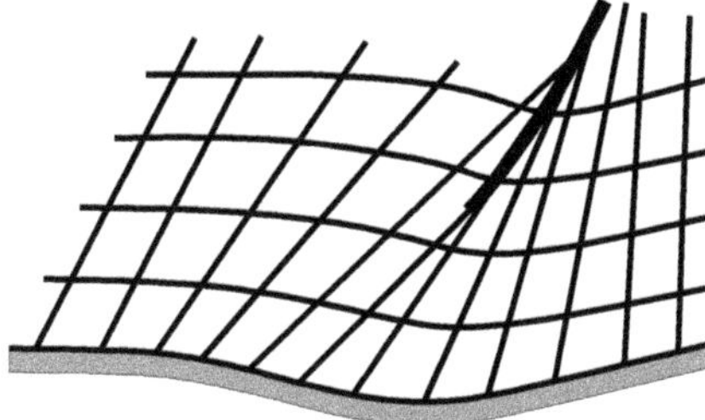

Fig. 4.116. Flow along a curved wall

pressure p_1 to p_2. After the pressure p_2 has been attained on characteristic 2, the flow continues uniformly and in a straight line. The flow direction forms the angle α_2 with characteristic 2, with which w_2 is associated.

If a wall with one or more convex corners is at hand, the flow here also takes place as a combination of linear flows and expansion regimes that always adjoin each other at the Mach angle. Even the flow along a continuously curved wall can be represented as a composition of individual elements. The wall can also be concavely curved for this approach. However, in this case, the solution of the potential equation is correct only as long as the jets forming each Mach angle do not intersect (Figure 4.116). If this does occur, the flow at this position becomes discontinuous, and a shock ensues.

In the case of a concave corner, associated with a pressure rise, as well as in the case of outgoing flow into a space with higher pressure, the flow always becomes unsteady. *Oblique shocks* (Figure 4.117) form. The characteristic 2 in Figure 4.117 would lie upstream from characteristic 1, and this is not possible. Instead, unsteady compression takes places, with the shock plane lying between directions 1 and 2. The equations for the velocity components perpendicular to the shock plane are the same as those for the normal shock wave in Section 4.3.4. The remaining transversal velocity component that is unchanged by the shock is simply superimposed. The three upper schlieren photographs in Figure 4.119 are examples of this theoretical superposition for the corner flow with expansion, or with oblique shocks at the outlet from a nozzle.

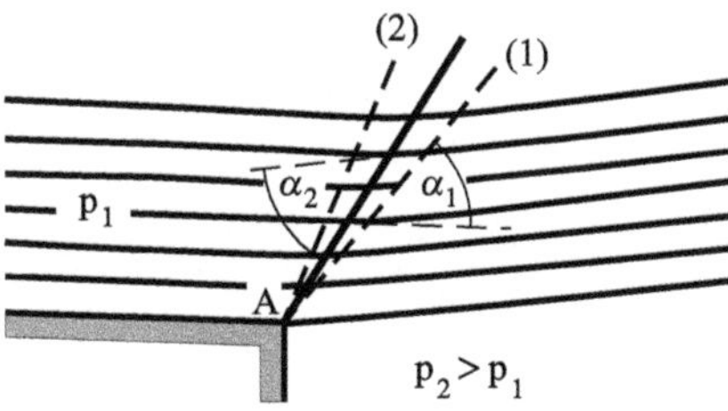

Fig. 4.117. Oblique shock wave

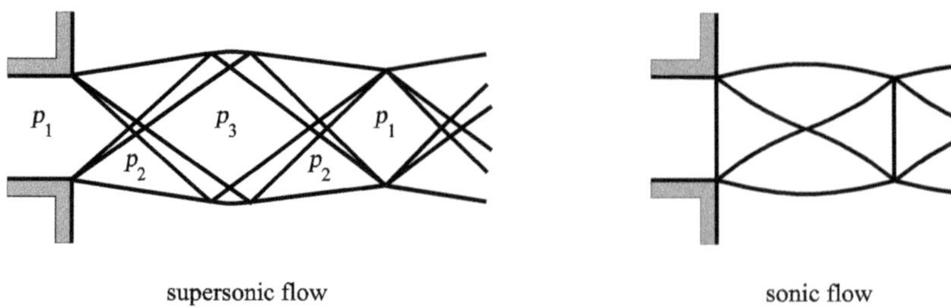

Fig. 4.118. Wave figures of free jets

Free Jets

In a supersonic free jet, a periodic structure of shock waves and expansion waves forms. The oblique expansion and compression waves penetrate each other without mutual disturbance. They are completely reflected at the boundaries of the free jet in such a manner that an expansion wave is reflected as a compression wave and vice versa. For a parallel supersonic flow where the pressure at the outlet is lower than in the jet, an expansion wave arises at every outlet edge, as seen in the left-hand picture of Figure 4.118. These intersect and are reflected as compression waves at the opposite jet boundaries. They propagate downstream and are again reflected at the opposite jet boundary as expansion waves. This process repeats itself periodically. The pressure p_3 in the middle of the wave is lower than the outer pressure p_2, by a similar amount as p_1 is larger than p_2.

If the outer pressure is larger than that in the jet, oblique shock waves initially occur (Figure 4.118). These are reflected at the edge of the jet as expansion waves, which then propagate as shown in the first picture of Figure 4.119. If the initial velocity is equal to the velocity of sound, the Mach angle at the outlet is $\alpha = 90°$, and the characteristic node structure with normal shock

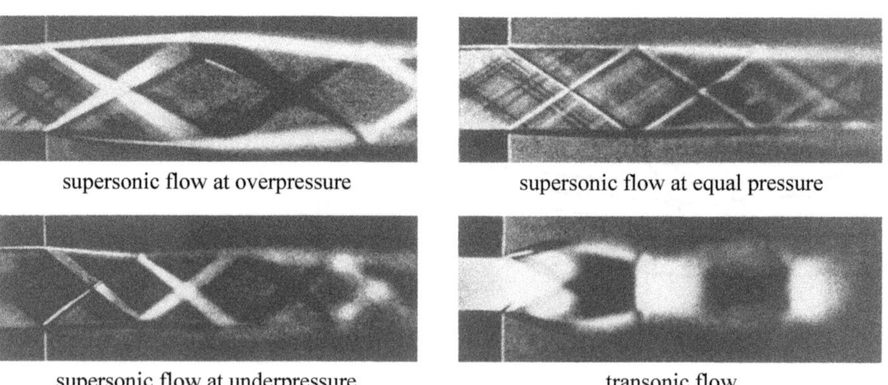

Fig. 4.119. Free jet structures at different outlet conditions, *L. Prandtl* (1907)

waves shown in the right-hand picture of Figure 4.118 occurs. The schlieren photographs of Figure 4.119 show the free jet structures for overpressure, equal pressure, and underpressure at the outlet. The outlet velocity is the same for the first three photographs. The fourth picture shows the case in which the outlet velocity is the same as the velocity of sound. In all these schlieren photographs, the light areas indicate expansion and the dark areas compression. If the jet is not a parallel jet as it departs the outlet, the wave pictures become more complex. The wavelength remains almost constant. Using (4.155), for two-dimensional motion it is

$$\lambda = 2 \cdot d_m \cdot \cot(\alpha_m) = 2 \cdot d_m \cdot \sqrt{\left(\frac{w}{c}\right)_m^2 - 1}.$$

Here d_m is the mean jet diameter, while α_m and $(w/c)_m$ are the mean values of α and w/c.

The node structure in round free jets, more complicated because of the conical intersection of the waves, is shown in Figure 4.106. The wavelength in these free jets, where the outlet velocity is equal to the velocity of sound, was determined experimentally by *R. Emden* for compressed air to be

$$\lambda = 0.89 \cdot d \cdot \sqrt{\frac{p_0 - 1.9 \cdot p_2}{p_2}}.$$

Here d is the diameter of the outlet, p_0 the reservoir pressure, and p_2 the outlet pressure.

4.3.6 Flows with Small Perturbations

In this section we treat inviscid steady flows in which both the magnitude and the direction of the velocity deviate only slightly from a given velocity u_0, which may be a subsonic or a supersonic velocity. The small deviations of the velocity from u_0 are denoted by u and v. All derivations are carried out only to first order in u and v. The magnitude of the velocity of the flow is $w = \sqrt{(u_0 + u)^2 + v^2}$.

The generalized *Bernoulli equation* (4.158) is valid:

$$\int \frac{\mathrm{d}p}{\rho} + \frac{w^2}{2} = \text{const.}$$

or else in differential form,

$$\frac{\mathrm{d}p}{\rho} + w \cdot \mathrm{d}w = 0.$$

With $\mathrm{d}p/\rho = (\mathrm{d}p/\mathrm{d}\rho) \cdot \mathrm{d}\rho/\rho = c^2 \cdot \mathrm{d}\rho/\rho$ we obtain the equation

$$\frac{\mathrm{d}\rho}{\rho} = -\frac{w^2}{c^2} \cdot \frac{\mathrm{d}w}{w} = -M^2 \cdot \frac{\mathrm{d}w}{w}, \tag{4.174}$$

which is valid for all flows with a unique *Bernoulli* constant, i.e. for irrotational flows. The relative change of the density $\mathrm{d}\rho/\rho$ vanishes for small Mach numbers. For Mach numbers smaller than 0.2 we can therefore calculate the flow as an incompressible flow. At $M = 1$ the relative change of the density is exactly the opposite of the relative change of the velocity. This means that $\rho \cdot w$ is approximately constant, or a constant stream filament cross-section.

The continuity equation yields

$$\frac{\partial}{\partial x}(\rho \cdot (u_0 + u)) + \frac{\partial}{\partial y}(\rho \cdot v) = 0.$$

Since $w^2 = (u_0 + u)^2 + v^2$, the velocity perturbation is given by the u perturbation to first order, and linearization of the continuity condition with (4.174) yields

$$(1 - M_0^2) \cdot \frac{\partial u}{\partial x} + \frac{\partial v}{\partial y} = 0. \tag{4.175}$$

This is the *linear gas-dynamic equation* with Mach number $M_0 = u_0/c_0$. This equation is not valid for the transonic regime $M \approx 1$, where the perturbations are no longer small and where linearization is not possible. Introducing a perturbation position φ (see Section 4.1.5), we set

$$u = \frac{\partial \varphi}{\partial x} \quad \text{and} \quad v = \frac{\partial \varphi}{\partial y},$$

and obtain from (4.175)

$$(1 - M_0^2) \cdot \frac{\partial^2 \varphi}{\partial x^2} + \frac{\partial^2 \varphi}{\partial y^2} = 0. \tag{4.176}$$

The factor in front of $\partial^2\varphi/\partial x^2$ alters its sign at $M_0 = 1$. For subsonic Mach numbers $M_0 < 1$, the differential equation, like the potential equation, is of *elliptical type*. For supersonic Mach numbers it has the form of the vibration differential equation; i.e. it is *hyperbolic*. For $M_0 > 1$ every continuous and twice-differentiable function F with argument $(y \pm x \cdot \tan(\overline{\alpha}))$ is a solution of (4.176), where $\overline{\alpha}$ is to be suitably determined. We obtain

$$\frac{\partial^2 \varphi}{\partial x^2} = \mathrm{F}'' \cdot \tan^2(\overline{\alpha}) \quad \text{and} \quad \frac{\partial^2 \varphi}{\partial y^2} = \mathrm{F}''.$$

In order to satisfy (4.176),

$$(M_0^2 - 1) \cdot \tan^2(\overline{\alpha}) = 1$$

must hold, i.e.

$$\tan(\overline{\alpha}) = \pm \frac{1}{\sqrt{M_0^2 - 1}}.$$

This yields

$$\sin(\overline{\alpha}) = \frac{\tan(\overline{\alpha})}{\sqrt{1 + \tan^2(\overline{\alpha})}} = \pm \frac{1}{M_0}.$$

The solution represents waves of arbitrary wave form, whose straight fronts ($y = \pm x \cdot \tan(\overline{\alpha}) + \text{const}$) are inclined to the left or the right toward the x axis in the entire flow field with the constant Mach angle $\overline{\alpha}$.

For *subsonic flows* we obtain characteristic solutions of the following form: The compressible flow with weak perturbations is compared with the corresponding incompressible flow under the same conditions. The small deviations of the velocity of u_0 of the incompressible flow are denoted by U and V and the associated coordinates by X and Y. According to Section 4.1.5, the incompressible flow with the associated potential Φ must satisfy the potential equation

$$\frac{\partial^2 \Phi}{\partial X^2} + \frac{\partial^2 \Phi}{\partial Y^2} = 0. \tag{4.177}$$

Comparison with the compressible flow is carried out by setting the potentials φ and Φ proportional to one another:

$$\varphi(x, y) = \mathrm{a} \cdot \Phi(X, Y), \tag{4.178}$$

where a is a numerical factor.

So that both φ can satisfy the differential equation (4.176) and Φ the equation (4.177), the ratios of x to X and y to Y must be different. If $Y/y = \mathrm{b} \cdot X/x$ is set to scale with the factor b, a suitable choice of b allows us to obtain the association of the potentials according to (4.178). For simplicity we arbitrarily set $x = X$, so that $Y = \mathrm{b} \cdot y$. With this relation and with (4.178), (4.176) leads to

$$\mathrm{a} \cdot \frac{\partial^2 \Phi}{\partial X^2} \cdot (1 - M_0^2) + \mathrm{a} \cdot \mathrm{b}^2 \cdot \frac{\partial^2 \Phi}{\partial Y^2} = 0. \tag{4.179}$$

This equation becomes identical to (4.177) if we set $\mathrm{b}^2 = 1 - M_0^2$.

The angle δ that forms a streamline with the x axis satisfies

$$\tan(\delta) = \frac{v}{u_0 + u}.$$

To first order this is also $\tan(\delta) = v/u_0 = (1/u_0) \cdot \partial\varphi/\partial y$. Similarly, we obtain the angle Δ between the streamline and the X axis of the incompressible flow:

$$\tan(\Delta) = \frac{V}{u_0} = \frac{1}{u_0} \cdot \frac{\partial \Phi}{\partial Y}.$$

If the same body is placed in both flows, $\tan(\delta) = \tan(\Delta)$ must be satisfied on the bounding streamline. This yields $\partial\varphi/\partial y = \partial\Phi/\partial Y$. With (4.178) and $Y = \mathrm{b} \cdot y$ we obtain $\mathrm{a} \cdot \mathrm{b} = 1$, i.e. the condition

$$\mathrm{a} = \frac{1}{\mathrm{b}} = \frac{1}{\sqrt{1 - M_0^2}}. \tag{4.180}$$

To compare the pressure distributions of both flows we merely need to consider the pressure gradient in the x direction. The finite pressure differences in both flows behave like their gradients. From the nonlinear term of the Euler equation $\rho \cdot (u_0 + u) \cdot \partial u / \partial x$, to first order we have $\rho \cdot u_0 \cdot \partial u / \partial x = \rho \cdot u_0 \cdot \partial^2 \varphi / \partial x^2$. This term is to be compared with the term $\rho \cdot u_0 \cdot \partial^2 \Phi / \partial X^2$ of the incompressible flow. The ratio is a. To first order in the Euler equation we have $\partial p / \partial x = -\rho \cdot u_0 \cdot \partial u / \partial x$. This means that the pressure differences of the compressible flow are, to first order, $1/\sqrt{1 - M_0^2}$ times larger than those in the incompressible comparison flow.

Flow past an Airfoil

This relation can be applied approximately for narrow wings at a small angle of attack, as long as the velocity of sound is not reached on the wing (Figure 4.120). The lift for compressible flow past a wing is in the same ratio as that in (4.180) compared to that for the incompressible flow (*Prandtl's rule*).

The question as to the value of a in (4.178) can also be formulated differently. How must a body be shaped so that the pressure differences in the compressible flow and in the incompressible comparison flow are equally large? This question is of importance for the case in which the pressure distribution for the incompressible comparison flow is at the limit of flow separation. In this case, $\mathrm{a} = 1$ must be selected. Then $\tan(\delta) = \mathrm{b} \cdot \tan(\Delta)$. The body in the compressible flow must be narrower, the closer u_0 approaches the velocity of sound if separation of the flow is to be avoided.

Wavy Wall

A flow with mean velocity u_0 flows along a slightly wavy wall. The contour of the wall is given by the equation

$$y_1 = a \cdot \sin(\mu \cdot x), \quad \text{with} \quad \mu = \frac{2 \cdot \pi}{\lambda}.$$

Here λ is the wavelength. From $v/u_0 = \mathrm{d}y_1/\mathrm{d}x$ and close to $y = 0$ we obtain

$$v_0 = u_0 \cdot a \cdot \mu \cdot \cos(\mu \cdot x).$$

In the incompressible comparison fluid, $V_0 = v_0$ at $Y = 0$. The associated potential is

$$\Phi = -u_0 \cdot a \cdot \cos(\mu \cdot X) \cdot e^{-\mu \cdot Y}.$$

This corresponds to the following potential in the incompressible flow:

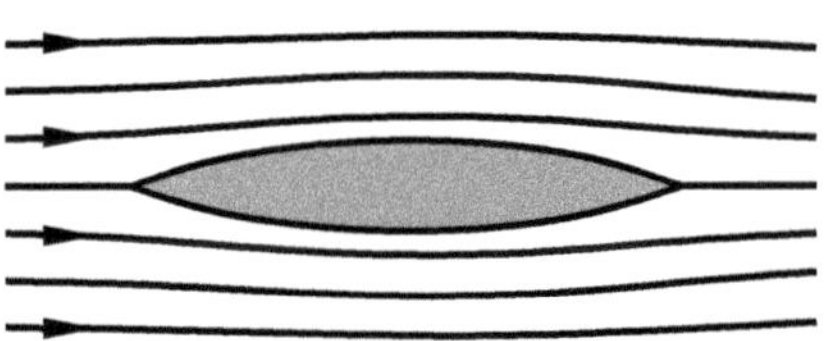

Fig. 4.120. Flow past a slender airfoil

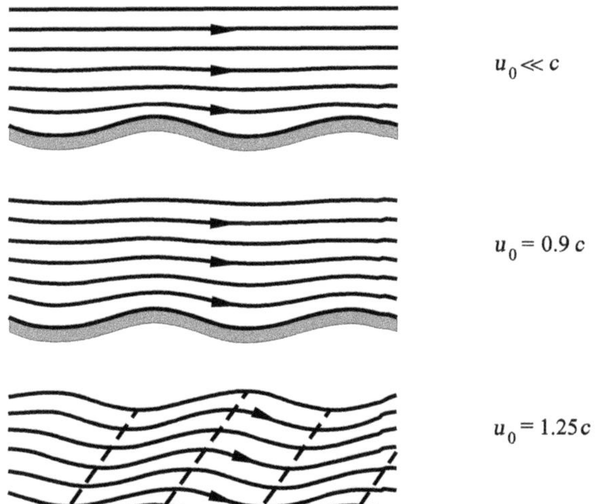

Fig. 4.121. Flow past a wavy wall

$$\varphi = -A \cdot \cos(\mu \cdot x) \cdot e^{-\mu \cdot y \cdot \sqrt{1-M_0^2}}. \tag{4.181}$$

For $y = 0$ we thus obtain $v_0 = \partial\varphi/\partial y = A\cdot\mu\cdot\sqrt{1-M_0^2}\cdot\cos(\mu\cdot x)$. Comparison with the incompressible flow leads to $A = u_0\cdot a/\sqrt{1-M_0^2}$. Figure 4.121 shows the flow past a wavy wall for incompressible flow ($u_0 \ll c$), compressible subsonic flow ($u_0 = 0.9 \cdot c$), and compressible supersonic flow ($u_0 = 1.25 \cdot c$).

4.3.7 Flows past Airfoils

Supersonic Flow

For airfoils that are sufficiently slender and peaked, the characteristic method can also be applied to two-dimensional *supersonic flow past airfoils.* The pressure on every surface element of the airfoil is given by the free-stream velocity and the inclination of the surface element, neglecting small losses of the front shock wave. For the airfoil shown in Figure 4.122, an oblique shock wave (*head wave*) occurs at the tip of the airfoil, generating an overpressure. The convex curvature of the surface of the airfoil causes expansion waves to be formed, through which the overpressure is reduced, until an underpressure occurs at the rear part of the airfoil. The two flows along the upper and lower sides meet at similar angles at the trailing edge of the airfoil. This leads to a further shock wave (*trail wave*). After this, the pressure is approximately the same as the unperturbed pressure of the free stream. The expansion waves travel divergently. The waves originating from the front part of the airfoil meet at the front shock wave, while those originating from the rear part meet at the trail wave. The strength of these shock waves therefore becomes gradually weaker in the flow field. This theoretically determined image is confirmed by the schlieren photograph in Figure 4.122. The schlieren aperture was positioned so that lighter areas indicate an increase in density, while darker areas imply a reduction in density.

In order to investigate the effect of the independence of the thickness of the airfoil of the incidence, we first consider a thin plate inclined at an angle, as shown in Figure 4.123.

A shock wave occurs on the pressure side, while an expansion wave occurs on the suction side. Both deviate the flow direction by the angle α, the angle of inclination of the plate toward the outgoing flow direction. As long as the direction of the flow remains constant, the pressure and the velocity do not change.The resulting force therefore acts on the center of the plate. At the trailing edge the pressure is equalized, leading to a shock wave on the suction side and an expansion wave on the pressure side. The resulting force at small angles of inclination is approximately proportional to the angle of inclination α and, for inviscid flow, is exactly perpendicular to the plate. The equivalent of the lift is included in the transverse velocities generated in both waves. The transverse velocity becomes smaller after a certain distance, because the expansion waves join with the shock waves. However, its magnitude in the perpendicular direction increases to the same degree, so that at every cross-section perpendicular to the direction of flow behind the plate, the lift is still imparted as momentum.

A thin plate, peaked at the front and the rear, possibly with a slightly arched suction side, is the most favorable airfoil profile for supersonic flows.

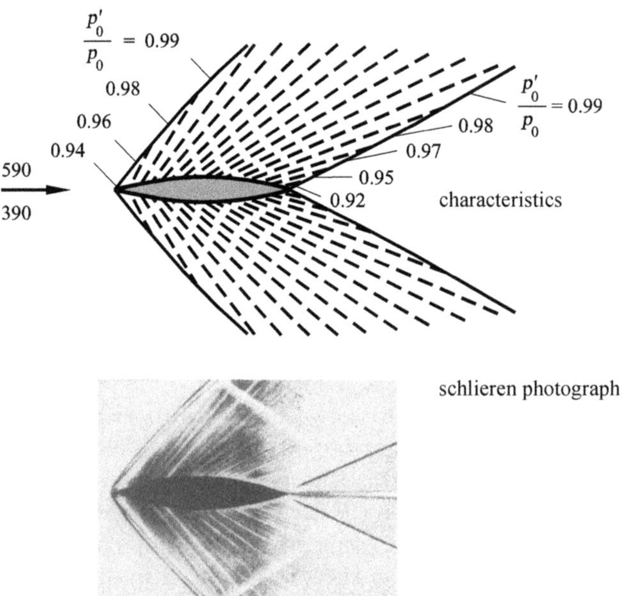

Fig. 4.122. Supersonic flow at a slender airfoil, p_0 reservoir pressure, p_0' reservoir pressure after the shock wave

The usual wing profiles that are thick at the front are not suitable for supersonic flows because of their large drag. The best ratio of drag to lift W/A is therefore, in contrast to subsonic flows, never smaller than $\tan(\alpha)$.

Approximate solutions of the supersonic flow past the profiles in Figures 4.122 and 4.123 can be determined with the differential equation (4.176). Each potential $\varphi = \mathrm{F}(x - y \cdot \cot(\alpha'))$ yields a possible perturbation flow to the basic flow u_0. Here α' is the Mach angle of the free stream. If F' is the derivative of the potential with respect to the argument $x - y \cdot \cot(\alpha')$, we obtain the perturbation components $u = \varphi_x = \mathrm{F}'$ and $v = \varphi_y = -\mathrm{F}' \cdot \cot(\alpha')$, or

$$u = -\frac{v}{\cot(\alpha')}. \tag{4.182}$$

Since the flow angle is approximately given by $\tan(\delta) = v/u_0$, and the pressure differences are proportional to u, the pressure coefficient is c_p:

$$c_p = \frac{p - p_0}{\frac{1}{2} \cdot \rho_0 \cdot u_0^2} = 2 \cdot \tan(\delta) \cdot \tan(\alpha'). \tag{4.183}$$

An overpressure occurs for airfoils with a positive angle of attack, while at negative angles of attack an underpressure occurs. Therefore, a wing in an inviscid supersonic flow also has a drag.

In order to obtain dimensionless coefficients of the forces, we divide the forces by the product of pressure and surface area. The stagnation pressure $\rho_0 \cdot u_0^2/2$ is used for the pressure. At higher Mach numbers the stagnation pressure $\rho_0 \cdot u_0^2/2$ is half the oncoming momentum associated with the pressure rise in the head wave. The surface A is chosen to be the largest projection surface area of the profile. Therefore we set

$$F_A = c_a \cdot A \cdot \frac{\rho_0 \cdot u_0^2}{2}, \qquad W = c_w \cdot A \cdot \frac{\rho_0 \cdot u_0^2}{2}. \tag{4.184}$$

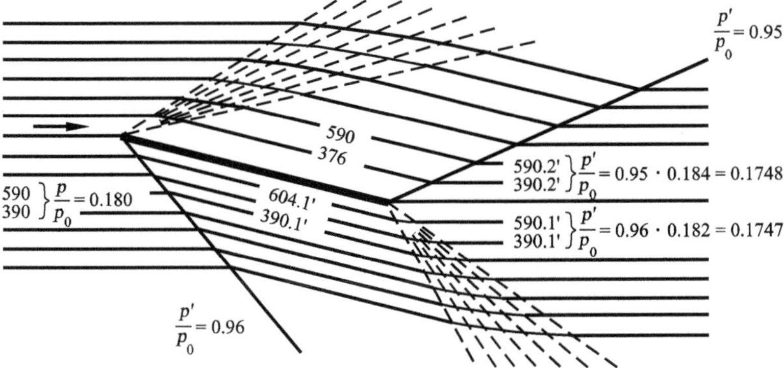

Fig. 4.123. Supersonic flow at an inclined plate, numbers by *A. Busemann* (1931)

The plate at an angle of incidence in supersonic flow in Figure 4.123 has a constant underpressure on the upper side and a constant overpressure on the lower side. The quantities c_a on the pressure and suction sides each correspond to (4.183) if the flow angle δ is replaced by the angle of attack α:

$$c_a = \frac{4 \cdot \alpha}{\sqrt{M_0^2 - 1}} . \tag{4.185}$$

Since the tangential force in the supersonic regime ($M_0 > 1$) vanishes, the drag coefficient is

$$c_w = c_a \cdot \tan(\alpha) = \frac{4 \cdot \alpha^2}{\sqrt{M_0^2 - 1}} . \tag{4.186}$$

Equations (4.185) and (4.186) were first presented by *J. Ackeret* (1925).

Transonic Flow

The balance of energy (4.167) can be used to derive the following exact relation for ideal gases of constant specific heat, after introducing the velocity of sound instead of the temperature in the enthalpy:

$$\frac{1}{M^2} - 1 = \frac{\kappa + 1}{2} \cdot \left(\frac{c'^2}{w^2} - 1 \right) . \tag{4.187}$$

Here c' is the critical velocity of sound. If w is only slightly different from c', as in the case of flow close to the speed of sound $M \approx 1$, then the perturbation component u approximately satisfies

$$\begin{aligned} 1 - M^2 &= (\kappa + 1) \cdot \left(\frac{c'}{u_0 + u} - 1 \right) + \cdots \\ &= (\kappa + 1) \cdot \left(1 - \frac{u_0 + u}{c'} \right) + \cdots . \end{aligned} \tag{4.188}$$

The difference $1 - M^2$ is therefore proportional to $c' - (u_0 + u)$.

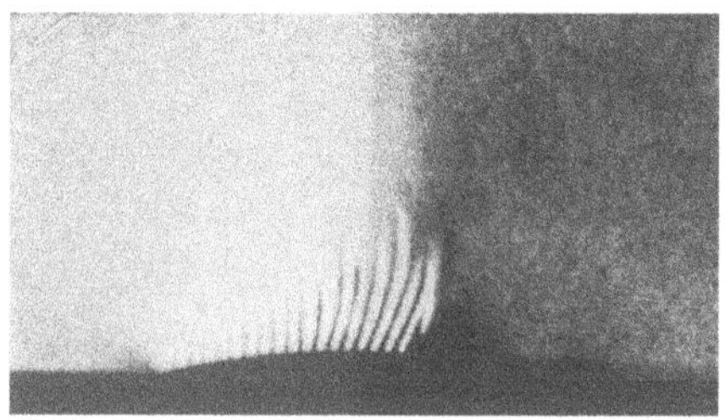

static pressure 1.6 *bar*

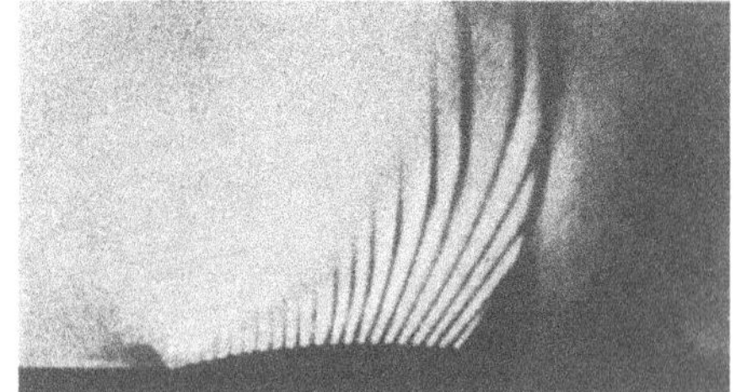

static pressure 1.89 *bar*

Fig. 4.124. Local supersonic regions

As we approach $M_0 = 1$ from a supersonic free-stream regime, the shock separates from the tip of the airfoil and moves upstream away from the body. The trail wave, on the other hand, remains in the supersonic region at the end of the airfoil. The closer M_0 is to 1, the weaker the head wave, until at $M_0 = 1$ it finally vanishes. The pressure distribution now has a subsonic character in the stagnation region at the nose of the airfoil and a supersonic character with a trail wave in the underpressure region on the body. This trail wave is still retained for $M_0 < 1$. The local supersonic regions that occur on airfoils for transonic subsonic free streams are shown in Figure 4.124. Here the supersonic characteristics in the local supersonic regimes have been made visible by disturbances on the surface of the airfoil. A flow drag is generated by the suction peak that has been displaced downstream and the final shock wave.

When the free stream is close to the velocity of sound, there is a *Mach number* distribution almost independent of M_0, particularly on the front part of the airfoil. This is because for a slightly supersonic free stream the shock is an almost perpendicular shock far in front of the airfoil that generates an approximately parallel subsonic flow. For this reason, the Mach number distributions on an airfoil differ only slightly, whether the free stream has $M_0 = 0.90$ or $M_0 = 1.10$. This effect is called *freezing* of the *Mach number distribution.*

Equation (4.188) yields the pressure coefficient to first order:

$$c_p = \frac{p - p_0}{\frac{1}{2} \cdot \rho_0 \cdot u_0^2} = -2 \cdot \frac{u}{u_0} = -2 \cdot \left(\frac{u_0 + u}{c'} - 1 \right) \tag{4.189}$$
$$= -\frac{2}{\kappa + 1} \cdot (M^2 - M_0^2).$$

Therefore, the change of c_p with M_0 at $M_0 = 1$ is

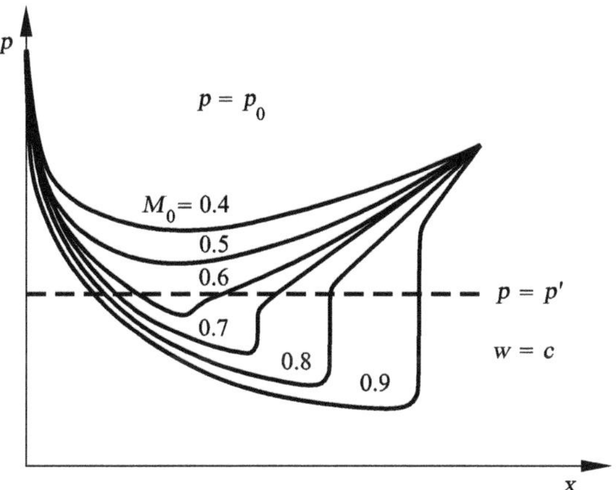

Fig. 4.125. Pressure on airfoils with a subsonic free stream

$$\left.\frac{\mathrm{d}c_p}{\mathrm{d}M_0}\right|_{M_0=1} = \frac{4}{\kappa+1}. \tag{4.190}$$

This allows us to determine the change in drag at $M_0 = 1$.

The pressure distributions on the airfoil for free streams in the linear and transonic subsonic regimes are shown in Figure 4.125. The appearance of shock waves that complete the local supersonic regime downstream cause the pressure drag to increase.

The flow past transonic wings will be treated in detail in Section 4.4.5.

4.3.8 Problems

4.17

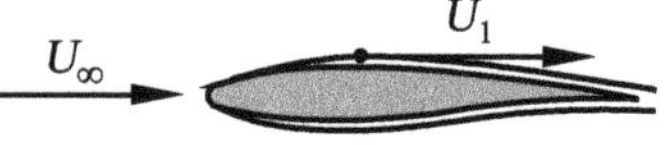

The maximum flow velocity U_1 at the edge of the boundary layer of a wing is 1.7 times the free-stream velocity U_∞.

How large is the local Mach number M_1 at the position of the largest velocity U_1 if the free-stream Mach number M is equal to 0.5 ($\kappa = 1.4$)? Treat the outer flow at the edge of the boundary layer as inviscid.

$$M_1 = \frac{1}{\sqrt{\frac{\kappa-1}{2}\cdot\left[\left(\frac{1}{1.7}\right)^2 - 1\right] + \left(\frac{1}{1.7}\right)^2 \cdot \frac{1}{M^2}}} = 0.893.$$

4.18

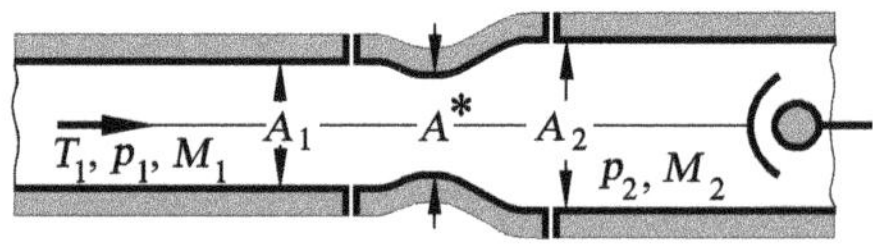

For the operation of a supersonic measuring track, an air flow with pressure p_1, temperature T_1, and Mach number M_1 is led through a pipe with cross-sectional area A_1 and a Laval nozzle.

This expands the flow to the pressure p_2 in the measuring track, so that the flow there is a supersonic parallel jet. The experiment in this parallel jet consists of a blunt displacer that causes a shock wave that can be considered in the region of interest in front of the stagnation point of the displacer to be a normal shock. The nozzle flow is steady, one-dimensional, and, apart from the shock, isentropic.

The following numerical values are given: $p_1 = 6.5$ bar, $T_1 = 440$ K, $M_1 = 0.5$, $A_1 = 160$ cm^2, $p_2 = 1.0$ bar, specific gas constant $R = 287$ m^2/(s$^2\cdot$K), isentropic exponent $\kappa = 1.4$.

Determine the following quantities for the experimental setup.

(a) What Mach number M_2 is reached in the measuring track?

$M_2 = 2.0.$

(b) How large must the areas A^* and A_2 be?

$$A^* = \frac{A_1 \cdot M_1}{\left[1 + \frac{\kappa - 1}{\kappa + 1} \cdot (M_1^2 - 1)\right]^{\frac{\kappa+1}{2\cdot(\kappa-1)}}} = 119.4 \text{ cm}^2,$$

$$A_2 = \frac{A^*}{M_2} \cdot \left[1 + \frac{\kappa - 1}{\kappa + 1} \cdot (M_1^2 - 1)\right]^{\frac{\kappa+1}{2\cdot(\kappa-1)}} = 201.5 \text{ cm}^2.$$

(c) How large is the mass flux through the experimental setup?

$\dot{m} = 17.33 \text{ kg/s}.$

(d) What are the values of the Mach number M_3, the pressure p_3, and the temperature T_3 directly downstream from the shock, and how large is the temperature T_s at the stagnation point of the displacer?

$$M_3 = 0.577\,, \quad p_3 = 4.5 \text{ bar}\,, \quad T_3 = 433.16 \text{ K}\,, \quad T_s = T_0 = 462 \text{ K}.$$

4.19

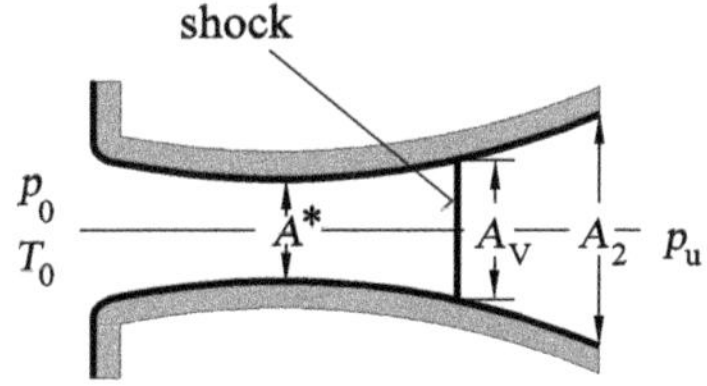

Air flows through a Laval nozzle out of a large container where the pressure is p_0 and the temperature T_0 into an atmosphere with pressure p_u. At the narrowest cross-section with area A^*, the velocity is that of sound, and further downstream, at the position with the cross-sectional area A_v, there is a normal, steady shock.

The following quantities are given: $p_0 = 5$ bar, $T_0 = 273.15$ K, $A^* = 2$ cm^2, $A_v = 3.1$ cm^2, $A_2 = 4.0$ cm^2, $\kappa = 1.4$, $R = 287$ m^2/(s$^2 \cdot$ K).

Determine the following quantities:

(a) the density ρ_0 in the container.

$$\rho_0 = \frac{p_0}{R \cdot T} = 6.378 \text{ kg/m}^3.$$

(b) the state quantities p_v, T_v, ρ_v of the air as well as the flow velocity c_v directly in front of the shock.

$$T_v = \frac{T_0}{1 + \frac{\kappa - 1}{2} \cdot M_v^2} = 158.9 \text{ K}, \; p_v = \frac{p_0}{\left(1 + \frac{\kappa - 1}{2} \cdot M_v^2\right)^{\frac{\kappa}{\kappa-1}}} = 0.75 \text{ bar},$$

$$\rho_v = \frac{\rho_0}{\left(1 + \frac{\kappa - 1}{2} \cdot M_v^2\right)^{\frac{1}{\kappa-1}}} = 1.646 \text{ kg/m}^3, \quad c_v = 479.1 \text{ m/s}.$$

(c) the total pressure $p_{0,v}$ and the total temperature $T_{0,v}$ directly in front of the shock.

$$p_{0,v} = p_0 = 0.5 \text{ bar}, \quad T_{0,v} = T_0 = 273.15 \text{ K}.$$

(d) the state quantities p'_v, T'_v, ρ'_v of the air and the flow velocity c'_v directly after the shock.

$$M_v^{2'} = \frac{1 + \frac{\kappa - 1}{\kappa + 1} \cdot (M_v^2 - 1)}{1 + \frac{2 \cdot \kappa}{\kappa + 1} \cdot (M_v^2 - 1)} = 0.3557,$$

$$\frac{p'_v}{p_v} = 1 + \frac{2 \cdot \kappa}{\kappa + 1} \cdot (M_v^2 + 1) = 4.03,$$

$$\frac{T'_v}{T_v} = \left[1 + \frac{2 \cdot \kappa}{\kappa + 1} \cdot (M_v^2 + 1)\right] \cdot \left[1 - \frac{2}{\kappa + 1} \cdot \left(1 - \frac{1}{M_v^2}\right)\right] = 1.605,$$

$$\frac{\rho'_v}{\rho_v} = \frac{p'_v}{p_v} \frac{T_v}{T'_v} = 2.51,$$

$$\frac{p'_{0,v}}{p_{0,v}} = \left[1 + \frac{2 \cdot \kappa}{\kappa + 1} \cdot (M_v^2 + 1)\right]^{-\frac{1}{\kappa-1}} \cdot \left[1 - \frac{2}{\kappa + 1} \cdot \left(1 - \frac{1}{M_v^2}\right)\right]^{-\frac{\kappa}{\kappa-1}} = 0.7692,$$

$M'_v = 0.596, \quad p'_v = 3.023$ bar, $\quad T'_v = 255$ K,
$\rho'_v = 4.13 \text{ kg/m}^3, \quad c'_v = 190.8$ m/s.

4.20

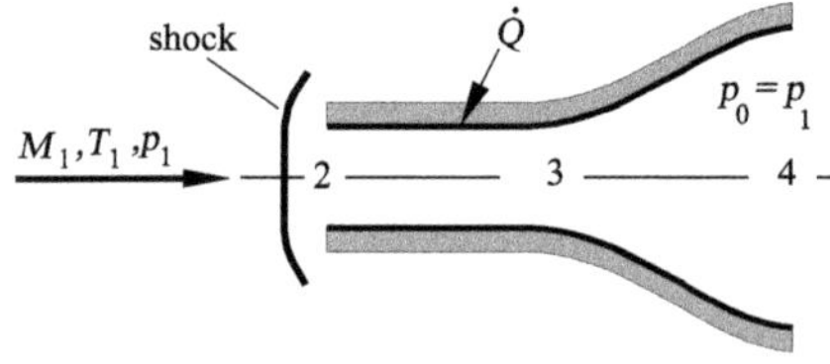

An approximately normal shock occurs in front of a supersonic propulsion ($M_1 = 2$, $p_1 = 0.3$ bar, $T_1 = 250$ K). Between states 2 and 3, the heat $\dot{Q}$ is supplied in the combustion chamber, so that the Mach number in state 3 becomes $M_3 = 1$.

In the divergent part of the drive nozzle an isentropic supersonic flow occurs with the outlet pressure $p_4 = p_1$ ($A_2 = A_3 = 0.4$ m^2, $A_4 = 0.56$ m^2, $\kappa = 1.4$, $R = 287$ J/kg K, $c_p = 104.5$ J/kg K).
(a) Calculate p_2, T_2, u_2 after the shock as well as the mass flux $\dot{m}$.

$$p_2 = 1.35 \text{ bar}\,,\; T_2 = 421.8 \text{ K}\,,\; u_2 = 237.5 \text{ m/s}\,,\; \dot{m} = 106 \text{ kg/s}.$$

(b) Calculate M_4, T_4, and u_4 at the outlet of the nozzle.

$$M_4 = 1.76\,,\; T_4 = 379.9 \text{ K}\,,\; u_4 = 687.6 \text{ m/s}.$$

(c) With the energy equation determine the heat transfer $\dot{Q}$ between 2 and 3.

$$T_3 = 512.9 \text{ K}\,,\; u_3 = 454 \text{ m/s}\,,\; \dot{Q} = 17630 \text{ kJ/s}.$$

4.21

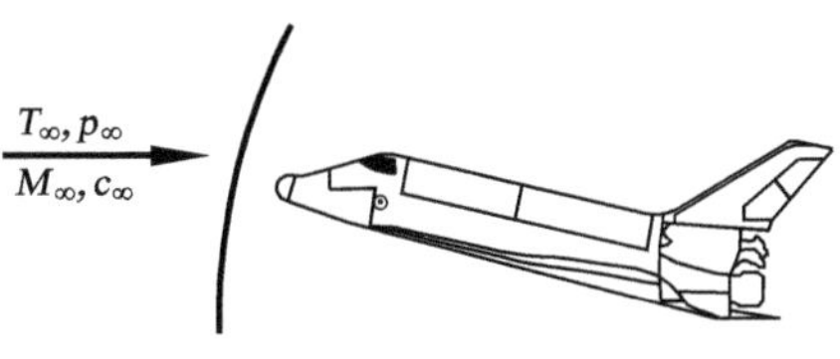

A shock forms in front of a reentry aircraft as it enters the atmosphere. This can be treated approximately as a normal shock. With the exception of the shock, the flow is an ideal, isentropic gas ($\kappa = 1.4$, $R = 287$ J/kg K).

(a) What are the maximum Mach number M_1, velocity u_1, and density ρ_1 allowed so that the maximum permissible temperature $T_{0,\max} = 840$ K in the stagnation point of the orbiter is not exceeded ($T_1 = 200$ K, $p_1 = 0.1$ bar)?

$$M_{1,\max} = \sqrt{\frac{2}{\kappa-1}\cdot\left(\frac{T_{0,1}}{T_1}-1\right)} = 4\,,\; u_{1,\max} = a_1 \cdot M_{1,\max} = 1133.9 \text{ m/s}.$$

(b) Using the above result, calculate the Mach number M_2, the velocity u_2, the pressure p_2, and the density ρ_2 for the flight state directly behind the shock. Determine the stagnation pressure $p_{0,2}$ at this position.

$$M_2 = \sqrt{\frac{1+\frac{\kappa-1}{\kappa+1}\cdot(M_{1,\max}^2-1)}{1+\frac{2\cdot\kappa}{\kappa+1}\cdot(M_{1,\max}^2-1)}} = 0.435,$$

$$p_2 = p_1 \cdot \left[1+\frac{2\cdot\kappa}{\kappa+1}\cdot(M_{1,\max}^2-1)\right] = 1.85 \text{ bar},$$

$$\rho_2 = \frac{p_1}{1-\frac{2\cdot\kappa}{\kappa+1}\cdot\left(1-\frac{1}{M_{1,\max}^2}\right)} = 0.08 \text{ kg/m}^3,$$

$$p_{0,2} = \left(1+\frac{\kappa-1}{2}\cdot M_2^2\right)^{\frac{\kappa}{\kappa-1}} \cdot p_1 \cdot \left[1+\frac{2\cdot\kappa}{\kappa+1}\cdot(M_{1,\max}^2-1)\right] = 2.11 \text{ bar}.$$

4.4 Aerodynamics

The aim of aerodynamics is to predict the forces and moments acting on bodies in a flow, such as airfoils, wings, fuselages, engines cells, or the whole airplane. Aerodynamics also includes prediction of wind forces on buildings, motor vehicles, and ships, as well as prediction of the aerodynamic heating of reentry vehicles on entry into Earth's or another planet's atmosphere. Further aims are the computation of losses and heat transfer in airplane engines, rocket engines, and pipelines.

In this section we restrict ourselves to the basics of the aerodynamics of airplanes and in particular to the *aerodynamics of wings*, which is greatly determined by the Mach number of the unperturbed free stream M_∞.

According to *D. Küchemann's* (1978) vision, any place in the world can be reached within the same flight time as long as the wing form is adapted to the flight Mach number M_∞ required. Figure 4.126 shows the dependence of different airplane shapes on M_∞ for distances D referred to the circumference of the Earth. For short distances, unswept wings at subsonic Mach numbers are used. Intermediate distances are covered using swept transonic wings. Supersonic flight is used for long distances. The vision of hypersonic flight could be realized with waveriders.

Eventually, it was the swept wing at transonic flight Mach numbers ($M_\infty = 0.8$) that prevailed for civil aircraft. This allows wide-bodied jets to transport a large number of passengers at flight times of up to 16 hours over distances of 14 000 km. In supersonic flight ($M_\infty = 2$), only possible over the sea or desert regions because of the supersonic boom from the head and tail shock waves, the flight time is halved.

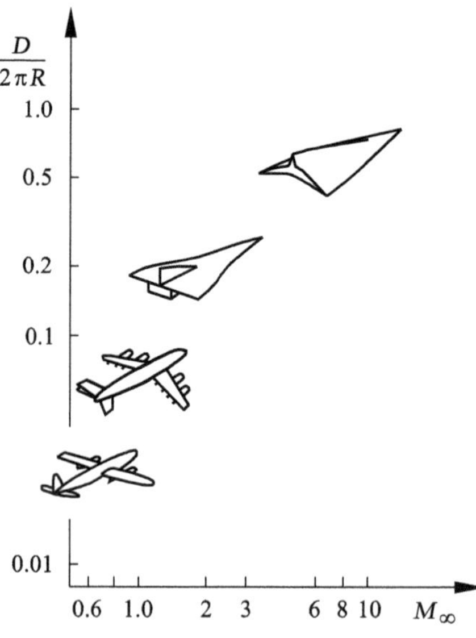

Fig. 4.126. Dependence of airplane forms on the flight Mach number M_∞

4.4.1 Bird Flight

In the last 10^8 years, evolution has developed flight in different ways in insects, bats, saurians, and birds. Because rotation about an axis is biologically impossible, the necessary lift and propulsion to fly are attained by the back and forth movement of the flap of a wing. The propulsion arises because the downward flap is carried out with great force while the upward flap takes place with as low a drag as possible. In birds, the largest part of the propulsion is due to the outer part of the wing, which completes the greatest vertical motion, as shown in Figure 4.127. The angle of inclination of different sections of the wing is changed during one period of oscillation by the deformation of the wing. The lift is essentially generated by the inner part of the wing. The functions of the wing and driving propeller of an airplane are integrated into a bird's wing, but this is paid for by the fact that the lift and propulsion vary in the course of one oscillation.

The stability problems related to this are counteracted by the aerodynamic forces on the tail surfaces that balance the oscillations as a horizontal rudder. The largest bird of passage, the albatros, has a span of 3.8 m, a top speed of up to 110 km/h, and a gliding number (lift to drag) of 20.

The qualitative dimensionless pressure distribution c_p (4.193) of a characteristic section of a bird wing is shown in Figure 4.128. Because of the different curvatures of the upper and lower sides of the wing, the flow is greatly accelerated, leading to a larger pressure drop on the upper suction side of the wing. Downstream from the point of suction the flow is decelerated, leading to a corresponding pressure increase. Because of the strong curvature of the profile, the flow tends to separate in the decelerated region, and this is prevented by the unsteady flap of the bird wing.

The first successful technical application of bird flight was carried out by *Otto Lilienthal* (1891) with his glider. Figure 4.129 shows the birdlike shape of the rigid wing with integrated vertical and horizontal surfaces, which ensured

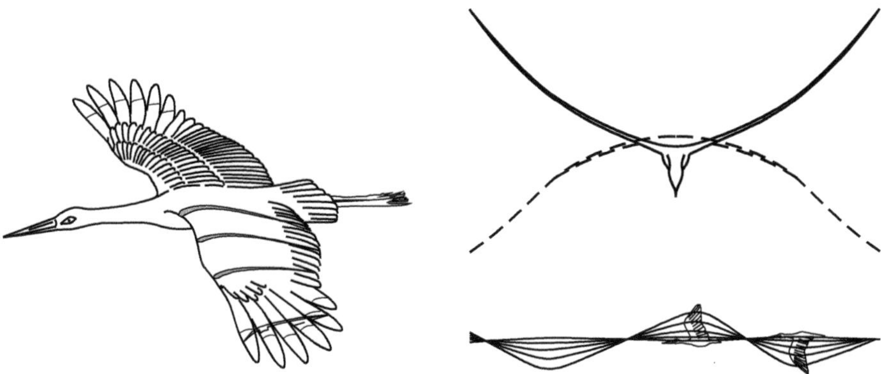

Fig. 4.127. Wing cross-sections and pathlines of bird flight

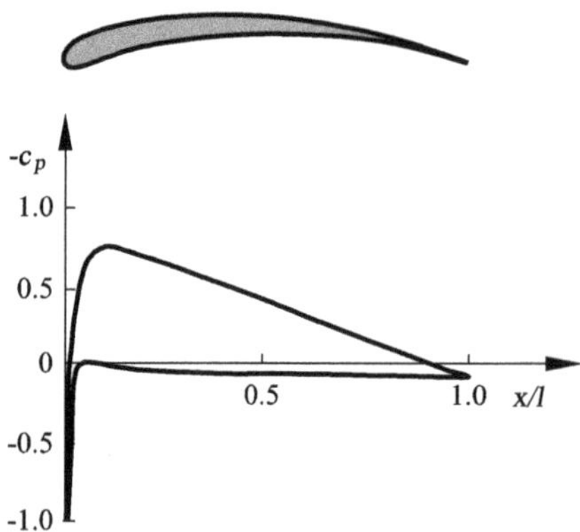

Fig. 4.128. Profile and pressure distribution of a bird's wing

stability. The hang-glider was controlled by shifting the weight of the body underneath the glider.

Prior to this, in 1889, Lilienthal had published a book with the title *Bird Flight as the Basis of the Art of Flying*, which contained all the aerodynamic data of that time. Even modern civil aircraft 100 years later still use rigid wings (see Figure 4.130). The flap of the bird's wing has been replaced by fan engines, which, because of their size, are placed under the wings. The fuselage holds the passengers, and the side and upper tails provide the required stability. What has changed compared to bird flight is the speed. The endeavor to fly from one place to the next as fast, comfortably, and economically as possible has led to transonic flight speeds of 950 km/h at a Mach number 0.8 and an altitude of 10 km. Flow losses are reduced at transonic flow Mach numbers using swept wings, to be treated in Section 4.4.5. The winglets at the ends of the wings are modeled on the tip of a bird's wing, and these reduce the strength of the edge vortex and thus the wing drag.

Fig. 4.129. Lilienthal's hang-glider

Fig. 4.130. Civil airplane

4.4.2 Airfoils and Wings

If an airplane moves with constant velocity V_∞, it experiences the resultant aerodynamic force R (Figure 4.131). The component of this force in the free-stream direction is the drag W, and the component perpendicular to this is the lift A. The inclination of the resultant R to the free-stream velocity and therefore the ratio of lift to drag essentially depends on the geometric shape of the wing and the free-stream direction. A large value of the ratio A/W is desirable. For steady gliding of an airplane without an engine, the resulting force R must be equal and opposite to the weight G. Thus the *gliding angle* α is defined by the relation

$$\tan(\alpha) = \frac{W}{A}. \tag{4.191}$$

A wing of a civil airplane swept with the angle ϕ is sketched in Figure 4.131. Each vertical cut through the *wing* is called a *profile.* The camber line, the average of the distance between the upper and lower sides of the wing, is a particular profile line that is needed in describing inviscid design methods. The angle of the profile to the unperturbed free stream V_∞ is denoted by α. As was explained in Section 4.2.10, the aerodynamic forces *lift* A, *drag* W, and the *resultant* R are caused by the pressure distribution and the distribution of the wall shear stresses on the surfaces of the wing. In addition, a *moment* M is also produced, and this is responsible for the rotation of the wing. The

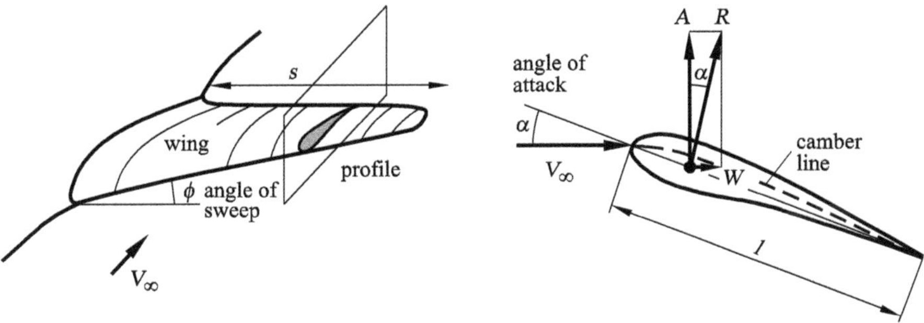

Fig. 4.131. Sketch of wing and airfoil

relevant dimensionless coefficients are

$$c_a = \frac{A}{q_\infty \cdot S}, \qquad c_w = \frac{W}{q_\infty \cdot S}, \qquad c_m = \frac{M}{q_\infty \cdot S \cdot l}, \tag{4.192}$$

with $q_\infty = 0.5 \cdot \rho \cdot V_\infty^2$ and the wing area S. The pressure and friction coefficients are

$$c_p = \frac{p - p_\infty}{q_\infty}, \qquad c_f = \frac{\tau}{q_\infty}, \tag{4.193}$$

with the pressure of the unperturbed free stream p_∞. All coefficients are functions of the free-stream Mach number M_∞, the Reynolds number Re_l, the angle of attack α, and the sweep angle ϕ.

Profile Flow

Typical profiles of different Mach number regimes are sketched in Figure 4.132. In contrast to the thin bird profiles in Figure 4.138, *L. Prandtl* showed in 1917 that subsonic profiles of thickness $d/l = 13\%$ (e.g. the Göttingen profile 298) have a larger lift coefficient c_a at a smaller drag coefficient c_w.

According to Figure 4.132 the profiles for transonic free streams have to be thinner, so that the transition to supersonic flow takes place as far downstream on the profile as possible. Oblique shock waves occur on profiles in supersonic flow, so that the drag can be kept small with sharp leading and trailing edges.

The various flow regimes are shown in Figure 4.133 for transonic subsonic and supersonic Mach numbers. Transonic subsonic Mach numbers are those for which, as in the first figure, the acceleration on the profile passes over to the supersonic regime. The supersonic regime is then concluded with a shock wave, which in turn leads to an additional pressure drag c_s. The shock waves are shown in bold in Figure 4.133, and the sonic lines $M = 1$ dashed. The deceleration of the flow on the profile causes a pressure increase up to the trailing edge. The pressure that occurs here is slightly above the pressure of the unperturbed free stream.

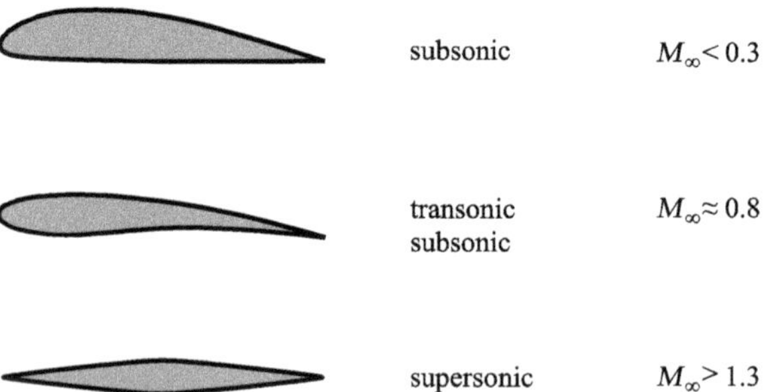

Fig. 4.132. Characteristic profile shapes for subsonic, transonic subsonic, and supersonic Mach numbers

If the transonic free-stream Mach number is increased to a value greater than 0.8, the supersonic regime extends over the entire upper side of the profile, as in the second picture. The shock wave moves to the trailing edge, while a local supersonic regime with a shock wave also occurs on the lower side. The shock at the trailing edge provides the necessary pressure increase, which is carried over into the pressure of the wake flow.

The limiting case of a free stream with Mach number $M_\infty = 1$ is sketched in the third drawing in Figure 4.133. The shock waves on the upper and lower sides of the profile move down as far as the wake and branch into two oblique shock waves and one vertical shock wave at the trailing edge. The sonic line extends through the entire flow field, and almost the entire profile is in a supersonic flow. If the free-stream Mach number is slightly higher than 1, a separated head wave forms far in front of the profile.

For a supersonic free stream $M_\infty \geq 1$, the head wave distance decreases. A subsonic regime forms between the shock and the profile. The oblique shock waves move out of the wake at the trailing edge of the profile. If the free-stream Mach number is increased further, attached oblique shock waves, corresponding to those at the trailing edge, form at the sharp leading edge. Figure 4.134 shows the dependence of the *lift and drag coefficients* on the Mach number for a given profile. At subsonic Mach numbers, the lift coefficient increases with increasing Mach number, corresponding to the *Prandtl–Glauert rule* (see Section 4.4.8):

$$c_a = \frac{2 \cdot \pi}{\sqrt{1 - M_\infty^2}}, \qquad M_\infty < 1. \tag{4.194}$$

Here the pressure coefficient of the profile computed with the linear theory is

$$c_p = \frac{c_{p0}}{\sqrt{1 - M_\infty^2}},$$

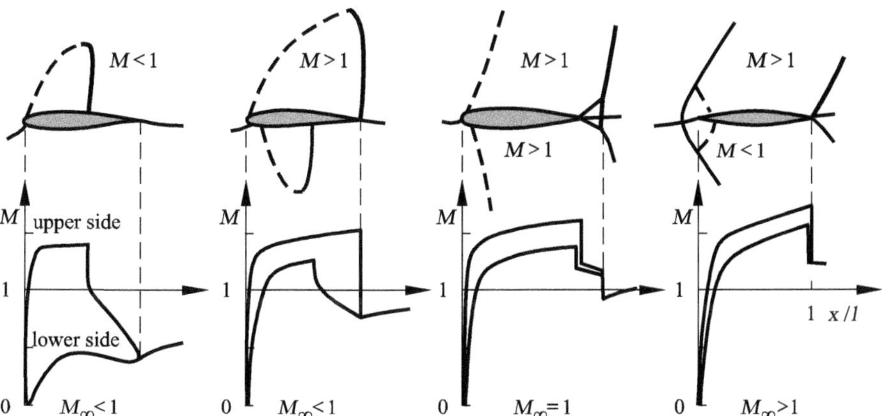

Fig. 4.133. Mach number distribution in transonic profile flows

where c_{p0} is the pressure coefficient of the incompressible flow.

A decrease in the lift coefficient is found with the linear supersonic theory corresponding to the *Ackeret rule*:

$$c_a = \frac{4}{\sqrt{M_\infty^2 - 1}}, \qquad M_\infty > 1. \tag{4.195}$$

The lift coefficient passes through a maximum in the transonic subsonic regime. The sudden drop in the lift coefficient is due to the appearance of the supersonic regime and the second shock wave on the lower side of the profile. The Mach number distribution shown in Figure 4.133 causes the lift coefficient to decrease drastically, only to increase again for Mach numbers greater than 0.9. The renewed increase in the lift coefficient occurs whenever the shock waves have moved from the wave to the trailing edge of the profile, to weaken because of the small shock angle. It is only with the appearance of the head wave and the subsonic regime between the shock wave and the profile that the lift coefficient in the supersonic regime decreases again, in accordance with the *Ackeret* equation (4.195).

In designing the profile of a civil airplane, the flight Mach number in the transonic subsonic regime is chosen to be around the maximum of about 0.8.

The drag coefficient c_w behaves similarly to the lift coefficient, except that the second maximum in the transonic subsonic Mach number does not appear. Until the point where the supersonic regime appears on the upper side of the profile, the drag coefficient remains essentially constant with increasing free-stream Mach number. When the shock wave on the lower side of the profile occurs, the drag coefficient increases considerably. Up until the drag maximum is reached at the Mach number $M_\infty = 1$, local Mach numbers up to $M = 2$ may be reached in the supersonic regimes. The shock waves on the profile are so strong that the pressure increase causes flow separation, with the drag increasing even further.

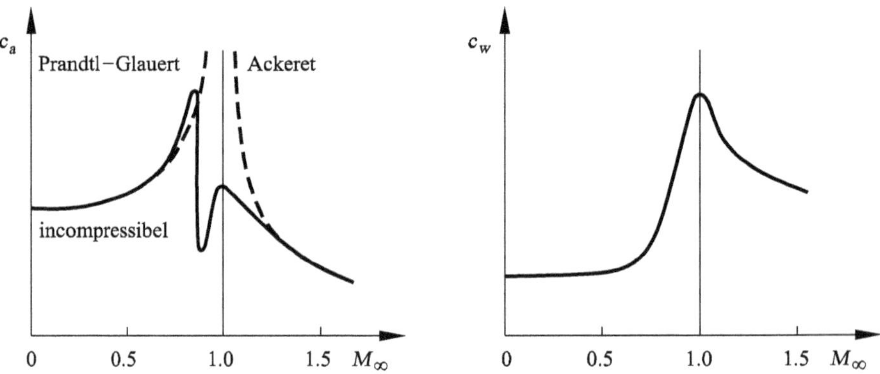

Fig. 4.134. Dependence of lift coefficient c_a and drag coefficient c_w on the free-stream Mach number M_∞

This leads to the design of supercritical profiles (Figure 4.135) with the aim of increasing the transonic flight Mach number at as low a drag as possible. Here the thickest point of the profile lies close to the leading edge, and the extended supersonic regime on the profile is concluded with a weak shock wave as far upstream as possible. In contrast to conventional transonic profiles, the suction peak in the front region of the profile is avoided. The dependence of the lift coefficient c_a on the *angle of attack* α is shown in Figure 4.136 for a given subsonic profile. The lift initially increases linearly with increasing angle of attack, as long as the flow remains attached. Even for the angle of attack $\alpha = 0°$, the asymmetry of the profile means that the lift coefficient is positive. The lift coefficient passes through a maximum at a critical angle of attack α_{crit} and then decreases sharply for larger α. The snapshot of the flow in Figure 4.136 shows that the flow on the entire upper side of the profile separates unsteadily. The collapse of the lift coefficient is accompanied by an increase in the drag of the profile.

In order to be able to take off and land with a wing, the surface area of the wing is increased at low velocities with front and rear flaps. This leads to the dashed lift curve in Figure 4.136, yielding higher lift values.

A tool that is useful in the design of profiles is the *polar diagram* (Figure 4.137), where the lift coefficient c_a is plotted against the drag coefficient c_w for different angles of attack α. The polar curves are so called because the forces acting on the profile can be directly read off from Figure 4.137. The vector from the origin to a point on the polar curve shows the resultant force $\boldsymbol{R}$. For the supercritical profile of Figure 4.135, the increase of the lift coefficient with increasing angle of attack is large, while the maximum value of c_a is small compared with subsonic profiles. The drag coefficient remains small for a large range of angles of attack. At the free-stream Mach number $M_\infty = 0.76$, the design yields a lift coefficient of $c_a = 0.57$.

In order to be able to analyze the effect of friction on the flow past a profile, consider the pressure distributions for different types of separation for inviscid and viscous flow for a subsonic profile set at an angle, as shown in Figure

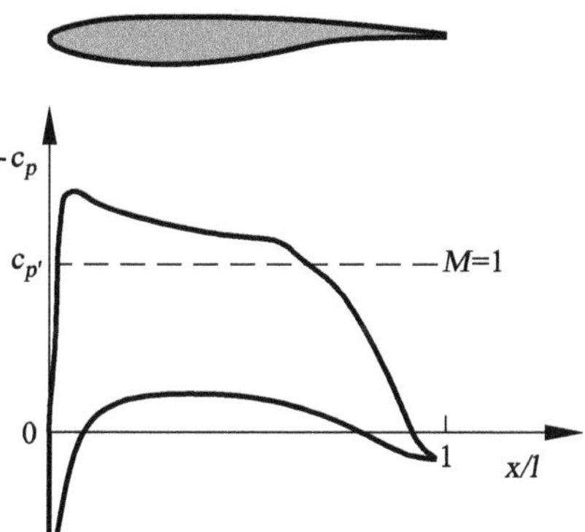

Fig. 4.135. Pressure distribution c_p on a supercritical profile

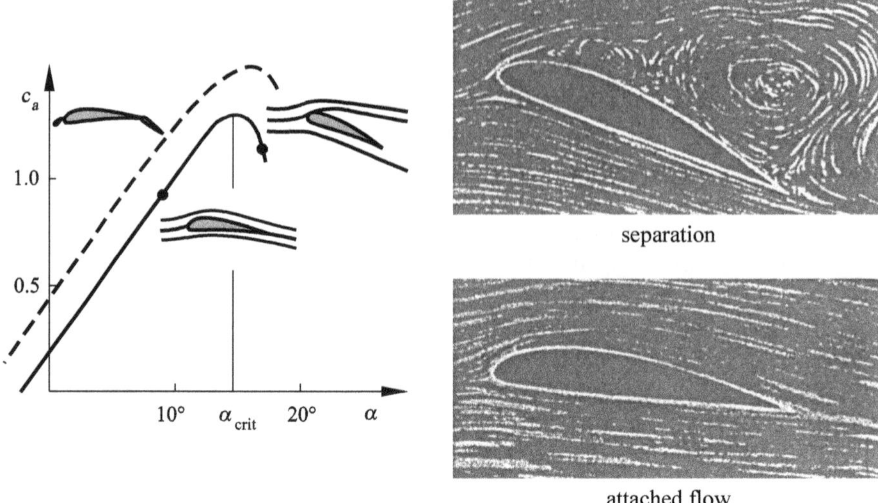

Fig. 4.136. Dependence of the lift coefficient c_a and flow portraits on the angle of attack α

4.138. As long as the boundary-layer flow remains attached to the profile, the displacement action of the viscous part of the pressure distribution causes the pressure to increase. If the flow separates, a time-averaged backflow region with constant pressure forms on the profile, causing this lift to be decreased.

If separation begins already at the leading edge, reattachment of the flow can occur on the profile. The region of constant pressure then lies in the suction peak of the profile, and the lift collapses. The flow is determined by the gray viscous part of the pressure distribution, so that the theory of inviscid flow past a profile treated in Section 4.4.3 remains restricted to the region of inviscid outer flow of the attached profile boundary layer.

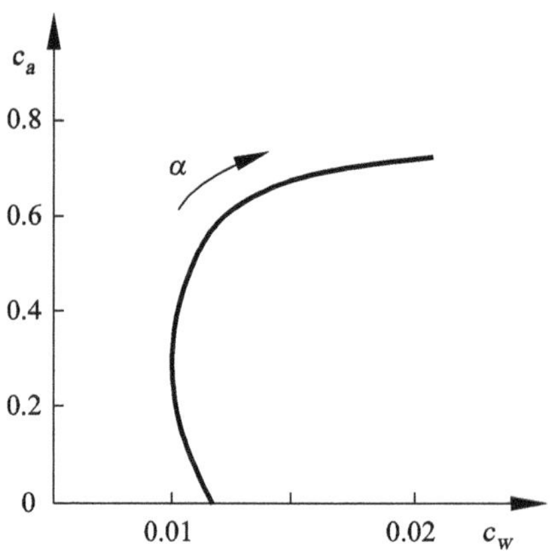

Fig. 4.137. Polar diagram of a transonic profile

Flow past Wings

In what follows we transfer the results of the flow past profiles to the finite *wing* in Figure 4.131. The flow past a wing is three-dimensional.

A third velocity component in the direction of the wing span is superimposed onto the two-dimensional flow past a profile. The explanation for this is to be found in Figure 4.139. On the upper side of the wing there is underpressure, and on the lower side, overpressure. This leads to a flow past the edges of the wing, which leads to a vortex in the wake in each case. These vortices cause a velocity component directed downward behind the wing. The additional vortex formation at the edges of the wing changes the pressure distribution such that an additional pressure drag arises, known as the *induced drag*. The drag balance (4.124), consisting of the pressure and friction drag, is extended in the case of a wing by the induced pressure drag c_i:

$$c_w = c_d + c_f + c_i + c_s. \tag{4.196}$$

In the case of a transonic wing, there is also the additional pressure drag of the shock wave on the upper side of the wing, and this is known as the *wave* or *shock drag* c_s. The contributions to the drag for a wing with supercritical profile are 51% the friction drag c_f, 35% the induced drag c_i, 10% the pressure drag c_d, and 4% the shock drag c_s (see Figure 4.155).

These are the figures for a swept transonic wing. It lowers the local free-stream Mach number of the profile in such a manner that the increase in drag in Figure 4.134 is postponed to higher Mach numbers. The fact that the sweep ϕ causes the effective profile Mach number to be lowered by $M_n = M_\infty \cdot \cos(\phi)$ was first noted by *A. Betz* (1939) (Figure 4.140). He considered that the free-stream pressure drag is generated only by the normal component $\boldsymbol{v}_n$. If the free stream is directed tangentially to the span of the wing with velocity $\boldsymbol{v}_t$,

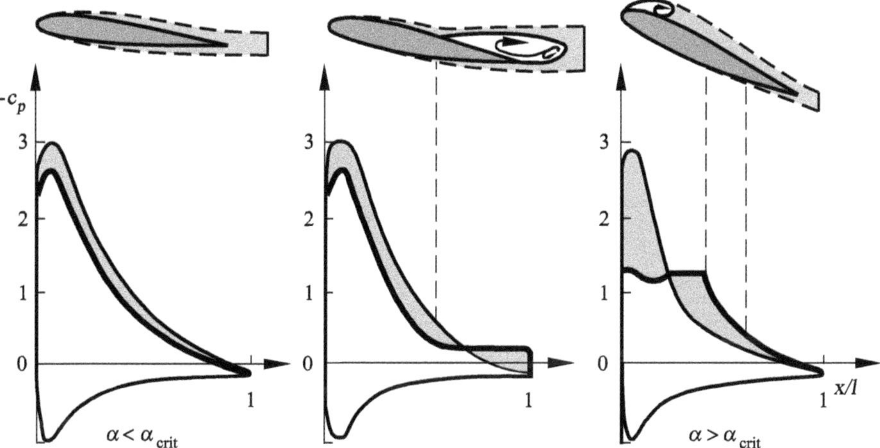

Fig. 4.138. Pressure distributions for inviscid and viscous flow past a profile

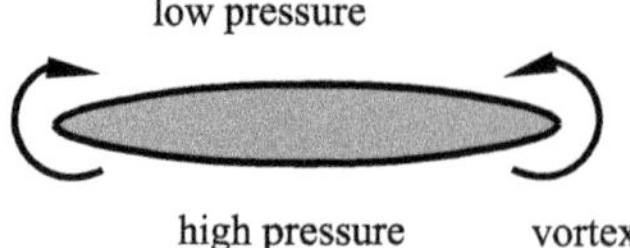

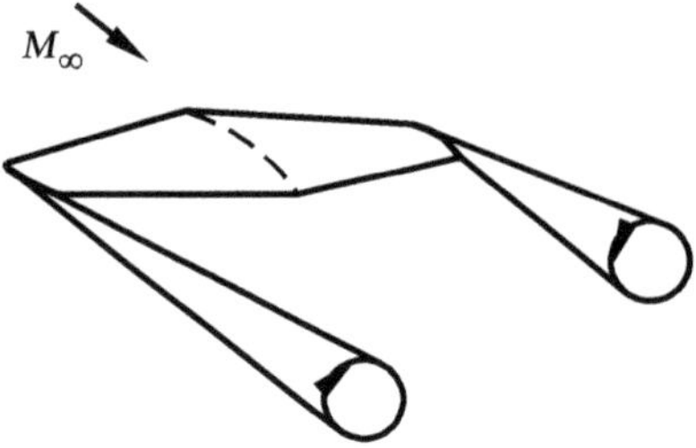

Fig. 4.139. Edge vortex at a finite wing

this flow cannot cause a pressure change at the wing, and only friction drag occurs.

4.4.3 Airfoil and Wing Theory

The basis of Prandtl's airfoil and wing theory was the discovery that aerodynamic lift is caused by the circulation distribution around the wing. For large Reynolds numbers it is assumed that the pressure and circulation distribution of the wing can be approximately computed with the potential equation $\Delta\Phi = 0$ (4.20) of inviscid flow.

There are two different mathematical ways of calculating the inviscid *flow past an airfoil*: The method of conformal mapping and the singularity method. In what follows we will discuss the singularity method, in particular with regard to the calculation of the three-dimensional flow past a wing.

We begin with the particular solution of the linear potential equation, as discussed in Section 4.1.5. The flow past a vaulted profile with finite thickness

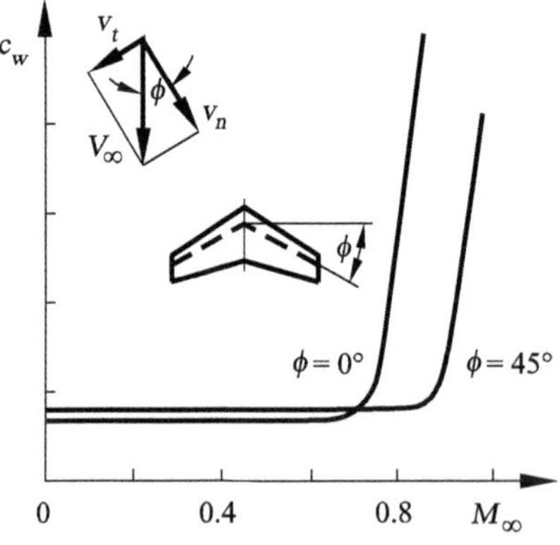

Fig. 4.140. Effect of the sweep angle ϕ on the drag coefficient c_w

at angle of attack α can be calculated using linear superposition of sources, sinks (thickness), vortices (angle of attack), and the additional superposition of a translation velocity (free stream). These are shown in Figure 4.141. With the circulation $\Gamma = \oint \boldsymbol{v} \cdot \mathrm{d}\boldsymbol{s}$, and the *Kutta–Joukowski condition* at the trailing edge and using linear superposition of individual solutions, a lift per unit length A can be computed for the inviscid flow past a profile in Figure 4.142:

$$A = \rho \cdot \Gamma \cdot V_{\infty}. \tag{4.197}$$

The onset of circulation at a wing is explained in Figure 4.143. As the wing begins to move, a startup vortex with negative circulation $-\Gamma$ forms at the trailing edge. According to Thomson's law (Section 4.1.5) the circulation must be conserved, and so the same circulation but with a positive sign forms around the wing. This is called the attached vortex. Combining the attached vortex, the startup vortex and the edge vortex of Figure 4.139 together, we have the closed vortex system shown in Figure 4.144, since according to Helmholtz's law, no vortex can end in free flow. The lift of the attached vortex is linked to the induced drag c_i of (4.196).

We now present the fundamentals of Prandtl's theory, since even today it is used to carry out the initial design of a wing for *subsonic flow. L. Prandtl's* theoretical starting point in 1920 was to assume that to compute the lift, the slender wing can be replaced by a lift line (camber line) superimposed with a circulation distribution. The simplest vortex system of a finite wing consists of the attached vortex of strength Γ and two edge vortices with the same vortex strength (Figure 4.145). Since the lift distribution decreases as we move toward the edges of the wing, this distribution can be approximately represented with a vortex system of infinitesimal strength across the span s of

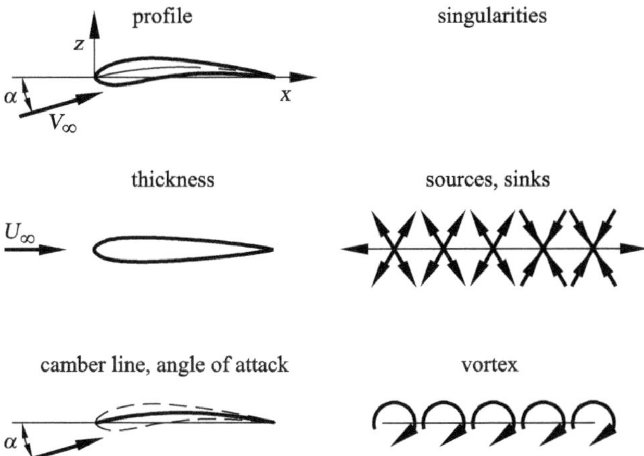

Fig. 4.141. Singularity distribution of a profile of finite thickness at an angle of attack

the wing. For the vortex system in Figure 4.145, there is in the center of the wing a vortex extending infinitely far forward and backward with strength Γ. At distance d we obtain the velocity $w = \Gamma/(2 \cdot \pi \cdot d)$. For symmetry reasons, a vortex extending only backward from the cutting plane has only half the velocity $\Gamma/(4 \cdot \pi \cdot d)$. In the middle of the wing, $d = s/2$, the velocity from the right and left vortices is combined, yielding

$$w_0 = 2 \cdot \frac{\Gamma}{4 \cdot \pi \cdot \frac{s}{2}} = \frac{\Gamma}{\pi \cdot s}.$$

With the Kutta–Joukowski condition $\Gamma = A/(\rho \cdot s \cdot V_\infty)$ for a wing with span s we obtain

$$w_0 = \frac{A}{\pi \cdot \rho \cdot V_\infty \cdot s^2}.$$

Around the center of the wing larger velocities are found, increasing to infinity close to the ends of the wing. This means that the assumption of a constant lift as far as the ends of the wing is inadmissable. Assuming the *elliptical lift distribution* shown in Figure 4.146, we obtain the constant vertical velocity w over the wing. In the center, the circulation is $4/\pi$ times larger than the average circulation, and so the individual vortex filaments are on average closer to the center, and w becomes larger than w_0. Integration over all vortex filaments yields

$$w = 2 \cdot w_0 = \frac{2 \cdot A}{\pi \cdot \rho \cdot V_\infty \cdot s^2}, \tag{4.198}$$

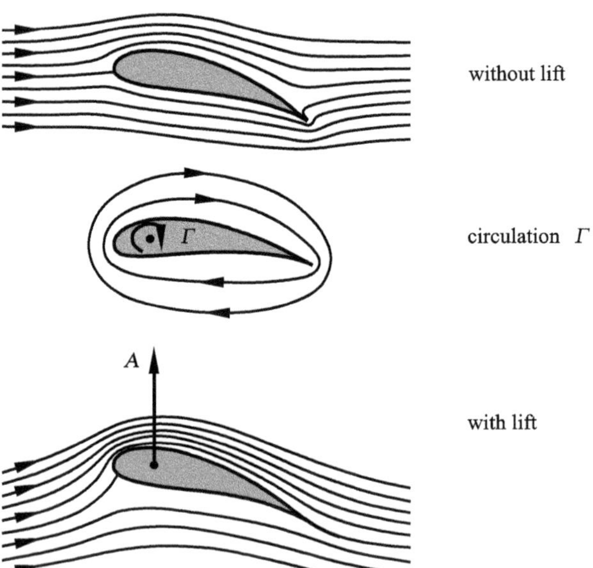

Fig. 4.142. Lift generation at a wing profile (potential flow, see Figure 4.30)

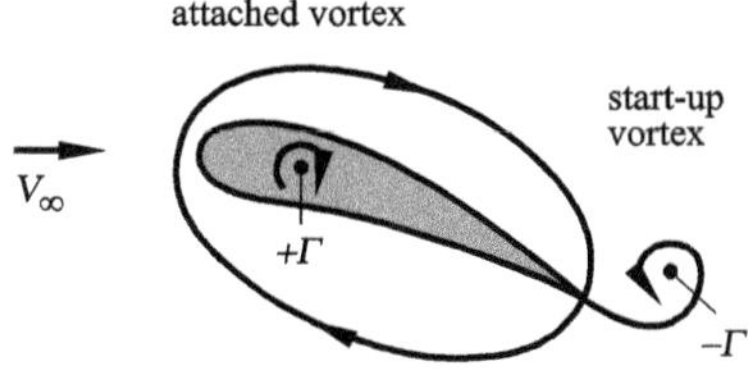

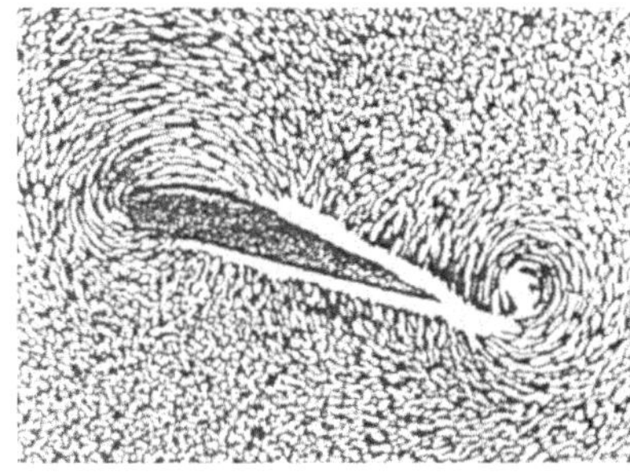

Fig. 4.143. Startup vortex and attached vortex of a wing profile, *L. Prandtl, O.G. Tietjens* (1934)

and so

$$\tan(\alpha) = \frac{w}{v_\infty} = \frac{2 \cdot A}{\pi \cdot \rho \cdot V_\infty^2 \cdot s^2} = \frac{A}{\pi \cdot p_s \cdot s^2},$$

with impact pressure p_s. Since w is constant over the span for an elliptical lift distribution, $\tan(\alpha)$ is also constant. Therefore, the induced drag $W_i = A \cdot \tan(\alpha)$ is

$$W_i = \frac{A^2}{\pi \cdot p_s \cdot s^2}. \tag{4.199}$$

Equation (4.199) shows that the induced drag becomes smaller, the larger the span over which the lift is distributed, and so for airplanes in a subsonic free stream the wings have a large span. The wing chord l does not appear in equation (4.199). Only the flow state behind the wing is of importance, not the distribution of circulation over the chord of the wing.

The distribution of the vortex strength along the camber line of a slender profile is obtained from the kinematic condition that the camber line must

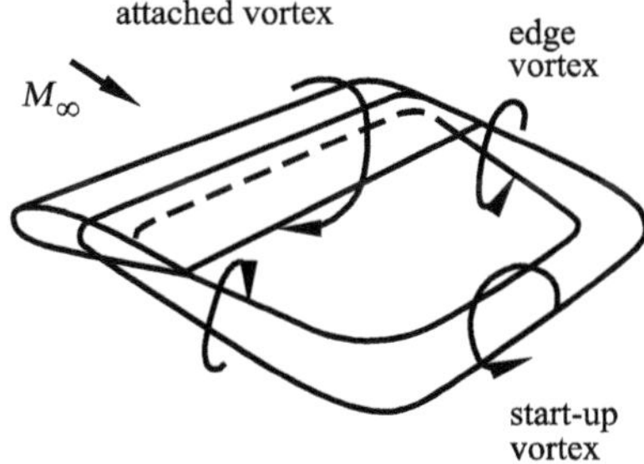

Fig. 4.144. Vortex system around a wing

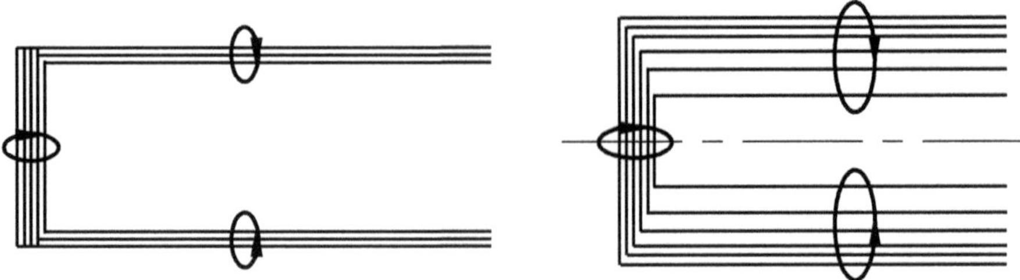

Fig. 4.145. Simplified vortex system of a wing

be a streamline. The translation velocity V_∞, which forms the angle of attack α with the chord, is superimposed onto the vortex distribution (Figure 4.147). At all points on a streamline the vertical velocity component must vanish. For a slender profile, the camber line can be replaced by the chord, by approximation, so that to first order we obtain

$$V_\infty \cdot \left(\alpha - \frac{\mathrm{d}z}{\mathrm{d}x}\right) + w(x) = 0. \tag{4.200}$$

The vortex strength per unit length (vortex density) is denoted by $\gamma(x)$. An infinitesimal vortex element of strength $\gamma(x') \cdot \mathrm{d}x'$ at position x' generates the infinitesimal velocity

$$\mathrm{d}w = -\frac{\gamma(x') \cdot \mathrm{d}x'}{4 \cdot \pi \cdot (x - x')}. \tag{4.201}$$

Integration over the chord of the wing l yields the vertical velocity

$$w(x) = -\frac{1}{4 \cdot \pi} \cdot \int\limits_0^l \frac{\gamma(x') \cdot \mathrm{d}x'}{x - x'}. \tag{4.202}$$

Equation (4.200) with vertical velocity (4.202) is the *fundamental equation for slender profiles*, which arises from the requirement that the camber line be a streamline. This can also be used to compute the increase of the lift coefficient c_a in Figure 4.136:

$$\frac{\mathrm{d}c_a}{\mathrm{d}\alpha} = 2 \cdot \pi. \tag{4.203}$$

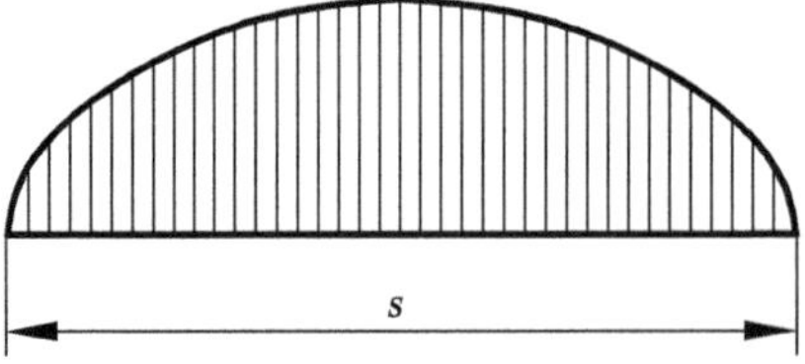

Fig. 4.146. Elliptical lift distribution

To transfer this result to the *wing*, we look again at the vortex filaments, the attached and free boundary vortices of Figure 4.145, also called *horseshoe vortices.*

A vortex filament extending to infinity in both directions, as in Figure 4.148, generates for each infinitesimal vortex element $\mathrm{d}\boldsymbol{l}$ the following velocity at the point P:

$$\mathrm{d}\boldsymbol{v} = \frac{\Gamma}{4 \cdot \pi} \cdot \frac{\mathrm{d}\boldsymbol{l} \times \boldsymbol{r}}{|\boldsymbol{r}^3|}. \tag{4.204}$$

The relation is known as the *Biot–Savart law.* Integration along the vortex filament yields

$$\boldsymbol{v} = \int\limits_{-\infty}^{\infty} \frac{\Gamma}{4 \cdot \pi} \cdot \frac{\mathrm{d}\boldsymbol{l} \times \boldsymbol{r}}{|\boldsymbol{r}^3|}. \tag{4.205}$$

Using the definition of the vector product, we see that the direction of the velocity vector $w = |\boldsymbol{v}|$ is downward, and

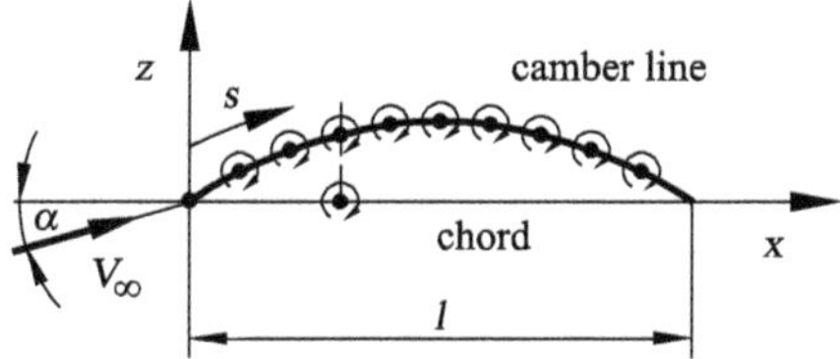

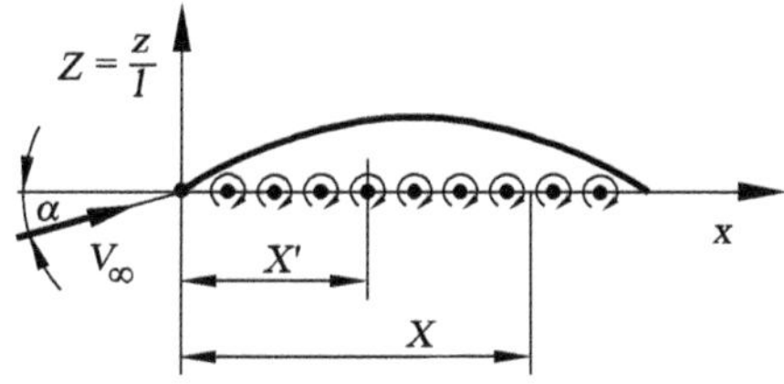

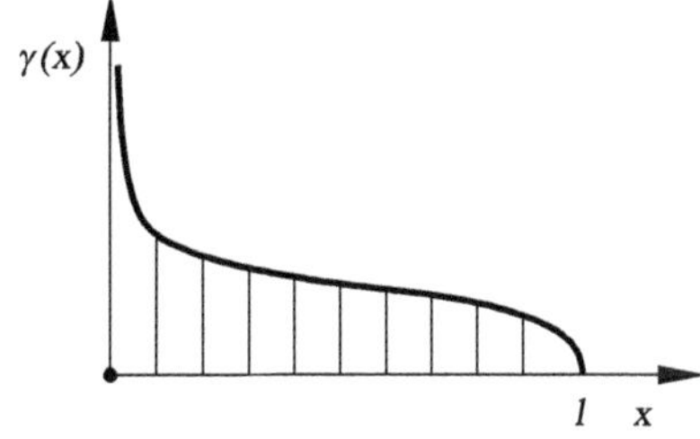

Fig. 4.147. Vortex strength distribution along the camber line and chord of a slender profile

$$w = \frac{\Gamma}{4 \cdot \pi} \cdot \int_{-\infty}^{\infty} \frac{\sin(\Theta)}{r^2} \cdot \mathrm{d}l. \tag{4.206}$$

With the vertical distance h to the vortex element $\mathrm{d}l$, integration for a semi-infinite vortex filament delivers

$$w = \frac{\Gamma}{4 \cdot \pi \cdot h}. \tag{4.207}$$

The concept of a vortex filament was first introduced by *H. Helmholtz* to compute inviscid incompressible flows. The *Helmholtz vortex laws* state that the vortex strength Γ along a vortex filament is constant and that a vortex filament may not end in the flow field. However, the end of a vortex filament may indeed lie at infinity, where closure with the startup vortex (Figure 4.139) takes place. As already discussed, *L. Prandtl* extended the concept of the horseshoe vortex with the attached vortex and two edge vortices extending to infinity to tackle the computation of induced lift on a wing. Here the circulation distribution over the finite wing is taken into account (see Figure 4.146).

If we consider the single horseshoe vortex in Figure 4.149, we see that the attached vortex of span s does not give rise to a velocity component along the vortex filament. The vertical component is $w(y)$. The edge vortices are also superimposed onto a vertical component of the velocity. With (4.207) we obtain the contribution of the semi-infinite edge vortex:

$$w = -\frac{\Gamma}{4 \cdot \pi \cdot \left(\frac{s}{2} + y\right)} - \frac{\Gamma}{4 \cdot \pi \cdot \left(\frac{s}{2} - y\right)} = -\frac{\Gamma}{4 \cdot \pi} \cdot \frac{s}{\frac{s^2}{4} - y^2}. \tag{4.208}$$

Note that w tends to $-\infty$ at the ends of the wing $\pm s/2$. Because of this, *L. Prandtl* considered not just a single horseshoe vortex on the wing, but rather a large number of horseshoe vortices of different lengths of the attached vortex. These are arranged along a line called the *lift line*. Figure 4.150 shows first the superposition of three horseshoe vortices. The first horseshoe vortex, of strength $\mathrm{d}\Gamma_1$, encompasses the entire attached vortex from point A ($y = -s/2$) to point F ($y = +s/2$). Superimposed on this is the second horseshoe

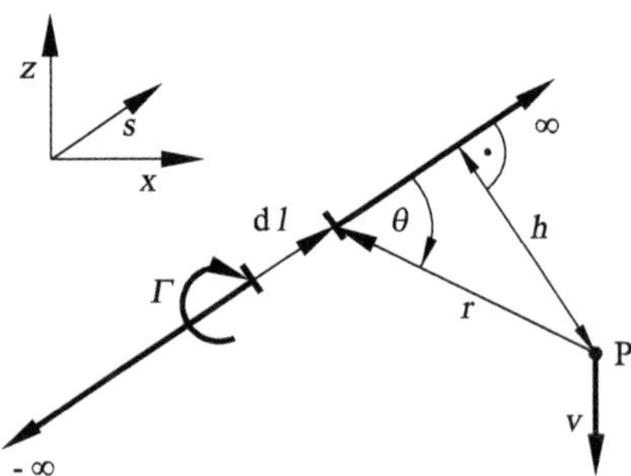

Fig. 4.148. Velocity $\boldsymbol{v}$ at the point P of a straight vortex filament

vortex, of strength $\mathrm{d}\Gamma_2$, from B to E, covering only a part of the attached vortex. The third horseshoe vortex $\mathrm{d}\Gamma_3$ is superimposed from C to D. This means that the vortex strength $\Gamma(y)$ varies along the attached vortex (lift line). Along $\overline{\mathrm{AB}}$ and $\overline{\mathrm{EF}}$ the strength is $\mathrm{d}\Gamma_1$, along $\overline{\mathrm{BC}}$ and $\overline{\mathrm{DE}}$ it is $\mathrm{d}\Gamma_1+\mathrm{d}\Gamma_2$, and along $\overline{\mathrm{CD}}$ it is $\mathrm{d}\Gamma_1 + \mathrm{d}\Gamma_2 + \mathrm{d}\Gamma_3$. Two edge vortices are assigned to each vortex element along the lift line. The vortex strength of each edge vortex is equal to the change in circulation along the lift line.

If we extrapolate the superposition to infinitely many horseshoe vortices of infinitesimal vortex strength $\mathrm{d}\Gamma$, we obtain a continuous distribution of the vortex strength $\Gamma(y)$ along the span of the wing. Let the maximum value of the circulation be Γ_0. The finite number of horseshoe vortices has become a continuous vortex street parallel to the free stream V_∞. Integration of the vortex strength perpendicular to the vortex street is zero, since the boundary vortices are paired with the same vortex strength but opposite signs.

If we consider an infinitesimal element $\mathrm{d}y$ on the lift line with vortex strength $\Gamma(y)$, the variation along the element is $\mathrm{d}\Gamma = (\mathrm{d}\Gamma/\mathrm{d}y) \cdot \mathrm{d}y$. The vortex strength of the edge vortex at position y is equal to the change in vortex strength $\mathrm{d}\Gamma$. At position y', following the Biot–Savart law (4.204), each element $\mathrm{d}x$ of the boundary vortex causes the vertical velocity

$$\mathrm{d}w = \frac{\frac{\mathrm{d}\Gamma}{\mathrm{d}y} \cdot \mathrm{d}y}{4 \cdot \pi \cdot (y' - y)}. \tag{4.209}$$

Integration along all edge vortices yields

$$w(y') = \frac{1}{4 \cdot \pi} \cdot \int\limits_{-\frac{s}{2}}^{\frac{s}{2}} \frac{\frac{\mathrm{d}\Gamma}{\mathrm{d}y}}{y' - y} \cdot \mathrm{d}y. \tag{4.210}$$

The circulation distribution $\Gamma(y)$ for a given wing and thus the induced lift and drag still have to be calculated. The notation for the derivation of the *Prandtl wing theory* is indicated in Figure 4.151. The geometric angle of

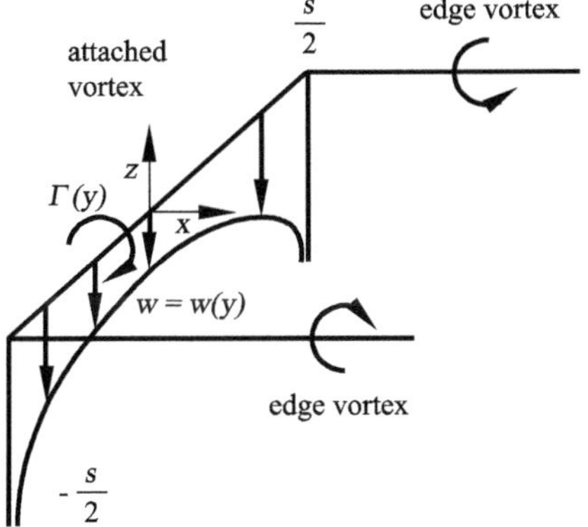

Fig. 4.149. Distribution of the vertical velocity $w(y)$ for a single horseshoe vortex

attack from Figure 4.131 is supplemented by the induced angle of attack α_i, with the free-stream velocity V_∞. This results in the effective angle of attack α_{eff} between the profile chord and the local free stream:

$$\alpha_{\text{eff}} = \alpha - \alpha_i. \tag{4.211}$$

From this we can compute one component of the local lift vector in the direction V_∞, called the induced drag W_i. Denoting the position of the profile by y', we see that the induced angle of attack is

$$\alpha_i(y') = \frac{1}{\tan\left(\frac{-w(y')}{V_\infty}\right)}. \tag{4.212}$$

In general, w is one order of magnitude smaller than V_∞, so that (4.212) yields

$$\alpha_i(y') = -\frac{w(y')}{V_\infty}. \tag{4.213}$$

Using (4.210) we obtain a relation between the induced angle of attack α_i and the circulation distribution $\Gamma(y)$:

$$\alpha_i(y') = \frac{1}{4 \cdot \pi \cdot V_\infty} \cdot \int\limits_{-\frac{s}{2}}^{\frac{s}{2}} \frac{\frac{\mathrm{d}\Gamma}{\mathrm{d}y}}{y' - y} \cdot \mathrm{d}y. \tag{4.214}$$

As shown in Figure 4.151, α_{eff} is the effective angle of attack for the local profile at position y'. As the downward directed vertical velocity varies over the wing span, the effective angle of attack α_{eff} also changes. Therefore, the lift coefficient at the position $y = y'$ is

$$c_a = a' \cdot (\alpha_{eff}(y') - \alpha_{A=0}) = 2 \cdot \pi \cdot (\alpha_{eff}(y') - \alpha_{A=0}). \tag{4.215}$$

Here the increase a' of the lift coefficient has been replaced by the value $2 \cdot \pi$, where the angle $\alpha_{A=0}$ at the lift $A = 0$ varies along a wing with wash with y'. For a wing without wash $\alpha_{A=0}$ is constant and is thus a known quantity for

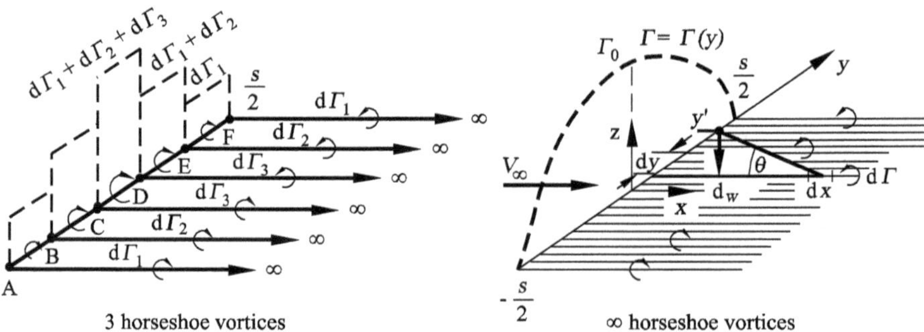

Fig. 4.150. Superposition of horseshoe vortices along the lift line

a given wing. With the Kutta–Joukowski condition we obtain the following lift for the local profile with length $l(y')$:

$$A' = \frac{1}{2} \cdot \rho_\infty \cdot V_\infty^2 \cdot l(y') \cdot c_a = \rho_\infty \cdot V_\infty \cdot \Gamma(y'). \tag{4.216}$$

Therefore, the lift coefficient is

$$c_a = \frac{2 \cdot \Gamma(y')}{V_\infty \cdot l(y')}. \tag{4.217}$$

The effective angle of attack is found using (4.215):

$$\alpha_{eff} = \frac{\Gamma(y')}{\pi \cdot V_\infty \cdot l(y')} + \alpha_{A=0}. \tag{4.218}$$

With $\alpha_{eff} = \alpha - \alpha_i$ and (4.214) we obtain the fundamental equation of the *Prandtl wing theory*:

$$\alpha(y') = \frac{\Gamma(y')}{\pi \cdot V_\infty \cdot l(y')} + \alpha_{A=0}(y') + \frac{1}{4 \cdot \pi \cdot V_\infty} \cdot \int_{-\frac{s}{2}}^{\frac{s}{2}} \frac{\frac{\mathrm{d}\Gamma}{\mathrm{d}y}}{y' - y} \cdot \mathrm{d}y. \tag{4.219}$$

This integrodifferential equation uses the fact that the geometric angle of attack is equal to the sum of the effective angle of attack and the induced angle of attack. The only unknown is the circulation distribution Γ. All other quantities α, l, V_∞, and $\alpha_{A=0}$ are known for a given wing. The solution of (4.219) yields $\Gamma = \Gamma(y')$, where y' varies across the wing span from $y = -s/2$ to $y = s/2$. Using the Kutta–Joukowski condition we can therefore obtain the induced lift:

$$A'(y') = \rho_\infty \cdot V_\infty \cdot \Gamma(y'), \tag{4.220}$$

and the total lift

$$A = \rho_\infty \cdot V_\infty \cdot \int_{-\frac{s}{2}}^{\frac{s}{2}} \Gamma(y) \cdot \mathrm{d}y. \tag{4.221}$$

Using (4.192) we obtain the lift coefficient:

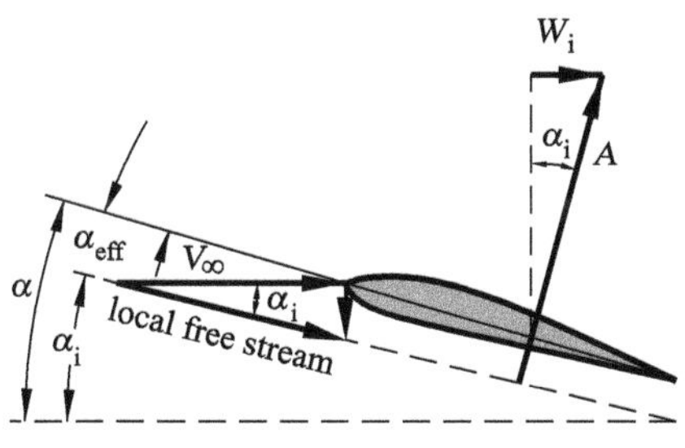

Fig. 4.151. Geometric α, induced α_i and effective α_{eff} angle of attack of a local wing profile

$$c_a = \frac{A}{\frac{1}{2} \cdot \rho_\infty \cdot V_\infty^2 \cdot S} = \frac{2}{V_\infty \cdot S} \cdot \int\limits_{-\frac{s}{2}}^{\frac{s}{2}} \Gamma(y) \cdot \mathrm{d}y, \tag{4.222}$$

with the wing surface area S.

Integration over the wingspan yields the induced total drag:

$$W_i = \rho_\infty \cdot V_\infty \cdot \int\limits_{-\frac{s}{2}}^{\frac{s}{2}} \Gamma(y) \cdot \alpha_i(y) \cdot \mathrm{d}y. \tag{4.223}$$

The drag coefficient is then

$$c_{w_i} = \frac{W_i}{\frac{1}{2} \cdot \rho_\infty \cdot V_\infty^2 \cdot S} = \frac{2}{V_\infty \cdot S} \cdot \int\limits_{-\frac{s}{2}}^{\frac{s}{2}} \Gamma(y) \cdot \alpha_i(y) \cdot \mathrm{d}y. \tag{4.224}$$

Prandtl's wing theory therefore delivers all aerodynamic properties of a given wing. The methods of solution of (4.219), such as the vortex-filament method (*J.D. Anderson Jr.* 1991), are treated in depth in the aerodynamic literature, and so will not be discussed further here.

The different shapes of subsonic wings are shown in Figure 4.152. The wing with elliptical area leads to a minimal induced drag. However, since elliptical wings are difficult to produce, in practice, tapered wings are used, which approximately realize an elliptical lift distribution.

An important result of wing theory is that the induced drag is inversely proportional to the wing span s. In order to keep the induced drag as low as possible, the span s must be chosen as large as possible when the wing is being designed. This was confirmed experimentally on rectangular wings of aspect ratio s/l from 1 to 7 by *L. Prandtl* (1915). The results are summarized

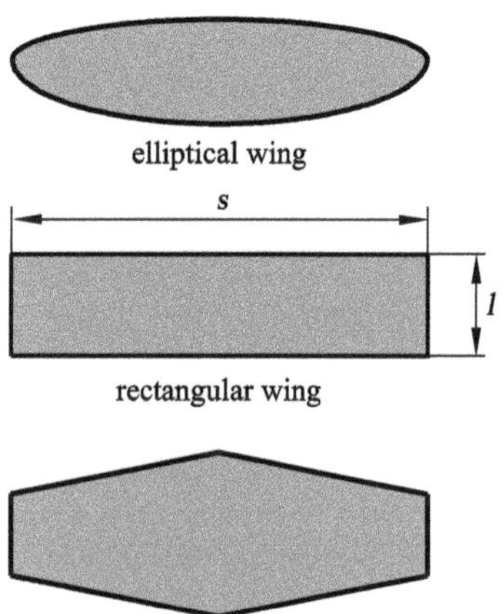

Fig. 4.152. Different shapes of a flat wing

in Figure 4.153. The lift and drag coefficients on the rectangular wing were scaled here with the aspect ratio $s/l = 5$.

Wing Computation

The extension of Prandtl's wing theory to *wings with finite thickness* and the computational methods of, for example, the pressure distribution, are described in the aerodynamics books of *J.D. Anderson jr.* (1991) and *D. Küchemann* (1987). Figure 4.154 shows typical pressure distributions over the surface of subsonic wings. The almost elliptical span distribution is due to the fact discussed above. The large acceleration downstream of the leading edge of the wing leads to different pressure peaks on the upper and lower sides. This is ultimately responsible for the lift of the wing. For the swept subsonic wing, treated in Section 4.4.5, the pressure distribution changes considerably over the wingspan. The pressure peaks are more distinctive at the ends of the wing, a fact that is undesirable in the design of wings.

Until now, we have treated only inviscid wing theory. We know from (4.130) that the total drag c_w and the lift c_a have a friction contribution c_f as well as pressure and induced contributions c_d and c_i. Figure 4.155 presents an overview of the different contributions along the span of a swept subsonic wing at the Reynolds number $\mathrm{Re}_l = 1.7 \cdot 10^6$ with a given lift coefficient $c_a = 0.56$ of a civil aircraft.

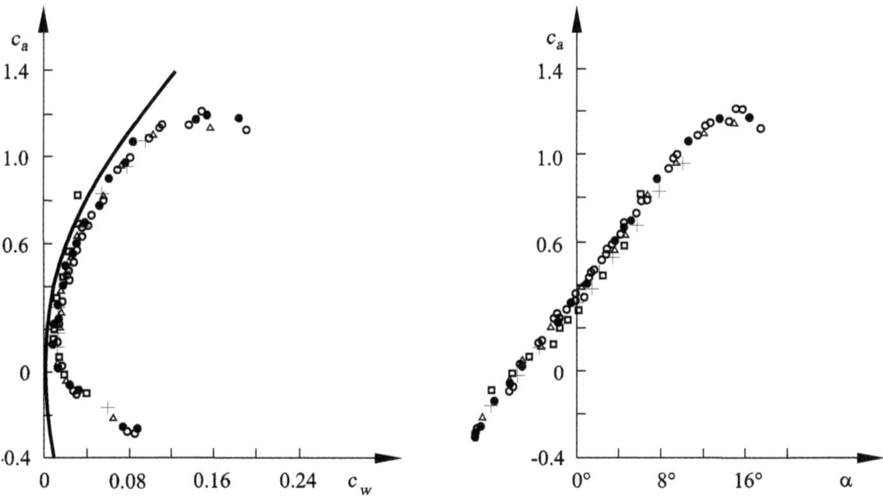

Fig. 4.153. Polar coefficients and lift coefficients of rectangular wings with aspect ratios of $s/l = 1$ to 7, *L. Prandtl* (1915)

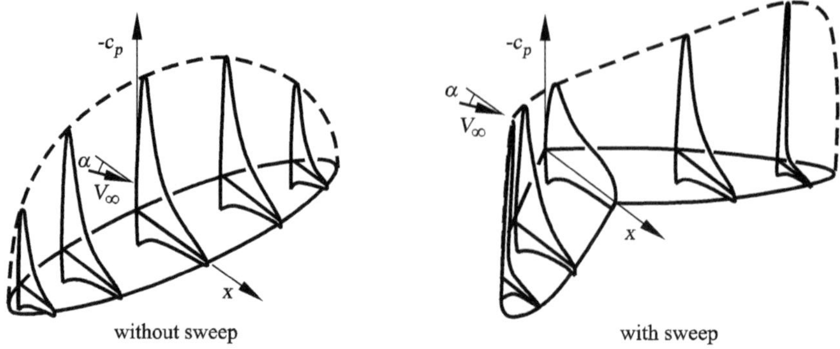

Fig. 4.154. Pressure distributions of long wings *D. Küchemann* (1978)

Numerical Wing Computation

These days, both educational and commercial fluid-mechanics software packages are available to calculate the viscous flow past a wing (in particular, see *H. Oertel Jr.* 2003). These numerically solve the Navier–Stokes equations (5.65) in the laminar flow regime and the Reynolds equations (5.95) in the turbulent regime. The development of numerical methods in fluid mechanics has ranged from finite difference methods (FDM), to finite volume methods (FVM), to adaptive finite element methods (FEM) for unsteady three-

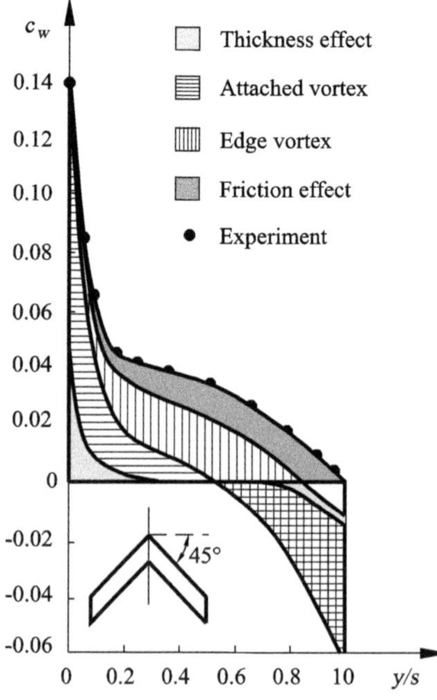

Fig. 4.155. Drag contributions along the span of a swept subsonic wing $\mathrm{Re}_l = 1.7 \cdot 10^6$, *D. Küchemann* (1978)

dimensional flow problems. In parallel to this, spectral methods (SM) have also been developed, to deal particularly with the solution of fluid-mechanical stability problems and direct flow simulation. Of the many different solution algorithms, we have selected the finite volume method (FVM) for the numerical calculation of the wing. As an example we take the transonic wing of a civil airplane. This is treated in detail in the next section. Figure 4.156 shows the procedure in designing and calculating a wing. The *preliminary design* of the wing is carried out using the inviscid Prandtl wing theory. The curvature of the wing profile, the aerodynamic coefficients, and the pressure distribution (as sketched) are fixed provisionally. The second step is the *calculation* of the designed wing, taking into account the sweeping and the warping of the transonic wing. The first calculation of the wing will generally not attain the desired lift coefficient c_a, or else the drag coefficient c_w may still be too large. A further iteration step is then needed, to permit an improved preliminary design with the calculated data. These design iterations are carried out in several steps.

When the required aerodynamic coefficients are satisfied, the third step of the design process takes place, namely the *verification* and *validation* of the wing design in the wind tunnel. Verification is the comparison of the experimental results with the numerical results, as well as the adaptation

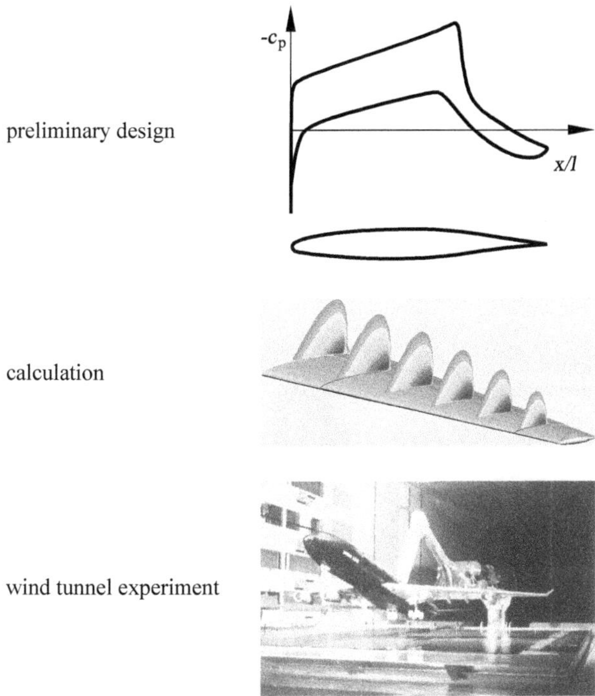

Fig. 4.156. Wing design

of the numerical methods and the instrumentation in the wind tunnel (see Section 4.4.4). Validation requires the further development of the physical models, in particular of turbulence models, in the fundamental equations. This is a very time-consuming process, which has a great influence on the development time of an airplane.

In the verification and validation phases, the calculation, or the preliminary design, is corrected in a few iteration steps, until the initial requirements are satisfied. In each iteration step, a new wind tunnel model has to be built, and the time-consuming measurements in the wind tunnels repeated. The fewer the number of iteration steps that have to be carried out, the more successful the design process. The more precise the numerical methods for calculation, the more efficient the design.

The result of the calculation for the Mach number $M_\infty = 0.78$, the Reynolds number $\mathrm{Re}_l = 26.6 \cdot 10^6$, and the sweep angle $\phi = 20°$ is shown as isobars in Figure 4.157. The numerical solution shows the supersonic field and the denser isobars in the region of the shock wave that concludes this supersonic regime upstream. For the given lift coefficient $c_a = 0.0506$ of a transonic model wing, we calculate a drag coefficient $c_w = 0.0184$. This small drag coefficient is obtained for a laminar wing, where the laminar–turbulent transition on the upper side of the wing is displaced into the shock–boundary-layer region while that on the lower side is moved to the thickest part of the wing. This is attained with a continuously accelerating pressure distribution

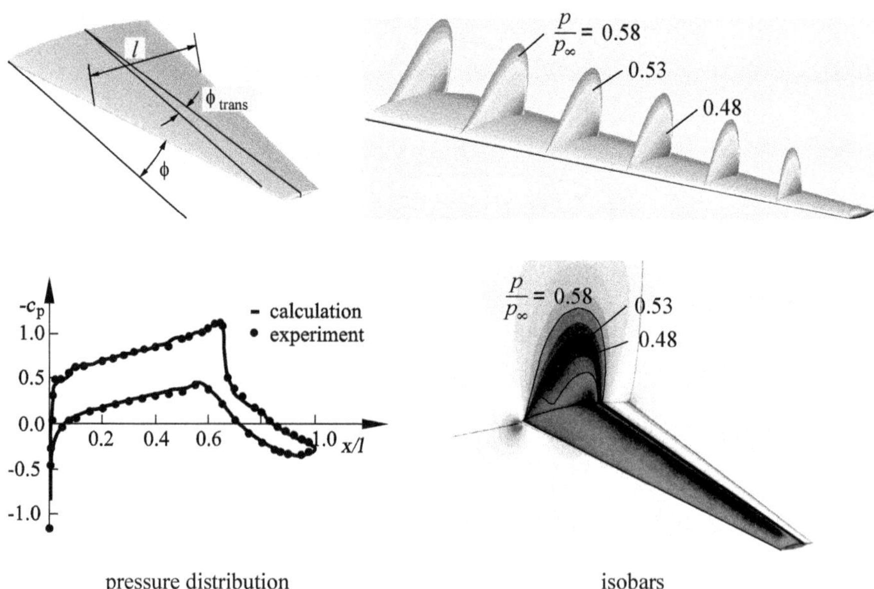

Fig. 4.157. Isobars in profile sections and on the surface of a swept transonic wing, $M_\infty = 0.78, \mathrm{Re}_l = 26.6 \cdot 10^6$, angle of attack $\alpha = 2°$, and angle of sweep $\phi = 20°$

and is associated with a reduction in the drag coefficient (see Figure 4.164). The isobars on the upper side of the wing can be obtained from the load distribution on the wing.

4.4.4 Aerodynamic Facilities

In this section we will not discuss the many different types of wind tunnel and methods of measurement, but rather we will present the *Prandtl-constructed wind tunnel.* Transonic, supersonic, and hypersonic wind tunnels, as well as the associated measuring techniques are treated in the books referred to at the end of the text. The Prandtl, or Göttingen, wind tunnel consists of a closed circuit with open test tracks, where the wing or airplane model to be measured is placed on a scale. Figure 4.158 shows a sketch of the Prandtl wind tunnel. The air is supplied from the ventilator in a continually expanding channel with deflectors of the nozzle of 2 m diameter. The accelerated air reaches the open measurement track, from there moves to the collecting funnel, is decelerated in the diffuser that follows, and is then led back to the blower. The wind tunnel was designed for a wind velocity of 40 m/s, the speed reached by airplanes of that time. The air is smoothed in front of the nozzle by means of a rectifier and screens. These are shown in Figure 4.159. In order to achieve a homogeneous air flow in the measurement track with uniform velocity over the cross-section, the contraction ratio of the nozzle must be chosen appropriately. The pressure drop $p_1 - p_2$ causes the same increase in kinetic energy in all air particles.

Relative fluctuations are essentially compensated by the nozzle contraction. If the ratio of velocities is 1 : 5, that of the stagnation pressure is 1 : 25. Any vortex strength of the free stream has to be reduced by *rectifiers*, systems of parallel channels. The angular velocity of a mass of air rotating around

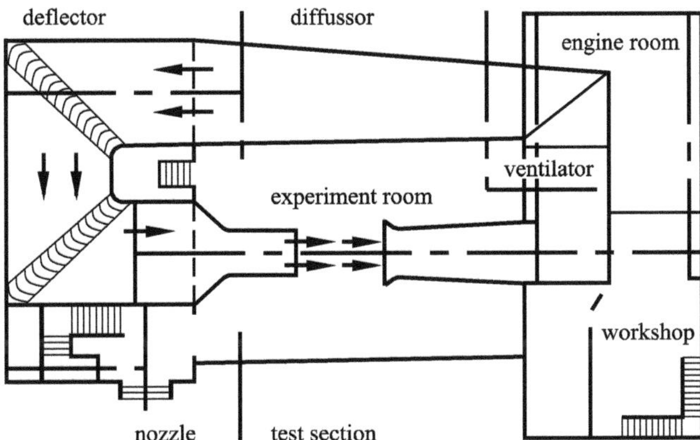

Fig. 4.158. Prandtl-constructed subsonic wind tunnel, *L. Prandtl* (1915)

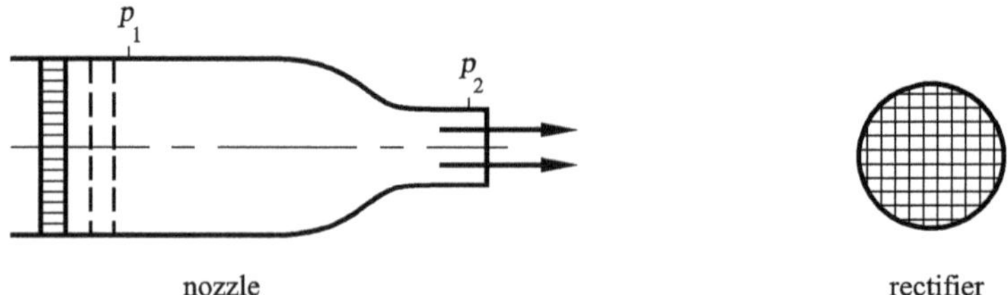

Fig. 4.159. Nozzle with rectifier and screens

an axis parallel to the direction of flow increases n times for a cross-section contraction to $1/n$ of its size.

Since the diameter perpendicular to the streamline is reduced in the ratio $1/\sqrt{n}$, there is an increase in the transverse velocity $(r \cdot \omega)$ in the ratio $\sqrt{n}$, while the longitudinal velocity increases in the ratio n. In contrast, a rotation about an axis perpendicular to the streamline yields a decrease in the angular velocity ω proportional to the decrease in radius r; i.e. it is reduced $1/\sqrt{n}$ times. The perturbation velocity $r \cdot \omega$ is reduced $1/\sqrt{n}$ times. To compensate the longitudinal velocity fluctuations, additional wire mesh screens are used. In addition to these local velocity fluctuations, the turbulent flow also causes velocity fluctuations in time. Uniform fine-meshed screens are placed behind the rectifier to dampen the approaching turbulence. Because of the contraction of the nozzle, the turbulence is also reduced by similar processes to those used in compensating the spatial velocity fluctuations. The longitudinal component of the fluctuation velocity $\sqrt{\overline{u'^2}}$ is reduced much more than the transverse components $\sqrt{\overline{v'^2}}$ and $\sqrt{\overline{w'^2}}$, so that anisotropic turbulence is present directly after the nozzle, although it becomes isotropic again downstream. It is to be noted that the damping screens themselves introduce turbulence into the flow again, although this decreases downstream. It can be reduced by providing a *calming track* between the last screen and the nozzle.

4.4.5 Transonic Aerodynamics, Swept Wings

Civil airplanes with jet engines fly in the transonic subsonic Mach number regime. A typical flight envelope of a civil airplane is shown in Figure 4.160. The flight speed v_∞ in climbing flight is limited by the breaking point of the airplane, although it may not fall below a certain minimum speed, the so-called stalling point. At high altitudes the flight speed is determined by the design Mach number $M_\infty = 0.8$. Below 11 km in altitude, the velocity of sound decreases, leading to higher flight speeds at constant Mach number. Above 11 km the speed of sound is constant. The upper limit of the altitude is set by the design of the pressure cabin.

At transonic subsonic Mach numbers of $M_\infty = 0.8$, the flow is compressible, and the supersonic regime on the wing is concluded by a shock wave.

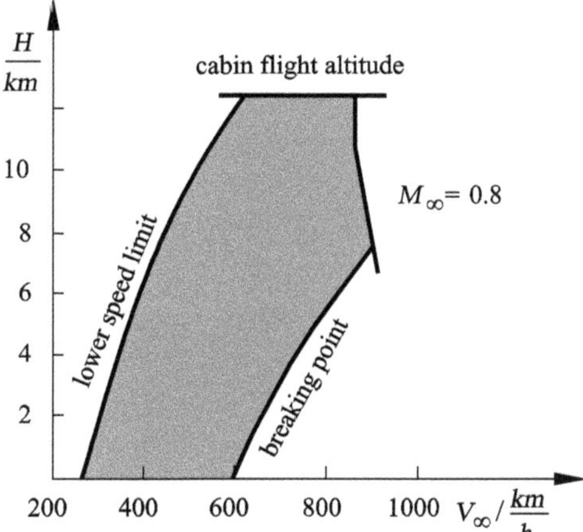

Fig. 4.160. Flight envelope of a civil airplane

The wings of civil aircraft are swept, for the reasons mentioned in Section 4.4.2. This leads to a reduction in the total drag c_w (Figure 4.140). The effect of the sweep was already known in 1939, as documented in Figure 4.162. Because of the sweep, the boundary layer becomes *three-dimensional*, which also affects the laminar–turbulent boundary-layer transition.

Transonic flows past wings are *nonlinear*. For example, the linear increase in the lift coefficient c_a with the angle of attack α for subsonic flows in (4.203) is replaced by a nonlinear progression. In addition, potential theory may no longer be applied to the nonlinear flow, so that numerical methods of solution of the Navier–Stokes and Reynolds equations have to be applied to compute the transonic flow past a wing. The shock wave associated with the shock–boundary-layer interaction on the wing changes the flow field compared to subsonic flow in such a way that the lift coefficient c_a can no longer be computed inviscidly. Figure 4.161 shows the comparison of an inviscidly and a viscously computed flow past a profile at the transonic Mach number $M_\infty =$

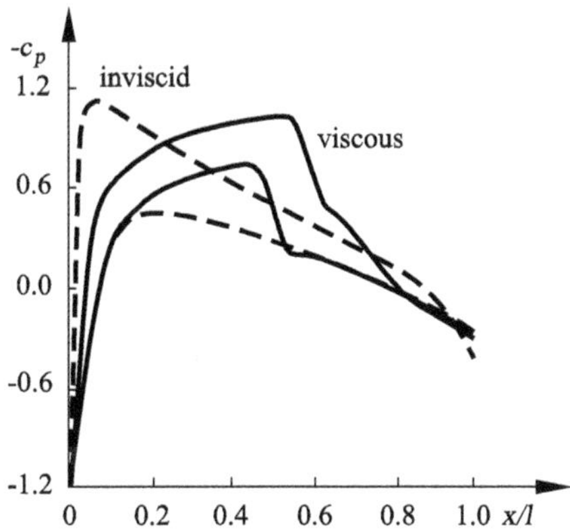

Fig. 4.161. Inviscid and viscous transonic flow past a profile, $M_\infty = 0.82$

0.82. The pressure distribution computed with the linear potential theory has nothing in common with the transonic pressure distribution. Figure 4.162 shows the polar curves of the swept wing in comparison to those of the unswept wing for the Mach number $M_\infty = 0.9$. Because of the sweep, the streamlines in the wing boundary layer are curved. Applying the Bernoulli equation transverse to the streamline at the edge of the boundary layer, we obtain approximately

$$\frac{\partial p}{\partial n} = \rho \cdot \frac{u_\delta^2}{R}, \tag{4.225}$$

with n the direction normal to the streamline, u_δ the velocity at the edge of the boundary layer, and R the local radius of curvature. Because of the no-slip condition we have $\boldsymbol{v} = 0$ at the wall. The pressure is imposed onto the boundary layer, yielding approximately

$$\left.\frac{\partial p}{\partial n}\right|_{z=\delta} = \left.\frac{\partial p}{\partial n}\right|_{z=0} .$$

This pressure gradient perpendicular to the streamline causes a cross-flow component $v(z)$, sketched in Figure 4.163. The laminar–turbulent transi-

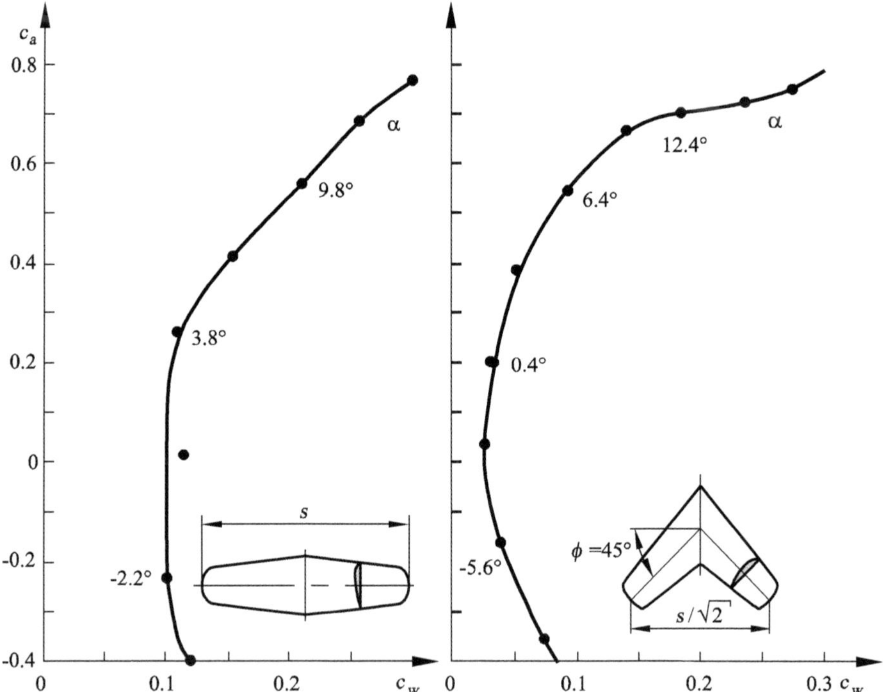

Fig. 4.162. Polar curves of the unswept and swept wing at the transonic Mach number $M_\infty = 0.9$, *H. Ludwieg* (1939)

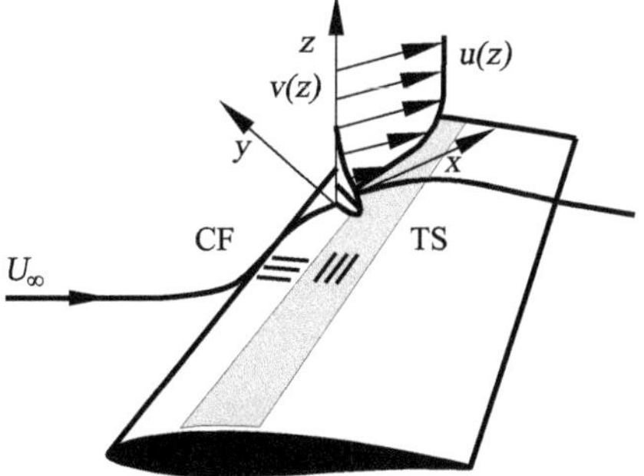

Fig. 4.163. Three-dimensional boundary layer profile of a swept wing, Tollmien–Schlichting waves TS and cross-flow waves CF

tion in the three-dimensional boundary layer is determined not only by the *Tollmien–Schlichting waves* TS. Because of the cross-flow component, additional *cross-flow instabilities* CF occur, to be treated in Section 6.2. The streamline curvature is greatest downstream in the streamlines, so that it can be assumed that the transition to turbulence in the boundary layer takes place in the front region of the transonic wing. The transition line has to be determined suitably when the flow past a wing is computed. At the Mach number $M_\infty = 0.8$ a *shock wave* occurs on the wing. The pressure distributions of a swept transonic wing are sketched in Figure 4.164 for a conventional transonic profile. A strong shock wave, known from Figure 4.133, occurs on the upper side of the wing. Further compression waves, or shock waves, occur from the wing tip or wing–fuselage region. These deflect the three-dimensional supersonic flow in the front region of the wing to a supersonic flow parallel to the camber line (Figure 4.165), and this flow becomes subsonic by means of an almost perpendicular shock wave on the wing. For the pressure distributions shown in Figure 4.164, the laminar–turbulent transition in the three-dimensional wing boundary layer is to be expected at the

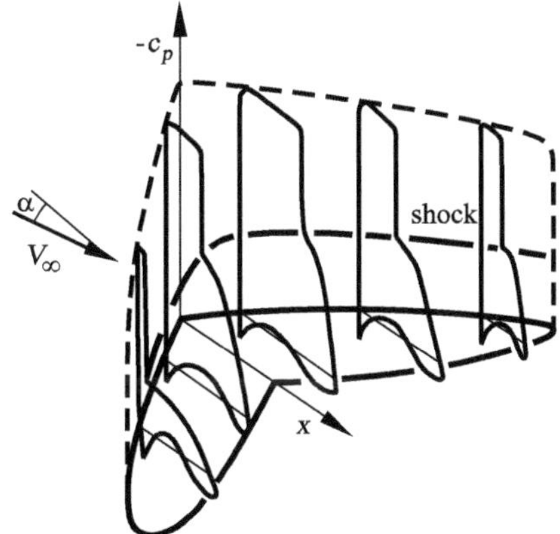

Fig. 4.164. Pressure distributions of a swept wing in a transonic free stream, *D. Küchemann* (1978)

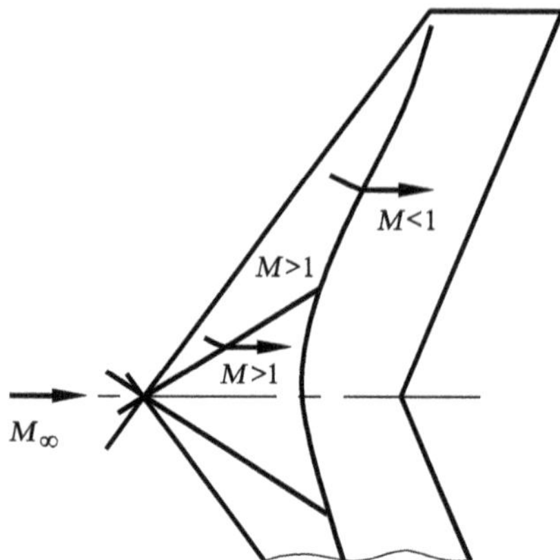

Fig. 4.165. Shock position on a swept wing in a transonic free flow

pressure peak on the lower side and in the shock–boundary-layer interaction region on the upper side.

Because of the high wave drag c_s of strong shock waves, the supercritical profile shown in Figure 4.135 was introduced. The shape of the front region of the wing was chosen so that the subsonic regime is extended downstream and a weakened shock wave occurs in the rear region of the wing. The resultant pressure distribution for a free-stream Mach number of 0.75 is shown as a dashed line in Figure 4.166. If the friction drag c_f of the wing is to be reduced, the wing has to be shaped so that the laminar–turbulent transition in the wing boundary layer is shifted downstream. In addition, the suction tip on the upper side of the wing has to be avoided and a continuous acceleration as far as the shock wave achieved. Such a pressure distribution is shown in Figure 4.166 as a heavy line. It leads to smaller leading-edge radii and steeper pressure increases at the trailing edge. The shape of the profile is chosen so that the onset of the Tollmien–Schlichting waves TS is shifted

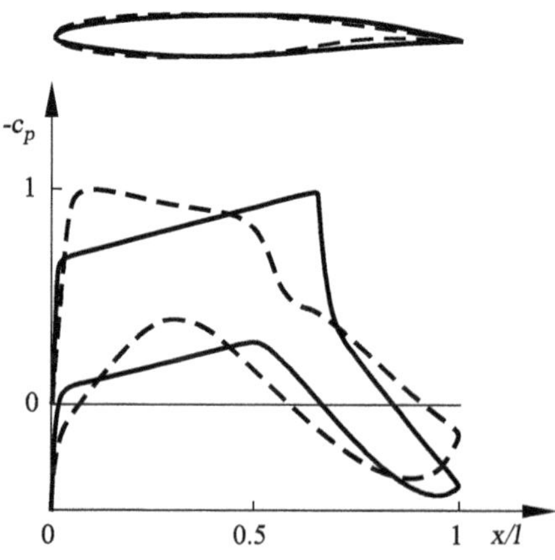

Fig. 4.166. Conventional supercritical profile and laminar profile, $M_\infty = 0.75$, $c_a = 0.5$, $\mathrm{Re}_l = 25 \cdot 10^6$

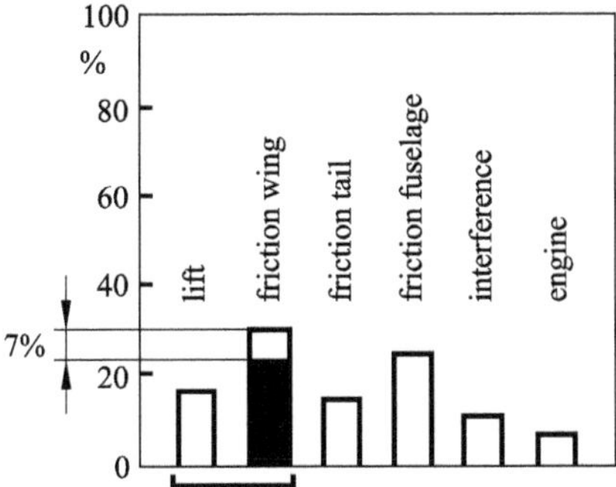

Fig. 4.167. Drag contributions of a civil airplane

downstream into the shock–boundary-layer interaction region. The sweep of the wing also has to be reduced so that no cross-flow instabilities occur at the leading edge. The solution of the Navier–Stokes and Reynolds equations (5.65) and (5.95) for such a *laminar wing* of the transonic free-stream Mach number 0.78 is shown in Figure 4.157. A sweep angle of $\phi = 20°$ is chosen, at which the amplification rate of the cross-flow instabilities close to the leading edge is considerably smaller than the amplification rate of the Tollmien–Schlichting instabilities. The laminar boundary-layer flow is retained right into the shock–boundary-layer interaction region. The extended supersonic region on the transonic wing is concluded by a weak shock wave, seen in Figure 4.157 as the compression of the isobars. The drag contributions for the entire airplane are summarized in Figure 4.167. The contribution due to the wing is 46%. By making the selected wing laminar, a decrease of 15% in the drag is attained. For the airplane this means a reduction potential of

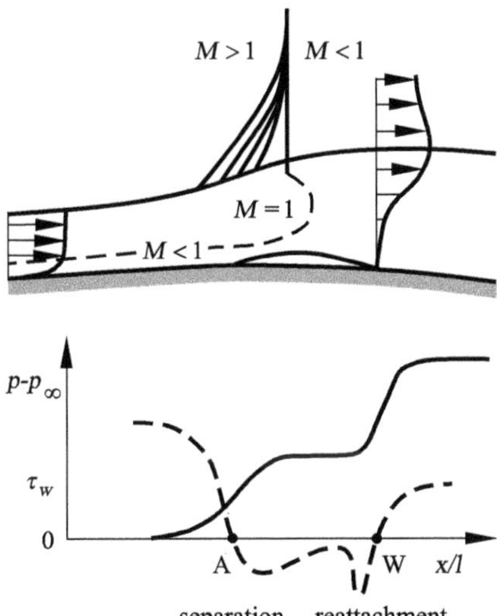

Fig. 4.168. shock–boundary-layer interaction with flow separation, pressure and wall shear stress distribution

about 7%. Further possibilities to reduce the drag are suction of the boundary layer on the wing, influencing the viscous sublayer of the turbulent wing boundary layer with so-called riblets, and influencing the shock–boundary-layer interaction region with a bulge on the wing, to be discussed in the following section.

4.4.6 Shock–Boundary-Layer Interaction

The interaction of the shock wave with the turbulent wing boundary layer leads to an increase in the boundary-layer thickness already in front of the shock wave (Figure 4.168). The thickening of the boundary layer causes pressure perturbations in front of the shock that can lead to an oblique shock wave and then to a branching of the shock. Behind the shock, the boundary layer grows further, which, because of the displacement effect, leads to an additional acceleration of the flow. In the interaction regime the pressure at the wing wall increases in front of the shock. This pressure increase is related to a decrease in the wall shear stress. If the shock wave is strong, the wall shear stress becomes negative, and the boundary layer separates. Because of the acceleration behind the shock and the compensation by the turbulent mixing, both due to the boundary-layer thickening, the separation bubble is reattached. The pressure at the wing wall in the separation bubble is almost constant. There are two fundamentally different ways of calculating the shock–boundary-layer interaction. On the one hand, we can use the numerical methods of wing calculation introduced in the previous section, with a fine resolution of the interaction region. Results based on the shock–boundary-layer interaction are presented at the end of the section. On the other hand, there is the possibility to derive approximate solutions of the shock–boundary-layer interaction using semianalytical methods and a zonal

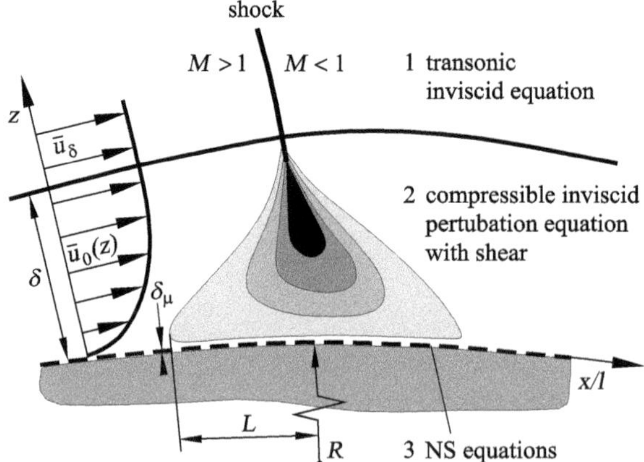

Fig. 4.169. Flow model in the shock–boundary-layer interaction regime

division of regions of the two-dimensional turbulent boundary layer of a transonic wing (see *R. Bohning* 1982).

In order to make the shock–boundary-layer interaction accessible to an analytical calculation, the zonal flow model in Figure 4.169 is used. The two-dimensional Navier–Stokes equations may then be simplified according to the physical properties of each region.

In the inviscid outer region of the turbulent boundary-layer flow, the nonlinear potential equation of transonic flow holds. The turbulent boundary layer is divided into two further subregions according to the discussions in Section 4.2.5. The outer part 2 of the boundary layer is modeled by a turbulent, compressible shear layer in which the effect of friction appears only via a given time-averaged velocity profile $\overline{u}_0(z)$ in the otherwise inviscid perturbation equations. Region 3 close to the wall is the viscous sublayer of thickness δ_μ. Friction acts in this layer, and so it is here that the complete Navier–Stokes equations must be solved. The name perturbation equation is due to the fact that the basic flow 0 is perturbed by a weak vertical shock wave. In what follows we treat the approximate solution of this inviscid perturbation equation in the boundary layer region 2, leading us ultimately to the application of the analytical method of separation.

We first determine the basic flow $u_0(z)$. The dependence of the basic flow quantities on the downstream coordinate x is neglected. This is permissible only if the curvature of the wing profile may be assumed to be suitably small and the region under discussion in the x direction of length L not too large. This leads to a two-dimensional discussion of the interaction region. The compressible steady basic flow profile is then given by the time-averaged turbulent quantities: the velocity $\overline{u}_0(z)$, the density $\overline{\rho}_0(z)$, the temperature $\overline{T}_0(z)$, as well as the pressure $\overline{p}_0$. Apart from the pressure, in this local discussion all quantities depend on the z coordinate normal to the wall. According to the boundary-layer approximation $\partial\overline{p}_0/\partial z$, the pressure of the basic flow $\overline{p}_0$ is a constant.

In deriving the perturbation differential equations, we start out from the two-dimensional compressible boundary-layer equations

$$\frac{\partial(\rho \cdot u)}{\partial x} + \frac{\partial(\rho \cdot w)}{\partial z} = 0, \tag{4.226}$$

$$\rho \cdot \left(u \cdot \frac{\partial u}{\partial x} + w \cdot \frac{\partial u}{\partial z} \right) = -\frac{\partial p}{\partial x} + \mu \cdot \frac{\partial^2 u}{\partial z^2}. \tag{4.227}$$

Because of the shock wave encroaching into the boundary layer, a pressure gradient $\partial p/\partial z$ in the z direction normal to the wall must be taken into account. Figure 4.169 shows, however, that the characteristic length region in the x direction and the boundary layer thickness δ in region 2 are of the same order of magnitude. Therefore, in the boundary-layer case at very large Reynolds numbers Re_l and for $\delta/L \approx 1$, the friction terms in the z direction vanish. The equation of motion in the z direction then becomes

$$\rho \cdot \left(u \cdot \frac{\partial w}{\partial x} + w \cdot \frac{\partial w}{\partial z} \right) = -\frac{\partial p}{\partial z}. \tag{4.228}$$

Using the energy equation

$$c_p \cdot T + \frac{u^2}{2} = \text{const} \tag{4.229}$$

and the equation of state of the ideal gas

$$\frac{p}{\rho} = R \cdot T, \tag{4.230}$$

we obtain five equations to determine the five dependent variables u, w, p, ρ, and T. With the perturbation ansatz

$$u = \overline{u}_0(z) + u', \qquad w = w', \\ p = \overline{p}_0 + p', \qquad \rho = \overline{\rho}_0(z) + \rho', \qquad T = \overline{T}_0(z) + T', \tag{4.231}$$

and neglecting the product of perturbation quantities (linearization), we obtain the perturbation differential equations

$$\overline{\rho}_0 \cdot \frac{\partial u'}{\partial x} + \overline{u}_0 \cdot \frac{\partial \rho'}{\partial x} + \frac{\partial(\overline{\rho}_0 \cdot w')}{\partial z} = 0, \tag{4.232}$$

$$\overline{\rho}_0 \cdot \overline{u}_0 \cdot \frac{\partial u'}{\partial x} + \overline{\rho}_0 \cdot w' \cdot \frac{\mathrm{d}\overline{u}_0}{\mathrm{d}z} = -\frac{\partial p'}{\partial x} + \mu \cdot \left(\frac{\mathrm{d}^2 \overline{u}_0}{\mathrm{d}z^2} + \frac{\partial^2 u'}{\partial z^2} \right), \tag{4.233}$$

$$\overline{\rho}_0 \cdot \overline{u}_0 \cdot \frac{\partial w'}{\partial x} = -\frac{\partial p'}{\partial z}. \tag{4.234}$$

The dashed flow quantities u', w', p', ρ', and T' are the perturbations in the flow field due to the shock. In contrast to the quantities of the basic flow, they are dependent on both spatial coordinates x and z.

After linearization, the energy equation and the equation of state yield the equations

$$\overline{u}_0 \cdot u' + c_p \cdot T' = 0, \tag{4.235}$$

$$\overline{\rho}_0 \cdot T' + \rho' \cdot \overline{T}_0 = p' \cdot \frac{\overline{\rho}_0 \cdot \overline{T}_0}{\overline{p}_0}. \tag{4.236}$$

Introducing the critical values of both Mach numbers $M = 1$ as reference values, with the critical speed of sound a_k and p_k, ρ_k, T_k, we obtain the dimensionless perturbation differential equations

$$\overline{\rho}_0 \cdot \frac{\partial u'}{\partial x} + M_k \cdot \frac{\partial \rho'}{\partial x} + \frac{L}{\delta} \cdot \frac{\partial(\overline{\rho}_0 \cdot w')}{\partial z} = 0, \tag{4.237}$$

$$\overline{\rho}_0 \cdot M_k \cdot \frac{\partial u'}{\partial x} + \overline{\rho}_0 \cdot w' \cdot \frac{L}{\delta} \cdot \frac{\mathrm{d}M_k}{\mathrm{d}z} = -\frac{1}{\kappa} \cdot \frac{\partial p'}{\partial x} + \frac{1}{\mathrm{Re}_\delta} \cdot \frac{L}{\delta} \cdot \left(\frac{\mathrm{d}^2 M_k}{\mathrm{d}z^2} + \frac{\partial^2 u'}{\partial z^2} \right), \tag{4.238}$$

$$\overline{\rho}_0 \cdot M_k \cdot \frac{\partial w'}{\partial x} = -\frac{1}{\kappa} \cdot \frac{L}{\delta} \cdot \frac{\partial p'}{\partial z}, \tag{4.239}$$

with the characteristic length L for the flow coordinate x and the boundary-layer thickness δ for the coordinate normal to the wall z. The dimensionless characteristic numbers $\mathrm{Re}_\delta = a_k \cdot \rho_k \cdot \delta/\mu$ and $M_k = \overline{u}_0/a_k$ with the critical velocity of sound $a_k^2 = \kappa \cdot p_k/\rho_k$ appear. In region 2, L and δ are of the same order of magnitude, so that $L/\delta = 1$ may be set. For $\mathrm{Re}_\delta \gg 1$, therefore, the frictional terms in the perturbation differential equations (4.237)–(4.239) may be neglected. The friction enters only indirectly via the velocity profile $u_0(z)$ of the given basic flow. This yields a simplified system of differential equations for region 2:

$$\overline{\rho}_0 \cdot \left(1 + \overline{\rho}_0 \cdot (\kappa - 1) \cdot M_k^2\right) \cdot \frac{\partial u'}{\partial x} + \overline{\rho}_0 \cdot M_k \cdot \frac{\partial p'}{\partial x} + \frac{\partial(\overline{\rho}_0 \cdot w')}{\partial z} = 0, \qquad (4.240)$$

$$\overline{\rho}_0 \cdot M_k \cdot \frac{\partial u'}{\partial x} + \overline{\rho}_0 \cdot w' \cdot \frac{\mathrm{d}M_k}{\mathrm{d}z} = -\frac{1}{\kappa} \cdot \frac{\partial p'}{\partial x}, \qquad (4.241)$$

$$\overline{\rho}_0 \cdot M_k \cdot \frac{\partial w'}{\partial x} = -\frac{1}{\kappa} \cdot \frac{\partial p'}{\partial z}. \qquad (4.242)$$

We also have the dimensionless energy equation and equation of state

$$T' + (\kappa - 1) \cdot M_k \cdot u' = 0, \qquad (4.243)$$

$$\overline{\rho}_0 \cdot T' + \rho' \cdot \overline{T}_0 = p'. \qquad (4.244)$$

By eliminating u', ρ', T', the system of equations (4.240)–(4.242) may be transformed to a system of two equations in the two unknowns p' and w', to which the analytical method of separation may be applied:

$$\frac{1}{\kappa} \cdot (M_0^2 - 1) \cdot \frac{\partial p'}{\partial x} - \overline{\rho}_0 \cdot w' \cdot \frac{\mathrm{d}M_k}{\mathrm{d}z} + \overline{\rho}_0 \cdot M_k \cdot \frac{\partial w'}{\partial z} = 0, \qquad (4.245)$$

$$\frac{1}{\kappa} \cdot \frac{\partial p'}{\partial z} + \overline{\rho}_0 \cdot M_k \cdot \frac{\partial w'}{\partial x} = 0. \qquad (4.246)$$

The boundary value problem for p' and w' still has to be formulated for (4.245), (4.246). On the one hand, this is because boundary values are given by the shock on the outer edge of the boundary layer of region 2, while on the other hand, the viscous sublayer of region 3 has boundary conditions

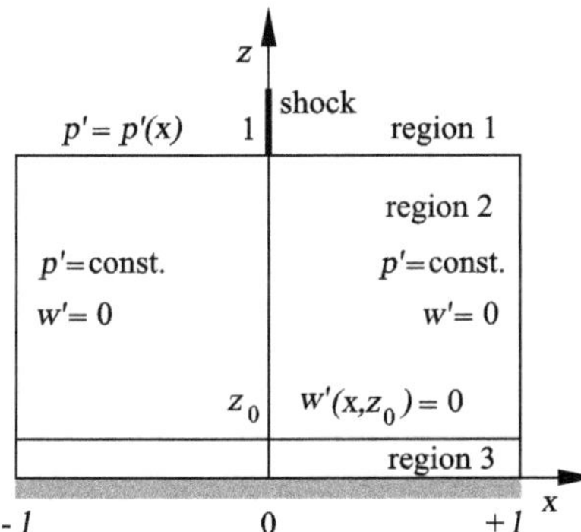

Fig. 4.170. Boundary conditions of the perturbation problem

at the wall that have to be satisfied. As derivatives of both perturbation quantities p' and w' with respect to x and to z appear, we need to formulate four boundary conditions, as shown in Figure 4.170. At the outer edge of the boundary layer between region 2 and region 1, the pressure distribution of the outer flow is imposed on region 1. The pressure perturbation p' is therefore given at position $z = 1$ for all x:

$$p' = p'(x, 1) \qquad \text{for} \qquad z = 1. \tag{4.247}$$

At a sufficiently large distance upstream and downstream of the shock, at the dimensionless coordinates $x = \pm l$, the perturbation velocity w' must vanish, to guarantee a continuous transition to the basic flow. We obtain the two boundary conditions

$$\begin{aligned} w' &= 0 \qquad \text{for} \qquad x = +l, \\ w' &= 0 \qquad \text{for} \qquad x = -l. \end{aligned} \tag{4.248}$$

For the viscous sublayer in region 3 we have the known boundary condition that the pressure along the wall coordinate z is constant for all x:

$$\frac{\partial p'(x, z_0)}{\partial z} = 0 \qquad \text{for} \qquad z = z_0 = \frac{\delta_\mu}{\delta}.$$

It then follows from (4.246) that $\partial w'/\partial x = 0$. Together with the condition (4.248), we obtain the fourth boundary condition:

$$w'(x, z_0) = 0 \qquad \text{with} \qquad z_0 = \frac{\delta_\mu}{\delta}. \tag{4.249}$$

Figure 4.171 shows the calculated pressure distribution, plotted against the downstream coordinate x/δ. The diagram shows the distribution of the wall pressure for $z = 0$ compared to experimental results. It can easily be seen how the pressure jump in the outer flow caused by the shock wave is spread out by the effect of friction. A weak shock wave was assumed, so that the flow separation sketched in Figure 4.168 does not occur.

Shock–Boundary-Layer Control

The thickening of the turbulent boundary layer in the interaction regime causes an increase in the total drag of the wing. In order to reduce this in-

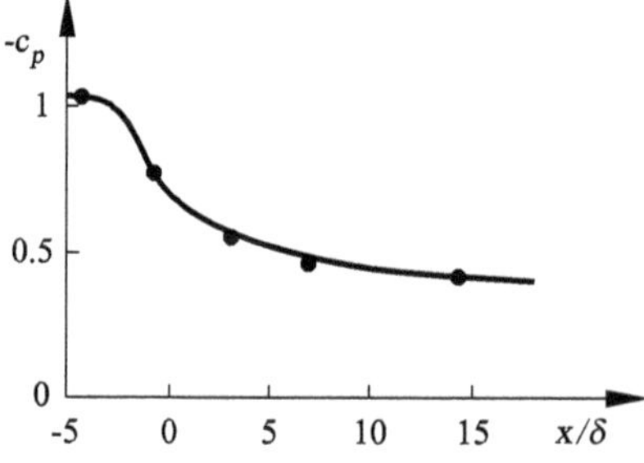

Fig. 4.171. Calculated pressure distribution $-c_p$ at the wall of a transonic profile compared to experimental results •

crease in drag, attempts were initially made to weaken the shock and thus reduce the wave drag using a pressure compensation chamber in the wing. Figure 4.172 shows the effect of the chamber. The passive pressure compensation takes place at a porous part of the wall in the shock region. Behind this wall is the compensation chamber, which permits partial pressure compensation in the shock–boundary-layer interaction region via self-induced ventilation flow. The ventilation affects the displacement effect of the boundary layer so that the structure of the shock wave is altered, and instead of a strong shock wave, a weakened shock wave is formed. As a consequence of the shock weakening, the wave drag and friction drag are reduced and separation bubbles close to the wall avoided.

The isomach lines in the shock regime computed with the Reynolds equations (5.95) show shock-induced thickening for the uninfluenced transonic profile, as well as the postexpansion regime already discussed. In front of the shock a pre-compression takes place, which leads to the shock branching described. This branching is stronger, the higher the chosen chamber pressure. The oblique shock wave occurs at the start of the ventilation chamber. This has the additional effect that the shock is fixed at the start of the influencing zone. Because of the pressure difference in front of and behind the shock branching, a secondary flow through the wall perforation and into the ventilation chamber occurs. This has the consequence that air is blown out of the front region of the compensation chamber, and the displacement thickness and turbulence intensity in the boundary layer increase at that point. Behind the oblique shocks, the flow is sucked, reducing the growth of the boundary layer downstream. Since the entropy and thus the wave drag increase with the third power of the shock strength, the wave drag across two weakened oblique shock waves is smaller than that across a single vertical shock. In this manner the pressure compensation chamber achieves the desired drag reduction.

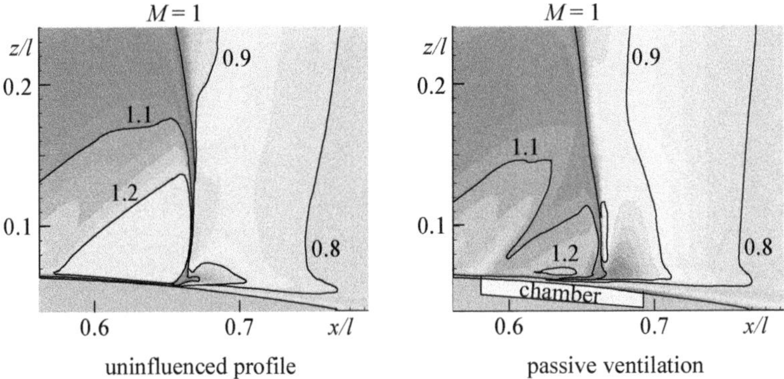

Fig. 4.172. Isomach lines of the shock–boundary-layer interaction, effect of a compensation chamber $M_\infty = 0.76$, $\mathrm{Re}_l = 6 \cdot 10^6$, $\alpha = 2°$

One further method to reduce the drag is a specific *change in the contour* in the shock regime. This is simpler to construct on a wing than a pressure compensation chamber. By means of a slight *arching, a streamline such as that formed by the passive ventilation* at a compensation chamber is copied.

Figure 4.173 shows the wing solution of Figure 4.157 with a bulge. Again shock-wave branching takes place. With the bulge the boundary layer is not perturbed by an additional flow out of the compensation chamber, and so the turbulence intensity in the interaction regime remains smaller and the boundary layer does not thicken so much. As with the compensation chamber, the bulge prevents shock-induced separation. The curvature increase at the bulge causes the postexpansion regime to be extended, further reducing the separation tendency. Altogether, a *reduction in the total drag of* 8% is achieved, and the lift of the wing additionally slightly improved.

4.4.7 Flow Separation

It has already been shown in Section 4.4.2 that the flow on the wing separates above a critical angle of attack α_{crit} (Figure 4.136). The increased displacement leads to an increase in the pressure and friction drag, while the lift simultaneously drops (Figure 4.138). As the angle of attack α increases, flow separation on the wing takes place with a separation bubble that is steady

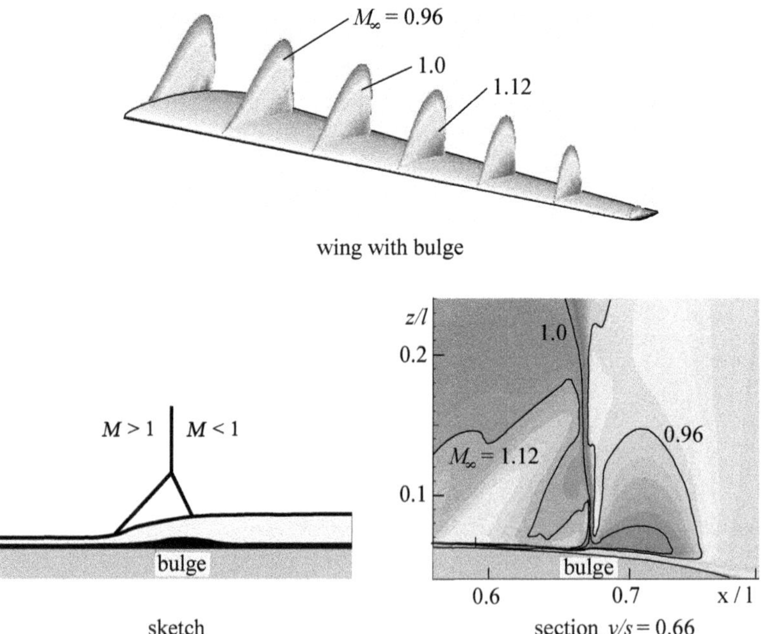

Fig. 4.173. Isomach lines of the shock–boundary-layer interaction, effect of a bulge $M_\infty = 0.78$, $\text{Re}_l = 27 \cdot 10^6$, $\Phi = 20°$, $\alpha = 2°$

in the time average. The separation line A and the reattachment line W are half-saddle lines S′ (Figure 4.174), following the notation from Section 3.3. As the angle of attack is increased, a secondary separation takes place, leading to two further half-saddles. At the front part of the wing, the separation initially remains steady, in the time average. However downstream, an open flow surface forms that is part of an unsteady three-dimensional flow separation, also called *buffeting*. The third illustration in Figure 4.174 shows all flow surfaces in the flow field. The separation surfaces roll up and form a vortex street. The secondary separation now leads to a second vortex street, since the flow close to the wall can no longer move against the pressure gradient caused by the primary vortex separation.

In earlier chapters we have already used *Prandtl's separation criterion*, according to which the wall shear stress τ_w is zero on both the separation and reattachment lines. This is related to a branching of the wall streamlines, leading to a singular half-saddle S′. However, this separation criterion is restricted to two-dimensional flow. For three-dimensional flow, the discussion of flow separation at a delta wing (Figure 3.9) has already shown that the wall streamlines on the wing converge to a separation line that forms a separation surface in the flow field. The Prandtl separation criterion $\tau_w = 0$ for three-dimensional flow separation is therefore replaced by the criterion of *convergence of the wall streamlines*.

Figure 4.175 shows two possibilities of three-dimensional separation. The first illustration shows the three-dimensional separation bubble, and the second the formation of a free shear surface that leads to a vortex street. In the separation bubble case, the backflow in the bubble is separated from the main flow by a three-dimensional shear layer. This shear layer leads to Kelvin–Helmholtz instabilities, which do not, however, change the position of the separation bubble in the time average. The free shear surface of the second figure leads to a flow surface bifurcation line at the wall and the separation surface that rolls up downstream as in Figure 4.174 and forms an unsteady vortex street. Prandtl's separation criterion $\tau_w = 0$ cannot be applied for three-dimensional flow separation, and an additional theory of flow surface bifurcation is necessary. Many three-dimensional separation criteria

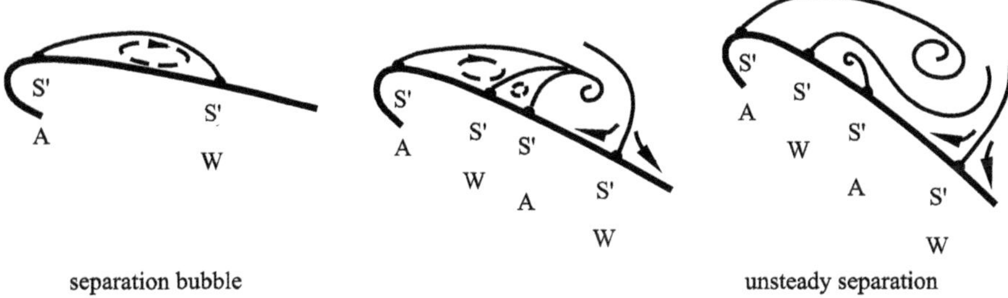

Fig. 4.174. Flow separation on a wing at increasing angle of attack α

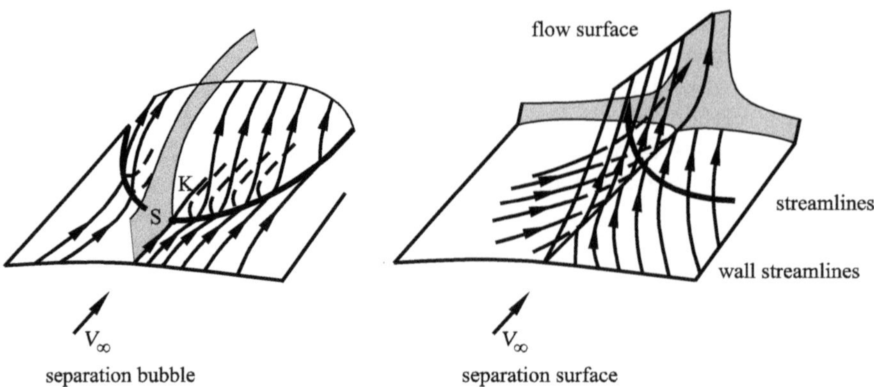

Fig. 4.175. Three-dimensional flow separation

are mentioned in the references, but they have not yet led to a conclusive theory.

4.4.8 Supersonic Aerodynamics, Delta Wings

The aerodynamics of supersonic flight is fundamentally different from that of subsonic flight. This is because of the shock waves at the tip and end of the profile, as discussed in Section 4.4.2. The shape of the wing for supersonic flight is to be chosen so that the wave drag and thus the shock strength are kept as small as possible. This can amount to up to half of the total drag. The oblique shocks in the head and tail waves are weaker, the smaller the sweep angle of the wing and the sharper the leading edge of the wing. In the supersonic case this leads to *delta wings*. Their aerodynamics are determined by the shock waves as well as the leading edge separation and the resulting vortex system on the wing (Figure 4.176). This causes the additional lift, which becomes larger with increasing angle of attack.

If we consider the dependence of lift-to-drag coefficient ratio c_a/c_w on the flight Mach number in Figure 4.177, three airplane shapes are seen. The civil airplane with swept wings in the transonic subsonic regime was considered in the previous section. At Mach number 0.7 the value of c_a/c_w is 16. At Mach number $M_\infty = 1$ the ratio of c_a/c_w drops, because of the increasing wave drag. A slender supersonic airplane with delta wings can reach c_a/c_w values of up to 8 at the Mach number $M_\infty = 2$.

In the *supersonic flow* of a delta wing (Figure 4.176), two different situations can occur. If the Mach line (see Section 4.3.1) lies in front of the wing edge, as in Figure 4.178, the normal component of the free-flow velocity v_n is smaller than the speed of sound a_∞. We then have a *subsonic leading edge* with $\alpha' > \phi$ and $v_n < a_\infty$. On the other hand, if the Mach line lies behind the wing edge, the situation is that of a *supersonic leading edge* with $\alpha' < \phi$ and $v_n > a_\infty$. This division into subsonic and supersonic is important not only for the leading edge, but also for the trailing edge of the wing. If there is a subsonic trailing edge, the Kutta condition can be applied, and pressure compensation occurs between the lower and upper sides of the wing. In the case of a supersonic trailing edge, oblique shock waves occur, and these lead to an unsteady change in the flow quantities. A finite pressure difference exists between the upper and lower sides of the wing. There is a sharp bend in the pressure distribution along the chord of the wing, as shown in Figure 4.178.

Assuming weak shock waves (small perturbations), the linearized potential equation (4.20) can be applied for inviscid supersonic flow, as for subsonic flow:

$$(1 - M_\infty^2) \cdot \frac{\partial^2 \Phi}{\partial x^2} + \frac{\partial^2 \Phi}{\partial y^2} + \frac{\partial^2 \Phi}{\partial z^2} = 0. \qquad (4.250)$$

Again the flow behaves linearly. We have already used the *Prandtl–Glauert rule* in the subsonic regime and the *Ackeret rule* in the supersonic regime in Section 4.4.2. In order to derive these similarity rules, we carry out a transformation of the potential equation (4.250). This transformation should be such that the Mach number of the free flow no longer appears explicitly in the transformed potential equation. We assume a transformed reference flow as follows:

$$x' = x, \qquad y' = C_1 \cdot y, \qquad z' = C_1 \cdot z, \qquad \Phi' = C_2 \cdot \Phi. \qquad (4.251)$$

The factor C_1 is determined so that the Mach number drops out. This yields $C_1 = \sqrt{1 - M_\infty^2}$ for subsonic velocities $M_\infty < 1$ and $C_1 = \sqrt{M_\infty^2 - 1}$ for

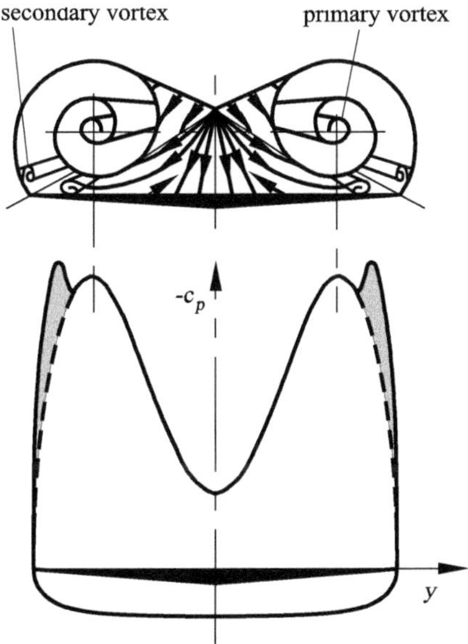

Fig. 4.176. Vortex formation and pressure distribution at a section of a delta wing

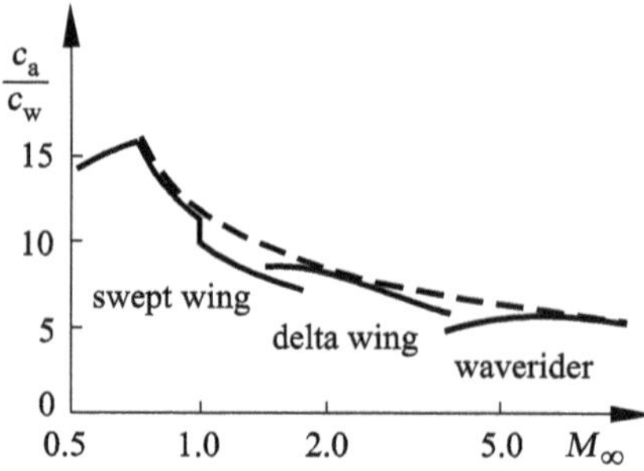

Fig. 4.177. Dependence of the lift to drag coefficient ratio c_a/c_w on the Mach number

supersonic velocities $M_\infty > 1$. The transformed potential equation of the reference flow for subsonic flow yields

$$\frac{\partial^2 \Phi'}{\partial x'^2} + \frac{\partial^2 \Phi'}{\partial y'^2} + \frac{\partial^2 \Phi'}{\partial z'^2} = 0, \tag{4.252}$$

and that for supersonic flow yields

$$\frac{\partial^2 \Phi'}{\partial x'^2} - \frac{\partial^2 \Phi'}{\partial y'^2} - \frac{\partial^2 \Phi'}{\partial z'^2} = 0. \tag{4.253}$$

The transformed equation of subsonic flow is identical to the potential equation for incompressible flow. The transformed equation for supersonic flow is identical to the linearized potential equation (4.250) for the Mach number $M_\infty = \sqrt{2}$. The transformation shows that the calculation of supersonic flows for arbitrary Mach numbers may be reduced to that for $M_\infty = \sqrt{2}$. The transformation (4.251) is called the *Prandtl–Glauert–Ackeret similarity rule* in wing theory.

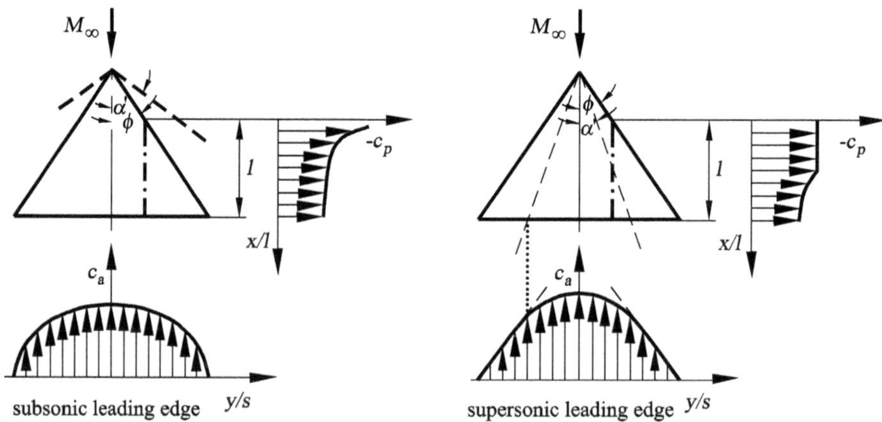

Fig. 4.178. Pressure distribution along the wing chord l and lift distribution along the wing span s of a delta wing

For a given delta wing, we obtain the transformed wing by shrinking or increasing its dimensions transverse to the free flow direction by the factor C_1 corresponding to (4.251). Figure 4.179 shows the transformation of a given delta wing for different Mach numbers. The transformed wings for subsonic Mach numbers $M_\infty < 1$ were computed for incompressible flow $M_\infty = 0$ and for supersonic Mach numbers $M_\infty > 1$ at the Mach number $M_\infty = \sqrt{2}$.

The *Prandtl–Glauert–Ackeret rule* may also be carried over to the profile section and the angle of attack. The transformed thickness ratio d'/l' and the transformed angle of attack α' are computed from

$$\frac{d'}{l'} = \frac{d}{l} \cdot \sqrt{|1 - M_\infty^2|}, \qquad \alpha' = \alpha \cdot \sqrt{|1 - M_\infty^2|}. \tag{4.254}$$

For $M_\infty < \sqrt{2}$, the transformed wing has a smaller thickness as well as a smaller angle of attack than the given wing. For $M_\infty > \sqrt{2}$ a greater thickness and angle of attack are found.

The transformation of the *pressure distribution* is obtained from (4.251) and

$$c_p = -2 \cdot \frac{u}{u_\infty} = -\frac{2}{u_\infty} \cdot \frac{\partial \Phi}{\partial x}, \qquad c_p' = -2 \cdot \frac{u'}{u_\infty} = -\frac{2}{u_\infty} \cdot \frac{\partial \Phi'}{\partial x'}, \tag{4.255}$$

where the free flow u_∞ is the same size for the given and the transformed wings. With (4.251) we obtain

$$c_p = C_2 \cdot c_p'. \tag{4.256}$$

The transformation factor C_2 is determined from the streamline analogy of both wings. From $w = \partial \Phi / \partial z$ and $w' = \partial \Phi' / \partial z'$, we obtain

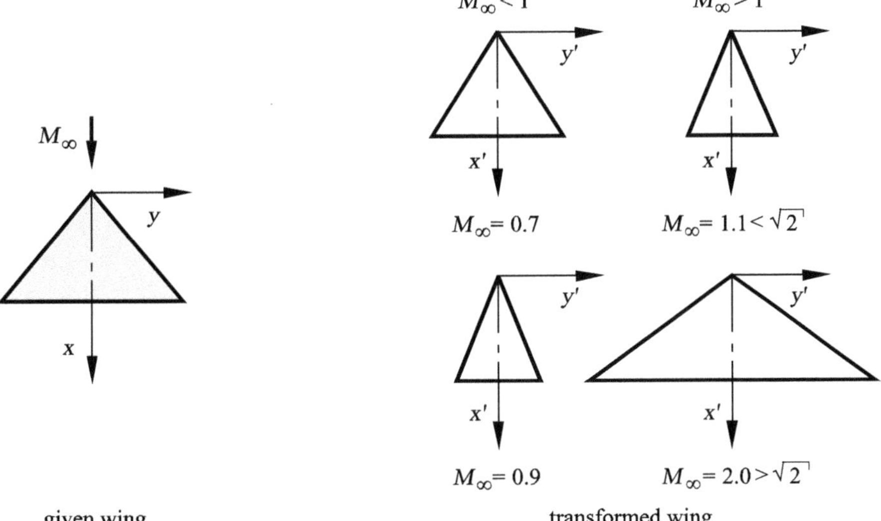

Fig. 4.179. Application of the *Prandtl–Glauert–Ackeret rule* at a delta wing

$$C_1^2 \cdot C_2 = 1,$$

and with $C_1 = \sqrt{|1 - M_\infty^2|}$ we obtain

$$C_2 = \frac{1}{|1 - M_\infty^2|}.$$

This leads to the pressure distribution

$$c_p = \frac{c_p'}{|1 - M_\infty^2|}. \tag{4.257}$$

If we carry out the transformation in such a way that only the dimensions in the y direction (wingspan) are distorted, while the dimensions in the z direction (profile and angle of attack) remain unchanged, the transformation in (4.254) is inverted. We then obtain the pressure coefficient

$$c_p = \frac{c_p'}{\sqrt{|1 - M_\infty^2|}}. \tag{4.258}$$

This relation, already used in Section 4.4.2, is shown in Figure 4.180.

A delta wing designed for supersonic flight also has to have good slow-flight properties for takeoff and landing in the subsonic regime. The vortex system on the delta wing, discussed in Section 3.3 (Figures 3.9 and 4.176), has to be stable in the entire Mach number range, in order to guarantee continuous lift. This requires a subsonic leading edge of the delta wing, for example at a flight Mach number of $M_\infty = 2$. The sweep angle of the wing is chosen so that an approximately conical flow forms (Figure 4.181), which causes a wave drag as small as possible. The angle of attack of the delta wing is restricted by the occurrence of unsteady vortex separation or the bursting of the wing. This limiting angle is reached at about $\alpha \approx 40°$, so that a stable vortex system occurs in a large angle of attack range, compared to subsonic wings. The stable vortex system in Figure 4.182 exists both in

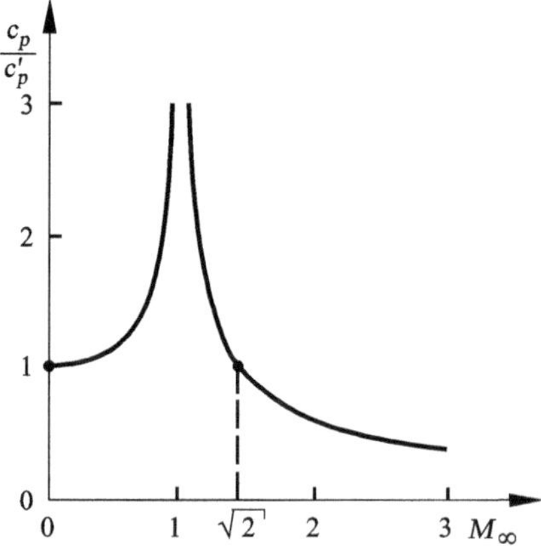

Fig. 4.180. Transformation of pressure coefficients

Fig. 4.181. Steady vortex separation at the leading edge of a delta wing

the subsonic regime and the supersonic regime, as long as a subsonic leading edge is realized. This occurs for a span-to-chord ratio of about $s/l \approx 0.5$.

Figure 4.182 shows the lift-to-drag ratio of a wing-fuselage configuration at a flight Mach number of $M_\infty = 2$ and a given angle of attack. The maximum value of c_a/c_w for this example is 7.4, at a lift coefficient of $c_a = 0.15$. For the subsonic flight at takeoff and landing, $c_a/c_w = 11.6$ at the same lift coefficient. In contrast to the swept wing of transonic subsonic flight, which requires high-lift flaps to sustain the lift at takeoff and landing, these high-lift aids are not needed for delta wings. Because of the stable vortex system, the values of c_a/c_w for the delta wing in subsonic flight are higher than those in supersonic flight.

The *supersonic airplane Concorde* was designed using the concept described for the flight Mach number $M_\infty = 2$. The airplane has a length of $l = 62$ m and a span of $s = 26$ m (Figure 4.183). This yields a ratio of $s/l = 0.42$, thus approximately realizing the slender, conical flow of the vortex trail with a conical head wave in supersonic flight, which was described above. The head wave heats up the flowing gas, so that temperatures of 128° C in the stagnation point and 105° C at the leading edge of the wing are attained in supersonic flight. As well as the mechanical strains, there are also additional thermal strains on the cell structure of a supersonic airplane.

The dependence of the drag contributions to the total drag c_w of the supersonic airplane on the flight Mach number are shown in Figure 4.184. In subsonic flight the friction drag c_f caused by the vortex trail dominates. At Mach number $M_\infty = 2$, the wave drag c_s and the wave drag of the vortex trail c_{si} dominate.

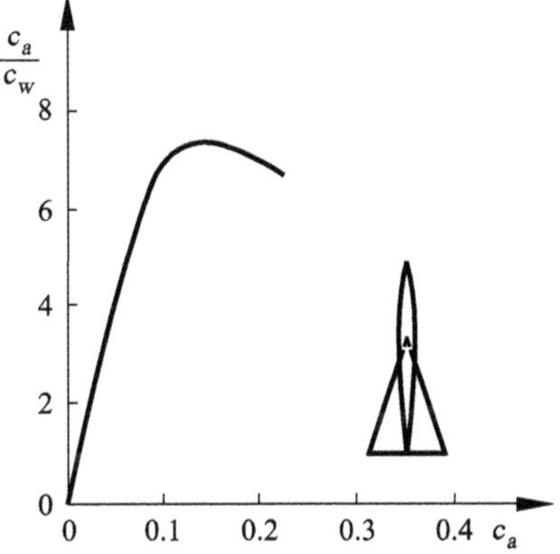

Fig. 4.182. Ratio of lift to drag coefficient c_a/c_w for a slender supersonic airplane with delta wings

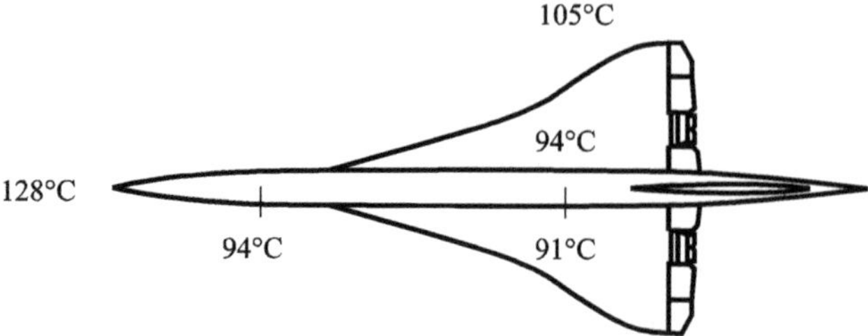

Fig. 4.183. Supersonic airplane Concorde, $M_\infty = 2$

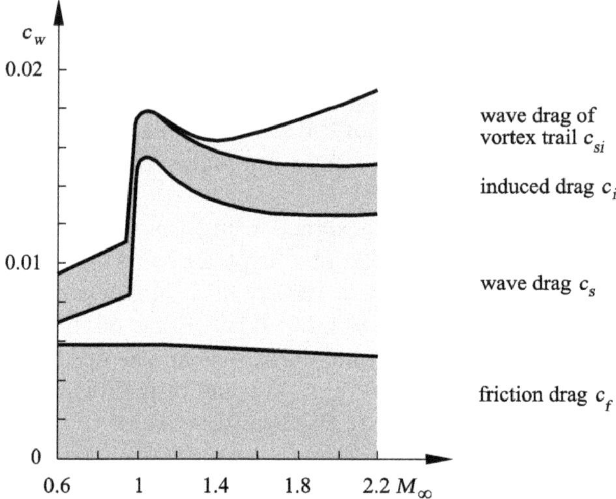

Fig. 4.184. Contributions to the total drag c_w of the supersonic airplane Concorde plotted against the Mach number M_∞

4.4.9 Problems

4.22

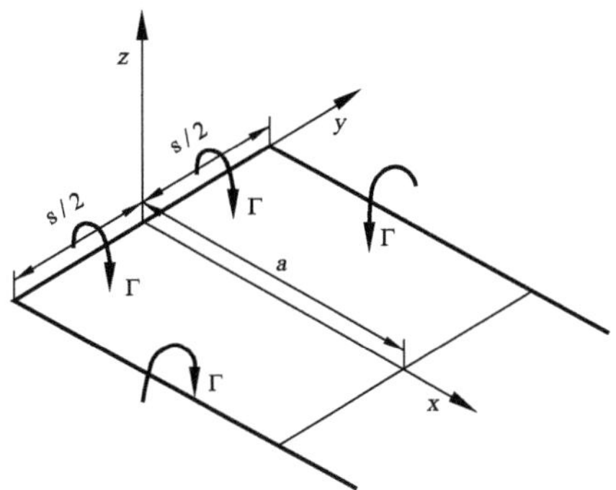

The simplest representation of the inviscid flow past a wing is a horseshoe vortex with circulation Γ.

(a) Using the Biot–Savart law, compute the vertical velocity component w along the line $x = a$, $-s/2 \leq y \leq s/2$. The contribution of each vortex filament is to be computed.

$$
\begin{aligned}
w &= -(w_1 + w_2 + w_3),\\
w_1 &= \frac{\Gamma}{4\Pi}\frac{1}{s/2+y}\left(1+\frac{a}{\sqrt{a^2+(s/2+y)^2}}\right),\\
w_2 &= \frac{\Gamma}{4\Pi}\frac{1}{a}\left(\frac{s/2-y}{\sqrt{a^2+(s/2-y)^2}}+\frac{s/2+y}{\sqrt{a^2+(s/2+y)^2}}\right),\\
w_3 &= \frac{\Gamma}{4\Pi}\frac{1}{s/2-y}\left(1+\frac{a}{\sqrt{a^2+(s/2-y)^2}}\right).
\end{aligned}
$$

(b) Compute the induced velocity along the x-axis for $a \to \infty$.

$$
\begin{aligned}
w(y=0,a) &= -\frac{\Gamma}{4\Pi}\left(\frac{4}{s}+\frac{4}{as}\sqrt{a^2+(s/2)^2}\right),\\
w(y=0,a\to\infty) &= -\frac{2\Gamma}{\Pi s}.
\end{aligned}
$$

4.23

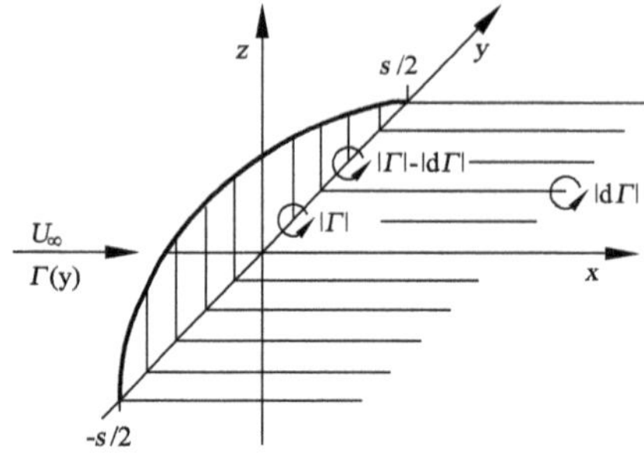

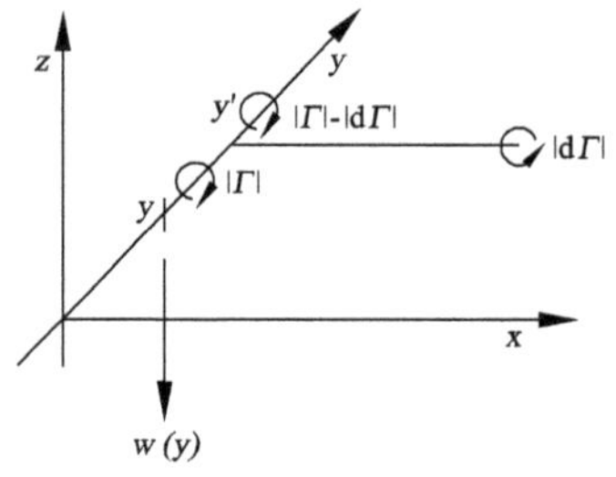

Assuming an elliptical circulation distribution

$$\Gamma(y) = -\Gamma_0 \sqrt{1 - \left(\frac{y}{s/2}\right)^2},$$

one obtains a wing flow with minimal induced drag. Because of the first Helmholtz vortex law, an infinitesimal free vortex with vortex strength $\mathrm{d}\Gamma = (\mathrm{d}\Gamma/\mathrm{d}y')\mathrm{d}y'$ is induced at every point y' in the flow field.

(a) Compute the induced vertical velocity $w(y)$ that is induced by the free vortex at the position of the attached vortex.

$$w(y) = \frac{\Gamma_0}{2s} = \text{const.}$$

(b) Compute the induced angle of attack $\alpha_i = w/U_\infty$ and the lift of the wing at the position of the attached vortex.

$$\alpha_i = \frac{\Gamma_0}{2sU_\infty}, \qquad A_i = \rho \Gamma_0 U_\infty s \frac{\Pi}{4},$$

$$c_a = \frac{\Pi}{2} \frac{\Gamma_0 s}{2U_\infty S}, \qquad S = \text{wing surface area.}$$

(c) Compute the induced drag w_i.

$$w_i = \frac{\Pi}{8} \rho \Gamma_0^2, \qquad c_w = \frac{c_a^2}{\Pi s^2/S}.$$

4.24

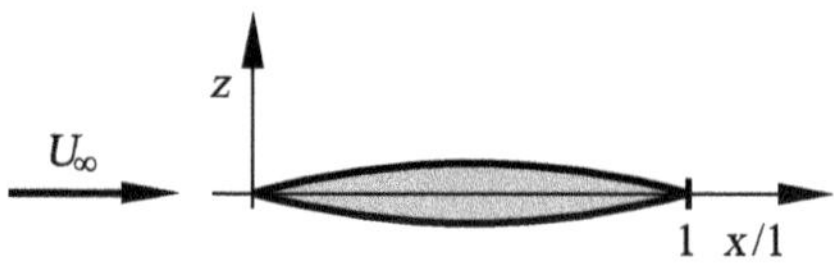

The camber line of a slender airfoil can be computed with a given vortex strength distribution $\gamma(x)$. This is called the inverse design method.

With the direct computation method, the vortex strength distribution can be computed from a given geometry of the camber line.

(a) Determine the equation of the camber line for the given vortex strength distribution $\gamma(x) = 2U_\infty C = \text{const.}$

$$f(x) = \alpha(x-1) + \frac{C}{\Pi}\left((x-1)\ln(1-x) - x\ln x\right).$$

(b) Calculate the vortex strength distribution and the lift coefficient for the given camber line $z = f(x) = \epsilon x(1 - x/\alpha)$.

$$\gamma(x) = 4\sqrt{x(1-x)}\left(\frac{\alpha}{2x} + \epsilon\right),$$

$$c_a = \Pi(2\alpha + \epsilon).$$

4.25

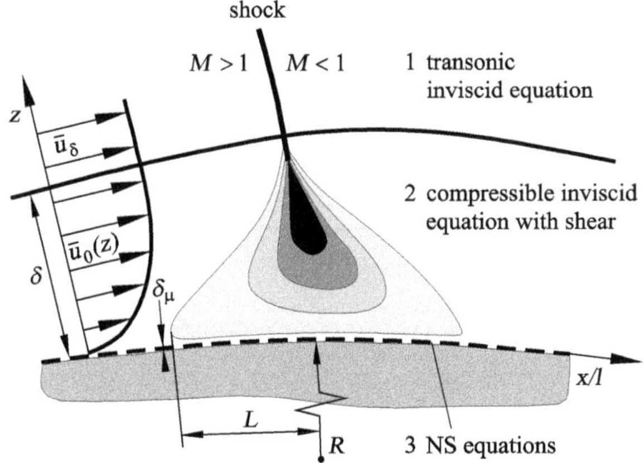

In order to calculate the turbulent shock–boundary-layer interaction of a transonic airfoil, the interaction regime is divided into three zones: (1) the inviscid outer flow, close to the speed of sound; (2) the boundary-layer region, with the inviscid flow differential equations with shearing; and (3) the viscous sublayer. In the boundary-layer zone 2, the effect of friction is taken into account only by a given time-averaged velocity profile $\overline{u}_0(z)$, which is perturbed by the pressure perturbation of the shock p'.

(a) Using the perturbation ansatz

$$u = \bar{u}_0(z) + u', \qquad w = w', \qquad p = \bar{p}_0 + p',$$

$$\rho = \bar{\rho}_0(z) + \rho', \qquad T = \bar{T}_0(z) + T'',$$

the linearized perturbation differential equations are to be determined from the two-dimensional boundary-layer equations

$$\frac{\partial(\rho \cdot u)}{\partial x} + \frac{\partial(\rho \cdot w)}{\partial z} = 0,$$

$$\rho\left(u\cdot\frac{\partial u}{\partial x}+w\cdot\frac{\partial u}{\partial z}\right)=-\frac{\partial p}{\partial x}+\mu\frac{\partial^2 u}{\partial z^2},$$

$$\rho\left(u\cdot\frac{\partial w}{\partial x}+w\cdot\frac{\partial w}{\partial z}\right)=-\frac{\partial p}{\partial z},$$

$$c_p\cdot T+\frac{u^2}{2}=\text{const},\qquad p=R\rho T.$$

Note that because of the shock, the term $\partial p/\partial z$ in the boundary-layer equation has to be taken into account, but because $\delta/L\propto 1$, the friction terms may be neglected.

$$\overline{\rho}_0\frac{\partial u'}{\partial x}+\overline{u}_0\frac{\partial \rho'}{\partial x}+\frac{\partial(\overline{\rho}_0\cdot w)}{\partial z}=0,$$

$$\overline{\rho}_0\overline{u}_0\cdot\frac{\partial u'}{\partial x}+\overline{\rho}_0\cdot w'\cdot\frac{\partial\overline{u}_0}{\partial z}=-\frac{\partial p'}{\partial x}+\mu\left(\frac{\partial^2\overline{u}_0}{\partial z^2}+\frac{\partial^2 u'}{\partial z^2}\right),$$

$$\overline{\rho}_0\cdot\overline{u}_0\cdot\frac{\partial w'}{\partial x}=-\frac{\partial p'}{\partial z},$$

$$\overline{u}_0\cdot u'+c_p\cdot T'=0,\overline{\rho}_0\cdot T'+\rho'\cdot\overline{T}_0=p'\frac{\overline{\rho}_0\overline{T}_0}{\overline{p}_0}.$$

(b) Make the perturbation differential equations dimensionless with the critical values k at $M=1$ and eliminate u' to obtain two perturbation differential equations for p' and w'. These can then be analytically solved using, for example, a separation trial solution.

$$\frac{1}{\kappa}(M_0^2-1)\frac{\partial p'}{\partial x}-\overline{\rho}_0\cdot w'\cdot\frac{\mathrm{d}M_k}{\mathrm{d}z}+\overline{\rho}_0\cdot M_k\cdot\frac{\partial w'}{\partial z}=0,$$

$$\frac{1}{\kappa}\cdot\frac{\partial p'}{\partial z}+\overline{\rho}_0\cdot M_k\cdot\frac{\partial w'}{\partial x}=0,$$

with $M_0(z)=M_k(z)\sqrt{\overline{\rho}_0(z)}$, $M_k=\overline{u}_0/a_k$, and $\mathrm{Re}_\delta\gg 1$.

4.26

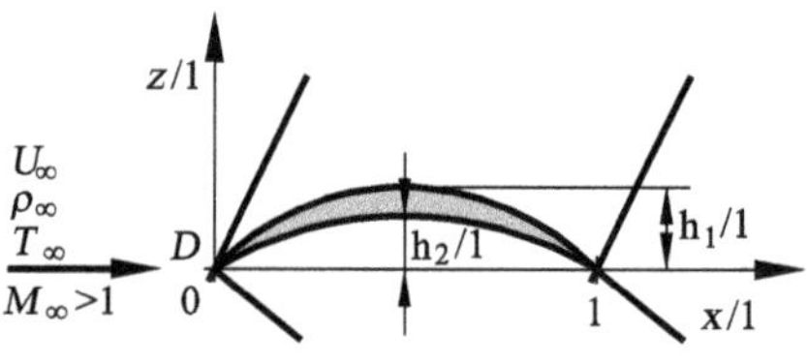

An airfoil of width b whose contours on the upper and lower sides are given by two parabolic equations is placed in supersonic free stream with Mach number M_∞:

$$\frac{z}{l}=4\cdot\frac{h_1}{l}\cdot\frac{x}{l}\cdot\left(1-\frac{x}{l}\right),\qquad\frac{z}{l}=4\cdot\frac{h_2}{l}\cdot\frac{x}{l}\cdot\left(1-\frac{x}{l}\right).$$

(a) Determine the x dependence of the c_p value along the upper and lower sides of the airfoil.

$$c_{p,u} = \frac{8 \cdot \frac{h_1}{l} \cdot \left(1 - \frac{2 \cdot x}{l}\right)}{\sqrt{M_\infty^2 - 1}}, \qquad c_{p,l} = \frac{8 \cdot \frac{h_2}{l} \cdot \left(1 - \frac{2 \cdot x}{l}\right)}{\sqrt{M_\infty^2 - 1}}.$$

(b) How large is the torque M_D acting on the point D that results on the pressure distributions on the upper and lower sides of the airfoil: $M_\infty = 1.4$, $l = 4\,\mathrm{m}$, $b = 15\,\mathrm{m}$, $h_1 = 0.1\,\mathrm{m}$, $h_2 = 0.05\,\mathrm{m}$, $\rho_\infty = 0.265\,\mathrm{kg/m^3}$, $U_\infty = 413\,\mathrm{m/s}$.

$$M_D = \frac{2 \cdot \rho_\infty \cdot U_\infty^2 \cdot b \cdot (h_1 + h_2) \cdot l}{3 \cdot \sqrt{M_\infty^2 - 1}} = 276.8\,\mathrm{kNm}.$$

5. Fundamental Equations of Fluid Mechanics

5.1 Continuity Equation

The *mass conservation at a volume element* $\mathrm{d}V = \mathrm{d}x \cdot \mathrm{d}y \cdot \mathrm{d}z$ for steady, incompressible flow

$$\frac{\partial u}{\partial x} + \frac{\partial v}{\partial y} + \frac{\partial w}{\partial z} = 0,$$

with velocity components u, v, w of the velocity vector $\boldsymbol{v}$, was introduced in Section 4.2.1. In this chapter we again consider the derivation of the continuity equation at a volume element $\mathrm{d}V$, but now extended to unsteady and compressible flows.

In general, the conservation of mass at a volume element may be formulated as follows:

The rate of change of mass in a volume element
$= \sum$ *the mass fluxes into the volume element*
$- \sum$ *the mass fluxes out of the volume element.*

Figure 5.1 shows the volume element $\mathrm{d}V$. Its edges have lengths $\mathrm{d}x$, $\mathrm{d}y$, and $\mathrm{d}z$. The mass flux $\rho \cdot u \cdot \mathrm{d}y \cdot \mathrm{d}z$ flows in through the left surface of the volume element with surface $\mathrm{d}y \cdot \mathrm{d}z$. The quantity $\rho \cdot u$ changes its value from position x to position $x + \mathrm{d}x$ in the x direction by $\partial(\rho \cdot u)/\partial x \cdot \mathrm{d}x$. Therefore,

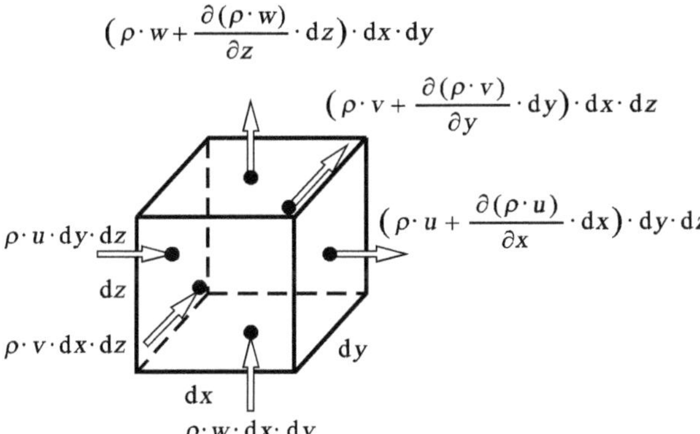

Fig. 5.1. Mass fluxes entering and exiting the volume element $\mathrm{d}V$

H. Oertel (ed.), *Prandtl-Essentials of Fluid Mechanics*,
Applied Mathematical Sciences 158, DOI 10.1007/978-1-4419-1564-1_5,

the mass flux exiting the volume element through the right surface $\mathrm{d}y \cdot \mathrm{d}z$ can be written as

$$\left(\rho \cdot u + \frac{\partial(\rho \cdot u)}{\partial x} \cdot \mathrm{d}x\right) \cdot \mathrm{d}y \cdot \mathrm{d}z.$$

In the y and z directions, analogous expressions can be computed for the surfaces $\mathrm{d}x \cdot \mathrm{d}z$ and $\mathrm{d}x \cdot \mathrm{d}y$.

According to the conservation of mass, the rate of change of mass inside the volume element under consideration corresponds to the difference between the mass fluxes entering and exiting. The term

$$\frac{\partial(\rho \cdot \mathrm{d}x \cdot \mathrm{d}y \cdot \mathrm{d}z)}{\partial t} = \frac{\partial \rho}{\partial t} \cdot \mathrm{d}x \cdot \mathrm{d}y \cdot \mathrm{d}z$$

is the mathematical expression for the rate of change of mass in the volume element. According to the discussion above, we have

$$\begin{aligned}\frac{\partial \rho}{\partial t} \cdot \mathrm{d}x \cdot \mathrm{d}y \cdot \mathrm{d}z = & \left(\rho \cdot u - (\rho \cdot u + \frac{\partial(\rho \cdot u)}{\partial x} \cdot \mathrm{d}x)\right) \cdot \mathrm{d}y \cdot \mathrm{d}z \\ & + \left(\rho \cdot v - (\rho \cdot v + \frac{\partial(\rho \cdot v)}{\partial y} \cdot \mathrm{d}y)\right) \cdot \mathrm{d}x \cdot \mathrm{d}z \\ & + \left(\rho \cdot w - (\rho \cdot w + \frac{\partial(\rho \cdot w)}{\partial z} \cdot \mathrm{d}z)\right) \cdot \mathrm{d}x \cdot \mathrm{d}y.\end{aligned}$$

This leads us to the *continuity equation for compressible flows*:

$$\frac{\partial \rho}{\partial t} + \frac{\partial(\rho \cdot u)}{\partial x} + \frac{\partial(\rho \cdot v)}{\partial y} + \frac{\partial(\rho \cdot w)}{\partial z} = 0. \tag{5.1}$$

For an *incompressible fluid*, this simplifies to

$$\frac{\partial u}{\partial x} + \frac{\partial v}{\partial y} + \frac{\partial w}{\partial z} = 0. \tag{5.2}$$

Using vector notation, these equations in a general coordinate system read

$$\frac{\partial \rho}{\partial t} + \nabla \cdot (\rho \cdot \boldsymbol{v}) = 0 \qquad \text{and} \qquad \nabla \cdot \boldsymbol{v} = 0, \tag{5.3}$$

where the operator $\nabla \cdot$ denotes the divergence of the vector. The Nabla operator ∇ has the following components:

$$\nabla = \left(\frac{\partial}{\partial x}, \frac{\partial}{\partial y}, \frac{\partial}{\partial z}\right)^T.$$

5.2 Navier–Stokes Equations

5.2.1 Laminar Flows

The Navier–Stokes equation results from the *conservation of momentum* at a volume element $\mathrm{d}V$. It was derived for viscous, incompressible flow in Section

4.2.1. We now consider the derivation for compressible flow. As with the derivation of the continuity equation at the volume element in Figure 5.1, we consider the rate of change of momentum in such a volume element. The momentum is the product of mass and velocity. The fluid inside the volume therefore has the momentum $\rho \cdot \mathrm{d}x \cdot \mathrm{d}y \cdot \mathrm{d}z \cdot \boldsymbol{v}$, and its rate of change can be written as

$$\frac{\partial(\rho \cdot \mathrm{d}x \cdot \mathrm{d}y \cdot \mathrm{d}z \cdot \boldsymbol{v})}{\partial t} = \frac{\partial(\rho \cdot \boldsymbol{v})}{\partial t} \cdot \mathrm{d}x \cdot \mathrm{d}y \cdot \mathrm{d}z. \tag{5.4}$$

In general, we can say:

The rate of change of momentum in a volume element
$= \sum$ *the momentum fluxes entering the volume element*
$- \sum$ *the momentum fluxes exiting the volume element*
$+ \sum$ *the shear and normal stresses acting on the volume element*
$+ \sum$ *the forces acting on the mass of the volume element.*

Initially, we consider only one component of the momentum vector $\rho \cdot \mathrm{d}x \cdot \mathrm{d}y \cdot \mathrm{d}z \cdot \boldsymbol{v}$, namely, the component that points in the x direction. Its rate of change can be expressed as follows:

$$\frac{\partial(\rho \cdot \mathrm{d}x \cdot \mathrm{d}y \cdot \mathrm{d}z \cdot u)}{\partial t} = \frac{\partial(\rho \cdot u)}{\partial t} \cdot \mathrm{d}x \cdot \mathrm{d}y \cdot \mathrm{d}z. \tag{5.5}$$

Just as in consideration of the mass fluxes, momentum enters or exits the volume through the surfaces of the volume element per unit time. In deriving the continuity equation, the quantity ρ (mass per unit volume) was used. Now we consider the quantity $(\rho \cdot u)$ (momentum per unit volume). As with the derivation of the continuity equation, we write down the momentum fluxes entering and exiting the volume element.

Again we consider the volume element shown in Figure 5.2 together with the momentum fluxes. Initially, we restrict ourselves to the x direction of the rate of change of the momentum $\rho \cdot \mathrm{d}x \cdot \mathrm{d}y \cdot \mathrm{d}z \cdot \boldsymbol{v}$.

The momentum flux

$$(\rho \cdot u) \cdot u \cdot \mathrm{d}y \cdot \mathrm{d}z = \rho \cdot u \cdot u \cdot \mathrm{d}y \cdot \mathrm{d}z \tag{5.6}$$

enters through the left surface $\mathrm{d}y \cdot \mathrm{d}z$ of the volume element. The quantity $\rho \cdot u \cdot u$ changes its value in the x direction by

$$\frac{\partial(\rho \cdot u \cdot u)}{\partial x} \cdot \mathrm{d}x, \tag{5.7}$$

so that the momentum flux exiting the volume element through the right surface $\mathrm{d}y \cdot \mathrm{d}z$ may be denoted by

$$\left(\rho \cdot u \cdot u + \frac{\partial(\rho \cdot u \cdot u)}{\partial x} \cdot \mathrm{d}x\right) \cdot \mathrm{d}y \cdot \mathrm{d}z. \tag{5.8}$$

The momentum $\rho \cdot u$ acting in the x direction also enters and exits the volume element through the remaining surfaces $\mathrm{d}x \cdot \mathrm{d}z$ and $\mathrm{d}x \cdot \mathrm{d}y$, but then with velocity components v and w, respectively.

Similar expressions can be written down for the y and z directions, so that in total three momentum fluxes can be given on each surface (Figure 5.2).

Now, the momentum fluxes entering and exiting are not the only cause of the rate of change of momentum within the volume element. The momentum inside the volume element is also changed by the forces acting on this volume element. These forces include the normal stresses and shear stresses, which are shown in Figure 5.3. These stresses vary in the x, y, and z directions, and the figure shows each quantity and its corresponding change at each of the positions $x + \mathrm{d}x$, $y + \mathrm{d}y$, and $z + \mathrm{d}z$.

The normal stresses and shear stresses are denoted in the same manner as in Section 4.2.1: The first index indicates on which surface the stress acts.

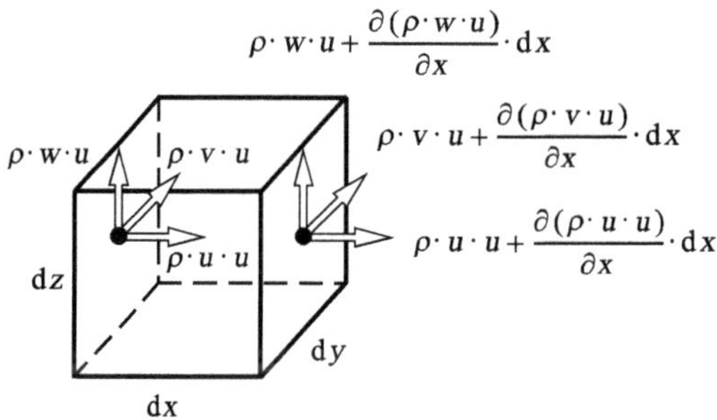

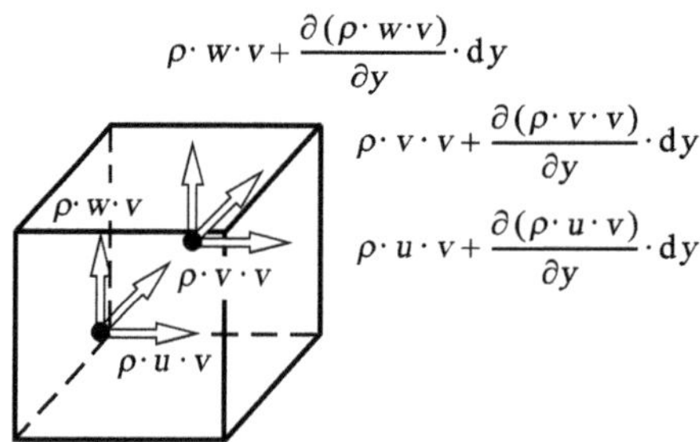

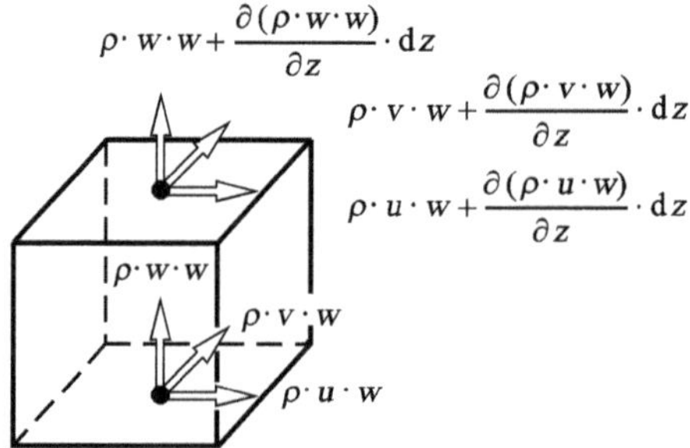

Fig. 5.2. Momentum fluxes entering and exiting the volume element $\mathrm{d}V$

For example, if the normal to the surface on which the stress acts points in the x direction, then this stress is given an x as its first index. The second index indicates in which coordinate direction the force resulting from the stress acts (Figure 5.3).

In deriving the equations, the signs of the stresses are determined as follows: A force is positive if the surface normal points in the positive coordinate direction, and is negative if the normal points in the negative coordinate direction.

The volume forces act on the mass of the volume element. They include gravity as well as the electric and magnetic forces that act on a flow, and are denoted by $\boldsymbol{f} = (f_x, f_y, f_z)^T$.

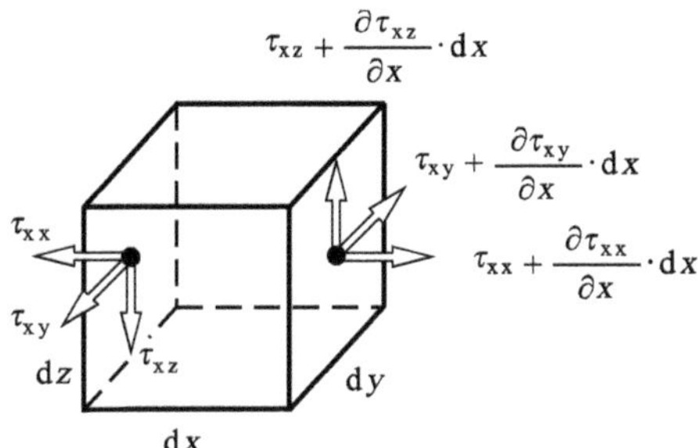

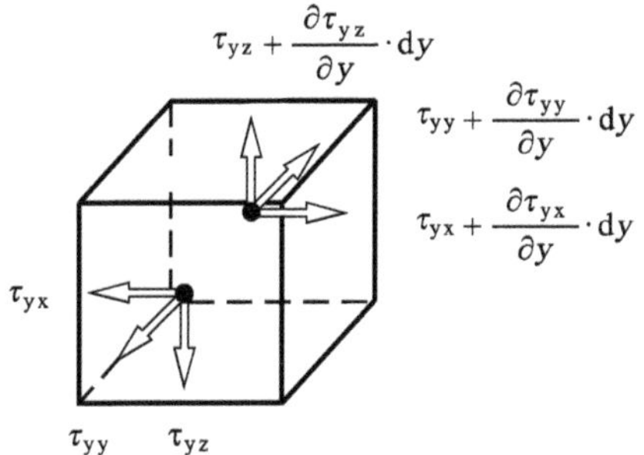

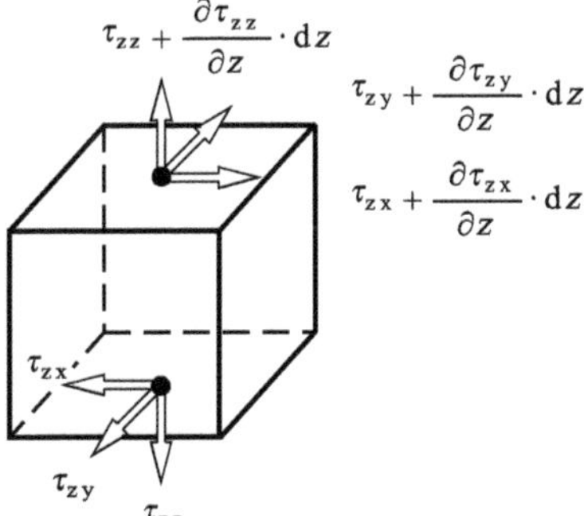

Fig. 5.3. Normal stresses and shear stresses at the volume element dV

Corresponding to the basic principle at the start of this chapter, the rate of change of momentum $\rho \cdot \mathrm{d}x \cdot \mathrm{d}y \cdot \mathrm{d}z \cdot u$ is

$$\begin{aligned}
\frac{\partial(\rho \cdot u)}{\partial t} \cdot \mathrm{d}x \cdot \mathrm{d}y \cdot \mathrm{d}z = & \left(\rho \cdot u \cdot u - (\rho \cdot u \cdot u + \frac{\partial(\rho \cdot u \cdot u)}{\partial x} \cdot \mathrm{d}x) \right) \cdot \mathrm{d}y \cdot \mathrm{d}z \\
& + \left(\rho \cdot u \cdot v - (\rho \cdot u \cdot v + \frac{\partial(\rho \cdot u \cdot v)}{\partial y} \cdot \mathrm{d}y) \right) \cdot \mathrm{d}x \cdot \mathrm{d}z \\
& + \left(\rho \cdot u \cdot w - (\rho \cdot u \cdot w + \frac{\partial(\rho \cdot u \cdot w)}{\partial z} \cdot \mathrm{d}z) \right) \cdot \mathrm{d}x \cdot \mathrm{d}y \\
& + f_x \cdot \mathrm{d}x \cdot \mathrm{d}y \cdot \mathrm{d}z \qquad (5.9) \\
& + \left(-\tau_{xx} + (\tau_{xx} + \frac{\partial \tau_{xx}}{\partial x} \cdot \mathrm{d}x) \right) \cdot \mathrm{d}y \cdot \mathrm{d}z \\
& + \left(-\tau_{yx} + (\tau_{yx} + \frac{\partial \tau_{yx}}{\partial y} \cdot \mathrm{d}y) \right) \cdot \mathrm{d}x \cdot \mathrm{d}z \\
& + \left(-\tau_{zx} + (\tau_{zx} + \frac{\partial \tau_{zx}}{\partial z} \cdot \mathrm{d}z) \right) \cdot \mathrm{d}x \cdot \mathrm{d}y.
\end{aligned}$$

This yields

$$\begin{aligned}
& \frac{\partial(\rho \cdot u)}{\partial t} + \frac{\partial(\rho \cdot u \cdot u)}{\partial x} + \frac{\partial(\rho \cdot u \cdot v)}{\partial y} + \frac{\partial(\rho \cdot u \cdot w)}{\partial z} \\
& \qquad = f_x + \frac{\partial \tau_{xx}}{\partial x} + \frac{\partial \tau_{yx}}{\partial y} + \frac{\partial \tau_{zx}}{\partial z}. \qquad (5.10)
\end{aligned}$$

The following equations are obtained for the y and z directions:

$$\begin{aligned}
& \frac{\partial(\rho \cdot v)}{\partial t} + \frac{\partial(\rho \cdot v \cdot u)}{\partial x} + \frac{\partial(\rho \cdot v \cdot v)}{\partial y} + \frac{\partial(\rho \cdot v \cdot w)}{\partial z} \\
& \qquad = f_y + \frac{\partial \tau_{xy}}{\partial x} + \frac{\partial \tau_{yy}}{\partial y} + \frac{\partial \tau_{zy}}{\partial z},
\end{aligned}$$

$$\begin{aligned}
& \frac{\partial(\rho \cdot w)}{\partial t} + \frac{\partial(\rho \cdot w \cdot u)}{\partial x} + \frac{\partial(\rho \cdot w \cdot v)}{\partial y} + \frac{\partial(\rho \cdot w \cdot w)}{\partial z} \\
& \qquad = f_z + \frac{\partial \tau_{xz}}{\partial x} + \frac{\partial \tau_{yz}}{\partial y} + \frac{\partial \tau_{zz}}{\partial z}.
\end{aligned}$$

The pressure p can be written as the trace of the stress tensor:

$$p = -\frac{\tau_{xx} + \tau_{yy} + \tau_{zz}}{3}. \qquad (5.11)$$

The minus sign takes into account the fact that the pressure acts as a negative normal stress.

The three normal stresses τ_{xx}, τ_{yy}, and τ_{zz} can each be split up into two parts, the pressure p and the contributions due to the friction of the fluid, σ_{xx}, σ_{yy}, and σ_{zz}:

$$\tau_{xx} = \sigma_{xx} - p, \qquad \tau_{yy} = \sigma_{yy} - p, \qquad \tau_{zz} = \sigma_{zz} - p. \qquad (5.12)$$

Inserting τ_{xx}, τ_{yy}, and τ_{zz} from (5.12) into (5.10), we obtain

$$\frac{\partial(\rho \cdot u)}{\partial t} + \frac{\partial(\rho \cdot u^2)}{\partial x} + \frac{\partial(\rho \cdot u \cdot v)}{\partial y} + \frac{\partial(\rho \cdot u \cdot w)}{\partial z} = f_x - \frac{\partial p}{\partial x} + \frac{\partial \sigma_{xx}}{\partial x} + \frac{\partial \tau_{yx}}{\partial y} + \frac{\partial \tau_{zx}}{\partial z}, \tag{5.13}$$

$$\frac{\partial(\rho \cdot v)}{\partial t} + \frac{\partial(\rho \cdot v \cdot u)}{\partial x} + \frac{\partial(\rho \cdot v^2)}{\partial y} + \frac{\partial(\rho \cdot v \cdot w)}{\partial z} = f_y - \frac{\partial p}{\partial y} + \frac{\partial \tau_{xy}}{\partial x} + \frac{\partial \sigma_{yy}}{\partial y} + \frac{\partial \tau_{zy}}{\partial z}, \tag{5.14}$$

$$\frac{\partial(\rho \cdot w)}{\partial t} + \frac{\partial(\rho \cdot w \cdot u)}{\partial x} + \frac{\partial(\rho \cdot w \cdot v)}{\partial y} + \frac{\partial(\rho \cdot w^2)}{\partial z} = f_z - \frac{\partial p}{\partial z} + \frac{\partial \tau_{xz}}{\partial x} + \frac{\partial \tau_{yz}}{\partial y} + \frac{\partial \sigma_{zz}}{\partial z}. \tag{5.15}$$

For Newtonian fluids the following relations hold:

$$\begin{aligned}
\sigma_{xx} &= 2 \cdot \mu \cdot \frac{\partial u}{\partial x} - \frac{2}{3} \cdot \mu \cdot \left(\frac{\partial u}{\partial x} + \frac{\partial v}{\partial y} + \frac{\partial w}{\partial z}\right), \\
\sigma_{yy} &= 2 \cdot \mu \cdot \frac{\partial v}{\partial y} - \frac{2}{3} \cdot \mu \cdot \left(\frac{\partial u}{\partial x} + \frac{\partial v}{\partial y} + \frac{\partial w}{\partial z}\right), \\
\sigma_{zz} &= 2 \cdot \mu \cdot \frac{\partial w}{\partial z} - \frac{2}{3} \cdot \mu \cdot \left(\frac{\partial u}{\partial x} + \frac{\partial v}{\partial y} + \frac{\partial w}{\partial z}\right), \\
\tau_{yx} &= \tau_{xy} = \mu \cdot \left(\frac{\partial v}{\partial x} + \frac{\partial u}{\partial y}\right), \qquad \tau_{yz} = \tau_{zy} = \mu \cdot \left(\frac{\partial w}{\partial y} + \frac{\partial v}{\partial z}\right), \\
\tau_{zx} &= \tau_{xz} = \mu \cdot \left(\frac{\partial u}{\partial z} + \frac{\partial w}{\partial x}\right),
\end{aligned} \tag{5.16}$$

with the symmetry condition

$$\tau_{yx} = \tau_{xy}, \qquad \tau_{yz} = \tau_{zy}, \qquad \tau_{zx} = \tau_{xz}. \tag{5.17}$$

Inserting the normal stresses and shear stresses according to equations (5.16) into the conservation of momentum equations (5.13), (5.14), and (5.15), we obtain the *Navier–Stokes equations*:

$$\begin{aligned}
&\frac{\partial(\rho \cdot u)}{\partial t} + \frac{\partial(\rho \cdot u^2)}{\partial x} + \frac{\partial(\rho \cdot u \cdot v)}{\partial y} + \frac{\partial(\rho \cdot u \cdot w)}{\partial z} \\
&= f_x - \frac{\partial p}{\partial x} + \frac{\partial}{\partial x}\left[\mu \cdot \left(2 \cdot \frac{\partial u}{\partial x} - \frac{2}{3} \cdot (\nabla \cdot \mathbf{v})\right)\right] \\
&\quad + \frac{\partial}{\partial y}\left[\mu \cdot \left(\frac{\partial u}{\partial y} + \frac{\partial v}{\partial x}\right)\right] + \frac{\partial}{\partial z}\left[\mu \cdot \left(\frac{\partial w}{\partial x} + \frac{\partial u}{\partial z}\right)\right],
\end{aligned}$$

$$\frac{\partial(\rho \cdot v)}{\partial t} + \frac{\partial(\rho \cdot v \cdot u)}{\partial x} + \frac{\partial(\rho \cdot v^2)}{\partial y} + \frac{\partial(\rho \cdot v \cdot w)}{\partial z}$$
$$= f_y - \frac{\partial p}{\partial y} + \frac{\partial}{\partial x}\left[\mu \cdot \left(\frac{\partial u}{\partial y} + \frac{\partial v}{\partial x}\right)\right]$$
$$+ \frac{\partial}{\partial y}\left[\mu \cdot \left(2 \cdot \frac{\partial v}{\partial y} - \frac{2}{3} \cdot (\nabla \cdot \mathbf{v})\right)\right] + \frac{\partial}{\partial z}\left[\mu \cdot \left(\frac{\partial v}{\partial z} + \frac{\partial w}{\partial y}\right)\right],$$

$$\frac{\partial(\rho \cdot w)}{\partial t} + \frac{\partial(\rho \cdot w \cdot u)}{\partial x} + \frac{\partial(\rho \cdot w \cdot v)}{\partial y} + \frac{\partial(\rho \cdot w^2)}{\partial z}$$
$$= f_z - \frac{\partial p}{\partial z} + \frac{\partial}{\partial x}\left[\mu \cdot \left(\frac{\partial w}{\partial x} + \frac{\partial u}{\partial z}\right)\right]$$
$$+ \frac{\partial}{\partial y}\left[\mu \cdot \left(\frac{\partial v}{\partial z} + \frac{\partial w}{\partial y}\right)\right] + \frac{\partial}{\partial z}\left[\mu \cdot \left(2 \cdot \frac{\partial w}{\partial z} - \frac{2}{3} \cdot (\nabla \cdot \mathbf{v})\right)\right].$$

For *incompressible flows*, we can use the continuity equation $\nabla \cdot \boldsymbol{v} = 0$ (5.2) to obtain the Navier–Stokes equations:

$$\rho \cdot \left(\frac{\partial u}{\partial t} + \frac{\partial(u \cdot u)}{\partial x} + \frac{\partial(v \cdot u)}{\partial y} + \frac{\partial(w \cdot u)}{\partial z}\right)$$
$$= f_x - \frac{\partial p}{\partial x} + \frac{\partial}{\partial x}\left[2 \cdot \mu \cdot \frac{\partial u}{\partial x}\right]$$
$$+ \frac{\partial}{\partial y}\left[\mu \cdot \left(\frac{\partial u}{\partial y} + \frac{\partial v}{\partial x}\right)\right] + \frac{\partial}{\partial z}\left[\mu \cdot \left(\frac{\partial w}{\partial x} + \frac{\partial u}{\partial z}\right)\right],$$

$$\rho \cdot \left(\frac{\partial v}{\partial t} + \frac{\partial(u \cdot v)}{\partial x} + \frac{\partial(v \cdot v)}{\partial y} + \frac{\partial(w \cdot v)}{\partial z}\right)$$
$$= f_y - \frac{\partial p}{\partial y} + \frac{\partial}{\partial x}\left[\mu \cdot \left(\frac{\partial u}{\partial y} + \frac{\partial v}{\partial x}\right)\right] \tag{5.18}$$
$$+ \frac{\partial}{\partial y}\left[2 \cdot \mu \cdot \frac{\partial v}{\partial y}\right] + \frac{\partial}{\partial z}\left[\mu \cdot \left(\frac{\partial v}{\partial z} + \frac{\partial w}{\partial y}\right)\right],$$

$$\rho \cdot \left(\frac{\partial w}{\partial t} + \frac{\partial(u \cdot w)}{\partial x} + \frac{\partial(v \cdot w)}{\partial y} + \frac{\partial(w \cdot w)}{\partial z}\right)$$
$$= f_z - \frac{\partial p}{\partial z} + \frac{\partial}{\partial x}\left[\mu \cdot \left(\frac{\partial w}{\partial x} + \frac{\partial u}{\partial z}\right)\right]$$
$$+ \frac{\partial}{\partial y}\left[\mu \cdot \left(\frac{\partial v}{\partial z} + \frac{\partial w}{\partial y}\right)\right] + \frac{\partial}{\partial z}\left[2 \cdot \mu \cdot \frac{\partial w}{\partial z}\right].$$

Using the continuity equation (5.2), these may be rewritten in nonconservative form, assuming constant viscosity:

$$\begin{aligned}
\rho\cdot \left(\frac{\partial u}{\partial t} + u\cdot\frac{\partial u}{\partial x} + v\cdot\frac{\partial u}{\partial y} + w\cdot\frac{\partial u}{\partial z}\right) &= f_x - \frac{\partial p}{\partial x} + \mu\cdot\left(\frac{\partial^2 u}{\partial x^2} + \frac{\partial^2 u}{\partial y^2} + \frac{\partial^2 u}{\partial z^2}\right),\\
\rho\cdot \left(\frac{\partial v}{\partial t} + u\cdot\frac{\partial v}{\partial x} + v\cdot\frac{\partial v}{\partial y} + w\cdot\frac{\partial v}{\partial z}\right) &= f_y - \frac{\partial p}{\partial y} + \mu\cdot\left(\frac{\partial^2 v}{\partial x^2} + \frac{\partial^2 v}{\partial y^2} + \frac{\partial^2 v}{\partial z^2}\right),\\
\rho\cdot \left(\frac{\partial w}{\partial t} + u\cdot\frac{\partial w}{\partial x} + v\cdot\frac{\partial w}{\partial y} + w\cdot\frac{\partial w}{\partial z}\right) &= f_z - \frac{\partial p}{\partial z} + \mu\cdot\left(\frac{\partial^2 w}{\partial x^2} + \frac{\partial^2 w}{\partial y^2} + \frac{\partial^2 w}{\partial z^2}\right).
\end{aligned} \tag{5.19}$$

These equations can be summarized using vector notation as follows

$$\rho\cdot\left(\frac{\partial \boldsymbol{v}}{\partial t} + (\boldsymbol{v}\cdot\nabla)\boldsymbol{v}\right) = \boldsymbol{f} - \nabla p + \mu\cdot\Delta\boldsymbol{v}, \tag{5.20}$$

where ∇p is the gradient of p, and $(\boldsymbol{v}\cdot\nabla)$ the scalar product of the velocity vector and the Nabla operator. This is a convection operator that can be applied to each component of the velocity vector $\boldsymbol{v}$. Here $\Delta\boldsymbol{v}$ denotes the Laplace operator applied to $\boldsymbol{v}$:

$$\begin{aligned}
\nabla p &= \left(\frac{\partial p}{\partial x}, \frac{\partial p}{\partial y}, \frac{\partial p}{\partial z}\right)^T, \qquad \boldsymbol{v}\cdot\nabla = u\cdot\frac{\partial}{\partial x} + v\cdot\frac{\partial}{\partial y} + w\cdot\frac{\partial}{\partial z},\\
\Delta\boldsymbol{v} &= \frac{\partial^2 \boldsymbol{v}}{\partial x^2} + \frac{\partial^2 \boldsymbol{v}}{\partial y^2} + \frac{\partial^2 \boldsymbol{v}}{\partial z^2}.
\end{aligned} \tag{5.21}$$

Together with the continuity equation (5.2), equations (5.19) form a system of *four nonlinear second-order partial differential equations* for the *four unknowns u, v, w, and p*. This system has to be solved for given initial and boundary conditions.

On the other hand, if we consider a compressible fluid, the density ρ has to be taken into account as an additional unknown. A further equation, the *energy equation*, is then needed. This is discussed for laminar flows in Section 5.3.1.

5.2.2 Reynolds Equations for Turbulent Flows

For turbulent flows, the Reynolds ansatz (4.63) introduced in Section 4.2.4 holds. In order to be able to apply this to turbulent compressible flows too, we introduce mass-averaged quantities:

$$\tilde{u} = \frac{\overline{\rho \cdot u}}{\overline{\rho}}, \qquad \tilde{v} = \frac{\overline{\rho \cdot v}}{\overline{\rho}}, \qquad \tilde{w} = \frac{\overline{\rho \cdot w}}{\overline{\rho}}. \tag{5.22}$$

The line over the products denotes time averaging according to equation (4.64):

$$\overline{\rho \cdot u} = \frac{1}{T} \cdot \int_0^T (\rho \cdot u) \cdot \mathrm{d}t, \tag{5.23}$$

also known as *Favre averaging.*

The velocity components u, v, etc., are now made up of the time-averaged values according to equations (5.22) and a fluctuating quantity, which will now be denoted by two dashes. The pressure p and the density ρ do not have to be mass averaged. Their fluctuating quantities are denoted by one dash only. Thus we have the *Reynolds ansatz* for compressible flows:

$$\begin{aligned} &\rho = \overline{\rho} + \rho', \qquad &&p = \overline{p} + p', \\ &u = \tilde{u} + u'', \qquad &&v = \tilde{v} + v'', \qquad w = \tilde{w} + w''. \end{aligned} \tag{5.24}$$

It is important to note that the time-averaged quantities $\overline{f''}$ (where f'' is one of the fluctuating quantities u'', v'', etc.) are nonzero. On the other hand, the quantity $\overline{\rho \cdot f''}$ is equal to zero.

The following computational rules hold for any two quantities f and g:

$$\overline{\frac{\partial f}{\partial s}} = \frac{\partial \overline{f}}{\partial s}, \qquad \overline{f + g} = \overline{f} + \overline{g}, \qquad \overline{\rho' \cdot \tilde{u}} = 0, \qquad \overline{\rho \cdot u''} = 0. \tag{5.25}$$

The time average of the continuity equation (5.1) is written as

$$\frac{1}{T} \cdot \int_0^T \left(\frac{\partial \rho}{\partial t} + \frac{\partial (\rho \cdot u)}{\partial x} + \frac{\partial (\rho \cdot v)}{\partial y} + \frac{\partial (\rho \cdot w)}{\partial z} \right) \cdot \mathrm{d}t = 0,$$

or

$$\overline{\frac{\partial \rho}{\partial t} + \frac{\partial (\rho \cdot u)}{\partial x} + \frac{\partial (\rho \cdot v)}{\partial y} + \frac{\partial (\rho \cdot w)}{\partial z}} = 0. \tag{5.26}$$

Inserting the quantities u, v, and w according to equations (5.24) into equation (5.26), we use the computational rules (5.25) and $\overline{\rho \cdot f''} = 0$ to obtain

$$\overline{\frac{\partial \rho}{\partial t}} + \overline{\frac{\partial [\rho \cdot (\tilde{u} + u'')]}{\partial x}} + \overline{\frac{\partial [\rho \cdot (\tilde{v} + v'')]}{\partial y}} + \overline{\frac{\partial [\rho \cdot (\tilde{w} + w'')]}{\partial z}} = 0,$$

$$\frac{\partial \overline{\rho}}{\partial t} + \frac{\partial \overline{[\rho \cdot (\tilde{u} + u'')]}}{\partial x} + \frac{\partial \overline{[\rho \cdot (\tilde{v} + v'')]}}{\partial y} + \frac{\partial \overline{[\rho \cdot (\tilde{w} + w'')]}}{\partial z} = 0,$$

$$\frac{\partial \overline{\rho}}{\partial t} + \frac{\partial \overline{[\rho \cdot (\tilde{u}_i + u_i'')]}}{\partial x_i} = 0.$$

The second term contains the abbreviated notation for the three coordinate and velocity directions ($i = 1, \ldots, 3$); i.e.

$$\frac{\partial\overline{[\rho\cdot(\tilde{u}_i+u_i'')]}}{\partial x_i}=\frac{\partial(\overline{\rho\cdot\tilde{u}_i})}{\partial x_i}+\frac{\partial(\overline{\rho\cdot u_i''})}{\partial x_i}=\frac{\partial(\overline{\rho}\cdot\tilde{u}_i)}{\partial x_i}.$$

Therefore, the time-averaged continuity equation for compressible flows reads

$$\frac{\partial\overline{\rho}}{\partial t}+\frac{\partial(\overline{\rho}\cdot\tilde{u})}{\partial x}+\frac{\partial(\overline{\rho}\cdot\tilde{v})}{\partial y}+\frac{\partial(\overline{\rho}\cdot\tilde{w})}{\partial z}=0. \tag{5.27}$$

It now no longer contains the quantities ρ and u_i, but rather $\overline{\rho}$ and $\tilde{u}_i$.

For incompressible flows the continuity equation reads

$$\frac{\partial(\overline{u})}{\partial x}+\frac{\partial(\overline{v})}{\partial y}+\frac{\partial(\overline{w})}{\partial z}=0. \tag{5.28}$$

The time averaging of the Navier–Stokes equations is carried out in the same manner as the averaging of the continuity equation. First, we consider the equation for the x direction. Equation (5.13) yields

$$\overline{\frac{\partial(\rho\cdot u)}{\partial t}+\frac{\partial(\rho\cdot u^2)}{\partial x}+\frac{\partial(\rho\cdot u\cdot v)}{\partial y}+\frac{\partial(\rho\cdot u\cdot w)}{\partial z}}$$
$$=\overline{f_x-\frac{\partial p}{\partial x}+\frac{\partial\sigma_{xx}}{\partial x}+\frac{\partial\tau_{yx}}{\partial y}+\frac{\partial\tau_{zx}}{\partial z}},$$

with (5.16)

$$\sigma_{xx}=\mu\cdot\left(2\cdot\frac{\partial u}{\partial x}-\frac{2}{3}\cdot(\nabla\cdot\mathbf{v})\right),\qquad \tau_{ij}=\mu\cdot\left(\frac{\partial u_i}{\partial x_j}+\frac{\partial u_j}{\partial x_i}\right).$$

Using the computational rules (5.25) we obtain

$$\frac{\partial(\overline{\rho\cdot u})}{\partial t}+\frac{\partial(\overline{\rho\cdot u^2})}{\partial x}+\frac{\partial(\overline{\rho\cdot u\cdot v})}{\partial y}+\frac{\partial(\overline{\rho\cdot u\cdot w})}{\partial z}$$
$$=f_x-\frac{\partial\overline{p}}{\partial x}+\frac{\partial\overline{\sigma}_{xx}}{\partial x}+\frac{\partial\overline{\tau}_{yx}}{\partial y}+\frac{\partial\overline{\tau}_{zx}}{\partial z}. \tag{5.29}$$

According to the definition of $\tilde{u}$, $\overline{\rho\cdot u}=\overline{\rho}\cdot\tilde{u}$. Therefore, all the time-averaged terms on the left- and right-hand sides of equation (5.29) are known, apart from the three terms on the left-hand side that contain the spatial partial derivatives. These will be considered further in what follows. By inserting the Reynolds ansatz (5.24) for u, v, and w, we obtain

$$\frac{\partial\overline{[\rho\cdot(\tilde{u}+u'')^2]}}{\partial x}+\frac{\partial\overline{[\rho\cdot(\tilde{u}+u'')\cdot(\tilde{v}+v'')]}}{\partial y}+\frac{\partial\overline{[\rho\cdot(\tilde{u}+u'')\cdot(\tilde{w}+w'')]}}{\partial z}$$
$$=\frac{\partial(\overline{\rho\cdot\tilde{u}^2})}{\partial x}+\frac{\partial(\overline{\rho\cdot u''^2})}{\partial x}+\frac{\partial(\overline{2\cdot\rho\cdot\tilde{u}\cdot u''})}{\partial x}$$
$$+\frac{\partial(\overline{\rho\cdot\tilde{u}\cdot\tilde{v}})}{\partial y}+\frac{\partial(\overline{\rho\cdot\tilde{u}\cdot v''})}{\partial y}+\frac{\partial(\overline{\rho\cdot u''\cdot\tilde{v}})}{\partial y}+\frac{\partial(\overline{\rho\cdot u''\cdot v''})}{\partial y}$$
$$+\frac{\partial(\overline{\rho\cdot\tilde{u}\cdot\tilde{w}})}{\partial z}+\frac{\partial(\overline{\rho\cdot\tilde{u}\cdot w''})}{\partial z}+\frac{\partial(\overline{\rho\cdot u''\cdot\tilde{w}})}{\partial z}+\frac{\partial(\overline{\rho\cdot u''\cdot w''})}{\partial z}$$

$$= \frac{\partial(\overline{\rho} \cdot \tilde{u}^2)}{\partial x} + \frac{\partial\overline{(\rho \cdot u''^2)}}{\partial x} + \frac{\partial(\overline{\rho} \cdot \tilde{u} \cdot \tilde{v})}{\partial y} + \frac{\partial\overline{(\rho \cdot u'' \cdot v'')}}{\partial y}$$
$$+ \frac{\partial(\overline{\rho} \cdot \tilde{u} \cdot \tilde{w})}{\partial z} + \frac{\partial\overline{(\rho \cdot u'' \cdot w'')}}{\partial z}.$$

Inserting this result into equation (5.29), we obtain the *Reynolds equation* for the x direction:

$$\frac{\partial(\overline{\rho} \cdot \tilde{u})}{\partial t} + \frac{\partial(\overline{\rho} \cdot \tilde{u}^2)}{\partial x} + \frac{\partial(\overline{\rho} \cdot \tilde{u} \cdot \tilde{v})}{\partial y} + \frac{\partial(\overline{\rho} \cdot \tilde{u} \cdot \tilde{w})}{\partial z}$$
$$= f_x - \frac{\partial \overline{p}}{\partial x} + \frac{\partial \overline{\sigma}_{xx}}{\partial x} + \frac{\partial \overline{\tau}_{yx}}{\partial y} + \frac{\partial \overline{\tau}_{zx}}{\partial z}$$
$$- \left(\frac{\partial\overline{(\rho \cdot u''^2)}}{\partial x} + \frac{\partial\overline{(\rho \cdot u'' \cdot v'')}}{\partial y} + \frac{\partial\overline{(\rho \cdot u'' \cdot w'')}}{\partial z} \right). \tag{5.30}$$

A simple additional calculation leads to the following equations for the time-averaged normal and shear stresses σ_{xx}, τ_{yx}, and τ_{zx}:

$$\overline{\sigma}_{xx} = \mu \cdot \left(2 \cdot \frac{\partial \tilde{u}}{\partial x} - \frac{2}{3} \cdot (\nabla \cdot \tilde{\boldsymbol{v}}) \right) + \mu \cdot \left(2 \cdot \frac{\partial \overline{u''}}{\partial x} - \frac{2}{3} \cdot (\nabla \cdot \overline{\boldsymbol{v}''}) \right), \tag{5.31}$$

$$\overline{\tau}_{ij} = \mu \cdot \left(\frac{\partial \tilde{u}_i}{\partial x_j} + \frac{\partial \tilde{u}_j}{\partial x_i} \right) + \mu \cdot \left(\frac{\partial \overline{u''_i}}{\partial x_j} + \frac{\partial \overline{u''_j}}{\partial x_i} \right). \tag{5.32}$$

The expressions $\nabla \cdot \tilde{\boldsymbol{v}}$ and $\nabla \cdot \overline{\boldsymbol{v}''}$ denote the divergences

$$\frac{\partial \tilde{u}}{\partial x} + \frac{\partial \tilde{v}}{\partial y} + \frac{\partial \tilde{w}}{\partial z}, \qquad \frac{\partial \overline{u''}}{\partial x} + \frac{\partial \overline{v''}}{\partial y} + \frac{\partial \overline{w''}}{\partial z}.$$

Compared to the Navier–Stokes equation for laminar flows (5.18), equation (5.30) contains additional terms on the right-hand side that take into account the fluctuating motion of the flow. The additional terms in (5.30) have to be modeled suitably, since no closed theory of turbulence modeling is known.

The same holds for the y and z directions, so that the *Reynolds equations* for turbulent compressible flows can be modeled suitably:

$$\frac{\partial(\overline{\rho} \cdot \tilde{u})}{\partial t} + \frac{\partial(\overline{\rho} \cdot \tilde{u}^2)}{\partial x} + \frac{\partial(\overline{\rho} \cdot \tilde{u} \cdot \tilde{v})}{\partial y} + \frac{\partial(\overline{\rho} \cdot \tilde{u} \cdot \tilde{w})}{\partial z}$$
$$= f_x - \frac{\partial \overline{p}}{\partial x} + \frac{\partial \overline{\sigma}_{xx}}{\partial x} + \frac{\partial \overline{\tau}_{yx}}{\partial y} + \frac{\partial \overline{\tau}_{zx}}{\partial z}$$
$$- \left(\frac{\partial\overline{(\rho \cdot u''^2)}}{\partial x} + \frac{\partial\overline{(\rho \cdot u'' \cdot v'')}}{\partial y} + \frac{\partial\overline{(\rho \cdot u'' \cdot w'')}}{\partial z} \right), \tag{5.33}$$

$$\frac{\partial(\overline{\rho}\cdot\tilde{v})}{\partial t}+\frac{\partial(\overline{\rho}\cdot\tilde{v}\cdot\tilde{u})}{\partial x}+\frac{\partial(\overline{\rho}\cdot\tilde{v}^2)}{\partial y}+\frac{\partial(\overline{\rho}\cdot\tilde{v}\cdot\tilde{w})}{\partial z}$$
$$=f_y-\frac{\partial\overline{p}}{\partial y}+\frac{\partial\overline{\tau}_{xy}}{\partial x}+\frac{\partial\overline{\sigma}_{yy}}{\partial y}+\frac{\partial\overline{\tau}_{zy}}{\partial z}$$
$$-\left(\frac{\partial\overline{(\rho\cdot v''\cdot u'')}}{\partial x}+\frac{\partial\overline{(\rho\cdot v''^2)}}{\partial y}+\frac{\partial\overline{(\rho\cdot v''\cdot w'')}}{\partial z}\right),\tag{5.34}$$

$$\frac{\partial(\overline{\rho}\cdot\tilde{w})}{\partial t}+\frac{\partial(\overline{\rho}\cdot\tilde{w}\cdot\tilde{u})}{\partial x}+\frac{\partial(\overline{\rho}\cdot\tilde{w}\cdot\tilde{v})}{\partial y}+\frac{\partial(\overline{\rho}\cdot\tilde{w}^2)}{\partial z}$$
$$=f_z-\frac{\partial\overline{p}}{\partial z}+\frac{\partial\overline{\tau}_{xz}}{\partial x}+\frac{\partial\overline{\tau}_{yz}}{\partial y}+\frac{\partial\overline{\sigma}_{zz}}{\partial z}$$
$$-\left(\frac{\partial\overline{(\rho\cdot w''\cdot u'')}}{\partial x}+\frac{\partial\overline{(\rho\cdot w''\cdot v'')}}{\partial y}+\frac{\partial\overline{(\rho\cdot w''^2)}}{\partial z}\right),\tag{5.35}$$

with

$$\overline{\sigma}_{ii}=\mu\cdot\left(2\cdot\frac{\partial\tilde{u}_i}{\partial x_i}-\frac{2}{3}\cdot(\nabla\cdot\tilde{\boldsymbol{v}})\right)+\mu\cdot\left(2\cdot\frac{\partial\overline{u_i''}}{\partial x_i}-\frac{2}{3}\cdot(\nabla\cdot\overline{\boldsymbol{v}''})\right),\tag{5.36}$$

$$\overline{\tau}_{ij}=\mu\cdot\left(\frac{\partial\tilde{u}_i}{\partial x_j}+\frac{\partial\tilde{u}_j}{\partial x_i}\right)+\mu\cdot\left(\frac{\partial\overline{u_i''}}{\partial x_j}+\frac{\partial\overline{u_j''}}{\partial x_i}\right).\tag{5.37}$$

For *incompressible flows*, equations (5.22) and (5.24) simplify to

$$\begin{aligned}&\tilde{u}=\overline{u}, &&\tilde{v}=\overline{v}, &&\tilde{w}=\overline{w},\\ &u=\overline{u}+u', &&v=\overline{v}+v', &&w=\overline{w}+w', \quad p=\overline{p}+p'.\end{aligned}\tag{5.38}$$

The *continuity equation* reads

$$\frac{\partial(\overline{u})}{\partial x}+\frac{\partial(\overline{v})}{\partial y}+\frac{\partial(\overline{w})}{\partial z}=0.\tag{5.39}$$

The time-averaged *Navier–Stokes equations* for incompressible flows are

$$\rho\cdot\left(\frac{\partial(\overline{u})}{\partial t}+\frac{\partial(\overline{u}^2)}{\partial x}+\frac{\partial(\overline{u}\cdot\overline{v})}{\partial y}+\frac{\partial(\overline{u}\cdot\overline{w})}{\partial z}\right)$$
$$=f_x-\frac{\partial\overline{p}}{\partial x}+\frac{\partial\overline{\sigma}_{xx}}{\partial x}+\frac{\partial\overline{\tau}_{yx}}{\partial y}+\frac{\partial\overline{\tau}_{zx}}{\partial z}$$
$$-\rho\left(\frac{\partial(\overline{u'^2})}{\partial x}+\frac{\partial(\overline{u'\cdot v'})}{\partial y}+\frac{\partial(\overline{u'\cdot w'})}{\partial z}\right),\tag{5.40}$$

$$\rho \cdot \left(\frac{\partial(\overline{v})}{\partial t} + \frac{\partial(\overline{v} \cdot \overline{u})}{\partial x} + \frac{\partial(\overline{v}^2)}{\partial y} + \frac{\partial(\overline{v} \cdot \overline{w})}{\partial z} \right)$$
$$= f_y - \frac{\partial \overline{p}}{\partial y} + \frac{\partial \overline{\tau}_{xy}}{\partial x} + \frac{\partial \overline{\sigma}_{yy}}{\partial y} + \frac{\partial \overline{\tau}_{zy}}{\partial z}$$
$$-\rho \left(\frac{\partial(\overline{v' \cdot u'})}{\partial x} + \frac{\partial(\overline{v'^2})}{\partial y} + \frac{\partial(\overline{v' \cdot w'})}{\partial z} \right), \tag{5.41}$$

$$\rho \cdot \left(\frac{\partial(\overline{w})}{\partial t} + \frac{\partial(\overline{w} \cdot \overline{u})}{\partial x} + \frac{\partial(\overline{w} \cdot \overline{v})}{\partial y} + \frac{\partial(\overline{w}^2)}{\partial z} \right)$$
$$= f_z - \frac{\partial \overline{p}}{\partial z} + \frac{\partial \overline{\tau}_{xz}}{\partial x} + \frac{\partial \overline{\tau}_{yz}}{\partial y} + \frac{\partial \overline{\sigma}_{zz}}{\partial z}$$
$$-\rho \left(\frac{\partial(\overline{w' \cdot u'})}{\partial x} + \frac{\partial(\overline{w' \cdot v'})}{\partial y} + \frac{\partial(\overline{w'^2})}{\partial z} \right). \tag{5.42}$$

5.3 Energy Equation

5.3.1 Laminar Flows

The energy equation for steady inviscid fluids has already been used in Section 4.3.3. The principle on which the three-dimensional energy balance at the volume element $\mathrm{d}V$ in Figure 5.4 is based is as follows:

The rate of change of the total energy in a volume element
$=\sum$ *the energy fluxes entering and exiting with the flow*
$+\sum$ *the energy fluxes entering and exiting by means of heat conduction*
$+\sum$ *the work done on the volume element per unit time due to the pressure forces, normal stress forces and shear stress forces*
$+$ *the energy supply from outside*
$+$ *the work done per unit time due to the effect of volume forces.*

The total energy E found within the volume element is made up of the internal energy $\rho \cdot e \cdot \mathrm{d}x \cdot \mathrm{d}y \cdot \mathrm{d}z$ and the kinetic energy $\rho \cdot (V^2/2) \cdot \mathrm{d}x \cdot \mathrm{d}y \cdot \mathrm{d}z = 0.5 \cdot \rho \cdot (u^2 + v^2 + w^2) \cdot \mathrm{d}x \cdot \mathrm{d}y \cdot \mathrm{d}z$ (where $\boldsymbol{v} \cdot \boldsymbol{v} = V^2$). The rate of change of energy in the volume element is

$$\frac{\partial[\rho \cdot \left(e + \frac{V^2}{2}\right) \cdot \mathrm{d}x \cdot \mathrm{d}y \cdot \mathrm{d}z]}{\partial t} = \frac{\partial[\rho \cdot \left(e + \frac{V^2}{2}\right)]}{\partial t} \cdot \mathrm{d}x \cdot \mathrm{d}y \cdot \mathrm{d}z. \tag{5.43}$$

The energy in the volume element is changed by the internal energy transported into and out of the volume element per unit time with the flow. This part is denoted by $\mathrm{d}\dot{E}$. Figure 5.4 shows the energy fluxes flowing inward and outwards. Using a similar approach to that used in deriving the Navier–Stokes equation, we consider the term $\mathrm{d}\dot{E}$:

$$\mathrm{d}\dot{E}=\left[\rho\cdot\left(e+\frac{V^2}{2}\right)\cdot u-\left(\rho\cdot\left(e+\frac{V^2}{2}\right)\cdot u+\frac{\partial(\rho\cdot\left(e+\frac{V^2}{2}\right)\cdot u)}{\partial x}\cdot \mathrm{d}x\right)\right]\cdot \mathrm{d}y\cdot \mathrm{d}z$$
$$+\left[\rho\cdot\left(e+\frac{V^2}{2}\right)\cdot v-\left(\rho\cdot\left(e+\frac{V^2}{2}\right)\cdot v+\frac{\partial(\rho\cdot\left(e+\frac{V^2}{2}\right)\cdot v)}{\partial y}\cdot \mathrm{d}y\right)\right]\cdot \mathrm{d}x\cdot \mathrm{d}z$$
$$+\left[\rho\cdot\left(e+\frac{V^2}{2}\right)\cdot w-\left(\rho\cdot\left(e+\frac{V^2}{2}\right)\cdot w+\frac{\partial\left(\rho\cdot(e+\frac{V^2}{2}\right)\cdot w)}{\partial z}\cdot \mathrm{d}z\right)\right]\cdot \mathrm{d}x\cdot \mathrm{d}y,$$

$$\mathrm{d}\dot{E}=-\left(\frac{\partial(\rho\cdot\left(e+\frac{V^2}{2}\right)\cdot u)}{\partial x}+\frac{\partial(\rho\cdot\left(e+\frac{V^2}{2}\right)\cdot v)}{\partial y}\right.$$
$$\left.+\frac{\partial(\rho\cdot\left(e+\frac{V^2}{2}\right)\cdot w)}{\partial z}\right)\cdot \mathrm{d}x\cdot \mathrm{d}y\cdot \mathrm{d}z. \tag{5.44}$$

The energy in the volume element is also changed by the transport of energy that enters or exits the volume per unit time by means of heat conduction. This part of the change in energy will be denoted by $\mathrm{d}\dot{Q}$ in what follows. According to the Fourier heat conduction law, heat energy flows in the direction of decreasing temperature. For example, the equation $\dot{q}=-\lambda\cdot(\mathrm{d}T/\mathrm{d}x)$ holds for the one-dimensional heat conduction problem, $\dot{q}$ stands for the heat

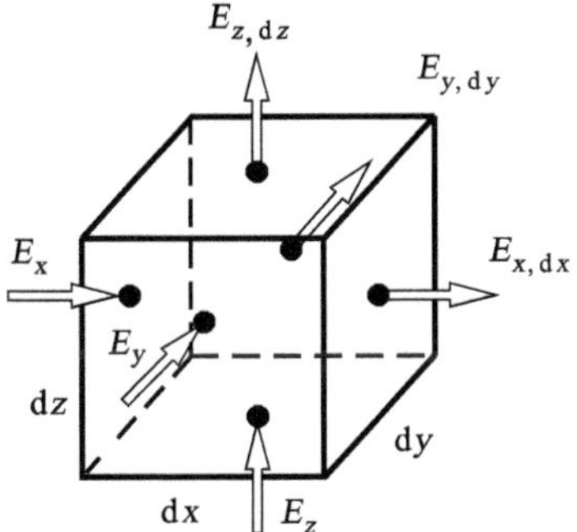

Fig. 5.4. Convective energy fluxes at the volume element $\mathrm{d}V$

flux per unit area, and λ for the thermal conductivity, which is in general dependent on the particular fluid the pressure, and the temperature. Using the Fourier heat conduction law to determine the term $\mathrm{d}\dot{Q}$, we obtain the following expression for the total energy flux due to heat conduction into or out of the volume element:

$$\begin{aligned}\mathrm{d}\dot{Q} = &\left(-\lambda\cdot\frac{\partial T}{\partial x} - \left[-\lambda\cdot\frac{\partial T}{\partial x} + \frac{\partial}{\partial x}\left(-\lambda\cdot\frac{\partial T}{\partial x}\right)\cdot \mathrm{d}x\right]\right)\cdot \mathrm{d}y\cdot \mathrm{d}z\\ &+\left(-\lambda\cdot\frac{\partial T}{\partial y} - \left[-\lambda\cdot\frac{\partial T}{\partial y} + \frac{\partial}{\partial y}\left(-\lambda\cdot\frac{\partial T}{\partial y}\right)\cdot \mathrm{d}y\right]\right)\cdot \mathrm{d}x\cdot \mathrm{d}z\\ &+\left(-\lambda\cdot\frac{\partial T}{\partial z} - \left[-\lambda\cdot\frac{\partial T}{\partial z} + \frac{\partial}{\partial z}\left(-\lambda\cdot\frac{\partial T}{\partial z}\right)\cdot \mathrm{d}z\right]\right)\cdot \mathrm{d}x\cdot \mathrm{d}y,\end{aligned} \tag{5.45}$$

$$\mathrm{d}\dot{Q} = \left(\frac{\partial}{\partial x}\left(\lambda\cdot\frac{\partial T}{\partial x}\right) + \frac{\partial}{\partial y}\left(\lambda\cdot\frac{\partial T}{\partial y}\right) + \frac{\partial}{\partial z}\left(\lambda\cdot\frac{\partial T}{\partial z}\right)\right)\cdot \mathrm{d}x\cdot \mathrm{d}y\cdot \mathrm{d}z. \tag{5.46}$$

In what follows we will determine the relations for the work done on the volume element by the pressure, normal stress, and shear stress forces. On each surface of the volume element three stresses that are due to the friction act, as does the static pressure. The forces resulting from the pressure and the stresses perform work on the volume element. The work per unit time, also called the power, is computed from the product of the velocity and the force that acts in the direction of the velocity component at hand. The work per unit time is given a positive sign when the velocity component points in the direction of the pressure, normal stress, or shear stress force. If this is not the case, the work per unit time is given a negative sign.

First, we consider the work per unit time $\mathrm{d}\dot{A}_x$ done on the volume element through the two surfaces with area $\mathrm{d}y\cdot \mathrm{d}z$:

$$\begin{aligned}\mathrm{d}\dot{A}_x = &\,p\cdot \mathrm{d}y\cdot \mathrm{d}z\cdot u - \left(p\cdot \mathrm{d}y\cdot \mathrm{d}z\cdot u + \frac{\partial(p\cdot \mathrm{d}y\cdot \mathrm{d}z\cdot u)}{\partial x}\cdot \mathrm{d}x\right)\\ &-\sigma_{xx}\cdot \mathrm{d}y\cdot \mathrm{d}z\cdot u + \left(\sigma_{xx}\cdot \mathrm{d}y\cdot \mathrm{d}z\cdot u + \frac{\partial(\sigma_{xx}\cdot \mathrm{d}y\cdot \mathrm{d}z\cdot u)}{\partial x}\cdot \mathrm{d}x\right)\\ &-\tau_{xy}\cdot \mathrm{d}y\cdot \mathrm{d}z\cdot v + \left(\tau_{xy}\cdot \mathrm{d}y\cdot \mathrm{d}z\cdot v + \frac{\partial(\tau_{xy}\cdot \mathrm{d}y\cdot \mathrm{d}z\cdot v)}{\partial x}\cdot \mathrm{d}x\right)\\ &-\tau_{xz}\cdot \mathrm{d}y\cdot \mathrm{d}z\cdot w + \left(\tau_{xz}\cdot \mathrm{d}y\cdot \mathrm{d}z\cdot w + \frac{\partial(\tau_{xz}\cdot \mathrm{d}y\cdot \mathrm{d}z\cdot w)}{\partial x}\cdot \mathrm{d}x\right),\end{aligned} \tag{5.47}$$

$$\mathrm{d}\dot{A}_x = \left(-\frac{\partial(p\cdot u)}{\partial x} + \frac{\partial(\sigma_{xx}\cdot u)}{\partial x} + \frac{\partial(\tau_{xy}\cdot v)}{\partial x} + \frac{\partial(\tau_{xz}\cdot w)}{\partial x}\right)\cdot \mathrm{d}x\cdot \mathrm{d}y\cdot \mathrm{d}z. \tag{5.48}$$

In the y and z directions, we obtain similar expressions for $\mathrm{d}\dot{A}_y$ and $\mathrm{d}\dot{A}_z$:

$$\mathrm{d}\dot{A}_y = \left(-\frac{\partial(p\cdot v)}{\partial y} + \frac{\partial(\tau_{yx}\cdot u)}{\partial y} + \frac{\partial(\sigma_{yy}\cdot v)}{\partial y} + \frac{\partial(\tau_{yz}\cdot w)}{\partial y}\right)\cdot \mathrm{d}x\cdot \mathrm{d}y\cdot \mathrm{d}z, \tag{5.49}$$

$$\mathrm{d}\dot{A}_z = \left(-\frac{\partial(p\cdot w)}{\partial z} + \frac{\partial(\tau_{zx}\cdot u)}{\partial z} + \frac{\partial(\tau_{zy}\cdot v)}{\partial z} + \frac{\partial(\sigma_{zz}\cdot w)}{\partial z}\right)\cdot \mathrm{d}x\cdot \mathrm{d}y\cdot \mathrm{d}z. \quad (5.50)$$

Then $\mathrm{d}\dot{A}$ is the sum of $\mathrm{d}\dot{A}_x$, $\mathrm{d}\dot{A}_y$, and $\mathrm{d}\dot{A}_z$.

Now, according to the guiding principle and the equations (5.43), (5.44), (5.46), (5.48), (5.49), (5.50) as well as $(\boldsymbol{f}\cdot\boldsymbol{v})\cdot \mathrm{d}x\cdot \mathrm{d}y\cdot \mathrm{d}z$ for the power of the volume forces, the *balance of energy* reads

$$\begin{aligned}
&\frac{\partial(\rho\cdot\left[e+\frac{V^2}{2}\right])}{\partial t}\\
&= -\left(\frac{\partial(\rho\cdot\left[e+\frac{V^2}{2}\right]\cdot u)}{\partial x} + \frac{\partial(\rho\cdot\left[[e+\frac{V^2}{2}\right]\cdot v)}{\partial y} + \frac{\partial(\rho\cdot\left[e+\frac{V^2}{2}\right]\cdot w)}{\partial z}\right)\\
&\quad + \left(\frac{\partial}{\partial x}\left[\lambda\cdot\frac{\partial T}{\partial x}\right] + \frac{\partial}{\partial y}\left[\lambda\cdot\frac{\partial T}{\partial y}\right] + \frac{\partial}{\partial z}\left[\lambda\cdot\frac{\partial T}{\partial z}\right]\right)\\
&\quad + \left(-\frac{\partial(p\cdot u)}{\partial x} + \frac{\partial(\sigma_{xx}\cdot u)}{\partial x} + \frac{\partial(\tau_{xy}\cdot v)}{\partial x} + \frac{\partial(\tau_{xz}\cdot w)}{\partial x}\right) \qquad (5.51)\\
&\quad + \left(-\frac{\partial(p\cdot v)}{\partial y} + \frac{\partial(\tau_{yx}\cdot u)}{\partial y} + \frac{\partial(\sigma_{yy}\cdot v)}{\partial y} + \frac{\partial(\tau_{yz}\cdot w)}{\partial y}\right)\\
&\quad + \left(-\frac{\partial(p\cdot w)}{\partial z} + \frac{\partial(\tau_{zx}\cdot u)}{\partial z} + \frac{\partial(\tau_{zy}\cdot v)}{\partial z} + \frac{\partial(\sigma_{zz}\cdot w)}{\partial z}\right) + \boldsymbol{f}\cdot\boldsymbol{v} + \rho\cdot\dot{q}_s.
\end{aligned}$$

Using the ansatz for normal and shear stresses (5.16) and the continuity equation (5.1), and neglecting the radiation, we can obtain the following expression:

$$\begin{aligned}
&\rho\cdot\left(\frac{\partial e}{\partial t} + u\cdot\frac{\partial e}{\partial x} + v\cdot\frac{\partial e}{\partial y} + w\cdot\frac{\partial e}{\partial z}\right)\\
&= \left(\frac{\partial}{\partial x}\left[\lambda\cdot\frac{\partial T}{\partial x}\right] + \frac{\partial}{\partial y}\left[\lambda\cdot\frac{\partial T}{\partial y}\right] + \frac{\partial}{\partial z}\left[\lambda\cdot\frac{\partial T}{\partial z}\right]\right) - p\cdot(\nabla\cdot\boldsymbol{v}) + \mu\cdot\Phi, \quad (5.52)
\end{aligned}$$

with the dissipation function Φ:

$$\begin{aligned}
\Phi = 2\cdot&\left[\left(\frac{\partial u}{\partial x}\right)^2 + \left(\frac{\partial v}{\partial y}\right)^2 + \left(\frac{\partial w}{\partial z}\right)^2\right] + \left(\frac{\partial v}{\partial x} + \frac{\partial u}{\partial y}\right)^2\\
&+ \left(\frac{\partial w}{\partial y} + \frac{\partial v}{\partial z}\right)^2 + \left(\frac{\partial u}{\partial z} + \frac{\partial w}{\partial x}\right)^2 - \frac{2}{3}\cdot\left(\frac{\partial u}{\partial x} + \frac{\partial v}{\partial y} + \frac{\partial w}{\partial z}\right)^2. \quad (5.53)
\end{aligned}$$

This contains only quadratic terms and is therefore greater than or equal to zero at all points in the flow field.

In deriving the energy equation no restrictions were made. This equation is valid in general, and describes the energy budget in a very small volume element, even for flows in which chemical or, equivalently, combustion processes take place. It was assumed that the flow is homogeneous and that the fluid is a Newtonian medium. In what follows we shall set down the energy equation for caloric ideal gases.

In a caloric ideal gas, the specific heat capacities c_p and c_v are temperature independent, and the following thermodynamic relations hold:

$$e = c_v \cdot T, \qquad h = e + \frac{p}{\rho} = c_p \cdot T, \tag{5.54}$$

or

$$e = c_p \cdot T - \frac{p}{\rho}. \tag{5.55}$$

Inserting the left side of equation (5.55) into equation (5.52) for e, and using the continuity equation (5.1), we obtain the *energy equation* for a *caloric ideal gas*:

$$\begin{aligned} \rho \cdot c_p \cdot \left(\frac{\partial T}{\partial t} + u \cdot \frac{\partial T}{\partial x} + v \cdot \frac{\partial T}{\partial y} + w \cdot \frac{\partial T}{\partial z} \right) \\ = \left(\frac{\partial p}{\partial t} + u \cdot \frac{\partial p}{\partial x} + v \cdot \frac{\partial p}{\partial y} + w \cdot \frac{\partial p}{\partial z} \right) \\ + \left(\frac{\partial}{\partial x} \left[\lambda \cdot \frac{\partial T}{\partial x} \right] + \frac{\partial}{\partial y} \left[\lambda \cdot \frac{\partial T}{\partial y} \right] + \frac{\partial}{\partial z} \left[\lambda \cdot \frac{\partial T}{\partial z} \right] \right) + \mu \cdot \Phi. \end{aligned} \tag{5.56}$$

5.3.2 Turbulent Flows

For the time average of the energy equation, the mass-averaged flow quantities (5.22) are extended by

$$\tilde{T} = \frac{\overline{\rho \cdot T}}{\overline{\rho}}, \qquad \tilde{e} = \frac{\overline{\rho \cdot e}}{\overline{\rho}}, \tag{5.57}$$

and the Reynolds ansatz (5.24) by

$$T = \tilde{T} + T'', \qquad e = \tilde{e} + e''. \tag{5.58}$$

This yields the energy equation, neglecting the dissipation:

$$\begin{aligned} \frac{\partial(\overline{\rho} \cdot \tilde{e}_{\text{tot}})}{\partial t} + \frac{\partial[\tilde{u} \cdot (\overline{\rho} \cdot \tilde{e}_{\text{tot}} + \overline{p})]}{\partial x} + \frac{\partial[\tilde{v} \cdot (\overline{\rho} \cdot \tilde{e}_{\text{tot}} + \overline{p})]}{\partial y} + \frac{\partial[\tilde{w} \cdot (\overline{\rho} \cdot \tilde{e}_{\text{tot}} + \overline{p})]}{\partial z} \\ = \frac{\partial(\overline{\tau}_{xx} \cdot \tilde{u} + \overline{\tau}_{xy} \cdot \tilde{v} + \overline{\tau}_{xz} \cdot \tilde{w})}{\partial x} \\ + \frac{\partial(\overline{\tau}_{yx} \cdot \tilde{u} + \overline{\tau}_{yy} \cdot \tilde{v} + \overline{\tau}_{yz} \cdot \tilde{w})}{\partial y} + \frac{\partial(\overline{\tau}_{zx} \cdot \tilde{u} + \overline{\tau}_{zy} \cdot \tilde{v} + \overline{\tau}_{zz} \cdot \tilde{w})}{\partial z} \\ - \sum_{\text{l}=1}^{3} \left[\overline{\tau_{\text{ml}} \cdot u_{\text{l}}''} - \tilde{u}_{\text{m}} \cdot \overline{\rho} \cdot \widetilde{u_{\text{l}}'' \cdot u_{\text{m}}''} - \frac{1}{2} \cdot \overline{\rho} \cdot u_{\text{l}}'' \cdot \widetilde{u_{\text{l}}'' \cdot u_{\text{m}}''} \right] \\ - \overline{p \cdot u_{\text{m}}''} - \overline{\rho \cdot e'' \cdot u_{\text{m}}''} - \overline{q}_x - \overline{q}_y - \overline{q}_z. \end{aligned} \tag{5.59}$$

The turbulent total energy $\tilde{e}_{\text{tot}}$ per volume is made up of the average internal energy $\tilde{e}$, the kinetic energy of the average flow, and the kinetic energy contained in the turbulent fluctuations:

$$\tilde{e}_{\text{tot}} = \tilde{e} + \frac{1}{2} \cdot (\tilde{u}^2 + \tilde{v}^2 + \tilde{w}^2) + \frac{1}{2} \cdot \left(\widetilde{u'' \cdot u''} + \widetilde{v'' \cdot v''} + \widetilde{w'' \cdot w''}\right) . \quad (5.60)$$

The Reynolds-averaged pressure can be calculated from the equation of state of the ideal gas:

$$\overline{p} = R \cdot \overline{\rho \cdot T} = R \cdot \overline{\rho} \cdot \tilde{T},$$

where the Favre-averaged temperature appearing on the right-hand side can be determined directly from the Favre-averaged internal energy:

$$\tilde{T} = \frac{\tilde{e}}{c_v}.$$

Since the heat flux is Reynolds-averaged, whereas the temperature is Favre-averaged, additional terms appear in the calculation:

$$\begin{aligned} \overline{q}_x &= -\lambda \cdot \left(\frac{\partial \tilde{T}}{\partial x} + \frac{\partial \overline{T''}}{\partial x}\right) , \\ \overline{q}_y &= -\lambda \cdot \left(\frac{\partial \tilde{T}}{\partial y} + \frac{\partial \overline{T''}}{\partial y}\right) , \\ \overline{q}_z &= -\lambda \cdot \left(\frac{\partial \tilde{T}}{\partial z} + \frac{\partial \overline{T''}}{\partial z}\right) . \end{aligned} \quad (5.61)$$

Similarly for the stresses:

$$\begin{aligned} \overline{\tau}_{\text{ml}} = \mu \cdot &\left(\frac{\partial \tilde{u}_{\text{l}}}{\partial x_{\text{m}}} + \frac{\partial \tilde{u}_{\text{m}}}{\partial x_{\text{l}}}\right) - \delta_{\text{ml}} \cdot \frac{2}{3} \cdot \nabla \cdot \tilde{\boldsymbol{v}} \\ + \mu \cdot &\left(\frac{\partial \overline{u''_{\text{l}}}}{\partial x_{\text{m}}} + \frac{\partial \overline{u''_{\text{m}}}}{\partial x_{\text{l}}}\right) - \delta_{\text{ml}} \cdot \frac{2}{3} \cdot \nabla \cdot \overline{\boldsymbol{v}''} \quad . \end{aligned} \quad (5.62)$$

The energy equation for an incompressible flow with $c = c_v$ and neglecting the dissipation reads

$$\begin{aligned} &\rho \cdot c \cdot \left(\frac{\partial(\overline{T})}{\partial t} + \frac{\partial(\overline{T} \cdot \overline{u})}{\partial x} + \frac{\partial(\overline{T} \cdot \overline{v})}{\partial y} + \frac{\partial(\overline{T} \cdot \overline{w})}{\partial z}\right) \\ &= \frac{\partial}{\partial x}\left(\lambda \cdot \frac{\partial \overline{T}}{\partial x} - \rho \cdot c \cdot \overline{T' \cdot u'}\right) + \frac{\partial}{\partial y}\left(\lambda \cdot \frac{\partial \overline{T}}{\partial y} - \rho \cdot c \cdot \overline{T' \cdot v'}\right) \\ &\quad + \frac{\partial}{\partial z}\left(\lambda \cdot \frac{\partial \overline{T}}{\partial z} - \rho \cdot c \cdot \overline{T' \cdot w'}\right) . \end{aligned} \quad (5.63)$$

In the calculation of *incompressible flows*, the energy equation is decoupled from the continuity equation and the Navier–Stokes equations; i.e. equations (5.40) to (5.42) may be solved first, and then used in the energy equation, together with the knowledge of $\overline{u}$, $\overline{v}$, $\overline{w}$, and $\overline{p}$, to determine the temperature field.

5.4 Fundamental Equations as Conservation Laws

5.4.1 Hierarchy of Fundamental Equations

The *continuum mechanical conservation equations* for mass, momentum, and energy, which were derived in Sections 5.1 to 5.3, are found as shown in Figure 5.5, by formation of the moments of the *Boltzmann equation* that describes the flow as a collection of fluid particles that move and collide with each other.

The *Navier–Stokes equations* are obtained for Newtonian media, and time averaging leads to the *Reynolds equations* for turbulent flows. The calculation of small perturbations in the flow field is carried out via a perturbation ansatz with the *perturbation differential equations.*

The *Boltzmann equation* is the transport equation of the distribution function f that describes the statistical distribution of the particles in velocity space $\boldsymbol{c} = c_m$ and in physical space $\boldsymbol{x} = x_m$:

$$\frac{\partial f}{\partial t} + \boldsymbol{c} \cdot \frac{\partial f}{\partial \boldsymbol{x}} + \frac{\boldsymbol{F}}{m} \cdot \frac{\partial f}{\partial \boldsymbol{c}} = \left(\frac{\partial f}{\partial t} \right)_{\text{coll}} \quad . \tag{5.64}$$

The left-hand side of the Boltzmann equation is the substantial derivative of the distribution function f with respect to time in six-dimensional phase space, where the term $\frac{\boldsymbol{F}}{m} \frac{df}{d\boldsymbol{c}}$ describes the change in the distribution function by the acceleration of the particles due to external force fields $\boldsymbol{F}$. The right-hand side represents the change in the distribution function as a consequence of the collisions between the particles.

In the *microscopic description* of a flow, the spatial coordinate system is specified. The velocities of the molecules inside the volume element $dV =$

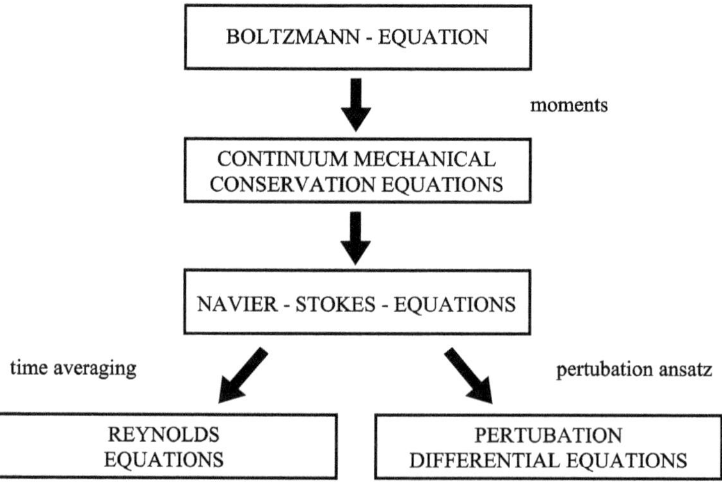

Fig. 5.5. Hierarchy of fluid-mechanical fundamental equations

$dx \cdot dy \cdot dz$ generally differ in their magnitude and direction. To characterize the velocities, velocity space is also introduced. Both spaces are shown in Figure 5.6. A point in this six-dimensional space representing a molecule is identified by specifying its Cartesian coordinates $\boldsymbol{x} = (x, y, z)$ and velocity $\boldsymbol{c} = (c_x, c_y, c_z)$.

A fluid with N particles is therefore represented by N points in the six-dimensional space. Therefore one mole of a fluid has $6 \cdot 10^{23}$ image points. In order to describe the particle density in six-dimensional space, the distribution function is defined as

$$f(\boldsymbol{x}, \boldsymbol{c}) = \frac{\mathrm{d}N}{\mathrm{d}\boldsymbol{x} \cdot \mathrm{d}\boldsymbol{c}} \quad . \tag{5.65}$$

This describes the statistical distribution of the particles in physical space and velocity space. dN is the number of image points in the volume element $dx \cdot dy \cdot dz \cdot dc_x \cdot dc_y \cdot dc_z$. Integrating the distribution function over all velocity and space coordinates yields the total number of particles as the sum of all image points:

$$N = \int_{\boldsymbol{c}} \int_{\boldsymbol{x}} f(\boldsymbol{x}, \boldsymbol{c}, t) \cdot \mathrm{d}\boldsymbol{x} \cdot \mathrm{d}\boldsymbol{c} \quad . \tag{5.66}$$

Knowing the microscopic structure of the flow in the form of the scalar distribution function $f(\boldsymbol{x}, \boldsymbol{c}, t)$, the dependence of all fluid properties on time can be derived. In velocity space, a distribution function can be defined by the relation

$$\mathrm{d}N = N \cdot f(\boldsymbol{c}) \cdot \mathrm{d}\boldsymbol{c} \tag{5.67}$$

Macroscopic quantities at a particular point in time are interpreted as averages of the molecular properties. The macroscopic quantities are obtained by averaging the molecular quantities Q, weighted according to the distribution function f:

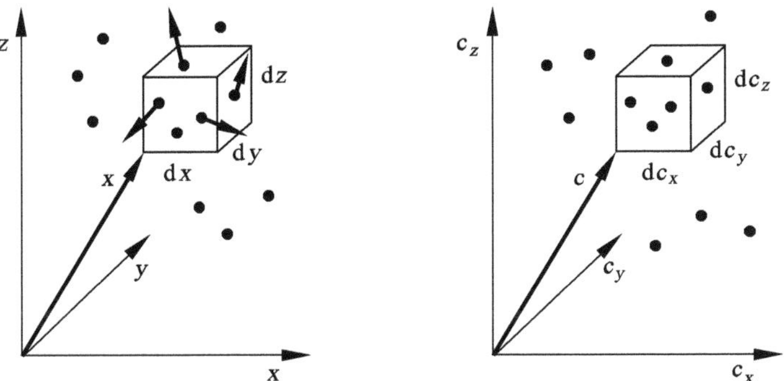

Fig. 5.6. Physical space $\boldsymbol{x}$ and velocity space $\boldsymbol{c}$

$$\bar{Q} = \frac{1}{N} \cdot \int_N Q \cdot \mathrm{d}N \quad \text{, with equation (5.66)}$$

$$\bar{Q} = \frac{1}{N} \cdot \int_{-\infty}^{+\infty} Q \cdot f(\mathbf{c}) \cdot N \cdot \mathrm{d}\mathbf{c} \quad , \tag{5.68}$$

$$\bar{Q} = \int_{-\infty}^{+\infty} Q \cdot f(\mathbf{c}) \cdot \mathrm{d}\mathbf{c} \quad .$$

The procedure described is called formation of the moments of the distribution function. The most important moments of the distribution function are the mean flow velocity

$$\overline{\mathbf{c}} = \int_{-\infty}^{+\infty} \mathbf{c} \cdot f(\mathbf{c}) \cdot \mathrm{d}\mathbf{c} \quad , \tag{5.69}$$

the pressure p

$$p = \int_{-\infty}^{+\infty} \frac{m}{3} \cdot \mathbf{c}^2 \cdot f(\mathbf{c}) \cdot \mathrm{d}\mathbf{c} \tag{5.70}$$

and the temperature T

$$T = \frac{2}{3 \cdot n \cdot k} \cdot \int_{-\infty}^{+\infty} \frac{m}{2} \cdot \mathbf{c}^2 \cdot f(\mathbf{c}) \cdot \mathrm{d}\mathbf{c} \quad , \tag{5.71}$$

with the particle density n (number of particles per unit volume), the particle mass m and the Boltzmann constant k. Equations (5.69) – (5.71) are used to establish the relationship between the microscopic and macroscopic approaches.

The *simplified model equations*, represented in Figure 5.7, can be derived from the Navier–Stokes equations. The *Euler equation* is obtained for inviscid flows. If the flow is also irrotational, it is the *potential equation* that holds. Flows at low Mach numbers lead to the *Navier–Stokes equations for incompressible fluids*. If the density of the fluid is dependent only on the

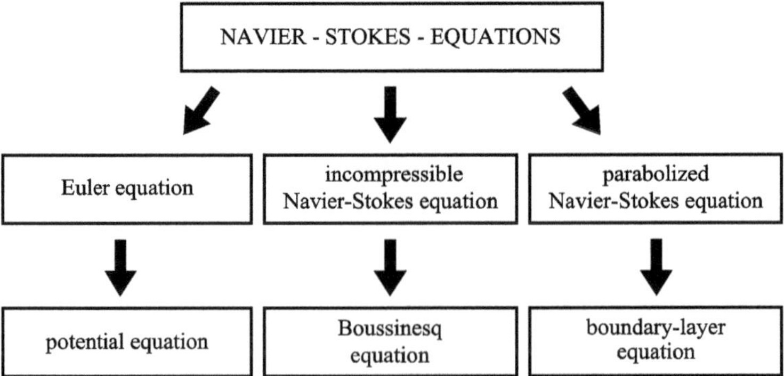

Fig. 5.7. Simplified model equations

temperature and not on the pressure, the buoyancy is taken into account and the *Boussinesq equation* obtained. For flows at large Reynolds numbers, the thickness of the boundary layer close to the wall is small compared to the geometric dimensions of the body, so that certain terms may be neglected inside the boundary layer. This leads to the *parabolized Navier–Stokes equations* and the *boundary-layer equations.*

5.4.2 Navier–Stokes Equations

In order to compute flow numerically, it is advantageous to rewrite the fundamental equations (5.1), (5.18), and (5.52) from the previous Sections in *conservative form.* This means that the conserved quantities mass, momentum, and energy are written as the divergence of these quantities. For example, the continuity equation then contains the divergence $\nabla \cdot (\rho \cdot \boldsymbol{v})$, the momentum equation contains the expression $\nabla \cdot (\rho \cdot \boldsymbol{v}\boldsymbol{v})$, and finally the energy equation contains the divergence $\nabla \cdot (\rho \cdot E \cdot \boldsymbol{v})$ with the total energy E.

Introducing dimensionless quantities $(^*)$, the dimensionless Cartesian coordinates become

$$x_m^* = \frac{x_m}{l}, \qquad m = 1, 2, 3,$$

where l is a *reference length* characteristic for the entire flow field.

Here x_m^* stands for

$$\boldsymbol{x}^* = \begin{pmatrix} x_1^* \\ x_2^* \\ x_3^* \end{pmatrix} = \begin{pmatrix} x^* \\ y^* \\ z^* \end{pmatrix},$$

and the dimensionless time is

$$t^* = \frac{t \cdot u_\infty}{l},$$

where u_∞ is a *reference velocity* characteristic for the entire flow field. The quantities x_m^* and t^* are the four *independent variables* in which the differential equations are formulated. The *dependent variables* are summarized in the *solution vector*

$$\boldsymbol{U}^*(x_{\mathrm{m}}^*, t^*) = \begin{pmatrix} \rho^* \\ \rho^* \cdot u_1^* \\ \rho^* \cdot u_2^* \\ \rho^* \cdot u_3^* \\ \rho^* \cdot E^* \end{pmatrix}, \tag{5.72}$$

with the dimensionless *density*

$$\rho^* = \frac{\rho}{\rho_\infty},$$

where ρ_∞ is a reference density characteristic for the entire flow field. The components $\rho^* \cdot u_m^*$ of the dimensionless *momentum vector* per unit volume are

$$\rho^* \boldsymbol{u}^* = \frac{\rho \cdot \boldsymbol{u}}{\rho_\infty \cdot u_\infty} = \begin{pmatrix} \rho^* \cdot u_1^* \\ \rho^* \cdot u_2^* \\ \rho^* \cdot u_3^* \end{pmatrix},$$

and with the dimensionless specific *total energy* of the fluid per unit volume E,

$$\rho^* \cdot E^* = \frac{\rho \cdot E}{\rho_\infty \cdot u_\infty^2}$$

The quantity $\boldsymbol{u}$ denotes the *velocity vector*, and E is the total energy per unit mass (internal energy + kinetic energy $1/2 \cdot \boldsymbol{u}^2$).

The dimensionless *Navier–Stokes equations* for a compressible fluid in *conservative form* (mass, momentum, and energy conservation) read

$$\frac{\partial \boldsymbol{U}^*}{\partial t^*} + \sum_{m=1}^{3} \frac{\partial \boldsymbol{F}_m^*}{\partial x_m^*} - \frac{1}{\mathrm{Re}_l} \cdot \sum_{m=1}^{3} \frac{\partial \boldsymbol{G}_m^*}{\partial x_m^*} = 0. \tag{5.73}$$

This conservative form of the equations is so called because the system of differential equations (5.73) was derived at a control volume fixed in space, so that each equation expresses the mass, momentum, and energy conservation directly. Each line of the solution vector (5.72) contains the *conservative variable*, referred to the volume, i.e. mass per unit volume ρ^*, momentum per unit volume $\rho^* \cdot \boldsymbol{u}^*$, and total energy per unit volume $\rho^* \cdot E^*$. In contrast to the conservative variables are the *primitive variables* velocity, pressure, and temperature, used in the previous sections.

In equation (5.73), $\boldsymbol{F}_{\mathrm{m}}^*$ is the vector of *convective fluxes* in the direction m,

$$\boldsymbol{F}_{\mathrm{m}}^* = \begin{pmatrix} \rho^* \cdot u^* \mathrm{m} \\ \rho^* \cdot u_{\mathrm{m}}^* \cdot u_1^* + \delta_{1m} \cdot p^* \\ \rho^* \cdot u_{\mathrm{m}}^* \cdot u_2^* + \delta_{2m} \cdot p^* \\ \rho^* \cdot u_{\mathrm{m}}^* \cdot u_3^* + \delta_{3m} \cdot p^* \\ u_{\mathrm{m}}^* \cdot (\rho^* \cdot E^* + p^*) \end{pmatrix}, \tag{5.74}$$

where ($\delta_{ij} = 1$ for $i = j$; $\delta_{ij} = 0$ for $i \neq j$), and $\boldsymbol{G}_m^*$ is the vector of *dissipative fluxes* in the coordinate direction m,

$$\boldsymbol{G}_m^* = \begin{pmatrix} 0 \\ \tau_{m1}^* \\ \tau_{m2}^* \\ \tau_{m3}^* \\ \sum_{l=1}^{3} u_l^* \cdot \tau_{lm}^* + \dot{q}_m^* \end{pmatrix}, \tag{5.75}$$

with the dimensionless *internal energy*

$$e^* = E^* - \frac{1}{2} \cdot \sum_{m=1}^{3} u_m^{*2},$$

the dimensionless *pressure*

$$p^* = (\kappa - 1) \cdot \rho^* \cdot e^* = \frac{p}{\rho_\infty \cdot u_\infty^2},$$

the dimensionless *temperature*

$$T^* = (\kappa - 1) \cdot \kappa \cdot M_\infty^2 \cdot e^* = \frac{T}{T_\infty},$$

the dimensionless *stresses*

$$\tau_{ij}^* = \mu^* \cdot \left(\frac{\partial u_i^*}{\partial x_j^*} + \frac{\partial u_j^*}{\partial x_i^*} \right) - \frac{2}{3} \cdot \mu^* \cdot \sum_{k=1}^{3} \frac{\partial u_k^*}{\partial x_k^*} \cdot \delta_{ij},$$

and the dimensionless *heat flux* in the direction m,

$$\dot{q}_m^* = \frac{\mu^*}{(\kappa - 1) \cdot M_\infty^2 \cdot \mathrm{Pr}_\infty} \cdot \frac{\partial T^*}{\partial x_m^*} = \frac{\mu^* \cdot \kappa}{\mathrm{Pr}} \cdot \frac{\partial e^*}{\partial x_m^*} = \frac{\lambda^*}{(\kappa - 1) \cdot M_\infty^2 \cdot \mathrm{Pr}_\infty} \cdot \frac{\partial T^*}{\partial x_m^*}.$$

These equations contain the following *material properties*: the Prandtl number $\mathrm{Pr}_\infty = \nu_\infty / k_\infty$, the ratio of specific heat capacities $\kappa = c_p / c_v$, the dimensionless dynamic viscosity μ^*. For air under atmosphere conditions these have the values $\mathrm{Pr}_\infty = 0.71$, $\kappa = 1.4$, and the *Sutherland formula*

$$\mu^* = (T^*)^{\frac{3}{2}} \cdot \frac{1 + S}{T^* + S}, \qquad S = \frac{110.4\ \mathrm{K}}{T_\infty}.$$

The reference quantity T_∞ is again characteristic for the flow. The following *dimensionless characteristic numbers*characterize the flow field:

$$M_\infty = \frac{u_\infty}{a_\infty} \qquad \text{Mach number,}$$

$$\mathrm{Re}_l = \frac{\rho_\infty \cdot u_\infty \cdot l}{\mu_\infty} \qquad \text{Reynolds number,}$$

$$\mathrm{Pr}_\infty = \frac{\nu_\infty}{k_\infty} \qquad \text{Prandtl number.}$$

Here a_∞ is a characteristic velocity of sound, and μ_∞ a characteristic viscosity.

The Navier–Stokes equations are a system of five coupled nonlinear second-order partial differential equations. Because they contain the time as an independent variable and describe spatially directed transport mechanisms, the equations are parabolic.

If steady flows are of interest, the time derivatives are neglected. The equations are then elliptic in subsonic regimes, and hyperbolic in supersonic regimes. For this reason they are also said to be of *mixed type*.

The following *boundary conditions* have to be taken into account:

At a *solid wall*, the *no-slip condition* holds,

$$\boldsymbol{u}^* = \boldsymbol{0}$$

as well as either the temperature boundary condition at an *isothermal wall*

$$T^* = T^*_{\mathrm{W}} \quad ,$$

with a given dimensionless wall temperature T^*_W, or the temperature boundary condition at an *adiabatic wall*,

$$\frac{\partial T^*}{\partial \boldsymbol{n}^*} = \frac{\partial T^*}{\partial x^*_1} \cdot n^*_1 + \frac{\partial T^*}{\partial x^*_2} \cdot n^*_2 + \frac{\partial T^*}{\partial x^*_3} \cdot n^*_3 = 0 \quad ,$$

with the dimensionless coordinate $\boldsymbol{n}^*$ in the direction normal to the wall.

A further boundary is the *far-field boundary*, the outer edge of the computational region in problems involving flow past bodies. If the far-field boundary is far enough away from the body, the flow there is the unperturbed outer flow $\boldsymbol{u}_\infty$, i.e. the boundary condition of inviscid flow from Section 5.4.3.

If it is not possible to determine the far-field boundary in this way so that the friction does not play a role, e.g. if a boundary layer, a separation bubble, or a wake flow leaves the region of integration, no mathematically exact boundary condition can be given. In this case, extrapolation is used to determine the flow quantities at the outer edge.

The solution vector $t = t_0 = 0$ is determined by the *initial condition*

$$\boldsymbol{U}^*(x^*_i, 0) = \boldsymbol{U}^*_0(x^*_i).$$

5.4.3 Derived Model Equations

As shown in Figure 5.7, by neglecting the term $\boldsymbol{G}^*$ in the Navier–Stokes equations (5.73), we obtain the dimensionless *Euler equation* in conservative form for laminar compressible flows

$$\frac{\partial \boldsymbol{U}^*}{\partial t^*} + \sum_{\mathrm{m}=1}^{3} \frac{\partial \boldsymbol{F}^*_{\mathrm{m}}}{\partial x^*_{\mathrm{m}}} = \boldsymbol{0} \quad , \tag{5.76}$$

where the solution vector $\boldsymbol{U}^*$ and the convective fluxes $\boldsymbol{F}^*_{\mathrm{m}}$ are as previously defined ((5.72) and (5.74), respectively).

We now have a system of five coupled nonlinear first-order differential equations. The Euler equations describe inviscid flows in which curved *shock waves* can occur. The flow field is characterized by the Mach number M_∞.

At a solid wall, the *slip condition* is the *boundary condition*

$$\boldsymbol{u}^* \cdot \boldsymbol{n} = 0, \tag{5.77}$$

with $\boldsymbol{n}$ the vector normal to the wall. This condition states that the flow cannot pass through the wall and that the velocity vector is directed parallel to the wall.

At the edge of the flow field, the expansion of information is vital in determining the boundary conditions. In order to do this, we have to distinguish between inflow and outflow boundaries (depending on the direction of the flow), and between supersonic and subsonic boundaries (depending on whether the local Mach number is larger or smaller than one). Neither too much nor too little information may be given at each boundary, because otherwise the problem would be mathematically overdetermined or underdetermined. The number of boundary conditions leads to the theory of characteristics.

	Inflow boundary		Outflow boundary	
	supersonic	subsonic	supersonic	subsonic
Number of variables to be specified	5	4	0	1
Number of variables to be computed	0	1	5	4

A further simplification is obtained if we assume that the flow is additionally *isentropic*. In this case, the flow may no longer contain any straight or curved shock waves. It can be shown that such flows are *irrotational*:

$$\boldsymbol{\omega}^* = \mathrm{rot}\boldsymbol{u}^* = \begin{pmatrix} \dfrac{\partial u_3^*}{\partial x_2^*} - \dfrac{\partial u_2^*}{\partial x_3^*} \\[2ex] \dfrac{\partial u_1^*}{\partial x_3^*} - \dfrac{\partial u_3^*}{\partial x_1^*} \\[2ex] \dfrac{\partial u_2^*}{\partial x_1^*} - \dfrac{\partial u_1^*}{\partial x_2^*} \end{pmatrix} = 0,$$

or in vector notation,

$$\boldsymbol{\omega}^* = \nabla \times \boldsymbol{u}^* = 0.$$

For irrotational flows it makes sense to introduce the *potential function* Φ^*:

$$\frac{\partial \Phi^*}{\partial x_1^*} = u_1^*, \qquad \frac{\partial \Phi^*}{\partial x_2^*} = u_2^*, \qquad \frac{\partial \Phi^*}{\partial x_3^*} = u_3^*. \tag{5.78}$$

Inserting this into the Euler equations and after simplifying, we obtain the dimensionless linearized *potential equation*

$$\frac{\partial^2 \Phi^*}{\partial x_1^{*2}} + \frac{\partial^2 \Phi^*}{\partial x_2^{*2}} + \frac{\partial^2 \Phi^*}{\partial x_3^{*2}} = 0, \qquad \Delta \Phi^* = 0. \tag{5.79}$$

This *scalar* equation is linear, of second order, and elliptic. Flows that can be described using the potential equation are also known as *potential flows*, already introduced in Section 4.1.5.

Conservation of momentum for an incompressible flow is automatically satisfied by the assumption that it is irrotational. This energy equation is an additional decoupled equation.

As in the case of the Euler equation, the slip condition is the *boundary condition* at a solid wall:

$$\frac{\partial \Phi^*}{\partial x_1^*} \cdot n_1^* + \frac{\partial \Phi^*}{\partial x_2^*} \cdot n_2^* + \frac{\partial \Phi^*}{\partial x_3^*} \cdot n_3^* = 0, \tag{5.80}$$

with the components of the vector normal to the wall n_1^*, n_2^*, and n_3^*. Each streamline can be considered to be a solid wall.

At the far-field boundary, any perturbations due to a body must have died away, i.e.

$$\frac{\partial \Phi^*}{\partial x_1^*} = \frac{\partial \Phi^*}{\partial x_2^*} = \frac{\partial \Phi^*}{\partial x_3^*} = 0. \tag{5.81}$$

These boundary conditions determine the solution only up to a constant, since only derivatives of the potential function appear in (5.75). Therefore, the value of Φ^* has also to be determined at some position in the flow field.

The advantage of the potential equation is that it is linear. This means that any linear combination of known solutions (e.g. parallel flow, source, sink, potential vortex) is also a solution.

For *incompressible laminar* flows, the *Navier–Stokes equations* (5.20) hold. These, together with the continuity equation (5.3), are given below:

$$\nabla \cdot \boldsymbol{u}^* = 0 \quad ,$$

$$\frac{\partial \boldsymbol{u}^*}{\partial t^*} + (\boldsymbol{u}^* \cdot \nabla)\boldsymbol{u}^* = -\nabla p^* + \frac{1}{Re_l} \cdot \Delta \boldsymbol{u}^* \quad . \tag{5.82}$$

The no-slip condition holds at solid walls:

$$\boldsymbol{u}^* = 0. \tag{5.83}$$

The pressure level has to be determined at some point (x_1^*, x_2^*, x_3^*):

$$p^*(x_1^*, y_1^*, z_1^*) = p_1^*.$$

The direction or magnitude can be prescribed at inflow or outflow boundaries, but in doing so we must ensure that the continuity equation is satisfied.

It may be desirable to prescribe the pressure at an inflow or outflow boundary, e.g. prescribing a certain pressure difference between two cross-sections of a pipe flow. Here it must be noted that the velocity profile can take on any shape at these cross-sections. It is only in exceptional cases that both the velocity and the pressure may be prescribed at the same boundary.

In Chapters 6 and 7 we will treat *flows with heat transfer*. In many applications of such flows, the density change as a result of pressure change can be neglected. However, because of heat expansion, the density does change with the temperature. For example, in *convection flows* this is the origin of a *buoyancy* $\rho^*(T) \cdot \boldsymbol{g}$.

Within the framework of the *Boussinesq approximation*, the density change will be taken into account only in the lift term and will be neglected in all other terms. The ansatz used for the density is then

$$\rho(T) = \rho_0 \cdot [1 - \alpha \cdot (T - T_0)], \tag{5.84}$$

where α is the *coefficient of heat expansion*, ρ_0 is a reference density, and T_0 a reference temperature. The viscosity is assumed to be constant, and in addition, the dissipation is neglected. Inserting these assumptions into the Navier–Stokes equations (5.18) and the energy equation (5.56) and taking into account the dimensionless quantities

$$x_m^* = \frac{x_m}{l}, \qquad t^* = \frac{k_\infty \cdot t}{l^2}, \qquad \boldsymbol{u}^* = \frac{l}{k_\infty} \cdot \boldsymbol{u},$$

$$T^* = \frac{T - T_\infty}{T_W - T_\infty}, \qquad p^* = (p + \rho_\infty \cdot g \cdot x_3) \cdot \frac{l^2}{\rho_\infty \cdot \nu_\infty \cdot k_\infty},$$

we obtain the dimensionless *Boussinesq equations*

$$\nabla \cdot \boldsymbol{u}^* = 0,$$

$$\frac{1}{\mathrm{Pr}_\infty} \cdot \left(\frac{\partial \boldsymbol{u}^*}{\partial t^*} + (\boldsymbol{u}^* \cdot \nabla)\boldsymbol{u}^* \right) = \mathrm{Ra}_\infty \cdot T^* \cdot \begin{pmatrix} 0 \\ 0 \\ 1 \end{pmatrix} - \nabla p^* + \Delta \boldsymbol{u}^*, \tag{5.85}$$

$$\frac{\partial T^*}{\partial t^*} + \boldsymbol{u}^* \cdot \nabla T^* = \Delta T^*,$$

with the dimensionless Rayleigh number

$$\mathrm{Ra}_\infty = \frac{g \cdot l^3}{k_\infty \cdot \nu_\infty} \cdot \alpha \cdot (T - T_\infty) \quad .$$

Different steady or unsteady behavior of the flow is expected depending on the size of the Prandtl number Pr_∞. If Pr_∞ is small (e.g. 0.71 for air, 10^{-2} for liquid metals), then the flow is unsteady. If Pr_∞ is large (7 for water, 10^3 for oil), we obtain a steady flow in the form of convection rolls. In this case, the unsteady term has only a minor effect, since it is multiplied by the small factor $1/\mathrm{Pr}_\infty$.

If we also take the mass diffusion in a *two-component liquid layer* (e.g. salt solution) into account, the concentration gradient means that we obtain a second part to the lift.

To describe the mass exchange inside a multicomponent mixture consisting of N species, N mass-balance equations can be written down. If m_k is the mass of species k, then the quantity ρ_k is called the *partial density* of species k in the mixture. The density of the mixture ρ is defined as

$$\rho = \sum_{k=1}^{N} \rho_k. \tag{5.86}$$

As well as this, each species has its own velocity $\boldsymbol{u}_k$. In analogy to a single-component fluid (N = 1) in the presence of mass sources or mass sinks, N mass balances can be formulated:

$$\frac{\partial \rho_k}{\partial t} + \nabla \cdot (\rho_k \cdot \boldsymbol{u}_k) = 0, \qquad k = 1, \ldots, N. \tag{5.87}$$

Summing these *component continuity equations* and introducing the *mass concentration* $c_k = \rho_k/\rho$ of the species k ($\sum_{k=1}^{N} c_k = 1$) we obtain the following dimensionless relation for the mixture density:

$$\frac{\partial \rho^*}{\partial t^*} + \nabla \cdot \left(\rho^* \cdot \sum_{k=1}^{N} \boldsymbol{u}^* \right) = 0, \tag{5.88}$$

with the dimensionless flow velocity of the mixture

$$\boldsymbol{u}^* = \sum_{k=1}^{N} c_k \cdot \boldsymbol{u}_k^*. \tag{5.89}$$

Together with the dimensionless total pressure $p^* = \sum_{k=1}^{N} p_k^*$ and the linear thermal equation of state $\rho^* = 1 - (\alpha_m \cdot (T^* - T_m)) \cdot (T^* - T_m) - (\beta_m \cdot (c - c_m)) \cdot (c - c_m)$ (β concentration expansion coefficient, m mean temperature or concentration) we obtain the following dimensionless Boussinesq equations for the two-component mixture:

$$\begin{aligned} \nabla \cdot \boldsymbol{u}^* &= 0, \\ \mathrm{Le}_\infty \cdot \left(\frac{\partial c}{\partial t^*} + (\boldsymbol{u}^* \cdot \nabla c) \right) &= \Delta c, \\ \frac{1}{\mathrm{Pr}_\infty} \cdot \left(\frac{\partial \boldsymbol{u}^*}{\partial t^*} + (\boldsymbol{u}^* \cdot \nabla)\boldsymbol{u}^* \right) &= \Delta \boldsymbol{u}^* - \nabla p^* \\ &\quad + (\mathrm{Ra}_\infty \cdot T^* + \mathrm{Ra}_{D\infty} \cdot c) \cdot \begin{pmatrix} 0 \\ 0 \\ 1 \end{pmatrix}, \\ \frac{\partial T^*}{\partial t^*} + \boldsymbol{u}^* \cdot \nabla T^* &= \Delta T^*, \end{aligned} \tag{5.90}$$

with the following dimensionless characteristic numbers: The diffusion Rayleigh number $\mathrm{Ra}_{D\infty} = -\beta_m \cdot (c - c_m) \cdot g \cdot l^3/(k_m \cdot \nu_m)$ and the Lewis number $\mathrm{Le} = k_m/D_m$ (coefficient of diffusion D).

If we look at a salt solution, it is easy to see that for $c = 0$ (pure water) or $c = 1$ (i.e. $\mathrm{Ra}_{D\infty} = 0$) (salt water at its solubility limit) the above system of equations goes over to the system describing Rayleigh–Bénard convection.

The Euler equation (5.76) forms the basis for some incompressible flow problems with *freely movable interfaces*, to be treated in (Chapter 8). Here it is assumed that the flow on both sides of the interface is irrotational. Integrating the Euler equation over the spatial coordinates and simultaneously introducing a velocity potential $\boldsymbol{u} = -\nabla\Phi$ leads to the generalized Bernoulli equation in the form

$$-\frac{\partial \Phi}{\partial t} + \frac{1}{2} \cdot (\nabla\Phi)^2 + \frac{p}{\rho} + \boldsymbol{g} \cdot \boldsymbol{x} = \mathrm{C}_k, \tag{5.91}$$

where C_k is a constant of integration that can vary on both side of the interface for different phases k. The Bernoulli equation (5.91) is the starting point for the description of wave processes in layered incompressible media. In addition, it can be used to describe the dynamics of starting phases of pressure-induced bubble growth.

For spherically shaped bubbles, a differential equation for the radius R_B of a single bubble under the effect of a pressure field was derived from the Bernoulli equation by *J. W. S. Rayleigh* (1917) and by *M. Plesset* and *S. A. Zwick* (1954). A spherically shaped bubble in an infinitely extended liquid was considered. Using a time-varying bubble volume, the mass balance yields the following relation for the velocity $\boldsymbol{u}(R_B, r, t)$ at a radius r outside the bubble:

$$\boldsymbol{u}(R_B, r, t) = \frac{R_B^2}{r^2} \cdot \frac{\mathrm{d}R_B}{\mathrm{d}t}. \tag{5.92}$$

The velocity can be assigned a potential in the form

$$\Phi = \frac{R_B^2}{r} \cdot \frac{\mathrm{d}R_B}{\mathrm{d}t}.$$

Inserting this relation into the Bernoulli equation (5.87), we obtain the *Rayleigh–Plesset equation* connecting the states at the edge of the bubble and at a great distance from the edge, i.e. for $r \to \infty$,

$$R_B \cdot \frac{\mathrm{d}^2 R_B}{\mathrm{d}t^2} + \frac{3}{2} \cdot \left(\frac{\mathrm{d}R_B}{\mathrm{d}t}\right)^2 = \frac{1}{\rho_k} \cdot (p_R - p_\infty). \tag{5.93}$$

Here the index k denotes the liquid phase, and the indices R and ∞ the pressure states at the edge of the bubble and at infinity, respectively. The equation has to be modified if phase transitions occur at the edge of the bubble, if surface stresses or viscous forces act, or if the gas and the liquid are not in thermodynamic equilibrium.

For flows at large Reynolds numbers, the fact that the factor $\boldsymbol{G}_m^*$ (5.73) in front of the $1/\mathrm{Re}_l$ term is small does not necessarily permit the dissipative fluxes to be neglected. For flows with boundary-layer character, the size of the viscous terms depends on whether velocity gradients parallel or perpendicular to the contour of the body are considered.

Since the contour of the body does not in general run parallel to one of the coordinate axes, the Navier–Stokes equations (5.73) first have to transformed to body-fitted curvilinear coordinates. The curvilinear coordinates ξ_1^*, ξ_2^*, ξ_3^* are given by the transformation equations

$$\xi_1^* = \xi_1^*(t^*, x_\mathrm{m}^*) \quad , \qquad \xi_2^* = \xi_2^*(t^*, x_\mathrm{m}^*) \quad , \quad \xi_3^* = \xi_3^*(t^*, x_\mathrm{m}^*) \quad , \qquad t^* = \tau^* \quad ,$$

$$\frac{\partial}{\partial t^*} = \frac{\partial}{\partial \tau^*} + \frac{\partial \xi_1^*}{\partial t^*} \cdot \frac{\partial}{\partial \xi_1^*} + \frac{\partial \xi_2^*}{\partial t^*} \cdot \frac{\partial}{\partial \xi_2^*} + \frac{\partial \xi_3^*}{\partial t^*} \cdot \frac{\partial}{\partial \xi_3^*} \quad ,$$

$$\frac{\partial}{\partial x_\mathrm{m}^*} = \frac{\partial \xi_1^*}{\partial x_\mathrm{m}^*} \cdot \frac{\partial}{\partial \xi_1^*} + \frac{\partial \xi_2^*}{\partial x_\mathrm{m}^*} \cdot \frac{\partial}{\partial \xi_2^*} + \frac{\partial \xi_3^*}{\partial x_\mathrm{m}^*} \cdot \frac{\partial}{\partial \xi_3^*} \quad .$$

The transformed equations read

$$\frac{\partial \hat{\boldsymbol{U}}^*}{\partial t^*} + \sum_{\mathrm{m}=1}^{3} \frac{\partial \hat{\boldsymbol{F}}_\mathrm{m}^*}{\partial \xi_\mathrm{m}^*} - \frac{1}{Re_l} \cdot \sum_{\mathrm{m}=1}^{3} \frac{\partial \hat{\boldsymbol{G}}_\mathrm{m}^*}{\partial \xi_\mathrm{m}^*} = \mathbf{0} \quad , \tag{5.94}$$

with

$$\hat{\boldsymbol{U}}^* = J \cdot \boldsymbol{U}^* \quad ,$$

$$\hat{\boldsymbol{F}}_\mathrm{m}^* = J \cdot \left(\frac{\partial \xi_\mathrm{m}^*}{\partial t^*} \cdot \boldsymbol{U}^* + \frac{\partial \xi_\mathrm{m}^*}{\partial x_1^*} \cdot \boldsymbol{F}_1^* + \frac{\partial \xi_\mathrm{m}^*}{\partial x_2^*} \cdot \boldsymbol{F}_2^* + \frac{\partial \xi_\mathrm{m}^*}{\partial x_3^*} \cdot \boldsymbol{F}_3^* \right) \quad ,$$

$$\hat{\boldsymbol{G}}_\mathrm{m}^* = J \cdot \left(\frac{\partial \xi_\mathrm{m}^*}{\partial x_1^*} \cdot \boldsymbol{G}_1^* + \frac{\partial \xi_\mathrm{m}^*}{\partial x_2^*} \cdot \boldsymbol{G}_2^* + \frac{\partial \xi_\mathrm{m}^*}{\partial x_3^*} \cdot \boldsymbol{G}_3^* \right) \quad ,$$

with the Jacobi determinant

$$J^{-1} = \frac{\partial x_1^*}{\partial \xi_1^*} \cdot \frac{\partial x_2^*}{\partial \xi_2^*} \cdot \frac{\partial x_3^*}{\partial \xi_3^*} + \frac{\partial x_1^*}{\partial \xi_3^*} \cdot \frac{\partial x_2^*}{\partial \xi_1^*} \cdot \frac{\partial x_3^*}{\partial \xi_2^*} + \frac{\partial x_1^*}{\partial \xi_2^*} \cdot \frac{\partial x_2^*}{\partial \xi_3^*} \cdot \frac{\partial x_3^*}{\partial \xi_1^*}$$
$$- \frac{\partial x_1^*}{\partial \xi_1^*} \cdot \frac{\partial x_2^*}{\partial \xi_3^*} \cdot \frac{\partial x_3^*}{\partial \xi_2^*} - \frac{\partial x_1^*}{\partial \xi_2^*} \cdot \frac{\partial x_2^*}{\partial \xi_1^*} \cdot \frac{\partial x_3^*}{\partial \xi_3^*} - \frac{\partial x_1^*}{\partial \xi_3^*} \cdot \frac{\partial x_2^*}{\partial \xi_2^*} \cdot \frac{\partial x_3^*}{\partial \xi_1^*} .$$

The terms in $\boldsymbol{G}^*$ that contain derivatives parallel to the body contour are in general small (except in the case of flow separation). This means that they may be neglected.

Perturbations generally expand downstream only, corresponding to a parabolic expansion mechanism. This property is passed on to the *steady* Navier–Stokes equations by impressing the pressure onto the boundary layer. The pressure gradient in the direction normal to the wall ξ_3^* is neglected.

This yields the dimensionless *parabolized Navier–Stokes equations* for steady boundary-layer flows:

$$\sum_{m=1}^{3} \frac{\partial \hat{\hat{\boldsymbol{f}}}_m^*}{\partial \xi_m^*} - \frac{1}{\mathrm{Re}_l} \cdot \frac{\partial \hat{\hat{\boldsymbol{G}}}_3^*}{\partial \xi_3^*} = 0, \tag{5.95}$$

with

$$\hat{\hat{\boldsymbol{f}}}_m^* = J^{-1} \cdot \begin{pmatrix} \rho^* \cdot \hat{u}_m^* \\ \rho^* \cdot \hat{u}_m^* \cdot u_1^* + \frac{\partial \xi_m^*}{\partial x_1^*} \cdot p_s^* \\ \rho^* \cdot \hat{u}_m^* \cdot u_2^* + \frac{\partial \xi_m^*}{\partial x_2^*} \cdot p_s^* \\ \rho^* \cdot \hat{u}_m^* u \cdot_3^* + \frac{\partial \xi_m^*}{\partial x_3^*} \cdot p_s^* \\ \hat{u}_m^* \cdot (\rho^* \cdot e_{\text{tot}}^* + p_s^*) \end{pmatrix} , \tag{5.96}$$

$$\hat{\hat{\boldsymbol{G}}}_3^* = J^{-1} \cdot \begin{pmatrix} 0 \\ \sum_{l=1}^{3} \frac{\partial \xi_3^*}{\partial x_l^*} \cdot \tau_{l1}^* \\ \sum_{l=1}^{3} \frac{\partial \xi_3^*}{\partial x_l^*} \cdot \tau_{l2}^* \\ \sum_{l=1}^{3} \frac{\partial \xi_3^*}{\partial x_l^*} \cdot \tau_{l3}^* \\ \sum_{l=1}^{3} \frac{\partial \xi_3^*}{\partial x_l^*} \cdot \left(\sum_{m=1}^{3} u_m^* \cdot \tau_{m3}^* + \dot{q}_3^* \right) \end{pmatrix} , \tag{5.97}$$

and

$$\hat{u}_m^* = \frac{\partial \xi_m^*}{\partial x_1^*} \cdot u_1^* + \frac{\partial \xi_m^*}{\partial x_2^*} \cdot u_2^* + \frac{\partial \xi_m^*}{\partial x_3^*} \cdot u_3^*, \qquad m = 1, 2, 3.$$

If the effect of the curvature of the contour is also neglected, the order of magnitude estimation carried out by *Prandtl* leads to all the derivatives with respect to x_1 and x_2 in the friction terms of (5.95) being neglected. This is justified for high Reynolds number flows if the boundary-layer thickness is small compared to the dimensions of the body. Since the pressure is impressed onto the boundary layer ($\partial p^*/\partial x_3^* = 0$), the third momentum conservation equation drops away, and we obtain the *Prandtl boundary-layer equations* in Cartesian coordinates x_m^*:

$$\frac{\partial(\rho^* \cdot u_1^*)}{\partial x_1^*} + \frac{\partial(\rho^* \cdot u_2^*)}{\partial x_2^*} + \frac{\partial(\rho^* \cdot u_3^*)}{\partial x_3^*} = 0,$$

$$\rho^* \cdot \left(u_1^* \cdot \frac{\partial u_1^*}{\partial x_1^*} + u_2^* \cdot \frac{\partial u_1^*}{\partial x_2^*} + u_3^* \cdot \frac{\partial u_1^*}{\partial x_3^*} \right) = -\frac{\partial p_s^*}{\partial x_1^*} + \frac{1}{\text{Re}_l} \cdot \frac{\partial}{\partial x_3^*} \left(\mu^* \cdot \frac{\partial u_1^*}{\partial x_3^*} \right),$$

$$\rho^* \cdot \left(u_1^* \cdot \frac{\partial u_2^*}{\partial x_1^*} + u_2^* \cdot \frac{\partial u_2^*}{\partial x_2^*} + u_3^* \cdot \frac{\partial u_2^*}{\partial x_3^*} \right) = -\frac{\partial p_s^*}{\partial x_2^*} + \frac{1}{\text{Re}_l} \cdot \frac{\partial}{\partial x_3^*} \left(\mu^* \cdot \frac{\partial u_2^*}{\partial x_3^*} \right),$$

$$\begin{aligned} \rho^* \cdot \left(u_1^* \cdot \frac{\partial T^*}{\partial x_1^*} + u_2^* \cdot \frac{\partial T^*}{\partial x_2^*} + u_3^* \cdot \frac{\partial T^*}{\partial x_3^*} \right) &= \frac{\mu^*}{(\kappa - 1) \cdot \text{Re}_l \cdot \text{Pr}_\infty} \cdot \frac{\partial^2 T^*}{\partial x_3^{*2}} \\ &+ \frac{\mu *}{\text{Re}_l} \cdot \left[\left(\frac{\partial u_1^*}{\partial x_3^*} \right)^2 + \left(\frac{\partial u_2^*}{\partial x_3^*} \right)^2 \right] \\ &+ u_1^* \cdot \frac{\partial p_s^*}{\partial x_1^*} + u_2^* \cdot \frac{\partial p_s^*}{\partial x_2^*}. \end{aligned} \tag{5.98}$$

5.4.4 Reynolds Equations for Turbulent Flows

Writing the *Reynolds equations* (5.27), (5.33)–(5.35), (5.59) in *conservative form* as in Section 5.4.2, and using mass-averaged flow quantities, we obtain the time-averaged fundamental equations for dimensional flow quantities:

$$\frac{\partial \overline{\boldsymbol{U}}}{\partial t} + \sum_{m=1}^{3} \frac{\partial \overline{\boldsymbol{f}}_m}{\partial x_m} - \frac{1}{\mathrm{Re}_l} \sum_{m=1}^{3} \frac{\partial \overline{\boldsymbol{G}}_m}{\partial x_m} + \sum_{m=1}^{3} \frac{\partial \boldsymbol{R}_m}{\partial x_m} = 0, \tag{5.99}$$

with the solution vector

$$\overline{\boldsymbol{U}}(x_m, t) = \begin{pmatrix} \overline{\rho} \\ \overline{\rho} \cdot \tilde{u}_1 \\ \overline{\rho} \cdot \tilde{u}_2 \\ \overline{\rho} \cdot \tilde{u}_3 \\ \overline{\rho} \cdot \tilde{E} \end{pmatrix} . \tag{5.100}$$

Compared to the Navier–Stokes equations in conservative form (5.73), the Reynolds ansatz and the time averaging have introduced the term $\boldsymbol{R}_m$.

The vector of the time-averaged convective fluxes is

$$\overline{\boldsymbol{f}}_m = \begin{pmatrix} \overline{\rho} \cdot \tilde{u}_m \\ \overline{\rho} \cdot \tilde{u}_m \cdot \tilde{u}_1 + \delta_{1m} \cdot \overline{p} \\ \overline{\rho} \cdot \tilde{u}_m \cdot \tilde{u}_2 + \delta_{2m} \cdot \overline{p} \\ \overline{\rho} \cdot \tilde{u}_m \cdot \tilde{u}_3 + \delta_{3m} \cdot \overline{p} \\ \tilde{u}_m \cdot (\overline{\rho} \cdot \tilde{e}_{\mathrm{tot}} + \overline{p}) \end{pmatrix} , \tag{5.101}$$

the vector of the average dissipative fluxes is

$$\overline{\boldsymbol{G}}_{\mathrm{m}} = \begin{pmatrix} 0 \\ \overline{\tau}_{\mathrm{m1}} \\ \overline{\tau}_{\mathrm{m2}} \\ \overline{\tau}_{\mathrm{m3}} \\ \sum_{\mathrm{l}=1}^{3} \tilde{u}_{\mathrm{l}} \cdot \overline{\tau}_{\mathrm{lm}} + \overline{\dot{q}}_{\mathrm{m}} \end{pmatrix} \tag{5.102}$$

and the additional vector of the turbulent fluxes is

$$\boldsymbol{R}_{\mathrm{m}} = \begin{pmatrix} 0 \\ \overline{\rho} \cdot \widetilde{u_1'' \cdot u_{\mathrm{m}}''} \\ \overline{\rho} \cdot \widetilde{u_2'' \cdot u_{\mathrm{m}}''} \\ \overline{\rho} \cdot \widetilde{u_3'' \cdot u_{\mathrm{m}}''} \\ R_{\mathrm{m,E}} \end{pmatrix} , \tag{5.103}$$

where

$$\mathrm{R}_{m,E} = \sum_{\mathrm{l}=1}^{3} \left(\overline{\tau_{\mathrm{ml}} \cdot u_{\mathrm{l}}''} + \tilde{u}_{\mathrm{m}} \cdot \bar{\rho} \cdot \widetilde{u_{\mathrm{l}}'' \cdot u_{\mathrm{m}}''} - \frac{1}{2} \cdot \bar{\rho} \cdot u_{\mathrm{l}}'' \cdot \widetilde{u_{\mathrm{l}}'' \cdot u_{\mathrm{m}}''} \right)$$
$$- \overline{p \cdot u_{\mathrm{m}}''} - \overline{\rho \cdot e'' \cdot u_{\mathrm{m}}''} \quad ,$$

with the total energy

$$\tilde{E} = \tilde{e} + \sum_{\mathrm{m}=1}^{3} \frac{\tilde{u}_{\mathrm{m}}^2}{2} + K \quad ,$$

$$K = \sum_{\mathrm{m}=1}^{3} \frac{\widetilde{u_{\mathrm{m}}'' \cdot u_{\mathrm{m}}''}}{2} \quad ,$$

where K is the *turbulent kinetic energy.*

The fluctuating quantities appearing in the additional term $\boldsymbol{R}_{\mathrm{m}}$ are unknown. It is clear that the system of equations has more unknowns than equations and is therefore *not closed.* The related closure problem of the Reynolds equations for turbulent flows means that the individual terms in $\boldsymbol{R}_{\mathrm{m}}$ have to be modeled using empirical assumptions for each flow problem.

5.4.5 Turbulence Models

Momentum and heat transport take place on a microscopic scale in all flows as a consequence of molecular diffusion processes. They are represented by molecular viscosity and thermal conductivity respectively. It is useful to model the exchange processes due to turbulence in an analogous manner and to introduce a turbulent viscosity and a turbulent thermal conductivity. For simple one-dimensional flows this is realized by means of the classical mixing length approach due to Prandtl (see Section 4.2.5).

Following this idea, we model the Reynolds stresses using the Boussinesq ansatz:

$$-\bar{\rho} \cdot \widetilde{u_{\mathrm{i}}'' \cdot u_{\mathrm{j}}''} = \mu_{\mathrm{t}} \cdot \left(\frac{\partial \tilde{u}_{\mathrm{i}}}{\partial x_{\mathrm{j}}} + \frac{\partial \tilde{u}_{\mathrm{j}}}{\partial x_{\mathrm{i}}} \right) - \frac{2}{3} \cdot \bar{\rho} \cdot K \cdot \delta_{\mathrm{ij}} \quad . \tag{5.104}$$

Here μ_{t} is the turbulent viscosity or *eddy viscosity.* The right-hand term in equation (5.104) represents the turbulent pressure (with $\delta_{\mathrm{ij}} = 1$ for $i = j$ and $\delta_{\mathrm{ij}} = 0$ for $i \neq j$), assumed proportional to the turbulent kinetic energy per unit mass

$$K = \frac{1}{2} \cdot \widetilde{u_{\mathrm{i}}'' \cdot u_{\mathrm{i}}''} = \frac{1}{2} \cdot \left(\widetilde{u_1''^2} + \widetilde{u_2''^2} + \widetilde{u_3''^2} \right) \tag{5.105}$$

This may be neglected in what follows.

The analogy to molecular exchange processes, as well as characteristic length scales of turbulent flows under normal conditions, are illustrated in Figure 5.8, where the left-hand picture shows the continuum mechanical velocity as the average of the molecular motion, while the right-hand picture sketches the mean velocity as the average of the turbulent instantaneous velocity. In both cases, the velocity to be considered is taken as the average value of numerous individual velocities (of molecules or of eddies). In each case a relevant length scale is observed; the mean free path or a turbulent length scale.

Turbulence models that apply the eddy viscosity approach described above are called *eddy viscosity models*. Compared to the independent modeling of all six Reynolds stresses, the effort involved is less for only one further quantity, the eddy viscosity. The eddy viscosity is not a material property of the fluid but rather is a property of the turbulence of each flow.

According to the analogy, we may also model the turbulent heat fluxes with the Fourier law of thermal conductivity, i.e. with the ansatz

$$-\rho \cdot c_v \cdot \overline{u'_\mathrm{i} \cdot T'} = \lambda_\mathrm{t} \cdot \frac{\partial \overline{T}}{\partial x_\mathrm{i}} \quad , \qquad -\overline{u'_\mathrm{i} \cdot T'} = a_\mathrm{t} \cdot \frac{\partial \overline{T}}{\partial x_\mathrm{i}} \quad . \tag{5.106}$$

Therefore, the turbulent heat fluxes are assumed proportional to the gradient of the mean temperature. The quantity λ_t is called the turbulent thermal conductivity and $a_\mathrm{t} = \lambda_\mathrm{t}/(\rho \cdot c)$ the turbulent thermal diffusivity. One of these quantities must be modeled. In most cases the turbulent heat transport is considerably larger than the molecular heat transport, and, independently of the material properties, it is this which determines the effect of the turbulence on the mean flow.

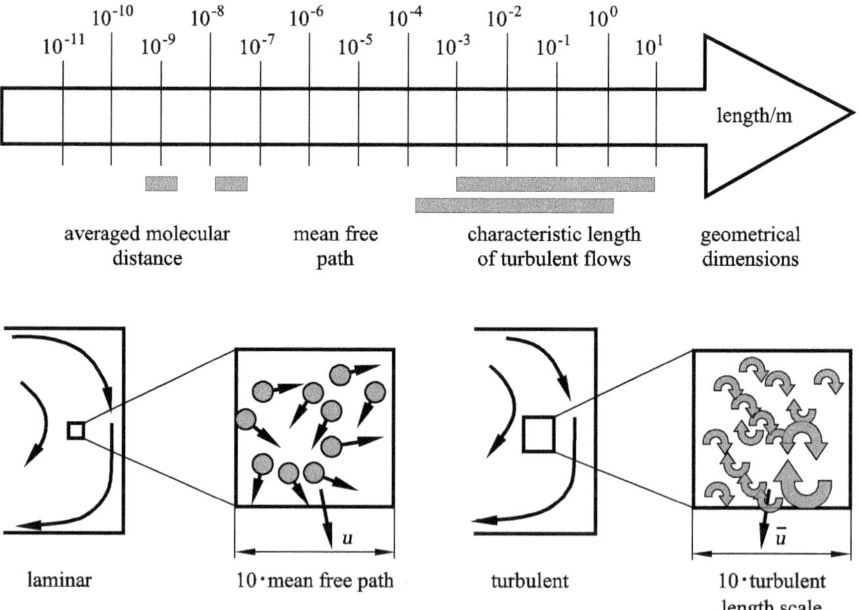

Fig. 5.8. Analogy of the detailed and averaged point of view in turbulence

The eddy viscosity and the turbulent thermal diffusivity are not independent of each other. In analogy to the molecular Prandtl number, we also define a turbulent Prandtl number as the ratio of the two transport coefficients:

$$Pr_{\mathrm{t}} = \frac{\nu_{\mathrm{t}}}{a_{\mathrm{t}}} \quad . \tag{5.107}$$

This has approximately the value one. In practice, for fluids of low thermal conductivity (air, water), $Pr_{\mathrm{t}} = 0.9$ is generally applied. Fluids with very high thermal conductivity compared to viscosity, i.e. fluids with very flow molecular Prandtl number (e.g. liquid metals), are an exception to this. Here the turbulent fluctuations of the velocity field have a lesser effect on the turbulent thermal conductivity and diffusivity than on the turbulent viscosity. Therefore a higher turbulent Prandtl number is to be selected for such fluids, with a value of about $Pr_{\mathrm{t}} = 3$ for $Pr = 0.01$.

This reduces turbulent modeling to modeling the dependence of the eddy viscosity on the mean flow. If ν_{t} is known, a_{t} may be computed using the assumed turbulent Prandtl number.

If we insert the ansatz for the eddy viscosity into the Reynolds equations for incompressible flows (5.40 - 5.42) and the energy equation, after dividing by the density we obtain

$$\frac{\partial \overline{u}_{\mathrm{i}}}{\partial t} + \frac{\partial}{\partial x_{\mathrm{j}}} \left(\overline{u}_{\mathrm{j}} \cdot \overline{u}_{\mathrm{i}} \right) = -\frac{\partial \overline{p}}{\partial x_{\mathrm{i}}} + \frac{\partial}{\partial x_{\mathrm{j}}} \left[(\nu + \nu_{\mathrm{t}}) \cdot \left(\frac{\partial \overline{u}_{\mathrm{i}}}{\partial x_{\mathrm{j}}} + \frac{\partial \overline{u}_{\mathrm{j}}}{\partial x_{\mathrm{i}}} \right) \right] \quad , \tag{5.108}$$

$$\frac{\partial \overline{T}}{\partial t} + \frac{\partial}{\partial x_{\mathrm{j}}} \left(\overline{u}_{\mathrm{j}} \cdot \overline{T} \right) = \frac{\partial}{\partial x_{\mathrm{j}}} \left[\left(a + \frac{\nu_{\mathrm{t}}}{Pr_{\mathrm{t}}} \right) \cdot \frac{\partial \overline{T}}{\partial x_{\mathrm{j}}} \right] \quad . \tag{5.109}$$

Therefore as well as the molecular transport coefficients we also have the turbulent transport coefficients; the turbulent pressure is neglected.

The idea that the intensity of the turbulent mixing is represented by a single quantity, the eddy viscosity, assumes that the turbulent fluctuations are the same in all spatial directions. The turbulence is hence isotropic. However, isotropic turbulence seldom occurs in practice, at best in the turbulent parallel flow behind a grid. Close to a wall and in free-shear layers, the turbulence is anisotropic to a greater or lesser degree. For example, in a turbulent boundary layer the fluctuations in the direction parallel to the wall are twice as large as those in the direction normal to the wall, as the wall suppresses normal motions.

The reproduction of the Fourier heat conduction law for the turbulent heat flux implies that temperature gradients once present in the flow are leveled out by the mixing processes. The turbulent heat flux is therefore orientated in the opposite direction to the temperature gradient. Only flows with greatly anisotropic turbulence permit anti-gradient heat transport, e.g. a fluid layer with internal heat sources. The turbulent heat flux then moves in the direction of the higher temperature. This cannot be modeled using the above approaches, as the thermal conductivity would be negative.

However, for many flows with heat transport, which will be considered in Chapter 7, the approach using the turbulent Prandtl number and the assumption of eddy viscosity has shown itself to be reliable in practice.

Algebraic Turbulence Models

The algebraic eddy viscosity models are the simplest class of turbulence models. In some shear flows along solid walls, e.g. in fully developed pipe flow or in boundary-layer flow along a flat plate, the spatial dependence of the eddy viscosity can be reduced to a single coordinate, namely the distance from the wall. This is because turbulent boundary layers, just as laminar boundary layers, are similar and the boundary-layer equation may be solved using a similarity transformation. Only the dependence of the eddy viscosity on the distance from the wall must be given.

This was carried out by Prandtl with the help of the *mixing length ansatz* (see Section 4.2.5):

$$-\rho \cdot \overline{u' \cdot v'} = -\rho \cdot \overline{l \cdot \frac{\partial \overline{u}}{\partial z} \cdot l \cdot \frac{\partial \overline{u}}{\partial z}} = \mu_t \cdot \frac{\partial \overline{u}}{\partial z} \quad , \tag{5.110}$$

from which, neglecting the sign of the eddy viscosity, it follows that:

$$\mu_t = \rho \cdot l^2 \cdot \left| \frac{\partial \overline{u}}{\partial z} \right| \quad . \tag{5.111}$$

Here $\overline{u}$ is the mean velocity component parallel to the wall and z the distance from the wall. l is called the Prandtl mixing length, and is thus the distance downstream covered by a turbulence ball until it has completely mixed with its surroundings.

Numerous measurements for different shear flows have shown that the mixing length may be assumed proportional to the distance from the wall to good accuracy:

$$l = 0.41 \cdot z \quad , \tag{5.112}$$

where the pre-factor 0.41 is called the von Kármán constant. Close to the wall (up to about 1/3 of the thickness of the boundary layer) this even holds for boundary layers with pressure gradient, for channel and pipe flows, as well as for other shear flows attached to a wall.

The Baldwin-Lomax turbulence model was developed for flows with boundary-layer character past bodies, e.g. the calculation of wing flows (Figure 5.9). The model assumes that the outer flow is inviscid. The flow is divided into two layers depending on the distance from the wall. In the inner layer, which includes the zone close to the wall and the viscous sublayer, a modified mixing length ansatz is used:

$$(\mu_t)_{\text{inner}} = \rho \cdot l_{\text{mod}}^2 \cdot |\overline{\omega}| \quad . \tag{5.113}$$

Instead of the velocity gradient, the magnitude of the rotation of the mean flow appears:

$$|\overline{\omega}| = |\nabla \times \overline{\boldsymbol{u}}| = \sqrt{\left(\frac{\partial \overline{u_2}}{\partial x_3} - \frac{\partial \overline{u_3}}{\partial x_2}\right)^2 + \left(\frac{\partial \overline{u_3}}{\partial x_1} - \frac{\partial \overline{u_1}}{\partial x_3}\right)^2 + \left(\frac{\partial \overline{u_1}}{\partial x_2} - \frac{\partial \overline{u_2}}{\partial x_1}\right)^2}\,, \qquad (5.114)$$

The modified mixing length reads:

$$l_{\text{mod}} = 0.41 \cdot z \cdot \left[1 - \exp\left(-\frac{z^+}{\mathrm{A}^+}\right)\right] \quad , \qquad (5.115)$$

with the van Driest damping factor (the expression in square brackets). $z^+ = u_\tau/\nu \cdot z$ is the dimensionless coordinate, with $u_\tau = \sqrt{\tau_{\text{w}}/\overline{\rho}_{\text{w}}}$. The model constant has the value $A^+ = 26$. Outside the viscous sublayer, the damping factor has approximately the value one, and so hardly changes the eddy viscosity in the zone close to the wall. In this viscous sublayer, however, this factor takes into account the changed conditions as it reduces l_{mod} and hence also the eddy viscosity.

In the outer layer the strength of the turbulence depends on the state of the outer flow. The ansatz for the eddy viscosity reads:

$$(\mu_{\text{t}})_{\text{outer}} = \rho \cdot \tilde{\mathrm{K}} \cdot \mathrm{C_{CP}} \cdot \mathrm{F_{wake}} \cdot \mathrm{F_{Kleb}} \quad , \qquad (5.116)$$

with the constant $\tilde{\mathrm{K}} = 0.0168$ and the Clauser parameter $\mathrm{C_{CP}} = 1.6$. Apart from F_{Kleb} at the position x, all quantities in this ansatz are constant.

In order to calculate the constant $\mathrm{F_{wake}}$, we consider the function

$$\mathrm{F}(z) = z \cdot |\overline{\omega}| \cdot \left[1 - \exp\left(-\frac{z^+}{\mathrm{A}^+}\right)\right] \quad , \qquad (5.117)$$

in which the van Driest damping factor is approximately one. Whereas the factor z increases with increasing distance from the wall, the mean rotation decreases to zero, and so this function has a maximum $F_{\max}$ at the position $x_{\max}$. For subsequent modeling, this may be written as follows:

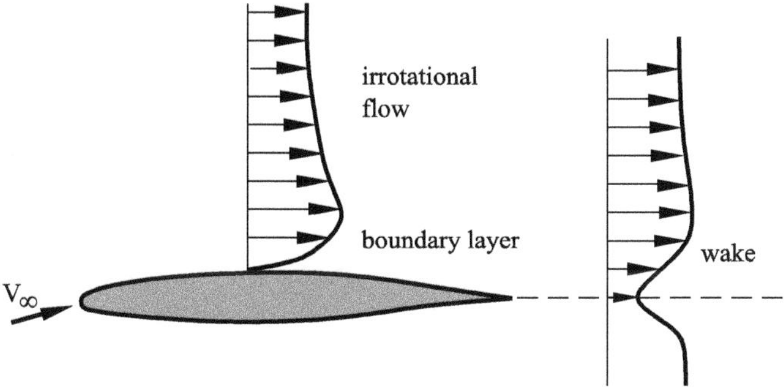

Fig. 5.9. Example of application of the Baldwin-Lomax turbulence model: flow past a wing with wake

$$\mathrm{F_{wake}} = \min(z_{\max} \cdot \mathrm{F_{max}}, z_{\max} \cdot \frac{u_{\mathrm{dif}}^2}{\mathrm{F_{max}}}) \quad , \tag{5.118}$$

with the maximum velocity difference

$$u_{\mathrm{dif}} = |\overline{\boldsymbol{u}}|_{\max} - |\overline{\boldsymbol{u}}|_{\min} \quad .$$

Because of the no-slip condition, the quantity $|\overline{u}|_{\min}$ has the value zero at the wall. Baldwin and Lomax also applied the turbulence model to wake flows of wings. The velocity in the center of the wake profile is then taken (Figure 5.8).

The Klebanoff intermittence factor

$$\mathrm{F_{Kleb}} = \left[1 + 5.5 \cdot \left(\frac{\mathrm{C_{Kleb}} \cdot z}{z_{\max}}\right)^6\right]^{-1} \quad , \qquad \mathrm{C_{Kleb}} = 0.3 \tag{5.119}$$

ensures that the eddy viscosity falls outwards to zero. In takes into account the fact that in the outer regime of a boundary layer laminar and turbulent phases alternate, as the laminar outer flow can briefly extend into the boundary layer, or the turbulent structures can move with spatial and time shifts into the outer flow and only there dissipate (intermittence). The data measured in a boundary layer without pressure gradient are assumed here for all boundary layers. The quantity $z_{\max}$ is used instead of the boundary-layer thickness.

In the inner layer, the eddy viscosity increases with distance from the wall, while in the outer layer it decreases with distance from the wall. The boundary between the inner and outer layers is at the point of intersection of these two progressions. In practice the eddy viscosity is calculated in both regions and the minimum is taken.

If the boundary layer is not turbulent from the leading edge onwards, but rather it starts off laminar and only becomes turbulent within a transition region, the turbulent model is only used from the end of the transition region. However, the laminar-turbulent transition cannot be determined using a turbulence model. Determining the transition is a stability problem of the laminar flow (see Section 4.2.4). If the end of the transition region is not known from experiment, a transition model is necessary to determine the position where the transition region ends (see *H. Oertel jr., J. Delfs* (1996), (2005)).

Transport Models

The assumption made above, that the turbulence at one position in the flow field is only dependent on the local conditions, is a considerable restriction. Frequently the mechanisms by which turbulence is transported must also be taken into account. It often happens in technical flows that turbulence arises in certain regions of the flow field and is then transported to other regions, where it affects the mean flow. It may then die away again in other regions of the flow field. Turbulence models that take these transport mechanism into account are generally known as transport models.

In the *Prandtl one-equation model*, the eddy viscosity is modeled with:

$$\mu_t = C_\mu \cdot \rho \cdot l \cdot \sqrt{K} \quad , \qquad C_\mu = 0.09 \tag{5.120}$$

where $l = 0.41 \cdot z$ is again the Prandtl mixing length, but the shearing is now replaced by the square root of the turbulent kinetic energy in accordance with equation (5.105).

The transport of the turbulent kinetic energy K can be derived from the Reynolds equations (5.40)–(5.42). The ith component of the Reynolds equation is multiplied by the velocity fluctuation u_i':

$$\rho \cdot \frac{\partial(\overline{u}_i + u_i')}{\partial t} \cdot u_i' + \rho \cdot (\overline{u}_j + u_j') \cdot \frac{\partial(\overline{u}_i + u_i')}{\partial x_j} \cdot u_i' = - \frac{\partial(\overline{p} + p')}{\partial x_i} \cdot u_i' + \mu \cdot \frac{\partial^2(\overline{u}_i + u_i')}{\partial x_j^2} \cdot u_i' \quad . \tag{5.121}$$

All terms are now multiplied out and then time-averaged. The first term in each case is identical with the corresponding term from the Reynolds equation and can be cancelled out. The transport equations are then added, and transformed using the identities:

$$\frac{\partial u_i'}{\partial t} \cdot u_i' = \frac{\partial}{\partial t} \left(\frac{1}{2} \cdot u_i'^2 \right) \quad , \qquad \frac{\partial u_i'}{\partial x_j} \cdot u_i' = \frac{\partial}{\partial x_j} \left(\frac{1}{2} \cdot u_i'^2 \right) \quad , \tag{5.122}$$

$$\frac{\partial^2 u_i'}{\partial x_j^2} \cdot u_i' = \frac{\partial}{\partial x_j} \left(\frac{\partial u_i'}{\partial x_j} \cdot u_i' \right) - \left(\frac{\partial u_i'}{\partial x_j} \right)^2 \tag{5.123}$$

The equation for K is then formulated:

$$\rho \cdot \frac{\partial K}{\partial t} + \rho \cdot \overline{u}_j \cdot \frac{\partial K}{\partial x_j} = - \frac{\partial \overline{u}_i}{\partial x_j} \cdot \rho \cdot \overline{u_i' \cdot u_j'} + \frac{\partial}{\partial x_j} \left(\mu \cdot \frac{\partial K}{\partial x_j} - \frac{1}{2} \cdot \rho \cdot \overline{u_i' \cdot u_i' \cdot u_j'} - \overline{p' \cdot u_j'} \right) - \mu \cdot \overline{\frac{\partial u_i'}{\partial x_j} \cdot \frac{\partial u_i'}{\partial x_j}} \quad . \tag{5.124}$$

The terms on the left-hand side are the convection terms. The first term on the right-hand side does not contain the transport quantity K, and is therefore called a production term (source term). The further terms in brackets are the molecular diffusion, the turbulent diffusion and the pressure diffusion. The last term is always negative, and so this represents a sink term. It describes the draining away and decay (dissipation) of the turbulence. Terms that contain unknown fluctuation quantities have to be modeled.

We begin with the production term. As both i and j appear twice, we have to sum over both indices (9 terms). Each term consists of the product of a shear component and a Reynolds stress. The Reynolds stresses have already been modeled in (5.104) using the eddy viscosity, and we employ this again:

$$- \frac{\partial \overline{u}_i}{\partial x_j} \cdot \rho \cdot \overline{u_i' \cdot u_j'} = \mu_t \cdot \frac{\partial \overline{u}_i}{\partial x_j} \cdot \left(\frac{\partial \overline{u}_i}{\partial x_j} + \frac{\partial \overline{u}_j}{\partial x_i} \right) \quad . \tag{5.125}$$

According to this model, turbulence is produced where the mean flow has a velocity gradient. This is in good agreement with the idea that shear layers generate turbulence because of their instability.

The turbulence diffusion has the form of a triple product, where we sum over i and j (9 terms). The expansion of turbulence takes place because of its own dynamics. These processes are very complex and can only be modeled for each geometrical class in a very simplified manner. It makes sense to consider the diffusion as a gradient transport. This means that differences in the turbulence intensity, i.e. gradients in K, are equalized. The necessary transport coefficient is proportional to the eddy viscosity:

$$-\frac{1}{2} \cdot \rho \cdot \overline{u_{\mathrm{i}}' \cdot u_{\mathrm{i}}' \cdot u_{\mathrm{j}}'} - \overline{p' \cdot u_{\mathrm{j}}'} = \frac{\mu_{\mathrm{t}}}{\sigma_{\mathrm{k}}} \cdot \frac{\partial K}{\partial x_{\mathrm{j}}} \quad , \tag{5.126}$$

where σ_{k} is the ratio between the eddy viscosity and the turbulent diffusion coefficient, in analogy to the Prandtl number. This model constant can be assumed to be one. The pressure diffusion is not modeled separately, but rather is included in the model of the turbulent diffusion.

Modeling the dissipation is done in an entirely empirical manner. The turbulence in a parallel flow behind a lattice decreases with increasing distance from the lattice because of the internal friction of the turbulent structures. Experiments have shown that the dissipation is proportional to $K^{3/2}$. In the one-equation model we therefore use

$$\mu \cdot \overline{\frac{\partial u_{\mathrm{i}}'}{\partial x_{\mathrm{j}}} \cdot \frac{\partial u_{\mathrm{i}}'}{\partial x_{\mathrm{j}}}} = \mathrm{C_D} \cdot \rho \cdot \frac{K^{\frac{3}{2}}}{l} \quad , \qquad \mathrm{C_D} = 0.09 \quad . \tag{5.127}$$

Here l is again the Prandtl mixing length, and is introduced to strengthen the dissipation close to solid walls.

Therefore the model equation for the turbulent kinetic energy reads:

$$\rho \cdot \frac{\partial K}{\partial t} + \rho \cdot \overline{u}_{\mathrm{j}} \cdot \frac{\partial K}{\partial x_{\mathrm{j}}} = \tag{5.128}$$

$$\mu_{\mathrm{t}} \cdot \frac{\partial \overline{u}_{\mathrm{i}}}{\partial x_{\mathrm{j}}} \cdot \left(\frac{\partial \overline{u}_{\mathrm{i}}}{\partial x_{\mathrm{j}}} + \frac{\partial \overline{u}_{\mathrm{j}}}{\partial x_{\mathrm{i}}} \right) + \frac{\partial}{\partial x_{\mathrm{j}}} \left(\mu \cdot \frac{\partial K}{\partial x_{\mathrm{j}}} + \frac{\mu_{\mathrm{t}}}{\sigma_{\mathrm{k}}} \cdot \frac{\partial K}{\partial x_{\mathrm{j}}} \right) - \mathrm{C_D} \cdot \rho \cdot \frac{K^{\frac{3}{2}}}{l} \quad .$$

In order to determine $K(x_1, x_2, x_3)$ boundary conditions are necessary. The turbulent kinetic energy vanishes at a solid wall because of the no-slip condition. K has to be specified at a boundary where a turbulent flow arises.

From the theoretical point of view there is some criticism of the application of the mixing length l. In the layer close to the wall the transport processes are not important. If the corresponding terms are neglected, the remaining equation makes clear that l must be modified in order that the ansatz used here passes over to the Prandtl mixing length model and hence, for a flat plate without pressure gradient, for example, to the logarithmic law of the wall. In addition it is to be noted that without neglecting the transport terms there is no crossover to the mixing length ansatz, and thus the use of a mixing length in connection with a one-equation model is also meaningless. Instead of this it is two-equation models that have gained acceptance for practical calculations.

With the *two-equation model (K-ε model)* we calculate the eddy viscosity using the ansatz:

$$\mu_t = \rho \cdot C_\mu \cdot \frac{K^2}{\varepsilon} \quad , \qquad C_\mu = 0.09 \quad . \tag{5.129}$$

We do not have the problem of having to fix a characteristic length.

$$\varepsilon = \nu \cdot \overline{\frac{\partial u_i'}{\partial x_k} \cdot \frac{\partial u_i'}{\partial x_k}} \tag{5.130}$$

is the dissipation, for which a transport equation also must be solved. This can be derived or modeled in a similar manner from the Reynolds equations. The two transport equations read:

$$\rho \cdot \frac{\partial K}{\partial t} + \rho \cdot \overline{u}_j \cdot \frac{\partial K}{\partial x_j} = \mu_t \cdot \frac{\partial \overline{u}_i}{\partial x_j} \cdot \left(\frac{\partial \overline{u}_i}{\partial x_j} + \frac{\partial \overline{u}_j}{\partial x_i} \right) + \frac{\partial}{\partial x_j} \left(\mu \cdot \frac{\partial K}{\partial x_j} + \frac{\mu_t}{\sigma_k} \cdot \frac{\partial K}{\partial x_j} \right) - \rho \cdot \varepsilon \quad , \tag{5.131}$$

$$\rho \cdot \frac{\partial \varepsilon}{\partial t} + \rho \cdot \overline{u}_j \cdot \frac{\partial \varepsilon}{\partial x_j} = C_{\varepsilon 1} \cdot \frac{\varepsilon}{K} \cdot \mu_t \cdot \frac{\partial \overline{u}_i}{\partial x_j} \cdot \left(\frac{\partial \overline{u}_i}{\partial x_j} + \frac{\partial \overline{u}_j}{\partial x_i} \right) + \frac{\partial}{\partial x_j} \left(\mu \cdot \frac{\partial \varepsilon}{\partial x_j} - \frac{\mu_t}{\sigma_\varepsilon} \cdot \frac{\partial \varepsilon}{\partial x_j} \right) - C_{\varepsilon 2} \cdot \rho \cdot \frac{\varepsilon^2}{K} \quad , \tag{5.132}$$

with further model constants of the ε-equation $C_{\varepsilon 1} = 1.44$, $C_{\varepsilon 2} = 1.92$ and $\sigma_\varepsilon = 1.3$. As a boundary condition, the derivative of ε perpendicular to the wall is set to zero. Calculation of the value of ε at the wall is then not necessary. In an intake cross-section, ε, just as K, must be prescribed.

The characterization of the turbulence by the two transport quantities K and ε can be understood if we consider the processes of the onset and decay of turbulent structures (eddies) as energy cascades (see Section 6.4.5). As a consequence of instability of the mean flow, large-scale structures initially arise. However, these are unstable and decay into smaller structures, which themselves decay, and so on. The largest part of the kinetic energy is associated with the large-scale eddies. In contrast, the dissipation mainly takes place at the smallest scales. The energy-carrying eddies may therefore be associated with the transport quantity K in the K-ε model, whereas the small eddies are connected to ε.

There are numerous variants of the K-ε model known. One example is the *low Reynolds number K-ε model.* If the wall shear stress, the wall heat flux or flow separation have to be calculated, the layer close to the wall and the viscous sublayer must be modeled. With increasing Reynolds number, these layers become even smaller and thus their solution ever more important. Hence we are restricted to flows with low Reynolds numbers. To contrast with the standard K-ε model, we speak of a *low Reynolds number K-ε model.* Modifications of the eddy viscosity ansatz and the transport equations must be employed to approximate the layer close to the wall.

The ansatz for eddy viscosity can also be extended by the damping function f_μ:

$$\mu_\mathrm{t} = \rho \cdot \mathrm{f}_\mu \cdot \mathrm{C}_\mu \cdot \frac{K^2}{\varepsilon} \quad . \tag{5.133}$$

We have already seen one way of doing this with the van Driest damping function. However, as this is a function of the distance from the wall, we need to look for alternatives, as the distance from the wall is not uniquely defined for complex geometries. Functions that depend only on K or ε are more suitable as damping functions.

Equations (5.131) and (5.132) may only be applied in the layer close to the wall with some modification. It can be shown that the most important Reynolds stress $\rho \cdot \overline{u_1' \cdot u_3'}$ must drop off at the wall with z^4. Yet the K-ε model yields a decay that goes as z^3. In addition, ε has a relative maximum at the edge of the viscous sublayer, and this would not be reproduced correctly without further modification. There are numerous low Reynolds number K-ε models that employ a modified damping function and an additional term in the ε-equation in order to remedy these deficiencies.

For example, the damping function

$$\mathrm{f}_\mu = \exp\left(\frac{-3.4}{\left(1+\frac{\mathrm{R_t}}{50}\right)^2}\right) \quad , \qquad \mathrm{R_t} = \frac{K^2}{\nu \cdot \varepsilon} \quad , \tag{5.134}$$

with the further term

$$D = 2 \cdot \nu \cdot \left(\frac{\partial \sqrt{K}}{\partial z}\right)^2 \tag{5.135}$$

as the additional dissipation on the right-hand side of the K equation (5.131) is used.

Reynolds Stress Models

In flows with strongly anisotropic turbulence, the ansatz of an eddy viscosity can no longer be used, as the turbulence, in both its structure and effect on the mean flow, is dependent on the direction. The turbulent kinetic energy is not suitable for turbulence modeling as this does not take into account the directional dependence.

Secondary flows in, for example, non-circular pipes, can be a direct consequence of the anisotropy of the turbulence. The onset of the secondary flow sketched in Figure 5.10 cannot be explained by an increase in the viscosity, as in the eddy viscosity ansatz. Rather, it is due to directionally dependent Reynolds stresses.

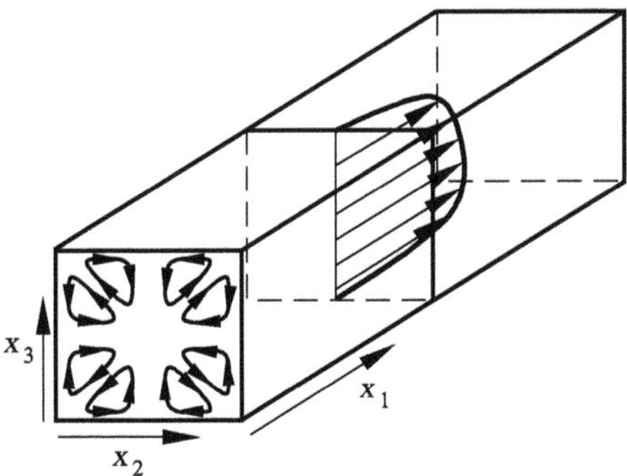

Fig. 5.10. Fully developed flow in a square pipe

Two-equation models are also no longer suitable if the curvature of the streamlines plays a role. This curvature can either strengthen or weaken the turbulence, depending on whether it is destabilizing or stabilizing (see Figure 5.11). This is taken into account in a turbulence model via the production term, a positive or negative term. In the K-ε model the positive parts generally dominate, so that the predicted eddy viscosity is therefore too large in the case of stabilizing curvature, e.g. in a rotating system or an eddy.

The effect of streamline curvature causes damping of the turbulent fluctuations along a convex surface. For flows along a concave surface, the fluctuations are amplified. Therefore, the streamline curvature can indeed reduce the Reynolds stresses, a point that is not represented in the eddy viscosity ansatz. For this reason eddy viscosity models break down if angular momentum is present.

The solution to this problem lies in the calculation of the individual components of the Reynolds stress tensor. We consider isothermal flows and ne-

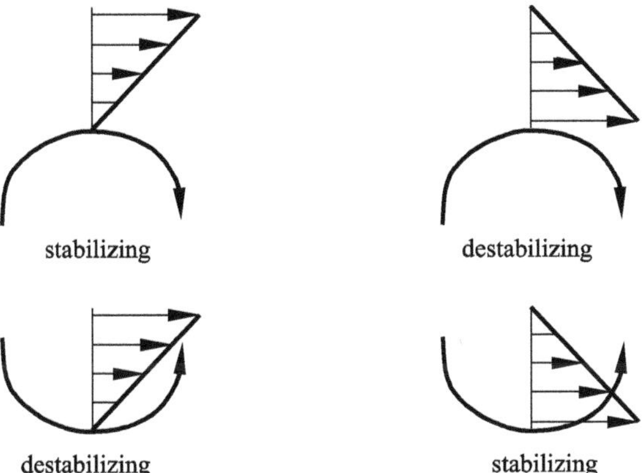

Fig. 5.11. Stabilizing and destabilizing effect of the streamline curvature

glect the energy equation. Thus, instead of an eddy viscosity, six Reynolds stresses, three normal stresses and three shear stresses have to be computed.

Among the Reynolds stress model are *algebraic models* like the eddy viscosity models, where all Reynolds stresses are modeled only as a function of the geometry. In addition there are *transport equation models* where each Reynolds stress is integrated into its own transport equation.

The *transport equation* of the *Reynolds stresses* is obtained from the Navier-Stokes equations

$$N(u_{\mathrm{i}}) = \rho \cdot \frac{\partial u_{\mathrm{i}}}{\partial t} + \rho \cdot u_{\mathrm{k}} \cdot \frac{\partial u_{\mathrm{i}}}{\partial x_{\mathrm{k}}} + \frac{\partial p}{\partial x_{\mathrm{i}}} - \mu \cdot \frac{\partial}{\partial x_{\mathrm{j}}} \left(\frac{\partial u_{\mathrm{i}}}{\partial x_{\mathrm{j}}} + \frac{\partial u_{\mathrm{j}}}{\partial x_{\mathrm{i}}} \right) = 0 \qquad (5.136)$$

by multiplication of the equation for the ith component with the fluctuation velocity u'_j and time averaging:

$$\overline{u'_{\mathrm{i}} \cdot N(u_{\mathrm{j}}) + u'_{\mathrm{j}} \cdot N(u_{\mathrm{i}})} = 0 \quad . \qquad (5.137)$$

The equation for the mean flow, the Reynolds equation, is subtracted. All terms in which the fluctuation velocity appears only once vanish because of the averaging. The remaining double products of the fluctuation velocity are the Reynolds stresses, thus the dependent variables of the associated equation.

Therefore the transport equations of the Reynolds stresses read

$$\frac{\partial \tau^{\mathrm{t}}_{\mathrm{ij}}}{\partial t} + \overline{u}_{\mathrm{k}} \cdot \frac{\partial \tau^{\mathrm{t}}_{\mathrm{ij}}}{\partial x_{\mathrm{k}}} = - \tau^{\mathrm{t}}_{\mathrm{ik}} \cdot \frac{\partial \overline{u}_{\mathrm{j}}}{\partial x_{\mathrm{k}}} - \tau^{\mathrm{t}}_{\mathrm{jk}} \cdot \frac{\partial \overline{u}_{\mathrm{i}}}{\partial x_{\mathrm{k}}} - \varepsilon_{\mathrm{ij}} + \Pi_{\mathrm{ij}} + \frac{\partial}{\partial x_{\mathrm{k}}} \left(\nu \cdot \frac{\partial \tau^{\mathrm{t}}_{\mathrm{ij}}}{\partial x_{\mathrm{k}}} + C_{\mathrm{ijk}} \right) \quad . \qquad (5.138)$$

These are nine combinations of the indices i and j; however for symmetry reasons there are only six different equations. The mathematical operations carried out above are known as the formation of the second moments of the Navier-Stokes equations. For this reason Reynolds stress transport equations are also called *second moment closures*.

As well as the double products that we have identified as the Reynolds stresses, there are further terms, namely the dissipation tensor:

$$\varepsilon_{\mathrm{ij}} = 2 \cdot \mu \cdot \overline{\frac{\partial u'_{\mathrm{i}}}{\partial x_{\mathrm{k}}} \cdot \frac{\partial u'_{\mathrm{j}}}{\partial x_{\mathrm{k}}}} \quad , \qquad (5.139)$$

the pressure-shear correlation or the pressure dilatation:

$$\Pi_{\mathrm{ij}} = \overline{p' \cdot \left(\frac{\partial u'_{\mathrm{i}}}{\partial x_{\mathrm{j}}} + \frac{\partial u'_{\mathrm{j}}}{\partial x_{\mathrm{i}}} \right)} \qquad (5.140)$$

and the turbulent diffusion correlation:

$$C_{\mathrm{ijk}} = \rho \cdot \overline{u'_{\mathrm{i}} \cdot u'_{\mathrm{j}} \cdot u'_{\mathrm{k}}} + \overline{p' \cdot u'_{\mathrm{i}}} \cdot \delta_{\mathrm{jk}} + \overline{p' \cdot u'_{\mathrm{j}}} \cdot \delta_{\mathrm{ik}} \quad , \qquad (5.141)$$

which consists of the turbulent diffusion and the pressure diffusion.

The convection of the Reynolds stresses with the mean flow (Figure 5.12) appears on the right-hand side of the transport equation. The first two terms on the right-hand side

$$P_{\mathrm{ij}} = -\tau^{\mathrm{t}}_{\mathrm{ik}} \cdot \frac{\partial \overline{u}_{\mathrm{j}}}{\partial x_{\mathrm{k}}} - \tau^{\mathrm{t}}_{\mathrm{jk}} \cdot \frac{\partial \overline{u}_{\mathrm{i}}}{\partial x_{\mathrm{k}}} \tag{5.142}$$

are source terms and represent the production, i.e. the amplified or damping effect of the mean flow on the individual Reynolds stresses. These terms can be positive or negative. The term $\varepsilon_{\mathrm{ij}}$ denotes the turbulent dissipation, i.e. the draining away of the turbulence. In contrast to most laminar flows, in turbulent flows the dissipation as a result of fluctuations must be taken into account. These two terms also appear in the K-equation.

According to (5.140), the pressure dilatation Π_{ij} is the interaction of the pressure with the velocity fluctuations. This term dropped out in the derivation of the K-equation, and so is not to be interpreted as a source or sink of the turbulence, but rather only describes a redistribution of the Reynolds stresses among one another. The redistribution can lead to certain Reynolds stresses increasing or decreasing at the expense of others. Redistribution takes place in all flows apart from those with homogeneous turbulence.

The last term on the right-hand side of equation (5.138) consists of the diffusion of the Reynolds stresses due to the molecular viscosity, as well as a term C_{ijk} that also contains triple products. This term describes the diffusion of the Reynolds stresses due to the turbulent mixing and can be separated into the so-called turbulent diffusion and the pressure diffusion.

It is now a matter of modeling the unknown terms in equations (5.139) - (5.141). The dissipation and the diffusion are already known from the K-equation. The difference here is that these quantities now must be formulated separately for each Reynolds stress.

In contrast to these, the pressure-shear correlation is new. The pressure can be eliminated from this term and it can be shown that it is made up of two parts with respect to the velocities: one part $(\Pi_{\mathrm{ij}})_1$ that contains only fluctuation velocities and another part $(\Pi_{\mathrm{ij}})_2$ that contains both fluctuations and the mean flow.

The first (slow) part, which is independent of the mean flow, is generally modeled:

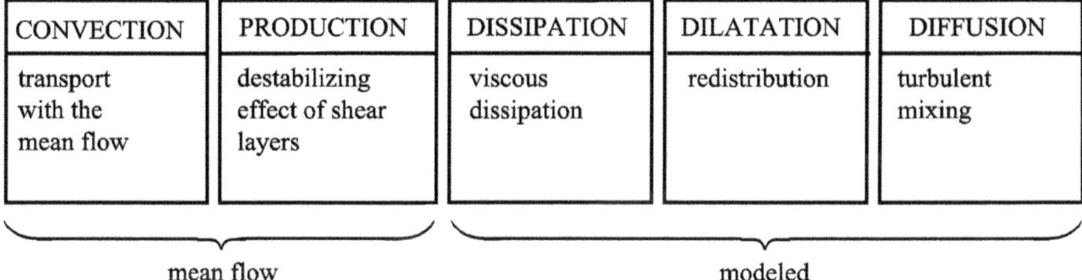

Fig. 5.12. Interpretation of the transport equations for the Reynolds stresses

$$(\Pi_{ij})_1 = -C_1 \cdot \frac{\varepsilon}{K} \cdot \left(\tau_{ij}^{Re} - \frac{2}{3} \cdot \delta_{ij} \cdot K \right) \quad , \qquad C_1 = 1.4 \quad , \tag{5.143}$$

where C_1 is a model constant and ε/K is the inverse of a decay time of the turbulence characteristic for the pressure dilatation. K and ε are defined in the usual manner. The signs are chosen so that the term always results in a return to isotropy. Therefore this means that we assume any deviation in the turbulence from the isotropic state is reduced and the turbulence will slowly become isotropic if left to itself. This is in good agreement with observations far from solid walls.

The second (fast) term depends on the mean flow and its dependence on this must be modeled, e.g. by means of

$$(\Pi_{ij})_2 = -C_2 \cdot \left(P_{ij} - \frac{2}{3} \cdot P_k \cdot \delta_{ij} \right) \quad , \qquad C_2 = 0.6 \quad , \tag{5.144}$$

where P_{ij} is the production tensor according to (5.142) and P_k the production term of the turbulent kinetic energy. Therefore the dependence of the redistribution on the Reynolds stresses is modeled. In particular close to a wall, this effect dominates the mean flow.

It is known of boundary-layer flows that the Reynolds normal stresses are approximately twice as large in the downstream direction compared to those in the direction normal to the wall, while the normal stress in the transverse direction lies roughly between these values. This is due to the fact that the wall inhibits most strongly the normal components of the fluctuations. Transport processes play only a minor role close to the wall, and so this anisotropy can influence the turbulence by modification of the pressure-shear correlation close to the wall, e.g. by mean of the wall-effect term:

$$(\Pi_{ij})_2^{w} = \left[0.125 \cdot \frac{\varepsilon}{K} \cdot \left(\overline{u_i \cdot u_j} - \frac{2}{3} \cdot K \cdot \delta_{ij} \right) + 0.015 \cdot (P_{ij} - D_{ij}) \right] \cdot f(z) \, , \tag{5.145}$$

with

$$D_{ij} = -\tau_{ik}^{t} \cdot \frac{\partial \overline{u}_k}{\partial x_j} - \tau_{jk}^{t} \cdot \frac{\partial \overline{u}_k}{\partial x_i} \quad , \tag{5.146}$$

where $f(z)$ is a function of the distance from the wall z (the weighting function), which drops off outwards from a value of one at the wall. The expression in brackets is constructed so that the non-isotropic turbulence close to the wall is reproduced as it is known from experiments.

The turbulent diffusion and the pressure diffusion of the Reynolds stresses read:

$$\frac{\partial C_{ijk}}{\partial x_k} = \frac{\partial (\rho \cdot \overline{u_i' \cdot u_j' \cdot u_k'})}{\partial x_k} + \frac{\partial (\overline{p' \cdot u_i'} \cdot \delta_{jk} + \overline{p' \cdot u_j'} \cdot \delta_{ik})}{\partial x_k} \quad . \tag{5.147}$$

There are hardly any known ways of modeling the second term, the pressure diffusion, and so this term is generally neglected. The first term consists of a triple correlation, whereby we sum over the index k. This can be modeled in many different manners, e.g. according to *C. C. Shir* (1973):

$$\rho \cdot \overline{u_i' \cdot u_j' \cdot u_k'} = -C_s \cdot \frac{K^2}{\varepsilon} \cdot \frac{\partial \overline{u_i \cdot u_j}}{\partial x_k} \quad , \tag{5.148}$$

according to *B. J. Daly and F, H. Harlow* (1970):

$$\rho \cdot \overline{u_i' \cdot u_j' \cdot u_k'} = -C_s \cdot \frac{K}{\varepsilon} \cdot \overline{u_k \cdot u_l} \cdot \frac{\partial \overline{u_i \cdot u_j}}{\partial x_l} \tag{5.149}$$

or according to *G. L. Mellor and H. J. Herring* (1973):

$$\rho \cdot \overline{u_i' \cdot u_j' \cdot u_k'} = -C_s \cdot \frac{K^2}{\varepsilon} \cdot \left(\frac{\partial \overline{u_j \cdot u_k}}{\partial x_i} + \frac{\partial \overline{u_k \cdot u_i}}{\partial x_j} + \frac{\partial \overline{u_i \cdot u_j}}{\partial x_k} \right) \quad , \tag{5.150}$$

where C_s is again a model constant and K/ε represents the timescale of the turbulent diffusion. The turbulent diffusion is therefore reduced to the Reynolds stresses themselves. The model by *Shir* is equivalent to the ansatz for the K-ε model. No one of this alternatives has as yet gained overall acceptance.

Although in modeling all the Reynolds stresses it is the directional dependence of the turbulence that is the most important property, in modeling the dissipation the assumption of isotropic turbulence is indeed sensible. This is in agreement with the idea that the directional dependence is lost as large structures decay into smaller structures. The modeling of the dissipation tensor using the scalar dissipation is therefore:

$$\varepsilon_{ij} = \frac{2}{3} \cdot \delta_{ij} \cdot \varepsilon \quad . \tag{5.151}$$

Such a model is known as a *τ-ε model.* The same transport equation as that in the K-ε model (5.132) can be used to calculated ε.

Each Reynolds stress is generated separately, dependent on the mean flow, and is transported by convection. At high Reynolds numbers transport by diffusion plays only a minor role. However, only the normal stresses are dissipated, so that the shear stresses are primarily reduced via redistribution. In Reynolds stress models the pressure-shear correlation is of particular importance, so that the various transport equation Reynolds stress models differ mainly in this term. Further variations are obtained by modeling mechanisms that cause precisely this redistribution and thus the deviation from isotropy, e.g. the effect of a wall in a three-dimensional flow.

Large-Eddy Simulation and Fine-Structure Models

If we divide turbulent structures in flows with high Reynolds numbers into two types, large-scale and fine-scale (Figure 5.13), we approach a different method of modeling. The temporal and spatial development of large-scale structures of a turbulent flow are directly calculated and only the fine structures are modeled. This method is known as *Large-Eddy Simulation* LES.

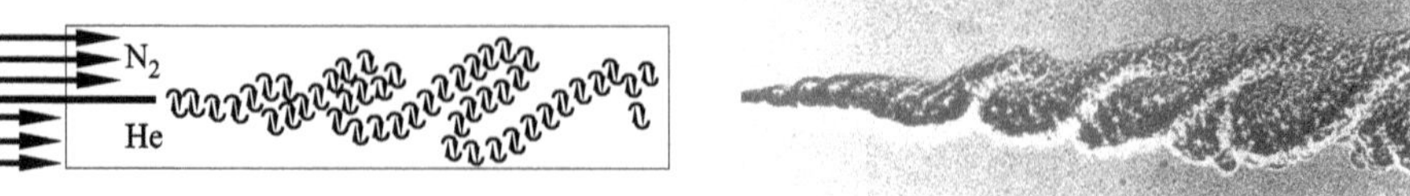

Fig. 5.13. Decomposition of the turbulence into large-scale and small-scale structures in the example of the mixing layer, *A. Roshko* (1976)

A typical logarithmic energy turbulence spectrum E at high Reynolds numbers (Figure 5.14) is subdivided into different regimes. The regime of low frequencies f or wave numbers a is generated by the large-scale energy-carrying eddies. This regime also contains the generation of turbulence. These structures have the strongest anisotropy, as at their onset they are strongly related to the geometry of the flow regime. These structures are simulated with large-scale eddy simulations, that is, without a turbulence model.

The regime of moderate frequencies or wave numbers is known as the inertial regime. In this regime further decay into ever small structures takes place. It can be shown that the nonlinear inertial terms are responsible for this behavior, while the friction plays a minor role. During decay, the turbulence becomes more and more isotropic and the geometry of the flow regime is less important. The theory of isotropic turbulence developed by Kolmogorov states that the energy E decreases with the wave number a as $E \sim a^{-5/3}$. This has been experimentally confirmed for numerous flows. The inertial regime

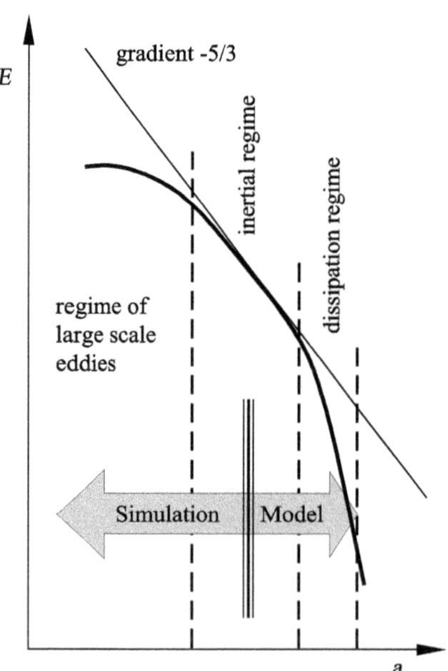

Fig. 5.14. Energy spectrum of turbulence

is thus greater the higher the Reynolds number. This regime also contains the boundary between large-scale and small-scale structures from the point of view of Large-Eddy Simulations.

In the high frequency or high wave number regime, the inertial regime passes gradually into the dissipation regime, where the loss of energy with wave number increases to $E \sim a^{-7/3}$ in magnitude. The turbulent dissipation also plays a role, as decreasing eddy size means the friction effects dominate the inertial effects more and more. This regime is modeled with respect to its effects on the large-scale structures using a fine-structure turbulence model.

To describe the method, we consider in Figure 5.15 the spatial distribution of an experimental signal along a coordinate x. We see from the sketch that both large-scale and fine-scale structures are present. In order to separate them, we employ mathematical filtering, i.e. at each position x we multiply the flow quantity f with a filter function $\mathrm{G}(x')$ and then integrate over Δx

$$\bar{\mathrm{f}}(x,t) = \frac{1}{\Delta x} \cdot \int\limits_{-\frac{\Delta x}{2}}^{\frac{\Delta x}{2}} \mathrm{f}(x - x', t) \cdot \mathrm{G}(x - x') \cdot \mathrm{d}x' \quad . \tag{5.152}$$

Here x' is the associated integration variable. The filtered signal corresponds to the dashed line. This is not a steady quantity, as in Reynolds averaging, but rather the filtered value is itself a function of time. Different filter functions have been suggested, of which we consider here the Gauß filter (other filter functions lead to similar results). Filtering is carried out in all three directions in space. The difference between filtering and averaging is that here we multiply with the filter function before carrying out the integration.

As in Reynolds averaging, we consider each local flow quantity as the sum of the filtered value and the fluctuation value. For example, for the velocity components we obtain:

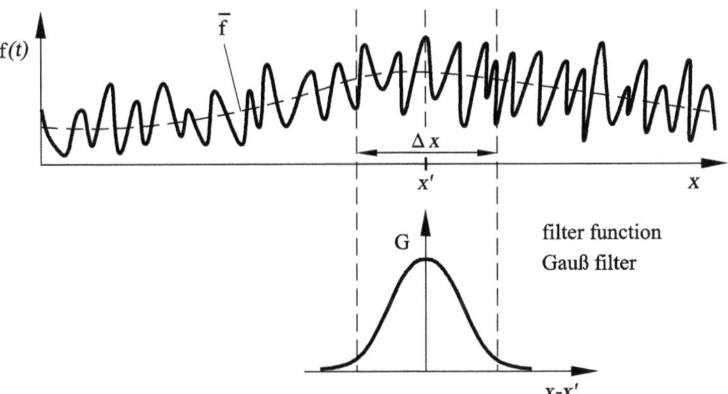

Fig. 5.15. Filtering of a flow quantity

$$u_{\mathrm{m}}(x,t) = \overline{u}_{\mathrm{m}}(x,t) + u'_{\mathrm{m}}(x,t) \quad , \tag{5.153}$$

where the filtered value is indicated with a bar. In contrast to averaging, the filtered fluctuation does not vanish:

$$\overline{u'}_{\mathrm{m}} \neq 0 \quad . \tag{5.154}$$

Taking this difference into account, the derivation of the fundamental equations of the large-scale simulation can now be carried out in analogy to the derivation of the Reynolds equations, e.g. for incompressible flows with heat transfer. The averaging operator implies filtering, and we obtain the filtered continuity equation:

$$\frac{\partial \overline{u}_{\mathrm{i}}}{\partial x_{\mathrm{i}}} = 0 \quad , \tag{5.155}$$

the filtered Navier-Stokes equations:

$$\rho \cdot \left(\frac{\partial \overline{u}_{\mathrm{i}}}{\partial t} + \frac{\partial (\overline{u}_{\mathrm{j}} \cdot \overline{u}_{\mathrm{i}})}{\partial x_{\mathrm{j}}} \right) = -\frac{\partial \overline{p}}{\partial x_{\mathrm{i}}} + \frac{\partial}{\partial x_{\mathrm{j}}} \left[\mu \cdot \left(\frac{\partial \overline{u}_{\mathrm{i}}}{\partial x_{\mathrm{j}}} + \frac{\partial \overline{u}_{\mathrm{j}}}{\partial x_{\mathrm{i}}} \right) - \rho \cdot \left(\overline{u'_{\mathrm{i}} \cdot u'_{\mathrm{j}}} + \overline{\overline{u}_{\mathrm{i}} \cdot u'_{\mathrm{j}}} + \overline{u'_{\mathrm{i}} \cdot \overline{u}_{\mathrm{j}}} \right) \right] \tag{5.156}$$

and the filtered energy equation:

$$\rho \cdot c \cdot \left(\frac{\partial \overline{T}}{\partial t} + \frac{\partial (\overline{u}_{\mathrm{j}} \cdot \overline{T})}{\partial x_{\mathrm{j}}} \right) = \frac{\partial}{\partial x_{\mathrm{j}}} \left[\lambda \cdot \frac{\partial \overline{T}}{\partial x_{\mathrm{j}}} - \rho \cdot \left(\overline{u'_{\mathrm{j}} \cdot T'} + \overline{\overline{u}_{\mathrm{j}} \cdot T'} + \overline{u'_{\mathrm{j}} \cdot \overline{T}} \right) \right] \quad . \tag{5.157}$$

The equations are formally the same as the Reynolds equations. However the fine-structure stresses appear as additional terms:

$$-\rho \cdot \overline{u'_{\mathrm{i}} \cdot u'_{\mathrm{j}}} \quad . \tag{5.158}$$

The additional terms of the energy equation are the fine-structure heat fluxes:

$$-\rho \cdot \overline{u'_{\mathrm{i}} \cdot T'} \quad , \tag{5.159}$$

which represent the effect of the fine-structure turbulence on the filtered model (large-scale). These quantities are unknown and must be modeled with a fine-structure turbulence model.

Further terms, the so-called cross terms

$$-\rho \cdot \left(\overline{\overline{u}_{\mathrm{i}} \cdot u'_{\mathrm{j}}} + \overline{u'_{\mathrm{i}} \cdot \overline{u}_{\mathrm{j}}} \right) \quad \text{and} \quad -\rho \cdot \left(\overline{\overline{u}_{\mathrm{j}} \cdot T'} + \overline{u'_{\mathrm{j}} \cdot \overline{T}} \right) \tag{5.160}$$

are also unknown. These quantities are neglected in most large-scale simulations. The remaining system of equations, without the cross terms, therefore corresponds formally to the Reynolds equations.

The simplest *fine-structure turbulence model* is the *Smagorinski model*, where the fine-structure stresses are modeled using the fine-structure eddy viscosity:

$$\overline{u_i' \cdot u_j'} = \nu_{SGS} \cdot 2 \cdot S_{ij} = \nu_{SGS} \cdot \left(\frac{\partial \overline{u}_i}{\partial x_j} + \frac{\partial \overline{u}_j}{\partial x_i} \right) \quad . \tag{5.161}$$

Similarly for the heat fluxes

$$\overline{u_i' \cdot T'} = a_{SGS} \cdot \frac{\partial \overline{T}}{\partial x_i} \quad , \qquad Pr_{SGS} \cdot \frac{\nu_{SGS}}{a_{SGS}} \approx 0.4 \quad . \tag{5.162}$$

The eddy viscosity is determined algebraically from the fine-structure shearing

$$\nu_{SGS} = (C_s \cdot h)^2 \cdot \sqrt{S_{ij} \cdot S_{ij}} \quad , \qquad h = \sqrt[3]{\Delta x \cdot \Delta y \cdot \Delta z} \quad . \tag{5.163}$$

Here $C_s = 0.17$ is the Smagorinski constant and h is a measure of the grid size of a structured numerical grid. It is assumed that structures not resolved by the numerical grid (size of sub-grid scale, index SGS) have to be modeled.

The theoretical value for the Smagorinski constant has been shown not to be universally applicable. Rather it has been seen that the value of this model constant can vary greatly from flow to flow, so that a suitable choice is indeed a problem. For this reason other models have been suggested where this parameter is adapted to each turbulence field (dynamic model). Fine-structure transport equation models have also been formulated.

An introduction to the theory of Large-Eddy Simulation is given in the book by *P. Sagaut* (2001).

5.4.6 Multiphase Flows

For multiphase flows, which will be treated in detail in Chapter 8, the conservation equations are formulated for each individual phase k. In a Euler representation, every quantity Ψ_k that is transported with velocity $\boldsymbol{u}_k$ satisfies the following conservation equation:

$$\frac{\partial(\rho_k \cdot \Psi_k)}{\partial t} + \nabla \cdot (\rho_k \cdot \boldsymbol{u}_k \cdot \Psi_k) = \rho_k \cdot \boldsymbol{f}_k + \nabla \boldsymbol{j}_k. \tag{5.164}$$

The rate of change of a volume specific conservation quantity $\rho_k \cdot \Psi_k$ with the convective flux $\rho_k \cdot \boldsymbol{u}_k \cdot \Psi_k$ is determined by a volume-specific source or sink $\rho_k \cdot \boldsymbol{f}_k$ and the dissipative fluxes $\boldsymbol{j}_k$. The meanings of $\boldsymbol{f}_k$ and $\boldsymbol{j}_k$ in equation (5.164) are listed below for the conservation quantities $\rho_k \cdot \Psi_k$ of mass, momentum and energy:

Φ_k	Quantity Ψ_k	Source/sink f_k	Flux j_k	Jump M_i
Mass	1	0	0	0
Momentum	u_k	g	$T_k = -p_k \cdot I + \tau_k$	m_i^σ
Energy	$E_k = e_k + \frac{1}{2} \cdot u_k^2$	$g \cdot u_k + Q_k$	$q_k = (-p_k \cdot I + \tau_k) \cdot u_k + q_w$	γ_i^σ

Here g is the gravity vector, I the unit tensor, p_k the hydrodynamic pressure, τ_k the shear stress tensor, Q_k a volume heat source and and q_k and q_w the energy and heat fluxes.

In addition to the usual conditions at the edges of the multiphase flow region, there are also further conditions at the interfaces between the individual phases, given in the form of discontinuity or jump relations between specific phase properties. Thus the fluxes at the interfaces must satisfy the following jump conditions:

$$[(\rho_k \cdot \Psi_k) \cdot (u_k - v_i) - j_k) \cdot n_k] = M_i \quad . \tag{5.165}$$

Here M_i denotes the jump that the conservation quantity on the left-hand side of the equation undergoes at the phase interface. v_i denotes the local velocity of a point on the phase interface, and n_k is the unit vector normal to the interface. The jump M_i depends on the local deformation of the interface and the liquid-specific interface stress σ. The terms m_i^σ and γ_i^σ describe the specific jumps of momentum and energy.

The analytical treatment of multiphase flows requires the introduction of averaged state quantities. In terms of a generally valid statistical approach, it is convenient to apply ensemble averaging. In fluid mechanics, however, experimental considerations mean that temporal, spatial or spatiotemporal averaging processes are introduced in modeling complex flows. In the following, these replace ensemble averaging in flows whose states are statistically independent in time and space and whose averaging intervals tend to ∞. This assumption is generally not satisfied. Therefore the degree of approximation must be checked in each case.

By introducing a weighting function $X_k(x, t)$ for each individual phase, the spatiotemporal average of a quantity Φ_k in the phase k can be defined as follows:

$$\overline{\Phi_{\mathrm{k}}(\boldsymbol{x},t)}^{\mathrm{k}} = \frac{\dfrac{1}{V}\cdot\displaystyle\int_V\left(\frac{1}{\Delta t}\cdot\int_0^{\Delta t}\mathrm{X}_{\mathrm{k}}\cdot\Phi_{\mathrm{k}}\cdot\mathrm{d}t\right)\cdot\mathrm{d}V}{\dfrac{1}{V}\cdot\displaystyle\int_V\left(\frac{1}{\Delta t}\cdot\int_0^{\Delta t}\mathrm{X}_{\mathrm{k}}\cdot\mathrm{d}t\right)\cdot\mathrm{d}V} \quad . \tag{5.166}$$

The integration is carried out over a control volume V that is small compared to the entire flow region, and over a time interval Δt that is small compared to the total time of the flow process under consideration. The length and time scales for the averaging process are to be selected according to the flow phenomena to be described. In multiphase flows it is difficult to determine average values for each phase at a certain position, as several phases may appear in different flow states at the same position at different times. Multiphase flows are generally unsteady to a great degree. It is therefore convenient to introduce a phase indicator function in the form of a Heaviside function:

$$X_{\mathrm{k}}(\boldsymbol{x},t) = \begin{cases} 1 \; ; & x \in V_{\mathrm{k}} \quad , \quad t \in t_{\mathrm{k}} \subset \Delta t \\ 0 \; ; & x \notin V_{\mathrm{k}} \quad , \quad t \notin t_{\mathrm{k}} \subset \Delta t \end{cases} \quad . \tag{5.167}$$

In particular, a volume fraction ϵ_{k} of the phase k may be defined as the spatiotemporal average of the phase indicator function as

$$\epsilon_{\mathrm{k}} = \overline{X}_{\mathrm{k}}^{\mathrm{k}} = \frac{1}{V}\cdot\int_V \frac{1}{\Delta t}\cdot\int_0^{\Delta t}\mathrm{X}_{\mathrm{k}}\cdot\mathrm{d}t\cdot\mathrm{d}V \quad . \tag{5.168}$$

For turbulent multiphase flows with specific turbulence time scales, for practical reasons spatial and temporal averaging are frequently applied in succession.

The general form of the conservation equations for the averaged state functions is derived by multiplying the conservation equation (5.164) term by term with the phase indicator function and then carrying out an averaging procedure corresponding to the definition in (5.166). Transformation laws between volume and surface integrals, known as the Leibnitz and Gauß relations, need to be applied. The averaged conservation equation (5.164) can then be brought to the following form:

$$\begin{aligned} &\frac{\partial(\overline{X\cdot\rho_{\mathrm{k}}\cdot\Psi_{\mathrm{k}}}^{\mathrm{k}})}{\partial t} + \nabla\cdot(\overline{X\cdot\rho_{\mathrm{k}}\cdot\boldsymbol{u}_{\mathrm{k}}\cdot\Psi_{\mathrm{k}}}^{\mathrm{k}}) - \overline{\frac{1}{V}\cdot\int_{A_{\mathrm{i}}}\rho_{\mathrm{k}}\cdot\Psi_{\mathrm{k}}\cdot((\boldsymbol{v}_{\mathrm{i}}-\boldsymbol{u}_{\mathrm{k}})\cdot\boldsymbol{n}_{\mathrm{k}})\cdot\mathrm{d}S}^{\Delta\mathrm{t}} \\ &\quad -\overline{\frac{1}{V}\cdot\int_{A_{\mathrm{Wk}}}\rho_{\mathrm{k}}\cdot\Psi_{\mathrm{k}}\cdot((\boldsymbol{v}_{\mathrm{i}}-\boldsymbol{u}_{\mathrm{k}})\cdot\boldsymbol{n}_{\mathrm{k}})\cdot\mathrm{d}S}^{\Delta\mathrm{t}} \\ &\quad = \overline{X\cdot\boldsymbol{f}_{\mathrm{k}}}^{\mathrm{k}} + \nabla\cdot\overline{X\cdot\boldsymbol{j}_{\mathrm{k}}}^{\mathrm{k}} - \overline{\frac{1}{V}\cdot\int_{A_{\mathrm{i}}+A_{\mathrm{wk}}}(\boldsymbol{j}_{\mathrm{k}}\cdot\boldsymbol{n}_{\mathrm{k}})\cdot\mathrm{d}S}^{\Delta t} \quad . \end{aligned} \tag{5.169}$$

Here the horizontal lines with the superscript Δt denote a time average of the surface integrals. The boundaries of the flow region of phase k consists of free interfaces between the phases in the core of the flow with an instantaneous total surface A_{i} and of liquid-solid boundaries between the phase k and the solid edges of the entire multiphase flow region with instantaneous total surface A_{wk}.

The free interfaces move in general with a velocity $\boldsymbol{v}_{\mathrm{i}}$ that is different from the velocity $\boldsymbol{u}_{\mathrm{k}}$ of phase k. The velocity $\boldsymbol{u}_{\mathrm{k}}$ vanishes at solid walls. The time averaged first surface integral on the left-hand side of equation (5.169) therefore reflects, with the factor $\boldsymbol{v}_{\mathrm{i}} - \boldsymbol{u}_{\mathrm{k}}$, the relative motion of phase k to the interface. The second surface integral vanishes at stationary solid walls. The right-hand side of the equation describes, in the order in which the terms appear, the phase-specific body force, the phase-specific divergence of the diffusion fluxes and the time averaged diffusion fluxes over all the interfaces of phase k, i.e. over the surfaces $A_{\mathrm{i}} + A_{\mathrm{wk}}$. In averaging the jump conditions (5.165) it must be noted that the sum of the mass fluxes over all interfaces A_{i} must cancel out to zero. However the jump values remain for the averaged values of the interface stresses and the interface energies.

In averaging certain conservation quantities, it has been found convenient to use the product $\rho_{\mathrm{k}} \cdot X_{\mathrm{k}}(\boldsymbol{x}, t)$, rather than $X_{\mathrm{k}}(\boldsymbol{x}, t)$, as a weighting function in the definition (5.166). This corresponds to mass-weighted or Favre averaging, known from the treatment of turbulent compressible flows. In the relevant literature the following set of definitions for the averages of state and constitutive quantities has become accepted:

$$\overline{\rho}_{\mathrm{k}}^{\mathrm{k}} = \frac{\overline{X_{\mathrm{k}} \cdot \rho_{\mathrm{k}}}^{\mathrm{k}}}{\epsilon_{\mathrm{k}}} \qquad \text{density,}$$

$$\overline{\boldsymbol{u}}_{\mathrm{k}}^{\rho \mathrm{k}} = \frac{\overline{X_{\mathrm{k}} \cdot \rho_{\mathrm{k}} \cdot \boldsymbol{u}_{\mathrm{k}}}^{\mathrm{k}}}{\epsilon_{\mathrm{k}} \cdot \overline{\rho}_{\mathrm{k}}^{\mathrm{k}}} \qquad \text{velocity,}$$

$$\overline{E}_{\mathrm{k}}^{\rho \mathrm{k}} = \frac{\overline{X_{\mathrm{k}} \cdot \rho_{\mathrm{k}} \cdot E_{\mathrm{k}}}^{\mathrm{k}}}{\epsilon_{\mathrm{k}} \cdot \overline{\rho}_{\mathrm{k}}^{\mathrm{k}}} \qquad \text{energy,}$$

$$-\overline{p}_{\mathrm{k}}^{\mathrm{k}} \cdot \boldsymbol{I} + \overline{\boldsymbol{\tau}}_{\mathrm{k}}^{\mathrm{k}} = \frac{\overline{X_{\mathrm{k}} \cdot (-p_{\mathrm{k}} \cdot \boldsymbol{I} + \boldsymbol{\tau}_{\mathrm{k}})}^{\mathrm{k}}}{\epsilon_{\mathrm{k}}} \qquad \text{pressure and molecular shear stresses,}$$

$$\overline{\boldsymbol{q}}_{\mathrm{wk}}^{\mathrm{k}} = \frac{\overline{X_{\mathrm{k}} \cdot \boldsymbol{q}_{\mathrm{wk}}}^{\mathrm{k}}}{\epsilon_{\mathrm{k}}} \qquad \text{energy flux,}$$

$$\overline{Q}_{\mathrm{k}}^{\rho \mathrm{k}} = \frac{\overline{X_{\mathrm{k}} \cdot \rho_{\mathrm{k}} \cdot Q_{\mathrm{k}}}^{\mathrm{k}}}{\epsilon_{\mathrm{k}} \cdot \overline{\rho}_{\mathrm{k}}^{\mathrm{k}}} \qquad \text{energy sources and energy sinks.}$$

With these definitions for the averages of the relevant quantities, the conservation equations for mass, momentum and energy can be written in the following form:

$$\frac{\partial(\epsilon_k \cdot \overline{\rho}_k^k)}{\partial t} + \nabla \cdot (\epsilon_k \cdot \overline{\rho}_k^k \cdot \overline{\boldsymbol{u}}_k^{\rho k}) = \Gamma_k \quad , \tag{5.170}$$

$$\frac{\partial(\epsilon_k \cdot \overline{\rho}_k^k \cdot \overline{\boldsymbol{u}}_k^{\rho k})}{\partial t} + \nabla \cdot (\epsilon_k \cdot (\overline{\rho_k \cdot \boldsymbol{u}_k \cdot \boldsymbol{u}_k}^k + \overline{p}_k^k \cdot \boldsymbol{I} - \overline{\boldsymbol{\tau}}_k^k))$$
$$= \epsilon_k \cdot \overline{\rho}_k^k \cdot \boldsymbol{g} + \boldsymbol{M}_k + \overline{\boldsymbol{u}_{ik} \cdot \Gamma_k}^k \quad , \tag{5.171}$$

$$\frac{\partial(\epsilon_k \cdot \overline{\rho}_k^k \cdot \overline{E}_k^{\rho k})}{\partial t} + \nabla \cdot (\epsilon_k \cdot (\overline{\rho_k \cdot E_k \cdot \boldsymbol{u}_k}^k + \overline{p_k \cdot \boldsymbol{u}_k}^k - \overline{\boldsymbol{\tau}_k \cdot \boldsymbol{u}_k}^k + \overline{\boldsymbol{q}}_{wk}^k))$$
$$= \epsilon_k \cdot \overline{\rho}_k^k \cdot \overline{\boldsymbol{u}}_k^{\rho k} \cdot \boldsymbol{g} + \epsilon_k \cdot \overline{\rho}_k^k \cdot \overline{Q}_k^{\rho k} + W_k + F_k + \overline{E_{ik} \cdot \Gamma_k}^k \ . \tag{5.172}$$

The following abbreviations have been introduced for surface integrals:

$$\Gamma_k = \overline{\frac{1}{V} \cdot \int_{A_i} \rho_k \cdot ((\boldsymbol{v}_i - \boldsymbol{u}_k) \cdot \boldsymbol{n}_k) \cdot dS}^{\Delta t}$$ interface mass sources,

$$\boldsymbol{M}_k = \overline{-\frac{1}{V} \cdot \int_{A_i} ((\boldsymbol{\tau}_i - p_k \cdot \boldsymbol{I}) \cdot \boldsymbol{n}_k) \cdot dS}^{\Delta t}$$ interface momentum sources,

$$\overline{\boldsymbol{u}_{ik} \cdot \Gamma_k}^k = \overline{\frac{1}{V} \cdot \int_{A_i} \rho_k \cdot (\boldsymbol{u}_k \cdot ((\boldsymbol{v}_i - \boldsymbol{u}_k) \cdot \boldsymbol{n}_k)) \cdot dS}^{\Delta t}$$ interface energy sources,

$$W_k = \overline{-\frac{1}{V} \cdot \int_{A_i} (((\boldsymbol{\tau}_i - p_k \cdot \boldsymbol{I}) \cdot \boldsymbol{u}_k) \cdot \boldsymbol{n}_k) \cdot dS}^{\Delta t}$$ work done by the interface stresses,

$$F_k = \overline{\frac{1}{V} \cdot \int_{A_i + A_{Wk}} (\boldsymbol{q}_k \cdot \boldsymbol{n}_k) \cdot dS}^{\Delta t}$$ heat sources at interfaces and edges,

$$\overline{E_{ik} \cdot \Gamma_k}^k = \overline{\frac{1}{V} \cdot \int_{A_i} \rho_k \cdot (E_k \cdot (\boldsymbol{v}_i - \boldsymbol{u}_k) \cdot \boldsymbol{n}_k) \cdot dS}^{\Delta t}$$ energy sources at interfaces during mass transfer.

By averaging the jump conditions (5.165) and using the above definition, we reach the following set of necessary conditions at the interfaces:

$$\sum_k \Gamma_k = 0 \quad ,$$
$$\sum_k (\boldsymbol{M}_k + \overline{\boldsymbol{u}_{ik} \cdot \Gamma_k}^k) = \boldsymbol{m}_i^\sigma \quad , \tag{5.173}$$
$$\sum_k (W_k + F_k + \overline{E_{ik} \cdot \Gamma_k}^k) = \gamma_i^\sigma \quad .$$

Here $\boldsymbol{m}_i^\sigma$ represents the jump in the stresses and γ_i^σ the jump in the energy at the interfaces.

In order to describe the exchange processes at phase boundaries, e.g. evaporation and condensation processes in single-component systems, it is has been found to be convenient, within the framework of the spatiotemporal averaging, to define an interface concentration using the integral

$$a_{\mathrm{i}} = \overline{\frac{1}{V} \cdot \int_{A_{\mathrm{i}}} \mathrm{d}S}^{\Delta t}$$

Locally averaged exchange fluxes, such as mass and heat fluxes at the interfaces, can then be represented as products of the average values of the flux densities and the interface concentration. For two-phase or multiphase flows that vary on microscopic scales, this approach is a significant simplification when modeling local exchange processes. However it implies that a conservation equation may also be set up for the interface concentration a_{i} and solved simultaneously with the other conservation equations for mass, momentum and energy for the problem at hand. This has the form:

$$\frac{\partial a_{\mathrm{i}}}{\partial t} + \nabla \cdot (a_{\mathrm{i}} \cdot \boldsymbol{v}_{\mathrm{i}}) = \sum_{\mathrm{j}=1}^{4} \Phi_{\mathrm{j}} + \Phi_{\mathrm{ph}} + \Phi_{\mathrm{n}} \quad . \tag{5.174}$$

Here $\boldsymbol{v}_{\mathrm{i}}$ are the locally averaged velocities of the interface, Φ_{j} are the rates of change of the interface concentration due to particle decay and coalescence and Φ_{ph} are the rates of change due to phase transitions. The relationship between the density averaged velocity of the particle phase $\boldsymbol{v}_{\mathrm{i}}$ and the interface concentration $\boldsymbol{a}_{\mathrm{i}}$ is $\boldsymbol{v}_{\mathrm{i}} = \boldsymbol{v}_{\mathrm{p}} \cdot \boldsymbol{a}_{\mathrm{i}} \cdot \rho_{\mathrm{i}}$. The final source term Φ_{n} represents possible nucleation processes.

In order to solve practical problems, the conservation equations (5.170) to (5.172) with the additional equation (5.174) must be supplemented by further constitutive relations for mass and heat transfer at the interfaces and edges. This represents a substantial part of the modeling of multiphase flows. This is now a closure problem, where the number of quantities to be determined must be aligned with the number of available equations.

Special models have been developed for highly dilute two-phase flows consisting of a continuous substrate phase and a small admixture of disperse particles, as found for example in technical spray systems, in the atmospheric transport of aerosols and in sandstorms. In these models the substrate is treated using the conservation equations for mass, momentum and energy in Eulerian form. The motion of the individual particles, in contrast, is described in the Lagrangian manner using Newton's force law that relates the local particle acceleration to the reaction forces between particle and substrate. The trajectories of all particles, or at least of a representative group, can be determined by double time integration of the particle's acceleration, starting with a defined initial state for each individual particle.

We forgo the representation of the conservation equations for the substrate, as this aspect is discussed in detail for single phase flows in Sections

5.1 to 5.4.5. Of course, they may also be derived by simplification of the above conservation equations (5.170) to (5.172). However, we expressly point out that the conservation equations for the substrate must be supplemented by source terms if the effect of the particle motion on the substrate is to be taken into account; this is the case for higher particle concentrations or where the effect of the particles on the turbulence of the substrate is significant. The balance of forces for an individual particle with mass m_{p}, volume V_{p} and velocity $\boldsymbol{v}_{\mathrm{p}}$ at position $x_{\mathrm{p}}(t)$ in a substrate flow with velocity $\boldsymbol{u}_{\mathrm{c}}$ and density ρ_{c} may be given in the following form:

$$m_{\mathrm{p}} \cdot \frac{\mathrm{d}\boldsymbol{v}_{\mathrm{p}}}{\mathrm{d}t} = -V_{\mathrm{p}} \cdot \nabla p_{\mathrm{c}} - \frac{1}{2} \cdot \rho_{\mathrm{c}} \cdot V_{\mathrm{p}} \cdot C_{\mathrm{A}} \cdot \frac{\mathrm{d}}{\mathrm{d}t}(\boldsymbol{v}_{\mathrm{p}} - \boldsymbol{u}_{\mathrm{c}})$$
$$-\frac{1}{2} \cdot \rho_{\mathrm{p}} \cdot V_{\mathrm{p}} \cdot C_{\mathrm{D}} \cdot |\boldsymbol{v}_{\mathrm{p}} - \boldsymbol{u}_{\mathrm{c}}| \cdot (\boldsymbol{v}_{\mathrm{p}} - \boldsymbol{u}_{\mathrm{c}}) - \rho_{\mathrm{c}} \cdot V_{\mathrm{p}} \cdot C_{\mathrm{L}} \cdot (\boldsymbol{v}_{\mathrm{p}} - \boldsymbol{u}_{\mathrm{c}}) \times \boldsymbol{\omega}$$
$$-\frac{3}{2} \cdot d_{\mathrm{p}}^2 \cdot \rho_{\mathrm{c}} \cdot \sqrt{\nu_{\mathrm{c}}} \cdot C_{\mathrm{H}} \cdot \int_{t_0}^{t} \frac{1}{\sqrt{t-t'}} \cdot \frac{\mathrm{d}(\boldsymbol{v}_{\mathrm{p}} - \boldsymbol{u}_{\mathrm{c}})}{\mathrm{d}t'} \cdot \mathrm{d}t' + \boldsymbol{F}_{\mathrm{i}} \quad . \tag{5.175}$$

∇p_{c} denotes the local pressure gradient in the continuous phase and $\boldsymbol{\omega}$ is a vortex strength that is generally made up of the local vortex strength in the continuous phase $\boldsymbol{\omega}_{\mathrm{c}} = \nabla \times \boldsymbol{u}_{\mathrm{c}}$ and a part of the innate rotation of the particle $\boldsymbol{\omega}_{\mathrm{p}}$. $\boldsymbol{F}_{\mathrm{i}}$ are potential forces under the effect of external force fields such as gravitational force and electromagnetic fields. The physical meaning of the different terms in equation (5.175) is as follows: The term on the left-hand side describes the acceleration force of the particle. The first term on the right-hand side describes the effect of the pressure gradient in the substrate on the particle. The second term represents a force that displaces the substrate and accelerates the *virtual additional mass* close the the particle relative to the core flow of the continuum. This additional mass is equal to half the mass of the continuous phase with a volume half that of the particle. The third term describes the resistance of the particle in the surrounding flow due to its relative velocity. The fourth term is a force transverse to the direction of motion of the particle. It is known as the Saffmann lift force. As already mentioned, it is caused by the vortex strength of the substrate and the innate rotation of the particle relative to the flow. The effect of the latter property is also known as the Magnus effect. The fifth term represents the so-called *Basset force*. It describes the effect of the viscosity when the particle is accelerated compared to the fluid and has a relaxing effect on the motion of the particle along the trajectory.

The coefficients C_{A}, C_{D}, C_{L} and C_{H} generally depend on the specific flow conditions in the continuous phase and on the size of the particles. This means that they are a functions of particle Reynolds number $Re_{\mathrm{p}} = |\boldsymbol{v}_{\mathrm{p}} - \boldsymbol{u}_{\mathrm{p}}| \cdot d_{\mathrm{p}}/\nu_{\mathrm{c}}$. The functional dependencies have to be modeled on the basis of the fluid mechanical conditions.

Equation (5.175) for the motion of the individual particle is of course subject to considerable constraints. First, for the dilute disperse phase $\epsilon_{\mathrm{p}} \ll$

1 must hold. Further, for the velocity $|\boldsymbol{v}_\mathrm{p} - \boldsymbol{u}_\mathrm{c}|/|\boldsymbol{u}_\mathrm{c}| \ll 1$ holds and the characteristic particle diameter d_p is small compared to the distance over which significant variations in state take place in the continuous phase. This means that it is smaller than the Kolmogorov microlength $d_\mathrm{p} < (\nu_\mathrm{c}^3/\varepsilon_\mathrm{c})^{1/4}$, with ε_c the rate of dissipation of the turbulent kinetic energy in the continuous phase.

The relation (5.175) also does not take into account any forces that are due to a direct or indirect interaction between individual particles. Even the dependence on the particle concentration ϵ_p is absent in this relation. Thus phenomena such as cluster formation of particles, particle coalescence or particle decay cannot be described by the above Euler-Lagrange model for a disperse two-phase flow.

In order to remedy such serious deficiencies in a Euler-Lagrange model, advanced statistical models are being developed based on the distribution functions for groups of particles and associated transport equations. Such models are principally suitable for the description of direct particle collisions and conglomerations and decays. Further details on such modeling is to be found in the text books by *C. T. Crowe et al.* (1998) and *W. A. Sirignano* (1999).

In many technical processes and in numerous geophysical events, turbulent motions determine the transport and exchange processes in multiphase flows. Examples of this are steam-water flows in thermal power plants, technical spray flows, bubble flows, dust storms and the transport and deposit of sediment in rivers and bays. The *modeling of turbulence* in single phase flows was described in Section 5.4.5. It is characterized by a large range of length and time scales of the flow vortices. In multiphase flows, the number of length and time scales multiplies, because of the number of possible phase distributions. A well-founded theoretical treatment of turbulent multiphase flows is in general possible if it is possible to separate the turbulent time and length scales from those of the phase distribution. This is true for dilute disperse two-phase flows with a turbulent continuous phase and with a volume fraction of the disperse phase of $\epsilon_\mathrm{p} \ll 1$. The above examples fall into this class of flows.

The turbulence in the substrate is generally due to strong shearing effects in inhomogeneous high-velocity flows or to pulsing flows through grids or past corners and edges. The particles of the disperse phase can, depending on their size and weight, have purely passive behavior in convective transport or can affect the substrate phase through their own dynamics and thence either increase or dampen the level of turbulence in the substrate flow. The effect that will prevail depends to a great extent on a set of characteristic parameters of the disperse two-phase flow. The most important particle parameters can be easily identified using the equation for *particle motion* (5.174). These are the mass-density ratio of the phases $\beta_\mathrm{p} = \rho_\mathrm{p}/\rho_\mathrm{c}$, a particle Reynolds number

$Re_{\rm p} = |\boldsymbol{v}_{\rm p} - \boldsymbol{u}_{\rm c}| \cdot d_{\rm p}/\nu_{\rm c}$ and a time measure for the particle, defined as the relaxation time of a particle in a viscous Stokes flow:

$$\tau_{\rm p} = \frac{\rho_{\rm p}}{\rho_{\rm c}} \cdot \frac{d_{\rm p}^2}{18 \cdot \nu_{\rm c}} \quad .$$

These quantities are to be related to the relevant parameters of the turbulent substrate flow. These are the flow Reynolds number $Re_{\rm c} = U_{0\rm c} \cdot D_{\rm c}/\nu_{\rm c}$ with a volume flux density $U_{0\rm c}$ and a characteristic hydraulic diameter $D_{\rm C}$ and the turbulence Reynolds number $Re_{\rm ct} = u' \cdot L/\nu_{\rm c}$ with u' the square root of the variance of the fluctuation velocities and L a specific length for the energy-filled vortex. This corresponds to an integral length measure of the turbulence. In addition there are two relevant time measures: the vortex circulation time $\tau_{\rm e} = L/u'$ and the dissipation time of the smallest vortices according according to Kolmogorov $\tau_{\rm k} = \nu_{\rm c}/\varepsilon_{\rm c}$, with $\varepsilon_{\rm c}$ the rate of dissipation of the continuous phase. For turbulent dilute disperse two-phase flows in general $Re_{\rm p} \ll Re_{\rm c}$ holds and $Re_{\rm ct} \geq 1$. In order to describe the interaction between the phases it is convenient to introduce the particle volume fraction $\epsilon_{\rm p}$ and the ratio of the above characteristic time scales $\tau_{\rm p}/\tau_{\rm k}$ and $\tau_{\rm p}/\tau_{\rm e}$. The latter time ratio is called the Stokes number. Figure 5.16 shows the main interaction mechanisms between the phases. For very small volume fractions $\epsilon_{\rm p} < 10^{-6}$ the particles behave passively. Any effect of the movement of the particles on the turbulence of the continuous phase may be neglected (one-

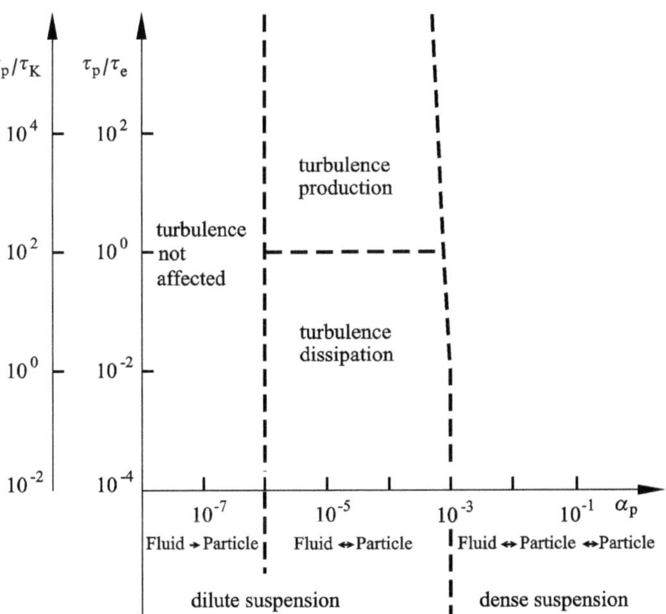

Fig. 5.16. Classification of the interaction of particles in a turbulent substrate, *S. E. Eglobashi* (1994)

sided coupling between the phases). In the region $10^{-6} < \epsilon_{\mathrm{p}} < 10^{-3}$ of the volume fraction, the particles affect the degree of turbulence of the continuous phase; it is increased for $\tau_{\mathrm{p}}/\tau_{\mathrm{e}} > 1$ or $\tau_{\mathrm{p}}/\tau_{\mathrm{k}} > 10^2$. For $\tau_{\mathrm{p}}/\tau_{\mathrm{e}} < 1$ or $\tau_{\mathrm{p}}/\tau_{\mathrm{k}} < 10^2$ the particles dampen the turbulence by increasing the rate of dissipation ε_{c} (mutual coupling). The region $\epsilon_{\mathrm{p}} < 10^{-3}$ is classified as a dilute suspension. Here there are no interactions between particles. Beyond this region with $\epsilon_{\mathrm{p}} > 10^{-3}$, the mean distance L_{p} between the particles, which may be defined with the relation $d_{\mathrm{p}} = L_{\mathrm{p}} \cdot (6 \cdot \epsilon_{\mathrm{p}}/\pi)^{1/3}$ decreases, so that finally the ratio becomes $d_{\mathrm{p}}/L_{\mathrm{p}} > 0.1$. Under such conditions the particles interact with each other, either indirectly via the spatial effect of their boundary layers and wake flows, or via direct contact between individual particles. This parameter regime is known as dense suspension. Modeling a two-phase flow in this regime requires four-fold coupling between particles and substrate, which can no longer be provided by a simple mechanical model such as that on which equation (5.174) is based.

One criterion for limiting the region of turbulence amplification and damping, as indicated in Figure 5.16 for $\tau_{\mathrm{p}}/\tau_{\mathrm{e}} = 1$, has been suggested by *S. Hosokawa* and *A. Tomiyama* (2004) on the basis of experimental investigations. They introduce the ratio of two turbulent viscosities ν_{tp} and ν_{tc} as a characteristic number, whereby the first describes the contribution of particle motion to the total turbulence and the second the contribution to the turbulence generated only by shearing in the continuous phase. This ratio $\nu_{\mathrm{tp}}/\nu_{\mathrm{tc}}$ is equivalent to the ratio of the Reynolds numbers for the particles and for the turbulence $Re_{\mathrm{p}}/Re_{\mathrm{ct}}$ as defined above. In the case of channel flows, the authors introduce quantities averaged over the cross-section and use the known relation between the integral turbulence length measure L_{c} and the channel diameter D in the form $L_{\mathrm{ct}} = 0.2 \cdot D$. The criterion has the form:

$$\frac{\nu_{\mathrm{tp}}}{\nu_{\mathrm{tc}}} = \frac{\overline{|\boldsymbol{v}_{\mathrm{p}} - \boldsymbol{u}_{\mathrm{c}}|}^{\mathrm{k}} \cdot \overline{\boldsymbol{d}_{\mathrm{p}}}^{\mathrm{k}}}{0.2 \cdot \overline{\boldsymbol{u}'_{\mathrm{c}}} \cdot D} = \begin{cases} \leq 1 & \text{damping} \\ \geq 1 & \text{amplification} \end{cases} .$$

In calculations of technical applications, methods such as those developed for the treatment of single phase turbulent flows in Sections 5.2.2 and 5.4.4 have also proved valuable for the description of turbulent disperse multiphase flows. The derivation of the equations is based on the averaged conservation equations (5.170) to (5.172). The Reynolds ansatz for multiphase flows may be written as

$$\rho_{\mathrm{k}} = \overline{\rho_{\mathrm{k}}}^{\mathrm{k}} + \rho'_{\mathrm{k}} \ , \ \boldsymbol{u}_{\mathrm{k}} = \overline{\boldsymbol{u}_{\mathrm{k}}}^{\rho\mathrm{k}} + \boldsymbol{u}'_{\mathrm{k}} \ , \ p_{\mathrm{k}} = \overline{p_{\mathrm{k}}}^{\mathrm{k}} + p'_{\mathrm{k}} \ , \ E_{\mathrm{k}} = \overline{E'_{\mathrm{k}}}^{\rho\mathrm{k}} + E'_{\mathrm{k}} \ . \quad (5.176)$$

Inserting these expressions into the averaged conservation equation (5.169), a new set of double and tripple correlations of the fluctuating quantities arises. Typical examples are $\overline{X_{\mathrm{k}} \cdot \rho'_{\mathrm{k}} \cdot \boldsymbol{u}'_{\mathrm{k}}}^{\mathrm{k}}$, $\overline{X_{\mathrm{k}} \cdot \rho'_{\mathrm{k}} \cdot (\boldsymbol{u}'_{\mathrm{k}} \cdot \boldsymbol{u}'_{\mathrm{k}})}^{\mathrm{k}}$ and $\overline{X_{\mathrm{k}} \cdot \rho'_{\mathrm{k}} \cdot p'_{\mathrm{k}}}^{\mathrm{k}}$. The turbulent density fluctuations ρ'_{k} may be replaced by the corresponding fluctuations in the volume fraction ϵ'_{k} via the relation $\rho'_{\mathrm{k}} = \overline{\rho_{\mathrm{k}}}^{\mathrm{k}} \cdot \epsilon'_{\mathrm{k}}$, where $\overline{\rho_{\mathrm{k}}}^{\mathrm{k}}$

is the material density of the component k. For turbulent disperse multiphase flows, we obtain the Reynolds equations in the following form:

$$\frac{\partial(\epsilon_k \cdot \overline{\rho}_k^k)}{\partial t} + \nabla \cdot (\epsilon_k \cdot \overline{\rho}_k^k \cdot \overline{\boldsymbol{u}}_k^{\rho k} + \boldsymbol{j}_k^{Re}) = \Gamma_k \quad , \tag{5.177}$$

$$\begin{aligned}\frac{\partial(\epsilon_k \cdot \overline{\rho}_k^k \cdot \overline{\boldsymbol{u}}_k^{\rho k})}{\partial t} &+ \nabla \cdot (\epsilon_k \cdot (\overline{\rho}_k^k \cdot \overline{\boldsymbol{u}}_k^{\rho k} \cdot \overline{\boldsymbol{u}}_k^{\rho k} + \overline{p}_k^k \cdot \boldsymbol{I} - \overline{\boldsymbol{\tau}}_k^k - \boldsymbol{\tau}_k^{Re})) \\ &= \epsilon_k \cdot \overline{\rho}_k^k \cdot \boldsymbol{g} + \boldsymbol{M}_k^* + \overline{\boldsymbol{u}_{ik} \cdot \Gamma_k}^k \quad ,\end{aligned} \tag{5.178}$$

$$\begin{aligned}\frac{\partial(\epsilon_k \cdot \overline{\rho}_k^k \cdot (\overline{E}_k^{\rho k} + E_k^{Re}))}{\partial t} &+ \nabla \cdot (\epsilon_k \cdot (\overline{\rho}_k^k \cdot \overline{\boldsymbol{u}}_k^{\rho k} \cdot (\overline{E}_k^{\rho k} + E_k^{Re}) + \overline{p}_k^k \cdot \overline{\boldsymbol{u}}_k^{\rho k} + \overline{p_k' \cdot \boldsymbol{u}_k'}^k \\ &- (\overline{\boldsymbol{\tau}}_k^k + \boldsymbol{\tau}_k^{Re}) \cdot \overline{\boldsymbol{u}}_k^{\rho k} + \overline{\boldsymbol{q}}_{wk}^k + \boldsymbol{q}_k^{Re})) \\ &= \epsilon_k \cdot \overline{\rho}_k^k \cdot (\overline{\boldsymbol{u}}_k^{\rho k} \cdot \boldsymbol{g} + \overline{Q}_k^{\rho k}) + W_k^* + F_k^* + \overline{E_{ik} \cdot \Gamma_k}^k .\end{aligned} \tag{5.179}$$

Here terms with a superscript Re denote contributions of the turbulence to the fluxes that are explicitly given by double or triple correlations. The additional turbulent mass flux $\boldsymbol{j}_k^{Re}$, the turbulent Reynolds shear stresses $\boldsymbol{\tau}_k^{Re}$, the turbulent kinetic energy E_k^{Re} and the turbulent heat fluxes $\boldsymbol{q}_k^{Re}$ are defined as

$$\boldsymbol{j}_k^{Re} = \overline{\rho_k' \cdot \boldsymbol{u}_k'}^{\rho k} = \overline{\rho_k}^k \cdot \overline{\epsilon_k' \cdot \boldsymbol{u}_k'}^{\rho k} \quad , \quad \boldsymbol{\tau}_k^{Re} = \frac{\overline{\rho_k \cdot \boldsymbol{u}_k' \cdot \boldsymbol{u}_k'}^k}{\epsilon_k} \quad ,$$

$$E_k^{Re} = \frac{\overline{\rho_k \cdot (\boldsymbol{u}_k')^2}^k}{2 \cdot \epsilon_k \cdot \overline{\rho_k}^k} \quad , \quad \boldsymbol{q}_k^{Re} = \frac{\overline{\rho_k \cdot E_k' \cdot \boldsymbol{u}_k'}^k + \overline{p_k' \cdot \boldsymbol{u}_k'}^k - \overline{\boldsymbol{\tau}_k' \cdot \boldsymbol{u}_k'}^k}{\epsilon_k} .$$

Further contributions of the turbulent fluctuations are present in the volume and surface source terms $\boldsymbol{M}_k^*$, F_k^*, W_k^*, $\overline{\boldsymbol{u}_{ik} \cdot \Gamma_k}^k$ and $\overline{E_{ik} \cdot \Gamma_k}^k$ on the right-hand side of the equations, but for simplicity are not stated here explicitly. Their relevance needs to be discussed in relation to the problem at hand and accordingly taken into account in the representation. Clearly the number of unknown quantities in the conservation equations exceeds the number of equations. Thus we have a closure problem. As in the procedure for single phase turbulence flows, to solve the problem closure relations must be provided for all constitutive quantities on the basis of physical deliberations and rational demands on the representation of functional dependencies. In analogy to single phase turbulence flows, for multiphase flows multiple correlations are also given in the form of algebraic relations using gradients of the primitive variables and dimensionless characteristic numbers. This means essentially that fluxes are modeled as diffusion processes. This will now be explained for a two-phase particle flow using the example of the mass flux $\boldsymbol{j}_k^{Re}$ in equation (5.177). Using a gradient ansatz, the particle flux in, for example, a dilute suspension may be represented as

$$\boldsymbol{j}_{\mathrm{k}}^{\mathrm{Re}} = \overline{\rho_{\mathrm{p}}}^{\mathrm{k}} \cdot \overline{\epsilon_{\mathrm{p}}' \cdot \boldsymbol{u}_{\mathrm{p}}'}^{\mathrm{k}} = \overline{\rho_{\mathrm{p}}}^{\mathrm{k}} \cdot D_{\mathrm{p}} \cdot \nabla \epsilon_{\mathrm{p}} \quad ,$$

where D_{p} denotes a turbulent particle diffusion coefficient. D_{p} will generally still depend on the relevant characteristic numbers introduced above, such as Re_{p}, St, Re_{c}. In the literature it is suggested that D_{p} be related to a turbulent vortex viscosity of the continuous phase ν_{ct} and a turbulent particle mass diffusion characteristic number, the Schmidt number Sc_{p}, as $D_{\mathrm{p}} = \nu_{\mathrm{ct}}/Sc_{\mathrm{p}}$ needs to be determined empirically. Now the above ansatz is incomplete in terms of a rational mechanical approach, as the necessary dependence on gradients of the velocity $\overline{\boldsymbol{u}_{\mathrm{k}}}^{\rho\mathrm{k}}$ has not been demonstrated. Therefore physical arguments must be used in each case to see if this latter dependence may be neglected.

In the relation for the diffusion coefficient D_{p}, the vortex viscosity ν_{ct} either may be expressed using a mixing path approach with gradients of the mean velocities, or may be related to the turbulent kinetic energy $E_{\mathrm{K}}^{\mathrm{Re}}$ and the turbulent dissipation ε_{C} in the continuous phase using the Prandtl-Kolmogorov relation $\nu_{\mathrm{ct}} \sim E_{\mathrm{K}}^{\mathrm{Re}}/\varepsilon_{\mathrm{C}}$.

An algebraic gradient ansatz of the kind described in the example is frequently insufficient to describe more complex flows. This is the case, for example, for recirculation flows behind steps and in regions where channels suddenly widen, where the dimensions of the energy-carrying vortices become comparable to distances over which the particle concentration changes significantly. In such situations a fundamentally different approach is needed, one that describes the spatiotemporal development of the particle mass flux with its own transport equation. This approach itself presents further complications, as new double correlations, such as the variance of the particle concentration $\overline{\epsilon_{\mathrm{p}}'^2}^{\mathrm{k}}$, appear in the corresponding transport equations in addition to the multiple correlations that already appear in the basis equations (5.177) – (5.179).

Following the description of turbulent single phase flows, a multiple equation method has also been developed to solve the closure problem for multiphase flows. This is based on the introduction of further transport equations for relevant turbulent quantities, such as the turbulent kinetic energy, the Reynolds shear stresses and the dissipation in the individual phases. Further details on this will be introduced in Section 8.6 on multiphase flows.

In principle, the multifluid model in the Eulerian description presented here is capable of treating turbulent dispersion flows with mutual coupling between the phases. Thus it may be applied for problems with particle concentrations $\epsilon_{\mathrm{p}} \sim 10^{-3}$ and beyond, according to Figure 5.16. Events such as coalescence or decay of particles, as in bubble or drop flows, require an extension of the model to groups of particle quantities with corresponding source and sink terms in the transport equations for the different groups. Such models are currently under development.

The fluid-particle model in the Euler-Lagrange form presented at the start of the section can also be adapted for turbulent flows, by using the Reynolds equations to describe the continuous phase. The damping or amplification of turbulence in the substrate flow by particle motion may be taken into account by additional source or sink terms in the Reynolds equations. These source or sink terms depend by nature on the relevant characteristic numbers of the suspension flow Re_{p}, Re_{c}, St and need to be formulated as closure conditions on the basis of physical deliberations.

Equation (5.175) for the particle motion in the substrate formulates only a dependence on the current velocity $\boldsymbol{u}_{\mathrm{c}}$ of the substrate. Within the framework of a turbulence model for the substrate, such as the K-ε model, in order to close the problem the current velocity introduced in equation (5.175) must be reconstructed from the mean velocity $\overline{\boldsymbol{u}_{\mathrm{c}}}$ and the turbulent kinetic energy $E_{\mathrm{c,kin}}^{\mathrm{Re}} = 0.5 \cdot \overline{\boldsymbol{u}_{\mathrm{c}}^{\prime 2}}$. This is achieved by introducing a suitable distribution function for the velocity fluctuations $\boldsymbol{u}_{\mathrm{c}}^{\prime}$, such as a Gauß distribution, with a variance that corresponds to the value of the turbulent kinetic energy. Application of this model to dilute suspension flows is restricted to weak coupling between the phases. In order to describe the interaction between the particles, the Euler-Lagrange model must be extended based on statistical methods with the introduction of distribution functions for particle groups and the preparation of the associated evolution equations.

To describe dense suspension flows with $\epsilon_{\mathrm{p}} > 10^{-3}$, the above *Euler-Lagrange models* should be used based on conservation equations extended by turbulence effects.

For the sake of completion, we note that direct numerical simulation (DNS) and Large-Eddy Simulation (LES), described in Section 5.4.5, are used to investigate basic phenomena of turbulent disperse multiphase flows. DNS is used to describe the motion of the phase boundary, while LES describes the transport of the vortex with an additional transport equation. Naturally these methods are only of restricted usefulness for technical applications because of the limit of numerical resolution for large spatial regions, but may be of great practical use in deriving and validating closure conditions for the model equations.

5.4.7 Reactive Flows

As well as the conservation equations for mass, momentum and total energy introduced in Sections 5.1 to 5.3, the description of *laminar* and *turbulent reactive flows* also requires balance equations for the *partial mass density* ρ_{i} of each reactive particle type i in the flow. The local flow velocity $\boldsymbol{u}_{\mathrm{i}}$ of particle type i is made up of the *mean flow velocity* $\boldsymbol{u}$ of the center of gravity and a so-called *diffusion velocity* $\boldsymbol{U}_{\mathrm{i}}$ for particle type i that represents the relative velocity of type i to the center of gravity of the system of particles.

Since species are changed into one another by chemical reaction, a *source term* appears. This is made up of the product of the molar masses M_{i} of the species and the rate of formation $\dot{\omega}_{\mathrm{i}}$ on a molar scale (e.g. in $mol/(m^3 \cdot s)$).

If we denote $\rho_i \cdot \boldsymbol{U}_i = \boldsymbol{j}_i$ as the *diffusion flow density* or *diffusion flux*, in analogy to relation (5.177) in Section 5.4.6 we obtain:

$$\frac{\partial \rho_i}{\partial t} + \nabla \cdot (\rho_i \cdot \boldsymbol{u}) + \nabla \cdot \boldsymbol{j}_i = M_i \cdot \dot{\omega}_i \quad . \tag{5.180}$$

From the conservation equation for the total energy $\rho \cdot E = \rho \cdot e + (1/2) \cdot \rho \cdot \boldsymbol{u}^2$ we can, using the momentum equation, derive the conservation equation for the *specific internal energy* e, neglecting gravity, in the form

$$\frac{\partial(\rho \cdot e)}{\partial t} + \nabla \cdot (\rho \cdot e \cdot \boldsymbol{u}) + \nabla \cdot \boldsymbol{j}_q + \tau : \nabla \boldsymbol{u} = 0 \tag{5.181}$$

where " : " is the double contraction of the two tensors τ und $\nabla \boldsymbol{u}$. Using the relation $\rho \cdot h = \rho \cdot e + p$ this relation can be transformed into a conservation equation for the *specific enthalpy*:

$$\frac{\partial(\rho \cdot h)}{\partial t} - \frac{\partial p}{\partial t} + \nabla \cdot (\rho \cdot h \cdot \boldsymbol{u}) + \nabla \cdot \boldsymbol{j}_q + \tau : \nabla \boldsymbol{u} - \nabla \cdot (p \cdot \boldsymbol{u}) = 0 \, . \tag{5.182}$$

Detailed models to calculate the diffusion flux $\boldsymbol{j}_i$, the heat flux $\boldsymbol{j}_q$ of the shear stress tensor τ and the viscosity μ for multi-component flows are presented in Chapter 9. The enthalpy and the internal energy are also determined there as functions of temperature and composition of the mixture.

For *turbulent* flows, if it is time averaged values rather than the fluctuations in time that are of interest, the averaged Reynolds equations can be derived. As in the treatment of compressible flows in Section 5.2.2 and multiphase flows in Section 5.4.6, for reactive flows it is also convenient to use density-weighted Favre averaging.

For the conservation of particle i, using the constitutive relation $\boldsymbol{j}_i = -D_i \cdot \rho \cdot \nabla \omega_i$ we obtain:

$$\frac{\partial(\overline{\rho} \cdot \tilde{\omega}_i)}{\partial t} + \nabla \cdot (\overline{\rho} \cdot \tilde{\boldsymbol{u}} \cdot \tilde{\omega}_i) + \nabla \cdot (-\overline{\rho \cdot D_i \cdot \nabla \omega_i} + \overline{\rho \cdot \boldsymbol{u}'' \cdot \omega_i''}) = \overline{M_i \cdot \dot{\omega}_i} \, . \tag{5.183}$$

For the conservation of energy (5.182) using the ansatz $\boldsymbol{j}_q = -\lambda \cdot \nabla T$ we obtain

$$\frac{\partial(\overline{\rho} \cdot \tilde{h})}{\partial t} - \frac{\partial \overline{p}}{\partial t} + \nabla \cdot (\overline{\rho} \cdot \tilde{\boldsymbol{u}} \cdot \tilde{h}) + \nabla \cdot (\overline{-\lambda \cdot \nabla T} + \overline{\rho \cdot \boldsymbol{u}'' \cdot h''}) = 0 \, . \tag{5.184}$$

Here the terms $\tau : \nabla \boldsymbol{u}$ and $\nabla \cdot (p \cdot \boldsymbol{u})$ have been neglected as they are only important when shock waves or detonations occur, i.e. at extreme pressure gradients. In analogy to the unaveraged equations, we need a thermal equation of state. From $p = \rho \cdot R \cdot T \cdot \sum_i (\omega_i / M_i)$, averaging yields:

$$\tilde{p} = R \cdot \sum_{i=1}^{N} \left((\overline{\rho} \cdot \tilde{T} \cdot \tilde{\omega}_i + \overline{\rho} \cdot T'' \cdot \omega_i'') \cdot \frac{1}{M_i} \right) \quad . \tag{5.185}$$

If the molar masses are similar, an approximate assumption is that the mean molar mass barely fluctuates. After averaging the ideal gas equation we obtain:

$$\tilde{p} = \frac{\overline{\rho} \cdot R \cdot \tilde{T}}{\overline{M}} \quad , \tag{5.186}$$

whereby in this equation $\overline{M}$ is the averaged mean molar mass of the mixture under consideration.

Source terms appear in the particle conservation equations, and their treatment is frequently very difficult. For this reason it is convenient to consider *element conservation equations*. In chemical reactions elements are neither created nor destroyed, and so the source terms in conservation equations for the elements vanish. We introduce the *element mass fraction*:

$$Z_{\mathrm{i}} = \sum_{\mathrm{j}=1}^{\mathrm{N}} (\mu_{\mathrm{ij}} \cdot \omega_{\mathrm{j}}) \quad , \qquad i = 1, ..., \mathrm{M} \tag{5.187}$$

where N is the number of substances, M the number of elements in the mixture under consideration and μ_{ij} denotes the mass fraction of element i in substance j.

If it is assumed approximately that all diffusion coefficients D_{i} in (5.183) are equal, the conservations equations can be multiplied by μ_{ij} and summed and we obtain the simple relation:

$$\frac{\partial(\rho \cdot Z_{\mathrm{i}})}{\partial t} + \nabla \cdot (\rho \cdot Z_{\mathrm{i}} \cdot \boldsymbol{u}) - \nabla \cdot (\rho \cdot D \cdot \nabla Z_{\mathrm{i}}) = 0 \quad . \tag{5.188}$$

Because of the conservation of elements $\sum(\mu_{\mathrm{ij}} \cdot M_{\mathrm{i}} \cdot \omega_{\mathrm{i}}) = 0$, this equation no longer contains any reaction terms; this is used to advantage in Chapter 9. Following time averaging, (5.188) also leads to a source free equation:

$$\frac{\partial(\overline{\rho} \cdot \tilde{Z}_{\mathrm{i}})}{\partial t} + \nabla \cdot (\overline{\rho} \cdot \tilde{\boldsymbol{u}} \cdot \tilde{Z}_{\mathrm{i}}) + \nabla \cdot (-\overline{\rho \cdot D \cdot \nabla Z_{\mathrm{i}}} + \overline{\rho \cdot \boldsymbol{u}'' \cdot Z_{\mathrm{i}}''}) = 0 \ . \tag{5.189}$$

Whereas the Navier-Stokes equations are closed when classical constitutive relations are used for the flux densities and so can be solved numerically, for the averaged conservation equations terms of the form $\overline{\rho \cdot \boldsymbol{v}'' \cdot q''}$ appear and these are not known explicitly as functions of the averaged values. Thus there are more unknowns than determining equations. This is the *closure problem* of turbulence that is described in Section 5.4.4.

In order to find a solution to the problem, models are used that describe the dependence of the Reynolds stress terms $\overline{\rho \cdot \boldsymbol{v}'' \cdot q''}$ on the averaged values. The turbulence models generally used today (see for example *B. E. Launder* and *D. B. Spalding* (1972), *W. P. Jones* and *J. H. Whitelaw* (1985)) interpret the term $\overline{\rho \cdot \boldsymbol{v}'' \cdot q''}$ with $q = w_{\mathrm{i}}, \boldsymbol{v}, h, Z_{\mathrm{i}})$ in (5.181) as *turbulent transport* and thus model it in the framework of the Boussinesq approximation in analogy with the laminar case by means of a *gradient approach*, according to which the term is proportional to the gradient of the averaged value of the quantity under consideration:

$$\overline{\rho \cdot \boldsymbol{v}'' \cdot q_{\mathrm{i}}''} = -\rho \cdot \nu_{\mathrm{T}} \cdot \nabla \tilde{q}_{\mathrm{i}} \quad , \tag{5.190}$$

where ν_T is called the turbulent exchange coefficient. However, for certain flow situations this approach breaks down. Indeed experiments indicate that turbulent transport can also take place against the gradient (*J. B. Moss* (1979)).

The turbulent transport is generally much faster than the molecular diffusive transport processes in laminar flow. For this reason, the averaged laminar transport terms may be neglected in very many cases.

The conservation equations for turbulent reactive flows can be solved numerically if the turbulent exchange coefficient ν_T is known. It can be assumed that it takes on different values for the different equations. In order to determine the exchange coefficient numerous models exist (see Section 5.4.5). Generally the *K-ε turbulence model* (*B. E. Launder* and *D. B. Spalding* (1972), *W. P. Jones* and *J. H. Whitelaw* (1985)) is used. This uses an equation for the turbulent kinetic energy K and the *rate of dissipation* ε of the kinetic energy. The turbulent exchange coefficient ν_T is then

$$\nu_T = C_\nu \cdot \frac{\tilde{K}^2}{\tilde{\varepsilon}} \quad . \tag{5.191}$$

If we neglect the laminar transport in the conservation equations, and use the gradient approach and assume that the turbulent exchange coefficient is the same for all transport quantities, together with the momentum equation and the equations for $\tilde{k}$ and $\tilde{\varepsilon}$, we obtain the averaged conservation equations for reactive flows:

$$\frac{\partial(\bar{\rho} \cdot \tilde{\omega}_i)}{\partial t} + \nabla \cdot (\bar{\rho} \cdot \tilde{\boldsymbol{u}} \cdot \tilde{\omega}_i) - \nabla \cdot (\bar{\rho} \cdot \nu_T \cdot \nabla \tilde{\omega}_i) = \overline{M_i \cdot \dot{\omega}_i} \quad , \tag{5.192}$$

$$\frac{\partial(\bar{\rho} \cdot \tilde{h})}{\partial t} - \frac{\partial \bar{p}}{\partial t} + \nabla \cdot (\bar{\rho} \cdot \tilde{\boldsymbol{u}} \cdot \tilde{h}) - \nabla \cdot (\bar{\rho} \cdot \nu_T \cdot \nabla \tilde{h}) = 0 \quad , \tag{5.193}$$

$$\frac{\partial(\bar{\rho} \cdot \tilde{Z}_i)}{\partial t} + \nabla \cdot (\bar{\rho} \cdot \tilde{\boldsymbol{u}} \cdot \tilde{Z}_i) - \nabla \cdot (\bar{\rho} \cdot \nu_T \cdot \nabla \tilde{Z}_i) = 0 \quad . \tag{5.194}$$

These equations are then closed in the framework of the model assumptions discussed above if the averaged source term for the individual species equations can be determined. For this there are again numerous models of differing complexity, which are presented in Section 5.4.5.

5.5 Differential Equations of Perturbations

Fluid mechanical instabilities are treated in Chapter 6. The necessary perturbation differential equations are obtained using the trial ansatz:

$$\boldsymbol{u}^*(x, y, z, t) = \boldsymbol{U}_0^*(x, y, z) + \epsilon \cdot \boldsymbol{u}^{*\prime}(x, y, z, t). \tag{5.195}$$

Here $\boldsymbol{U}_0^*$ is the dimensionless basic flow, which is perturbed by the small disturbance $\boldsymbol{u}'$ (fluid mechanical instability), and $\epsilon \ll 1$ is the expansion parameter, a measure for the small perturbing quantity. The initial perturbation at time $t^* = 0$ is normalized to 1:

$$|\epsilon \cdot \boldsymbol{u}^{*\prime}|_{t^*=0} = \epsilon \quad \Rightarrow \quad |\boldsymbol{u}^{*\prime}|_{t^*=0} = 1.$$

The dimensionless flow quantities $\mathbf{u}^*, p^*, \rho^*, T^*$ are then written using the perturbation ansatz:

$$\begin{aligned} \boldsymbol{u}^* &= \boldsymbol{U}_0^* + \epsilon \cdot \boldsymbol{u}^{*\prime}, \qquad & p^* &= p_0^* + \epsilon \cdot p^{*\prime}, \\ \rho^* &= \rho_0^* + \epsilon \cdot \rho^{*\prime}, \qquad & T^* &= T_0^* + \epsilon \cdot T^{*\prime}. \end{aligned} \tag{5.196}$$

Inserting these into the dimensionless fundamental equations of *compressible flows* (5.1), (5.18), (5.56)

$$\frac{\partial \rho^*}{\partial t} + \boldsymbol{u}^* \cdot \nabla \rho^* = -\rho^* \cdot \nabla \cdot \boldsymbol{u}^*, \tag{5.197}$$

$$\begin{aligned} \rho^* \cdot \left(\frac{\partial \boldsymbol{u}^*}{\partial t} + \boldsymbol{u}^* \cdot \nabla \boldsymbol{u}^* \right) &= -\frac{1}{\kappa \cdot M_\infty^2} \cdot \nabla p^* \\ &+ \frac{1}{\mathrm{Re}_l} \cdot \left(\nabla \cdot (\mu [\nabla \boldsymbol{u}^* + {}^t\nabla \boldsymbol{u}^*]) - \frac{2}{3} \cdot \nabla (\mu \cdot \nabla \cdot \boldsymbol{u}^*) \right), \end{aligned} \tag{5.198}$$

$$\begin{aligned} \rho^* \cdot \left(\frac{\partial T^*}{\partial t} + \boldsymbol{u}^* \cdot \nabla T^* \right) &= -(\kappa - 1) \cdot p^* \cdot \nabla \cdot \boldsymbol{u}^* \\ &+ \frac{\kappa}{\mathrm{Re}_l} \cdot \left(\frac{1}{\mathrm{Pr}_\infty} \cdot \nabla \cdot (\lambda \cdot \nabla T^*) - (\kappa - 1) \cdot M_\infty^2 \cdot \Phi^* \right), \\ \Phi^* &= \mu \cdot \left(\frac{1}{2} \cdot (\nabla \boldsymbol{u}^* + {}^t\nabla \boldsymbol{u}^*)^2 - \frac{2}{3} \cdot (\nabla \cdot \boldsymbol{u}^*)^2 \right), \end{aligned} \tag{5.199}$$

we obtain the *perturbation differential equations* (see also *H. Oertel, J. Delfs* (1996), (2005)):

$$\begin{aligned} &\frac{\partial \rho'}{\partial t} + \boldsymbol{u}' \cdot \nabla \rho_0 + \boldsymbol{U}_0 \cdot \nabla \rho' + \rho' \cdot \nabla \cdot \boldsymbol{U}_0 + \rho_0 \cdot \nabla \cdot \boldsymbol{u}' \\ &\quad = -\epsilon \cdot [\nabla \cdot (\rho' \cdot \boldsymbol{u}')], \end{aligned} \tag{5.200}$$

$$\begin{aligned} &\rho_0 \cdot \left(\frac{\partial \mathbf{u}'}{\partial t} + \boldsymbol{u}' \cdot \nabla \boldsymbol{U}_0 + \boldsymbol{U}_0 \cdot \nabla \boldsymbol{u}' \right) + \rho' \cdot (\boldsymbol{U}_0 \cdot \boldsymbol{U}_0) \\ &\quad + \frac{1}{\kappa \cdot M_\infty^2} \cdot \nabla (\rho_0 \cdot T' + T_0 \cdot \rho') - \frac{1}{\mathrm{Re}_l} \cdot \Big[\nabla \cdot (\mu_0 \cdot [\nabla \boldsymbol{u}' + {}^t\nabla \boldsymbol{u}'] \\ &\quad + {\mu_\epsilon}' \cdot [\nabla \boldsymbol{U}_0 + {}^t\nabla \boldsymbol{U}_0]) - \frac{2}{3} \cdot \nabla (\mu_0 \cdot \nabla \cdot \boldsymbol{u}' + {\mu_\epsilon}' \cdot \nabla \cdot \boldsymbol{u}') \Big] \\ &\quad = \epsilon \cdot \Bigg(- \rho' \cdot \left(\frac{\partial \boldsymbol{u}'}{\partial t} + \boldsymbol{u}' \cdot \nabla \boldsymbol{U}_0 + \boldsymbol{U}_0 \cdot \nabla \boldsymbol{u}' \right) - \rho_0 \cdot \boldsymbol{u}' \cdot \nabla \cdot \boldsymbol{u}' \\ &\quad - \frac{1}{\kappa \cdot M_\infty^2} \cdot \nabla (\rho' \cdot T') + \frac{1}{\mathrm{Re}_l} \cdot \Big[\nabla \cdot ({\mu_\epsilon}' \cdot [\nabla \boldsymbol{u}' + {}^t\nabla \boldsymbol{u}'] \\ &\quad + {\mu_{\epsilon\epsilon}}' \cdot [\nabla \boldsymbol{U}_0 + {}^t\nabla \boldsymbol{U}_0]) - \frac{2}{3} \cdot \nabla ({\mu_\epsilon}' \nabla \cdot \boldsymbol{u}' + {\mu_{\epsilon\epsilon}}' \cdot \boldsymbol{U}_0) \Big] \Bigg), \end{aligned} \tag{5.201}$$

$$
\begin{aligned}
&\rho_0 \cdot \left(\frac{\partial T'}{\partial t} + \boldsymbol{u}' \cdot \nabla T_0 + \boldsymbol{U}_0 \cdot \nabla T' \right) \\
&\quad + (\kappa - 1) \cdot [(T_0 \cdot \rho' + \rho_0 \cdot T') \cdot \nabla \cdot \boldsymbol{U}_0 + T_0 \cdot \rho_0 \cdot \nabla \cdot \boldsymbol{u}'] \\
&\quad + \frac{\kappa}{\mathrm{Re}_l} \cdot \left[(\kappa - 1) \cdot M_\infty^2 \cdot {\Phi_\epsilon}' - \frac{1}{\mathrm{Pr}_\infty} \cdot \nabla \cdot (\lambda_0 \cdot \nabla T' + {\lambda_\epsilon}' \cdot \nabla T_0) \right] \\
&= \epsilon \cdot \Bigg(- \rho' \cdot \left(\frac{\partial T'}{\partial t} + \boldsymbol{u}' \cdot \nabla T_0 + \boldsymbol{U}_0 \cdot \nabla T' \right) - \rho_0 \cdot \boldsymbol{u}' \cdot \nabla T' \qquad (5.202) \\
&\quad - (\kappa - 1) \cdot [(T_0 \cdot \rho' + \rho_0 \cdot T') \cdot \nabla \cdot \boldsymbol{u}' + T' \cdot \rho' \cdot \nabla \cdot \boldsymbol{U}_0] \\
&\quad - \frac{\kappa}{\mathrm{Re}_l} \cdot \left[(\kappa - 1) \cdot M_\infty^2 \cdot {\Phi_{\epsilon\epsilon}}' - \frac{1}{\mathrm{Pr}_\infty} \cdot \nabla \cdot ({\lambda_\epsilon}' \cdot \nabla T' + {\lambda_{\epsilon\epsilon}}' \cdot \nabla T_0) \right] \Bigg) .
\end{aligned}
$$

The index $(*)$ for the dimensionless perturbation quantities has been omitted, and the temperature dependence of the viscosity $\mu(T)$ and the thermal conductivity $\lambda(T)$ have been taken into account according to the *Sutherland equation*

$$\mu = \lambda = T^{\frac{2}{3}} \cdot \frac{1+S}{T+S}, \qquad S = \frac{110.4\ K}{T_\infty} .$$

Perturbations in the density ρ and the temperature T also lead to perturbations in these functions. The functions μ and λ are expanded in a Taylor series about the ground state μ_0, λ_0:

$$
\begin{aligned}
(\mu, \lambda) &= (\mu, \lambda)_0 + \left(\frac{\mathrm{d}(\mu, \lambda)}{\mathrm{d}T} \right)_0 \cdot (T - T_0) + \frac{1}{2!} \cdot \left(\frac{\mathrm{d}^2(\mu, \lambda)}{\mathrm{d}T^2} \right)_0 \cdot (T - T_0)^2 + \cdots \\
&= (\mu, \lambda)_0 + \epsilon \cdot \left(\frac{\mathrm{d}(\mu, \lambda)}{\mathrm{d}T} \right)_0 \cdot T' + \epsilon^2 \cdot \frac{1}{2!} \cdot \left(\frac{\mathrm{d}^2(\mu, \lambda)}{\mathrm{d}T^2} \right)_0 \cdot {T'}^2 + \cdots . \qquad (5.203)
\end{aligned}
$$

Here it can be seen that deviations of the transport coefficient from the ground state $(\mu - \mu_0)$ or $(\lambda - \lambda_0)$ contain not only terms of order of magnitude ϵ. They also contain terms of higher orders of ϵ. We introduce the notation

$$(\mu - \mu_0, \lambda - \lambda_0) = \epsilon \cdot (\mu'_\epsilon, \lambda'_\epsilon) + \epsilon^2 \cdot (\mu'_{\epsilon\epsilon}, \lambda'_{\epsilon\epsilon}) + \cdots ,$$

where

$$
\begin{aligned}
(\mu'_\epsilon, \lambda'_\epsilon) &:= \frac{1}{1!} \cdot \left(\frac{\mathrm{d}(\mu, \lambda)}{\mathrm{d}T} \right)_0 \cdot T' , \\
(\mu'_{\epsilon\epsilon}, \lambda'_{\epsilon\epsilon}) &:= \frac{1}{2!} \cdot \left(\frac{\mathrm{d}^2(\mu, \lambda)}{\mathrm{d}T^2} \right)_0 \cdot {T'}^2 .
\end{aligned}
$$

Similarly, deviations in the dissipation function Φ due to the perturbation $\epsilon \cdot \boldsymbol{u}'$ are defined as

$$\Phi - \Phi_0 = \epsilon \cdot \Phi'_\epsilon + \epsilon^2 \cdot \Phi'_{\epsilon\epsilon} + \cdots . \qquad (5.204)$$

We insert the perturbed flow state $\mathbf{u} = \boldsymbol{U}_0 + \epsilon \cdot \boldsymbol{u}'$ and the perturbed viscosity into the dissipation function (5.199) and sort for powers of ϵ:

$$\Phi - \Phi_0 = \epsilon \Bigg[\mu_\epsilon{}' \cdot \left(\frac{1}{2} \cdot (\nabla \boldsymbol{U}_0 + {}^t\nabla \boldsymbol{U}_0)^2 - \frac{2}{3} \cdot (\nabla \cdot \boldsymbol{U}_0)^2 \right) \tag{5.205}$$
$$+ 2 \cdot \mu_0 \cdot \left(\frac{1}{2} \cdot (\nabla \boldsymbol{U}_0 + {}^t\nabla \boldsymbol{U}_0) \cdot (\nabla \boldsymbol{u}' + {}^t\nabla \boldsymbol{u}') - \frac{2}{3} \cdot (\nabla \cdot \boldsymbol{U}_0) \cdot (\nabla \cdot \boldsymbol{u}') \right) \Bigg]$$
$$+ \epsilon^2 \cdot \Bigg[\mu_{\epsilon\epsilon}{}' \cdot \left(\frac{1}{2} \cdot (\nabla \boldsymbol{U}_0 + {}^t\nabla \boldsymbol{U}_0)^2 - \frac{2}{3} \cdot (\nabla \cdot \boldsymbol{U}_0)^2 \right)$$
$$+ 2 \cdot \mu_\epsilon{}' \cdot \left(\frac{1}{2} (\nabla \boldsymbol{U}_0 + {}^t\nabla \boldsymbol{U}_0) \cdot (\nabla \boldsymbol{u}' + {}^t\nabla \boldsymbol{u}') - \frac{2}{3} \cdot (\nabla \cdot \boldsymbol{U}_0) \cdot (\nabla \cdot \boldsymbol{u}') \right)$$
$$+ \mu_0 \cdot \left(\frac{1}{2} \cdot (\nabla \boldsymbol{u}' + {}^t\nabla \boldsymbol{u}') - \frac{2}{3} \cdot \nabla (\nabla \cdot \boldsymbol{u}')^2 \right) \Bigg],$$

with

$$\Phi_\epsilon{}' = \mu_\epsilon{}' \cdot \left(\frac{1}{2} \cdot (\nabla \boldsymbol{U}_0 + {}^t\nabla \boldsymbol{U}_0)^2 - \frac{2}{3} \cdot (\nabla \cdot \boldsymbol{U}_0)^2 \right)$$
$$+ 2 \cdot \mu_0 \cdot \left(\frac{1}{2} \cdot (\nabla \boldsymbol{U}_0 + {}^t\nabla \boldsymbol{U}_0) \cdot (\nabla \boldsymbol{u}' + {}^t\nabla \boldsymbol{u}') - \frac{2}{3} \cdot (\nabla \cdot \boldsymbol{U}_0) \cdot (\nabla \cdot \boldsymbol{u}') \right),$$
$$\Phi_{\epsilon\epsilon}{}' = \mu_{\epsilon\epsilon}{}' \cdot \left(\frac{1}{2} \cdot (\nabla \boldsymbol{U}_0 + {}^t\nabla \boldsymbol{U}_0)^2 - \frac{2}{3} \cdot (\nabla \cdot \boldsymbol{U}_0)^2 \right)$$
$$+ 2 \cdot \mu_\epsilon{}' \cdot \left(\frac{1}{2} (\nabla \boldsymbol{U}_0 + {}^t\nabla \boldsymbol{U}_0) \cdot (\nabla \boldsymbol{u}' + {}^t\nabla \boldsymbol{u}') - \frac{2}{3} \cdot (\nabla \cdot \boldsymbol{U}_0) \cdot (\nabla \cdot \boldsymbol{u}') \right)$$
$$+ \mu_0 \cdot \left(\frac{1}{2} \cdot (\nabla \boldsymbol{u}' + {}^t\nabla \boldsymbol{u}') - \frac{2}{3} \cdot \nabla (\nabla \cdot \boldsymbol{u}')^2 \right).$$

After inserting the perturbation ansatz $\boldsymbol{u} = \boldsymbol{U}_0 + \epsilon \cdot \boldsymbol{u}'$, the pressure terms of the fundamental equations also generate an expression that is of second order in ϵ. After Taylor expanding, only one exact term with a simple product-like dependence of the pressure on the density and pressure, as in the ideal gas law, remains:

$$\begin{aligned} p - p_0 &= \epsilon \cdot \left(\frac{\partial p}{\partial \rho} \right)_0 \cdot \rho' + \epsilon \cdot \left(\frac{\partial p}{\partial T} \right)_0 \cdot T' + \epsilon^2 \cdot \left(\frac{\partial^2}{\partial \rho \cdot \partial T} \right)_0 \cdot (\rho' \cdot T') \\ &= \epsilon \cdot (T_0 \cdot \rho' + \rho_0 \cdot T') + \epsilon^2 \cdot \rho' \cdot T' \quad . \end{aligned} \tag{5.206}$$

The system of perturbation differential equations (5.200)–(5.202) describes the behavior of an arbitrary perturbation $\boldsymbol{u}'(x, y, z, t)$ of the steady ground state flow $\boldsymbol{U}_0(x, y, z)$. The nonlinear terms are on the right-hand sides. If small but finite perturbations are assumed, the powers of ϵ may be interpreted as an order-of-magnitude division of the nonlinear effects on the perturbation expansion. If we consider infinitesimally small perturbations, i.e. $\epsilon \to 0$, the right-hand sides vanish in the limit, and we obtain linear differential equations. If we increase ϵ as a measure of the size of the perturbations, these terms increase in importance, and nonlinear effects affect the expansion of the perturbation.

We note that the third- and higher-order terms in the momentum equations (5.201) and the fourth- and higher-order terms in the energy equation (5.202) (independent of the size of ϵ) are only a consequence of the generally very weak second- and higher-order derivatives of the transport coefficients μ and λ with respect to the temperature. They may be justifiably neglected, even for moderate perturbations.

In Chapter 6 the fluid-mechanical instabilities of *infinitesimally small perturbations* with $\epsilon \to 0$ are treated. The *linear perturbation differential equations* of compressible flows that describe such perturbations are obtained by neglecting the right-hand sides of (5.200)–(5.202).

As in every flow, the perturbing flow $\boldsymbol{u}'$ also has to satisfy boundary conditions. First, the no-slip condition has to be satisfied at solid walls. The boundary condition that is additionally required for the temperature perturbation will be briefly discussed here. For simplicity, we begin with the case of an isothermal wall. According to the perturbation ansatz, $T_0(x_w, y_w, z_w) + \epsilon \cdot {T_w}'(x_w, y_w, z_w) = T_w$ must hold. This yields the temperature condition

$$ {T_w}' = 0 $$

for arbitrary ϵ. The calculation for adiabatic walls is carried out in a similar manner, yielding

$$ \boldsymbol{n} \cdot \nabla T' = 0. $$

No explicit boundary condition may be demanded of the density perturbation, since only its second derivative appears in the equations. Instead, the density is determined from the continuity equation (5.200) evaluated at the boundary.

In the treatment of problems involving flows past a body, we also require that all perturbations die away to zero in the far field, i.e. at infinite distances from the walls.

For *incompressible flows*, the fundamental equations at constant density ρ and constant dynamic viscosity μ as in (5.82) simplify to

$$ \nabla \cdot \boldsymbol{u}^* = 0, \tag{5.207} $$

$$ \frac{\partial \boldsymbol{u}^*}{\partial t} + \boldsymbol{u}^* \cdot \nabla \boldsymbol{u}^* = -\nabla p^* + \frac{1}{\mathrm{Re}_l} \cdot \Delta \boldsymbol{u}^*. \tag{5.208} $$

The perturbation ansatz

$$ \boldsymbol{u} = \mathbf{U}_0 + \epsilon \cdot \boldsymbol{u}', \qquad p = p_0 + \epsilon \cdot p' $$

leads to the *perturbation differential equations* for incompressible fluids

$$ \nabla \cdot \boldsymbol{u}' = 0, \tag{5.209} $$

$$ \frac{\partial \boldsymbol{u}'}{\partial t} + \boldsymbol{U}_0 \cdot \nabla \boldsymbol{u}' + \boldsymbol{u}' \cdot \nabla \boldsymbol{U}_0 + \epsilon \cdot \boldsymbol{u}' \cdot \nabla \boldsymbol{u}' = -\nabla p' + \frac{1}{\mathrm{Re}_l} \cdot \Delta \boldsymbol{u}'. \tag{5.210} $$

For *small perturbations* with $\epsilon \to 0$, the left-hand side of (5.210), with the factor ϵ, is neglected, and so the *linear perturbation differential equations* are written

$$\nabla \cdot \boldsymbol{u}' = 0, \tag{5.211}$$

$$\frac{\partial \boldsymbol{u}'}{\partial t} + \boldsymbol{U}_0 \cdot \nabla \boldsymbol{u}' + \boldsymbol{u}' \cdot \nabla \boldsymbol{U}_0 = -\nabla p' + \frac{1}{\mathrm{Re}_l} \cdot \Delta \boldsymbol{u}'. \tag{5.212}$$

The perturbation differential equations for small perturbations of the *Boussinesq equation* (5.89) are similarly found to be

$$\nabla \cdot \boldsymbol{u}' = 0, \tag{5.213}$$

$$\frac{1}{\mathrm{Pr}_\infty} \cdot \left(\frac{\partial \mathbf{u}'}{\partial t} + \boldsymbol{U}_0 \cdot \nabla \boldsymbol{u}' + \boldsymbol{u}' \cdot \nabla \boldsymbol{U}_0 \right) =$$

$$-\nabla p' + \Delta \boldsymbol{u}' + \mathrm{Ra}_\infty \cdot T' \cdot \begin{pmatrix} 0 \\ 0 \\ 1 \end{pmatrix}, \tag{5.214}$$

$$\frac{\partial T'}{\partial t} + \boldsymbol{U}_0 \cdot \nabla T' + T' \cdot \nabla \boldsymbol{U}_0 = \Delta T'. \tag{5.215}$$

For double diffusion–convection (temperature and concentration gradients), equation (5.90) is used to write the perturbation differential equations as

$$\nabla \cdot \boldsymbol{u}' = 0, \tag{5.216}$$

$$\mathrm{Le}_\infty \cdot \frac{\partial c'}{\partial t} = \Delta c' + \mathrm{Le}_\infty \cdot \omega', \tag{5.217}$$

$$\frac{1}{\mathrm{Pr}_\infty} \cdot \left(\frac{\partial \mathbf{u}'}{\partial t} + \boldsymbol{U}_0 \cdot \nabla \boldsymbol{u}' + \boldsymbol{u}' \cdot \nabla \boldsymbol{U}_0 \right) =$$

$$-\nabla p' + \Delta \boldsymbol{u}' + (\mathrm{Ra}_\infty \cdot T' + \mathrm{Ra}_{D\infty} \cdot c') \cdot \begin{pmatrix} 0 \\ 0 \\ 1 \end{pmatrix}, \tag{5.218}$$

$$\frac{\partial T'}{\partial t} + \boldsymbol{U}_0 \cdot \nabla T' + T' \cdot \nabla \boldsymbol{U}_0 = \Delta T'. \tag{5.219}$$

5.6 Problems

5.1

Given an ideal gas ($p = \rho RT, R = \text{const}$), as well as the continuity equation

$$\frac{\mathrm{d}\rho}{\mathrm{d}t} + \rho \cdot (\nabla \cdot \boldsymbol{v}) = 0,$$

(a) Show that the following relation for the total time derivative of the pressure can be derived from the continuity equation:

$$\frac{1}{p} \cdot \frac{\mathrm{d}p}{\mathrm{d}t} = \frac{1}{T} \cdot \frac{\mathrm{d}T}{\mathrm{d}t} - \nabla \cdot \boldsymbol{v}.$$

(b) For the dimensionless velocity field

$$\boldsymbol{v}(x, y, z) = \begin{pmatrix} u \\ v \end{pmatrix} = V_0 \cdot \sqrt{x^2 + y^2} \cdot \begin{pmatrix} \sin(\omega \cdot t) \\ \cos(\omega \cdot t) \end{pmatrix},$$

with the constant V_0 as well as the constant angular velocity ω and the dimensionless temperature distribution

$$T(x, y) = A_0 \cdot \sqrt{x^2 + y^2} + T_0$$

with the constants A_0 and T_0, determine the relative substantial temperature change $(1/T) \cdot (\mathrm{d}T/\mathrm{d}t)$ as well as the divergence $(\nabla \cdot \boldsymbol{v})$ of the velocity field. In doing so, first transform the substantial change in T into the local change and the convective part.

$$\frac{1}{T} \cdot \frac{\mathrm{d}T}{\mathrm{d}t} = \frac{V_0 \cdot A_0}{A_0 \cdot \sqrt{x^2 + y^2} + T_0} \cdot [x \cdot \sin(\omega \cdot t) + y \cdot \cos(\omega \cdot t)],$$

$$\frac{\partial u}{\partial x} + \frac{\partial v}{\partial y} = \frac{V_0}{\sqrt{x^2 + y^2}} \cdot [x \cdot \sin(\omega \cdot t) + y \cdot \cos(\omega \cdot t)].$$

5.2

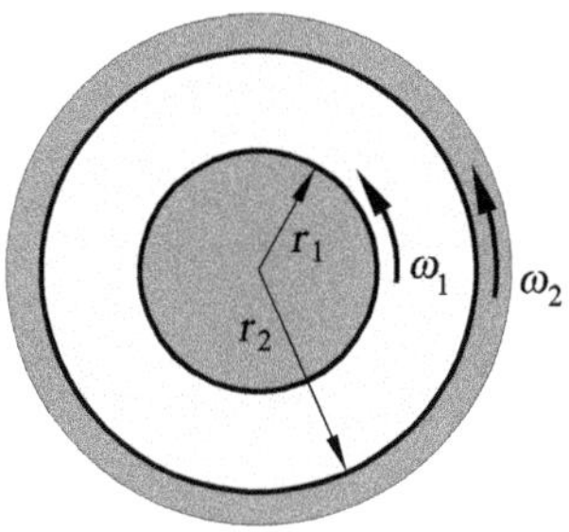

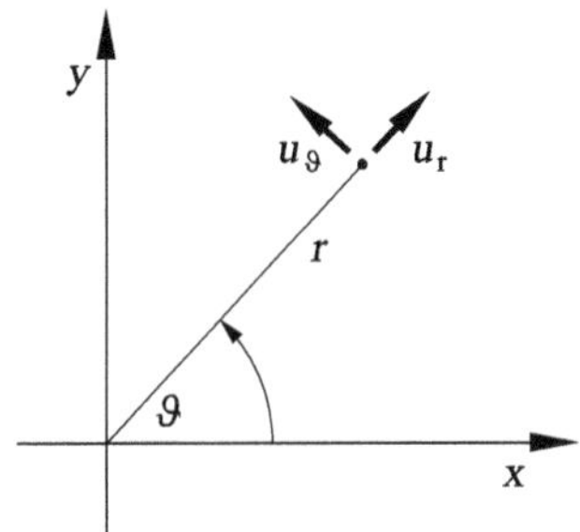

A cylinder with radius r_1 is surrounded by an outer cylinder with radius r_2. The inner cylinder rotates with angular velocity ω_1, and the outer cylinder with angular velocity ω_2. There is a fluid between the cylinders. Determine the laminar incompressible velocity profile of the fluid between the two cylinders. Use the Navier–Stokes equations in polar coordinates:

Continuity equation:

$$\frac{\partial u_r}{\partial r} + \frac{u_r}{r} + \frac{1}{r} \cdot \frac{\partial u_\vartheta}{\partial \vartheta} = 0.$$

Navier–Stokes equations:

$$\rho \cdot \left(u_r \cdot \frac{\partial u_r}{\partial r} + \frac{u_\vartheta}{r} \cdot \frac{\partial u_r}{\partial \vartheta} + \frac{u_\vartheta^2}{r} \right)$$
$$= -\frac{\partial p}{\partial r} + \mu \cdot \left(\frac{\partial^2 u_r}{\partial r^2} + \frac{1}{r} \cdot \frac{\partial u_r}{\partial r} - \frac{u_r}{r^2} + \frac{1}{r^2} \cdot \frac{\partial^2 u_r}{\partial \vartheta^2} - \frac{2}{r^2} \cdot \frac{\partial u_\vartheta}{\partial \vartheta} \right) + f_r,$$

$$\rho \cdot \left(u_r \cdot \frac{\partial u_\vartheta}{\partial r} + \frac{u_\vartheta}{r} \cdot \frac{\partial u_\vartheta}{\partial \vartheta} + \frac{u_r \cdot u_\vartheta}{r} \right)$$
$$= -\frac{1}{r} \cdot \frac{\partial p}{\partial \vartheta} + \mu \cdot \left(\frac{\partial^2 u_\vartheta}{\partial r^2} + \frac{1}{r} \cdot \frac{\partial u_\vartheta}{\partial r} - \frac{u_\vartheta}{r^2} + \frac{1}{r^2} \cdot \frac{\partial^2 u_\vartheta}{\partial \vartheta^2} - \frac{2}{r^2} \cdot \frac{\partial u_r}{\partial \vartheta} \right) + f_\vartheta.$$

Simplify the Navier–Stokes equations and determine $u(r)$.

$$\frac{\mathrm{d}u_r}{\mathrm{d}r}+\frac{u_r}{r}=0,$$

$$\rho\cdot\frac{u^2}{r}=-\frac{\mathrm{d}p}{\mathrm{d}r},$$

$$\frac{\mathrm{d}^2u}{\mathrm{d}r^2}+\frac{1}{r}\cdot\frac{\mathrm{d}u}{\mathrm{d}r}-\frac{u}{r^2}=0,$$

$$u(r)=\frac{1}{r_2^2-r_1^2}\cdot\left[r\cdot(\omega_2\cdot r_2^2-\omega_1\cdot r_1^2)-\frac{r_1^2\cdot r_2^2}{r}\cdot(\omega_2-\omega_1)\right].$$

5.3

The Reynolds equations for compressible media with constant material values μ and c_v contain terms of the following form:

$$\frac{\partial(\rho\cdot u\cdot w)}{\partial z},$$

$$\sigma_{zz}=2\cdot\mu\cdot\frac{\partial w}{\partial z}-\frac{2}{3}\cdot\mu\cdot\left(\frac{\partial u}{\partial x}+\frac{\partial v}{\partial y}+\frac{\partial w}{\partial z}\right),$$

$$\rho\cdot c_v\cdot\left(\frac{\partial T}{\partial t}+w\frac{\partial T}{\partial z}\right),$$

$$\left(\frac{\partial u}{\partial x}+\frac{\partial v}{\partial y}\right)^2.$$

The turbulent flow under consideration is quasi-steady. The turbulent fluctuation quantities, the velocity components, and the temperature are to be mass-averaged (Le Favre averaging) and the density simply averaged. The following assumptions hold:

$$u=\tilde{u}+u'',\qquad v=\tilde{v}+v'',\qquad w=\tilde{w}+w'',$$
$$T=\tilde{T}+T'',\qquad \rho=\bar{\rho}+\rho'.$$

Insert these assumptions into the terms of the Reynolds equations and carry out the time-averaging.

$$\overline{\frac{\partial(\rho\cdot u\cdot w)}{\partial z}}=\frac{\partial(\overline{\rho\cdot\tilde{u}\cdot\tilde{w}})}{\partial z}+\frac{\partial(\overline{\rho\cdot u''\cdot w''})}{\partial z}$$
$$=\frac{\partial(\overline{\rho}\cdot\tilde{u}\cdot\tilde{w})}{\partial z}+\frac{\partial(\overline{\rho}\cdot\overline{u''\cdot w''})}{\partial z}+\frac{\partial(\overline{\rho'\cdot u''\cdot w''})}{\partial z},$$

$$\overline{\sigma_{zz}}=2\cdot\mu\cdot\frac{\partial\tilde{w}}{\partial z}-\frac{2}{3}\cdot\mu\cdot\left(\frac{\partial\tilde{u}}{\partial x}+\frac{\partial\tilde{v}}{\partial y}+\frac{\partial\tilde{w}}{\partial z}\right)$$
$$+2\cdot\mu\cdot\frac{\partial w''}{\partial z}-\frac{2}{3}\cdot\mu\cdot\left(\frac{\partial\overline{u''}}{\partial x}+\frac{\partial\overline{v''}}{\partial y}+\frac{\partial\overline{w''}}{\partial z}\right),$$

$$\overline{\rho \cdot c_v \cdot \left(\frac{\partial T}{\partial t} + w \frac{\partial T}{\partial z} \right)} = c_v \cdot \left(\overline{\rho \cdot \tilde{w} \cdot \frac{\partial \overline{T}}{\partial z} + \tilde{w} \cdot \rho \cdot \frac{\partial T''}{\partial z} + \rho \cdot w'' \cdot \frac{\partial T''}{\partial z}} \right),$$

$$\left(\frac{\partial u}{\partial x} + \frac{\partial v}{\partial y} \right)^2$$

$$= \left(\frac{\partial \tilde{u}}{\partial x} \right)^2 + 2 \cdot \frac{\partial \tilde{u}}{\partial x} \cdot \frac{\partial \tilde{v}}{\partial y} + \left(\frac{\partial \tilde{v}}{\partial y} \right)^2 + \overline{\left(\frac{\partial u''}{\partial y} \right)^2} + 2 \cdot \overline{\frac{\partial u''}{\partial x} \cdot \frac{\partial v''}{\partial y}} + \overline{\left(\frac{\partial v''}{\partial y} \right)^2}$$

$$+ 2 \cdot \frac{\partial \tilde{u}}{\partial x} \cdot \frac{\partial \overline{u''}}{\partial x} + 2 \cdot \frac{\partial \tilde{u}}{\partial x} \cdot \frac{\partial \overline{v''}}{\partial y} + 2 \cdot \frac{\partial \tilde{v}}{\partial y} \cdot \frac{\partial \overline{u''}}{\partial x} + 2 \cdot \frac{\partial \tilde{v}}{\partial y} \cdot \frac{\partial \overline{v''}}{\partial y}.$$

5.4

Consider the energy equation for the mass-specific internal energy e with $e = c_v T$ for a compressible medium with constant material properties c_v and λ, neglecting the effects of radiation:

$$\rho \cdot c_v \cdot \left(\frac{\partial T}{\partial t} + u \cdot \frac{\partial T}{\partial x} + v \cdot \frac{\partial T}{\partial y} + w \cdot \frac{\partial T}{\partial z} \right)$$

$$= \lambda \cdot \left(\frac{\partial^2 T}{\partial x^2} + \frac{\partial^2 T}{\partial y^2} + \frac{\partial^2 T}{\partial z^2} \right) - p \cdot \left(\frac{\partial u}{\partial x} + \frac{\partial v}{\partial y} + \frac{\partial w}{\partial z} \right) + \mu \cdot \Phi.$$

Write down the Favre-averaged energy equation for a quasi-steady turbulent flow. For simplicity, the time average of the dissipation term may be written as $\overline{\mu \cdot \Phi}$ and need not be decomposed into its individual terms.

$$c_v \cdot \overline{\rho} \cdot \left(\tilde{u} \cdot \frac{\partial \tilde{T}}{\partial x} + \tilde{v} \cdot \frac{\partial \tilde{T}}{\partial y} + \tilde{w} \cdot \frac{\partial \tilde{T}}{\partial z} \right)$$

$$+ c_v \cdot \left(\overline{\rho \cdot u'' \cdot \frac{\partial T''}{\partial x}} + \overline{\rho \cdot v'' \cdot \frac{\partial T''}{\partial y}} + \overline{\rho \cdot w'' \cdot \frac{\partial T''}{\partial z}} \right)$$

$$= \lambda \cdot \left(\frac{\partial^2 \tilde{T}}{\partial x^2} + \frac{\partial^2 \tilde{T}}{\partial y^2} + \frac{\partial^2 \tilde{T}}{\partial z^2} \right) + \lambda \cdot \left(\frac{\partial^2 \overline{T''}}{\partial x^2} + \frac{\partial^2 \overline{T''}}{\partial y^2} + \frac{\partial^2 \overline{T''}}{\partial z^2} \right)$$

$$- \overline{p} \cdot \left(\frac{\partial \tilde{u}}{\partial x} + \frac{\partial \tilde{v}}{\partial y} + \frac{\partial \tilde{w}}{\partial z} \right) + \overline{\mu \cdot \Phi}$$

$$- \overline{p} \cdot \left(\frac{\partial \overline{u''}}{\partial x} + \frac{\partial \overline{v''}}{\partial y} + \frac{\partial \overline{w''}}{\partial z} \right) - \left(\overline{p' \cdot \frac{\partial u''}{\partial x}} + \overline{p' \cdot \frac{\partial v''}{\partial y}} + \overline{p' \cdot \frac{\partial w''}{\partial z}} \right).$$

5.5

For the numerical computation of compressible turbulent flow fields, it is useful to write down the fundamental equations in dimensionless conservative form for the mass-averaged flow quantities:

$$\frac{\partial \overline{\boldsymbol{U}^*}}{\partial t^*} + \sum_{m=1}^{3} \frac{\partial \overline{\boldsymbol{f}_m^*}}{\partial x_m^*} - \frac{1}{\mathrm{Re}_l} \cdot \sum_{m=1}^{3} \frac{\partial \overline{\boldsymbol{G}_m^*}}{\partial x_m^*} + \sum_{m=1}^{3} \frac{\partial \boldsymbol{R}_m^*}{\partial x_m^*} = 0.$$

Explain the difference between these equations and the laminar form of the conservation equations

$$\frac{\partial \boldsymbol{U}^*}{\partial t^*} + \sum_{m=1}^{3} \frac{\partial \boldsymbol{f}_m^*}{\partial x_m^*} - \frac{1}{\mathrm{Re}_l} \cdot \sum_{m=1}^{3} \frac{\partial \boldsymbol{G}_m^*}{\partial x_m^*} = 0,$$

and explain the necessity of the turbulence modeling.

5.6

Starting from the general form of the multiphase flow equations (5.169), derive the one-dimensional two-fluid model equations (8.14)–(8.19). In particular, express the constitutive variables in these equations using the area and line-averaged primitive variables. Using the table in Section 5.4.6, reduce the surface integrals to line integrals in a cross-section.

5.7

What is the rate of formation (change in concentration per unit time) of species H for each of the following reactions? Assume that the rate constants $(k_1, k_2, \ldots)$ are known:

$$\begin{aligned}
\mathrm{H} + \mathrm{H} + \mathrm{M} &\xrightarrow{k_1} \mathrm{H_2} + \mathrm{M}, \\
\mathrm{H_2} + \mathrm{M} &\xrightarrow{k_2} \mathrm{H} + \mathrm{H} + \mathrm{M}, \\
\mathrm{H} + \mathrm{H} + \mathrm{H} &\xrightarrow{k_3} \mathrm{H_2} + \mathrm{H}, \\
\mathrm{H} + \mathrm{O_2} + \mathrm{M} &\xrightarrow{k_4} \mathrm{OH} + \mathrm{O}, \\
\mathrm{H} + \mathrm{O_2} &\xrightarrow{k_5} \mathrm{HO_2} + \mathrm{M}.
\end{aligned}$$

For an elementary reaction r given by

$$\sum_{s=1}^{S} \nu_{rs}^{(a)} A_s \xrightarrow{k_r} \sum_{s=1}^{S} \nu_{rs}^{(p)} A_s,$$

the rate of formation of species i in the reaction r is

$$\left(\frac{\partial c_i}{\partial t}\right)_{\mathrm{chem},r} = k_r \left(\nu_{ri}^{(p)} - \nu_{ri}^{(a)}\right) \prod_{s=1}^{S} c_s^{\nu_{rs}^{(a)}},$$

where $\nu_{rs}^{(a)}$ and $\nu_{rs}^{(p)}$ are the stoichiometric coefficients of reactants and products, and c_s are the concentrations of the species s $(s = 1, \ldots, S)$.

This leads to the rate of formation

$$\frac{\mathrm{d[H]}}{\mathrm{d}t} = k_1(0 - 2)\mathrm{[H][H][M]} = -2k_1\mathrm{[H]}^2\mathrm{[M]}.$$

5.8

Formulate the flow differential equations of the stability analysis in general form, so that they can also be applied to multiphase flows and flows with chemical reactions.
Write the conservation equations in the form

$$\boldsymbol{N}_I\left(\frac{\partial}{\partial t}\boldsymbol{U}\right) + \boldsymbol{N}_S(\boldsymbol{U}) = 0.$$

Here $\boldsymbol{U}$ is the solution vector, $\boldsymbol{N}_I$ acts on the unsteady terms in the conservation equations, $\boldsymbol{N}_S$ represents the nonlinear differential expression of the steady terms.
(a) With the flow ansatz

$$\boldsymbol{U} = \boldsymbol{U}_0 + \epsilon \cdot \boldsymbol{u}',$$

transform the conservation equations into the flow-differential equations by expanding $\boldsymbol{N}_I$ and $\boldsymbol{N}_S$ as Taylor series in ϵ.
Using the linear differential expressions

$$\boldsymbol{L}_I(\frac{\partial}{\partial t}\boldsymbol{u}')=\boldsymbol{N}_I(\frac{\partial}{\partial t}\boldsymbol{u}')_{\epsilon=0},$$

$$\boldsymbol{L}_S(\boldsymbol{u}')=\left(\frac{\mathrm{d}}{\mathrm{d}\epsilon}\boldsymbol{N}_S\right)_{\epsilon=0},$$

note that for infinitesimal perturbations ($\epsilon \rightarrow 0$),

$$\boldsymbol{L}_I(\frac{\partial}{\partial t}\boldsymbol{u}')+\boldsymbol{L}_S(\boldsymbol{u}') = 0.$$

(b) Apply this formal procedure of stability analysis to the dissipation function $\Phi - \Phi_0$ from Section 5.5. The same result is obtained

$$\Phi_\epsilon{}' = \mu_\epsilon{}' \cdot \left(\frac{1}{2} \cdot (\nabla \boldsymbol{U}_0 + {}^t\nabla \boldsymbol{U}_0)^2 - \frac{2}{3} \cdot (\nabla \cdot \boldsymbol{U}_0)^2\right)$$

$$+ 2 \cdot \mu_0 \cdot \left(\frac{1}{2} \cdot (\nabla \boldsymbol{U}_0 + {}^t\nabla \boldsymbol{U}_0) \cdot (\nabla \boldsymbol{u}' + {}^t\nabla \boldsymbol{u}') - \frac{2}{3} \cdot (\nabla \cdot \boldsymbol{U}_0) \cdot (\nabla \cdot \boldsymbol{u}')\right).$$

6. Instabilities and Turbulent Flows

6.1 Fundamentals of Turbulent Flows

When a viscous fluid flows through long straight tubes at reasonably high speeds, the Hagen–Poiseuille law (4.45), according to which the pressure drop is linearly proportional to the volume of fluid flowing through the pipe, is replaced by another law, in which the pressure drop is significantly greater, and almost proportional to the square of the volume flow rate of fluid. At the same time it is found that the flow field, which is smooth and straight (or *laminar*) in the Hagen–Poiseuille regime, becomes at higher velocities full of irregular eddying motions (or *turbulent*). This may be seen clearly in the case of a fluid flowing through glass tubes if a dye is introduced through a small injector at the inlet (Figures 6.1, 4.52). The colored filament is straight and smooth for low speeds but breaks off and disperses almost uniformly when turbulence develops. As a second example, introduced in Chapter 1, consider a jet of water that emerges from a circular orifice into a tank of still water. At

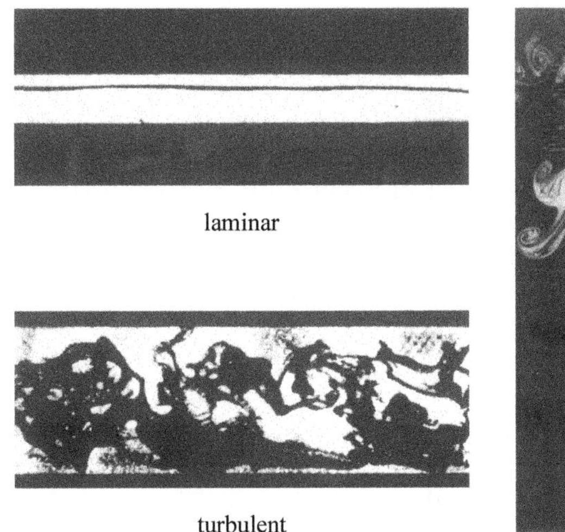

Laminar and turbulent pipe flow

Turbulent jet of water

Fig. 6.1. Laminar and turbulent flows

H. Oertel (ed.), *Prandtl-Essentials of Fluid Mechanics*,
Applied Mathematical Sciences 158, DOI 10.1007/978-1-4419-1564-1_6,

very low speeds of the fluid the jet is smooth and steady. For higher speeds, it develops swirls of various sizes amidst avalanches of complexity.

The two figures, being static, do not do justice to the dynamical interactions occurring within the flow. Observation suggests that parcels of fluid get stretched, folded, and tilted as they evolve, in turn losing shape by agglomeration and breakup, while new ones are constantly being created. This evolution and development of the flow does not repeat itself in full detail. Together, these features have a profound influence on the ability of the turbulent flow to transport heat, mass, and momentum. Under suitable conditions, turbulence occurs in such varied flow configurations as boundary-layers, wakes behind objects, thermal convection, and geophysical and astrophysical flows. The turbulence in each of these contexts is different in detail but similar in its function.

As a practical matter, turbulence plays an important role in technology and control phenomena such as weather and climate that have a large effect on human activities. Without turbulence, the mixing of air and fuel in an automobile engine would not occur on useful time scales. The transport and dispersion of heat, pollutants, and momentum in the atmosphere and oceans would be far weaker. In short, life as we know it would not be possible on Earth. Turbulence also has undesirable consequences. It increases energy consumption of pipelines, aircraft, ships, and automobiles and is an aspect to be reckoned with in air-travel safety, and it distorts the propagation of electromagnetic signals, and so forth. A major goal of a turbulence practitioner is the prediction and control of the effects of turbulence in various applications such as industrial mixers and burners, nuclear reactors, aircraft intakes, around ships, and inside of rocket nozzles. A major goal of a physicist working in turbulence is to understand the dynamical origin of this complexity, describe and quantify its features, and understand the universal properties embedded in features that are specific to a flow. A larger goal is to understand whether the statistical complexity of turbulence is shared in a serious way by other phenomena such as granular flows, fracture, and earthquakes.

In summary, then, turbulence is a rich problem both as a paradigm of spatiotemporal complexity and as a matter of practical importance. There are three major aspects to be considered: the origin of turbulence, the phenomena of flows in which turbulence is already developed, and the control of turbulence in a given situation.

6.2 Onset of Turbulence

During the last 120 years or so a great deal of ingenuity has been expended, on both mathematical and experimental fronts, on answering the question of how turbulence arises, and a reasonable picture has emerged, at least in some instances (see Section 4.2.4). Qualitatively, the transition from the laminar to the turbulent state occurs if the momentum exchange by molecular transport cannot compete sufficiently effectively with the transport due to macroscopic fluctuations in flow velocity. Making use of the ideas of dynamic similarity, *O. Reynolds* (1883), (1894) argued that the transition from the laminar to the turbulent state occurs when a dimensionless parameter, now bearing his name, exceeds a certain critical value. The Reynolds number (4.51) is defined as Ul/ν, where U is a characteristic velocity of the flow, l its characteristic size, and ν the kinematic viscosity of the fluid.

The situation is more complex than was originally presumed by Reynolds. For instance, the numerical value of the critical Reynolds number depends on the flow and a number of other factors such as the initial disturbance level (besides the obvious dependence on the precise definitions selected for the velocity and length scales). The notion that flows are laminar and stable up to a certain critical Reynolds number, becoming turbulent thereafter, turns out to be somewhat naive in practice.

6.2.1 Fluid-Mechanical Instabilities

A generic case of instability to consider in a carefully prepared experiment is one in which the perturbations are small. This idea has prompted a vast development of linear stability theory, the theory that calculates the Reynolds number at which laminar motion becomes unstable to small perturbations. Starting with *Lord Rayleigh* in the 1880s, *O. Reynolds* (1883), *W. M. F. Orr* (1907), *A. Sommerfeld* (1908), *G. I. Taylor* (1923), *W. Heisenberg* (1924), *C. C. Lin* (1955), *S. Chandrasekhar* (1961), and others (see, for example, *P. G. Drazin and W. H. Reid* (1981), *H. Oertel Jr. and J. Delfs* (1996), (2005) for details) have made lasting contributions to the subject.

Since the instabilities grow only at relatively high Reynolds numbers (or equivalently, small viscosities), it appears reasonable at first to treat the problem as essentially inviscid. Indeed, inviscid instability is often able to explain certain observations concerning the behavior of fluids with finite viscosity. This turns out to be the case particularly for flows for which the maximum vorticity occurs within the bulk of the fluid instead of on the boundaries. An excellent example is the so-called mixing layer, the flow formed when two parallel streams with different velocities come together (see *A. Michalke* (1970)).

Inviscid instability yields implausible answers for certain other flows. For instance, the theory yields the result that the flow between two parallel plates, one of which is stationary while the other moves with finite velocity, called plane Couette flow, is stable at all Reynolds numbers. Experiments, on the other hand, show that the flow does indeed become unstable at some finite Reynolds number on the order of a thousand (when based on the velocity of the moving plate and the distance between the plates). This phenomenon is puzzling at first sight because, if a flow is stable in the absence of viscosity, the additional damping provided by viscosity may be thought reasonably to make it even more stable, not less so. However, viscosity plays a subtle role, as explained by *W. Tollmien* (1929), and more fully by *C. C. Lin* (1955), and can promote instability (see *P. G. Drazin and W. H. Reid* (1981)).

These issues are best explained for the case of a boundary-layer on a thin flat plate, for which extensive literature is available (see Section 4.2.4). This is an important flow in practice because it will be seen that turbulence often arises within a boundary-layer. To study the initial growth of the perturbation in the boundary-layer of a viscous fluid, *W. M. F. Orr* (1907) and *A. Sommerfeld* (1908) derived from the Navier–Stokes equations a linear differential equation (4.73) that is now named after them. The solutions of this equation are of the form shown in Figures 6.2 and 4.57. Inside the neutral curve ($\omega_\mathrm{i} = 0$), the two-dimensional wave perturbations are unstable ($\omega_\mathrm{i} > 0$), and outside, they are stable ($\omega_\mathrm{i} < 0$). In regions of instability, the perturbations grow exponentially with time if they are spatially homogeneous. The perturbations grow exponentially with space if introduced at some point in space and allowed to grow as they propagate, or in both space and time if the perturbations are in the form of a wave packet.

Further investigation shows that a second characteristic layer is formed at the position in the flow where the velocity of the main flow is the same as the phase velocity of the oscillation. In the absence of friction this would lead to singularities in the motion of fluid particles, since they are subject to the same pressure gradient for a very long time. However, if viscosity is postulated in this second layer also, then the disturbance is free from singularities. With the presence of viscosity, the phase displacement of longitudinal motion produces a damping effect, which, in conjunction with the amplification due to the secondary boundary-layer, gives a critical value for the Reynolds number. Here we have only hinted at the basic physics, but it was the notable achievement of *W. Tollmien* (1929) to carry out the calculation needed to compute the critical Reynolds number.

The so-called *Tollmien–Schlichting waves* are spatially amplified downstream. Via several intermediate states in the transition regime, the state of fully developed turbulence is reached, as described in Section 4.2.5. Above a second critical Reynolds number, plane Tollmien–Schlichting waves initially become unstable to cross-wave perturbations. Downstream, they form the so-called lambda structures with local shear layers in the boundary-layer. It is

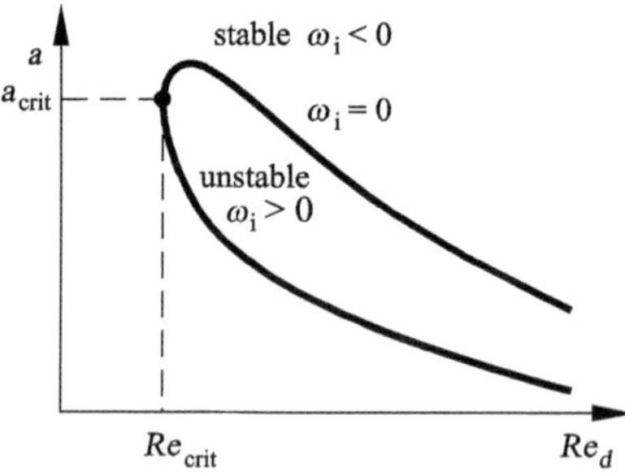

Fig. 6.2. Stability diagram of Tollmien-Schlichting waves in a flat plate boundary-layer

only when these shear layers lose their identity that the turbulent boundary-layer flow is fully developed (see Figures 6.3 and 4.56).

In *three-dimensional boundary-layers*, such as those that occur on a swept wing of a civil aircraft, the cross flow along the wing can also lead to further instabilities as well as the Tollmien–Schlichting transition. These *cross-flow instabilities* occur downstream from the stagnation line of the wing. They form traveling waves and a steady vortex pattern along the cross-flow component of the three-dimensional boundary-layer. This pattern decays with the same mechanisms as the Tollmien–Schlichting transition and passes over to the turbulent boundary layer close to the stagnation line.

There are other flows for which the linear stability theory gives excellent results for the loss of stability. This loss of stability is often expressed in terms of dimensionless parameters that are related to a suitably defined Reynolds number. For instance, the theory predicts rather well the so-called Taylor number at which the flow between concentrically rotating cylinders loses stability and begins to form toroidal vortices (*G. I. Taylor* (1923)). The Taylor number is the square of the Reynolds number based on the angular velocity of the rotating cylinder, the gap between the two cylinders, and the viscosity.

The theory similarly predicts well the so-called Rayleigh number (*Lord Rayleigh* (1916)) at which the heat transfer changes from a steady conductive case to a structured form involving hexagonal or roll patterns (Figure 1.5, see also Section 7.2.1). The Rayleigh number is a measure of the ratio of the effect of buoyancy, which tends to accelerate a fluid parcel against gravity, to the viscous and diffusive effects that tend to slow it down. For the fluid between a pair of infinitely extended horizontal plates, with the bottom plate heated and the top plate cooled, the heat transport ceases to be purely conductive at $\mathrm{Ra} = 1708$. In engineering literature on so-called free convection problems, the Grashof number $\mathrm{Gr} = \mathrm{Ra} \cdot \mathrm{Pr}$ is often used, where the Prandtl number is $\mathrm{Pr} = \nu/\kappa$, κ being the thermal diffusivity of the fluid.

In a broad class of flows, a few of which were just mentioned, the loss of linear stability of the laminar state is a significant first step in the formation of turbulence. The next step in the process of complexity is the nonlinear stage, at which the perturbations have grown to a sufficiently large amplitude at which they begin to interact with the mean flow and cease to grow exponentially as a result.

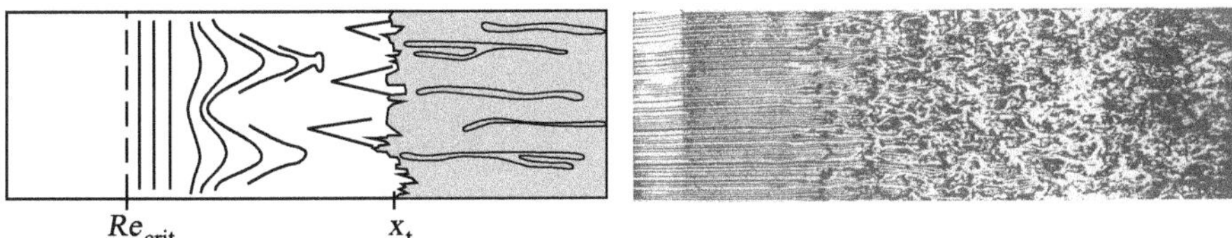

Fig. 6.3. Transition process in a flat plate boundary-layer

From the variety of fluid mechanical instabilities the *boundary-layer instabilities* are considered in some detail here. The classical linear stability analysis of two- and three-dimensional boundary-layer follows. A more detailed summary of the theory of fluid mechanical instabilities is given in the second edition of Prandtl-Essentials of Fluid Mechanics (*H. Oertel* (2004)) and in the 11th German edition (*H. Oertel* (2002)).

6.2.2 Linear Stability Analysis

The definition of *fluid mechanical instability* depends on whether one considers *temporal* or *spatial perturbation development.* Figure 6.4 shows the steady laminar convection flow as an example on a vertical, heated plate. The flow field is perturbed with a harmonic periodic perturbation wave w' with small amplitude:

$$w'(x,z,t) = \hat{w}(x) \cdot \exp(\mathrm{i} \cdot a \cdot z - \mathrm{i} \cdot \omega \cdot t). \tag{6.1}$$

For a given wavelength $\lambda = 2 \cdot \pi / a$, the laminar initial state is regarded as *temporally unstable* with respect to this wavelength if the flow causes the wave amplitude to be amplified in time ($\mathrm{Im}(\omega(a)) > 0$). If the perturbing wave is damped in time ($\mathrm{Im}(\omega(a)) < 0$), the laminar initial flow is *temporally stable* with respect to the given wavelength. The *temporally neutral* state is the limiting case of a temporally constant perturbation amplitude. Instead of temporal perturbation development, the concept of stability can also be

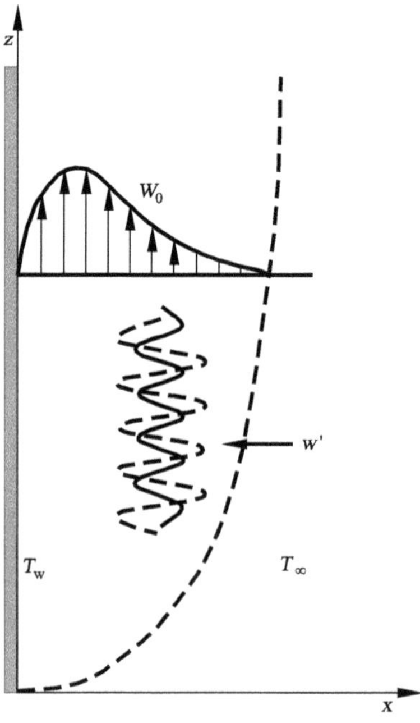

Fig. 6.4. The concept of stability in fluid mechanics: thermal convection at a vertical wall boundary-layer with $T_\mathrm{w} > T_\infty$

defined with respect to the purely spatial (ω real, a complex), or, more generally, the spatiotemporal (ω, a complex) development of perturbations. In the latter case, the division into so-called absolute and convective instabilities is convenient. A *convective instability* is present when the temporally amplifying perturbation energy moves downstream with the flow. On the other hand, if the perturbation remains in one place, the instability is *absolutely unstable.*

In the mathematical definition of stability, we assume a steady flow state $\boldsymbol{U}_0 = (x, y, z)$, which is completely defined by, for example, its dimensionless density distribution ρ_0, temperature distribution T_0, and the three components of the velocity vector (u_0, v_0, w_0) at each spatial position (x, y, z). The state $\boldsymbol{U}_0 = (\rho_0, u_0, v_0, w_0, T_0)$ satisfies the fluid-mechanical equations. The question is whether further solutions, that is, additional equilibrium states of the system, exist. In order to be able to answer this question, we disturb the flow state $\boldsymbol{U}_0$ out of its basic state with a small perturbation $\boldsymbol{u}'(x, y, z, t)$. This disturbance must be physically possible; i.e., the new flow state $\boldsymbol{u}(x, y, z, t)$ occurring at the time $t = 0$ must satisfy the boundary conditions of the flow problem.

We obtain the ansatz introduced in Section 4.2.4, equation (4.67):

$$\boldsymbol{u}(x, y, z, t) = \boldsymbol{U}_0(x, y, z) + \boldsymbol{u}'(x, y, z, t). \tag{6.2}$$

The size of the perturbation is introduced with

$$|\boldsymbol{u}'| = \int_V |\boldsymbol{u}'(x, y, z)^2| \cdot \mathrm{d}V. \tag{6.3}$$

This is a measure of the deviation of the perturbed flow $\boldsymbol{u}$ from the basic flow $\boldsymbol{U}_0$ in the entire flow field V. The quantity $|\boldsymbol{u}'|$ will be called the *perturbation energy* in the flow field.

The basic flow is stable as long as the size of a perturbation remains smaller than a given number ϵ for all times $t \geq 0$:

$$|\boldsymbol{u}'|_t < \epsilon \quad \text{with} \qquad t \geq 0, \tag{6.4}$$

for all initial perturbations $\boldsymbol{u}'(x, y, z, t = 0)$ with perturbation energy smaller than a constant. Otherwise, the basic flow is unstable. Figure 6.5 shows examples that can be divided into stable and unstable flows by applying the above definition to the temporal behavior of the perturbation energy in the flow. On the fundamental flows $\boldsymbol{U}_0$ we superimpose various initial perturbations, e.g. $\boldsymbol{u}_1'(t = 0)$, $\boldsymbol{u}_2'(t = 0)$, $\boldsymbol{u}_3'(t = 0)$, $\boldsymbol{u}_4'(t = 0)$. We note that, of the infinitely many possible perturbations, there are those excited in unstable flow that then die away over time, such as the perturbation $\boldsymbol{u}_3'(t = 0)$. In general, flows $\boldsymbol{U}_0$ are investigated for *asymptotic stability*, which is then present when any given perturbation dies away over time:

$$\lim_{t \to \infty} |\boldsymbol{u}'(t)| = 0. \tag{6.5}$$

In this case the perturbed system again takes on its temporally asymptotic initial state $\boldsymbol{U}_0$. This case is sketched in Figure 6.6.

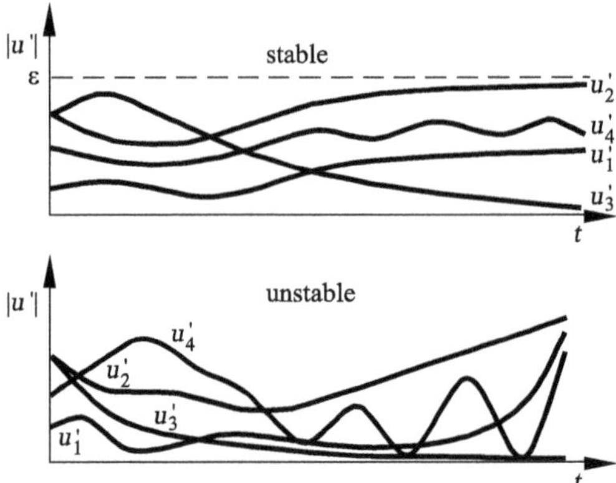

Fig. 6.5. The definition of stability

Note that the definition of stable and unstable flow is not a statement about the spatiotemporal expansion of unstable perturbations.

To clarify the problem, we compare two unstable basic flows $\boldsymbol{U}_0$, which have qualitatively different behaviors after the onset of the perturbation. Under the idealizing assumption of freedom from perturbations, a steady wake behind a body in the flow could be generated even for supercritical Reynolds numbers, so that no Kármán vortex street would occur, in contrast to the situation in Figure 6.7. Similarly, an ideal perturbation-free longitudinal flow past a flat plate would be laminar, although unstable, even at supercritical Reynolds numbers.

In the example of the wake flow, if a local perturbation is quickly placed close to the steady wake region of the body at time t_0, over time, a Kármán vortex street will form. Such a perturbation in the unstable plate boundary-layer flow behaves qualitatively quite differently. The size of the perturbation also grows here, but the perturbation simultaneously moves downstream, as in the sketch. Clearly, the instability in the wake flow leads to a self-excited oscillation of the system at a fixed position while, in the boundary-layer flow, perturbations at a fixed position vanish over time. Perturbation energy at a fixed position can be observed here only if continuous perturbation energy is introduced upstream, from outside the system.

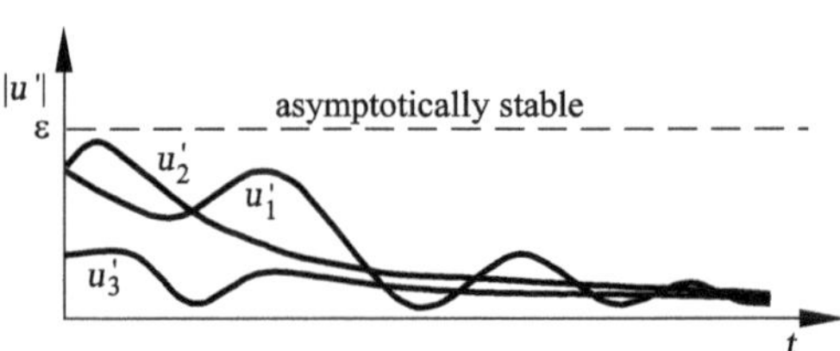

Fig. 6.6. Behavior of perturbations in asymptotically stable flow

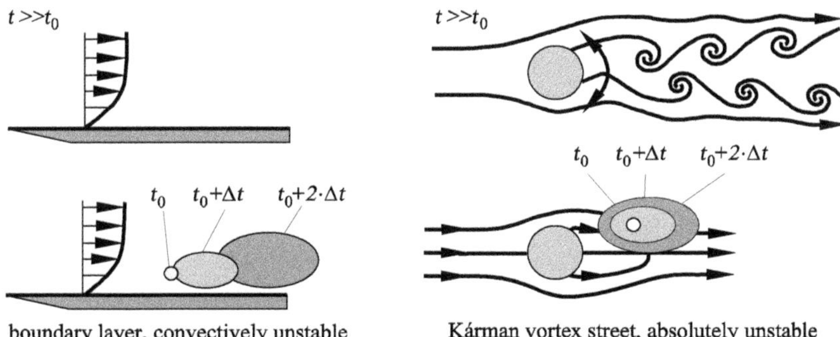

Fig. 6.7. Expansion of unstable perturbations in convective and absolute instability

In order to be able to make a statement regarding the spatial behavior of the perturbation, we clearly have to introduce a measure for the local size of the perturbations. To do this, we shrink the region of integration V to one small region. This shrinking is carried out until the region of integration has been reduced to an infinitesimally small size $\mathrm{d}V$. From (6.3) we then have

$$\mathrm{d}|\boldsymbol{u}'| = |\boldsymbol{u}'|^2 \mathrm{d}V.$$

Dividing by the volume element $\mathrm{d}V$, we obtain a perturbation energy density A with

$$A(x, y, z, t) = \frac{\mathrm{d}|\boldsymbol{u}'|}{\mathrm{d}V} = |\boldsymbol{u}'|^2, \tag{6.6}$$

which will be defined in what follows as a measure of the size of the perturbation at position x, y, z at time t. If the perturbation energy density A in an initially perturbation-free unstable flow dies away time-asymptotically at the position where the perturbation was introduced, this flow is *convectively unstable*. Otherwise, the flow is *absolutely unstable*. The wake flow shown in Figure 6.7 is therefore absolutely unstable, while the plate boundary-layer is convectively unstable.

Boundary-Layer Instabilities

The description of the laminar–turbulent transition in boundary-layers was first met in Section 4.2.4. In the *plate boundary-layer*, the instability occurs with two-dimensional *Tollmien–Schlichting waves* at the critical Reynolds number $\mathrm{Re}_{x,\mathrm{crit}} = 5 \cdot 10^5$ or, with $d = \sqrt{\nu \cdot x/U_\delta}$ at $\mathrm{Re}_{d,\mathrm{crit}} = 302$, which corresponds to the displacement thickness Reynolds number $\mathrm{Re}_{\delta^*,\mathrm{crit}} = 520$. The wave fronts are shown in Figure 4.55 and 6.3. The primary perturbation amplitudes grow downstream, and so the flow in this region becomes unstable to three-dimensional *secondary perturbations* (region (2) in Figure 6.8). The vortex lines are deformed in a wavelike shape. Further downstream, the vortex tubes deformed with the vortex lines are stretched and form the *lambda* *structures* (3). The subsequent decay of these structures and the spatial and temporal irregular appearance of quickly growing *turbulent spots* (4) completes the transition process at position x_t, called the *position of complete transition.* Following this is the developed turbulent state (5). Even fully developed turbulence is not without structure, since longitudinal stripe-shaped regions with greatly reduced downstream components of the velocity (*streaks*) are observed close to the wall. Other structures also exist.

Throughout the entire transition process (1)–(4), there is a significant increase in the thickness of the boundary-layer. This is because the ever growing perturbation amplitudes, particularly the vertical oscillations, result in distributing the time average of the downstream momentum more evenly within the boundary-layer. The greatest oscillation intensity initially takes place directly at the surface, causing the time-averaged wall shear stress in the transition regime to take on an even higher value than that in the region of full turbulence. Note particularly that the transition described does not take place at one position, but rather over an extended distance $x_{\text{crit}} < x < x_t$.

The unstable primary perturbation (1) of the laminar flow (0) causes lasting change to the flow field only downstream of the critical position x_{crit}. Upstream of this position, the flow remains laminar. If a local perturbation is introduced into the boundary-layer at a point $x > x_{\text{crit}}$, the perturbing wave packet expands downstream with a characteristic velocity and simultaneously disintegrates, while the perturbation intensity due to the instability grows. If such an unstable wave packet does not continue to affect the original position of the perturbation, the instability is convectively unstable (see Figure 6.7). Thus, the primary perturbation of the boundary-layer is convectively unstable. Therefore, the turbulence does not occur abruptly, as for the Taylor instabilities, but rather develops within a transition region that extends downstream. The onset of the Tollmien–Schlichting waves in the two-dimensional plate boundary-layer was described in Section 4.2.4 as an eigenvalue problem of the Orr–Sommerfeld equation (4.73). The stability diagram and the critical Reynolds number Re_{crit} and wave number a_{crit} are shown in Figure 4.2.4 for the Blasius boundary-layer flow. In the following section we extend the stability analysis to three-dimensional perturbations.

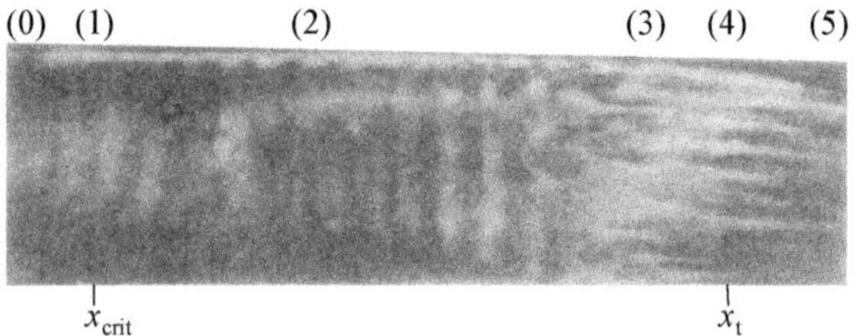

Fig. 6.8. Transition process in the boundary-layer of a rotationally symmetric body, *F. N. M. Brown* (1957)

The *stability analysis* begins with the determination of the *basic flow*. Usually, it consists of solving the Navier–Stokes or boundary-layer equations by numerical methods.

In the stability analysis of boundary-layers the increase of the boundary-layer thickness δ in the downstream direction x has to be considered. The flow quantities are therefore not only dependent on the position z in the normal direction on the boundary-layer, but also on x and y. Therefore, as well as z, x and y are also inhomogeneous directions.

However, if boundary-layer flows are considered in the large Reynolds number regime, the boundary-layer thickness $\delta(x, y)$ typically varies only moderately (for example, in the case of a plate, $\delta \sim x/\sqrt{\mathrm{Re}_x}$), and so the dependence of the flow velocity on x, y is considerably weaker than that on z.

It is known from experimental results that the dependence of the *perturbations* on the parallel directions x, y is, in contrast to the basic flow, not at all weak. All perturbation velocities are referred to the free stream velocity U_∞, the lengths to δ, and the perturbation pressure to ρU_∞^2. The perturbation ansatz for the *incompressible boundary-layer* reads

$$
\begin{aligned}
u &= U_\infty \cdot (u_0(\overline{x}, \overline{y}, z) + \epsilon \cdot u') \quad , \\
v &= U_\infty \cdot (v_0(\overline{x}, \overline{y}, z) + \epsilon \cdot v') \quad , \\
w &= U_\infty \cdot (\epsilon \cdot w_0(\overline{x}, \overline{y}, z) + \epsilon \cdot w') \quad , \\
p &= \rho \cdot U_\infty^2 \cdot (p_0(\overline{x}, \overline{y}, z) + \epsilon \cdot p') \quad .
\end{aligned}
\tag{6.7}
$$

Here, ϵ is a suitable expansion parameter, chosen for the boundary layer after careful consideration to be $\epsilon = 1/\mathrm{Re}_\delta$. The problem depends on two different length scales, namely, a long scale $\overline{d} = \delta/\epsilon$ and the much shorter scale δ. Because these scales are so different, it is appropriate to formulate the general dependence of the solution on x or y as separate dependencies on both long-scale variables $\overline{x}$ and $\overline{y}$ and on short-scale variables $\tilde{x}$ and $\tilde{y}$. This method is called the *method of multiple scales*. Their relation to the original variable x or y is found as follows:

$$
\begin{aligned}
\tilde{x} &= x \quad , \qquad \overline{x} = \epsilon \cdot x \quad , \\
\tilde{y} &= y \quad , \qquad \overline{y} = \epsilon \cdot y \quad .
\end{aligned}
\tag{6.8}
$$

It is understood that all perturbation quantities are functions of both variables, for example $u' = u'(t, x, y, z) = u'(t, \tilde{x}, \overline{x}, \tilde{y}, \overline{y}, z)$. Derivatives with respect to x are then written in the form $\partial u'/\partial x = (\partial u'/\partial \tilde{x}) \cdot \mathrm{d}\tilde{x}/\mathrm{d}x + (\partial u'/\partial \overline{x}) \cdot \mathrm{d}\overline{x}/\mathrm{d}x = \partial u'/\partial \tilde{x} + \epsilon \cdot \partial u'/\partial \overline{x}$.

This yields the linearized *perturbation differential equations*

$$
\frac{\partial u'}{\partial x} + \frac{\partial v'}{\partial y} + \frac{\partial w'}{\partial z} = 0 \quad , \tag{6.9}
$$

$$\frac{\partial u'}{\partial t} + u_0 \cdot \frac{\partial u'}{\partial x} + v_0 \cdot \frac{\partial u'}{\partial y} + \frac{\mathrm{d}u_0}{\mathrm{d}z} \cdot w' + \frac{\partial p'}{\partial x} - \frac{1}{\mathrm{Re}_d} \cdot \left(\frac{\partial^2 u'}{\partial x^2} + \frac{\partial^2 u'}{\partial y^2} + \frac{\partial^2 u'}{\partial z^2} \right) = 0 \quad , \tag{6.10}$$

$$\frac{\partial v'}{\partial t} + u_0 \cdot \frac{\partial v'}{\partial x} + v_0 \cdot \frac{\partial v'}{\partial y} + \frac{\mathrm{d}v_0}{\mathrm{d}z} \cdot w' + \frac{\partial p'}{\partial y} - \frac{1}{\mathrm{Re}_d} \cdot \left(\frac{\partial^2 v'}{\partial x^2} + \frac{\partial^2 v'}{\partial y^2} + \frac{\partial^2 v'}{\partial z^2} \right) = 0 \quad , \tag{6.11}$$

$$\frac{\partial w'}{\partial t} + u_0 \cdot \frac{\partial w'}{\partial x} + v_0 \cdot \frac{\partial w'}{\partial y} + \frac{\partial p'}{\partial z} - \frac{1}{\mathrm{Re}_d} \cdot \left(\frac{\partial^2 w'}{\partial x^2} + \frac{\partial^2 w'}{\partial y^2} + \frac{\partial^2 w'}{\partial z^2} \right) = 0 \quad . \tag{6.12}$$

It is essential that the coefficients, e.g. $u_0(\overline{x}, \overline{y}, z)$, of this homogeneous linear system of partial differential equations in the variables t, $\tilde{x}$, $\tilde{y}$, z depend only on the variables $\overline{x}$, $\overline{y}$, z and not on the small-scale variables $\tilde{x}$, $\tilde{y}$. It can be seen that no explicit derivatives with respect to $\overline{x}$ or $\overline{y}$ appear in (6.9)–(6.12). Within the framework of the above approximation, the solution of the system of differential equations is therefore only algebraically dependent on the spatial variables $\overline{x}$, $\overline{y}$ and not differentially dependent. This stability analysis is then called a *local stability analysis.* The constant basic solution with respect to the short-scale parallel coordinates $\tilde{x}$, $\tilde{y}$ is given at the fixed selected position $\overline{x}$, $\overline{y}$, and the stability analysis is carried out locally here. We also note that the perturbation differential equation is homogeneous in t, $\tilde{x}$, and $\tilde{y}$.

In deriving the perturbation differential equations, the dependence on the normal component w_0 of the basic flow drops away. This is called the *parallel flow assumption.* Its validity was confirmed by *T. Herbert* and *F. P. Bertolotti* (1987) for the plate boundary-layer flow.

The perturbations satisfy the *boundary conditions*

$$u'(x, y, z = z_\mathrm{w}, t) = v'(x, y, z = z_\mathrm{w}, t) = w'(x, y, z = z_\mathrm{w}, t) = 0 \quad , \tag{6.13}$$

at the wall $z = z_\mathrm{w}$, and additionally, the far-field boundary conditions

$$v'(x, y, z \to \infty, t) = 0 \quad , \qquad p'(x, y, z \to \infty, t) = 0 \quad . \tag{6.14}$$

The system of perturbation differential equations (6.9)–(6.12) is homogeneous in $\tilde{x}$, $\tilde{y}$, and t. We can carry out a separation trial solution (wave ansatz)

$$\begin{pmatrix} \tilde{u}'(\tilde{x}, \tilde{y}, z, t; \overline{x}, \overline{y}) \\ \tilde{v}'(\tilde{x}, \tilde{y}, z, t; \overline{x}, \overline{y}) \\ \tilde{w}'(\tilde{x}, \tilde{y}, z, t; \overline{x}, \overline{y}) \\ \tilde{p}'(\tilde{x}, \tilde{y}, z, t; \overline{x}, \overline{y}) \end{pmatrix} = \mathrm{F}_x(\tilde{x}; \overline{x}, \overline{y}) \cdot \mathrm{F}_y(\tilde{y}; \overline{x}, \overline{y}) \cdot \mathrm{F}_t(t; \overline{x}, \overline{y}) \cdot \begin{pmatrix} \hat{u}(z; \overline{x}, \overline{y}) \\ \hat{v}(z; \overline{x}, \overline{y}) \\ \hat{w}(z; \overline{x}, \overline{y}) \\ \hat{p}(z; \overline{x}, \overline{y}) \end{pmatrix} , \tag{6.15}$$

because the boundary conditions depend only on z. Inserting (6.15) into the continuity equation (6.9), we obtain

$$\left(\frac{1}{\mathrm{F}_x}\cdot\frac{\mathrm{dF}_x}{\mathrm{d}\tilde{x}}\right)\cdot\hat{u}+\frac{\mathrm{d}\hat{w}}{\mathrm{d}z}+\left(\frac{1}{\mathrm{F}_y}\frac{\mathrm{dF}_y}{\mathrm{d}\tilde{y}}\right)\cdot\hat{v}=0\quad,$$

where the two terms on the right are independent of $\tilde{x}$, and the two terms on the left are independent of $\tilde{y}$, so that the expressions in parentheses are each constants with respect to $\tilde{x}$ and $\tilde{y}$. The same procedure can be carried out with the function F_t. Inserting the separation ansatz into equation (6.12), we obtain

$$\frac{1}{\mathrm{F}_x}\cdot\frac{\mathrm{dF}_x}{\mathrm{d}\tilde{x}}=\mathrm{i}\cdot a(\overline{x},\overline{y}),\quad\frac{1}{\mathrm{F}_y}\cdot\frac{\mathrm{dF}_y}{\mathrm{d}\tilde{y}}=\mathrm{i}\cdot b(\overline{x},\overline{y}),\quad\frac{1}{\mathrm{F}_t}\cdot\frac{\mathrm{dF}_t}{\mathrm{d}t}=-\mathrm{i}\cdot\omega(\overline{x},\overline{y})\quad,$$

where the three separation parameters a, b, and ω have been introduced, and these are still functions of the long-scale variables. From the equations for F_x, F_y, and F_t it follows that

$$\begin{pmatrix}\tilde{u}'(\tilde{x},\tilde{y},z,t)\\\tilde{v}'(\tilde{x},\tilde{y},z,t)\\\tilde{w}'(\tilde{x},\tilde{y},z,t)\\\tilde{p}'(\tilde{x},\tilde{y},z,t)\end{pmatrix}=\exp(\mathrm{i}\cdot a\cdot\tilde{x}+\mathrm{i}\cdot b\cdot\tilde{y}-\mathrm{i}\cdot\omega\cdot t)\begin{pmatrix}\hat{u}(z)\\\hat{v}(z)\\\hat{w}(z)\\\hat{p}(z)\end{pmatrix}\quad,\tag{6.16}$$

where the dependence of the functions on $\overline{x}$ and $\overline{y}$ has not been indicated here. The exponent $a(\overline{x},\overline{y})\cdot\tilde{x}+b(\overline{x},\overline{y})\cdot\tilde{y}-\omega(\overline{x},\overline{y})\cdot t$ is also called the *phase*. The separation parameters a, b, and ω are initially any, generally complex, numbers.

Inserting the wave ansatz (6.16) into the system of equations (6.9)–(6.12), we obtain

$$a\cdot\hat{u}+b\cdot\hat{v}=\mathrm{i}\cdot\frac{\mathrm{d}\hat{w}}{\mathrm{d}z}\quad,\tag{6.17}$$

$$(a\cdot u_0+b\cdot v_0-\omega)\cdot\hat{u}-\mathrm{i}\cdot\frac{\mathrm{d}u_0}{\mathrm{d}z}\cdot\hat{w}=-a\cdot\hat{p}+\frac{i}{\mathrm{Re}_d}\cdot\left(a^2+b^2-\frac{\mathrm{d}^2}{\mathrm{d}z^2}\right)\hat{u},\tag{6.18}$$

$$(a\cdot u_0+b\cdot v_0-\omega)\cdot\hat{v}-\mathrm{i}\cdot\frac{\mathrm{d}v_0}{\mathrm{d}z}\cdot\hat{w}=b\cdot\hat{p}+\frac{i}{\mathrm{Re}_d}\cdot\left(a^2+b^2-\frac{\mathrm{d}^2}{\mathrm{d}z^2}\right)\hat{v},\tag{6.19}$$

$$(a\cdot u_0+b\cdot v_0-\omega)\cdot\hat{w}=\mathrm{i}\cdot\frac{\mathrm{d}\hat{p}}{\mathrm{d}z}+\frac{i}{\mathrm{Re}_d}\cdot\left(a^2+b^2-\frac{\mathrm{d}^2}{\mathrm{d}z^2}\right)\hat{w}.\tag{6.20}$$

With the boundary conditions (6.13) and (6.14),

$$\hat{u}(z=z_\mathrm{w})=\hat{v}(z=z_\mathrm{w})=0\quad,\qquad\hat{w}(z=z_\mathrm{w})=0\quad,\tag{6.21}$$

$$\hat{\boldsymbol{v}}(z\rightarrow\infty)=0\quad,\qquad\hat{p}(z\rightarrow\infty)=0\quad,\tag{6.22}$$

where we have formulated the eigenvalue problem for the *wave instabilities*. It is a linear system of homogeneous differential equations that contains the four parameters Re_d, a, b, and ω. The Reynolds number is given as a real number. Apart from the trivial solution, the system of equations is solvable only for certain a, b, and ω. This defines a mutual relation among these three relations, called the *dispersion relation*:

$$D(a, b, \omega) = 0 \quad . \tag{6.23}$$

In the *eigenvalue problem* described, two of the quantities a, b, and ω are given, and the remaining one is to be computed as an eigenvalue from the equations.

The *stability analysis* is concerned with the variation of the perturbation amplitude $|\boldsymbol{u}'|$ of a perturbation $\boldsymbol{u}'$ introduced into a flow $\boldsymbol{U}_0$. As seen in the beginning of the section, the stability is defined via the temporal amplification of the perturbation amplitudes. In boundary-layers the perturbations are represented as waves that run along the directions x and y:

$$\boldsymbol{u}'(x, y, z, t) = \boldsymbol{u}(z) \cdot \exp(\mathrm{i} \cdot a \cdot x + \mathrm{i} \cdot b \cdot y - \mathrm{i} \cdot \omega \cdot t). \tag{6.24}$$

The tilde above the x and y has again been left out for clarity. According to the definition of stability, an eigenform is given by the wave number components a and b, and the associated value $\omega = \omega_r + \mathrm{i} \cdot \omega_\mathrm{i}$ is computed from the eigenvalue problem. If spatially periodic waves (i.e. real $a = a_r$ and $b = b_r$) are given, the problem concerns the *temporal stability analysis.* Since the system can develop further only in the positive time direction, a wave perturbation with given $a = a_r$ and $b = b_r$ is then temporally unstable only if its amplitude is amplified in time, i.e. if $\omega_\mathrm{i} > 0$. Here ω_i is the *temporal amplification rate.* A perturbation for which $\omega_\mathrm{i} = 0$ holds is called an indifferent or neutral perturbation. The quantity ω may also be given and the associated eigenform (represented by a and b) computed. The problem becomes one of *spatial stability analysis* if $\omega = \omega_r$ is given as a real value (i.e. consideration of all possible waves with a given frequency), and, for example, a is computed for a given b. The real part a_r of the computed number a is then the wave number, and the imaginary part a_i is a measure for the spatial amplification in x. An explicit definition for spatial amplification is clearly obtained only when a direction of consideration is given. Let it be represented by the unit vector $\boldsymbol{e}_\phi = \boldsymbol{e}_x \cdot \cos(\phi) + \boldsymbol{e}_y \cdot \sin(\phi)$ (Figure 6.9).

The variation in the amplitude $|\boldsymbol{u}'| = |\hat{\boldsymbol{u}}| \cdot \exp(-a_\mathrm{i} \cdot x - b_\mathrm{i} \cdot y + \omega_\mathrm{i} \cdot t)$ of the wave is determined along the given direction ϕ as $\mathrm{d}|\boldsymbol{u}'|/\mathrm{d}x_\phi = \boldsymbol{e}_\phi \cdot \boldsymbol{\nabla}|\boldsymbol{u}'|$. It is found that $\mathrm{d}|\boldsymbol{u}'|/\mathrm{d}x_\phi = -(a_\mathrm{i} \cdot \cos(\phi) + b_\mathrm{i} \cdot \sin(\phi)) \cdot |\boldsymbol{u}'|$. The amplitude grows along $\boldsymbol{e}_\phi$ if $\mathrm{d}|\boldsymbol{u}'|/\mathrm{d}x_\phi$ is positive. The wave is amplified with respect to the direction ϕ if

$$a_\mathrm{i} \cdot \cos(\phi) + b_\mathrm{i} \cdot \sin(\phi) < 0 \quad .$$

The quantities a_i and b_i are also called *spatial amplification rates.* It is noted that the necessity to specify a direction ϕ is to a certain degree arbitrary. For this reason it is necessary to check whether the wave with the phase velocity vector $\boldsymbol{c} = (c_x, c_y, 0) = \omega_r/(a_r^2 + b_r^2) \cdot (a_r, b_r, 0)$ moves in the direction of increasing amplitude. The direction of consideration $\boldsymbol{e}_\phi$ is allowed to lie along the direction of motion of the wave $\boldsymbol{e}_{\text{crit}} = (a_r, b_r, 0) \cdot \text{sgn}(\omega_r)/\sqrt{a_r^2 + b_r^2}$, where $\text{sgn}(\omega_r) = \omega_r/|\omega_r|$ (cf. Figure 6.9). A temporally periodic wave experiences an increase in amplitude along its direction of motion if

$$\omega_r \cdot (a_r \cdot a_\text{i} + b_r \cdot b_\text{i}) < 0 \quad .$$

A two-dimensional wave ($b = 0$) can be called spatially amplified if for $\omega_r > 0$, the imaginary part satisfies $a_\text{i} < 0$. However, which wave actually contributes to the spatial amplification of perturbations can be answered precisely within the framework of the concept of the stability analysis of local perturbations for convective instabilities.

The eigenvalue problem can deliver either a for a given $b = b_r + \text{i} \cdot b_r$ and $\omega = \omega_r$ or b for a given $a = a_r + \text{i} \cdot a_r$ and $\omega = \omega_r$. Rather than specifying a complex wave number, it is clearer in the spatial analysis to determine, for example, the amplification $\phi = 1/\tan(b_\text{i}/a_\text{i})$. This corresponds to determining the ratio of the imaginary parts a_i and b_i of a and b.

We note that the temporal stability analysis is simpler to carry out than the spatial stability analysis. In the eigenvalue problem (6.17)–(6.20), ω appears only linearly, whereas a and b appear quadratically. The solution of a quadratic eigenvalue problem requires considerably more computational effort than the solution of a linear problem. Therefore, a method by which temporal amplifications could be transformed into spatial amplifications was examined. Such a relation was given by *M. Gaster* (1962) for $b = 0$. The transformation of the temporal amplification ω_i of a spatially periodic wave with given real wave number a_r and associated frequency ω_r to a temporally periodic wave (i.e. $\omega_\text{i} = 0$) with the same wave number a_r and frequency ω_r is performed using

$$a_\text{i} \approx -\frac{1}{\dfrac{\partial \omega_r}{\partial a_r}} \cdot \omega_\text{i} \quad .$$

This yields the spatial amplification of the wave from the temporal amplification of the associated wave using the *group velocity* $\partial \omega_r / \partial a_r$. The above relation is called the *Gaster transformation*. It is valid only for small ampli-

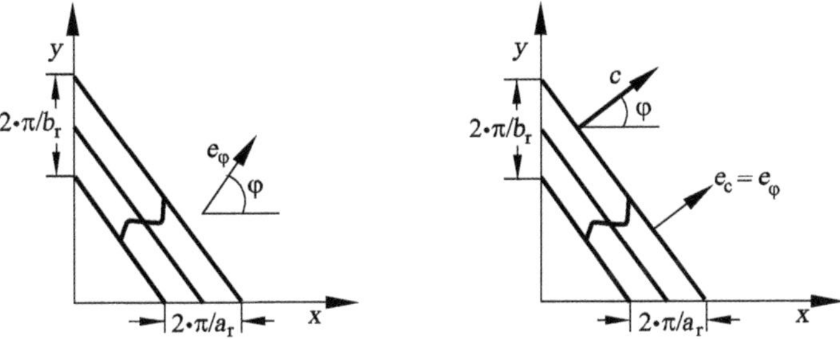

Fig. 6.9. Expansion of a wave perturbation

fication rates a_{i}, ω_{i}, since it is based on a Taylor expansion of the dispersion relation $D(a,\omega)=0$ about the neutral state $a_{\mathrm{i}}=0$, $\omega_{\mathrm{i}}=0$.

The system of perturbation differential equations (6.17)–(6.20) has a remarkable property. It can be summarized by a single fourth-order differential equation that represents an extension to the Orr–Sommerfeld equation (4.73) for obliquely traveling waves, with $\hat{u}$, $\hat{v}$, and $\hat{p}$ eliminated. Using the *Squire transformation*

$$a_\varphi \cdot u_{0,\varphi} = a \cdot u_0 + b \cdot v_0 \quad , \qquad a_\varphi^2 = a^2 + b^2 \quad ,$$

which represents a coordination rotation in the direction of expansion, we obtain the *Orr–Sommerfeld equation*

$$\left[(a_\varphi \cdot u_{0,\varphi} - \omega) \cdot \left(\frac{\mathrm{d}^2}{\mathrm{d}z^2} - a_\varphi^2\right) - a_\varphi \cdot \frac{\mathrm{d}^2 u_{0,\varphi}}{\mathrm{d}z^2} + \mathrm{i} \cdot \frac{1}{\mathrm{Re}_d} \cdot \left(\frac{\mathrm{d}^2}{\mathrm{d}z^2} - a_\varphi^2\right)^2\right] \hat{w} = 0 \quad , \tag{6.25}$$

with the following boundary conditions for $\hat{w}$:

$$\hat{w} = 0 \quad , \qquad \frac{\mathrm{d}\hat{w}}{\mathrm{d}z} = 0 \quad \text{for} \qquad z = z_{\mathrm{w}}, \tag{6.26}$$

$$\hat{w} = 0 \quad , \qquad \frac{\mathrm{d}\hat{w}}{\mathrm{d}z} = 0 \quad \text{for} \qquad z \to \infty \quad . \tag{6.27}$$

If in equation (6.25), a_φ is replaced by a, and $a_\varphi \cdot u_{0,\varphi}$ by $a \cdot u_0$, this represents the two-dimensional case (4.73). In Figure 6.10, the stability diagram has been supplemented by a typical eigenfunction. We point out that the vertical component $|\hat{w}|$ of the perturbation velocity has been enlarged 10 times. It is very small compared to the amplitude of the downstream component $|\hat{u}|$. The largest perturbation amplitudes for $\hat{u}$ are assumed to be largest directly at the wall. Now, the perturbations have not died away when the boundary-layer thickness is reached. They extend far out of the boundary-layer. The sharp minimum of $|\hat{u}|$ at a distance from the wall of about 2/3 of the boundary-layer thickness δ is only a consequence of forming the magnitude of $\hat{u}$. In fact, the

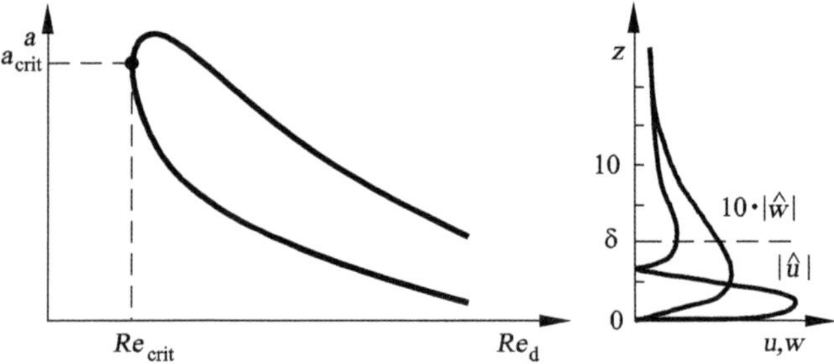

Fig. 6.10. Stability diagram for real a, $b=0$ for the flat plate and eigenfunction for $a=0,16$, $b=0$, $\mathrm{Re}_{d,\mathrm{crit}}=302$

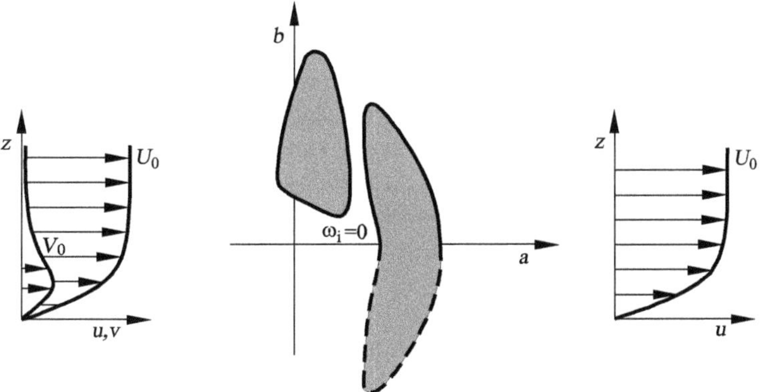

Fig. 6.11. Unstable waves for boundary-layers with and without cross-flow component $\boldsymbol{V}_0(z)$

function $\hat{u}$ passes through zero at this position, a fact that is related to a phase change of the wave of 180°.

In three-dimensional boundary-layers, *Tollmien–Schlichting waves* occur and also, because of the cross-flow component of the basic profile, do the *cross-flow instabilities*. Which waves have cross-flow instabilities is shown in the wave number diagram of Figure 6.11, using the instability region for fixed Reynolds number. Tollmien–Schlichting waves occur downstream only when the critical Reynolds number is exceeded. Note, however, that the Reynolds number in this regime is very small, and therefore there is a strong viscous effect, in this case damping. For comparison, an instability region for the two-dimensional velocity profile $U_0(z)$ is also included. It is typical that in-

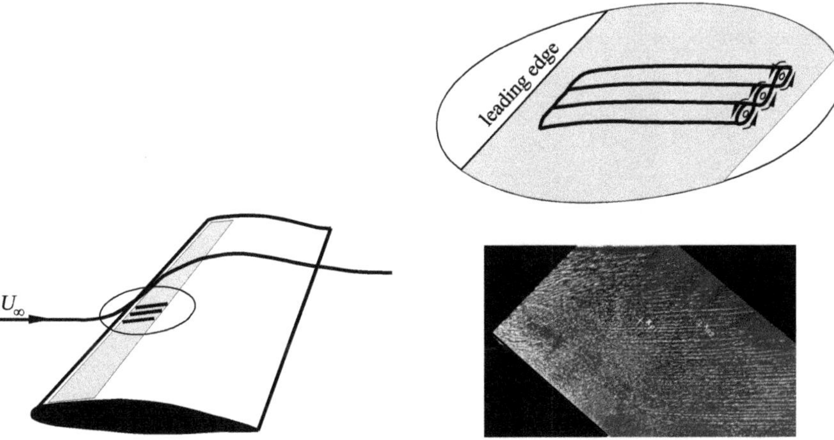

Fig. 6.12. Unstable cross-flow vortex in a three-dimensional boundary-layer, *Y. Kohama* (1989)

stability waves with considerably larger king pin inclinations $\varphi = 1/\tan(b/a)$ exist than in the three-dimensional boundary-layer. Because of its characteristic form, the neutral curve $\omega_{\rm i} = 0$ in the wave number diagram for two-dimensional boundary-layers is also called a kidney curve.

Equally typical for cross-flow instabilities is the appearance of standing perturbation vortices. Since the angular frequency of these standing perturbation waves is $\omega_r = 0$, they are also called *0-Hertz modes.* Their wave normal is almost perpendicular to the downstream direction at the edge of the boundary-layer. These standing waves can be made visible in experiment, with, for example, smoke introduced into the flow, and then have a clear structure in the downstream direction (see Figure 6.12). The perturbation waves that are amplified the most are, however, generally unsteady and travel at a large angle φ transverse to the downstream direction x.

Secondary Instabilities

Until now we have considered *primary instabilities.* The ground state $\boldsymbol{U}_0$ was replaced by the instability, denoted by $\boldsymbol{U}_1$. The new ground state for the *secondary instability* is $\boldsymbol{U}_1$, which can in turn become unstable to perturbations. The perturbation ansatz for the secondary instabilities is $\boldsymbol{u} = \boldsymbol{U}_1 + \varepsilon \cdot \boldsymbol{u}''$. In the plate boundary-layer, the two-dimensional Tollmien–Schlichting wave is replaced by the three-dimensional Λ-structures. The vortex lines, still straight lines in the case of the primary perturbation, are deformed to a wave form in the span direction y. This curvature of the vortex lines is the origin of an abrupt onset of vortex-dynamic induction and self induction, which further deforms and stretches the vortex lines. During this process the characteristic

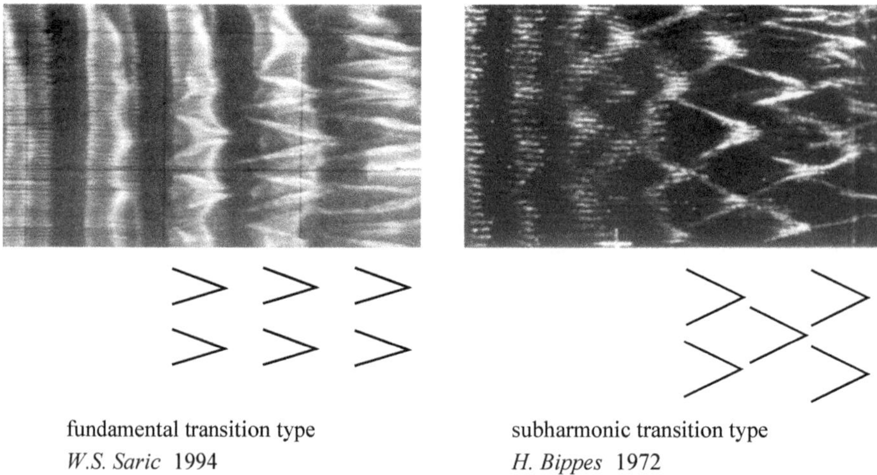

Fig. 6.13. Secondary instabilities in the transitional plate boundary-layer

Λ structures form (Figure 6.13). The secondary instabilities can be analysed with the *Floquet analysis.*

Just as was done for the primary stability analysis, the first step in a secondary stability analysis is to compute the basic flow $\boldsymbol{U}_1(x,y,t)$. In order to be accessible to a secondary stability analysis, $\boldsymbol{U}_1$ must be periodic with respect to a spatial direction parallel to the wall $\boldsymbol{e}_\varphi = \boldsymbol{e}_x \cdot \cos(\varphi) + \boldsymbol{e}_y \cdot \sin(\varphi) = \boldsymbol{e}_{\xi'}$ with the coordinate ξ', and homogeneous to the second parallel direction $\boldsymbol{e}_{\varphi+90^\circ} = -\boldsymbol{e}_x \cdot \sin(\varphi) + \boldsymbol{e}_y \cdot \cos(\varphi) = \boldsymbol{e}_\eta$, i.e. $\boldsymbol{U}(\xi',\eta,t) = \boldsymbol{U}(\xi'+\lambda,t)$. In addition, the basic flow must be able to be written down as a steady flow in a suitable coordinate system $\xi = \xi' - c \cdot t$ (Figure 6.14), i.e. $\boldsymbol{U}_1(\xi',t) = \boldsymbol{U}_1(\xi) = \boldsymbol{U}_1(\xi+\lambda)$. In this way, such basic flows $\boldsymbol{U}_1(z) = \langle \boldsymbol{U}_1 \rangle(z) + \boldsymbol{U}_1^p(\xi,z)$, consisting of a parallel boundary-layer flow, spatially averaged with respect to ξ, $\langle \boldsymbol{U}_1 \rangle(z) = 1/\lambda \cdot \int_\xi^{\xi+\lambda} \boldsymbol{U}_1(\xi,z) \cdot \mathrm{d}\xi$, and a spatially periodic part $\boldsymbol{U}_1^p(\xi,z)$, can be investigated for secondary instabilities. The periodic part does not have a spatial average, but rather has a finite amplitude $A(z) = (1/\lambda \cdot \int_\xi^{\xi+\lambda} |\boldsymbol{U}_1^p(\xi,z)|^2 \cdot \mathrm{d}\xi)^{0.5}$, i.e. it is not assumed that A is infinitesimally small. The basic flow is given in a coordinate system (x,y,z) in which, as usual, the x axis points along the direction of the main flow $\langle \boldsymbol{U}_1 \rangle(z)$ (in three-dimensional boundary-layer flows, typically at the edge of the boundary-layer). A coordinate system adapted to the periodic direction $\boldsymbol{e}_\varphi = \boldsymbol{e}_\xi$ is then chosen. This is obtained from the transformation

$$\begin{pmatrix} \xi \\ \eta \\ z \end{pmatrix} = \begin{bmatrix} \cos(\varphi) & \sin(\varphi) & 0 \\ -\sin(\varphi) & \cos(\varphi) & 0 \\ 0 & 0 & 1 \end{bmatrix} \cdot \begin{pmatrix} x \\ y \\ z \end{pmatrix} - \underbrace{\begin{pmatrix} c \cdot t \\ 0 \\ 0 \end{pmatrix}}_{= \boldsymbol{c} \cdot t} . \tag{6.28}$$

Therefore, in the (ξ,η,z) coordinate system, $\boldsymbol{U}_1(x,y,z,t)$ appears as a steady flow $\boldsymbol{U}_1(\xi,z)$. In contrast, in a two-dimensional boundary layer $\boldsymbol{U}_0(z)$, $\boldsymbol{c} = (c_{\mathrm{TS}},0,0)$ is a downstream traveling wave perturbation with phase veloc-

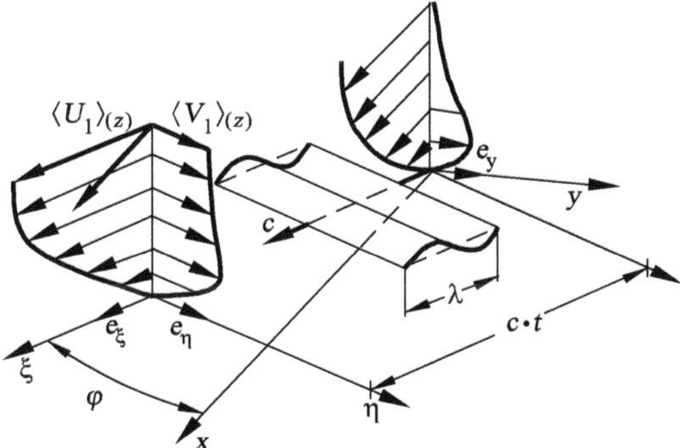

Fig. 6.14. Coordinate system used to describe secondary instabilities

ity c_{TS} . Such a wave perturbation can occur in the course of the amplitude growth of a Tollmien–Schlichting wave (Figure 6.8). Although the basic flow $\boldsymbol{U}_1$ is not actually periodic (weak growth in the boundary-layer thickness downstream, weak spatial amplitude growth of perturbation waves), periodicity is assumed.

The *perturbation differential equation* of the secondary instabilities will not be explicitly given here. They may be found in, for example, *H. Oertel, J. Delfs* (1996), (2005). They are inhomogeneous in t and η, and so exponential trial solutions may be assumed in these directions:

$$\boldsymbol{u}'' = \boldsymbol{V}(\xi, z) \cdot \exp(\mathrm{i} \cdot \beta \cdot \eta) \cdot \exp(\sigma \cdot t) \quad . \tag{6.29}$$

Here $\beta = \beta_r$ is a given real number. This determines the period of the perturbation to be computed with respect to η, i.e. perpendicular to the wave normal of the primary instability (Figure 6.15). The value $\beta = 0$ indicates the special case of a two-dimensional secondary instability. The constant $\sigma = \sigma_r + \mathrm{i} \cdot \sigma_\mathrm{i}$ is in general complex. In analogy to the primary stability analysis, the real part σ_r denotes a temporal amplification rate.

Characteristic of the perturbation differential equations of the secondary instability is the ξ periodicity of the coefficients. The period is $\lambda = 2 \cdot \pi / a_\varphi$ with $a_\varphi = \sqrt{a_r^2 + b_r^2}$. In analogy to linear differential equations with constant coefficients, linear differential equations with periodic coefficients can be solved using a general *Floquet ansatz*:

$$\boldsymbol{V}(\xi, z) = \exp(\mathrm{i} \cdot \alpha \cdot \xi) \cdot \tilde{\boldsymbol{V}}(\xi, z) \quad , \qquad \tilde{\boldsymbol{V}}(\xi, z) = \tilde{\boldsymbol{V}}(\xi + \lambda, z) \quad . \tag{6.30}$$

The solution clearly consists of a function $\tilde{\boldsymbol{V}}(\xi, z)$, to be determined, with the same period as the coefficients of the differential equation, multiplied by an exponential ansatz $\exp(\mathrm{i} \cdot \alpha \cdot \xi)$, in which a generally complex constant α appears. The function $\tilde{\boldsymbol{V}}(\xi, z)$ is expanded as a Fourier series, and the perturbation flow written as

$$\boldsymbol{u}'' = \exp(\mathrm{i} \cdot \alpha \cdot \xi + \mathrm{i} \cdot \beta \cdot \eta) \cdot \exp(\sigma \cdot t) \cdot \sum_{\mathrm{j}=-\infty}^{\infty} \hat{\boldsymbol{V}}_\mathrm{j}(z) \cdot \exp(\mathrm{i} \cdot j \cdot a_\varphi \cdot \xi). \tag{6.31}$$

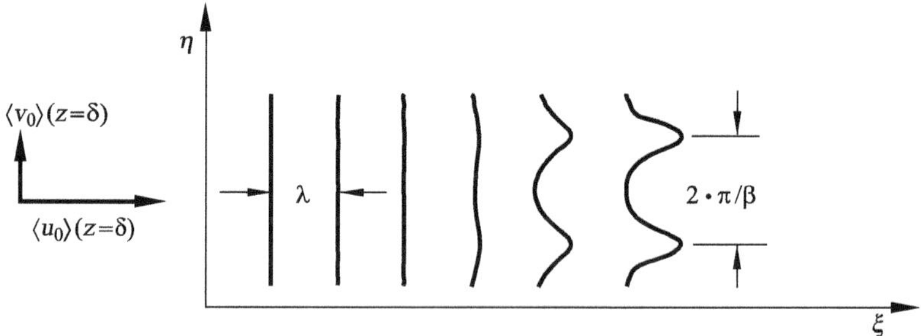

Fig. 6.15. The parameter β in secondary instabilities in the boundary-layer

Inserting the components (u'', w'') of $\boldsymbol{u}''$ into the system of perturbation differential equations and sorting according to the different exponential terms $\exp(\mathrm{i} \cdot (\mathrm{j} \cdot a_\varphi + \alpha) \cdot \xi)$, we obtain a system of infinitely many homogeneous ordinary differential equations in z for the Fourier coefficients $\hat{V}_\mathrm{j}(z)$. This system of equations has nontrivial solutions only for certain combinations $(\alpha,\ \beta,\ \sigma)$, again called eigenfunctions of the secondary stability theory. To actually compute this eigenvalue problem of the secondary stability theory, the Fourier series in (6.31) is interrupted after finitely many terms N. Numerical investigations have shown that for $\varphi = 0$ only two terms $\mathrm{j} = 0, 1$ deliver results that are sufficiently accurate. In the case of oblique primary waves, in particular in cross-flow waves, several modes have to be used.

In analogy to the primary stability theory, we distinguish between temporal and spatial analysis. A temporal stability calculation is carried out by specifying real α and β and determining σ as a generally complex number from the eigenvalue problem. The real part σ_r of the temporal eigenvalue σ denotes the temporal amplification rate. The basic flow $\boldsymbol{U}_1$ is unstable to secondary perturbations if the eigenvalue problem of the secondary stability analysis delivers a value $\sigma_r > 0$. The imaginary part is the total angular frequency of all modes of the secondary eigenfunction $\boldsymbol{u}''$ in the moving reference frame (ξ, η, z). For $\sigma_\mathrm{i} = 0$, all modes of the secondary eigenfunction are standing waves with respect to (ξ, η, z). They do not move relative to the primary wave.

A spatial stability analysis is carried out when no temporal amplification is permitted in the system at rest $(\xi + c \cdot t, \eta, z)$, but rather a temporally periodic process is assumed. In the moving system, σ_r is not set to zero, but rather $\sigma_r = \alpha_\mathrm{i} \cdot c$. The frequency Ω that occurs in the coordinate system at rest appears in the moving coordinate system as $\sigma_\mathrm{i} = \Omega - \alpha_r \cdot c$ and is inserted into the equations as such.

Fundamental modes of secondary instabilities (Figure 6.13) are present in the following Fourier series ansatz:

$$\boldsymbol{u}_f'' = \exp(-\alpha_\mathrm{i} \cdot \xi + \mathrm{i} \cdot \beta \cdot \eta) \cdot \exp(\sigma \cdot t) \cdot \sum_{\mathrm{j}=-\infty}^{\infty} \hat{V}_\mathrm{j}(z) \cdot \exp(\mathrm{i} \cdot \mathrm{j} \cdot a_\varphi \cdot \xi). \qquad (6.32)$$

It is typical for this form of instabilities that they have the same period with respect to ξ as the basic flow.

With the ansatz below we see the subharmonic transition type

$$\boldsymbol{u}_s'' = \exp(-\alpha_\mathrm{i} \cdot \xi + \mathrm{i} \cdot \beta \cdot \eta) \cdot \exp(\sigma \cdot t) \cdot \sum_{\mathrm{j}=-\infty}^{\infty} \hat{V}_\mathrm{j}(z) \cdot \exp\left(\mathrm{i} \cdot \left(\mathrm{j} + \frac{1}{2}\right) \cdot a_\varphi \cdot \xi\right). \qquad (6.33)$$

This secondary instability has double the period of the primary.

The temporal secondary eigenvalue analysis shows that the largest rate of amplification and therefore the dominant eigensolution occurs in both cases for $\sigma_\mathrm{i} = 0$. The entire system of waves given by the modes $\hat{V}_\mathrm{j}$ of the secondary eigenfunction is steady with respect to the primary Tollmien–Schlichting wave

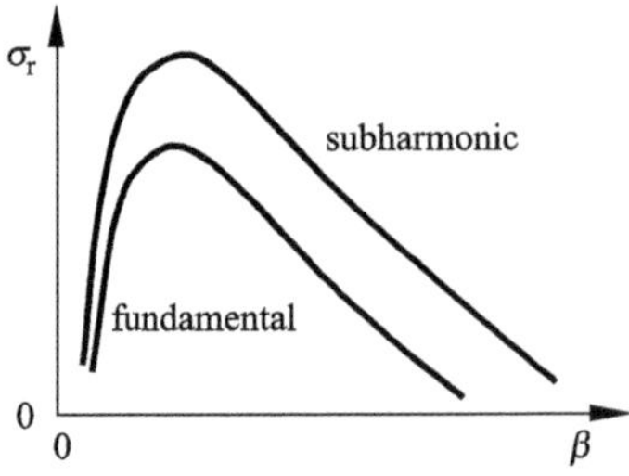

Fig. 6.16. Amplification rate at fundamental and subharmonic resonance of a two-dimensional boundary-layer

of finite amplitude. The secondary modes are coupled with the motion of the primary waves, where they clearly can take up the most perturbation e5nergy. This state is also called *phase-coupled.* Which of the eigenforms is actually taken on at the start of the transition process depends greatly on the initial perturbation spectrum. For small amplitudes $A \lesssim 2\%$ of the Tollmien–Schlichting wave, the amplification rates of the subharmonic secondary instability are largest and those of the fundamental type smallest (Figure 6.16). These proportions change as soon as the amplitudes of the primary perturbation become large $A \gtrsim 2\%$. The fundamental resonance then dominates over the other forms.

The typical maximal amplification rates of secondary instabilities are considerably larger than primary amplification rates, even at small amplitudes $A \approx 1\%$. Therefore, it is justified to consider the primary perturbation to be locally periodic with frozen amplitude A, since A varies only a little, while the secondary modes are greatly amplified. What is important is the size of the primary amplitude, and less so its variation.

According to Figure 6.17, the secondary instability exists for an entire band of the transverse wave number β, whose width grows with increasing primary amplitude A. The width of the transitional flow structures determined by β is therefore in no way uniquely determined; rather, it can be completely different depending on the perturbation. It can clearly be seen that for too small β, the secondary amplification rates for the Blasius plate boundary-layer flow fall drastically to zero.

In contrast to the other modes, the fundamental modes from (6.32) contain an aperiodic part. This partial wave is independent of ξ, and its wave normal points in the direction of the η coordinate. This means that it repre-

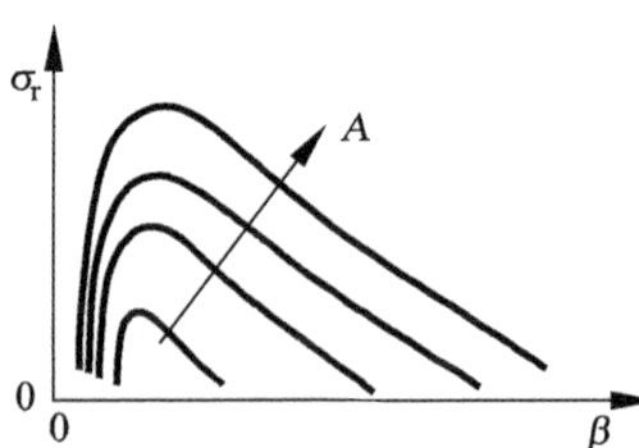

Fig. 6.17. Growth of the secondary amplification rate with the primary amplitude

sents a periodic longitudinal vortex in η. These vortices are ordered in pairs rotating in opposite directions, as follows from the symmetry of the flow field $\boldsymbol{U}_1$ with respect to the ξ-z plane. The structure of the longitudinal vortex is also called a *peak–valley structure.* In the planes $\eta = \eta_p$, in which the vortices induce upward velocities, slow-moving fluid close to the wall is transported in high layers z with relatively large mean velocities. This leads to strong shearing, favoring the perturbation development. For this reason, the $\eta = \eta_p$ plane is called the *peak plane.* The planes displaced from the peak planes by half a width π/β at $\eta = \eta_v = \eta_p \pm \pi/\beta$ are called *valley planes*, to indicate that the perturbation development here is much weaker than in the peak plane.

The secondary stability analysis in *three-dimensional boundary-layers* shows that in the case of *cross-flow vortices* in the boundary-layer of a moving wing, the temporal secondary amplification σ_r is of the same order of magnitude as the primary amplification rate. In addition, the boundary-layer thickening and the wall curvature have a great influence on the stability properties of this flow close to the leading edge, so that the corresponding results are essentially only qualitative. Figure 6.18 shows the instantaneous streamlines of a sequence of the oscillatory secondary instability in the direction of expansion of the cross-flow waves. It is seen that the secondary perturbation waves oscillating about the primary cross-flow angle weaken and strengthen periodically.

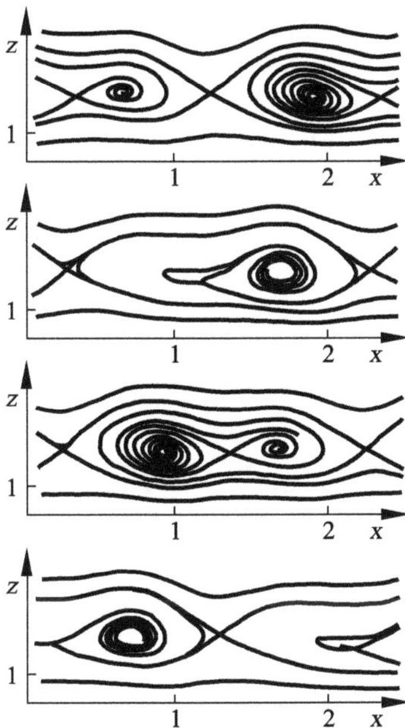

Fig. 6.18. Sequence of a period of instantaneous streamlines of the secondary cross-flow waves in section along the expansion direction of the primary perturbation waves and perpendicular to the wall, *T.M. Fischer* (1987)

Stability of Nonparallel Boundary-Layer Flows

Until now we have described the local stability analysis. The change in the boundary-layer in the direction of the flow was neglected. In this section, the stability analysis will be extended to nonparallel flows. Note that the effect of the flow relations that change in the direction parallel to the layer on the perturbation development strongly depends on the type of perturbation. What is important in the effect of the change in the basic flow on the perturbation is how great a change occurs in one perturbation wavelength. For example, consider the Blasius plate boundary-layer flow in Figure 6.19 whose boundary-layer $\delta(x)$ thickens in the flow direction x. For a given wavelength $\lambda = 2 \cdot \pi / \sqrt{a_r^2 + b_r^2}$, the boundary-layer thickening has a greater effect on the perturbation wave, the larger the king pin inclination, $\varphi = 1/\tan(b_r/a_r)$, of the wave with respect to x. This is because the wavelength section $\lambda_x = 2 \cdot \pi / a_r = \lambda / \cos(\varphi)$ in the flow direction x increases greatly with φ. In particular, in the limiting case of transverse traveling perturbation waves, i.e. $\varphi = 90°$, the parallel flow assumption of the local analysis infringes greatly the actual physical facts.

Two fundamentally different procedures to investigate the *stability of nonparallel flows* have been developed. One of these approaches is a direct extension of the local stability analysis with analytical methods. It builds on the previously described method of multiple scales and yields correction terms from taking the nonparallel effects only at the position under consideration. The second approach involves parabolizing the fundamental equations (5.91) and the perturbation differential equations derived from them. This approach has the advantage that the history of the perturbation development upstream of the position under consideration is taken into account. Both procedures contain the special case of the local analysis for parallel basic flows.

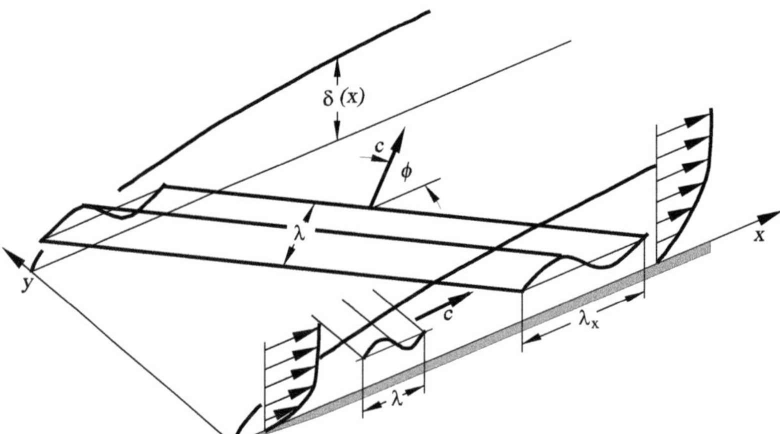

Fig. 6.19. Thickening effect with inclined waves

The analysis shows that the boundary-layer thickening has a generally destabilizing effect on perturbation waves especially in the low Reynolds-Number regime (see Figure 6.20). This means that for a given frequency, the spatial amplification is greater when the basic flow is taken to be nonparallel than when the parallel flow assumption is used. This is particularly true for obliquely traveling waves moving opposite to the main flow direction, where the wavelength component in the downstream direction is large. The effect is strong for perturbation waves whose wave normal is perpendicular to the main flow in the span direction y. A nonparallel basic flow also has a greatly amplifying effect on the cross-flow instability. The amplification rate of unstable perturbation waves in compressible boundary-layer flows is similarly greatly increased, since compressible boundary-layers thicken more than incompressible boundary-layers, due to the heating of the medium close to the wall and the consequent volume expansion.

It can be shown that effects due to wall curvature and curvature of the wave fronts (divergence or convergence of the wave normals) frequently affect the spatial amplification rate just as much as a nonparallel basic flow. For example, the curvature has a considerable effect on the cross-flow instabilities close to the leading edge of a moving wing, where a strong convex wall curvature is present. The convex wall curvature stabilizes such perturbation waves and, in this case, acts to oppose the effect of the nonparallel basic flow that strengthens the amplification. For a consistent theory, both effects have to be taken into account simultaneously.

In addition to the stability analysis, direct simulation of the transition process up to turbulent boundary-layer flow by numerical solution of the complete Navier–Stokes equations (5.65) has also been also performed. Figure 6.21 shows the simulation results of the *Tollmien–Schlichting transition* and

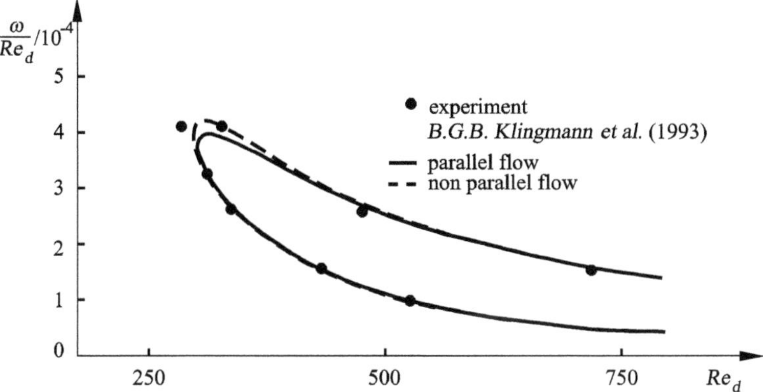

Fig. 6.20. Stability diagrams of Tollmien-Schlichting waves for parallel and non-parallel boundary-layer flows

the *transition of the cross-flow vortices* in a three-dimensional wing boundary-layer at Mach number $M_\infty = 0.62$ and Reynolds number $\mathrm{Re}_l = 26 \cdot 10^6$. Contour surfaces of the rotation $\boldsymbol{\omega} = \boldsymbol{\nabla} \times \boldsymbol{u}$ are shown. The transition process of the Tollmien–Schlichting waves begins with downstream traveling plane waves. As in Figure 6.8, three-dimensional perturbations are superimposed, and Λ-structures (fundamental transition type) form. The Λ-structures are regions of local shearing and excess velocity in the peaks. They are lined up periodically in the span and form several rows periodically ordered behind each other. The occurrence of the Λ-structures is associated with the appearance of strong free-shear layers. These are prominent local maxima of the shear stress far from the wall. As the transition proceeds, the high shear rates decay into increasingly smaller structures, leading eventually to

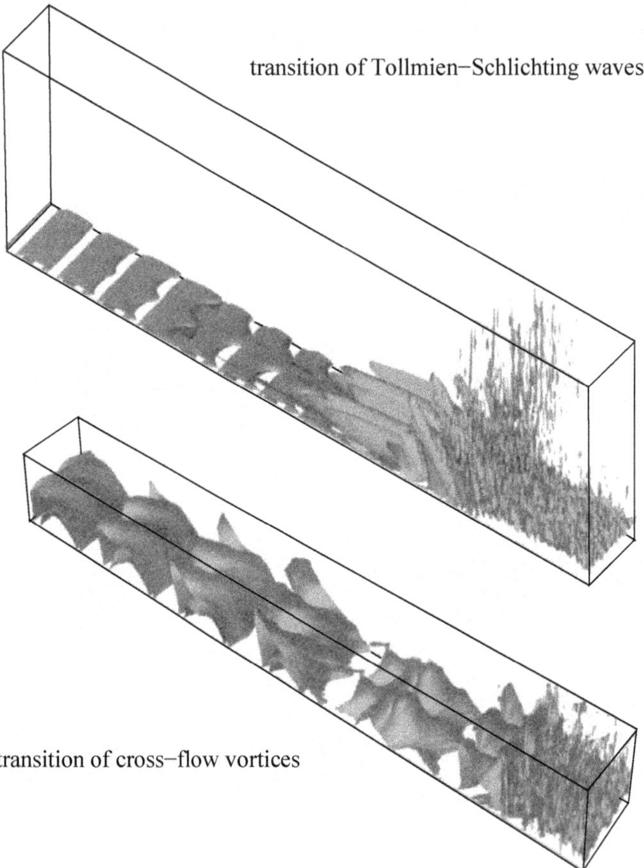

Fig. 6.21. Laminar–turbulent transition in the compressible wing boundary-layer, $M_\infty = 0.62$, $\mathrm{Re}_l = 26 \cdot 10^6$

the turbulent state. The decay of the shear layers takes place within a few wavelengths of the Tollmien–Schlichting waves.

The mechanisms of the transition process of cross-flow vortices are similar. Figure 6.21 shows the formation of the Λ-structures associated with high shear rates and fluctuation in the perturbation quantities in the peaks. In the final state of the transition they decay within a short distance into the turbulent boundary-layer flow.

Local Perturbations

Figure 6.22 shows a sketch of local perturbations of the Tollmien–Schlichting transition and the transition of cross-flow instabilities in the three-dimensional boundary-layer of a swept transonic wing. Both instabilities are convectively unstable in the boundary-layer.

In what follows we will briefly analyze the behavior of three-dimensional wave packets in a three-dimensional compressible boundary-layer. In contrast to the investigation into two-dimensional perturbations, the transverse wave number b now also appears in the dispersion relation function $D(\omega, a, b)$, whose roots are indeed given by those combinations (ω, a, b), representing the solutions of the stability eigenvalue problem for complex ω, a, b. We consider the change in amplitude of a perturbation wave packet in the plane reference frame, moving with the group velocity (U, V). The frequency observed is then

$$\omega' = \omega - a \cdot U - b \cdot V \quad . \tag{6.34}$$

As in the two-dimensional case, we again have to find those waves whose group velocity vector $(\partial\Omega/\partial a, \partial\Omega/\partial b)$ is real. The complex frequency function $\Omega(a, b)$ is then defined by $D(\Omega(a, b), a, b) \equiv 0$. The relative temporal amplification ω_i' is then plotted, not just as a function of $U = \partial\Omega/\partial a$, but also against the group velocity plane (U, V). The line of height $\omega_i' = 0$ is of particular interest, since it encloses the region in the (U, V) plane in which $\omega_i' > 0$. Therefore, this region represents the parts of the perturbation that

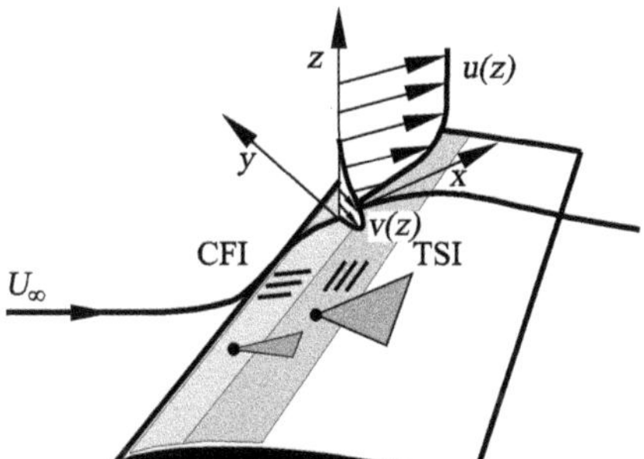

Fig. 6.22. Local cross-flow (CFI) and Tollmien–Schlichting instabilities (TSI) in the three-dimensional boundary-layer of a swept wing

contribute time-asymptotically to the wave packet. Figure 6.23 contains diagrams with the regions of temporal amplification at two representative positions on a swept wing. The lower diagram in the figure shows a typical curve $\omega_i' = 0$, which is computed for a position close to the leading edge of the swept wing, i.e. in the cross-flow instability region. The upper diagram shows the same curve at a position further downstream on the wing, where Tollmien–Schlichting instabilities are present. We see that both instabilities have convective character, since in both cases the origin $(U, V) = (0, 0)$ is not contained in the $\omega_i' > 0$ region. The growing perturbation energy is transported downstream in both cases. The tangents at the curves $\omega_i' = 0$ determine the angular region within which these amplified perturbations remain. In the case of the cross-flow instabilities, the angular range is very narrow and lies essentially downstream. Note that the associated instabilities are waves that travel practically perpendicular to the downstream direction. This clearly indicates the fundamental difference between group velocity and phase velocity.

Now that we have determined that the cross-flow instabilities are convective in nature and that they induce a spatially extended transition process downstream, the associated spatial wave packet amplification rates ($g_{\max} = [(\omega_i - a_i \cdot U - b_i \cdot V)/\sqrt{U^2 + V^2}]_{\max}$) for the transonic swept-wing boundary-layer have been computed. Figure 6.24 shows the eigenvalues, eigenfunctions, and unstable regions of wave packet perturbations for angles of sweep from 15° to 25°. In developing a swept laminar wing, it is essential to avoid cross-flow instabilities, since they induce a transition process already directly at the leading edge. Using the methods of stability analysis, the region of the design parameters of a swept wing can be determined within which active influencing measures are not needed (corresponding to the natural laminar behavior). One of these parameters is the angle of sweep.

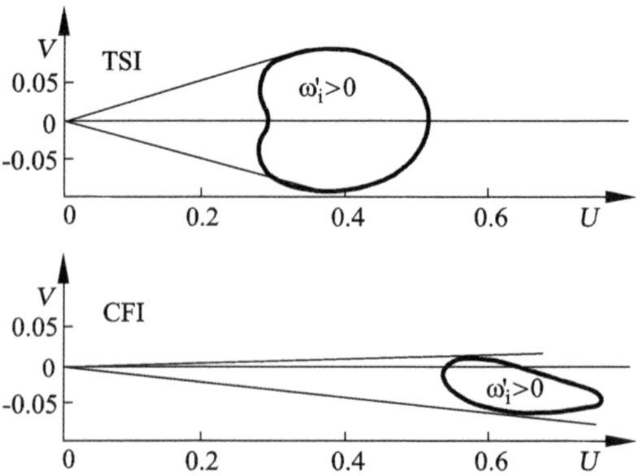

Fig. 6.23. Regions of relative temporal amplification of the Tollmien–Schlichting instabilities (TSI) and cross-flow instabilities (CFI) in the group velocity plane (U, V)

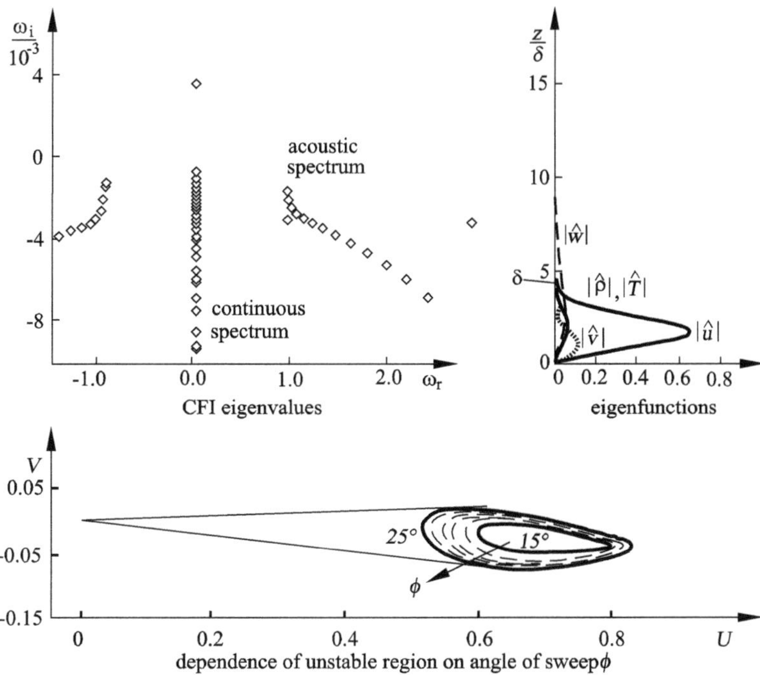

Fig. 6.24. Eigenvalues, eigenfunctions, and unstable regions of the cross-flow instability in the compressible boundary-layers of swept wings, $M_\infty = 0.78$, $\mathrm{Re}_l = 26 \cdot 10^6$

In an otherwise identical free stream, there is a critical range of angle of sweep within which the transition process changes from TSI-dominated to CFI-dominated (Figure 6.22). Stability theory therefore leads us to a limit for the angle of sweep of a laminar wing.

6.2.3 Transition to Turbulence

Other fascinating advances have been made with respect to successive instabilities potentially leading to turbulence. *L. D. Landau* proposed a quasi-periodic route to turbulence (see *L. D. Landau, E. M. Lifschitz* (1991)) in which successive instabilities occur at ever faster rates and culminate in turbulence at their accumulation point. Possibilities such as the few-step route (*D. Ruelle, F. Takens* (1971)) and the period-doubling route (e.g. *M. J. Feigenbaum* (1978)) have been proposed for the generation of temporal complexity (or chaos) in a variety of nonlinear systems. Indeed, these scenarios have been observed in many nonlinear systems including a restricted class of fluid flows (and are thus believed to be universal in scope), but the appearance of turbulence is an issue of both temporal and spatial complexity. Here, progress is attained more or less on a case-by-case basis, although some generality of concepts does exist.

In particular, the route to turbulence is not unique, because, among other things, the process is not merely one of successive instabilities but also one of flow receptivity to a variety of background fluctuations that are invariably present. For instance, for the flat plate boundary-layer, unless the disturbance level is carefully controlled, some steps described earlier in the process of transition may be bypassed altogether, and pointlike disturbances may evolve into three-dimensional wave packets that grow quickly into spots of turbulence. These spots coalesce to form turbulence as we know it.

Let us summarize some aspects of what we have already covered in earlier sections. For the boundary-layer, when the background noise level and initial conditions are carefully controlled, a variety of details can be reproduced, and the following sequence of events occurs (Figures 4.55, 6.8). Once the modes of primary instability, the Tollmien–Schlichting waves, grow to finite amplitudes, the flow develops spanwise variations. These spanwise variations appear rather slowly in wind tunnels, and are better studied when induced artificially, as was done by *P. S. Klebanoff et al.* (1962), who attached small strips of tape at equal intervals across the plate (Figure 6.25). Their measurements revealed the appearance of counterrotating vortices, and the development of definite *peaks* and *valleys* in the fluctuation velocity. As spanwise variation intensifies, a thin layer of high shear appears, especially at the peak, consistent with the observations of *L. S. G. Kovasznay et al.* (1962). *J. T. Stuart* (1963) has shown that the convection and vortex-stretching in the presence of large spanwise variations produce small layers of high intensity, resembling those observed experimentally. These layers possess inflection points and are inviscidly unstable, thus leading to further high-frequency modes and the formation of new vorticity in both longitudinal and spanwise directions. The passage of the vortex structures results in spikes in velocity signals, as seen in the extensive studies of *M. Nishioka et al.* (1990) for the case of a two-

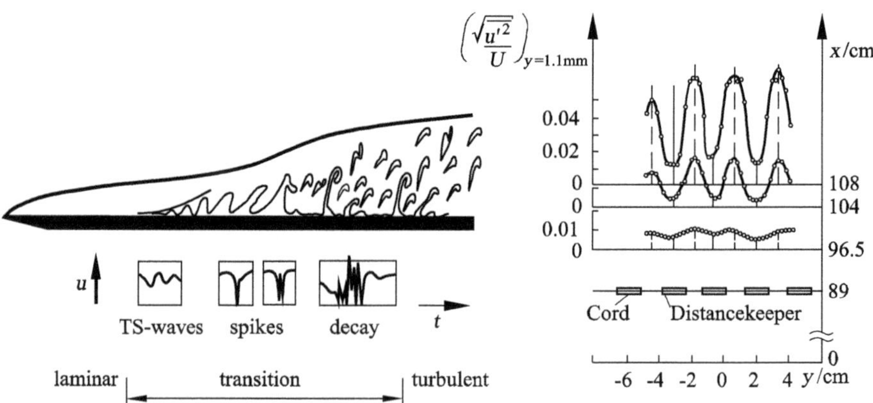

Fig. 6.25. Transition in the plate boundary-layer, *M. Nishioka et al.* (1990) and *P. S. Klebanoff et al.* (1962)

dimensional boundary-layer (Figure 6.25). Near where the spikes occur, spots of turbulence are born. Turbulent spots (*H. W. Emmons* (1951)) have a well-defined shape within which the fluid is in nearly turbulent motion, and are surrounded by essentially laminar flow (Figure 6.26). The spots grow as they propagate and merge with other spots to become fully developed turbulent flow. The growth rate of isolated spots is proportional to the square root of the difference between the Reynolds number of the flow and the Reynolds number at which spots are born.

For a more detailed description of laminar–turbulent transition in the boundary layer, reference may also be made to *R. Narasimha* (1985) and *A. V. Boiko et al.* (2002).

The combination of stability theory and experiment has been able to advance our understanding of the origin of turbulence in certain broad classes of flows. However, there are other circumstances for which linear stability is an unsuitable starting point for understanding the onset of turbulence. In those instances the onset of turbulence is sudden, and a fundamentally different sequence of events is involved. In particular, the many scales of turbulence appear more or less at the same time. Flow through pipes is an excellent example of this kind of transition. Typically, flows of this kind are stable to all linear perturbations, and one of their strong characteristics is that the transition has no reproducible *critical* Reynolds number, as would be characteristic of linear instability. The Reynolds number at which the transition to turbulence occurs depends on the type, form, and magnitude of the disturbance. For the onset of turbulence, the initial disturbance and the Reynolds number need to be large enough, and play complementary roles, where a smaller disturbance level is needed at larger Reynolds numbers, and vice versa. If the pipe is joined to a smooth-walled vessel by a sharp edge, the critical Reynolds number is about 2800. If the inlet is well rounded and the flow there is prepared to be relatively free of disturbances, transition values as high as 10^5 can be observed. If the inflow is very irregular, it may fall to about 2300 (see Section 4.2.4). In fact, in the last case, the transition Reynolds number is representative of the conditions at which large initial disturbances *just* manage to regenerate continually. In contrast to pipe flow, which is linearly stable for asisymmetric perturbations and for all Reynolds numbers, channel flow is expected to become linearly unstable at a finite critical Reynolds number of 5772 (*C. C. Lin* (1945), *S. A. Orszag* (1971)).

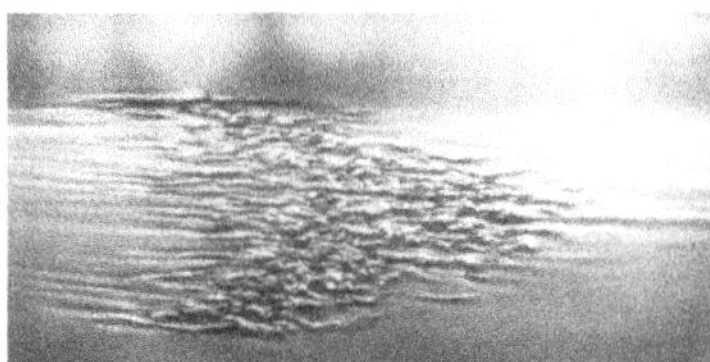

Fig. 6.26. Turbulent spot, *R. Falco* (1980)

However, experiments show that the transition does not usually wait until that Reynolds number is reached, but occurs at lower Reynolds numbers.

The mechanism of transition in these cases is called subcritical because it occurs below the linear stability value. *W. M. F. Orr* (1907) knew that linear disturbances of a shear flow could grow for some time even if they are stable (since the concept of stability is related to the asymptotic growth of perturbations, see Section 6.2.2). Many later authors have expanded on this theme (for a summary see *S. Grossmann* (2000)).

Figure 6.27 shows a schematic plot of subcritical transition. With increasing initial disturbance amplitudes A the transition to turbulence occurs at smaller Reynolds numbers Re_l. The transition line should be interpreted as the envelope of all stability lines for possible types of disturbances.

It is now clear that the nonnormality of eigenfunctions of the linear operator for the perturbation equation is the essential property responsible for the transient growth of disturbances. This, together with the proper action of the nonlinear interactions between finite disturbances of sufficient amplitude, leads to the onset of turbulence. The nonnormality of the linear dynamics quite generally implies a bunching of the eigendirections. Those disturbances that fit into the bundle decay with time, while those that do not do so first grow algebraically at a rate that depends on the nonnormality and the Reynolds number. Only after this transient increase do they decay. But if there is sufficient transient amplification, the nonlinearity, which can no longer be neglected, drastically modifies the dynamics, and the appearance of an irregularly fluctuating velocity field can be expected.

The *transition process* can be divided qualitatively into *different stages.* The first is usually the *receptivity stage*, which is associated with disturbances in the flow. Receptivity is often the most difficult process to conclude the transition prediction for realistic flow situations. It entails knowledge about the ambient disturbance environment and the mechanisms by which disturbances are projected into growing eigenmodes.

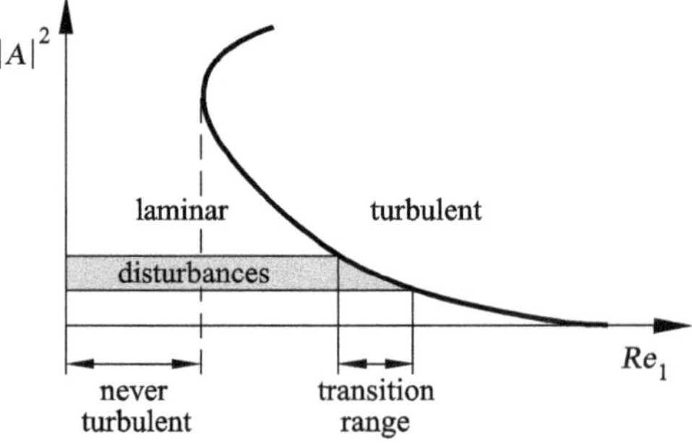

Fig. 6.27. Subcritical transition

The next stage is the *linear growth stage* of primary instabilities as Tollmien-Schlichting waves in the boundary-layer, where small disturbances are amplified until they reach a size where nonlinear interactions become important. This amplification can be in the form of exponential growth of eigenmodes, nonmodal growth of optimal disturbances, or nonmodal response to forcing.

Once a disturbance has reached a finite amplitude, it often *saturates* and transforms the flow into a new state. Only in a few cases does the primary instability lead the flow directly into a turbulent state. Instead, the new unstable flow becomes a base flow on which secondary instabilities can grow, as was shown in Section 6.2.2.

The *secondary instability stage* can be viewed as a new instability of a more complicated flow. This stage of the transition process is in many cases more rapid than the stage where primary instabilities prevail.

The last stage is the *breakdown stage* where nonlinearities and higher instabilities excite an increasing number of scales and frequencies in the flow. This stage is often more rapid than both the linear stage and the secondary instability stage.

Dividing the transition process into these five stages, receptivity, linear growth, nonlinear saturation, secondary instability, and breakdown certainly idealizes the transition process, because all stages cannot always be expected to occur in an unambiguous manner. However, they often provide a good framework to view transition even for complicated flows.

Figure 6.28 shows one example of a transition scenario at high free-stream turbulence level in comparison with the flat plate transition process described in Figure 6.3 of Section 6.2.1. In the first stage the formation of streaks by free-stream-localized vortical disturbances in the boundary-layer can be observed in the experiments of *P. H. Alfredson* et al. (1996). The streaks modulate the boundary-layer in the spanwise direction. The second stage

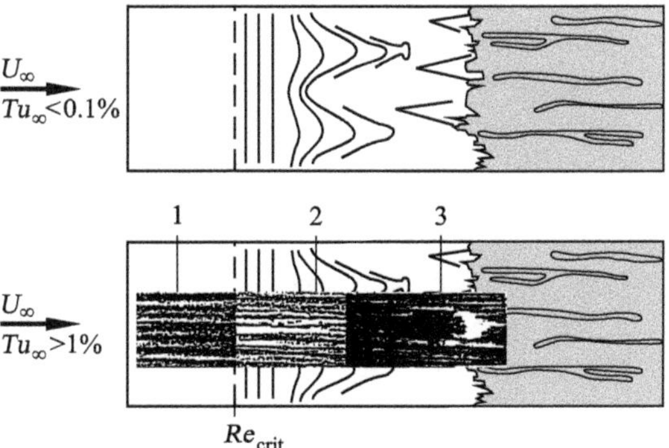

Fig. 6.28. Transition scenario at low and high free-stream turbulence, *P. H. Alfredson* et al. (1996)

includes the following streak development accompanied by the generation of high-frequency wave packets and incipient spots due to different nonlinear mechanisms including the interaction with Tollmien-Schlichting waves and secondary instability. The third stage of the transition includes development and interaction of the turbulent spots which completes the laminar-turbulent transition in the boundary-layer.

6.3 Developed Turbulence

6.3.1 The Notion of a Mixing Length

The two flows with which we started this chapter are examples of developed turbulence. In practice, we do not need to know all the details of turbulent flows, but are often content to obtain answers, for example, to questions such as, how fast does a jet grow on the average? How much power is required to pump a fluid at a certain rate through a pipe? How much power is required to fly an aircraft? How much fuel is consumed in providing a required amount of thermal energy in a combustion chamber?

It is useful for these purposes to decompose the velocity into mean and fluctuating parts (called *Reynolds decomposition* after Osborne Reynolds (4.63); see Section 4.2.4), and to obtain the *Reynolds* equation (5.95) for the mean part. The Reynolds shear stress term appears in addition to the viscous stress. Mathematically, the source of this new term is the nonlinearity of the advection term in the Navier–Stokes equations. Physically, turbulent fluctuations give rise, on average, to increased transport of momentum by transporting it from place to place in the flow. The turbulent, or the Reynolds shear stress has the form $\tau_t = -\rho\overline{u'v'}$, where u', v' are deviations of the velocity components from their average values $\overline{u}$ and $\overline{v}$, respectively, and the overline indicates the average over time. In order to solve the Reynolds equations and obtain formulas of practical use, we must express τ in terms of quantities related to the mean velocity. The situation, called the *closure problem*, is analogous to that in kinetic theory in which the momentum transport of molecular theory is seen as a macroscopic viscosity, which must be prescribed. However, viscosity is a property of a fluid that can be measured once and for all. Such simplicity does not exist in turbulence for reasons that we shall mention presently, and so a variety of methods has been devised to express the Reynolds stresses in terms of the distribution of the mean velocity (see Section 5.4.5). The methods developed have varying levels of success but are not universally applicable to all turbulent flows. Approaches vary from the application of sophisticated statistical mechanical principles or hypotheses, whose physical content is not immediately apparent, to the use of more or less transparent physical ideas, which cannot always be justified.

The simplest intuitive physical picture of *L. Prandtl* has historically allowed us to make some progress by assuming that fluid parcels (or *eddies*) of a certain size transport momentum through the fluid by means of their seemingly random motion (see Section 4.2.5). If so, it is appropriate to associate one length scale with the "diameter" of these eddies, and another for the distance through which they remain intact as they propagate relative to the rest of the fluid. We cannot say a priori that these two lengths are the same, but we might speculate that they would be proportional to each other. We now assume that the flow is such that the mean velocity varies in a direction at right angles to the streamlines (as in pipe flows). If, as shown in Figure 4.61, a fluid parcel is displaced from a position y where the mean velocity is $\overline{u}(y)$ by a distance l in a direction transverse to the flow, the difference between its old and new velocities is $\overline{u}(y+l) - \overline{u}(y)$. As a first approximation, we may write this as $l\partial\overline{u}/\partial y$. This gives an estimate of the order of magnitude of the fluctuation u'. The value of v' is found from the assumption that two parcels of fluid, which enter the layer in question from opposite sides and subsequently move on after each other, approach or recede from one another with relative velocity $2l\partial\overline{u}/\partial y$. This gives rise to transverse velocities of the same order of magnitude as u'. Thus, in forming the average value $\overline{u'v'}$, we have still to consider the signs of the corresponding u and v components. It is easy to see, however, that in crossing a control surface parallel to the boundary, the fluid particles moving away from the boundary are relatively slow compared to those moving toward the boundary. Therefore, in general, negative values of u' are associated with positive values of v', and positive values of u' with negative values of v'. Thus the product $u'v'$ tends to be negative in both cases, and the new shearing stress is positive and of order $\rho(l\partial\overline{u}/\partial y)^2$. If we arbitrarily take the unknown factor of proportionality as unity, we merely make a slight change in the meaning of l. To make the formula accurately express the fact that a positive shearing stress corresponds to positive values of $\partial\overline{u}/\partial y$ and a negative shearing stress to negative values of $\partial\overline{u}/\partial y$, we must write

$$\tau' = \rho\, l^2 \left|\frac{\partial\overline{u}}{\partial y}\right| \frac{\partial\overline{u}}{\partial y}. \tag{6.35}$$

From this approximate expression, we infer that the Reynolds stresses due to turbulent motion are proportional to the square of velocity increments, leading to the notion that fluid resistance varies roughly as the square of the velocity in a turbulent flow. The length l, called *Prandtl's mixing length*, is not unlike the molecular mean free path $\overline{\lambda}$ in the kinetic theory of gases. In the latter, the transfer of momentum due to motion of molecules is discussed in a way similar to our present account of the transfer of momentum by the large-scale motion of fluid parcels. As in the present case, the deviation from the mean velocity of particles, moving upward or downward, is given by $u' = \pm\overline{\lambda}\,\partial\overline{u}/\partial y$. The transverse velocity v', however, is not proportional to u', but is equal to the molecular velocity, effectively a constant. Thus, the shearing stresses due to molecular motion (the viscous stresses) are linearly proportional to $\partial\overline{u}/\partial y$. In gases, the mean free path $\overline{\lambda}$ is inversely proportional to the density ρ, so that the factor $\rho\overline{\lambda}$ present in the definition of viscosity is independent of the density.

If we insert $\mu_t = \rho l^2|\partial\overline{u}/\partial y|$ into (6.35), we obtain the equation $\tau' = \mu_t\partial\overline{u}/\partial y$. This is of the same type as the equation for the viscous shearing stress $\tau = \mu\partial u/\partial y$, and μ_t has the dimensions of viscosity. Unlike the molecular viscosity coefficient, however, μ_t, known as the *eddy viscosity coefficient*, depends on the details of the flow and its Reynolds number. Another important difference from ordinary viscosity is that μ_t is not a unique property of the fluid and varies from point to point in the flow. For example, it tends to zero as the boundary wall is approached. In practice, these attributes limit the usefulness of the concept of eddy viscosity. Neither is the notion as compelling as in the molecular case, where there is a large separation of scales between the molecular mean free path and the length scale characterizing the mean flow gradient. Indeed, in turbulence, the mixing length is often not a negligibly small fraction of the flow size. In spite of these basic limitations, the notion of mixing length is qualitatively ingrained even in sophisticated theories of turbulence.

6.3.2 Turbulent Mixing

The effects of turbulence include not only increased momentum transport, but also the transport by convection of all the properties of moving matter (heat content, quantity of admixed matter, etc). With some exceptions, the transport of a given property will occur, on average, from regions rich in that property to those that are lacking in that property. In the case of temperature differences, this means that some type of turbulent heat conduction; in the case of differences in concentration, a type of turbulent diffusion, will result. Thus, since the quantity of heat contained in unit mass of a fluid is c_pT, where T is the temperature and c_p the specific heat at constant pressure, the net quantity of heat flowing across unit area per unit time is given by

$$Q = -c_p\kappa_t\frac{\partial\overline{T}}{\partial y} = -c_p\rho l^2\left|\frac{\partial\overline{u}}{\partial y}\right|\frac{\partial\overline{T}}{\partial y}. \tag{6.36}$$

That is, $c_p\kappa_t$ is the effective thermal diffusivity ($\lambda_t = c_p \cdot \rho\kappa_t$). In the case of a chemical or mechanical admixture of concentration c, the mass of admixed substance transferred across unit area in unit time is given by

$$M = -\rho D_t\frac{\partial\overline{c}}{\partial y}. \tag{6.37}$$

There remains the question of whether κ_t and D_t agree numerically with $\nu_t = \mu_t/\rho$, considering that the mechanism of transfer of a property of matter, or of an admixed substance, is not quite the same as that of transfer of momentum. The ratios ν_t/κ_t and ν_t/D_t are known as the turbulent Prandtl number and turbulent Schmidt number, respectively (see Section 7.4). Their numerical values depend on whether one is considering turbulence near a solid boundary, or in regions away from it (the so-called free-shear flow).

This difference between free shear flows and wall-bounded flows is connected with differences in the eddy structure between the two classes of flows. Loosely speaking, eddies with their axes parallel to the direction of flow predominate near a solid boundary, whereas the eddies with their axes nearly at right angles to the flow direction predominate in free-shear flows. Eddies of the first kind make no contribution to the transport of momentum, whereas eddies of the second kind make a very considerable contribution. Therefore, the distributions of mean velocity and of mean temperature or concentration exhibit marked differences. That the heat exchange is more dominant than the momentum exchange in the case of free turbulence has also been shown by experiments on the smoothing out of temperature and velocity distributions in the rear of lattices of heated rods, where the temperature differences vanishes much more rapidly than the velocity differences.

In general, since turbulent transport and mixing depend largely on the motion of parcels of fluid, one may imagine that they become essentially independent of molecular properties. It may be expected that the momentum transport far away from the wall becomes asymptotically independent of the fluid viscosity. The situation very near the wall is that the viscosity always plays an important role because turbulent fluctuations are small there. Turbulent mixing of admixtures does seem to retain a weak dependence on the molecular Prandtl or Schmidt number (as appropriate). This seems to be the result of the fact that parcels moving through the turbulent background develop transient boundary-layers on their front side, and reintroduce the molecular Prandtl or Schmidt number effects indirectly.

6.3.3 Energy Relations in Turbulent Flows

Work is done on a fluid element by the Reynolds stresses and the corresponding pressure differences. This work serves to maintain the turbulent motion within the element. In the very simple picture considered above, the work done on unit volume per second is $\tau' \partial \overline{u} / \partial y$. This work enables the eddying motion to maintain itself against the resistance that it encounters in its motion. The initial forward motion of the individual eddy, relative to its surroundings, is itself a turbulent motion, which, if its Reynolds number is sufficiently high, gives rise to a turbulence of the second-order with smaller eddies of turbulence. These, in turn, produce turbulence of the third-order. This process continues until the final eddies are so small that they cannot become turbulent. What is left of the kinetic energy of these smallest eddies is transformed into heat as a result of viscosity. This suggests that a large range of scales can be created in turbulence, and that this range is larger if we start out with a larger Reynolds number. This notion has been formalized by *L. F. Richardson* (1920) and, particularly, *A. N. Kolmogorov* (1941). In describing this work, it is customary to speak loosely of scales of turbulence,

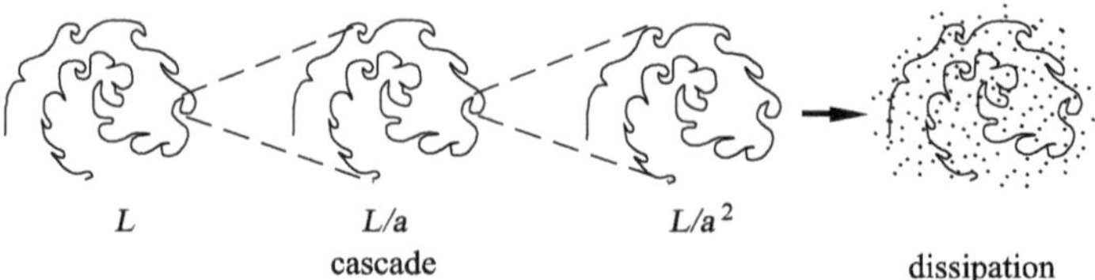

Fig. 6.29. Turbulent cascade

which, while being another word for sizes of turbulent eddies, conveys a far less specific picture than balls of fluid moving about in a cohesive manner. For instance, in a Fourier representation of the turbulent velocity, the scale size would be the wavelength of a given mode. The Kolmogorov picture is that the turbulent energy is introduced at the largest scale, say L, which then cascades down to smaller and smaller scales without dissipation (see Figures 6.29 and Figure 6.30, neither of which should be taken too literally) until a certain smallest scale is reached, where the velocity gradients are so large that dissipation is large enough to damp out the generation of further smaller scales. While the velocity gradients of the small scale are very large, their amplitudes are rather small.

Figure 6.31 shows the spectrum of turbulence energy E connected with wave numbers, a, of turbulent eddies with the cascade regime at intermediate wave numbers and the dissipation regime at large wave numbers corresponding to such wavelengths of the turbulent eddies. The energy associated with the motion of eddies is dissipated and converted to thermal energy. This permanent dissipation process results in a consecutive kinetic energy loss. The stretching work on large eddies, done by the mean flow, provides the energy to maintain turbulence. In a stationary state, it is precisely the energy produced at the large scale that is dissipated at small scales. Thus, though viscosity is responsible for dissipating the energy, it does not control the amount of

Fig. 6.30. Cascading decay of a vortex ring, *H.J. Lugt* (1983)

energy dissipated; this is set instead by the action of the large scales. This important property is characteristic of turbulence away from the boundary.

The amount of energy transformed into heat per unit volume per unit time, denoted by ϵ, is made up of the mean values of the squares and products of the partial derivatives of u', v', and w' with respect to x, y, and z. One can use ϵ and ν to define the characteristic length and velocity scales of these smallest scales as

$$l_{\mathrm{k}} = (\nu^3/\epsilon)^{1/4} \quad , \quad v_{\mathrm{k}} = (\nu \cdot \epsilon)^{1/2} \quad . \tag{6.38}$$

These are known as *Kolmogorov length* and *velocity* scales, respectively. It is easy to verify that the Reynolds number based on these scales is exactly unity, consistent with the idea that their order of magnitude corresponds to the smallest dynamical scale in turbulence.

The energy of the intermediate scales between L and l_k, which form a hierarchy, is given entirely by the consideration that their function is simply to transmit energy to the next smallest scales. Their amplitudes adjust themselves to the requirement that the rate of energy transmission be independent of the scale. Since the time scales associated with smaller length scales is shorter, the energy accordingly diminishes with decreasing scale size in a self-similar manner. Kolmogorov also postulated that the scales will become increasingly isotropic (i.e. direction-independent) as their size becomes smaller. This picture of turbulence is the basis for Large Eddy Simulation (LES) in turbulence modeling, Chapter 5.4.5.

Following the discussion above, the standard understanding is that there are only two length scales of intrinsic interest in turbulence, namely L and l_k. This is not expected to be true near the boundary or if there are multiple

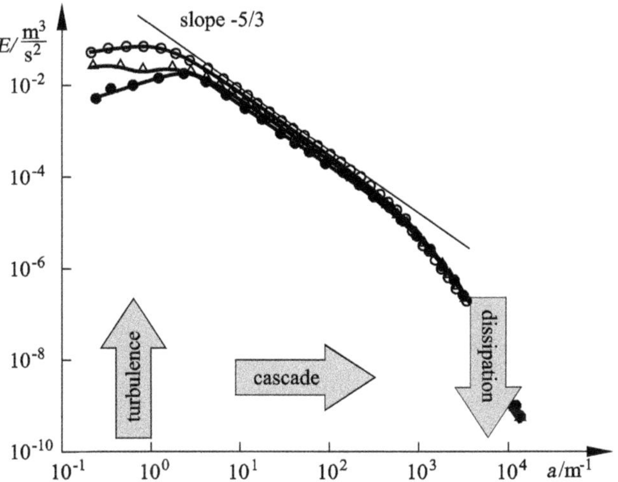

Fig. 6.31. Energy spectrum of turbulence, *F. H. Champagne* (1978)

mechanisms for the generation of turbulence. Even if the simple picture might be true, one can define other length scales. The most popular one is the so-called Taylor microscale λ:

$$\overline{\left(\left(\frac{\partial u'}{\partial x}\right)^2\right)} = \frac{\overline{(u'^2)}}{\lambda^2} \text{ const.} \tag{6.39}$$

In the case of isotropic turbulence (described in Section 6.4.4), *G.I. Taylor* showed that ϵ is given by the expression

$$\epsilon = 7.5 \cdot \mu \cdot \overline{(\partial u'/\partial y)^2}.$$

For other forms of turbulence (wall turbulence, free-shear flow turbulence), it is not clear that the dissipation can be related to the gradient of a single velocity gradient through a universal numerical coefficient as above, but the proportionality is still quite frequently used. If, for brevity, we write u' instead of $\sqrt{\overline{(u')^2}}$, we may put $\epsilon = \text{konst.} \cdot \mu \cdot (u'/\lambda)^2$. Since $u' = l \cdot |\partial \overline{u}/\partial y|$, we can put u'/l as an approximation for $|\partial \overline{u}/\partial y|$ and replace $|\tau|$ by $\rho \cdot u'^2$ in the equation $\epsilon = \tau' \cdot |\partial \overline{u}/\partial y|$. We thus have

$$\mu \cdot \left(\frac{u'}{\lambda}\right)^2 \cdot \text{const.} = \rho \cdot \frac{u'^3}{l} \quad ,$$

and so

$$\lambda = \text{const} \cdot \sqrt{(\nu l/u')}.$$

If $Re_l = u' \cdot l/\nu$ is introduced as the Reynolds number for the motion of an eddy, we have

$$\lambda \approx \frac{l}{\sqrt{Re_l}} \quad .$$

Thus, the Taylor microscale Reynolds number $u'\lambda/\nu$ is proportional to the square root of the large-scale Reynolds number $u'l/\nu$.

6.4 Classification of Turbulent Flows

The *mixing length* l in turbulent motion in general varies from place to place. As yet, no general theory is available regarding its magnitude, although in a number of particular cases it has been found possible to make assumptions leading to results in good agreement with experiment. In many cases it is permissible to neglect the actual shearing stresses arising from the viscosity in comparison with the apparent shearing stresses (see remarks above on turbulent transport and mixing). In other instances, the more far-reaching assumption is made that the effect of viscosity on the magnitude of l is negligible. In these cases, therefore, one may well be inclined to deal with the turbulence of an ideal fluid with zero viscosity. If the Reynolds number is sufficiently large, this point of view may well justified for certain purposes away from the walls.

We shall first discuss two cases in more detail, the so-called free turbulence and the turbulence that arises along a smooth boundary (Sections 6.4.1 and 6.4.2). The effect of viscosity in the latter case, the flow along a rough boundary and the flow past a plate, are discussed in Section 6.4.2. Section 6.4.3 deals with stratified fluid and the flow in curved flows, Section 6.4.4 with turbulence in wind tunnels (including some mention of isotropic turbulence), and Section 6.4.5 deals with two-dimensional turbulence. Finally, Section 6.4.6 contains a few elementary comments on the role of structures in turbulent flows.

6.4.1 Free Turbulence

In cases such as the mixing of a free jet (see Figure 4.62) having a sufficiently high value of Reynolds number with the fluid surrounding it at rest, it seems reasonable to take the mixing length for every cross-section as being proportional to the width of the jet there ($l = \alpha \cdot b$). By b we may, for example, mean half the base of a parabolic or paraboloidal distribution of velocity, in which the maximum velocity and quantity of fluid moving coincide with those of the actual flow considered. Some such assumption is necessary, since the actual flow passes, in an average sense, smoothly into the external fluid without any perceptible boundary. Making an assumption of this kind, we get values for α of approximately $1/8$ far away from the origin of the jet.

Observation shows that free round jets, for example of Figure 6.1, in a sufficiently large space full of fluid at rest spread out in such a way that except in the immediate neighborhood of the outflow, the width of the jet is proportional to the distance from the point of outflow, while the velocity is inversely proportional to that distance. Throughout the jet the pressure is nearly the same as in the surrounding fluid.

In discussions of an ordinary liquid spray, the assumption is sometimes made that there is a rise of pressure in the air jet as the velocity decreases, by Bernoulli's theorem, and that the pressure at the point of outflow is therefore reduced, thus causing fluid to be sucked up. This is incorrect: Bernoulli's theorem is true only when frictional stresses are absent, which is certainly not the case here. On the contrary, the suction is due to the flow around the edge of the tube that projects into the jet at right angles. In the spreading jet the pressure is practically the same as in the surrounding air at rest.

The decrease in velocity with increase of distance from the point of outflow is therefore due to the frictional stresses alone. Further, the decrease in velocity does not take place in such a way that the same quantity of fluid flows across all cross-sections. That this cannot be the case is clear because, during the advance of the flow, fresh masses of fluid at rest are carried along with it. This is called entrainment of the outer fluid into the jet. On the other hand, the momentum of the jet, $I = \rho \cdot \iint u^2 \, da$, is constant on account of the constant pressure. We have $I = \rho \cdot u_1^2 \cdot \pi \cdot b^2 \cdot \text{const}$, where u_1 is the maximum velocity in the cross-section. It follows from the fact that I is constant that u_1 is proportional to $1/b$, i.e. to $1/x$.

Another important case is that of the spread of the edge of a jet (Figure 6.32) as it exits from, say, a two-dimensional orifice; this is the so-called mixing layer. Here $u_1 = \text{const}$. If we put $l = \alpha \cdot b$, we have, as before, $\overline{\tau'} \sim \alpha^2 \cdot \rho \cdot u_1^2$; i.e. $\overline{\tau'}$ is also constant. The loss of momentum of the part of the flow coming from the orifice is proportional to $\rho \cdot u_1^2 \cdot b$, and the corresponding resistance is proportional to $\overline{\tau} \cdot x$, so that $b \sim \alpha^2 \cdot x$, as in the previous example. The loss of momentum and the resistance are calculated for a cross-section of unit depth in the direction perpendicular to the plane of the paper. The fluid entrained from the surrounding region at rest shows an equal gain of momentum. The slope of the boundary between the undisturbed portion of the jet and the turbulent zone is of practical importance. It may be taken as $1 : 10$.

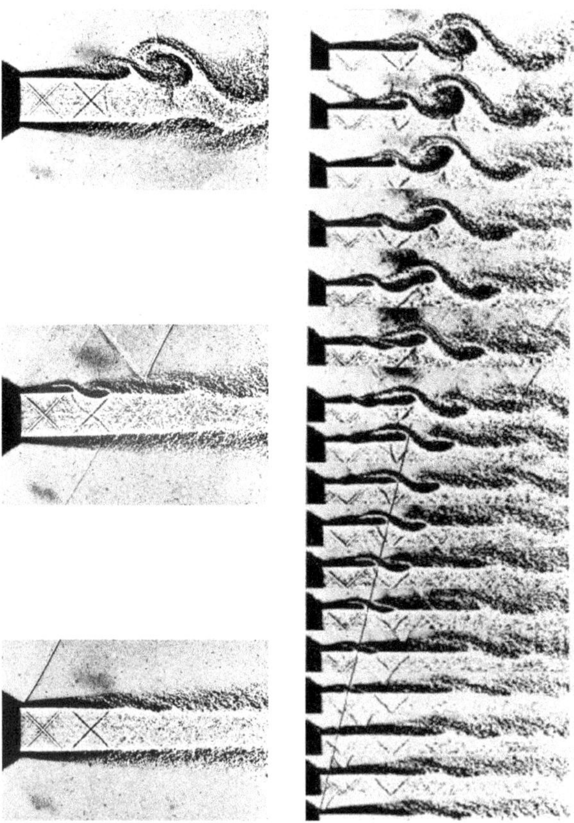

Fig. 6.32. Jet perturbation, *H. Oertel sr. and H. Oertel jr.* (1989)

Yet another case is that of the wake in the rear of a moving body (Figures 4.94 and 4.95). These and other canonical flows have been studied in detail, and a summary can be found in books such as *J.O. Hinze* (1975).

An important development to which we should draw attention is that the instantaneous boundary between the turbulent and nonturbulent parts of free-shear flows (see, for example, Figure 6.1) is quite well defined and relatively sharp at high Reynolds numbers. The instantaneous boundary is dynamic and is distinct from the average boundary of the flow, no matter how one defines the latter. This is also true of wall-bounded flows on the side exposed to the free stream (Figure 6.33). The turbulent transport is reasonably constant within this dynamic boundary, whereas it makes a sharp transition to zero as one cross it into the outside stream. Such boundaries, or interfaces, also exist for admixtures. In a given flow, the interfaces for turbulence itself and those for admixtures of various kinds do not necessarily match, either on the average or instantaneously. But all these interfaces have convolutions on many scales, from the largest possible to the smallest allowed by viscous or diffusive effects. The stochastic geometry of these boundaries in a range of scales can be described in terms of fractals (see, for example, *K.R. Sreenivasan* (1991)).

An observer placed near the free boundary of a turbulent flow will find himself sometimes immersed within the turbulent region and sometimes outside of it. If he stays close to the solid surface, he may be expected to remain within the turbulent region nearly all the time, whereas the fraction of time that this happens becomes increasingly smaller as one moves away from the surface. The average fraction of time one encounters the turbulent region is called the outer intermittency factor γ. Its behavior is shown on the right of Figure 6.38. δ_R is an effective thickness of the boundary layer obtained by equalizing the shaded areas above and below the intermittency factor. It is quite distinct from the intermittency of small scales to which we will return in Section 6.5.

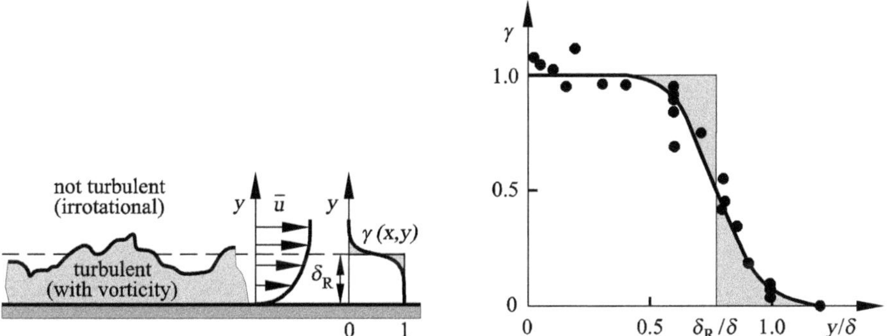

Fig. 6.33. Intermittency factor γ in the turbulent boundary-layer, *P. S. Klebanoff* (1955), y is now the wall-normal distance.

6.4.2 Turbulence near Solid Boundaries

In cases of flow along solid boundaries, which represents an important class, the *mixing length* must tend to zero as the boundaries are approached, as is clear from the definition of the mixing length. It follows that $\partial \overline{u}/\partial y$ reaches large values in the neighborhood of the boundaries, in relation to that in the interior of the flow. Figure 4.53 shows the differences between the distributions of velocity for turbulent flow and laminar flow in a tube.

Following Section 4.2.5, we might consider that a layer of fluid immediately next to the boundary lacks in its ability to transport mass, momentum and heat by turbulent mechanisms, even if the overall boundary layer is turbulent, because of the damping of wall-normal fluctuations imposed by the boundaries. This thin sublayer is formed according the rough definition $\partial \overline{u}/\partial y = \tau_{\text{wall}}/\mu$, provided that the boundary is smooth. It should be stressed that the viscous sublayer is highly disturbed and is far from being laminar, as it once was thought to be. For large values of the Reynolds number, the value of τ_{wall} is quite considerable, owing to the vigorous mixing in the interior of the flow, so that the rate of increase $\partial \overline{u}/\partial y$ is extremely rapid near the boundary of the viscous sublayer, which is accordingly very thin. Because the sublayer is so thin, to a superficial and global observation it might appear as if, in turbulent motion, the velocity has a finite value at the boundary itself, which is often used for technical turbulence modeling (Section 5.4.5).

From the theoretical point of view, a general idea may be obtained simply if we assume that the shearing stress is constant throughout the region outside the viscous sublayer. In reality, τ decreases continuously as the distance from the boundary increases beyond a point. (For the pipe, τ becomes zero on the axis.) Nevertheless, the formulas obtained by putting $\tau = \text{const} = \tau_{\text{wall}}$ give very useful approximations at least up to a wall-normal distance that is not directly affected by the outer intermittency. For pipes, the formulas below hold nearly to the axis. The total shearing stress ($\tau = \overline{\tau} + \tau'$, the mean value of the viscous stress plus the apparent shearing stress due to turbulence) is then given by

$$\tau = \mu \cdot \frac{\partial \overline{u}}{\partial y} + \rho \cdot l^2 \cdot \left(\frac{\partial \overline{u}}{\partial y} \right)^2 . \tag{6.40}$$

The first term is important only for very small distances from the boundary. Outside this region, the second term is so much greater than the first if the Reynolds number is large, that the first may be neglected in comparison with it. Taking the square root of the resulting simplified form of equation (6.40), we have

$$\sqrt{\frac{\tau}{\rho}} = l \cdot \frac{\partial \overline{u}}{\partial y}. \tag{6.41}$$

From the right-hand side we readily see that $\sqrt{(\tau/\rho)}$ has the dimensions of a velocity. For simplicity, we introduce the symbol u_τ and call it the friction velocity. It is of the same order of magnitude as the velocities u', v'

due to turbulence (or, more accurately, u_τ is of the order of $\sqrt{\overline{(u'v')}}$). With the assumption we have made here, however, u_τ is a constant for a given streamwise position along the flow.

We shall now suppose that $y = 0$ represents a smooth wall, and, for simplicity, regard it as extending to an infinite distance in both horizontal directions. We shall assume that another wall is at an infinite distance in y away from the first wall. Then $\overline{u}$ depends on y only. In what follows, therefore, we shall write $d\overline{u}/dy$ for $\partial\overline{u}/\partial y$, and since we shall momentarily not be concerned with fluctuations, we shall also drop the averaging symbol.

We have now to find a reasonable law for the mixing length l, i.e. one that gives the correct dimensions. If we make the further assumption (suggested by observation) that l is unaffected by fluid viscosity, the only length we have at our disposal is the distance from the wall y. The only dimensionally correct formula for l is then

$$l = \kappa \cdot y. \tag{6.42}$$

The numerical factor κ is essentially a universal constant of this problem in turbulent flow. It is known as the Kármán constant, due to *Th. von Kármán.* From equation (6.41) we then have

$$u_\tau = \kappa \cdot y \cdot \frac{du}{dy} \quad . \tag{6.43}$$

Since u_τ is constant, this can be immediately solved, giving (4.82)

$$u = u_\tau \left(\frac{1}{\kappa} \cdot \ln y + C \right) \quad . \tag{6.44}$$

For large values of the Reynolds number this expression is in reasonable agreement with observation, with 0.41 as the accepted value of the Kármán constant. It is true that for $y = 0$ the formula gives the value $-\infty$ instead of the value 0, but we know already that our simplified calculation will not apply at or near $y = 0$; instead, we have to use the more accurate equation (6.40) and set up a modified formula for l involving the second length ν/u_τ. We shall discuss the role of this second length scale later.

We can also obtain an expression for C, the constant of integration in equation (6.44), from the fact that the viscosity becomes important in the immediate neighborhood of the wall. The expression in parentheses in equation (6.44) must be a pure number and must not depend on the units employed. This is achieved if we subtract from $\ln y$ the logarithm of the length ν/u_τ mentioned above, i.e. if we put

$$C = C_1 - \frac{1}{\kappa} \cdot \ln \frac{\nu}{u_\tau}. \tag{6.45}$$

Then C_1 is a second universal number, and we have (4.83)

$$u = u_\tau \left(\frac{1}{\kappa} \cdot \ln \frac{y \cdot u_\tau}{\nu} + C_1 \right) \quad . \tag{6.46}$$

Since the greatest velocity differences occur in the immediate neighborhood of the wall, equation (6.46) may also be used as a good approximation in cases in which the shearing stress τ depends mildly on y. We have merely to set $u_\tau = \sqrt{(\tau_{\text{wall}}/\rho)}$, and obtain values of the velocity that are found to lie very close to the observed values. For these cases that deviate from the theory, e.g. for flow in pipes, the observed values of u/u_τ can be plotted against $\log_{10} \frac{yu_\tau}{\nu}$. The curve obtained is almost a straight line. If equation (6.46) is used in this way as an approximation to the distribution of velocity in smooth-walled straight pipes, Nikuradse's experiments (*J. Nikuradse* (1932)) give $\kappa = 0.40$ and $C_1 = 5.5$. Passing from natural logarithms to ordinary logarithms, we obtain

$$u = u_\tau \cdot \left(5.75 \cdot \log_{10}(\frac{y \cdot u_\tau}{\nu}) + 5.5\right) \quad . \tag{6.47}$$

M.V. Zagarola and *A.J. Smits* (1998) have extended the range of pressure drop measurements in a pipe up to about 36 million in the Reynolds number based on the pipe diameter, thus extending Nikuradse's range by a factor of about 10. They confirm the existence of a logarithmic region (though the Kármán constant in these measurements is a few percent different).

It should be mentioned that there is a different scheme of describing the velocity distribution in pipe flows (and, in general, in wall-bounded flows). This scheme, in its modern form, is due primarily to *G.I. Barenblatt* (1993). It proposes that (6.43) is not strictly valid because the influence of the second length scale, namely ν/u_τ, never strictly disappears but remains intact, though perhaps only weakly. Loosely speaking, this expectation is in keeping with the spirit of the behavior of condensed matter near the critical point. Instead of equation (6.43), one then has

$$\frac{\mathrm{d}u}{\mathrm{d}y} = \frac{1}{\kappa} \cdot \frac{u_\tau}{y} \cdot \left(\frac{y \cdot u_\tau}{\nu}\right)^\beta \quad , \tag{6.48}$$

where β is an undetermined constant. Integrating the equation, one can see that a power law emerges for the velocity distribution. *Barenblatt* and his collaborators have examined the data of *Nikuradse*, and also those of *Zagarola and Smits* in the lower range of Reynolds number, and concluded that the power law provides a better fit to the velocity distribution than the logarithmic law. They have determined the constants in the power-law velocity distribution by empirical fit to the data.

The question is not merely one of which of the two forms fits the data better, but is one of principle. Even at high Reynolds numbers, and not too close to the wall, does the influence of the second length scale ν/u_τ disappear altogether, or remain weakly present? A firm answer to this question will seriously influence our thinking on how one quantity scales with another in wall-bounded flows.

6.4.3 Rotating and Stratified Flows

So far, we have not considered the effects of *rotation* and *density stratification* that are evident in most natural flows (see Chapter 10). Such effects are particularly important in geophysical flows. Large-scale flows such as hurricanes are clearly affected by both Earth's rotation and density stratification. These effects are sometimes important even on the laboratory scale. One need only consider the bathtub vortex and the direction of rotation of the fluid as it nears the drain.

The main effect of *rotation* is to introduce *centrifugal* and *Coriolis forces.* The centrifugal force always acts perpendicular to the axis of rotation, and is similar in structure to the pressure gradient, with which it is often considered together. In the case of the flow past curved objects, turbulence is diminished or increased as a result of the centrifugal forces, according to whether the velocity increases or decreases from the center of curvature outwards. Here the variation in magnitude of the centrifugal forces plays the same role as that played by variation of the force of gravity in the flow of layers differing in density.

The Coriolis force, which acts perpendicular to the axis of rotation and is perpendicular to the relative velocity, may be explained as follows. If a fluid mass moves from Earth's equator to the north, it crosses latitudes with decreasing radius. To preserve its angular momentum, the fluid parcel has to spin faster and thus move to the right. A fluid parcel moving toward the equator will have to slow down and move, relative to Earth, to the left. The movements in the Southern Hemisphere are just the opposite. The Coriolis force, which thus depends on the latitude, is proportional in magnitude to the sine of the latitude, and is a source of additional vorticity and turbulence in rotating systems.

The precise circumstances in which the Coriolis force is important depend on the relative magnitude of other forces. The ratio of inertial to Coriolis forces is called the *Rossby number.* A second parameter, called the *Ekman number*, is the ratio of frictional forces to the Coriolis force. In most geophysical flows, which include atmospheric and oceanic motions, the inertial force is by far stronger than the frictional force, so it is often the Rossby number that is important. In the boundary-layers, of course, the *Ekman number* is also of consequence.

An additional effect is due to *density stratification.* In a flow that is predominantly horizontal, if the density of the medium diminishes rapidly upward as, for example, in a mass of air with the temperature increasing upward, or where there is a layer of fresh water superimposed on salt water, the process of turbulent mixing must cause heavier layers to be moved above the lighter, and lighter layers to be pushed down below the heavier. That is, part of the work available for the maintenance of turbulence, derived from the main flow, is used up against gravity. This may cause the turbulent motion to be diminished and possibly die out altogether. This is the explanation of the cessation of turbulence and dying down of the wind at night in the lower layers of the atmosphere, with the wind still continuing unabated at a higher level. Conversely, turbulence is increased by irradiation from the ground, which causes a reversal of the stratification, resulting in dense layers higher than less dense ones. This is what happens, for example, in so-called Rayleigh–Bénard convection, in which a fluid layer contained between two horizontal plates is heated at the lower plate and cooled at the top plate (see Section 7.2.1 and Figure 1.5).

6.4.4 Turbulence in Wind Tunnels

Much attention has been devoted to turbulence in wind tunnels, in which turbulence is undesirable, since one purpose of experiments in wind tunnels is to simulate the state of the flow past a body moving at a uniform speed through air at rest. Such tests are rather important for the development of new designs of automobiles to aircraft, as well as stationary objects such as bridges and towers that are exposed to winds. Turbulence, however, cannot be entirely avoided in wind tunnels. Residual turbulence exists even after the air has passed through a honeycomb and screens at the entrance section of the tunnel (see Figure 6.34). This particularly affects the occurrence of turbulence in the boundary-layers on bodies under investigation, and also the separation of the flow from the bodies. Separation changes the character of the flow near the wall and affects transport properties immensely. Needless to say, controlling the wind-tunnel turbulence is especially important in studies of laminar–turbulent transition in boundary-layers and other flows.

The earliest way of measuring turbulence in a wind tunnel was by the fall in the drag of a sphere due to the onset of turbulence in the boundary-layers. Later, *G. B. Schubauer, H. K. Skramstad* (1947), and *H. L. Dryden* (1948) worked out methods using hot-wire anemometers, by which numerical values for the small fluctuations of velocity could be obtained quite reliably. It was found that wind-tunnel turbulence (or, more generally, any turbulence arising from flow through a grid of bars) has simple properties at sufficient downstream distances from it. It is found to be nearly *homogeneous and isotropic*; that is, the fluctuations of velocity are of the same magnitude across

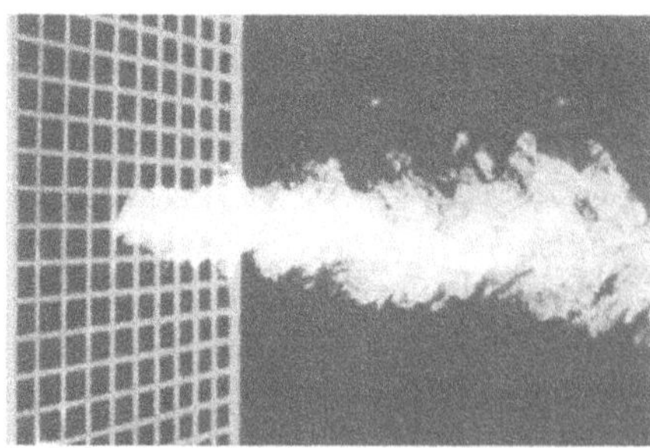

Fig. 6.34. Turbulent flow behind a wind-tunnel honeycomb, *M. Lesieur* (1997), picture by *J. L. Balint et al.*

the wind tunnel cross-section (except very close to the wind tunnel walls), and their average measures are the same in all directions as well.

The simplest statistical quantity is the mean energy of fluctuation

$$E = \frac{1}{2} \cdot \rho \cdot \left(\overline{(u'^2)} + \overline{(v'^2)} + \overline{(w'^2)} \right) = \frac{1}{2} \cdot \rho \cdot q^2 \quad . \tag{6.49}$$

From a series of measurements made for a grid of mesh-width m through which fluid moves with mean velocity $\overline{U}$, it is now known that q decays, over some intermediate distance from the grid, as a power of that distance. Equivalently, in situations where the turbulence is generated by sweeping a grid of bars at velocity $\overline{U}$ through a fluid medium at rest, the decay of the energy follows a power law in time. The power-law exponent is roughly -1.25. It is not clear whether this exponent is universal (experiments yield a value roughly between 1 and 1.4), or depends weakly on a number of features such as m, the diameter of the rod, the geometry of the rod itself, and on whether the grid is passive or has some moving elements in it. The constant of proportionality in the formula is indeed nonuniversal and depends strongly on the details just mentioned.

Isotropic Turbulence

As the name suggests, *isotropic turbulence* (see Figure 1.4) has no directional preference and is a mathematical construct. In fact, turbulence can be generated only in the presence of local shear or near boundaries, and the process of generation of turbulence tends to maintain a preferred direction. However, the turbulence that is found far enough away from the boundary where the mean velocity gradients are small is often approximately isotropic. The turbulence generated behind grids is roughly isotropic sufficiently away from the grid. Further, small scales of turbulence in all flows tend to be *statistically* isotropic though individual structures do show deviations from isotropy. For all these reasons, isotropic turbulence is of some interest. In any case, this is the form of turbulence most accessible to theoretical development, and has consequently assumed an importance in its own right. Isotropic turbulence is also homogeneous, though the mention of the latter is often omitted for brevity.

Homogeneous and isotropic turbulence is a paradigm that can be dealt with up to a point by statistical theory and by experiments suggested by theoretical work. Special reference should be made to the work by *G. I. Taylor* (1935, 1936), who introduced the concept, and *Th. von Kármán* (1948), who was responsible for deriving an important equation for statistical quantities from the Navier–Stokes equations. A detailed discussion can be found in *A. S. Monin, A. M. Yaglom* (1975).

An indication of the spatial character of velocity fluctuations may be obtained by studying the *correlation* between the velocities at neighboring points 1 and 2. For isotropic turbulence there is only one independent correlation function, which is a function of the distance r. In Figure 6.35, R is the correlation between the components of velocity at 1 and at 2 parallel to r:

$$R(r) = \frac{\overline{u'_1 \cdot u'_2}}{\sqrt{\overline{u'^2_1}} \cdot \sqrt{\overline{u'^2_2}}} \quad . \tag{6.50}$$

From the graph of R the characteristic length of turbulence can be defined as

$$\int_0^\infty R(r) \cdot \mathrm{d}r = L \quad .$$

It is closely related to the mixing length l. The value of L in Figure 6.33 is a measure of the large eddies in the turbulence motion, in which the energy of turbulence is controlled by the manner in which turbulence is produced. According to *G. I. Taylor* (1936), the statistical mean value of the dissipation is proportional to

$$\mu \cdot q^2 \cdot \left(\frac{\mathrm{d}^2 R}{\mathrm{d}r^2}\right)_{r=0} \quad , \tag{6.51}$$

where

$$\left(\frac{\mathrm{d}^2 R}{\mathrm{d}r^2}\right)_{r=0} = \frac{1}{\lambda^2} \quad .$$

The Reynolds number based on λ, the Taylor microscale introduced in Section 6.3.3, and the root-mean-square fluctuation velocity u', is often used to compare properties among different flows for which the characteristic large scale depends on the geometry, and is thus not a useful scale of comparison.

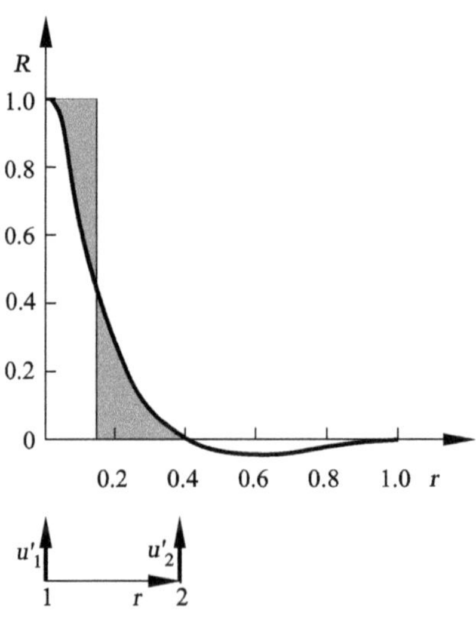

Fig. 6.35. Correlation of velocity fluctuations. The integral in the figure extends only up to the first crossing of the correlation, as is often done experimentally, and the shaded areas above and below the correlation function are equal.

It should be remembered that λ does not represent the smallest scales of turbulence. That function is served by the *A. N. Kolmogorov* scale l_k (6.58).

The main dynamical problems in isotropic turbulence are the nonlinear transfer of energy from one scale to another, and its eventual dissipation to heat. On average, the energy transfer occurs from large scales to small scales, though instantaneously, there is some two-way transfer, whose details are not fully understood. The average transfer is assumed to proceed from one scale to a neighboring smaller scale in the form of an energy cascade. When the scales involved are large, that is, their characteristic Reynolds numbers, based on their own size, are sufficiently high, it is assumed that the scales merely transmit energy to the next smaller ones without dissipating any part of the energy. When the energy reaches the smallest scales, it is presumed to be dissipated there. If the cascade picture holds for any type of turbulence at all, isotropic turbulence is the most likely candidate.

One consequence of the energy cascade is that when the scales that contain most of the energy (of order L) and the scales that dissipate most of the energy (of order l_k) are significantly disparate, the energy dissipation rate is the same as the rate at which energy is being pumped into turbulence at large scales—as already discussed. This equality has been verified both experimentally in grid turbulence and by solving the equations of motion on a high-speed computer, as long as the Reynolds number of turbulence is sufficiently high for the said scale separation to exist. Thus arises the notion that the energy dissipation rate in high-Reynolds-number turbulence is independent of fluid viscosity. This seemingly anomalous behavior is of great consequence, and shows that the limit of high Reynolds number (or vanishing viscosity) is not the same as the case of zero viscosity. It may be recalled that this feature is common to all singular perturbation problems including boundary-layers.

A significant contribution related to energy cascades is due to *A. N. Kolmogorov* (1941), which is the basis of turbulence modeling in Chapter 5.4.5. Since this work is related strongly to scales that are considerably smaller than that at which turbulence is produced, a somewhat more detailed description is postponed to the later section on small scales of turbulence. One result may, however, be worth presenting here. This result is thought to hold even for general anisotropic turbulence for the so-called inertial range of scales, which is smaller than the energy-containing scales L and larger than the dissipating scales l_k. In this range, the energy transfer process adjusts itself so that the spectral distribution of energy is given by (see Figure 6.31)

$$E(a) = C_k \cdot \epsilon^{2/3} \cdot a^{-5/3} \quad , \tag{6.52}$$

where C_k is the so-called Kolmogorov constant and ϵ is the rate of energy dissipation. The integral of $E(a)$ over all wave numbers a gives the total turbulent kinetic energy. Here, the wave number a takes the role of distinguishing different scales of turbulence: Small values of a correspond to large scales, and large a represent small scales. The constant C_k cannot be deduced theoretically but is known from experiment to be a constant of about 1.5 at high Reynolds numbers. Note that Figure 6.31 indeed corresponds to a flow that is anisotropic on the large scale.

As already remarked, isotropic turbulence has been studied in wind tunnels behind a grid of bars, or by pulling a grid of bars through a stationary mass of fluid. Recently, as computer power has increased, the Navier–Stokes equations of Section 5.2 have been solved numerically in periodic boxes, starting with an initial realization of a prescribed random field. In due course, the computer solutions attain properties that are essentially independent of the initial conditions and replicate those of measured turbulence. Such simulations have provided a powerful tool for understanding turbulence in general, and isotropic turbulence in particular. An interesting result to emerge is that the structure at small scales is in the form of intense vortex tubes that are long compared to their diameter. The vortex tubes form mosaics of several different scales. It is not yet clear whether this observation is of fundamental consequence to the theory of turbulence.

6.4.5 Two-Dimensional Turbulence

A study of the appropriate equations shows that, as a rule, the components of turbulent fluctuating velocities in all three directions tend to be of the same order of magnitude except close to solid surfaces (where there is a preferential damping of the wall-normal velocity), or when certain types of body forces act on the flow. This is true even in flows that are two-dimensional on average such as boundary-layers on extended flat plates, or wakes behind long cylinders, which do not have significant average variations along the span. There are circumstances, however, in which the turbulence fluctuations are close to being two-dimensional (i.e. fluctuations are largely planar).

Examples are atmospheric and oceanic flows (see Chapter 10), which often have a very large spatial extent in two directions and a relatively short extent in the direction of their depth. Such flows occur in a stratified, often rotating, environment and are central to understanding and predicting weather, dispersion of particles and chemicals in the atmosphere and oceans, and other natural phenomena. A laboratory realization of two-dimensional turbulence is the turbulent flow on a soap film, which is shown in Figure 6.36.

While these examples are not purely two-dimensional, there is promising evidence that the strictly two-dimensional mathematical approximation will allow us to make some headway. On the experimental front, there has been some success in generating in the laboratory close approximations to two-dimensional flows that compare well to both natural flows and the mathematical ideal. Two-dimensional turbulence is also studied with the expectation that it could provide insight into the three-dimensional problem. For instance, the two problems have in common fundamental properties such as energy transfer between scales, dissipation mechanisms, and structure formation and evolution.

Major Theoretical Results

The relative simplicity of the two-dimensional Navier–Stokes equation allows several fundamental properties to be derived. The first result can be derived in a straightforward manner by taking the curl of the Navier–Stokes equation for an incompressible fluid, and taking the inviscid (Euler) limit. We obtain the *Helmholtz theorem*

$$\frac{\partial \omega}{\partial t} + \boldsymbol{v} \cdot \nabla \omega = 0 \quad , \tag{6.53}$$

where the vorticity $\omega = \nabla \times \boldsymbol{v}$ is always along the axis normal to the plane. Here arises a fundamental difference from the situation in three dimensions: The Helmholtz equation means that vorticity of a fluid parcel is conserved through the lifetime of turbulence. In contrast, three-dimensional turbulence permits an additional vortex-stretching term $(\omega \nabla \boldsymbol{v})$, which is nonzero due to the presence of the additional degree of freedom in the third dimension. Furthermore, the restriction of the flow to the plane results in the following equations for energy $E = \frac{1}{2}\langle \boldsymbol{v}^2 \rangle$ and enstrophy $\Omega = \langle \omega^2 \rangle$ in two-dimensional homogeneous turbulence:

$$\frac{\mathrm{d}E}{\mathrm{d}t} = -\nu \cdot \Omega \quad , \qquad \frac{\mathrm{d}\Omega}{\mathrm{d}t} = -\nu \cdot \langle (\nabla \omega)^2 \rangle \quad . \tag{6.54}$$

Here, the angular brackets imply suitable averaging and their distinction from the overbar used in earlier sections is not important for present purposes. For three-dimensional turbulence, the zero-viscosity limit is known to lead to an increase of enstrophy, because, in that limit, viscous diffusion of vorticity decreases and stretching of vortex lines is less restrained. Thus, as already mentioned, the rate of energy dissipation for three-dimensional turbulence remains finite even in the inviscid limit. In two dimensions, however, the enstrophy changes only due to viscous effects, and thus can only decrease. This leads to zero rate of energy dissipation in the inviscid limit (6.54). *G. K.*

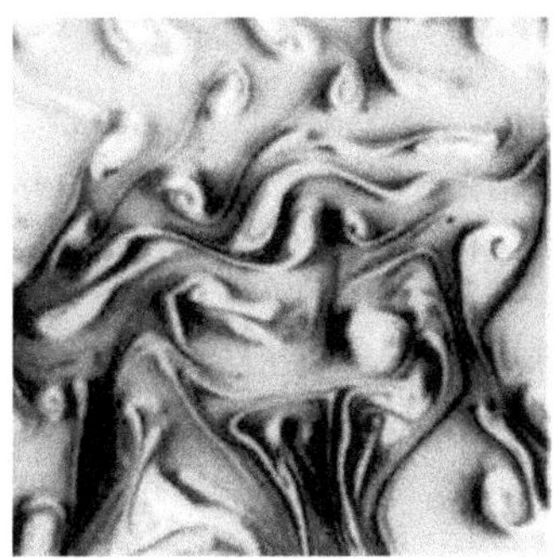

Fig. 6.36. Turbulent flow on a soap film, *P. Vorobieff, R. E. Ecke* (2003)

Batchelor (1948) provided arguments that the rate of dissipation of enstrophy is nonzero in the inviscid limit in the two-dimensional case; this is the so-called enstrophy dissipation anomaly.

The final picture, then, is that two-dimensional systems do not dissipate energy in small scales. The energy is transported to larger scales and eventually gets dissipated by friction at the boundaries of a finite system. On the other hand, enstrophy is allowed to cascade down the scales to be dissipated in the small scales. Therefore, there appears to be some value to casting the two-dimensional enstrophy (vorticity) as analogous to three-dimensional energy (velocity). This was the approach of *R. H. Kraichnan* (1967).

The Energy and Enstrophy Cascades

R.H. Kraichnan (1967) recognized that the enstrophy and the energy cascades can exist simultaneously in two dimensions. From the study of the conservation equations and triadic wave number interactions, it can be shown that energy is transferred, on average, toward small wave numbers (large scales), while the enstrophy is transferred toward large wave numbers (small scales). The prediction for the energy spectrum in the inverse cascade is a scaling law $E(a) \sim a^{-5/3}$, which has been verified in numerical simulations (*G. Boffetta* (2007)) and experiments (*J. Somméria* (1986), *P. Tabeling* (1997)). The inverse energy cascade implies a mechanism by which large eddies are created from small eddies instead of the other way around, as in three-dimensional turbulence.

The phenomenological picture is that the initial vortices, formed by the forcing, get conjoined to other vortices to form larger ones during their lifetime, i.e. in the time it takes for friction at the boundary to damp them out by depleting all their energy. The three-dimensional Richardson cascade of the breakup of eddies is replaced by an aggregation process among vortices in two dimensions. The Kraichnan conjecture for a (stationary) inverse cascade seems to hold only if there is a sink for energy at large scales. While the fluid itself has no such property, the boundary conditions in both simulations and experiments provide the artificial sink for energy, for example, the friction at the walls. This allows for observation of a sustained (stationary) inverse energy cascade.

The enstrophy Ω, as already mentioned, is dissipated in the small scales in the inviscid limit. In a forced two-dimensional system the enstrophy cascades from the energy injection scale down to small scales. The enstrophy spectrum in the inertial range, according to the theories of *G. K. Batchelor* and *R. H. Kraichnan*, has the behavior $\Omega(a) \sim a^{-1}$. The corresponding energy spectrum in the inertial range follows to be $E(a) \sim a^{-3}$. Experimental observations of the decaying energy spectrum have yielded slopes ranging from -3 to -4 over varying times and ranges of initial conditions. A full description of these aspects is summarized by *P. Tabeling* (2002).

6.4.6 Structures and Statistics

Both forced and decaying two-dimensional turbulence have a well-documented tendency to form coherent structures. The remarkable feature of the two-dimensional coherent structures observed in both numerical and experimental work is their long life-times. Generally speaking, much energy has been put into identifying the coherence in vortical structures, determining their stability properties, and analyzing the dynamics of vortex interactions including merging. The goal is to provide a satisfactory link between statistical theories and coherent structures of turbulence. This approach has been more successful in two dimensions than in three.

G. K. Batchelor (1969) was the first to propose self-similarity in time of the decay process in two dimensions. A dimensional argument led him to the following estimate of the decay rate of the vortex density ρ,

$$\rho \sim E^{-1} \cdot t^{-2}, \tag{6.55}$$

where E is the kinetic energy density. The same dimensional analysis shows that both vortex size and intervortex spacing grow at a rate linear in time t. This was the initial attempt at a statistical description. It was soon discovered in numerical simulations (*J. C. McWilliams* (1990)) that although power laws seem to hold, the exponents of the decay deviated from Batchelor's prediction. The vortex density decayed more slowly, as did the growth of their size and spacing. These numerical observations have more recently been supported by experimental data. *G. F. Carneve et al.* (1991) proposed that another invariant must be present in the system in addition to E. This invariant is the global maximum vorticity of the system. While the physical justification for this quantity as an invariant of a decaying system is not rigorous, it seems to derive reasonable numerical support. On recalculation of the scaling exponents , good agreement with empirical evidence is achieved, supporting this framework, known as the *universal decay theory*.

The subject of structures in three-dimensional turbulence is less well developed, despite considerable effort. That large scale structures do exist is clear even to a casual observer but what is not clear is the degree of their temporal coherence at very high Reynolds numbers. They are, of course, far

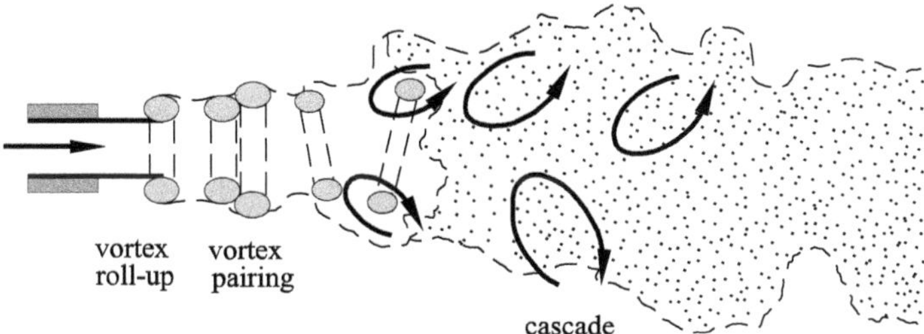

Fig. 6.37. Coherent structures near the exit of an axisymmetric turbulent jet

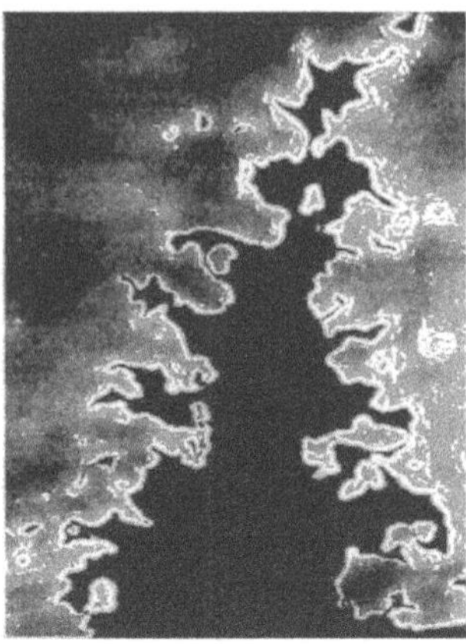

Fig. 6.38. Coherent structures in a turbulent flame (see Section 9.3.6)

more evident in special cases such as the mixing regions near the jet exit (Figure 6.37) and in low-Reynolds-number flames (see Figure 6.38 which shows a laser-induced fluorescence sheet (LIF) of the OH concentration of a turbulent premixed air–gas flame). There is considerable discussion as to whether the largest scales, which give the shape to a flow, are efficient in transporting heat, mass and momentum. The same questions apply for reacting flows, and examples of coherent structures in turbulent flames are discussed in detail in Chapter 9.

On the other hand, it is quite clear that the small-scale and intense vortical linking structures to statistical theories do tend to arrange themselves coherently, with a diameter that scales on the Kolmogorov scale and length somewhere between the Taylor microscale and the integral scale. While one would imagine that they, too, play a role in turbulence dynamics, we have to emphasize that the relation between the observed structure and the dynamical quantities that one usually measures is not fully clear. We should reemphasize, that a central problem in turbulence is to connect the structure and dynamics in some fundamental and systematic way. In this context, concepts such as helicity (which is the volume integral of the scalar product of the velocity and vorticity, see *H. K. Moffatt* (1969)) and reconnections (see *S. Kida, M. Takayoka* (1994)) play an important role.

6.5 Some New Developments in Turbulence

6.5.1 Decomposition into small and large scales

The past few decades have seen an increased interest in the statistical descriptions of turbulence, and the desire to incorporate the observed structure in such descriptions. While turbulence involves the creation and interaction of structures and patterns of different length scales, some of them coherent, its vast spatial and temporal complexity necessitates a stochastic description. It is hoped that a probabilistic description will yield a simplified picture of its universal properties. The length scales within which universality may be applicable are smaller than the large scale L that characterizes the size of the system, or of the manner in which turbulence is generated. The focus on small scales, while offering plenty of promise, tends to gloss over large-scale phenomena such as structure formation and coherence, and the sweeping effects on the small structures. Certain properties of the large-scale motion have their origin in flow instability, but they are nonuniversal in that their shapes, onset, and precise manifestation differ from flow to flow. The two regimes of turbulence, namely the small and large scales, have often been examined independently of each other, based on the assumption that sufficient separation between them offers independence from each other. In reality, of course, this independence is to be regarded only as a convenient model.

We first present a summary of the experimental methods in use, and then discuss some recent work.

Experimental Methods

The measurement of small-scale, rapidly fluctuating quantities such as velocity and velocity derivatives is still most successfully done using thermal anemometry and hotwire probes (see, e.g. *H. H. Brunn* (1995) for a survey of the methods). Data from such measurements are used to calculate statistics of flows ranging from mean properties to high-order moments such as Reynolds stress and structure functions (which are moments of velocity differences between two neighboring points in space). A limitation of hotwire data is that their spatial information is often obtained by some means of surrogation, for instance the use of Taylor's hypothesis, which assumes that the flow is swept past a probe without any distortion, at the local flow speed. Of course, multiple probes can be, and have been, used to transcend this limitation but there is a limit beyond which this escalation becomes both cumbersome and invasive. In its simplest form, laser Doppler velocimetry (LDV) again yields single-point measurements. The advantage of LDV is that it is noninvasive (see *F. Durst* (1980)) and can be used in hostile environments such as flames. The need for full spatial information has led to the development of particle image velocimetry (PIV); see *Raffel et al.* (2007). However, the advantage of PIV over hotwire (or LDV) is sometimes constrained by the present technology, which places limits on the temporal resolution attainable, and hence on resolution of the fluctuations at high Reynolds numbers. A recent effort to remedy this constraint of classical PIV has been made. High-energy particle detectors have been modified to serve as optical imaging devices for tracking particles in a high-Reynolds-number flow (*G. A. Voth et al.* (1998)). Finally, the incentive to create very high Reynolds number flows under controlled laboratory conditions has motivated the use of low-viscosity cryogenic helium as a test fluid (see, e.g. *K. R. Sreenivasan, R. J. Donnelly* (2000)). (Helium below the so-called λ-point, roughly 2.3 K, is bestowed with the property of *superfluidity* and generates *quantized vortices.* Tangles of quantized vortices, called *quantum or superfluid turbulence*, possess properties which are analogous to those of classical turbulence; see *C. F. Barenghi et al.* (2001).) Facilities based on highly compressed air *M. V. Zagarola, A. J. Smits* (1998) and that being built in Göttingen using compressed sulfur-hexafluoride are also noteworthy for their versatility. In all these methods, very high Reynolds numbers can be achieved in a moderately sized apparatus but their quantitative measurements are still not satisfactory because of limitations of the instrumentation.

Small-Scale Turbulence

To study small-scale turbulence, one needs measures that are independent of the large-scale motion on which small scales are thought to be superimposed. A simple such measure is the velocity difference between two points separated by a distance r that is small compared to the large scale L. It is generally assumed that such quantities, for $r \ll L$, behave as in isotropic turbulence. This is the assumption of *local isotropy*. The rate at which anisotropic effects of the large scale diminish with the reduction in scale is a subject of much study and practical interest, and a survey can be found in *S. Kurien, K. R. Sreenivasan* (2001).

One exact relation valid at high Reynolds numbers is the so-called Kolmogorov's law, according to which the following relation holds in the inertial range $l_k \ll r \ll L$:

$$\left\langle (u(x+r) - u(x))^3 \right\rangle = -\frac{4}{5} \cdot \langle \epsilon \rangle \cdot r \quad . \tag{6.56}$$

This law has provided the basis for an enormous volume of work. The classical interpretation of equation (6.56) (e.g. *A. S. Monin, A. M. Yaglom* (1975)) is that the energy flux from large to small scales is unidirectional on average. Other attempts have been made to extract more information from this equation. The equation fixes the extent of the inertial range in experiments and estimates $\langle \epsilon \rangle$ with less ambiguity than the local isotropy relation $\langle \epsilon \rangle = 15 \cdot \nu \cdot \left\langle (\partial u / \partial x)^2 \right\rangle$.

Extrapolating the implications of Kolmogorov's arguments for higher-order moments of velocity increments, we have

$$\langle (u(x+r) - u(x))^{\mathrm{n}} \rangle = \mathrm{C_n} \cdot (\epsilon \cdot r)^{\frac{\mathrm{n}}{3}} \quad , \tag{6.57}$$

The spectral equivalent of (6.57) for the special case with $n = 2$ is the one-dimensional version of (6.52) which we have already encountered. It can be written as

$$\phi(a_1) = C_K \cdot \langle \epsilon \rangle^{2/3} \cdot a_1^{-5/3} \quad , \tag{6.58}$$

where $\phi(a_1)$ is the one-dimensional spectrum (We have already encountered the value of the Kolmogorov constant for three-dimensional spectrum in

(6.52)) of the wave number component a_1, and C is the Kolmogorov constant in one-dimensional spectrum. *H. L. Grant et al.* (1962) verified equation (6.58) adequately for the first time. Subsequent investigators have also found the spectral slope to be close to 5/3. The value of C, determined empirically, is about 0.5.

In the dissipation range, Kolmogorov's arguments yield the following result for the spectral density:

$$\phi(a_1) = \mathrm{f}(A) \cdot \langle\epsilon\rangle^{2/3} \cdot a_1^{-5/3} \quad , \tag{6.59}$$

where $A = a_1 \cdot l_\mathrm{k}$ is the wave number normalized by the Kolmogorov length scale $l_k = (\nu^3/\langle\epsilon\rangle)^{1/4}$, and the universal function $f(A)$ is unknown (except that it approaches C for small A). From numerical simulations at low Reynolds numbers, it appears that the spectral density is of the form $A^{a'} \exp(-g \cdot a_{1\eta})$, where $a' \approx 3.3$ and $g \approx 7.1$, though it appears to be smaller at higher Reynolds numbers. Experimental data support equation (6.59) to some extent, but the situation is not fully satisfactory because of data scatter. A different type of spectral universality in the dissipation region has been proposed on the basis of multifractality of the small scale. For a discussion of this approach, see *U. Frisch* (1995).

The present situation is such that it is not possible to state that (6.57) works exactly, even for second-order statistics. There certainly appear to be departures from (6.57) for large enough n. In atmospheric boundary-layers, in high-Reynolds-number air and helium flows, the probability density functions of the velocity increments in the inertial range vary continuously with scale separation r. If fitted by stretched exponentials exp $[\Delta u_r^m]$, the stretching exponent m varies smoothly with r, from about 0.5 in the dissipative range to about 2 as r approaches integral-scale separations (i.e. the distribution becomes Gaussian). If Kolmogorov's arguments were right, m would be a constant independent of r. Given the empirical evidence, one is forced to modify the Kolmogorov universality in its broadest sense, though it remains of considerable value in making estimates at most finite Reynolds numbers.

Intermittency in the Inertial and Dissipation Ranges

It is now believed, following *A. M. Obukhov* (1962), that the reason for the inadequacy of Kolmogorov's universality is the strong spatial variation of the energy dissipation rate, a phenomenon known as *intermittency.* Note that this intermittency of the small scales is quite distinct from the outer intermittency we encountered in Section 6.4.2. Obukhov suggested replacing the (global) mean energy dissipation rate $\langle\epsilon\rangle$ in Kolmogorov's formulas by the local average value ϵ_r defined over a ball of radius r. For $r \ll L$, where L is a characteristic large scale, the variable $\epsilon_r/\langle\epsilon\rangle$ is a fluctuating quantity and, according to *Obukhov's* suggestion, a function of the ratio r/L. In this way, whenever averages are taken over regions containing varying levels of energy dissipation rate, the large scale enters inertial-range statistics explicitly, in contrast to the previously held understanding that L would not appear explicitly in inertial range properties. *A. N. Kolmogorov* (1962) made *Obukhov's* suggestion more explicit by assuming that the dissipation rate is lognormally distributed. He also refined his original hypotheses in an essential way by taking note of *Obukhov's* suggestion. This gave rise to the so-called refined similarity hypothesis. The resulting modification is that one may expect power laws of the form

$$\langle \Delta u_r^n \rangle \sim \cdot (r/L)^{\zeta_n} \quad , \tag{6.60}$$

where the factors of proportionality, omitted here, are nonuniversal, but the exponents ζ_n, although different from $n/3$, are presumed to be universal. The deviation of the exponents ζ_n from $n/3$ is the hallmark of inertial-range intermittency. Inertial range intermittency is also inferred from the empirical fact that the probability density functions of wave number bands show increasingly flattened tails for increasing midband wave numbers.

G. K. Batchelor and A. A. Townsend (1949) showed that the non-Gaussian behavior of the probability density of dissipation quantities increases with decreasing scale. In a complementary sense, dissipation quantities become increasingly non-Gaussian as the Reynolds number increases. These are the two hallmarks of dissipation-scale intermittency. The scaling exponents ν_q for the energy dissipation are defined as

$$\langle \epsilon_r^q \rangle \sim (r/L)^{-\nu_q} . \tag{6.61}$$

The proportionality constants omitted here are not expected to be universal. The rationale for writing this power law can be explained in terms of the so-called breakdown coefficients or multipliers, which are supposed to represent the fractions in which the energy dissipation is shared when an eddy of size r is broken into two eddies, say, of size $r/2$. It is not clear that the multipliers, although quite useful, are fundamental to turbulence. Nontrivial scaling implies that ν_q is a nonlinear function of q.

Indeed, there exist a broad class of models that attempt to explain the observed intermittency of the dissipative and inertial scales. These models are cast best in terms of *multifractals* (see *M. S. Borgas* (1992) for a summary), which provide a convenient superstructure. *Kolmogorov's* original model is a degenerate case, as are other later models described in *A. S. Monin and A. M. Yaglom* (1975). The connection of these models to the Navier–Stokes equations is tenuous, and since the detailed physics of the models cannot be tested directly, their success should be evaluated chiefly on the basis of how well they agree with experiments.

Several efforts have been made to measure the exponents ν_q in (6.61), in both high and low Reynolds number flows. Given the difficulties in measuring them, the agreement among various data sets is surprisingly good.

Some other measures of the dissipation range intermittency include the scaling exponents for vorticity and circulation. The conclusion is that enstrophy is more intermittent than the energy dissipation rate, at least at moderate Reynolds numbers. Similarly, the dissipation rate exponents for the passive scalar appear to also be more intermittent than the energy dissipation field in the inertial–convective range (between L and l_k). By contrast, in the viscous–convective range, it has been found that the scaling exponents are trivial (that is, there is no intermittency, and all intermittency exponents are essentially zero). A summary of this discussion can be found in *K. R. Sreenivasan and R. A. Antonia* (1997).

Computation of Turbulent Flows

Computing power has increased exponentially with time in the last few decades. One can in principle start with suitable initial conditions and compute the evolution of a turbulent flow, subject to appropriate boundary conditions, by solving the Navier–Stokes equations without any further physical approximations. These are called direct numerical simulations (DNS) (see, e.g. *P. Moin, K. Mahesh* (1998)). The DNS has been used in Section 6.2 by simulating the transition process in three-dimensional boundary-layers. Another field of application of DNS is combustion, which is described in Chapter 9.

The hope is that it will be possible to compute many of the technologically important flows by DNS, though it is clear that some others, such as the flow around an entire aircraft or ship, or in the ocean and the atmosphere, will remain out of bounds for many years to come, if they ever become amenable to direct numerical simulations. Note that the range of scales needing to be resolved increases nominally as the third power of the Reynolds number. Thus, some inventiveness in our ability to calculate flows will be needed. It is also clear that the physics of turbulence cannot be understood merely by computing, though that step will help immensely if combined with organizing principles of the sort illustrated in this chapter. In one sense, we are still in the early stages of organization of our knowledge of turbulence. Vortex *methods* (see, e.g. *A. J. Chorin* (1994)), based on the representation of the turbulence by means of the vorticity field, offer an alternative in some cases, especially in two dimensions.

On the other end of the spectrum, since we are interested quite often in the mean characteristics of turbulent flow, one can write down the Reynolds equations (5.33)–(5.35) for the mean quantities of interest by averaging the Navier–Stokes equations. It is clear from the discussion of Section 6.3 that additional terms will appear. For the equations describing the mean velocity these terms are the standard Reynolds stress terms, which need to be modeled suitably (see Section 5.4.5, Section 5.4.6 for multiphase flows and Section 5.4.7 for combustion). This aspect of research has been important in practice, and is motivated by the need to adapt our partial understanding of turbulence dynamics to obtain predictions of acceptable accuracy in engineering problems. One account of these models can be found in *C. G. Speziale* (1991).

In between these two extremes lies the scheme that computes only the large scale in temporal and spatial detail, while not resolving the small scales. The notion is that the large scales (not just the largest ones that give shape and form to a flow) carry larger share of burden in the transport of heat, mass, and momentum, while the effects of the unresolved small scales (or subgrid scales), which need not be known in detail for most purposes, can be modeled by suitable parameterizations. A sensible modeling of small scales is in principle attainable because of their nearly universal properties. This scheme of computation is known as *Large Eddy Simulations* (LES) of turbulence which is introduced in Section 5.4.5. Here, one writes down the equations for large scales only, and models the new terms that appear. These new terms are similar to the Reynolds stress terms in the mean flow equations. Part of the reason for studying small-scale structure is indeed the understanding of its universal properties, so it can be suitably modeled and parameterized, thus allowing the computation of the large scale correctly. The biggest bottleneck in using the LES methods extensively is the complexity of turbulence near the wall. For reviews of these methods, see *M. Lesieur, O. Metais* (1996) and *S. B. Pope* (2000).

In recent years, a numerical scheme based on microscopic models and mesoscopic kinetic equations has been successfully employed to compute several turbulent flows. The models are based on what is now called the lattice Boltzmann methods (LBM), which has been used in Chapter 11 for microflows. In conventional computational methods of fluid dynamics, one discretizes the macroscopic continuum equations on a suitably defined fine mesh before solving them. In LBM, on the other hand, one constructs simplified microscopic models that incorporate the essential physics. The basic premise is that the macroscopic dynamics, which are the result of a collective behavior of microscopic particles, are insensitive to the precise details of the microscopic physics, as long as one satisfies certain conservation properties. These methods are particularly suitable for fluid flows involving interfacial dynamics in Chapter 8 and 11 and complex boundaries as airplanes, ships and cars. A summary of the methods can be found in *S. Chen, G. D. Doolen* (1998).

6.5.2 Lagrangian Investigations of Turbulence

Since the transport properties of turbulence are dominated by the advection of infinitesimal fluid elements, it is natural to resort to the Lagrangian viewpoint of Section 3.2, following the motion of the fluid elements. Lagrangian stochastic models have become important for the prediction of turbulent mixing and dispersion, with a particular emphasis on reacting flows in Chapter 9, see *S. B. Pope* (2000). A convenient reference for early theoretical development in Lagrangian methods is *A. S. Monin, A. M. Yaglom* (1975), and an idea of the recent work can be had from *P. K. Yeung* (2002).

Among the activities currently being pursued, one of the important ones is the use of DNS data, obtained in the Eulerian frame, to construct Lagrangian trajectories and compute selected properties, including velocity, acceleration, time scales, velocity gradients, dissipation of energy, properties of the scalar passively carried along Lagrangian trajectories, and so forth. Lagrangian concepts have been usefully employed in subgrid scale modeling. At a fundamental level, they have been used to solve aspects of a model for passive scalars (see Section 6.5.3), and also to study the influence of geometry on scaling considerations by following Lagrangian clusters. There is, of course, the thought that Lagrangian studies may be more natural for studying the properties of coherent structures in turbulence. Finally, using some clever experimental methods initially developed for data acquisition in high-energy physics, *G. A. Voth et al.* (1998) have measured Lagrangian acceleration of particles and shown that the distributions have tails that spread to many standard deviations.

6.5.3 Field-Theoretic Methods

The turbulence problem, more than once described as the last unsolved problem in classical physics, perhaps no longer appears to be as *exceptional* as it once did, for other important strong-coupling problems have since been faced in theoretical physics. Some of these, such as color confinement in quantum chromodynamics, are still with us. For others, such as critical phenomena in three spatial dimensions, the critical scaling exponents have been calculated successfully by several methods, although other nonuniversal quantities of significant interest, such as critical temperatures, cannot yet be readily calculated for physical systems found in nature or realized in laboratories.

It is only natural to attempt to use these methods, employed with some success in similar problems, to address the basic problem of nonlinear coupling among scales of turbulence. Unfortunately, none of these methods that enabled breakthrough successes in the theory of critical phenomena have yet yielded results of comparable significance in understanding or predicting turbulent flows. Nevertheless, considerable progress has been made, and the application of such methods to turbulence has yielded some important insights. In particular, field-theoretic techniques have scored a significant success in calculating turbulent scaling exponents in a simplified model of a white-noise advected passive scalar (for a review, see *G. Falkovich et al.* (2001)).

6.5.4 Outlook

Turbulence is perhaps the most complex form of motion that fluid flows take. It contains structures and strong fluctuations, one embedded in the other. Consideration of one, and the neglect of the other, does not provide a full picture valid in all instances.

Our understanding of turbulence is still imperfect. In attempting to understand it, one needs to employ a combination of tools that consist of novel experiments, advanced computations and theoretical understanding of the behavior of the equations of motion. This is a long-drawn process and no fast success can be expected. On the other hand, one now knows many general features of turbulence and compute some of them. Since one cannot wait for a full understanding has been attained, practical flows are being computed by various turbulence modeling. Modeling and fundamental studies will no doubt continue to exist side by side.

Many fascinating aspects of turbulence come to the fore when combined with some other physical aspect. We briefly drew attention to the effects of rotation and stratification. To these two topics one can add the effects of turbulence in the presence of magnetic fields, particle and bubble loading, polymers, complex boundaries including rough walls, combustion, and so forth. There is a gold mine of problems to be explored in this vast domain.

We have implied that it is convenient to think of a scale separation between the large scales that provide the shape and form for a given turbulent flow and the dissipative small scales, and that the interaction between them is weak. This feature renders the small scales nearly universal, and amenable to a study independent of too many details of the flow. However, this is merely a working model of turbulence, whose elucidation has taken much work. Details are emerging slowly.

It is often said that each turbulent flow is different. The large scales are indeed different. There is a varying degree of coherence in the large-scale motion, depending on initial and boundary conditions. The effects of this coherence can (and should) be captured eventually by appropriate statistics, but it is not clear that the statistics one uses, and constructs for reasons of mathematical convenience, are necessarily best adopted for taking faithful account of this observed coherence.

Another remark often made is that turbulence has nothing to do with stability. It is indeed the case that the instability caused by linear disturbances of negligible amplitude plays very little role in maintaining a turbulent flow, but stability arguments have been used consistently and often successfully to describe the observed coherent structures. The nature of this instability, which is described at the beginning of the chapter for laminar flows, remains unclear in the turbulent context. However, it is clear that a good student of turbulence ought to be versed in different aspects of hydrodynamic stability, and the variety of structures that can be generated by this mechanism.

Ultimately, turbulence dynamics consists of incorporating stability and multi-scale structure into the framework of statistical theories and universality.

7. Convective Heat and Mass Transfer

This chapter on *convective heat and mass transfer* starts out from Prandtl's original chapter *Heat Transfer in Flowing Liquids.* We will treat *free convection flows,* caused by the density changes in the fluid due to temperature and concentration gradients. These cause a lift in the gravitational field, which in turn causes convection flows. Examples of free convection flows at heated cylinders and plates were shown in the introductory chapter in Figure 1.6. The Rayleigh–Bénard convection of Figure 1.5 and diffusion convection are also examples of free convection flows.

Forced convection flows occur when an external force, such as a pressure gradient, also acts on the flow. Forced convection flows occur, for example, in heated or cooled pipes such as those used in heat exchangers.

Heat and mass exchange processes are found in the ocean and in many different processes in chemical technology, such as absorption, adsorption, extraction, and distillation. When water evaporates on the surface of the oceans, a high salt concentration remains, and an unstable density layer with diffusion instabilities occurs. The expansion of substances in solvents and the separation of substances in centrifuges are further examples. Examples of biological mass exchange processes are the supply of oxygen to the blood and absorption of food in the body.

Fig. 7.1. Basalt columns caused by cellular convection at the solidification contour

H. Oertel (ed.), *Prandtl-Essentials of Fluid Mechanics*,
Applied Mathematical Sciences 158, DOI 10.1007/978-1-4419-1564-1_7,

7.1 Fundamentals of Heat and Mass Transfer

7.1.1 Free and Forced Convection

Instabilities occur in free convection flow with unstable thermal straification. In *Rayleigh–Bénard convection*, the ground state is given as heat conduction. It is replaced by thermal cellular convection at the critical Rayleigh number. The convection flow causes the heat flow in a horizontal liquid layer to increase.

A *Rayleigh–Bénard instability* is also observed in the cooling process of molten magma. The surface cools, and an unstable thermal boundary layer forms in the magma. In the region of the unstable thermal boundary layer, gravity causes a convection flow structured in hexagonal cells to occur. After solidification, these leave typical basalt columns (Figure 7.1).

In *diffusion convection* the ground state is a concentration profile that is caused by diffusion and heat conduction in a horizontal fluid layer with several components. At the critical diffusion Rayleigh number, the mass and heat flux increase because of the free convection flow.

The *density differences* causing the convection flow can also be due to concentration gradients in the fluid. Just as with the Bénard convection, hexagonal flow cells also form on free surfaces. An example of such a situation is the drying up of a salt lake. The water evaporating from the surface leaves high salt concentrations with corresponding density increases. Heavy unstable fluid is therefore layered over lighter fluid. When a critical concentration difference is exceeded, convection flow forms and lifts up sand and dust particles from the ground, where the flow is in the direction of the cell centers. These particles are then carried through the lift zone in the center of the cell and are distributed at the edges of the cell, according to the flow sketched in Figure 1.5. Here the convection motion causes the particles eventually to sink to the ground, where they are finally deposited. In this manner

Salt lake

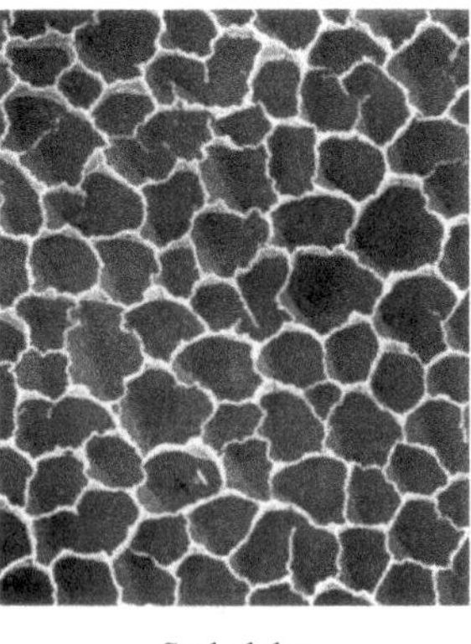

Soda lake

Fig. 7.2. Cellular convection due to concentration gradients

the structures shown in Figure 7.2 are formed at the bottom of dried-up salt lakes.

A further example and introduction to the chapter on heat transfer is the free convection flow at a *heated vertical plate* (Figure 1.6), which will be treated in Section 7.2.1. Figure 7.3 shows the velocity and temperature profiles in the air with Prandtl number $\mathrm{Pr} = 0.71$ for an isothermal wall. The wall temperature T_w is higher than the ambient temperature T_∞. The heat transferred from the plate to the fluid causes a temperature increase in the fluid close to the wall and, because of the temperature dependence of the density, to a change in the density. If the density decreases with increasing temperature, lift forces occur close to the wall and warmer fluid rises along the plate. The effect of the plate is restricted to the wall boundary layer. The ratio of the thickness of the viscous boundary layer δ to the thickness of the thermal boundary layer δ_T behaves like $\sqrt{\mathrm{Pr}}$. Now in the boundary layer of the perpendicular plate, the laminar–turbulent transition takes place above a critical dimensionless characteristic number. Since the heat transport also has to be taken into account, the transition to turbulent boundary-layer flow is initiated at the critical Rayleigh number.

Forced convection is the cause of other external forces in addition to the lift forces. An example of this is the pipe flow of Section 4.2.1 with heat transport, which will be discussed in Section 7.3.1. Figure 7.4 shows the parabolic velocity profile in the intake of laminar pipe flow and also the formation of the temperature profile for an isothermal pipe wall.

In the intake region, the velocity and temperature distributions depend on the radial coordinate r and on x. For viscous intake and uniform flow, we can assume $l \approx 0.05 \cdot \mathrm{Re}_D$. The ratio of the thermal intake length to the

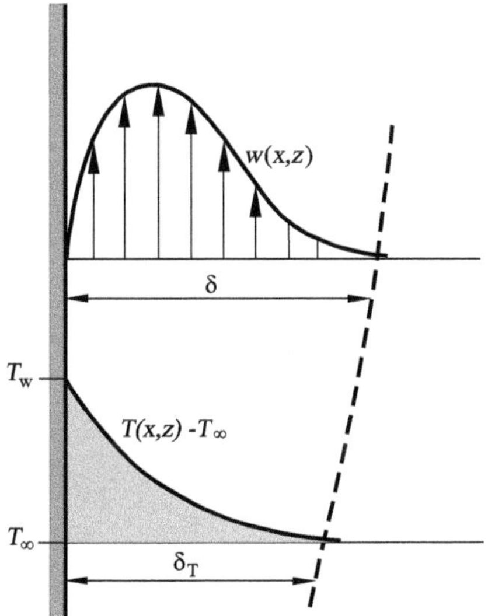

Fig. 7.3. Convection flow at a heated vertical plate

viscous intake length again depends on the Prandtl number of the fluid. For liquid metals, because $\delta_T \gg \delta$, the thermal intake can be neglected compared to the viscous intake. This is the other way round for highly viscous oils with $\delta_T \ll \delta$.

7.1.2 Heat Conduction and Convection

Energy transport at temperature gradients that do not act parallel to the gravitational field is due to heat conduction and superimposed thermal convection flow. A critical Rayleigh number for the onset of cellular convection exists only for horizontal fluid layers heated from below. The heat radiation will be neglected in what follows. The amount of heat transferred to a wall per unit area and time is

$$q_\mathrm{w} = h \cdot (T_\mathrm{m} - T_\mathrm{w}), \tag{7.1}$$

where h is the coefficient of heat transfer, T_w the wall temperature, and T_m the mean temperature of the flowing medium. In the case of a body in a flow, the temperature of the unperturbed free flow T_∞ is chosen. The dimensionless number that characterizes the heat transport is the *Nusselt number*

$$\mathrm{Nu}_l = \frac{q_\mathrm{w} \cdot l}{\lambda \cdot (T_\mathrm{m} - T_\mathrm{w})} = \frac{h \cdot l}{\lambda} \quad . \tag{7.2}$$

It describes the ratio of the heat transfer due to heat conduction and convection to the heat conduction of the fluid at rest.

Since we have initially no given reference velocity for free convection flow, we have to find a characteristic number for convection flow instead of the Reynolds number, namely, the Grashof number

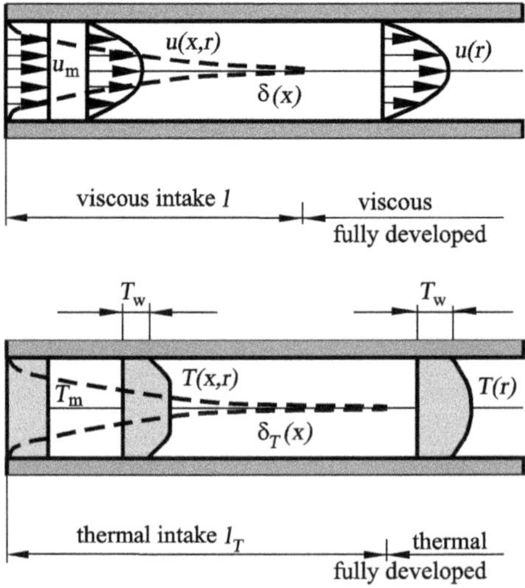

Fig. 7.4. Development of the velocity and temperature profiles of cooled pipe flow

$$\mathrm{Gr_l} = \frac{\alpha \cdot g \cdot (T_\mathrm{m} - T_\infty) \cdot l^3}{\nu^2} \quad . \tag{7.3}$$

Comparing this with the square of the Reynolds number, $\mathrm{Re}_l^2 = w^2 \cdot l^2/\nu^2$, we obtain the following characteristic velocity for free convection flow:

$$w = \sqrt{\alpha \cdot g \cdot (T_\mathrm{m} - T_\infty) \cdot l} \quad . \tag{7.4}$$

The relation with the Prandtl number $\mathrm{Pr} = c_p \cdot \mu/\lambda = \nu/k$ yields the *Rayleigh number* for free convection flow.

$$\mathrm{Ra} = \mathrm{Pr} \cdot Gr \quad . \tag{7.5}$$

If the heat flux into or from the wall is given, the *Grashof number* can be written as

$$\mathrm{Gr_q} = \frac{\alpha \cdot g \cdot q_\mathrm{w} \cdot l^4}{\nu^2 \cdot \lambda} \quad . \tag{7.6}$$

At the heated vertical plate, the thickening of the thermal boundary layer causes the heat flux q_w and the coefficient of heat transfer h to vary in proportion to $l^{-1/4}$.

In *forced convection*, a further independent characteristic number is the Eckert number

$$\mathrm{Ec} = \frac{w^2}{c_p \cdot (T_\mathrm{m} - T_\mathrm{w})} \quad . \tag{7.7}$$

Here the kinetic energy of the flowing medium is referred to the thermal enthalpy difference in the fluid.

For a given heat transport problem we therefore have the dimensionless relation

$$\mathrm{Nu} = \mathrm{f}(\mathrm{Re}, \mathrm{Pr}, \mathrm{Ec}) \tag{7.8}$$

and this can be determined either numerically by solving the fluid-mechanical fundamental equations of Section 5.4 or experimentally. For flow velocities that are not too large, the Eckert number is so small that the relation (7.8) reduces to $\mathrm{Nu} = \mathrm{f}(\mathrm{Re}, \mathrm{Pr})$.

There is no characteristic velocity given for *free convection flow*, so that (7.8) is to be replaced by

$$\mathrm{Nu} = \mathrm{f}(\mathrm{Gr}, \mathrm{Pr}) \tag{7.9}$$

Therefore, the Grashof number in free convection corresponds to the Reynolds number in forced convection flow. Whereas the Reynolds number, the Eckert number, and the Grashof number all depend on the geometric, dynamic, and thermodynamic parameters of the heat transport problem, the Prandtl number is a characteristic number that is substance-specific.

7.1.3 Diffusion and Convection

In convection flows caused by diffusion processes, there are similar laws to those that describe heat transport. The mass transport takes place along the largest concentration gradients. The coefficient of thermal expansion $\alpha = (1/\rho)\cdot \mathrm{d}\rho/\mathrm{d}T$ is now replaced by the coefficient of concentration expansion $\beta = (1/\rho) \cdot \mathrm{d}\rho/\mathrm{d}c$, and the heat conduction number k by the diffusion coefficient D. Similarly, the Rayleigh number for free diffusion convection is replaced by the *diffusion Rayleigh number*

$$\mathrm{Ra_D} = \frac{\beta \cdot g \cdot (c_\mathrm{m} - c_\infty) \cdot l^3}{\nu \cdot D} \quad , \tag{7.10}$$

with a mean mass concentration c_m and the reference concentration c_∞. The Prandtl number is replaced by the *Schmidt number*

$$\mathrm{Sc} = \frac{\nu}{D} \quad . \tag{7.11}$$

In analogy to the heat flux, we specify the diffusion flux at the wall $j_\mathrm{w} = D \cdot \partial c_\mathrm{w}/\partial n$ with wall normal n, and we obtain the relation

$$\mathrm{Gr_D} = \frac{\beta \cdot g \cdot j_\mathrm{w} \cdot l^4}{\nu^2 \cdot D} \tag{7.12}$$

for the diffusion Grashof number in diffusion-caused free convection, and the Nusselt number

$$\mathrm{Nu_D} = \frac{j_\mathrm{w} \cdot l}{D \cdot (c_\mathrm{m} - c_\mathrm{w})} \quad . \tag{7.13}$$

for mass transfer. For a given diffusion problem we have to determine the relation

$$\mathrm{Nu_D} = \mathrm{f}(\mathrm{Gr_D}, \mathrm{Sc}) \tag{7.14}$$

for *free convection flow*, and for *forced convection flow*, the relation

$$\mathrm{Nu_D} = \mathrm{f}(\mathrm{Re}, \mathrm{Sc}) \tag{7.15}$$

The question arises of how large the Schmidt number Sc is compared to the thermal Prandtl number. For gases, the Schmidt number, like the Prandtl number, has order of magnitude 1, since k and D are only slightly different. For the diffusion of steam in air, the value is $\mathrm{Sc} \approx 0.62$ at a mean temperature of 8° C. For the diffusion of CO_2 in air at 0° C $\mathrm{Sc} \approx 1.1$. Therefore, in gases with the same Reynolds or Grashof numbers, the Nusselt numbers have the same order of magnitude for heat and mass transport. On the other hand, in aqueous solutions, the Schmidt numbers are considerably larger than the Prandtl numbers. For the diffusion of macromolecules in aqueous solutions we obtain Schmidt numbers of order of magnitude 10^4, while the Prandtl number of water is 7. Mass exchange in aqueous solutions is therefore related to heat exchange in viscous oils.

7.2 Free Convection

7.2.1 Rayleigh–Bénard Convection

We now consider thermal unstable Rayleigh–Bénard convection in a horizontal liquid layer under the effect of gravity, heated from below. Let the layer be infinitely extended in the horizontal plane and have height h. Its lower side is heated to temperature T_1 and its upper side kept at the temperature $T_2 < T_1$ (Figure 7.5). When a critical temperature difference $\Delta T = (T_1 - T_2)$ between the upper and lower boundaries of the liquid layer is exceeded, straight convection rolls form in the horizontal liquid layer. The longitudinal axes of these steady convection rolls are horizontal and ordered periodically next to one another. This process is known as *thermal cellular convection.*

Because of the additional thermal exchange processes, convection results in a increase in the heat flux $\dot{q}$, compared to the case of heat conduction. The *Nusselt number* Nu equal to $(\dot{q}_{\text{conduction}} + \dot{q}_{\text{convection}})/\dot{q}_{\text{conduction}}$, as the dimensionless heat flux is plotted against the *Rayleigh number* $\text{Ra} = \alpha \cdot \Delta T \cdot g \cdot l^3/(\nu \cdot k)$ in Figure 7.6, with the coefficient of thermal expansion α, the kinematic viscosity ν, and the thermal conductivity number k of the medium. The diagram shows that the Nusselt number remains constant ($\text{Nu} = 1$) up to a critical Rayleigh number of $\text{Ra}_{\text{crit}} = 1708$. Clearly, it is pure heat conduction at play in this regime. When this critical Rayleigh number is exceeded, the dimensionless heat flux branches, and there is a strong dependence on the Rayleigh number and the Prandtl number $\text{Pr} = \nu/k$ of the medium. This sudden process is clearly connected with a fluid-mechanical instability. The original state (pure heat conduction, medium at rest) can no longer be retained. It becomes unstable and is replaced by a new state (heat conduction + convection, medium in motion). The critical Rayleigh number is indepen-

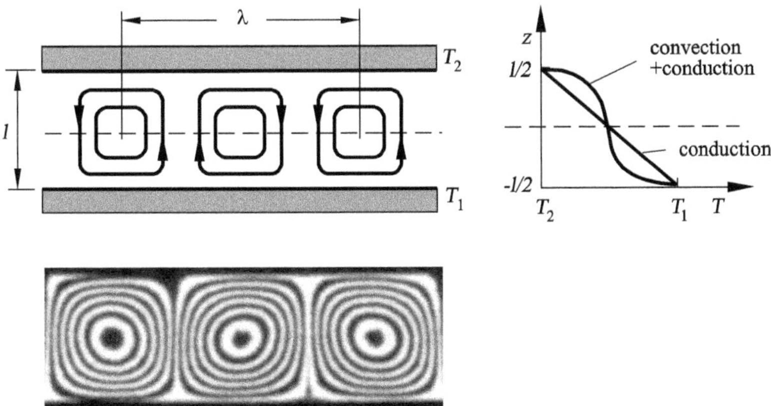

Fig. 7.5. Thermal cellular convection

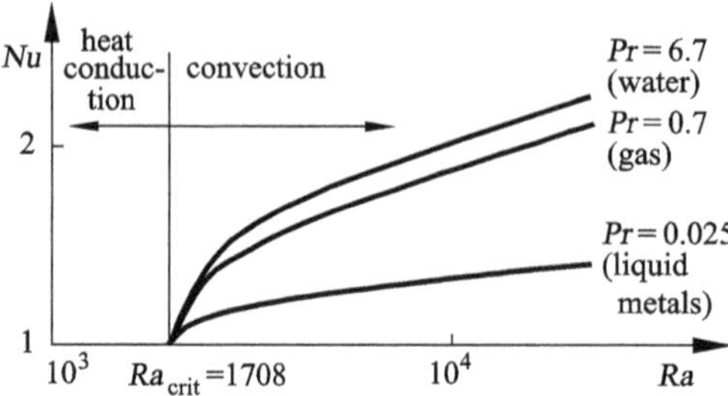

Fig. 7.6. Branching diagram of the dimensionless heat flux Nu against the Rayleigh number Ra

dent of the medium, since the branching point $(\mathrm{Nu}, \mathrm{Ra}_{\mathrm{crit}}) = (1, 1708)$ is independent of the Prandtl number Pr.

Thermal cellular convection is important in many technical problems. On the one hand, an engineer endeavors to design heat insulation out of air layers (e.g. thermopane layers) so that thermal cellular convection is prevented. On the other hand, the construction of a heat regenerator requires convection processes that are as strong as possible.

Let us briefly look at the origin of the instability because of the higher temperature a liquid particle from a lower layer z_1, has a lower density than a particle in a higher layer $z_2 > z_1$. This is called an *unstable stratification.* If the particle in z_1 is relocated to a layer above, it experiences a lift force in the new surroundings of less dense fluid and is accelerated upwards. Frictional forces and heat conduction act against this tendency, which tries to compensate the driving temperature difference and density difference of the particle.

Let the fluid element under consideration have size d (Figure 7.7). Let the element move with a perturbation velocity $\boldsymbol{v}$ from $z = z_0$ to a layer above at $z_0 + d$. This takes place within the time interval $\Delta t = d/\boldsymbol{v}$. The density

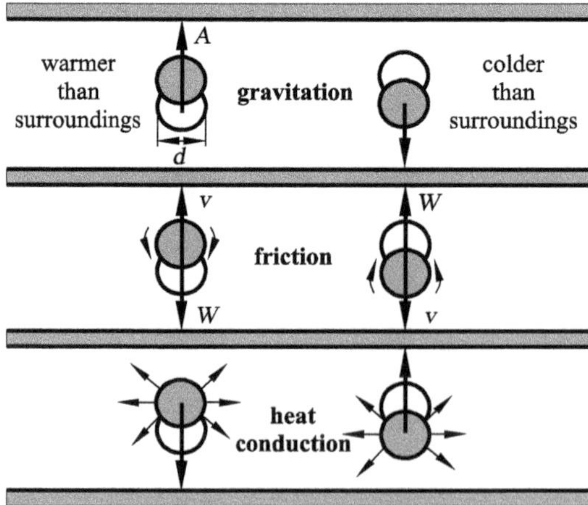

Fig. 7.7. The physical interpretation of thermal cellular convection

difference $\Delta\rho \sim \rho_m \cdot \alpha_m \cdot \Delta T$ causes a lift $A = \Delta\rho_m \cdot g \cdot V_k \sim \rho_m \cdot \alpha_m \cdot \Delta T \cdot g \cdot d^3$. Simultaneously, at a small perturbation velocity, Stokes's law states that the drag $W \sim \mu \cdot d^2 \cdot \boldsymbol{v}/d = \mu \cdot d^2/\Delta t$ occurs. What is important is the degree to which the heat conduction in time interval Δt balances out the driving temperature difference between the fluid element and its new surroundings. The difference in internal energy of $E_k \sim \rho \cdot c_v \cdot \Delta T \cdot d^3$ is transferred to the surroundings through a cross-sectional area $\sim d^2$ by means of the heat conduction $\dot{q} \sim \lambda \cdot \Delta T/d$. The time scale for this process is therefore $\Delta t = E_k/\dot{q} \cdot d^2 \sim d^2/k$, and it can be inserted into the proportionality considerations above.

The system clearly becomes unstable when the lift dominates the drag:

$$A \geq W \Longleftrightarrow \rho \cdot \alpha \cdot \Delta T \cdot g \cdot d^3 \geq \mu \cdot d^2 \cdot \frac{k}{d^2} \cdot \mathrm{C} \quad ,$$

or, with $d = l$,

$$\frac{\alpha \cdot \Delta T \cdot g \cdot l^3}{k \cdot \nu} = Ra \geq \mathrm{C} = Ra_{\mathrm{crit}} \quad . \tag{7.16}$$

The Rayleigh number is clearly the ratio of lift force to frictional force.

Stability Analysis

The *fundamental equations* of thermal cellular convection (5.85), assuming the *Boussinesq approximation*, were introduced in Section 5.4.3 The perturbation ansatz (5.196) yields the perturbation differential equations (5.213)–(5.215). The *ground state* $\boldsymbol{U}_0$, p_0, T_0, whose stability is to be investigated, is the state of rest with $\boldsymbol{U}_0 = 0$. The energy equation (5.215) then yields

$$\Delta T_0 = 0 \quad . \tag{7.17}$$

This is the steady heat conduction problem. For the state of rest, it is necessary that the temperature gradient be parallel to $\boldsymbol{e}_z = (0, 0, 1)$. The boundary condition for Rayleigh–Bénard convection is

$$T_0(x, y, z = -\frac{1}{2}) = T_1 \quad , \qquad T_0(x, y, z = \frac{1}{2}) = T_2 \quad . \tag{7.18}$$

The ground state is dependent only on the vertical direction z:

$$\frac{\mathrm{d}^2 T_0}{\mathrm{d}z^2} = 0 \quad , \qquad T_0(z) = \mathrm{C}_1 \cdot z + \mathrm{C}_0 \quad . \tag{7.19}$$

The constants $(\mathrm{C}_1, \mathrm{C}_2)$ follow from the boundary conditions (7.18), yielding $\mathrm{C}_1 = -1$, $\mathrm{C}_0 = (T_1 + T_2 - 2 \cdot T_m)/\Delta T$, with $T_m = (T_1 + T_2)/2$. For the heat conduction ground state, we obtain

$$T_0 = -z \quad . \tag{7.20}$$

The momentum equations (5.214) yield

$$0 = -\frac{\mathrm{d}p_0}{\mathrm{d}z} + Ra \cdot T_0 \quad ,$$

with (7.20) and hence the pressure

$$p_0 = -\frac{1}{2} \cdot Ra \cdot z^2 + p_\infty \quad , \qquad (7.21)$$

with the ambient pressure p_∞. The temperature distribution determined above, and therefore the entire heat conduction problem, is independent of p_∞. It is not the pressure p_∞ itself that affects the stability problem, but rather its gradient alone.

For the *boundary conditions* for the perturbation quantities, we distinguish between free and fixed horizontal boundaries of the liquid layer. On *free* boundaries (liquid surface), $z = \pm 1/2$ is the kinematic flow condition of impermeability of the surface, with

$$w'(x, y, \pm\frac{1}{2}) = 0 \quad , \qquad (7.22)$$

assuming that the deformation due to small perturbations may be neglected. On *fixed* boundaries, the no-slip conditions holds:

$$\boldsymbol{u}'(x, y, \pm\frac{1}{2}) = 0 \quad . \qquad (7.23)$$

We also distinguish between isothermal (5.70) and adiabatic (5.71) boundaries. If the horizontal boundary has a large thermal conductivity, it behaves *isothermally*, and the temperature perturbations vanish:

$$T'(x, y, \pm\frac{1}{2}) = 0 \quad . \qquad (7.24)$$

At *adiabatic boundaries*, there is a constant heat flux. Changes $\dot{q}' = -\lambda \cdot \partial T'/\partial z$ to this heat flux by means of temperature perturbations are zero if the thermal conductivity of the bounding medium is very small.

Let the local changes in the temperature of the bounding medium associated with the fixed heat flux also be small, and effects on the basic solution negligible, as usual. This leads to the thermal boundary condition

$$\frac{\partial T'}{\partial z}(x, y, z_r) = 0 \quad . \qquad (7.25)$$

The perturbation differential equations (5.213)–(5.215) with their boundary conditions lead to an eigenvalue problem that allows the critical Rayleigh number $\mathrm{Ra}_{\mathrm{crit}}$ and wave number a_{crit} of the periodic cell structures of the Rayleigh–Bénard convection to be computed. Summarizing the variables of the solutions vector $\boldsymbol{u}' = (u', v', w', p', T')$ in the perturbation differential equations, the separation trial solution

$$\boldsymbol{u}' = \boldsymbol{u}'_x(x, y, z, \omega) \cdot \exp(-\mathrm{i} \cdot \omega \cdot t) \qquad (7.26)$$

can be used to separate the time and space dependence. If we further eliminate u', v', and p' (see, for example *H. Oertel* and *J. Delfs* (1996)), we obtain the perturbation differential equations as

$$\left\{ \begin{bmatrix} -\Delta^2 - \text{Ra} \cdot \left(\dfrac{\partial^2}{\partial x^2} + \dfrac{\partial^2}{\partial y^2} \right) \\ -1 \qquad -\Delta \end{bmatrix} - \text{i} \cdot \omega \cdot \begin{bmatrix} \frac{1}{\text{Pr}} \cdot \Delta 0 \\ 0 \quad 1 \end{bmatrix} \right\} \begin{pmatrix} w' \\ T' \end{pmatrix}_x = \begin{pmatrix} 0 \\ 0 \end{pmatrix} . \qquad (7.27)$$

Eliminating w' and with $\boldsymbol{u}' = T'$, we obtain the eigenvalue problem

$$\left\{ \left[-\Delta^3 + \text{Ra} \cdot \left(\frac{\partial^2}{\partial x^2} + \frac{\partial^2}{\partial y^2} \right) \right] - \text{i} \cdot \omega \cdot \left[\left(1 + \frac{1}{\text{Pr}} \right) \cdot \Delta^2 \right] \right.$$
$$\left. + \ \omega^2 \cdot \left[\frac{1}{\text{Pr}} \cdot \Delta \right] \right\} T'_x = 0. \qquad (7.28)$$

In the eigenvalue problem for T'_x, the eigenvalue ω appears quadratically.

For an infinitely extended liquid layer of the Rayleigh–Bénard stability problem, the periodic cell structure permits the following separation ansatz:

$$(u', v', w', p', T')(x, y, z) = \text{F}(x, y) \cdot \left(\hat{u}(z), \hat{v}(z), \hat{w}(z), \hat{p}(z), \hat{T}(z) \right) . \qquad (7.29)$$

We note that this ansatz is no longer possible if there are boundaries at the sides of the container, since in this case explicit boundary conditions at the side walls are required. Substituting ansatz (7.29) into the steady energy equation (5.215) initially yields the relation of the function $\text{F}(x, y)$ to the arbitrarily chosen separation parameter a^2:

$$\frac{\dfrac{\text{d}^2 \hat{T}}{\text{d}z^2} + \hat{w}}{\hat{T}} = - \frac{\dfrac{\partial^2 \text{F}}{\partial x^2} + \dfrac{\partial^2 \text{F}}{\partial y^2}}{F} = a^2 = \text{const.} \qquad (7.30)$$

In the separated differential equation for T' (7.28), a^2 then appears in relation to the assumption $\omega = 0$ (neutral state):

$$\left(\frac{\text{d}^2}{\text{d}z^2} - a^2 \right)^3 \hat{T}(z) + \text{Ra} \cdot a^2 \cdot \hat{T}(z) = 0. \qquad (7.31)$$

With the boundary conditions we again define an eigenvalue problem in which, for a given a of the periodic cell structure, the Rayleigh number Ra appears as an eigenvalue. The eigenvalue problem (7.31) describes the onset of thermal cellular convection of a fluid. For a given wave number a, the

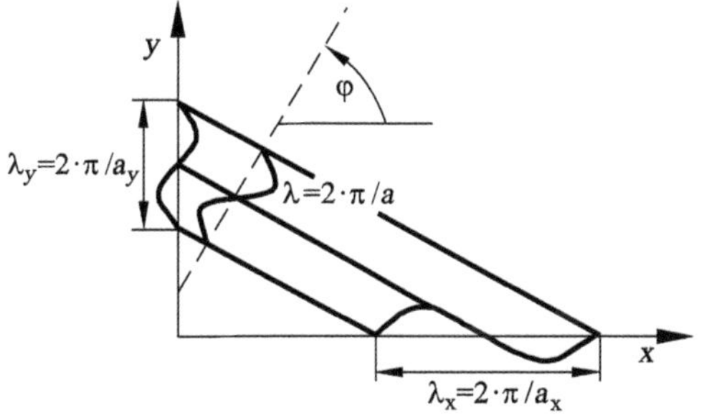

Fig. 7.8. Interpretation of the separation parameter a as wave numbers

associated Rayleigh number $\mathrm{Ra}(a)$ is determined. The spatially periodic cell structure then satisfies

$$\mathrm{F}(x, y) = \exp(\mathrm{i} \cdot a_x \cdot x + \mathrm{i} \cdot a_y \cdot y), \tag{7.32}$$

with

$$a^2 = a_x^2 + a_y^2. \tag{7.33}$$

For real numbers a_x, a_y, expression (7.32) describes a spatially periodic plane wave with the partial wave numbers $a_x = 2 \cdot \pi/\lambda_x$ and $a_y = 2 \cdot \pi/\lambda_y$ (Figure 7.8). It can be seen that the choice of a partial wave number a_x (or a_y) is restricted only by the condition $a_x^2 \leq a^2$ (or $a_y^2 \leq a^2$). The other partial wave number then follows from (7.33). The separation parameter a is clearly a characteristic wave number. The stability problem is determined only by the wavelength $\lambda = 2 \cdot \pi/a$ of the associated characteristic perturbation wave, and not by the orientation of its wave normal $\varphi = 1/\tan(a_y/a_x)$ in the x-y plane.

Because there is no characteristic direction, we may choose, for example, $a_x \in [0, a/\sqrt{2}]$ without any loss of generality. The x-y structure of the solution is also independent of the particular solution $\hat{T}(z)$ determined from (7.31). If we determine, for example, the critical wave number a_{crit} from the eigenvalue problem (7.31), there are infinitely many possibilities to construct this out of partial waves using (7.33).

Therefore one-dimensional (e.g. $a_x = 0$, $a_y = a$) roll structures are just as likely as two-dimensional hexagonally shaped cell structures. An example is shown in Figure 7.9, where the function $\mathrm{f}(x, y) = \cos(a \cdot y) + \cos\left(\sqrt{3}/2 \cdot a \cdot x + 0.5 \cdot a \cdot y\right) + \cos\left(\sqrt{3}/2 \cdot a \cdot x - 0.5 \cdot a \cdot y\right)$ with $a = 2 \cdot \pi$ is plotted.

According to the linear theory, it is solely the initial conditions that determine which of the possible structures forms. In reality, however, it is seen that the hexagonal cells are preferred for free boundaries, even for different initial perturbations, while for fixed boundaries it is roll structures that are observed.

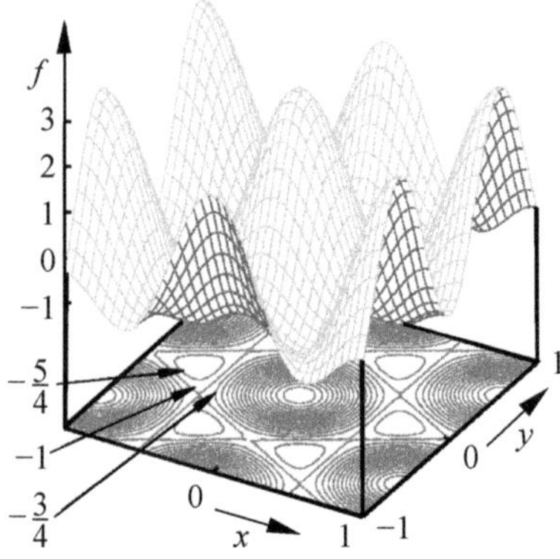

Fig. 7.9. Occurrence of hexagonal cell structures (contours) by superposition of three eigensolutions

Stability Diagram

In this section we discuss three solutions of the eigenvalue problem (7.31) for different boundary conditions (7.22)–(7.25).

In the case of two *free isothermal boundaries*, the solution of the eigenvalue problem may be written down in closed form. We have the boundary conditions

$$\hat{T}\left(z=\pm\frac{1}{2}\right)=0, \quad \frac{\mathrm{d}^2\hat{T}}{\mathrm{d}z^2}\left(z=\pm\frac{1}{2}\right)=0, \quad \frac{\mathrm{d}^4\hat{T}}{\mathrm{d}z^4}\left(z=\pm\frac{1}{2}\right)=0. \tag{7.34}$$

Every even function $\hat{T}^e(z) = \cos((2 \cdot n + 1) \cdot \pi \cdot z)$ satisfies these boundary conditions. The same holds for the odd functions $\hat{T}^o(z) = \sin(2 \cdot n \cdot \pi \cdot z)$. Inserting $\hat{T}^e$ into the eigenvalue problem (7.31) leads to the eigensolution

$$\mathrm{Ra}(a) = \frac{((2 \cdot n + 1)^2 \cdot \pi^2 + a^2)^3}{a^2}. \tag{7.35}$$

This is the desired relation between the Rayleigh number Ra and the wave number a on the indifference curve $\mathrm{Ra}(a)$. Looking closely at (7.35), we see that there is an infinite number of such indifference curves, because the order n can be given arbitrarily. It is easy to see that the lowest (and therefore the most relevant) Rayleigh numbers for all a are those for the fundamental mode $n = 0$. The critical Rayleigh number $\mathrm{Ra}_{\mathrm{crit}}$ is obtained from the condition that the derivative of the function $\mathrm{Ra}(a)$ must vanish at its minimum:

$$\mathrm{Ra}_{\mathrm{crit}} = \frac{27}{4} \cdot \pi^4 = 658, \text{ for} \qquad a_{\mathrm{crit}} = \frac{\pi}{\sqrt{2}} = 2.22. \tag{7.36}$$

Inserting the odd eigenfunctions $\hat{T}^o$, we see that the lowest lying $\mathrm{Ra}(a)$ curve is far above that for the even eigenfunctions. It has a critical Rayleigh number of $\mathrm{Ra}_{\mathrm{crit}} = 108 \cdot \pi^4 \approx 10520$ at $a = \sqrt{2} \cdot \pi \approx 4.44$. It can be seen from this that the odd solution is physically irrelevant, because an even eigensolution will always become unstable first. The lowest-order indifference curve for even and odd eigensolutions is shown in Figure 7.10. At the critical Rayleigh number, long *convection rolls* or hexagonal cells occur. Their appearance in meteorology will be treated in Chapter 10.

The boundary conditions of thermal cellular convection at two *fixed isothermal boundaries* are

$$\hat{T}\left(z=\pm\frac{1}{2}\right)=0, \qquad \frac{\mathrm{d}^2\hat{T}}{\mathrm{d}z^2}\left(z=\pm\frac{1}{2}\right)=0, \tag{7.37}$$

$$\left(\frac{\mathrm{d}^2}{\mathrm{d}z^2}-a^2\right)\frac{\mathrm{d}\hat{T}}{\mathrm{d}z}\left(z=\pm\frac{1}{2}\right)=0.$$

The eigenvalue problem is given by the linear sixth-order ordinary differential equation in z with constant coefficients (7.31). Using an $e^{\lambda \cdot z}$-ansatz, this equation is reduced to the characteristic equation

$$\left(\lambda^2 - a^2\right)^3 + \mathrm{Ra} \cdot a^2 = 0. \tag{7.38}$$

The constants $\mathrm{C_i}$ of the general solution $\hat{T}(z) = \sum_{\mathrm{i}=1}^{6} \mathrm{C_i} \cdot e^{(\lambda_\mathrm{i} \cdot z)}$ have to be fitted nontrivially to the six homogeneous boundary conditions (7.37). Nontrivial solutions $\hat{T}(z) \neq 0$ exist only if the determinant of the corresponding 6×6 matrix vanishes. This condition leads to the desired relation between the Rayleigh number Ra and the wave number a, which is solved numerically.

The indifference curve $\omega_\mathrm{i} = 0$ is shown in the stability diagram in Figure 7.10. A negative rate of amplification $\omega_\mathrm{i} < 0$ indicates that the perturbations die away in time. The heat conduction ground state remains stable. Positive amplification rates $\omega_\mathrm{i} > 0$ lead to instability. The indifference curve has a minimum Rayleigh number $\mathrm{Ra_{crit}}$ below which perturbations of all wavelengths die away. This limit is computed as the minimum of the function $\mathrm{Ra}(a)$ as

$$\mathrm{Ra_{crit}} = 1708 \quad , \qquad a_{\mathrm{crit}} = 3.12 \quad . \tag{7.39}$$

The eigensolutions are longitudinal *convection rolls*, as already shown in the introductory chapter in Figure 1.5. The solution for the odd (asymmetric) eigenfunctions $\mathrm{f}_i^u(z)$ would also yield an indifference curve as in Figure 7.10. However, the critical Rayleigh number in this case is about 10 times higher than in the case of even (symmetric) perturbation functions ($\mathrm{Ra_{crit}} \approx 17610$ at $a_{\mathrm{crit}} \approx 5.37$). The odd perturbation functions are therefore amplified in time only when the even functions are already unstable.

For the case of one free and one fixed isothermal boundary, the condition (7.34) is to be satisfied at $z = 0.5$, while the condition (7.37) is to be satisfied at $z = -0.5$. This problem can be reduced to the preceding problem of two fixed boundaries. Because an odd function always vanishes with all its even derivatives at $z = 0$, the odd eigensolution satisfies precisely the conditions of the free isothermal boundary at $z = 0$. Therefore, the upper half $0 < z \leq 0.5$ of the Rayleigh–Bénard convection with doubly fixed boundaries can be removed from the problem. The Rayleigh number and the dimensionless wave number a merely have to be referred to a layer of thickness l that has been reduced by half. To do this, we halve the temperature difference ΔT and in

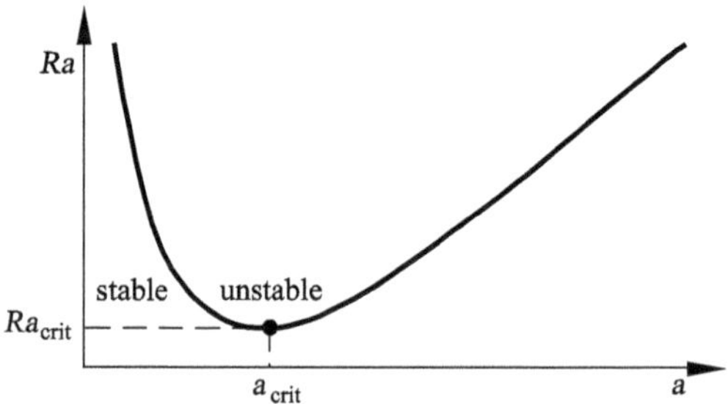

Fig. 7.10. Indifference curve of Rayleigh–Bénard cellular convection

the definition of the Rayleigh number substitute l by $l/2$: $\mathrm{Ra}(\Delta T/2, l/2) = 2^{-4} \cdot \mathrm{Ra}$. Since the wave number a was made dimensionless by multiplication by l, it has to be halved: $a(l/2) = 0.5 \cdot a$. This yields

$$\mathrm{Ra}_{\mathrm{crit}} = \frac{17610}{2^4} = 1101 \quad \text{at} \qquad a_{\mathrm{crit}} = \frac{5.37}{2} = 2.68. \tag{7.40}$$

At the critical Rayleigh number *hexagonal convection cells* are observed. On the free liquid surface these are caused by the temperature dependence of the surface stresses. Examples are shown in Figures 1.5 and 7.1.

Effect of Container Boundaries

In the stability problems treated until now, the basic flow was inhomogeneous in only one spatial direction (z). It was only in this direction that explicit boundary conditions were required. In the homogeneous directions (without explicit boundary conditions), wave trial solutions (separation ansatz) could be used, leading to ordinary homogeneous differential equations. However, if the Rayleigh–Bénard instability is observed in containers with finite cross-sections, explicit boundary conditions have to be satisfied at all walls, and separate consideration of given wave perturbations is no longer permitted. The numerical solution of the eigenvalue problem (7.28) for $\omega_{\mathrm{i}} = 0$ becomes more difficult.

Results of the numerical solution of the eigenvalue problem show that the vertical boundaries act to stabilize the onset of cellular convection, because the no-slip condition introduces additional friction. This is clear from Figure 7.11, where the critical Rayleigh number is plotted against the ratio of container length l_x to container height l. For a given ratio $l_y/l = 4$ the critical Rayleigh number tends toward the value 1815 for large l_x/l. It can also be seen from Figure 7.11 that the asymptotic value of the critical Rayleigh number is reached already at relatively low values of the ratio l_x/l. As the container length l_x/l is reduced to very small values, the critical Rayleigh number increases greatly. The frictional force due to the no-slip condition on the sides acts in the entire flow field and completely prevents the formation

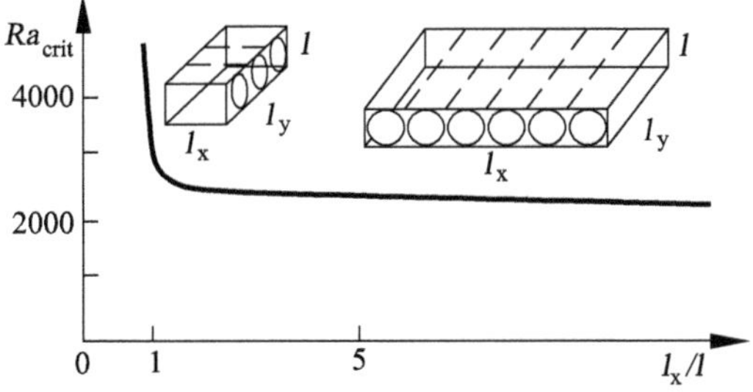

Fig. 7.11. Critical Rayleigh numbers of rectangular containers with finite size $l_y/l = 4$

of convection rolls. In general, the longitudinal axes of the convection rolls orient themselves parallel to the shorter sides of the container. The flow field is fundamentally three-dimensional. The effect of the boundaries acts as far as a depth of about one characteristic length l of the flow field. The inner part of the flow field can be computed as if no boundaries were present. This leads to the surprising result that even in the middle of a circular container time-asymptotically straight roll structures form, and not, as had been suspected earlier, concentric ring cells.

Secondary Instabilities

Until now, we have treated the onset of thermal cellular convection. For supercritical Rayleigh numbers, a great number of different branch solutions occur, depending on the initial and boundary conditions. (see Section 6.2.1). Steady three-dimensional and time-dependent oscillatory cell structures, as well as turbulent cellular convection, occur.

The theory of these secondary instabilities is described in *H. Oertel* and *J. Delfs* (1996). Here it is assumed that the ground state $\boldsymbol{U}_0$ is replaced by the unstable steady cellular convection at the critical Rayleigh number. This cellular convection is now denoted by $\boldsymbol{U}_1$ and is taken to be the new ground state. In analogy to the primary stability analysis, a small perturbation $\varepsilon \cdot \boldsymbol{u}''$ is superimposed on the periodic basic flow $\boldsymbol{U}_1$. This leads to the perturbation ansatz

$$\boldsymbol{u} = \boldsymbol{U}_1 + \varepsilon \cdot \boldsymbol{u}'' \tag{7.41}$$

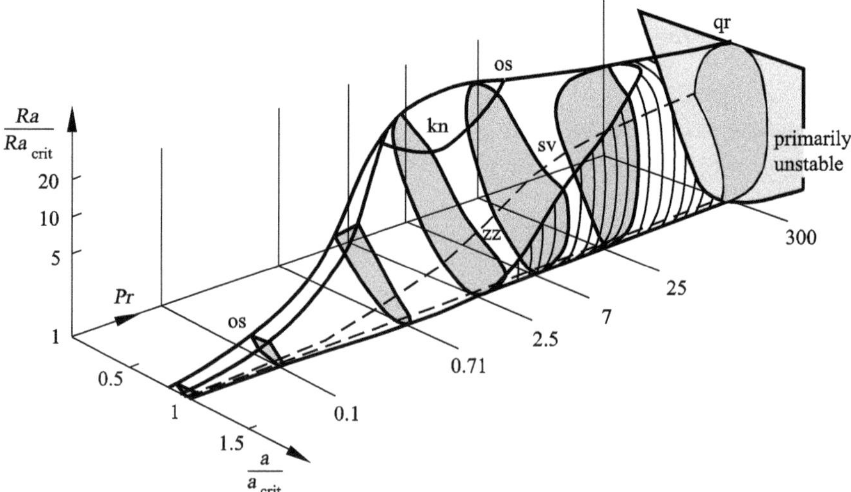

Fig. 7.12. Stability regime for convection rolls between two fixed horizontal boundaries. Secondary eigensolutions: *os*: oscillatory, *sv*: oblique–varicose, *zz*: zigzag, *qr*: cross-rolls, *kn*: nodes, $\mathrm{Ra}_{\mathrm{crit}}$: critical Rayleigh number of primary instability

and the perturbation differential equations for the secondary instabilities $\boldsymbol{u}''$.

F. H. Busse's (1978) theoretical and experimental results are summarized in Figures 7.12 and 7.13. Extensive parameter variations have shown that for given Rayleigh and Prandtl numbers and the same fundamental wavelength λ of the Bénard cells, several different unstable secondary eigenforms $\boldsymbol{u}''$ can exist. The appear secondary instabilities different depending on the combination Pr, Ra, $a = 2 \cdot \pi / \lambda$. In the very small Prandtl number regime, convection rolls, for example, are unstable to unsteady oscillatory perturbation forms. This has the following plausible explanation. The local acceleration $\partial \boldsymbol{u} / \partial t$ is divided by the Prandtl number in the fundamental Boussinesq equations (5.85), and therefore, the smaller the Prandtl number, the greater the effect of the unsteady terms.

The states in which this oscillatory instability occurs on the convection rolls are also shown in Figure 7.12, for the case of an infinitely extending fluid layer with fixed horizontal boundaries. It shows the three-dimensional regime in $(a, \mathrm{Pr}, \mathrm{Ra})$ space for which all secondary perturbations die away in time. The convection rolls characterized by the parameter inside the stability region are therefore stable to small perturbations. The shape of this region is indicated with five cross-sections, each at a constant Prandtl number. Depending on where we depart the region of stability, the convection rolls will become unstable to different perturbation forms. The entire insta-

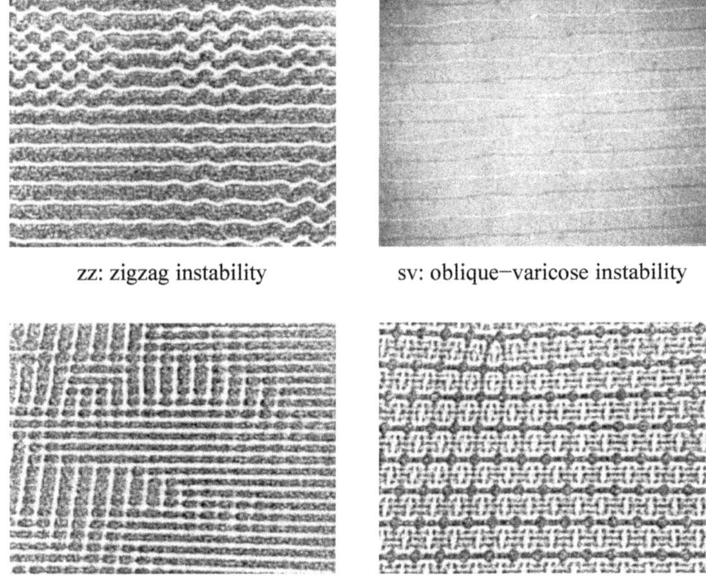

Fig. 7.13. Snapshots of the time-dependent cellular convection, original state: convection rolls with given wavelength, *F. H. Busse* (1978)

bility regime touches the line $\mathrm{Ra} = \mathrm{Ra}_{\mathrm{crit}}$, $a = a_{\mathrm{crit}}$, Pr, which represents the critical state $\mathrm{Ra}_{\mathrm{crit}} = 1708$, $a_{\mathrm{crit}} = 3.12$ of the primary instability. The critical Rayleigh number $\mathrm{Ra}_{\mathrm{crit}}$ becomes independent of the Prandtl number Pr. In the $\mathrm{Pr} = 300$ cross-section, the instability regime of the primary stability analysis (Figure 7.10) is also indicated in Figure 7.12, to show that the stability regime for secondary instabilities is embedded within it. The stability diagram is unable to say whether a secondary instability eventually will form a flow state that corresponds to the eigenform of this instability in the course of a perturbation development. It can say only that the convection rolls become unstable to infinitesimally small perturbations as the critical surface surrounding the stability regime is passed over. The secondary stability analysis also indicates the spatiotemporal character of the amplified perturbation forms, as long as they still have an infinitesimally small amplitude.

In addition to the oscillatory instabilities *os*, we also distinguish between three types of time-dependent secondary instabilities (Figure 7.13), the zigzag instability *zz*, the oblique–varicose instability *sv*, and the cross-roll instability *qr*. The zigzag instability occurs when the given wavelength of the convection rolls is too large at that Reynolds number and the wavelength is reduced by the formation of zigzags. The oblique–varicose instability forms a spatially periodic variation in both horizontal directions with a periodic displacement of the rolls from one roll to the next. The cross-roll instability eventually leads to a complete displacement of the convection rolls, which are then oriented at right angles to each other with different wavelengths.

For liquids with Prandtl numbers greater than 7, the three-dimensional flow at Rayleigh numbers greater than $2 \cdot 10^4$ is steady. This instability is called a node (bimodal) instability (fourth picture in Figure 7.13). In gases with Prandtl number 0.71, the convection rolls begin to oscillate at the Rayleigh number $1 \cdot 10^4$, and no steady node instabilities are observed. A further increase in the Rayleigh number leads to an increase in the oscillation amplitude. The time-dependent structure of the convection cells becomes increasingly irregular until the transition to turbulent convection flow is eventually complete. In liquid metals with Prandtl numbers of order of magnitude 10^{-2}, the Rayleigh number regime of steady convection flow is very small, and turbulent flow is reached already for a Rayleigh number of 2500.

7.2.2 Convection at a Vertical Plate

Figure 7.14 shows the velocity and temperature profiles of *laminar convection flow* of a heated vertical plate. From the Boussinesq equations (5.85), estimation of the orders of magnitude yields the two-dimensional boundary-layer equations. With the boundary-layer transformation

$$
\begin{aligned}
x^* &= \frac{x}{l} \cdot \mathrm{Gr}_z^{\frac{1}{4}}, \qquad z^* = \frac{z}{l}, \\
u^* &= \frac{u}{\sqrt{g \cdot \alpha \cdot l \cdot (T_\mathrm{m} - T_\infty)}} \cdot \mathrm{Gr}_z^{\frac{1}{4}}, \\
w^* &= \frac{w}{\sqrt{g \cdot \alpha \cdot l \cdot (T_\mathrm{m} - T_\infty)}}, \qquad\qquad (7.42) \\
T^* &= \frac{T - T_\infty}{T_\mathrm{m} - T_\infty},
\end{aligned}
$$

the *boundary-layer equations* are made independent of the Rayleigh and Grashof numbers. Dropping the $*$ denoting dimensionless quantities, we obtain the following system of equations:

$$
\frac{\partial u}{\partial x} + \frac{\partial w}{\partial z} = 0, \qquad (7.43)
$$

$$
u \cdot \frac{\partial w}{\partial x} + w \cdot \frac{\partial w}{\partial z} = \frac{\partial^2 w}{\partial x^2} + T, \qquad (7.44)
$$

$$
u \cdot \frac{\partial T}{\partial x} + w \cdot \frac{\partial T}{\partial z} = \frac{1}{\mathrm{Pr}} \cdot \frac{\partial^2 T}{\partial x^2}. \qquad (7.45)
$$

The energy and momentum balances are coupled via the temperature in the buoyancy term. The temperature distribution of the free convection flow therefore induces a velocity distribution.

The velocity and temperature profiles of the heated vertical plate are similar, so that they may be transformed into one another with a suitable coordinate transformation. The system of equations (7.43)–(7.45) yields two ordinary differential equations for the velocity w and the temperature T, which have to be solved numerically.

The computed velocity and temperature profiles for different Prandtl numbers are shown in Figure 7.14 for an isothermal boundary at constant

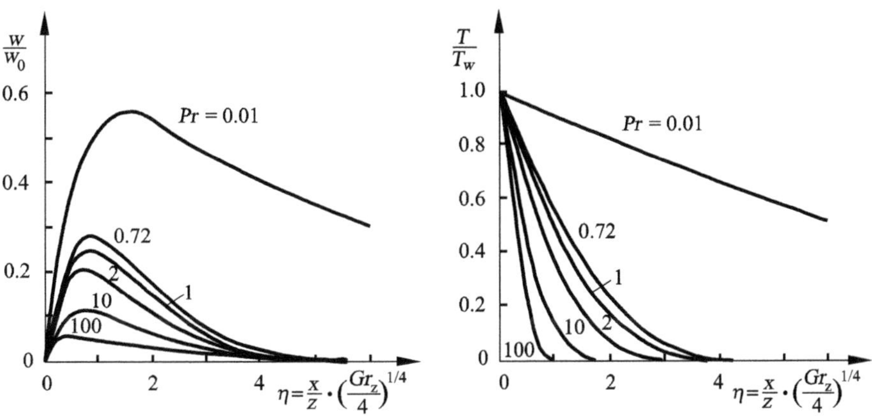

Fig. 7.14. Velocity and temperature profiles at a vertical heated plate at constant wall temperature T_w

wall temperature T_w. The characteristic velocity $w_0 = \sqrt{g \cdot \alpha \cdot l \cdot (T_\mathrm{m} - T_\infty)}$ corresponds to the transformation equation (7.42). For $\mathrm{Pr} \leq 1$ the viscous boundary-layer thickness δ and the thermal boundary-layer thickness δ_T are about the same size. For $\mathrm{Pr} \gg 1$ the thermal boundary-layer is restricted to a layer close to the wall. The heat transfer at the wall follows from

$$q_\mathrm{w} = -\lambda \cdot \left(\frac{\partial T}{\partial x}\right)_\mathrm{w} = -\lambda \cdot (T_\mathrm{w} - T_\infty) \cdot \frac{\mathrm{C}}{z^{\frac{1}{4}}} \cdot \left(\frac{\mathrm{d}T}{\mathrm{d}\eta}\right)_\mathrm{w}, \tag{7.46}$$

with the dimensionless vertical coordinate

$$\eta = -\frac{x}{z} \cdot \left(\frac{\mathrm{Gr}_z}{4}\right)^{\frac{1}{4}}$$

and the constant C. Here

$$\mathrm{Gr}_z = \frac{\alpha \cdot g \cdot (T_\mathrm{w} - T_\infty) \cdot z^3}{\nu^2} \tag{7.47}$$

is the local Grashof number formed with the z coordinate.

The local Nusselt number

$$\mathrm{Nu}_z = \frac{h \cdot z}{\lambda} = -\left(\frac{\mathrm{Gr}_z}{4}\right)^{\frac{1}{4}} \cdot \left(\frac{\mathrm{d}T}{\mathrm{d}\eta}\right)_\mathrm{w} \tag{7.48}$$

is shown plotted against the Prandtl number in Figure 7.15. The numerical solution can be approximated by the relation

$$\frac{\mathrm{Nu}_z}{\left(\frac{\mathrm{Gr}_z}{4}\right)^{\frac{1}{4}}} = \frac{0.676 \cdot \mathrm{Pr}^{\frac{1}{2}}}{(0.861 + \mathrm{Pr})^{\frac{1}{4}}}. \tag{7.49}$$

As well as the local Nusselt number, the mean Nusselt number is also of interest:

$$\frac{\mathrm{Nu}_l}{\left(\frac{\mathrm{Gr}_l}{4}\right)^{\frac{1}{4}}} = \frac{0.902 \cdot \mathrm{Pr}^{\frac{1}{2}}}{(0.861 + \mathrm{Pr})^{\frac{1}{4}}}. \tag{7.50}$$

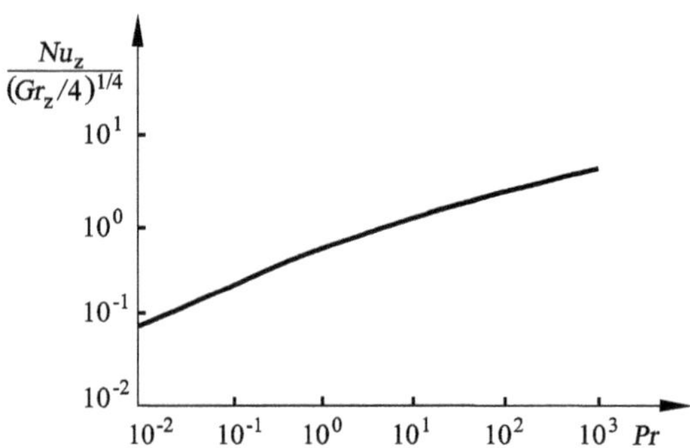

Fig. 7.15. Local Nusselt number at the vertical heated plate at constant wall temperature T_w

The solution functions for the velocities yield the friction coefficient

$$c_f = 2 \cdot \left(\frac{\mathrm{Gr}_z}{4}\right)^{-\frac{1}{4}} \cdot \left(\frac{\mathrm{d}w}{\mathrm{d}\eta}\right)_{\mathrm{w}} . \tag{7.51}$$

If the *heat flux* q_{w} is prescribed, rather than the wall temperature T_{w}, we obtain the Grashof number (7.6). The system of equations (7.43)–(7.45) remains unchanged, and it can be solved with the boundary condition $(\partial T/\partial x) = q_{\mathrm{w}}(z)/\lambda$ (heat conduction at the position $x = 0$). For the boundary layer thickness δ we obtain $\delta \sim \nu^{2/5}$, rather than $\delta \sim \sqrt{\nu}$ for a given wall temperature T_{w}.

The *region of validity* of the laminar boundary-layer flow with heat transport described until now is restricted to $10^4 < \mathrm{Ra}_l = \mathrm{Gr}_l \cdot \mathrm{Pr} < 10^8$. For Rayleigh numbers smaller than 10^4, the boundary-layer approximation is no long valid, and for Rayleigh numbers greater than 10^8 the transition to turbulence-free convection flow takes place.

Using the linear stability theory of Section 6.2.2, we use the basic profiles of the system of equations (7.43)–(7.45) to obtain a critical Grashof number $\mathrm{Gr}_{\mathrm{crit}}$ of $3 \cdot 10^6$ for air with $\mathrm{Pr} = 0.71$. This is considerably smaller than that found in experiment at the end of the transition process. This indicates that the small-amplitude perturbation waves are not recognized in experiment and are measured only upstream from the completion of the transition process. Figure 7.16 shows a differential interferogram in air of the laminar convection flow at the vertical plate at constant wall temperature T_{w} for the Grashof number $8 \cdot 10^6$, which is stable in experiment. The interference stripes show approximately lines of equal temperature gradient.

For the region of *turbulence-free convection flow* the Reynolds equations (5.40)–(5.42) and the energy equation (5.63) have to be solved numerically

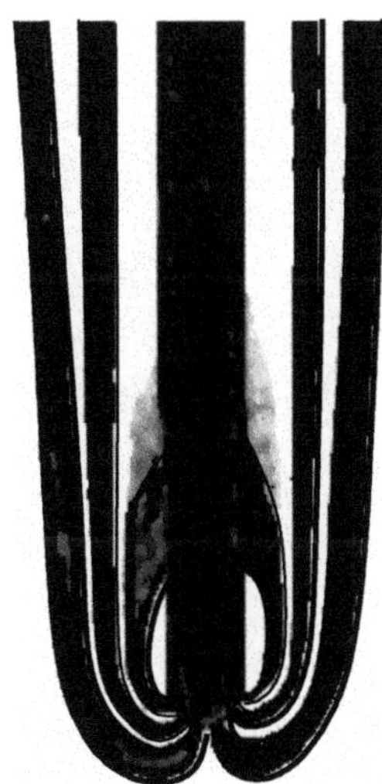

Fig. 7.16. Differential interferogram of the vertical heated plate, $\mathrm{Gr}_z = 8 \cdot 10^6$

with the buoyancy term and the Boussinesq approximation (5.85) in the boundary-layer approximation.

The dimensional system of equations of the two-dimensional turbulent boundary layer yields

$$\overline{u} \cdot \frac{\partial \overline{u}}{\partial x} + \overline{w} \cdot \frac{\partial \overline{u}}{\partial z} = \nu \cdot \left(\frac{\partial^2 \overline{u}}{\partial x^2} + \frac{\partial^2 \overline{u}}{\partial z^2} \right) - \frac{\overline{\partial u'^2}}{\partial x} - \frac{\overline{\partial (u' \cdot w')}}{\partial z}, \tag{7.52}$$

$$\overline{u} \cdot \frac{\partial \overline{w}}{\partial x} + \overline{w} \cdot \frac{\partial \overline{w}}{\partial z} = \nu \cdot \left(\frac{\partial^2 \overline{w}}{\partial x^2} + \frac{\partial^2 \overline{w}}{\partial z^2} \right) - \frac{\overline{\partial (u' \cdot w')}}{\partial x} - \frac{\overline{\partial w'^2}}{\partial z} + \alpha \cdot z \cdot (\overline{T} - T_\infty), \tag{7.53}$$

$$\overline{u} \cdot \frac{\overline{\partial T}}{\partial x} + \overline{w} \cdot \frac{\overline{\partial T}}{\partial z} = k \cdot \left(\frac{\partial^2 \overline{T}}{\partial x^2} + \frac{\partial^2 \overline{T}}{\partial z^2} \right) - \frac{\overline{\partial (u' \cdot T')}}{\partial x} - \frac{\overline{\partial (w' \cdot T')}}{\partial z}, \tag{7.54}$$

with the turbulent oscillation quantities u', w', T' of the Reynolds ansatz.

The turbulent velocity profile at the vertical heated plate is sketched in Figure 7.17. It can be divided into three regions. At a large enough distance from the wall we find the region of fully developed turbulent flow. Directly at the wall is the region of the viscous sublayer, introduced in Section 4.2.5. Between these is a transition region where the velocity changes only very little.

We use the Boussinesq ansatz to compute the wall shear stress as

$$\tau_{\mathrm{w}} = (\mu + \mu_{\mathrm{t}}) \cdot \left(\frac{\partial \overline{w}}{\partial x} \right)_{x=0} \tag{7.55}$$

and the heat flux at the wall

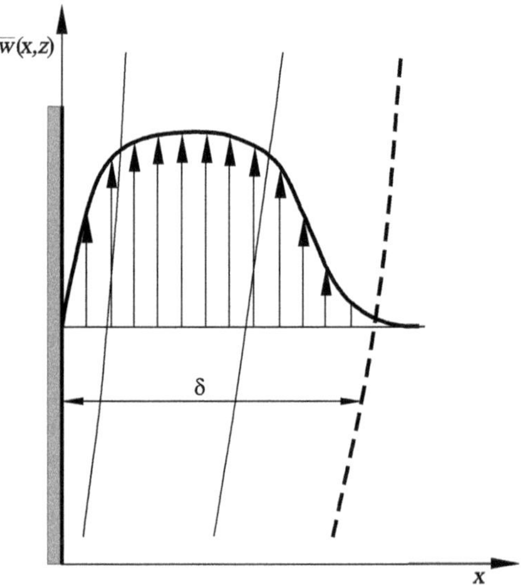

Fig. 7.17. Turbulent velocity profile at the vertical heated plate

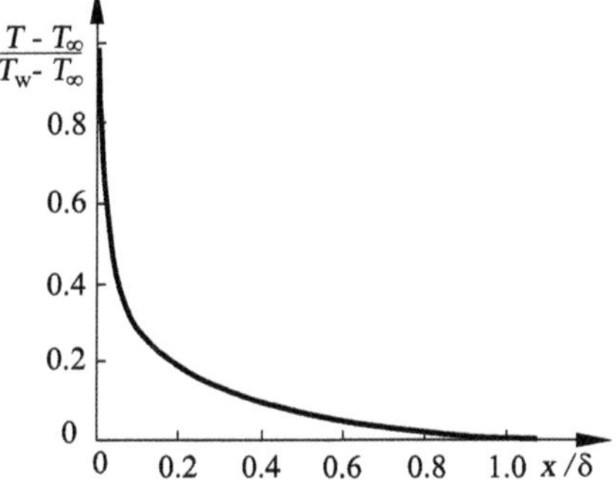

Fig. 7.18. Turbulent temperature profile at the vertical heated plate in air $\mathrm{Pr} = 0.71$ at a given wall temperature T_w

$$q_\mathrm{w} = (\lambda + \lambda_\mathrm{t}) \cdot \left(\frac{\partial \overline{T}}{\partial x}\right)_{x=0} . \tag{7.56}$$

The time-averaged temperature profile in air is shown in Figure 7.18. For the averaged heat flux we obtain the correlation

$$\overline{\mathrm{Nu}}_z \sim (\mathrm{Pr} \cdot \mathrm{Gr}_z)^{\frac{1}{3}} \tag{7.57}$$

for large values of $\mathrm{Pr} \cdot \mathrm{Gr}_z$.

The turbulence production due to buoyancy leads to considerably improved heat transfer. This is true for fluids with large Prandtl numbers. For media with small Prandtl numbers such as air, the turbulence production due to lift may be approximately neglected. The dependence of the local heat transfer for air and water is shown in Figure 7.19.

In practice, interpolation formulas are used to estimate the heat transfer of the heated vertical plate. For the mean heat flux in the region $0 < \mathrm{Pr} \cdot \mathrm{Gr}_z < 10^{12}$ we obtain

$$\sqrt{\overline{\mathrm{Nu}}_z} = 0.825 + \frac{0.387 \cdot (\mathrm{Pr} \cdot \mathrm{Gr}_z)^{\frac{1}{6}}}{\left(1 + \left(\frac{0.492}{\mathrm{Pr}}\right)^{\frac{9}{16}}\right)^{\frac{8}{27}}} . \tag{7.58}$$

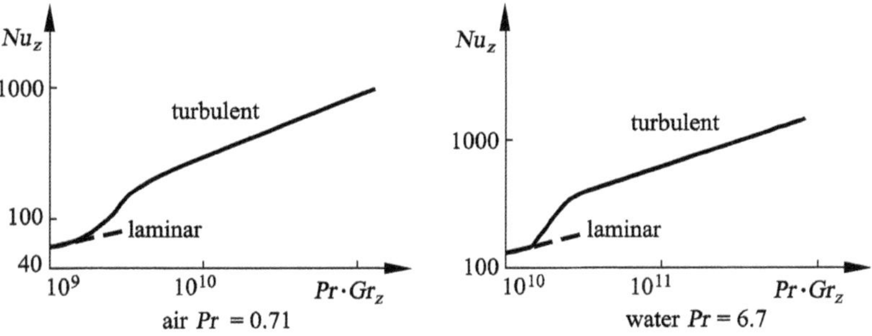

Fig. 7.19. Local heat transfer at the vertical heated plate

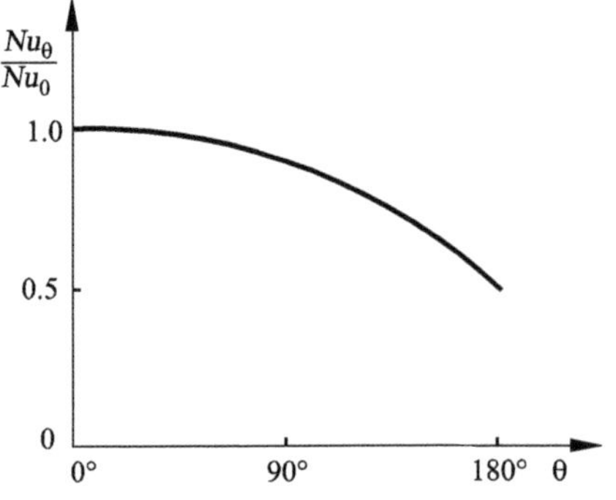

Fig. 7.20. Local heat transfer on the circumference of a horizontal circular cylinder in air Pr = 0.71 at a given wall temperature T_w

7.2.3 Convection at a Horizontal Cylinder

The free convection flow around a heated horizontal cylinder is shown in Figure 1.6. The system of equations (7.43)–(7.45) again leads to similar solutions for the velocity and temperature distributions, so that all the conclusions of the previous section may be used here, too. The laminar–turbulent transition is also completed here at a critical Grashof number of 10^8.

Figure 7.20 shows the local Nusselt number Nu_θ over the circumference of the horizontal circular cylinder for air at a given wall temperature T_w. Here Nu_0 denotes the heat transfer at the stagnation point. Integrating the Nusselt number Nu_θ over the circumference yields the mean Nusselt number $\overline{\mathrm{Nu}} \cdot \mathrm{Gr}^{(-1/4)} = 0.372$. Figure 7.21 shows a plot of the mean Nusselt number against the Rayleigh number $\mathrm{Ra} = \mathrm{Pr} \cdot \mathrm{Gr}$. For large Grashof numbers this behaves like $\overline{\mathrm{Nu}} \sim \mathrm{Ra}^{(1/4)}$, where the dependence on the Prandtl number for $\mathrm{Pr} > 0.71$ is small.

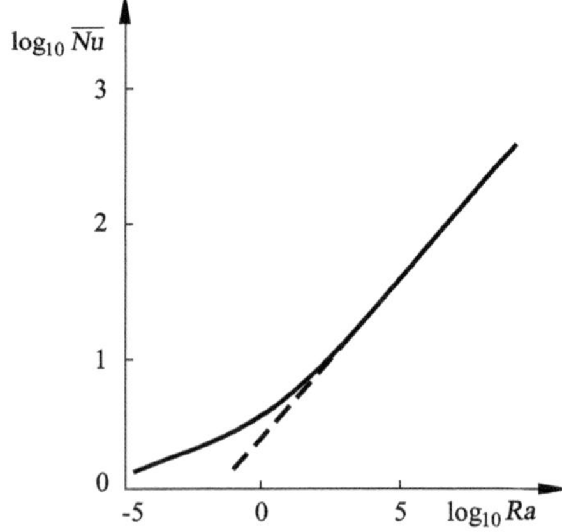

Fig. 7.21. Mean Nusselt number of a heated horizontal circular cylinder in air Pr = 0.71 for a given wall temperature T_w

7.3 Forced Convection

7.3.1 Pipe Flows

Fully developed pipe flow (Figure 7.4) has the following parabolic velocity profile, as treated in Section 4.2.1:

$$\frac{u}{u_{\max}} = 1 - \left(\frac{r}{R}\right)^2, \tag{7.59}$$

with the pipe radius R, the maximum velocity $u_{\max}$ is equal to $\Delta p \cdot R^2/(4 \cdot \mu \cdot l) = 2 \cdot u_{\mathrm{m}}$, and the constant pressure gradient is $\Delta p/l$. The thermal fully developed temperature profile is computed with the energy equation

$$u \cdot \frac{\partial T}{\partial x} = k \cdot \frac{1}{r} \cdot \frac{\partial}{\partial r}\left(r \cdot \frac{\partial T}{\partial r}\right). \tag{7.60}$$

The mean velocity u_{m} and the mean temperature T_{m} are found from

$$u_{\mathrm{m}} = \frac{1}{\pi \cdot R^2} \cdot \int_0^R 2 \cdot \pi \cdot r \cdot u \cdot \mathrm{d}r,$$

$$T_{\mathrm{m}} = \frac{1}{u_{\mathrm{m}} \cdot \pi \cdot R^2} \cdot \int_0^R 2 \cdot \pi \cdot r \cdot u \cdot T \cdot \mathrm{d}r.$$

We will compute the temperature profile for the two cases of constant heat transfer q_{w} and constant wall temperature T_{w}.

In the case of *constant heat transfer* $q_{\mathrm{w}} = h \cdot (T_{\mathrm{w}} - T_{\mathrm{m}})$, for thermally fully developed pipe flow the coefficient of heat transfer h is constant:

$$h = \frac{q_{\mathrm{w}}}{T_{\mathrm{w}} - T_{\mathrm{m}}} = \frac{\lambda}{R} \cdot \left(\frac{\partial}{\partial\left(\frac{z}{R}\right)} \cdot \left(\frac{T_{\mathrm{w}} - T}{T_{\mathrm{w}} - T_{\mathrm{m}}}\right)\right)_{\mathrm{w}}. \tag{7.61}$$

The $(T_{\mathrm{w}} - T_{\mathrm{m}})$ is constant, leading to

$$\frac{\partial T}{\partial x} = \frac{\mathrm{d}T_{\mathrm{w}}}{\mathrm{d}x} = \frac{\mathrm{d}T_{\mathrm{m}}}{\mathrm{d}x}.$$

Inserting this into the energy equation (7.60), we obtain

$$\frac{u}{k} \cdot \frac{\mathrm{d}T_{\mathrm{m}}}{\mathrm{d}x} = \frac{1}{r} \cdot \frac{\partial}{\partial r} \cdot \left(r \cdot \frac{\partial T}{\partial r}\right) \quad \text{for} \quad q_{\mathrm{w}} = \text{const.} \tag{7.62}$$

The case of constant heat flux density is found in many technical applications, such as electrical heating, nuclear heating, and heat exchangers.

For the thermally fully developed pipe flow with a *given wall temperature* T_{w}, we have

$$\frac{\partial T}{\partial x} = \frac{T_{\mathrm{w}} - T}{T_{\mathrm{w}} - T_{\mathrm{m}}} \cdot \frac{\mathrm{d}T_{\mathrm{m}}}{\mathrm{d}x}.$$

Therefore, the energy equation (7.60) becomes

$$\frac{u}{k} \cdot \left(\frac{T_{\mathrm{w}} - T}{T_{\mathrm{w}} - T_{\mathrm{m}}} \right) \cdot \frac{\mathrm{d}T_{\mathrm{m}}}{\mathrm{d}x} = \frac{1}{r} \cdot \frac{\partial}{\partial r} \cdot \left(r \cdot \frac{\partial T}{\partial r} \right) \quad \text{for} \quad T_{\mathrm{w}} = \text{const.} \tag{7.63}$$

The solutions of (7.62) and (7.63) are shown in Figure 7.22. In the case in which $q_{\mathrm{w}} = \text{const}$, the temperature difference is $(T_{\mathrm{w}} - T_{\mathrm{m}}) = \text{const}$. In the case $T_{\mathrm{w}} = \text{const}$, $(T_{\mathrm{w}} - T_{\mathrm{m}}(x))$ decreases with the pipe length x, since $T_{\mathrm{m}}(x)$ increases because of the energy supply. For $q_{\mathrm{w}} = \text{const}$ we obtain the Nusselt number $\mathrm{Nu} = 4.36$, and for $T_{\mathrm{w}} = \text{const}$, the value is $\mathrm{Nu} = 3.66$.

If we take the intake flow (Figure 7.4) into account, we obtain the local Nusselt number along the pipe with diameter $D = 2 \cdot R$. Figure 7.23 shows Nu_l for $q_{\mathrm{w}} = \text{const}$ and $T_{\mathrm{w}} = \text{const}$, together with the limiting cases for hydrodynamically and thermally fully developed pipe flow of air with $\mathrm{Pr} = 0.71$. It is seen that the thermal intake stretch l can be approximated by

$$\frac{l_T}{D} \approx 0.05 \cdot \mathrm{Re}_D \cdot \mathrm{Pr}. \tag{7.64}$$

The ratio of the intake stretches is $l_T/l \approx \mathrm{Pr}$. Highly viscous oils therefore have large thermal intake stretches.

The heat transfer coefficient is larger in the intake stretch than in the fully developed region. This is understandable because the boundary layer grows in the intake region and therefore, the local heat transfer drops.

For practical application, the mean Nusselt number is of interest:

$$\overline{\mathrm{Nu}} = \frac{1}{l} \cdot \int_0^l \mathrm{Nu}_x \cdot \mathrm{d}x. \tag{7.65}$$

Comparison with experiment yields variation at large temperature differences. These originate in the material constants that have been assumed constant until now. At large temperature differences, the viscosity and the thermal conductivity vary across the radius of the pipe. Figure 7.24 shows

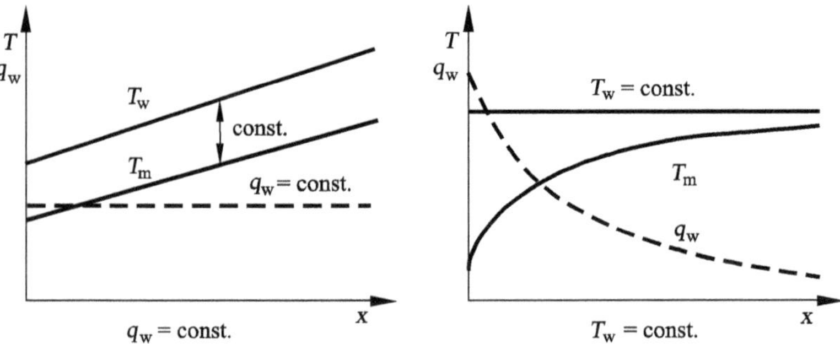

Fig. 7.22. Mean temperature T_{m} and wall temperature T_{w}, and heat flux q_{w} at a heated pipe wall

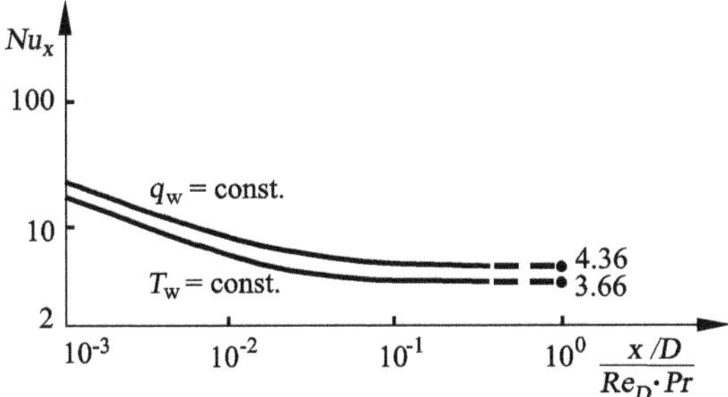

Fig. 7.23. Local Nusselt number in the intake of a pipe flow

the effect of varying viscosity on the velocity profile. For $\mu_w > \mu_m$ the increase in viscosity close to the wall caused by cooling a liquid or heating a gas leads to a more slender velocity profile. For $\mu_w < \mu_m$ the viscosity close to the wall is smaller for heated liquids or cooled gases, so that the velocity profile becomes broader.

Similar results are obtained for pipe cross-sections that are not circular or that vary. The rotational symmetry is then lost, and the complete system of equations for laminar incompressible flow (5.85) has to be solved numerically.

Turbulent pipe flow without heat supply has already been described in Section 4.2.5. With heat transport, the Reynolds equations (5.40)–(5.42) and (5.63) have to be solved numerically. The following simplifications can be applied to rotationally symmetric pipe flow with constant cross-section. For the shear stress $\tau(r)$ of turbulent pipe flow we obtain

$$\tau(r) = \tau_w \cdot \frac{r}{R} = -\mu \cdot \frac{\partial \overline{u}}{\partial r} + \rho \cdot \overline{u' \cdot v'} = -(\mu + \rho \cdot \epsilon_\tau) \cdot \frac{\partial \overline{u}}{\partial r}, \tag{7.66}$$

with $\tau_w = -(\mathrm{d}p/\mathrm{d}x) \cdot R/2$, and for the heat flux we obtain

$$q(r) = \frac{2 \cdot q_w}{u_m \cdot r \cdot R} \cdot \int_0^r \overline{u} \cdot r \cdot \mathrm{d}r = \lambda \cdot \frac{\partial \overline{T}}{\partial r} - \rho \cdot c_p \cdot \overline{T' \cdot v'} \tag{7.67}$$

$$= (\lambda + \rho \cdot c_p \cdot \epsilon_q) \cdot \frac{\partial \overline{T}}{\partial r},$$

with the turbulent exchange quantities ϵ_τ and ϵ_q.

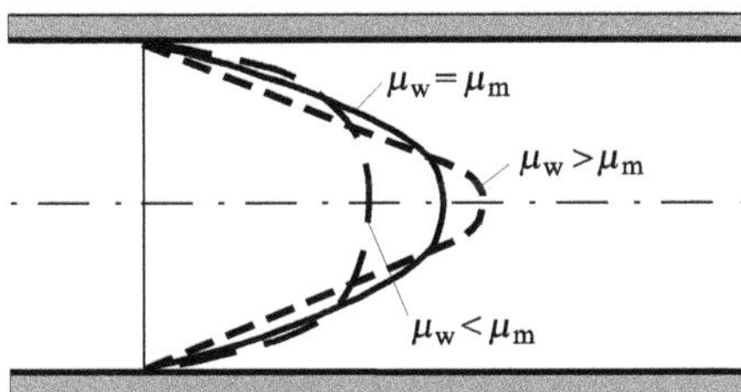

Fig. 7.24. Effect of varying viscosity on the parabolic velocity profile

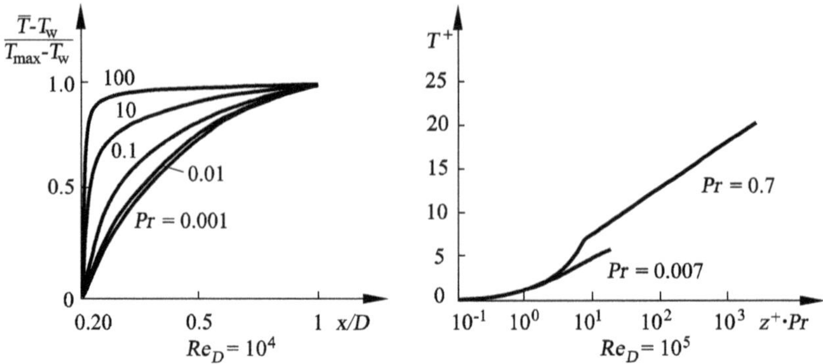

Fig. 7.25. Temperature profiles of fully developed turbulent pipe flow for $q_{\mathrm{w}} =$ const

Using the simplifying assumption of a given heat flux $q_{\mathrm{w}} =$ const at the pipe wall and thus neglecting the convective terms in the energy equation (5.63), no information about the time-average velocity profile is needed. We obtain the solution of the simplified energy equation:

$$-(\lambda + \rho \cdot c_p \cdot \epsilon_q) \cdot \frac{\mathrm{d}\overline{T}}{\mathrm{d}r} = -\mu \cdot c_p \cdot \left(\frac{1}{\mathrm{Pr}} + \frac{\epsilon_\tau}{\nu \cdot \mathrm{Pr_t}} \right) \cdot \frac{\mathrm{d}\overline{T}}{\mathrm{d}r} . \tag{7.68}$$

With the dimensionless variables

$$z^+ = \frac{r \cdot u_\tau}{\nu} , \quad T^+ = \frac{(T_{\mathrm{w}} - \overline{T}) \cdot \rho \cdot c_p \cdot u_\tau}{q_{\mathrm{w}}} , \quad u_\tau = \sqrt{\frac{\tau_{\mathrm{w}}}{\rho}} , \tag{7.69}$$

and empirical trial solutions for $\mathrm{Pr_t}$ and ϵ_τ, we obtain the temperature distributions of fully developed pipe flow (Figure 7.25) for a given heat flux

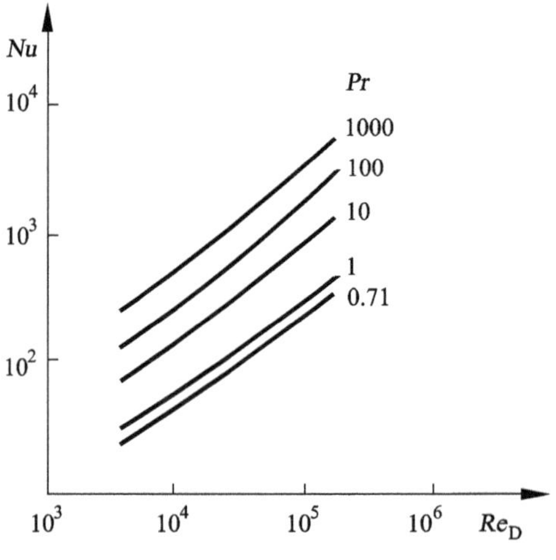

Fig. 7.26. Nusselt number for fully developed turbulent pipe flow for $q_{\mathrm{w}} =$ const.

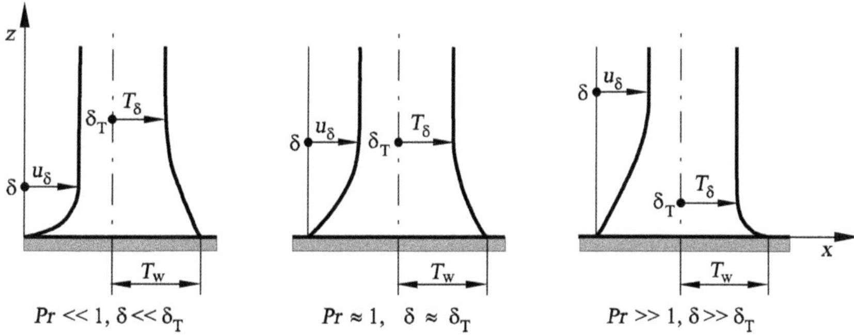

Fig. 7.27. Ratio of boundary-layer thicknesses δ_T, δ for different Prandtl numbers

$q_{\rm w}$ = const. In the logarithmic regime of the time-averaged velocity profile, molecular exchange can be approximately neglected compared to turbulent exchange. With increasing Prandtl number, this regime moves ever closer to the pipe wall. This viscous sublayer becomes thinner. The drag increases compared to the heat conduction, and the temperature profiles become broader, leading to an increase in the heat transfer. The dependence of the Nusselt number Nu on the Reynolds number Re_D and the Prandtl number Pr is shown in Figure 7.26.

There is a series of empirical relations for the Nusselt number to be found in the literature. These may be used both for constant heat flux $q_{\rm w}$ and for constant wall temperature $T_{\rm w}$. An example of such a relation is

$$\mathrm{Nu} = \frac{(\mathrm{Re}_D - 1000)\cdot \mathrm{Pr}\cdot \frac{\tau_{\rm w}}{\rho\cdot u_{\rm m}^2}}{1 + 12.7\cdot\sqrt{\frac{\tau_{\rm w}}{\rho\cdot u_{\rm m}^2}}\cdot(\mathrm{Pr}^{\frac{2}{3}} - 1)}\cdot\left(1 + \left(\frac{D}{l}\right)^{\frac{2}{3}}\right), \tag{7.70}$$

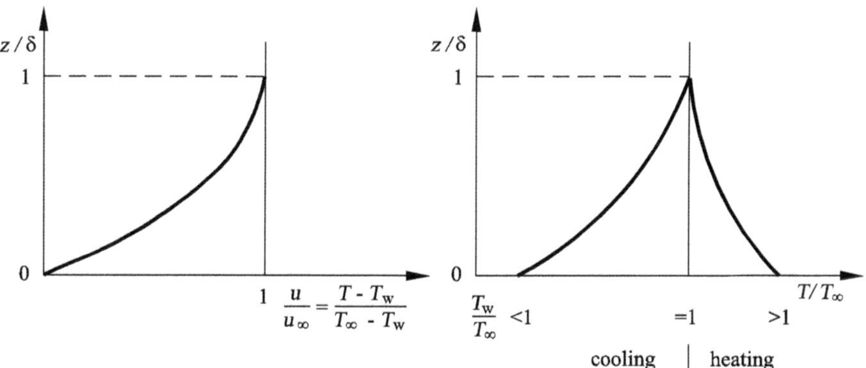

Fig. 7.28. Velocity and temperature profiles of the flat plate flow for Pr = 1 and given wall temperature $T_{\rm w}$ = const

with $\tau_\mathrm{w} = (\mathrm{d}p/\mathrm{d}x) \cdot R/2$.

7.3.2 Boundary-Layer Flows

In *forced convection flow in the boundary layer* of a plate placed longitudinally in a flow, the pressure gradient in the boundary-layer equation for free convection flow (7.44) has to be extended. The pressure work is not taken into account in the energy equation (7.45). Similarly, in what follows, the dissipation will be neglected, a fact that is approximately satisfied for incompressible flows. The ratio of thermal to flow boundary-layer thicknesses is that of free convection flow:

$$\frac{\delta_T}{\delta} \sim \frac{1}{\sqrt{\mathrm{Pr}}}. \tag{7.71}$$

If the convective heat transport and the heat conduction are of the same order of magnitude, we obtain

$$\frac{\delta_T}{\delta} \sim \frac{1}{\sqrt{\mathrm{Re}_x \cdot \mathrm{Pr}}}. \tag{7.72}$$

For different Prandtl numbers, we obtain the ratios sketched in Figure 7.27. Liquid metals, with $\mathrm{Pr} \ll 1$, have very good thermal conductivity at small viscosity. Gases, with $\mathrm{Pr} \approx 1$, have a comparably small viscosity and thermal conductivity, whereas oils with $\mathrm{Pr} \gg 1$ conduct heat badly but have a high viscosity.

For liquid metals, the flow boundary layer may be neglected. In order to compute the thermal boundary layer, the velocity profile can be approximately determined on the edge of the boundary layer $U_\delta(x)$. For gas flows, the thicknesses of the thermal and flow boundary layers are of the same order of magnitude, and the complete boundary-layer equations have to be solved. Corresponding to (7.43)–(7.45), these are written for incompressible forced convection flow as

$$\frac{\partial u}{\partial x} + \frac{\partial w}{\partial z} = 0, \tag{7.73}$$

$$u \cdot \frac{\partial u}{\partial x} + w \cdot \frac{\partial u}{\partial z} = -\frac{\mathrm{d}p}{\mathrm{d}x} + \frac{1}{\mathrm{Re}_l} \cdot \frac{\partial^2 u}{\partial z^2}, \tag{7.74}$$

$$u \cdot \frac{\partial T}{\partial x} + w \cdot \frac{\partial T}{\partial z} = \frac{1}{\mathrm{Pr} \cdot \mathrm{Re}_l} \cdot \frac{\partial^2 T}{\partial z^2}. \tag{7.75}$$

In order for the boundary-layer equations to be valid, and because of the requirement $\mathrm{Re}_l \gg 1$, we also demand that $\mathrm{Re}_l \cdot \mathrm{Pr} \gg 1$. The continuity and momentum equations (7.73) and (7.74) are now decoupled from the energy equation (7.75), and so the flow boundary layer can be computed independently of the thermal boundary layer.

For $\mathrm{Pr} = 1$, the boundary-layer equations (7.73)–(7.75) can be solved exactly. As well as the Blasius flow, the solution of the energy equation is also given. Figure 7.28 shows the computed velocity and temperature profiles for a given wall temperature T_w. Since the temperature and velocity profiles are identical, there is a direct proportionality between the heat transfer and the wall shear stress. This is called the *Reynolds analogy* between momentum and heat exchange, with the Stanton number

$$\mathrm{St} = \frac{c_f}{2}, \tag{7.76}$$

$$\mathrm{St} = \frac{q_\mathrm{w}}{\rho \cdot c_p \cdot (T_\mathrm{w} - T_\infty) \cdot u_\infty}$$

and the coefficient of friction $c_f = 2 \cdot \tau_\mathrm{w} / (\rho \cdot u_\infty^2)$. For the heat transfer, we obtain the exact solution

$$\mathrm{St} \cdot \sqrt{\mathrm{Re}_x} = \frac{\mathrm{Nu}_x}{\sqrt{\mathrm{Re}_x}} = 0.332, \tag{7.77}$$

and for the mean Nusselt number,

$$\overline{\mathrm{Nu}} = \frac{\overline{h} \cdot l}{R} = 0.664 \cdot \sqrt{\mathrm{Re}_l}. \tag{7.78}$$

The numerical solutions of the system of equations with the Stanton number (7.73)–(7.75) show that, in contrast to previous estimates, the ratio of the boundary-layer thicknesses for $\mathrm{Pr} > 1$ is proportional to $\mathrm{Pr}^{(-1/3)}$:

$$\frac{\delta_T}{\delta} = \frac{0.975}{\mathrm{Pr}^{\frac{1}{3}}}. \tag{7.79}$$

Therefore, the local Nusselt number is

$$\frac{\mathrm{Nu}_x}{\sqrt{\mathrm{Re}_x}} = 0.332 \cdot \mathrm{Pr}^{\frac{1}{3}}. \tag{7.80}$$

For liquid metals with $\mathrm{Pr} \ll 1$ we again obtain the dependence on $\sqrt{\mathrm{Pr}}$,

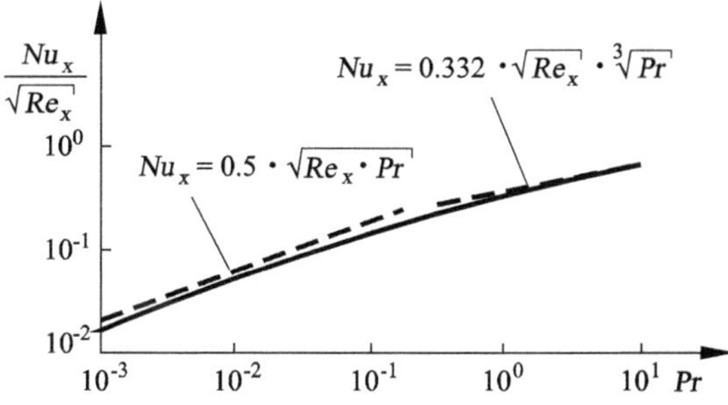

Fig. 7.29. Dependence of the local Nusselt number on the Prandtl number for the flat plate with constant wall temperature T_w

$$\frac{\delta_T}{\delta} = \frac{0.58}{\sqrt{\mathrm{Pr}}}, \tag{7.81}$$

and the local Nusselt number

$$\frac{\mathrm{Nu}_x}{\sqrt{\mathrm{Re}_x}} = 0.5 \cdot \sqrt{\mathrm{Pr}}. \tag{7.82}$$

Figure 7.29 summarizes the results of the dependence of the local heat transfer on the Prandtl number.

Dissipation

At *high flow velocities* the dissipation

$$\Phi = \mu \cdot \left(\frac{\partial u}{\partial z}\right)^2 \tag{7.83}$$

of two-dimensional boundary-layer flow cannot be neglected. For the case of an adiabatic wall with $q_\mathrm{w} = 0$ the temperature profile in Figure 7.30 is expected. The dissipation is largest close to the wall. Therefore, the temperature T_{q_w} will have a maximum at the wall, called the *recovery temperature.* Figure 7.31 shows the temperature profiles for different Prandtl numbers for the adiabatic wall. The dissipation causes the temperature profiles to become broader. The adiabatic wall temperature (recovery factor) shows that for $\mathrm{Pr} > 1$ a temperature T_{q_w} is obtained that is larger than the adiabatic stagnation temperature. The adiabatic wall temperature can be approximated by

$$\frac{c_p \cdot (T_{q_\mathrm{w}} - T_\infty)}{\frac{1}{2} \cdot u_\infty^2} \approx \begin{cases} \sqrt{\mathrm{Pr}} & \text{for} \quad 0.6 < \mathrm{Pr} < 1.5, \\ 1.9 \cdot \mathrm{Pr}^{\frac{1}{3}} & \text{for} \quad \mathrm{Pr} \gg 1. \end{cases} \tag{7.84}$$

For the case of constant wall temperature T_w, the temperature T_∞ is replaced by T_{q_w} in the definition of the heat transfer coefficient h. Therefore, we have, also with dissipation, the relation (7.80):

$$\frac{\mathrm{Nu}_x}{\sqrt{\mathrm{Re}_x}} = 0.332 \cdot \mathrm{Pr}^{\frac{1}{3}} \quad \text{for} \quad 0.6 < \mathrm{Pr} < 10.$$

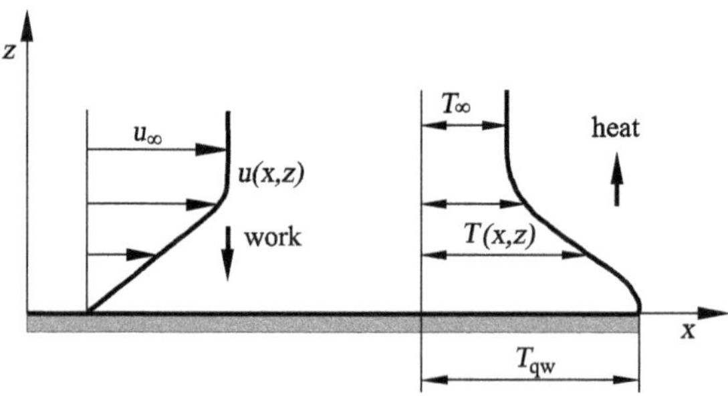

Fig. 7.30. Temperature profile due to dissipation at an adiabatic wall

Compressibility

Because the stagnation temperature grows quadratically with the Mach number in gases, compressibility effects very soon have to be taken into account. As well as the Mach number, the Eckert number

$$\mathrm{Ec} = \frac{u_\infty^2}{c_p \cdot (T_\mathrm{m} - T_\mathrm{w})} \tag{7.85}$$

is also a measure for the compressibility, since $\mathrm{Ec} \sim M^2$. The Reynolds analogy (7.76) is also valid for compressible flow, in the form

$$\mathrm{St} = \frac{c_f}{2 \cdot \mathrm{Pr}}. \tag{7.86}$$

For compressible boundary-layer flows, there is a coupling between temperature and velocity

$$\frac{T}{T_\infty} = \frac{T_\mathrm{w}}{T_\infty} + \frac{T_\infty - T_\mathrm{w}}{T_\infty} \cdot \frac{u}{u_\infty} + \mathrm{Pr} \cdot \frac{\kappa - 1}{2} \cdot M_\infty^2 \cdot \left(1 - \frac{u}{u_\infty}\right), \tag{7.87}$$

as given by *L. Crocco* (1932) and *A. Busemann* (1935). The effect of compressibility is seen in the third term and the effect of heat transfer in the second term of the relation (7.87).

Figure 7.32 shows the velocity and temperature profiles at the adiabatic flat plate for $\mathrm{Pr} = 1$. The boundary-layer thickness grows with increasing Mach number, and the velocity profile takes on an almost linear form for large Mach numbers. For strong cooling of the wall, the thickening and therefore the displacement effect of the boundary layer is reduced, and the velocity profiles become broader. Heating the wall increases the displacement effect of the compressible plate boundary layer.

Turbulent Boundary-Layer Flow

Our knowledge of turbulent pipe flow can also be applied to the flat plate boundary-layer flow. The starting point is the boundary-layer equations

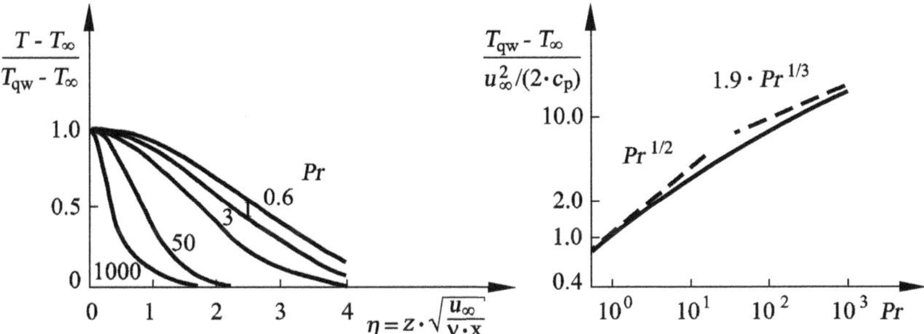

Fig. 7.31. Temperature profile and adiabatic wall temperature at the flat plate for constant fluid properties

(7.73)–(7.75). The Reynolds ansatz yields the time-average boundary-layer equations, neglecting the pressure work and the dissipation:

$$\frac{\partial \overline{u}}{\partial x}+\frac{\partial \overline{w}}{\partial z}=0, \tag{7.88}$$

$$\overline{u}\cdot\frac{\partial \overline{u}}{\partial x}+\overline{w}\cdot\frac{\partial \overline{u}}{\partial z}=-\frac{1}{\rho}\cdot\frac{\mathrm{d}\overline{p}}{\mathrm{d}x}+\frac{\partial^2 \overline{u}}{\partial z^2}-\frac{\partial(\overline{u'\cdot w'})}{\partial z}, \tag{7.89}$$

$$\overline{u}\cdot\frac{\partial \overline{T}}{\partial x}+\overline{w}\cdot\frac{\partial \overline{T}}{\partial z}=k\cdot\frac{\partial^2 \overline{T}}{\partial z^2}-\frac{\partial(\overline{w'\cdot T'})}{\partial z}, \tag{7.90}$$

with the Reynolds heat flux

$$q_\mathrm{t}=\rho\cdot c_p\cdot\overline{w'\cdot T'}.$$

For $\mathrm{Pr}=1$ we also have the *Reynolds analogy* for plate boundary-layer flow:

$$\mathrm{St}=\frac{\mathrm{Nu}_x}{\mathrm{Re}_x\cdot\mathrm{Pr}}=\frac{c_f}{2}. \tag{7.91}$$

In the *Prandtl analogy*, the flow field is divided into the viscous sublayer and the fully turbulent region. This yields

$$\mathrm{St}=\frac{\frac{c_f}{2}}{1+5\cdot\sqrt{\frac{c_f}{2}}\cdot(\mathrm{Pr}-1)}. \tag{7.92}$$

For $\mathrm{Pr}=1$, the Prandtl analogy is identical to the Reynolds analogy (7.91).

Von Kármán built on Prandtl's ideas and divided the boundary layer into three regimes. Between the viscous sublayer and the fully turbulent regime, he considered a transition layer ($5<z^+<30$) in which the molecular and turbulent exchange quantities are of the same order of magnitude. He formulated the *von Kármán analogy*

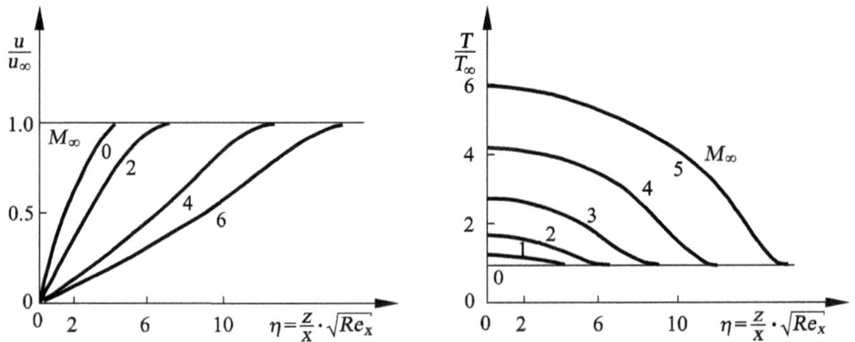

Fig. 7.32. Velocity and temperature distributions at the adiabatic flat plate for $\mathrm{Pr}=1$

$$St = \frac{\frac{c_f}{2}}{1 + 5 \cdot \sqrt{\frac{c_f}{2}} \cdot \left(\mathrm{Pr} - 1 + \ln \left(\frac{5 \cdot \mathrm{Pr} + 1}{6} \right) \right)}, \tag{7.93}$$

which is again identical to the Reynolds analogy for $\mathrm{Pr} = 1$.

The local coefficient of friction for the flat plate is

$$c_f = \mathrm{const} \cdot \mathrm{Re}_x^{-\frac{1}{5}}. \tag{7.94}$$

The local Nusselt numbers for the turbulent boundary layer of the flat plate are shown in Figure 7.33. For all analogies it is assumed that the turbulent Prandtl number is set to $\mathrm{Pr_t} = 1$. For this reason, they cannot be applied for liquid metals where $\mathrm{Pr} \ll 1$.

For the fully developed turbulent pipe flow it was assumed that the ratio of heat flux density to shear stress in the central flow is approximately constant over the cross-section of the pipe. This is also approximately true for plate flow. With the dimensionless quantities $u^+ = \overline{u}/u_\tau$ and $T^+ = (T_\mathrm{w} - \overline{T}) \cdot \rho \cdot c_p \cdot u_\tau / q_\mathrm{w}$ as well as $z^+ = z \cdot u_\tau/\nu$ we obtain the temperature profiles shown in Figure 7.34.

7.3.3 Bodies in Flows

The simplest case of a body with heat transfer in a flow is the circular cylinder in a transverse flow. In a large range of Reynolds numbers, the heat transfer takes place mainly in the boundary layer, so that the relations from the previous section can be applied to the cylinder boundary layer.

According to experiments by *R. Hilpert* (1933), the dependence of the mean Nusselt number in air and at constant cylinder temperature T_w can be divided into different Reynolds number regimes:

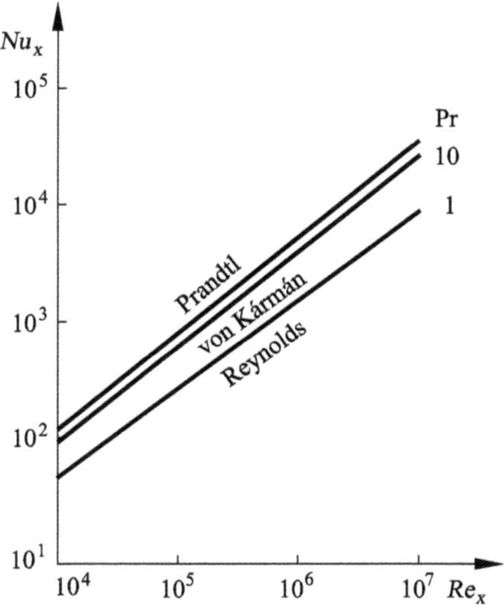

Fig. 7.33. Local Nusselt number of the turbulent flat plate boundary-layer flow

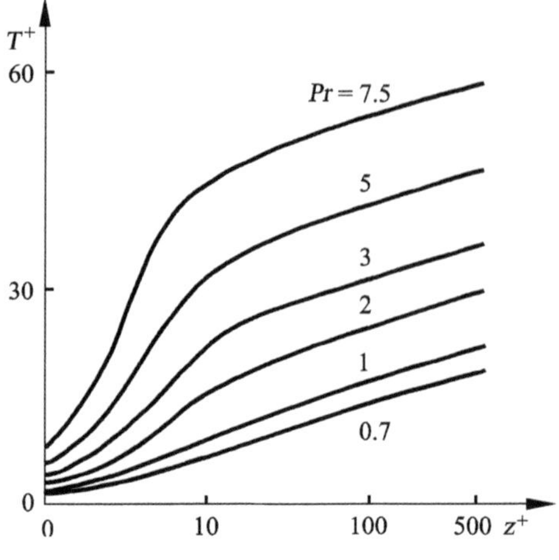

Fig. 7.34. Temperature profile of the turbulent flat plate boundary-layer flow

$$
\begin{aligned}
40 < \mathrm{Re}_D < 4000, \qquad & \overline{\mathrm{Nu}}=0.615 \cdot \mathrm{Re}_D^{0.466}, \\
4000 < \mathrm{Re}_D < 4 \cdot 10^4, \qquad & \overline{\mathrm{Nu}}=0.174 \cdot \mathrm{Re}_D^{0.618}, \\
4 \cdot 10^4 < \mathrm{Re}_D < 2.5 \cdot 10^5, \qquad & \overline{\mathrm{Nu}}=0.0239 \cdot \mathrm{Re}_D^{0.805}.
\end{aligned}
\tag{7.95}
$$

The exponent increases from 0.46 to 0.8 at Reynolds numbers larger than $4 \cdot 10^4$. This indicates that the Kármán vortex street in the wake flow contributes ever more to the heat transfer for increasing Reynolds number. Relations (7.95) hold for turbulence-free streams. If the turbulence intensity in the free flow is increased to 2.5%, the mean Nusselt number rises by up to 80%. This explains why the Nusselt numbers measured in wind tunnels are generally higher than those given in (7.95).

7.4 Heat and Mass Exchange

Heat and mass exchange processes occur in boundary layers if, for example, coolant gas is supplied to the boundary-layer flow. Blowing a light gas from the wall reduces the heat transfer. Evaporated liquid layers at the wall assist the cooling. As well as momentum and heat exchange, there is also mass exchange due to diffusion. As well as the velocity and thermal boundary layers, there are also concentration boundary layers.

7.4.1 Diffusion Convection

In analogy to Rayleigh–Bénard convection, a concentration gradient can be responsible for an unstable density layering in a mixture, even at constant temperature. For example, in a salt solution the density increases with the concentration. If water evaporates at the free surface of a salt solution (Figure 7.35), a high salt concentration remains, and an unstable density layering arises. We will see that the treatment of a convection flow in a binary mixture driven by concentration differences is identical to the analysis of Rayleigh–Bénard convection. Only the characteristic temperature difference ΔT has to be replaced by the concentration difference Δc, the heat expansion coefficient $\alpha = \rho^{-1} \cdot \mathrm{d}\rho/\mathrm{d}T$ by the concentration expansion coefficient $\beta = \rho^{-1} \cdot \mathrm{d}\rho/\mathrm{d}c$, and the thermal conductivity k by the diffusion coefficient D. Similarly, the diffusion Rayleigh number $\mathrm{Ra}_D = \beta_m \cdot \Delta \cdot c_m \cdot g \cdot l^3/(\nu \cdot D)$ replaces the Rayleigh number Ra, and the Schmidt number $\mathrm{Sc} = \nu/D$ replaces the Prandtl number Pr, where g is again the gravitational acceleration, l the thickness of the liquid layer, and ν the kinematic viscosity. All results from thermal cellular convection may therefore be directly carried over to diffusion convection.

In what follows we will therefore treat the *double diffusion instability*. The single diffusion instability is then a special case. Double diffusion phenomena are processes in which two diffusion effects occur simultaneously: mass diffusion and heat diffusion (heat conduction). We will treat the stability of a double diffusion system that is due to the superposition of mass diffusion (e.g. saltwater solution in the ocean) with heat conduction. Depending on the case, these two different diffusion processes may interact either to introduce an instability or to stabilize a liquid layer.

The upper side of the liquid layer is kept at a higher temperature T_2 than the base T_1 (see Figure 7.35). Let the salt concentration c_2 on the surface also be greater than that on the base c_1. Here c indicates the mass concentration $c = \rho_s/\rho$ with the partial density of the salt ρ_s and the total density of the solution ρ. Both a stable and an unstable density layering can exist in this arrangement.

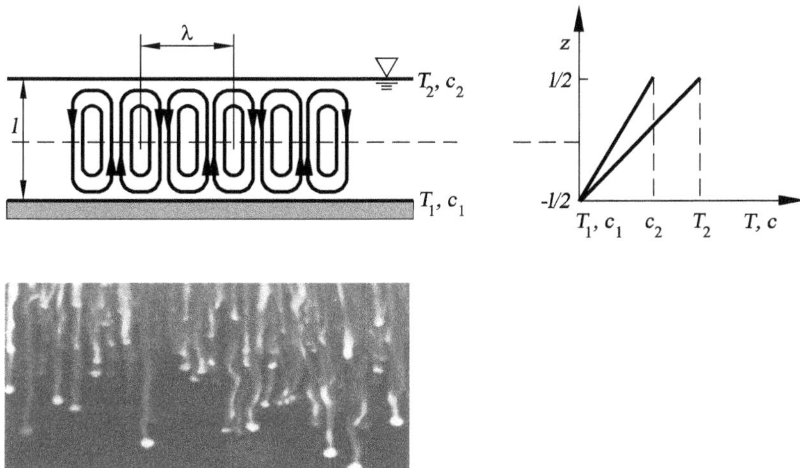

Fig. 7.35. Double diffusion instability, *J. Turner* (1985)

If the unstable concentration difference $\Delta c = c_2 - c_1$ exceeds a critical value, the horizontal liquid layer will become unstable, even if the thermal stratification is stable. Salt fingers form, which evolve into narrow high convection cells.

In analogy to Rayleigh–Bénard convection, we consider a liquid element of characteristic size l that rises with a small vertical velocity w because of a small perturbation (Figure 7.36). In the new layer it has a lower temperature and a lower salt content compared to the surroundings. It rises with velocity w and passes along the temperature gradient $\Delta T/l$ in the surrounding fluid layer. The associated change in internal energy in the volume l^3 of the particle is $\dot{E}_k = \rho_m \cdot c_v \cdot w \cdot (\Delta T/l) \cdot l^3$. This change is achieved by the energy flow through the particle surface $\sim l^2$ due to heat conduction $\dot{q} \sim \lambda \cdot \Delta T_w/l$. The effective temperature gradient ΔT_w was introduced to indicate that it is generally not the entire temperature gradient ΔT in the layer that acts during the process. If the upward velocity w of the particle is large, the particle does not have enough time to adapt itself to the ambient temperature. The balance $\dot{E}_k = \dot{q} \cdot l^2$ is an estimate for the effective $\Delta T_w = w \cdot l \cdot \Delta T/k$, with the thermal conductivity $k = \lambda/(\rho_m \cdot c_v)$. If the particle velocity were such that temperature equilibrium is just achieved, ΔT and ΔT_w would be the same. The associated thermal diffusion velocity w_T would then be $w_T = k/d$.

While the particle is exposed to the concentration gradient of the layer $\Delta c/l$ it accumulates salt. The change in concentration that it experiences while rising with velocity w is therefore $w \cdot \Delta c/l$. This corresponds to a mass change of $\dot{m} = \rho_m \cdot w \cdot (\Delta c/l) \cdot l^3$. The salt accumulation takes place as a diffusion flux $j = \rho_m \cdot D \cdot (\Delta c_w/l) \cdot l^2$ flowing through the surface l^2 of the particle. Here D denotes the diffusion coefficient. Again we have introduced the effective concentration difference Δc_w, since the speed of the fluid particle means that it does not have enough time to experience the full concentration difference Δc. Setting $\dot{m} = j$ we obtain the effective concentration gradient

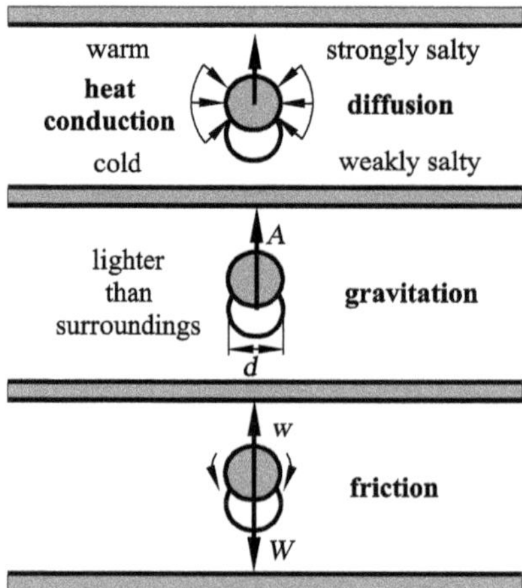

Fig. 7.36. Physical explanation of the double diffusion instability

as $\Delta c_w = w \cdot l \cdot \Delta c/D$. If the particle velocity were such that the concentration compensation is just attained, Δc and Δc_w would be equal. The associated mass diffusion velocity w_D would then be $w_D = D/l$.

A statement about the instability of the state is reached in the same way as with Rayleigh–Bénard convection, by comparing the lift force A acting on the fluid element to the drag force W. The lift force $A = A_T + A_D$ is made up of a thermal part A_T and a diffusion part A_D. The density change of the fluid element due to the temperature change is $\Delta \rho_T \sim \rho_m \cdot \alpha_m \cdot \Delta T_w$. The part of the lift force due to this effect is $A_T \sim \rho_m \cdot \alpha_m \cdot \Delta T_w \cdot g \cdot l^3$, with the mean coefficient of thermal expansion α_m. The density change due to diffusion is $\Delta \rho_D \sim \rho_m \cdot \beta_m \cdot \Delta c_w$, which leads to the lift force $A_D \sim -\rho_m \cdot \beta_m \cdot \Delta c_w \cdot g \cdot l^3$. Here β_m denotes the mean coefficient of concentration expansion. The minus sign was introduced so that A_T and A_D point in the same direction when Δc_w and ΔT_w have the same sign. The motion of the particle acts against the drag force W. For creeping flows (small perturbation velocities w), Stokes's law states that $W \sim \mu \cdot w \cdot l = \mu \cdot l^2/\Delta t$. The condition for instability is clearly given by the domination of lift over drag:

$$A = A_T + A_D \geq W,$$

$$\rho_m \cdot \alpha_m \cdot \Delta T_w \cdot g \cdot l^3 - \rho_m \cdot \beta_m \cdot \Delta c_w \cdot g \cdot l^3 \geq \mu \cdot w \cdot l \cdot \mathrm{C}.$$

The constant C summarizes all the factors of proportionality used in the above estimates. Using the above relations for ΔT_w and Δc_w and dividing by $\mu \cdot w \cdot l$, we obtain

$$\underbrace{\frac{\alpha_m \cdot \Delta T \cdot g \cdot l^3}{k \cdot \nu}}_{\mathrm{Ra}} - \underbrace{\frac{\beta_m \cdot \Delta c \cdot g \cdot l^3}{D \cdot \nu}}_{\mathrm{Le} \cdot \mathrm{Ra}_D} \geq \mathrm{C}. \tag{7.96}$$

The first dimensionless collection of quantities on the left-hand side is simply the Rayleigh number again. The second dimensionless collection is generally written as the product of the diffusion Rayleigh number $\mathrm{Ra}_D = \beta_m \cdot \Delta c \cdot g \cdot l^3/(k \cdot \nu)$ and the *Lewis number* $\mathrm{Le} = k/D$. The *Lewis number* is the ratio of the characteristic thermal diffusion velocity w_T and the material diffusion velocity w_D: $\mathrm{Le} = w_T/w_D$.

We note that the onset of Rayleigh–Bénard convection is a special case of the above stability criterion. Without the effect of diffusion, $\mathrm{Ra}_D = 0$, and we obtain the stability criterion (7.16). We also note that the constant C, which has the meaning of the value of a critical characteristic number, can simply be read off the analysis of Rayleigh–Bénard convection ($\mathrm{Ra}_D = 0$), i.e. $\mathrm{C} = \mathrm{Ra}_{\mathrm{crit}}$. From this phenomenological discussion we can write

$$\mathrm{Ra} - \mathrm{Le} \cdot \mathrm{Ra}_D \geq \mathrm{Ra}_{\mathrm{crit}}. \tag{7.97}$$

The relation (7.97) for $\mathrm{Ra} - \mathrm{Le} \cdot \mathrm{Ra}_D = \mathrm{Ra}_{\mathrm{crit}}$ is the equation for a straight line in the $\mathrm{Ra}(\mathrm{Ra}_D)$ diagram. This critical straight line has slope Le (see Figure 7.37).

For positive Ra there is thermally unstable density stratification, and for negative Ra_D, diffusively unstable density stratification. For example, for a given $\mathrm{Ra}_D < 0$, the density stratification becomes unstable already at values of $\mathrm{Ra} < \mathrm{Ra}_{\text{crit}}$.

Also note that the thermal and diffusion density gradients $\Delta\rho_T = \rho_m \cdot \alpha_m \cdot \Delta T$ and $\Delta\rho_D = \rho_m \cdot \beta_m \cdot \Delta c$ of the layer cancel each other out for $\mathrm{Ra} = \mathrm{Ra}_D$. The condition $\mathrm{Ra} > \mathrm{Ra}_D$ states that the denser medium lies above the lighter medium. Beyond the point where the critical straight line from (7.97) cuts the straight line $\mathrm{Ra} = \mathrm{Ra}_D$ (such a point exists for $\mathrm{Le} \neq 1$), instability is also possible for stable density stratification.

Although (7.97) is an exact stability criterion, we also mention that, in the regime of very large positive diffusion Rayleigh numbers Ra_D, this equation is no longer valid. The stratification is already unstable at smaller thermal Rayleigh numbers Ra than predicted by (7.97). The strong density changes of the particle at relatively strong concentration and temperature gradients are responsible for this. These density changes ensure that the inertial force, as well as the lift and frictional forces, also affects the equilibrium. The instabilities that then occur are unsteady. A further dimensionless characteristic number that then appears is the Prandtl number $\mathrm{Pr} = \nu/k$. Until now, the inertial forces have been neglected in Rayleigh–Bénard convection. These forces are necessary to describe the onset of the steady instabilities correctly. They occur in the form of narrow high convection cells, and are generally known as *finger instabilities* (cf. Figure 7.35).

Stability Analysis

The *fundamental equations of double diffusion convection* (5.90) (concentration and temperature gradients) and the *perturbation differential equations* (5.216)–(5.219) are given in Section 5.4.3 and 5.5.

The *ground state* of the double diffusion convection instability $\boldsymbol{U}_0 = (c_0, \boldsymbol{u}_0, p_0, T_0)$ is obtained from the continuity equation and energy equation (5.90):

$$\boldsymbol{\Delta} c_0 = 0, \qquad \boldsymbol{\Delta} T_0 = 0. \tag{7.98}$$

A state of rest $\boldsymbol{u} = 0$ is also possible here if the temperature gradient $\boldsymbol{\nabla} T_0$ is not parallel to the direction of the force of gravity $\boldsymbol{e}_z$. Taking the curl of the momentum equation (5.90) and inserting $\boldsymbol{u} = 0$, we obtain the condition $(\boldsymbol{\nabla} T_0 - \mathrm{Ra}_D/\mathrm{Ra} \cdot \boldsymbol{\nabla} c_0) \times \boldsymbol{e}_z = 0$. The requirement that these vectors be parallel is now more generally valid for the vector sum of the temperature and concentration gradients. Here $\mathrm{Ra}_D/\mathrm{Ra} = -\boldsymbol{\Delta}\rho_D/\boldsymbol{\Delta}\rho_T$ is to be interpreted as the ratio of the density change due to temperature gradients $\Delta\rho_T = -\rho_m \cdot \alpha_m \cdot \boldsymbol{\Delta} T$ to the density change due to concentration gradients $\boldsymbol{\Delta}\rho_D = \rho_m \cdot \beta_m \cdot \boldsymbol{\Delta} c$. For $\boldsymbol{\Delta}\rho_D/\boldsymbol{\Delta}\rho_T = 1$, the density is the same at every position, since in this case the density changes due to temperature and concentration just cancel each other out. The situation of neutral density stratification is therefore given by $\mathrm{Ra} = \mathrm{Ra}_D$.

For a layer that is infinitely extended in the horizontal directions x and y, the ground state is independent of x and y. Let the temperature and the concentrations at the two horizontal boundaries of the liquid layer be constant and given by

$$T_0\left(x, y, z = -\frac{1}{2}\right) = T_1, \qquad T_0\left(x, y, z = \frac{1}{2}\right) = T_2,$$
$$c_0\left(x, y, z = -\frac{1}{2}\right) = c_1, \qquad c_0\left(x, y, z = \frac{1}{2}\right) = c_2.$$

Along the homogeneous parallel directions x, y, the ground state is dependent only on the vertical direction z. From the above Laplace equations for T_0 and c_0 we obtain

$$T_0(z) = \mathrm{C}_1^T \cdot z + \mathrm{C}_0^T, \qquad c_0(z) = \mathrm{C}_1^c \cdot z + \mathrm{C}_0^c.$$

The constants (C_0^T, C_1^T) and (C_0^c, C_1^c) follow from the boundary conditions, since $\mathrm{C}_1^T = -1$, $\mathrm{C}_0^T = (T_1 + T_2 - 2 \cdot T_m)/\Delta T$ and $\mathrm{C}_1^c = -1$, $\mathrm{C}_0^c = (c_1 + c_2 - 2 \cdot c_m)/\Delta c$. With $T_m = 0.5 \cdot (T_1 + T_2)$ as in Rayleigh–Bénard convection, and $c_m = 0.5 \cdot (c_1 + c_2)$ we obtain the fundamental solution

$$T_0 = c_0 = -z. \tag{7.99}$$

From the first two Boussinesq equations (5.85), we obtain $p_0 = p_0(z)$. The z Boussinesq equation yields

$$0 = -\frac{\mathrm{d}p_0}{\mathrm{d}z} + (\mathrm{Ra} \cdot T_0 - \mathrm{Ra}_D \cdot c_0),$$

or, using (7.98) for the pressure,

$$p_0 = -\frac{1}{2} \cdot (\mathrm{Ra} - \mathrm{Ra}_D) \cdot z^2 + p_\infty. \tag{7.100}$$

Here p_∞ is the ambient pressure. The temperature and concentration distributions, and thus also the entire heat conduction diffusion problem, are independent of p_∞. It is not the magnitude of the pressure p_0 that influences the problem, but only its gradient $\mathrm{d}p_0/\mathrm{d}z$.

The linear stability analysis again yields a *stability diagram* (Figure 7.37). The procedure corresponds to that of Rayleigh–Bénard covection in Section 7.2.1. For a horizontal two-component layer with *free boundaries*, the *finger instability* may be either steady or oscillatory. The indifference curve for the steady double diffusion instability is calculated with (see *H. Oertel* and *J. Delfs* (1996))

$$\overline{\Pi}(a) = \mathrm{Ra} - \mathrm{Le} \cdot \mathrm{Ra}_D = \frac{(a^2 + \pi^2)^3}{a^2}. \tag{7.101}$$

Here $\overline{\Pi}(a)$ describes the same curve as $\mathrm{Ra}(a)$ in Rayleigh–Bénard convection. For the indifference curve for oscillatory finger instabilities we obtain

$$\tilde{\Pi}(a) = \frac{\mathrm{Pr} \cdot \mathrm{Le}^2 \cdot \mathrm{Ra} - \mathrm{Pr} \cdot \mathrm{Le} \cdot \dfrac{1 + \mathrm{Pr} \cdot \mathrm{Le}}{1 + \mathrm{Pr}} \cdot \mathrm{Ra}_D}{\mathrm{Pr} \cdot \mathrm{Le}^2 + \mathrm{Le} \cdot (1 + \mathrm{Pr}) + 1} = \frac{(a^2 + \pi^2)^3}{a^2}. \tag{7.102}$$

The characteristic number $\tilde{\Pi}$ in the oscillatory case corresponds to the characteristic number $\overline{\Pi}$ for the steady instability. Further, $\tilde{\Pi}$ has the same form as $\mathrm{Ra}(a)$ in Rayleigh–Bénard convection. In noting this we have reduced the double diffusion problem for a liquid layer with free boundaries without temperature and concentration perturbations to the much simpler steady Rayleigh–Bénard convection.

The minimum of the function $\overline{\Pi}(a) = \tilde{\Pi}(a)$ yields the critical values (cf. free liquid layer in Rayleigh–Bénard convection) $\Pi_{\mathrm{crit}} = (27/4) \cdot \pi^4 = 658$ and the critical wave number $a_{\mathrm{crit}} = \pi/\sqrt{2} = 2.22$.

In doing this we have determined the critical states of the liquid layer. Because the Lewis number Le and the Prandtl number Pr can be taken to be constant, fixed material properties, it makes sense to depict the critical states in a diagram of the Rayleigh numbers $\mathrm{Ra}(\mathrm{Ra}_D)$. From (7.101) we obtain the linear equation

$$\overline{\mathrm{Ra}} = \Pi_{\mathrm{crit}} + \mathrm{Le} \cdot \mathrm{Ra}_D, \tag{7.103}$$

and from (7.102) then

$$\tilde{\mathrm{Ra}} = \Pi_{\mathrm{crit}} \cdot \left(1 + \frac{1}{\mathrm{Le}}\right) \cdot \left(1 + \frac{1}{\mathrm{Le} \cdot \mathrm{Pr}}\right) + \frac{\dfrac{1}{\mathrm{Le}} + \mathrm{Pr}}{1 + \mathrm{Pr}} \cdot \mathrm{Ra}_D. \tag{7.104}$$

Both straight lines are shown in Figure 7.37. The diagram also shows that the limits of stability $\overline{\mathrm{Ra}}$ and $\tilde{\mathrm{Ra}}$ generally cut the median line $\mathrm{Ra} = \mathrm{Ra}_D$ that is the left-hand boundary of the Rayleigh number regime in which a stable density stratification (lighter fluid above heavier) exists. This shows that double diffusion instabilities are also possible in stable density stratification.

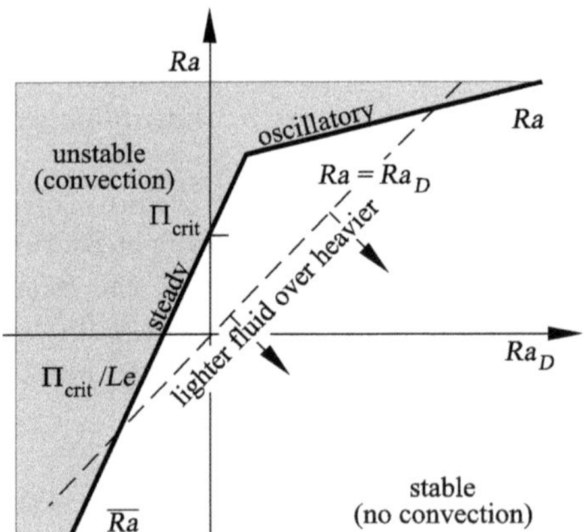

Fig. 7.37. Critical states of a liquid layer of a two-component mixture with free boundaries

Even if heavier fluid is lying above lighter fluid, the state of the liquid layer can still be stable.

7.4.2 Mass Exchange at a Flat Plate

The forced convection flow of the *flat plate with mass exchange* in a longitudinal flow is sketched in Figure 7.38. Cool air with velocity component $w(x)$ in the direction normal to the wall is superimposed on a hot gas flow. In this section the simplest case of a flat incompressible boundary-layer flow of an inert binary mixture will be treated. The cool gas 1 with mass concentration $c_1 = c$ diffuses through the porous surface into the flowing gas 2 with mass concentration $c_2 = 1 - c_1$. The two-dimensional boundary-layer equations with heat transport (7.73)–(7.75) discussed until now are extended by the mass transport equation:

$$\frac{\partial u}{\partial x} + \frac{\partial w}{\partial z} = 0, \tag{7.105}$$

$$u \cdot \frac{\partial u}{\partial x} + w \cdot \frac{\partial u}{\partial z} = -\frac{\partial p}{\partial x} + \frac{1}{\mathrm{Re}_l} \cdot \frac{\partial^2 u}{\partial z^2}, \tag{7.106}$$

$$u \cdot \frac{\partial T}{\partial x} + w \cdot \frac{\partial T}{\partial z} = \frac{1}{\mathrm{Pr} \cdot \mathrm{Re}_l} \cdot \frac{\partial^2 T}{\partial z^2}, \tag{7.107}$$

$$u \cdot \frac{\partial c}{\partial x} + w \cdot \frac{\partial c}{\partial z} = \frac{1}{\mathrm{Le}} \cdot \frac{\partial^2 c}{\partial z^2}, \tag{7.108}$$

with the Lewis number $\mathrm{Le} = D/k = \mathrm{Pr}/\mathrm{Sc}$. Here coupling effects such as thermodiffusion, as is used in the separation of isotopes, are neglected. The physical values of a binary mixture are dependent not only on the temperature and pressure, but also on the concentration. This dependence, however, is small and is neglected, as is the pressure dependence. Within the Boussinesq approximation, the physical values are assumed to be constant at the mean temperature T_{m}. In the energy equation (7.107), the energy diffusion term has been neglected compared to the heat conduction term, a fact that is approximately satisfied for inert gas mixtures. Therefore, the mass exchange influences the velocity profile only via the boundary condition $w_{\mathrm{w}}(x)$. The continuity equation (7.105) and the momentum equation (7.106) remain unchanged.

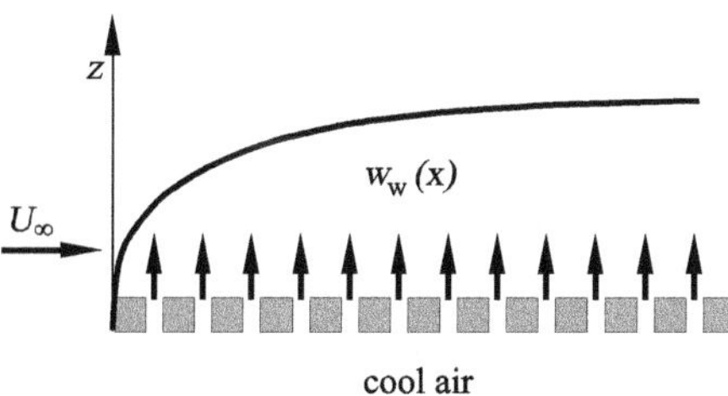

Fig. 7.38. Mass exchange in the flat plate boundary layer

The system of equations for the flat plate boundary layer has the boundary conditions at the wall $w = w_\mathrm{w}$, the given wall temperature T_w, and the wall concentration $c = c_\mathrm{w}$. At the edge of the far field we have $T = T_\infty$ and $c = c_\infty$. Figure 7.39 shows the computed temperature and concentration profiles at different blowing rates for the Prandtl and Schmidt numbers 0.7. Because of the mass transport to the wall with $w_\mathrm{w} < 0$, the profiles become broader. Suction is generally used in practice to prevent flow separation in boundary layers with pressure gradients. The mass transport in blowing allows the profile to become flatter, causing flow separation to be favored. The profiles have a turning point, which also causes the laminar–turbulent transition in the boundary layer.

The ratios of the flow and thermal boundary layers can also be applied to mass transport. The statement $\delta/\delta_T \approx \mathrm{Pr}^{(1/3)}$ for $\mathrm{Pr} \geq 1$ corresponds to

$$\frac{\delta}{\delta_D} \approx \mathrm{Sc}^{\frac{1}{3}} \quad \text{for} \quad \mathrm{Sc} \geq 1. \tag{7.109}$$

For the diffusion Nusselt number Nu_D we have

$$\frac{\mathrm{Nu}_D}{\sqrt{\mathrm{Re}_x}} = 0.332 \cdot \mathrm{Sc}^{\frac{1}{3}} \quad \text{for} \quad \mathrm{Sc} \geq 1, \tag{7.110}$$

as long as the blowing velocity is very small. The constant in (7.110) has to be suitably adapted for arbitrary blowing rates.

For turbulent mass transport, the Reynolds equations (7.88)–(7.90) of the flat plate boundary layer are extended by the Reynolds transport equation

$$\overline{u} \cdot \frac{\partial \overline{c}}{\partial x} + \overline{w} \cdot \frac{\partial \overline{c}}{\partial z} = D \cdot \frac{\partial^2 \overline{c}}{\partial z^2} - \frac{\partial (\overline{w' \cdot c'})}{\partial z}, \tag{7.111}$$

with the Reynolds mass flux

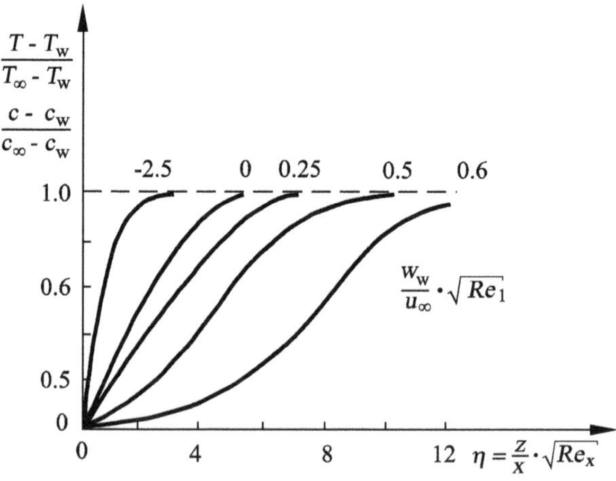

Fig. 7.39. Temperature and concentration profiles at the flat plate with mass transport, $\mathrm{Pr} = \mathrm{Sc} = 0.7$

$$j_z = \rho \cdot \overline{w' \cdot c'} = -\rho \cdot \epsilon_D \cdot \frac{\partial \overline{c}}{\partial z}$$

and the turbulent exchange quantity ϵ_D for the mass exchange. Following on from the molecular Prandtl and Schmidt numbers, we define the turbulent Prandtl and Schmidt numbers as

$$\mathrm{Pr_t} = \frac{\epsilon_\tau}{\epsilon_q} = \frac{\overline{u' \cdot w'}}{\overline{w' \cdot T'}} \cdot \frac{\frac{\partial \overline{T}}{\partial z}}{\frac{\partial \overline{u}}{\partial z}}, \tag{7.112}$$

$$\mathrm{Sc_t} = \frac{\epsilon_\tau}{\epsilon_D} = \frac{\overline{u' \cdot w'}}{\overline{w' \cdot c'}} \cdot \frac{\frac{\partial \overline{c}}{\partial z}}{\frac{\partial \overline{u}}{\partial z}}. \tag{7.113}$$

This yields the turbulent Lewis number

$$\mathrm{Le_t} = \frac{\mathrm{Sc_t}}{\mathrm{Pr_t}} = \frac{\epsilon_q}{\epsilon_D} = \frac{\overline{w' \cdot T'}}{\overline{w' \cdot c'}} \cdot \frac{\frac{\partial \overline{c}}{\partial z}}{\frac{\partial \overline{T}}{\partial z}}. \tag{7.114}$$

The value $\mathrm{Le_t} = 1$ may be set approximately in shear layers, and so all statements on turbulent momentum exchange can be applied to turbulent mass exchange.

Laminar and turbulent mass transport with chemical reactions will be discussed in Chapter 9.

8. Multiphase Flows

8.1 Fundamentals of Multiphase Flows

Multiphase flow is the kind of flow that occurs most frequently in nature and technology. The concept of a phase is to be understood in the thermodynamic sense as a solid, liquid, or gaslike state that can occur simultaneously in one-component or many-component systems. Storm clouds drifting with rain drops and hailstones, a raging current in the mountains, a snow-dust avalanche, and the cloud of a volcano are impressive examples of multiphase flows in nature.

In power station and chemical technology, multiphase flows are often an important method of heat and material transport. Two-phase flows determine the processes in steam generators, condensators, and cooling towers of steam power stations. Multiphase multicomponent flows are used in the extraction, transport, and treatment of oil and natural gas. These types of flows are also greatly involved in distillation and rectification processes in chemical industry.

The importance of these flow processes for the environment and technology demands that we have a fundamental physical understanding of transport processes and interactions in flowing multiphase multicomponent mixtures.

Multiphase flows generally manifest themselves as unsteady processes with a chaotic character. Therefore, to a much greater extent than for turbulent flows, a formal description requires the use of average states and statistical methods, as well as scaling laws, to be able to make quantitative statements about the expected phenomena, such as pressure drops and phase distributions.

The very different forms and structures that are seen even in the simplest geometries such as pipes and channels of constant cross-section in gas–liquid or gas–solid flows make a consistent mathematical physical description of multiphase flows difficult. The effect of gravity is considerable. In addition, interfacial tensions and electrostatic forces in solids are of central importance. Examples of such typical, repeatedly observed flow forms for a gas–liquid flow in the horizontal pipe are shown schematically in Figure 8.1.

Multiphase flows can fundamentally be described in two different ways. On the one hand, a multiphase flow can be considered as a moving continuum of phases penetrating into each other, whereby each phase is present at every

H. Oertel (ed.), *Prandtl-Essentials of Fluid Mechanics*,
Applied Mathematical Sciences 158, DOI 10.1007/978-1-4419-1564-1_8,

location to a certain extent. This model is useful if the large-scale behavior of a multiphase fluid is to be described. On the other hand, the motion in each phase can be described separately, with the coupling between the phases at the interfaces of particular importance. This is expressed mathematically by computing the motion of the interfaces in detail by specific mathematical methods (for details, see *W. Shyy et al.* (1996)). This kind of consideration is advantageous if the processes are governed by interactions at the interfaces, such as mass fluxes. The attention here is on small-scale effects.

In Section 5.4.6 the fundamental equations of *two-phase flow* and simplified models in the sense of the first method of consideration are presented. The second method of consideration, which was presented by way of example in Section 5.4.3 with the Rayleigh–Plesset equation, will not be applied in this chapter.

8.1.1 Definitions

Using the definition of average values introduced in Section 5.4.6, some fundamental quantities and concepts of multiphase flows will now be introduced.

The *void* or the volume fraction $\epsilon_k(x,t)$ of the phase k in the flow denotes the amount of the volume of the flow channel in space and time that is occupied by the phase k (gas or liquid). The volume fraction ϵ can be defined as a local quantity averaged over the time, over a chord length L, over the channel cross-sectional area A, or over a channel volume ΔV. Accordingly, the *time fraction*, *surface fraction*, and *volume fraction* of the phase k are defined by the simple relations

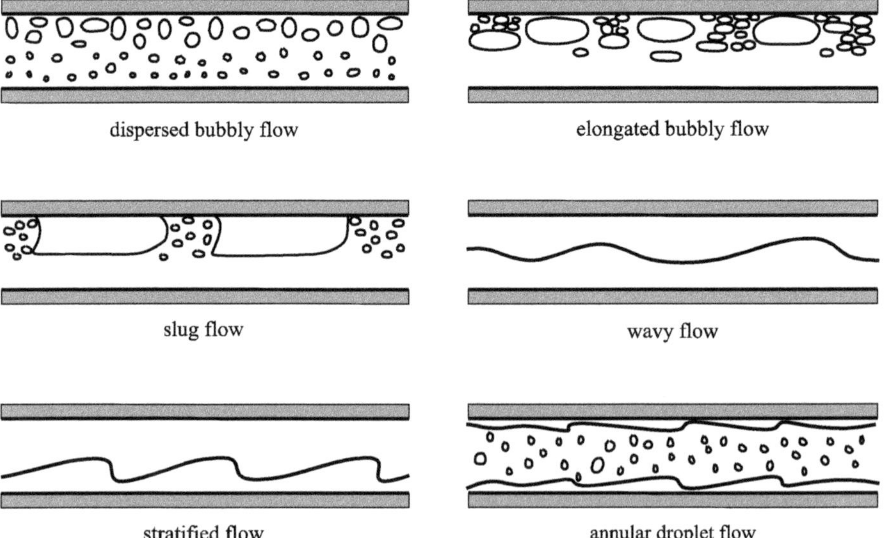

Fig. 8.1. Flow patterns in horizontal pipes

$$\epsilon_{k,t} = \frac{\Delta t_k}{\Delta t}, \qquad \epsilon_{k,A} = \frac{A_k}{A}, \qquad \epsilon_{k,V} = \frac{V_k}{\Delta V}, \tag{8.1}$$

where Δt_k, A_k, and V_k are to be understood as the corresponding averages of the phase indicator function $X(x,t)$. The surface and volume fractions can additionally be time-averaged.

For the velocities of the phases, time-averaged flow values $\overline{u_k(x,t)}^k$, cross-sectionally averaged $\langle u_k(x,t)\rangle_{k,A}$, and spatially averaged values $\langle U_k(x,t)\rangle_{k,V}$ are similarly introduced. For simplicity, further discussion will be for one velocity component only.

A *superficial velocity* is the product of the *phase fraction* ϵ_k and the *phase velocity* u_k, and is defined as

$$U_k = \epsilon_k \cdot u_k.$$

The averaged values are then

$$\overline{U_k}^k = \overline{\epsilon_k \cdot u_k}^k, \quad \langle U_k\rangle_{k,A} = \langle \epsilon_k \cdot u_k\rangle_{k,A}, \quad \langle U_k\rangle_{k,V} = \langle \epsilon_k \cdot u_k\rangle_{k,V}. \tag{8.2}$$

In particular, the cross-sectionally averaged velocity can then also be interpreted as the mean volumetric flux of the phase k and written in the form

$$\langle U_k\rangle = \frac{\dot{V}_k}{A}. \tag{8.3}$$

The velocity defined in this manner is called the *volumetric flux* or *superficial velocity*. Here $\dot{V}_k$ is the volumetric flow rate of the phase k. The volumetric flux of the phase k is therefore to be physically interpreted as if the phase k were flowing alone in the channel. At this point we also define the total velocity by the relation

$$U = \sum_k U_k, \tag{8.4}$$

which can be used in the local form as here, or in the cross-sectionally averaged form $\langle U\rangle = \sum_k \langle U_k\rangle$. The nature of averaging is such that the following relation holds between the mean quantities $\langle U_k\rangle$, $\langle u_k\rangle$, and $\langle \epsilon_k\rangle$:

$$\langle U_k\rangle = \langle \epsilon_k \cdot u_k\rangle = \mathrm{C} \cdot \langle \epsilon_k\rangle \cdot \langle u_k\rangle, \tag{8.5}$$

with C as the correlation factor. This permits the cross-sectionally averaged phase velocities $\langle U_k\rangle$ to be represented by $\langle \epsilon_k\rangle$ and $\langle u_k\rangle$. It is useful to introduce the ratio of the phase velocities:

$$S = \frac{\langle u_i\rangle}{\langle u_k\rangle} = \frac{\langle \epsilon_i\rangle \cdot \langle U_k\rangle}{\langle \epsilon_k\rangle \cdot \langle U_i\rangle} \cdot \frac{C_{0,i}}{C_{0,k}}. \tag{8.6}$$

This ratio is frequently called the *slip*. This is misleading, and we shall call it rather the *velocity ratio*. Using the velocities and the densities ρ_k of the phases as defined, we can write down the mass flux densities $\dot{m}_k$ and the mass flow rates $\dot{M}_k$. We have the relations

$$\dot{m}_k = \rho_k \cdot \langle u_k\rangle, \qquad \dot{M}_k = \rho_k \cdot \langle U_k\rangle \cdot A. \tag{8.7}$$

The mass flux density is also known as the *mass velocity*. For the total mass flow rate, the balance of mass implies $\dot{M} = \sum \dot{M}_k$.

In order to characterize multiphase flows, as well as the volume fraction (void) ϵ_k, we also use a *mass fraction* χ as the ratio of the mass flow rate of the phase k to the total mass flux:

$$\chi_k = \frac{\dot{M}_k}{\dot{M}}, \quad \text{with} \quad \dot{M} = \sum \dot{M}_k. \tag{8.8}$$

For gas–liquid flow this ratio is called the *quality*. It is thermodynamically determined by the enthalpy of the phases for one-component two-phase flows such as water-water vapor (cf. Section 8.4). There is a functional dependence between the phase velocities, mass flow rates, and densities. For two-phase flows, gas–liquid, the dependence can be given in the form

$$\frac{\chi_G}{\chi_L} = \frac{\rho_G}{\rho_L} \cdot S \cdot \frac{\epsilon_G}{\epsilon_L}, \tag{8.9}$$

whereby, because of mass conservation, for any type of averaging, $\chi_L = 1 - \chi_G$ and $\epsilon_L = 1 - \epsilon_G$ hold. In particular, it is clear from this relation that the volume fraction ϵ quite generally depends on the velocity ratio, on the density ratio, and on the quality of a two-phase flow.

Because two-phase flows are of particular importance in applications, further expressions for velocities have been introduced to describe transport processes. The *drift velocity* is the deviation of the actual phase velocity u_k from the total *volumetric flux* $U = U_G + U_L$. For example, for a gas–liquid flow it is defined as

$$u_{G,U} = (u_G - U), \qquad u_{L,U} = (u_L - U). \tag{8.10}$$

These *drift velocities* are simply related to the *relative velocity* $u_G - u_L$ between the two phases. We have

$$u_{G,U} = (1 - \epsilon_G) \cdot (u_G - u_L), \qquad u_{L,U} = -\epsilon_G \cdot (u_G - u_L). \tag{8.11}$$

This relation is also valid for all types of averaging. For this reason, we do not denote the averaging here with a particular symbol. In analogy to the volumetric fluxes U_k, in deriving some models describing two-phase flows, we also introduce referred drifts, called *drift fluxes*. They are defined as

$$U_{\mathrm{G,U}} = \epsilon_{\mathrm{G}} \cdot u_{\mathrm{G,U}} \quad , \qquad U_{\mathrm{L,U}} = (1 - \epsilon_{\mathrm{G}}) \cdot u_{\mathrm{L,U}} \quad . \tag{8.12}$$

A further characteristic parameter for two-phase flows was introduced by *R. Lockhart* and *R. Martinelli* (1949). It is given by the ratio of the friction pressure losses, for the cases in which gas and liquid each flow alone through the channel. If the pressure losses of the averaged liquid and gas volumetric fluxes are given by $(\mathrm{d}p/\mathrm{d}z)_L$ and $(\mathrm{d}p/\mathrm{d}z)_G$, respectively, then the *Martinelli parameter* is defined as

$$X^2 = \frac{\left(\frac{\mathrm{d}p}{\mathrm{d}z}\right)_L}{\left(\frac{\mathrm{d}p}{\mathrm{d}z}\right)_G}. \tag{8.13}$$

This parameter is in general a measure for the volume fraction of the flow. For $X^2 \gg 1$ the two-phase flow consists essentially of liquid, and for $X^2 \ll 1$, of gas.

8.1.2 Flow Patterns

Two-phase flows can take different forms, depending on the different types of interface interaction at different volumetric fluxes of the phases. At very different densities of the phases, gravity has a considerable effect. In order to characterize the effect of gravity, it has been useful to classify the flow forms for horizontal and vertical pipe flows. The typical flow patterns are sketched for both cases in Figures 8.1 and 8.2.

The flow patterns shown occur in this order as the gas fraction ϵ and the gas velocity are increased. The transitions between the patterns are not sharp and are influenced by the flow turbulence in each phase, the volume fraction, and the stability of the interfaces.

8.1.3 Flow Pattern Maps

In order to distinguish between different flow patterns, flow charts have been developed using experimental observations. A greatly simplified representation of the states can be obtained for a given gas–liquid mixture, such as air and water, using the volumetric fluxes of liquid and gas as control parameters. Such a *flow chart* or *flow map* was produced from a large database of experimental results by *J. M. Mandhane et al.* (1974). It was derived from a variation of liquid and gas volumetric fluxes in a horizontal test pipe. Figure 8.3 shows the Mandhane flow map.

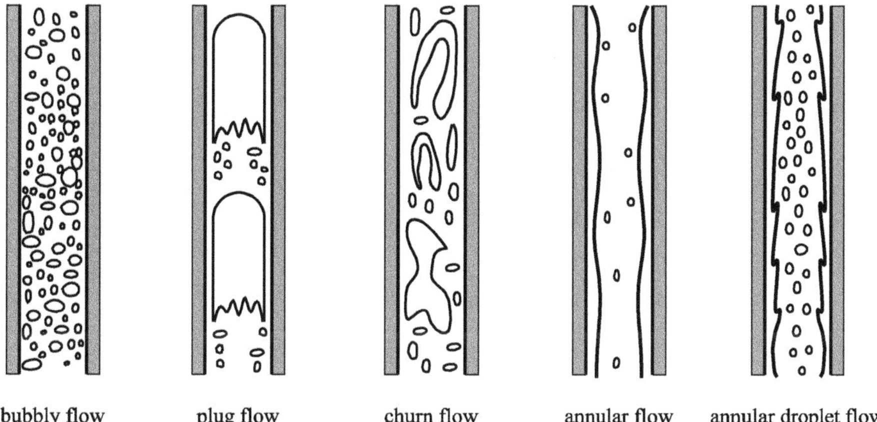

Fig. 8.2. Flow patterns in vertical pipes, cocurrent upward flows

Similar maps have also been given by *G. Govier* and *K. Aziz* (1972) and *Y. Taitel et al.* (1980) for air–water flows in vertical pipes. The boundaries between the flow patterns are not sharply marked and in some cases show hysteresis. In vertical pipe flows, the pipe length and the flow intake, for example, considerably affect the transition between plug flows and churn flows. Using a linear scaling of the volumetric fluxes for air–water mixtures with property parameters that take into account the density and the surface tension, *G. Govier* and *K. Aziz* (1972) and *J. M. Mandhane et al.* (1974) were able to generalize their flow maps for other gas–liquid mixtures. They introduced the density ratios between gas and air ρ_G/ρ_{air}, liquid and water $\rho_L/\rho_{\text{water}}$, and the ratio of the surface tensions $\sigma/\sigma_{\text{air/water}}$ for alternative mixtures and for air and water. They define

$$Y = \left(\frac{\rho_L}{\rho_{\text{water}}} \cdot \frac{\sigma_{\text{air/water}}}{\sigma}\right)^{\frac{1}{4}} \quad \text{and} \quad X = \left(\frac{\rho_G}{\rho_{\text{air}}}\right)^{\frac{1}{3}} \cdot Y.$$

As modified volumetric fluxes for the alternative mixture they set $U_G^* = X \cdot U_{\text{air}}$ and $U_L^* = Y \cdot U_{\text{water}}$.

Y. Taitel and *A. Dukler* (1976) and *Y. Taitel* (1990) derive flow maps from purely theoretical considerations. They distinguish between three classes of flow: *stratified flows* in smooth or wavelike form, *intermittent flows* in the form of *slug and plug flows*, and *dispersed flows* in the form of *bubbly* or *annular-droplet flows.* A condition for the transition from *stratified* to *intermittent flow* is derived from the instability condition for a soliton wave. The limiting

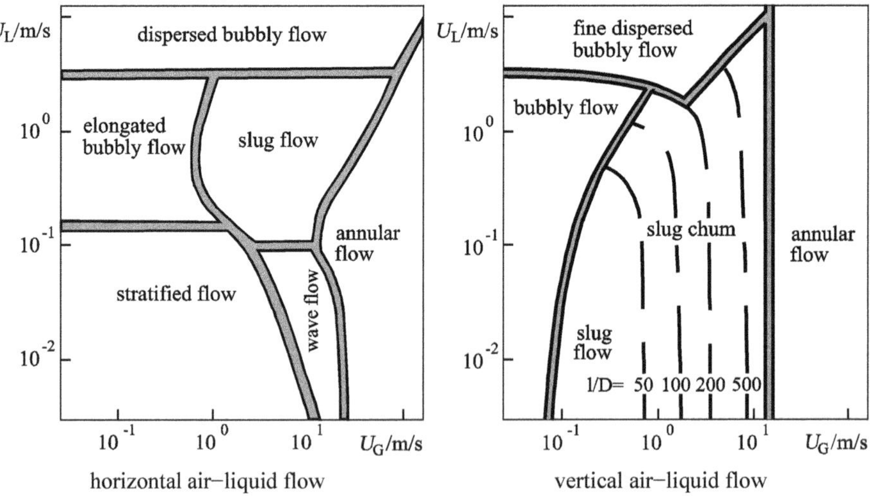

Fig. 8.3. Flow map by *J. M. Mandhane et al.* (1974) for a horizontal air–water flow with experimental data: pressure 0.1 MPa, pipe diameter 2.5 cm. Flow map by *Y. Taitel et al.* (1980) for vertical air–water flow with experimental data: pressure 0.1 MPa, pipe diameter 5.1 cm, $l/_D$ denotes the ratio of pipe length to pipe diameter

curve is given by a modified Froude number,

$$F = \sqrt{\frac{\rho_G}{\rho_L - \rho_G}} \cdot \frac{U_G}{\sqrt{D \cdot g \cdot \cos(\beta)}},$$

as a function of the Martinelli parameter X. Here β is the angle of inclination of the pipe, and D the diameter of the pipe. The transition from smooth to wavelike stratified flow is determined by the *Kelvin–Helmholtz instability condition.* After some simplifying assumptions it can be given the form

$$K = \frac{U_\mathrm{G}}{\sqrt{D \cdot g \cdot \cos(\beta)}} \cdot \sqrt{\frac{\rho_\mathrm{G}}{\rho_\mathrm{L} - \rho_\mathrm{G}}} \cdot \sqrt{\frac{U_\mathrm{L} \cdot D}{\nu_\mathrm{L}}} \gtrsim 20 \cdot \sqrt{\epsilon_\mathrm{L}} \cdot \epsilon_\mathrm{G}$$

where ν_L is the kinematic viscosity of the liquid. A correlation to the Martinelli parameter can be determined for the volume fractions, so that the limiting curve can be obtained in the form $K(X)$.

The transition from annular flow to intermittent flow in not too strongly inclined pipes is given by the minimum possible liquid fraction in a slug interspersed with gas bubbles. According to *Y. Taitel* and *A. Dukler* (1976), intermittent slug flows occur for $\epsilon_\mathrm{L} \gtrsim 0.24$. This corresponds approximately to the value $X \approx 1.6$ for the Martinelli parameter. The state of dispersed bubbly flow and intermittent flow is determined by turbulent agitation, by gravity, and by the collapse and coalescence of bubbles. The turbulence intensity in the liquid phase may be characterized by the pressure drop in the liquid phase. *Y. Taitel* and *A. Dukler* (1976) introduce the ratio of the *superficial* pressure drop of the liquid phase and the hydrostatic lift of the bubbles as a characteristic number, in the form

$$T^2 = \left| \left(\frac{\mathrm{d}p}{\mathrm{d}z} \right) \right|_\mathrm{L} \cdot \frac{1}{(\rho_\mathrm{L} - \rho_\mathrm{G}) \cdot g \cdot \cos(\beta)}$$

The transition between the two regimes can be given as the function T of the Martinelli parameter X. After evaluating the functional relations, the flow map of *Y. Taitel* and *A. Dukler* (1976) has the form shown in Figure 8.4, where each of the limiting curves $K(X)$, $F(X)$, and $T(X)$ is assigned separately to an axis.

Y. Taitel (1990) generalized the theory of the flow regime boundaries so that two-phase flows in pipes can be classified with arbitrary angles of inclination. The transition conditions are then either given graphically or can be numerically determined point by point.

The different state quantities of the two-phase flow as well as their derivatives, such as the pressure gradient, the volume fraction, and the heat transfer coefficient are greatly dependent on the flow pattern. Therefore, in general, computational methods for two-phase flows have to be developed individually for each characteristic flow pattern. This is a major and complex task. The different computational models are then linked via flow maps or computer-supported transition conditions, in order to determine sufficient accurately the two-phase states for technical systems, such as steam generators. The computation of two-phase flows with respect to the different regimes has still not been satisfactorily solved. Currently, for certain technically relevant quantities, such as pressure drop and heat flux, correlations are still being used that were derived based on extensive experimental data. The modeling of two-phase flows will be discussed in the following sections.

8.2 Flow Models

For any model development it is very useful to divide two-phase flows according to the scheme of *Y. Taitel* and *A. Dukler* (1976) into three classes: separate flows, such as stratified flows, wavy flows and annular flows; intermittent or transition flows in the form of elongated bubble flows, slug flows, and plug flows; and dispersed flows like bubble flows, churn flows, and droplet or mist flows. In order to describe two-phase flow, the mechanical coupling of the state variables velocity, pressure, and temperature is usually carried out in a Euler form of the conservation equations for mass, momentum, and energy. In the general case, the balance for each phase is taken separately, and the description of the two-phase flow is then called a two-fluid model. This procedure can generally also be applied to describe a flow with N fluids, and it yields an N-fluid model. This has already been presented in general form in Section 5.4.6. Next, we discuss the one-dimensional form of the two-fluid model.

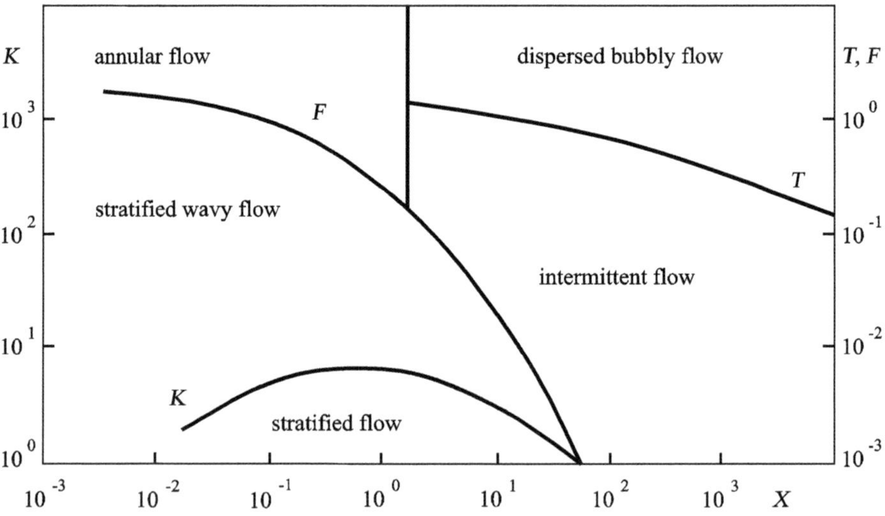

Fig. 8.4. Flow map in horizontal and slightly inclined pipe, after *Y. Taitel* and *A. Dukler* (1976), with the characteristic numbers K, F, and T as a function of the Martinelli parameter X

8.2.1 The One-Dimensional Two-Fluid Model

It is expedient to derive the one-dimensional two-fluid model by utilizing the idea of a stratified flow in a pipe, as shown in Figure 8.1 in Section 8.1.2, and here schematically depicted in Figure 8.5 to clarify some specific quantities.

In order to obtain the one-dimensional two-fluid equations from the general fundamental equations for multiphase flows, some fundamental assumptions are made with regard to the spatial (cross-sectional) averaging and the time averaging.

In the spatial or time averaging of a product of state variables f and g we have in general $\langle f \cdot g \rangle = c \cdot \langle f \rangle \cdot \langle g \rangle$ or $\overline{f \cdot g} = c \cdot \overline{f} \cdot \overline{g}$. In the simplest representation of the two-fluid model we set $c = 1$.

The thermodynamic equations of state and the constitutive relations for local quantities also hold for the averaged quantities.

The heat conduction and the change of the shear stresses in the direction of flow, as well as the dissipation of the frictional forces in each phase, are neglected.

Local thermodynamic equilibrium in each phase is assumed. However, the phases do not have to be in thermodynamic equilibrium with each other.

In vertical two-phase flows the pressure over the cross-section of the pipe is assumed to be constant. In many applications this is also valid to good approximation for horizontal flows.

The effect of the interfacial stresses can initially be neglected. However, it frequently reappears when interfacial stresses enter the constitutive relations needed to close the system of model equations.

In order to present the one-dimensional balance equations, we use the following notation for the averaged quantities:

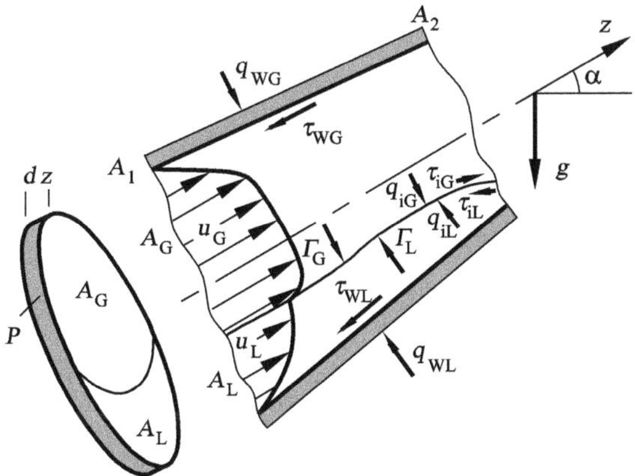

Fig. 8.5. Control space for a separate two-phase flow

geometry:	A_G, A_L, A	cross-sectional areas, $A_G + A_L = A$,
	P_G, P_L, P	circumferential segments, $P_G + P_L = P$,
	P_i	circumferential sections,
	$\epsilon_G = A_G/A$,	cross-sectional fractions, $\epsilon_G + \epsilon_L = 1$,
	$\epsilon_L = A_L/A$	
state variables:	u_G, u_L	velocities,
	p_G, p_L	pressures,
	ρ_G, ρ_L	densities,
	$\dot{m}_G = \rho_G \cdot u_G$,	mass flux densities
	$\dot{m}_L = \rho_L \cdot u_L$	(or mass velocities),
	e_G, e_L	specific internal energy,
	h_G, h_L	specific enthalpy,
constitutive variables:	$\tau_{w,G}$, $\tau_{w,L}$	wall shear stress,
	$q_{w,G}$, $q_{w,L}$	wall heat flux density,
	$\tau_{i,G}$, $\tau_{i,L}$	shear stresses at phase interface,
	$q_{i,G}$, $q_{i,L}$	heat flux densities at phase interface,
	Γ_G, Γ_L	mass source densities at phase interface,
	u_i	velocity at phase interface,
	p_i	pressure at phase interface,
in phase transitions:	$M_G^{(\Gamma_G)}$, $M_L^{(\Gamma_L)}$	momentum source term due to mass exchange at phase interface,
	$L_{\tau,G}$, $L_{\tau,L}$	power source densities due to wall shear stresses at interface,
	$L_G^{(\Gamma_G)}$, $L_L^{(\Gamma_L)}$	power source densities due to mass exchange at interface.
	L_g, L_q	power source densities due to gravity and heat supply.

Because of the local equilibrium we have the following relations between the constitutive variables at the interface:

$$\begin{aligned}
\Gamma_G - \Gamma_L &= 0,\\
\tau_{i,G} - \tau_{i,L} &= 0,\\
q_{i,G} - q_{i,L} &= 0,\\
M_G^{(\Gamma_G)} - M_L^{(\Gamma_L)} &= 0,\\
L_G^{(\Gamma_G)} - L_L^{(\Gamma_L)} &= 0.
\end{aligned}$$

The momentum source densities $M_G^{(\Gamma_G)}$ and $M_L^{(\Gamma_L)}$ and the power densities can be specified further (cf. *M. Ishii* (1975), *M. Ishii* und *T. Hibiki* (2006) and *J. Delhaye et al.* (1981)). Those parts of the quantities that are due to mass exchange sum to zero, as already stated above. However, if surface stresses play a role, there are further terms in the balance equations for the phase mixture that take into account the surface stresses. For simplicity, the surface stress effects will not be taken into account here. Using the assumptions discussed here, the one-dimensional balance equations for the two-fluid model can be written in the following form:

Mass:

$$\frac{\partial}{\partial t}(\rho_G \cdot \epsilon \cdot A) + \frac{\partial}{\partial z}(\dot{m}_G \cdot \epsilon \cdot A) = \Gamma_G, \tag{8.14}$$

$$\frac{\partial}{\partial t}[\rho_L \cdot (1-\epsilon) \cdot A] + \frac{\partial}{\partial z}[\dot{m}_L \cdot (1-\epsilon) \cdot A] = \Gamma_L. \tag{8.15}$$

Momentum:

$$\begin{aligned}
\frac{\partial}{\partial t}(\dot{m}_G \cdot \epsilon \cdot A) + \frac{\partial}{\partial z}(\dot{m}_G \cdot u_G \cdot \epsilon \cdot A) &= -\epsilon \cdot A \cdot \frac{\partial}{\partial z} p_G - \tau_{w,G} \cdot P_G\\
&\quad - \tau_{i,G} \cdot P_i - \epsilon \cdot A \cdot \rho_G \cdot g \cdot \sin(\alpha) + M_G^{(\Gamma_G)},
\end{aligned} \tag{8.16}$$

$$\begin{aligned}
\frac{\partial}{\partial t}[\dot{m}_L \cdot (1-\epsilon) \cdot A] + \frac{\partial}{\partial z}[\dot{m}_L \cdot (1-\epsilon) \cdot A \cdot u_L] &= -(1-\epsilon) \cdot A \cdot \frac{\partial}{\partial z} \cdot p_L\\
-\tau_{w,L} \cdot P_L - \tau_{i,L} \cdot P_i - (1-\epsilon) \cdot A \cdot \rho_L \cdot g \cdot \sin(\alpha) + M_L^{(\Gamma_L)},
\end{aligned} \tag{8.17}$$

with $\epsilon_G = \epsilon$. Here the assumption $p_L = p_G = p$ can be introduced.
Energy:

$$\begin{aligned}
\frac{\partial}{\partial t}(\rho_G \cdot E_G \cdot \epsilon \cdot A) + \frac{\partial}{\partial z}\left[\dot{m}_G \cdot \left(h_G + \frac{u_G^2}{2}\right) \cdot \epsilon \cdot A\right]\\
= L_{\tau,G} + L_{g,G} + L_G^{(\Gamma_G)} + L_{q,G}.
\end{aligned} \tag{8.18}$$

Here $E_G = e_G + u_G^2/2$ is the energy density, h_G the specific enthalpy of the gas, and on the right-hand side of the equation we have written down, without closer specification, the power contributions of shear stress, of gravity, of momentum exchange due to mass exchange between the phases, and of heat fluxes. Similarly, the energy equation for the liquid phase can be written as follows:

$$\frac{\partial}{\partial t}(\rho_L \cdot E_L \cdot (1-\epsilon) \cdot A) + \frac{\partial}{\partial z}\left[\dot{m}_L \cdot \left(h_L + \frac{u_L^2}{2}\right) \cdot (1-\epsilon) \cdot A\right]$$
$$= L_{\tau,L} + L_{g,L} + L_L^{(\Gamma_L)} + L_{q,L}. \tag{8.19}$$

For the simplest case of an incompressible flow, the equations contain the six variables of state u_G, u_L, e_G, e_L, p, ϵ. As well as these, there is a great number of constitutive variables, which, in stringent derivation of the momentum and energy equations, may be reduced to the following independent variables: Γ, $(\tau_{w,G} \cdot P_G)$, $(\tau_{w,L} \cdot P_L)$, $(\tau_{i,G} \cdot P_G)$, $(q_{w,G} \cdot P_G)$, $(q_{w,L} \cdot P_L)$, $(q_{i,L} \cdot P_L)$ (cf. *G. Yadigaroglou* and *J. R. T. Lahey* (1976)). Correlations between the constitutive variables and the state variables have to be set up, based on experimental evidence or theoretical ideas, in order to close the balance equations and to apply them to solving two-phase flow problems. The model correlations for the constitutive variables are to be developed individually for the different flow regimes. A set of specific correlations has been given by *M. Ishii* and *K. Mishima* (1984) and in a more general form by *M. Ishii* and *T. Hibiki* (2006).

8.2.2 Mixing Models

The two-fluid model can be simplified if, for technical reasons, only integral states of the two-phase flow, such as total mass flow rate, total pressure drop, and total heat transport, are of interest. By adding the same kind of balance equations for the individual phases, we obtain three balance equations for the total mass flow rate, the total momentum, and the total energy of the two-phase mixture. These equations can be written in a form that corresponds to the form of the one-dimensional fluid-mechanical equations for compressible media if the densities, the wall shear stress, and the wall heat flux of the mixture are introduced as weighted quantities. Using the general definitions from Section 8.1, simple algebraic manipulation leads to the following three conservation equations for the two-phase mixture:

Mass:

$$\frac{\partial \rho_H}{\partial t} + \frac{1}{A} \cdot \frac{\partial \dot{M}}{\partial z} = 0, \tag{8.20}$$

with

$$\rho_H = \epsilon \cdot \rho_G + (1-\epsilon) \cdot \rho_L. \tag{8.21}$$

Momentum:

$$\frac{\partial}{\partial t}\dot{M} + \frac{\partial}{\partial z}\left(\frac{1}{\rho_I} \cdot \frac{\dot{M}^2}{A}\right) = -A \cdot \frac{\partial p}{\partial z} - \langle \tau_w \rangle \cdot P - A \cdot \rho_H \cdot g \cdot \sin(\alpha), \tag{8.22}$$

with $\chi = \chi_G$ and

$$\frac{1}{\rho_I} = \frac{\chi^2}{\epsilon \cdot \rho_G} + \frac{(1-\chi)^2}{(1-\epsilon) \cdot \rho_L}, \qquad \langle \tau_w \rangle = \tau_{w,G} \cdot P_G + \tau_{w,L} \cdot P_L. \tag{8.23}$$

Energy:

$$\frac{\partial}{\partial t} E + \frac{1}{A} \cdot \frac{\partial}{\partial z} \left[\dot{M} \left(h + \frac{1}{\rho_E^2} \cdot \frac{\dot{M}^2}{2 \cdot A^2} \right) \right] = L_{\tau,w} + L_G + L_{q,w}, \tag{8.24}$$

with the total energy $E = \rho_G \cdot E_G + \rho_L \cdot E_L$ and the total enthalpy

$$h = \chi \cdot h_G + (1-\chi) \cdot h_L \quad \text{and} \quad \frac{1}{\rho_E^2} = \frac{\chi^3}{\epsilon^2 \cdot \rho_G^2} + \frac{(1-\chi)^3}{(1-\epsilon)^2 \cdot \rho_L^2}. \tag{8.25}$$

It can be seen that the densities of the individual phases are differently weighted in the different balance equations and therefore have different effects compared to single-phase flows. The mixture equations can be used to define mixture densities, which are occasionally used to evaluate signals of a two-phase flow instrumentation. The definitions are set out in the following:

Definitions of mixture densities based on mixture balances		
Mass	$\rho_H = \epsilon \cdot \rho_G + (1-\epsilon) \cdot \rho_L$	homogeneous density
Momentum	$\rho_I = \left(\frac{\chi^2}{\epsilon \cdot \rho_G} + \frac{(1-\chi)^2}{(1-\epsilon) \cdot \rho_L} \right)^{-1}$	momentum density
Energy	$\rho_E = \left(\frac{\chi^3}{\epsilon^2 \cdot \rho_G^2} + \frac{(1-\chi^3)}{(1-\epsilon)^2 \cdot \rho_L^2} \right)^{-1/2}$	energy density

Although the model for the mixture flow has been simplified by a reduction of the number of equations, we now encounter the new problem that the volume fraction ϵ has to be correlated with the steam quality χ by an empirical relation in order to be able to use the simplified model for the solution of problems. In general, ϵ is correlated to χ via the velocity ratio $S = u_G/u_L$ of the phases (cf. (8.9)). Therefore, empirical relations are sometimes determined for the velocity ratio S, and the $\epsilon(S, \chi)$-relation is inserted into the mixture equations.

The mixture equations are naturally suited to applied calculations if dispersed flow patterns such as bubble or droplet flows are present. They can be even further simplified if we assume mechanical equilibrium between the phases, i.e. if the dispersed phase has the same velocity as the homogeneous phase. A mixing model simplified in this manner is also called a *homogeneous flow model*. Because of its simplicity, it can be conveniently used and can be applied to dispersed flows with very small volume fractions of the dispersed phase. With the assumption $S = 1$ we obtain a unique relation between ϵ and χ according to (8.9). It reads

$$\epsilon_H = \frac{1}{1 + \frac{1-\chi}{\chi} \cdot \frac{\rho_G}{\rho_L}}. \tag{8.26}$$

The subscript H denotes the *homogeneous model*. Using this relation, all remaining density definitions ρ_I and ρ_E can be simply transformed algebraically to the expression $\rho_H = \epsilon \cdot \rho_G + (1-\epsilon) \cdot \rho_L$ in the mixture balance; i.e. $\rho_H = \rho_I = \rho_E$ holds. The one-dimensional homogeneous flow model is therefore described by the equations

$$\frac{\partial}{\partial t}\rho_H + \frac{1}{A} \cdot \frac{\partial \dot{M}}{\partial z} = 0, \tag{8.27}$$

$$\frac{\partial \dot{M}}{\partial t} + \frac{\partial}{\partial z}\left(\frac{1}{\rho_H} \cdot \frac{\dot{M}^2}{A}\right) = -A \cdot \frac{\partial p}{\partial z} - \overline{\tau} \cdot P - A \cdot \rho_H \cdot g \cdot \sin(\alpha), \tag{8.28}$$

if the phases are in thermodynamic equilibrium, i.e. $T_G = T_L$. The homogeneous model is the simplest of all two-phase flow models. It can be extended to dispersed flows with evaporation and condensation processes that are not in thermal equilibrium.

Dispersed two-phase flows with larger fractions of the dispersed phase are in general not in mechanical equilibrium. In order to take this fact into account and still use the simplifying idea of well-mixed phases, the *drift-flow model* has been developed. This will be described in what follows.

8.2.3 The Drift-Flow Model

The drift-flow model was suggested by *H. Zuber* and *J. A. Findley* (1965); it is based on the fundamental idea that both phases are well mixed together, but move relative to each other and, in general, have different thermodynamic states. The range of application of the model is therefore mainly for dispersed flows, that is, bubble, churn, and droplet flows. However, attempts have been made to extend the model to plug and annular flows.

The model is based on a *density-weighted mixture velocity*

$$u_M = \frac{\epsilon \cdot \rho_G \cdot u_G + (1-\epsilon) \cdot \rho_L \cdot u_L}{\epsilon \rho_G + (1-\epsilon) \cdot \rho_L} \tag{8.29}$$

and takes into account the relative velocities by means of so-called drift velocities, which are initially introduced as a local property in the form

$$u_{G,U}^{(l)} = u_G^{(l)} - U^{(l)}, \tag{8.30}$$

$$u_{L,U}^{(l)} = u_L^{(l)} - U^{(l)}, \tag{8.31}$$

where $U^{(l)}$ is the local total volumetric flux, given by the relative local velocities $U^{(l)} = U_G^{(l)} + U_L^{(l)}$. By cross-sectional averaging, these definitions can be used to relate the average values of the volume fraction $\langle \epsilon \rangle$, the total volumetric flux $\langle U \rangle$, and a still to be defined mean drift velocity. This is carried out by multiplication of the relation (8.30) by the local volume fraction ϵ, and subsequent cross-sectional averaging. It is to be noted that, as discussed in Section 8.1, in general we have $\langle \epsilon \cdot U \rangle = \mathrm{C}_0 \cdot \langle \epsilon \rangle \cdot \langle U \rangle$. After some algebraic transformations we obtain the relation.

$$\langle \epsilon \cdot u_{G,U} \rangle = \langle \epsilon \cdot u_G \rangle - \mathrm{C}_0 \cdot \langle \epsilon \rangle \cdot \langle U \rangle. \tag{8.32}$$

Using the definitions $\overline{u}_{G,U} = \langle \epsilon \cdot u_{G,U} \rangle / \langle \epsilon \rangle$ and $U_G = \langle \epsilon \cdot u_G \rangle$ as the cross-sectionally weighted drift velocity and average volumetric gas flow, equation (8.32) yields the cross-sectionally averaged volume fraction of gas as

$$\langle \epsilon \rangle = \frac{U_G}{\mathrm{C}_0 \cdot U + \overline{u}_{G,U}}. \tag{8.33}$$

In order to define a mean drift velocity, we now introduce the volume-weighted quantities $\overline{u}_{G,U}$ and $\overline{u}_G = \langle \epsilon \cdot u_G \rangle / \langle \epsilon \rangle$. To write down the balance equations, we use the relations

$$\overline{u}_{G,U} = \overline{u}_G - U, \qquad \overline{u}_{L,U} = \overline{u}_L - U. \tag{8.34}$$

For simplicity, we will drop the averaging symbol '-' in what follows. The drift velocity can therefore be considered as the velocity of a phase relative to a surface moving with the *mixture velocity* U (the total volumetric flux). With the relation for the mixture velocity $U = \epsilon \cdot u_G + (1-\epsilon) \cdot u_L$, the relation of the drift velocity and relative velocity can immediately be given as

$$u_{G,U} = (1-\epsilon) \cdot (u_G - u_L), \qquad u_{L,U} = \epsilon \cdot (u_G - u_L). \tag{8.35}$$

From this relation and the defining equation (8.29), we obtain a relation between u_G, u_M, and $u_{g,U}$ in the form

$$u_G = u_M + \frac{\rho_L}{\rho_H} \cdot u_{G,U}, \qquad u_L = u_M - \frac{\rho_L}{\rho_H} \cdot \frac{\epsilon}{1-\epsilon} \cdot u_{G,U}. \tag{8.36}$$

Similar relations interrelate u_G and u_L with u_M and $u_{L,U}$.

The expressions (8.36) are inserted into the balance equation for the mixtures, and the mass balance equation for the gas phase, which is also retained to describe phase transitions. After some algebraic transformations we obtain four balance equations for the state variables mean velocity u_M, pressure p, mean enthalpy h_M and volume fraction ϵ.

The equations have the following form:
Mass:

$$\frac{\partial}{\partial t} \cdot \rho_H + \frac{1}{A} \cdot \frac{\partial}{\partial z}(\rho_H \cdot u_M \cdot A) = 0, \tag{8.37}$$

$$\frac{\partial}{\partial t}(\epsilon \cdot \rho_G) + \frac{1}{A} \cdot \frac{\partial}{\partial z}(\epsilon \cdot A \cdot \rho_G \cdot u_M) + \frac{1}{A} \cdot \frac{\partial}{\partial z}\left(\epsilon \cdot A \cdot \frac{\rho_G \cdot \rho_L}{\rho_H} \cdot u_{G,U}\right) = \frac{\Gamma_G}{A}, \tag{8.38}$$

Momentum:

$$\frac{\partial}{\partial t}(\rho_H \cdot u_M) + \frac{1}{A} \cdot \frac{\partial}{\partial z}(A \cdot \rho_H \cdot u_M^2) + \frac{1}{A} \cdot \frac{\partial}{\partial z}\left(A \cdot \frac{\epsilon}{1-\epsilon} \cdot \frac{\rho_G \cdot \rho_L}{\rho_H} \cdot u_{G,U}^2\right)$$
$$= -\frac{\partial p}{\partial z} - \tau_w \cdot \frac{P}{A} - \rho_H \cdot g \cdot \sin(\alpha). \tag{8.39}$$

After some further transformations and using the momentum balance equations for the individual phases, we obtain the energy equation

$$\frac{\partial}{\partial t}(\rho_H \cdot h_M) + \frac{1}{A} \cdot \frac{\partial}{\partial z}(A \cdot \rho_H \cdot h_M \cdot u_M)$$
$$= \frac{1}{A} \cdot (q_{w,G} \cdot P_G + q_{w,L} \cdot P_L) + \frac{\partial p}{\partial t} + u_M \cdot \frac{\partial p}{\partial z} + u_{G,U} \cdot \frac{\rho_L - \rho_G}{\rho_H} \cdot \frac{\partial p}{\partial z}$$
$$-\frac{1}{A} \cdot \frac{\partial}{\partial z}\left(A \cdot \frac{\epsilon \cdot \rho_L \cdot \rho_G}{\rho_H} \cdot u_{G,U} \cdot \Delta h_{LG}\right) + \frac{1}{A} \cdot L_{\text{diss}}. \tag{8.40}$$

Here h_M is the *density-averaged enthalpy.* It is defined as

$$h_M = \frac{\epsilon \cdot \rho_G \cdot h_G + (1-\epsilon) \cdot \rho_L \cdot h_L}{\epsilon \cdot \rho_G + (1-\epsilon) \cdot \rho_L}. \tag{8.41}$$

Here Δh_{LG} is the heat of evaporation in phase transitions.

Successful application of the simple drift-flow model depends essentially on whether constitutive relations can be developed for the drift velocity $u_{G,U}$.

8.2.4 Bubbles and Drops

The motion of bubbles and drops in a moving liquid or gas is a fundamental element for the modeling of two-phase flows and the representation of the drift velocity. Extensive investigations into this topic have been carried out, and these are presented in great detail in the book by *R. Clift et al.* (1978). Heuristic considerations show that the relative velocity of bubbles and drops in a continuum depends on the type of interaction between the phases, their interaction with the boundaries, and the effect of the external field forces, e.g. the gravitational force. This fact can be expressed by the following functional relation:

$$u_r = u_G - u_L = \mathrm{f}\left(\frac{\mu_G}{\mu_L}, \frac{\rho_G}{\rho_L}, \sigma, \epsilon, \frac{\rho_L - \rho_G}{\rho_L} \cdot g, \frac{D_B}{d}\right). \tag{8.42}$$

Here μ_G and μ_L are the viscosities of the gas and the liquid, ρ_G and ρ_L their densities, σ the surface tension, g the gravitational acceleration, D_B the bubble or drop diameter, and d a typical container dimension.

The statements that follow concentrate on the behavior of bubbles. With certain modifications they are also true for drops.

In quasi-steady dispersed two-phase flows, the effect of acceleration forces on single bubbles can frequently be neglected. The equilibrium velocity u_∞ of a single bubble in the fluid continuum is then determined by the balance of the drag and field forces. In the case of the lift force we have

$$u_\infty^2 = \frac{4 \cdot (\rho_L - \rho_G) \cdot g}{3 \cdot \rho_L} \cdot \frac{D_B}{c_w}. \tag{8.43}$$

Here c_w is the drag coefficient in the definition $c_w = W/(0.5 \cdot \rho_G \cdot u_\infty^2 \cdot A)$ with W the drag force and A the cross-sectional area of a sphere with the equivalent volume of the bubble. The equivalent bubble radius is found from the relation $D_B = 2 \cdot (3/(4 \cdot \pi) \cdot V_B)^{(1/3)}$ with V_B the bubble volume. The introduction of the equivalent bubble radius permits us also to consider deformed bubbles and to associate a drag coefficient with them from experiments on a solid sphere in a single-phase flow. However, the deformation of the bubbles under the effect of relative motion can be so great that comparison with a moving solid sphere yields incorrect results. Therefore, in many experiments the terminal rise velocity of single bubbles in the gravitational field has been measured, where in particular, the shape of the bubble was investigated as an additional effect. *R. Clift et al.* (1978) presented the results in a graph, ordered by dimensionless characteristic numbers. They introduced the following bubble characteristic numbers:

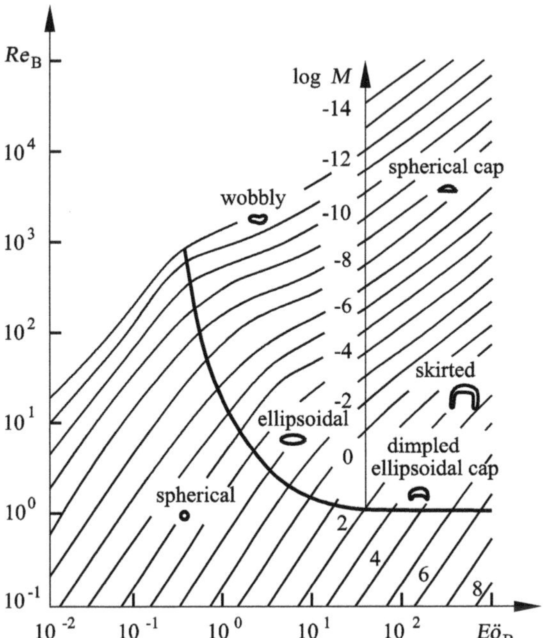

Fig. 8.6. Form of ascending bubbles in a liquid, *R. Clift et al.* (1978)

$$\mathrm{Re}_B = \frac{u_\infty \cdot D_B \cdot \rho_L}{\mu_L} \qquad \textit{Reynolds number,}$$
$$\mathrm{E\ddot{o}}_B = \frac{g \cdot (\rho_L - \rho_G) \cdot D_B^2}{\sigma} \qquad \textit{Eötvös number,}$$
$$\mathrm{Mo}_B = \frac{g \cdot \mu_L^4 \cdot (\rho_L - \rho_G)}{\rho_L^2 \cdot \sigma^3} \qquad \textit{Morton number.}$$

The Eötvös number describes the interaction between gravitational and capillary forces, while the Morton number essentially relates viscous, capillary, and gravitational forces. The graph is shown in Figure 8.6. It allows the dependence of the equilibrium velocity u_∞ to be determined as a function of all other quantities appearing in the characteristic numbers, and it also gives qualitative insight into the form of the bubble. Some experimentally observed bubble shapes are sketched in Figure 8.7.

The effect of finite containers and neighboring bubbles on the equilibrium velocity u_B of a single bubble is frequently modeled with a power product ansatz of the influencing quantities ϵ, $1 - \epsilon$, D_B/d in the form

$$u_B = u_\infty \cdot \left(1 + \alpha \cdot \frac{D_B}{d}\right)^m \cdot (1 - \epsilon)^n \cdot \epsilon^p \tag{8.44}$$

(see *R. Collins* (1967), *G. Wallis* (1969)), with the parameters α, m, n, and p to be determined from experiment. For example, *R. Collins* corrected the bubble ascent velocity of single bubbles in a vertical pipe with diameter d in the form

$$u_B = u_\infty \cdot \left(1 + \alpha \cdot \frac{D_B}{d}\right)^{-1},$$

with $\alpha = 1.6$ for deformed soft gas bubbles and $\alpha = 2.4$ for spherical hard gas bubbles.

G. Wallis (1969) states a fundamental relation for the representation of the drift velocity $u_{G,U}$ as a function of the equilibrium velocity and the gas volume fraction in the form

$$u_{G,U} = u_\infty \cdot (1 - \epsilon)^n.$$

He determines the exponent n using the experimental data of *F. N. Peebles* and *H. J. Garber* (1953) for different bubble shapes and bubble Reynolds

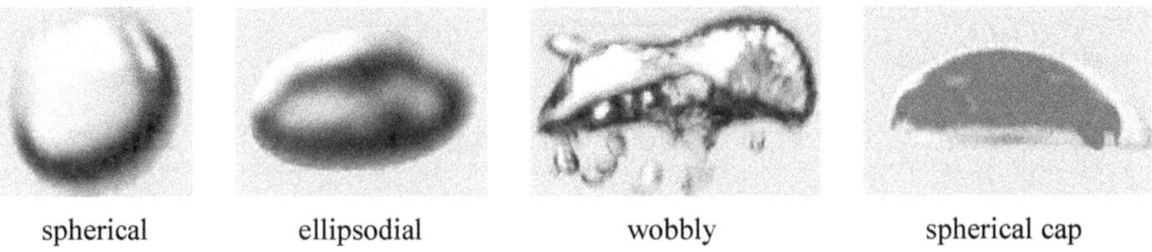

Fig. 8.7. Dependence of different bubble shapes on the bubble volume and bubble characteristic numbers, see Figure 8.6

numbers. In what follows we present the results of his investigations into the terminal rise velocity of single bubbles:

u_∞	n	Region of validity	bubble form
$\frac{D_B^2 \cdot (\rho_L - \rho_G) \cdot g}{18 \cdot \mu_L}$	2	$\mathrm{Re}_B < 2$	rigid spherical bubble
$0.33 \cdot \frac{2 \cdot \nu_L}{D_B} \cdot \left(\frac{g \cdot D_B^3}{8 \cdot \nu_L^2}\right)^{0.76}$	1.75	$2 < \mathrm{Re}_B < 4.02 \cdot G_1^{-0.214}$	spherical bubble with inner flow
$1.18 \cdot \left(\frac{g \cdot \sigma}{\rho_L}\right)^{0.25}$	1.5	$3.10 \cdot G_1^{-0.25} < \mathrm{Re}_B$ $5.75 < G_2$	oscillating elliptical bubble
$1.00 \cdot \sqrt{g \cdot D_B}$	$0 < n < 1$	$\sqrt{\frac{g \cdot \rho_L \cdot D_B^2}{\sigma}} > 4$	cap bubble

The Galileo number G_1 is

$$G_1 = \frac{g \cdot \mu_L^4}{\rho_L \cdot \sigma^3}, \tag{8.45}$$

and G_2 is defined as

$$G_2 = \frac{1}{16} \cdot G_1 \cdot \mathrm{Re}_B^4 = \frac{g \cdot u_\infty^4 \cdot \rho_L^4 \cdot D_B^4}{16 \cdot \sigma^3}.$$

Further details have been outlined by *R. Clift* et. al. (1978).

Using the representation of the drift velocity for different bubble shapes according to *G. Wallis* (1969), the drift flow model for dispersed two-phase flows is closed.

H. Zuber and *J. A. Findley* (1965) and *M. Ishii* (1977) have applied the drift-flow model to other flow forms such as plug, annular, and turbulent churn flows. Their papers contain details of the relevant constitutive parameters.

A computationally supported description of the mass and heat exchange processes of multidimensional two-phase flows frequently requires a precise

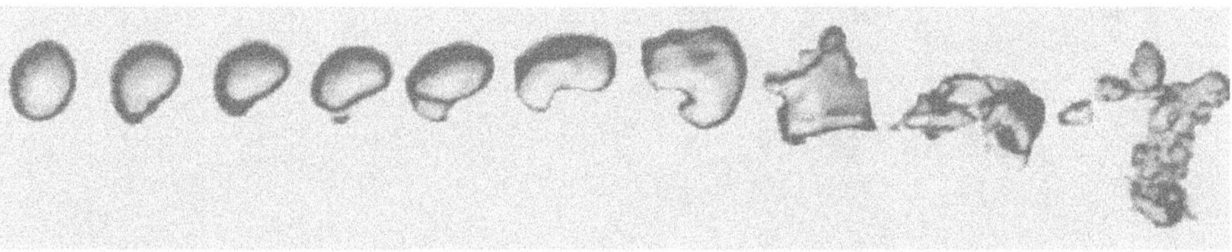

Fig. 8.8. Sequence of different bubble forms until bubble decay, under the effect of a turbulent liquid flow, F. Risso and J. Fabre (1998)

examination of the development of the phase interfaces in the flow. In the framework of a two-fluid model, this takes place on the basis of a balance equation for the interface concentration $\alpha_{\rm i}$, as shown briefly in Section 5.4.6. The development of the phase interface in disperse bubble or drop flows is then essentially determined by the coalescence and decay processes of individual particles as they interact. The physical understanding of these two processes is therefore centrally important and is has been the subject of many individual investigations. The complexity of these processes can be seen in Figures 8.8 and 8.9 from the sequence of deformation states of the particles in the decay of a bubble and the coalescence of a drop.

Observations show that the coalescence and decay of bubbles and drops essentially takes place in three steps.

In bubble flows, the bubbles move towards each other and collide, and a thin liquid film forms between them. The liquid film must be displaced by the motion of the bubbles until the film thickness reaches a critical size. The film then tears and coalescence begins. The process can be characterized by three time scales. There is a mean collision time of the bubbles, an effective contact time in which the liquid film thins, and the opening time of the tear in the film.

The collision time is essentially determined by the convective motion and concentration of the bubbles in the continuous phase. The motion is apparent in relative velocities of the bubbles in laminar shear flows, in fluctuation velocities of turbulent flows or in the different lift velocities of bubbles of different shapes and sizes. Bubble concentration, relative bubble velocity and an effective collision cross-section fixed by the effective bubble cross-section determine the collision frequency of the process. Similar considerations hold for drops in gas flows.

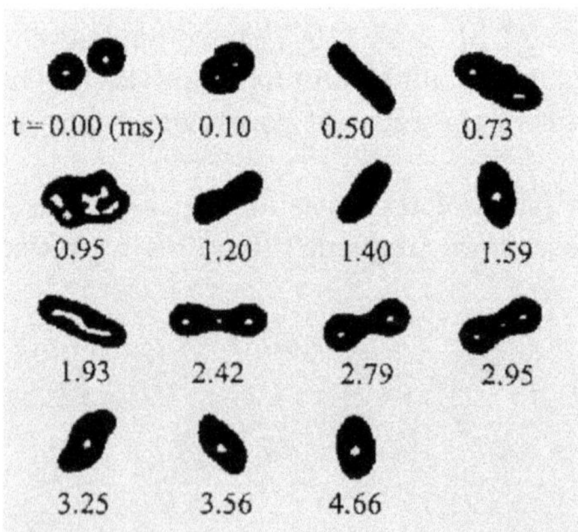

Fig. 8.9. Sequence of different drop forms after a binary, non-central drop collision, until stable drop coalescence, J. Quian und C. R. Law (1997)

Not every collision process ends in the coalescence or decay of the particles. Rather it depends on whether and in which topological drop form an equilibrium between interface and motion energies of the particles can be reached after the collision.

The energy due to friction dissipation or lifting work is generally less important. As a measure of the ratio of kinetic to surface energy of a bubble or a drop, we can use a Weber number in the form $We = u_p^2 \cdot d_p \cdot \rho_L / \sigma$, where u_p is the particle velocity, d_p is an equivalent particle diameter, ρ_L the density of the liquid phase and σ the interface stress. A rough approximation for upper and lower limits for coalescence and decay can be introduced with the static equilibrium condition $We = 1$, leading to a critical particle velocity $u_{cp} = \sqrt{\sigma / (d_p \cdot \rho_L)}$. For the technically relevant turbulent two-phase flows, a stochastic relative motion of the particle is essentially produced by the turbulent fluctuation velocities $|u_c'|$ of the continuous phase filled with vortices. These velocities are determined by the total energy supply and the viscous dissipation in the fluid system (see *H. Tennekes* and *J. L. Lumley* (1972)). The critical particle diameter d_{cp} for coalescence and decay can therefore be directly associated with these quantities using the condition $We = 1$. This simplified criterion is extended by a further characteristic quantity if the fluids involved have significant viscosity. *J. O. Hinze* (1955) adds a further characteristic number, the so-called Ohnesorge number, in the form $Oh = \mu_L / \sqrt{\rho_p \cdot \sigma \cdot d_p}$ or sets up an empirical relation $Oh(We)$ as a process criterion. The coalescence or decay is essentially determined by whether the particle with a diameter d_p reaches the region of influence of a sufficiently energetic turbulent vortex with comparable diameter with a certain probability, i.e. if it "collides" with it. If the kinetic energy transferred in the interaction $0.5 \cdot \rho_L \cdot |u_c'|^2$ exceeds the critical amount $0.5 \cdot \sigma / (\rho_L \cdot d_{cp})$, decay will occur, or coalescence if the transferred energy remains smaller. The ratio of the energies $E_i = |u_i'|^2 / u_{cp}^2$ may be seen as a measure of the efficiency of the processes. The literature contains distribution functions for the efficiency to quantify the measure of decay or coalescence. A particularly simple distribution function was suggested by *C. A. Coulaloglon* and *L. L. Tavlarides* (1977) in the form $\beta_i = \exp(-u_{cp} / |u_i'|^2)$ (for more recent developments see *C. Tsouris* and *L. L. Tavlarides* (1994); *H. A. Jakobsen et al.* (2005)).

The product of the number of collisions of the particles and the efficiency of the collision can serve as a rate for the generation and destruction of bubbles and drops with a certain diameter. Suitable summing over all drop sizes in the flow space that are available to decay or coalesce can be used to derive a measure of the source or sink rates of the interface concentration (cf. *G. Kocamustafaogullari* and *M. Ishii* (1995), *H. A. Jakobsen et al.* (2005)).

8.2.5 Spray Flows

Dispersed droplet flows with very low liquid fraction are frequently called spray flows. They are of great technical importance for the generation of optimal combustion mixtures in motors, gas turbines, and furnaces. Other important areas of application are the cooling of thermally stressed surfaces and containments by evaporation and gas scrubbers in process plants. There are two crucial questions in the treatment of spray flows: the generation of drop clusters and the transport of the drops in the flowing carrier gas. Both aspects have recently been presented in overview articles and books, such as those by *S. P. Lin* and *R. D. Reitz* (1998), *L. Bolle* und *J. C. Moureau* (1982), *C. Crowe et al.* (1998). Here only some basic facts will be presented.

The decomposition of liquid into drops requires energy. This is proportional to the surface tension and to the increase in surface area during the formation of drops. The necessary supply of energy frequently comes from pressure drop in nozzle flow, from centrifugal forces in film flow on rotating disks, or from shear forces that act on liquid films or jets via shearing gas flows. In the generation of drops in nozzels, the nozzle geometry, the liquid properties, viscosity, surface stress, and the flow velocity affect the size of the drops. The dependencies can be formulated using the dimensionless characteristic Weber and Reynolds numbers that are based on a characteristic nozzle diameter D. The shape of the nozzle exit is generally circular, annular, or slit-shaped. Consequently, either liquid free jets or thin conelike or fan-shaped liquid disks initially form at the nozzle exit. The objects then decay into drops in further steps by the action of flow instabilities. If the decay of the liquid jet or disk is determined by inertial forces and surface tension, i.e. by capillary waves, the liquid is said to splatter. If turbulent shear forces determine the drop formation inside the liquid jet and at its edge at high velocities, it

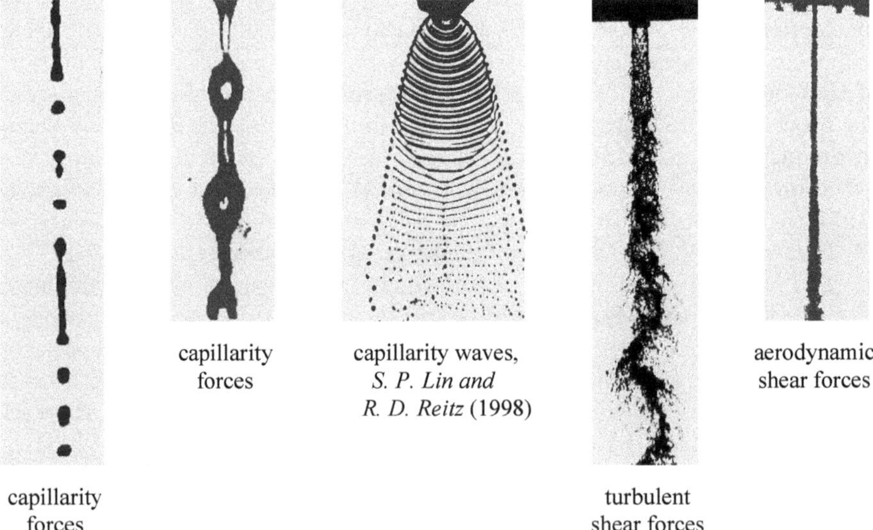

Fig. 8.10. Drop formation in the decay of liquid jets and lamellae via different flow instabilities

is called atomization. In the relevant literature the different forms of drop formation are frequently represented in graphical form using the Reynolds number and the Ohnesorge number $Oh = \mu_L/\sqrt{(\rho_L \cdot \sigma \cdot d_p)}$. Typical phenomena of drop formation in free jets are shown in Figure 8.10. Figure 8.11 shows an ordering chart for the different drop formation processes.

For specific technical applications, the nozzle geometry and the hydraulic characteristic data, i.e. the driving pressure drop, have to be selected such that atomization occurs at the desired mean drop diameter and into a required spatial angle. The Sauter diameter is usually taken as a measure for the drop diameter. It is defined as the ratio of the total volume of the drop to its total surface area.

The development of spray flows from the drop formation stage to the fully developed, weakly concentrated, dispersed droplet flow is extremely complex and inaccessible to a simple general description. One key question is the representation of further drop disintegration under the effect of the flow forces until at a certain drop size distribution an equilibrium is achieved. Different decay mechanisms of free drops in shear flows have been observed in experiments. Either simple drop oscillations, Rayleigh inertial instabilities, Kelvin–Helmholtz wave instabilities, or pure shear flow instabilities play a role. The observed decay phenomena have been characterized and summarized by *S. P. Pilch* and *R. D. Erdmann* (1987), *L. Bolle* and *J. C. Moureau* (1982), *C. Crowe et. al.* (1998) and are displayed in Figure 8.12, ordered by increasing Weber numbers.

The mathematical description of fully developed spray flows starts from the concept of a dilute droplet distribution in the carrier gas. It is assumed that there is a relevant interaction only between the gas and the droplets and not between the drops themselves. The quality of the interaction is determined by the so-called Stokes number St. It characterizes the temporal reaction of the droplet to a change in the gas flow, and is therefore defined as the time ratio $\mathrm{St} = \tau_p/\tau_e$ (where the subscript d indicates disperse, c

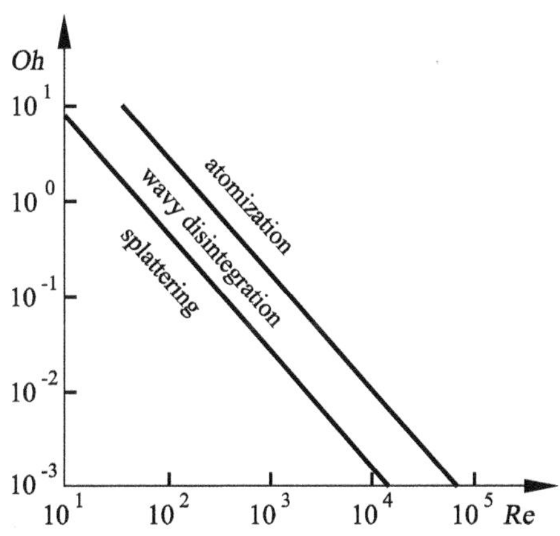

Fig. 8.11. Graph of drop decay processes according to *W. von Ohnesorge (1936)*

continuous) with $\tau_p = \rho_p \cdot d_p^2/(18 \cdot \mu_c)$, where d is the drop diameter and $\tau_e = L_{tc}/U_c$. Here L_{tc} is the characteristic length of the flow domain and U_c the characteristic velocity of the continuous phase. Therefore, very small Stokes numbers indicate an almost inertia-free motion of the droplets with the gas, whereas values of order one imply a considerable interaction between the phases.

In constructing the model, the gas flow, a continuum in the Euler representation, is treated as an inviscid, viscous, or fully turbulent fluid, with additional locally acting flow forces and mass sources or sinks from the drop–gas interaction (see Section 5.4.6.). The droplet motion takes place along

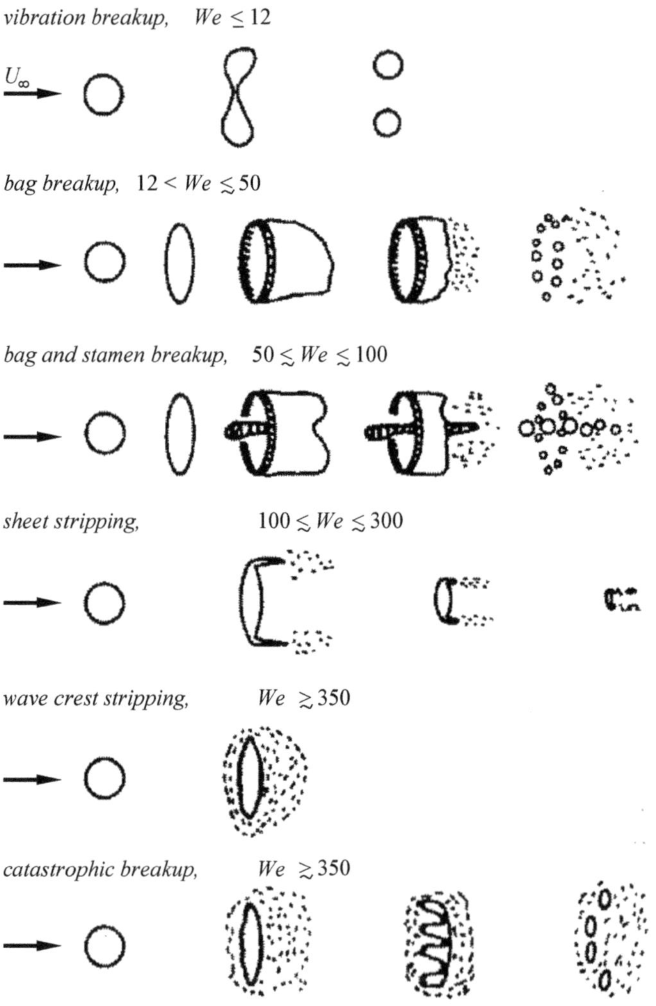

Fig. 8.12. Breakup mechanisms, according to *M. Pilch* and *C. A. Erdmann* (1987)

trajectories in the Lagrangian description, and is determined by the acting flow forces: drag, lift and gravity. The key problem of the modeling is the representation of the interaction between the phases by simple physical models in algebraic form. This is an area of active research, in particular with a view to the interaction of turbulence and droplet dispersion. Here we refer to the recent relevant literature, e.g. *C. Crowe*, et. al. (1998), *W. A. Sirignano* (1999). The hybrid Euler–Lagrange model is a significant alternative to the two-fluid model treated earlier (see Section 5.4.6 and 8.2) for the case of dilute, dispersed two-phase or two-component flows.

8.2.6 Liquid–Solid Transport

The transport of solid particles in gases or liquids occurs in various technical applications and in a series of geological phenomena. Examples are the pneumatic or hydraulic transportation of bulk goods in pipes, the transport of sediment in rivers and processes in mudslides and powder and snow avalanches.

Solid–liquid mixtures behave like a fluid if the particles appear in suspended form in the flow of the mixture and interact essentially indirectly via viscous friction forces or through turbulent shear forces in the liquid phase. Such mixture flows are called dilute suspension flows. If the mean distance between the solid particles becomes comparable with a mean particle diameter d_p as the volume fraction of the solids ϵ_p grows, the mixture flow is increasingly determined by the direct momentum exchange of colliding particles, whereby the solid friction as the particles touch each other and the surrounding edges also significantly affects the processes. This situation is also called a dense suspension flow. Ultimately an increase of the volume fraction of the solid above the value $\epsilon_p = 0.4$ leads to locking of the solid particles in the transport channel as a consequence of the excessively high solid friction between the particles and the channel wall. An immobile particle bed forms, through which the gas or liquid seeps if there is a pressure drop present.

The essential phenomena will be briefly clarified using the example of transport of solids in horizontal pipelines. The design of such a transport channel requires that a maximum particle transport efficiency is achieved at a propulsion power for the substrate that is as low as possible.

As with liquid–gas flows, patterns also develop in solid suspension flows, and these depend on the relevant system parameters such as the mean particle diameter d_p, the diameter of the pipe D, the mass density ratio of the solid to the fluid phase ρ_p/ρ_f, the mass flux density ratio $\dot{m}^* = \rho_p \cdot u_p/(\rho_f \cdot u_f)$, the particle volume fraction ϵ_p, the pressure difference Δp for the mixture propulsion, the turbulent fluctuation velocity in the substrate u'_f and finally, for the fluid dynamic characterization of the particles, the characteristic sink rate $w_{s\infty}$ of the particles in the stationary substrate. A progression of typical flow patterns as well as schematic distributions of the velocity and the particle fraction is shown in Figure 8.13, following *M. Weber* (1974).

For the homogeneous and saltation transport states, the velocity of the substrate is large enough to keep the particles in suspension and well mixed, due the effect of the turbulence, the lift forces and the elastic collisions with the wall. If the gas velocity decreases, some of the particles settle on the base of the channel. First moving particle strands form and, as the velocity is further decreased, stationary deposits form. On the surface of the deposits the gas flow causes wave-like grooves and eventually dune-like clusters, if the mass flux ratio exceeds limits of $\dot{m}^* > 30$. At even greater particle concentrations, with values $\dot{m}^* \gg 30$, particle plugs form and finally a compact, immobile bed of particles. Such states are to be avoided in the use of transport systems.

The driving pressure drop in the transport channel has a characteristic dependence on the volume flux U_f of the substrate and on the mass flux ratio $\dot{M}_s = \dot{M}_p/\dot{M}$, with $\dot{M} = \dot{M}_p + \dot{M}_f$. This dependence is shown schematically in Figure 8.14, whereby the volume flux density and the pressure are each made dimensionless with the relevant fall velocity $v_g = \sqrt{g \cdot D}$.

For finite loads $\dot{M}_s$, the pressure drop in the substrate increases, both for high and for low volume flux densities, and becomes a minimum for a certain intermediate value. The higher the particle load, the more this minimum

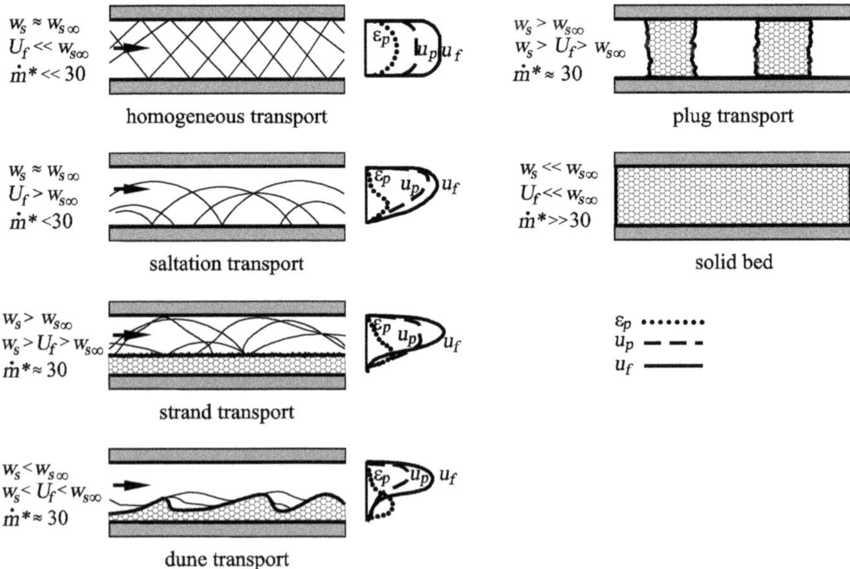

Fig. 8.13. Sketch of observed states in pneumatic solid transport in horizontal pipes. Parameters used: U_p, U_f relative particle and fluid velocities, $w_{s\infty}$ sink rate of the particles in the stationary substrate, ϵ_p particle volume fraction, $\dot{m}^*$ mass flux density fraction

pressure drop is displaced to higher volume flux densities, and observations show the formation of strands and depositions on the base of the transport pipelines. Therefore the associated volume flux of the substrate can be taken to be a conservative critical condition for stable transport. The increase of the pressure drop at low volume fluxes and increasing particle loads reflects the increasing effect of strand formation to the point of development of particle plugs and a solid bed of particles, and is a feature of the hydraulics of seeping flows in porous containers.

Pressure loss diagrams and correlations have been empirically developed for different classes of granular and powder-like materials such as sand, coal, corn, flour and ores and can be found in the relevant textbooks (see for example *G. Govier* and *K. Aziz* (1972), *M. Weber* (1974), *O. Molerus* (1982). Because of the great variation in the geometric and mechanical properties of the different transport goods, until now there is no generally valid relation for the transport pressure drop. The minimum pressure drop can be determined easily, and roughly marks the transition from suspended particle transport to strand formation. In order to make this pressure criterion more exact, another significant quantity for the formation of strands, namely the sink rate of the particles in the substrate, has been investigated more exactly. We discuss the example of one of many empirical relations used as a design criterion for the onset of strand formation in transport channels. According to the observations and measurements of *R. A. Duckworth* (1971), the minimum velocity $U_{f_{min}}$ for suspended particle transport under the effect of gravity depends essentially on the mass load $\dot{m}^*$, the ratio of mean particle diameter to pipe diameter d_p/D and the settling velocity of a particle in the stationary substrate.
As a measure of the latter, we have the relation

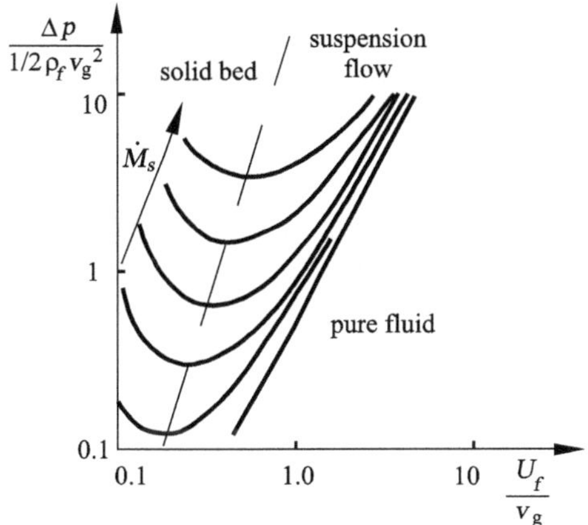

Fig. 8.14. Pressure loss as a function of the velocity of the substrate and the particle load

$$w_{s\infty} = \sqrt{\frac{4 \cdot g \cdot d_p \cdot (\rho^* - 1)}{3 \cdot C_D}} \quad ,$$

with $\rho^* = \rho_p/\rho_f$, C_D the hydrodynamic drag coefficient of the particles in the flow and g the acceleration due to gravity. Taking the free fall velocity as a measure of the velocity, the critical, minimal transport velocity $Fc_{min} = Uc_{min}/\sqrt{g \cdot D}$ can be represented generally as a dimensionless function of the form $Fc_{min} = f(d_p/D, \dot{m}^*, U_{p\infty}/\sqrt{\rho \cdot g})$. *R. A. Duckworth* (1971) suggests the following empirical relation:

$$F_{c\,min} = C \cdot (\dot{m}^*)^{0.2} \cdot \left(\frac{d_p}{D}\right)^{-0.6} \cdot \sqrt{\frac{w_{s\infty}}{\sqrt{g \cdot D}}} \quad ,$$

where C as a constant of proportionality includes other specific material properties of the particles, such as particle shape and surface properties.

In the case of vertical transport pipes, similar patterns are found in the form of suspension, strand and plug flows to the point of solid particle bed flows. However, as gravity only acts in the direction of transport, the velocities and particle concentrations are symmetrically distributed over the flow cross-section.

8.2.7 Fluidization of Particle Beds

In process engineering, aerated or vented particle beds are frequently used in chemical reactors and separating apparatus. These are generally particle fills in a container into which gas or fluid can be introduced from below through a grid of nozzels. Depending on the magnitude of the volume flux of the injected gas or liquid, different flow forms can form in the fill, and these are important for the efficiency of the apparatus concerned. Figure 8.15 shows some frequently observed states at increasing volume fluxes.

For very small volume fluxes, the particle bed remains compact and its weight is mainly carried by the perforated base wall. If the volume flux increases, a flow state occurs where the weight of the fill is compensated by the resisting force of the flow in the collection of particles. At even larger volume fluxes, the particle bed becomes ever looser, until the particles are finally freely suspended in the substrate and as the volume flux is increased even further, the particles are finally carried out of the container as a suspension. The floating of the particle bed is called fluidization. As sketched in Figure 8.15, the fluidization state can occur in different forms that depend mainly on the shape of the container, the type of injection of the carrier fluid, and the geometric and material properties of the particles. A crucial control quantity for the use of fluidization systems is the critical pressure drop at the onset of the fluidization process. This can be determined from measurements of the dependence of the pressure drop on the volume flux density in the fill. Figure 8.16 shows this relationship schematically.

For the flow in a compact particle bed, the Ergun linear relation (*S. Ergun* (1952)) holds between the pressure gradient and the local volume flux density, i.e. the relative velocity, in the form $\nabla p = k \cdot U_f$. The pressure drop remains constant over a fully fluidized particle bed without particle removal as the volume flux density of the substrate increases. However, a small transition zone with hysteresis behavior occurs, with slightly increased or reduced pressure values for increasing or decreasing volume fluxes. This effect is caused by shape effects of the particles and segregation effects among the particle quantities. The critical quantities for the onset of fluidization can be determined from the intersection of the extrapolated linear pressure–volume flux curves for the regions of compact and fluidized beds.

Each fluidization state sketched in Figure 8.15 can be observed in a certain region of volume flux densities. Empirical relations have been developed for the transitions between the different states and these can be found in the relevant textbooks (*D. Kunii* and *O. Levenspiel* (1991), *J. R. Grace* (1982)).

D. Geldert (1973) investigated in detail the effect of particle and fluid properties, such as the mean diameter and the density of the particles and the viscosity of the substrate, on the fluidization states. He introduced four size classes of particles: very fine powder with mean particle diameters $\overline{d}_p < 20\,\mu m$, two sand groups with $20\,\mu m \leq \overline{d}_p \leq 90\,\mu m$ and $90\,\mu m \leq \overline{d}_p \leq 650\,\mu m$, and granular fills with $\overline{d}_p > 650\,\mu m$.

He discovered that controlled fluidization for fine powders is very difficult to achieve, as electrostatic and van der Waals forces and, in the case of liquid substrate flows, interface stresses greatly influence the pure mechanical interaction between the particles. In contrast, sands can easily be fluidized into all different states. In rough particle beds, a state with continuous flow channels tends to occur, in which particles are transported to the surface, while in the regions between the channels the particle bed sinks down. For the sake of completion, we mention that it is possible to reproduce the fluidization states with newly developed numerical methods of multiphase fluid dynamics

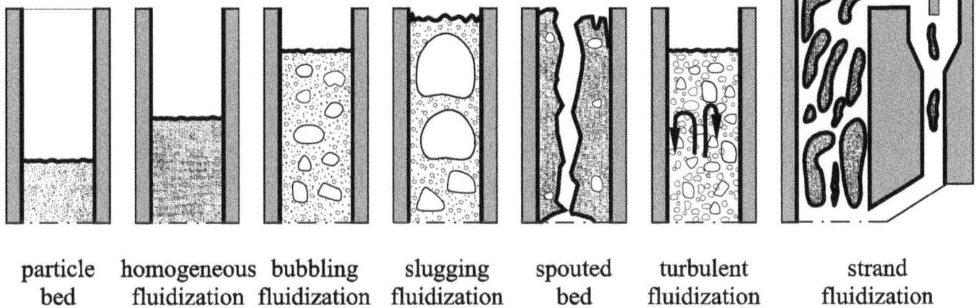

Fig. 8.15. Schematic representation of typical phenomena in the gaseous aeration of a particle bed with increasing volume flux of the substrate

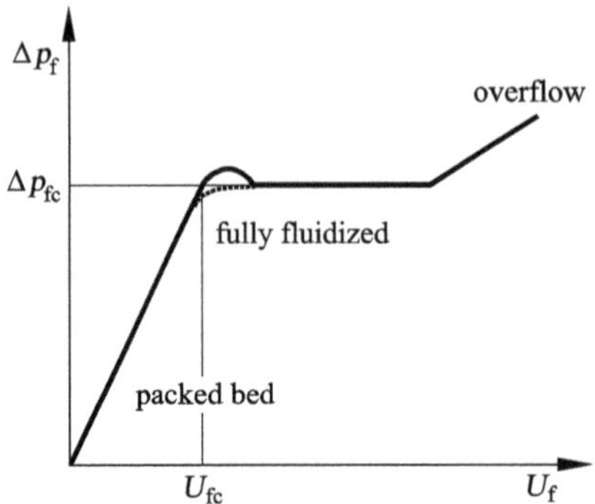

Fig. 8.16. Schematic representation of the dependence on pressure drop and volume flux density in compact and fluidized particle beds, definition of the critical quantities Δp_{fc} and U_{fc} for the onset of fluidization

(see Section 5.4.6). A critical evaluation of this possibility has been presented by *J. R. Grace* and *F. Taghipour* (2004).

8.3 Pressure Loss and Volume Fraction in Hydraulic Components

Pressure losses in two-phase flows are of great importance in power and process engineering. Therefore, robust empirical pressure loss correlations, like those for single-phase flow hydraulics, have been developed on the basis of measurements and simple models. These relations do not distinguish between specific flow regimes. Yet the concept of a dispersed flow on the one hand and that of a separate two-phase flow on the other hand has led to two different variants for pressure drop relations. The total pressure drop in a pipe or channel consists quite generally of the losses due to friction, due to acceleration, and due to gravity. Symbolically, we can write

$$\left(\frac{\mathrm{d}p}{\mathrm{d}z}\right)_{\mathrm{tot}} = \left(\frac{\mathrm{d}p}{\mathrm{d}z}\right)_f + \left(\frac{\mathrm{d}p}{\mathrm{d}z}\right)_a + \left(\frac{\mathrm{d}p}{\mathrm{d}z}\right)_g , \tag{8.46}$$

with the subscripts f for friction, a for acceleration, and g for gravity. While in horizontal straight pipes of constant cross-section, only the friction acts, in contracting or expanding elements, such as valves or junctions and bends, the acceleration predominates as in single-phase flow, but to a much greater degree. First we consider here the horizontal straight pipe.

8.3.1 Friction Loss in Horizontal Straight Pipes

The Homogeneous Model

Assuming a horizontal pipe with constant cross-section, we obtain the following representation, as in the single-phase flow:

$$-\left(\frac{\mathrm{d}p}{\mathrm{d}z}\right)_f = -\tau_w \cdot \frac{4}{d}, \qquad \tau_w = \frac{1}{4} \cdot c_{f,2\mathrm{Ph}} \cdot \frac{1}{2} \cdot \rho_H \cdot u^2 \tag{8.47}$$

with d the hydraulic diameter, $c_{f,2\mathrm{Ph}}$ the friction coefficient of the two-phase flow, and the homogeneous density ρ_H. The friction coefficient is given as a function of a still to be defined Reynolds number for the two-phase flow. Frequently, the known relations for the friction coefficient of single-phase flow are taken, such as the Stokes law for laminar flow and the Blasius law for turbulent flow. The friction coefficient can then be chosen depending on the flow form. For an annular-droplet flow, the value for rough pipe walls is chosen as $c_{f,2\mathrm{Ph}} \simeq 0.02$. If the volume fraction of the gas is large, i.e. $(1-\epsilon) \ll 1$, the single-phase value $c_f = c_{f,G}$ can be selected. If $\epsilon \ll 1$, then $c_f = c_{f,L}$, at which the viscosity of the gas or the liquid is selected for the Reynolds number. Frequently, following the classical implicit relation of Prandtl for single-phase fully turbulent flows, the following relation is also used:

$$\frac{1}{4} \cdot c_{f,2\mathrm{Ph}} = 0.0014 + 0.125 \cdot \mathrm{Re}_{2\mathrm{Ph}}^{-0.32} . \tag{8.48}$$

In order to form the Reynolds number, the total mass flux density and a mixture viscosity are used. The simplest relations for the weighted viscosities are

$$\begin{aligned}
\mu_{2\mathrm{Ph}} &= \frac{U_G}{U} \cdot \mu_G + \frac{U_L}{U} \cdot \mu_L , \\
\mu_{2\mathrm{Ph}} &= \chi \cdot \mu_G + (1-\chi) \cdot \mu_L , \\
\frac{1}{\mu_{2\mathrm{Ph}}} &= \frac{\chi}{\mu_\mathrm{G}} + \frac{1-\chi}{\mu_\mathrm{L}} .
\end{aligned} \tag{8.49}$$

For mixtures such as water and vapor that can be condensed, a so-called two-phase multiplier is frequently used for practical calculations. This multiplier is defined as the ratio of the pressure drop in the actual two-phase flow with mass-flow rate $\dot{M}$ to the pressure drop of the overall condensed liquid flow through the same pipe cross-section and with the same mass-flow rate:

$$\Phi_{L0}^2 = \frac{\left(\frac{\mathrm{d}p}{\mathrm{d}z}\right)_{2\mathrm{Ph}}}{\left(\frac{\mathrm{d}p}{\mathrm{d}z}\right)_{L0}} . \tag{8.50}$$

The subscript $L0$ indicates that the pressure drop of a pure liquid flow with the same mass flux as that of the two-phase flow was chosen as the reference measure. Correlations are given that are essentially a function of the gas–liquid properties and of the vapor content, and so have the general form

$\Phi_{L0}^2 = f(\mu_G/\mu_L, \rho_G/\rho_L, \chi)$. *W. Idsinga et al.* (1977) give a relation of the form

$$\Phi_{L0}^2 = \frac{\rho_L}{\rho_H} \cdot \left[1 + \chi \cdot \left(\frac{\rho_L - \rho_H}{\rho_G}\right)\right] . \tag{8.51}$$

It is valid for flows with very small gas volume fraction $\epsilon \ll 1$ for which $\mu_{2Ph} = \mu_L$ can be assumed. Similar relations are found when other relations for the mixture viscosity are used.

The Separate Model

The separate model is based on the idea that both phases flow in two separate regions of the pipe, as in stratified flow or annular flow, but that both phases are in a pressure equilibrium independent of the flow pattern. The total pressure drop is then described by the momentum equation for the two-phase mixture according to (8.22). For the horizontal pipe of constant cross-section and for steady flows we have

$$-\frac{\mathrm{d}p}{\mathrm{d}z} = \langle\tau_w\rangle \cdot \frac{P}{A} + \frac{\dot{M}^2}{A^2} \cdot \frac{\mathrm{d}}{\mathrm{d}z}\left(\frac{\chi^2}{\epsilon \cdot \rho_G} + \frac{(1-\chi)^2}{(1-\epsilon) \cdot \rho_L}\right) . \tag{8.52}$$

For the case in which no heat is supplied across the edge of the pipe, χ and ϵ do not change along the pipe. The wall shear stress is in equilibrium with the pressure force. *R. Lockhart* and *R. Martinelli* (1949) introduce two-phase multipliers to form the ratio of the pressure drop in the two-phase flow and the pressure drop in the gas or liquid phase when either of each flows alone in the pipe.

The multipliers of *R. Lockhart* and *R. Martinelli* (1949) are defined as follows:

$$\Phi_G^2 = \frac{\left(\frac{\mathrm{d}p}{\mathrm{d}z}\right)_{2\mathrm{Ph}}}{\left(\frac{\mathrm{d}p}{\mathrm{d}z}\right)_G} , \qquad \Phi_L^2 = \frac{\left(\frac{\mathrm{d}p}{\mathrm{d}z}\right)_{2\mathrm{Ph}}}{\left(\frac{\mathrm{d}p}{\mathrm{d}z}\right)_L} . \tag{8.53}$$

Based on experimental data, the authors place them in a relation with the so-called Martinelli parameter (cf. definition (8.13)), the pressure drop ratio. It can be computed explicitly if the volumetric fluxes of the two-phase flow and their laminar or turbulent flow states are known. The turbulent states are determined by the Reynolds number of the gas or liquid flow. For $\mathrm{Re}_{G,L} > 2000$ a turbulent state is assumed, and for $\mathrm{Re}_{G,L} < 2000$ the flow is assumed to be laminar. Therefore, there are four possible forms of the Martinelli parameter, depending on whether the gas and liquid phases are laminar or turbulent. The dependencies in the classical representation of *R. Lockhart* and *R. Martinelli* (1949) are shown in Figure 8.17. An analytical representation of the graphs was given by *D. Chisholm* (1967). He states the relations

$$\Phi_G^2 = 1 + \mathrm{C} \cdot X + X^2, \qquad \Phi_L^2 = 1 + \frac{\mathrm{C}}{X} + \frac{1}{X^2}, \tag{8.54}$$

where the parameter C takes on the values 20, 12, 10, and 5 for these four cases, and with 20 determines the doubly turbulent case, and with 5 the doubly laminar case.

The pressure loss relation of *R. Lockhart* and *R. Martinelli* is based on a relatively limited set of data of system pressures ($p < 0.4$ MPa) and pipe diameters ($d < 3 \cdot 10^{-2}$ m). Therefore, calculations using this model may be affected by an uncertainty up to an order of 40%. However, the relation is very simple and is therefore frequently used for initial estimations.

R. Martinelli and *D. Nelson* (1948) extended the pressure correlation of *R. Lockhart* and *R. Martinelli* to flows with higher system pressures, up to critical system pressures. They represent the two-phase multiplier in the form Φ_{L0} (see (8.50)), which utilizes the single-phase liquid flow as the reference state and displays it as a function of the vapor quality χ. This relation is still used to compute pressure losses in liquid–vapor flows.

For more exact calculations, empirical pressure loss correlations have been developed by *D. Chisholm* (1973) and *L. Friedel* (1978), based on a large amount of data ($2 \cdot 10^4$ experimental measurements). These relationships take into account further specific dependencies on the two-phase mass flux (Reynolds number), the surface tension (Weber number) and gravity (Froude number). They are complex functional relations between dimensionless characteristic numbers that represent different physical phenomena. For example, we discuss here the Friedel correlation. *L. Friedel* (1978) chooses the form in the definition (8.50) for the two-phase multiplier with dependence on different characteristic numbers. His correlation reads

$$\Phi_{L0}^2 = \mathrm{E} + \frac{3.24 \cdot \mathrm{F} \cdot \mathrm{H}}{\mathrm{Fr}^{0.045} \cdot \mathrm{We}^{0.035}}, \tag{8.55}$$

with the expressions

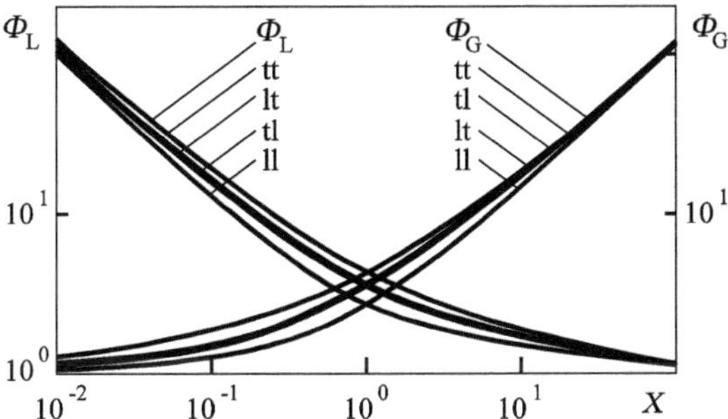

Fig. 8.17. Correlations according to *R. Lockhart* and *R. Martinelli* (1949): tt both phases turbulent; lt fluid laminar, gas turbulent; ll both phases laminar; tl fluid turbulent, gas laminar

$$E = (1-\chi)^2 + \chi^2 \cdot \frac{\rho_L}{\rho_G} \cdot \frac{c_{f,G0}}{c_{f,L0}},$$

$$F = \chi^{0.78} \cdot (1-\chi)^{0.24},$$

$$H = \left(\frac{\rho_L}{\rho_G}\right)^{0.91} \cdot \left(\frac{\mu_G}{\mu_L}\right)^{0.19} \cdot \left(1 - \frac{\mu_G}{\mu_L}\right)^{0.7},$$

$$Fr = \frac{\dot{m}^2}{g \cdot D \cdot \rho_H^2}, \qquad We = \frac{\dot{m}^2 \cdot D}{\rho_H \cdot \sigma}.$$

For application of the different empirical pressure loss relations, *P. B. Whalley et al.* (1981) have carried out extensive comparative calculations and have given recommendations.

Correlations for the Volume Fraction

Of equal importance for the calculation of the pressure drop is a quantitative estimation of the volume fraction in a two-phase flow. Independently of the possibility to calculate this from the two-fluid model, models and correlations were developed from experimental data to determine this quantity. Within the framework of the drift-flow model, the volume fraction was referred to the correlation coefficient C_0, the drift velocity $u_{G,U}$, and the volumetric fluxes (8.3), where C_0 and $u_{G,U}$ are determined according to experiments and physical relations for each flow regime.

R. Lockhart and *R. Martinelli* (1949) developed an empirical relation for volume fractions associated with their pressure drop measurements that is independent of the flow patterns. This is shown graphically in Figure 8.18. *D. Chisholm* (1967) gives a simple algebraic relation for the graph in Figure 8.18 in the form

$$1 - \epsilon = \frac{\chi}{\sqrt{\chi^2 + 20 \cdot \chi + 1}}. \tag{8.56}$$

A. Premoli et al. (1970) derived a correlation (called the CISE correlation) for ϵ that is based on an empirical relation for the velocity ratio $S = u_G/u_L$.

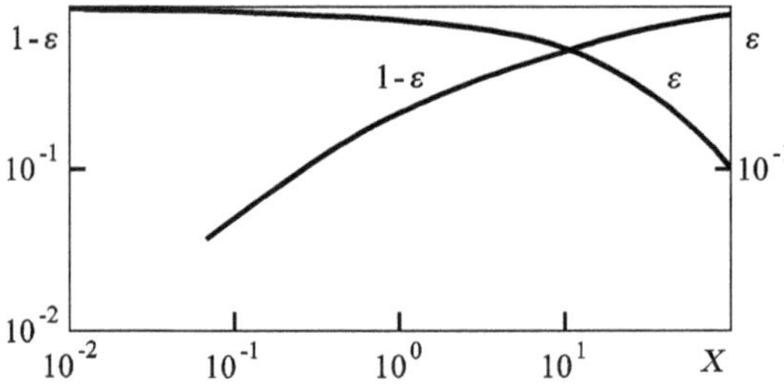

Fig. 8.18. Liquid volume fraction, according to *R. Lockhart* and *R. Martinelli* (1949)

According to (8.9), ϵ can be written in the form

$$\epsilon = \frac{1}{1 + S \cdot \frac{1-\chi}{\chi} \cdot \frac{\rho_G}{\rho_L}}. \tag{8.57}$$

They developed the following empirical relation for the velocity ratio:

$$S = 1 + \mathrm{E}_1 \cdot \sqrt{\frac{Y}{1 + Y \cdot \mathrm{E}_2} - Y \cdot \mathrm{E}_2}, \tag{8.58}$$

with the expressions

$$Y = \frac{\dot{V}_G}{\dot{V}_L},$$

$$\mathrm{E}_1 = 1.578 \cdot \mathrm{Re}^{-0.19} \left(\frac{\rho_L}{\rho_G}\right)^{0.22},$$

$$\mathrm{E}_2 = 0.0273 \cdot \mathrm{We} \cdot \mathrm{Re}^{-0.51} \left(\frac{\rho_L}{\rho_G}\right)^{-0.08},$$

$$\mathrm{Re} = \frac{\dot{m} \cdot D}{\mu_L}, \qquad \mathrm{We} = \frac{\dot{m}^2 \cdot D}{\rho_L \cdot \sigma}.$$

Here $\dot{V}_G$ and $\dot{V}_L$ are the volumetric flow rates of the phases. This correlation has also been developed independently of flow regimes. We also mention here a more complex relation by *B. Chexal et al.* (1997), which was developed according to the ideas of the drift flow model and which can also be applied to two-phase flows moving in opposite directions. Because of its complex form, we do not further outline this relation here.

8.3.2 Acceleration Losses

To a much greater degree than in single-phase flow, acceleration losses occur in pipe expansions or contractions, in pipe bends, and in pipe junctions. In designing apparatus, the pressure loss relations have to be given, in analogy to similar correlations for single-phase flows neglecting the wall friction. In pipe expansions there is generally a deceleration of the flow, and so in two-phase flows a separation of the phases is to be expected. Therefore, we use

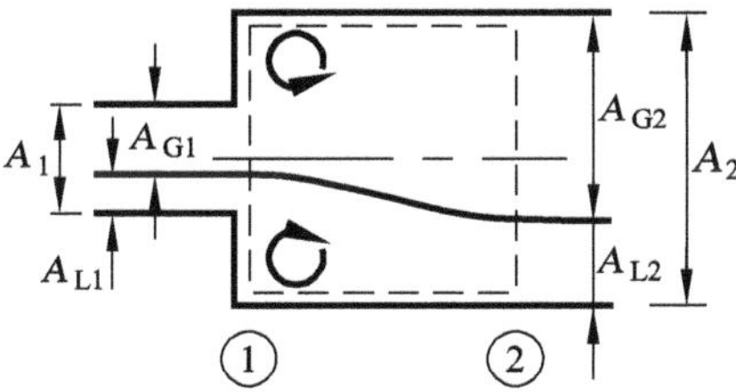

Fig. 8.19. The separate flow in the Carnot diffusor

the separate model in the form of the mixture balance equations to calculate a pressure change. Using the mixture equations for the separate model, a relation for the pressure gain in the Carnot diffusor can easily be written down neglecting the wall friction. For the control volume shown in Figure 8.19, *P. A. Lottes* (1961) derived the relation.

$$p_2 - p_1 = \dot{m}_1^2 \cdot \frac{A_1}{A_2} \cdot \left(\frac{1}{(\rho_I)_1} - \frac{A_1}{A_2} \cdot \frac{1}{(\rho_I)_2} \right) . \tag{8.59}$$

Here $(\rho_I)_{1,2}$ is the momentum density defined in (8.23). If we go over to single-phase flow, i.e. we choose $\chi = 0$ or $\chi = 1$, the momentum density changes to the density of the single-phase flow, as does the expression for the pressure recovery. Of course, expression (8.59) can be evaluated only if the volume fraction and the steam qualities of the cross-sections 1 and 2 can be related to each other. At low system pressures, if $p \ll p_{\text{crit}}$, and there are no phase transitions by evaporation, the volume fraction of the gas ϵ and the quality χ essentially do not change (cf. *B. Richardson* (1958), *L. Velasco* (1975)). In this case the analogy to single-phase flows is evident, since $\rho_{I1} \equiv \rho_{I2}$ holds. Observations show that downstream an equilibrium of the two-phase flow is attained only after a relatively long distance of about $30 - 70$ pipe diameters. This fact requires a pressure-dependent correlation for the change of the vapor content (cf. *M. Patric* and *B. Swanson* (1950)) for more exact calculations. In vapor–liquid flows with phase transitions, empirical relations between the volume fraction ϵ and the vapor quality χ are used (*J. Weisman et al.* (1976)).

In pipe contractions, there is an acceleration of the two-phase flow that leads to an improved mixing, and therefore a pressure loss calculation can be carried out using the *homogeneous flow model* to good approximation. Since the laws of the single-phase flow are valid for the homogeneous model, with the homogeneous density ρ_H as the only characteristic quantity, we obtain the known relation of the single-phase flow, which is easily confirmed using the schematic representation in Figure 8.20. We have

$$p_2 - p_1 = \frac{\dot{m}_2^2}{2 \cdot \rho_H} \cdot \left[1 - \left(\frac{1}{\sigma_c} \right)^2 + \left(\frac{1}{\sigma_c} - 1 \right)^2 \right] , \tag{8.60}$$

with $\sigma_c = A_c/A_2$ as the contraction ratio. Here the σ_c values for single-phase

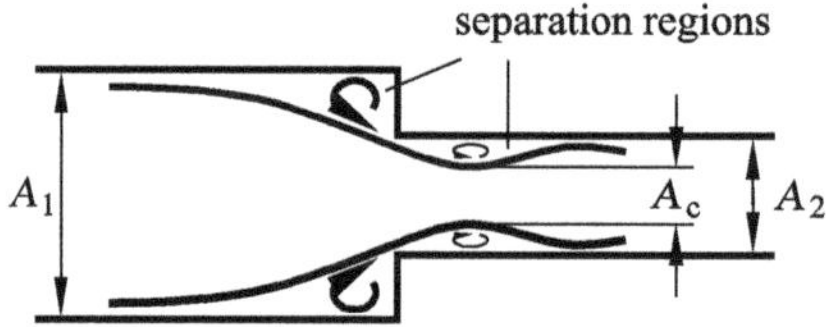

Fig. 8.20. The flow through a pipe contraction

flow according to *W. Archer* (1913) are used. The two terms in the square brackets can be identified as the irreversible and reversible contributions to the pressure loss. The contraction number reflects the local narrowing of the flow as a consequence of the separation of the flow at the edge (see Figure 8.20).

A separation bubble can also be observed in pipe bends at high-speed flows. In case of two-phase flow, a demixing of the phases occurs due to centrifugal forces. The effect is sketched in Figure 8.21. The gas phase collects on the inner side, while the liquid flows in the outer region of the bend. Dispersed flows change locally to stratified flows.

A new equilibrium between the phases, corresponding to the stationary intake conditions, is achieved only after 30 – 70 pipe diameters. This means that in the design of apparatus, only seldom can fully developed two-phase flows be assumed. Relations for the pressure drop in pipe bends have been developed by *D. Chisholm* (1967) on the basis of empirical two-phase multipliers.

The behavior of two-phase flows in pipe junctions is essentially determined by the branching angle and the orientation of the branching and the intake section with regard to the gravity vector. For any asymmetric orientation of the outlet and the branching to the intake section or to gravity, there is a redistribution of the phases that in particular cases can lead to complete phase separation. Because of the weaker inertial force, the gas phase follows curved trajectories more readily. Depending on the orientation of the branching to the gravity vector, this trend may be supported or compensated. This separation phenomenon is shown schematically for the example of a bend with horizontal intake and run but differently orientated branch in the graphs of Figure 8.22. The phase redistribution is here represented by the ratio of the steam quality χ_3 in the branch to that in the intake χ_1 plotted against the mass flux density ratio $\dot{m}_3/\dot{m}_1$, for three different orientations of the branch to the direction of gravity: opposite, in the direction of gravity, and perpendicular.

The graph shows that an almost complete separation of the phases occurs for a branch directed vertically upward. In case of a horizontal branch, the

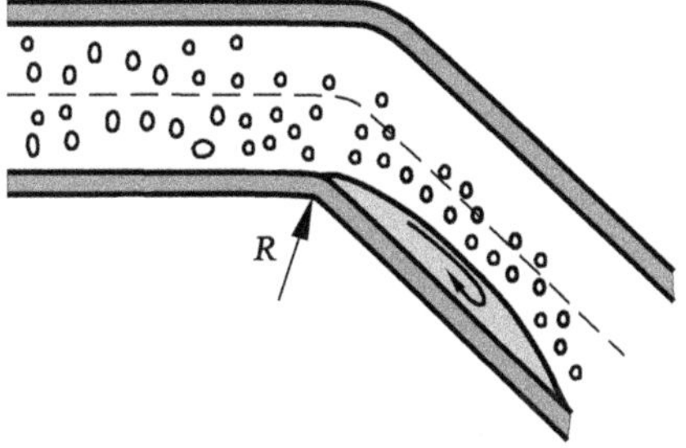

Fig. 8.21. Flow separation and phase separation in a pipe bend

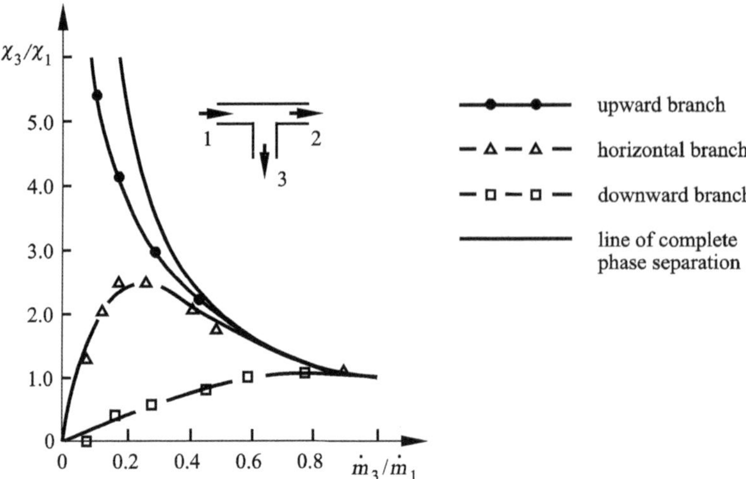

Fig. 8.22. Phase redistribution in a T-junction with different orientations to gravity

gas phase accumulates more strongly in the branch in the entire parameter regime $\dot{m}_3/\dot{m}_1$, namely, with a maximum at $\dot{m}_3/\dot{m}_1 \sim 0.25$. For the case of a downward branch and for small branching mass fluxes, gravity causes the liquid to follow the branch. Only when the inertial forces of the denser phase dominate gravity, in the example for $\dot{m}_3/\dot{m}_1 \sim 0.6$, does the steam quality in the branch become greater than that in the intake. The gas accumulates in the branch.

With regard to the pressure change, the branching behaves from the intake to the run like the cross-section expansion in the diffusor. From the intake to the branch, a flow acceleration occurs, as in a flow contraction. This is qualitatively in agreement with the observations for single-phase flows. In T-junctions extensive separation zones are observed which lead to a local cross-section contraction of the active two-phase flow. This is indicated in the sketch of Figure 8.23. Based on visual observations, a pseudoseparation line is frequently introduced to mark transmitted and branched mass flux densities.

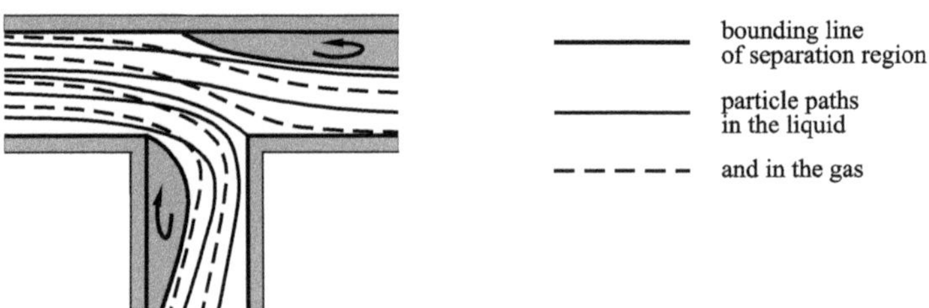

Fig. 8.23. Schematic representation of the phase redistribution in a T-junction. Denoted by pseudostreamlines and obvious separation regions

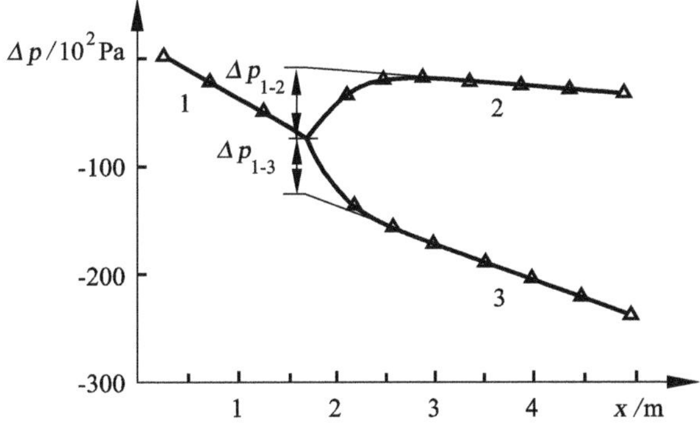

Fig. 8.24. Pressure variation in a T-junction, 1 intake, 2 outlet, 3 branch. The branch is horizontal. The following control values hold: $p_1 = 0.6$ Pa, $U_{L1} = 1.5$ m/s, $U_{G1} = 14.5$ m/s, $\dot{m}_3/\dot{m}_1 = 0.51$, air–water flow

With this assumption, pressure loss calculations for each partial mass flux can be carried out according to the separate model or the homogeneous two-phase flow model. Adaptation parameters in the pressure correlations are taken into account by means of experimental data. In the two-phase flow, the absolute pressure changes are significantly larger. An example of this behavior is shown in Figure 8.24. Further details of two-phase flows in pipe junctions are summarized by *B. J. Azzopardi* and *E. Hervieu* (1994).

For a detailed analysis of the phase redistribution in pipe-branchings three-dimensional computer calculations based on a two-fluid model have been carried out more recently (cf. *R. T. Lahey* (1990)). The calculations show good agreement with experimental observations.

8.4 Propagation Velocity of Density Waves and Critical Mass Fluxes

8.4.1 Density Waves

When a two-phase mixture flows out of a pressure reservoir through an exit of narrow cross-section, above a certain critical pressure difference of container and ambient pressure $p_1 - p_0$, there occurs a limitation of the mass flow rate. A further reduction of the ambient pressure leads to no further increase in the mass flux. A similar phenomenon is to be seen in compressible flows. The compressible flow in the Laval nozzle is the classical example for mass flow rate limitation as a consequence of the compressibility of the gas. The physical cause of this phenomenon is the same in both cases. Above the critical pressure difference, wavelike pressure and density perturbations in the liquid can move only downstream, because the flow velocity has become greater than the propagation velocity of small perturbations. Influencing the flow regime upstream by a change of state further downstream is not possible. In the case of compressible flow, the small perturbations are sound waves or Mach waves, whereas in the two-phase flow, they are small changes in the volume fraction of the gas phase. In both cases, the propagation velocity of these small perturbations can be represented by the same thermodynamic change of state, namely, by an isotropic variation of the density with the pressure. In the approximation of small wave amplitudes we get for the velocity of the wave propagation

$$a^2 = \left(\frac{\partial p}{\partial \rho}\right)_s . \tag{8.61}$$

In gas dynamics this is the propagation velocity of sound waves. In two-phase flows this is the propagation of density perturbations, primarily as a consequence of changes in the steam quality, and secondly due to changes in the densities of each phase with pressure. The term *velocity of sound* in connection with the propagation of small density perturbations in two-phase flows is therefore misleading.

To compute the critical mass flux in a pipe contraction, as in gas dynamics, we use the balance equations for mass and momentum, with certain equations of state for the gas and liquid phases. We begin with the equations for the two-phase mixture, (8.20)–(8.25), or with the simplified form (8.27) and (8.28). A simple calculation, analogous to that in gas dynamics, leads to the statement that the critical mass flux is given by the propagation velocity of the *density wave* at the narrowest point of the flow constriction, and is written in the form

$$\dot{m}_{\text{crit}} = A^* \cdot a^* \cdot \rho^* , \tag{8.62}$$

with a in the definition (8.61). The symbol * denotes the narrowest cross-section, which in some cases such as orifices, because of flow separation, is not the same as the geometrically narrowest cross-section. Its precise determination may be difficult. However, this notation immediately indicates the typical problem in two-phase flow. The critical mass flux depends on the definition of the two-phase density ρ_{2Ph} which, depending on the mixing model (separate or homogeneous model), can have different forms. Since the flow is always accelerating when it flows out of or past a body, it is generally assumed that the phases are well mixed and that the homogeneous density $\rho_H(\chi)$ describes the mixture well. A formal derivation of the expression for the density at constant entropy, i.e. in the approximation in which the changes of state are adiabatic in each phase, but where phase changes occur at the phase boundaries, then yields

$$\left(\frac{1}{a_{\text{2Ph}}^2}\right)_H = \left(\frac{\partial \rho_H}{\partial p}\right)_s$$
$$= \rho_H^2 \cdot \left[\frac{1}{\rho_L^2 \cdot a_L^2} + \chi \cdot \left(\frac{1}{\rho_G^2 \cdot a_G^2} - \frac{1}{\rho_L^2 \cdot a_L^2}\right) - \left(\frac{\partial \chi}{\partial p}\right)_s \cdot \left(\frac{1}{\rho_G} - \frac{1}{\rho_L}\right)\right] . \tag{8.63}$$

Here the index H indicates that the result refers to the homogeneous flow model. The velocities of sound in the gas $a_G^2 = (\partial p/\partial \rho_G)_s$ and in the liquid $a_L^2 = (\partial p/\partial \rho_L)_s$ were also introduced. Here we may also assume that the velocity of sound in the liquid is considerably larger than that in the gas ($a_L^2 \gg a_G^2$). After some algebraic manipulation, and using the definitions, this leads to a relation of the form

$$(a_{2\mathrm{Ph}}^2)_H = a_G^2 \cdot \left(\chi + \frac{\rho_G}{\rho_L} \cdot (1-\chi)\right)^2 \cdot \left[\chi - \left(\frac{\partial \chi}{\partial p}\right)_s \cdot a_G^2 \cdot \frac{\rho_G}{\rho_L} \cdot (\rho_L - \rho_G)\right]^{-1} . \tag{8.64}$$

It can clearly be seen that the propagation velocity of the density wave essentially depends on the steam quality and its change under isentropic thermodynamic equilibrium conditions. In many technically relevant flows, an evaporation process in thermodynamic equilibrium does not take place by pressure reduction at the narrowest flow cross-section, because the pressure drop occurs too fast and it is too small. This means that the relaxation time for the evaporation is considerably larger than the time the flow takes to pass through the constriction. Cases such as these are called metastable or frozen thermodynamic equilibrium. This occurs when $(\partial \chi/\partial p)_s = 0$. If we replace in the further simplified expression (8.64) the steam quality χ by the volume fraction ϵ, we obtain the following expression for the frozen density wave propagation velocity in a homogeneous two-phase flow:

$$(a_{2\mathrm{Ph}}^2)_H = a_G^2 \cdot \frac{1}{\epsilon \cdot \left(\epsilon + \frac{\rho_L}{\rho_G} \cdot (1-\epsilon)\right)} \approx a_G^2 \cdot \frac{\rho_G}{\rho_L} \cdot \frac{1}{\epsilon \cdot (1-\epsilon)} . \tag{8.65}$$

The final relation holds with the assumption $\rho_G/\rho_L \ll 1$. We note that the velocity of the density wave is considerably smaller than the velocity of sound in the gas and has its smallest value for $\epsilon = 0.5$. This behavior has been confirmed experimentally for air–water flows. Figure 8.25 shows that experiments confirm this behavior well at system pressures that are not too high. The low value of the wave propagation velocity compared to the velocity

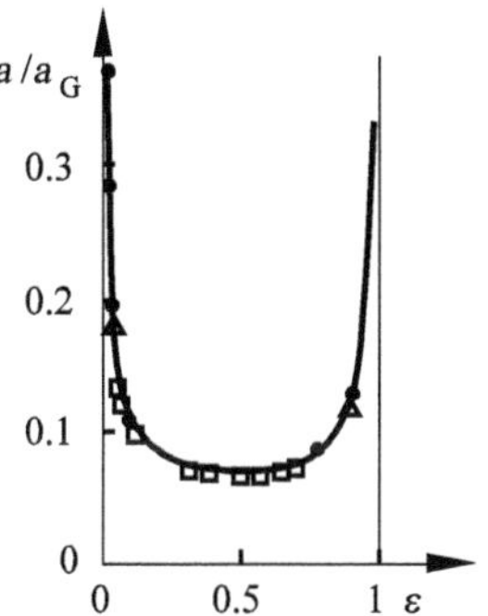

Fig. 8.25. Wave velocity of small perturbations in water–air mixtures, assuming a homogeneous mixture, compared to experiments by *P. von Böckh* (1975)

of sound in the gas is surprising. Its minimum is less than 10% of the velocity of sound in the gas. As a consequence of this and relation (8.62), the mass flow limitation at the geometrically narrowest points of the flow occurs already at very low two-phase mass fluxes. This is technically extraordinarily important with regard to the release of gas–liquid mixtures from pressurized containers.

If we take the momentum density of the separate model as the characteristic density for the two-phase mixture, after a tedious derivation process we obtain a complicated expression for the density wave velocity. However, this is dependent on both variables of state, the vapor quality χ and the volume fraction ϵ. However, since these quantities are coupled together via the velocity ratio, the wave propagation velocity is not only dependent on thermodynamic changes of state, but also on the kinematic quantity $S = u_G/u_L$ and its change with pressure. Thus in general, one gets

$$(a_{2\mathrm{Ph}}^2)_{\mathrm{sep}} = f\left(a_G, a_F, \chi, \left(\frac{\partial \chi}{\partial p}\right)_s, S, \left(\frac{\partial S}{\partial p}\right)_s\right). \tag{8.66}$$

The derivative $(\partial S/\partial p)_s$ expresses the momentum transfer between the phases. Several authors have attempted to develop model relationships for $(\partial S/\partial p)_s$ in bubbly flows with different bubble shapes; see, for example, *R. Henry et al.* (1971). However, these complex models have not endured. The general analytic relations are therefore more likely to be based on the simpler homogeneous density model.

8.4.2 Critical Mass Fluxes

Analytical models to compute critical mass fluxes can be roughly classified according to whether they assume thermodynamic and mechanical equilibrium

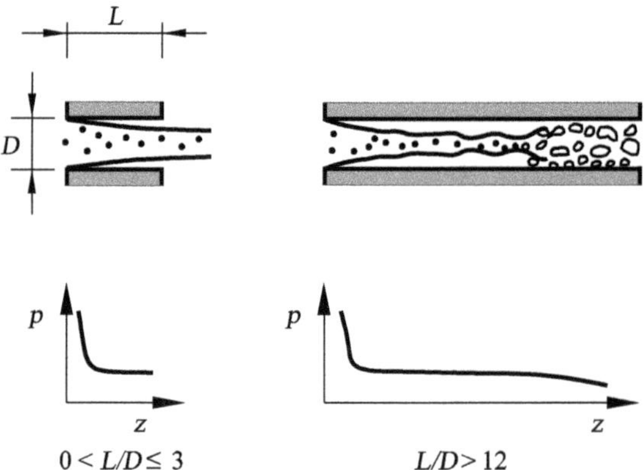

Fig. 8.26. Schematic representation of the outflow of a two-phase mixture from a pressure container, on the left under frozen thermodynamic equilibrium and on the right under complete thermodynamic equilibrium

between the phases. Flow experiments show that in general there is neither thermodynamic nor mechanical equilibrium in a flow process through a pipe. In flashing evaporation, temperature differences form between the phases that do not equilibriate via heat transfer at the interfaces during the short expansion process. Simultaneously, there is a difference velocity between the phases. This may be illustrated by a flow out of a reservoir with low vapor quality ($\chi_0 < 0.05$) through short ($L/D < 3$) or long ($L/D \leq 12$) pipe studs.

Figure 8.26 illustrates the process. In the flow out of a reservoir through a short stud (left sketch), there is in general no thermodynamic equilibrium in the separated free jet, and there is no significant vapor formation in the center of the free jet. This leads to a sharp pressure drop at the pipe inlet with a subsequent pressure plateau in the free jet regime.

Long pipes between pressure reservoirs lead, however, after a certain distance, to reattachment of the free jet to the wall, and independently of this, after a certain relaxation distance, to thermodynamic equilibrium with considerable vapor formation rates. The great increase of the vapor quality in the flow then leads to a significant two-phase friction and acceleration pressure drop. This is shown schematically in the right sketch of Figure 8.26.

The two situations can each be roughly described by a one-dimensional homogeneous two-phase flow that is thermodynamically fully frozen or that is in complete equilibrium. Because thermodynamic effects in the outlet flow can essentially be determined via the evaporation, we briefly outline the thermodynamic changes of state using a real gas equation.

In a $p-(1/\rho)$ diagram for real gases and liquids (see Figure 8.27), the two-phase regime is separated from the liquid state by the so-called boiling-point curve and from the vapor state by the so-called dew-point curve. The boiling-point and dew-point curves meet in the critical point T_K, which limits the

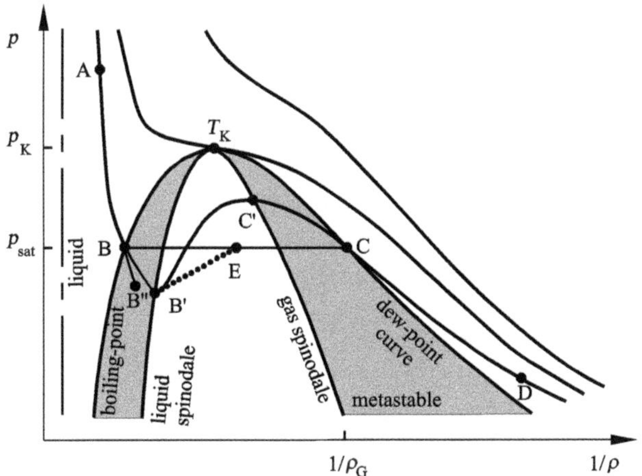

Fig. 8.27. Schematic representation of a state diagram for a real gas. The dotted line shows the apparent change in nonequilibrium. The two-phase regime is shaded.

two-phase regime to higher pressures. The isothermal lines in the two-phase regime typically have a maximum and a minimum. The line that joins all the minima is called the liquid spinodale, while the line that joins all maxima is called the vapor spinodale. When the pressure is reduced, the liquid changes its state from, for example, point A on an isothermal line to the boiling-point curve at point B. There it attains the saturation value of the pressure p_{sat} and the temperature T_{sat}. If thermodynamic equilibrium is guaranteed by small, slow changes of state, evaporation of the liquid phase at constant pressure takes place by increasing the volume of the mixture until the dew-point curve is reached (point C). The straight line that joins these points $\overline{\text{BC}}$ is the equilibrium isothermal line in the two-phase regime. On the other hand, if the pressure reduction is large and sudden, the boiling-point curve is not in thermodynamic equilibrium, and the pressure reduction follows the real isothermal line into the two-phase regime without evaporation occurring. This is a change of state in a metastable or fully frozen thermal equilibrium. An isothermal expansion of the liquid phase can take only place as far as the spinodale, point B′. If it is attained, or almost attained in a real system, the system passes discontinuously through an explosion-like evaporation to an equilibrium state along the two-phase isothermal line, for example, to point E. The thermodynamic nonequilibrium on the isothermal line between the boiling-point curve and the spinodale can be characterized by comparison with the corresponding equilibrium state on the boiling-point curve. This liquid has been overheated by the sudden expansion around the temperature range $T - T_{\text{sat}}$. Overheating typically occurs in boiling processes in liquids with heat supply. Overheating is necessary to activate boiling centers in the formation of vapor bubbles. Therefore, in evaporation processes via pressure reduction or heat supply there are a series of comparable phenomena; for details see the relevant literature (e.g. *J. G. Collier* and *J. R. Thome* (1994), *C. E. Brennen* (1995)). Of course, in nonisothermal expansion with partial evaporation, other nonequilibrium states can be reached in the regime between the boiling curve and the liquid spinodale. However, in reality, it is difficult to control such nonequilibrium transients or to describe them by physical models. This is an area of current research. Therefore, in what follows we discuss the limiting cases. We do mention that in the transition from the vapor phase to the liquid phase in the regime between the dew-point curve and the gas spinodale, namely, via vapor condensation, similar phenomena of thermal nonequilibrium can occur. They are not important in what follows and so will not be discussed.

Both limiting cases of complete equilibrium and complete nonequilibrium are described using a simplification of (8.64) for the critical propagation velocity of a homogeneous two-phase flow. Using the assumption that for the velocities of sound in the vapor and liquid phases we have $a_G^2 \ll a_L^2$, we obtain the following simplified expression for the critical or maximum mass flux:

$$\dot{m}^*_{\text{crit}} = \frac{1}{\sqrt{\left(\frac{\chi}{\rho_G^2 \cdot a_G^2} - \left(\frac{1}{\rho_G} - \frac{1}{\rho_L}\right) \cdot \left(\frac{\partial \chi}{\partial p}\right)_s\right)^*}} \quad . \tag{8.67}$$

The outflow at complete *frozen thermal equilibrium* is characterized by $(\partial\chi/\partial p)_s = 0$. Complete thermal equilibrium is given by the total expression of (8.67). The interrelation of the states at the narrowest flow cross-section to the stagnation values in the pressurized reservoir is obtained by using the equation of state of the corresponding vapor–liquid mixture (e.g. from the steam tables). Comparison with experimental data for short pipes shows that mass fluxes computed according to the *homogeneous equilibrium model* (HEM) are in general far too low, while those values computed according to the *frozen equilibrium model* (FEM) yield better results (see Figure 8.30).

In order to compute systematically the maximum mass flux, we need to integrate the balance equations for the two-phase mixture, taking into account the friction losses from the intake to the narrowest cross-section or to the exit of the connecting channel. Integration of the momentum balance equation for the mixture in (8.22), assuming mechanical equilibrium with $S = 1$ and the additional condition $(\partial\dot{m}/\partial p)_s = 0$, leads to a tabulated or graphical representation of the critical mass flux. Figure 8.28 shows the graphical representation of the critical mass fluxes for the case of thermodynamical and mechanical equilibrium dependent on the reservoir stagnation values for pressure p_0 and enthalpy h_0.

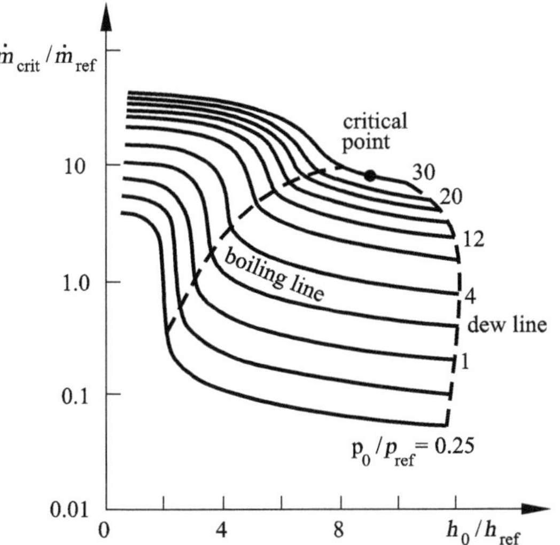

Fig. 8.28. Critical mass fluxes according to the homogeneous equilibrium model (HEM), dependent on the stagnation values,. $p_{\text{ref}} = 689.5\ kN/m^2$, $h_{\text{ref}} = 232.6\ kJ/kg$, $\dot{m}_{\text{ref}} = 4.882\ kg/(m^2 s)$

If we drop the assumption of mechanical equilibrium, a maximum mass flux density can be determined by integrating the momentum or energy balance equation for the mixture ((8.22) and (8.24)) under the additional conditions $(\partial \dot{m}/\partial S)_s = 0$.

This procedure yields relations for the critical mass flux with given velocity ratios, which have the value $S = \sqrt{\rho_L/\rho_G}$ for the integrated momentum balance equation, and the value $S = (\rho_L/\rho_G)^{1/3}$ on integration of the energy balance equation. These simple models for critical two-phase mass fluxes were first derived by *K. Fauske* (1963) and *F. J. Moody* (1965) and presented in the form of diagrams. As an example, the graph computed by *F. J. Moody* is shown in Figure 8.29. Comparison with experiment has shown that the critical mass flux computed by *F. J. Moody* (1965) from an energy balance is considerably higher than the values obtained from experiments. For this reason, the Moody model is frequently used in safety analyses for conservative estimates of leakage rates.

For further illustration, Figure 8.30 shows the model calculations discussed here compared to experimental data obtained with short-outlet pipe studs.

In summary, we have seen that the assumptions made in the simple models concerning thermodynamical and mechanical equilibrium give insufficient accuracy for a quantitative comparison of experimental data and model calculations.

In order to describe the actual processes in two-phase flow through nozzles, apertures, or pipes under high pressures, the local and temporal deviations from thermodynamic equilibrium and the mechanical interaction between the phases have to be taken into account.

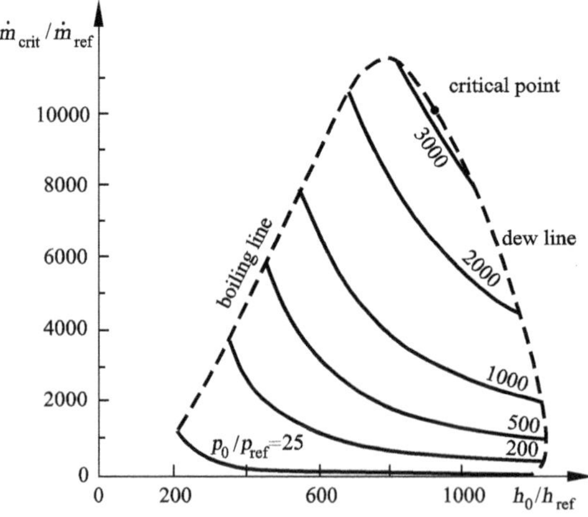

Fig. 8.29. Critical mass fluxes according to *F. J. Moody* (1965), dependent on the stagnation values

Based on the relation for the homogeneous equilibrium model, (8.67), *H. Henry* and *H. Fauske* (1971) developed an empirical nonequilibrium model supported by experiment. In this relationship they replace the equilibrium vapor fraction χ_{eq} by a real vapor fraction χ, which depends nonlinearly on χ_{eq} and on the velocity ratio S. They succeed in finding a function that satisfactorily describes the experimental data in a certain parameter range. This is, however, a more formal adaptation of the relationship (8.67) to the experimental facts.

For a model of nonequilibrium processes that is physically better founded, the full set of steady one-dimensional conservation equations (8.14)–(8.19) of the two-fluid model has to be integrated in principle along the outlet path.

In particular, in modeling the source terms in the mass balance equations, the deviation from thermodynamic equilibrium has to be taken into account. This has still not been achieved satisfactorily for deviations that are very large, as in the case of very strong depressurizations. However, it is generally

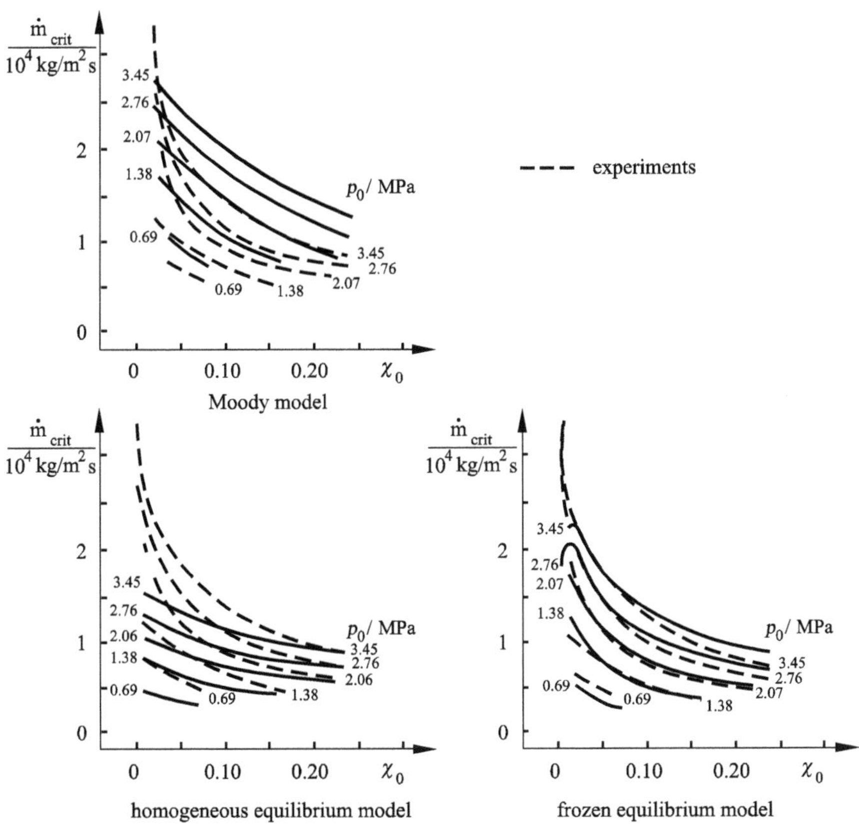

Fig. 8.30. Comparison between model calculations and experimental data according to different models (*G. B. Wallis* (1980))

observed that the phenomena at high pressures are predominantly determined by the effects of the thermodynamic nonequilibrium and less by the changing relative velocity between the phases. In this case, knowledge of vapor nuclei within the fluid that may be activated is of great importance.

In order to consider thermodynamic nonequilibrium in phase transitions quantitatively, *H. Lemonnier* und *Z. Bilicki* (1994), among other authors, have suggested an evolution equation for the actual steam quality χ compared to the thermodynamic equilibrium steam quality χ_{eq} as a supplement to the balance equations and the equation of state for the system. The relationship links the vapor production rate $\mathrm{d}\chi/\mathrm{d}t$ linearly to the deviation of the actual steam quality from its equilibrium value $\chi - \chi_{\mathrm{eq}}$ via a *relaxation time parameter* Θ. The steam quality difference depends directly on the superheating of the liquid. The evolution equation has the form

$$\frac{\mathrm{d}\chi}{\mathrm{d}t} = \frac{\partial \chi}{\partial t} + \frac{\dot{m}_G}{\rho_G} \cdot \frac{\partial \chi}{\partial z} = \frac{\chi - \chi_{\mathrm{eq}}}{\Theta}. \tag{8.68}$$

Starting out from the stagnation states, simultaneous integration of the conservation equations (8.14)–(8.19) and the relaxation differential (8.68) then yields the actual states in the pressure reduction channel. The remaining difficulty now concerns the determination of the relaxation parameter Θ for a specific arrangement and a specific fluid. In principle, Θ embodies the physics of a real homogeneous or heterogeneous vapor formation process. In the absence of any general known physical interrelation for Θ, *H. Lemonnier* and *Z. Bilicki* (1994) take a pragmatic view and determine Θ simultaneously with the calculation of the pressure and the mass flux along the integration path from a prescribed measured gas volume fraction and the liquid overheat. In their model equations they assume mechanical equilibrium, that is, homo-

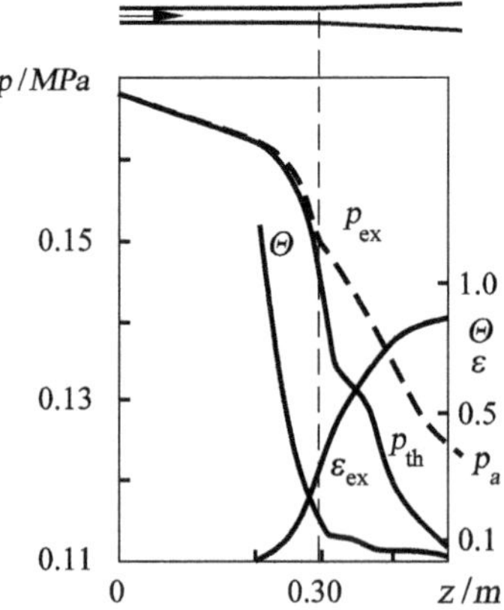

Fig. 8.31. Pressure p and relaxation coefficient Θ as a two-phase mixture passes through a narrow nozzle with supercritical pressure difference. For example $\dot{m} = 6526\ kg/(m^2 s)$, $p_{\mathrm{a}} = 0.123\ MPa$, p_{ex} measured pressure, p_{th} calculated pressure

geneous flow conditions. Compared to simple analytical models, they find good agreement with the measured pressure distribution in narrow expansion nozzles (cf. Figure 8.31). The calculation of Θ offers a new approach to the understanding of fundamental nonequilibrium processes. Independently of the particular requirements in the physical modeling, in the numerical integration of the differential equations there are difficulties related to the singular behavior at the narrowest cross-section. Further details are given by *H. Lemonnier* and *Z. Bilicki* (1994).

8.4.3 Cavitation

In liquid flows at high velocity, a pressure drop in the flow past bodies and corners can lead to locally bounded vapor or gas formation. This phenomenon is called cavitation. It is a locally bounded flashing vaporization with subsequent condensation, or otherwise the release of dissolved foreign gases from the liquid by pressure reduction. It occurs occasionally in hydraulic flow machinery such as pumps and turbines and in other hydraulic components like valves or injection nozzles of combustion engines. Undesirable side effects such as deterioration of operation control, noise development, mechanical oscillations and local material wear and tear are observed. Avoidance and control of cavitation is therefore of great importance in hydraulic engineering. Cavitation processes have been investigated intensively for many years with the aim of deriving criteria for its onset and extent. Many articles and books summarize the area of cavitation, such as those by *T. Knapp et. al.* (1970), *A. J. Acosta* and *B. R. Parkin* (1975), *C. E. Brennen* (1995), *Y. Lecoffre* (1999), *J. P. Franc* and *J. M. Michel* (2004).

In single-component flows, cavitation may occur if the local static pressure in the flow reaches and falls below the thermodynamic saturation pressure p_{sat} of the fluid. This necessary condition for the onset of cavitation is characterized in an inviscid, incompressible flow by a dimensionless *cavitation number* σ. This characteristic number is defined as

$$\sigma = 2 \cdot \frac{(p - p_{\text{sat}})}{\rho \cdot u^2} \quad , \tag{8.69}$$

where ρ is the density and u the local velocity. For values of $\sigma \leq 0$, vaporization of the fluid can occur. Because of mechanical and thermodynamic real effects, the "ideal" critical value of the cavitation number $\sigma_i = 0$ for the onset of cavitation can shift to positive or negative values. The relevant quantities that influence such deviations will be discussed in what follows.

The types of cavitation observed are as diverse as the flow patterns in two-phase flows in pipes (see Section 8.2). They are always highly unsteady. They may be classified according to increasing vapor content, as cloud cavitation, bubble cavitation, stratified cavitation, and supercavitation. Collections of small bubbles in the micron and submicron region, which can form in greatly sheared boundary layers, are called cloud cavitation. Extended bubble crowds develop in the low-pressure regimes of separation flows behind body edges, on the suction side of airfoils at an angle of attack, and in regions of separated flows. This is then called bubble cavitation. The formation of coherent vapor or gas films occurs in some parts of an airfoil or the edge of a body when bubbles coalesce, if the vapor or gas volume fraction exceeds a critical amount of $\epsilon \sim 0.5$. This form is called stratified cavitation. If the body, at very high flow velocities and low local pressures, is covered by a vapor film on all or part of its contour surface, this is called supercavitation. Such extreme conditions are occasionally attained on the propellers of speedboats. Figure 8.32 shows these cavitation forms, in experiments by *J. P. Franc* and *J. M. Michel* (1985) on an NACA-foil in a water channel.

These investigations have shown that, as well as the cavitation coefficient σ already introduced, the following properties have considerable influence on

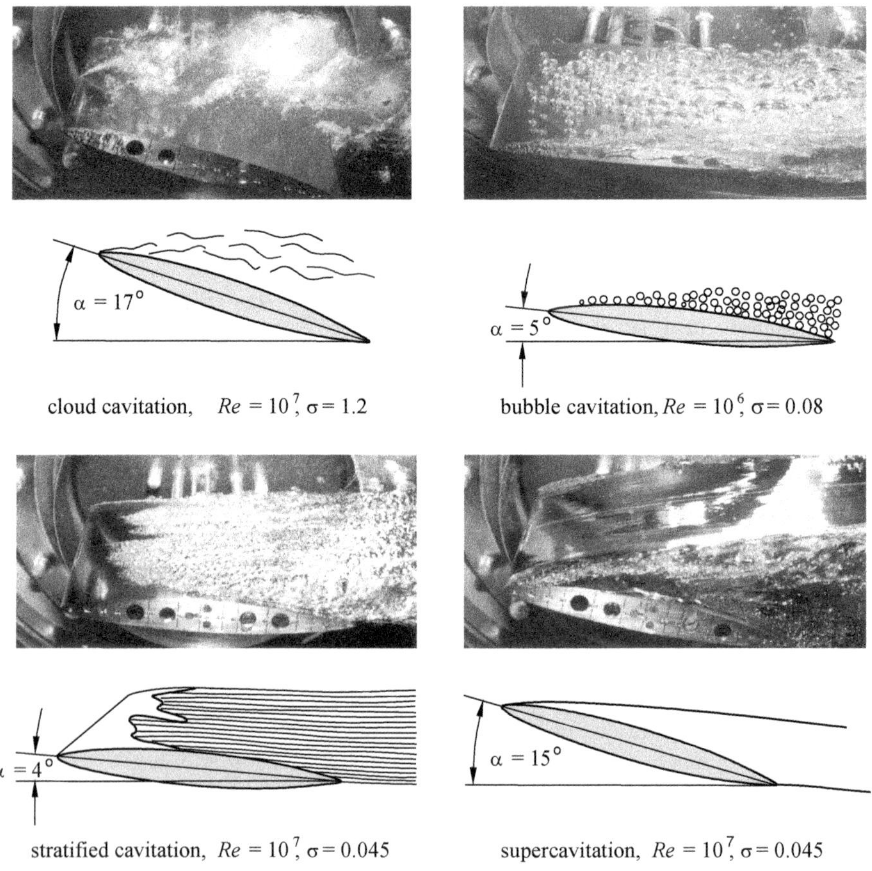

Fig. 8.32. Different cavitation patterns on an airfoil of type NACA 16012 at an angle of attack in a water channel, according to *J. P. Franc* and *J. M. Michel* (1985)

the cavitation: the shape of the body, such as its degree of slenderness and angle of attack, as well as fluid properties such as viscosity, surface stress, parameters of the real gas equations, thermal conductivity, heat capacity, latent heat, and concentration of foreign material in the fluid in the form of foreign gases or particles.

The onset and the dynamic behavior of the cavitation process are also greatly affected by thermal microprocesses in the activation of the nucleation centers, and by the degree of turbulence of the flow.

The effects of the different parameters can in principle be described in the form of dimensionless fluid-mechanical and thermodynamic characteristic numbers, such as the Reynolds, Weber, and Stefan numbers. In hydraulic engineering, attempts have been made to develop a relationship that is as simple as possible between the cavitation coefficient and a normalized volumetric flux. In shipbuilding, relations between cavitation coefficient, angle of attack, propulsion, and drag coefficient have been derived for certain classes of hydrofoils and propellers. Experimental investigations show, however, that the metrological determination of these correlations depends greatly on the quality of the test liquid. The quality of the liquid is characterized by the concentration of dissolved foreign substances, and the concentration and size distribution of finely distributed undissolved foreign particles, since these determine the tensile load limit of a fluid without vapor formation and thus the onset of cavitation. Therefore, in recent cavitation experiments, the effect of the water quality is taken into account by a specified addition of gas or solid particles to the test liquid.

Coherent structures in turbulent flow may significantly promote the generation of individual vapor bubbles and, moreover, lead to the formation of bubble collectives. Bubble growth and collapse is influenced by pressure fluctuations associated with small scale high intensity vortices. Bubble accumulation occurs preferentially in the core region of vortices of major extent where even coherent vapor tubes may be formed by bubble coalescence.

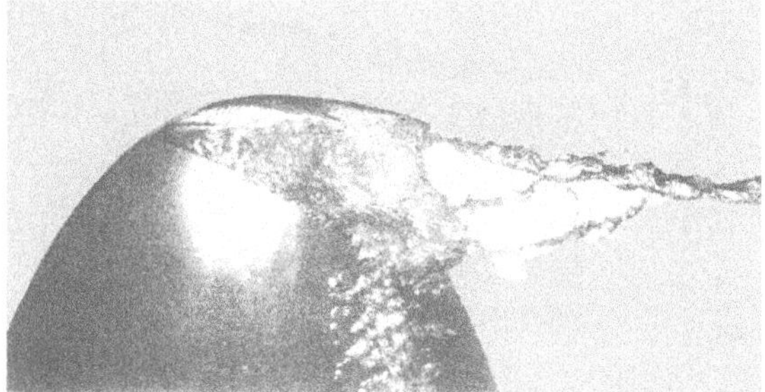

Fig. 8.33. Free cavitation tube at the tip of a wing, ending in vapor intake from a wing-cavitation layer (*R. E. A. Arndt* and *V. H. Arakeri* (1991))

Figure 8.33 demonstrates such a process in form of a *wing tip caviation tube*. The same mechanism may also explain the observed initiation of cavitations and formation of cavitation clouds within boundary layers and free shear layers, contrasting the expectations that cavitation originates from rigid boundaries of fluid flow domains.

Unsatisfactory results in the description of cavitation by power law correlations between simple flow characteristic numbers have recently increased efforts for the mathematical–physical modeling of the two-phase processes. The same basic physical concepts are used as for the computation of critical mass fluxes. Cavitation generally occurs in bounded regions of the flow, and therefore, two- or three-dimensional computations are certainly necessary. A starting point is given by the general equations for two-phase flows in Section 5.4.6. For simplification, a homogeneous two-phase flow is frequently assumed.

In order to compute the vapor volume fraction ϵ, the balance equation for the vapor phase, (8.14), with a source term for evaporation and condensation, is used. The central point of this modeling is the representation of the source term. For a given distribution of nucleation centers in the fluid, the space term can be computed from the growth of individual bubbles in the pressure field of the homogeneous two-phase flow. This can be performed using the Rayleigh–Plesset equation (5.89) of Chapter 5, or other similar descriptions of individual bubble dynamics. One method has been presented by *Y. Chen* and *S. D. Heister* (1994), and independently by *J. Sauer* and *G. H. Schnerr* (2000). Figure 8.34 shows an example of a computation of cavitation in an injection nozzle.

In spite of remarkable advances in the numerical computation of cavitation processes in channels, there are several significant effects, such as mechanical nonequilibrium between the phases, turbulence effects in the homogeneous phase, and thermodynamic nonequilibrium during the formation

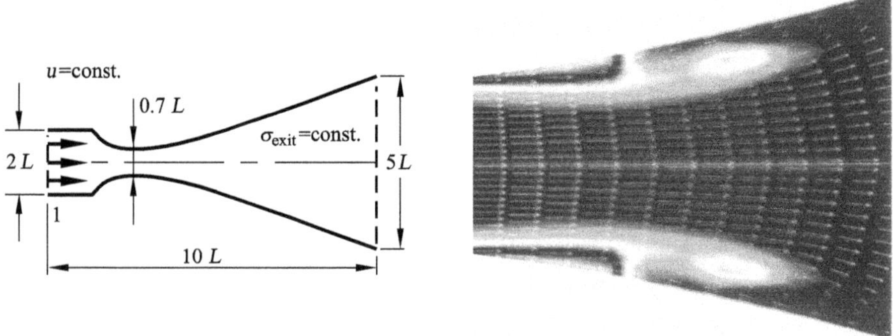

Fig. 8.34. Numerically computed cavitation regime in the contraction region of a nozzle, according to *J. Sauer* and *G. H. Schnerr* (2000)

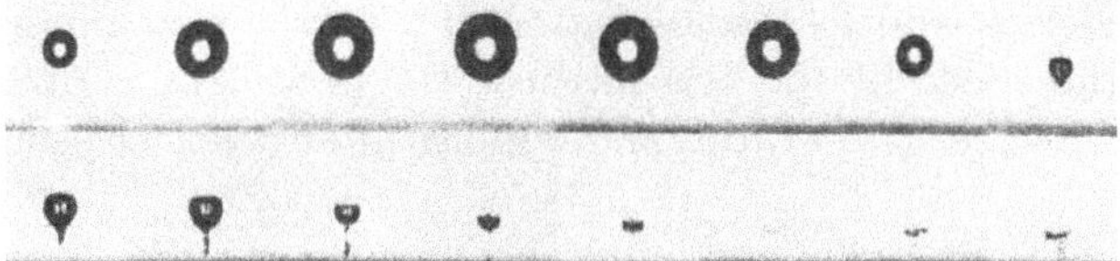

experiment computed bubble contours

Fig. 8.35. Collapse of a vapor bubble close to a solid wall in the phase of jet formation toward the wall: left, experiment; right, computed sequence of bubble contours by *J. R. Blake et al.* (1986)

of bubbles and their collapse by condensation, that have not yet been described adequately, since relevant physical models are lacking.

In contrast, the modeling of cavitation in the form of the growth and collapse of individual vapor bubbles is presently well understood. Experimental investigations (such as those by *W. Lauterborn* and *H. Bolle* (1975)) show that cavitation bubbles close to walls collapse asymmetrically with the formation of a high-velocity liquid jet directed toward the wall (Figure 8.35) associated with an intensive pressure wave. It has been shown experimentally by *A. Phillip* and *W. Lauterborn* (1998) that wall material damage is caused by high-frequency pressure shock waves and high-velocity jets. The computation of this process was first performed by *S. Plesset* and *R. B. Chapman* (1971). Their calculations were later completed by, among others, *J. R. Blake et. al.* (1986). Figure 8.35 shows the time development of a bubble collapse with bubble contour lines.

8.5 Instabilities in Two-Phase Flows

Two-phase flows may occur in different patterns in pipe and channel flows (see Section 8.1). Each of these patterns exists in a certain range of the control parameters such as mass fluxes of the phases and volume fraction. On variation of the control parameters there are transitions between these patterns. The transitions are frequently triggered by flow instabilities at the interfaces. They are essentially wave instabilities caused by the effect of the relative velocity between the phases, surface tension, and acceleration forces. They are known as *Kelvin–Helmoltz* and *Rayleigh–Taylor instabilites.* These instabilities also play a central role in the deformation and disintegration of bubbles, drops, liquid lamellae, and liquid films. Figure 8.36 shows the situation. If the gas and the liquid move with velocity $u_G - u_L$ relative to one another at a smooth interface, small wavelike perturbations are amplified in time under the effect of surface tension and acceleration forces, such as gravity, acting from outside on the interface. The rate of amplification generally depends on the wavelength of the perturbation. The perturbation wavelength λ_m with the largest amplification rate leads to the decay of the smooth interface, such as that of gas and liquid jets, and to the formation of a new spatial phase distribution that is characterized by the new intrinsic length scale λ_m. This length scale also determines the size of drops and bubbles after the transition process.

A linear stability analysis for an inviscid infinitely extended two-layer flow yields an amplification rate c_i of the form

$$c_i = \frac{1}{a} \cdot \sqrt{\frac{\rho_1 \cdot \rho_2 \cdot (u_1 - u_2)^2 \cdot a^2}{\rho_1 + \rho_2} - \frac{\sigma \cdot a^2 - g \cdot (\rho_1 - \rho_2) \cdot a}{\rho_1 + \rho_2}}. \tag{8.70}$$

Here u_1 and u_2 are the velocities of each phase, a is the wave number defined as $a = 2\pi/\lambda$, and the other quantities have the same meanings as introduced earlier. The acceleration is the acceleration due to gravity, although any other acceleration with corresponding effect could take its place. The maximum of the amplification rate as a function of the wave number can be determined from the condition $\partial c_i / \partial a = 0$. For the following discussion, the wavelength $\lambda = 2 \cdot \pi / a$ is introduced. For the case in which there is no velocity difference between the phases and $u_1 - u_2 = 0$, the maximally amplified perturbation has wavelength

$$\lambda_m = 2 \cdot \pi \cdot \sqrt{\frac{3 \cdot \sigma}{g \cdot (\rho_1 - \rho_2)}}. \tag{8.71}$$

Here it is assumed that the denser fluid is accelerated in the direction of the less-dense fluid. Otherwise, only a damped oscillation will occur. This is called the *Rayleigh–Taylor instability.* The critical wavelength for amplified perturbations is given by a vanishing amplification rate c_i:

$$\lambda_c = 2 \cdot \pi \cdot \sqrt{\frac{\sigma}{g \cdot (\rho_1 - \rho_2)}}. \tag{8.72}$$

From this we conclude that the interface is stable to small perturbations with small wavelength and experiences no lasting deformation, since the surface tension is in equilibrium with the acceleration forces. For perturbations with larger wavelengths, the interface will be deformed permanently. A visible example of this interface instability is the prevention of an outflow of liquids from containers with small enough apertures at the bottom, or the decay-free rise of gas bubbles with diameter $d_B \leq \lambda_c$ in a liquid.

If the acceleration is not strong, but nevertheless significant velocity differences $u_1 - u_2$ occur between the phases and surface tension is present, the amplified perturbations of the wavelength λ are bounded from below by the critical wavelength

Fig. 8.36. Stability of an interface between two layered fluids, velocity of the phases u_1, u_2

$$\lambda_c = 2 \cdot \pi \cdot \left(\frac{\sigma}{\rho_1 \cdot (u_1 - u_2)^2} \right). \tag{8.73}$$

Here it is assumed that the densities of the phases are very different at normal conditions, i.e. $\rho_2 \ll \rho_1$, as in the case of gas and liquid. The perturbation with the greatest amplification has wavelength $\lambda_m = 1.5 \cdot \lambda_c$.

For two-phase flows, the consequence of this capillary instability is that the shear velocities at the interface cause undamped capillary waves that lead to the decay of the interface. For example, large drops break up into smaller components if the drop diameter is larger than the critical wavelength λ_c. This instability of interfaces to shearing motion is a *Kelvin–Helmholtz instability.* It substantiates the well-known Weber number criterion for the decay of liquid jets and drops (see Section 8.2.4). This empirical criterion claims that a decay of moving volumes of liquid with free surfaces occurs if the *Weber number* We_L formed with the characteristic length L of the drop volume exceeds the value one:

$$\mathrm{We}_L = \frac{\rho_1 \cdot (u_1 - u_2)^2 \cdot L}{\sigma} > 1. \tag{8.74}$$

The phases 1 and 2 are identified with a gas and a liquid, respectively.

Both instabilities introduced here have a great effect on the bubble and drop formation in two-phase flows and explain several significant phenomena in boiling and condensation processes. For example, the breakdown of nucleate boiling and film boiling are due to the Rayleigh–Taylor instability. The formation of wavy stratified flows and of homogeneous bubbly and droplet flows is directly related to the Kelvin–Helmholtz instability, and its initiation can be described by using a Weber number criterion.

The simple representation of the instabilities discussed here can be completed by taking further effects into account, such as the viscosity of the phases and geometric dimensions of the containers and channels. This can be looked up in detail in the textbooks by *C. H. Yih* (1980) or *S. Chandrasekhar* (1968).

Apart from the small-scale interface instabilities, which are significant for the phase distribution, there are further fundamental large-scale instability mechanisms that determine the temporal behavior of the two-phase flow in hydraulic systems with phase transitions. Since such instabilities can lead to uncontrolled mechanical pressure and shock loads and moreover to thermally induced stresses in the channel walls of the system, the stability bounds of such processes represent practical design and operation criteria for such systems. These include chemical reactors, nuclear steam generators, refrigerating sets, fluid flow engines etc.

A typical two-phase instability can occur in a system that consists of two pressure reservoirs and two hydraulic components arranged in series between them, namely, a centrifugal pump and an evaporator tube with constant heat supply. The system is shown in Figure 8.37.

The stability behavior of this system is determined by the different pressure to mass-flux dependencies of the pump and evaporator tube. Whereas the pressure head of a centrifugal pump typically drops monotonically with increasing mass-flux, the two-phase pressure drop in the evaporator tube typically shows nonmonotonic cubic behavior in its dependence on the mean mass-flux. The nonmonotonic behavior is essentially due to the different contributions of the frictional and accelerational pressure drop to the overall pressure loss of the two-phase flow in the evaporator tube. The monotonic branches of the pressure loss curve are determined by the high vapor fraction at low mass-fluxes, and by the high liquid volume fraction at high mass-fluxes. Therefore, in general, the system can assume three steady states of operation as the mass-flux is varied. They are given by the intersecting points P, P′, P″ of the two pressure–mass-flux curves for the centrifugal pump and the evaporator tube. Assuming a small variation of the quantities Δp and $\dot{m}$ in a neighborhood of the steady states, it can easily be shown that the states P′ and P″ are stable, whereas a pressure and mass-flux variation close to P results in a change in the power of the pump and thence to a change in pressure loss with exactly opposite sign. This leads to a transition to one of the two stable states of operation P′ or P″.

The simple approaches show that the condition $\partial \Delta p / \partial \dot{m} > 0$ is necessary for stability of an operating point, where Δp is the driving system pressure. This temporal behavior is caused by the instability and is characterized by a simple transient from an unstable to a steady stable state; hence it is called statically stable or unstable. *M. Ledinegg* (1938) first investigated this *instability*, and it is named after him. We do not present a precise analytical model for this instability at this point, but instead refer to the literature (*G. Yadigaroglou* (1981, 2006), *M. Ozawa* (1999)).

The static behavior of the system may lead to dynamic, i.e. oscillatory, behavior if a pressurizer is added to the component chain consisting of reservoirs, pump and evaporator tube (see the dashed component in Figure 8.37). When the two-phase flow pressure loss decreases, the pressurizer temporarily stores the excess pump power as compression energy and passes it back to

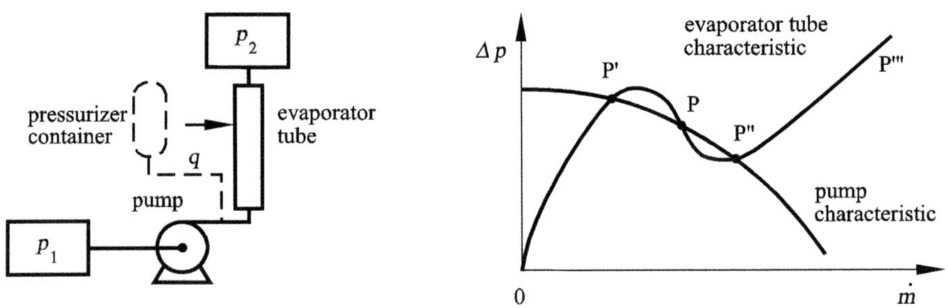

Fig. 8.37. A two-phase system with possible instability. Left: flow system; right: pressure–mass-flux diagram for the radial pump and evaporator tube

the system after a time offset. The buffer effect of the pressurizer means that the characteristic line of the driving pressure in the diagram in Figure 8.37 oscillates in time. This leads to a displacement of the points of intersection P', P" on the characteristic line of operation. In the case of strong oscillations, such a system can in principle carry out a relaxation oscillation between all three original static states. This is undesirable for the operation of a technical system. Such operative fluctuations are prevented by the addition of throttling components, in the form of screens or other flow constrictions.

A typical two-phase instability driven by pressure dependent evaporation processes is found in nature in geysers. Heat is supplied to a cavern filled with supercooled water that is connected to a higher-lying water reservoir via a narrow channel. The water heats to saturation temperature and begins to boil. The steam formed leaves the cavern in the form of a two-phase mixture via the connecting channel. Because of the increasing steam content, the hydrostatic pressure in the cavern is reduced. This leads to further amplification of the evaporation and to further water discharge out of the cavern. As the water deficit increases, the steam development is reduced and supercooled water can penetrate into the cavern from the upper reservoir against a reduced flow of steam. This completely cuts off the boiling process. The continuing heat supply to the cavern again heats to saturation temperature the water that has penetrated and the process repeats itself. If the external conditions, the heat supply to the cavern and the heat removal from the reservoir remain the same, a periodic process with a typical relaxation character is set up. Similar phenomena can also occur in technical steam generators if certain operative failures occur (cf. *M. Ozawa* (1999)).

Similar flow instabilities also occur if steam is introduced to supercooled water reservoirs via pipe chambers. The instabilities are generated by the condensation of large bubbles of steam in the reservoir and in the entry

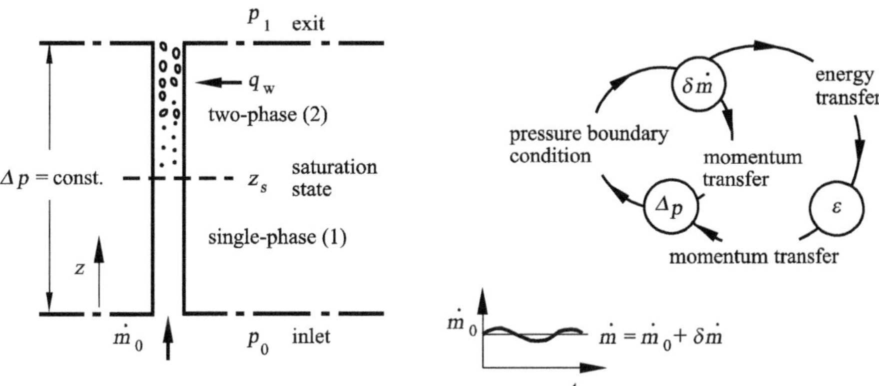

Fig. 8.38. Left: sketch of the formation of the *density-wave instability*; right: feedback effects

chambers, and can lead to unacceptable material strain on the pipe and container walls.

Another system-covering two-phase instability that can be observed in the form of oscillations of the gas volume fraction is the so-called *density wave instability*. This instability occurs frequently in systems with phase transitions and is based on a feedback mechanism between mass flux, vapor formation or condensation rate, and pressure drop in the boiling or condensation region. This instability can be analytically described for a system with constant heat supply or removal using the one-dimensional two-fluid model (8.14)–(8.18) and a classical linear stability analysis (see *G. Yadigaroglou* (1981)). Here we do not present the analytical model, but describe the essential mechanism using a simple evaporator tube connecting two pressure reservoirs with constant pressures p_0 and p_1. Figure 8.38 shows a sketch of the situation.

Adding a small periodic perturbation to the mass flux $\partial \dot{m}$ at the inlet of the evaporator tube, the location of the saturation temperature and therefore the location of the onset of evaporation inside the tube will follow the fluctuation, since mass-flux oscillations in single-phase flow regimes include enthalpy oscillations. Changes in the mass flux and changes in the length of the single-phase flow section in the pipe cause pressure fluctuations $\partial \Delta p_1$ in the single-phase flow regime. In the two-phase regime, an enthalpy perturbation acts as a perturbation of the vapor volume fraction ϵ, which moves in the direction of flow as a density wave. The change in the vapor content, together with the mass-flux and length perturbations, leads to an increased pressure perturbation $\partial \Delta p_2$ in the two-phase regime. Since the total pressure difference of the system acting on the evaporation tube is constant, the individual pressure fluctuations $\partial \Delta p_1$ and $\partial \Delta p_2$ have to cancel each other. This implies a feedback between the two-phase and single-phase regimes, which, for a suitable phase relation between the perturbations, leads to a resonant amplification of the small initial perturbation. The consequence is massive oscillations that affect the whole boiling regime and in particular, change the vapor fraction in the evaporator tube. In technical systems this must be prevented in order to avoid uncontrolled thermal stresses at the heated wall. Therefore, the prediction of limits for the *density-wave instability* is important in the design of tube-type steam generators. The tendency to unstable behavior increases if several evaporator tubes are arranged in parallel.

A similar instability may be observed in a cavitating radial fluid flow engine. Under certain conditions, local cavitation regimes in the blade channels of the rotor move from one channel to the next at a particular frequency. The situation is analogous to the system of parallel evaporator tubes. In the radial machine there is a certain pressure difference between the inlet and exit of the blade channel, determined by the rotation rate. The phase transition in the blade channel takes place as explained in Section 8.4.3, at the position where the hydrodynamic pressure drops below the evaporation pressure, i.e. the saturation vapor pressure of the liquid. Small perturbations of the mass flux in the individual blade channels can therefore cause the same resonant feedback mechanism as in evaporator tubes, and lead to inadmissible oscillations and a deterioration of the efficiency of the fluid flow engine.

8.6 Turbulence in Dispersed Two-Phase Flows

8.6.1 General Aspects

A cruical issue for the analysis of dispersed two-phase flow in technical systems is the spatio-temporal distribution of particles such as bubbles, drops and solid granules in turbulent carrier flows. Experiments have shown that even for the simple situation of two-phase flows in vertical ducts distinctly different particle distributions occur depending on the overall flow direction and the phase density ratio ρ_p/ρ_c. Two typical examples from measurements of *T. L. Liu* and *S. G. Bankoff* (1992) and *X. Sun et. al.* (2004) are shown in Figures 8.39 and 8.40. For instance, for turbulent upward bubble flow a concentration peak of the volume fraction ϵ_p is observed in the vicinity of the duct walls, while for downward flow the maximum of the particle concentration is generally located in the pipe center.

Observations and measurements indicate that bubbles do not enter the viscous wall region in turbulent pipe flow. *E. Moursali et. al.* (1995) have observed this behavior also in turbulent bubbly flow along vertical flat plates. Furthermore, non-uniformities in the distribution of particles have been realized in the free shear layers of two-phase flow jets occuring in spray or aerating systems. A typical vortex induced structural phenomenon in a dilute particle jet flow is demonstrated in Figure 8.41 according to investigations of *E. K. Longmire* and *J. K. Eaton* (1992).

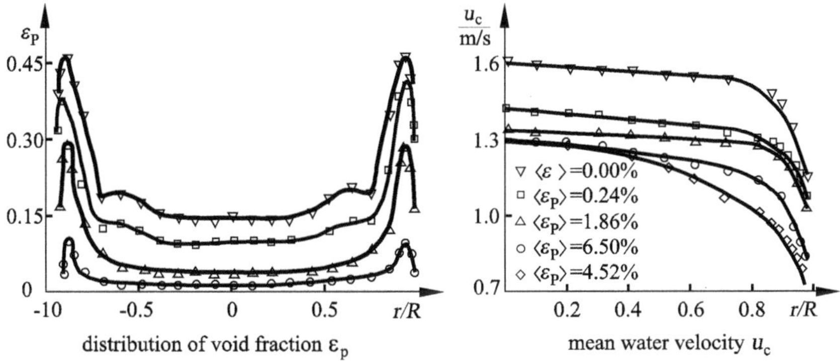

Fig. 8.39. Upward bubbly air-water flow in a vertical pipe after *T. J. Liu* and *S. G. Bankoff* (1993), pipe diameter $D = 38\ mm$, bubble diameter $d_b \leq 5\ mm$, volumetric water flux $\bar{u}_c = 1.087\ m/s$, volumetric gas flux $\bar{u}_p = 0.027, 0.112, 0.230, 0.347\ m/s$

The vortex structures in the gas jet are visualized by smoke tracers. Obviously there are particle accumulations in zones of low vorticity between vortex centers. Characteristic features are summarized in the schematic sketch of Figure 8.41.

Preferential concentration and rarefaction of particles within specific regions of a homogenous turbulent flow field is a general feature of high Reynolds-number dilute particle flows. These phenomena have also been realized and predicted by direct numerical simulations using the model equations for dilute two-phase flow as presented in Section 5.4.6 (see *J. D. Kulick et. al.* (1994), *K. D. Squires* and *J. D. Eaton* (1990)).

The outlined phenomena can be explained by forces acting on an individual particle moving in a fluid continuum. Under the influence of centrifugal, Coriolis, lift and body forces the particle trajectories generally cross the path and streamlines lines of the carrier flow. This is sketched for an axisymmetric rotation in Figure 8.42. Particles heavier than the carrier fluid ($\rho_p/\rho_c > 1$) are removed from the vortex center by centrifugal and Coriolis forces and, as a consequence, lead to a local thinning of particles. In a system of several vortices the particles will accumulate near flow stagnation points or lines, regions with high shear and moderate vorticity. This situation is sketched for a two-dimensional stagnation flow in Figure 8.42. For particle slighter than the carrier fluid ($\rho_p/\rho_c < 1$) the situation is reversed. Bubbles accumulate near vortex centers and escape from stagnation zones. Regions of preferential concentration thus originate from vortex formation and interaction of vortices in the continuous phase.

In one-dimensional pipe flow mainly gravity and lift forces are responsible for non-uniformities of the particle distribution (see 5.4.6). The elevated volume fraction ϵ_p near the wall in upward vertical pipe flow and the increased level of ϵ_p in the center of the pipe in downward flow can indeed be explained

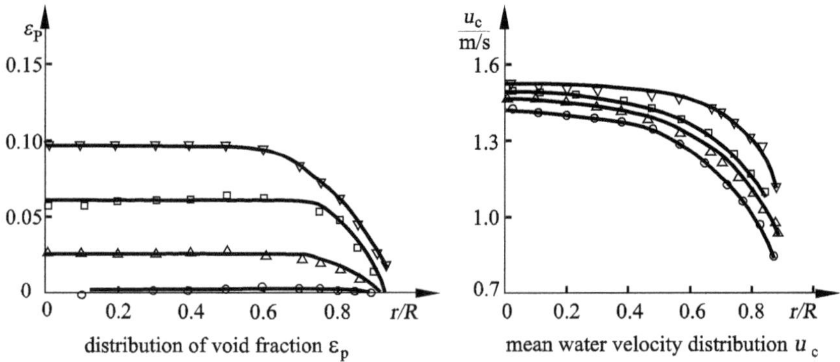

Fig. 8.40. Downward bubbly air-water flow in a vertical pipe after *X. Sun et al.* (2004), pipe diameter $D = 25.4\ mm$, water flux velocity $\bar{u}_c = 1.25\ m/s$, air flux velocity $\bar{u}_p = 0.02; 0.10; 0.29\ m/s$

by the acting lift force. According to (5.111) this force is proportional to the relative velocity $\boldsymbol{v}_{\mathrm{p}} - \boldsymbol{u}_{\mathrm{c}}$ between particle and the carrier phase and to the local vorticity of the carrier flow expressed by the transversal gradient of the mean velocity $\omega_{\mathrm{c}} = -\partial \boldsymbol{u}_{\mathrm{c}}/\partial \mathrm{n}$ with n as a coordinate perpendicular to the direction of flow. For particles lighter than the carrier fluid with $\rho_{\mathrm{p}}/\rho_{\mathrm{c}} < 1$ and for upward bubbly flow $\boldsymbol{v}_{\mathrm{p}} - \boldsymbol{u}_{\mathrm{c}} > 0$ holds because of buoyancy forces. Thus, the lift force is directed away from the center and towards the wall, resulting in an accumulation of bubbles near the pipe wall. The direction of the hydrodynamic lift force is reversed for downward bubbly flow with $\boldsymbol{v}_{\mathrm{p}} - \boldsymbol{u}_{\mathrm{c}} < 0$, as buoyancy effects reduce the particle velocity $\boldsymbol{v}_{\mathrm{p}}$. As a consequence, the bubbles migrate towards the pipe center forming an enhanced concentration at this location. For particles heavier than the carrier fluid, i. e. $\rho_{\mathrm{p}}/\rho_{\mathrm{c}} > 1$ the scenario is reversed with concentration peaks of particles near walls for downward flow and accumulations near the center for upward flow.

There is yet another vorticity related force that inhibits particles from touching the wall. When rising bubbles approach the wall, the flow around the individual bubble becomes non-uniform. Due to the non-slip condition at the wall the drainage rate on the wall side of the bubble is much smaller than on the opposite side. This gives rise to an asymmetric relative incident flow to the bubble creating a lift force away from the wall. Utilizing analytical considerations together with numerical simulations *S. P. Antal et. al.* (1991) derived the repulsive wall force on the particle in the form:

$$F_{\mathrm{p}}^{\mathrm{c}} = \epsilon_{\mathrm{p}} \cdot \rho_{\mathrm{c}} \cdot \frac{2}{d_{\mathrm{p}}} \cdot (\boldsymbol{v}_{\mathrm{p}} - \boldsymbol{u}_{\mathrm{c}})^2 \cdot \left(c_{\mathrm{w1}} + c_{\mathrm{w2}} \cdot \frac{d_{\mathrm{p}}}{2 \cdot y} \right) \quad . \tag{8.75}$$

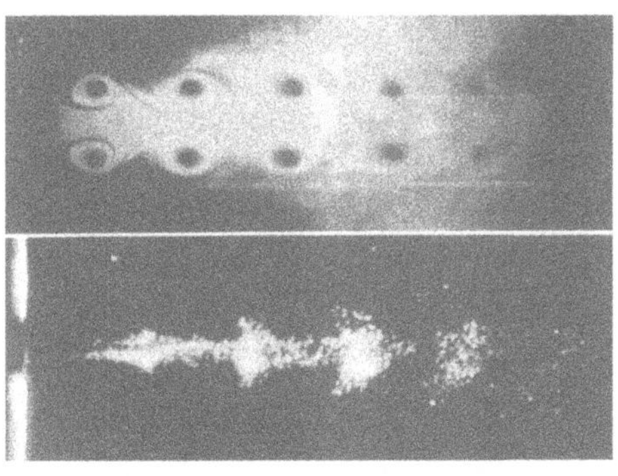

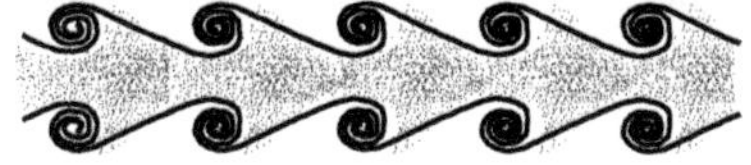

Fig. 8.41. Above: Flow visualization of a single phase pulsed gas-jet by smoke tracers at $Re_{\mathrm{c}} = 23000$; middle: visualization of the dilute concentration of glass particles carried by the jet for the same carrier gas flow rate, particle diameter $d_{\mathrm{p}} = 0.55\ \mu m$; below: schematic sketch of the phenomena, after *E. K. Longmire* and *J. K. Eaton* (1992)

Here y is the distance from the wall, $\boldsymbol{v}_\mathrm{p} - \boldsymbol{u}_\mathrm{c}$ the relative velocity parallel to the wall and $c_{\mathrm{w}1}$ and $c_{\mathrm{w}2}$ are coefficients obtained from numerical adjustments to experimental observations with $c_{\mathrm{w}1} = -0.104 - 0.06 \cdot |\boldsymbol{v}_\mathrm{p} - \boldsymbol{u}_\mathrm{c}|$ and $c_{\mathrm{w}2} = 0.147$. The effect of wall repulsive forces is evident in Figure 8.39 as the measured volume fraction decreases steeply towards zero close to the pipe wall. It should be mentioned here that any deformation of the particle by acting fluid dynamic forces feeds back on the quantity of the particle drag and lift force and has to be taken into account if necessary (see Section 8.2.4).

There is the obvious question of what is the kinematic and dynamic structure of a dispersed two-phase flow in the proximity of walls. Measurements of the wall shear stress and velocity of the continuous phase have shown that similar to single-phase fully developed boundary-layer flow three characteristic zones exist: a viscous sublayer, an intermediate inertial layer with a logarithmic velocity distribution and an outer region with wake flow character, as described in Section 4.2.5. This can be seen in Figure 8.43. The graph shows the dependence of measured wall-parallel mean velocities in bubbly flows along a vertical plate on the wall distance, displayed in standard variables $U^* = u/v^*$ and $y^* = v^* \cdot y/\nu_\mathrm{c}$ with $v^* = \tau_\mathrm{w}/\rho_\mathrm{c}$ the shear-stress-velocity scale, ν_c and ρ_c the kinematic viscosity and density of the continuous phase respectively (for details see *J. L. Marie et. al.* (1997)).

It is obvious from Figure 8.43 that parameters in a generalized logarithmic law of the wall for dispersed particle flow should depend on the particle fraction and also take into account body forces such as gravity. *J. L. Marie et. al.* (1997) have derived a generalized law of the wall for bubbly flow from their measured data.

It was outlined already in Section 5.4.6 by Figure 5.16 that particle motion and turbulent fluctuations in the continuous phase may interact in a complex way. In dispersed two-phase flows the dynamics of particles may either surpress or enhance turbulence at all turbulence length scales (wavelengths) of energetic vortices. They even may do this selectively in a limited

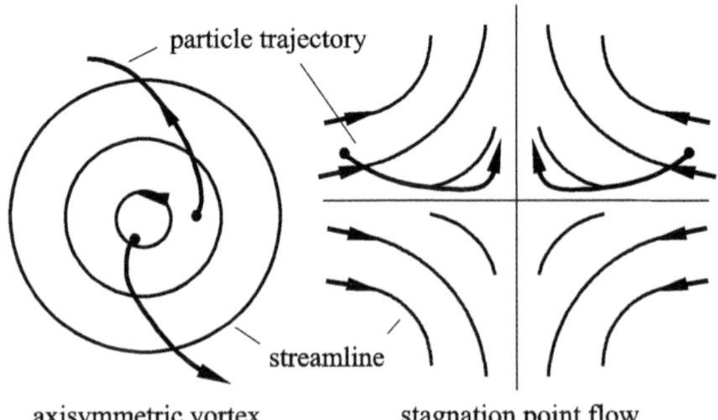

Fig. 8.42. Particle trajectories in simple two-dimensional flows, after *K. D. Squires* and *K. J. Eaton* (1990)

range of wavelengths. It has been repeatedly observed in experiments under various turbulent flow conditions that collections of small particles extract energy from large energetic vortices and transfer this dispersively to smaller turbulence vortices which finally dissipate all energy by friction. This becomes evident from evaluated power density spectra of the fluid velocity of turbulent bubbly flow. Figure 8.44 shows such spectra evaluated by *M. Lance* and *J. Battaille* (1991). The measurements were taken in bubbly flows behind grids.

As outlined in Section 5.4.6, essentially four parameters govern the features of turbulent two-phase flow: the Reynolds number of the continuous phase $Re_c = \bar{u}_c \cdot D_c/\nu_c$, the particle Reynolds number $Re_p = \bar{u}_r \cdot d_p/\nu_c$ with $\bar{u}_r$ the relative particle velocity, the Stokes number $St = \tau_p/\tau_e$ and the mass loading ratio $\phi = \dot{m}_p/\dot{m}_c$. The fluid Reynolds number determines mainly the turbulence level induced by the macro-scale shearing character of the fluid velocity profiles. The particle Reynolds number may serve as an indicator for either additional viscous damping in the continuous phase or for particle induced small-scale turbulence production owing to vortex formation and shedding in the wakes of moving particles. These effects generally occur for $Re_p \leq 1$ and $Re_p > 100$ respectively. The Stokes number describes the available relative time for momentum exchange when a particle crosses the domains of energetic vortices in a turbulent flow field, and thus may be used to qualify the interaction process between particles and continuous phase as depicted in Figure 5.16, Section 5.4.6. The mass loading ratio ϕ quantifies the mass portion of the dispersive phase and quantifies the integral interaction process regarding its spatial distribution and its dynamic intensity.

After the discussion of several characteristic phenomena in turbulent dispersed two-phase flow in the following two sections some principle ideas will

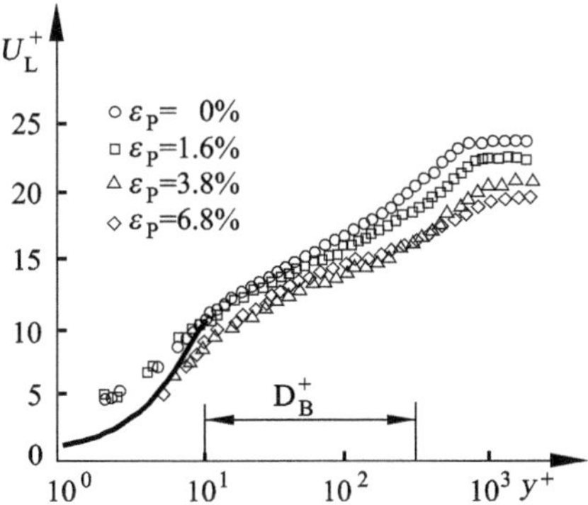

Fig. 8.43. Velocity profiles displayed in standard inner variables for different volume fractions ϵ as indicated and for a volumetric fluid flux $\bar{u}_c = 1\ m/s$. For comparison the logarithmic law of the wall for single-phase flow ($\epsilon = 0$) is shown in Figure 4.64: $u^+ = 2.51 \cdot ln(y^+) + 5.5$ after *J. L. Marie et al.* (1997)

be outlined on how to describe these complex flows by physico-analytical modeling which extends the considerations of Section 5.4.6.

8.6.2 The Mixing Length Concept

It is reasonable to apply Prandtl's mixing length concept also to dispersed two-phase flow in order to describe the relevant integral quantities in turbulent two-phase flow, such as pressure drop $\Delta\rho_{\mathrm{TP}}$, shear-stresses τ, mean velocities $\bar{u}_{\mathrm{c}}$, and void-fraction ϵ_p in a most simple way. *Y. Sato* and *K. Sekoguchi* (1975) proposed that the shear stress in fully developed two-phase pipe flow is composed of three independent constituents, a viscous Newtonion part, a shear-flow induced turbulent part, and a particle induced pseudo-turbulent part. Using a gradient-diffusion ansatz and introducing eddy diffusivities ν' and ν'' for the two turbulence portions they write for the shear stress in the continuous phase

$$\tau_{\mathrm{c}} = (1-\epsilon) \cdot \rho_{\mathrm{c}} \cdot (\nu_{\mathrm{c}} + \nu' + \nu'') \cdot \frac{\mathrm{d}\bar{u}_{\mathrm{c}}}{\mathrm{d}y} \quad . \tag{8.76}$$

Here ν_{c} is the kinematic viscosity and $\bar{u}_{\mathrm{c}}$ the local mean velocity of the continuous phase, y is a coordinate perpendicular to the flow direction. They next utilize Prandtl's mixing length concept for shear flows and for turbulent wake flows and correlate the eddy diffusivities ν' and ν'' to mean velocitiy $\bar{u}_{\mathrm{c}}$ of pipe or boundary layer flows. They propose

$$\nu' = \kappa \cdot y^2 \cdot \frac{\mathrm{d}\bar{u}_{\mathrm{c}}}{\mathrm{d}y} \quad , \qquad \nu'' = \mathrm{K}_1 \cdot \epsilon_{\mathrm{p}} \cdot \frac{d_{\mathrm{p}}}{2} \cdot \bar{u}_{\mathrm{r}} \quad . \tag{8.77}$$

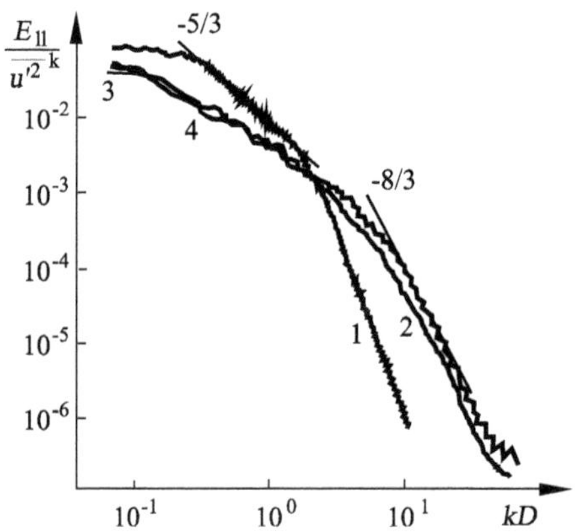

Fig. 8.44. Influence of the gas fraction on the one-dimensional spectrum of the velocity fluctuations in the liquid phase after *M. Lance* and *J. Battaille* (1991). Liquid mean velocity $\bar{u}_{\mathrm{c}} = 0.9\ m/s$; spectrum 1, $\epsilon_{\mathrm{p}} = 0$ %; spectrum 2, $\epsilon_{\mathrm{p}} = 1$ %; spectrum 3, $\epsilon_{\mathrm{p}} = 2.5$ %; spectrum 4, $\epsilon_{\mathrm{p}} = 4$ %. Ratio of bubble diameter to grid spacing: $d_{\mathrm{b}}/D = 0.125$; probe-distance from the grid: $L = 36 \cdot D$

Here $\kappa = 0.4$ is the Karmán mixing length constant, d_{p} is a mean particle diameter, $\bar{u}_{\mathrm{r}} = \bar{u}_{\mathrm{p}} - \bar{u}_{\mathrm{c}}$ the relative particle velocity and ϵ_{p} the particle volume fraction which accounts for the bulk effect of the particle induced turbulence. K_1 is an empirial constant. This has been determined as $\mathrm{K}_1 = 1.2$ by integrating (8.76) for $\bar{u}_{\mathrm{c}}$ using measured values of the volume fraction in an upward bubbly pipe flow and adapting the calculated values $\bar{u}_{\mathrm{c}}$ to the measurements. A similar approach was taken by *D. A. Drew* and *R. T. Lahey* (1980) when they derived a consistent predictive model for the velocity and volume fraction distribution in fully developed upward and downward channel and pipe flow. The proximity of rigid walls results in low velocities in the continuous phase due to repulsion forces in a depleted particle concentration near walls. Thus, the particle induced pseudo-turbulence is negligible in this region and for the flow close to vertical flat walls a balance between shear stress and particle related buoyancy forces exists. This is expressed as

$$\tau = \tau_{\mathrm{w}} - g \cdot \rho_{\mathrm{c}} \cdot \int_0^y (\epsilon - \epsilon_{\mathrm{E}}) \cdot \mathrm{d}y \quad . \tag{8.78}$$

Here τ_{w} is the wall shear stress, ϵ the local void fraction and ϵ_{E} the equilibrium free stream void fraction. *J. L. Marie et. al.* (1997) proposed that the buoyancy term may be approximated by introducing the bubble diameter d_{p} as the relevant length scale and the peak void fraction ϵ_{max} as the relevant concentration measure in the wall region and obtained the relation

$$\tau_{\mathrm{w}}^* = \tau_{\mathrm{w}} - g \cdot \rho_{\mathrm{c}} \cdot (\epsilon_{\mathrm{p}} - \epsilon_{\mathrm{max}}) \cdot d_{\mathrm{p}} \quad . \tag{8.79}$$

Using the mixing length model, the velocity distribution in the wall region can be evaluated. In the particle free viscous sublayer the velocity varies linearly with the distance from the wall. It is matched by a logarithmic distribution in the turbulent boundary layer region, which is still influenced by void and buoyancy effects. *J. L. Marie et. al.* (1997) have shown that the velocity of continuous phase near the wall can be represented in the generalized form

$$\frac{\bar{u}_{\mathrm{c}}}{u_\tau^*} = \frac{1}{\mathrm{K}^*} \cdot \log\left(\frac{y \cdot u_\tau^*}{\nu_{\mathrm{c}}}\right) + \mathrm{C}^* \quad . \tag{8.80}$$

Here $u_\tau^* = \sqrt{\tau_{\mathrm{w}}^*/\rho_{\mathrm{c}}}$ is the wall shear stress velocity and K^* and C^* are constants of integration which themselves depend weakly on τ_w and the free steam velocity or, in case of channel flows, the center line velocity. Moreover, in bubbly flow they also depended on buoyancy effects due to non-uniform void distributions. This is validated by the experimental data displayed in Figure 8.43. *J. L. Marie et. al.* (1997) have also derived analytical expressions for K^* and C^* in terms a Froude number, accounting for buoyancy effects, and a ratio of wall shear stresses related to a two-phase and a single-phase flow respectively. Details are omitted here, but it is emphasized that these findings are fundamental for satisfying wall boundary conditions in multi-dimensional two-phase flow calculations, using the Eulerian approach outlined in Section 5.4.6.

8.6.3 Transport Equation Models for Turbulence

For more complex, non-homogeneous situations such as two-phase flows in pipe branchings, container intake flows or wake flows simple algebraic closure relationships such as those in (8.76) – (8.78) for turbulence quantities based on a mixing length concept and defined eddy viscosities are insufficient for a realistic simulation of phase redistribution and separation effects. As for single-phase flow, the associated three-dimensional phenomena can be understood properly only by solving transport equations for the crucial turbulence properties together with the conservation equations for mass, momentum and energy as formulated in Section 5.4.6. In this context the relevant turbulent quantities in the momentum equation are the Reynolds shear stresses $\boldsymbol{\tau}_{\mathrm{t}}^{\mathrm{Re}} = \overline{\boldsymbol{u}' \cdot \boldsymbol{u}'}$. Another relevant quantity, the turbulent dissipation $\epsilon_{\mathrm{diss}}^{\mathrm{Re}}$ appears in the energy equation. For two-phase flows the transport equations for these quantties can be derived by the very same procedure as used for single-phase flow (see *B. E. Launder et al.* (1975)). However, one has to start from the balance of momentum (5.170) – (5.173) for two-phase flow. For a general discussion, details are omitted here and the two transport equation for the Reynolds stresses $\boldsymbol{\tau}_{\mathrm{t}}^{\mathrm{Re}} = \overline{\boldsymbol{u}' \cdot \boldsymbol{u}'}$ and the dissipation $\epsilon_{\mathrm{diss}}^{\mathrm{Re}} = \nu_{\mathrm{c}} \cdot (\overline{\nabla \boldsymbol{u}' : \nabla \boldsymbol{u}'})$ are given here in symbolic forms for the continuous phase as derived by *L.Bertodano et. al.* (1990) and *R. T. Lahey* (1990). The equations read as:

$$
\begin{aligned}
(1-\epsilon) \cdot \left(\frac{\partial}{\partial t} \boldsymbol{\tau}_{\mathrm{c}}^{\mathrm{Re}} + \bar{\boldsymbol{u}}_{\mathrm{c}} \cdot \nabla \boldsymbol{\tau}_{\mathrm{c}}^{\mathrm{Re}} \right) =& \nabla \left[(1-\epsilon) \cdot \left(\nu_{\mathrm{c}} \cdot \nabla \boldsymbol{\tau}_{\mathrm{c}}^{\mathrm{Re}} - \overline{\boldsymbol{u}' \cdot \boldsymbol{u}' \cdot \boldsymbol{u}'} \right) \right] \\
& + (1-\epsilon) \cdot \left(\boldsymbol{P} + \boldsymbol{\Phi} - 2 \cdot \epsilon_{\mathrm{diss}}^{\mathrm{Re}} \cdot \boldsymbol{I} + \boldsymbol{S}_{\mathrm{i}} \right) \quad ,
\end{aligned} \tag{8.81}
$$

$$
\begin{aligned}
(1-\epsilon) \cdot \left(\frac{\partial}{\partial t} \epsilon_{\mathrm{diss}}^{\mathrm{Re}} + \bar{\boldsymbol{u}}_{\mathrm{c}} \cdot \nabla \epsilon_{\mathrm{diss}}^{\mathrm{Re}} \right) =& \nabla \left[(1-\epsilon) \cdot \left(\nu_{\mathrm{c}} \cdot \nabla \epsilon_{\mathrm{diss}}^{\mathrm{Re}} - \overline{\boldsymbol{u}' \cdot \epsilon'^{\mathrm{Re}}_{\mathrm{diss}}} \right) \right] \\
& + (1-\epsilon) \cdot \left(P_{\mathrm{E}} - \epsilon_{\mathrm{Ediss}}^{\mathrm{Re}} + S_{\mathrm{Ei}} \right) \quad .
\end{aligned} \tag{8.82}
$$

The characters $\boldsymbol{P}$ and P_{E} represent the production of stresses and dissipation, $\boldsymbol{\Phi}$ symbolizes a momentum source originating from pressure–velocity correlations. $\epsilon_{\mathrm{diss}}^{\mathrm{Re}}$ and $\epsilon_{\mathrm{Ediss}}^{\mathrm{Re}}$ describe the dissipation of momentum and energy. The terms $\boldsymbol{S}_{\mathrm{i}}$ and S_{Ei} are specific for an activation of turbulence by the particles of the dispersed phase. In order to solve the extended set of model equations for turbulent two-phase flow for specific conditions, additional model correlations are required for all terms on the right side of (8.81) and (8.82). Most of these terms are of the same structure as in the case of single-phase flow. Therefore, closure relationships similar to the ones derived by *B. E. Launder et al.* (1975) for single-phase flow may be utilized (see also Section 5.4.5). However, the source terms, $\boldsymbol{S}_{\mathrm{i}}$ und S_{Ei}, associated with contributions from the dispersed phase, have to be specifically modeled. Based on physical and mathematical deliberations *L. Bertodano et al.* (1990) proposed the following relationships:

$$S_{\mathrm{i}} = \mathrm{C_i} \cdot \begin{pmatrix} \frac{4}{5} & 0 & 0 \\ 0 & \frac{3}{5} & 0 \\ 0 & 0 & \frac{3}{5} \end{pmatrix} \cdot \frac{3}{4} \cdot C_{\mathrm{D}} \cdot \frac{\rho_{\mathrm{c}} \cdot \epsilon_{\mathrm{p}}}{d_{\mathrm{p}}} \cdot |\bar{\boldsymbol{u}}_{\mathrm{r}}|^3 \quad , \tag{8.83}$$

$$S_{\mathrm{Ei}} = \mathrm{C_{Ei}} \cdot \frac{\epsilon_{\mathrm{diss}}^{\mathrm{Re}}}{\frac{1}{2} \cdot \overline{u' \cdot u'}} \cdot \frac{3}{4} \cdot C_{\mathrm{D}} \cdot \frac{\rho_{\mathrm{c}} \cdot \epsilon_{\mathrm{p}}}{d_{\mathrm{p}}} \cdot |\bar{\boldsymbol{u}}_{\mathrm{r}}|^3 \quad . \tag{8.84}$$

Here C_{D} is suitable drag coefficient for the particle moving with the relative velocity $\bar{\boldsymbol{u}}_{\mathrm{r}} = \boldsymbol{v}_{\mathrm{p}} - \boldsymbol{u}_{\mathrm{c}}$ in the continuous phase, d_{p} is a representative particle diameter and $\mathrm{C_i}$ is an empirical small constant set as $\mathrm{C_i} = 0.02$ to account for the small contribution of the particle induced pseudo-turbulence. $\mathrm{K} = 0.5 \cdot \overline{\boldsymbol{u}' \cdot \boldsymbol{u}'}$ is the turbulent kinetic energy and $\mathrm{C_{Ei}}$ another empirical constant set as $\mathrm{C_{Ei}} = 1.92$ according to *L. Bertodano et al.* (1990).

Together with properly defined boundary conditions (5.113) – (5.115) in Section 5.4.6 and the transport equations (8.81) and (8.82) supplemented by the closure relationship, (8.83) and (8.84) form a complete set of differential equations to calculate three-dimensional turbulent, dispersed two-phase flows. For rigid walls the same approximation as in single-phase flow may be applied in determining the conditions at the inertial sublayer where a modified logarithmic law of the wall (8.80) holds for the tangential component of the velocity while the normal component can be assumed to approach zero.

9. Reactive Flows

9.1 Fundamentals of Reactive Flows

The use of *combustion processes*, the most important example of a chemically reactive flow, is one of the oldest and at the same time one of the most successful technologies to serve humans. In spite of all efforts made in the development of alternative sources of energy, currently more than 80 % of the energy supplies of the world still rely on combustion processes. Because of their broad spectrum of application (heat, electricity, transport and chemistry), the fossil fuels that are currently used annually worldwide have taken about one million years of the Earth's history to form. The pollutants that are produced through this, such as CO_2, nitrogen oxides and soot, lead to undesirable changes in the atmosphere and biosphere of the Earth, as will be described in Section 10.4.

Reactive flows and thus also combustion processes are determined by a complex multi-dimensional and time-dependent interaction between a large number of chemical elementary reactions and transport processes for mass, momentum and energy, as well as phase boundary effects. Empirical methods to develop or improve environmentally friendly and efficient new processes have been largely exhausted. Rather a new approach is necessary. This approach no longer consists of describing reactive flows summarily, rather of assembling the microscopic processes and thereby deriving the visible macroscopic processes. In this manner it is possible, for example, to explain the origin of the formation of pollutants, the incomplete progression of combustion or the mode of functioning of catalysts.

Both non-intrusive analysis of combustion processes with the assistance of optical spectroscopy, and mathematical modeling and simulation play a central role. By means of advances in laser technology it has become possible quantitatively to record the chemically unstable particles that frequently appear only momentarily in combustion with laser light. Thereby insights are obtained into the microscopic progression of reaction in the flame. There are many reasons for the increasing interest in a mathematical description of combustion processes (modeling) and the solution on computers of the model equations developed (simulation). Simulations reduce the effort of experimental investigations by providing indications of possible advantageous conditions and thereby permit targeted design of experiments and targeted

H. Oertel (ed.), *Prandtl-Essentials of Fluid Mechanics*,
Applied Mathematical Sciences 158, DOI 10.1007/978-1-4419-1564-1_9,

experimentation. On the basis of reliable simulations, systems in which experiments are very difficult or impossible can then be optimized. In addition, simulations permit the recognition of systematic errors and the interpretation of indirect test results (parameter identification).

Modeling and simulation yield a detailed view into the physical-chemical processes on which combustion is based. Simulation yields distributions of all system quantities resolved in space and time, such as the temperature and concentration of the species undergoing the combustion process. In addition, comparison of detailed and simplified models allows the effect of certain simplifications to be understood, by *switching on and off* physical-chemical effects.

The interaction of flow, diffusion and release of heat by reaction that is typical of reactive flows can be illustrated in a simple manner with the example of a Bunsen burner (Figure 9.1). Fuel streams out of a nozzle into air that is at rest. By means of molecular transport (diffusion), the fuel and air mix and combust in the reaction zone. For this simple geometry the height of a flame can easily be estimating using a simplified approach.

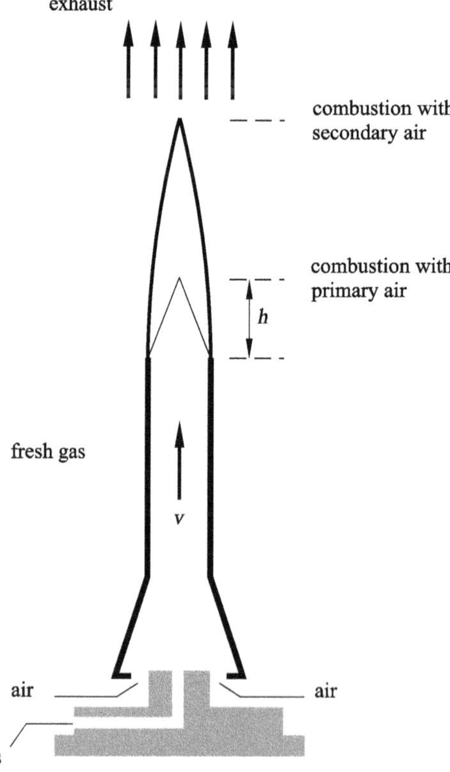

Fig. 9.1. Bunsen burner flame

Let the radius of the jet be r, the height of the flame z and the velocity in the direction of the jet v. At the center of the cylinder, the time required for the fuel to reach the tip of the jet can be estimated from the height of the nonpremixed flame and the intake velocity ($t = z/v$). This time corresponds to the time required for the fuel and air to mix. This mixing time can be estimated from Einstein's displacement law for the depth of penetration by diffusion ($r^2 = 2 \cdot D \cdot t$, D = coefficient of diffusion). Setting the time t in both expressions equal we obtain the equation $z = r^2 \cdot v/(2 \cdot D)$. Now if we replace the velocity v by the volume flux $\dot{V} = \pi \cdot r^2 \cdot v$, we obtain $z = \dot{V}/(2 \cdot \pi \cdot D)$. From this it follows that the flame height z depends only on the volume flux $\dot{V}$ and not on the dimensions of the nozzle r. The height is inversely proportional to the coefficient of diffusion, which is why, for example, a hydrogen flame is about 2.5 times lower than a carbon monoxide flame.

The general aim of this chapter on chemically reactive flows is to describe the coupling between chemical reactions and flows. It is divided into sections on the fundamentals of reaction kinetics, laminar and turbulent flows and hypersonic flows. For each of these classes of reactive flows, typical applications will be presented, with the development of models in the foreground, supported by experimental observations.

The focus is on specific aspects of the fluid mechanics of reactive flows, such as the change in density from reactions and the release of heat. These are complemented by specific questions of reaction kinetics, such as the oxidation of hydrocarbons, the analysis of reaction mechanisms and heterogeneous chemical reactions.

Beyond a purely phenomenological description, in all parts of the chapter indications are given of how the different flows can be modeled and how these models can be translated into equations.

9.1.1 Rate Laws and Reaction Orders

The rate law for a chemical reaction, given in general notation as

$$\mathrm{A} + \mathrm{B} + \mathrm{C} + \cdots \quad \xrightarrow{k^{(\mathrm{f})}} \quad \mathrm{D} + \mathrm{E} + \mathrm{F} + \cdots , \tag{9.1}$$

where A, B, C, ... are different species involved in the reaction, is understood as an empirical ansatz for the *reaction rate*, i.e. the rate at which a species involved in the reaction is formed or consumed. Considering species A, for example, the reaction rate can be written in the form

$$\frac{\mathrm{d}\,[\mathrm{A}]}{\mathrm{d}t} = -k^{(\mathrm{f})} \cdot [\mathrm{A}]^a \cdot [\mathrm{B}]^b \cdot [\mathrm{C}]^c \cdots . \tag{9.2}$$

Here a, b, c, ... are the *reaction orders* with respect to the species A, B, C, ..., and $k^{(\mathrm{f})}$ is the *rate coefficient* of the chemical reaction. The sum of all exponents is the *overall reaction order* of this reaction.

Frequently, there is an excess of some species. In this case, their concentrations change only imperceptibly. For example, if [B], [C], ... remain approximately constant during the reaction, the rate coefficients and the concentrations of the excess species can be used to define an effective rate coefficient. With, for example, $k = k^{(\mathrm{f})} \cdot [\mathrm{B}]^b \cdot [\mathrm{C}]^c \cdots$ we obtain

$$\frac{\mathrm{d}\,[\mathrm{A}]}{\mathrm{d}t} = -k \cdot [\mathrm{A}]^a . \tag{9.3}$$

Integrating the rate law (solving the differential equation), we can determine the temporal change of the concentration of species A.

For *first-order reactions* ($a = 1$) integration of (9.3) yields the first-order rate law

$$\frac{[\mathrm{A}]_t}{[\mathrm{A}]_0} = -k \cdot (t - t_0) , \tag{9.4}$$

where $[\mathrm{A}]_0$ and $[\mathrm{A}]_t$ denote the concentrations of species A at times t_0 and t, respectively.

Similarly, *second-order reactions* ($a = 2$) yield the rate law

$$\frac{1}{[\mathrm{A}]_t} - \frac{1}{[\mathrm{A}]_0} = k \cdot (t - t_0) , \tag{9.5}$$

and for *third-order reactions* ($a = 3$) we obtain the rate law

$$\frac{1}{[\mathrm{A}]_t^2} - \frac{1}{[\mathrm{A}]_0^2} = 2 \cdot k \cdot (t - t_0) . \tag{9.6}$$

If the temporal change of the concentration during a chemical reaction is experimentally determined, we can obtain the reaction orders. A logarithmic plot of the concentration against time for first-order reactions, or a plot of $1/[\mathrm{A}]_t$ against time for second-order reactions, is linear (Figure 9.2).

9.1.2 Relation Between Forward and Reverse Reactions

The reverse reaction of reaction (9.1) has, in analogy to (9.2), the rate law

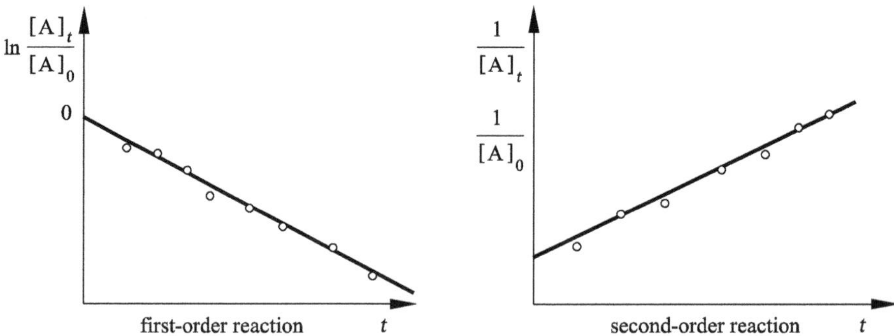

Fig. 9.2. Time histories of the concentration for first- and second-order reactions

$$\frac{\mathrm{d}\,[\mathrm{A}]}{\mathrm{d}t} = -k^{(\mathrm{r})} \cdot [\mathrm{D}]^d \cdot [\mathrm{E}]^e \cdot [\mathrm{F}]^f \cdots . \tag{9.7}$$

In chemical equilibrium, microscopic forward and reverse reactions have the same rate (the forward reaction is denoted by the superscript (f), the reverse reaction by the superscript (r)). Macroscopically, no conversion can be observed. For this reaction in chemical equilibrium the following holds:

$$k^{(\mathrm{f})} \cdot [\mathrm{A}]^a \cdot [\mathrm{B}]^b \cdot [\mathrm{C}]^c \cdots = k^{(\mathrm{r})} \cdot [\mathrm{D}]^d \cdot [\mathrm{E}]^e \cdot [\mathrm{F}]^f \cdots , \tag{9.8}$$

or

$$\frac{[\mathrm{D}]^d \cdot [\mathrm{E}]^e \cdot [\mathrm{F}]^f \cdots}{[\mathrm{A}]^a \cdot [\mathrm{B}]^b \cdot [\mathrm{C}]^c \cdots} = \frac{k^{(\mathrm{f})}}{k^{(\mathrm{r})}} . \tag{9.9}$$

The expression on the left-hand side corresponds to the equilibrium constant K_c of the reaction, which can be determined from thermodynamic data. Therefore, the relation between the rate coefficients for the forward and reverse reaction is

$$K_c = \frac{k^{(\mathrm{f})}}{k^{(\mathrm{r})}} = \exp\left(-\frac{\Delta_{\mathrm{R}}\overline{F}^0}{R \cdot T} \right) . \tag{9.10}$$

9.1.3 Elementary Reactions and Reaction Molecularity

An *elementary reaction* is a reaction that occurs on the molecular level precisely as described by the reaction equation. For example, the essential reaction involved in the combustion of hydrogen, the reaction of hydroxy radicals (OH) with hydrogen molecules (H_2) to form water and hydrogen atoms,

$$\mathrm{OH} + \mathrm{H_2} \longrightarrow \mathrm{H_2O} + \mathrm{H}, \tag{9.11}$$

is such an elementary reaction. Through the motion of the molecules in the gas, hydroxy radicals collide with hydrogen molecules. If they collide nonreactively, the molecules collide and then fly apart again. In reactive collisions, however, the molecules react and the products H_2O and H are formed. On the other hand, the reaction

$$2\mathrm{H_2} + \mathrm{O_2} \longrightarrow 2\mathrm{H_2O} \tag{9.12}$$

is not an elementary reaction, since on detailed investigation it is seen that the reactive particles H, O, and OH are formed as intermediate products, in addition to traces of end products other than H_2O. Such reactions are called *net reactions* or *overall reactions*. These global reactions generally have quite complicated rate laws of the form (9.2) or of an even more complex form. The reaction orders a, b, c, ... are generally not whole numbers, can also have negative values (*inhibition*), and depend on the time and on the experimental conditions. Moreover, extrapolation of the results to regimes where no measurements have been made is extremely unreliable or even wrong. A reaction-kinetic interpretation of these rate laws is normally impossible.

However, in all cases, global reactions may be decomposed into a number of elementary reactions, at least in principle. This is generally very difficult. For example, the formation of water (9.12) can be described by 38 elementary reactions, shown below in the H_2–O_2 system at $p = 1$ bar:

Reaction	A	β	E
	[cm · mol · s]	[-]	[kJ/mol]
H_2–O_2 reactions (excluding HO_2, H_2O_2)			
$O_2 + H = OH + O$	$2.00 \cdot 10^{14}$	0.00	70.30
$H_2 + O = OH + H$	$5.06 \cdot 10^{04}$	2.67	26.30
$H_2 + OH = H_2O + H$	$1.00 \cdot 10^{08}$	1.60	13.80
$OH + OH = H_2O + O$	$1.50 \cdot 10^{09}$	1.14	0.42
$H + H + M^\star = H_2 + M^\star$	$1.80 \cdot 10^{18}$	−1.00	0.00
$O + O + M^\star = O_2 + M^\star$	$2.90 \cdot 10^{17}$	−1.00	0.00
$H + OH + M^\star = H_2O + M^\star$	$2.20 \cdot 10^{22}$	−2.00	0.00
HO_2 formation/consumption			
$H + O_2 + M^\star = HO_2 + M^\star$	$2.30 \cdot 10^{18}$	−0.80	0.00
$HO_2 + H = OH + OH$	$1.50 \cdot 10^{14}$	0.00	4.20
$HO_2 + H = H_2 + O_2$	$2.50 \cdot 10^{13}$	0.00	2.90
$HO_2 + H = H_2O + O$	$3.00 \cdot 10^{13}$	0.00	7.20
$HO_2 + O = OH + O_2$	$1.80 \cdot 10^{13}$	0.00	−1.70
$HO_2 + OH = H_2O + O_2$	$6.00 \cdot 10^{13}$	0.00	0.00
H_2O_2 formation/consumption			
$HO_2 + HO_2 = H_2O_2 + O_2$	$2.50 \cdot 10^{11}$	0.00	−5.20
$OH + OH + M^\star = H_2O_2 + M^\star$	$3.25 \cdot 10^{22}$	−2.00	0.00
$H_2O_2 + H = H_2 + HO_2$	$1.70 \cdot 10^{12}$	0.00	15.7
$H_2O_2 + H = H_2O + OH$	$1.00 \cdot 10^{13}$	0.00	15.0
$H_2O_2 + O = OH + HO_2$	$2.80 \cdot 10^{13}$	0.00	26.8
$H_2O_2 + OH = H_2O + HO_2$	$5.40 \cdot 10^{12}$	0.00	4.20

The rate coefficients are given in the form $k = A \cdot T^\beta \cdot \exp(-E/R \cdot T)$, while $[M^\star] = [H_2] + 6.5 \cdot [H_2O] + 0.4 \cdot [O_2] + 0.4 \cdot [N_2]$ and the rate coefficient of the reverse reaction is calculated with (9.10).

The concept of using elementary reactions has great advantages. The reaction order of elementary reactions is always the same (in particular, it is independent of the time and of any experimental conditions), and it is easy to determine. We consider the *molecularity* of a reaction as the number of species leading to the reaction complex, that is, the transition state of the molecules during the reaction. In practice, there are only three essential values of the reaction molecularity:

Unimolecular reactions describe the decay or the dissociation of a molecule

$$\mathrm{A} \quad \longrightarrow \quad \text{products.} \tag{9.13}$$

They have a first-order rate law. When the initial concentration is doubled, so too is the reaction rate.

Bimolecular reactions are the type of reaction most frequently encountered. They proceed according to the reaction equations

$$\mathrm{A}+\mathrm{B} \quad \longrightarrow \quad \text{products} \tag{9.14}$$

or

$$\mathrm{A}+\mathrm{A} \quad \longrightarrow \quad \text{products.}$$

Bimolecular reactions always have a second-order rate law. Doubling the concentration of any one of the reaction partners causes the reaction rate to double.

Trimolecular reactions are generally recombination reactions. They basically satisfy a third-order rate law,

$$\mathrm{A}+\mathrm{B}+\mathrm{C} \quad \longrightarrow \quad \text{products} \tag{9.15}$$

or

$$\mathrm{A}+\mathrm{A}+\mathrm{B} \quad \longrightarrow \quad \text{products}$$

or

$$\mathrm{A}+\mathrm{A}+\mathrm{A} \quad \longrightarrow \quad \text{products.}$$

In general, for elementary reactions the reaction order is equal to the reaction molecularity. The rate laws can be derived from this. Let the equation of an elementary reaction r be given by

$$\sum_{s=1}^{S} \nu_{rs}^{(\mathrm{a})} \cdot \mathrm{A}_s \quad \xrightarrow{k_r} \quad \sum_{s=1}^{S} \nu_{rs}^{(\mathrm{p})} \cdot \mathrm{A}_s . \tag{9.16}$$

The rate of formation of species i in reaction r is then

$$\left(\frac{\partial c_i}{\partial t}\right)_{\mathrm{chem},r} = k_r \cdot \left(\nu_{ri}^{(\mathrm{p})} - \nu_{ri}^{(\mathrm{a})}\right) \cdot \prod_{s=1}^{S} c_s^{\nu_{rs}^{(\mathrm{a})}} . \tag{9.17}$$

Here $\nu_{rs}^{(a)}$ and $\nu_{rs}^{(p)}$ are stoichiometric coefficients for the initial reactants and products, and c_s the concentrations of the S different species s.

For example, we consider the elementary reaction $\mathrm{H} + \mathrm{O}_2 \longrightarrow \mathrm{OH} + \mathrm{O}$, and obtain the rate laws

$$\frac{\mathrm{d}\,[\mathrm{H}]}{\mathrm{d}t} = -k \cdot [\mathrm{H}] \cdot [\mathrm{O_2}]\,, \qquad \frac{\mathrm{d}\,[\mathrm{O_2}]}{\mathrm{d}t} = -k \cdot [\mathrm{H}] \cdot [\mathrm{O_2}]\,,$$
$$\frac{\mathrm{d}\,[\mathrm{OH}]}{\mathrm{d}t} = k \cdot [\mathrm{H}] \cdot [\mathrm{O_2}]\,, \qquad \frac{\mathrm{d}\,[\mathrm{O}]}{\mathrm{d}t} = k \cdot [\mathrm{H}] \cdot [\mathrm{O_2}]\,.$$

For the elementary reaction $\mathrm{OH} + \mathrm{OH} \longrightarrow \mathrm{H_2O} + \mathrm{O}$ (or $2\mathrm{OH} \longrightarrow \mathrm{H_2O} + \mathrm{O}$) we obtain

$$\frac{\mathrm{d}\,[\mathrm{OH}]}{\mathrm{d}t} = -2 \cdot k \cdot [\mathrm{OH}]^2\,, \quad \frac{\mathrm{d}\,[\mathrm{H_2O}]}{\mathrm{d}t} = k \cdot [\mathrm{OH}]^2\,, \quad \frac{\mathrm{d}\,[\mathrm{O}]}{\mathrm{d}t} = k \cdot [\mathrm{OH}]^2\,.$$

For *reaction mechanisms* consisting of sets of elementary reactions, the rate laws can then always be determined. If the mechanism covers all possible elementary reactions of the system (complete mechanism), then it is valid for all possible conditions, i.e. for all temperatures and compositions. For a mechanism consisting of R reactions of S species given by

$$\sum_{s=1}^{S} \nu_{rs}^{(a)} \cdot \mathrm{A}_s \quad \xrightarrow{k_r} \quad \sum_{s=1}^{S} \nu_{rs}^{(p)} \cdot \mathrm{A}_s \quad \text{with} \quad r = 1, \ldots, R, \tag{9.18}$$

we obtain the rate of formation of a species i by summation over the rate of formation (9.17) in the individual elementary reactions:

$$\left(\frac{\partial c_i}{\partial t}\right)_{\mathrm{chem},r} = \sum_{r=1}^{R} k_r \cdot \left(\nu_{ri}^{(\mathrm{p})} - \nu_{ri}^{(\mathrm{a})}\right) \cdot \prod_{s=1}^{S} c_s^{\nu_{rs}^{(\mathrm{a})}} \quad \text{with} \quad i = 1, \ldots, S. \tag{9.19}$$

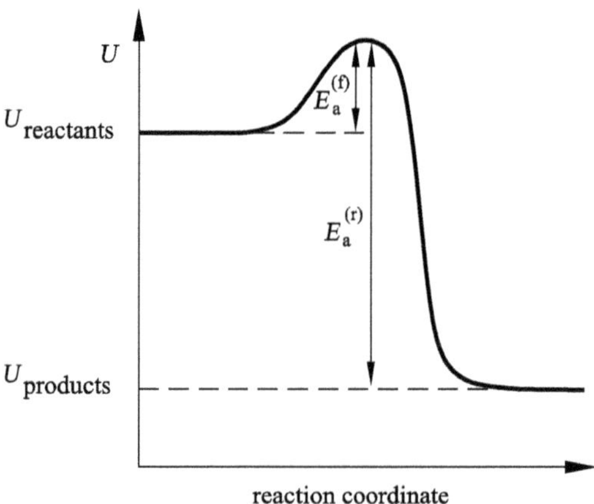

Fig. 9.3. Energy diagram for a chemical elementary reaction. The relation $E_{\mathrm{a}}^{(\mathrm{f})} - E_{\mathrm{a}}^{(\mathrm{r})} = U_{\mathrm{products}} - U_{\mathrm{reactants}}$ is a consequence of (9.10). The reaction coordinate is the path of minimal potential energy between reactants and products with respect to the varying interatomic distances (see, e.g. *W. P. Atkins* (1990))

9.1.4 Temperature Dependence of Rate Coefficients

One very important and typical characteristic of chemical reactions is that their rate coefficients depend very strongly and nonlinearly on the temperature. In this manner they determine the typical abrupt course of combustion processes. According to *S. A. Arrhenius* (1889), this temperature dependence can be described relatively simply with the *Arrhenius equation*:

$$k = A \cdot \exp\left(-\frac{E_{\mathrm{a}}}{R \cdot T}\right). \tag{9.20}$$

In precise measurements, a temperature dependence of the *pre-exponential factor* A that is small compared to the exponential dependence is frequently observed:

$$k = A' \cdot T^b \cdot \exp\left(-\frac{E'_{\mathrm{a}}}{R \cdot T}\right). \tag{9.21}$$

The *activation energy* E_a corresponds to an energy threshold that must be exceeded during the course of the reaction (see Figure 9.3). Its highest value corresponds to the binding energies involved (e.g. the activation energy in dissociation reactions is approximately equal to the binding energy of the chemical bond split), but can also be considerably smaller (or zero) if new bonds are formed at the same time as bonds are broken.

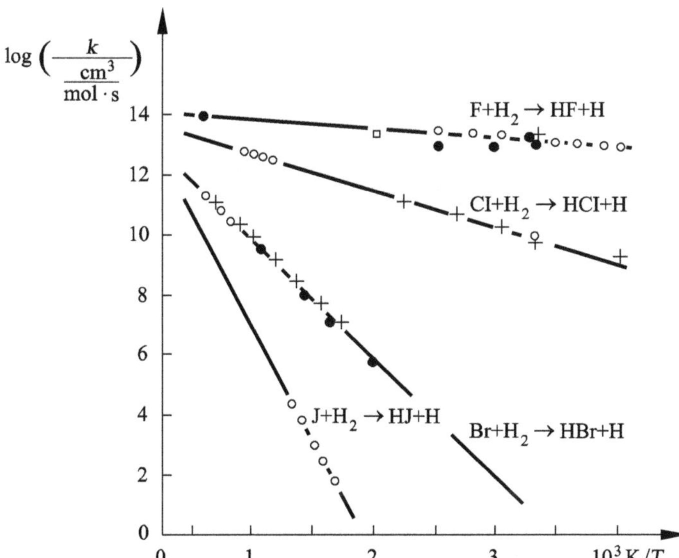

Fig. 9.4. Temperature dependence $k(T)$ for the reaction of halogen atoms with H_2, *K. H. Homann et al.* (1970)

Figure 9.4 shows an example of the temperature dependence of some elementary reactions (reactions of halogen atoms with hydrogen molecules). The logarithms of the rate coefficients k are plotted against the inverse of the temperature. According to (9.20), we find a linear dependence ($\log(k) = \log(A) - \text{const}/T$). Any temperature dependence of the pre-exponential factor is hidden by experimental errors.

When the activation energy vanishes, or at very high temperatures, the exponential term in (9.20) approaches the value 1. The reaction rate is then determined only by the pre-exponential factor A, or $A' \cdot T^b$. This factor has different physical interpretations for unimolecular, bimolecular, and trimolecular reactions.

For unimolecular reactions the inverse of A corresponds to a mean lifetime of a reactive (activated) molecule. In dissociation reactions this lifetime is determined by the frequency with which the atoms involved in the molecular bond vibrate. The pre-exponential factor is thus given by twice the oscillation frequency of the bond involved. From usual oscillation frequencies in molecules we find that $A \approx 10^{14} - 10^{15}\ \text{s}^{-1}$.

For bimolecular reactions the pre-exponential factor A corresponds to a *collision number*, i.e. the number of collisions between two molecules per unit time and volume. This is because the collision number fixes an upper limit to the reaction rate when there is no activation threshold or at very large temperatures. Kinetic gas theory yields numerical values for A of between 10^{13} and $10^{14}\ \text{cm}^3/(\text{mol} \cdot \text{s})$.

In trimolecular reactions, a third partner must meet a bimolecular collision complex. This third partner takes on the energy set free by the reaction (*collision partner*). For example, if two hydrogen atoms collide, the momentarily formed hydrogen molecule will immediately decay because of the large energy present. Since it is very difficult to define when the collision of three molecules occurs simultaneously, numerical values can be calculated only with great difficulty.

9.1.5 Pressure Dependence of Rate Coefficients

The pressure dependence of reaction rate coefficients of dissociation and recombination reactions is based on the fact that complex sequences of reactions are treated as elementary reactions. In the simplest case, the relations can be understood using the *Lindemann model* (1922). Unimolecular decay of a molecule is possible only if the molecule has enough energy to break a bond. For this reason it is necessary that energy be supplied to the molecule by another particle before the actual breaking of the molecular bond. The internal oscillation of the molecule, for example, can serve as an excitation. The excited molecule can then decay into the reaction products, with collision partner M:

$$\text{A} \quad + \text{M} \quad \xrightarrow{k_a} \text{A}^* \quad + \text{M} \qquad \text{(activation)},$$

$$\begin{array}{llll} A^* + M & \xrightarrow{k_{-a}} A + M & \text{(deactivation)}, & (9.22) \\ A^* & \xrightarrow{k_u} \text{products} & \text{(unimolecular reaction)}, & \end{array}$$

According to Section 9.1.3, the rate equations for this reaction mechanism are

$$\frac{d[P]}{dt} = -k_u \cdot [A^\star], \tag{9.23}$$

$$\frac{d[A^\star]}{dt} = k_a \cdot [A] \cdot [M] - k_{-a} \cdot [A^\star] \cdot [M] - k_u \cdot [A^\star]. \tag{9.24}$$

Assuming that the concentration of the reactive intermediate product $[A^\star]$ is quasi-steady,

$$\frac{d[A^\star]}{dt} \approx 0, \tag{9.25}$$

we obtain the following expressions for the concentration of the activated species $A^\star$ and the formation of the reaction product P:

$$[A^\star] = \frac{k_a \cdot [A] \cdot [M]}{k_{-a} \cdot [M] + k_u}, \tag{9.26}$$

$$\frac{d[P]}{dt} = \frac{k_u k_a \cdot [A] \cdot [M]}{k_{-a} \cdot [M] + k_u}. \tag{9.27}$$

We can now pick out two extreme cases: reactions at very low pressure and reactions at very high pressure.

In the *low pressure regime* the concentration of collision partner M is very small. With $k_{-a} \ll k_u$ we obtain the simplified second-order rate law

$$\frac{d[P]}{dt} = k_a \cdot [A] \cdot [M]. \tag{9.28}$$

The reaction rate is therefore proportional to the concentrations of species A and the collision partner M, since at low pressure the activation of the molecule is slow and hence determines the rate.

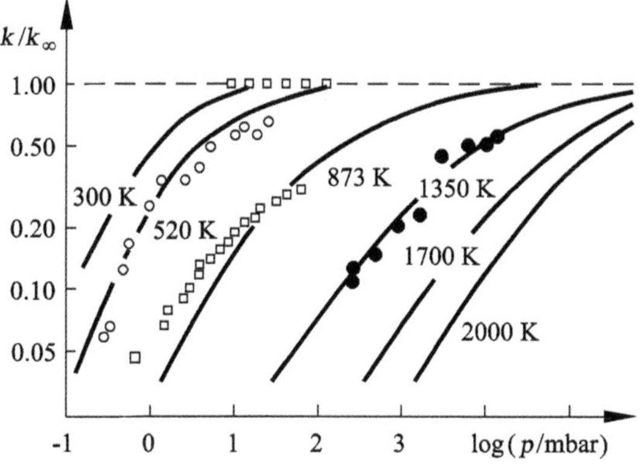

Fig. 9.5. "Falloff" curves for the unimolecular decay $C_2H_6 \longrightarrow CH_3 + CH_3$

In the *high pressure regime* the concentration of collision partner M is very high, and with $k_{-\mathrm{a}} \gg k_{\mathrm{u}}$ we obtain the simplified second-order rate law

$$\frac{\mathrm{d}\,[\mathrm{P}]}{\mathrm{d}t} = \frac{k_{\mathrm{u}} \cdot k_{\mathrm{a}}}{k_{-\mathrm{a}}} \cdot [\mathrm{A}] = k_{\infty} \cdot [\mathrm{A}]\,. \tag{9.29}$$

The reaction rate here is independent of the concentration of the collision partner, since at high pressure, collisions take place frequently, and so it is not the activation but rather the decay of the activated particle $\mathrm{A}^{\star}$ that determines the rate.

The Lindemann mechanism is a simple example of a case in which the reaction order of a complex reaction depends on the current conditions. However, the Lindemann mechanism itself is a simplified model. Precise results for the pressure dependence of unimolecular reactions can be obtained using the *theory of unimolecular reactions* (see, e.g. *P. J. Robinson, K. A. Holbrook* (1972), *H. Homann* (1975)). This theory takes into account the fact that in reality, it is not only a single activated particle $\mathrm{A}^{\star}$ that is present, but rather, depending on the energy transfer in activation, different degrees of activation result. Writing the rate law of a unimolecular reaction as $\mathrm{d}\,[\mathrm{P}]\,/\mathrm{d}t = k \cdot [\mathrm{A}]$, we see that the rate coefficient k is dependent on the pressure and the temperature. The *theory of unimolecular reactions* yields so-called *falloff curves* that describe the dependence of the rate coefficient k on the pressure for different temperatures. Generally, the logarithm of k is plotted against the logarithm of p. Figure 9.5 shows typical falloff curves. For $p \to \infty$, k approaches the limiting value k_{∞}; i.e. the rate coefficient becomes independent

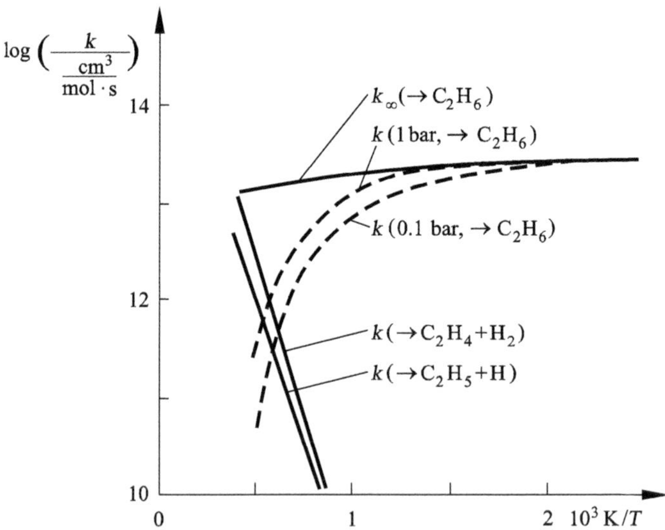

Fig. 9.6. Temperature dependence of the rate coefficient for the pressure-dependent reaction $CH_3 + CH_3 \longrightarrow$ products

of the pressure (9.29). At low pressures the rate coefficient k is proportional to the pressure (9.28) with a linear dependence. As can be seen in Figure 9.5, the falloff curves are greatly dependent on the temperature. For this reason the rate coefficients of unimolecular reactions often have greatly different temperature dependence for different values of the pressure (see Figure 9.6).

9.1.6 Characteristics of Reaction Mechanisms

Reaction mechanisms have some characteristic properties that are independent of the particular problem. Knowledge of these characteristics contributes to an understanding of the chemical reactions and can deliver extremely valuable indications for the subsequent simplification of reaction mechanisms. Of particular note in combustion process are *quasi-steady states* and *partial equilibria*, treated in detail in what follows.

Quasi-Steady States

We consider a simple reaction consisting of a sequence of two steps, which will also be used as an example in the following sections:

$$\mathrm{S_1} \quad \xrightarrow{k_{12}} \quad \mathrm{S_2} \quad \xrightarrow{k_{23}} \quad \mathrm{S_3}. \tag{9.30}$$

The rate laws for the species are then given by the expressions

$$\frac{\mathrm{d}\,[\mathrm{S_1}]}{dt} = -k_{12} \cdot [\mathrm{S_1}]\,, \tag{9.31}$$

$$\frac{\mathrm{d}\,[\mathrm{S_2}]}{dt} = k_{12} \cdot [\mathrm{S_1}] - k_{23} \cdot [\mathrm{S_2}]\,, \tag{9.32}$$

$$\frac{\mathrm{d}\,[\mathrm{S_3}]}{dt} = k_{23} \cdot [\mathrm{S_2}]\,. \tag{9.33}$$

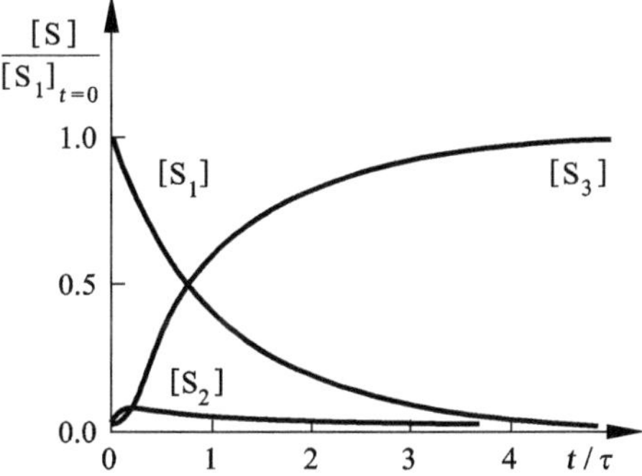

Fig. 9.7. Precise temporal behavior of the reaction $\mathrm{S_1} \longrightarrow \mathrm{S_2} \longrightarrow \mathrm{S_3}$; τ = lifetime of $\mathrm{S_1}$ (time for the decay of $[\mathrm{S_1}]$ to $[\mathrm{S_1}]$ / e)

We assume that S_2 is a very reactive species that therefore has a very short lifetime ($k_{23} \gg k_{12}$). Figure 9.7 shows the change in the concentrations for the ratio $k_{12}/k_{23} = 0.1$. The initial species S_1 decreases with time, while the final product S_3 is formed. Since $k_{23} \gg k_{12}$, the intermediate product S_2 occurs only at a very small concentration. As soon as it is formed in the slow first step of the reaction sequence, it is consumed by the very fast secondary reaction. This leads to a *quasi-steady state* of the intermediate product.

Since S_2 is very reactive, the consumption rate of S_2 must be approximately the same as the formation rate of S_2 (*quasi-steady state assumption*), so that we can write approximately

$$\frac{\mathrm{d}\,[S_2]}{\mathrm{d}t} = k_{12} \cdot [S_1] - k_{23} \cdot [S_2] \approx 0. \tag{9.34}$$

The temporal behavior of the concentration of S_1 can be determined, since (9.31) is integrable. We obtain

$$[S_1] = [S_1]_0 \cdot \exp(-k_{12} \cdot t). \tag{9.35}$$

If we are interested in the rate of formation of the final product S_3, (9.33) yields a statement of only limited usage, as only the concentration of the intermediate product S_2 appears in the rate law for S_3. Using the quasi-steady state assumption (9.34), however, we obtain a relationship that is easy to apply:

$$\frac{\mathrm{d}\,[S_3]}{\mathrm{d}t} = k_{12} \cdot [S_1]\,. \tag{9.36}$$

Inserting (9.35) into this expression, we obtain the differential equation

$$\frac{\mathrm{d}\,[S_3]}{\mathrm{d}t} = k_{12} \cdot [S_1]_0 \cdot \exp(-k_{12} \cdot t), \tag{9.37}$$

which can be integrated. The solution of this equation is

$$[S_3] = [S_1]_0 \cdot [1 - \exp(-k_{12} \cdot t)]\,. \tag{9.38}$$

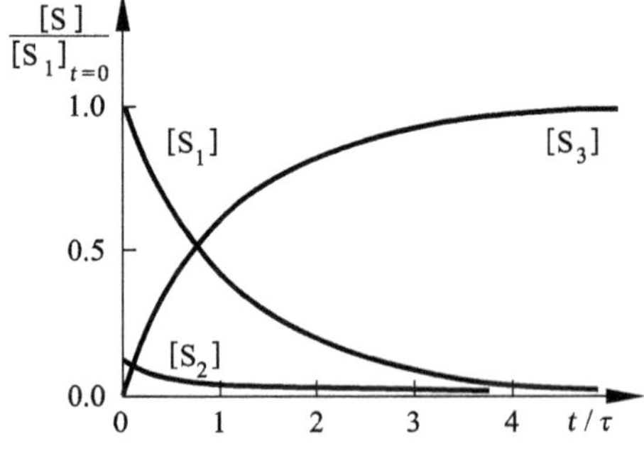

Fig. 9.8. Time development of the reaction $S_1 \rightarrow S_2 \rightarrow S_3$ when $[S_2]$ is quasi-steady

The results for the example above are shown in Figure 9.8. Comparing Figures 9.7 and 9.8, we see that the assumption of quasi-steadiness is a good approximation for the process. It is only at the beginning of the reaction that small deviations are present.

Partial Equilibrium

We consider the mechanism for the combustion of hydrogen discussed in Section 9.1.3. Analysis of experiments or simulations show that for high temperatures ($T > 1800$ K at $p = 1$ bar) the reaction rates for forward and reverse reactions are so fast that for the reactions

$$
\begin{array}{lcccccc}
OH & + & H_2 & = & H_2O & + & H, \\
H & + & O_2 & = & OH & + & O, \\
O & + & H_2 & = & OH & + & H,
\end{array}
$$

a so-called *partial equilibrium* is established, in which each individual reaction pair is in equilibrium. Forward and reverse reaction rates are therefore equally fast. Setting the reaction rates equal, we obtain

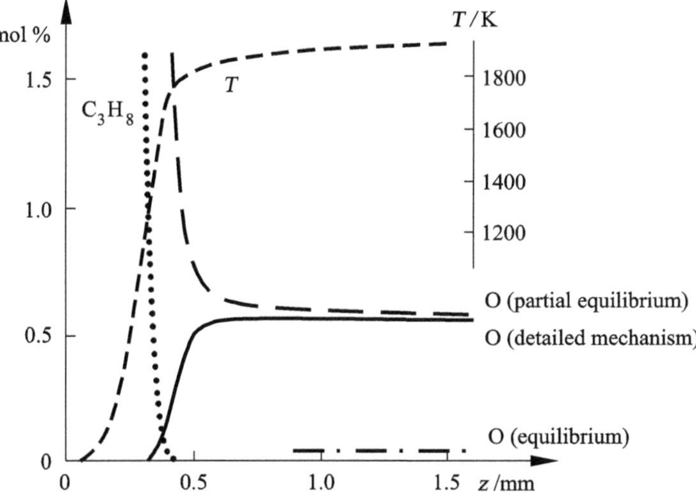

Fig. 9.9. Mole fractions of O in a premixed stoichiometric C_3H_8-air flame at $p =$ 1 bar, $T_u = 298$ K, calculated with a detailed mechanism, with the assumption of partial equilibrium and with the assumption of complete equilibrium

$$[\mathrm{H}] = \left(\frac{k_1^2 \cdot k_3 \cdot k_5 \cdot [\mathrm{O_2}] \cdot [\mathrm{H_2}]^3}{k_2 \cdot k_4 \cdot k_6 \cdot [\mathrm{H_2O}]^2} \right)^{\frac{1}{2}}, \tag{9.39}$$

$$[\mathrm{O}] = \frac{k_1 \cdot k_3 \cdot [\mathrm{O_2}] \cdot [\mathrm{H_2}]}{k_2 \cdot k_4 \cdot [\mathrm{H_2O}]}, \tag{9.40}$$

$$[\mathrm{OH}] = \left(\frac{k_3 \cdot k_5}{k_4 \cdot k_6} \cdot [\mathrm{O_2}] \cdot [\mathrm{H_2}] \right)^{\frac{1}{2}}. \tag{9.41}$$

The concentrations of the unstable species (which are hard to measure, since calibration is difficult) can therefore be reduced to those of the stable species H_2, O_2, and H_2O, which are easy to measure.

Finally, Figure 9.9 shows spatial profiles of the mole fractions of oxygen atoms in a premixed stoichiometric C_3H_8-air flame at $p = 1$ bar, $T_u = 298$ K, calculated with a detailed mechanism, with the assumption of partial equilibrium and with the assumption of complete equilibrium. Whereas the assumption of complete equilibrium leads to unsatisfactory results at all temperatures, partial equilibrium describes the mole fractions of oxygen atoms well, at least for sufficiently high temperatures. We note that the amount of oxygen atoms considered here greatly affects the formation of nitrogen oxides in a reaction system.

Sensitivity Analysis

The rate laws for a reaction mechanism of R reactions with S species involved can be written in the form of a system of ordinary differential equations (compare Section 9.1.3):

$$\frac{\mathrm{d}c_i}{\mathrm{d}t} = F_i(c_1, \ldots, c_s; k_1, \ldots, k_R)\,, \quad c_i(t = t_0) = c_i^0 \quad (i = 1, 2, \ldots, S). \tag{9.42}$$

The time t is the *independent variable*, the concentrations c_i of the species i are the *dependent variables*, and the k_r are the *parameters of the system.* The

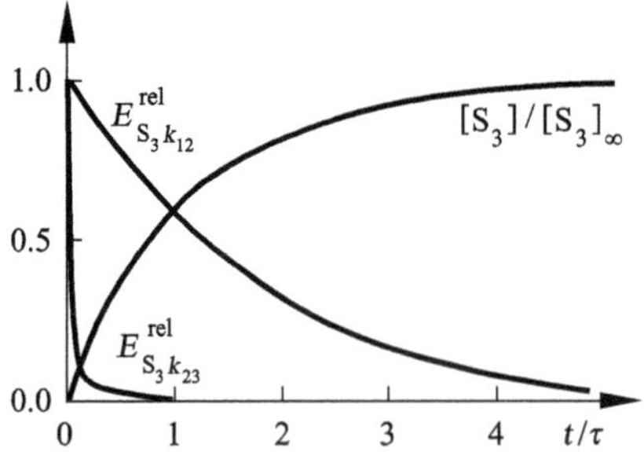

Fig. 9.10. Time development of the relative sensitivity coefficients for the reaction $S_1 \rightarrow S_2 \rightarrow S_3$

c_i^0 denote the initial conditions. Here only the rate coefficients of the chemical reactions are considered as the parameters of the system. However, in complete analogy we could identify the initial conditions, the pressure, etc., as the parameters of the system. The solution of the system of differential equations (9.42) depends both on the initial conditions and on the parameters. The question now arises of how the solution (i.e. the concentrations at time t) changes when the system parameters, i.e. the rate coefficients of the chemical reactions, are varied. The answer to this question delivers information about the rate-determining reaction steps and indicates what effect inaccuracy in the rate coefficients has on the total reaction (some of the elementary reactions that take place in reactive flows are known only to their order of magnitude).

The *sensitivity* of a reaction is the dependence of the solution c_i on the parameters k_r. We distinguish between absolute and relative (normalized) sensitivities:

$$E_{i,r} = \frac{\partial c_i}{k_r} \quad \text{or} \quad E_{i,r}^{\text{rel}} = \frac{k_r}{c_i} \cdot \frac{\partial c_i}{\partial k_r} = \frac{\partial \ln c_i}{\partial \ln k_r}. \tag{9.43}$$

Again we consider the simple reaction made up of a sequence of two steps (9.30). The time development of the relative sensitivity coefficients are plotted in dimensionless form together with the concentration of the final product in Figure 9.10, where $k_{12} = \tau^{-1}$, $k_{23} = 100 \cdot \tau^{-1}$ and τ = lifetime (see Figure 9.7). The result of the sensitivity analysis is that with respect to the slow (i.e. rate-determining) reaction ($S_1 \xrightarrow{k_{12}} S_2$) there is a large relative sensitivity of the formation of S_3, whereas for the fast reaction (which is

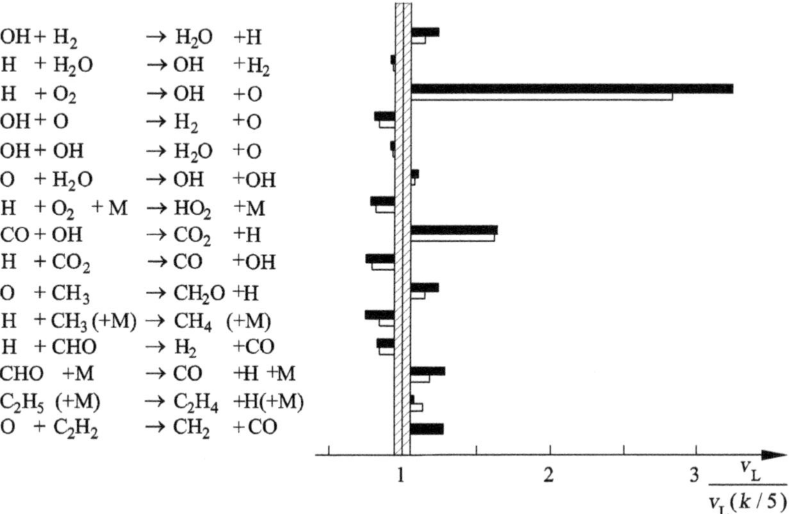

Fig. 9.11. Sensitivity analysis for the flame velocity v_L in premixed stoichiometric CH_4-air (black) and C_2H_6-air flames (white) at $p = 1$ bar, $T_u = 298$ K

not rate-limiting) ($S_2 \overset{k_{23}}{\rightarrow} S_3$) there is a small relative sensitivity. A sensitivity analysis can therefore identify the rate-determining reactions. Such analyses are therefore valuable tools in understanding complex reaction mechanisms.

Figure 9.11 shows an example of a sensitivity analysis for the flame velocity v_L in premixed stoichiometric CH_4-air and C_2H_6-air flames. The elementary reactions not shown in the diagram have negligibly small sensitivity. It can be seen that only a few of the many elementary reactions are sensitive. In addition, very different systems (CH_4 and C_2H_6) give qualitatively the same picture, indicating that in combustion processes some elementary reactions in the H_2-O_2-CO system are always rate-determining, independent of the fuel under consideration.

9.2 Laminar Reactive Flows

9.2.1 Structure of Premixed Flames

We now present a comparison of experimental (when available) and calculated data on the structure of laminar flat flames. The numerical simulations are based on a detailed mechanism, solving the Navier–Stokes equations.

It turns out that at flame conditions ($T > 1100$ K) the oxidation of large aliphatics R-H (such as octane C_8H_{18}, see Figure 9.12) begins with the attack of H, O, or OH on a C-H bond with the formation of a radical R•,

$$\text{H, O, OH} + \text{RH} \quad \longrightarrow \quad \text{H}_2\text{, OH, H}_2\text{O} + \text{R}\bullet \quad \text{(H-atom abstraction)}, \tag{9.44}$$

that then leads to an alkene and a smaller radical R′ by thermal decomposition,

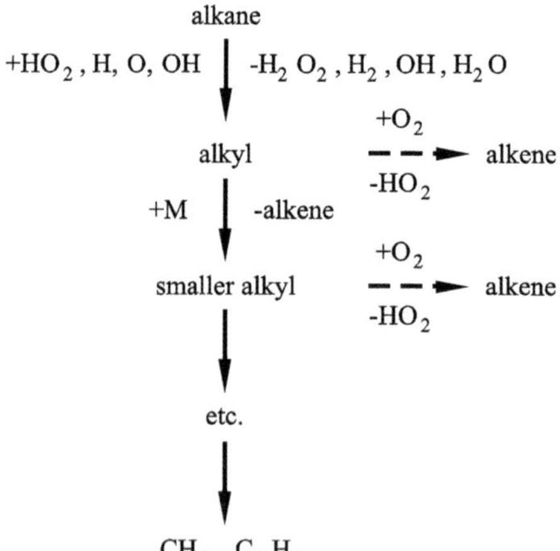

Fig. 9.12. Schematic reaction mechanism for the radical pyrolysis of large aliphatic hydrocarbons to form CH_3 and C_2H_5

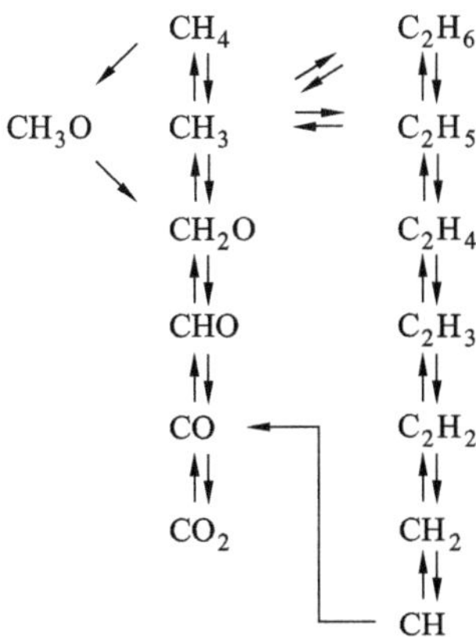

Fig. 9.13. Schematic mechanism of the oxidation of C_1 and C_2 hydrocarbons

$$R'-CH_2-H-C-R'' \quad \rightarrow \quad \bullet R' + CH_2{=}CHR'' \quad (\beta \text{ decay}), \qquad (9.45)$$

until the relatively stable radicals methyl (CH_3) and ethyl (C_2H_5) are formed; these are then slowly oxidized. In this way, the problem of alkane oxidation can be reduced to the relatively well understood oxidation of methyl and ethyl radicals (see Figure 9.13).

CH_3 mainly reacts with O atoms with the formation of formaldehyde (the role of the oxidation of CH_3 by OH is not yet fully understood). The CHO radical is then formed by H-atom abstraction. CHO can decompose thermally to CO and H, or the H atom can by abstracted from H or O_2.

This result, which is quite simple until this point, is then made complicated by the recombination of CH_3 radicals. In stoichiometric CH_4-air flames, this reaction path consumes about 30% of the CH_3 (neglecting recombination with H atoms). In fuel-rich flames, the proportion of recombination increases to about 80%.

The oxidation of CH_3 and C_2H_5 is the rate-determining (i.e. the slowest) step in this oxidation mechanism (see Figure 9.18) and is therefore the reason

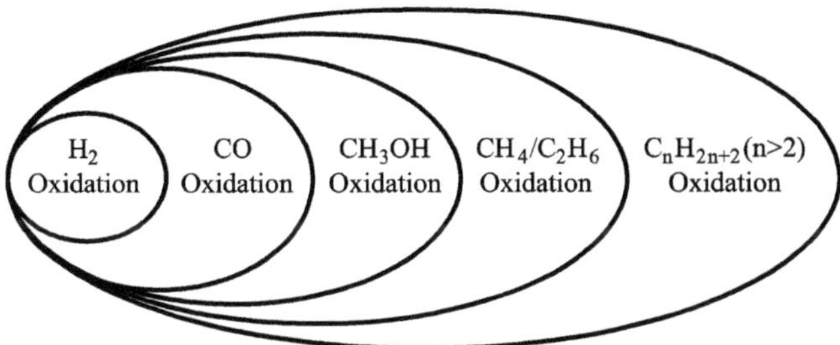

Fig. 9.14. Hierarchical structure of the reaction mechanism describing the combustion of aliphatic hydrocarbons

for the similarity of the combustion of all alkanes and alkenes. Related to this is the fact that the reaction mechanism for the combustion of hydrocarbons has a hierarchical structure, as shown in Figure 9.14 (*C. K. Westbrook, F. L. Dryer* (1981)).

Figure 9.15 shows an example of the flame structure of a propane–oxygen flame diluted with argon to reduce the temperature (*H. Bockhorn et al.* (1990)) at pressure $p = 100$ mbar. For other hydrocarbons the results are similar. The concentration profiles are determined with mass spectrometry (except for OH, which is determined with UV-light absorption measurements), while the temperature is measured using Na-D line inversion.

Another example is an ethyne (acetylene)–oxygen flame at sooting conditions. The appearance of CO and H_2 as stable products and the formation of higher hydrocarbons in connection with the formation of soot precursors (e.g. C_4H_2) are typical.

9.2.2 Flame Velocity of Premixed Flames

The pressure and temperature dependence in the case of a single-step reaction (*Y. B. Zeldovich, D. A. Frank-Kamenetskii* (1938)) is

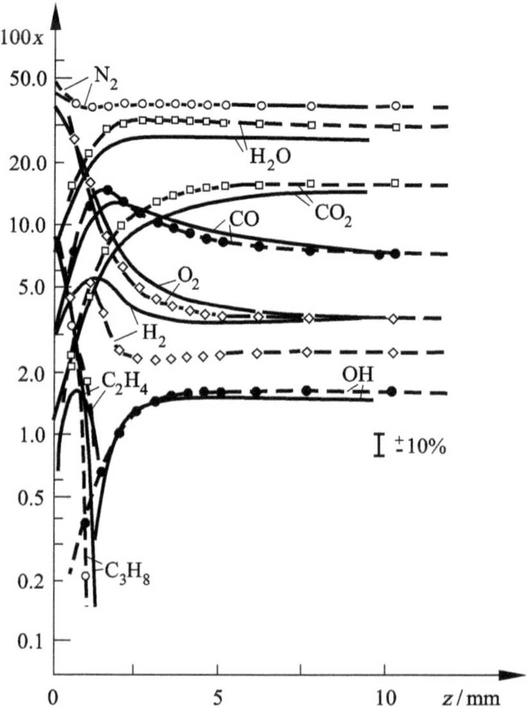

Fig. 9.15. Structure of a laminar premixed propane–oxygen flame (diluted with Ar) at $p = 100$ mbar, *H. Bockhorn* (1990). Points: experiments; lines: simulations

$$v_L \approx p^{\frac{n}{2}-1} \cdot \exp\left(-\frac{E}{2 \cdot R \cdot T_b}\right). \tag{9.46}$$

Here n is the reaction order, E is the activation energy of the single-step reaction, and T_b is the burnt gas temperature.

Figure 9.16 shows the dependence of the flame velocity on the pressure and temperature T_u for the example of a methane–air mixture. In addition, Figure 9.17 shows the dependence of the flame velocity on the composition for different fuels.

Figure 9.16 clearly indicates the weaknesses of the single-step model (T_u is the temperature of the unburned gas). For the rate-determining steps (see next section), the reaction order is 2 or 3, and the simplified model predicts either pressure independence or even a positive pressure dependence. The numerical results, on the other hand, indicate a negative pressure dependence of the flame velocity.

9.2.3 Sensitivity Analysis

Sensitivity analyses (see Section 9.1.6) yield quite similar results for all hydrocarbon–air mixtures for the flame velocity, *U. Nowak* (1988) (see Figures 9.18 and 9.19). In addition, the results are reasonably independent of the equivalence ratio. We note in particular that the number of reactions with sensitivity is low.

In all cases, the elementary reaction $H + O_2 \longrightarrow OH + O$ is greatly rate-determining as the slowest chain-branching reaction, while $H + O_2 + M \longrightarrow HO_2 + M$ has a negative sensitivity because of its chain-terminating character.

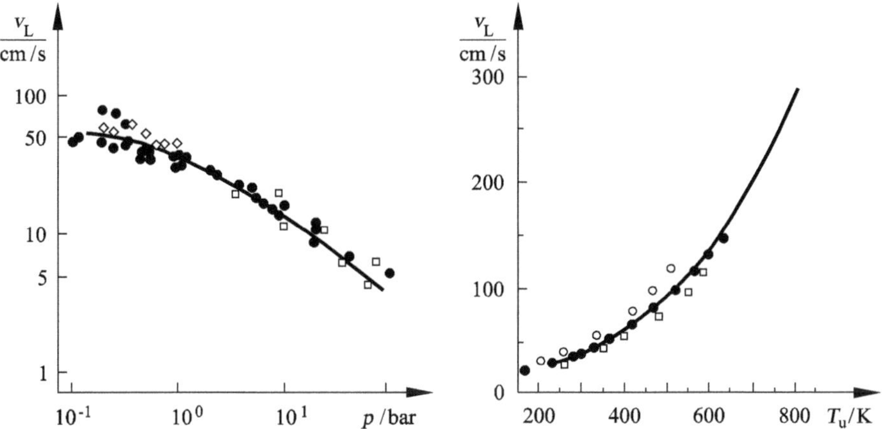

Fig. 9.16. Pressure dependence of v_L for T_u = 298 K (left) and temperature dependence of v_L for $p = 1$ bar (right) in stoichiometric CH_4-air mixtures. Points: experiments; lines: simulations

The reaction $CO + OH \longrightarrow CO_2 + H$ determines a large part of the release of heat and for this reason is also rate-determining.

In a similar way, in the combustion of large aliphatic hydrocarbons, the reactions $H+O_2 \rightarrow OH+O$, $H+O_2+M \rightarrow HO_2+M$ and $CO+OH \rightarrow CO_2+H$ are rate-determining, as demonstrated in Figure 9.20. Again it is seen that the fuel-specific reactions have essentially no sensitivity.

9.2.4 Nonpremixed Counterflow Flames

Nonpremixed flames are flames in which fuel and oxidizer are mixed together only in the combustion region. In practical devices, fuel and air are brought together by convection and then mixed as a result of a diffusion process. In general, this is a three-dimensional problem.

Deeper understanding of nonpremixed flames has therefore come from experiments in which the processes can be considered to be spatially one-dimensional. An example of a suitable simple burner *counterflow* is generated by two burners, in which a directed laminar fuel flow encounters a laminar

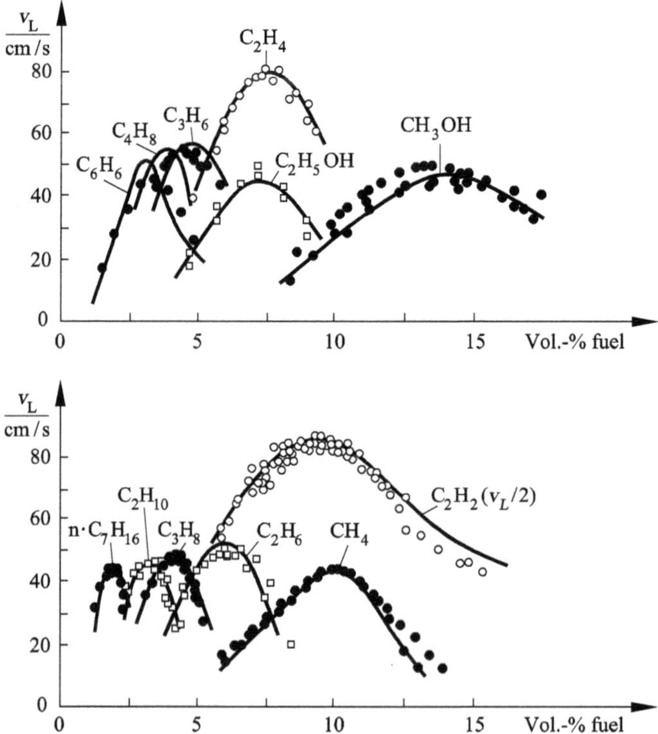

Fig. 9.17. Mixture composition dependence (at $p = 1$ bar, $T_u = 298$ K) of v_L in different fuel–air mixtures. Points: experiments; lines: simulations

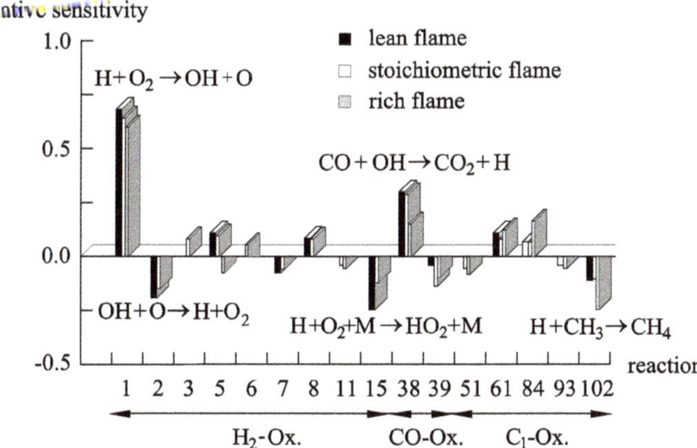

Fig. 9.18. Sensitivity analysis with respect to the rate coefficients of the elementary reactions involved for the laminar flame velocity of a methane–air flame

counterflow of the oxidizer in the opposite direction (see Figure 9.21). The mathematical treatment can be greatly simplified by restricting oneself to the flow properties in the stagnation point plane (see Figure 9.21). Using the *boundary-layer approximation* of Prandtl (i.e. neglecting the diffusion in the direction orthogonal to the stream line, in Figure 9.21 in the x direction), the problem is reduced to one spatial coordinate, namely, the distance from the stagnation point. In this manner the tangential gradients of the temperature and the mass fractions and the velocity components v_x can be eliminated.

If we consider solutions only along the y axis, the symmetry axis determined by the stagnation point, we obtain a system of equations that is

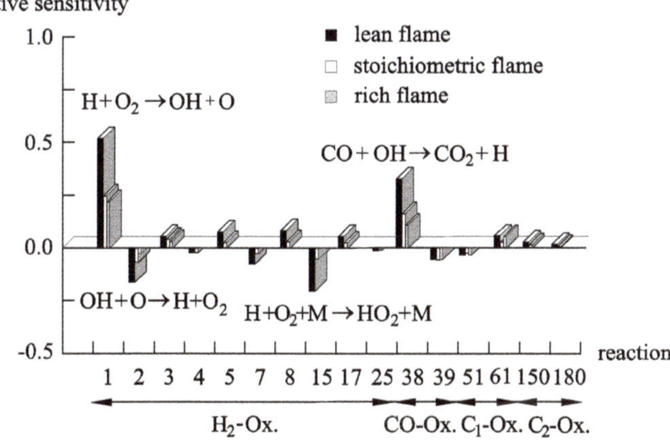

Fig. 9.19. Sensitivity analysis with respect to the rate coefficients of the elementary reactions involved for the laminar flame velocity of a propane–air flame

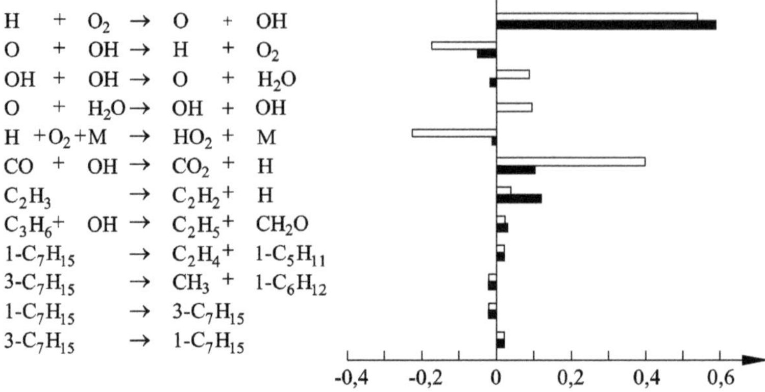

Fig. 9.20. Sensitivity analysis with respect to the rate coefficients of the elementary reactions involved for the laminar flame velocity of a stoichiometric n-heptane–air flame at $p = 1$ bar, $T_u = 298$ K

dependent only on the time t and the spatial coordinate y as independent variables. The pressure gradient J is an eigenvalue of the system; i.e. for given boundary conditions, J must have a value such that a solution of the problem exists. This permits the profiles of temperature, concentration, and velocity in laminar nonpremixed counterflow flames to be calculated and compared to experimental results. Figure 9.22 shows an example of calculated and experimentally determined temperature and concentration profiles (determined using CARS spectroscopy) in nonpremixed methane–air counterflow flames at a pressure of $p = 1$ bar. In the experiment the temperature of the inflowing air (right in the figure) is 300 K.

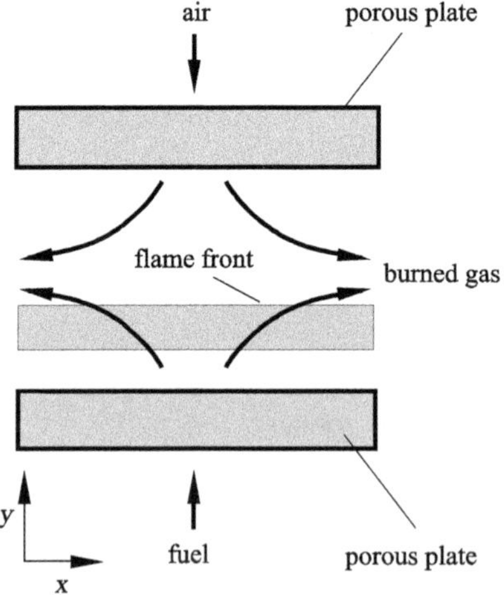

Fig. 9.21. Schematic depiction of a counterflow burner setup

As an example we consider a comparison between measured and calculated velocity profiles as shown in Figure 9.23. The velocities are determined experimentally from the tracking of added MgO particles. The shape of the velocity profile can easily be explained. A nonreactive flow is characterized by a monotonic transition between the velocities at the two boundaries. However, in combustion a strong change in density also occurs (caused by the high temperature of the burnt gas), and close to the flame front (around $y = 3$ mm) this causes a deviation from the monotonic behavior.

9.2.5 Nonpremixed Jet Flames

In order to describe this type of flame correctly, a treatment that is at least two-dimensional is necessary. This is very important, since such flames are widely used (*Bunsen burner*). The fuel streams out of a nozzle into air at rest. By molecular transport (diffusion) the fuel and the air mix and burn in the reaction zone.

The structure of such a nonpremixed Bunsen flame is shown in Figures 9.24 and 9.25 in examples. The results were calculated by complete numerical solution of the spatially two-dimensional conservation equations. The diameter of the fuel nozzle is 1.26 cm in this example, while the height of the flame shown is 30 cm. Temperature and concentration scales each start with the lowest of the grey scales. The maximum temperature is about 2000 K, while the maximum OH concentration corresponds to a mole fraction of 0.35%.

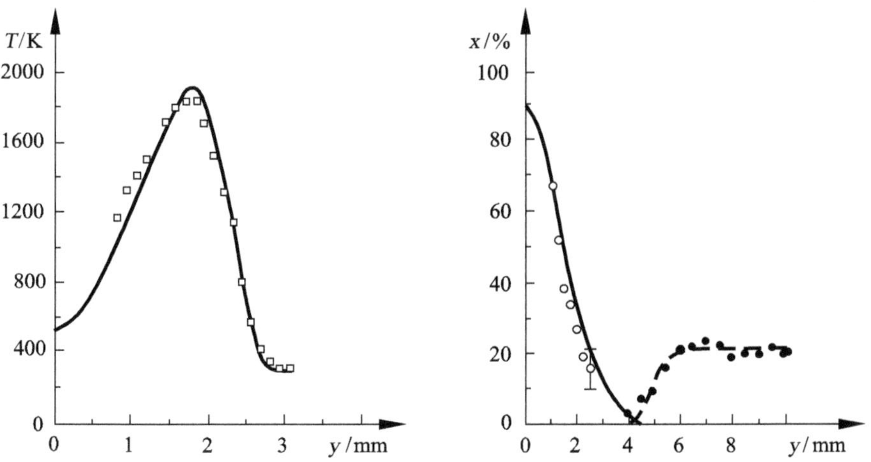

Fig. 9.22. Left: calculated (line) and experimentally determined (points) temperature profiles in a nonpremixed methane–air counterflow flame at a pressure of $p = 1$ bar; y denotes the distance from the burner (*V. Sick et al.* (1991)). Right: calculated (line) and experimentally determined (points) mole fraction profiles of methane and oxygen in a nonpremixed methane–air counterflow flame at a pressure of $p = 1$ bar, y denotes the distance from the burner (*T. Dreier et al.* (1987))

9.2.6 Nonpremixed Flames with Fast Chemistry

When the chemistry is infinitely fast (in practice, when it is very fast), the reaction can be written in the form of a single-step reaction of fuel and oxidizer to the reaction products:

$$\mathrm{F} + \mathrm{Ox} \quad \longrightarrow \quad \mathrm{P}. \tag{9.47}$$

This corresponds to the simplification "mixed = burnt," suggested in the thirties by *K. Rummel* (1937). In analogy to the mass fractions w_i, an element mass fraction Z_i can be defined that gives the mass fraction of a chemical element i and the total mass as (see (5.187))

$$Z_i = \sum_{j=1}^{S} \mu_{ij} \cdot w_j, \qquad i = 1, \ldots, M. \tag{9.48}$$

Here S is the number of species, and M is the number of elements in the mixture under consideration. The coefficients μ_{ij} denote the mass proportion of the element i in the species j.

The element mass fractions are of particular importance, since in a reactive flow they can be altered by neither convective nor chemical processes.

For simple nonpremixed flames, which can be treated as a *two-flow problem*, where one flow is the fuel (F) and the other the oxidizer (Ox), the element mass fractions Z_i can be used to define a *mixture fraction* ξ (the subscripts 1 and 2 denote the two flows):

$$\xi = \frac{Z_i - Z_{i2}}{Z_{i1} - Z_{i2}}. \tag{9.49}$$

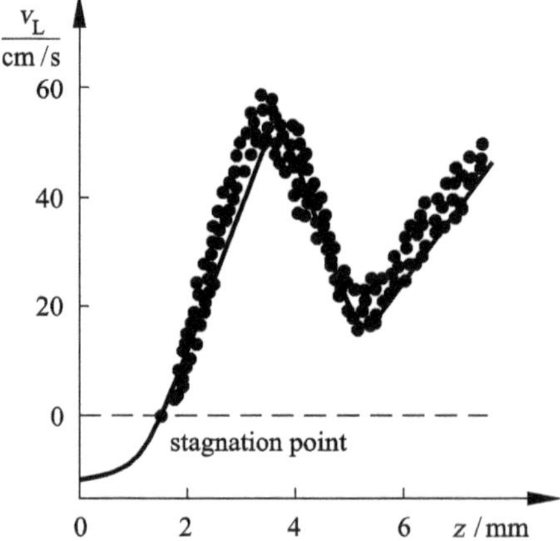

Fig. 9.23. Calculated (line) and experimentally determined (points) velocity profiles in a nonpremixed methane–air counterflow flame; y denotes the distance from the burner

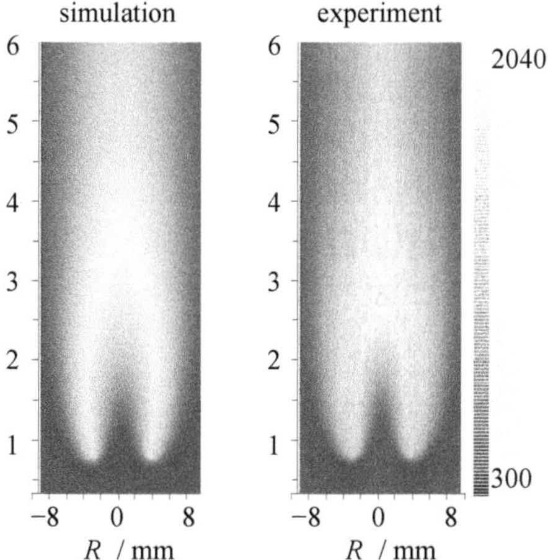

Fig. 9.24. Calculated temperature field (left) in a nonpremixed jet flame. The results can be directly compared with the corresponding results from LIF-experiments (right) (*M. D. Smooke et al.* (1989))

The advantage of this new concept is that because of (9.48) and (9.49), ξ is related in a linear manner to the mass fractions (see Figure 9.26). If the diffusion coefficients of the different chemical species are the same (often approximately true except for a few exceptions), then the mixture fraction defined in this manner is also independent of the choice of the element under consideration i ($i = 1, \ldots, M$).

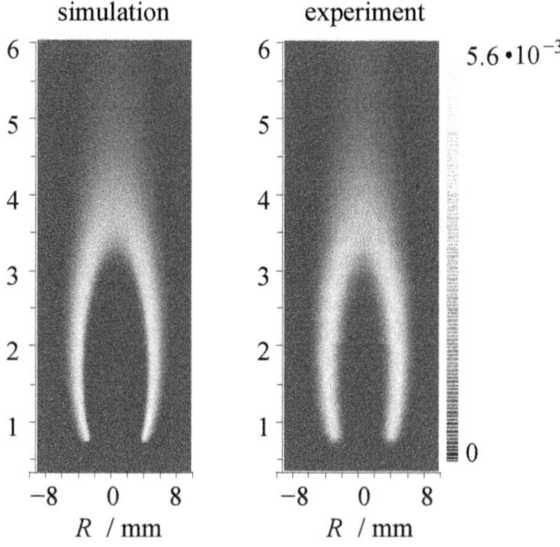

Fig. 9.25. Calculated hydroxyl-radical concentration (left) in a nonpremixed jet flame. The results can be directly compared with the corresponding results from LIF experiments (right) (*M. D. Smooke et al.* (1989))

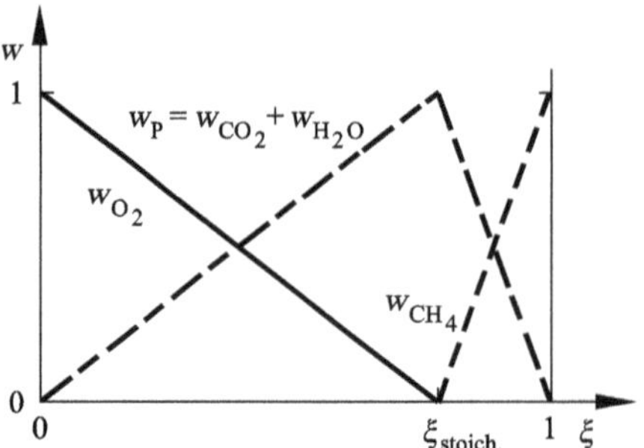

Fig. 9.26. Linear relations between mixture fraction and mass fractions for a simple reaction system

9.2.7 Exhaust Gas Cleaning with Plasma Sources

Increased efforts to protect the climate and use resources more efficiently has meant that the upper emission limits of internal combustion engines in street traffic are becoming ever stricter. To attain these statutory limits, in addition to primary measures to avoid pollutants during combustion, the after-treatment of exhausts is also being investigated intensively.

Three-way catalytic converters that simultaneously reduce NO_x, unburnt hydrocarbons, and CO have an efficiency of about 90% for an Otto engine if it is driven by a stoichiometric fuel–air mixture. Diesel engines and also *direct injection* Otto engines burn lean fuel–air mixtures and generate exhaust gas with an oxygen content of typically 5% for the Otto engine, and up to 20% for the Diesel engine. In noble-metal catalysts, oxidation processes take place only under these conditions with O_2, whereby only CO and unburned hydrocarbons (HC) are reduced, while there is no reduction of NO_x. *Plasma-chemical processes* are therefore increasingly used to supplement three-way catalytic converters in the treatment of exhaust gas. One plasma-chemical process that has low operational expense is the exhaust after-treatment with dielectric barrier discharges.

Figure 9.27 shows a sketch of a reactor. Such plasma reactors have been intensively investigated in recent times in order to determine their potential

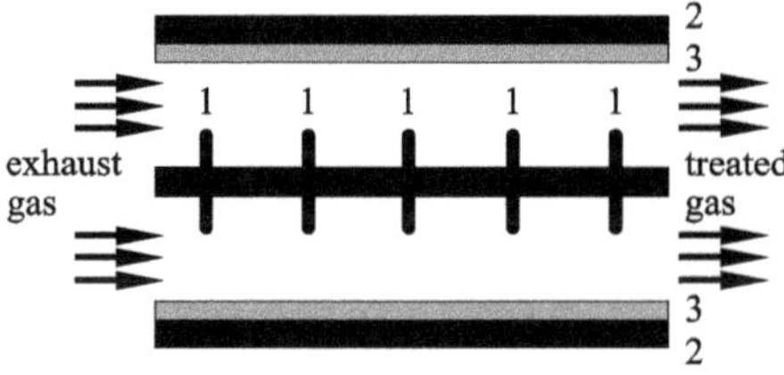

Fig. 9.27. Sketch of a plasma reactor for exhaust gas cleaning (1 inner electrode, 2 outer electrodes, 3 dielectric)

with regard to the oxidation of unburned hydrocarbons and the reduction of NO_x in the exhaust. It has been observed that the reduction of hydrocarbons is possible and depends specifically on the unburned hydrocarbon (*I. Orlandini, U. Riedel* (2000)). Figure 9.28 shows the attainable decrease for a model exhaust gas consisting of 72% N_2, 18% O_2, 10% H_2O, and 440 – 540 ppm unburned hydrocarbons (depending on the experimental conditions). The hydrocarbons investigated are ethane (C_2H_6), propane (C_3H_8), and ethene (C_2H_4).

The differing levels of reduction are due to reaction-kinetic effects in the flow, explained by reaction flux analyses and sensitivity analyses. Furthermore, it has been seen that in oxygen-rich exhausts NO is mainly oxidized to NO_2. Less than 10% of the NO initially present is reduced to N_2. Further measures are therefore necessary to remove the NO_2, such as catalytic reduction or reduction with ammonia.

Figure 9.29 shows the reduction of C_2H_4 by the plasma source for an exhaust composition of 72% N_2, 18% O_2, 10% H_2O, and 500 ppm of unburnt ethene and an exhaust flow rate of 500 liters per minute. The distribution after the first four pulses in the reactor is shown. The reduction of ethene is inhomogeneous perpendicularly to the direction of flow, since the radicals generated in the plasma discharge needed for reduction are also distributed nonuniformly.

9.2.8 Flows in Etching Reactors

In the manufacture of semiconductors, etching processes are used in a large number of production steps. Wet-etching with liquid chemicals is being replaced more end more by dry-etching processes with reactive gases. The etch-

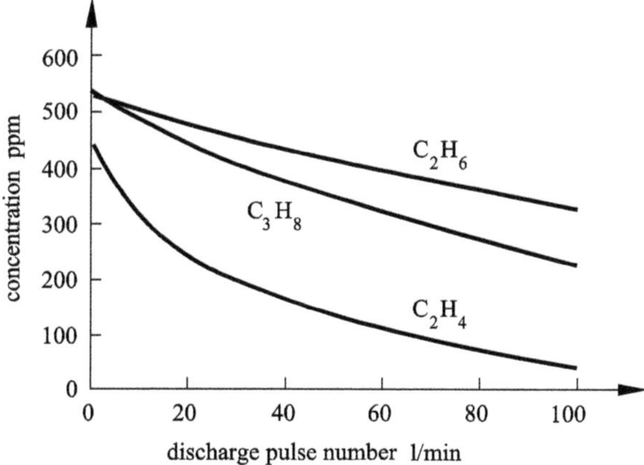

Fig. 9.28. Simulation of the decomposition of C_2H_6, C_3H_8, and C_2H_4 in a plasma reaction as a function of the number of discharge pulses with which the exhaust is treated

ing gases are frequently generated upstream from the actual reactor by means of a plasma source and are then fed to the reactor. Figure 9.30 shows the basic construction of a reactor. In order to attain uniform etching over the wafer, the reactor is operated at low pressures and with small flow velocities, since diffusion prevails over convection and reaction in these conditions and ensures an almost uniform distribution of the reactants.

The processes on the surface are closely coupled to the flow and diffusion from the gas phase. Some of the particles reaching the wafer are adsorbed and can react with other species from the gas phase or with other particles already adsorbed onto the surface. The reaction products formed in this way can then desorb and return to the gas phase.

Figure 9.31 shows the reaction product SiF_4 for an etching gas composition of 70% F atoms and 30% N_2 molecules in an axially symmetric reactor. The SiF_4 forms in surface reactions and, because of the low pressure in the reactor, rapidly diffuses away from the surface.

The etching rate is increased by about 3.5% at the edge of the wafer, which has a diameter of 200 mm (Figure 9.32). The acceleration of the flow close to the edges of the wafer toward the outlet of the reactor causes the convective flux of fluorine atoms here to be greater than close to the symmetry axis.

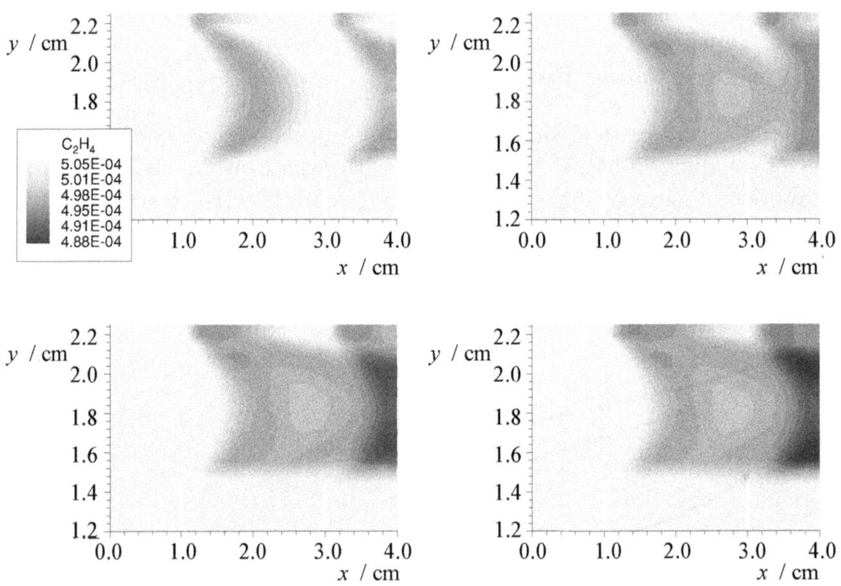

Fig. 9.29. Decomposition of C_2H_4 in a plasma reactor after $t = 0.5$ ms (above left), $t = 1.0$ ms (above right), $t = 1.5$ ms (below left), and $t = 2.0$ ms (below right)

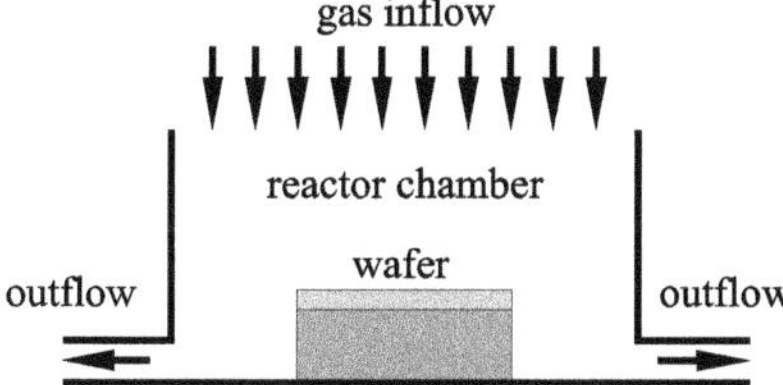

Fig. 9.30. Sketch of an etching reactor

9.2.9 Heterogeneous Catalysis

In heterogeneously catalyzed gas phase reactions the reactants and products are gases, while the reaction takes place on the surface of a solid, the catalyst. The catalyst enhances the rate of the reaction. The principle of catalytic reactions is based on the reduction in the activation energy necessary for a certain reaction, as shown schematically in Figure 9.33. Many reactions have reaction rates on surfaces that are orders of magnitude faster than in the gas phase. This permits the reaction to be carried out at considerably lower temperatures.

A catalyst has no effect on the thermodynamic equilibrium. However, the selectivity of the products can be changed by a catalyst. This is done by suitably selecting the time spent by the mixture in the chemical reaction or by isolating intermediate products. A large number of chemical synthesis methods are based on this procedure.

Heterogeneous-catalytic reactions can be divided into five steps:

1. Diffusion of the reactants to the catalyst,
2. Adsorption of the reactants onto the catalyst surface,
3. Reaction between the reactants,
4. Desorption of the products from the catalyst surface,
5. Diffusion of the products away from the catalyst.

The concentration of the reactants and products on the surface depends on those in the gas phase via the adsorption and desorption equilibria. On the other hand, these depend on changes due to chemical gas-phase reactions

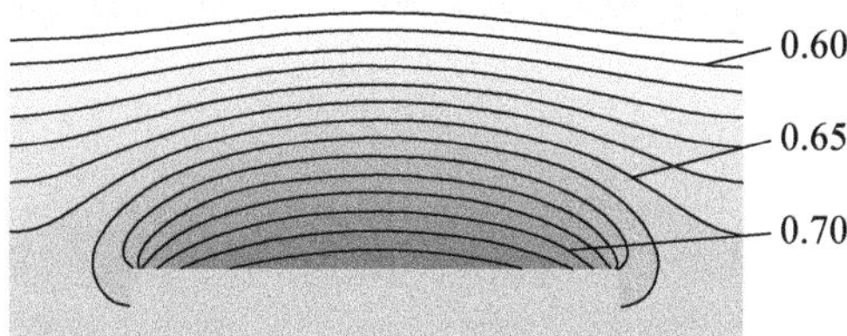

Fig. 9.31. Distribution of the reaction product SiF_4 in the reactor for the etching of silicon with fluorine at a pressure of 40 Pa

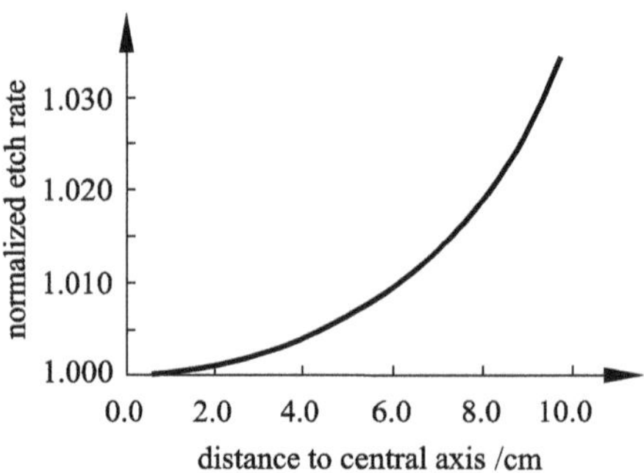

Fig. 9.32. Normalized etch rate as a function of the distance to the center of the wafer

and transport processes. Because of this, depending on the external conditions (temperature, pressure, concentration, flow conditions), different partial processes (mass transport, reaction-kinetic) are rate-determining for the global reaction system. In order to achieve a quantitative understanding of heterogeneous reaction systems it is therefore necessary to couple all the partial processes taking place and to describe the process by detailed models. In analogy to the gas phase, heterogeneous reactions can also be modeled with detailed reaction mechanisms consisting of molecular processes (*M. E. Coltrin et al.* (1990)). The *mean-field* approximation is used, where the catalytic surfaces are described by the temperature and the average degree of coverage with the adsorbed species.

In contrast to gas-phase reactions, only a few of the mechanisms of surface reactions are understood. In the past, a great number of spectroscopic and microscopic investigations into the interactions of molecules with single-crystal surface at low pressures have been used to investigate various elementary processes. Since the direct application of these results to higher pressures (*pressure gap*) and polycrystalline catalyst species (*material gap*) is not without its difficulties, recently, nonlinear optical methods such as sum-frequency

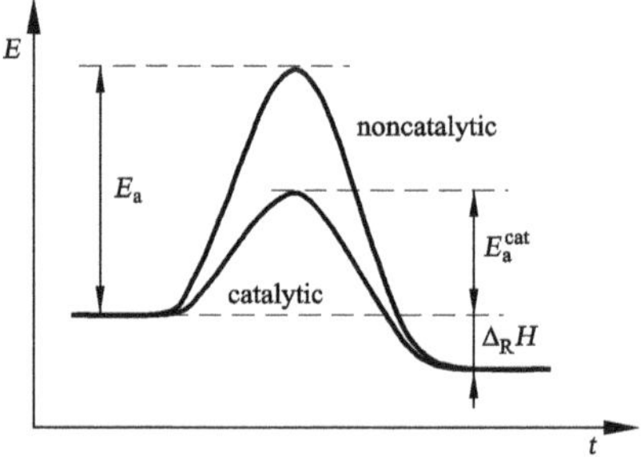

Fig. 9.33. Principle of catalytic reactions. E_a is the activation energy for the noncatalytic reaction and E_a^{cat} that for the catalytic reaction

spectroscopy (*U. Metka et al.* (2000)) have been used to investigate the catalytic surface under these technically relevant conditions. The first detailed heterogeneous reaction mechanisms have been put together, such as those describing catalytic combustion processes (*O. Deutschmann et al.* (1996)) and the partial oxidation of lower alkanes (*D. K. Zerkle et al.* (2000)).

These reaction mechanisms are coupled to the reactive flow via the balance equations at the gas surface interface (*M. E. Coltrin et al.* (1990), *O. Deutschmann et al.* (1996)). In particular, this concept has been applied successfully to the description of the laminar flow behavior in monolithic catalysts (*D. K. Zerkle et al.* (2000)).

9.3 Turbulent Reactive Flows

9.3.1 Overview and Concepts

Turbulent reactive flows play an important role in many industrial combustion processes. In contrast to laminar flow, turbulent processes are characterized by rapid fluctuations of velocity, density, temperature, and composition. This *chaotic* nature of turbulence is due to the high nonlinearity of the underlying physical-chemical processes. Even small variations in the parameters of a flow field can lead to instabilities and hence to the formation of turbulence.

The complexity of turbulent combustion processes (as a standard example for turbulent reactive flows) is the reason why the mathematical models describing them are not as highly developed as the models describing laminar systems. In the following sections we present general patterns of turbulent reactive flows as well as some methods to describe them mathematically. These methods have recently become established in commercial computer programs.

Turbulent nonpremixed flames (see Section 9.3.5) are of great importance in practical applications. They are found in jet engines, diesel engines, steam generators, furnaces, and hydrogen–oxygen rocket engines. Since the fuel and oxidizer mix only in the combustion zone, nonpremixed flames are easier to handle than premixed flames from a safety perspective. It is precisely their practical importance that is the reason why many mathematical models have been developed to simulate these combustion processes.

As will be shown below, the understanding of laminar nonpremixed flames forms the basis for the understanding of turbulent nonpremixed flames. Such flames were previously called *diffusion flames*, since the diffusion of fuel and oxidizer takes place slowly (and is thus rate determining) compared to the chemical reaction. However, since diffusion is also a requirement for the combustion of premixed flames, we use the more precise terms "premixed" and "nonpremixed" flames.

In (ideal) *turbulent premixed flames* (see Section 9.3.6) the unburnt gas is mixed thoroughly before the chemical reaction begins. The chemical reaction causes a rapid transition from unburnt to burnt gas at an interface. This interface moves with velocity v_L.

The motion of a premixed flame is a superposition of flame propagation and (possibly turbulent) flow. In short, this means that quantitative understanding of turbulent premixed flames presents a far greater challenge than the modeling of nonpremixed flames.

Frequently, it is not possible to distinguish clearly between premixed and nonpremixed flames if the time scales of mixing and chemical reaction are of the same order of magnitude. Local flame quenching in nonpremixed flames, for example, causes the fuel and air to mix before they are *ignited* by the surrounding nonpremixed combustion zone (leading to *partial premixed combustion*).

9.3.2 Direct Numerical Simulation

There are no indications that the Navier–Stokes equations are invalid even for turbulent reactive flows, as long as the turbulent length scale (see below) is large compared to the intermolecular distances. This is generally satisfied in combustion processes at atmospheric pressure, so that a turbulent flow could in principle be described by the solution of the Navier–Stokes equations. However, in *direct numerical simulation* (DNS, *W. C. Reynolds* (1989)) even the smallest length scales have to be resolved in the spatial discretization. Therefore, the problem lies in the computational effort required. At the current rate of development, a solution is to be expected only in 20 or 30 years. This can be demonstrated simply as follows. The ratio of the largest to the smallest turbulent length is given by

$$\frac{l_0}{l_K} \approx R_l^{\frac{3}{4}}, \tag{9.50}$$

where R_l is a *turbulence Reynolds number* for which in general $R_l < \mathrm{Re}$ holds, l_0 is the *integral length scale* that describes the largest length scale and is determined by the dimensions of the system, and l_K is the *Kolmogorov length scale*, which represents the length scale of the smallest turbulent structures.

For a typical turbulent flow with $R_l = 500$, $l_0/l_K \approx 100$, so that in order to resolve the smallest structures, we require a grid with ≈ 1000 grid points in each dimension, and so for three-dimensional problems we need 10^9 grid points. Taking into account that the description of an unsteady combustion processes requires at least 10 000 time steps, we obtain the number of floating point operations of order 10^{15}. Another problem lies in the fact that the computational time required for direct simulation is determined by the relation (9.50) and by the fact that the time steps must be reduced in inverse proportion to the square of the grid point distance. This means that the computational time for direct simulation increases with approximately the fourth power of the Reynolds number.

In spite of these problems, direct numerical simulations (DNS) are possible for small Reynolds numbers and simple chemical systems. These simulations are far from practical interest, but can still deliver very useful information on the character of turbulent combustion processes. Direct solution of the Navier–Stokes equations (5.180) and (5.181) for practical applications is not yet possible.

The formation of *closed regions* of unburnt gas that penetrate into the burnt gas is an interesting phenomenon in turbulent premixed flames. This transient process can be investigated using DNS. This process is of importance in determining the regimes of validity of current models and in developing new models to describe turbulent combustion. Figure 9.34 shows the concentration distribution of OH and CO radicals, as well as the vortex strength in a turbulent premixed methane flame, which we have already seen in the introductory chapter (Figure 1.9).

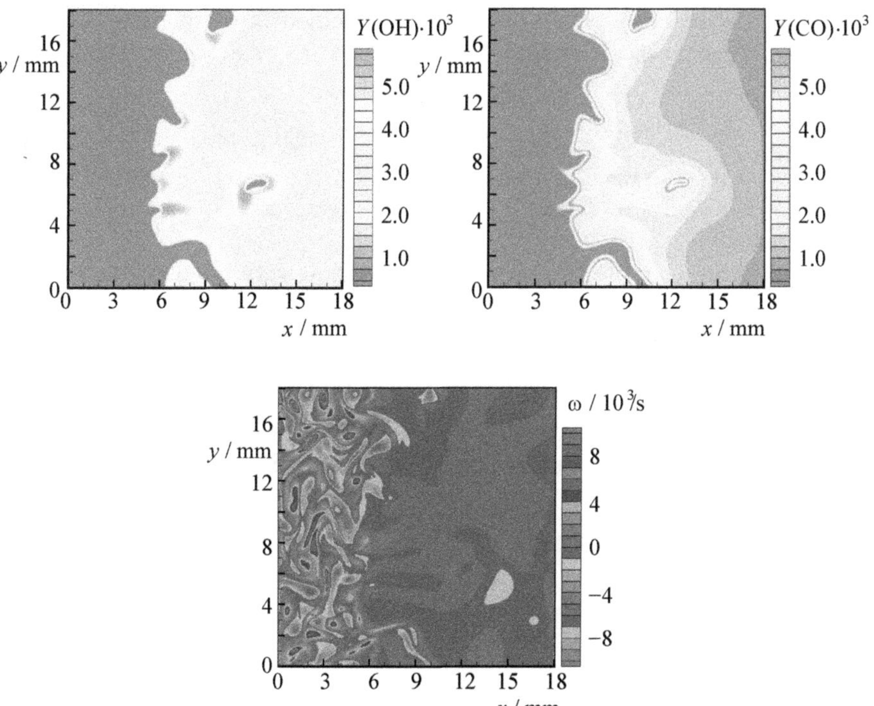

Fig. 9.34. Mass fraction of OH (above left), CO (above right), and vortex strength (below) in a turbulent premixed methane flame

9.3.3 Mean Reaction Rates

For the solution of the averaged conservation equations (5.183) and (5.184) in addition to the turbulence transport model of Section 5.4.7, the determination of the mean reaction rates $\tilde{\omega}_i$ is necessary. In order to demonstrate the problems caused here, we look at two examples (*P. A. Libby, F. A. Williams* (1980)).

In the first example we consider the reaction A + B $\longrightarrow$ products at constant temperature, but with variable concentrations. We assume a hypothetical (but similar to the character of turbulent nonpremixed combustion) time development of the concentration as shown in Figure 9.35, where c_A and c_B are never both simultaneously nonzero. In order to avoid confusion with the turbulent kinetic energy, the rate coefficient k is denoted with the subscript R. The reaction rate is

$$\omega_\mathrm{A} = -k_\mathrm{R} \cdot c_\mathrm{A} \cdot c_\mathrm{B} \quad \text{and} \quad \overline{\omega}_\mathrm{A} = 0;$$

i.e. the mean reaction rate cannot be determined directly from the mean values of the concentrations. Rather we have the relation for the average values:

$$\overline{\omega}_\mathrm{A} = -k_\mathrm{R} \cdot \overline{c_\mathrm{A} \cdot c_\mathrm{B}} = -k_\mathrm{R} \cdot \overline{c}_\mathrm{A} \cdot \overline{c}_\mathrm{B} - k_\mathrm{R} \cdot \overline{c'_\mathrm{A} \cdot c'_\mathrm{B}}. \tag{9.51}$$

Therefore, it is in no way permissible to calculate the reaction rates simply (or even approximately) by replacing the current concentrations by the averaged concentrations.

As a second example we consider a reaction at variable temperature (but constant concentrations), where the temperature development in time is assumed to be sinusoidal (see Figure 9.36). As a result of the strong nonlinearity of the rate coefficient $k_\mathrm{R} = A \cdot \exp(-T_\mathrm{a}/T)$, $\overline{k}_\mathrm{R}$ is entirely different from $k_\mathrm{R}(\overline{T})$. This is clarified with a numerical example. For $T_\mathrm{min} = 500$ K and $T_\mathrm{max} = 2000$ K we obtain $\overline{T} = 1250$ K. Calculating the reaction rate for an activation temperature of $T_\mathrm{a} = 5 \cdot 10^4$ K ($T_\mathrm{a} = E_\mathrm{a}/R$), we obtain

$$\begin{aligned} k_\mathrm{R}(T_\mathrm{max}) &= 1.4 \cdot 10^{-11} \cdot A, \\ k_\mathrm{R}(T_\mathrm{min}) &= 3.7 \cdot 10^{-44} \cdot A, \\ k_\mathrm{R}(\overline{T}) &= 4.3 \cdot 10^{-18} \cdot A, \end{aligned}$$

and after calculating the time average (by, for example, numerical integration) we obtain

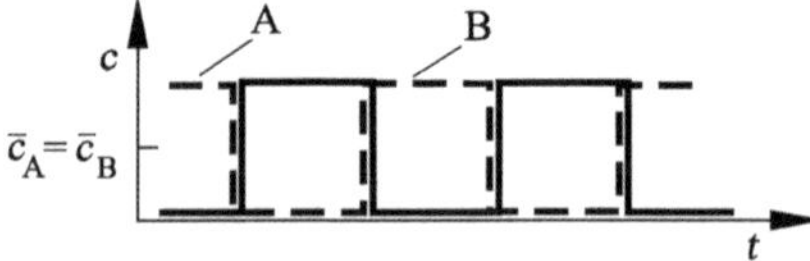

Fig. 9.35. Hypothetical time development of the concentrations in a reaction A + B $\longrightarrow$ products

$$\overline{k}_{\mathrm{R}} = 7.0 \cdot 10^{-12} \cdot A.$$

This fact is of particular interest in, for example, the treatment of nitrogen oxide formation, which is strongly temperature dependent because of its high activation temperature ($T_{\mathrm{a}} = 3.8 \cdot 10^4$ K). NO is therefore mainly formed at peak values of the temperature. Determining the amount of NO at the temperature average is therefore meaningless. Temperature fluctuations must be included in the investigation.

One way of formulating average reaction rates is the statistical treatment using probability density functions (PDFs). If the PDF is known, the average reaction term can be determined by integration. For the example $\mathrm{A} + \mathrm{B} \longrightarrow$ products it is found that (*P. A. Libby, F. A. Williams* (1980))

$$\begin{aligned}\overline{\omega} &= -\int_0^1 \cdots \int_0^1 \int_0^\infty \int_0^\infty k_{\mathrm{R}} \cdot c_{\mathrm{A}} \cdot c_{\mathrm{B}} \cdot P(\rho, T, w_1, \ldots, w_S, \boldsymbol{r}) \cdot \mathrm{d}\rho \cdot \mathrm{d}T \cdot \mathrm{d}w_1 \cdots \mathrm{d}w_S \\ &= -\frac{1}{M_{\mathrm{A}} \cdot M_{\mathrm{B}}} \int_0^1 \cdots \int_0^1 \int_0^\infty \int_0^\infty k_{\mathrm{R}}(T) \cdot \rho^2 \cdot w_{\mathrm{A}} \cdot w_{\mathrm{B}} \cdot P(\rho, T, w_1, \ldots, w_S, \boldsymbol{r}) \\ &\qquad \cdot \mathrm{d}\rho \cdot \mathrm{d}T \cdot \mathrm{d}w_1 \cdots \mathrm{d}w_S. \end{aligned} \tag{9.52}$$

The main problem in this method is that the probability density function P must be known. There are many methods of determining it, which can be used depending on the specific requirements of the case at hand.

PDF Transport Equations

(On this subject see, for example *B. C. Dopazo, E. E. O'Brien* (1974), *S. B. Pope* (1991)). The solution of PDF transport equations is the most general path. Transport equations for the time development of the PDF can be derived from the conservation equations for the particle masses. The great advantage of this procedure is that the chemical reaction is treated exactly (whereas here the molecular transport still has to be modeled empirically).

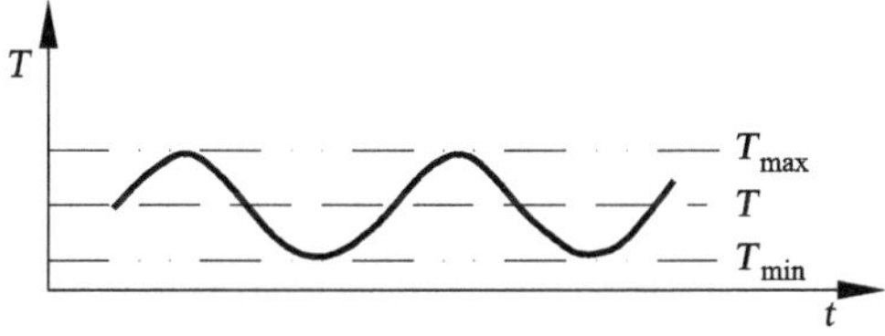

Fig. 9.36. Hypothetical time development of the temperature in a reaction $\mathrm{A} + \mathrm{B} \longrightarrow$ products

For the numerical solution of the transport equations we approximate the probability density function by a large number of different so-called stochastic particles, which represent individual realizations of the flow. The solution of the PDF transport equations is carried out using a Monte Carlo method. This is very complicated and at present confined to small chemical systems with at most four species, so that a reduced mechanism will certainly have to be used.

Empirical Construction of PDFs

In this method, probability density functions are constructed from empirical data. Consistent use is made of the fact that results of the simulation of turbulent flames generally depend only little on the precise shape of the PDF.

One simple way of constructing a multidimensional probability density function consists of assuming statistical independence of the individual variables. In this case the PDF can be decomposed into a product of one-dimensional PDFs (*Gutheil, Bockhorn* (1987)):

$$P(\rho, T, w_1, \ldots, w_S) = P(\rho) \cdot P(T) \cdot P(w_1) \cdots P(w_S). \tag{9.53}$$

Of course, this separation is not correct, since, for example, the mass fractions $w_1, w_2, \ldots, w_S$ are not independent of each other (since $\sum w_i = 1$). For this reason, additional correlations between the individual variables have to be taken into account.

One-dimensional PDFs can be empirically determined from experiment. In what follows we sketch some of these results for simple geometries (*P. A. Libby, F. A. Williams* (1994)).

Figure 9.37 shows a sketch of PDFs for the mass fraction of the fuel for different points of a turbulent mixing layer. At the edge of the mixing layer

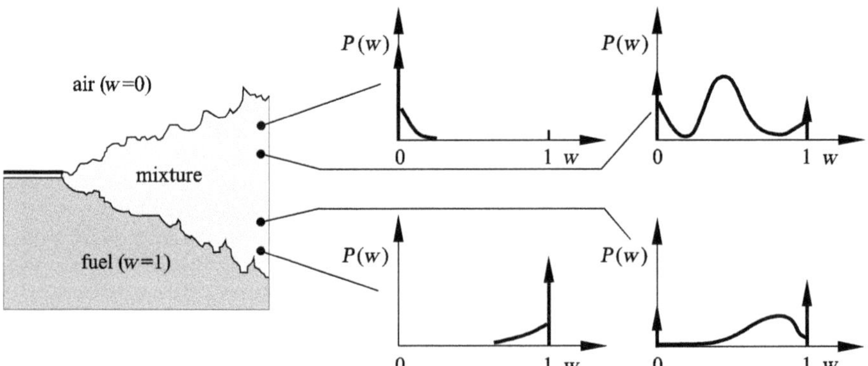

Fig. 9.37. Schematic representation of probability density functions for the mass fraction of the fuel in a turbulent mixing layer

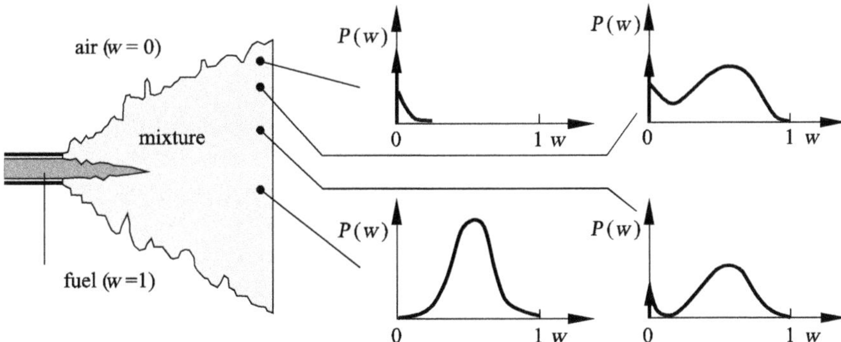

Fig. 9.38. Schematic representation of probability density functions for the mass fraction of the fuel in a turbulent jet

the probability of encountering pure fuel or pure air is very high (indicated by arrows), while the probability of encountering a mixture of fuel and air is only very small. Inside the mixing layer the probability of encountering a mixture of fuel and air is high. The PDF has a maximum for a certain mixture fraction. In spite of this, the probability here of encountering pure fuel or pure air is high (again indicated by arrows). The reason for this is *intermittence*, a phenomenon caused by the fact that the local boundaries between fuel, mixture, and air are constantly shifting. At a certain time a point will be in a pure fuel flow or in a pure air flow (see, e.g. *P. A. Libby, F. A. Williams* (1976, 1994)). Similar results are obtained for a turbulent jet, which can be considered as a combination of two mixing layers (see Figure 9.38).

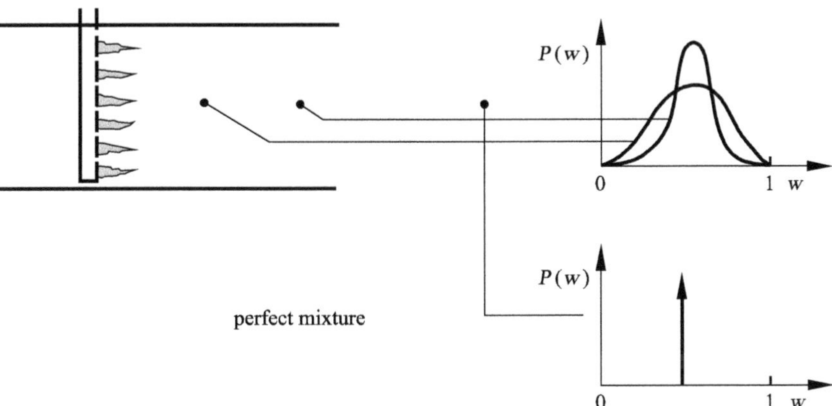

Fig. 9.39. Schematic representation of probability density functions for the mass fraction of the fuel in a turbulent reactor

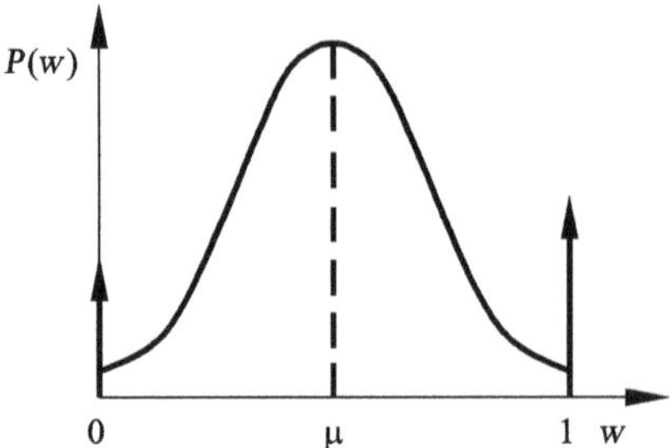

Fig. 9.40. Cut-off Gauß function

In a turbulent reactor (Figure 9.39) the probability density function is approximately a Gauß distribution. The further one is away from the inflow boundary, the greater the probability of encountering a complete mixture. The width of the Gauß function becomes ever smaller until it eventually becomes a *Dirac delta function* (the probability of encountering a complete mixture tends to one).

In order to describe one-dimensional PDFs analytically, a *cut-off Gauß function* or a β *function* may be used. The *cut-off Gauß function* (Figure 9.40) consists of a Gauß distribution and two Dirac δ functions to describe the *intermittence peaks* (*E. Gutheil, H. Bockhorn* (1987)).

An analytic representation of this frequently used function is given by (*F.A. Williams* (1985))

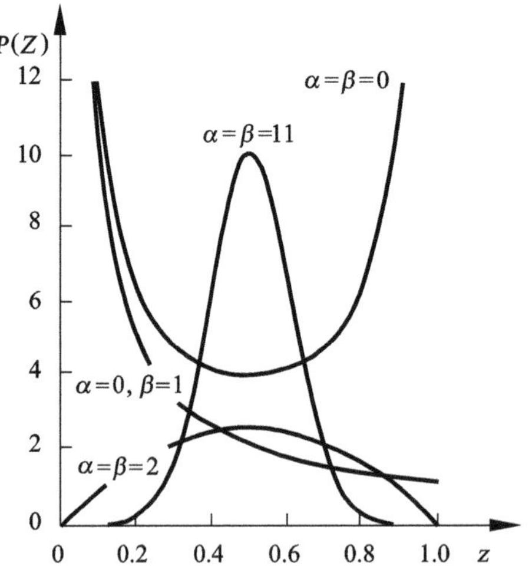

Fig. 9.41. β function for different sets of parameters α and β. For simplicity the normalization constant $\gamma = 1$ has been assumed

$$P(Z) = \alpha \cdot \delta(Z) + \beta \cdot \delta(1 - Z) + \gamma \cdot \exp\left(-\frac{(Z - \zeta)^2}{2 \cdot \sigma^2}\right). \tag{9.54}$$

Here ζ and σ characterize the position and the width of the Gauß function ($Z = w_i, T, \ldots$). The normalization constant γ for given α and β is

$$\gamma = \frac{(1 - \alpha - \beta) \cdot \sqrt{\frac{2 \cdot \sigma}{\pi}}}{\operatorname{erf}\left(\frac{1 - \zeta}{\sqrt{2} \cdot \sigma}\right) + \operatorname{erf}\left(\frac{\zeta}{\sqrt{2} \cdot \sigma}\right)}, \tag{9.55}$$

where the abbreviation "erf" denotes the *error function.*

The *β function* (Figure 9.41) has the great advantage that it contains only two parameters (α, β) but can still describe a great range of different shapes (*R. P. Rhodes* (1979)):

$$P(Z) = \gamma \cdot Z^{\alpha - 1} \cdot (1 - Z)^{\beta - 1} \quad \text{with} \quad \gamma = \frac{\Gamma(\alpha + \beta)}{\Gamma(\alpha) \cdot \Gamma(\beta)}. \tag{9.56}$$

The third parameter γ is obtained from the normalization condition $\int P(Z) \cdot \mathrm{d}Z = 1$. (Note that in mathematics the integral $B(\alpha, \beta) = \int_0^1 t^{\alpha - 1} \cdot (1 - t)^{\beta - 1} \cdot \mathrm{d}t$ is generally called the β function). The constants α and β can be determined from the average and variance of Z as

$$\overline{Z} = \frac{\alpha}{\alpha + \beta} \quad \text{and} \quad \overline{Z'^2} = \frac{\overline{Z} \cdot (1 - \overline{Z})}{1 + \alpha + \beta}. \tag{9.57}$$

9.3.4 Eddy-Break-Up Models

Eddy-break-up models are empirical models for the mean reaction rate at very fast chemistry. In this case the reaction rate is controlled by the rate of the turbulent dissipation (*"mixed is burnt"*). This model describes the reaction zone as a mixture of unburnt and almost completely burnt regions.

A formulation due to *D. B. Spalding* (1970) describes the rate with which regions of unburned gas break up into smaller fragments that have sufficient contact to gas that has already been combusted. They have therefore a sufficiently high temperature and hence react, in analogy to the reduction in turbulent energy. For the reaction rate (F= fuel, C_F is an empirical constant of order of magnitude 1) it is found that

$$\overline{\omega}_F = -\frac{\overline{\rho} \cdot C_F}{\overline{M}} \cdot \sqrt{\overline{w_F''^2}} \cdot \frac{\tilde{\varepsilon}}{\tilde{k}}. \tag{9.58}$$

9.3.5 Turbulent Nonpremixed Flames

Nonpremixed Flames with Equilibrium Chemistry

Insight into the character of nonpremixed turbulent flames is obtained by making the simplifying assumption that fuel and oxidizer react infinitely fast as soon as they have mixed. Using this assumption we have only to determine how fast the mixing takes place. An example of such a *turbulent mixing process* is shown in Figure 9.42. Fuel streams into the oxidizer (oxygen, air). Turbulent mixing causes the fuel and oxidizer to form a combustible mixture that reacts immediately under the above assumption of infinitely fast chemistry. As well as regions where the fuel predominates (rich mixture) and regions where there is a surplus of the oxidizer (lean mixture), there is a stoichiometric surface along which there is a stoichiometric mixture. The upper part of the figure shows an example of the mole fraction a certain distance from the burner. In many cases of turbulent nonpremixed flames, flame fronts appear in the region very close to the stoichiometric mixture. This can be identified by the intensive luminosity at this point.

As well as the assumption of infinitely fast chemistry, we now additionally simplify the description of the mixing process by assuming that the diffusion coefficients are the same. All species then mix equally fast, and we have only to consider the behavior of a single variable. As chemical species are formed or consumed during chemical reactions, it is easier to follow the mixing process for the elements. We introduce the *mixture fraction* ξ:

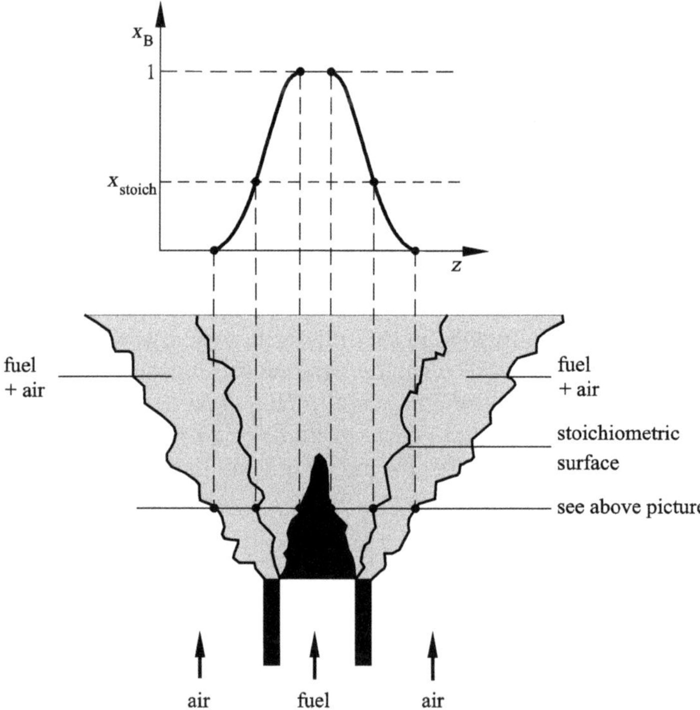

Fig. 9.42. Schematic representation of the momentary picture of a turbulent nonpremixed jet flame

$$\xi = \frac{Z_i - Z_{i2}}{Z_{i1} - Z_{i2}}, \tag{9.59}$$

where the Z_i are element mass fractions. We now consider a two-stream problem with the element mass fractions Z_{i1} and Z_{i2} in the two flows (e.g. in a jet flame). When the diffusivities are the same, ξ is independent of the choice of the element under consideration i ($i = 1, \ldots, M$) and, because of (9.59) and $Z_i = \sum \mu_{ij} \cdot w_j$ (9.48), it is linearly related to the mass fractions w_j. $\xi = 1$ in flow 1 and $\xi = 0$ in flow 2. The fraction ξ can be taken to describe the mass fraction of the species coming from flow 1, with $1 - \xi$ the mass fraction of the species coming from flow 2.

Because of the linear dependence, (9.59) and (5.188) can be used to derive a conservation equation for the mixture fraction ξ:

$$\frac{\partial(\rho \cdot \xi)}{\partial t} + \nabla \cdot (\rho \cdot \boldsymbol{v} \cdot \xi) - \nabla \cdot (\rho \cdot D \cdot \nabla \xi) = 0. \tag{9.60}$$

It is worth noting that there is no chemical source term for ξ in the conservation equation. Therefore, ξ is frequently called a *conserved scalar*. If we also assume that the Lewis number $\mathrm{Le} = \lambda/(D \cdot \rho \cdot c_p)$ is equal to 1 and that there are no heat losses, then the enthalpy or temperature field can also be described using ξ (the kinetic energy of the flow may be neglected and so the pressure is constant):

$$\xi = \frac{h - h_2}{h_1 - h_2}. \tag{9.61}$$

With the assumptions of (a) infinitely fast chemistry (equilibrium chemistry), (b) identical diffusitivities and $\mathrm{Le} = 1$, and (c) no heat losses, all scalar variables (temperature, mass fractions, and density) are well-defined functions of the mixture fraction. These functions are given directly by the equilibrium composition.

The problem of describing turbulent nonpremixed flames has been reduced to the problem of the description of a turbulent mixing process for the mixture fraction ξ. There are numerous approaches to this problem, such as DNS (*W. C. Reynolds* (1989)), LES (*A. McMurtry et al.* (1992)), the Lagrange integral method (LIM) (*W. J. A. Dahm et al.* (1995)), and the PDF method (*S. B. Pope* (1991)).

After forming the averages and using the gradient ansatz for the steady case, we obtain (compare (5.189))

$$\nabla \cdot (\overline{\rho} \cdot \tilde{\boldsymbol{v}} \cdot \tilde{\xi}) - \nabla \cdot (\overline{\rho} \cdot \nu_{\mathrm{T}} \cdot \nabla \tilde{\xi}) = 0. \tag{9.62}$$

If the PDF of the mixture fraction is known, we can calculate the averages of the scalar quantities. Since the average density enters equations (5.183) and (5.184), in this manner the system of averaged conservation equations can be closed. In the ideal case, the PDF should be calculated via its transport equation (*S. B. Pope* (1991)).

A simpler method of determining the probability density function of the mixture fraction consists of assuming that the distribution has a certain shape (such as a Gauß function or a β-function) characterized by the average and variance of ξ. Instead of the transport equation for the PDF, it is only balance equations for the average and variance of ξ that have to be solved. From equation (9.62) we can derive a conservation equation for the Favre variance $\widetilde{\xi''^2} = \overline{\rho \cdot \xi''^2}/\overline{\rho}$ (multiplication of (9.62) by ξ and subsequent formation of the average). We obtain (*R. W. Bilger* (1980))

$$\nabla \cdot (\overline{\rho} \cdot \tilde{\boldsymbol{v}} \cdot \widetilde{\xi''^2}) - \nabla \cdot (\overline{\rho} \cdot \nu_{\mathrm{T}} \cdot \widetilde{\xi''^2}) = 2 \cdot \overline{\rho} \cdot \nu_{\mathrm{T}} \cdot \nabla^2 \tilde{\xi} - 2 \cdot \overline{\rho \cdot D \cdot \nabla^2 \cdot \xi''}, \quad (9.63)$$

where $\nabla^2 \xi$ denotes the square of the absolute value of the gradient $(\nabla \xi)^{\mathrm{T}} \cdot \nabla \xi$. The last term of this equation is called the *scalar dissipation rate* χ. The dependence of this term χ on known quantities must also be modeled, using, for example, the simple gradient transport ansatz

$$\tilde{\chi} = 2 \cdot \frac{\overline{\rho \cdot D \cdot \nabla^2 \xi''}}{\overline{\rho}} \approx 2 \cdot D \cdot \nabla^2 \tilde{\xi}. \quad (9.64)$$

From $\tilde{\xi}$ and $\widetilde{\xi''^2}$ we can now determine the probability density function $P(\xi, \boldsymbol{r})$ (e.g. a β-function, see Section 9.3.4). With the help of the PDF we can calculate the average values of interest, since ρ, w_i, and T are all known as functions of ξ:

$$\begin{aligned}
\tilde{w}_i(\boldsymbol{r}) &= \int_0^1 w_i(\xi) \cdot \tilde{P}(\xi; \boldsymbol{r}) \cdot \mathrm{d}\xi, \\
\tilde{T}(\boldsymbol{r}) &= \int_0^1 T(\xi) \cdot \tilde{P}(\xi; \boldsymbol{r}) \cdot \mathrm{d}\xi, \\
\widetilde{w_i''^2}(\boldsymbol{r}) &= \int_0^1 [w_i(\xi) - \tilde{w}_i(\boldsymbol{r})]^2 \cdot \tilde{P}(\xi; \boldsymbol{r}) \cdot \mathrm{d}\xi, \\
\widetilde{T''^2}(\boldsymbol{r}) &= \int_0^1 \Big[T(\xi) - \tilde{T}(\boldsymbol{r})\Big]^2 \cdot \tilde{P}(\xi; \boldsymbol{r}) \cdot \mathrm{d}\xi, \qquad (9.65)
\end{aligned}$$

where $\tilde{P}$ is a Favre-averaged probability density function that can be calculated from the probability density function by integration over the density:

$$\tilde{P}(\xi; \boldsymbol{r}) = \frac{1}{\overline{\rho}} \cdot \int_0^{\infty} \rho \cdot P(\rho, \xi; \boldsymbol{r}) \cdot \mathrm{d}\rho. \quad (9.66)$$

The system of equations now consists of the conservation equations for the density and velocity fields (e.g. using the equations of the K-ϵ model) as well as the balance equations for the Favre average $\tilde{\xi}$ and the Favre variance $\widetilde{\xi''^2}$ of the mixture fraction ξ. From $\tilde{\xi}$ and $\widetilde{\xi''^2}$ we can determine the probability density function $P(\xi)$. Because of the well-defined relation between ξ and all scalar quantities (i.e. the equilibrium compositions), the statistics of every scalar can be calculated. These equations can be used to calculate flame lengths, temperature fields, and the concentration fields of the main components (fuel, oxygen, water, carbon dioxide).

However, the model will never be able to simulate flame quenching, since infinitely fast chemistry is assumed. The formation of soot and nitrogen oxides can also not be described by the model. Model improvements will therefore be treated that take into account the effect of finite rate chemistry.

Nonpremixed Flames with Finite Rate Chemistry

The complete conservation equations have to be considered in the case of finite rate chemistry; i.e. as well as the balance equations for total mass, energy and momentum, we also have to take into account all conservation equations for the individual species of the reaction system with the source terms $M_i \cdot \omega_i$:

$$\frac{\partial(\rho \cdot w_i)}{\partial t} + \nabla \cdot (\rho \cdot \boldsymbol{v} \cdot w_i) + \nabla \cdot (\rho \cdot D \cdot \nabla w_i) = M_i \cdot \omega_i, \quad i = 1, \ldots, S. \quad (9.67)$$

As described in Section 9.3.3, problems occur in the averaging of the source terms, since these depend nonlinearly on both the temperature and the concentrations.

Averaging is possible in principle if the PDFs of the mass fractions w_i are known. The equations can then be averaged and solved (*E. Gutheil, H. Bockhorn* (1987)). However, problems occur because the PDF is generally not well known, and in addition, the great number of different species means that the computational cost is too high.

One chemical process is first brought out of equilibrium with increasing mixing rate. If the mixing rate continues to increase, a second process then deviates from equilibrium. The chemical processes deviate from equilibrium one after another until the reactions that make up the main part of the energy balance take place on time scales comparable with that of the mixing process.

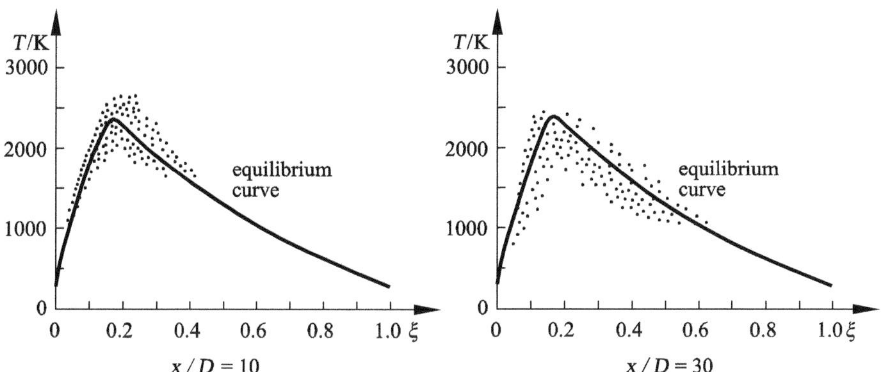

Fig. 9.43. Laser Raman scattering diagram of simultaneous measurements of the mixture fraction and the temperature in a turbulent nonpremixed hydrogen jet flame. The jet velocity in the right-hand picture is a factor of 3 larger (*P. Magre, R. W. Dibble* (1988))

If the mixing rate is then further increased, the temperature deviates from its equilibrium value.

This is shown in Figure 9.43. The temperature deviates only slightly from its equilibrium values. The diagrams on the left and on the right show the same experiment, where only the velocity of the hydrogen jet in the right picture has been increased by a factor of three. The laser Raman scattering experiment simultaneously measures the mixture fraction and the temperature. Each microsecond pulse is indicated by one point on the diagram.

In the left-hand picture, the measurements aggregate around the equilibrium line. The right-hand picture shows the decrease in the temperature where the mixing process, corresponding to a horizontal displacement in the diagram, competes with the release of heat by chemical reaction, corresponding to a vertical displacement in the diagram. The measurements are clearly below the equilibrium line. A further increase of the jet velocity leads to global flame quenching.

Figure 9.44 shows a different behavior. These scattering diagrams show local flame quenching. On the left is a nonpremixed methane–air flame with a small mixing rate. The right-hand picture shows measurements in the same flame but at a different position in the flame, where air mixes rapidly with the fuel. Local flame quenching is seen by the fact that numerous experimental

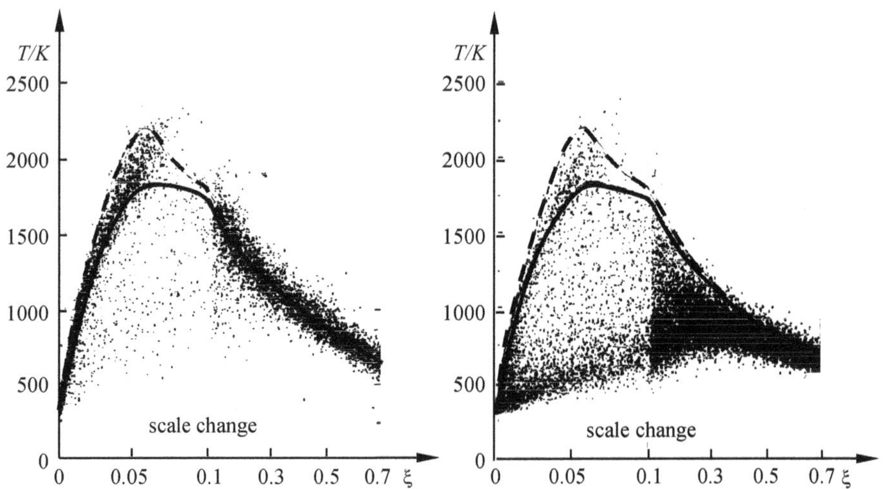

Fig. 9.44. Laser-Raman scattering diagram of simultaneous measurements of the mixing fraction ξ and the temperature T in a turbulent nonpremixed methane jet flame at different distances from the burner (*R. W. Dibble et al.* (1987)); the lines depict flamelet calculations for $a = 1\ \mathrm{s}^{-1}$ (dashed) and $a = 320\ \mathrm{s}^{-1}$

points are far from the equilibrium line. If the jet velocity is further increased, global flame quenching is observed here, too.

An improvement in the equilibrium model presented in the last section is obtained by calculating the rate of the first nonequilibrium process and assuming that the remaining (fast) chemical processes are in equilibrium. The faster the mixing takes place, the more this slow process will deviate from equilibrium. One parameter is required to describe this deviation from equilibrium.

The laminar counterflow flames from Section 9.2.4 have solutions that increasingly deviate from equilibrium. The crucial parameter here is the strain a with which the scalar dissipation rate $\chi = 2 \cdot D \cdot \nabla^2 \xi$ is connected by the relation (*W. J. A. Dahm* and *E. S. Bish* (1993))

$$a = 2 \cdot \pi \cdot D \cdot \left[\frac{\nabla^2 \xi}{(\xi^+ - \xi^-)^2} \right] \cdot \exp \left\{ 2 \cdot \mathrm{erf}^{-1} \left[\frac{\xi - \frac{1}{2} \cdot (\xi^+ + \xi^-)}{\frac{1}{2} \cdot (\xi^+ - \xi^-)} \right] \right\}^2 \tag{9.68}$$

for a locally two-dimensional flow. (For the Tsuji geometry, e.g. in Figure 9.21, the strain rate is generally approximated by the solution of the potential flow $a = 2 \cdot V/R$.) This equation correctly describes the fact that the scalar dissipation for each strain a can be large or small depending on whether the difference between ξ^+ and ξ^- is large or small.

The scalar rate of dissipation is therefore a suitable parameter that can describe the deviation from equilibrium. The scalar quantities in the flame are then again well-defined functions of the mixing fraction, where, however, not the equilibrium values are used, but rather the values of a strained flame. This means that the turbulent flame is taken to be an ensemble of many small laminar *flamelets* that all have the same scalar dissipation rate ξ. This model is a great improvement. Non-equilibrium concentrations of CO, NO, and other species are predicted. The model is further improved by permitting the ensemble of flamelets to have a distribution of the scalar dissipation rate, since the velocity field in the flame changes due to the motion of the vortex. Such a model will now be presented.

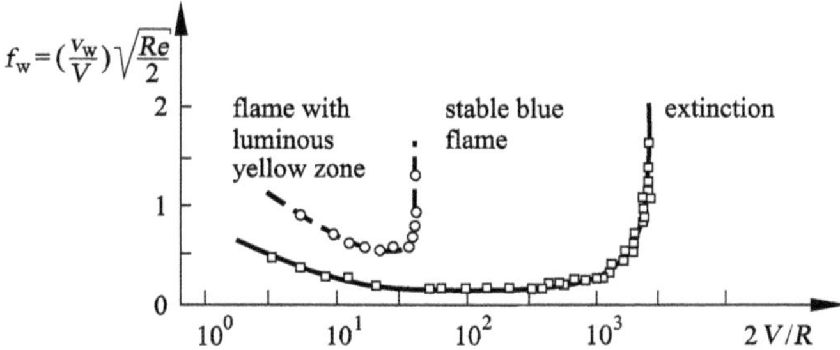

Fig. 9.45. Stability diagram of a laminar nonpremixed counterflow flame, *H. Tsuji*, *I. Yamaoka* (1967)

Flame Quenching

Laminar nonpremixed counterflow flames have already been described in Section 9.2.4. It turns out that characteristic parameters such as the flame temperature depend very strongly on the strain. The strain (characterized by the strain rate parameter a) here describes the velocity gradient along the flame surface.

For sufficiently large strain, laminar nonpremixed flames are extinguished. This behavior is shown in Figure 9.45. Above a critical strain parameter (corresponding to a critical free-stream velocity V of the air) the flame is "blown out." Here f_W is a dimensionless outflow parameter that can be calculated from the velocity V of the incoming air, the exit velocity v_W of the fuel from the porous cylinder, the Reynolds number Re, and the cylinder radius R. The strain is then given by $a = 2 \cdot V/R$.

Figure 9.46 shows calculated temperature profiles for different scalar dissipation rates χ, i.e. for different strains a, in a nonpremixed counterflow flame. Above a certain dissipation rate χ_q (here for $\chi_q = 20.6\ \mathrm{s}^{-1}$, where the subscript q stands for "quenching"), flame quenching finally occurs (*B. Rogg et al.* (1987)).

The temperature drops as convective–diffusive heat transport increases, while heat generation by chemical reaction simultaneously decreases due to the shorter time of direct contact. Flames close to quenching are sensitive to

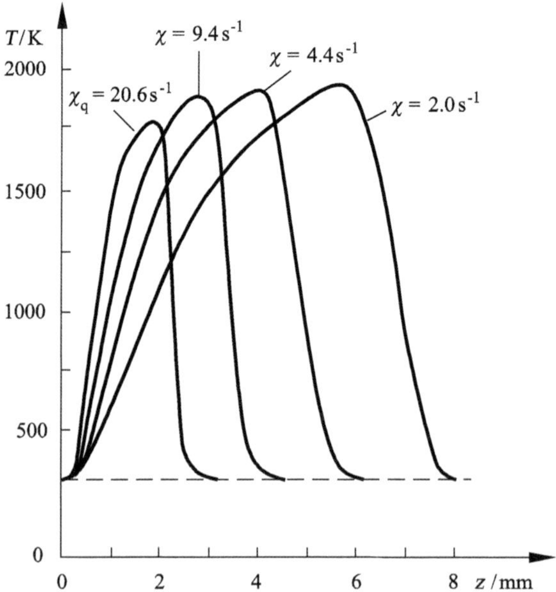

Fig. 9.46. Calculated temperature profiles of a nonpremixed CH_4-air counterflow flame for different scalar dissipation rates χ; flame quenching occurs for $\chi > 20.6\ \mathrm{s}^{-1}$; unburnt gas temperature $T = 298$ K on both sides; pressure 1 bar

the Lewis number $\mathrm{Le} = \lambda/(D \cdot \rho \cdot c_p)$, i.e. to the ratio of thermal diffusivity to mass diffusivity (*H. Tsuji*, *I. Yamaoka* (1967), *N. Peters*, *J. Warnatz* (1982)). In turbulent flames the strain of the laminar flamelet is determined by the scalar dissipation rate at the position of the stoichiometric mixing. The scalar dissipation rate is therefore a direct measure of the strain. If it exceeds a critical value, local quenching of the flamelet occurs. In this manner we can understand quenching processes in turbulent nonpremixed flames.

The flamelet model can also be used to explain the *lift-off* of turbulent flames to quenching through the high strain. This is shown schematically in Figure 9.47. At the exit of the nozzle the strain of the flame front is largest, and so it is here that quenching most frequently occurs. The mean luminous flame contour shows a lift-off that increases with increasing jet velocity. The practical importance of this approach via the lifting process lies in the possibility of carrying out quenching processes (e.g. on burning oil wells) optimally, namely, at the foot of the flame, where the sensitivity of the flame to quench is largest because of its high strain at this point.

In modeling turbulent nonpremixed flames, quenching processes are taken into account by integrating only over that region of the scalar dissipation rate in which no flame quenching occurs when the average values of the density, temperature, and mass fractions are being determined:

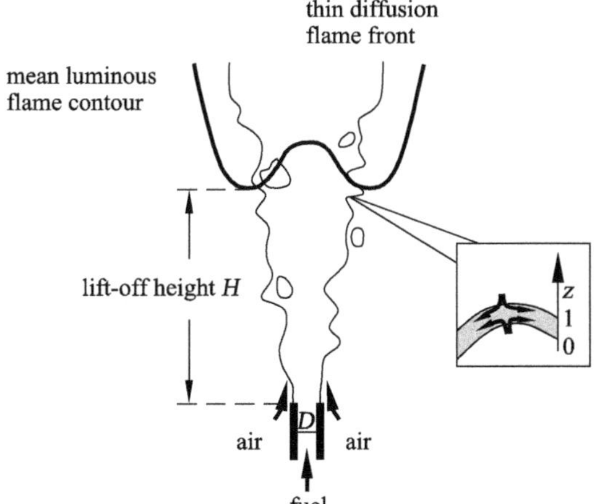

Fig. 9.47. Schematic representation of the lift-off behavior of a turbulent nonpremixed free jet flame

$$\tilde{T}(\boldsymbol{r}) = \int_0^1 \int_0^{\chi_\mathrm{q}} T^{(\mathrm{F})}(\chi, \xi) \cdot \tilde{P}^{(\mathrm{F})}(\chi, \xi; \boldsymbol{r}) \cdot \mathrm{d}\chi \cdot \mathrm{d}\xi + \int_0^1 \int_{\chi_\mathrm{q}}^{\infty} T_\mathrm{u}(\chi, \xi) \cdot \tilde{P}^{(\mathrm{F})}(\chi, \xi; \boldsymbol{r}) \cdot \mathrm{d}\chi \cdot \mathrm{d}\xi. \quad (9.69)$$

Analogous expressions are obtained for the other averages in (9.65). After the local quenching in nonpremixed flames, the reactants mix. This leads to local regions of partially premixed flames, and a further parameter is required to describe this premixing (*B. Rogg et al.* (1987)). The processes in turbulent premixed flames will be treated in Section 9.3.6.

PDF Simulations of Turbulent Nonpremixed Flames

In this section we noted that the *closure problem* of the chemical source terms is solved if the joint probability density function (PDF) of the scalar is known. Some of the methods used assume certain analytical expressions for the PDF (e.g. cut-off Gauß functions or β-functions). These functions are determined by the average and variance of one variable. The balance equations for these two variables can be derived from the Navier–Stokes equations.

Although great progress has been made with this process (see, e.g. *P. A. Libby* and *F. A. Williams* (1994)), the fact cannot be avoided that the actual PDFs often have properties that are only insufficiently reproduced by analytical functions. In principle, every PDF can be described by its (infinitely many) moments. However, the derivation of the balance equations for the higher moments and their solution is not viable from a practical point of view.

The form of the joint probability density function of the scalar comes from the mixing processes and the chemical reaction and is thus determined by the Navier–Stokes equations together with the species conservation equations. Starting out from these equations we can derive a transport equation for the *joint probability density function of velocity and scalars* (*S. B. Pope* (1991)). The single-point probability density function

$$f(v_x, v_y, v_z, \psi_1, \ldots, \psi_n; x, y, z, t) \cdot \mathrm{d}v_x \cdot \mathrm{d}v_y \cdot \mathrm{d}v_z \cdot \mathrm{d}\psi_1 \cdots \mathrm{d}\psi_n \quad (9.70)$$

indicates the probability that the fluid has velocity components in the range v_i and $v_i + \mathrm{d}v_i$ and values of the scalars (mass fractions, density, enthalpy) between ψ_α and $\psi_\alpha + \mathrm{d}\psi_\alpha$ at time t and at position x, y, z. The transport equation that describes the development of the PDF then reads (*S. B. Pope* (1991))

$$\rho(\boldsymbol{\Psi}) \cdot \frac{\partial f}{\partial t} + \rho(\boldsymbol{\Psi}) \cdot \sum_{j=1}^{3} \left(v_j \cdot \frac{\partial f}{\partial x_i} \right) + \sum_{j=1}^{3} \left(\left[\rho(\boldsymbol{\Psi}) \cdot g_j - \frac{\partial \overline{p}}{\partial x_j} \right] \cdot \frac{\partial f}{\partial v_j} \right)$$
$$+ \sum_{\alpha=1}^{n} \left(\frac{\partial}{\partial \boldsymbol{\Psi}_\alpha} [\rho(\boldsymbol{\Psi}) \cdot S_\alpha(\boldsymbol{\Psi}) \cdot f] \right)$$
$$= \sum_{j=1}^{3} \left(\frac{\partial}{\partial v_j} \left[\left\langle \frac{\partial p\prime}{\partial x_j} - \sum_{i=1}^{3} \frac{\partial \tau_{ij}}{\partial x_i} \middle| \boldsymbol{v}, \boldsymbol{\Psi} \right\rangle \cdot f \right] \right)$$
$$+ \sum_{\alpha=1}^{n} \left(\frac{\partial}{\partial \boldsymbol{\Psi}_\alpha} \left[\sum_{i=1}^{3} \left\langle \frac{\partial J_i^\alpha}{\partial x_i} \middle| \boldsymbol{v}, \boldsymbol{\Psi} \right\rangle \cdot f \right] \right), \tag{9.71}$$

where x_i denotes x, y, and z coordinates, g_i the gravitational acceleration in the x, y, and z directions, $\boldsymbol{\Psi}$ the n-dimensional vector of the scalars, v_j the components of the velocity vector $\boldsymbol{v}$, S_α the source terms for the scalars (e.g. chemical source terms), τ_{ij} the components of the shear stress tensor, and J_i^α the components of the molecular flux (e.g. diffusion or heat flux density) of the scalar α in the i direction. The terms $\langle q|\boldsymbol{v}, \boldsymbol{\Psi}\rangle$ denote *conditional expectation values* of the variable q. Thus $\langle q|\boldsymbol{v}, \boldsymbol{\Psi}\rangle$ is the average value of q under the condition that the velocity and composition take on the values $\boldsymbol{v}$ and $\boldsymbol{\Psi}$, respectively. Physically, this means that the conditional expectation values describe the average values of the molecular fluxes for certain values of the velocity and the scalars.

The first term on the left-hand side describes the rate of change of the PDF, the second describes the convection (transport in physical space), the third the transport in velocity space due to gravitation and pressure gradients, and the fourth the transport in state space due to source terms (e.g. chemical reactions). It is important to note that here all terms on the left-hand side of the equation appear in closed form. Therefore, the chemical reaction is treated exactly, the great advantage of this method.

However, the conditional expectation value $\langle q|\boldsymbol{v}, \boldsymbol{\Psi}\rangle$ of the molecular flux terms on the right-hand side of the equation have to be modeled, since they do not appear in closed form. This means that a dependence of these terms on known (e.g. calculated) quantities has to be formulated. Such models are necessary because of the fact that we are using only a single-point PDF to describe the flow and therefore have no information about spatial correlations.

The transport equation (9.71) for the single-point PDF cannot be solved simply with today's computers. The problem is its high dimensionality. Whereas in the Navier–Stokes equations the only independent variables are the time and the spatial coordinates, in the transport equation (9.71) the velocity components and the scalar variables are also independent variables. Use of the *Monte-Carlo* method is one way out of this problem. Here the PDF is approximated by a very large number (e.g. 10^5 in spatially two-dimensional systems) of *stochastic particles*. The properties of these particles vary in time, depending on the convection, chemical reaction, molecular transport, and external forces, and hence they mimic the development of the PDF (*S. B. Pope* (1991)).

In practical applications the joint probability density function of velocities and scalars $f(\boldsymbol{v}, T, w_i, \rho)$ is reduced to a PDF for the scalars (which describes the chemical reaction exactly) and the velocity field is calculated by means of a turbulence model (e.g. the K-ϵ model) that is based on the averaged Navier–Stokes equations. The two models are coupled via the density ρ. The PDF model yields a density field that is inserted into the turbulence model. From this a new flow field is computed, and the information is passed back to the PDF model. This process is repeated until a convergent solution is obtained. Such hybrid PDF/turbulence model simulations permit realistic treatment of turbulent flames. Figure 9.48 shows exemplarily a comparison between experimental results in a recirculating nonpremixed methane–air flame with a simulation. The simulation is based on a hybrid method in combination with simplified chemical kinetics (ILDM, *J. Warnatz et al.* (2001)). Agreement between the results is seen to be very good. The model is considerably better than an *eddy dissipation* model (improved eddy-break-up model; see Section 9.3.4), which assumes that the chemical reaction takes place much faster than the molecular mixing. The assumption of fast chemistry overestimates

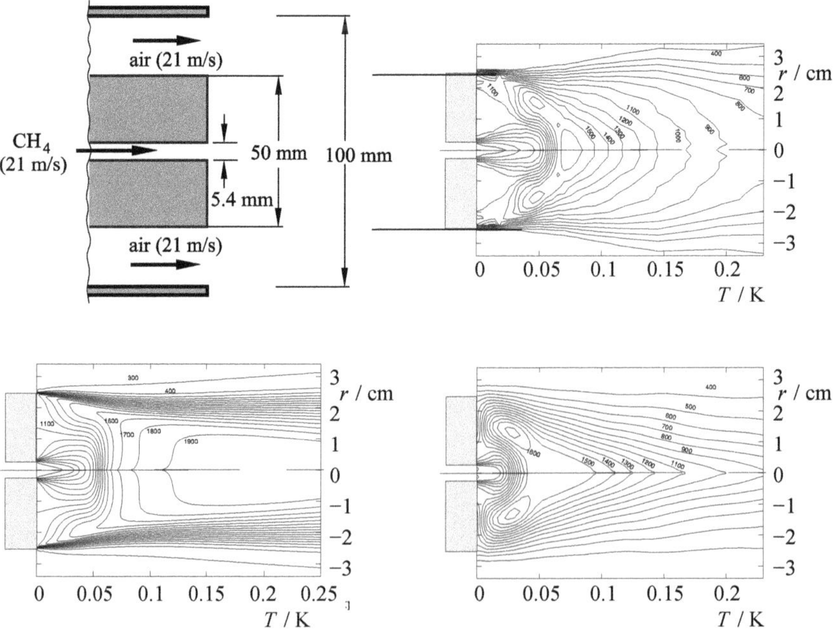

Fig. 9.48. Simulation of a nonpremixed CH_4-air jet flame, (above left) configuration, (above right) measured temperature profile, $T_{max} \approx 1600$ K, (below left) eddy dissipation model, $T_{max} \approx 1900$ K, (below right) combined PDF/turbulent flow model, $T_{max} \approx 1600$ K

the formation of products and, thus, the temperature rises. In consequence, the predicted values for the NO formation will be far too large.

9.3.6 Turbulent Premixed Flames

Figure 9.49 shows a premixed flame in a turbulent flow field. A mixture of fuel and oxidizer flows from above and a premixed flame is stabilized by the recirculation of hot gases behind a stagnation body. The flame propagates from the stagnation body into the unburnt mixture. If the flow were laminar, the flame would have a "V" shape. However, since the flow is turbulent, the angle of the flame changes constantly, depending on the local free-stream velocity, and the flame has the shape shown in Figure 9.49.

The three-dimensional structure increases with increasing degree of turbulence. This can be understood using the *Borghi diagram* (*R. Borghi* (1984), *S. Candel et al.* (1994), and *T. Poinsot et al.* (1991)), shown in Figure 9.50 in a double-logarithmic plot. Here v'/v_{L}, the turbulence intensity v' normalized by the laminar flame velocity v_{L}, is plotted against l_0/l_{L}, i.e. the largest length scale l_0 of the vortices normalized by the laminar flame thickness l_{L}.

The diagram is partitioned into various regions by different straight lines. If the turbulence Reynolds number $R_l = v' \cdot l_0/\nu$ is smaller than one, $R_l < 1$, laminar combustion takes place. The domain of turbulent combustion ($R_l > 1$) can be further subdivided. It is useful to introduce two new dimensionless quantities, namely, the *turbulent Karlovitz number* Ka and also the *turbulent Damköhler number* Da.

The turbulent Karlovitz number Ka describes the ratio of the time scale t_{L} of the laminar flame ($t_{\mathrm{L}} = l_{\mathrm{L}}/v_{\mathrm{L}}$) to the Kolmogorov time scale t_{K}:

$$\mathrm{Ka} = \frac{t_{\mathrm{L}}}{t_{\mathrm{K}}} \quad \text{with} \quad t_{\mathrm{K}} = \sqrt{\frac{\nu}{\tilde{\epsilon}}}, \tag{9.72}$$

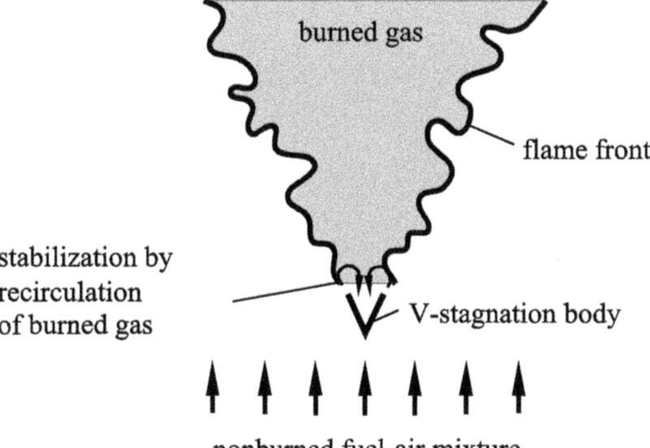

Fig. 9.49. Schematic representation of the instantaneous state of a turbulent premixed flame stabilized with a stagnation body

where ν is a characteristic kinematic viscosity ($\nu = \mu/\rho$), and $\tilde{\epsilon}$ is the dissipation rate of the turbulent kinetic energy. At the Kolmogorov scale, the time that a vortex of size l_{K} requires for one revolution is the same as the time required for diffusion through the vortex. If the time scale of the laminar flame is smaller than the Kolmogorov scale, local laminar premixed flames occur, embedded in the turbulent flow. In the Borghi diagram this *flamelet regime* lies below the straight line Ka = 1.

The turbulent Damköhler number Da describes the ratio between the macroscopic time scales and the time scale of the chemical reaction:

$$\mathrm{Da} = \frac{t_0}{t_{\mathrm{L}}} = \frac{l_0 \cdot v_{\mathrm{L}}}{v' \cdot l_{\mathrm{L}}}. \tag{9.73}$$

For Da < 1 the time needed for the chemical reaction is longer than the time needed for physical processes. In this regime, the vortices interact directly with the flame structure, which is spread out so much that it can hardly still be described as a "flame front." In the Borghi diagram, this regime lies above the straight line Da = 1. This reaction is also called a *homogeneous reactor*, *perfect mixing reactor*, or *ideal reactor*.

In between the ideal reactor regime and the flamelet regime is the distributed *reaction zone*, where some of the vortices are in the flame front (vortices whose length scales l_{K} are smaller than l_{L}). There is a wide spectrum of different dissipation rates $\tilde{\epsilon}$ in each turbulent flow, which probably have a logarithmic-normal distribution (*J. Warnatz et al.* (2001), *W. J. A. Dahm* and *E. S. Bish* (1993), *W. J. A. Dahm et al.* (1995)). For this reason, the conditions in a turbulent flame cannot be described as a point in the

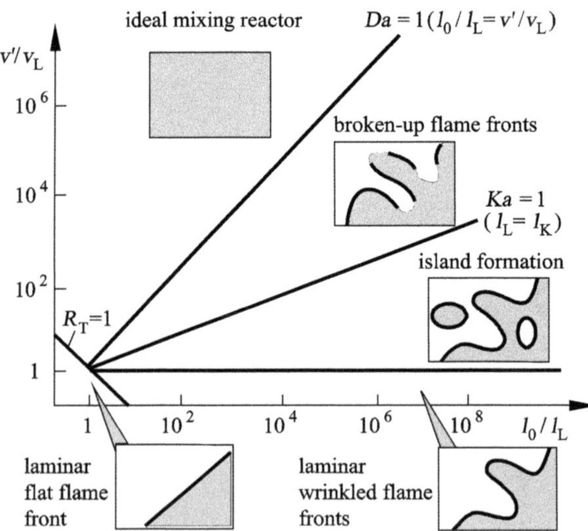

Fig. 9.50. Borghi diagram

Borghi diagram but rather as a zone that can extend across different domains of the diagram.

Flamelet Models

The methods described above permit the calculation of laminar premixed flames, e.g. the profiles of temperature and concentration (including pollutants) as well as the flame velocity. However, turbulent flames are three-dimensional and unsteady. Therefore, direct numerical simulation (DNS) (see Section 9.3.2) greatly exceeds the computational capacity available today. The practical alternative is to develop models that permit the most important properties of the turbulent flames to be described.

The *flamelet model* of turbulent premixed flames is analogous to the flamelet model of nonpremixed flames. The turbulent flame is considered as an ensemble of many small laminar flames in the turbulent flow field. If the turbulence Reynolds number R_l tends to zero, the model passes over correctly to the model of a laminar flame. It is generally agreed that the flamelet concept can be applied in the region of large Damköhler numbers where the turbulent time scales are larger than the time scale of laminar flames. This region is in the lower right part of the Borghi diagram (Figure 9.50).

In turbulent nonpremixed flames it was possible (at least in the case of fast chemistry) to fully describe the concentration field through the mixture

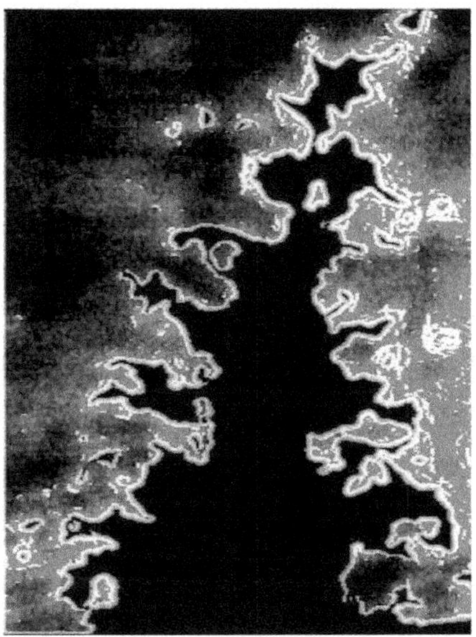

Fig. 9.51. Laser–light-sheet LIF measurement of the OH concentration in a turbulent premixed natural gas–air jet flame stabilized on a nozzle of 3 cm diameter; the black interior shows the region of the inflowing unburnt mixture ($\Phi = 0.8$, $R_l = 857$, $\mathrm{Ka} = 0.07$)

fraction. For turbulent premixed flames this concept has no meaning, since the fuel and oxidizer are already mixed together before the reaction. Therefore, another variable must be chosen to describe the combustion process. The use of a *reaction progress variable* c has gained acceptance. This describes the progress of the combustion in a premixed flame front and, like the mixture fraction, has values between zero and one (*K. N. C. Bray* (1980)). For example, the percentage of the formation of a final product such as

$$w_{CO_2} = c \cdot w_{CO_2,b} \tag{9.74}$$

can be used, where the index b indicates the burnt gas. The profile used may not have a maximum, since there would then be no way to uniquely determine c. The scalars such as OH, O_2, CO, CO_2, etc., are then uniquely determined at each point in the flow by the reaction progress variable c and, if necessary, by the local dissipation of c.

Laminar premixed flames with given values of the dissipation rate can, in the case of a counterflow arrangement, be obtained experimentally (*C. K. Law* (1989)) and numerically (*G. Stahl* and *J. Warnatz* (1991)).

Justification of the application of the flamelet model in premixed turbulent combustion at temporal resolution has been observed in laser–light-sheet experiments. An example is shown in Figure 9.51. In this turbulent Bunsen flame, the flamelet assumption seems justified. The figure shows an LIF-OH snapshot of a turbulent natural gas–air free-jet premixed flame at a burner in a semi-industrial scale. Again the wrinkled laminar flame structures can be seen. In order to apply the flamelet model, a model is required to describe the transport and the evolution of c. The flamelet model can then by used to determine from c the temperature, species concentrations, and density, which are then inserted into the turbulence flame model. There are many ways of coupling the flamelet and turbulence models, described in, for example, *T. Ashurst* (1995), *S. Candel et al.* (1994), *S. B. Pope* (1991), *P. A. Libby* and *F. A. Williams* (1994), *N. Peters* (1987).

Turbulent Flame Velocity

We also attempt to describe the progress of a turbulent premixed flame front (in analogy to the laminar case) by a turbulent flame velocity v_T. In the

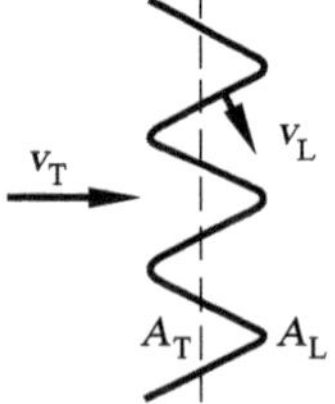

Fig. 9.52. Schematic representation of the propagation of a turbulent premixed flame front

simplest case, the turbulent flame front is considered to be a wrinkled laminar flame front (*G. Damköhler* (1940)), using the ansatz

$$\rho_\mathrm{u} \cdot v_\mathrm{T} \cdot A_\mathrm{T} = \rho_\mathrm{u} \cdot v_\mathrm{L} \cdot A_\mathrm{L}, \tag{9.75}$$

where A_L denotes the total surface area of the wrinkled laminar flame fronts, A_T the area of the mean turbulent flame front, and v_L the laminar flame velocity (see Figure 9.52). We then obtain the basic relation

$$v_\mathrm{T} = v_\mathrm{L} \cdot \frac{A_\mathrm{L}}{A_\mathrm{T}}. \tag{9.76}$$

The ratio of v_T and v_L is therefore given by the area ratio of laminar and (mean) turbulent flame surfaces. For example, Damköhler used the ansatz $A_\mathrm{L}/A_\mathrm{T} = 1 + v'/v_\mathrm{L}$, where v' indicates the turbulent fluctuation velocity (compare Section 9.3.6). Thus we obtain the expression

$$v_\mathrm{T} = v_\mathrm{L} + v' = v_\mathrm{L} \cdot \left(1 + \frac{v'}{v_\mathrm{L}}\right). \tag{9.77}$$

This result is in agreement with experimental results as long as the turbulence intensity is not too large (appearance of flame quenching). In particular, the model describes the fact that in automotive combustion engines, an increase in the piston speed (v' is approximately proportional to the rpm) leads to an increase in the combustion rate. Without this relation, effective automotive combustion would be restricted to low rpm (*J. B. Heywood* (1988)).

Also in agreement with experiment (*Y. Liu* and *B. Lenze* (1988)) is the fact that (9.75) indicates no dependence on the turbulent length scale (e.g. on the integral length scale l_0). This can be understood using a simple schematic diagram (Figure 9.53). Although both flame fronts shown have different length scales, the total areas of the laminar flame fronts are the same, and hence the turbulent flame velocities are the same.

Problems occur in this simple model if the mixture is too rich or too lean (outside the limits of combustibility, to be determined from Figure 9.55 by extrapolation). Then the laminar flame velocity v_L is zero, and, thus there is no flame, although the model incorrectly predicts $v_\mathrm{T} = v'$.

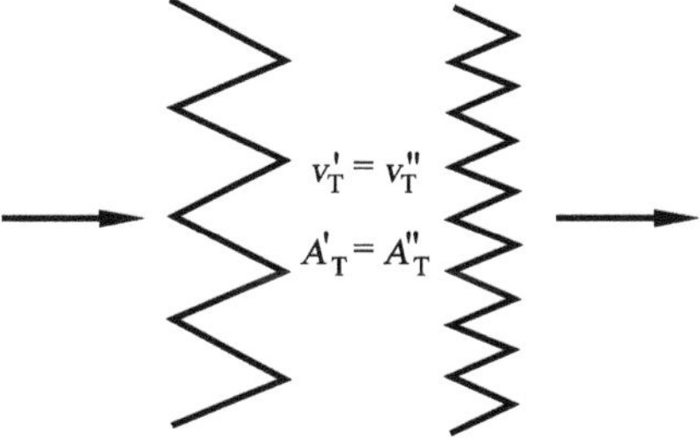

Fig. 9.53. Schematic representation of two flame fronts with different length scales but the same area

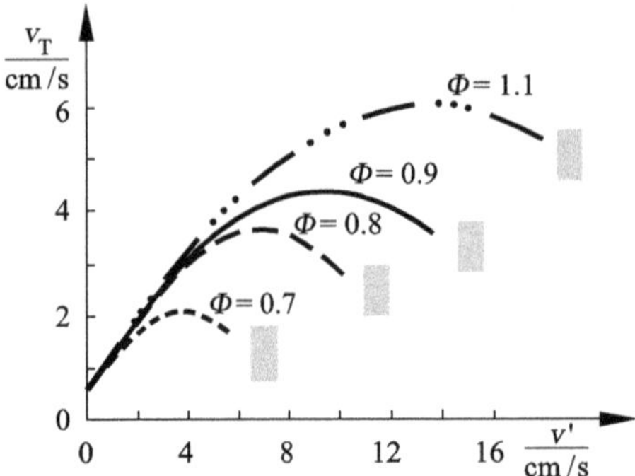

Fig. 9.54. Dependence of the turbulent flame velocity on the turbulence intensity, combustion of a C_3H_8-air mixture; gray areas: quenching region

Flame Quenching

With increasing turbulence intensity v' we observe a maximum of the turbulent flame velocity v_T that is caused by local flame quenching. This has been shown by *D. Bradley* and coworkers (1984), (1993) in a combustion vessel with C_3H_8-air at intensive turbulence generation by many strong ventilators (Figure 9.54). An explanation of this behavior is immediately obtained by recalling the flamelet idea (quenching at sufficiently high strain).

Figure 9.55 shows the strain necessary for quenching as a function of the equivalence ratio Φ for a pair of counterflow methane–air premixed flames. Different reaction mechanisms are investigated to ensure that the discrepancy between measurement and simulation is not due to the chemistry model. Experience shows that the small energy losses, which are difficult to quantify in experiment, may be responsible for the discrepancy (*G. Stahl* and *J. Warnatz* (1991)).

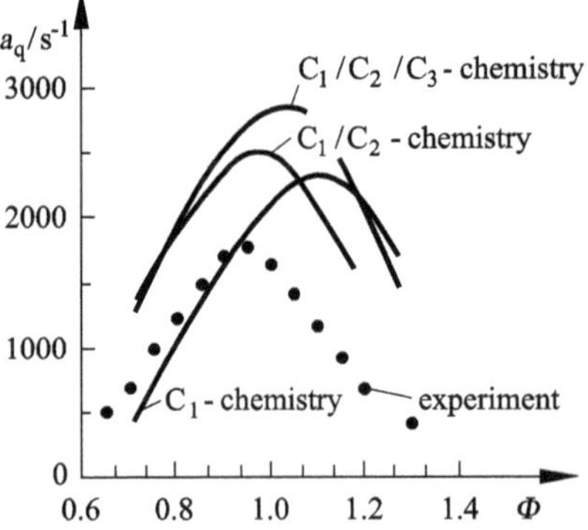

Fig. 9.55. Dependence of the necessary strain rate a_q for flame quenching on the mixture composition for propane-air flames

These measurements and simulations in laminar conditions together with a flamelet model permit the quenching observed in turbulent premixed flames to be explained.

Furthermore, calculations show that the characteristic time for flame quenching is only a fraction of a millisecond. The contraction of the gas caused by sudden quenching is considered to be the source of flame noise (together with the resonance conditions corresponding to the geometry) (*G. Stahl* and *J. Warnatz* (1991)).

As Figure 9.55 shows, lean (and also rich) mixtures are quenched particularly easily. This is one of the reasons why unexpectedly strong hydrocarbon emissions are observed in lean combustion engines. One might have assumed naively that the excess of oxygen would lead to complete combustion of the fuel.

9.4 Hypersonic Flows

9.4.1 Physical-Chemical Phenomena in Re-Entry Flight

A re-entry flight begins in the outermost layers of the atmosphere (Figure 9.56). In this part of the flight, the low densities mean that we are in the *free-molecular-flow* regime. In this gas kinetic regime the Boltzmann equation (5.64) with the distribution function (5.65) must be solved. With decreasing flight altitude and increasing air density, the mean free path of the gas particles is reduced. It is only at lower air layers that we can speak of a *continuum flow*. In this regime of the re-entry trajectory, the maximum heat transfer of the re-entry flight occurs and then the laminar-turbulent transition in the boundary layer of the vehicle (see *H. Oertel* (1994), (2005)).

The characteristic number that is decisive in determining the region of validity of the continuum-mechanical description is the *Knudsen number* $\mathrm{Kn} = \overline{\lambda}/L$, the ratio of the mean free path $\overline{\lambda}$ in the gas to a characteristic length L of the vehicle. The continuum is characterized by $\mathrm{Kn} \leq 10^{-2}$. For example, the mean free path at an altitude of 90 km is about 10^{-3} m, so

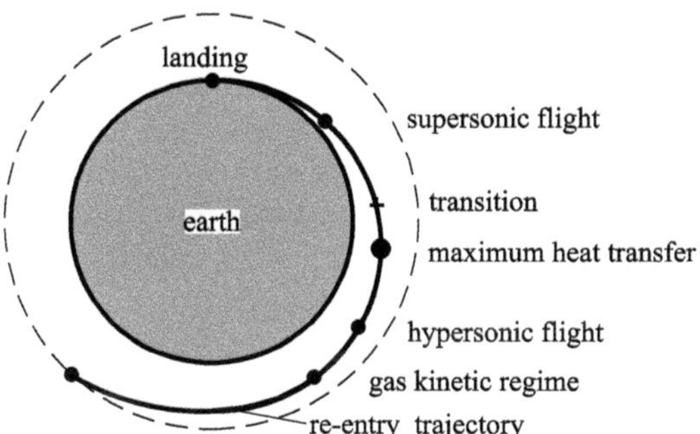

Fig. 9.56. Re-entry trajectory

that the Navier–Stokes equations can be used to describe the flow field past a body with a length of 0.1 m or more.

These conditions are very difficult to attain experimentally. Therefore, numerical aerothermodynamics, based on the Navier–Stokes equations of continuum mechanics in combination with detailed physical-chemical models, is a useful tool in predicting characteristic flow quantities such as heat flux, pressure distribution, and friction coefficient.

We can classify the physical-chemical models by considering the phenomena along the stagnation streamline in front of the vehicle. The supersonic flow causes a shock wave to form, resulting in a sharp increase in pressure, density, and temperature, related to a reduction in the flow velocity. This increase takes place on a length scale comparable with the mean free path of the molecules. In the shock wave, the high-velocity flow with Mach numbers $M \gg 1$ passes over to a high-enthalpy flow with $M < 1$.

In contrast to the translational degrees of freedom, the vibrational and rotational degrees of freedom of the molecule and the composition of the air past the shock initially experience no change. However, directly behind the shock wave, triggered by collisions of particles with higher translational temperature, the vibrational and rotational degrees of freedom are excited, and chemical reactions occur. Typical translational temperatures behind the shock wave that trigger these real gas phenomena are detailed below:

Temperature range	Effect
$T <$ 400 K	no real gas effects
400 K $< T <$ 2000 K	O_2-vibrational excitation
600 K $< T <$ 3000 K	N_2-vibrational excitation
2000 K $< T <$ 5000 K	O_2-dissociation
4000 K $< T <$ 10000 K	N_2-dissociation
1000 K $< T <$ 5000 K	NO-formation
3000 K $< T <$ 8000 K	NO-dissociation

Which of the above effects will actually occur depends on the actual re-entry trajectory of the vehicle. This determines the maximum translational temperature. As well as the processes presented already in the flow past re-entry bodies, there are additionally heterogeneous physical-chemical processes on the surface of the body that come under the collective term gas–wall interaction. In flows that have a high degree of dissociation close to the surface of the body, reactive interactions such as erosion and catalytic reactions are important. In order to predict the maximum heat load occurring on a re-entry body, models of the gas–wall interaction are essential.

In addition to these reaction-kinetic phenomena, mass, momentum, and energy transport occur in the entire flow, in particular in the boundary layer directly on the surface of the re-entry body. These processes are described by transport models with which the diffusion coefficient, viscosity, and thermal conductivity can be determined. If the velocity in re-entry flight is so high that temperatures considerably above 5000 K occur in the shock wave, ionization of the air also occurs.

9.4.2 Chemical Nonequilibrium

The increase of the translational temperature behind the shock wave leads to chemical reactions. The simplest reaction model is a five-component model common in the literature (*U. Riedel et al.* (1993)). The onflowing air contains oxygen and nitrogen molecules. These are dissociated by the temperature rise behind the shock wave, and O and N atoms appear, which then form NO molecules. The following reaction scheme for hot air is obtained. It consists of three dissociation reactions (R1), (R3), (R5) and two exchange reactions (R7) and (R9), as well as the associated reverse reactions (R2), (R4), (R6), and (R8) and (R10):

reaction	A	β	E_a	
$O_2 + M' \longrightarrow O + O + M'$	$2.70 \cdot 10^{19}$	-1.0	494.0	(R1),(R2)
$N_2 + M'' \longrightarrow N + N + M''$	$3.70 \cdot 10^{21}$	-1.6	941.0	(R3),(R4)
$NO + M''' \longrightarrow N + O + M'''$	$2.90 \cdot 10^{15}$	0.0	621.0	(R5),(R6)
$O + N_2 \longrightarrow NO + N$	$1.82 \cdot 10^{14}$	0.0	319.0	(R7),(R8)
$NO + O \longrightarrow O_2 + N$	$3.80 \cdot 10^{9}$	1.0	173.1	(R9),(R10)

	O_2	N_2	O	N	NO
M′	1.00	0.10	2.80	0.10	0.10
M″	0.10	1.00	0.10	2.80	0.10
M‴	0.05	0.05	1.00	1.00	1.00

with $k = A \cdot T^{\beta} \cdot \exp(-E_a/(R \cdot T))$. The Arrhenius parameter A is stated in cm · mol · s and the activation energy E_a in kJ/mol. The symbol M stands for an arbitrary species present in the reaction system that is involved in the reaction as a collision partner but that does not itself react.

The dissociation of oxygen takes place directly via the reaction (R1). Atomic nitrogen, on the other hand, has a triple bond with high bond energy and associated high activation energy and therefore is mainly formed via the two exchange reactions (R7) and (R9) and only to a very small degree via the dissociation reaction (R3). The NO formed in (R7) then decays via the NO dissociation reaction (R5) into N and O.

How fast and to what extent these reactions take place is a question of time scales. If the time scale prescribed by the flow is large compared with the time scale of the chemical reactions, the chemical reactions take place so fast that chemical equilibrium occurs. Formation and consumption of the individual species in this limiting case of infinitely fast reactions are no longer time dependent.

If the pressure and temperature are given, the equilibrium species concentrations are obtained by minimizing the *free enthalpy* of the system. The reaction system contains five species and is described according to *Gibbs's phase law* by three linearly independent reaction pairs ((R1),(R2); (R3),(R4);(R5),(R6)) and two components (N and O). However, we do not further treat this case here, since in hypersonic flows the flow velocities are typically so large that the above assumption about the time scales no longer leads to sufficiently precise predictions of the flow state.

The typical situation in hypersonic re-entry flights is rather that the time scale prescribed by the flow is comparable with the time scale of the chemical reactions. Therefore, the flow is in *chemical nonequilibrium*, which has to be taken into account in the physical-chemical model of the flow. In addition, such flows are frequently also in *thermal nonequilibrium* (see Section 9.4.3).

The rates of formation $(\partial Z_i/\partial t)_{\mathrm{chem}}$ (rate of change of the mass fraction) of each species in a fluid element is therefore time dependent. Assuming that the temperature and the pressure of the fluid element are known, they can be calculated from the rate equations presented in Section 9.1.3, a set of coupled ordinary differential equations. In general, when the temperature and pressure in the flow field are not only dependent on the chemical reactions but also change because of the mass, momentum, and energy transport, the mass fractions must be calculated coupled with the flow. However, when this coupling is neglected, useful insights into the rate-determining reactions and

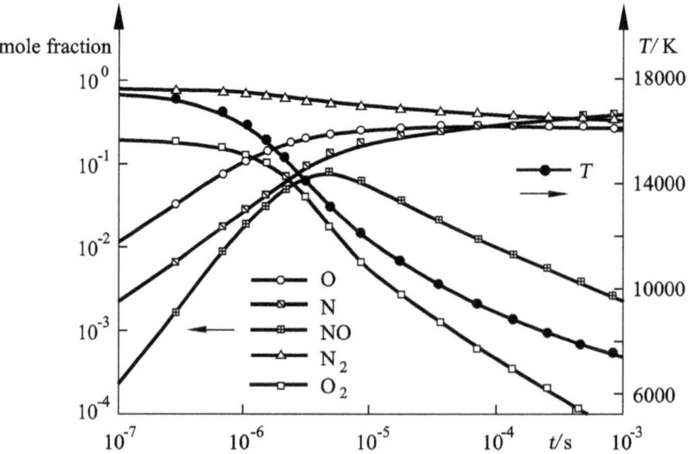

Fig. 9.57. Development of the mole fractions and the temperature for the reaction mechanism of hot air

the typical time scales of such a system are found, a technique known in the literature as a volume-averaged model or a zero-dimensional model.

In order to set up the equations for the rate of change of the mass fractions of the five components, Arrhenius parameters have to be specified for each reaction from (9.21). The parameters as evaluated by *U. Riedel* (1993) are the experimentally determined values for the reaction rates as functions of the temperature, as available in the literature.

Figure 9.57 shows the change of the mole fractions of O_2, O, N_2, N, and NO as functions of the time calculated with this reaction mechanism for an initial translational temperature behind the shock wave of $T = 17500\ K$. It is assumed that the initial composition of the air is 79% N_2 and 21% O_2. Trace gases are not taken into account. The total dissociation of the oxygen molecules, completed after about 0.5 ms, is clearly seen. In contrast, only a small part of N_2 has dissociated. NO passes through a maximum at 4.1 μs. The temperature then decreases most sharply if the change of the O and N mole fractions has a maximum, since then the greatest amount of energy is required for dissociation.

9.4.3 Thermal Nonequilibrium

As well as chemical reactions, there are also vibrational and rotational excitations of the molecules by collisions with particles of higher translational energy behind the shock wave. This can result in completely different behavior of the physics and chemistry of the dissociated air than that predicted by thermal models. Possible processes are the energy exchange between vibration and translation (known as V-T energy transfer), vibration and rotation (V-R energy transfer), or vibration levels (V-V energy transfer).

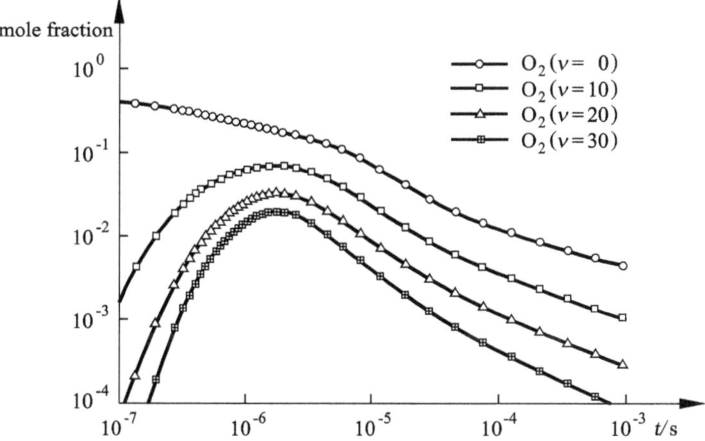

Fig. 9.58. Population of ground state and selected excited states of the oxygen molecule as a function of time

All these possible interactions can be described in principle by so-called *master equations*, which relate the change of the population number of an internal degree of freedom to the transition probability of the V-T, V-R, and V-V energy transfer of the O_2, N_2, and NO molecules. Since the rotational degrees of freedom attain thermal equilibrium quite rapidly, on average only three to four collisions are necessary, we can assume a common translational–rotational temperature. On the other hand, vibrational equilibrium can take considerably longer to be attained, and this (thermal) nonequilibrium situation must be taken into account in the description of the hypersonic flow.

One possible approach in modeling the vibrational excitation is to select the molecules according to their excited state. All excitation and relaxation processes associated with thermal nonequilibrium are thus mapped onto a detailed reaction mechanism of state-selected molecules $O_2(v)$ and $N_2(v)$, where v stands for the possible excited vibrational states and is counted from the ground state with $v = 0$ up to $v = v_{\max}$ directly below the dissociation limit of the molecule. If we describe the energy level that is occupied by vibrational excitation with the model of the *anharmonic oscillator*

$$E(v) = h \cdot c \cdot \left[\nu_0 \cdot (v + \frac{1}{2}) - \nu_0 \cdot x_e \cdot (v + \frac{1}{2})^2 \right] ,$$

for nitrogen we obtain $v_{\max}^{N_2} = 46$ and for oxygen $v_{\max}^{O_2} = 36$. Here ν_0 is the frequency of the ground oscillation, $\nu_0 \cdot x_e$ the anharmonicity constant, c the speed of light, and h Planck's constant. This splitting is not carried out for NO, since it appears only at low concentrations. Similarly, the rotational excitation is not taken into account (assumption of a common rotational–translational temperature).

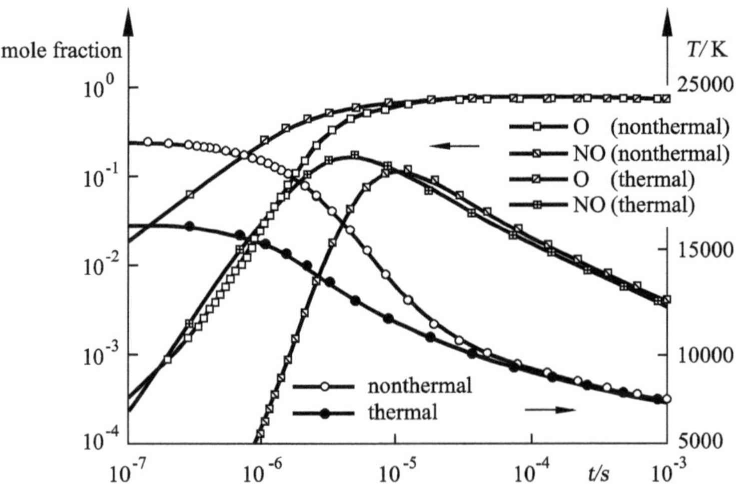

Fig. 9.59. Comparison of the temporal behavior of the mole fraction of the O atoms, NO molecules, and the temperature

Within the framework of this approach the difficulty arises of determining all detailed rate coefficients for the state-selected species. Experimental results in the literature are generally based only on the lower excited states. Model assumptions are necessary to be able to state the Arrhenius parameter for dissociation and exchange reactions, as well as for the dependence of the V-T energy transfer on the vibrational quantum number v. In total, using this approach described in detail in *U. Riedel et al.* (1993), a reaction mechanism of 87 species and 502 reactions is obtained.

Figure 9.58 shows the onset of selected vibrational states and the simultaneous reduction in occupation of the ground state. All excited states pass through a maximum at a time between 1 and 10 μs, which is later, the higher the excited state. It is assumed that initially, all oxygen and nitrogen molecules are in the ground state, since at a typical free-stream temperature of air of about 200 K only a small fraction of all molecules are in vibrationally excited states. At this temperature the ratio of the O_2 molecules in the first excited state to those in the ground state is about 10^{-5}.

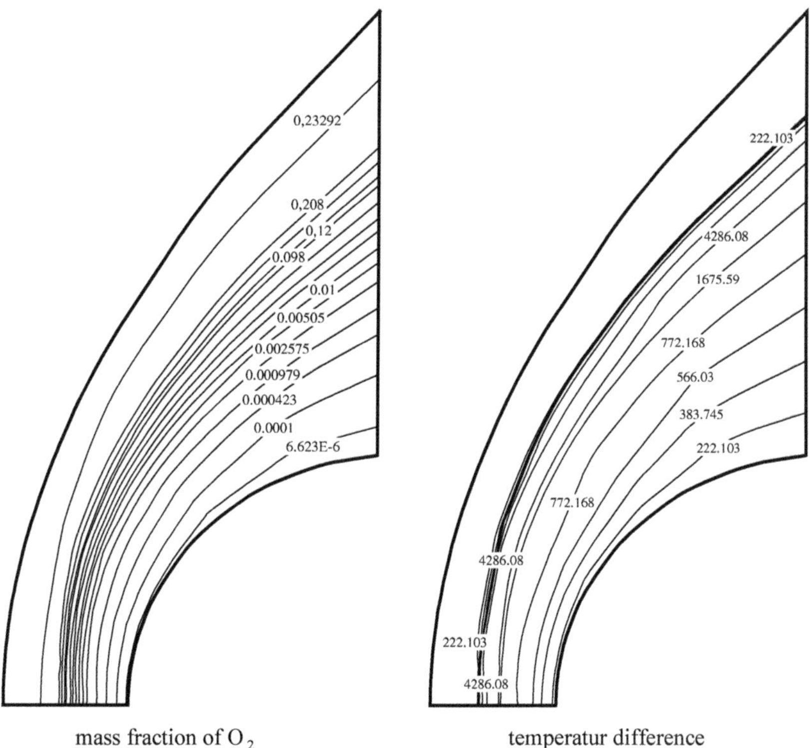

Fig. 9.60. Left: mass fraction of O_2 calculated with the nonthermal reaction mechanism for the flow past a circular cylinder of radius 1 m with a Mach number $M = 25$ (*U. Riedel et al.* (1993))

Figure 9.59 shows a comparison of the models. The heavy lines show the development of the mole fraction of O_2 and the temperature, calculated with the model in *U. Riedel et al.* (1993). The initial temperature for this case is 22000 K at a pressure of 25.6 hPa. After 1 ms the system has cooled to a temperature of 7300 K because of the energy used for the reaction and excitation processes. For comparison, the concentration and temperature development from Figure 9.57, obtained with the assumption of thermal equilibrium, is also shown. The two different initial temperatures were selected for this comparison so that for $t \to \infty$ the same translational temperature is attained. In thermal nonequilibrium the development of the concentration is slowed down. This is indicated clearly by the later rise in the O mole fraction due to delayed dissociation of the oxygen molecules and by the later maximum in the development of the NO mole fraction. In a thermal chemistry model this occurs at 3 μs and in a nonthermal chemistry model only at 9 μs.

Up until now, in the discussion of chemical and thermal nonequilibrium in hypersonic flows it was assumed that the temperature and the pressure of each fluid element are known. However, these two quantities are obtained only from a coupled investigation of the species masses, momentum, and energy balances based on the Navier–Stokes equations.

The left-hand side of Figure 9.60 shows the distribution of O_2 in the flow past a circular cylinder with radius 1 m at Mach number $M = 25$, calculated with the nonthermal model. The Navier–Stokes equations (see Section 5.4.7) are solved. The free-stream velocity is 7200 m/s at a temperature of 205 K. The mass fraction is determined by summation of all vibrationally excited states $O_2(v)$. In both the O_2 and the N_2 molecules the occupation of excited states is lower than that of a Boltzmann distribution in the presence of thermal equilibrium. This means that the lower occupation of vibrational degrees of freedom assumes less energy, and the translational temperature in the shock front rises to 15330 K, which is about 15% higher than the temperature calculated with the thermal model. In addition, the shock stand-off distance is larger. The right-hand side of the figure shows the difference in temperature distributions calculated with the nonthermal model and the thermal model. When an assumption of thermal equilibrium is made, a mass fraction of $6.1 \cdot 10^{-2}$ leads to the formation of about twice as many NO molecules as in the thermal reaction scheme. A consequence of this in the assumption of thermal equilibrium is the steeper rise in N atoms in the region behind the shock wave. However, the values in the region directly in front of the re-entry body no longer differ.

9.4.4 Surface Reactions on Re-entry Vehicles

In recent years, much research has been carried out on materials that could be used as heat shields for re-entry vehicles. These species include RCG (reaction cured glass), investigated for the American Space Shuttle. This consists of 94% SiO_2, 4% B_2O_3, and 2% SiB_4 and has a strong temperature dependence. At low temperatures the recombination probability is small, but it increases greatly with rising temperature (*O. Deutschmann et al.* (1995)). At the temperatures occurring on re-entry, the recombination of O and N atoms on the surface contributes considerably to the heat load of the vehicle.

In order to predict the heat transfer on the vehicle it is necessary to model the gas–wall interaction in addition to the pure gas phase reactions. The concept of elementary reactions is an adequate description of the recombination and the associated heat release. This will now be applied for both reactions in the gas phase as well as for reactions on the surface. The gas phase reaction mechanism is to be extended by the following surface reactions on the vehicle surface:

reaction	A	E_a	S^0
$O + (s) \longrightarrow O(s)$			0.1
$N + (s) \longrightarrow N(s)$			0.1
$O(s) \longrightarrow O + (s)$	$5.0 \cdot 10^{11}$	200.0	
$N(s) \longrightarrow N + (s)$	$7.3 \cdot 10^{11}$	215.0	
$O_2(s) \longrightarrow O_2 + (s)$	$1.0 \cdot 10^{12}$	10.0	
$N_2(s) \longrightarrow N_2 + (s)$	$1.0 \cdot 10^{12}$	10.0	
$O + O(s) \longrightarrow O_2 + (s)$	$6.0 \cdot 10^{13}$	60.0	
$N + N(s) \longrightarrow N_2 + (s)$	$6.0 \cdot 10^{13}$	60.0	
$O(s) + O(s) \longrightarrow O_2(s) + (s)$	$2.0 \cdot 10^{19}$	160.0	
$N(s) + N(s) \longrightarrow N_2(s) + (s)$	$7.0 \cdot 10^{17}$	160.0	

Here A is given in cm · mol · s, E_a in kJ/mol, and the sticking coefficient S^0 is dimensionless. This detailed reaction scheme of the surface processes comprises the adsorption and desorption of nitrogen and oxygen atoms as well as the desorption of the O and N atoms and the O_2 and N_2 molecules attached to the surface. The adsorption of the molecular oxygen and nitrogen is neglected, since the high temperatures mean that these atoms immediately desorb again. The interaction of NO with the surface is not taken into account in the model, since no experimental data are available.

The adsorption of O and N atoms is described via sticking coefficients, which state the probability that a particle will be adsorbed out of the gas phase onto the surface. With the formalism described by *O. Deutschmann et al.* (1995), the sticking coefficients can be transformed into Arrhenius form, leading to a pre-exponential factor dependent on the coverage of the surface.

The actual recombination step of O and N atoms to the associated molecule can take place via two possible reaction paths:

1. Reaction of an O atom (N atom) of the gas phase with an O atom (N atom) adsorbed onto the surface and subsequent desorption of the O_2 molecule (N_2 molecule). This path is called the Eley–Rideal reaction in the literature and is associated with a *lower energy accommodation.*

2. Reaction of two O atoms (N atoms) adsorbed on the surface together and subsequent desorption of the O_2 molecule (N_2 molecule). This path is called the Langmuir–Hinshelwood reaction in the literature and is associated with a *higher energy accommodation.*

The energy that contributes to heating of the surface depends both on the number of recombined atoms and on the amount of energy released that is accommodated by the surface. In order to quantify these two effects we define the recombination coefficient as the ratio

$$\gamma = \frac{j_{\text{reactive}}}{j_{\text{total}}}$$

of the total mass flux of the atoms striking the surface j_{total} and the mass flux of the recombined molecules j_{reactive}. The energy accommodation coefficient β is defined by

$$\beta = \frac{j_{\text{q}}}{j_{\text{reactive}} \cdot \Delta_{\text{D}} h},$$

with the heat flux to the surface j_{q} and $\Delta_{\text{D}} h$ the specific dissociation enthalpy. The energy flux to the surface caused by recombination processes therefore depends on the product $\gamma \cdot \beta$. Both γ and β are temperature dependent. Both coefficients can be calculated using the reaction-kinetic model described above. Comparison with experimental heat flux measurements (*O. Deutschmann et al.* (1995)) indicates the validity of the surface reaction scheme.

Figure 9.61 shows the development of O_2, N_2, and the temperature in the shock wave and along the surface of a semicylinder with radius 1 m in a flow (free-stream velocity 7200 m/s, free-stream temperature 205 K). The x values in the range -1.5 m $\leq x \leq -1.0$ m correspond to the line of symmetry in front of the body. Points on the surface have an x coordinate in the range $-1.0 \leq x \leq 0.0$. The influence of the surface reactions on the wall temperature and the species concentrations in the region close to the wall can clearly be seen.

Recombination of nitrogen atoms at the wall causes an increase in N_2 molecules in the region directly in front of the surface ($x = -1.0$ m). The same effect can be observed for O_2, but at a lower level, since the degree of dissociation of O_2 is much higher than that of N_2. If the gas–wall interaction is neglected, we obtain the conditions shown in the lower part of Figure 9.59, with a decrease in nitrogen and oxygen molecules close to the wall. In both models an increase in oxygen and nitrogen molecules is observed downstream along the body. This is due to the decreasing temperature and increased recombination of the atoms, independent of wall effects.

The temperature in the stagnation point rises due to the released heat of recombination to 1920 K and thus lies about 80 K above the result obtained when the gas–wall interaction is neglected.

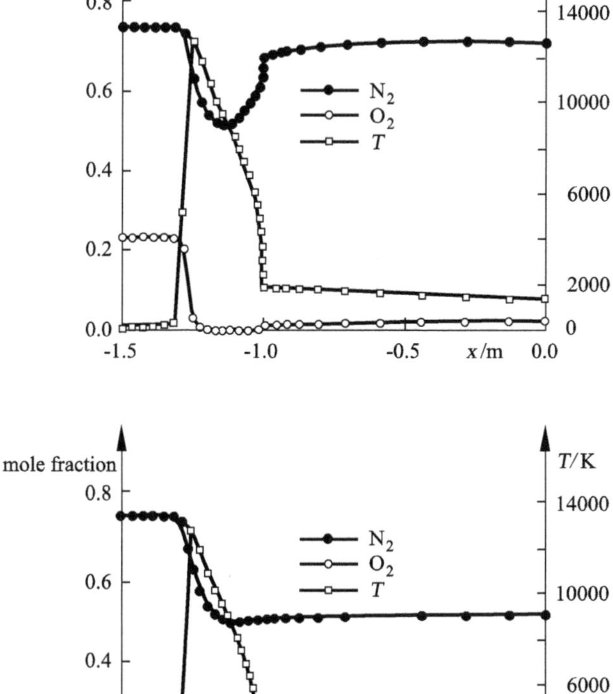

Fig. 9.61. Mass fractions of O_2 and N_2 and temperature along the symmetry line and along the surface. Above: with detailed reaction model of the gas–wall interaction; below: without gas–wall interaction

10. Flows in the Atmosphere and in the Ocean

10.1 Fundamentals of Flows in the Atmosphere and in the Ocean

10.1.1 Introduction

The flows in Earth's atmosphere (air flows) and in the oceans (oceanic currents) do not differ in principle from those flows in technical areas treated in the previous chapters. These are motions of gases (atmosphere) and liquids (ocean) that are acted on by gravity and are determined by pressure and frictional forces. The atmosphere and the oceans are part of a rotating system in which *Coriolis* and *centrifugal forces* also act.

From this point of view, a common treatment of flows in the atmosphere and in the ocean seems sensible. In fact, the name *geophysical fluid dynamics* has come to describe these flows. In particular, we refer to the textbooks of *B. Cushman-Roisin* (1994), *A. E. Gill* (1982), and *J. Pedlosky* (1994). Indeed, *Ludwig Prandtl*, in the original *Essentials of Fluid Mechanics*, considered flows in the atmosphere and oceans at several points throughout the book.

In this chapter we point out the essential elements of the geophysical flow processes in the atmosphere and oceans as discussed by Prandtl. Naturally, we can present only some elements of this topic, and this chapter in no way replaces the specialized literature from the areas of meteorology and oceanography.

10.1.2 Fundamental Equations in Rotating Systems

The fundamental equations for liquids and gases set up in Chapter 5 are also valid for the fluids in the ocean and the atmosphere. It is only the material-specific properties such as density, thermal conductivity, and viscosity of the medium under investigation that have to be set into the relevant equations. A new aspect that must be taken into account is that the Earth, with its atmosphere and oceans, rotates around an axis. The Navier-Stokes equations (5.18) and Reynolds equations (5.33) - (5.37) in Chapter 5.2 are valid for an inertial reference frame without acceleration. In meteorology and oceanography it has become established to refer the equations of motion to the rotating reference frame of the Earth.

H. Oertel (ed.), *Prandtl-Essentials of Fluid Mechanics*,
Applied Mathematical Sciences 158, DOI 10.1007/978-1-4419-1564-1_10,

The coordinate transformations necessary to do this are described in detail in textbooks (e.g. *D. Etling* (2002), *J. Pedolsky* (1994)). We now present a short derivation. We refer to Figure 10.1, that shows a rotating coordinate system. Here $\boldsymbol{\Omega}$ is the vector of rotation of the Earth and $\boldsymbol{r}$ the distance of the mass point. The magnitude of the rotation vector is defined as the angular frequency of the Earth's rotation $\Omega = 2 \cdot \pi / T$ where T is the period of rotation and so $\Omega = 2 \cdot \pi / 24\ h = 0.727 \cdot 10^{-4}\ s^{-1}$.

A fixed point on the surface of the Earth, whose position is given by the vector $\boldsymbol{r}$, has a velocity relative to the inertial system (indicated in the following by the index i) of

$$\left(\frac{\mathrm{d}\boldsymbol{r}}{\mathrm{d}t}\right)_{\mathrm{i}} = \boldsymbol{v}_{\mathrm{f}} = \boldsymbol{\Omega} \times \boldsymbol{r} \quad . \tag{10.1}$$

This velocity $\boldsymbol{v}_{\mathrm{f}}$, also called the peripheral velocity, is perpendicular to the vector of the Earth's rotation and to the radius vector, and is directed to the East.

If a particle of air has velocity $\boldsymbol{v}_{\mathrm{e}}$ relativ to the surface of the Earth (index e for coordinate system rotating with the Earth), in the inertial system it has velocity $\boldsymbol{v}_{\mathrm{i}}$ with

$$\boldsymbol{v}_{\mathrm{i}} = \boldsymbol{v}_{\mathrm{e}} + \boldsymbol{v}_{\mathrm{f}} = \boldsymbol{v}_{\mathrm{e}} + \boldsymbol{\Omega} \times \boldsymbol{r} \quad . \tag{10.2}$$

The relation between the rate of change of the velocity vector $\boldsymbol{v}_{\mathrm{e}}$ in the inertial system and in the Earth's system is:

$$\left(\frac{\mathrm{d}\boldsymbol{v}_{\mathrm{e}}}{\mathrm{d}t}\right)_{\mathrm{i}} = \left(\frac{\mathrm{d}\boldsymbol{v}_{\mathrm{e}}}{\mathrm{d}t}\right)_{\mathrm{e}} + \boldsymbol{\Omega} \times \boldsymbol{v}_{\mathrm{e}} \quad . \tag{10.3}$$

In order to obtain the relation between the accelerations in the inertial system and in the system rotating with the Earth, we apply (10.1) to (10.3)

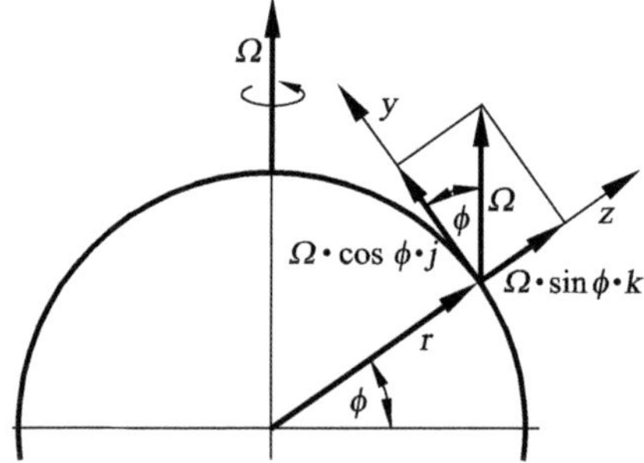

Fig. 10.1. Components of the Earth's rotation $\boldsymbol{\Omega}$ in a tangential phase (the x component and the unit vector $\boldsymbol{i}$ into the figure)

$$\left(\frac{\mathrm{d}\boldsymbol{v}_{\mathrm{i}}}{\mathrm{d}t}\right)_{\mathrm{i}} = \underbrace{\left(\frac{\mathrm{d}\boldsymbol{v}_{\mathrm{e}}}{\mathrm{d}t}\right)_{\mathrm{i}}} + \underbrace{\left(\frac{\mathrm{d}}{\mathrm{d}t}(\boldsymbol{\Omega}\times\boldsymbol{r})\right)_{\mathrm{i}}}$$

$$= \left(\frac{\mathrm{d}\boldsymbol{v}_{\mathrm{e}}}{\mathrm{d}t}\right)_{\mathrm{e}} + \boldsymbol{\Omega}\times\boldsymbol{v}_{\mathrm{e}} + \boldsymbol{\Omega}\times\underbrace{\left(\frac{\mathrm{d}\boldsymbol{r}}{\mathrm{d}t}\right)_{\mathrm{e}}}_{\boldsymbol{v}_{\mathrm{e}}} + \boldsymbol{\Omega}\times\boldsymbol{\Omega}\times\boldsymbol{r} \quad .$$

Now, dropping the index e for the coordinate system rotating with the Earth, the relation between the two accelerations reads

$$\left(\frac{\mathrm{d}\boldsymbol{v}_{\mathrm{i}}}{\mathrm{d}t}\right)_{\mathrm{i}} = \frac{\mathrm{d}\boldsymbol{v}}{\mathrm{d}t} + 2\cdot\boldsymbol{\Omega}\times\boldsymbol{v} + \boldsymbol{\Omega}\times\boldsymbol{\Omega}\times\boldsymbol{r} \quad . \tag{10.4}$$

The additional terms on the right-hand side of the equation are called the Coriolis acceleration $(2\cdot\boldsymbol{\Omega}\times\boldsymbol{v})$ and centrifugal acceleration $(\boldsymbol{\Omega}\times\boldsymbol{\Omega}\times\boldsymbol{r})$.

In total, the Navier-Stokes equations (5.20) for an incompressible flow, and coordinate system rotating with the Earth may be written as

$$\rho\cdot\left(\frac{\partial\boldsymbol{v}}{\partial t} + (\boldsymbol{v}\cdot\nabla)\boldsymbol{v} + 2\cdot\boldsymbol{\Omega}\times\boldsymbol{v} + \boldsymbol{\Omega}\times\boldsymbol{\Omega}\times\boldsymbol{r}\right) = \boldsymbol{f} - \nabla p + \mu\cdot\Delta\boldsymbol{v}. \tag{10.5}$$

For many problems in meteorology and oceanography, description of the motion in spherical polar coordinates, as would be suitable for the Earth, is not necessary. Rather, a Cartesian coordinate system is placed on the Earth's surface, so that its horizontal coordinates $(x,\ y)$, with the unit vectors $\boldsymbol{i}$, $\boldsymbol{j}$, form a *tangential plane* at a certain geographical latitude ϕ. The vertical coordinate z is then perpendicular to this plane, as shown in Figure 10.1. The rotation vector Ω can be decomposed in this coordinate system into

$$\boldsymbol{\Omega} = \boldsymbol{\Omega}\cdot\cos(\phi\cdot\boldsymbol{j}) + \boldsymbol{\Omega}\cdot\sin(\phi\cdot\boldsymbol{k}) = f^*\cdot\boldsymbol{j} + f\cdot\boldsymbol{k},$$

where $f^* = \boldsymbol{\Omega}\cdot\cos(\phi)$ and $f = \boldsymbol{\Omega}\cdot\sin(\phi)$ is called the *Coriolis parameter.*

Gravity acts in the atmosphere and oceans as a conservative force $\boldsymbol{f}$ in (10.5):

$$\boldsymbol{f} = -\rho\cdot g\cdot\boldsymbol{k} = -\rho\cdot\nabla\Phi,$$

where g is the acceleration due to gravity $g = 9.81\ \mathrm{m/s^2}$. This can also be given by the gradient of the gravitational potential Φ $(\Phi = g\cdot z)$. Gravity acts toward the center of the Earth, and so its component $g\cdot\cos(\phi)$ points in the direction of the axis of rotation. The centrifugal force $\boldsymbol{\Omega}\times\boldsymbol{\Omega}\times\boldsymbol{r}$ points outward from the axis of rotation and so opposes gravity. Because the magnitude of the centrifugal force $(\Omega^2\cdot r \approx 3\cdot 10^{-2}\ \mathrm{m/s})$ is so small, this is generally neglected in meteorology and oceanography compared to the force of gravity.

Using the above simplifications, the Navier–Stokes equations (5.20) can be written for a tangential plane as follows:

$$\frac{\partial\boldsymbol{v}}{\partial t} + (\boldsymbol{v}\cdot\nabla)\boldsymbol{v} + f\cdot\boldsymbol{k}\times\boldsymbol{v} = -\frac{1}{\rho}\cdot\nabla p - \nabla\Phi + \nu\cdot\Delta\boldsymbol{v}, \tag{10.6}$$

where $\nu = \mu/\rho$ is the kinematic viscosity.

The additional Coriolis force in the rotating reference frame actually ought to appear in all equations of motion, since all flow processes, including those in technology, take place on the rotating Earth.

The *Rossby number* has become established as a measure for the relative weight of the Coriolis force. This describes physically the ratio of inertial force $\boldsymbol{v} \cdot \nabla \boldsymbol{v}$ and Coriolis force $f \cdot \boldsymbol{k} \times \boldsymbol{v}$. For a flow with a typical dimension L and a typical velocity U, the orders of magnitude can be estimated:

$$\text{inertial force} \sim \frac{U^2}{L} \quad ,$$
$$\text{Coriolis force} \sim f \cdot U \quad .$$

Therefore, the Rossby number is:

$$\text{Ro} = \frac{\text{inertial force}}{\text{Coriolis force}} = \frac{U}{f \cdot L} . \tag{10.7}$$

A large Rossby number (Ro $\gg$ 1) therefore means that the Coriolis force may be neglected compared to the inertial force in the equations of motion (10.6). In the opposite case (Ro $\ll$ 1) the Coriolis force dominates and may not be neglected.

The following examples clarify this. Here the value $f = 10^{-4}\ \mathrm{s}^{-1}$ was used for the Coriolis force, as holds for a latitude of about 45°:

flow	dimension	velocity	Ro
cyclone	10^3 km	10 m/s	0.1
land–sea wind	50 km	5 m/s	1
dust devil	50 m	5 m/s	10^3
bathtub vortex	50 cm	5 cm/s	10^3

It can be seen from this table that the Coriolis force is not important for small-scale atmospheric flows and for technical flows. However, it must be taken into account in processes with large spatial dimensions (cyclones). This will become clear in the following sections, where some atmospheric and oceanic flow processes are described.

Equation (10.6) forms the basis for the formal description of the flow processes in the atmosphere and oceans. In addition, the continuity equation and the energy equation, as have already been treated in Chapters 5.1 and 5.3, are required. A summary of the equations is found in Section 10.4.1.

In addition, we now discuss the concept of potential temperature frequently used in meteorology. This is obtained from the first law of thermodynamics for an adiabatic process. In this case, the pressure p and temperature T are related by

$$\frac{T(p_0)}{T(p)} = \left(\frac{p_0}{p}\right)^{\frac{R}{c_p}} . \tag{10.8}$$

Here p_0 is a reference pressure, generally set to $p_0 = 1000$ hPa. The *potential temperature* θ is the temperature $T(p_0)$ that an air packet with temperature T and air pressure p assumes when its pressure is changed to p_0 by an adiabatic process:

$$\theta = T \cdot \left(\frac{p_0}{p}\right)^{\frac{R}{c_p}}. \tag{10.9}$$

Here $R = 287$ J/kg/K is the gas constant for dry air, and $c_p = 1005$ J/kg/K is the specific heat at constant pressure. The potential temperature is a conserved quantity for adiabatic processes, i.e. $\mathrm{d}\theta/\mathrm{d}t = 0$, which is why it is frequently used in the description of atmospheric processes.

10.1.3 Geostrophic Flow

In this section we consider the effect of the Coriolis force on flows in the atmosphere and oceans. We assume an inviscid, horizontal flow (subscript h). Therefore, (10.6) yields

$$\frac{\mathrm{d}\boldsymbol{v}_h}{\mathrm{d}t} + f \cdot \boldsymbol{k} \times \boldsymbol{v}_h = -\frac{1}{\rho} \cdot \nabla_h p. \tag{10.10}$$

A fluid particle can be accelerated in the horizontal plane by the Coriolis force and the pressure force. However, if the flow has no acceleration, i.e. $\mathrm{d}\boldsymbol{v}_h/\mathrm{d}t = 0$, in equilibrium we obtain

$$f \cdot \boldsymbol{k} \times \boldsymbol{v}_h = -\frac{1}{\rho} \cdot \nabla_h p.$$

Transforming this, we obtain the velocity $\boldsymbol{v}_h$:

$$\boldsymbol{v}_h = \frac{1}{\rho \cdot f} \cdot \boldsymbol{k} \times \nabla_h p. \tag{10.11}$$

This is called the *geostrophic* velocity (or geostrophic wind in the atmosphere) and is denoted with the subscript g ($\boldsymbol{v}_g$). As can be seen from (10.11), the flow direction is parallel to the lines of equal pressure (isobars) or perpendicular to the pressure gradient. This is shown in Figure 10.2. The initially surprising

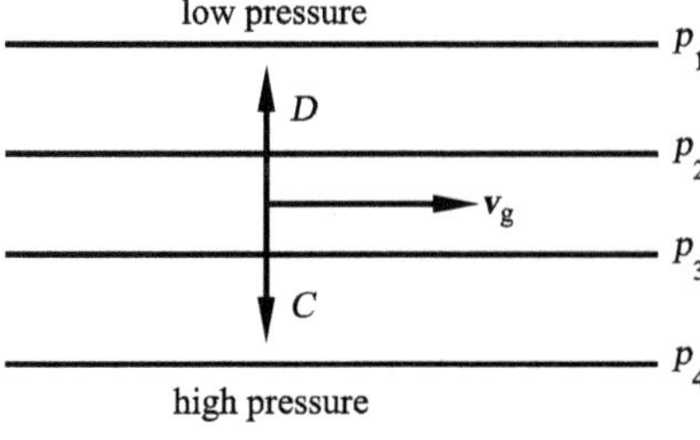

Fig. 10.2. Relation between the pressure field p, the pressure force $\boldsymbol{D}$, Coriolis force $\boldsymbol{C}$, and the geostrophic velocity $\boldsymbol{v}_g$

fact that a flow takes place perpendicular to the (pressure) force acting is due to the fact that in a rotating reference frame the Coriolis force causes a compensating force to appear that can lead to equilibrium (so-called geostrophic equilibrium). According to the discussion in Section 10.1.2, this is possible when the Rossby number satisfies $\mathrm{Ro} \to 0$. Because $\mathrm{Ro} = U/(f \cdot L)$, the flow process must take place in large spatial dimensions. In fact, an approximately geostrophic flow is observed in, for example, atmospheric high-pressure and low-pressure regions. Here the wind blows approximately parallel to the isobars, counter-clockwise in a low-pressure region and clockwise in a high pressure region (Figure 10.18), as can easily be checked on a weather map of the northern hemisphere. In the southern hemisphere the Coriolis parameter $f = 2 \cdot \boldsymbol{\Omega} \cdot \sin(\phi)$ becomes negative. For this reason the wind blows clockwise around a low. Therefore, the low air pressure lies to the right of the wind direction.

The geostrophic flow law (10.11) is valid for all layers of the atmosphere and of the ocean. Now observations have shown that the geostrophic wind in the atmosphere changes with altitude (typically, there is an increase in the wind velocity with height, see also Figure 10.21 in Section 10.2.5). Differentiating the wind law (10.11) with respect to the vertical coordinate z and using the equation of state for gases and the static fundamental equation (see Sections 10.2.4 and 10.2.5) leads to the following relation:

$$\frac{\partial \boldsymbol{v}_g}{\partial z} = \frac{g}{f \cdot T} \cdot \boldsymbol{k} \times \nabla T. \tag{10.12}$$

In (10.12), g is the gravitational acceleration, f is the Coriolis parameter, and T is the air temperature. This equation is also known as the *thermal wind relation*, since the change of the geostrophic wind with altitude depends on the horizontal temperature gradient. The integral of (10.12) is called the *thermal wind* $\boldsymbol{v}_T$:

$$\boldsymbol{v}_T = \boldsymbol{v}_g(z_2) - \boldsymbol{v}_g(z_1) = \int_{z_1}^{z_2} \left(\frac{g}{f \cdot T} \cdot \boldsymbol{k} \times \nabla T \right) \cdot \partial z.$$

The thermal wind $\boldsymbol{v}_T$ can be calculated with knowledge of the temperature field $T(x, y, z)$. Generally, a temperature is assumed that is independent of height between two altitudes z_1 and z_2 for simplicity, so that the following relation for the thermal wind is obtained:

$$\boldsymbol{v}_T = \frac{g}{f \cdot T} \cdot \boldsymbol{k} \times \nabla T \cdot (z_2 - z_1). \tag{10.13}$$

This thermal wind relation is very important for global atmospheric circulation (see Section 10.2.5). Among other things, it explains that between regions close to the equator and polar regions the temperature contrast practically always causes westerly winds in higher atmospheric layers.

On the other hand, the thermal wind arises from the equilibrium between the horizontal pressure force and the Coriolis force. An equilibrium can be both stable and unstable, as will be seen for convection and gravity waves in Sections 10.2.3 and 10.2.4. In the case of the thermal wind, in certain circumstances (e.g. large horizontal temperature gradient) the pressure and Coriolis forces can be brought out of equilibrium leading to a so-called *baroclinic instability*, with the consequence that low-pressure regions (cyclones) occur at Western European latitudes (see Section 10.2.5).

10.1.4 Vorticity

The flow processes in the atmosphere and oceans can be described using the equations of motion (10.6) and the continuity equation and energy equation introduced in Chapters 5.1 and 5.3. On large spatial scales, the flow processes are dominated by cyclonic and anticyclonic vortices with vertical axes, the low- and high-pressure regions (see Section 10.2.5). For this reason, in addition to the above equations, other equations are also used that describe the vortex strength of the geophysical flows. These are related to the concepts of *vorticity* and *potential vorticity.*

The *vorticity* (denoted by ω) is defined as the vertical component of the velocity rotation:

$$\omega = \boldsymbol{k} \cdot (\nabla \times \boldsymbol{v}) \quad . \tag{10.14}$$

Since the vorticity describes the vortex strength of flows with respect to an Earth-fixed reference frame, it is also called the relative vorticity. Considered from the inertial reference frame, the rotating Earth also has a vorticity, namely $2 \cdot \Omega$ or, perpendicular to the tangential plane, the value $2 \cdot \Omega \cdot \sin(\phi)$. The latter is precisely the Coriolis parameter f. The sum of the relative vorticity ω and the Coriolis parameter f is called the *absolute vorticity* and is denoted by η:

$$\eta = \omega + f \quad . \tag{10.15}$$

An equation for the rate of change of the absolute vorticity is obtained by applying the operator $\boldsymbol{k}\cdot\nabla\times$ to the equation of motion (10.6). After neglecting some terms and using the assumption of inviscid flow, we obtain the following equation for the absolute or relative vorticity for large-scale atmospheric and oceanic flow processes:

$$\frac{\mathrm{d}\eta}{\mathrm{d}t} = \frac{\mathrm{d}\omega}{\mathrm{d}t} + \beta \cdot v = -\eta \cdot \nabla_\mathrm{h} \cdot \boldsymbol{v}_\mathrm{h} \quad . \tag{10.16}$$

Here β is the so-called *beta parameter*

$$\beta = \frac{1}{R} \cdot \frac{\partial f}{\partial \phi} \quad , \tag{10.17}$$

with the radius of the Earth r, which states the variation of the Coriolis parameter f with latitude. In the special case of a two-dimensional, incompressible flow in the x-y plane, instead of (10.16) we obtain

$$\frac{\mathrm{d}\eta}{\mathrm{d}t} = \frac{\mathrm{d}\omega}{\mathrm{d}t} + \beta \cdot v = 0 \quad . \tag{10.18}$$

With these assumptions, the absolute vorticity is a conserved quantity. Thus

$$\eta = \omega + f = \text{const.}$$

Because of the variation of the Coriolis parameter f with latitude, according to (10.18) the relative vorticity of a flow must increase or decrease as we move north or south respectively. This effect of the spherical shape of the Earth on the Coriolis force is also called the beta (β) effect (from 10.17). This leads to the formation of large-scale oscillations in the meridional direction (Figure 10.3), and is observed in the atmosphere and oceans as so-called *Rossby waves.* This is formally obtained from the linearized form of the vorticity equation (10.18) with the assumption of a constant basic flow $\overline{u}$ (in the west–east direction) for the simplified perturbation vorticity $\omega' = \partial v'/\partial x$. With a wave ansatz of the form

$$v'(x,t) = v_0 \cdot \cos(a \cdot (x - c \cdot t))$$

(wave number $a = 2 \cdot \pi/\lambda$, wavelength λ, phase velocity c), the linearized vorticity equation leads to the following dispersion relation for the phase velocity of the *Rossby waves*:

$$c = \overline{u} - \frac{\beta}{a^2} = \overline{u} - \frac{\beta \cdot \lambda^2}{4 \cdot \pi^2} \quad . \tag{10.19}$$

The wavelength of steady waves (i.e. $c = 0$) can be estimated from the dispersion relation (10.19). For typical flow velocities in the atmosphere and oceans we obtain approximately

$$\text{atmosphere:} \quad \overline{u} \approx 15\ m/s \quad , \quad \lambda \approx 6000\ km \quad ,$$

$$\text{ocean:} \quad \overline{u} \approx 0.5\ m/s \quad , \quad \lambda \approx 1100\ km \quad .$$

Rossby waves are therefore large-scale flow processes in the atmosphere and in the ocean. Steady Rossby waves are stimulated by mountain ranges stretching in the north–south direction. An example of atmospheric Rossby waves is shown in Figure 10.4.

As well as vorticity and the vorticity equation, in recent years the concept of *potential vorticity* has come to be widely used in the areas of meteorology

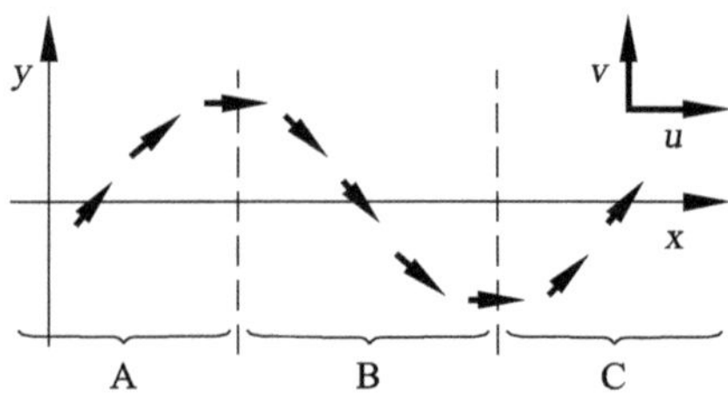

Fig. 10.3. Path of a fluid particle under the β effect. Regions A and C: $v > 0$, B: $v < 0$

and oceanography. The *potential vorticity* is generally denoted by PV and is defined as the product of the absolute vorticity η and the vertical gradient of the potential temperature $\partial\theta/\partial z$:

$$PV = \frac{1}{\rho} \cdot \frac{\partial \theta}{\partial z} \cdot \eta \quad . \tag{10.20}$$

For the case of adiabatic processes, we can derive an equation for the potential vorticity from the vorticity equation (10.16) and the first law of thermodynamics (*H. Ertel* (1942)).

$$\frac{\mathrm{d}}{\mathrm{d}t}\left(\frac{1}{\rho} \cdot \frac{\partial \theta}{\partial z} \cdot \eta\right) = 0 \quad . \tag{10.21}$$

This equation is valid for large-scale three-dimensional adiabatic flow processes. Here the potential vorticity is a conserved quantity.

In oceanography a modified form of the potential vorticity is frequently used, valid for *barotropic flows*:

$$PV = \frac{\eta}{H} \quad , \tag{10.22}$$

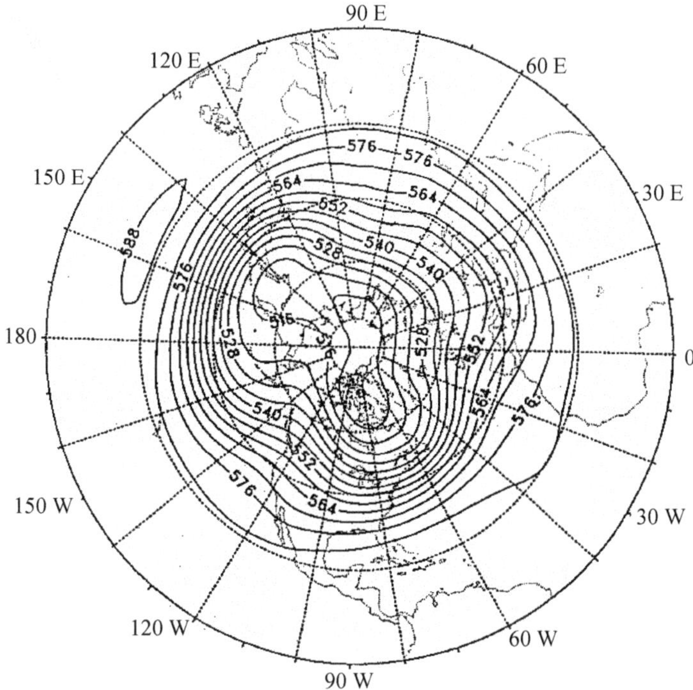

Fig. 10.4. Large-scale steady waves at an altitude of about 5 km in the northern hemisphere. The formation of a lee-side trough east of the Rocky Mountains can easily be seen

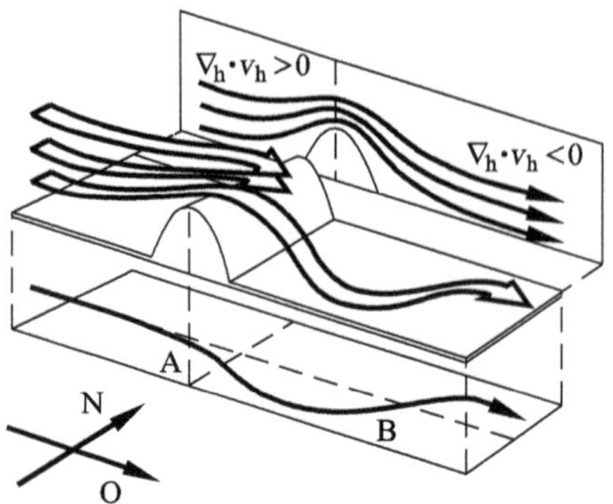

Fig. 10.5. Trajectories of fluid particles in the flow over a two-dimensional mountain (A). The influence of the divergence effect and the β-effect on the vorticity (10.12) causes the formation of a trough in the lee (B)

where H is the depth of a column of water. The corresponding vorticity equation reads

$$\frac{\mathrm{d}}{\mathrm{d}t}\left(\frac{\eta}{H}\right) = 0 \quad . \tag{10.23}$$

Conservation of the potential vorticity in the form (10.21) or (10.23) causes another large-scale phenomenon besides the formation of Rossby waves, namely, the *lee-side trough.* This is the formation of a trough (region of relatively low pressure) on the lee side of mountains running in a north–south direction. This is sketched in Figure 10.5. Because of the cross-sectional contraction close to the mountains, there are velocity divergences and convergences. As stated in the vorticity equation (10.16), this causes a change in the relative vorticity ω, which, together with the β-effect (equation (10.17) and Figure 10.3), leads to the formation of the lee-side trough.

10.1.5 Ekman Layer

In Section 10.1.3 geostrophic flow was derived as the equilibrium between pressure force and Coriolis force. In this section we retain the restrictions to horizontal acceleration-free flow but permit the action of a third force, the frictional force. The equations of motion now read

$$f \cdot \boldsymbol{k} \times \boldsymbol{v}_\mathrm{h} = -\frac{1}{\rho} \cdot \nabla_\mathrm{h} p + \nu \cdot \frac{\partial^2 \boldsymbol{v}_\mathrm{h}}{\partial z^2} \quad . \tag{10.24}$$

With the geostrophic wind relation (10.11) the pressure force may be replaced with $f \cdot \boldsymbol{k} \times \boldsymbol{v}_\mathrm{g}$, and hence

$$f \cdot \boldsymbol{k} \times (\boldsymbol{v}_\mathrm{h} - \boldsymbol{v}_\mathrm{g}) = \nu \cdot \frac{\partial^2 \boldsymbol{v}_\mathrm{h}}{\partial z^2} ,$$

or, with the velocity components u and v,

$$-f \cdot (v - v_{\mathrm{g}}) = \nu \cdot \frac{\partial^2 u}{\partial z^2} \quad , \tag{10.25}$$

$$f \cdot (u - u_{\mathrm{g}}) = \nu \cdot \frac{\partial^2 v}{\partial z^2} \quad . \tag{10.26}$$

From the relations (10.25) and (10.26) and with suitable boundary conditions, we can calculate the vertical dependence of the flow components $u(z)$ and $v(z)$. We first consider an example for the atmosphere. On the ground, the standard no-slip condition holds:

$$z = 0 \quad : \quad u = v = 0 \quad . \tag{10.27}$$

There is no fixed edge to the top of the atmosphere. However, it can be assumed plausibly that the effect of the fixed ground compared to the friction term becomes smaller, the further one is from the edge. For great heights (formally for $z \to \infty$), the flow is to match the geostrophic flow:

$$z \to \infty \quad : \quad u = u_{\mathrm{g}} \quad , \quad v = v_{\mathrm{g}} \quad , \tag{10.28}$$

Equations (10.25) and (10.26) together with the boundary conditions (10.27) and (10.28) can be solved analytically. This was first done by the Swedish oceanographer *V. W. Ekman* in the year 1905. For simplicity, we orient the coordinate system so that the geostrophic wind is in the x direction, and hence $\boldsymbol{v}_g = (u_g, 0)$. The solution reads

$$u(z) = u_{\mathrm{g}} \cdot \left(1 - \exp\left(-\frac{z}{D}\right) \cdot \cos\left(\frac{z}{D}\right)\right) \quad , \tag{10.29}$$

$$v(z) = u_{\mathrm{g}} \cdot \exp\left(-\frac{z}{D}\right) \cdot \sin\left(\frac{z}{D}\right) \quad , \tag{10.30}$$

with

$$D = \sqrt{\frac{2 \cdot \nu}{f}} \quad . \tag{10.31}$$

The length D (10.31) is called the *Ekman length*.

In order to better represent the change of direction of the wind with altitude, a so-called hodograph is frequently used, where the velocity vectors are projected onto the u-v plane. This is shown in Figure 10.6. As well as the increase in the magnitude of the velocity, a rotation of the wind with altitude toward the geostrophic wind direction is also seen. For altitudes $z > \pi \cdot D$ the hodograph is a spiral, also known as the *Ekman spiral*. Clearly, the deviations of the true wind $\boldsymbol{v}_h$ from the geostrophic wind $\boldsymbol{v}_g$ for $z > \pi \cdot D$ are quite small, which is why the layer below $\pi \cdot D$ is also known as the *Ekman boundary layer* or, for short, *Ekman layer*. This is a viscous boundary layer above a fixed base in a rotating system.

The situation is slightly different for the ocean. Here we consider the question of which flow occurs in the sea when no large-scale pressure gradient is present (geostrophic flow) under the effect of the wind shear stress on the surface. The fundamental equations then read

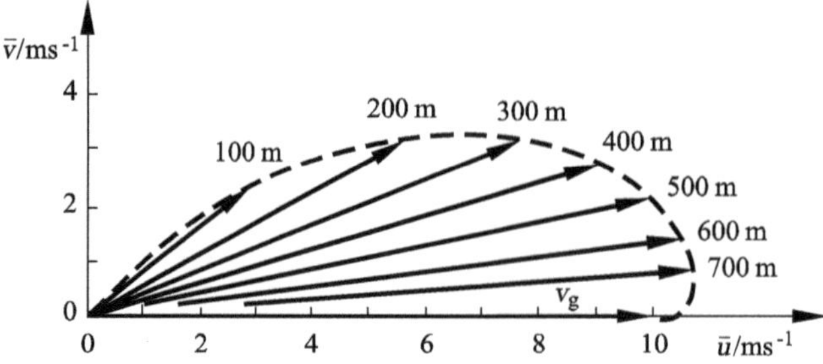

Fig. 10.6. Hodograph representation of the wind distribution in the atmospheric Ekman layer according to (10.29)–(10.31) for the case $|\boldsymbol{v}_g| = 10$ m/s, $f = 10^{-4}$ s^{-1}, $\nu = 10$ m^2/s, $\boldsymbol{v}_g$ = geostrophic wind

$$-f \cdot v = \nu \cdot \frac{\partial^2 u}{\partial z^2} \quad , \qquad f \cdot u = \nu \cdot \frac{\partial^2 v}{\partial z^2} \quad , \tag{10.32}$$

For the boundary conditions we set

$$z = 0 \quad : \quad \frac{\tau_x}{\rho} = \nu \cdot \frac{\partial u}{\partial z} \quad , \qquad \frac{\tau_y}{\rho} = \nu \cdot \frac{\partial v}{\partial z} \quad , \tag{10.33}$$

$$z \to -\infty \quad : \quad u = v = 0 \quad . \tag{10.34}$$

Here τ_x and τ_y are the components of the shear stress caused by the wind acting on the surface of the water.

The solution of this equation is again due to *V. W. Ekman*. For the case in which the wind shear stress acts in the y direction (i.e. $\tau_x = 0$), it reads

$$u(z) = \frac{\tau_y}{\rho \cdot \sqrt{\nu \cdot f}} \cdot \exp\left(\frac{z}{D}\right) \cdot \cos\left(\frac{z}{D} + \frac{\pi}{4}\right) \quad , \tag{10.35}$$

$$v(z) = \frac{\tau_y}{\rho \cdot \sqrt{\nu \cdot f}} \cdot \exp\left(\frac{z}{D}\right) \cdot \sin\left(\frac{z}{D} + \frac{\pi}{4}\right) \quad , \tag{10.36}$$

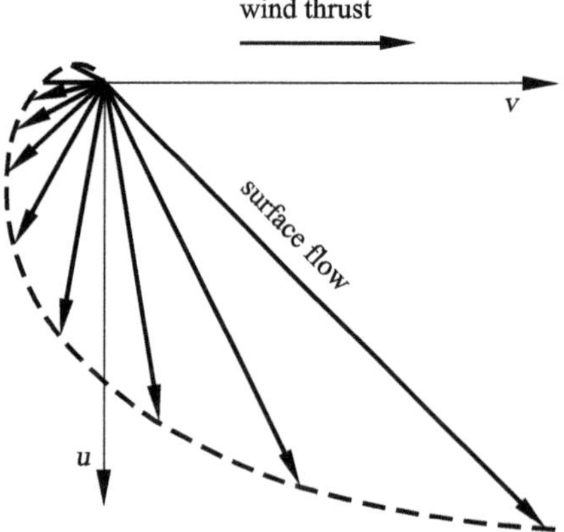

Fig. 10.7. Hodograph of the flow relations in the oceanic Ekman layer

Here z is downward negative.

The hodograph of the flow in the sea according to (10.35) and (10.36) is shown in Figure 10.7. Again an Ekman spiral is seen, similar to that in Figure 10.6. What is surprising in this solution is that the oceanic flow close to the surface is oriented 45° to the right of the wind shear stress acting. (*V. W. Ekman* observed this phenomenon on board the research ship *Frahm*, and this led him to derive the equations.)

In order to determine the vertical extension $H = \pi \cdot D$ of the Ekman boundary layer in the atmosphere and ocean, we insert the respective values of the kinematic viscosity (air: $\nu = 0.15\ \mathrm{cm}^2/\mathrm{s}$, water: $\nu = 0.01\ \mathrm{cm}^2/\mathrm{s}$), and obtain for $f = 10^{-4}\ \mathrm{s}^{-1}$

$$\begin{aligned} &\text{atmosphere}: \quad H = 55\ cm \quad , \\ &\text{ocean} \qquad\ \ : \quad H = 15\ cm \quad . \end{aligned}$$

These values are far smaller than those observed. In fact, the vertical extension of the Ekman layer in the atmosphere is about 1000 m and in the ocean about 50 m. The reason for the apparent discrepancy lies in the fact that the atmospheric and oceanic boundary layers are turbulent, and therefore the turbulent diffusion coefficient ν_t instead of the molecular kinematic viscosity ν must be used in equations (10.25), (10.26), and (10.32), where $\nu_t \approx 10\ \mathrm{m}^2/\mathrm{s}$ for the atmosphere and $\nu_t \approx 0.1\ \mathrm{m}^2/\mathrm{s}$ for the ocean.

A further deficiency of the Ekman solution compared to observations is the large angle of deviation of 45° between the ground wind and the geostrophic wind, or between the wind shear stress and the surface flow. This is because the turbulent diffusion coefficient is not a material constant, but rather changes with height. Numerical solutions of the Ekman equations with variable diffusion coefficient yield angles of deviations around 20°, in agreement with observations. In fact, *L. Prandtl* (1949) in the third edition of his book presented an analytical solution for a turbulent Ekman layer (see Section V.9 of the original work) that describes the observations quite closely.

10.1.6 Prandtl Layer

The Ekman layer described in the previous section describes the vertical dependence of the flow in the entire atmospheric or oceanic boundary layer. Close to the surface of the Earth, conditions have been investigated closely in numerous field experiments, and the concept of a surface boundary layer, also called the *Prandtl layer*, has been introduced. In short, it is assumed that the turbulent shear stress in the Prandtl layer is constant with respect to changes in altitude:

$$\tau(z) = \tau_0 = -\overline{\rho \cdot w' \cdot \boldsymbol{v}_{\mathrm{h}}'} = \overline{\rho} \cdot u_*^2 \quad . \tag{10.37}$$

The quantity $u_*^2 = |\overline{w' \cdot \boldsymbol{v}_n'}|$ is called the friction velocity. With the usual gradient ansatz for turbulent flows

$$\overline{w' \cdot \boldsymbol{v}_\mathrm{h}'} = -\nu_\mathrm{t} \cdot \frac{\partial \overline{\boldsymbol{v}}_\mathrm{h}}{\partial z}$$

and the Prandtl mixing length ansatz for the turbulent diffusion coefficient $\nu_t = k \cdot u_* \cdot z$, with the von Kármán constant $k = 0.4$, and with the choice of coordinate system $\overline{\boldsymbol{v}} = (u, 0)$, on integrating the wind profile we obtain

$$\overline{u}(z) = \frac{u_*}{k} \cdot \ln\left(\frac{z}{z_0}\right) \quad . \tag{10.38}$$

This relation is known by the name *logarithmic wind law* (4.82). The height z_0 occurring in the logarithm is called the roughness length and is a measure of the roughness of the ground (e.g. sand surface: $z_0 = 0.1$ mm, grass: $z_0 = 5$ cm). The logarithmic wind profile has been confirmed by numerous measurements and is valid for approximately the lowest $20 - 50$ m of the atmosphere. A similar law is also found in technical flows past rough plates or in pipes (wall law, Section 4.2.5).

One feature of the atmospheric Prandtl layer is the general simultaneous appearance of a vertical temperature gradient that leads to a stratified shear flow. In order to determine the effect of the temperature layering on the wind profile, the so-called similarity theory of *A. S. Monin* and *A. M. Obukhov* (1954) has come into use in the last decades. This states that suitable normalized velocity gradients in the Prandtl layer are universal:

$$\frac{k \cdot z}{u_*} \cdot \frac{\partial \overline{u}}{\partial z} = \phi\left(\frac{z}{L}\right) \quad . \tag{10.39}$$

The dimensionless similarity function ϕ depends only on the normalized height z/L. Here L is the *Monin–Obukhov length*, which determines the effect of the temperature layering via the turbulent temperature flux $\overline{w' \cdot \theta'}$:

$$L = -\frac{u_*^3}{\frac{k \cdot g}{\theta_0} \cdot \overline{w' \cdot \theta'}} \quad . \tag{10.40}$$

The function ϕ has been determined from numerous field measurements. A usual form is, for example,

$$\begin{aligned}
\phi &= 1 + 5 \cdot \frac{z}{L} \quad , && \frac{z}{L} > 0 \quad \text{stable stratification} \quad , \\
\phi &= 1 \quad , && \frac{z}{L} = 0 \quad \text{neutral stratification} \quad , \\
\phi &= \left(1 - 15 \cdot \frac{z}{L}\right)^{-\frac{1}{4}} \quad , && \frac{z}{L} < 0 \quad \text{unstable stratification} \quad .
\end{aligned}$$

The vertical wind profile in a thermally stratified Prandtl layer is obtained by integrating (10.39) using the empirical trial solutions for the similarity function $\phi(z/L)$.

In principle, a Prandtl layer with a logarithmic flow profile is also to be expected in the ocean. However, the layer of the ocean close to the surface is generally so greatly disturbed by waves that a velocity law analogous to (10.38) can generally be found only in particularly favorable circumstances. More information on boundary layers in the atmosphere and ocean and their interaction can be found in the monograph of *E. B. Kraus* and *J. Businger* (1994).

10.2 Flows in the Atmosphere

In the following sections we describe examples of individual flow forms in the atmosphere more closely. Of course, our selection must be restricted to a few typical phenomena, but it still covers all scales from small-scale dust vortices up to global atmospheric circulation.

Since the air itself is invisible, the question arises of how to make atmospheric flows visible. Clouds can help, since they naturally trace the flow. Examples are seen in the photo of the Earth taken by a weather satellite (Figure 10.8). The different atmospheric phenomena seen in the satellite photograph are explained in the interpretation aid in Figure 10.9. In the following sections we will discuss these flow forms more closely. Details of the different small-scale atmospheric phenomena are to be found in the monograph by *B. W. Atkinson* (1981), while numerous examples of satellite photographs and their interpretation with respect to atmospheric flows are given in *R. S. Scorer* (1986).

Extensive discussions of the forms of motion in the atmosphere may also be found in the German textbooks by *H. Häckel* (1999), *H. Kraus* (2000). The theoretical aspect of atmospheric flows is treated in detail in, for example, the monographs by *D. Etling* (2002) and *H. Pichler* (1997).

In the above books some detail of the treatment of the flow in the atmospheric boundary layer may also be found. In this instance we restrict ourselves to the remarks on the Ekman layer (Section 10.1.5) and the Prandtl layer (Section 10.1.6) and so do not treat boundary-layer flows in what follows. We recommend the special monographs on the atmospheric boundary layer by *J. R. Garratt* (1992) and *J. C. Kaimal* and *J. J. Finnigan* (1994) to the interested reader.

10.2.1 Thermal Wind Systems

As already discussed in Section 10.1.2, the motion of the air is caused or influenced by the pressure force, Coriolis force, viscous force, and gravitational force. The friction acts to weaken the flow, while the Coriolis force merely causes a change in direction. Since the gravitational force acts just on vertical motion, it is only the pressure force that is the actual driving force for horizontal atmospheric motion. We must consider where the pressure forces come from. Since the atmosphere is an almost ideal gas, the air pressure depends on the air temperature.

In what follows, such flow processes as occur from horizontal temperature differences will be briefly sketched. These are generally known as *thermal wind systems*. To this end we consider the circulation in an x-z plane. The circulation Γ along a closed curve S is defined as

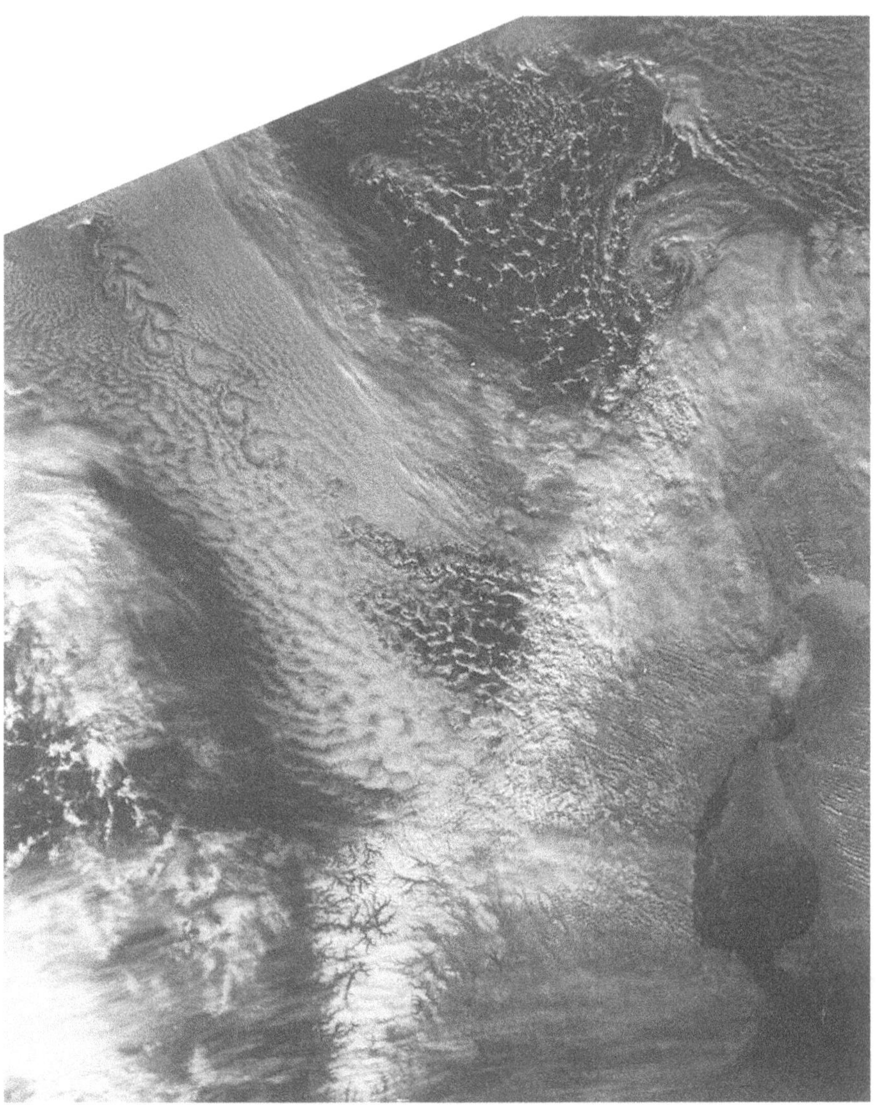

Fig. 10.8. Satellite photograph with different atmospheric phenomena (e.g. thermal convection in the form of rolls and cells, vortex street, gravity waves, cyclones) made visible by clouds

$$\Gamma = \oint_s \boldsymbol{v} \cdot \mathrm{d}\boldsymbol{s} = \oint_s v_s \cdot \mathrm{d}s. \tag{10.41}$$

Here v_s is the velocity component in the direction of the curve vector $\boldsymbol{s}$. If the evaluation of the integral in (10.41) yields a positive Γ, then the circulation is cyclonic (counter-clockwise), while in the case in which $\Gamma < 0$, the circulation is anticyclonic.

The rate of change of the circulation is obtained from the equation of motion (10.6). In what follows we consider small-scale phenomena, so that the Coriolis force may be neglected. Friction effects will also not be taken into account. We obtain

$$\frac{\mathrm{d}\Gamma}{\mathrm{d}t} = \frac{\mathrm{d}}{\mathrm{d}t} \oint_s v_s \cdot \mathrm{d}s = - \oint_s \frac{\mathrm{d}p}{\rho}, \tag{10.42}$$

or with the equation of state for ideal gases $p = \rho \cdot R \cdot T$,

$$\frac{\mathrm{d}\Gamma}{\mathrm{d}t} = - \oint_s \frac{R \cdot T}{p} \cdot \mathrm{d}p. \tag{10.43}$$

In the case of constant air temperature, integration along a closed curve in (10.43) yields $\mathrm{d}\Gamma/\mathrm{d}t = 0$. Thus, in order that there circulation occurs, spatial temperature variations must be present. This will be clarified for the case of the so-called *land–sea wind*. On sunny days the air over land heats up more

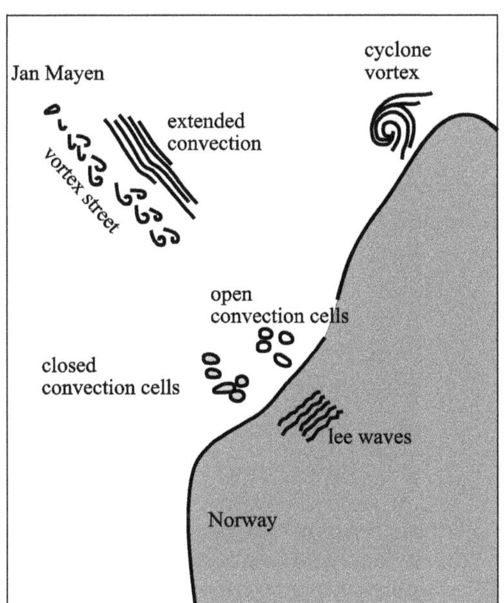

Fig. 10.9. Interpretation aid for the different flow types seen in the satellite photograph in Figure 10.8

than that over water. This is due to the different heat conductions and heat capacities of the ground (water or solid Earth). Since the air pressure always decreases with altitude, as already explained in Section 2.5, the distribution of isobars and isotherms shown in Figure 10.10 arises.

If we select the integration curve S such that it runs along the isobars and isotherms, in Figure 10.10 we obtain

$$\frac{\mathrm{d}\Gamma}{\mathrm{d}t} = R \cdot (T_3 - T_1) \cdot \ln\left(\frac{p_3}{p_1}\right) > 0. \tag{10.44}$$

A cyclonic circulation occurs that is directed from the cold water to the warm land close to the ground, i.e. under high air pressure. This is called a sea wind. At large altitudes (low air pressure) a compensating flow takes place from the land to the sea (land wind).

At night, the situation is reversed. The air over the land then cools more than that over the water. In Figure 10.10 the isotherms T_1 and T_3 would be swapped, and similarly, we would obtain an anticyclonic circulation, i.e. from the land to the sea close to the ground. In general, it may be said that thermal circulation in the atmosphere occurs to balance out the temperature between warmer and colder areas.

Of importance in this simple example is that atmospheric motion is caused by horizontal temperature difference. Land–sea wind circulation is an example of this. It occurs at practically all coasts and has a horizontal extension of $10 - 100$ km. The strength of the land–sea wind depends on the land–water temperature contrast, according to (10.44), and also on the large-scale wind flow that is generally superimposed. A further example is the so-called *slope current.* Close to slopes that are warmed during the day, an up-current occurs, becoming a down-current at night when the air close to the slope cools. Since it is colder air that flows down the slope at night, this is also called a cold-air drain.

In the simplest case, slope currents may also be explained with the circulation equation (10.42)–(10.44). In reality, the relations are somewhat more complex, since the atmosphere already has a vertical temperature layering,

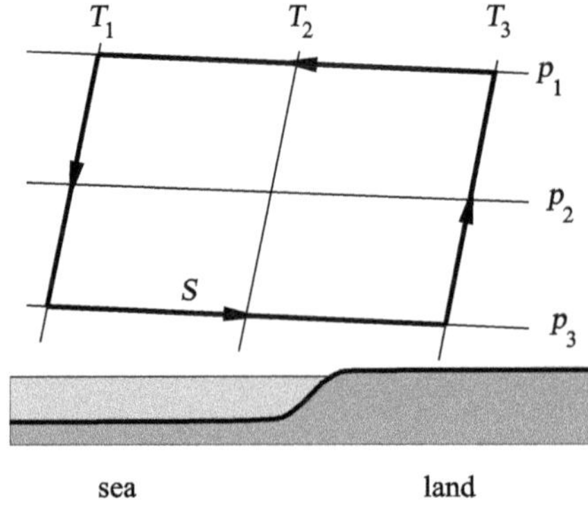

Fig. 10.10. Schematic representation of the land–sea wind circulation for the case where the surface of the land is warmer than the water (sea wind). Isotherms $T_1 < T_2 < T_3$, isobars $p_1 < p_2 < p_3$

and the isotherms therefore cannot run parallel to the slope. This was already noted by *L. Prandtl*, who gave an analytical solution of the slope current in an earlier edition of this book (third edition (1949), Section V.16). Further details on thermal wind systems may be found in the monographs by *B. W. Atkinson* (1981), and by *J. E. Simpson* (1994), (1997), *C. D. Whitemann* (2000).

10.2.2 Thermal Convection

In Chapter 7 the phenomenon of natural convection was also considered within the framework of heat transfer. In particular, the conditions for *cellular convection* in horizontal layers (Section 7.2.1) are also to be found in the atmosphere. Cellular convection is an instability in a thermally unstably stratified medium. The basic idea may be simply explained as follows.

For an atmosphere with a constant vertical temperature gradient, the equations of motion and the first law of thermodynamics can be used to derive the following relation for the vertical deviation z of a particle from its rest position:

$$\frac{\mathrm{d}^2 z}{\mathrm{d}t^2} + N^2 \cdot z = 0, \tag{10.45}$$

where N is the Brunt–Väisälä frequency, named after the English and Finnish meteorologists *B. Brunt* and *V. Väisälä*. It is defined via the vertical gradient of the air temperature T or the potential temperature θ (10.9) as

$$N = \sqrt{\frac{g}{T_0} \cdot \left(\frac{\partial T}{\partial z} + \Gamma' \right)} = \sqrt{\frac{g}{\theta_0} \cdot \frac{\partial \theta}{\partial z}}. \tag{10.46}$$

Here Γ' is the dry-adiabatic temperature gradient with a value of $\Gamma' = g/c_p = 9.8 \cdot 10^{-3}$ K/m, or about 1 K/100 m. If a packet of air is deviated vertically by the height Z_a from its equilibrium position, we obtain as a solution of (10.45)

$$Z(t) = Z_a \cdot \exp(N \cdot t) \quad \text{for} \quad \frac{\partial T}{\partial z} < -\Gamma' \quad \text{or} \quad \frac{\partial \theta}{\partial z} < 0,$$

$$Z(t) = Z_a \cdot \cos(N \cdot t) \quad \text{for} \quad \frac{\partial T}{\partial z} > -\Gamma' \quad \text{or} \quad \frac{\partial \theta}{\partial z} > 0.$$

In the first case, the air packet moves ever further away from its equilibrium position, and we have an unstable equilibrium. In this case, thermal convection can occur in the air. This condition is that the air temperature decreases faster with altitude that for the case of adiabatic (neutral) stratification.

In the second case, the air particle oscillates about its equilibrium position. This case, known as stable, will be discussed in Section 10.2.4 in the treatment of gravity waves.

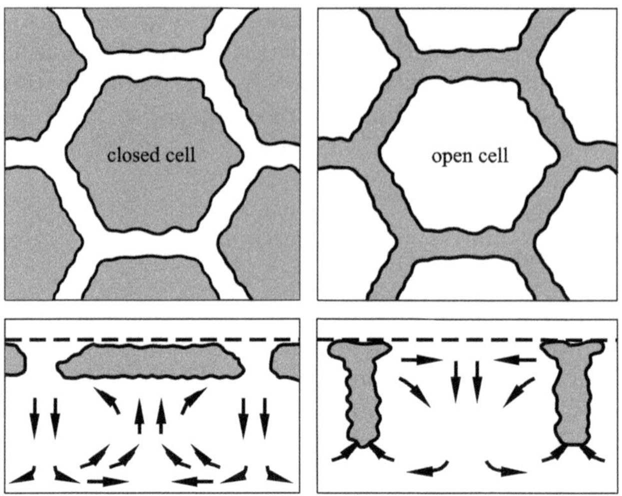

Fig. 10.11. Aspects and cross-sections through closed and open convection cells. Gray: clouds; arrows: circulation

A condition for the occurrence of thermal convection in the atmosphere is therefore heating of air layers from the ground. This form of cellular convection (rolls or hexagons) already discussed in Section 7.2.1 is also found in the atmosphere. It is made visible by clouds that form in the upper part of the convection cell by adiabatic cooling of the rising humid air. Figure 10.8 shows the different types of atmospheric convection on a satellite photograph: longitudinal convection rolls (cloud streets), as well as open and closed cells. The convection flow causing this cloud pattern is shown schematically in Figure 10.11, with a satellite photograph of the same in Figure 10.12.

Fig. 10.12. Satellite photograph of open convection cells

Although the patterns in Figure 10.8 and Figure 10.11 look similar to the experimental flows (Section 7.2.1), there are some principal differences. These concern both the dimensions and the physical origins. Atmospheric convection is generally restricted to a 1 – 2 km high layer above the surface of the Earth. The convection forms seen in Figures 10.8 and 10.9 have the following horizontal wavelengths: cloud streets 3 – 15 km, cells 10 – 30 km. The ratio of height to width is therefore between 1 : 3 and 1 : 10 in linear convection patterns and from 1 : 10 to 1 : 20 in cells. In the laboratory this ratio is about 1 : 3.

Many explanations have been given for the origin of the small aspect ratios in atmospheric convection, such as those in the overview articles by *B. W. Atkinson* and *J. W. Zhang* (1996) on cells and by *D. Etling* and *R. A. Brown* (1993) on rolls. In particular, the release of latent heat in the formation of clouds seems to play a role, an effect that of course does not occur in the laboratory. Further aspects of thermal convection in the atmosphere can be found in the monograph by *K. A. Emanuel* (1994).

10.2.3 Gravity Waves

In the previous section we treated the onset of thermal convection in an unstably stratified atmosphere. This arose due to warming of the air from the surface of the Earth. However, frequently the case occurs in which the atmosphere close to the ground cools down via long-wavelength radiation (e.g. at night) so that the air temperature increases with height. Such an atmosphere is then stably layered.

In the solution of (10.45) for stable stratification (second case), an oscillation was obtained:

$$Z(t) = Z_a \cdot \cos(N \cdot t), \tag{10.47}$$

with the Brunt–Väisälä frequency N as in (10.46). The period of oscillation of a vertically deviated air packet is $\tau = 2 \cdot \pi / N$. Below are some numerical values:

$\frac{\partial T}{\partial z}\left[\frac{\mathrm{K}}{100\ \mathrm{m}}\right]$	$\frac{\partial \theta}{\partial z}\left[\frac{\mathrm{K}}{100\ \mathrm{m}}\right]$	$N[\mathrm{s}^{-1}]$	$\tau[\mathrm{s}]$
−0.65	0.35	0.011	570
0	1.0	0.018	350
+1.0	2.0	0.026	240

A stably stratified atmosphere is a continuum that can carry out oscillations and thus permits the expansion of waves whose retroactive force is gravity. These waves are therefore also known as *gravity waves*. We now briefly derive the wave equation. We start out from the Boussinesq approximation of the equations of motion, as given in (5.85) in Section 5.4.3. With the usual linearization we obtain the form of the perturbation equations (5.213)–(5.215).

For simplicity we consider an atmosphere at rest. After bringing these equations together, we finally obtain a differential equation for the vertical velocity w:

$$\frac{\partial^2}{\partial t^2}\left(\frac{\partial^2 w}{\partial x^2} + \frac{\partial^2 w}{\partial z^2}\right) + N^2 \cdot \frac{\partial^2 w}{\partial x^2} = 0. \tag{10.48}$$

To solve this equation we consider a wave ansatz of the form

$$w(x, z, t) = w_0 \cdot \cos(a \cdot x + b \cdot z - \omega \cdot t). \tag{10.49}$$

Here a and b are the horizontal and vertical components ($a = 2 \cdot \pi/\lambda_x$, $b = 2 \cdot \pi/\lambda_z$) of the wave number vector $\boldsymbol{m}$, as shown in Figure 10.13, and ω is the eigenfrequency of the wave. Inserting the wave ansatz (10.49) into the wave equation (10.48), we obtain the following frequency conditions:

$$\omega = N \cdot \frac{a}{m} = N \cdot \cos(\alpha). \tag{10.50}$$

The frequency of oscillation of the gravity waves can therefore maximally attain the Brunt–Väisälä frequency N (for $\alpha = 0$, i.e. entirely horizontal wave expansion).

The phase velocity in the direction of expansion $\boldsymbol{m}$ is found using $\boldsymbol{m}{\cdot}c = \omega$ as

$$c = \pm\frac{N}{m} \cdot \cos(\alpha), \tag{10.51}$$

or, in the case of a horizontally expanding wave ($\alpha = 0$, $m = a$),

$$c = \pm\frac{N}{a} = \pm\frac{\lambda \cdot N}{2 \cdot \pi}.$$

Some numerical values: setting $\partial T/\partial z = 0$ ($\partial\theta/\partial z = 1$ K/100 m), i.e. $N = 0.018\ \mathrm{s}^{-1}$, we obtain

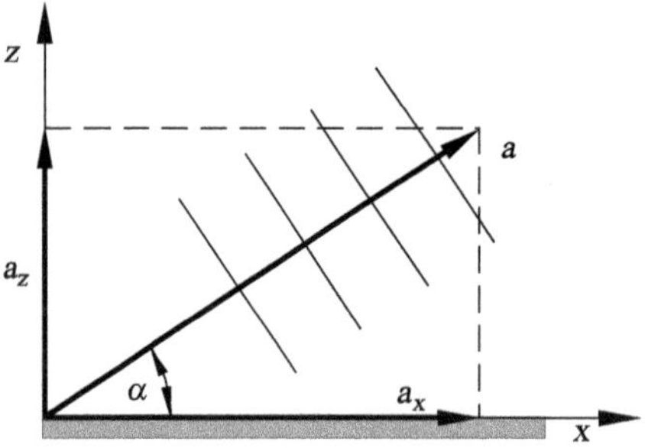

Fig. 10.13. The wave number vector $\boldsymbol{m}$ for internal gravity waves. The oscillation planes perpendicular to $\boldsymbol{m}$ are shown

$$\lambda = 1 \text{ km} \quad \rightarrow \quad c \approx 3 \text{ m/s},$$
$$\lambda = 3 \text{ km} \quad \rightarrow \quad c \approx 10 \text{ m/s}.$$

The phase velocity of gravity waves is therefore the order of magnitude of the wind velocity occurring in the atmosphere. If the direction of the wind and the phase velocity are opposed, steady gravity waves can occur. This is the case when the following relation is satisfied between wind velocity U, wavelength λ, and Brunt–Väisälä frequency N:

$$\lambda = \frac{2 \cdot \pi}{N} \cdot U. \tag{10.52}$$

Steady waves are found particularly in the lee of mountains that force the air to rise and so cause continuous vertical excitation of the gravity waves. The waves that occur in this case are also called *lee waves.*

If the air humidity is suitable, adiabatic cooling can occur in the upwind regions of the wave and so lead to cloud formation. Therefore, lee waves (and also gravity waves in general) become visible via periodic arrangement of clouds transverse to the wind direction (Figure 10.14). This can frequently be seen on satellite photographs of the type in Figure 10.15.

Since the free atmosphere is practically always stably stratified, gravity waves are a form of motion that is more or less continually to be observed in the atmosphere. Further details are to be found in the monographs by *E. E. Gossard* and *W. H. Hooke* (1975), *C. Nappo* (2002) and in the review article by *M. G. Wurtele et al.* (1996).

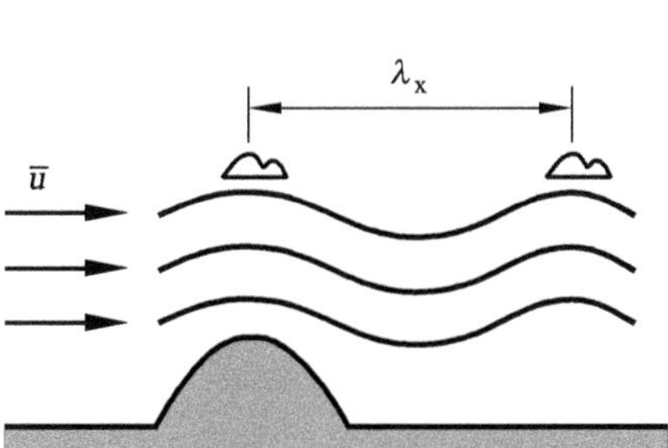

Fig. 10.14 Schematic representation of lee waves made visible by cloud formation

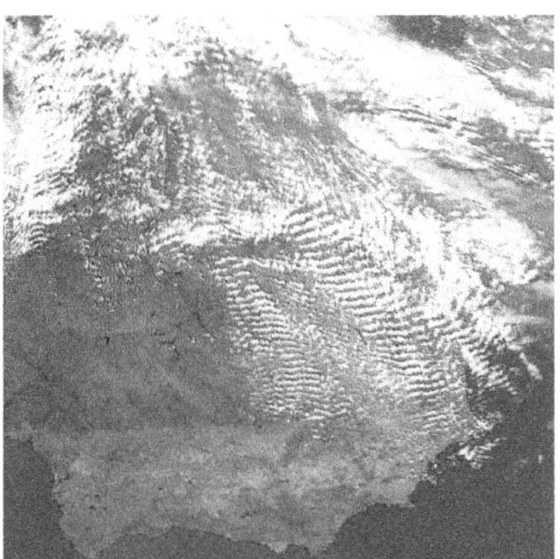

Fig. 10.15 Satellite photograph of gravity waves causing bandlike cloud formation (center of figure)

10.2.4 Vortices

As well as the processes described above, such as land–sea wind, thermal convection, and gravity waves, the dynamics of the atmosphere are greatly affected by *vortices* of many different sizes. These extend from the low-pressure regions with a horizontal extension of a few thousand kilometers, down to small-scale dust devils of 50 m diameter. In the following table we list examples of different vortex phenomena, with typical values of their diameter, wind velocity, and duration.

name	diameter [km]	duration	velocity [m/s]	rotation
low	2000	4 d	20	cyclonic
hurricane	500	10 d	80	cyclonic
orographic vortex	50	1 d	5	cycl. & anticycl.
tornado	1	1 h	100	cycl. & anticycl.
dust devil	0.1	1 min	10	cycl. & anticycl.

The examples above represent vortices with vertical axes, with the direction of rotation of the large-scale low-pressure region and hurricane always cyclonic, while the small-scale phenomena may be both cyclonic and anticyclonic. This difference is due to the effect of the Coriolis force, as seen in the force diagram in Figure 10.16.

In the ideal case of a rotationally symmetric vortex, neglecting the frictional forces in the equations of motion (10.6) yields the following balance of forces (Figure 10.16):

$$\frac{v^2}{r} + f \cdot v = \frac{1}{\rho} \cdot \frac{\partial p}{\partial r}, \tag{10.53}$$

centrifugal force + Coriolis force = pressure force,

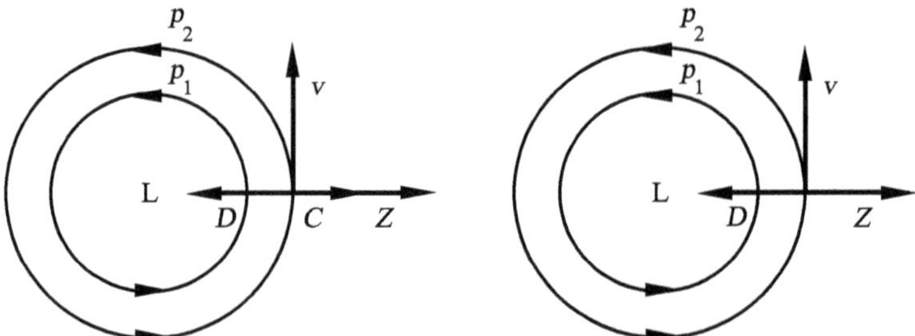

Fig. 10.16. Balance of forces in rotationally symmetric vortices. The air pressure in the center of each vortex is low (L). The flow velocity V, isobars P, Coriolis force C, pressure force D, and centrifugal force Z are shown. Left: large-scale vortex (low-pressure region), right: small-scale vortex (e.g. tornado)

where r is the distance from the center of the vortex.

In each case, the pressure force acts toward the center of the vortex, while the centrifugal force is directed away from the center of the vortex. In the case of a large-scale cyclone (Figure 10.16), the centrifugal force has the same direction as the Coriolis force, while in the case of an anticyclone (high-pressure region, not shown here) it has the direction of the pressure force.

We estimate the balance of forces in two examples:

	$\frac{v^2}{r}$	$f \cdot v$	$\frac{1}{\rho} \cdot \frac{\partial p}{\partial r}$	unit
low-pressure region	1	3	5	$\cdot 10^{-3}$ m/s
tornado	5000	5	5000	$\cdot 10^{-3}$ m/s

Whereas in the case of a large-scale vortex (low-pressure region) the Coriolis force plays an important role, it may be neglected for a small-scale vortex (tornado). This is sketched in the right-hand picture of Figure 10.16.

The causes of the onset of the vortex phenomena mentioned above are complex and will be briefly mentioned in the discussion of each example. However, all the examples have a vortex strengthening mechanism in common that can be obtained from the vorticity equation (10.16). If the latitudinal dependence of the Coriolis parameter f is neglected, this equation for the relative vorticity ω may be written as

$$\frac{d\omega}{dt} = -(f + \omega) \cdot \nabla_h \cdot \boldsymbol{v}_h. \tag{10.54}$$

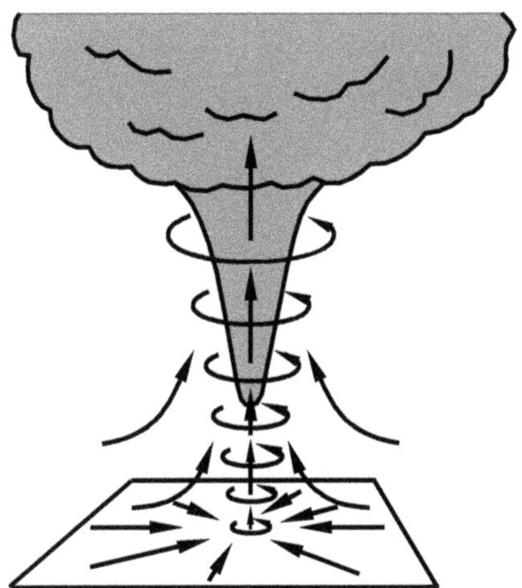

Fig. 10.17. Schematic representation of the vortex strength increase by means of horizontal flow convergence. The funnel-like protrusion of the cloud (shaded) corresponds approximately to the visible part of a *tornado* (right picture)

According to this equation, vortex strengthening (or weakening) occurs if a convergence (or divergence) is present in the horizontal flow field. We take the tornado as an example. In this case, $|\omega| \gg f$, and therefore

$$\frac{\mathrm{d}\omega}{\mathrm{d}t} = -\omega \cdot \nabla_h \cdot \boldsymbol{v}_h. \tag{10.55}$$

In these vortices a flow into the vortex core close to the ground (convergence) is always observed, so that $\nabla_h \cdot \boldsymbol{v}_h < 0$. If the initially weak vortex has a cyclonic vorticity ($\omega > 0$), the fact that $\nabla_h \cdot \boldsymbol{v}_h < 0$ means that this will increase with time. This is sketched in Figure 10.17. In the case of an initially anticyclonic direction of rotation ($\omega < 0$) the same effect occurs. The magnitude of the rotation is increased, and the direction of rotation remains anticyclonic.

We now briefly discuss individual types of vortices.

Low-Pressure Region

Low-pressure regions are large-scale atmospheric vortices with relatively low air pressure in the vortex core (hence the expression low-pressure region or, in short, low). The wind blows according to the geostrophic equilibrium (see Section 10.1.3) in the mathematically positive sense (counter-clockwise) around the low core, which is why a low is also called a *cyclone*. Because of its large spatial extension, the low can be recognized only through spiral cloud patterns on satellite photos (as in Figure 10.18) or in the ground pressure field on a weather map. Such a weather map with isobars characterizing the low is shown in Figure 10.26. The low-pressure regions with their cloud and

Fig. 10.18. Satellite photograph of a cyclone (low-pressure region) characterized by a spiral cloud pattern

rain regions essentially determine the weather at moderate latitudes in both hemispheres and thus are the most important type of large-scale vortex.

The cause of the onset of low-pressure regions can be explained by a certain type of instability, the so-called *baroclinic instability.* The theory of the baroclinic instability is discussed in detail in the monographs of *B. Cushman-Roisin*, *D. Etling*, *A. E. Gill*, *J. Pedlosky*, and *H. Pichler* mentioned earlier. At this point we indicate only the basic idea. As explained in Section 10.1.3, in ideal conditions ($\mathrm{Ro} \to 0$) there may be a balance between the pressure force and the Coriolis force. This leads to the geostrophic wind (10.11) and on to the thermal wind relation (10.12). The latter states that in a baroclinic atmosphere a horizontal temperature gradient, as well as the pressure gradient, leads to a variation of the geostrophic wind with altitude. Therefore, in this case, warm air masses are situated alongside cold air masses. However, this equilibrium is not stable but leads to a vertical displacement of the air when a critical horizontal temperature gradient is exceeded, i.e. the cold masses of air move below the warm air. Because of the great effect of the rotation of the Earth on large-scale motion, this leads to the formation of cyclonic horizontal motions that eventually become cyclones.

Tropical Cyclone

The *tropical cyclone*, as the name implies, is a low-pressure region in tropical areas of the atmosphere. In the western Atlantic region they are known as *hurricanes* and in Asia as *typhoons.* A satellite photo of such a hurricane was already shown in the introductory chapter, in Figure 1.10. The somewhat harmless expression *tropical cyclone* belies one of the most powerful wind systems of the atmosphere, which is why they are also known as tropical vortex storms. The high wind velocities (up to 300 km/h), together with the waves in the sea that they excite, regularly lead to massive destruction when such a storm comes on land. As example of such tropical storm Figure 10.19 shows the Hurricane Katrina (2005).

The origin of the onset of tropical vortex storms can only be indicated at this point: They form above the warm tropical oceans, where the air's humidity is very high. In the extensive thermal convection of these regions (storm clouds with altitudes of up to 15 km) the water vapor condenses and so releases latent heat energy. The rotation of the Earth eventually causes the formation of a cyclone, which strengthens in its passage westward over the wet, warm ocean, until it eventually becomes a vortex storm. Further information on the structure and appearance of tropical cyclones and on their onset may be found in the monograph of *J. B. Elsner* and *A. B. Kara* (1999) or in a video of natural catastrophes (Discovery Channel (1997)) that impressively documents the power of destruction of tropical vortex storms.

Orographic Vortex

It is known from fluid mechanics that vortices can form behind bodies in a flow. Such bodies are realized in the atmosphere as orographic obstacles (hills, mountains, mountain ranges). Of the many different types of orographically induced vortices, we discuss only the *Kármán vortex street* that arises in the wake of bodies and is well known from fluid mechanics. Such vortex streets occur in the atmosphere in the lee of large islands, as seen in Figure 10.8 in the example of Jan Mayen. The individual cyclonic and anticyclonic vortices have diameters of $10 - 30$ km, while the total length of the vortex street can be up to 400 km.

Since islands are quite flat obstacles (height : width $\approx 1 : 10$), the air is forced to flow around the obstacle. This takes place via thermal inversion below the height of the summit, which, because of the Archimedes lift forces, acts as a kind of lid on the low atmosphere.

Fig. 10.19. Satellite photograph of Hurricane Katrina (2005), Gulf of Mexico

Tornado and Dust Devil

A *tornado* is the general name in the USA for extremely strong tubelike vortices that arise in connection with large storm clouds (Figure 10.17). In contrast to the tropical vortex storms, their diameters are only a few hundred meters. Wind velocities of up to 400 km/h and the strong underpressure in the core of the vortex (up to 50 hPa compared to the surroundings) regularly cause great destruction. This power of destruction has been impressively documented on video (e.g. Discovery Channel (1997)).

The direction of rotation of the tornado can be both cyclonic and anticyclonic. However, in very strong tornados the cyclonic sense of rotation dominates, as the mother-cloud generally already has cyclonic rotation because of its large-scale wind conditions. The strengthening of this initial rotation to a tornado is very complex and is not completely understood. Again, part of the process may be explained using the rotation strengthening mechanism (10.55). Because of continuity, strong upward and downward currents of air (up to 40 m/s) in the storm cloud lead to strong horizontally divergent and convergent flows, which lead to an increase in the vorticity, according to (10.55). This is shown schematically in Figure 10.17.

Tornados also occur in Europe, but there they are generally much weaker and are also known as *wind spouts* (or water spouts above the sea). Extensive material on the appearance of tornados and their origin may be found in *C. Church et al.* (1993).

The *dust devil* is also a tubelike vortex with a vertical axis, but one that is not related to the presence of a cloud. Rather, it appears in connection with thermal convection and is therefore a fair-weather phenomenon. Because of its small size of about 10 – 100 m horizontally and 100 – 500 m vertically, as well as its moderate wind velocities of 10 m/s, it is sometimes also known as the younger brother of the tornado. Its name, dust devil, is due to the fact that it stirs up loose ground material and transports it upward in the vortex core. It is because of this that the vortex is visible at all.

The onset mechanism for vortices like the dust devil is not yet fully understood. The main effect is certainly again the vortex strengthening mechanism (10.55). Close to the heated ground, warm air rises in the form of thermic tubes, causing a horizontal slipstream of air. If a certain initial rotation is present, e.g. via an obstacle, the wind convergence close to the upwind tube leads to rotation strengthening according to (10.55).

10.2.5 Global Atmospheric Circulation

In the previous sections we presented various individual phenomena of atmospheric flows. To round off these discussions we now briefly sketch the behavior of the atmosphere on global scales. There are in particular two things that affect the large-scale dynamics: the rotation of the Earth and the radiation from the Sun. Because of the spherical shape of the Earth, the latter means that the atmosphere close to the equator receives more radiation energy than that in polar regions. This means that the air temperature in equatorial regions is higher than that in regions at greater latitudes.

In Section 10.2.1 we discussed thermal circulation and noted that a horizontal temperature gradient causes a vertical circulation, where the air close to the ground flows from colder to warmer regions. For this reason, such a circulation must also arise between the polar regions and the equator. This is indeed the case, if only in a restricted region between the equator and about 30° latitude north and south. This circulation is named after the English meteorologist *G. Hadley* and is called *Hadley circulation* (Figure 10.20).

This large-scale thermal circulation occurs on the rotating Earth, and so the effect of the Coriolis force must be taken into account. In Section 10.1.3, in connection with geostrophic equilibrium, we noted that a horizontal temperature gradient leads to a change with altitude of the geostrophic wind. The wind blows so that the warm air lies to the right of the direction of flow (thermal wind relation, (10.13)). In the case of the large-scale temperature gradient between the equator and the poles, with the effect of the Earth's rotation a westerly wind must form, i.e. a flow to the east more or less parallel to the lines of latitude. This is in fact observed, as can be seen in the zonal mean of the horizontal wind velocity in Figure 10.21. The figure also shows a wind maximum at an altitude of about 10 km, called the *jet stream*. Here the wind velocities can be 100 – 300 km/h, a fact that is of importance in air travel.

In the regions between the equator and about 30° latitude, the air pressure contrasts close to the ground (equatorial low-pressure trough, subtropical high) mean that easterly winds corresponding to the geostrophic wind relation (10.11) are observed. Together with the lower part of the *Hadley*

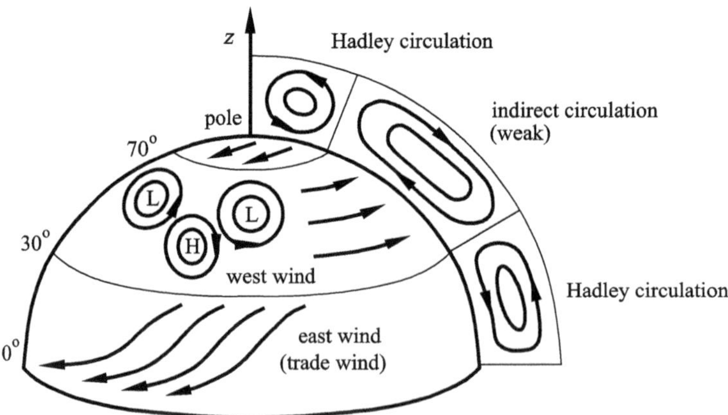

Fig. 10.20. Schematic representation of the global atmospheric circulation

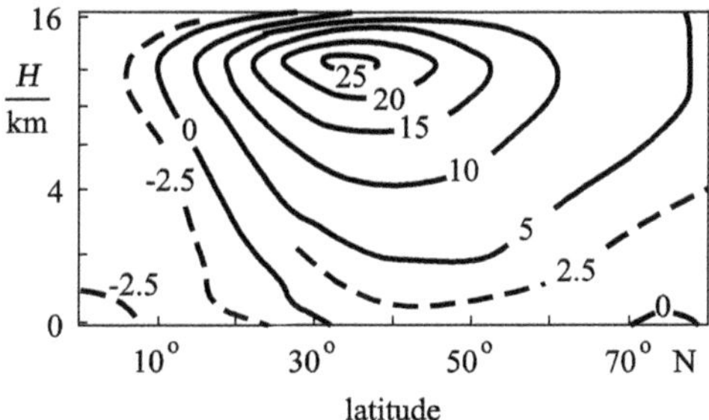

Fig. 10.21. Dependence of mean zonal wind velocity in m/s on the latitude and altitude. Positive values: west winds; negative values: east winds

circulation, these form the *trade winds* (Figure 10.20), the most permanent wind system of the atmosphere and one that was extremely important for mariners in earlier times.

At mid-latitudes, the large-scale atmospheric dynamics are determined by the formation and decay of low-pressure regions (cyclones) and high-pressure regions (anticyclones). As already discussed in brief in Section 10.2.5, the low-pressure regions arise via the *baroclinic instability* of the basic flow in the west-wind zone. This assumes that a critical horizontal temperature gradient is exceeded, as is frequently the case at moderate latitudes. These cyclones, together with their clouds and precipitation, essentially determine the weather at moderate latitudes.

Somewhat simplified, we may determine that the global atmospheric circulation causes temperature compensation between polar and equatorial regions in the atmosphere. Because of the spherical shape of the Earth and hence the uneven distribution of solar radiation, this temperature compensation cannot take place fully. However, the global air flows mean that the temperature difference between lower and higher altitudes is considerably more moderate than would be the case for just the pure balance of radiation.

The discussions above can be summarized in a simple scheme of the global atmospheric circulation, as shown in Figure 10.20. More extensive descriptions of global circulation may be found in the monographs of *R. Grotjahn* (1993) and of *J. P. Peixoto* and *A. H. Oort* (1992).

10.3 Flows in the Ocean

In Section 10.1 we considered different aspects that the flows in the atmosphere and in the ocean have in common. Examples of these were geostrophic flow (10.1.3), Rossby waves (10.1.4), and the Ekman layer (10.1.5). These flow forms will not be further treated in what follows. Although the basic fluid mechanics for both media are the same, there are several differences that cause particular flow conditions in the ocean. One main difference is that the air has essentially no side boundaries; it can flow around the Earth without restriction. However, the lateral motion of the oceans on our planet is restricted by the land masses. This means that the large-scale oceanic circulation is arranged in large anticyclonic vortices in the different basins (e.g. North Atlantic, Figure 10.22).

A further difference is the free, mobile surface of the sea that forms the upper boundary of the oceans. The atmosphere above causes a force to act on the water's surface via the wind shear stress. This force is the main driving force for currents in the sea. With respect to the vertical structure we note that not only does the water density ρ determine the pressure p and the temperature T as in the atmosphere, but in addition, the salt content c of the water has a great effect on the water's density.

For the reasons mentioned above, in what follows we can present only a brief discussion of oceanic flows, and this is in no way a replacement for the extensive research area of oceanography. Of the numerous monographs on oceanography, we merely mention those by *J. Pedlosky* (1996) and by *S. Pond* and *G. L. Pickard* (1991), where it is more the dynamic aspects of the currents in the sea that are treated.

10.3.1 Wind-Driven Flows

In Section 10.1.6 on the surface atmospheric boundary layer (Prandtl layer) we explained that the wind causes a tangential force to act on the surface of the Earth, called a tangential stress or shear stress per unit area. Experiments have shown that the shear stress, generally denoted by τ (see (10.37)), acts in the direction of the wind close to the ground and has a magnitude proportional to the square of the wind's velocity:

$$\tau = \rho \cdot c_w \cdot |\boldsymbol{v}| \cdot \boldsymbol{v}. \tag{10.56}$$

The coefficient c_w is called the *drag coefficient.* A typical value is about

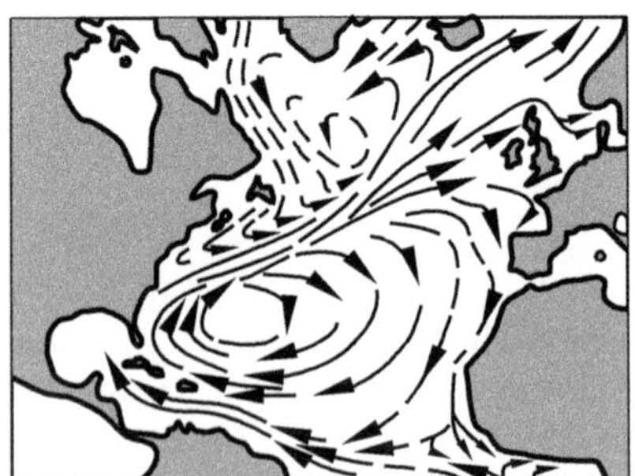

Fig. 10.22. Observed mean surface current conditions in the Atlantic Ocean. Full lines: warm ocean current; dashed: cold ocean current

$c_w \approx 1.5 \cdot 10^{-3}$. The ground shear stress (10.56) also acts on the surface of the sea, and since the sea is movable, a current is caused close to the surface. In the simple case of horizontal homogeneous conditions, we saw that an Ekman spiral arises, corresponding to (10.35) and (10.36) (see Section 10.1.5), where the flow on the surface of the sea is directed 45° to the right of the direction of the ground shear stress τ_w or of the ground wind $\boldsymbol{v}$. In the real ocean, however, this angle is considerably less and is about 20°. If we depart from this local approach and ask how the currents close to the surface of the sea are spatially distributed, clearly, the large-scale wind distribution must be considered as a driving force, according to (10.35), (10.36), and (10.56).

Considering the large-scale mean state of the atmosphere and the ocean, we noted in Section 10.2.5 that between the equatorial regions and about 30° N/S the trade winds dominate from easterly directions, while between 30° and about 70° the cyclones with mainly westerly winds dominate. Stated simply, at low latitudes a wind shear stress acts in a westerly direction and at higher latitudes in an easterly direction. This is shown in Figure 10.23. Hence ocean currents will develop from east to west or vice versa respectively.

As already mentioned in the introduction, the zonal extension of the oceans is obstructed by the continents running from north to south. Therefore, at low latitudes, more water mass will be supplied to an east coast, while at higher latitudes more water will be removed from the east coast. For reasons of continuity, at east coasts a flow from south to north must occur. The opposite is the case on west coasts. Therefore, in the ideal case, the wind shear stress causes a closed anticyclonic circulation in each oceanic basin, as shown schematically in Figure 10.23.

We note that the streamlines in the west of the basin (that is, at an east coast) are stronger than those in the east. Therefore, the current northward at the westerly edge of the basin is stronger than the southward, current at the easterly edge. The exact theoretical explanation for this initially unusual result is to be found in the monographs by *J. Pedlosky* and by *S. Pond* and *G. L. Pickard.* Stated simply, the current close to the surface of the sea is

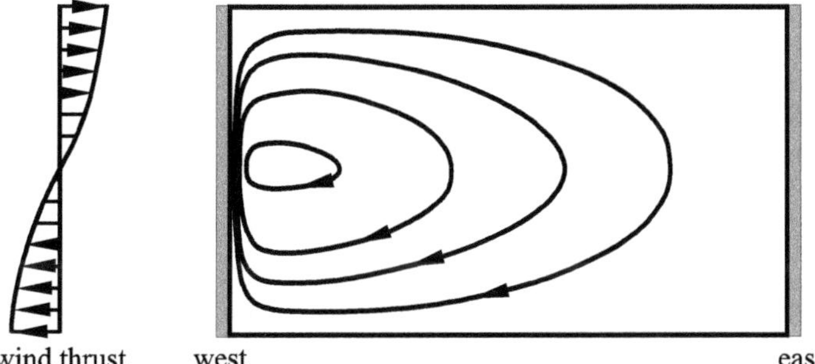

Fig. 10.23. Schematic representation of the zonal wind thrust and the surface current caused in an oceanic basin

in equilibrium between the frictional force and the Coriolis force. Now as in (10.6), the Coriolis force is the product of the Coriolis parameter f and the velocity $\boldsymbol{v}$. However, the Coriolis parameter changes with the latitude. For masses of water that are transported with a current from south to north, the Coriolis force increases northward. In equilibrium the frictional force must also increase. Since this latter is proportional to the velocity shearing action, the zonal gradient of the meridional velocity $(\partial v/\partial x)$ must increase. In the case of a flow from north to south, the exact opposite is true. Eventually, then, we find a strong velocity shearing action in the westerly part of a wind-driven oceanic circulation, and a lesser shearing action in the easterly part, as can be seen in Figure 10.23. This theoretical explanation of the idealized oceanic circulation can also be found in real oceans, as is shown in Figure 10.24 of the North Atlantic. The strong oceanic current along the east coast of the USA is known as the *Gulf Stream* (see also Figure 1.11 in Chapter 1).

Since the large-scale oceanic vortices have a large meridional extension, the water masses have different temperatures: warm in the south and cold in the north. The sea currents driven by the wind therefore transport warm water in the western region to the north and cold water in the eastern part to the south, as can be seen in Figure 10.23. This has huge consequences for the climate of the Earth. Comparing the air temperature in January along the 60th degree of latitude, we find in Ireland $+6°$ C, and close to Labrador $-10°$ C. Therefore, the mild winter climate of Western Europe exists thanks to the warm oceanic current of the *Gulf Stream*.

10.3.2 Water Waves

Like the atmosphere, the ocean is a medium that can carry out oscillations, and these appear in many different kinds of waves (*J. Pedlosky* (2003)). On a large scale, we have already treated *Rossby waves* in Section 10.1.4. These occur because of the latitude dependence of the Coriolis parameter (β effect). The dispersion relation (10.19) $c = \overline{u} - \beta/a^2$ is therefore valid for both the atmosphere and the ocean. We will not further discuss Rossby waves at this point.

In the case of a stably layered atmosphere, we considered *gravity waves* in Section 10.2.4. Now, the oceans, like the atmosphere, are also stably layered in large regions, so that the effect of the Archimedes lift force leads to the formation of internal gravity waves. The formal treatment of these waves for the ocean also takes place using (10.48) set up for atmospheric gravity waves. It is only in the *Brunt–Väisälä frequency* N (10.46) that the vertical density gradient in the ocean must be inserted, where this is determined by the salt content as well as the pressure and temperature. The dispersion relation for oceanic gravity waves is therefore identical to (10.51). The value of the Brunt–Väisälä frequency for the ocean is typically $N \approx 0.5 \cdot 10^{-2}\ \mathrm{s}^{-1}$, and so the waves have a period of about 30 minutes.

Neither Rossby waves nor internal gravity waves are generally accessible to the normal observer. The waves on the surface of the sea, one of the most common forms of motion of the ocean anywhere, are well known. We now look more closely at these *surface waves.*

Let us briefly consider the derivation of a dispersion relation for linear surface waves. The water mass is assumed to be incompressible and irrotational; i.e. $\nabla \cdot \boldsymbol{v} = 0$ and $\nabla \times \boldsymbol{v} = 0$. Therefore, the waves can be described by a potential flow, with the relations $\boldsymbol{v} = \nabla \phi$ and $\nabla^2 \phi = \Delta \phi = 0$ applying for the velocity potential ϕ.

In contrast to classical potential flow theory (see Section 4.1.5), the upper edge of the fluid consists of a movable surface whose height is variable; i.e. $\eta(x, y, z, t)$. In the simplified two-dimensional case, we select a wave ansatz of the form

$$\eta(x, t) = \eta_0 \cdot \cos(a \cdot (x - c \cdot t))$$

(wave number $a = 2 \cdot \pi / \lambda$, phase velocity c). With the physical boundary conditions

$$w(\eta) = \frac{\mathrm{d}\eta}{\mathrm{d}t}, \qquad w(z = -h) = 0,$$

and the depth of the base of the sea h, the linearized inviscid equations of motion (10.6), neglecting the Coriolis force, yield a relation for the phase velocity:

$$c = \sqrt{\frac{g}{a} \cdot \tanh(a \cdot h)}. \tag{10.57}$$

The phase velocity of the water waves clearly depends on the wavelength ($\lambda = 2 \cdot \pi / a$) and the water depth h. Depending on the values of λ and h, we may consider the following simplified limiting cases $h/\lambda > 0.5$ and $h/\lambda \leq 0.05$.

For

$$h/\lambda > 0.5 \rightarrow \tanh(a \cdot h) \approx 1,$$

(10.57) yields

$$c = \sqrt{\frac{g}{a}} = \sqrt{\frac{g \cdot \lambda}{2 \cdot \pi}}. \tag{10.58}$$

Because of the condition $h/\lambda \geq 1$, these waves are called *short waves* or *deep-water waves.* This does not mean that the water depth h must be large, only that it must be larger than the wavelength λ. Deep-water waves behave dispersively. According to (10.58), long waves move faster than short waves. For example, the occurrence of swell on the strand means that we can conclude that water waves have been excited by the wind of a storm that is still far offshore. Some numerical examples for the expansion velocity of deep-water waves are found below:

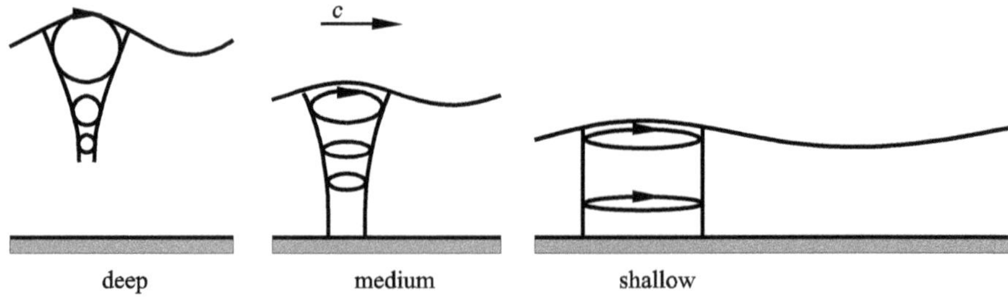

Fig. 10.24. Waves on the surface of the sea and resulting orbital motion of water particles for different water depths. Phase velocity c

$$\lambda = 10 \text{ m} \quad \rightarrow \quad c \approx 4 \text{ m/s},$$
$$\lambda = 100 \text{ m} \quad \rightarrow \quad c \approx 12 \text{ m/s}.$$

For

$$h/\lambda \leq 0.05 \rightarrow \tanh(a \cdot h) \approx a \cdot h,$$

(10.57) yields

$$c = \sqrt{g \cdot h}. \tag{10.59}$$

Because $h/\lambda \ll 1$, these waves are known as *long waves* or *shallow-water waves*. In contrast to the deep-water waves, these are not dispersive, and the phase velocity depends only on the water depth. Examples are

$$h = 10 \text{ m}, \qquad \lambda = 200 \text{ m} \quad \rightarrow \quad c \approx 10 \text{ m/s},$$
$$h = 1000 \text{ m}, \qquad \lambda = 20 \text{ km} \quad \rightarrow \quad c \approx 200 \text{ m/s}.$$

In the free ocean (large water depths), the relation (10.59) describes the phase velocity for very long waves, such as those caused by seaquakes (tsunamis).

In addition to the phase velocity (10.57), the solution of the potential flow equations also yields the velocity field in the water that is induced by the

Fig. 10.25. Water waves on the surface of the sea

surface waves. The precise analytical solutions may be found in *J. Lighthill* (1987). At this juncture we merely sketch the orbital motion of the water particles (Figure 10.24) that arises from the velocity fields.

The water waves described above are found for the idealized case of an inviscid liquid. In particular, if we include the surface stress as a further force, this dominates over the gravity effect in the case of very short waves (about $\lambda < 0.2$ m). These waves are visible as small ripples on the surface of the water and are also called capillary waves.

Now, the observer rarely sees the real surface of the sea in the form of one harmonic wave with a fixed wavelength. Rather, the superposition of many waves of different amplitudes, wavelengths, and phases is observed (Figure 10.25). At this point, we do not further discuss this so-called *sea spectrum* but refer to the more extensive literature on water waves (e.g. *J. Lighthill* (1987), *F. R. Young* (1999)).

10.4 Application to Atmospheric and Oceanic Flows

In addition to the fluid-mechanical phenomena in the atmosphere and the ocean, in recent times, the problems of weather forecasting, anthropogenic climate change, and ozone loss (ozone hole) have gained in importance. In order to take these into account, in addition to the actual fluid-mechanical topics we dedicate one short section to these long-term problems.

10.4.1 Weather Forecast

Even those who are not experts in the area of fluid mechanics are confronted with atmospheric flows almost daily via weather reports in the media. In the foreground of weather forecasting is the time development of the air temperature and the air pressure, as well as of clouds and precipitation. The prediction of wind strength and wind direction are additional fluid-mechanical components of the problem.

Weather forecasting has developed in the last 100 years from a somewhat empirical approach to the application of mathematical-physical methods based on the the dynamic and thermodynamic laws of fluid mechanics. The equations as set up in Chapter 5 are the basis for this. In a form common for the description of flows in the atmosphere and ocean, these equations read as follows:

Structure of equations describing the atmospheric–oceanic system

local change in time		advection		forces/ sources		diffusion	
$\partial \boldsymbol{v}/\partial t$	+	$\boldsymbol{v} \cdot \nabla \boldsymbol{v}$	=	$\boldsymbol{f}_i$	+	$K_\mu \cdot \nabla^2 \boldsymbol{v},$	(10.60)
$\partial \rho/\partial t$	+	$\boldsymbol{v} \cdot \nabla \rho$	=	$Q_\rho,$			(10.61)
$\partial T/\partial t$	+	$\boldsymbol{v} \cdot \nabla T$	=	Q_T	+	$K_T \cdot \nabla^2 T,$	(10.62)
$\partial q_i/\partial t$	+	$\boldsymbol{v} \cdot \nabla q_i$	=	Q_{q_i}	+	$K_q \cdot \nabla^2 q_i,$	(10.63)
$\partial c_n/\partial t$	+	$\boldsymbol{v} \cdot \nabla c_n$	=	Q_{c_n}	+	$K_c \cdot \nabla^2 c_n,$	(10.64)

with the velocity $\boldsymbol{v}$ and $\boldsymbol{f}_\mathrm{i} = -\rho \cdot f \cdot \boldsymbol{k} \times \boldsymbol{v} - \nabla p - \rho \cdot \nabla \phi$ in the *equation of motion* (10.60); the density ρ and $Q_\rho = -\rho \cdot \nabla \cdot \boldsymbol{v}$ for a compressible medium in the *continuity equation* (10.61); the temperature T and heat sources and sinks Q_T (e.g. adiabatic compression ($Q = -(1/(\rho \cdot c_p)) \cdot \mathrm{d}p/\mathrm{d}t$), divergence of long-wave and short-wave flows of radiation, phase change of water (latent heat) in the *equation for the internal energy* (10.62); the phase water vapor q_1, liquid water q_2, and ice q_3 and the phase changes Q_q (e.g. condensation, evaporation, freezing) in the *balance equation for water phases* q_i (10.63); and the gases, e.g. $c_1 = CO_2$, $c_2 = NO$, $c_3 = O_3$, etc., the salt content c in the ocean and the sources and sinks as well as the chemical transition of trace elements Q_c in the *balance equation for material* c_n ($\mathrm{n} = 1, 2, 3, \cdots$) (10.64). In the diffusion terms of the equations, K_μ, K_T, K_q, K_c are the turbulent diffusion coefficients for each flow property. The thermodynamic variables pressure p, density ρ, and temperature T are still included via equations of state. In the atmosphere, $p = R \cdot \rho \cdot T$, and in the ocean, $\rho = \rho(p, T, c)$ with the salt content c.

In addition to the equation of motion (10.60), the continuity equation (10.61) and the energy equation (10.62), there are also transport equations for water vapor and liquid water (clouds and raindrops) (10.63) as well as for atmospheric trace elements (10.64). It can be seen from the structure of the equations that atmospheric flows distribute air admixtures via both large-scale wind (advection) and small-scale turbulence (diffusion).

With equations (10.60)–(10.64), in principle, the time development of the variables wind, temperature, and precipitation in space can be predicted if the initial values are known. In practice, the initial values must be obtained from simultaneous worldwide measurements of the atmospheric variables. The solution of the equations cannot be obtained analytically because of their nonlinearity. Instead of this, numerical methods of solution are applied that are usual in other areas of fluid mechanics. In the science of the atmosphere, a specialist area with the name *numerical weather prediction* has developed, where the physical equations (10.60)–(10.64) are solved using the methods of numerical mathematics.

In fact, current weather prediction in the media is based on the results of the numerical solution of the fluid-mechanical equations (10.60)–(10.64). An example is the calculated prediction of the air pressure field on the ground shown in Figure 10.26. In summary, modern weather forecasting can be considered as the practical application of the laws of fluid mechanics to the atmosphere. The description of the fundamentals of weather forecasting and examples of its practical realization may be found in the monographs of *K. Balzer* et al. (1998), and *E. Kalnay* (2003).

10.4.2 Greenhouse Effect and Climate Prediction

Equations (10.60)–(10.64) for the atmosphere and ocean system can in principle be integrated over longer periods of time into the future than the few days taken for weather forecasting. Because of the nonlinearity and the known chaotic behavior of the system of equations, predictions over longer periods of time are no longer exact. The results of the numerical integration may

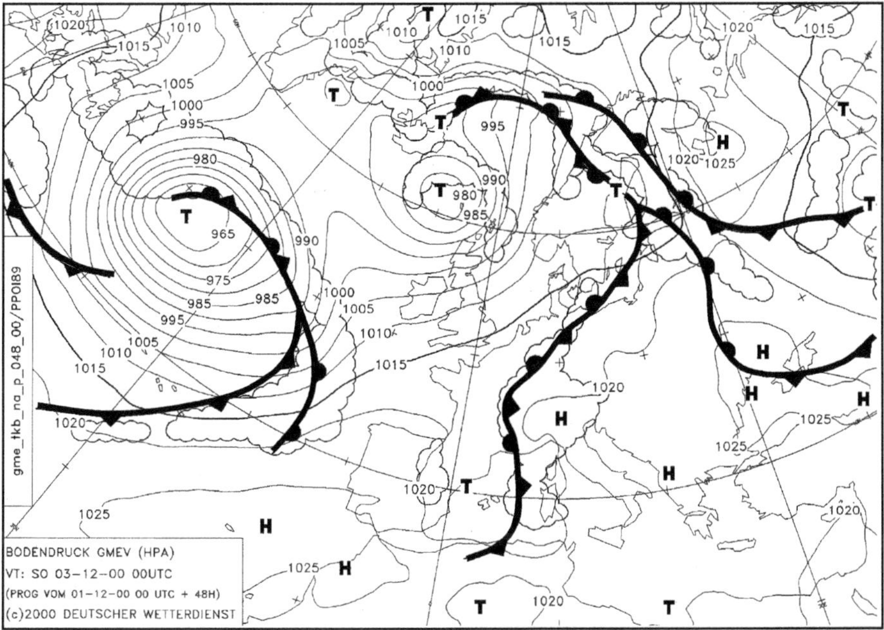

Fig. 10.26. Example of a 48-hour ground pressure prediction for the region Atlantic–Central Europe

therefore be interpreted only as spatial or temporal mean values for the different variables (e.g. mean air temperature in January). For observers, this corresponds to the mean conditions of the atmosphere, called the *climate.*

The fluid-mechanical equations (10.60)–(10.64) are therefore suitable for the prediction of the climate on the Earth. In the periods of time now considered (months, years, decades), the thermodynamic effects in the energy equation (10.62) dominate, in particular, the divergences of short-wave and long-wave radiation. The latter is greatly dependent on the spatial and temporal distribution of the radiation-affecting air admixtures (e.g. water vapor, carbon dioxide). The transport equations for these substances ((10.63) and (10.64)) therefore become more important in climate prediction. The simplest example here is the so-called *greenhouse effect*, which plays a large role in the discussion of a future change in climate.

In Section 10.2.5, on the global atmospheric circulation, we noted that the main origin of the large-scale motion can be seen in the different heating of the surface of the Earth by the short-wave solar radiation at different latitudes. These air flows, together with the temperature and water vapor distribution, determine the climate on our planet. The mean temperature of the surface of the earth T_0 is determined in the case of no atmosphere from the equilibrium between solar radiation S_o and long-wave blackbody radiation $\sigma \cdot T_0^4$:

$$\frac{S_o}{4} \cdot (1 - \alpha) = \sigma \cdot T_0^4. \tag{10.65}$$

Here $S_o = 1360$ W/m^2 is the solar constant, α the albedo of the Earth (amount of sun radiation reflected), and $\sigma = 5.67 \cdot 10^{-8}$ W/m^2/K^4 the Stefan–Boltzmann constant.

If we set the mean albedo of the Earth at $\alpha = 0.3$, we obtain from (10.65) $T_0 \approx 255$ K, corresponding to $-18°$ C. However, the observed mean air temperature close to the ground is about $+15°$ C, or 288 K. This is because the long-wave radiation does not come from solid bodies (such as the surface of the Earth) but also from certain gases. Of the gases present in the Earth's atmosphere, particularly water vapor H_2O, carbon dioxide CO_2, and ozone O_3 are known as absorbers and emitters of long-wave radiation. Depending on their temperature, these gases radiate both in the direction of space and in the opposite direction, i.e. toward the surface of the Earth. This part of the long-wave radiation is also called the *counter-radiation.* It reduces the effective long-wave radiation away from the surface of the Earth, so that instead of (10.65) we have

$$\frac{S_o}{4} \cdot (1 - \alpha) = \sigma \cdot T_0^4 - \lambda_g. \tag{10.66}$$

Here λ_g symbolizes the long-wave counter-radiation of the atmosphere. Finally, this counter-radiation is added to the solar radiation, so that a higher temperature T_0 must be found for (10.66) than that found from (10.65). This effect of the atmospheric counter-radiation on the global mean air temperature, that makes up about $+33°$C, is also called the *greenhouse effect.* In fact, it is the gases with their radiation that are present in the atmosphere that permit the climate necessary for life.

The anthropogenic change of the climate can firstly be described via the radiation as the driving force for the conversion of energy in the atmosphere–ocean–earth system. If additional amounts of greenhouse gases (e.g. CO_2, methane) are added to the atmosphere and if the long-wave atmospheric counter-radiation is strengthened (λ_g in (10.66)), according to the simple balance of radiation (10.65) the mean global ground temperature T_0 will rise. This effect of the greenhouse gases is partially compensated by the radiation effect of the aerosols (small droplets and particles, e.g. mineral dust or volcano ash, of a few μm diameter). These reflect a part of the short-wave solar radiation, meaning that the albedo α in equation (10.65) is increased. This causes the ground air temperature T_0 to be somewhat reduced. In spite of the cooling effect of the aerosols, currently, an increase of the global air temperature T_0 of about $1^\circ - 3^\circ$ C is assumed in the next 50 years.

Two aspects are emphasized in this section: the role of the atmosphere as a transport medium of trace gases and aerosols, and the estimation of the anthropogenic greenhouse effect by numerical simulation models.

We consider as an example the greenhouse gas CO_2. Since the beginning of industrialization, the combustion of fossil fuels has emitted additional carbon dioxide (as well as the CO_2 naturally present) into the atmosphere. Initially, CO_2 is carried from the sources close to the ground into the higher layers of air, and there it is distributed more or less uniformly over the entire atmosphere with the large-scale air flows (Figure 10.27). The formal description

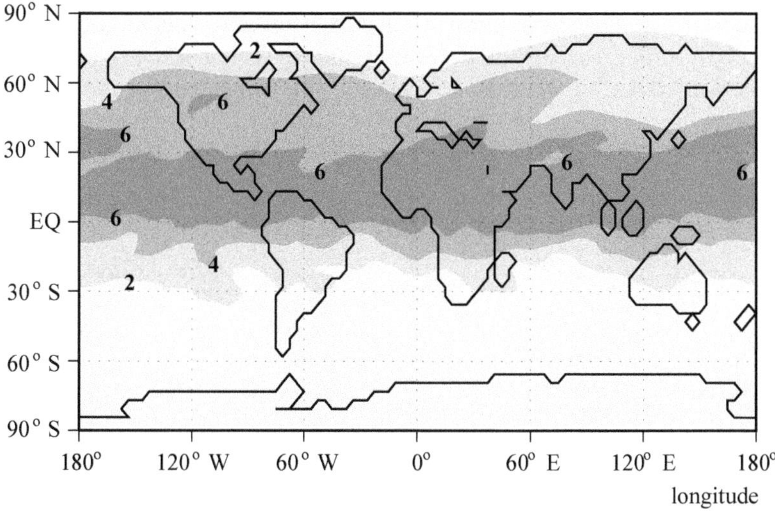

Fig. 10.27. Calculated global concentration distribution of aerosol (in $\mu g/m^3$) at an altitude of 20 km for November 15, 1991, five months after the eruption of the volcano Mt. Pinatubo

of transport and diffusion of CO_2 (and other greenhouse gases) in the atmosphere can be carried out using (10.64). The global atmospheric circulation (see Section 10.2.5) therefore causes a mixing of the atmosphere with CO_2 and other released radiating trace gases that contribute to the greenhouse effect.

Volcano eruptions are an example of the natural introduction of aerosols into the atmosphere. One of the greatest events in this century took place on June 15, 1991, with the eruption of Mt. Pinatubo. This volcano, situated in the Philippines (15.14°N, 120.35°E) shot sulphur aerosols up into the lower stratosphere at altitudes of between 20 and 25 km. There they dispersed rapidly over the globe with the atmospheric flows, and several months later were spread over the entire northern hemisphere and even in regions south of the equator. The dispersion of the volcano aerosols was calculated with a global transport model that used equations (10.60)–(10.64) (*C. Timmreck et al.* (1999)). The calculated aerosol concentration at an altitude of about 20 km is shown in Figure 10.27 for the date November 15, 1991, five months after the volcano erupted. As mentioned for the radiation properties of aerosols (reflection of solar radiation), in fact, in the one to two years after the eruption there was a reduction in the ground air temperature in the northern hemisphere of about 0.5° C.

The equations (10.60)–(10.64) of geophysical fluid mechanics yield information about the consequences of the greenhouse effect on the global flows in the atmosphere and also in the ocean. In the energy equation (the first law of thermodynamics), the divergences of the short-wave and long-wave radiation currents appear as heat sources, whose effect again depends on concentration and spatial distribution of trace gases and aerosols. If the latter is known, statements about the global temperature distribution and thus about the air flows caused by temperature gradients can be made. As has already been discussed in Section 10.4.1 on weather forecasting, the equations for the system

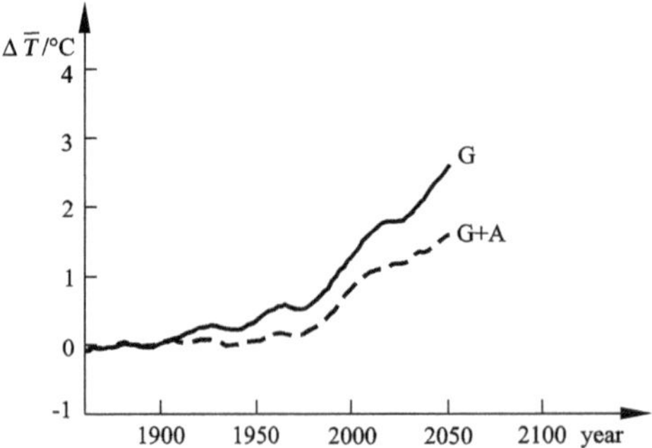

Fig. 10.28. Change in time of the globally averaged ground air temperature calculated with a climate model compared to that of an atmosphere without anthropogenic load with greenhouse gases and aerosols. Simulation **G**: only greenhouse gases; simulation **G+A**: greenhouse gases and aerosols

atmosphere–ocean can be solved only numerically. The prognosis of future climate changes caused by the anthropogenic greenhouse effect can therefore be carried out consistently only by discretizing equations (10.60)–(10.64) on a mesh and subsequently solving the initial–boundary value problem using numerical methods.

An example of the results of such a climate model is the rate of change of the globally averaged ground air temperature due to the anthropogenic increase of greenhouse gases and aerosols, shown in Figure 10.28. With the coupled models of atmosphere and ocean of the Max-Planck Institute for Meteorology in Hamburg, two scenarios were calculated (*E. Roeckner et al.* (1999)). In one case, **G** only, the natural and anthropogenic greenhouse gases were taken into account. In a second case, **G+A**, the radiation effect of the natural and anthropogenic aerosols was also calculated. An anthropogenically caused increase in the global air temperature of about 2.6° C in the next 50 years can be seen for the case of the pure greenhouse effect. This temperature increase is reduced to 1.6° C when the anthropogenic emission of aerosols (generally sulphur compounds) is taken into account.

Of the vast number of publications on climate issues, as well as the fundamentals of climate and climate change we mention only the monograph of *J. Houghton* (1997) and the report of the *Intergovernmental Panel on Climate Change (IPCC)* (2007). The latter also contains results from calculations with climate models. The principles of modeling the different parts of the climate system (atmosphere, ocean, biosphere, etc.) are given the the collected works by *K. E. Trenberth* (1992).

10.4.3 Ozone Hole

Besides the *greenhouse effect*, the *ozone hole* also plays a role in global climate change. This is a phenomenon in the *stratosphere* above the North Pole and South Pole. As winter passes, in spring there is a considerable reduction in concentration of the gas ozone O_3 at altitudes between 20 and 30 km. This is not a hole in the sense that the ozone has completely vanished. However, the reduction over the South Pole from typically 400 DU (*Dobson units*, a measure of the total ozone content in one column of air) in the year 1979 to 180 DU in 1992 is very obvious. The ozone hole denotes the more or less circular region with greatly reduced ozone concentration around the South Pole (Figure 10.29).

In the lower and middle stratosphere between 15 and 30 km altitude, there is a layer of maximal ozone concentration. Because of the ability of ozone to absorb short-wave solar radiation (ultraviolet (UV) radiation), this ozone layer protects life on Earth from harmful UV rays. The ozone O_3 is formed from molecular O_2 and atomic oxygen O via the absorption of ultraviolet solar radiation with wavelengths < 242 nm.

The ozone is then destroyed by short-wave solar radiation with wavelengths less than 1200 nm and split into molecular and atomic oxygen.

In total, these reactions form a photochemical equilibrium, and neither permits a loss mechanism for ozone. This occurs only via a further catalytic degradative reaction:

$$X + O_3 \quad \rightarrow \quad O + O_2, \tag{10.67}$$

$$OX + O \quad \rightarrow \quad X + O_2. \tag{10.68}$$

The catalyst X (e.g. chlorine, hydrogen, nitrogen oxides) is set free in this reaction and can destroy more ozone.

The decomposition of ozone in the polar stratosphere is due to such catalytic reactions. In particular, those materials that are in part of anthropogenic origin, e.g. nitrogen oxides (NO, NO_2), hydrogen radicals (OH, HO_2), chlorine Cl, chlorofluorocarbons (CFCs), seem to play a role. In the literature more than 30 different reaction mechanisms may be found that lead to a net reduction in the ozone of the atmosphere.

The reference to the previous sections lies in the mechanisms that transport the chemical substances into the middle stratosphere via the Antarctic and the North Pole. With the synoptic systems (low-pressure regions) and via the Hadley circulation, the particles are distributed more or less uniformly over the northern hemisphere. They must then overcome the barrier of the tropopause, which greatly obstructs the vertical exchange. By means of vertically extended thermal convection in the tropics and fronts of the low-pressure regions, at certain points tropospheric air reaches the stratosphere (Figure 10.30). The latter flows as (zonal) wind systems more or less parallel to the lines of latitude that permit particle distribution in the east–west direction. However, in order that the anthropogenic trace elements can enter the polar stratosphere, a meridional circulation is necessary. Such a mechanism indeed exists and is called *Brewer–Dobson circulation* after those who discovered it. The scheme of the meridional circulation in the troposphere

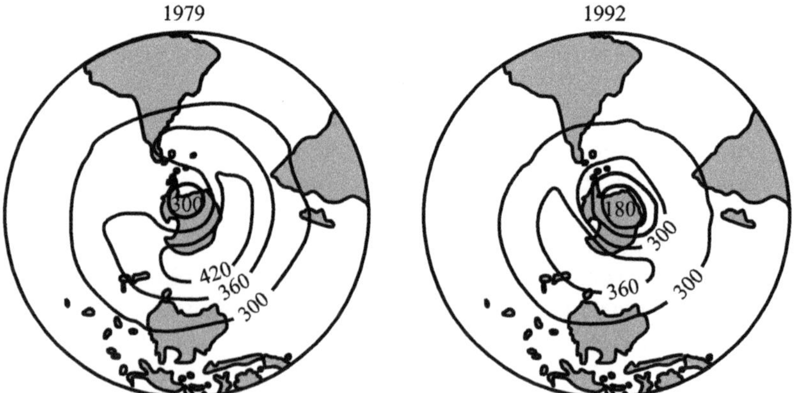

Fig. 10.29. Total ozone content of the atmosphere in Dobson units over the southern hemisphere in October 1979 and 1992

and stratosphere is shown in Figure 10.30. The period of revolution of this circulation, i.e. the transport of tropospheric air to the polar regions of the stratosphere, is several months. Therefore, only those chemical substances that have a long lifetime can contribute to the destruction of ozone. Indeed, the CFCs with their lifetimes of several years are among the candidates. Although the destruction of ozone is a purely photochemical process, the atmospheric transport methods are needed to explain the ozone hole in the Antarctic stratosphere.

In connection with the ozone hole, a further fluid-mechanical effect comes into play, which we briefly discuss here. In the processes of destruction of ozone the air temperature also plays an important role, in particular in relation to reactions via hydrogen. A lowest possible temperature (e.g. $-90°$ C) favors the reaction times of different processes that play a role in ozone chemistry. The stratospheric air above the South Pole must have the chance to cool down. This is guaranteed when it is not mixed with the relatively warm air from moderate latitudes, and in fact, this mixing is prevented by the very stable polar cyclonic vortex that forms in the winter months above the Antarctic. This vortex is characterized by high values of potential vorticity (see Section 10.1.4). Recent theoretical and numerical studies have shown that such a vortex permits essentially no mixing of external air masses (here from moderate latitudes). A *potential-vorticity barrier* is present. This property of the polar vortex also explains why the ozone layer appears less obviously above the North Pole than above the South Pole. The northern hemisphere polar vortex is variable, and there the polar air mixes more easily with air from middle latitudes.

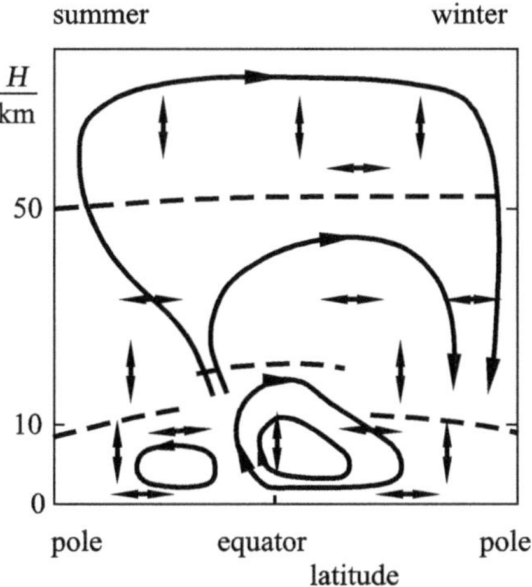

Fig. 10.30. Schematic representation of the meridional circulation in the troposphere ($\sim 0 - 10$ km) and stratosphere ($\sim 10 - 50$ km), as well as main transport paths (thick arrows) of atmospheric trace elements

In total, we can determine that the destruction of ozone in the polar stratosphere is a photochemical process that is caused by anthropogenic trace substances. Without the different transport processes in the atmosphere, from small-scale turbulent diffusion to vertically extensive thermal convection to stratospheric Brewer–Dobson circulation that first carry the substances to their reaction point, and the polar stratospheric vortex that essentially encloses these substances in winter, the destruction of ozone would not even be possible.

Further information on the ozone hole may be found in the monographs by *P. Fabian* (1992), *T. E. Graedel* and *P. J. Crutzen* (1994), *K. Labitzke* (1999) or in the review article by *S. Solomon* (1999).

11. Microflows

11.1 Fundamentals of Microflows

With advances in manufacturing technology, flow and transport processes in microchannels or past micro-objects have become relevant for technical applications. Modern manufacturing methods permit the construction of tiny structures of considerably less than one millimeter in different materials such as silicon, glass, metal or plastic. This results in microchannels and micro-objects, through which and past which flow and transport processes take place, thus realizing complex functions in tiny spaces. It is found that, depending on the fluid, a continuum-mechanical treatment of flows through and past very small geometries is no longer necessarily possible in many cases. To correctly represent the physics of the flow at such small length scales, corrections to the continuum-mechanical equations or even molecular methods are sometimes necessary.

On the one hand, the term *microflow* can be defined quite formally as a flow through a microchannel of width d or past a micro-object of dimension d, where $1 < d < 1000\,\mu m$. On the other hand, depending on the fluid, the flow at such length scales possibly can be described perfectly well by a continuum model. The term microflow is only then justified when the boundaries of the continuum-mechanical treatment are reached, or when certain effects, which play a less important role in macroscopic flows, become important. It is this physically-based concept of a microflow that we will discuss in this chapter.

11.1.1 Application of Microflows

In the Introduction (Chapter 1), we described the examples of the print head of an inkjet printer and a micro heat exchanger that have already been realized. We will now discuss two applications of microflows that point towards the future. The discussion of these examples is not meant to be complete; rather we wish to indicate the extensive application possibilities of microflows.

In physical, biological and chemical analysis a vision has developed of the assembly of a complete analysis laboratory on one chip in the not too distant future (cf. *A. Manz* and *H. Becker* (1999)): a *lab on a chip* or a *micro-total-analysis system, μTAS*. There are many advantages of miniaturization: (i)

H. Oertel (ed.), *Prandtl-Essentials of Fluid Mechanics*,
Applied Mathematical Sciences 158, DOI 10.1007/978-1-4419-1564-1_11,

only small specimen volumes are needed; (ii) the favorable ratio of fluid surface and volume permits efficient heat and mass transfer, catalytic reactions, and detection and separation processes; (iii) mechanical, optical and electrical components can be integrated; and (iv) the cost-effective mass production of such chips in bio-compatible and chemically compatible materials permits disposability. All these advantages lead us to expect that analysis, providing it is sensitive and reproducible, will be considerably more cost-effective and will be able to take place to a great degree in parallel. Such a miniaturized analysis lab requires the integration of a series of building blocks that provide fluid preparation, fluid transport, mixing of fluids, biological and/or chemical reactions, as well as separation and detection processes. A great number of these functions are closely connected to the flow and transport processes in the corresponding building blocks.

The combination of electronic and mechanical components in systems of considerably less than one millimeter in size (so-called micro-electro-mechanical systems, MEMS) are also opening up new possibilities in flow

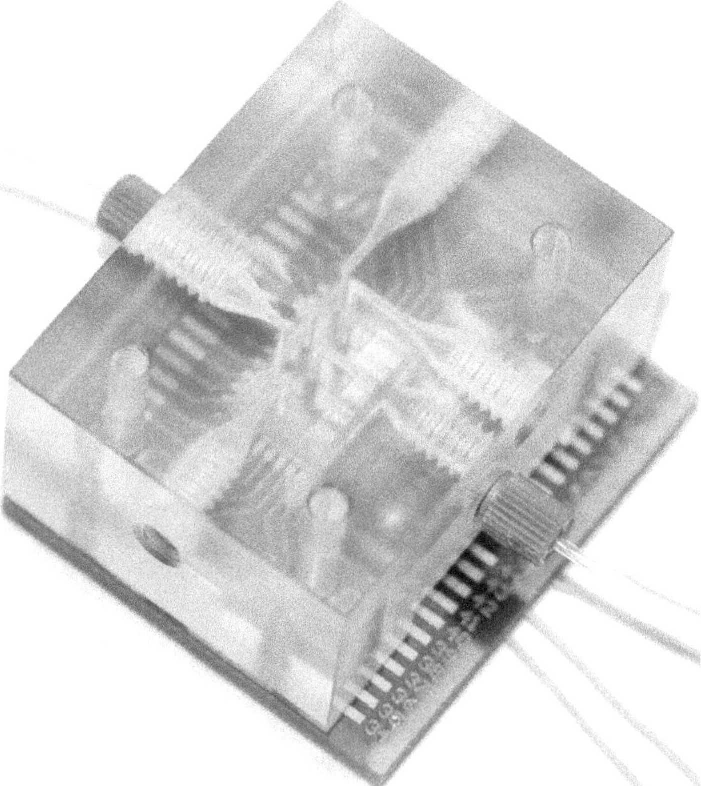

Fig. 11.1. Analysis laboratory for blood testing, *S. Zheng* and *C. Tai* (2006)

instrumentation (cf. *L. Löfdahl* and *M. Gad-el-Hak* (2002)). First, this is based on sensors for the flow velocity. The miniaturization of sensors based on the thermal principle (e.g. hot-wire probes) for one or more components of the velocity, or, further, the application of many such sensors in arrays, opens up new possibilities in turbulence research. Here, as well as the improved time resolution of small sensors and their cost-effective production, a fundamental improvement in the quality of the data is to be expected, because measurement is possible in many places at fine spatial resolution. Second, indirect or direct measurement of the shear stress is possible at similarly good spatial and time resolution. Indirect measurement of the shear stress is possible via similar thermal methods as those for the velocity. Direct measurement of the shear stress, through inserts embedded elastically into the wall, is now also possible at high spatial resolution with new (micro) production technologies. Third, it is possible to embed small, mainly capacitive pressure sensors in the wall. Here too, miniaturization promises high time and especially spatial resolution. These are the smallest structures in turbulent flows whose measurement resolution is possible with the help of micro-sensor-arrays. Accordingly, at least at moderate Reynolds numbers, the aim is for sensor surfaces with the dimensions of the Kolmogorov length.

A spatial collection of coherent structures in turbulent boundary-layers close to the wall can also serve as a starting point for the local control and suppression of turbulence. Additional arrays of miniaturized actors are necessary to induce the necessary reaction at the wall at the correct place and time. Such actors can also be manufactured as MEMS according to different principles and embedded cost-effectively as arrays in the wall. The perspectives of local turbulence control are qualitatively more far-reaching than those of global methods (such as surface suction) and may even yield an economical method of active turbulence control.

11.1.2 Fluid Models

The continuum mechanical description of the motion of fluids is the basis for conventional fluid mechanics (see Section 5.4.1). In conventional fluid mechanics the detailed character of the fluid as a collective of molecules is ignored and instead averaged flow quantities are used. The individual molecules have, for example, a random statistical Brownian motion which is superimposed by a translation. If we average the motion over a volume element with many molecules, only the translation remains; it is this that determines the continuum mechanical velocity vector. The Brownian motion of many molecules, whose amplitude incidentally is related to the temperature, is statistically independent and vanishes on averaging. Continuum mechanics thus describes the time and spatial change of flow quantities such as density, velocity, pressure and temperature averaged over a volume element. Such a representation is of course only reasonable if the averaging can be carried out over a sufficient number of molecules. In other words, the diameter of the averaging volume must be much larger than the mean distance of the molecules. At the same time, the diameter of the averaging volume must be small compared to the dimensions of the flow region. Thus there are clearly limits to continuum mechanics, such as when the number of molecules per averaging volume becomes very small (rarefied gases) or when the dimension of the flow region and thus the averaging volume becomes very small (microflows).

A further assumption in continuum mechanics is usually the following: in order to obtain an approximately linear relation between the stresses and the shear rates (Newtonian fluid), the fluid must be close to thermodynamic equilibrium. This assumption incidentally also means there is a linear relation between the heat fluxes and the temperature gradients (Fourier fluid) and between the mass flows and the concentration gradients (Fick's law). At the molecular level, thermodynamic equilibrium is present if there are enough interactions between the molecules per unit time. Simultaneously, this time interval must remain small compared to the time scale of the flow. If the fluid is far from thermodynamic equilibrium, discontinuous behaviour of the flow quantities may arise within the fluid. At the walls thermodynamic equilibrium is also responsible for there being no discontinuities in the velocity (no-slip condition) or temperature. For this, sufficient interactions between the fluid and wall molecules are necessary. If there is no thermodynamic equilibrium at the wall, discontinuities of the velocity (slip condition) and the temperature are possible there. The breakdown in the continuum mechanical treatment under certain (extreme) conditions makes it necessary to extend the spectrum of fluid models.

An overview of the basic equations and fluid models is given in Section 5.4.1 and Figures 5.5 and 5.7. Depending on the effect of friction, continuum models can be classified into the Euler equations, the Navier-Stokes equations for Newtonian fluids and the Burnett equations. The Burnett equations result from an approximation solution of the Boltzmann equation by introducing an expansion for very small Knudsen numbers Kn (the Chapman-Enskog expansion). Therefore they are also valid close to thermodynamic equilibrium. Formally, the Burnett equations have the appearance of the compressible momentum equations (5.13) – (5.15). However, the normal and shear stresses contain nonlinear terms of the velocity gradient as well as an inherent coupling with the energy equation (see *S. Chapman* and *T.G. Cowling* (1970)). For very small values of Kn the Burnett equations pass over to the compressible Navier Stokes equations (5.18). On the other hand there are molecular models, which use deterministic or statistical methods. Molecular dynamic simulation (MDS) is a deterministic method based on Newton's law, while Monte Carlo simulation (MCS) is of statistical nature and is based on the Boltzmann equation (5.64) (see *H. Oertel* (1994), (2005)).

In general continuum models are preferred over molecular models, as long as the Knudsen number range allows for this. This is because the mathematical or numerical effort needed for the treatment of the partial differential equations of continuum models is considerably less than the effort needed for molecular models: (i) molecular dynamic simulations and Monte-Carlo simulations require a description of the interaction and motion of all or of many selected molecules; (ii) the Boltzmann equation is an integro-differential equation whose (numerical) solution is difficult, particularly when the collision integral is present. Molecular dynamic simulations for technically relevant microflows, simply because of the number of molecules, remain highly extensive. Therefore is it important to define the limits of validity of the fluid models as precisely as possible, in order to be able to make substantiated decisions for any fluid model.

Gases and liquids differ essentially in the mean distance between their molecules. In gases the bond to neighboring molecules is broken and the large molecular distance permits quite free motion of the molecules, interrupted by collisions with other molecules. Therefore in gases the mean free path of the molecules between two collision is an important parameter. The theoretical treatment of gases within the framework of kinetic gas theory is relatively well developed. In contrast, in liquids the distance between the molecules is considerable smaller, so that the molecules are in constant interaction with neighboring molecules. Because the validity of the continuum models is essentially based on the molecular properties, in the following sections we will discuss gases and liquids separately.

11.1.3 Microflows of Gases

The mean free path in a gas is related to the frequency of collisions and thus also to the question of whether thermodynamic equilibrium has been attained in the gas (cf. *S. A. Schaaf* and *P. L. Chambré* (1961)). For an ideal gas of spherical molecules, for example, the mean free path has the form

$$\bar{\lambda} = \frac{k_\mathrm{B} \cdot T}{\sqrt{2} \cdot \pi \cdot p \cdot \sigma^2} \tag{11.1}$$

and so is dependent on pressure p and temperature T; k_B is the Boltzmann constant ($k_\mathrm{B} = 1.38 \cdot 10^{-23}\ J/K$) and σ is the scattering cross section, which is identical with the diameter of the molecules for the case of elastic spheres. Thermodynamic equilibrium is reached when the mean free path $\overline{\lambda}$ remains considerably smaller than the length scale L of the flow. The length scale of the flow may be the width of the channel; in general it is the length scale over which there are gradients in (macroscopic) flow quantities such as pressure, density, velocity or temperature. From the velocity profile $u(z)$ in a plane shear flow, say, the length scale is

$$L \sim \frac{u}{|du/dz|} \quad . \tag{11.2}$$

The ratio of the mean free path to the length scale of the flow is a dimensionless number called the Knudsen number

$$\mathrm{Kn} = \frac{\bar{\lambda}}{L} \quad . \tag{11.3}$$

For $\mathrm{Kn} \ll 1$ the flow is clearly in thermodynamic equilibrium. The mean free path, and therefore the Knudsen number, also characterizes the number of molecules (per averaging volume) to which a continuum-mechanical description may be applied. This may even be a large number of molecules (per averaging volume) if the mean free path is small, i.e. if $\mathrm{Kn} \ll 1$. Therefore the Knudsen number permits a gas flow to be characterized according to thermodynamic equilibrium and according to the continuum assumption.

The definition of the Knudsen number makes it clear that deviations from $\mathrm{Kn} \ll 1$ can occur for both large mean free paths and small length scales of the flow. Large $\bar{\lambda}$ occurs for flows of rarefied gases, while small L is found in microchannels. The flow of rarefied gases and the flow of gases past or through small geometries are therefore similar with respect to the Knudsen number. Therefore we may use the well-founded literature on flows of dilute gases to characterize the flow through or past small geometries. Using the Knudsen number, we find the following regimes (cf. *H. Oertel* (1994), (2005) *M. Gad-el-Hak* (1999)):

$\mathrm{Kn} \to 0$ $(Re \to \infty)$	Euler equations
$\mathrm{Kn} \le 10^{-2}$	Navier-Stokes equations with no-slip condition
$10^{-2} < \mathrm{Kn} \le 10^{-1}$	Navier-Stokes equations with slip condition
$10^{-1} < \mathrm{Kn} \le 10$	transition region
$10 < \mathrm{Kn}$	free molecular flow

The conventional continuum-mechanical equations and boundary conditions may be used for Knudsen numbers $\mathrm{Kn} \le 10^{-2}$. Increasing the Knudsen number means that a correction to the kinematic (and thermal) boundary conditions becomes necessary: discontinuities occur in the velocity (and temperature) at the wall. The region $10^{-1} \le \mathrm{Kn} \le 10$ is a transition region. In this region, for small Knudsen numbers the Burnett equations may still be used, and otherwise Monte-Carlo simulations are suitable. Finally, the region $10 < \mathrm{Kn}$ is characterized by free molecular flow, which is described by the Boltzmann equation.

We now clarify the Knudsen number regimes with a concrete example: consider air under normal conditions ($288\ K, 1\ bar$). The mean free path is then $\bar{\lambda} = 65\ nm$. A flow in a microchannel of $L = 1\,\mu m$ width therefore has $\mathrm{Kn} = 0.065$. This is already a flow where the slip of the gas at the wall has to be taken into account. If the pressure in the same channel were $0.1\ bar$, then $\bar{\lambda} = 650\ nm$ and so $\mathrm{Kn} = 0.65$. Therefore the flow could no longer be described using a continuum model.

In summary we may state that for gases with increasing Knudsen number, first thermodynamic equilibrium between the gas and the wall is lost and the gas slips at the wall. A further increase of the Knudsen number leads to loss of thermodynamic equilibrium inside the gas – the gas no longer behaves as a Newtonian fluid and the Navier-Stokes equations may no longer be used. Finally, at very large Knudsen numbers the continuum models completely break down and the gas must be treated as a collection of molecules.

11.1.4 Microflows of Liquids

Whereas for gases there is kinetic gas theory, a well-established (molecular) model that allows the regime beyond the limits of continuum mechanics to be characterized, for liquids the regime beyond the limits of continuum mechanics is much more difficult to deal with. The concept of the mean free path and the Knudsen number is not useful for liquids. As the molecules in a liquid are in constant interaction with neighboring molecules, molecular dynamic simulations (MDS) are preferable as a molecular model. In addition, experiments may be used to characterize the limits of the continuum-mechanical treatment of liquids.

From experiments with extremely thin liquid films between plates that are smooth to a molecular level (cf. *D. Y. C. Chan* and *R. G. Horn* (1985), *M. L. Gee et al.* (1990)), it is known that it is only at film thicknesses below about 10 molecule layers ($\sim 5\,nm$) that the liquid may no longer be treated as a continuum. Non-smooth changes in the normal and shear stresses are then observed — clear indications that the number of molecule layers is having an effect on the behaviour of the liquid. Furthermore, these experiments already show changes in the viscosity for liquid films below 100 molecule layers ($\sim 50\,nm$) — an indication that the liquid no longer has a Newtonian character. Similar indications are found in the molecular dynamic simulations of *W. Loose* and *S. Hess* (1989). These authors observe a shear layer of about 10 molecule layers of idealized, spherical molecules and for shear rates satisfying

$$\dot{\gamma} \geq 1.4 \cdot \sqrt{\frac{\epsilon}{\sigma^2 \cdot m}} \tag{11.4}$$

they find non-smooth behaviour of the flow quantities over the shear layer, while smooth behaviour is observed for small shear rates. The shear rate in a two-dimensional problem is related to the gradient of the velocity by $\dot{\gamma} = du/dz$. In (11.4) ϵ is the binding energy, m is the mass and σ is the diameter of molecules. For water molecules under normal conditions, the binding energy is $\epsilon \sim 3.5 \cdot 10^{-21} J$, the mass of a molecule is $m \sim 3 \cdot 10^{-26} kg$ and the diameter of a molecule is $\sigma \sim 3 \cdot 10^{-10} m$. The estimation according to (11.4) is therefore $\dot{\gamma} \geq 1.6 \cdot 10^{12} s^{-1}$. The water molecule is of course a non-spherical complex molecule, for which the simple model of *W. Loose* and

S. Hess (1989) cannot necessarily be directly applied. Nevertheless, it may serve as an estimate of the order of magnitude. Although the shear rates for simple molecules with small molecule mass and small molecule diameter may seem exorbitantly large, complex (heavy, large) molecules yield smaller critical shear rates, according to (11.4), that can indeed be reached in technical systems.

As well as the question of the continuum-mechanical treatment and the thermodynamical equilibrium of the liquid, there is also the question, as with gases, of whether discontinuities in the velocity (or the temperature) can occur at the wall ($z = 0$). For this it is useful to formulate the Navier slip condition in the form

$$u(z=0) - u_{\mathrm{w}} = L_{\mathrm{S}} \cdot \dot{\gamma}\ (z=0) \tag{11.5}$$

In (11.5) the discontinuity in the velocity tangential to the wall is proportional to the shear rate. The constant of proportionality L_S has the dimensions of a length and is called the slip length. A vanishing slip length ($L_S \to 0$) gives of course the no-slip condition. In the literature we find contradicting statements from experiments on the slipping of liquids at extremely smooth walls: (i) slip lengths of only $L_S \leq 20\,nm$ are found by *V. S. J. Craig et al.* (2001) in aqueous Newtonian liquid, and the slip length is observed to be dependent on the shear rate and the viscosity; (ii) slip lengths of up to $L_S \sim 1\,\mu m$ are found by *D. C. Tretheway* and *C. D. Meinhart* (2002) at hydrophobic walls of channels through which water is flowing, while at hydrophilic walls no noticeable slipping is observed; (iii) slip lengths of $L_S < 100\,nm$ are found by *P. Joseph* and *P. Tabeling* (2005) for water at hydrophilic and hydrophobic channel walls, whereby the experimental accuracy is given in the same range as $\pm 100\,nm$. Although the values are considerably scattered, it remains true that in all cases a (small) slipping of the liquid is observed at the wall. The experimental results can be supported by molecular dynamic simulations, at least at moderate shear rates. In an isothermal Couette shear flow, *P. A. Thompson* and *C. M. Troian* (1997) find a constant slip length of $L_S \leq 17\sigma$ for simple spherical molecules and for $\dot{\gamma} < \dot{\gamma}_c$. Therefore, slip lengths of up to $L_S \sim 5\,nm$ are to be expected for water, although the value depends in detail on the interaction and compatibility of the wall and liquid molecules. For $\dot{\gamma} < \dot{\gamma}_c$ the Navier slip condition is therefore confirmed. The critical shear rate $\dot{\gamma}_c$ is in the range

$$\dot{\gamma}_{\mathrm{c}} = 0.025 \ldots 0.4 \cdot \sqrt{\frac{\varepsilon}{\sigma^2 \cdot m}}, \tag{11.6}$$

where again there is a dependence on the interaction and compatibility of wall and liquid molecules. For water molecules (with the restrictions discussed above), this leads to shear rates of $\dot{\gamma}_c = 0.3 \ldots 4.5 \cdot 10^{11} s^{-1}$. Such exorbitantly large shear rates are hardly to be expected in technical systems. Here too it is to be noted that heavy and large liquid molecules lead to smaller critical shear rates $\dot{\gamma}_c$. For $\dot{\gamma} > \dot{\gamma}_c$ the slip length increases dramatically, and so the molecules slip freely. In this range the Navier slip condition is clearly no longer valid.

In summary, we may state that simple liquids at small shear rates slip according to the Navier slip model with small slip lengths. In this range their rheological behaviour corresponds to that of a Newtonian fluid. If a (very large) critical shear rate is exceeded, complete decoupling of the wall and liquid velocities can occur. The continuum assumption remains valid as long as there are at least 10 molecule layers and as long as exorbitantly high shear rates are not reached. For liquid films with fewer than 100 molecule layers, gradual deviation from Newtonian behaviour can occur.

11.2 Molecular Models

11.2.1 Fundamentals of Molecular Models

With a view to the methods of calculation still to be described for carrying out direct numerical *simulation of the distribution function* f, the gas-kinetic equations of the single particle collisions will be treated. In rarefied gases it is mainly collisions between two molecules that occur. For the gas-kinetic approach it is therefore generally sufficient to consider only two-particle collisions. The description of the process consists of calculation of the velocity vectors $\boldsymbol{c}_1'$ and $\boldsymbol{c}_2'$ and the internal energies $\epsilon\mathrm{i}, 1'$ and $\epsilon\mathrm{i}, 2'$ after collision between two individual particles.

The simplest case of a collision between particles is an elastic collision. Only translational energies are exchanged between the molecules; there is no exchange between translational energies and internal energies of the particles. Therefore this collision can be treated with the balance equations of mechanics.

In Figure 11.2 and in the following equations, the indices 1 and 2 indicate the two partners of the collision. Variables after the collision are indicated with a dash. Conservation of mass holds

$$m_1 + m_2 = m_1' + m_2' \quad . \tag{11.7}$$

Conservation of momentum yields

$$m_1 \cdot \boldsymbol{c}_1 + m_2 \cdot \boldsymbol{c}_2 = m_1 \cdot \boldsymbol{c}_1' + m_2 \cdot \boldsymbol{c}_2' = (m_1 + m_2) \cdot \boldsymbol{c}_\mathrm{m} \quad , \tag{11.8}$$

with the velocity of the center of gravity $\boldsymbol{c}_\mathrm{m}$. Conservation of energy may be written as

$$m_1 \cdot c_1^2 + m_2 \cdot c_2^2 = m_1 \cdot {c_1'}^2 + m_2 \cdot {c_2'}^2 \quad . \tag{11.9}$$

Defininining the reative velocities

$$\boldsymbol{c}_\mathrm{r} = \boldsymbol{c}_1 - \boldsymbol{c}_2 \quad \text{and} \qquad \boldsymbol{c}_\mathrm{r}' = \boldsymbol{c}_1' - \boldsymbol{c}_2' \quad ,$$

it follows from conservation of momentum and energy that

$$\boldsymbol{c}_1 = \boldsymbol{c}_\mathrm{m} + \frac{m_2}{m_1 + m_2} \cdot \boldsymbol{c}_\mathrm{r} \quad ,$$
$$\boldsymbol{c}_2 = \boldsymbol{c}_\mathrm{m} - \frac{m_1}{m_1 + m_2} \cdot \boldsymbol{c}_\mathrm{r} \quad ,$$
$$\boldsymbol{c}_1' = \boldsymbol{c}_\mathrm{m} + \frac{m_2}{m_1 + m_2} \cdot \boldsymbol{c}_\mathrm{r}' \quad ,$$
$$\boldsymbol{c}_2' = \boldsymbol{c}_\mathrm{m} - \frac{m_1}{m_1 + m_2} \cdot \boldsymbol{c}_\mathrm{r}' \quad .$$

Introducing these relations into the conservation equations and using the reduced mass

$$m_\mathrm{r} = \frac{m_1 \cdot m_2}{m_1 + m_2} \quad ,$$

we obtain the equations

$$m_1 \cdot \boldsymbol{c}_1^2 + m_2 \cdot \boldsymbol{c}_2^2 = (m_1 + m_2) \cdot \boldsymbol{c}_\mathrm{m}^2 + m_\mathrm{r} \cdot \boldsymbol{c}_\mathrm{r}^2 \quad ,$$
$$m_1 \cdot \boldsymbol{c}_1'^2 + m_2 \cdot \boldsymbol{c}_2'^2 = (m_1 + m_2) \cdot \boldsymbol{c}_\mathrm{m}^2 + m_\mathrm{r} \cdot \boldsymbol{c}_\mathrm{r}'^2 \quad ,$$

from which it follows immediately that the magnitude of the relative velocity does not change in the collision.

The direction of the relative velocities after collision is given by the two collision parameters χ and ε. We consider two particles and introduce a collision plane that passes through the center of particle 1 and is perpendicular to the relative velocity vector $\boldsymbol{c}_\mathrm{r}$ before the collision (see Figure 11.2). The position at which particle 2 strikes particle 1 is given by the polar coordinates b and ε. χ denotes the angle of deflection lying in the plane spanned by the vectors $\boldsymbol{c}_\mathrm{r}$ and $\boldsymbol{c}_\mathrm{r}'$.

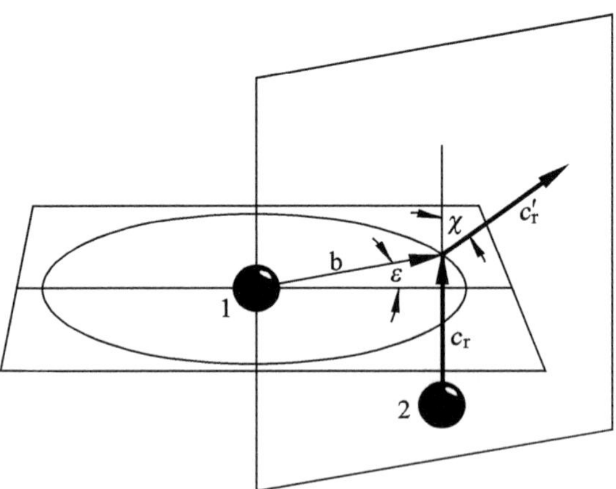

Fig. 11.2. Geometry of the two-particle collision in the center of gravity reference frame

The description of the transport properties of a gas, such as, for example, the viscosity μ or the thermal conductivity λ, is essentially determined by the potential used to describe the interaction between the particles. Figure 11.3 shows different models of the interaction potentials.

The classic interaction potential of gas kinetics is that of rigid elastic spheres, where an interaction occurs between the molecules only when they touch. This model yields the following results for the temperature dependence of the dynamic viscosity and the thermal conductivity

$$\mu(T) \sim T^{0.5} \quad \text{and} \quad \lambda(T) \sim T^{0.5} \quad .$$

These are independent of the type of gas. Further interaction potentials are the purely repulsive interaction potential and the Lennard-Jones potential. The purely repulsive interaction potential considers the electrostatic repulsion between two particles with the same electrical charge. The interaction potential force is given by $\boldsymbol{K} = -\nabla \boldsymbol{\Phi}$. As well as the electrostatic repulsion at small relative distances r between the colliding particles, the Lennard-Jones potential also considers the attractive van der Waals multipole interaction, which dominates at larger relative distances through the deformation of the electron shells of the colliding molecules or atoms. For our applications the interaction energies are so high ($> 1eV$) that the so-called *variable-hard-spheres* (VHS) model, which was derived from the hard-sphere model, provides a good approximation to describe the transport processes. In the variable-hard-spheres model the total scattering cross section is assumed to be a function of the relative kinetic energies in the form

$$\sigma_{\mathrm{T}} \sim \left(\frac{1}{2} \cdot m_{\mathrm{r}} \cdot c_{\mathrm{r}}^2\right)^{-\omega} \quad .$$

The exponent ω represents a gas-specific quantity. For $\omega = 0$ the VHS model describes the hard-sphere model, and for $\omega = 0.5$ the so-called Maxwell

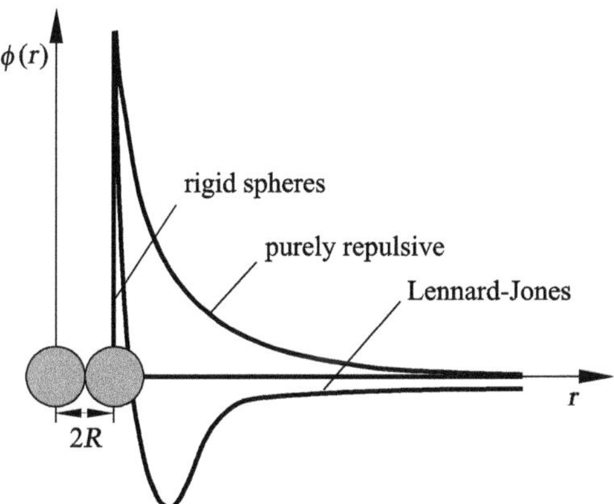

Fig. 11.3. Interaction potentials

molecules. The collision probability of Maxwell molecules is independent of the relative velocities of the molecules. In the following sections we use only this simplified interaction model. For air typically $\omega = 0.25$ is used.

11.2.2 Monte-Carlo-Simulation

We now consider the actual motion and the elastic and inelastic collisions of several hundred thousand molecular particles in a given region of simulation.

The approach to the gas-kinetic simulation yields the Boltzmann equation, made dimensionless with $x^* = x/L$, $\boldsymbol{c}^* = \boldsymbol{c}/\bar{c}$, $f^* \cdot \mathrm{d}c_\mathrm{i}^* = f \cdot \mathrm{d}c_\mathrm{i}/n$, $b^* \cdot \mathrm{d}b^* = b \cdot \mathrm{d}b/(\sqrt{2}\pi \cdot d^2)$ and $t^* = t/(L/\bar{c})$

$$\left(\frac{\partial}{\partial t^*} + \boldsymbol{c}^* \cdot \frac{\partial}{\partial \boldsymbol{r}^*}\right) f^* =$$
$$\frac{1}{\mathrm{Kn}} \cdot \int\int\int (f'^* \cdot f_1'^* - f^* \cdot f_1^*) \cdot c_\mathrm{rel}^* \cdot b^* \cdot \mathrm{d}b^* \cdot \mathrm{d}\varepsilon \cdot \mathrm{d}c_1^* \ , \quad (11.10)$$

that was introduced in Section 5.4.1. The dimensionless Boltzmann equation yields identical solutions for problems with the same Knudsen number

$$\mathrm{Kn} = \frac{\bar{\lambda}}{L} = \frac{1}{n \cdot \frac{\overline{\sigma \cdot c_\mathrm{r}}}{\bar{c}} \cdot L} \quad ,$$

i.e. for a given characteristic length L, the product of scattering cross section and particle density $\sigma \cdot n$ must be kept constant to obtain an identical solution. The actual number of molecules in a flow can thereby be replaced by several tens of thousands of model particles with artificially increased scattering cross section (theorem of *N. A. Derzko* (1972)). However a sufficient number of particles must be available to enable local averaging of the macroscopic quantities.

Of the multitude of numerical simulation methods, we have selected the direct simulation Monte-Carlo (DSMC) method and the molecular dynamics (MD) method. In the DSMC method the particles move as free molecules and the collision particles are statistically selected. In contrast to this, in the MD method the trajectories of the particles are traced exactly in time. For gases, a collision occurs only when two particles approach each other to their scattering cross section. For liquids, a permanent interaction with the neighboring particles is present. Because of the relatively high computational effort for the MD method, the heuristic DSMC method is recommended for gases. For further literature for the section on computational methods the review article of *J. N. Moss* and *G. A. Bird* (1984) is recommended.

The direct simulation Monte-Carlo (DSMC) method was developed by *G. A. Bird* (1976) and represents a powerful, heuristic method of investigating rarefied gas flows. The essential difference to the molecular dynamics (MD) method lies in the decoupled statistical treatment of the motion and the collisions of the model particles.

In the DSMC method the molecules that are actually present in the flow field are replaced by model particles. Several hundred thousand model particles are used. The initial state is selected randomly (Figure 11.4), as for the molecular dynamics method, and changes with the simulation time through the motion and collisions of the particles. To determine the macroscopic quantities and to guarantee correct local collision rates, the flow field is divided into cells. This grid can be matched to any body or can be rectangular (Figure 11.5).

In the Monte-Carlo method the particle motion and the collisions are decoupled. The central iteration step of the simulation is as follows (see Figure 11.4). The particles are moved according to a given time step Δt_{m}. Particles that leave the calculation area are removed and collisions of particles with the surface of the wall are calculated. For this, wall interaction models, as already described, must be taken into account. At the boundaries of the flow field new particles are generated for reasons of continuity. It is determined to which cell each particle belongs. Conversely, for each cell it is now determined which particle belongs in it. For each cell a number of collisions is carried out in accordance with the time step Δt_{m}. The positions of the particles remain unchanged.

In the manner in which the number of collisions per cell is determined and carried out, the methods of *G. A. Bird* (1976), *K. Nanbu* (1992) and *M.S. Ivanov* and *S. V. Rogasinsky* (1991) differ.

According to *G. A. Bird* (1976), the number of collisions per cell in the time step Δt_{m} is given by

$$N_{\mathrm{t}} = \frac{1}{2} \cdot N_{\mathrm{m}} \cdot n \cdot \Delta t_{\mathrm{m}} \cdot \overline{\sigma \cdot c_{\mathrm{r}}} \quad , \tag{11.11}$$

with the number of particles N_{m} per cell, the particle density n, the relative velocity c_{r} and the collision cross section σ of the collision partners.

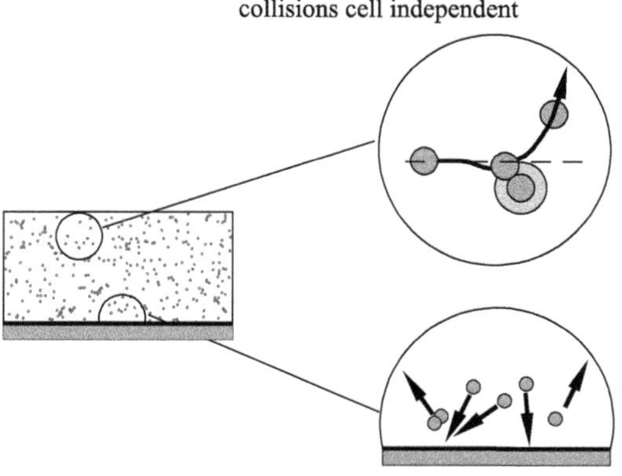

Fig. 11.4. Calculation with the DMSC method

Calculation of the product $\overline{\sigma \cdot c_{\mathrm{r}}}$ requires considerable effort, as all possible particle combinations in a cell must be included to form the average. *G. A. Bird* (1976) therefore introduced a collision time counter t_{C} that is increased after every collision using the collision cross section σ and the relative velocity c_{r} of each collision partner by

$$\Delta t_{\mathrm{C}} = \frac{2}{N_{\mathrm{r}} \cdot n \cdot \sigma \cdot c_{\mathrm{r}}}$$

until this counter is equal to the simulation time. In doing this the collision number N_{t} required by equation (11.11) is reached on average during time step Δt_{m}. The collision partners are selected randomly within the cells. From this we find that a collision between two particles becomes more probable the larger their collision cross sections and their relative velocity.

If a suitable pair is found, the six unknown velocity components of the selected collision partners are calculated. For this the balance equations for momentum and energy (11.8) – (11.9) are available. The direction of the relative velocity vectors after the collision is determined using random numbers, and thus, in contrast to direct simulation methods, this method is not deterministic.

The method of *K. Nanbu* (1992) differs from Bird's method only in the treatment of the collision process. The purely phenomenological model of *G. A. Bird* (1976) is replaced by *K. Nanbu* (1992) by a collision mechanism derived from the Boltzmann equation. In this method, only one particle changes its state in a collision. The number of collisions during the time step Δt_{m} is

$$N_{\mathrm{t}} = N_{\mathrm{m}} \cdot n \cdot \Delta t_{\mathrm{m}} \cdot \overline{\sigma \cdot c_{\mathrm{r}}} \quad .$$

The collisions occur with a probability

$$P_{\mathrm{i}} = \sum_{\mathrm{j}=1}^{N_{\mathrm{m}}} \frac{n \cdot \Delta t_m}{N_{\mathrm{m}}} \cdot (\sigma \cdot c_{\mathrm{r}})_{\mathrm{ij}} \quad .$$

For each particle the probability P_{i} is calculated in a cell and then a random number is used to decide whether a collision takes place in the time interval Δt_{m}. If a collision takes place, a collision partner is sought for this particle.

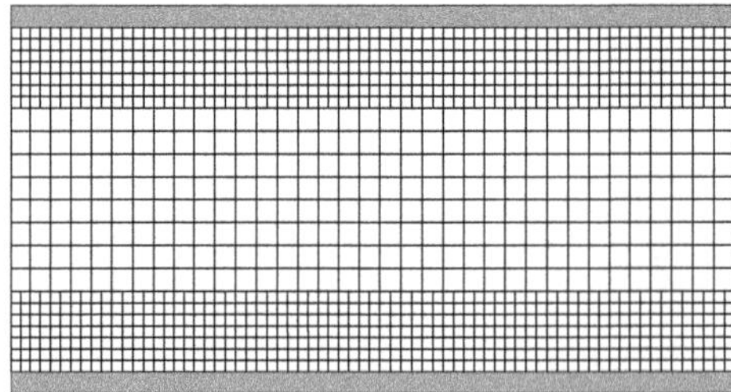

Fig. 11.5. Grid for the Monte-Carlo simulation

The calculation differs from the method of *G. A. Bird* (1976) in that only one collision partner experiences a change of velocity.

The method of *M. S. Ivanov* and *S. V. Rogasinsky* (1991) is known as the majorant frequency scheme. In contrast to *G. A. Bird* (1976), for each cell an upper estimate of the number of collisions is calculated:

$$N_{\mathrm{t,maj}} = \frac{1}{2} \cdot N_{\mathrm{m}} \cdot n \cdot \Delta t_{\mathrm{m}} \cdot (\sigma \cdot c_{\mathrm{r}})_{\max} \quad .$$

Instead of the average value $\overline{\sigma \cdot c_{\mathrm{r}}}$, a maximum value $(\sigma \cdot c_{\mathrm{r}})_{\max}$ that is simple to determine is used. Now for each cell, $N_{\mathrm{t,maj}}$ collision pairs are determined. The collisions occur with a probability $\sigma \cdot c_{\mathrm{r}} / (\sigma \cdot c_{\mathrm{r}})_{\max}$. The collisions that are accepted are called real collisions, while those that are rejected are called fictitious collisions. For the real collision the new velocities are determined according to the methods of *G. A. Bird* (1976). The number of real collisions yields the value required in (11.11).

The method of Ivanov has two advantages over that of Bird. The collisions can be calculated in a computationally more efficient manner. This considerably reduces the computational time. Statistically better results are found for small numbers of particles, as in the method of *G. A. Bird* (1976) improbable collisions (with small $\sigma \cdot c_{\mathrm{r}}$) are allowed to push the collision time counter far onwards. Thus over a long stretch of time no further collisions occur. The method of *M.S. Ivanov* and *S. V. Rogasinsky* (1991) was derived mathematically from the Boltzmann equation. Similar methods have since been presented by other authors and are called in the literature no time counter (NTC) schemes.

Conservation of angular momentum has not been ensured from the start in the methods presented in this section. However it has been shown in examples that angular momentum is conserved if enough particles are present in a cell.

11.2.3 Molecular Dynamic Simulation

The *molecular dynamics method* is characterised by the fact that only the initial state is determined by statistical methods. The continued progression is strictly deterministic, i.e. at each later point in time the state of the system can be derived from the initial state. At the start of the calculation a given number of model particles is positioned in the computational space with account taken of the geometrical boundary conditions. The components of the thermal velocity are associated with each particle. After superimposing the macroscopic velocity, the initial state of the flow field is then determined. These model particles are now moved with the associated velocity.

Molecular dynamic simulations are preferred when there is a constant (not collision-like) interaction between the molecules. This is generally the case for liquids and dense gases. An overview of molecular dynamic simulations may be found in *P. J. Koplik* and *J. R. Banavar* (1995). A molecular dynamic simulation calculates explicitly the motion of a large number of fluid molecules that may be interacting with their neighboring molecules (liquid, solid). Therefore the forces between the same and different fluid molecules, as well as between fluid and solid molecules, must be formulated. This can be done with the help of the Lennard-Jones potential, which describes the interaction of inert, non-ionized, non-polar, spherical atoms. According to *H. Green* (1960), for distances between the atoms r that are not too small, we obtain the approximate potential

$$\phi(r) \sim -\frac{\mathrm{C}_1}{r^6} + \frac{\mathrm{C}_2}{r^{12}} \quad . \tag{11.12}$$

According to (11.12), at large distances the attractive forces ($\phi \sim r^{-6}$) dominate. These are due to the mutual polarization of the atoms (van der Waals forces). At small distances the repulsive forces ($\phi \sim r^{-12}$) dominate. These come from the interaction between the electron shells. The constants C_1, C_2 have been determined for many atoms by the method of *J. E. Lennard-Jones* (1931). The interaction of complex fluid molecules, such as dipole molecules or chain molecules, can be realized by the elastic bond of several atoms. This leads to similar (generally more complicated) potentials for their interactions.

In the interaction between liquid atoms and the solid atoms of the wall, it must be taken into account that the solid atoms are embedded in an elastic lattice. Integrating the elastic forces, according to *F. F. Abraham* (1978), yields in the simplest case a potential of the form

$$\phi(r) \sim -\frac{\mathrm{C}_3}{r^4} + \frac{\mathrm{C}_4}{r^{10}} \quad . \tag{11.13}$$

Here too there are more realistic models for molecules available in the literature. The derivative $\partial\phi/\partial r$ of the potential is related to the force on the molecules. In practice we restrict ourselves to the nearest neighbor molecules. If the force on a single molecule is known, using Newton's law and integrating the acceleration with respect to time, the position of the molecule can be determined numerically and its path traced. Kinematic boundary conditions are now not necessary; at the molecular level the effect of solid boundaries on the fluid is completely described by the interaction between the solid and fluid molecules. Thermal boundary conditions can be realized by e.g. prescribing the Brownian motion of the solid molecules at the boundary. The kinematic and thermal motion of the fluid molecules close to the boundaries then permits conclusions to be drawn about the macroscopic boundary conditions.

Therefore, with considerable numerical effort, it is possible to simulate the motion of the fluid molecules, possibly with transitions between the liquid and gas phases, as well as including the interaction with solids or other fluids in detail. If the motion of the single molecules is known, in a natural way the behavior of macroscopic portions of fluid becomes accessible. The motion of the continuum is then determined by averaging over a large number of molecules, whereby the number determines the spatial resolution. A further averaging in time eliminates the thermal statistical motion of the molecules. Simulations of several hundred thousand molecules are needed in conjunction with a strict limitation of simulated real time, whereby all this is limited by the computational time required; typically regions of several hundred angstroms dimension are simulated for several nanoseconds. Molecular dynamic simulation is therefore particularly suitable for investigating the limit range between molecular processes and the continuum-mechanical approach. This is of particular interest at solid walls or at (moving) phase boundaries.

11.3 Continuum Models

The continuum-mechanical conservation equations may serve as a basis, provided that the continuum assumption holds. Compared to conventional flows, at small length scales modifications or additional effects may need to be taken into account. These are discussed in the following.

11.3.1 Similarity Discussion

The effect of small length scales may be formally answered using similarity analysis. It is sensible, first of all, to estimate the dimensional quantities. We follow the procedure of *H. Herwig* (2002). For microflows, we may distinguish two typical applications: A miniaturized analysis laboratory and a micro heat exchanger (see Figure 1.12). We first discuss the miniaturized analysis laboratory, e.g. given in Figure 11.1. In the table below, these estimates are given for macroflows and for microflows.

		macro	micro	micro-macro ratio
channel width	d	$\approx 10^{-2}\ m$	$\approx 10^{-5}\ m$	$\approx 10^{-3}$
channel length	l	$\approx 1\ m$	$\approx 10^{-2}\ m$	$\approx 10^{-2}$
velocity	$\bar{u}$	$\approx 1\ m/s$	$\approx 10^{-3}\ m/s$ to $1\ m/s$	$\approx 10^{-3}$ to 1

Channels or pipes in macroflows typically have a width of several cm, while we assume a channel width of several $10\,\mu m$ for microflows. Similarly we expect the channel length in macroflows to be several m and in microflows several cm. The average velocity in macroflows is generally several m/s, while in microflows several mm/s is present in a miniaturized analysis laboratory. The estimates of the quantities $d, l, \bar{u}$ may, if desired, be shifted by a factor of ten. This changes only the order of magnitude, but not the tendency of the following statements. Based on these estimated quantities $d, l, \bar{u}$, it is possible to estimate other quantities such as pressure drop, mass and volumetric flow rates, force ratios and thermal quantities. We always assume that the fluid properties in the macroflow and in the microflow are of the same order of magnitude. Furthermore, it is sufficient to consider the ratio of microflow quantity to the macroflow quantity in the table below.

		micro-macro ratio
pressure drop	Δp	≈ 10 to 10^4
volume, mass flux	$\dot{m}$, $\dot{V}$	$\approx 10^{-9}$ to 10^{-6}
Reynolds number, $F_{\mathrm{i}}/F_{\mathrm{f}}$	Re	$\approx 10^{-6}$ to 10^{-3}
$F_{\mathrm{g}}/F_{\mathrm{f}}$	Re/Fr	$\approx 10^{-3}$ to 10^{-6}
$F_{\mathrm{c}}/F_{\mathrm{f}}$	Re/We	$\approx 10^3$ to 1

The laminar pressure drop in a channel follows the relation $\Delta p \sim \bar{u} \cdot l/d^2$. Thus in microchannels a higher pressure drop is to be expected than in macrochannels. The volumetric and mass fluxes in channels behave according to $\dot{m}, \dot{V} \sim \bar{u} \cdot d^2$, so that drastically smaller volumetric and mass fluxes occur in microchannels.

Dimensionless groups based on force ratios are useful to evaluate the relevant forces in the flow. The Reynolds number characterizes the ratio of inertial forces F_i to friction forces F_f. It behaves according to $\mathrm{Re} \sim \bar{u} \cdot d$ and is considerably smaller in microchannels than in macrochannels. Generally, therefore, the friction forces dominate, and a transition to turbulence is not expected because of the weak inertial forces. Simultaneously, it is useful to compare the other forces with the friction force. The Stokes number, which gives the ratio of pressure forces F_p to friction forces F_f, is therefore always $\mathrm{Sto} = F_p/F_f \approx 1$. The ratio of gravitational force F_g to friction force F_f can be expressed using $\mathrm{Re}/\mathrm{Fr} \sim d^2/\bar{u}$ using the Reynolds number and the Froude number $\mathrm{Fr} = F_i/F_g$. Therefore, in microchannels the gravitational force, and in general all volumetric forces, are much weaker than in macrochannels. The ratio of capillary forces F_c and friction forces F_f is expressed in the ratio $\mathrm{Re}/\mathrm{We} \sim 1/\bar{u}$. This expression includes the Weber number $\mathrm{We} = F_i/F_c$. In microchannels, therefore, a greater effect of the capillary forces, and in general all interfacial forces, is to be expected. In single phase flows, after the initial wetting there are no fluid/fluid interfaces, so in such cases there are no capillary forces. On the other hand, at fluid/wall interfaces forces can occur, due for example to electrical fields. Such interfacial forces also follow the dependence $\sim \mathrm{Re}/\mathrm{We}$. The strong influence of all interfacial forces depends on the extremely large ratio of surface to volume in microchannels. This leads to the conclusion that heat and mass transfer in microchannels are also possible in a very efficient manner, as these too are determined by the transfer surface.

Now that we have looked at the mechanical aspects, we consider the aspect of heat transport and heat transfer in more detail. In the table below the corresponding ratios of microflow and macroflow are summarized.

		micro-macro ratio
temperature increase	ΔT	$\approx (10^4)$ to 10
heat conduction part	q_d/q_c	$\approx (10^5)$ to 10^2

Because of the analogy between heat and mass transport, these ideas may be directly transferred to mass transport. We now heat up the channel wall and ask by what temperature ΔT the fluid is heated as it passes through the channel. Here we find the dependence $\Delta T \sim l/(d \cdot \bar{u})$, according to the above table, and therefore there is a drastically higher temperature increase ΔT in the microchannel. Such a large temperature increase is of course not sensible, because the fluid takes on the temperature of the wall in a very short time and therefore no heat transfer takes place in the rest of the channel. It is therefore advisable to select considerable higher flow velocities of several m/s in a microchannel for heat transfer. With such parameters, a moderate temperature increase is achieved, as is more sensible for a heat exchanger. Unfortunately the large flow velocity causes the pressure to increase considerably. The tendency of the statements about the inertial force F_i and the gravitational force F_g remains the same while the order of magnitude is changed. The capillary and interfacial forces F_c, on the other hand, play a lesser role with this choice of parameters. In order to characterize the heat transport, ratios of heat fluxes may be used. The ratio of diffusively transported heat q_d to convectively transported heat q_c behaves according to $q_d/q_c \sim 1/(l \cdot \bar{u})$. Therefore, we may assume that the heat transported axially through the fluid by heat conduction is considerably more important in microchannels than in macrochannels. The same is also true for axial heat conduction in the wall.

11.3.2 Modifications of Boundary Conditions

We have already seen in Section 11.1.3 that gases, in particular, slip at solid walls as the Knudsen number increases. Therefore, in this section we restrict ourselves to the details of gas flow through a microchannel. Irrespective of this, Section 11.1.4 shows that liquids also slip at walls at high shear rates and the slip condition (11.5) may be used for modeling. However, first the slip length for liquids is very small; second there is little firm information in the literature on the slipping of liquids.

For the isothermal behavior of gas molecules at solid walls, we first discuss two idealized limiting cases. *J. Maxwell* (1879), in the kinetic theory of rarefied gases, characterized the behavior of spherical gas molecules at molecularly smooth solids. According to this theory, on collision with the wall ($z = 0$) each gas molecule retains its tangential momentum through symmetric reflection (cf. Figure 11.6). Thus only the normal momentum of the gas molecules changes. The absent exchange of tangential momentum between gas molecule and wall is equivalent to the perfect slipping of the gas molecule; the gas transfers no shear stress to the wall. If we assume a very rough wall, the situation changes fundamentally. Because of the roughness, the reflection of the gas molecules is statistically distributed in all directions. On average, therefore, these gas molecules no longer have any tangential momentum after reflection. The transfer of the tangential momentum to the wall corresponds to a finite shear stress. Balancing the forces leads to the slip law

$$u(z = 0) - u_{\mathrm{w}} = \bar{\lambda} \cdot \frac{\partial u}{\partial z}(z = 0) \quad , \tag{11.14}$$

with the mean free path $\bar{\lambda}$. This reflection is called diffuse reflection, because there is no correlation between the direction of arrival and the random direction of reflection.

Real walls are characterized by the fact that a small number of gas molecules undergo symmetric reflection, while a large number undergo diffuse reflection. In order to characterize the wall, we introduce a collision coefficient σ_{v} that gives the fraction of the diffusively reflected molecules from the total number of reflections. σ_{v} for real walls is in the range $\sigma_{\mathrm{v}} = 0.2 - 0.8$, whereby the value 0.2 is for exceedingly smooth walls, and the value 0.8 for technically relevant rough walls. Using σ_{v} the slip law can be generalized as

$$u(z = 0) - u_{\mathrm{w}} = \frac{2 - \sigma_{\mathrm{v}}}{\sigma_{\mathrm{v}}} \cdot \bar{\lambda} \cdot \frac{\partial u}{\partial z}(z = 0) \quad , \tag{11.15}$$

For $\sigma_{\mathrm{v}} \to 1$, (11.15) becomes (11.14), describing the perfect diffusively reflecting wall. The limiting case $\sigma \to 0$ in (11.15) leads to complete decoupling of the slip velocity and the shear rate, implying perfect slipping.

Frequently the situation is not isothermal, so that the effect of the fluid temperature T and the wall temperature T_{w} also must be taken into account. For $T \neq T_{\mathrm{w}}$ the slip law (11.15) must be modified, and a condition for the temperature jump must be included. According to *S. A. Schaaf* and *P. L. Chambré* (1961) we then have

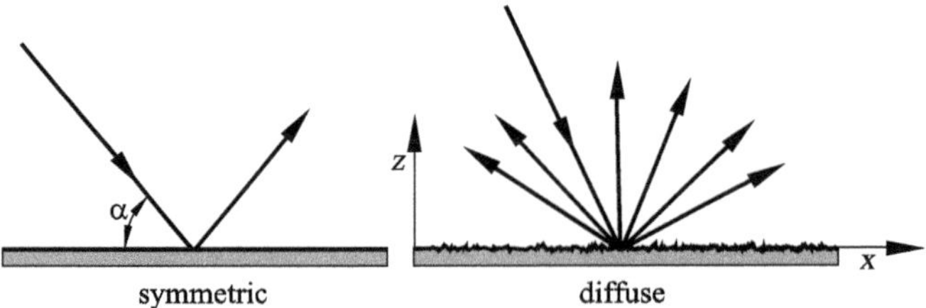

Fig. 11.6. Symmetric and diffuse reflection of gas molecules

$$u(z=0)-u_\mathrm{w} =$$

$$\frac{2-\sigma_\mathrm{v}}{\sigma_\mathrm{v}} \cdot \bar{\lambda} \cdot \frac{\partial u}{\partial z}(z=0) + \frac{3}{4} \cdot \frac{\mu}{\rho \cdot T(z=0)} \cdot \frac{\partial T}{\partial x}(z=0) \ , \quad (11.16)$$

$$T(z=0) - T_\mathrm{w} = \frac{2-\sigma_\mathrm{t}}{\sigma_\mathrm{t}} \cdot \frac{2 \cdot \kappa}{\kappa + 1} \cdot \frac{\bar{\lambda}}{Pr} \cdot \frac{\partial T}{\partial z}(z=0) \quad . \quad (11.17)$$

In analogy to σ_v, the thermal collision coefficient σ_t also appears in (11.17). Furthermore, the ratio of the specific heats $\kappa = c_p/c_v$ and the Prandtl number $Pr = \nu/k$ appear. Here ν is the kinematic viscosity and k is the thermal diffusivity of the gas. The second term on the right-hand side of the slip law (11.16) is called thermal creeping. If, because of the wall temperature $T_\mathrm{w}(x)$, a gas at the wall has a temperature gradient tangential to the wall $\partial T/\partial x > 0$, a velocity $u > u_\mathrm{w}$ occurs close to the wall. This fact is exploited in a Knudsen pump, which can be used to pump rarefied gases and which, without moving parts, delivers gas in a pipe from a cold zone to a warm zone. We now discuss the effect of the slip law (11.15) using a simple example. We consider an isothermal gas flow through a microgap, as sketched in Figure 11.7. The gap has a height $2 \cdot d$ and the length l; to reduce the mathematical difficulty and so that we can concentrate on the physics, in the y-direction the gap is infinite. The flow is driven by the pressure difference $(p_1 - p_0) > 0$, and we also assume, for simplicity, a steady, plane and essentially parallel flow $(u \gg w)$. These assumptions are generally satisfied to good approximation for a narrow gap $(d \ll l)$. Of course, the assumption of an isothermal flow is also an approximation, as the dissipation is neglected. For the mathematical description of the problem, we begin with the continuity equation (5.1) and the Navier-Stokes equations (5.18). Using the above approximations we obtain

$$\frac{\partial(\rho \cdot u)}{\partial x} \approx 0 \quad , \quad (11.18)$$

$$\rho \cdot u \cdot \frac{\partial u}{\partial x} \approx -\frac{\partial p}{\partial x} + \mu \cdot \frac{\partial^2 u}{\partial z^2} \quad . \quad (11.19)$$

In the case of a creeping flow ($\mathrm{Re}_d \ll 1$), which occurs frequently in microchannels, the momentum equation (11.19) loses the convective term and we

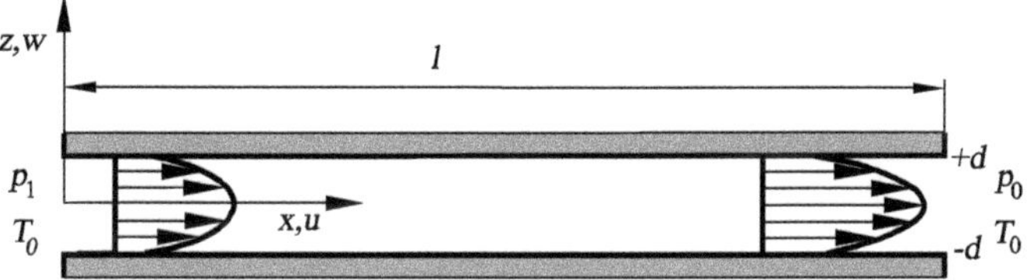

Fig. 11.7. Gas flow through a microgap

can integrate the equation using the boundary conditions (11.15) at $z = \pm d$, with the result

$$u(x,z) = -\frac{d^2}{2 \cdot \mu} \cdot \frac{\partial p}{\partial x} \cdot \left[1 - \left(\frac{z}{d}\right)^2 + 4 \cdot \mathrm{Kn} \cdot \frac{2 - \sigma_\mathrm{v}}{\sigma_\mathrm{v}} \right] \quad . \tag{11.20}$$

The first two terms in equation (11.20) describe the well-known Poiseuille flow between two plates, while the third term is responsible for the slipping of gas. The local Knudsen number $\mathrm{Kn}(x) = \bar{\lambda}(x)/2 \cdot d$ in equation (11.20) depends on the pressure $p(x)$ via the mean free path $\bar{\lambda}$. For an ideal gas at a given temperature T_0 we have the relation

$$\mathrm{Kn} = \frac{\mu}{2 \cdot d} \cdot \sqrt{\frac{\pi \cdot R \cdot T_0}{2}} \cdot \frac{1}{p} \quad , \tag{11.21}$$

with the specific gas constant R. Because the pressure along the gap drops continuously ($\partial p/\partial x < 0$), the Knudsen number along the gap grows correspondingly. According to equation (11.21), at the entrance to the gap ($x = 0$) the slipping is weak, while it is greater at the exit from the gap ($x = l$). The slipping of the gas at the wall therefore increases along the gap. The corresponding velocity profiles are shown qualitatively in Figure 11.7. The effects of the modified slip boundary condition on the gas flow in a gap can be confirmed by experiment. *J. C. Shih et al.* (1995) carried out experiments with helium in a microgap of $2 \cdot d = 1.2\,\mu m$ height and $l = 4000\,\mu m$ length; the gap width in the experiment is $\Delta y = 40\,\mu m$. The authors vary the entry pressure p_1, the exit is into the atmosphere, and the pressure inside the microgap is measured using sensors integrated into the wall. The local Knudsen number varies for an entry pressure of $p_1 = 1.5 \cdot 10^5\,Pa$ in the region $0.1 \leq \mathrm{Kn} \leq 0.16$. Figure 11.8 shows an example of measurements of the mass flux $\dot{m}$ as a function of the pressure difference $p_1 - p_0$. The mass flux can also be calculated by integration of the solution of (11.20), as

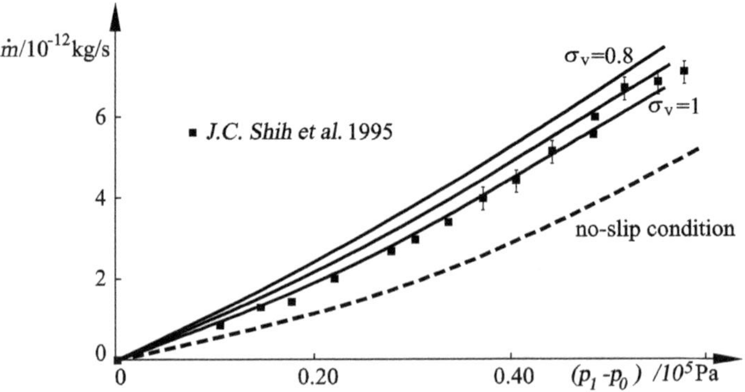

Fig. 11.8. Gas mass flux in a microgap, according to *J. C. Shih et al.* (1995)

$$\dot{m} \approx \rho \cdot \Delta y \cdot \int_{-d}^{d} u \cdot \mathrm{d}z \quad , \tag{11.22}$$

with $\Delta y = 40\,\mu m$. Figure 11.8 shows the results of the theoretical model for comparison, using the no-slip condition and using different slip conditions $\sigma_\mathrm{V} = 0.8, 0.9, 1$. We see on the one hand that the model with the no-slip condition considerably underestimates the mass flux. On the other hand, the experimental data and the model with slipping agree reasonably for $\sigma_\mathrm{V} = 0.9$ or 1. A collision coefficient of $\sigma \approx 1$ is equivalent to a molecularly rough wall that causes predominantly diffuse reflection of the gas molecules. These experimental findings in a gas flowing through a microgap prove that the use of modified boundary conditions is necessary for Knudsen numbers in the region of $\mathrm{Kn} \approx 0.1$.

11.3.3 Electrokinetic Effects

Electrokinetic effects are characterized by the interaction of electric fields with charges within fluids. Charges may occur in aqueous solutions in the form of free ions; gases can also be ionized under certain conditions. In the following discussion, we restrict ourselves to liquids with free ions as they may occur in a miniaturized analysis laboratory. Electrokinetic effects include electro-osmosis, electrophoresis, the flow potential and the sedimentation potential. While electro-osmosis and electrophoresis involve the effect of electric fields on the flow and the material transport, the flow potential and the sedimentation potential are the reverse of these effects. In these the flow and material transport via the charge transport affect the electric field. Here we consider only the first two effects; in parts we follow the review article of *K. V. Sharp et al.* (2002).

Electrical double layer

If we have a liquid with free ions, it usually contains an equal number of negative and positive ion charges, thus it is electrically neutral. The application of an electric field then, because of the Coulomb forces, causes both sorts of ion to move to the oppositely charged electrode. Generally this causes no motion of the liquid because the differently charged ions move in opposite directions. In other words: inside an electrically neutral liquid there are no resultant forces on the liquid.

The situation at interfaces (liquid/solid, liquid/liquid, liquid/gas) is completely different from that inside the liquid. Here an interaction between different types of molecules (atoms) is possible, so that in general there is no longer electrical neutrality. As an example, we consider the conditions at a liquid/solid boundary. Depending on the chemical composition of wall and liquid, adsorption of ions from the liquid or dissolution of molecules (atoms) from the wall can occur (cf. *R. Hunter* (1981)). Both lead to surface charges at the wall and these attract oppositely charged ions out of the liquid and repel identically charged ions. Thus there appears in the liquid an electrically non-neutral layer, the so-called electrical double layer. It is within this electrical double layer that electric forces act on the liquid if we apply a wall-tangential electric field from outside. The motion of excess ions leads via viscous effects to motion of the liquid. This motion of the liquid in the wall-tangential direction is termed electro-osmosis. Electro-osmosis occurs of course in both microchannels and large channels. While in microchannels the electric forces can become important, in large channels they are generally less important compared to the other forces. This has also to do with the fact that the electrical double layer is very thin. In order to model the electric forces in the Navier-Stokes equation, we need a model for the charge distribution in the electrical double layer; an overview of the models may be found in *R. Hunter* (1981). The Gouy-Chapman-Stern model assumes an immobile layer of counter ions directly at the wall (Stern layer). In Figure 11.9 we see in the direction normal to the wall z furthermore a shear layer with restricted mobility and the freely moveable diffuse layer (Gouy-Chapman layer), before the electrically neutral interior part of the liquid is reached. Conceptually the Stern layer and the shear layer, i.e. the first two layers of molecules, are to be associated with the immobile wall.

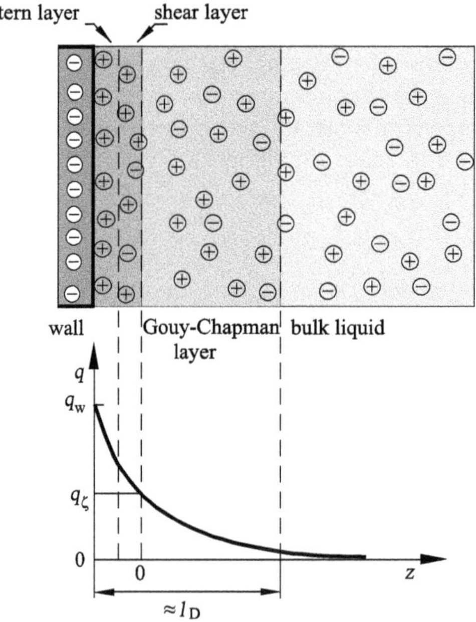

Fig. 11.9. Structure and charge density of the electrical double layer

In the following we assume a plane wall and a liquid with constant properties. Within the Gouy-Chapman layer is an interplay of electrostatic forces and diffuse thermal motion. The electric charge density q may therefore be summed from the Boltzmann distributions of the different species of ions, according to

$$q = e \cdot \sum_{\mathrm{i}} z_{\mathrm{i}} \cdot n_{\mathrm{i},\infty} \cdot \exp\left(\frac{-z_{\mathrm{i}} \cdot e \cdot \varphi}{k_{\mathrm{B}} \cdot T}\right) \quad . \tag{11.23}$$

Here e is the electron charge, z_i the valence and $n_{i,\infty}$ the ion density of the species i in the electrically neutral interior of the liquid; k_B is the Boltzmann constant, φ the electric potential and T the temperature. Furthermore, between the electric charge density q and the electric potential φ we have the relation

$$\nabla \cdot (\epsilon\mathrm{r} \cdot \nabla\varphi) = -\frac{q}{\epsilon 0} \quad , \tag{11.24}$$

where $\epsilon\mathrm{r} \cdot \epsilon 0$ characterizes the dielectric property of the fluid. Connecting equations (11.23) and (11.24) leads to a second order nonlinear differential equation to determine φ. If the energy of the thermal motion is much larger than that of the electrostatic forces, i.e. for $|z_i e\varphi| \ll |k_B T|$, the exponential function may be linearized and we obtain the so-called Debye-Hückel approximation (cf. *P. Debye* and *E. Hückel* (1923)). Within this approximation we obtain the solution for the charge density in the diffuse Gouy-Chapman layer as

$$q(z) \simeq \frac{q_\zeta}{l_{\mathrm{D}}} \cdot \exp\left(\frac{-z}{l_{\mathrm{D}}}\right) \quad . \tag{11.25}$$

In equation (11.25), q_ζ is the apparent wall charge density, which is related by $q_\zeta = -\zeta \cdot \epsilon\mathrm{r} \cdot \epsilon 0 / l_D$ to the zeta potential ζ, the potential at the boundary between the shear layer and the Gouy-Chapman layer. l_D is the so-called Debye length, which is a measure for the thickness of the electrical double layer. We have

$$l_{\mathrm{D}} = \sqrt{\frac{\epsilon\mathrm{r} \cdot \epsilon 0 \cdot k_{\mathrm{B}} \cdot T}{e^2 \cdot \sum_{\mathrm{i}} z_{\mathrm{i}}^2 \cdot n_{\mathrm{i},\infty}}} \quad . \tag{11.26}$$

Electro-osmosis

An electro-osmotic flow can be modeled if the electric charge densities in the electrical double layer and the interior of the liquid ($q = 0$) are known, by taking an electric volumetric force into account in the Navier-Stokes equations (cf. *R. Hunter* (1981)). We then have as a basis the system of equations

$$\nabla \cdot \boldsymbol{v} = 0 \quad , \tag{11.27}$$

$$\rho \cdot \left(\frac{\partial \boldsymbol{v}}{\partial t} + (\boldsymbol{v} \cdot \nabla)\, \boldsymbol{v}\right) = -\nabla p + \mu \cdot \Delta \boldsymbol{v} - q \cdot \nabla\varphi \quad , \tag{11.28}$$

and the electric potential φ remains to be determined. In general, because of the linear type of the underlying equations, the electric potential can be calculated from a superposition of the externally applied field and the self-induced field in the electrical double layer. Within the Debye-Hückel approximation, the self-induced electric field is given by the equations (11.24), (11.25). Depending on the electrical conductivity of the liquid, the externally applied field may be calculated electrostatically using the Gauß law or electrodynamically using Ohm's law. In an electrically neutral region with constant material properties, both laws lead to a Laplace equation for the electric potential φ. The boundary conditions may be the potentials at electrodes or spatial derivatives of the potential. An extensive discussion of these aspects can be found in *P. P. J. Barz* (2005).

We now discuss an example of the electro-osmotic flow in a microgap. Figure 11.10 shows this microgap with height $2 \cdot d$ and length l, and with an infinite extension in the y-direction. In order to simplify the mathematical treatment, we assume a steady, plane, fully-developed flow, driven by a pressure difference $(p_1 - p_0) > 0$ and a homogeneous external electric field with $\partial\varphi/\partial x = constant$, $\partial\varphi/\partial z = 0$. Taking the electrical double layers at both walls ($z = \pm d$) into account, with the above assumptions we obtain the x-component of the Navier-Stokes equation as

$$0 \approx -\frac{\partial p}{\partial x} + \mu \cdot \frac{\partial^2 u}{\partial z^2} - \frac{q_\zeta}{l_\mathrm{D}} \cdot \frac{\partial \varphi}{\partial x} \cdot \left[\exp\left(-\frac{z+d}{l_\mathrm{D}}\right) + \exp\left(\frac{z-d}{l_\mathrm{D}}\right)\right] . \qquad (11.29)$$

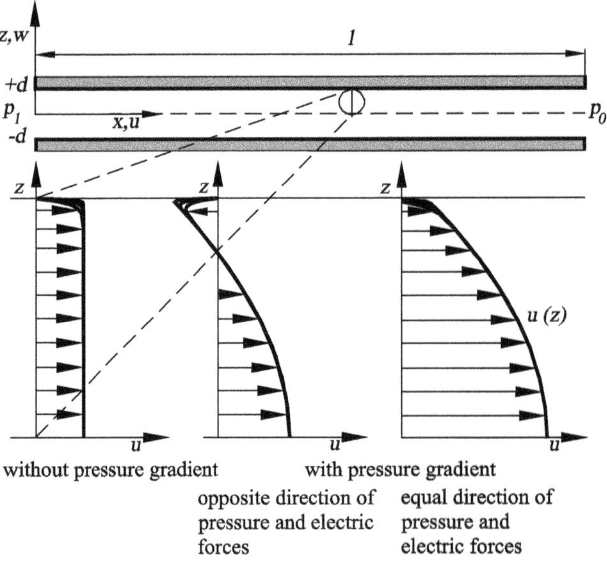

Fig. 11.10. Electro-osmotic flow in a microgap

In (11.29) only the external field $\partial\varphi/\partial x$ appears, because the self-induced fields only have components in the z-direction. Equation (11.29) can be solved by observing the no-slip condition at the walls, leading to the result (cf. *D. Burgreen* and *F. Nakache* (1964)):

$$\begin{aligned} u(z) =& -\frac{d^2}{2\cdot\mu}\cdot\frac{\partial p}{\partial x}\cdot\left[1-\left(\frac{z}{d}\right)^2\right] \\ & -\frac{2\cdot q_\zeta\cdot l_\mathrm{D}}{\mu}\cdot\exp\left(-\frac{d}{l_\mathrm{D}}\right)\cdot\frac{\partial\varphi}{\partial x}\cdot\left[\cosh\left(\frac{d}{l_\mathrm{D}}\right)-\cosh\left(\frac{z}{l_\mathrm{D}}\right)\right] . \end{aligned} \quad (11.30)$$

The first term in (11.30) corresponds to Poiseuille flow between two plates, while the second term represents the electro-osmotic part. Figure 11.10 shows the velocity profile $u(z)$ according to (11.30) for different pressure gradients $\partial p/\partial x$, different electric fields $\partial\varphi/\partial x$ and different values of l_D. If there is no pressure gradient ($\partial p/\partial x = 0$), for $\partial\varphi/\partial x < 0$ we obtain a plug-like velocity profile, with a sharp increase in the velocity in a thin layer close to the wall (cf. Figure 11.10). This is because the charged liquid close to the wall is drawn towards the oppositely charged electrode at the exit of the channel. This is a pure electro-osmotic flow. If the flow is additionally acted on by a pressure gradient $\partial p/\partial x < 0$, we obtain a superposition of pressure-driven and electro-osmotic flow. For $\partial\varphi/\partial x > 0$, the liquid in the electrical double layer moves to the left, against the pressure-driven flow. We obtain a velocity profile with a backflow close to the wall. Inside the liquid we obtain a parabolic velocity profile that is due to the pressure-driven Poiseuille flow. For $\partial\varphi/\partial x < 0$ we obtain the opposite effect of the electric forces close to the wall. The thickness of the kinematic boundary layer close to the wall is linked to the thickness of the electrical double layer l_D. The kinematic boundary layer is determined by a balance of viscous and electric forces. The profiles for $l_D/d = 0.004, 0.01, 0.02$ in Figure 11.10 differ from one another only within this kinematic boundary layer. The rotationally symmetric solution for a capillary with circular cross section may be found in *C. L. Rice* and *R. Whitehead* (1965).

According to (11.26), the Debye length l_D depends, inter alia, on the ion densities of the species involved. In the above discussion, l_D is assumed for simplicity to be constant. The same may also be stated for the zeta potential ζ and for q_ζ. Finally, the electrical conductivity of a liquid depends with great sensitivity on the ion densities, thus influencing the calculation of the electrical potential φ using Ohm's law. It is reasonable to assume constant material properties as long as the solutions are dilute and thus the concentration fields homogeneous. This couples the electric field into the flow equations in a simple manner. For a complete treatment of electro-osmotic flows, however, the concentration fields of all species involved must be calculated, resulting in complete coupling of all equations. This generally takes place via the solution of the additional corresponding transport equations using numerical methods.

The Debye length is generally in the regime $l_D < 100\,nm$, so that even in microchannels we have $l_D \ll d$. According to this, a one-dimensional wall-normal charge distribution, if necessary using the Debye-Hückel approximation, remains a useful approximation, as long as there is no curvature of the wall on the length scale l_D. Such strong curvatures can only occur at corners. Thus we have the following options for treating the electrical double layer: (1) The electrical double layer is solved numerically and no further assumptions are made about the charge distribution; because of the widely different length scales d and l_D, this can necessitate considerable numerical effort. (2) The charge distribution in the electrical double layer is assumed to be one-dimensional if necessary using the Debye-Hückel approximation and the solution for the flow close to the wall is asymptotically matched to the solution in the electrically-neutral interior of the flow; the superposition of both partial solutions yields an approximate solution for the flow in the entire region (cf. *I. Meisel* and *P. Ehrhard* (2006)). (3) The extension of the electrical double layer is ignored and instead a modified boundary condition is used for the tangential velocity at the wall. According to *R. Probstein* (1994) this so-called Helmholtz-Smouluchowski boundary condition is given by:

$$u(z=0) - u_w = -\frac{\epsilon r \cdot \epsilon 0 \cdot \zeta}{\mu} \cdot \frac{\partial \varphi}{\partial x} \quad . \tag{11.31}$$

Here x is the coordinate tangential to the wall and z that normal to the wall. The wall is located at $z = 0$.

Electrophoresis

If we have free ions in a liquid at rest, an externally applied electric field will cause the ions to move towards the oppositely-charged electrode (see Figure 11.11). This effect is called electrophoresis. If we consider a particle in a liquid at rest, the situation is similar. Irrespective of whether the particle is initially charged, surface charges arise on the particle by interaction with the liquid. Thus an electrical double layer is formed around the particle. The particle with the electrical double layer does not appear electrically neutral to the environment and is therefore also subject to electrophoresis.

On the one hand, Coulomb forces act on the ion or particle, causing motion. On the other hand, the motion of the ion or particle leads to viscous friction with the surrounding liquid. Because of the small ion or particle diameter d_0 and the small ion or particle velocity $\boldsymbol{v}_0$, we may assume $Re_0 = |\boldsymbol{v}_0| d_0/\nu \ll 1$. Therefore there is a creeping flow past the ion or particle. Now ions are small compared to the thickness of the electrical double layer, i.e. for ions $d_0 \ll l_D$ for various measurement techniques. The solid particles may be single cells or plastic spheres for various measurements, with typical diameters of several hundred nanometers. Thus for such particles $d_0 \gg l_D$.

The balance of forces at an ion yields

$$3 \cdot \pi \cdot \mu \cdot d_S \cdot \boldsymbol{v}_0 = -Q_{\text{eff}} \cdot \nabla\varphi \quad . \tag{11.32}$$

Here we note that a deformed non-spherical diffuse ion-atmosphere forms around the ion, whose shielding effect leads to $|Q_{\text{eff}}| \leq |z_i \cdot e|$. Simultaneously, in the friction force the Stokes diameter $d_S \geq d_0$ is to be used, i.e. the diameter of a sphere with identical friction. Thus for an ion we find the electrophoretic mobility as

$$\lambda' \equiv -\frac{\boldsymbol{v}_0}{\nabla\varphi} = \frac{Q_{\text{eff}}}{3 \cdot \pi \cdot \mu \cdot d_S} \quad . \tag{11.33}$$

The electrophoretic mobility of the ion is therefore directly proportional to its charge and inversely proportional to its size.

For a solid particle with $d_0 \gg l_D$, we may apply our knowledge of the electrical double layer at a plane wall. The force balance between viscous and electric forces leads to

$$\mu \cdot \frac{\pi \cdot d_0^2}{4 \cdot l_D} \cdot \boldsymbol{v}_0 = \epsilon r \cdot \epsilon 0 \cdot \zeta \cdot \frac{\pi \cdot d_0^2}{4 \cdot l_D} \cdot \nabla\varphi \quad . \tag{11.34}$$

The electrophoretic mobility therefore becomes

$$\lambda' = -\frac{\epsilon r \cdot \epsilon 0 \cdot \zeta}{\mu} \quad . \tag{11.35}$$

Thus the electrophoretic mobility of a solid particle in this limiting case is independent of the particle size and is directly proportional to the zeta potential.

The cases $d_0/l_D \ll 1$ und $d_0/l_D \gg 1$ lead to relatively simple relations, but it must be noted that for $d_0 \approx l_D$ or for non-spherical ions or particles, it is difficult to determine the electrophoretic mobility theoretically. Further approaches may be found in *R. Hunter* (1981). In addition, the quantities Q_{eff}, d_S and ζ in equations (11.33) and (11.35) are not well known. Therefore it is simpler to determine the electrophoretic mobility directly by experiment. According to the definition in (11.33), we measure the applied electric field $\nabla\varphi$ and the velocity of the ion or particle $\boldsymbol{v}_0$ in the solution. Both quantities appear accessible with a suitable experimental set up. It must be noted that as well as the electrophoresis for a quasi-steady electric field, high-frequency and inhomogeneous fields also cause forces on particles. An overview of this so-called dielectrophoresis is found in *A. Ramos et al.* (1998). In order to describe electrophoretic processes, it is desirable to have a suitable transport equation. On the one hand, the motion of an ion in solution in a liquid represents material transport, with the motion caused by electrophoresis, by convection and by diffusion. On the other hand, the transport of the charged ion causes an electric current through the liquid. The combination of material transport and charge transport allows a transport equation to be inferred (cf. *R. Probstein* (1994)). For dilute, conducting solutions, the electric current density may be expressed by the Nernst-Plank equation. Using conservation of charge in the volume element and a relation between the electric conductivity and the ion densities, we finally obtain i transport equations for the ion species in the form

$$\frac{\partial c_\mathrm{i}}{\partial t} + (\boldsymbol{v} - \lambda_\mathrm{i}' \cdot \nabla\varphi) \cdot \nabla c_\mathrm{i} = D_\mathrm{i} \cdot \Delta c_\mathrm{i} + \lambda_\mathrm{i}' \cdot c_\mathrm{i} \cdot \Delta\varphi + r \quad . \tag{11.36}$$

In (11.36), c_i is the concentration, λ_i' is the electrophoretic mobility and D_i is the diffusion coefficient of the species i. $\boldsymbol{v}$ is the velocity of the liquid, φ is the electric potential and r is a source or sink term that takes into account the change in the ion concentration due to chemical reactions. In an electrically-neutral region the second term on the right-hand side vanishes, because the electric potential always satisfies the Laplace equation. Equation (11.36) makes it clear that the electrophoretic motion modifies the convective transport. There is a superposition of flow velocity and electrophoretic velocity. In addition, for $\lambda_\mathrm{i}' \to 0$, (11.36) becomes a standard transport equation, such as is used for uncharged species. Although the derivation of this transport equation assumes free ions in a dilute solution, a similar equation can be used for particles. In this case there is no source or sink ($r = 0$) and for increasing particle size diffusion vanishes ($D_\mathrm{i} \to 0$). This describes the transport of particles taking into account their electrophoretic mobility.

Application to a separation channel

We now look at the example of the separation of a mixture of potassium ions (K^+), sodium ions (Na^+) and lithium ions (Li^+) in a micro-separation channel. *D. P. J. Barz* (2005) presents both numerical (FEM) simulations and experiments for validation. The equations (11.27) and (11.28) are used as a basis for the flow and (11.36) for the material transport of the species at constant material properties. The separation channel has a square cross section of 50 μm × 50 μm and a length of 72 mm from the junction in the channel to the detector (see Figure 11.12). The mixture of the ions in dilute solution is initially in the form of a plug in the region of the junction, surrounded by an aqueous solution without these ions. By application of a potential difference of 3 kV along the separation channel, corresponding to a

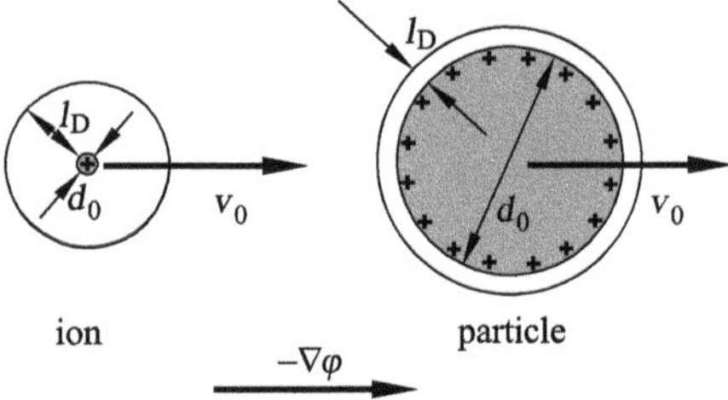

Fig. 11.11. The effect of an electric field on an ion and on a particle in a liquid at rest.

field strength of $\partial\varphi/\partial x = -35.3\ V/mm$, an electro-osmotic flow is generated in the positive x-direction. In addition the ions in the electric field are subject to electrophoresis.

From the concentration fields in Figure 11.12, we see that the plug has left the junction and the maxima of the concentration fields (arrows) are already at different positions. The potassium ions are clearly further than the sodium ions, which again are further than the lithium ions. Although all ions have the same number of charges, it is the size of the ion that causes the different mobility. The potassium ion is relatively small and therefore has a high electrophoretic mobility. A detector downstream captures the electric conductivity σ averaged over the cross section, which increases in the presence of ions. As the concentration fields pass through in time, we therefore see an increase in the conductivity. The simulated conductivities are in good agreement with the measured conductivities. The systematic deviations are not surprising, given the uncertain data for the zeta potential and the Debye length, as well as to a lesser degree for the electrophoretic mobility.

Similarity of electric field and flow

Under certain conditions it can be shown, according to *J. T. G. Overbeck* (1952), that the electric field and the electro-osmotic flow are similar. We then have

$$\boldsymbol{v}(x,y,z,t) = \frac{\epsilon\mathrm{r} \cdot \epsilon 0 \cdot \zeta}{\mu} \cdot \nabla\varphi(x,y,z,t) \quad . \tag{11.37}$$

This means that the streamlines and the electric field lines are parallel at all times and at all places, even for complex channel geometries. The assumptions for this similarity to hold are a thin electrical double layer with $l_\mathrm{D} \ll d$, constant liquid properties, a constant zeta potential, electrically isolating

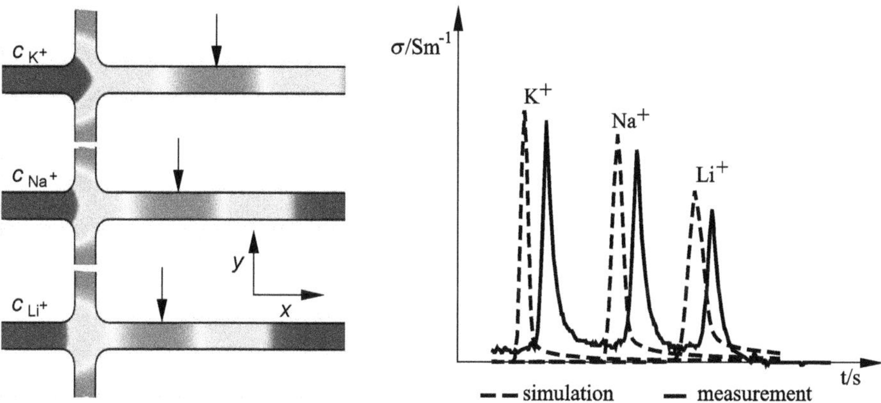

Fig. 11.12. Electrophoretic separation of an ion mixture, after *D. P. J. Barz* (2005)

channel walls, no external pressure gradient, $Re_d \ll 1$ and $Re_d \cdot Str \ll 1$. Here $Re_d = |\boldsymbol{v}| \cdot d/\nu$ is the Reynolds number in the channel and $Str = d/(|\boldsymbol{v}| \cdot \tau)$ is the Strouhal number, formed with the time scale τ of the time dependent electric field. These assumptions are frequently satisfied for electro-osmotic flows of dilute solutions under quasi-steady conditions. It is then sufficient to solve the Laplace equation for the external electric field, and the velocity field is known by (11.37). As the electrophoretic velocity of ions or particles $\boldsymbol{v}_{\mathrm{I,P}}$ also becomes similar to the electric field for similar conditions, it may be calculated by superposition according to

$$\boldsymbol{v}_{\mathrm{I,P}}(x, y, z, t) = \left(\frac{\epsilon \mathrm{r} \cdot \epsilon 0 \cdot \zeta}{\mu} - \lambda' \right) \cdot \nabla \varphi(x, y, z, t) \quad . \tag{11.38}$$

11.3.4 Wetting and Thin Films

Because of their small thickness or when wetting occurs, thin liquid films on solids can touch the limits of conventional continuum mechanics. In these cases possibly boundary conditions must be modified or the molecular forces must be taken into account. A liquid-gas interface, for example, is treated in continuum mechanics as a infinitely thin layer, across which there is a jump in the fluid properties. However, it is in fact a zone of finite thickness, across which these changes take place continuously. The integral properties of this zone are taken into account in continuum mechanics via the interfacial tension. Of course, we cannot expect that this model remains correct when, for example, two interfaces lie close together.

Wetting

When a liquid touches a solid, different types of wetting can be observed (cf. Figure 11.13).

On the one hand there may be partial wetting. For $t \to \infty$, a steady equilibrium arises between the liquid (l), the gas (g) and the solid (s). The contact line (cl) and the contact angle α_{s} are steady. The liquid, gas and solid meet at the contact line. Well-wetting systems are characterized by small contact angles, while poorly-wetting systems are characterized by large contact angles.

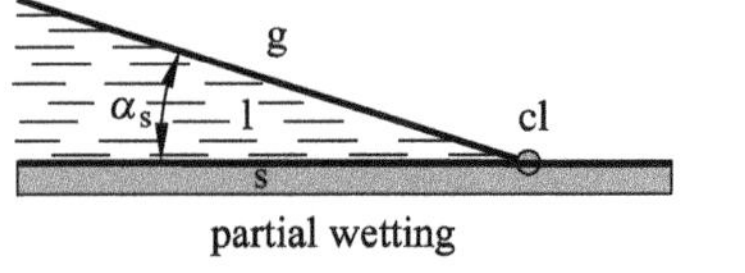

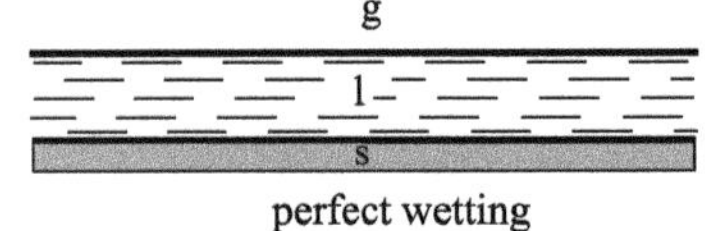

Fig. 11.13. Wetting of a solid

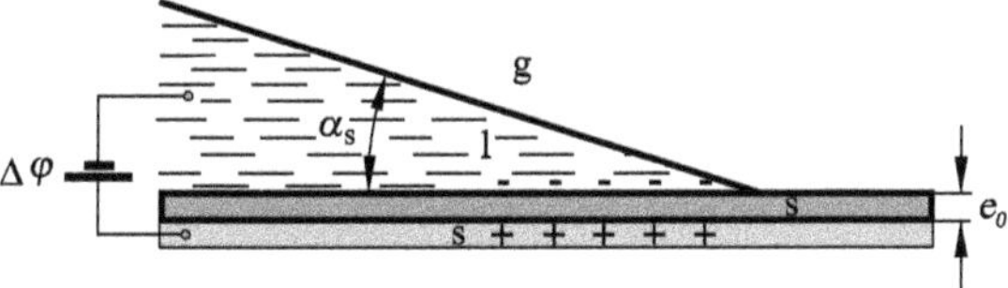

Fig. 11.14. Set up for electro-wetting

On the other hand there may be perfect wetting. In this case, for $t \to \infty$ the liquid spreads out without limit, until eventually a thin liquid film forms on the solid. This behaviour can be considered to be the limiting case $\alpha_s \to 0$.

At the molecular level there is a system of solid, liquid and gas molecules. It is the interaction between the solid and liquid molecules that can drive the wetting. As the wetting front progresses, however, gas molecules must be displaced from the solid. The interaction between gas and solid molecules can therefore inhibit the wetting. If the attraction between the molecules of the liquid and the solid is greater than that between the molecules of the gas and the solid, the wetting will progress. In contrast, the partial wetting in Figure 11.13 is an expression of a balance of these forces. *T. Young* (1805) derived the equation

$$\sigma_{sg} - \sigma_{sl} = \sigma_{lg} \cos \alpha_s \quad , \tag{11.39}$$

which relates the interfacial tension at the solid-gas (σ_{sg}), solid-liquid (σ_{sl}) and liquid-gas (σ_{lg}) interfaces (cf. Section 2.8).

The static contact angle in partial wetting can be changed by electric forces. According to *H. Vallet et al.* (1996), to achieve this an electrically conducting liquid and an electrically insulating solid must be present, into which an electrode is embedded (cf. Figure 11.14). If an electric potential difference $\Delta\varphi$ is now applied between the liquid and the electrode, the attractive electric forces modify Young's equation (11.39), and, according to

$$\cos \alpha_s = \frac{1}{\sigma_{lg}} \cdot \left(\sigma_{sg} - \sigma_{sl} + \frac{\epsilon r \cdot \epsilon 0}{2 \cdot e_0} \cdot \Delta\varphi^2 \right) \quad , \tag{11.40}$$

we obtain a smaller contact angle α_s. In this expression, $\epsilon r \cdot \epsilon 0$ characterizes the dielectric properties and e_0 the thickness of the insulating solid. This phenomenon is called electro-wetting and is applied technically e.g. for variable liquid lenses. A transition from partial to perfect wetting can also be

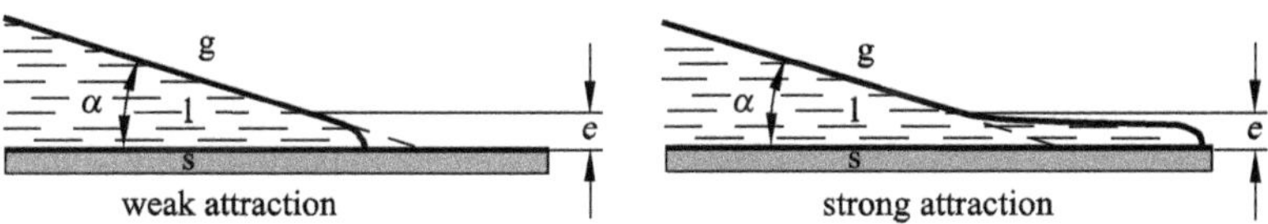

Fig. 11.15. Wetting interface shape on the molecular length scale

observed for a given set of materials for changes in the temperature (see *P. G. de Gennes* (1985)).

The straight line progression of the liquid-gas interface to the contact line is of course an idealization. On molecular length scales, according to *P. G. de Gennes* (1985), considerable deviations from this can occur, for both static and dynamic situations. Long-range attractive van der Waals forces lead to interface contours, as shown in Figure 11.15. These forces are caused by the mutual polarization of the molecules. Electrostatic forces have similar effects, as they act between charged or polar molecules. If there is a strong attraction between liquid and solid molecules, and at the same time a weak attraction between gas and solid molecules, a precursor liquid film is found. In all cases, the length scale e at which such deviations occur is of the order of several molecule diameters. These deviations are practically invisible for continuum mechanics. Figure 11.15 makes it clear that a macroscopic length scale is useful to define the contact angle. The macroscopic contact angle α is thus defined on a length scale of several micrometers. It can be accessed optically using standard methods. On the molecular length scale the contact angle is barely accessible and also, because of the motion of the molecules, not sharp.

Moving and contact line

To calculate the flow near a moving contact line, we must know the macroscopic situation. In the continuum mechanical treatment, these are taken account of via the boundary conditions. The kinematics of the flow on both sides of a moving contact line have been characterized by *E. B. Dussan* and *S. H. Davis* (1974). Without making any assumptions with respect to the geometric form of the interface and contact line, as well as with no assumptions on the rheology of the fluid, the authors obtain flow fields that are kinematically compatible with the interface, with the moving contact line and with the no-slip condition at the wall. These flow fields are shown in Figure 11.16 for the two basic cases. In each case the relative motion with respect to the moving contact line is shown. In the first picture a liquid element moves along the interface towards the contact line, so that the liquid propagates in a rolling motion on the solid. This generates a complex flow in the gas, which flows away along an inverse stagnation streamline. In the second picture a liquid element moves along the interface away from the contact line. A stagnation streamline is therefore formed inside the liquid, leading towards to the contact line. The gas is thus rolled away from the solid.

If we consider the velocity of the liquid on the one hand along the interface and on the other hand along the wall, as we approach the contact line two different limiting values are found. It is only for the contact angle $\alpha = 90°$ that this discrepancy vanishes. Therefore the velocity field is not continuous for general contact angles, rather there is an infinite velocity gradient at the contact line. Depending on the constitutive equations of the liquid, infinite stresses can therefore occur. This singularity is not a consequence of simplifying assumptions, but rather is due to the continuum mechanical formulation.

The question remains of how this singularity is to be treated. *E. B. Dussan* and *S. H. Davis* (1974) show that a change in the conditions at the interface does not eliminate this singularity. One possibility is to use suitable non-Newtonian constitutive equations. Although this does not eliminate the discontinuity of the velocity field, it ensures finite stresses. A further possibility is to relax the no-slip condition precisely at the contact line using a slip condition. This leads to a continuous velocity field and permits the use of Newtonian constitutive equations. The stresses remain finite. To be precise, the Navier slip law (11.5) can be used at the wall with a slip length L_S of several molecular diameters, so that appreciable slip only occurs directly at the moving contact line. The small slip of the liquid at the wall is not surprising if we consider the statements in Section 11.1.4. In addition, the appreciable slip directly at the moving contact line can be understood, as macroscopic wetting often takes place on top of an advancing microscopic liquid film. In the microscopic picture, therefore, the liquid slips along a microscopic liquid film (cf. Figure 11.15).

From two series of experiments by *W. Rose* and *R. W. Heins* (1962) for perfectly wetting material systems in a capillary tube, and using similarity arguments, *G. Friz* (1965) derives the empirical expression

$$\tan\alpha \approx 3.4 \cdot \left(\frac{\mu \cdot U}{\sigma_{\mathrm{lg}}}\right)^{\frac{1}{3}} \tag{11.41}$$

for the dynamic contact angle α. In this expression U is the velocity of the contact line, μ is the dynamic viscosity of the liquid and σ_{lg} is the interfacial tension between liquid and gas. The relation (11.41) was confirmed by *A. M. Schwartz* and *S. B. Tajeda* (1972) for further material systems with different wetting geometries. We can see a dependence of the macroscopic contact angle α on the velocity U of the contact line and on the liquid properties μ/σ_{lg}. For perfecly wetting systems the static contact angle $\alpha_{\mathrm{s}} \approx 0.$ holds. *E. B. Dussan* (1979) summarizes measurements of different authors taking into account partial wetting ($\alpha_{\mathrm{s}} \neq 0$) and finds, for advancing ($U > 0$) and retreating ($U < 0$) contact lines, the model laws

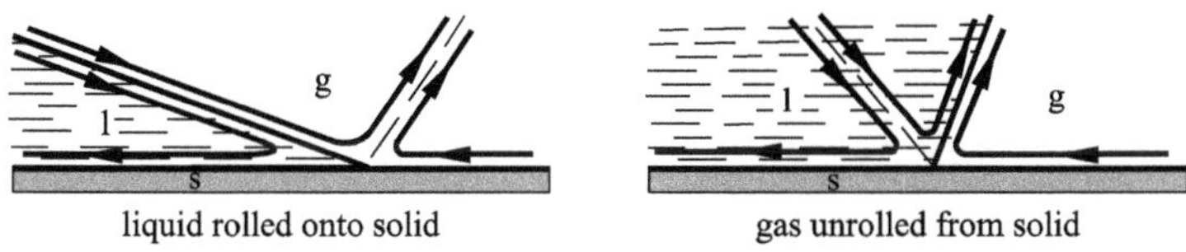

Fig. 11.16. Kinematics of the flow at a moving contact line, according to *E. B. Dussan* and *S. H. Davis* (1974), simplified for a plane wall

$$U > 0: \quad U = \kappa_{\mathrm{A}} \cdot (\alpha - \alpha_{\mathrm{A}})^{\mathrm{m}} \quad , \tag{11.42}$$

$$U < 0: \quad U = \kappa_{\mathrm{R}} \cdot (\alpha - \alpha_{\mathrm{R}})^{\mathrm{m}} \quad . \tag{11.43}$$

In the equations (11.42) and (11.43), κ_{A} and κ_{R} are empirical constants and α_{A} und α_{R} are the static contact angles after the contact line has advanced or retreated. As $\alpha_{\mathrm{A}} > \alpha_{\mathrm{R}}$, there is hysteresis in the contact angle. Experiments by *R. L. Hoffman* (1975) and *L. H. Tanner* (1979) for advancing contact lines suggest an exponent of $\mathrm{m} = 3$ (cf. Figure 11.17). For small contact angles and $\alpha_{\mathrm{A}}, \alpha_{\mathrm{R}} \to 0$ the model laws (11.42) and (11.43) are also consistent with (11.41).

In summary, we may say that the macroscopic contact angle in the dynamic case takes on different values from that in the static case. For advancing contact lines larger contact angles are observed, while for retreating contact lines smaller contact angles occur. The dependence according to the model laws (11.42) and (11.43) with $\mathrm{m} \approx 3$ seems to be confirmed experimentally, at least for advancing contact lines.

Thin films

The stability of thin liquid films on horizontal solids can serve to clarify their physics and particularly to indicate the limits of the continuum mechanical treatment. We follow the description in the review article by *A. Oron et al.* (1997).

A thin film is given by a liquid layer that lies between a horizontal solid plate at $z = 0$ and a liquid/gas interface at $z = h$ (cf. Figure 11.18). The extension of the film is infinite in the horizontal directions x and y. For simpler mathematical representation, we restrict ourselves to the two-dimensional problem in the x-z plane. A generalization to the three-dimensional problem may be found in *A. Oron* (2002). The two-dimensional problem is characterized by the vanishing velocity component $v = 0$ and vanishing gradients

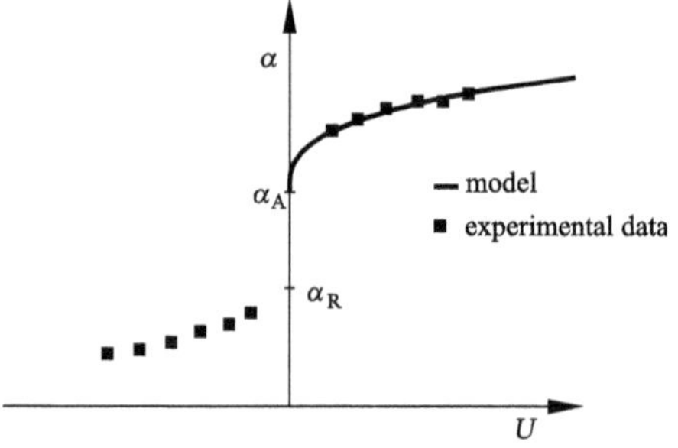

Fig. 11.17. Behavior of the dynamic contact angle, according to *E. B. Dussan* (1979)

$\partial/\partial y$, so that, from the incompressible continuity equation (5.3) and the incompressible Navier-Stokes equations (5.20), we obtain

$$\frac{\partial u}{\partial x} + \frac{\partial w}{\partial z} = 0 \quad , \tag{11.44}$$

$$\rho \cdot \left(\frac{\partial u}{\partial t} + u \cdot \frac{\partial u}{\partial x} + w \cdot \frac{\partial u}{\partial z} \right) = -\frac{\partial p}{\partial x} + \mu \cdot \left(\frac{\partial^2 u}{\partial x^2} + \frac{\partial^2 u}{\partial z^2} \right) - \frac{\partial \phi}{\partial x} \, , \tag{11.45}$$

$$\rho \cdot \left(\frac{\partial w}{\partial t} + u \cdot \frac{\partial w}{\partial x} + w \cdot \frac{\partial w}{\partial z} \right) = -\frac{\partial p}{\partial z} + \mu \cdot \left(\frac{\partial^2 w}{\partial x^2} + \frac{\partial^2 w}{\partial z^2} \right) - \frac{\partial \phi}{\partial z} - \rho \cdot g \, , \tag{11.46}$$

for the liquid. In addition we have the boundary conditions on the solid and at the liquid-gas interface

$$z = 0 : u = w = 0 \quad , \tag{11.47}$$

$$z = h : w = h_t + u \cdot h_x \quad , \tag{11.48}$$

$$\boldsymbol{n} \cdot \boldsymbol{T} \cdot \boldsymbol{n} = 2 \cdot H \cdot \sigma_{\mathrm{lg}} \quad , \tag{11.49}$$

$$\boldsymbol{t} \cdot \boldsymbol{T} \cdot \boldsymbol{n} = \boldsymbol{t} \cdot \nabla \sigma_{\mathrm{lg}} \quad . \tag{11.50}$$

Here $\boldsymbol{T}$ is the stress tensor in the liquid, $\boldsymbol{n}$ and $\boldsymbol{t}$ are the unit vectors in the normal and tangential directions (cf. Figure 11.18). The no-slip condition is satisfied on the solid. In this section we do not consider any moving contact lines. On the interface we use the kinematic boundary condition (11.48) to ensure a tangential flow. Further, the continuity of the stresses in the normal and tangential directions is retained by the dynamic boundary conditions (11.49) and (11.50). In the normal direction, for the simplest case of vanishing viscosity we obtain the Laplace pressure jump $\Delta p = 2 \cdot H \cdot \sigma_{\mathrm{lg}}$, whereby the mean curvature of the interface is given by

$$2 \cdot H = \frac{h_{xx}}{(1 + h_x^2)^{\frac{3}{2}}} \quad . \tag{11.51}$$

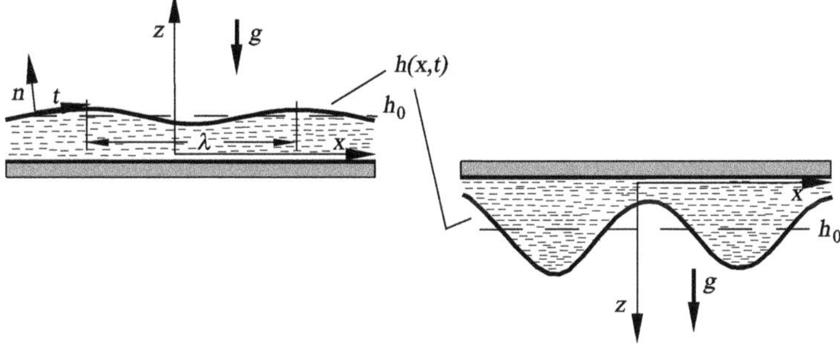

Fig. 11.18. Thin liquid films on and under horizontal plates

In the tangential direction we obtain shear stresses if surface tension σ_{lg} is not constant in space. This may be due to the temperature or to dissolved species. Further, because of $\mu_{\mathrm{gas}} \ll \mu$, we assume implicitely in (11.50) that the shear stresses in the surrounding gas may be neglected. This assumption permits the flow in the liquid to be treated separately from the flow in the gas. To simplify the notation, the partial derivatives of the function $h(x,t)$ are denoted by indices, i.e. $h_x = \partial h/\partial x$, $h_t = \partial h/\partial t$ etc. In the Navier-Stokes equations (11.45) and (11.46), as well as gravity we have introduced a further volumetric force in the form of a potential ϕ, which will be useful to model molecular forces.

By applying the so-called thin film approximation, the system (11.44) - (11.51) can be considerably simplified. We leave out the mathematical details, as they can be found in *A. Oron et al.* (1997). Basically it is assumed that the mean film thickness h_0 is very small compared to the wavelength λ of the perturbations of the interface. This leads to separate length and velocity scales in both directions, so that as well as $h_0 \ll \lambda$ also $w \ll u$ and $\partial/\partial x \ll \partial/\partial z$ can be utilized. If the equations are integrated over the film thickness in the region $0 \leq z \leq h$, taking account of the boundary conditions, a so-called evolution equation for the film thickness $h(x,t)$ can be derived.

We first consider a case in which the film thickness remains large compared to the diameter of the liquid molecules, so that we may expect to arrive at a continuum mechanical treatment. For constant surface tension σ_{lg} and for $\phi = 0$ we obtain the evolution equation

$$\mu \cdot h_t - \frac{1}{3} \cdot \rho \cdot g \cdot \left(h^3 \cdot h_x\right)_x + \frac{1}{3} \cdot \sigma_{\mathrm{lg}} \cdot \left(h^3 \cdot h_{xxx}\right)_x = 0 \quad . \tag{11.52}$$

In the evolution equation (11.52) the second term expresses the effect of gravity and the third term the effect of the capillary force. We first consider the effect of gravity. For a film on a plate a local deflection of the interface, as shown in Figure 11.19, leads hydrostatically to an increased pressure underneath the deflection. The consequence is a horizontal pressure gradient, which transports the liquid underneath the deflection to either side and thus reduces the deflection. Gravity therefore acts to stabilize the system. The capillary force has a similar effect. Because of the convex curvature of the interface, the pressure underneath the deflection is larger than that under-

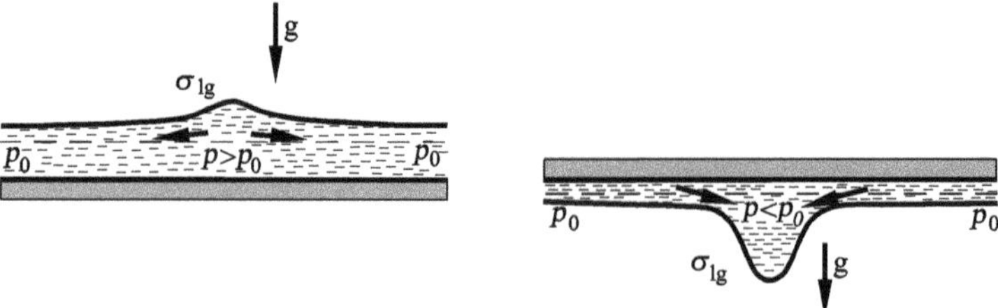

Fig. 11.19. Stability of a liquid film on and under a plate

neath the interface that is parallel to the wall. The capillary force therefore also acts to stabilize the system. If we subject the basic solution $h = h_0$ of equation (11.52) to a stability analysis against small perturbations that are periodic in x, we indeed find that all perturbations are damped in time for $\rho \cdot g > 0$ und $\sigma_{\mathrm{lg}} > 0$. Thus the basic solution remains stable. The condition for temporal growth of the pertubations of the film on the plate is

$$\rho g < -\sigma_{\mathrm{lg}} \cdot \left(\frac{2 \cdot \pi}{\lambda} \right)^2 \quad . \tag{11.53}$$

If the film is below the plate, as shown in Figure 11.19, the condiction for temporal growth is

$$\rho g > \sigma_{\mathrm{lg}} \cdot \left(\frac{2 \cdot \pi}{\lambda} \right)^2 \quad . \tag{11.54}$$

In this case the deflection reduces the pressure above the deflection hydrostatically, while the concave curved interface again causes a capillary pressure increase. Because of the change in direction of g, gravity now acts to destabilize the system. If (11.54) is satisfied, the destabilizing gravitational force prevails and the liquid is transported into the deflection, as in Figure 11.19. This situation is called the Rayleigh-Taylor instability, whereby equation (11.55) shows that large wavelengths λ are particularly critical. Evaluation yields

$$\lambda > 2 \cdot \pi \cdot \sqrt{\frac{\sigma_{\mathrm{lg}}}{\rho \cdot g}} \tag{11.55}$$

for the unstable region of the wavelengths. It is interesting to note that all results (11.53)-(11.55) do not depend on the mean film thickness h_0.

We now consider a case in which the liquid film is only 10 - 100 molecular layers thick, which corresponds, depending on the liquid, to several hundred angstroms. Under such conditions long-range molecular forces can play an important role. Inside the liquid, molecular forces are in principle be taken into account by the properties of the continuum. At individual interfaces the different molecular forces in both continua are also taken into account through surface tension. However, if two interfaces occur close together, as in the present case with the liquid-gas interface and the liquid-solid interface, the molecular forces additionally lead to an interaction of these interfaces. In the simplest case of parallel interfaces without the presence of ions, the force potential

$$\phi = \phi_0 + \frac{A}{6 \cdot \pi \cdot h^3} \tag{11.56}$$

may be used to model the system. Here ϕ_0 is a reference potential whose value is not relevant for the forces (cf. (11.45) and (11.46)) and A is the so-called Hamaker constant. If $A > 0$ the two interfaces attract each other, while for $A < 0$ they repel each other. For constant surface tension σ_{lg} and neglected gravity $g = 0$, in this case we obtain the evolution equation

$$\mu \cdot h_t + \frac{A}{6 \cdot \pi} \cdot \left(\frac{h_x}{h}\right)_x + \frac{1}{3} \cdot \sigma_{\text{lg}} \cdot \left(h^3 \cdot h_{xxx}\right)_x = 0 \quad . \tag{11.57}$$

Instead of the potential (11.56), to model the system a normal stress within the liquid can also be superimposed, the so-called disjoining pressure. Under the same conditions this leads to the same evolution equation.

A linear stability analysis of the basic solution h_0 for spatially periodic perturbations of the evolution equation (11.57) leads to amplification in time for

$$\frac{A}{6 \cdot \pi \cdot h_0} > \frac{1}{3} \cdot \sigma_{\text{lg}} \cdot \left(\frac{2 \cdot \pi}{\lambda}\right)^2 \cdot h_0^3 \quad . \tag{11.58}$$

For $A > 0$ therefore an instability is possible, which can be visualized as being due to a self-strengthening attraction between the two interfaces, leading to a local rupture of the liquid film as $h(x, t) \to 0$ occurs. Again it is large wavelengths that are critical. The stability analysis yields

$$\lambda > 2 \cdot \pi \cdot h_0^2 \cdot \sqrt{\frac{2 \cdot \pi \cdot \sigma_{\text{lg}}}{A}} \tag{11.59}$$

for the region of unstable wavelengths. For $A < 0$, on the other hand, because of the repulsive interaction between the two interfaces, no instability can occur. In both conditions (11.58) and (11.59) we can see a dependence on the mean film thickness h_0. In particular equation (11.58) indicates that for increasing h_0 the left side of the equation vanishes and the right side increases greatly. Thus for increasing film thicknesses the long-range molecular forces are no longer important.

The potential (11.56) is one possibility to take into account the effect of long-range van der Waals forces between the molecules. However, if we are to take into account ions in solution or polar molecules, or electric forces or forces at small molecular distances, a large number of suggestions for the potential $\phi(h)$ may be found in the literature. In addition the conditions of the liquid film may be generalized. The literature contains a great number of generalizations, including evaporation and condensation at the liquid-gas interface, surface tension σ_{lg} that is not constant in space, temperature or concentration dependent material properties, volumetric forces due to rotation of the films, and wetting and dewetting processes. A discussion of these aspects may be found in *A. Oron et al.* (1997) and *A. Oron* (2002).

11.4 Experiments

In recent years numerous experimental investigations of pressure loss, of the laminar-turbulent transition and of heat transfer in microchannels have been published in the literature. The sometimes surprising results deviate in part considerably from the conventional macroscopic correlations, although the conditions would suggest continuum mechanical behaviour. Now it has turned out that many of these deviations are due to erroneous interpretation, because experiments in microchannels frequently permit no access to local information. Thus, in general pressures and temperatures are measured in the entry and exit chambers, because local measurements inside the microchannels are hardly possible without causing disturbances. However, the derivation of the pressure loss and heat transfer correlations based on this integral information remains problematic. It is similarly difficult to detect a laminar-turbulent transition using integral data. Even if advances are made in some areas with local measurements, critical discussion will be necessary.

The data available in the literature can be divided into laminar flows and turbulent flows, whereby in each case circular, rectangular and trapezoid-shaped flow cross sections have been investigated. Here we restrict ourselves to circular cross sections.

11.4.1 Pressure Drop

In the discussion of experimental results on pressure loss in microchannels, we follow in part the review articles by *C. Sobhan* and *S. V. Garimella* (2001) and *G. Hetsroni et al.* (2005). The pressure drop in a pipe may be given by

$$\Delta p = \frac{\rho}{2} \cdot \bar{u}^2 \cdot \lambda \cdot \frac{l}{d} \quad . \tag{11.60}$$

Thus each measurement aims to determine the friction factor λ, generally represented as a function of the Reynolds number $Re_d = \bar{u} \cdot d/\nu$. Figure 11.20 shows the friction factor λ in the form of a so-called Nikuradse diagram as a function of the Reynolds number. Supplementary to Figure 4.84, data for micro pipes from literature are included. In conventional macroscopic pipes, the friction factor for Newtonian fluids in laminar, fully-developed flow is given by $\lambda = 64/Re_d$ (cf. equation (4.120)). For turbulent flows in smooth conventional pipes, we have the Blasius relation $\lambda = 0.316 \cdot Re_d^{-1/4}$ (cf. equation (4.121)). These curves are shown as solid lines in the laminar and turbulent regimes. Furthermore we find a family of selected results in the literature, given by various lines. These lines are drawn through the experimental values of the various authors.

We first turn to the laminar regime. The first curve is obtained by *S. B. Choi et al.* (1991) using nitrogen in pipes with diameters in the range $d = 3 - 81\,\mu m$. The curve confirms the dependence $\lambda \sim Re^{-1}$ that is valid for conventional pipes. However the curve lies about 17 % below the conventional curve in Figure 4.84. The second curve is determined by *D. Yu et al.* (1995) with nitrogen and water in pipes of $d = 19 - 102\,\mu m$ diameter. The curve also shows the dependence $\lambda \sim Re_d^{-1}$ and lies about 20 % below the conventional curve. The curves by *J. Judy et al.* (2002) are obtained for glass pipes with diameters $d = 52 - 149\,\mu m$ with water, isopropanol and methanol. Both curves confirm the dependence $\lambda \sim Re_d^{-1}$, with the curve for water lying about 3 % below the conventional curve. The curves for isopropanol and methanol cannot be differentiated and lie barely 3 % above the conventional curve. Incidentally *J. Judy et al.* (2002) use pipes of different lengths, to ensure that the entrance length, as well as the inlet and outlet effects, do not falsify their results. The curves of *D. Brutin et al.* (2003) are obtained for water in circular glass capillaries of diameter $d = 321\,\mu m$ and $540\,\mu m$. The authors use steady and transient experimental methods and analyze carefully the errors due to the entrance length. Both curves confirm the dependence $\lambda \sim Re_d^{-1}$, and they lie about 4 % and 5 % above the conventional curve.

The experiments by *S. B. Choi et al.* (1991) with nitrogen in the turbulent regime yield the dependence $\lambda \sim Re_d^{-0.182}$ and lie about 55 % below the Blasius correlation. The data of *D. Yu et al.* (1995) using nitrogen and water in the turbulent regime confirm the relation $\lambda \sim Re_d^{-0.25}$, but lie about 5 % below the Blasius correlation. For conventional flows the measurements for real rough pipes generally lie above the Blasius correlation, as may be seen for example from the Colebrook correlation. The location of curves below the Blasius correlation is therefore rather strange.

All measurements above are based on pressure measurements in the entry and exit chambers. Because of this no direct measurement of the pressure drop in the fully-developed flow is obtained. Effects due to the transition between the chambers and the channel, as well as to the entrance length, are always present. These effects are only corrected in a few experiments. Furthermore, inaccuracies in determining the pipe diameter ($\lambda \sim d^3$) and in measuring the integral mass flux ($\lambda \sim \dot{m}^{-2}$) seem to be particularly critical. The exact measurement of small mass fluxes is not trivial, particularly for gases. In the turbulent case there is also a considerable effect from the wall roughness to be expected, as the roughness depth is to be evaluated relative

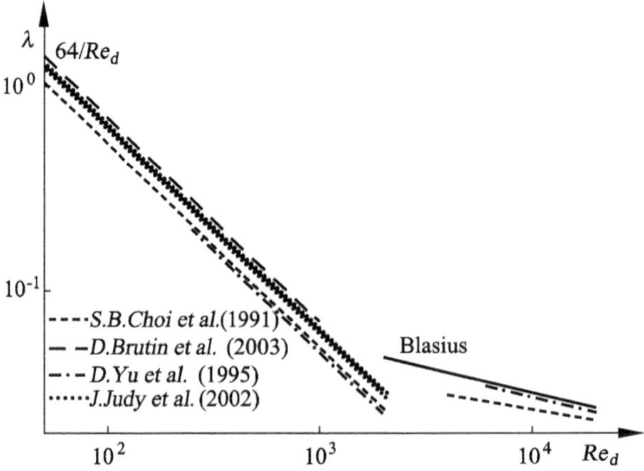

Fig. 11.20. Pressure loss in smooth circular micropipes

to the small diameters. Each inaccuracy in determining the pipe diameter and roughness depth is therefore critical. Finally, the effects of charge transport in the presence of ions (flow potential), of dissipative heating of the fluid, and of loss of thermodynamic equilibrium in gases (Kn $> 10^{-3}$) must be examined critically. In summary, comparison of measurements in micropipes with conventional correlations, at least in the turbulent regime, shows marked discrepancies whose origins remain unexplained.

11.4.2 Laminar-Turbulent Transition

The transition from laminar to turbulent flow in micropipes has been evaluated in the literature using, on the one hand, integral pressure loss data. On the other hand, more recent work uses so-called *micro particle image velocimetry (μPIV)* to obtain local information about the velocity field and from this to determine the transition regime. Figure 11.21 summarizes selected results on the laminar-turbulent transition regime for smooth circular micropipes.

D. Yu et al. (1995) use their pressure loss measurements with water and nitrogen in micropipes with diameters $d = 19 - 102\,\mu m$ to derive a transition range of $Re_d \approx 2000 - 6000$. *Z. X. Li et al.* (2003) carry out measurements with water in micropipes with diameters $d = 79.9 - 166.3\,\mu m$ and give the transition range based on their pressure loss measurements as $Re_d \approx 1535 - 2630$. *K. V. Sharp* and *R. J. Adrian* (2004) carry out their measurements with water and a 1-propanol-glycerol mixture in glass pipes with diameters $50 - 247\,\mu m$. In addition to pressure loss measurements, the authors draw on measurements of the velocity on the pipe axis (μPIV). They give the transition range as $Re_d \approx 1800 - 2300$. The measurements of *K. V. Sharp* and *R. J. Adrian* (2004) are particularly valuable as they drawn on an objective criterium for the velocity fluctuations on the pipe axis and thus on local information to determine the transition regime. Comparison of the transition regime according to *K. V. Sharp* and *R. J. Adrian* (2004) for micropipes with those for conventional macroscopic pipe flow does not indicate that the transition occurs in microflows at considerably smaller Reynolds numbers. Such statements of earlier authors (cf. review articles by *C. Sobhan* and *S. V. Garimella* (2001) or *G. Hetsroni et al.* (2005)) are to be evaluated very critically.

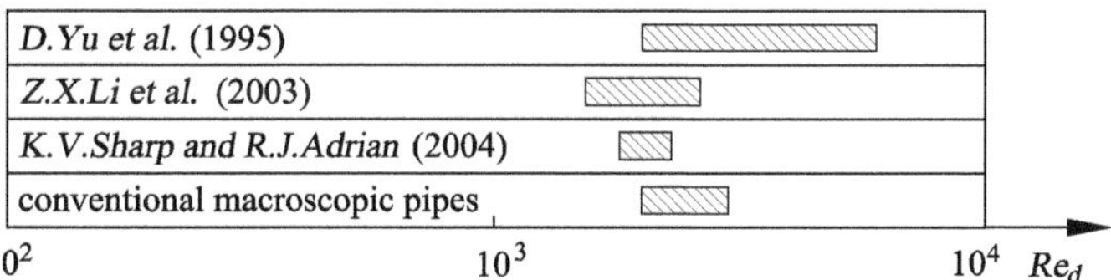

Fig. 11.21. Laminar-turbulent transition region for circular micropipes

11.4.3 Heat Transfer

In the discussion of experimental results on heat transfer in microchannels, we will follow in part the review articles by *C. Sobhan* and *S. V. Garimella* (2001) and *G. Hetsroni et al.* (2005)). For laminar flow in conventional circular pipes, we have in the literature the so-called Sieder-Tate correlation for the heat transfer

$$\frac{\mathrm{Nu}}{\mathrm{Pr}^{\frac{1}{3}}} \approx 1.86 \cdot Re_d^{\frac{1}{3}} \cdot \left(\frac{d}{l}\right)^{\frac{1}{3}} \quad . \tag{11.61}$$

Here the Nusselt number $\mathrm{Nu} = h \cdot d/\lambda$ may be understood as a dimensionless form of the heat transfer coefficient h. λ is the thermal conductivity of the fluid, d the diameter and l the length of the pipe. The Prandtl number $\mathrm{Pr} = \nu/k$ contains the material properties of the fluid in the form of the kinematic viscosity ν and the thermal diffusivity k. Figure 11.22 shows $\mathrm{Nu}/\mathrm{Pr}^{1/3}$ as a function of the Reynolds number. The Sieder-Tate correlation for the reference values of the parameter $d/l = 0.001$ and 0.005 from the experiments is shown as two solid curves in the laminar regime in Figure 11.22. Further, a family of experimental results is found in the literature for the laminar and turbulent regimes, and is shown in the form of various lines. Finally the conventional Dittus-Boelter correlation for turbulent flow

$$\frac{\mathrm{Nu}}{\mathrm{Pr}^{\frac{1}{3}}} \approx 0.023 \cdot Re_d^{0.8} \tag{11.62}$$

is shown as a solid line.

The curves by *S. B. Choi et al.* (1991) with nitrogen ($\mathrm{Pr} \approx 0.7$) in circular pipes with diameters $d = 9.7 - 81.2\,\mu m$. The curve in the laminar regime

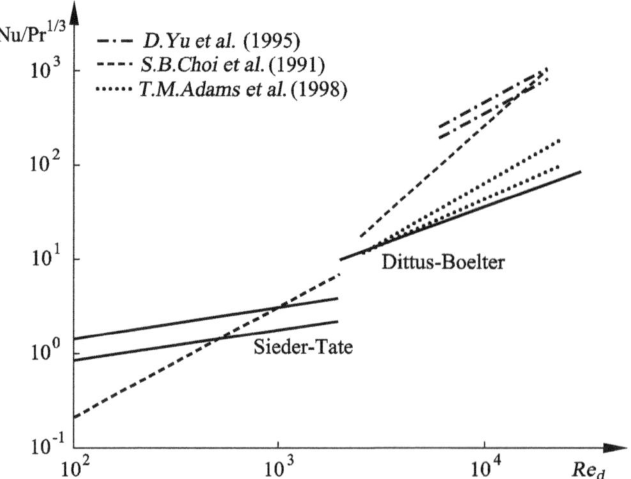

Fig. 11.22. Heat transfer in smooth circular micropipes

shows the relation $\mathrm{Nu} \sim \mathrm{Re_d}^{1.17}$ and is thus considerably steeper than the conventional Sieder-Tate correlation ($\mathrm{Nu} \sim \mathrm{Re_d}^{1/3}$). In the turbulent regime *S. B. Choi et al.* (1991) obtain the relation $\mathrm{Nu} \sim \mathrm{Re_d}^{1.96}$. This curve is also considerably steeper than the Dittus-Boelter correlation ($\mathrm{Nu} \sim \mathrm{Re_d}^{0.8}$). In particular, for large Reynolds numbers *S. B. Choi et al.* (1991) obtain considerably larger Nusselt numbers in micropipes than for macroscopic pipes. The results of *D. Yu et al.* (1995) for the heat transfer in turbulent flows of water ($\mathrm{Pr} \approx 5$) and nitrogen ($\mathrm{Pr} \approx 0.7$) show the relation $\mathrm{Nu} \sim \mathrm{Re_d}^{1.2} Pr^{0.2}$. On the one hand, this yields curves that are steeper than the Dittus-Boelter correlation. On the other hand, the dependence on the Prandtl number $\mathrm{Nu} \sim \mathrm{Pr}^{0.2}$ deviates from the Dittus-Boelter correlation ($\mathrm{Nu} \sim \mathrm{Pr}^{1/3}$). For this reason the two curves for $\mathrm{Pr} \approx 0.7$ and $\mathrm{Pr} \approx 5$ do not coincide. The experiments of *T. M. Adams et al.* (1998) on turbulent heat transfer of water in circular pipes of diameters $d = 760$, $1090\,\mu m$ confirm, on the one hand, the relation $\mathrm{Nu} \sim \mathrm{Pr}^{1/3}$. On the other hand, an explicit dependence on d/l remains, and both curves are steeper than the Dittus-Boelter correlation. The upper curve is valid for $d = 760\,\mu m$ and the lower curve for $d = 1090\,\mu m$.

As well as the problems already discussed for measurements on pressure loss (cf. Section 11.4.1), additional difficulties arise in measuring the heat transfer. All measurements are based on the temperatures in the entry and exit chambers. Thus effects due to the thermal entrance and the axial heat conduction in the fluid and in the wall are superimposed. In a few cases a correction has been carried out using numerical simulations. Comparison of the measured heat transfer correlations with the conventional macroscopic correlation shows considerable discrepancies, both in the laminar and in the turbulent regimes. Measurements based on local temperature fields would be necessary to yield precise data for pure the heat transfer.

12. Biofluid Mechanics

12.1 Fundamentals of Biofluid Mechanics

In contrast to the topics discussed in previous chapters, biofluid mechanics is concerned with flows that are influenced by flexible biological surfaces. We distinguish between *flows past living bodies* in air or in water, such as bird flight or the swimming of fish, and *internal flows*, such as the closed blood circulation of living beings. In the previous millions of years, evolution has developed crawling, running, swimming, gliding, and flying as methods of motion of living beings, depending on their size and weight.

The necessary propulsion for altering position requires flow control adapted to the Reynolds number of the motion. The motion of bacteria and amoebae takes place at very small Reynolds numbers, where friction dominates, with cilia and flagella. Tadpoles and octopuses use the inertia of jet propulsion for motion. Eels move in a wavelike manner; whales use vortex separation at the tail fin for motion at Reynolds numbers of up to 10^8. Fast-swimming fish such as the shark have longitudinal grooves on their scales that affect the viscous sublayer of the flow boundary layer so that the flow drag is reduced.

Heat and mass transport in living beings takes place in circulatory systems. These include respiration, the circulatory system of the blood, and the lymph, as well as that of water. All biological flows have in common that the motion is affected by external or internal highly flexible and structured surfaces. This leads to an actively controlled flow whose losses are kept small.

Of the many different biological flows, in this chapter we will treat the *flying of birds*, *the swimming of fishes* as well as looking at the *blood circulation of the human body* in depth.

Bird flight has already been introduced in Section 4.4.1. The lift and propulsion necessary for flight are generated by the flapping of the wings. The downward flap of the bird's wing is carried out with great force and the upward flap is performed with aerodynamic resistance that is as small as possible. The outer parts of the wing that carry out the greatest part of the vertical motion produce the greatest part of the propulsion. The orientation and form of the wing changes continuously during one flapping period. The inner part of the bird's wing generates the lift.

The propulsion of swimming *fish* is generated by up-and-down motion of the tail fin, or, in the case of small fish, by their side-to-side motion. The tail

H. Oertel (ed.), *Prandtl-Essentials of Fluid Mechanics*,
Applied Mathematical Sciences 158, DOI 10.1007/978-1-4419-1564-1_12,

fin has a symmetric profile, so that during one flapping period it generates continuous propulsion. Most fish control lift with a swim bladder located in their stomach. Large fish such as the whale and the dolphin also generate lift with the vertical motion of the tail fin. This is made possible by the horizontal positioning of the tail fin.

When friction is present, the motion of *bacteria* and *protozoa* takes place with cilia and flagella. Tadpoles and octopuses use the inertial force of jet propulsion to move, while eels move in a wavelike motion.

The *blood circulation of the human body* is driven by the heart. A sketch of the human circulatory system is shown in Figure 12.1. Every minute, the heart pumps about 5 l of blood into the circulation. If the body is under duress, the pump efficiency can increase to 20 to 30 l per minute. The blood circulation consists of two separated partial systems, which are connected via the heart. The first is called the systemic circulation and the second the pulmonary circulation. The complete circulation ensures the exchange of gas between the metabolism in the human tissue and the air in the atmosphere.

The systemic circulation begins with the aorta, which branches into large arteries. Also belonging to the circulation are the capillaries, where the blood gives up its oxygen and takes in carbon dioxide. The blood flows out of the capillaries into the veins, where it is led back again to the heart. From the heart, the blood is pumped into the pulmonary circulation, which consists of

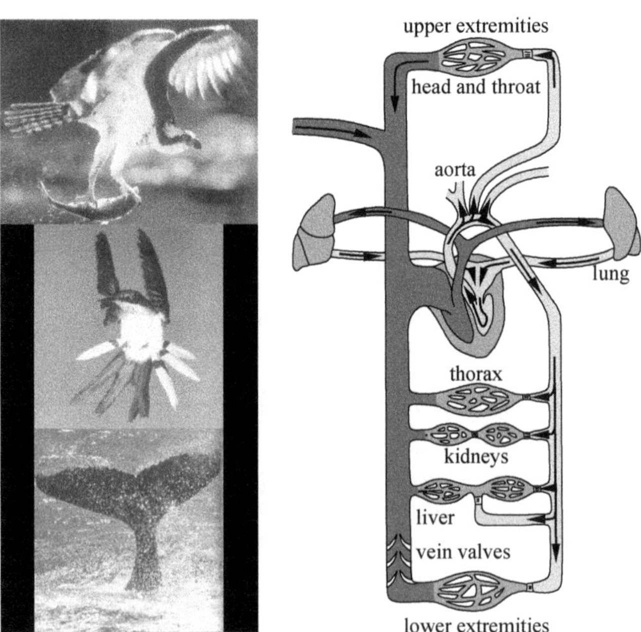

Fig. 12.1. Flight and swimming motion of animals and the blood circulation in the human body

pulmonary arteries, capillaries, and veins. In the pulmonary circulation, the blood gives up a part of its carbon dioxide and takes in as much oxygen as it had previously given up to the body tissues.

12.1.1 Biofluid Mechanics of Animals

Three quarters of all animals can swim or fly. In $6 \cdot 10^8$ years, evolution has developed a great variety of forms of motion in water and air. The Reynolds numbers achieved range from 10^{-3} for bacteria and protozoa to 10^8 for fast-swimming whales. In flight in the atmosphere, Reynolds numbers of 10^{-1} for the smallest insects and up to 10^7 for brief periods for fast-flying birds are reached. According to the discussion in Section 4.2, the flow past animals at Reynolds numbers $Re_l \ll 1$ is dominated by the friction of the fluid, while at Reynolds numbers $Re_l \gg 1$ the inertia of the fluid dominates. In the transition region $10^{-1} < Re_l < 10$ both friction and inertial effects determine the flow past the animals.

Depending on the Reynolds number regime, different forms of propulsion and lift have developed in nature. Bacteria and flagellates move with cilia and flagella. The oscillating motion of the cilia propels the protozoa forwards. In fish, this wavelike motion takes place only in the last third of the fish's body, so that their swimming using this method is slow. The greatest part of the propulsion of fast-swimming fish is achieved with the periodic flap of their tail fin. Sharks reach top speeds of up to $90 km/h$ by switching off the wavelike locomotion mode in the rear part of their bodies by means of pressure-controlled stiffening of their skin. The lift of fish in water is usually compensated by the swim bladder. Fast-swimming fish like sharks and whales compensate the lift with pectoral fins and with a vertical fin flap (see Figure 12.1) that generates the necessary lift as well as the propulsion.

The flight of insects with flapping wings of up to 1000 times per second developed already $3 \cdot 10^8$ years ago on earth, while the flight and gliding of birds appeared first $0.5 \cdot 10^8$ years ago. Compared to swimming, the flap of the wing in flight must generate both propulsion and lift at stable flight positions. This led to arched wing surfaces, whose continually varying pressure distribution on the upper and lower sides of the wing ensures the necessary lift. The wing flapping necessary for the propulsion of birds is stronger and at a higher frequency than the flapping of the fish's fin. As well as flying and gliding, most flying animals can also hover. This is used particularly during take-off and landing. The frequency of oscillation of the wings of the smallest mosquito is $1000 Hz$, that of the bee is $200 Hz$, of the colibri $45 Hz$, while that of the *condor* is $1.2 Hz$. In comparison, the tail fin of the blue whale oscillates at $0.25 Hz$.

Figure 12.2, with Figure 4.127, shows the flap of the wing of a seagull. At the beginning of the downward flap, the wing is at full stretch and moves without any forward component relative to the bird. In the middle of the downward flap, the tip of the wing is slightly *turned* and it generates the forward propulsion component. At the end of the downward flap, the wing is at full stretch and it generates lift along the entire span of the wing. At the beginning of the upward flap, the wing is bent, and at the same time the angle of attack increases to compensate for the loss of lift in the outer part of the wing. The wing moves slightly backwards and the tips of the wing are slightly splayed. The main feathers of the wing are in resting position. In the middle of the upward flap, the feathers are folded over each other. The backwards motion continues and the angle of attack is increased further. At the end of the upward flap, the wing is again stretched and the main feathers again oscillate forwards, as the downward flap begins.

The stability of bird flight is achieved by the tail feathers. The splaying of the tail feathers also permits abrupt flight manoeuvres such as braking, hovering, and gliding. The wings of birds are constructed for flight at high Reynolds numbers. The penetrability of the feathers, the slits in the front wings and the spreading of feathers at the back of the wing control the boundary layer and ensure that flow separation is avoided and the induced drag is kept small. By suitable shaping of the surfaces, such as crests at the leading edge and feather down, the friction drag is reduced and, in the case of the night owl for example, aerodynamic noise is reduced. A summary of the swimming and flight of animals is given in the book by *M. J. Lighthill* (1986).

The locomotion of protozoa takes place at a Reynolds number of 10^{-2} by means of transversal waves along the cilia, with increasing amplitude towards the end of the cilia (Figure 12.3). For a wave velocity V, the wave

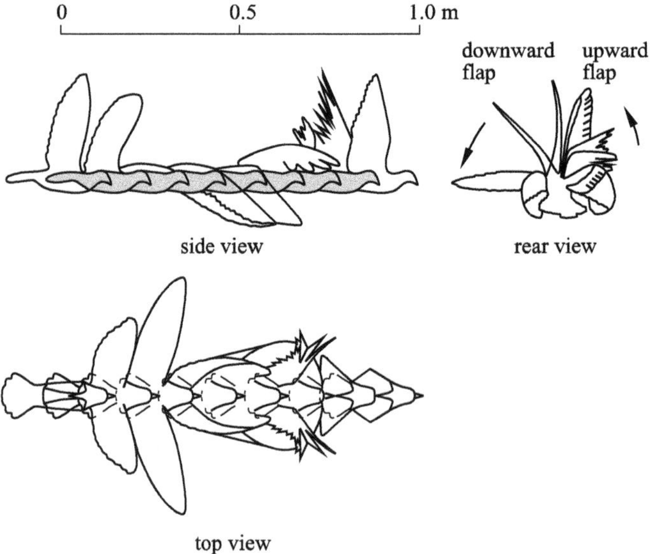

Fig. 12.2. Flap of the wing of a seagull, *J. Gray* (1968)

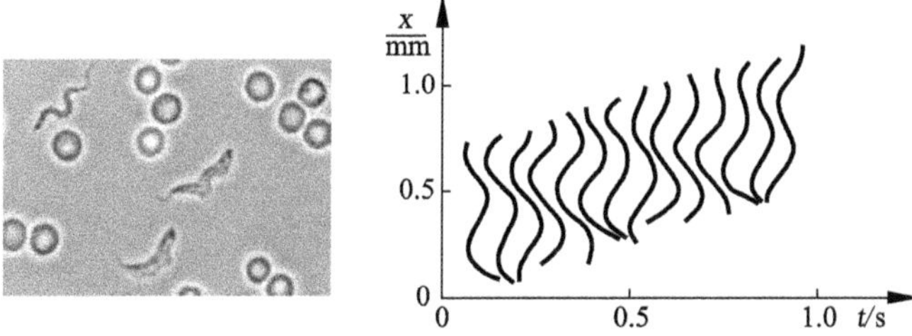

Fig. 12.3. Locomotion of protozoa and threadworms

motion induces a velocity of motion of the protozoa of the order $U = 0.2V$. Threadworms move in the same manner; with a length of about 1 mm, their Reynolds number is 1. The velocity of the wave along the body is about $V = 1\ mm/s$ and the resulting velocity of the threadworm is $U = 0.4V$. The reason for the higher velocity of the threadworm compared to that of the protozoa is that no additional head cell has to be moved (Figure 12.14). The wave amplitude at the end of the worm is much greater than at its head. Bigger worms with a length of 10 cm swim with Reynolds numbers of up to 10^3 and velocities of 10 mm/s. It is the transversal waves along the body that generate the propulsion. Eels use the transversal motion of their dorsal fin to swim slowly, while to swim at a greater velocity, they move their entire body in a wave form. Round single-celled organisms can also propel themselves forward by periodic thickening and thinning of their bodies. Changing the direction of the wave along the body enables the organism to move forwards and backwards.

At larger Reynolds numbers, the dominating inertial force means that wave motion of the entire body becomes inefficient. Therefore for *swimming* (Figure 12.4), only the last third of the body carries out wave motion, while the greatest part of the propulsion is generated by the periodic motion of

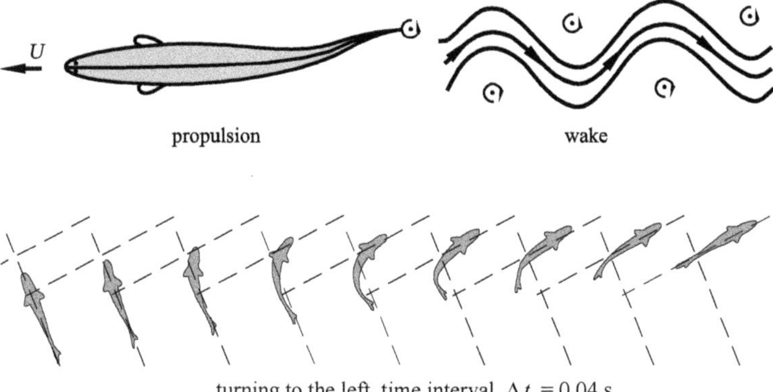

Fig. 12.4. Propulsion of fish

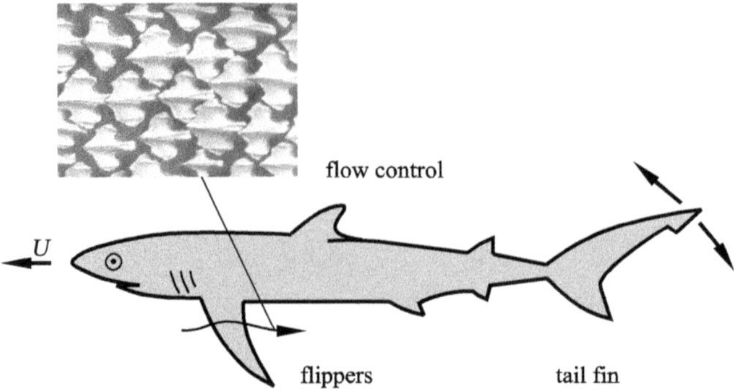

Fig. 12.5. Flippers and tail fin of the shark

the tail fin. This causes periodically separating vortices in the wake and thus flow losses. Therefore, depending on the Reynolds number, evolution has optimized motion in water by suitable shaping of the body to reduce the pressure drag, by adaptation of the texture of the surface of the fish's skin to reduce the friction drag and by suitable profiling of the tail fin to reduce the induced drag. Thus dolphins and penguins have an optimal body shape relative to the total drag; for the gentoo penguin the value is $c_w = 0.07$ at a Reynolds number of $Re_l = 10^6$. In spite of the *bulbous* body and the stabilizing back legs of the penguin, this value is very close to the value of the technical streamline body of a spindle of rotation, namely $c_w = 0.04$ (*W. Nachtigall* (2001)). The plumage of the penguin affects the viscous sublayer of the boundary layer so that the friction drag is reduced.

Dolphins achieve the same effect with a mucilaginous surface that dampens the laminar-turbulent transition in the boundary layer and, by the addition of tiny amounts of polymers into the water flowing past, reduces the friction drag.

Fast-swimming fish like the shark (Figure 12.5) impede the transverse component of the oscillating velocity in the viscous sublayer of the boundary layer by longitudinal grooves in their scales and thus achieve momentary top speeds of up to $90 km/h$.

Fish have additional flippers that equalize the rolling and yawing moments generated by the flapping of the fin. It is these that make swimming in one direction at all possible. In spite of the dominant inertial force at large Reynolds numbers, they also permit braking and abrupt changes of direction during swimming.

12.1.2 Biofluid Mechanics of Humans

The human body has several circulatory systems for which fluid mechanics is relevant. There is the respiratory system, the circulation of the blood and lymph and the water supply. We consider the cardiovascular circulatory system that ensures the transport of nutrients, gases, metabolic products and hormones between the organs of the human body. The most interesting part of the blood circulation from a fluid mechanical viewpoint is the circulation through the arteries, characterized by the pulsing intake flow, the secondary flows in the curves and branches of arteries, and the quasi-steady flow in the arterioles and capillaries with gas and material exchange.

The Reynolds numbers of the blood flow in the arteries in Figure 12.1 are between 100 and several thousand. The flow pulse of the heart causes a periodic laminar flow in the smaller arteries and a *transitional flow* in the larger arteries. The transition to the turbulent artery flow is introduced by temporary turning-point profiles. These occur in the unsteady backflow close to the wall of the arteries during the relaxation phase of the heart. The time of one cardiac cycle is however too short to form a fully developed turbulent flow. The smaller the branching of the arteries, the less the pulsing flow is noticeable.

In the curved arteries and in particular in the aorta, because of the centrifugal force *secondary flows* form. These were discussed in Section 4.2.7. A velocity component perpendicular to the streamlines develops and this causes a rotating flow in the direction of the outer wall. This acts to stabilize the transition process. The critical Reynolds number of the time-averaged velocity profile increases from 2300 for a straight pipe to up to 6000 for a curved pipe. The peak Reynolds numbers of the flow pulses of healthy humans are such that the secondary flow in the curves of the aorta prevents the onset of turbulence. In reality, the unsteady transitional flow described takes place in the boundary layer close to the wall during the braking phase of the pump cycle. However, the instabilities that occur are dampened after a short time by the change in time of the velocity profile.

The blood flow that leaves the heart is subdivided in up to 30 *branchings*, until we reach the *microcirculation* of several hundred million individual flows in arterioles with diameters of a few hundred micrometers and in capillaries with diameters of less than 10 micrometers.

From the exit from the ventricle into the aorta, as well as after every branching, an *intake flow* forms. The intake flow in a straight pipe of diameter D is about $0.03 \cdot Re_D \cdot D$. This means that the greatest part of the arteries after the branchings are characterized by intake flows and no averaged Poiseuille flow forms. If we consider the large bend in the aorta in Figure 12.1, the intake flow means that in spite of the large curvature we can expect no fully developed secondary flow.

The pressure pulse of the heart causes *expansion of the arteries* by about 2 %. The velocity of propagation of the pressure wave in the visco-elastic artery walls is about five times larger than the maximum blood velocity.

The first Fourier component of the pressure pulse (see equation (12.67)) with angular frequency ω is critically dependent on the ratio of the artery radius R and the oscillating *boundary layer thickness* $\sqrt{\nu/\omega}$. If we take the viscosity of the blood to be $4 \cdot 10^{-6} m^2/s$ and the angular frequency of the blood pulse to be $\omega = 8\ s^{-1}$, the boundary layer thickness ρ is about $0.7\ mm$. For the n-th Fourier mode of the blood pulse, the boundary layer thickness is to be multiplied by $\sqrt{n}$. For the large arteries, the ratio of the artery radius R to the boundary layer thickness ρ is of the order of magnitude 10. This implies that the velocity distribution over the cross-section of the arteries is almost uniform. Changes in the velocity distribution only cause changes in the wall boundary layer that makes up 10% of the artery radius. This means that, according to the Euler equation (5.72), almost the entire pressure gradient of the blood pulse is transformed into acceleration. Compared to the pressure gradient, the flow has a phase shift of almost 90°. In the boundary layer the phase shift of the wall shear stress is only 45°.

The *blood pulse* has an expansion velocity of $5\ m/s$ in the aorta. It is not simply a traveling wave going out from the heart. Every artery branching causes reflected waves, which are superimposed on the original pressure and velocity pulses. In the arteries this leads to an intermittent character between a traveling and a standing wave. The consequence of this is that the aorta acts as a *volume reservoir* for the heart output and ensures an almost continuous volume flux of the blood circulation.

The blood circulation of the human body is driven by the heart. The heart consists of two separate pump chambers, the left and right *ventricles* and *atria*, which are made out of cardiac muscle (Figure 12.6). The right

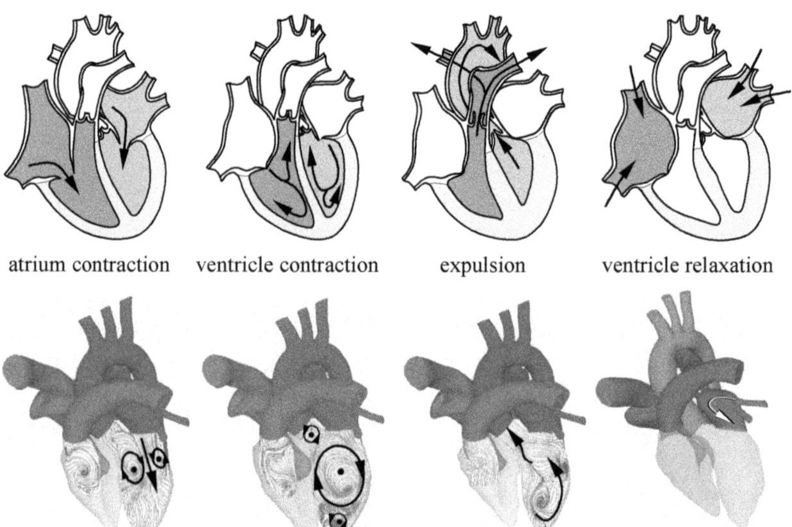

Fig. 12.6. Cross-sections and flow calculation of the heart during the four phases of the cardiac cycle

atrium contains blood that is weak in oxygen following systemic circulation. The right ventricle subsequently fills with blood from the right atrium, and then empties again into the pulmonary circulation on contraction. The re-oxygenated blood from the pulmonary circulation reaches the left atrium and is passed along from the left ventricle into the systemic circulation. The atria and ventricles are separated by atrioventricular valves, which regulate the filling of the heart ventricles. The right valve has three flaps and is therefore called the *tricuspid valve.* The left bicuspid valve has two flaps and is known as the *mitral valve.* The flaps ensure that the atria can fill with blood between the heart beats and prevent backflow of the blood when the ventricles contract. While the ventricles relax, the *pulmonary valve* prevents backflow of blood out of the aorta into the left ventricle.

During one cardiac cycle, the ventricles carry out a periodic contraction and relaxation. This pump cycle is associated with changes in the ventricle and arterial pressures. Figure 12.7 shows the pressure in the left chamber of the heart. The entire cycle can be split into four phases. The isovolumetric ventricle contraction is called the filling phase (1) and contraction phase (2), and the isovolumetric ventricle relaxation is called the evacuation phase (3)

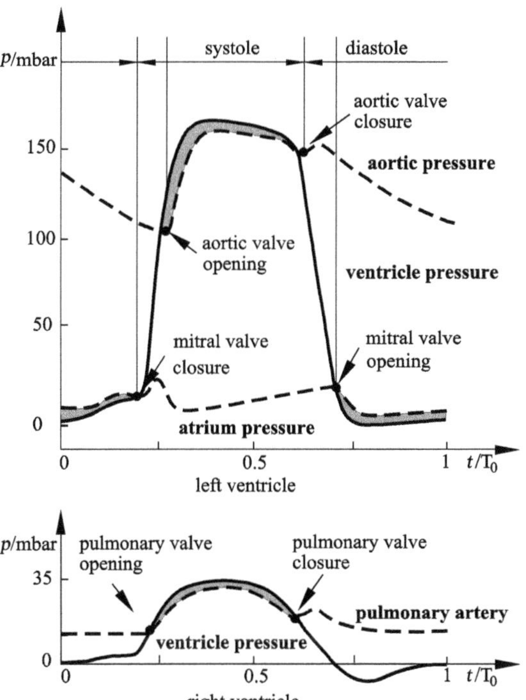

Fig. 12.7. Pressure in the aorta and pulmonary artery, left and right ventricle during one cardiac cycle, $T_0 = 0.8\ s$

and relaxation phase (4). Phases (2) and (3) of the ventricle contraction are called *systole*, and phases (4) and (1) of the ventricle relaxation are called *diastole*. The ventricle is filled during phase (4). The pressure at this point is only slightly higher in the left atrium than in the left ventricle. The mitral valve is therefore open, and the blood flows out of the pulmonary veins into the atrium and on into the left ventricle. As the filling volume increases and the ventricle expands, the ventricle pressure increases. The pressure in the aorta is considerably larger, so the aortic valve remains closed. The arterial pressure sinks continuously during the subsequent diastole, corresponding to the blood flow into the arterial vascular system. With the start of the ventricular contraction, the ventricle pressure increases above that of the atrium, so that the mitral valve closes. When the valve is closed, the ventricle contracts to retain a constant blood volume. While this increases the ventricle pressure to 166 mbar, the pressure decrease in the arteries continues. The aortic valve is opened when the ventricle pressure exceeds that in the aorta. Now a constant quantity of blood is forced out into the aorta. While the constant blood volume in the aorta is forced, the aortic pressure increases from its minimum value of 107 mbar to its maximum value of 160 mbar. After ventricle relaxation has begun, the ventricle pressure drops below that of the arteries, and the aortic and pulmonary valves close. The phase of isovolumetric relaxation now follows. The first phase of the diastole goes

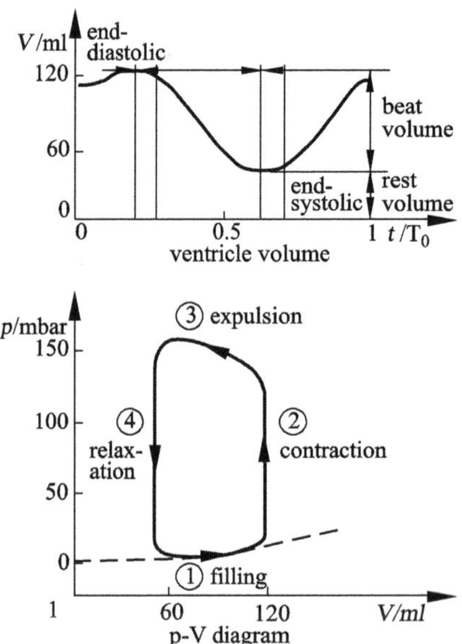

Fig. 12.8. Pressure-volume and volume diagram in the left ventricle during one cardiac cycle

on as long as the ventricle pressure is below the atrium pressure. Then the bicuspid valve opens, and the cardiac cycle starts the next filling phase.

The pressure–volume diagram in Figure 12.8 shows the filling of the left ventricle (1) along the rest expansion curve, the isovolumetric contraction (2), as well as the evacuation (3) and the isovolumetric relaxation (4). The enclosed surface is the systolic work done by the left heart ventricle of $1W$. When the body is under duress, the work diagram shifts along the rest expansion curve to large ventricle volumes and higher pressure. Increasing the heart filling leads to an increase in the cardiac work. When the aortic pressure is increased, the aortic valve opens later, so that the isovolumetric contraction phase reaches higher pressures.

The *systemic circulation* can be split into the blood distribution system, consisting of the *aorta*, large and small *arteries* and *arterioles*. These further divide into the *capillaries*, where gas and material exchange takes place in the microcirculation by diffusion. The blood flows back to the heart via the venules, small and large *veins*, and the *vena cava*.

The mean blood pressure on leaving the left ventricle is about 133 mbar. This drops to 13 mbar when the blood returns to the right ventricle. Figure 12.9 shows the mean pressure and pressure variations in the different artery regimes. Because of the elastic properties of the aorta, the pressure pulses between 120 mbar and 160 mbar around the mean value. In the large arteries, the amplitude of the pulsation initially increases, because of the wave reflection. It then sinks drastically to a mean value of 40 mbar in the arteriole region over a distance of a few millimeters. In the capillaries and venules, the pressure drop continues less sharply. Eventually, there is a pressure of 13 mbar to push the blood back to the right ventricle. In the large veins and the vena cava, there is no pulse and no considerable pressure drop. Simultaneously, pressure waves occur that are due to the pulsation of the right ventricle and move in the opposite direction to the flow of blood. The systolic pressure in the pulmonary arteries is quite small, about 20 mbar. A pressure drop of only 13 to 7 mbar is needed in order to overcome the flow drag in the lung volume, and so 13 to 7 mbar filling pressure remains for the left ventricle.

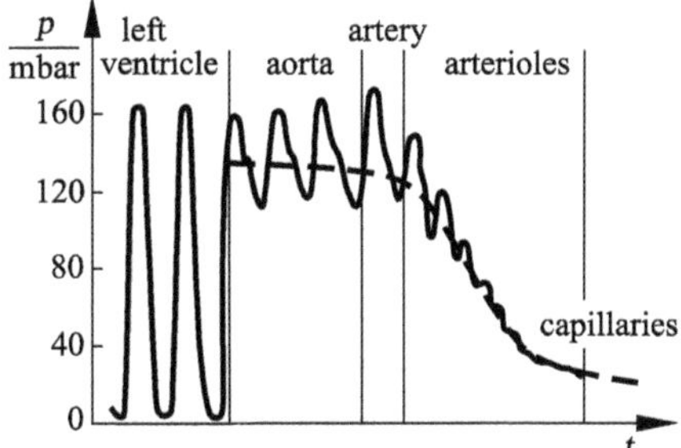

Fig. 12.9. Pressure in the arterial circulation

Because of their elasticity, the aorta and the large arteries act as a *volume reservoir*. The acceleration part of the blood pulse is reduced, and a higher pressure level is retained during the diastole and systole. This means that the flow in the arterial branches is smoother.

The wave form of the pressure and velocity pulse in the arterial branches is shown in Figure 12.10. Between each pressure pulse, the arteries contract by about 5% and so maintain the blood transport. The pressure pulse in the arteries is positive, even during the diastole of the heart. In contrast, a backflow occurs in the large arteries for a short time. The flow velocity is zero as the aortic valve is closed. The amplitude of the flow pulse decreases with increasing arterial branches, and the pulse width increases while a smaller backflow occurs. The forward motion of the pressure pulse through the arterial branches is initially associated with an increase in the pressure amplitude, which is caused by the arterial branches and also by the decrease in elasticity of the arterial walls. The flow profile in the branched arteries becomes more uniform.

The Reynolds numbers formed with the mean velocity are 3600 for the aorta, 500 for the large arteries, 0.7 for the arterioles, $2 \cdot 10^{-3}$ in the capillaries, 0.01 in the venules, 140 in the large veins, and 600 in the vena cava. Because of the unsteady intake flows and the secondary flows in the curves of the

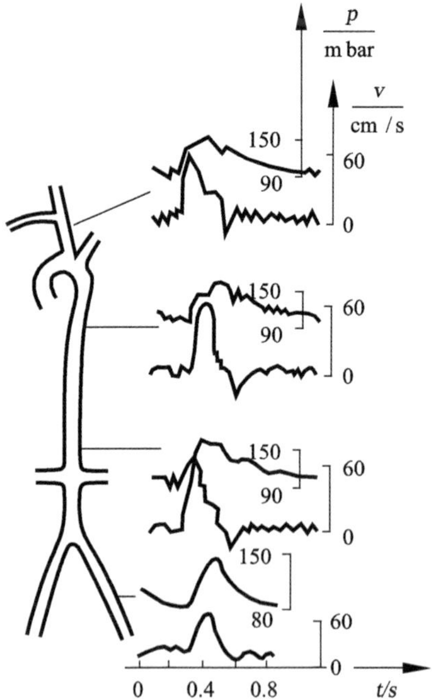

Fig. 12.10. Pressure and velocity waves in the arterial branches, *C. J. Mills et al.* (1970)

vessels, as described in the beginning of the chapter, a transitional laminar flow occurs in the vessel branches. The transition to turbulent flow takes place over a short time in the turning points in the velocity profile, close to the walls of the arteries.

12.1.3 Blood Rheology

The blood consists of the *blood plasma*, and suspended in it the red blood corpuscles (*erythrocytes*), white blood corpuscles (*leucocytes*), and platelets (*thrombocytes*). The blood plasma is the carrier fluid and is made up of 90% water, proteins, antibodies, and fibrinogens. The task of the blood is to ensure the supply to the cells and collection from the cells of nutrients, respiratory gases, minerals, enzymes, hormones, metabolic products, waste products, water, and heat. It is a transport system for the blood corpuscles that guarantees the immune reaction of the body and safeguards the circulatory system from injury. The mean volume of blood in a man is about 5 liters and in a woman about 4 liters. Of the entire circulatory system about 84% is essentially in the systemic vascular system, only 9% in the pulmonary circulation and 7% in the heart.

The behavior of blood flow is important for the flow in the heart and in the circulatory system. In particular, it has to be determined in which flow regimes and at which shear rates the Newtonian properties of the plasma or the non-Newtonian properties of the suspension have to be taken into account. These then determine the drag of the circulation that has to be compensated by the pump energy of the heart.

The viscosity of the blood can be considered only if the suspension occurs as a homogeneous liquid. This is true for blood in the large vessels. In the small vessels and particularly in the capillaries, the elastic erythrocytes with their diameter of 8 μm have to be considered as an inhomogeneity.

While the plasma can be treated to good approximation as a Newtonian fluid, the blood as a whole is a pseudoelastic thixotropic suspension. The viscosity of the suspension depends on the relative volume of all the suspended particles. The erythrocytes make up the largest part, with 99% by volume of all the particles and 40 – 45% by volume of the blood (hemato-value). The

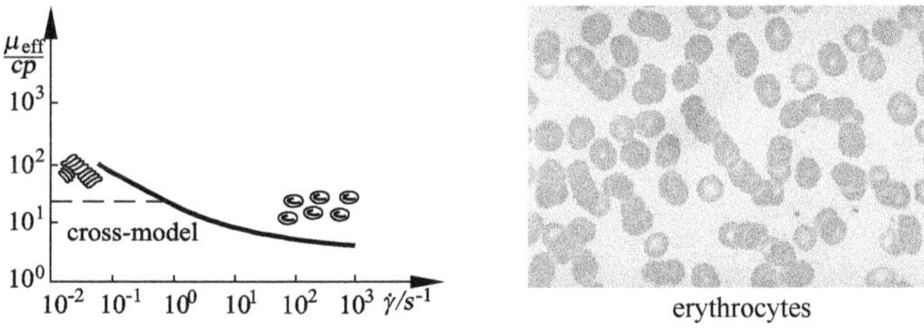

Fig. 12.11. Dependence of the viscosity of the blood μ on the shear rate $\dot{\gamma}$

thrombocytes and leucocytes make up less than 1% of the volume and have no effect on the rheology of the blood.

Figure 12.11 shows the dependence of the viscosity μ of the blood on the shear rate $\dot{\gamma}$. For the flow in vessels, the shear rate is $\dot{\gamma} = \partial u/\partial r$. In vascular branches, and in the aorta and ventricles, the dominant component of the shear rate tensor has to be chosen for $\dot{\gamma}$. In a wide range of varying velocity gradients, a drop in the viscosity of up to two orders of magnitude is noted. The region of velocity gradient in a healthy circulation varies between 8000 s^{-1} (arterioles) and 100 s^{-1} (vena cava). Therefore, in the asymptotic region, the velocity is almost constant. In the region of very high velocity gradients and therefore very large shear stresses, there is a deformation of the erythrocytes, which itself affects the viscosity of the blood suspension. At shear stresses over 50 N/m^2, the erythrocytes begin to pull apart in a spindle-like manner.

At shear rates of less than 1, such as those that occur in the backflow regions of an unhealthy circulation, aggregation of erythrocytes occurs. The cells pile up onto one another and form connected cell stacks that are linked together. However, in a healthy circulation system no aggregation can take place in the largest arteries. This is because the aggregation time is about 10 s, while the pulse is a factor of 10 shorter.

The dependence of the shear stress of the blood τ on the shear rate $\dot{\gamma}$ can be described to good approximation with the *Casson equation*

$$\sqrt{\tau} = \sqrt{\mu_{\text{eff}} \cdot \dot{\gamma}} = K \cdot \sqrt{\dot{\gamma}} + \sqrt{C} \tag{12.1}$$

Here K is the Casson viscosity and C the deformation stress of the blood. Fitting this equation to experimental results leads to the equation

$$\sqrt{\frac{\tau}{\mu_p}} = 1.53 \cdot \sqrt{\dot{\gamma}} + 2 \quad , \tag{12.2}$$

with the plasma viscosity $\mu_p = 0.012$ p. For shear rates larger than 100 the blood behaves as a Newtonian medium.

The non-Newtonian properties of the blood as it flows through the vessels lead to a reduction of the erythrocytes close to the vessel walls and therefore to a reduction in the viscosity. This alters the velocity profile close to the

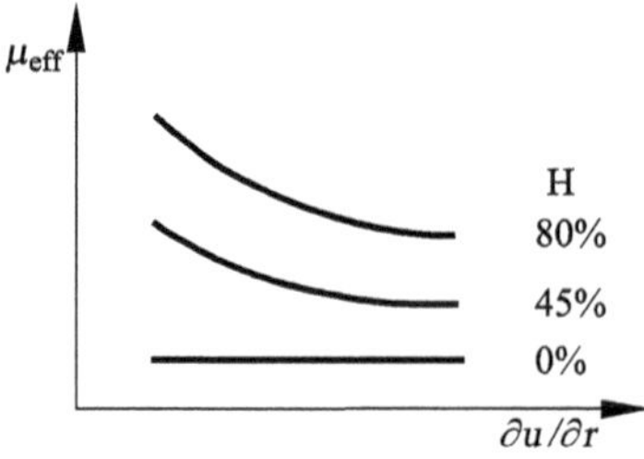

Fig. 12.12. Effect of the hematocrit value H on the viscosity μ_{eff} of the blood

wall and so also the drag of the blood. Segregation close to the wall leads to a plasma zone that is almost cell-free, which can be computed with the plasma viscosity μ_p. For steady Poiseuille flow, this leads to a velocity profile as has already been described in Figure 4.98.

For shear rates $1 < \dot{\gamma} < 50$ the slope $n = -0.28$ in (4.50) may be used to calculate approximately, and for $\dot{\gamma} > 100$ the slope $n = 1$ (Newtonian medium) may be used.

The Casson equation (12.1) leads to a modified ansatz for the viscosity:

$$\mu_{\text{eff}} = \frac{(\mathrm{K} \cdot \sqrt{\dot{\gamma}} + \sqrt{\mathrm{C}})^2}{\dot{\gamma}} \quad . \tag{12.3}$$

The modified *Cross model* is also used for the numerical calculation of the pulsating blood flow:

$$\mu_{\text{eff}} = \mu_\infty + \frac{\mu_0 - \mu_\infty}{(1 + (t_0 \cdot \dot{\gamma})^{\mathrm{b}})^{\mathrm{a}}} \quad . \tag{12.4}$$

The constants $\mu_\infty = 0.03$ p, $\mu_0 = 0.1315$ p, $t_0 = 0.5$ s, $a = 0.3$, $b = 1.7$ were determined with experiments by *Liepsch et al.* (1991). Here μ_∞ is a limiting viscosity for high shear rates $\dot{\gamma}$ and μ_0 a limiting viscosity for small values of $\dot{\gamma}$.

The viscosity of blood μ_{eff} changes with the *hematocrit value* H of human blood. The hematocrit value is defined as the ratio of the volume fraction of the red blood corpuscles to the total volume of the blood. For $H = 0$ the constant viscosity of the Newtonian blood plasma is found (Figure 12.12). For the hematocrit value $H = 45\%$ we obtain the viscosity shown in Figure 12.11. The viscosity of the blood increases further for larger values of the hematocrit value.

Nature optimizes the transport of oxygen in the circulation and in doing so has to accommodate two opposing demands. On the one hand a large hematocrit value is necessary to transport as much oxygen as possible. On the other hand a small value is required so that the blood viscosity sinks and

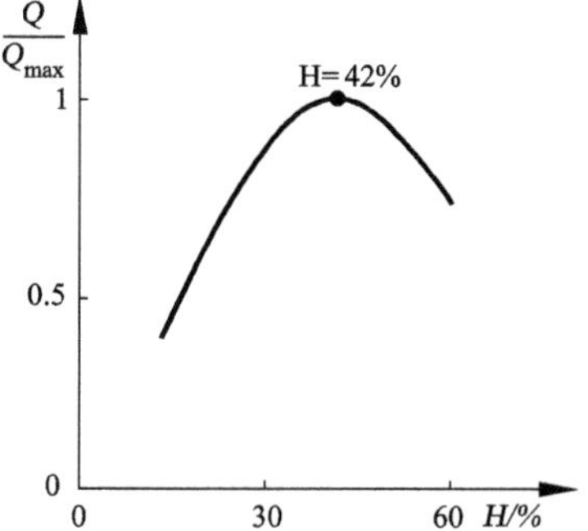

Fig. 12.13. Dependence of the particle flux $Q/Q_{\max}$ on the hematocrit value H of the blood

the volume flux in the vessels increases. Therefore it is not the overriding goal to achieve oxygen bonding by as many red blood corpuscles as possible. It is more important to optimize the flowing behavior of the blood, whereby it is essential to transport a sufficiently large amount of oxygen without impacting the other blood functions too much. As seen in Figure 12.13, in human circulation the maximum volume flux takes place at a hematocrit value of $H = 42\%$.

12.2 Swimming and Flight

The swimming and flight of animals was introduced in Sections 12.1.1 and 4.4.1. In this section, we treat the fluid mechanical fundamentals needed to determine biofluid mechanics at small Reynolds numbers. As examples we consider the motion of protozoa, the swimming and flow control of fish at large Reynolds numbers, and the aerodynamics of bird flight.

12.2.1 Motion of Protozoa

The motion of swimming animals of $1mm$ in size or smaller at Reynolds numbers $Re_l \leq 1$ is determined by friction, with the inertial forces playing a less important role. Thus the change in momentum or angular momentum caused by the animal may be neglected compared to the friction forces. The relative forward motion of the center of gravity of the animal has a translational velocity U due to the periodic curvature of the animal's body with a wave propagation velocity V.

In every fluid element there is a pressure force $-\nabla p$ in equilibrium with the friction force $\mu \cdot \Delta u$. The continuity equation of incompressible flow (5.3) holds:

$$\nabla \cdot \boldsymbol{u} = 0 \quad . \tag{12.5}$$

Neglecting the inertial forces, the Navier-Stokes equation (5.20) yields

$$-\nabla p + \mu \cdot \Delta \boldsymbol{u} = 0 \quad . \tag{12.6}$$

With the assumptions stated above, the dimensionless Navier-Stokes equation (5.82) yields

$$-\nabla p + \frac{1}{Re_l} \cdot \Delta \boldsymbol{u} = 0 \quad , \tag{12.7}$$

where the Reynolds number is formed with the characteristic length of the animal l.

Equations (12.5), (12.6) and (12.7) lead to the conclusion that the Laplace equation is valid for the pressure:

$$\Delta p = 0 \quad . \tag{12.8}$$

Therefore the pressure is a harmonic function for every inertia-free flow. From (12.5) and (12.6) it can be derived that the velocity is a biharmonic function that solves the equation

$$\nabla^4 \boldsymbol{u} = 0 \tag{12.9}$$

The wave motion of the animal's body can be represented by superposition of point forces:

$$\boldsymbol{f} \cdot \delta(\boldsymbol{r}) \quad , \tag{12.10}$$

where $\boldsymbol{F}$ is the force per unit volume, δ is the delta function and $\boldsymbol{r}$ is the position vector from the position of the action of the force. The force equilibrium on the fluid element for the force distribution in (12.10) yields the Navier-Stokes equation (12.6):

$$-\nabla p + \mu \cdot \Delta \boldsymbol{u} + \boldsymbol{f} \cdot \delta(\boldsymbol{r}) = 0 \quad . \tag{12.11}$$

With the continuity equation (12.3) we obtain

$$\Delta p = \nabla \cdot (\boldsymbol{f} \cdot \delta(\boldsymbol{r})) = 0 \quad , \tag{12.12}$$

whose solution is the classical dipole field

$$p = -\nabla \cdot \left(\frac{\boldsymbol{F}}{4 \cdot \pi \cdot \boldsymbol{r}} \right) \quad . \tag{12.13}$$

If we consider a protozoa with a flagellum of length l (Figure 12.14), we can write the form of the motion wave along the flagellum as

$$(x, y, z) = (X(s), Y(s), Z(s)) \tag{12.14}$$

Here s is the longitudinal coordinate along the flagellum with

$$X'^2(s) + Y'^2(s) + Z'^2(s) = 1 \quad . \tag{12.15}$$

The locomotion wave has the wavelength

$$X(s + \Lambda) = X(s) + \lambda \quad , \quad Y(s + \Lambda) = Y(s) \quad , \quad Z(s + \Lambda) = Z(s) \; , \tag{12.16}$$

with Λ the wavelength along the curved flagellum, $\lambda = \alpha \Lambda$ the wavelength in the direction of propulsion and $\alpha < 1$ the length contraction of the flagellum due to the wave motion.

In a reference system moving with the propulsion wave, the flagellate moves tangentially along the wavefront (12.14) with velocity c. At time t we have

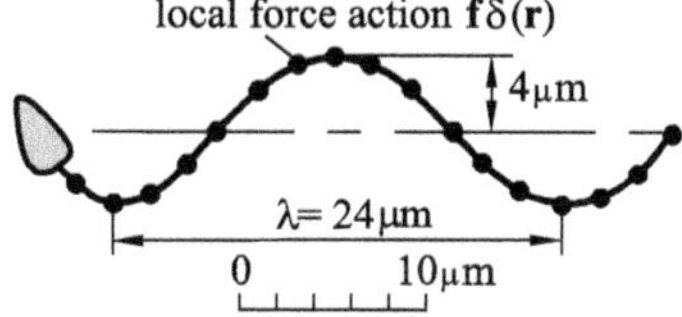

Fig. 12.14. Locomotion with flagella

$$(x,y,z) = (X(s-c\cdot t), Y(s-c\cdot t), Z(s-c\cdot t)) \quad . \tag{12.17}$$

c is the velocity along the curved body of the flagellate. The relationship with V, the wave velocity in the moving reference system U, is:

$$V = \alpha \cdot c \quad , \tag{12.18}$$

as the wave period can be described with Λ/c or $\lambda/V = \alpha\Lambda/V$. The propulsion wave moves downstream with the relative velocity $V - U$. This yields the velocity of the flagellate relative to the fluid as the vector sum of the velocity c along the forward tangent and the velocity $(V - U, 0, 0)$. The component along the backward tangent is

$$(V-U)\cdot X'(s-c\cdot t) - c \quad , \tag{12.19}$$

while the component along the backward normal is

$$(V-U)\cdot \sqrt{1-X'^2(s-c\cdot t)} \quad , \tag{12.20}$$

Here X' and $\sqrt{1-X'^2}$ denote the cosine directions of the tangent and the normal.

The propulsion P of the protozoan can be written as the x-component of the sum of the tangential forces F_t (12.19) and the normal forces F_n:

$$P = \int_0^l (F_\mathrm{t}((V-U)\cdot X'(s-c\cdot t) - c)\cdot X'(s-c\cdot t) + F_\mathrm{n}((V-U)\cdot(1-X'^2(s-c\cdot t))))\cdot \mathrm{d}s \quad , \tag{12.21}$$

with

$$\int_0^l X'(s-c\cdot t)\cdot \mathrm{d}s = \alpha\cdot l = V\cdot\frac{l}{c} \quad ,$$

where αl is the length of the wave motion in the direction of the propulsion. Using the definition

$$\int_0^l X'^2(s-c\cdot t)\cdot \mathrm{d}s = \beta\cdot l$$

we obtain

$$P = F_\mathrm{t}\cdot l\cdot((V-U)\cdot\beta - V) + F_\mathrm{n}\cdot l\cdot(V-U)\cdot(1-\beta) \quad . \tag{12.22}$$

This propulsion P must be in equilibrium with the drag of the head of the protozoan moving with U. For the drag force we write

$$F_\mathrm{n}\cdot l\cdot U\cdot\delta \quad , \tag{12.23}$$

with δ the ratio of drag of the head and drag of the normal motion of the flagellate.

Assuming that the head is a sphere with radius R, equation (12.22) with (12.23) yields

$$\frac{U}{V} = \frac{(1-\beta)\cdot(1-\frac{F_\mathrm{t}}{F_\mathrm{n}})}{1-\beta+\frac{F_\mathrm{t}}{F_\mathrm{n}}\cdot\beta+\delta} \quad , \tag{12.24}$$

for the ratio of the velocity of motion U to the wave velocity V. As $\beta < 1$ (with no motion $\beta = 1$), U/V varies between 0 and the maximum value

$$\left(\frac{U}{V}\right)_\mathrm{max} = \frac{1-\frac{F_\mathrm{t}}{F_\mathrm{n}}}{1+\delta} \quad . \tag{12.25}$$

The maximum value is obtained for $\beta \to 0$. For a spherical head with radius $R = 1\ \mu m$, the wavelength $\lambda = 45\ \mu m$, the amplitude $4\ \mu m$ and Stokes drag law $F = 6\cdot\pi\cdot\mu\cdot R\cdot U$ we obtain the values $\delta = 0.11$ and $\beta = 0.65$.

12.2.2 Swimming of Fish

Fish swim in the Reynolds number regime $10^4 < Re_l < 10^8$. Compared to the previous section, now the inertial force dominates the friction force and, for fast-swimming fish, a turbulent boundary layer forms downstream along the body. As explained in the introductory section, the propulsion of a fish, when it swims slowly, is due to the *wave motion* of the last third of its body, and, when it swims quickly, to the *flap of the tail* (see Figure 12.15).

In a simplified view of the *wave motion* of the fish in a reference frame moving with the propulsion velocity U, the propulsion force P is in equilibrium with the drag W of the fish. The efficiency η is formed with the average values

$$\eta = \frac{U\cdot\bar{P}}{\bar{A}} \quad , \tag{12.26}$$

where $\bar{A}$ is the mean power of the fish.

The deflection of the cross-sectional area in the z-direction about the equilibrium position of the fish is given by $h(x,t)$. Because of the wave motion of the end of the fish, a vertical velocity w arises:

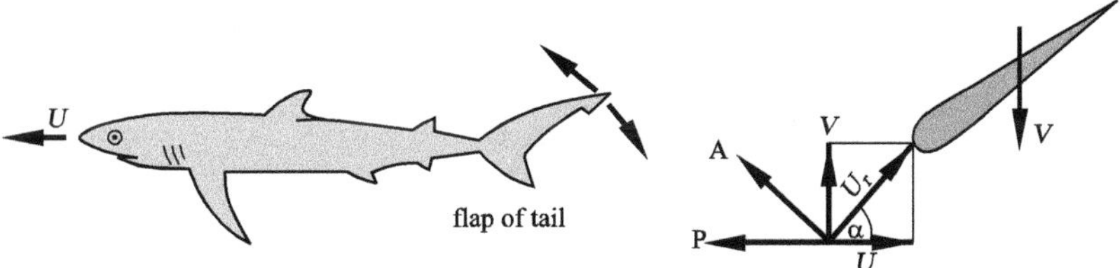

Fig. 12.15. Flap of a fish's tail fin

$$w = \frac{\partial h}{\partial t} + U \cdot \frac{\partial h}{\partial x} \quad , \tag{12.27}$$

with transverse velocity $\partial h/\partial t$. If the wave velocity V is only slightly larger than the propagation velocity U, for a constant amplitude of the propagation wave, we obtain

$$\frac{\partial h}{\partial t} + V \cdot \frac{\partial h}{\partial x} = 0 \quad .$$

This yields

$$w = \frac{\partial h}{\partial t} \cdot \frac{V - U}{V} \quad . \tag{12.28}$$

This means that to obtain a high efficiency η, the forward motion $w/(\partial h/\partial t)$ must be small. However, the value must be large enough to overcome the drag of the fish W.

On the basis of this simplified view of the wave motion of the fish, *M. J. Lighthill* (1960) developed the linear theory of the longitudinally stretched body. He assumed that the changes in the cross-section of the fish are small and that the wave motion perturbs the flow only slightly. However these assumptions are only fulfilled under certain conditions for fast-swimming fish.

In order to take account of the *flap of the tail fin* and the turbulent wake flow, the continuity equation (5.39) and the Reynolds equations (5.40) - (5.42) for incompressible flow have to be solved:

$$\nabla \bar{\boldsymbol{u}} = 0 \quad , \tag{12.29}$$

$$\frac{\partial \bar{\boldsymbol{u}}}{\partial t} + (\bar{\boldsymbol{u}} \cdot \nabla)\bar{\boldsymbol{u}} = -\frac{1}{\rho} \cdot \nabla \bar{p} + \nu \cdot \Delta \bar{\boldsymbol{u}} - \frac{\partial \overline{u'_{\mathrm{i}} \cdot u'_{\mathrm{j}}}}{\partial x_{\mathrm{j}}} \quad . \tag{12.30}$$

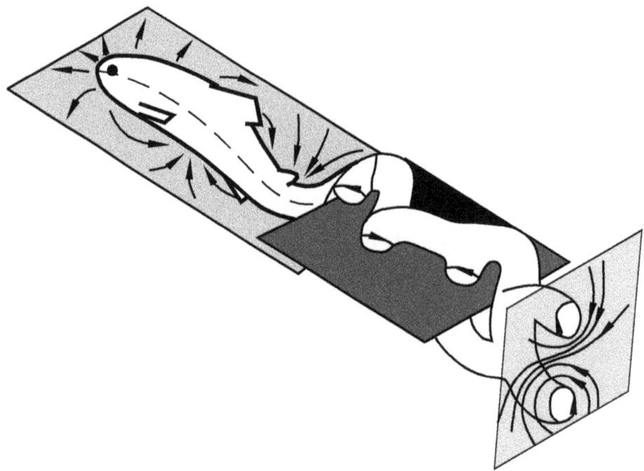

Fig. 12.16. Wake flow of a fish, *W. Nachtigall* (2001)

To model the Reynolds stresses (5.104), in Section 5.4.5 a nonlinear two-equation turbulence model or a large-eddy simulation with a fine-structure model is selected.

The tail fin with symmetric profile in Figure 12.15 moves with velocity V, and generates a propulsion velocity U. The fish controls the forward motion such that the resulting outward flow relative to the tail fin U_r has a positive angle of attack α compared to the body. The resulting lift A, whose component in the direction of swimming $A \cdot \sin\alpha$ generates the propulsion, is perpendicular to the relative velocity U_r. The propulsion is therefore $P = U \cdot A \cdot \sin\alpha$. In order to generate this propulsion, the tail fin moves sideways against the force $A \cdot \cos\alpha$, requiring a power of $V \cdot A \cdot \cos\alpha$. For slow-swimming fish this is only about 0.6 mW.

The vortices in the three-dimensional wake of the tail fin (Figure 12.16) are generated in both fulcrums of the tail fin. Alternating vortices of positive and negative directions of rotation arise. Between the fulcrums a propulsion jet occurs that acts against the drag of the wake. The tail fin of the fish is optimized so that the induced drag and friction drag are as small as possible, but large enough that propulsion can be generated.

12.2.3 Flow Control

Investigations on sharks have shown that fast-swimming fish have scales with longitudinal grooves. These prevent the onset of transverse turbulence in the viscous sublayer of the boundary layer and lead to a relaminization of the boundary layer and thus to a reduction of the friction drag.

Figure 12.17 shows the *longitudinal grooves* of the *shark scales* as well as their technical application in the form of riblet foil. The riblets are triangular ridges spaced 60 μm apart that reduce the wall shear stress τ_w by up to 8 %. Figure 12.18 shows the result of the numerical solution of the Navier-Stokes equation (5.20) directly at the surface. The dark regions have high velocity fluctuations and the lighter regions have lower fluctuations. A picture appears of longitudinal structures that cause an increased wall shear stress τ_w. The longitudinal rills suppress the v'-component of the velocity fluctuations and

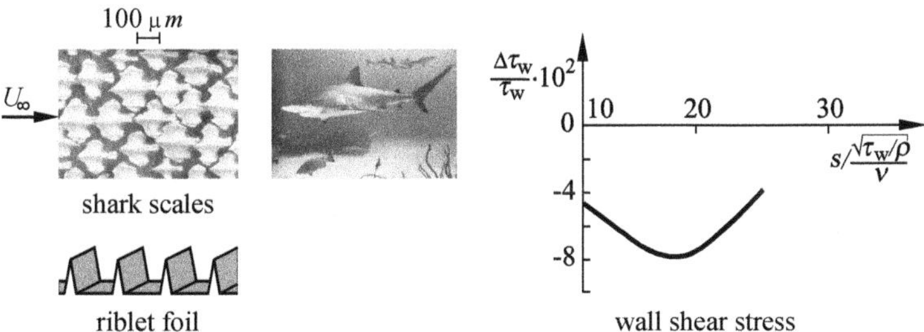

Fig. 12.17. Shark scales and riblet foil, *D. W. Bechert et al.* (2000)

thus the region of high shear rate in the viscous sublayer. This leads to relaminization of the turbulent boundary layer, permitting the shark to reach momentarily peak speeds of up to $90km/h$.

There is another method of relaminization in nature where particular surface properties delay the onset of the laminar-turbulent transition (see Section 4.2.4). Dolphins swim much faster than they ought to be able to due to their muscle mass. The *mucous membrane* of the *dolphin* consists of peg-shaped ridges in the outer skin, as shown in Figure 12.19, interspersed with a flexible fluid-containing lower skin. This *damping skin* dampens all perturbations in the boundary layer. Thus perturbations that introduce the laminar-turbulent transition are delayed.

In technically realizing such a damping layer, liquid with an adjustable rigidity is enclosed between a smooth and a knobbed plate. Relaminization by an oscillating surface has been demonstrated experimentally. Figure 12.19 shows how the reduction of the drag coefficient c_w of the artificial damping skin depends on the Reynolds number. By suitable adjustment of the amplitude and phase of the damping, particularly at large Reynolds numbers, the drag coefficient of laminar flow can be attained. As would be expected, if the artificial damping skin is placed in a turbulent flow the relaminization effect breaks down.

Dolphins are thought to use another method to relaminize the viscous sublayer too. By adding high-molecular polymers to a liquid, drag reductions of over 50% can be reached. Such polymers are secreted in small amounts from the mucous membrane of the dolphin. This relaminization effect is exploited in, for example, the Alaska pipeline where only a few parts per million of such polymers are added to viscous cold oil in order to reduce the pump power by 30%.

A further possibility to reduce the drag is with the introduction of *air bubbles* into the boundary-layer flow. The air bubbles reduce the mean density close to the surface and thus generate a drag reduction of over 50%. The *penguin* exploits this as he swims on the surface of the water and carries air bubbles with him in his coat, which are emitted as he swims. He reaches top

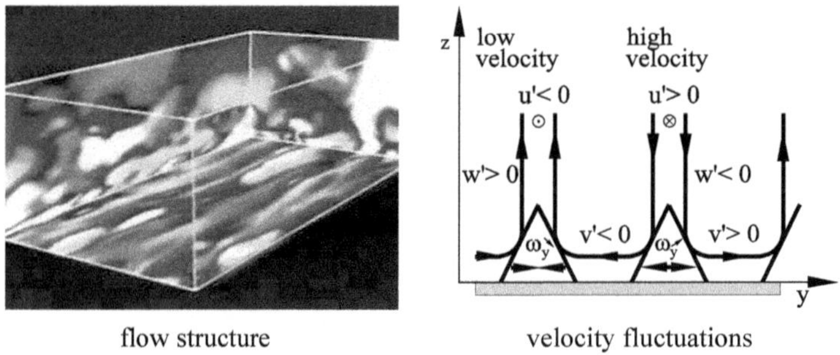

Fig. 12.18. Transverse turbulence of the viscous sublayer

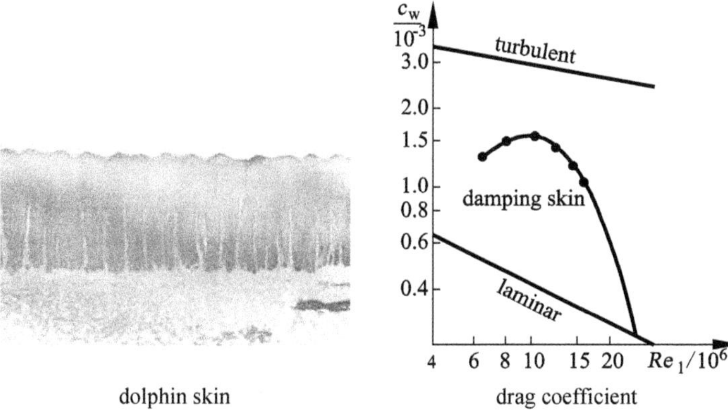

Fig. 12.19. Damping skin of the dolphin, *W. Nachtigall* (2001)

speeds of up to 25 km/h with a drag coefficient of $c_w = 0.03$ and Reynolds number $1.6 \cdot 10^6$.

12.2.4 Bird Flight

Bird flight has already been described in Section 4.4.1 and in the introduction 12.1.1. In this section we expand on the basic relations of *forward flight, gliding flight* and *hovering*.

The flapping of a wing during forward flight is shown in Figure 12.2 with the example of the seagull. A slice along the profile of a pigeon's wing is shown in Figure 12.20. The profile of the middle part of the wing is greatly arched. This is to generate a large lift when the wing is flapping and when the bird is gliding. Towards the tip of the wing the profiling and the arching of the profile of the wing decrease continuously. This promotes efficient propulsion. Gliding is performed at an angle of attack between 3° and 5°. During the flap of the wing and during maneuvers, the entire polar diagram of Figure 12.20 is passed through (see also Figure 4.137). The bird continuously optimizes the necessary lift with the necessary propulsion for the flight, always avoiding exceeding the limiting angle of attack of about 25°. Figure 12.21 shows two characteristic pressure distributions at different angles of attack. They show the typical form of the subsonic profile as known from Section 4.4.2.

For horizontal flight the bird needs power P to compensate the weight G at flight velocity V. With the rate of descent V_s and the drag W, the propulsion power is

$$P = G \cdot V_{\mathrm{s}} = W \cdot V \quad . \tag{12.31}$$

The drag W of the bird and the lift A can be calculated from the coefficients c_w and c_a:

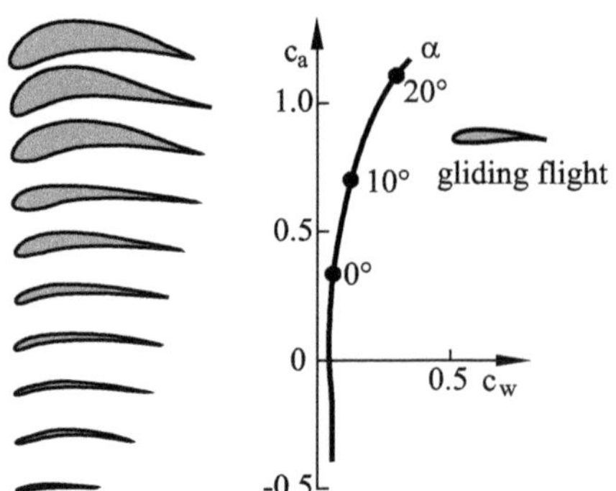

Fig. 12.20. Slices through the wing profile and polar diagram of a pigeon in gliding flight

$$
\begin{aligned}
W &= \frac{1}{2} \cdot c_\mathrm{w} \cdot \rho \cdot S \cdot V^2 \quad , \\
A &= \frac{1}{2} \cdot c_\mathrm{a} \cdot \rho \cdot S \cdot V^2 \quad ,
\end{aligned}
\tag{12.32}
$$

with the wing surface S.

This leads to

$$
W = \frac{c_\mathrm{w}}{c_\mathrm{a}} \cdot A \quad , \tag{12.33}
$$

with the gliding number c_a/c_w introduced in Section 4.4.1. In horizontal flight the weight G is compensated by the lift A. This yields

$$
W = \frac{c_\mathrm{w}}{c_\mathrm{a}} \cdot G \tag{12.34}
$$

and the propulsion power

$$
P = \frac{c_\mathrm{w}}{c_\mathrm{a}} \cdot G \cdot V \quad , \tag{12.35}
$$

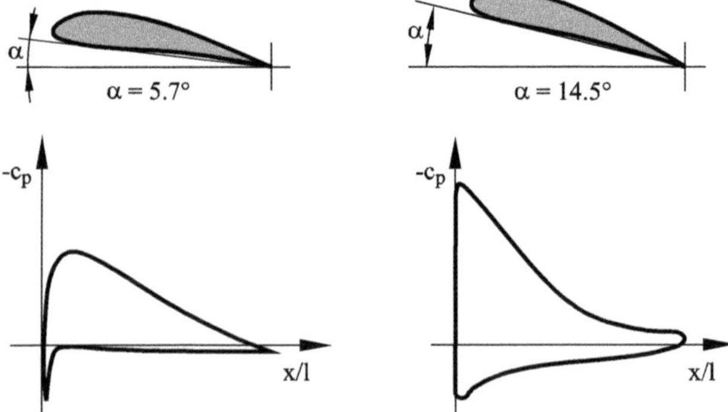

Fig. 12.21. Dependence of the pressure distribution on the angle of attack

Estimations of the necessary power for the bird lead to the conclusion that the available muscular force cannot be sufficient to sustain flight. The same issue is found for the swimming of the dolphin. In Section 12.2.3 this led to the conclusion that by suitable flow control by the dolphin's mucous membrane, the drag is reduced so that swimming at high speeds is made possible even at low muscular force. It is the same with bird flight. By spreading the end feathers of the wing, the induced drag of the boundary vortex (see Figure 4.139) is reduced during flight. When flapping the wing, the bird's moveable coat permits optimal flight control that minimizes flow separation and, because of the partial porosity of the feathers during, for example, the upward flap, also reduces the drag. This a bird can reach the flight velocity V with a smaller muscular force than that predicted by (12.35).

In *horizontal flight* the lift of the wing is in equilibrium with the weight of the bird. Therefore the flight velocity V can be expressed as a function of the weight G and the surface of the wing S

$$\frac{1}{2} \cdot c_{\mathrm{a}} \cdot \rho \cdot S \cdot V^2 = G \quad ,$$
$$V = \sqrt{\frac{2}{c_{\mathrm{a}} \cdot \rho} \cdot \frac{G}{S}} = \mathrm{K} \cdot \sqrt{\frac{G}{S}} \quad , \tag{12.36}$$

with the constant $K = \sqrt{2/c_a \rho}$.

The power necessary for flight $P = W \cdot V$ (12.35) is therefore proportional to the product of the drag W and the square root of the surface load G/S.

The force balance in *gliding flight* is shown in Figure 12.22. The gliding line is inclined at an angle α to the horizontal and the bird glides with outstretched wings. The weight of the bird G acts vertically downwards and has the components $P = G \cdot \sin(\alpha)$ and $N = G \cdot \cos(\alpha)$. Steady gliding is found when the drag W is equal and opposite to the part of the weight that propels the bird forwards $G \cdot \sin(\alpha)$, and the lift A is equal to $G \cdot \cos \alpha$. At smaller gliding angles α the lift can be set equal to the weight:

$$W = G \cdot \sin(\alpha) = \frac{1}{2} \cdot c_{\mathrm{w}} \cdot \rho \cdot S \cdot V^2 \quad ,$$
$$A = G = \frac{1}{2} \cdot c_{\mathrm{a}} \cdot \rho \cdot S \cdot V^2 \quad . \tag{12.37}$$

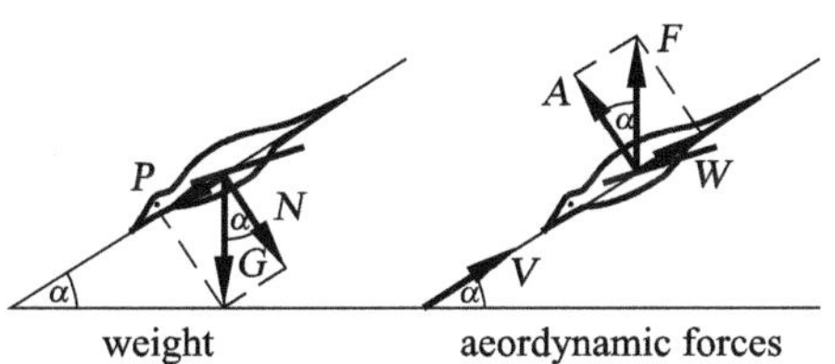

Fig. 12.22. Equilibrium of forces during gliding flight

The vector sum of the lift and the drag, F, is opposite in direction to the weight. Therefore the gliding angle satisfies

$$\tan(\alpha) = \frac{c_\mathrm{w}}{c_\mathrm{a}} \quad . \tag{12.38}$$

The gliding angle is therefore independent of the weight of the bird and of the surface area of the wing. It is only dependent on the profile of the wing. With (12.34) we obtain the velocity of the bird

$$V = \sqrt{\frac{2 \cdot G \cdot \cos(\alpha)}{c_\mathrm{a} \cdot \rho \cdot S}} = \mathrm{K} \cdot \sqrt{\frac{G}{S}} \quad , \tag{12.39}$$

Therefore birds with a high weight and small wings glide faster than light birds with big wings. At small gliding angles, the lift is equal to the weight and the lift coefficient is inversely proportional to the velocity

$$c_\mathrm{a} = \frac{2 \cdot G}{S \cdot \rho \cdot V^2} \quad . \tag{12.40}$$

Therefore every change in the gliding angle causes a change in the flight velocity. If we take into account the separating behavior of the bird's wing in Figure 12.20, there is a maximum lift coefficient and a corresponding minimum flight velocity V_min at which gliding is possible.

Figure 12.23 shows the dependence of the rate of descent V_s on the flight velocity V for a buzzard. The minimum rate of descent is 0.8 m/s at a flight velocity of 15 m/s. The buzzard then has a lift coefficient of $c_a = 1.8$ and a drag coefficient of $c_w = 0.06$ at the flight Reynolds number $Re_l = 2 \cdot 10^5$.

When *soaring* in the rising air of a thermal, the rate of descent of the bird is compensated by the vertical component of the wind. The circling of the thermal permits the eagle, for example, to cover great distances. Seagulls use the luff of a cliff to rise without muscular effort. Further examples are lee-wave soaring in updrafts behind mountain ridges and the dynamic soaring of the albatross who uses the updrafts in front of waves in the sea.

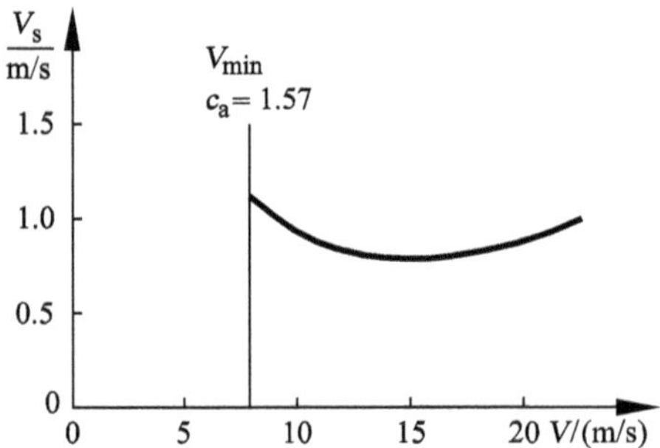

Fig. 12.23. Rate of descent V_s of the buzzard

In soaring and in forward flight, the bird has a considerable range of maneuverability, to retain the flight direction and to compensate for air turbulence. The inclination and motion of the wing relative to the wind velocity are constantly being adapted. Changes in direction in the flight demand a turning motion of the body, so that a centrifugal force Z acts. This can be written as a relation between the weight G and the flight velocity V:

$$Z = \frac{G \cdot V^2}{r} \quad , \tag{12.41}$$

where r is the radius of the curve.

If Φ denotes the angle at which the wing is inclined, we have

$$\tan(\Phi) = \frac{Z}{G} = \frac{V^2}{r} \quad . \tag{12.42}$$

However, in all flight positions, the vertical component of the lift must be equal to the weight of the bird. Thus the total lift of the wing must be larger when the bird is flying in a curve than when it is flying horizontally. For angle of inclination Φ, the total lift is $A = G/\cos\Phi$. If the angle is $\Phi = 60°$, the total lift of the wing must double in order to ensure flight in a straight line. Therefore the surface load of the wing increases considerably during curved flights, in order to achieve the necessary lift and increased angle of attack of the wing.

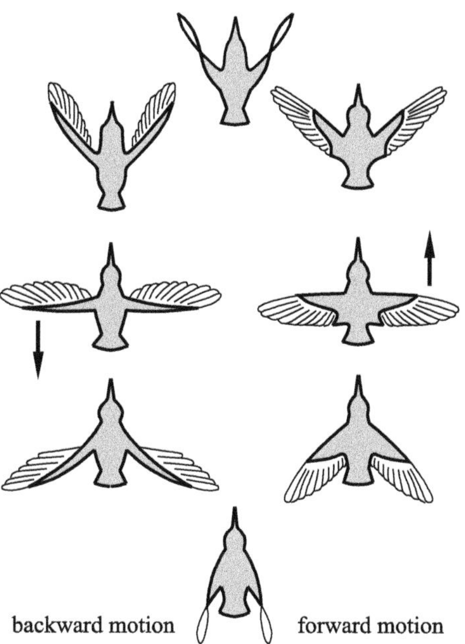

Fig. 12.24. Hovering

In *formation flight* over large distances, birds fly in a characteristic V-formation. Each bird uses the lift zone of the boundary vortex of the preceding bird, as in Figure 4.139, in order to fly large distances while saving energy. The displacement of each bird relative to the next is 1/4 of the wing span. In formation flight birds can theoretically extend their flight distance by 70 % of the single flight distance for the same muscular effort.

As well as flying forwards and gliding, most birds are able to *hover*. Figure 12.24 shows the forward and backward flap of the wing during one hover cycle. At the correct angle of attack, this forward and backward flapping ensures the lift necessary for hovering. The entire wing rotates along the longitudinal axis. The main feathers of the wing are tilted backwards and downwards during the upward flap, while during the downward flap they are tilted forwards. The action of the main feathers corresponds to the horizontal oscillating tail fin of, for example, the whale, that was treated in Section 12.2.2.

12.3 Human Heart Flow

Following the overview of the heart functions in the circulatory system in Section 12.1.2, this section will describe in detail the physiology and anatomy of the human heart with the interaction of electrical impulses, the electro-mechanical coupling, and the pulsing, three-dimensional flow.

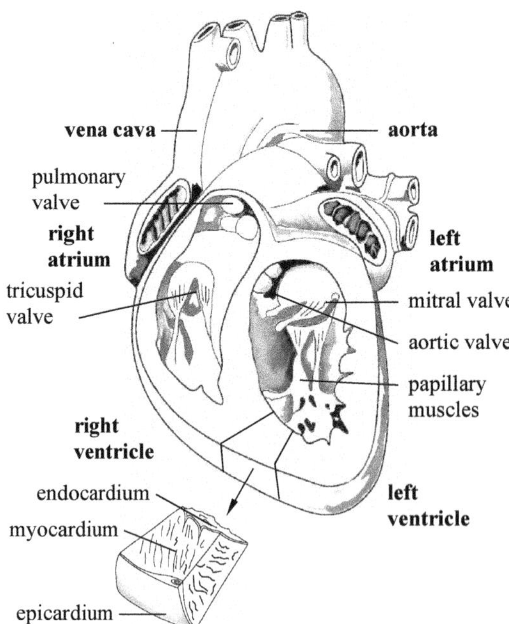

Fig. 12.25. Inner view of the heart

12.3.1 Physiology and Anatomy of the Heart

Figure 12.25 shows an *internal view of the heart* as it is shown in textbooks on medical physiology. The left and right atria of the heart are separated by the atrium septum, while the ventricle septum divides the two ventricles of the heart. The muscular heart wall is called the myocardium. It is surrounded by the endocardium on the inside and by the epicardium on the outside. The heart is in a sack of tissue called the pericardium. As shown in Figure 12.26, three groups of *muscle fibers* wind around both ventricles, while a further group of muscle fibers is wound around only the left ventricle. The cardiac muscles cells are oriented tangentially to the the heart rather than radially. Since the electrical resistance is lower along a muscle fiber, this effects the electrical impulses of the heart muscles.

Filling of the left and right ventricles from the atria is controlled by the mitral valve with two flaps and the tricuspid valve with three flaps. The flaps of the valves are very thin, so that they close quickly at the start of the ventricular contraction. They are held by tendon threads, which use papillary muscles to stop the valves turning inside out at high pressure. As the ventricle relaxes, the pulmonary valve prevents backflow of blood out of the pulmonary artery, and the aortic valve prevents backflow out of the aorta. Both valves are made of three semilunar sacks of connective tissue. These are more stable than the atrioventricular flap valves, because of the higher pressure that acts on the semilunar valves during the longest period of the heart beat.

The vena cava and the coronary sinus carry deoxygenated blood from the systemic circulation into the right atrium. At their openings into the right atrium there are two further valves, the Eustacian valve and the Thebesian valve, respectively. As the atrium contracts, these prevent backflow into the lower pressure vein circulation. Four pulmonary veins carrying oxygenated blood from the lungs to the left ventricle empty into the left atrium. In contrast to the right atrium, the left atrium has no backflow valves.

The total volume of the heart is about 750 ml for a man and 550 ml for a woman. With cardiovascular fitness training and the associated increased intake of oxygen during the load on the heart, the volume of the heart can increase to 1400 ml to 1700 ml.

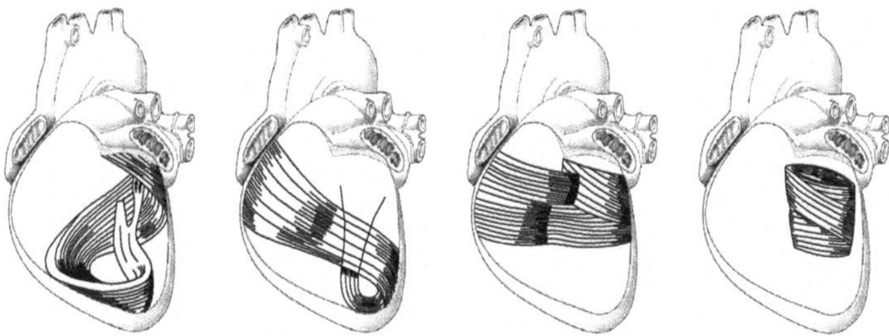

Fig. 12.26. Orientation of the cardiac muscle fibers

This is shown in the pressure-volume diagram of Figure 12.27. With increasing performance of the ventricle, the p - V curves are shifted to higher pressures and expulsion volumes. They are bounded by the end-diastolic ED and end-systolic ES pressure-volume curves. The area within each p - V curve gives the work done by the ventricle. The *Frank-Sterling law* states that the work done by the ventricle increases with increased filling volume of the ventricle. This has to do with the mechanical properties of the cardiac muscle and permits the heart to continuously adapt to different positions of the body, different efforts and different frequencies of breathing.

The mechanical contraction of the cardiac muscle is controlled by *periodic electrical impulses*. It starts with the excitation of the sinoatrial node (Figure 12.28), which carries out cyclical electrical depolarization and polarization, and therefore has the function of the primary pacemaker. During the depolarization phase, the discharge extends across the conduction paths with a velocity of 1 m/s into the surrounding muscles of the atria, which then contract. The electrical impulse of the sinoatrial node is delayed in the ventricular node. This delay permits optimal filling of the ventricles during contraction of the atria. The impulse passes along the His nerve fibers and the sides of the chamber with a velocity of $1-4$ m/s and reaches the ventricle muscles after about 110 ms. In the direction of the ventricle, the bundle of His divides into the left and right sides of the chamber.

These branch out in Purkinje fibers, which run right below the epicardium in each heart chamber. They first pass along the septum toward the apex of the heart and from there along the ventricle walls to the base of the heart. As the ventricles begin to contract, the contraction in the atria is finished, thanks to the delay of the conduction in the ventricular nodes. At this point all nerve cells in the impulse conduction system, apart from the impulse-forming cells in the sinoatrial nodes and ventricular nodes, can be spontaneously depolar-

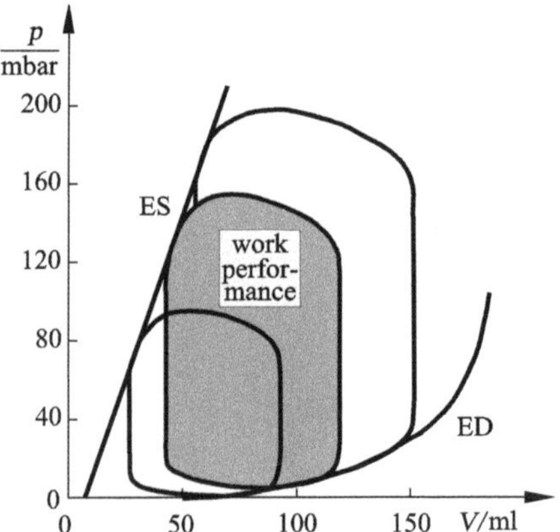

Fig. 12.27. Pressure-volume diagram for the left ventricle

ized. The ventricular depolarization in the electrocardiogram in Figure 12.28 takes less than 0.1 s.

Nerve cells and hormonal influences outside the heart affect the electrical impulses and cause different pulse frequencies. They modify the electrical conductivity and therefore the velocity of the depolarization wave through the heart. The cycle of depolarization and polarization generates a small electrical potential, which can be measured on the surface of the body. Figure 12.28 shows a typical electrocardiogram. The depolarization of the atria causes a small deflection, called the P-wave. After a pause of about 0.2 s, this is followed by a strong deflection due to the depolarization of both ventricles (QRS). The T-wave then follows, caused by renewed polarization of the ventricles.

As the mitral valve closes, the pressure in the left ventricle rises. This is associated with a sound wave that is detected as the first *heart beat.* This induces the systole, the contraction of the ventricle. At the second heart beat the diastole, the phase of ventricle relaxation, begins.

12.3.2 Structure of the Heart

In order to calculate the flow in the heart, a model of the geometry of the ventricles and of the cardiac valves during one cardiac cycle is needed. This is worked out using the methods of structural mechanics.

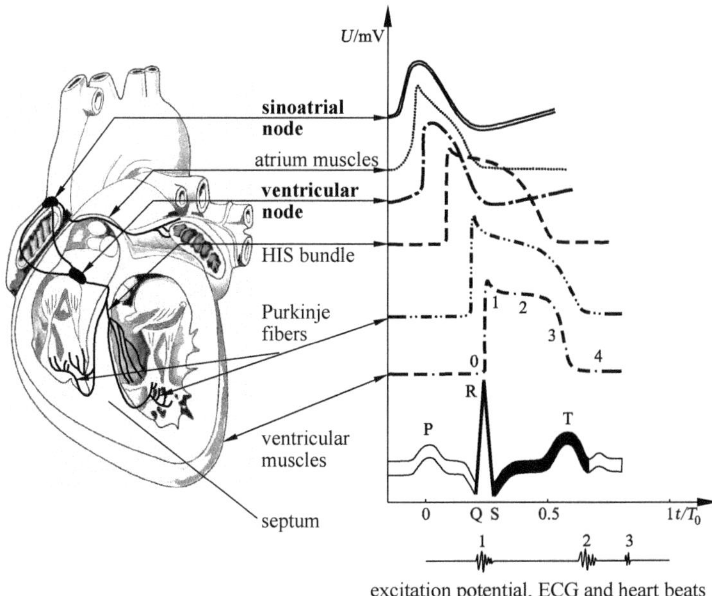

Fig. 12.28. Electrical impulse conductors exitation potential and echocardiogram (ECG) in the heart

Figure 12.29 shows a simplified model of the ventricular motion as well as real horizontal sections of human heart. During the contraction phase, the mitral and tricuspid valves are closed. The aortic and pulmonary valves are open. The muscle fibers of both ventricles contract. The left ventricle pumps blood low in oxygen into the lung. The pressure in the left ventricle is much larger than that in the right ventricle, as shown in Figure 12.7. For this reason, the left ventricle retains an almost elliptical cross-section during the contraction phase, while the right ventricle arranges itself around the left ventricle.

The motion of the ventricle walls is mainly radial, and because of the higher pressure in the left ventricle, it is greater in the left than in the right ventricle. The radial motion is accompanied by a shortening of the heart in the longitudinal direction. Because of the *spiral-like* arrangement of some of the muscle fibers (Figure 12.26), a rotating motion is superimposed onto the longitudinal motion. The shear stress distribution in the ventricles is therefore inhomogeneous and anisotropic (Figure 12.31).

The basis of the mathematical description of the ventricular motion is the *equation of motion* of structural mechanics. This is then numerically solved with finite element methods. The deformation velocity vector $\boldsymbol{v} = (v_1, v_2, v_3)$ is

$$\rho \left(\frac{\partial v_{\mathrm{i}}}{\partial t} + v_i \cdot \frac{\partial v_{\mathrm{i}}}{\partial x_{\mathrm{i}}} \right) = \frac{\partial \sigma_{\mathrm{ij}}}{\partial x_{\mathrm{j}}} + f_{\mathrm{i}} \quad , \tag{12.43}$$

with the stress tensor σ_{ij}, the external volume specific force f_{i}, and density ρ of the myocardium. The stress tensor σ_{ij} can be written down for an elastic body, assuming small deformations as a linear function of the rate of deformation tensor e_{kl}:

$$\boldsymbol{\sigma} = \boldsymbol{\sigma}_{\mathrm{ij}} = c_{\mathrm{ijkl}} \cdot e_{\mathrm{kl}} \quad . \tag{12.44}$$

Here c_{ijkl} is the tensor of the elastic constants that have to be determined for the heart.

For biological bodies, including the heart and the blood vessels, the stress tensor (12.44) can be taken to be approximately quasi-linear. The elastic deviation of each point in the body can then be determined relative to a

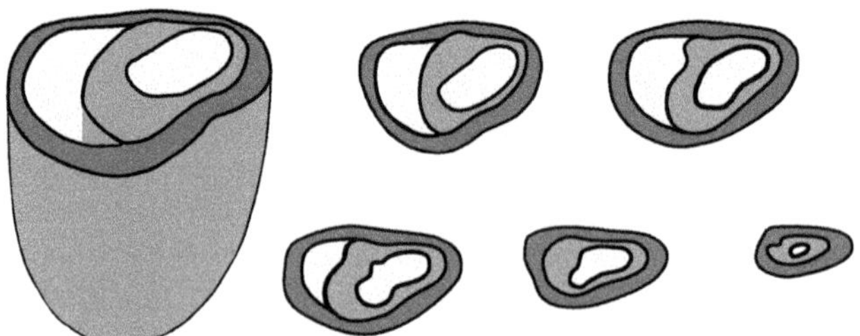

Fig. 12.29. Forms of ventricular contraction

ground state. For the details of tensor notation, we refer to the book by *Y. C. Fung* (1990).

For an elastic biological body that is undergoing a finite deformation $u_i = x_i - a_i$ (coordinate a_i before and x_i after the deformation), a strain–energy function $\rho_0 \cdot W(E_{11}, E_{12}, \cdots)$ exists. Its derivative leads to the Kirchhoff stress tensor

$$S_{\mathrm{ij}} = \frac{\partial\left(\rho_0 \cdot W\right)}{\partial E_{\mathrm{ij}}} \quad , \tag{12.45}$$

with the Green's strain

$$E_{\mathrm{ij}} = \frac{1}{2} \cdot \left(\delta_{\alpha\beta} \cdot \frac{\partial x_\alpha}{\partial a_{\mathrm{i}}} \cdot \frac{\partial x_\beta}{\partial a_{\mathrm{j}}} - \delta_{\mathrm{ij}}\right) = \frac{1}{2} \cdot \left(\frac{\partial u_{\mathrm{j}}}{\partial a_{\mathrm{i}}} + \frac{\partial u_{\mathrm{i}}}{\partial a_{\mathrm{j}}} + \frac{\partial u_\alpha}{\partial a_{\mathrm{i}}} \cdot \frac{\partial u_\beta}{\partial a_{\mathrm{j}}}\right) \tag{12.46}$$

The Kirchhoff stress tensor S_{ij} can be transformed into the Cauchy stress tensor σ_{ij} using the following relation:

$$\sigma_{\mathrm{ij}} = \frac{\rho}{\rho_0} \cdot \left(S_{\mathrm{ij}} \cdot \left(\delta_{\mathrm{i}\beta} \cdot \frac{\partial u_{\mathrm{j}}}{\partial a_\alpha} + \delta_{\mathrm{j}\alpha} \cdot \frac{\partial u_{\mathrm{i}}}{\partial a_\beta} \cdot \frac{\partial u_{\mathrm{j}}}{\partial a_\alpha}\right) \cdot S_{\alpha\beta}\right) \quad . \tag{12.47}$$

Here ρ and ρ_0 are the material densities in the deformed state and in the ground state.

According to *Y. J. Fung* (1993), the following simplified strain–energy function can be used for the heart during the filling phase:

$$\rho_0 \cdot W = \frac{c}{2} \cdot \left(e^Q - Q - 1\right) + \frac{q}{2} \tag{12.48}$$

Here c is a constant, and q, Q are quadratic forms of the Green's tensor:

$$\begin{aligned} Q =& k_{11} \cdot E_{11}^2 + k_{22} \cdot E_{22}^2 + k_{33} \cdot E_{33}^2 + 2 \cdot k'_{12} \cdot E_{11} \cdot E_{22} + 2 \cdot k'_{23} \cdot E_{22} \cdot E_{33} \\ &+ 2 \cdot k'_{31} \cdot E_{33} \cdot E_{11} + k_{12} \cdot E_{12}^2 + k_{23} \cdot E_{23}^2 + k_{31} \cdot E_{31}^2 \quad , \\ q =& b_{11} \cdot E_{11}^2 + b_{22} \cdot E_{22}^2 + b_{33} \cdot E_{33}^2 + 2 \cdot b'_{12} \cdot E_{11} \cdot E_{22} + 2 \cdot b'_{23} \cdot E_{22} \cdot E_{33} \\ &+ 2 \cdot b'_{31} \cdot E_{33} \cdot E_{11} + b_{12} \cdot E_{12}^2 + b_{23} \cdot E_{23}^2 + b_{31} \cdot E_{31}^2 \quad , \end{aligned}$$

with the material constants $k_{\mathrm{ij}}, b_{\mathrm{ij}}$. The units of c and b_{ij} are those of stress; k_{ij} are dimensionless. For $c = 0$ the linear Hooke's law is obtained.

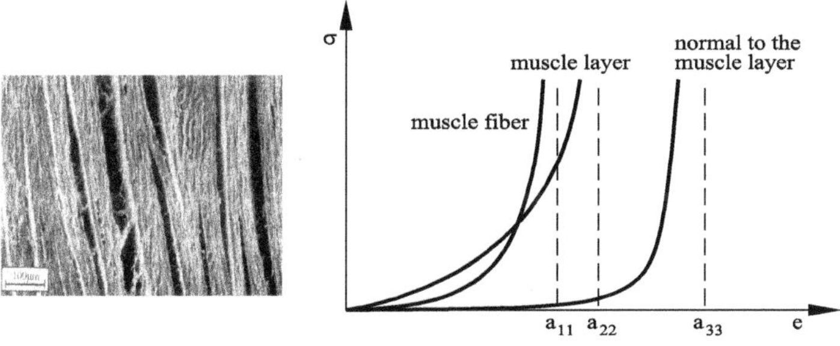

Fig. 12.30. Muscle layer in the ventricle, and axial stress-strain diagram

Structure-mechanical modeling of the myocardium is based on the stress measurements on thin muscle fiber layers of animal hearts. The myocardium has nonlinear and anisotropic stress–strain behavior. Figure 12.30 shows the axial stress–strain curves in a thin layer of muscle along the muscle fiber, in the muscle layer and normal to the muscle layer. The greatest variation in material properties of the myocardium is in the maximum strain a_{ii} along the selected axes. If the myocardium probe along the muscle fibers is extended, the limiting value of the strain is 1.3. In the direction perpendicular to the muscle fibers in the muscle layer, the limiting value 1.5 is found. The stress values perpendicular to the muscle layer are considerably smaller than those along the horizontal axis. Allowance must be made for these nonlinear anisotropic material properties of the myocardium in modeling the structure of the heart.

Numerous simplifications for the heart have been published. *J. P. Hunter et al.* (1997) and *J. P. Hunter* and *B. H. Smaill* (2000) used the following simplified ansatz of the strain–energy function for their simulations:

$$\begin{aligned} W = & k_{11} \cdot \frac{E_{11}^2}{|a_{11} - E_{11}|^{b_{11}}} + k_{22} \cdot \frac{E_{22}^2}{|a_{22} - E_{22}|^{b_{22}}} + k_{33} \cdot \frac{E_{33}^2}{|a_{33} - E_{33}|^{b_{33}}} \\ & k_{12} \cdot \frac{E_{12}^2}{|a_{12} - E_{12}|^{b_{12}}} + k_{13} \cdot \frac{E_{13}^2}{|a_{13} - E_{13}|^{b_{13}}} + k_{23} \cdot \frac{E_{23}^2}{|a_{23} - E_{23}|^{b_{23}}} \quad . \end{aligned} \tag{12.49}$$

Here the strain–energy function is separated into the individual parts of the stresses along each material axis. The a_{ij} are the poles of the limiting strains, the b_{ij} the curvature of the stress–strain curve for each deformation axis, and the k_{ij} the weighting factors of each deformation mode. Equation (12.49) consists of the six parts of the deformation modes of the Green's strain E_{ij}. The first three terms are the axial modes of the deformation and the remaining terms are the shear deformations between the material axes.

The strain-energy function (12.49) is the first order of an expansion about the poles of the limiting strains, where the vector product between the different modes of the axial and shear deformations is neglected. Further expansion of the myocardium structure model where the vector product is taken into account requires more extensive measurements of the myocardium.

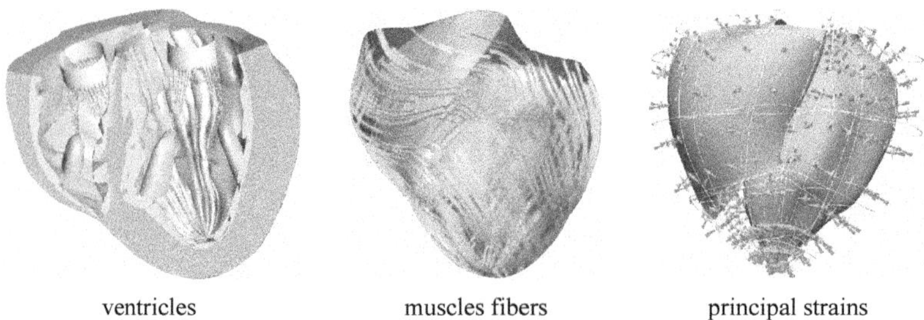

Fig. 12.31. Finite element model of the heart structure, *J.P. Hunter* (2001)

Figure 12.32 shows the finite element model of the heart ventricles, the orientation of the muscle fibers, and the principal strains at the end of diastole. The finite element model was developed by *J. P. Hunter et al.* (1993, 1997, 2000) on the basis of finite elasticity theory and the energy function (12.49). Stress measurements on active muscles fibers have shown that the muscle fiber forces behave more isotropically in the orthogonal direction than in the transverse direction.

Using the further developed model of the extension–energy function (12.49), the stress distribution on the surface of the heart was numerically computed with finite element discretization. The regions of larger and smaller muscle fiber stress are shown in Figure 12.32 with isolines for different phases of the cardiac cycle. Solid lines show large extension stresses, dashed lines large compression stresses. Initially, the heart muscle is relaxed, and the stresses are small. The progress of the compression stress from the atria during the filling phase can easily be seen. The stress then passes into the ventricles during the expulsion phase.

12.3.3 Excitation Physiology of the Heart

In addition to the description of the electrophysiology of the heart in Section 12.3.1, Figure 12.28 also shows the electric excitation potential in the individual region of the heart. The action potential U inside and outside the muscle cells was measured with microelectrodes.

At the start of the electrical excitation (0), the cardiac muscle cells are depolarized, and the potential difference across the cell membranes increases from −90 mV to +20 mV (1). The depolarization of the cardiac muscle cells is based on the opening of ion channels in the cell membrane (see, for example, *J. Malmivno, R. Plonsey* (1995)). The activation of the depolarization takes place within 1 ms. The mechanical contraction of the cardiac muscle cells is time delayed. There is a rapid drop of the activation potential, and repolarization is induced. This is delayed in phase (2) and reaches the original value via the drop (3). In this phase the action potential in the muscles is initiated, and the maximum of the muscular contraction is reached in phase (3). Repolarization takes place within 0.3 s, while the depolarization pulse lasts only 1 ms.

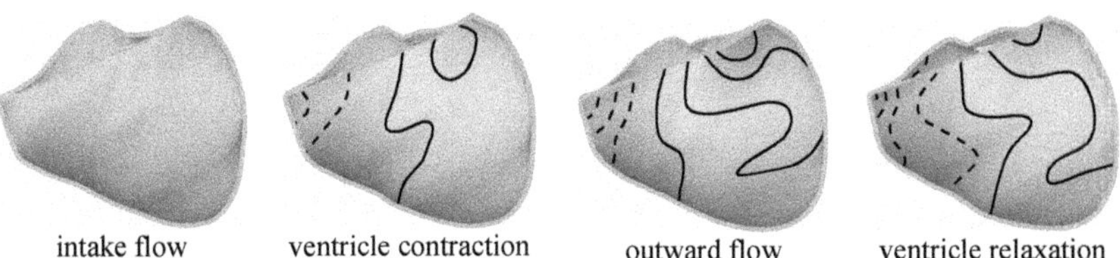

Fig. 12.32. Stress distribution in the muscle fibers on the surface of the heart, *J.P. Hunter* (2000)

Figure 12.33 shows a longitudinal section of the excitation in the heart in relation to an echocardiogram ECG. The excitation of the heart begins at the sinoatrial node and then propagates over the atria. This is associated with the T-wave in the ECG. The PQ time interval follows with the delayed excitation of the His conduction system. The excitation of the ventricles begins at the left side of the ventricular septum with the negative Q-deflection in the ECG. A short time later the walls of the right and left ventricle are excited from the inside out, including the apex of the heart. This leads to the R-spike in the ECG with positive polarity. The propagation of the excitation in the ventricles ends at the base of the left ventricle with the negative S-spike. After the excitation of the ventricles is complete, the entire surface of the heart is negatively charged. This phase in the excitation cycle is associated with the ST-stretch in the ECG. The repolarization phase of the heart begins in the subendocardial layers of the myocardium and progresses towards the endocardium and there is a field strength component that points away from the negative endocardial layers that are still excited towards the already positive regions that are no longer excited. The positive T-wave is associated with the repolarization phase.

Electrochemical investigations of the cardiac muscle cells show that the different regimes of the action potential are related to sodium Na^+ and potassium Ka^+ ion channels in the cell. Calcium Ca^{2+} ions in the cell membranes cause stimulation of the contraction in the muscle cells. In this manner, the form of the action potential affects the contraction behavior of the cardiac

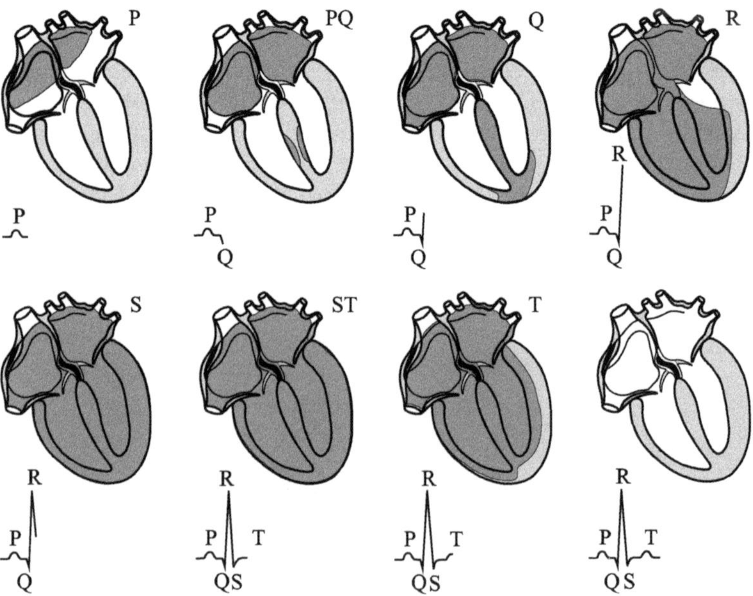

Fig. 12.33. Excitation of the heart and echocardiogram ECG

muscle cells in the different regions in the heart. The depolarization wave moves from the endocardium to the epicardium. The repolarization wave moves in the opposite direction.

Mathematical modeling of the depolarization wave and its expansion in the cardiac muscle cells requires modeling the nonlinear coupling of the depolarization excitation model with a model for the excitation expansion. Expansion with velocities between 0.03 (sinoatrial nodes) and 0.6 m/s (atrium and ventricle) can be computed either via a system of individual coupled cells or as a continuum.

The mathematical description of the excitation expansion in the heart is carried out with a system of nonlinear partial differential equations:

$$\frac{\partial u_{\mathrm{i}}}{\partial t} = f_{\mathrm{i}}(u_1, \cdots, u_{\mathrm{n}}) + D_{\mathrm{i}} \cdot \Delta u_{\mathrm{i}} \quad , \qquad \mathrm{i} = 1, \cdots, \mathrm{n} \quad . \tag{12.50}$$

The u_i are the n variables, $f_i(u_1, \ldots, u_n)$ the nonlinear stimulation function, and $D_i \cdot \Delta u_i$ the diffusion term.

The *FitzHugh–Nagumo equations* are a simple model with two variables:

$$\begin{aligned} \frac{\partial u_1}{\partial t} &= \frac{u_1 - \frac{u_1^3}{3} - u_2}{\varepsilon} + D_1 \cdot \Delta u_1 \quad , \\ \frac{\partial u_2}{\partial t} &= \varepsilon \cdot (u_1 + \beta - \gamma \cdot u_2) \quad , \end{aligned} \tag{12.51}$$

with the parameters $0 < \beta < \sqrt{3}$, $0 < \gamma < 1$, and $\varepsilon \ll 1$.

In order to determine the excitation function f_i, suitable model equations of the ion fluxes into the individual muscle cells have to be found. A selection of these model equations can be found in, for example, *A. V. Panfilov, A. V. Holden* (1997).

Figure 12.34 shows the result of a simulation of the expansion of the excitation potential on the surface of the heart. As in Figure 12.28, the potential expands outward from the inner heart wall. On the surface of the heart, this is expressed by a large excitation potential (dark). This moves from the sinoatrial nodes (1), past the two atria (2), and excites the ventricles, while the

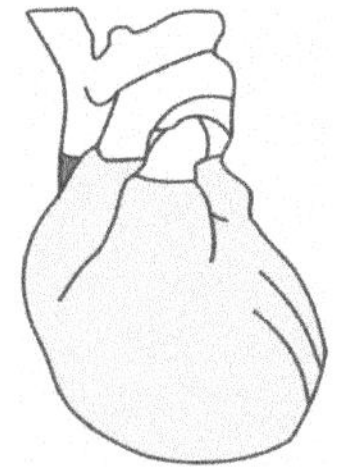

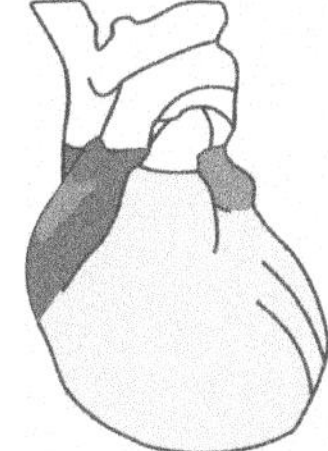

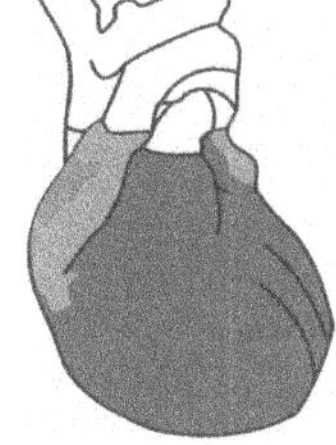

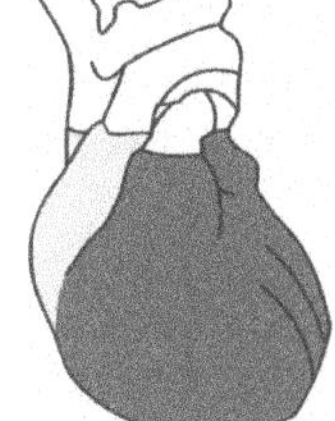

Fig. 12.34. Excitation potential distribution on the surface of the heart, *C. D. Werner et al.* (2000)

atria are again repolarized (3). At the end of the cardiac cycle, the ventricles are repolarized again.

12.3.4 Flow in the Heart

Calculation of the incompressible flow in the heart is carried out using the *continuity equation* (5.2)

$$\boldsymbol{\nabla} \cdot \boldsymbol{v} = 0 \tag{12.52}$$

and the *Navier–Stokes equation* for laminar, transitional flow (5.20)

$$\rho \cdot \left(\frac{\partial \boldsymbol{v}}{\partial t} + (\boldsymbol{v} \cdot \boldsymbol{\nabla})\boldsymbol{v} \right) = -\nabla p + \mu_{\text{eff}} \cdot \boldsymbol{\Delta v} + \boldsymbol{f} \quad . \tag{12.53}$$

Here $\boldsymbol{f}$ is the volume force that acts on the flow from the inner walls of the heart, $\boldsymbol{v}$ the velocity vector an p the pressure. The non-Newtonian properties of blood are taken into account approximately with the *cross model* (12.4).

The volume force $\boldsymbol{f}$ is computed from the shear stress distribution inside the heart, determined by the structure program of Section 12.3.2.

Following equation (12.43), the equation of motion of the structure mechanics for the velocity of deformation v_{i} and the stress tensor σ_{ij} is:

$$\rho \cdot \frac{\mathrm{d}v_{\text{i}}}{\mathrm{d}t} = \rho \cdot \left(\frac{\partial v_{\text{i}}}{\partial t} + v_{\text{j}} \cdot \frac{\partial v_{\text{i}}}{\partial x_{\text{j}}} \right) = \frac{\partial \sigma_{\text{ij}}}{\partial x_{\text{j}}} + f_{\text{i}} \quad , \tag{12.54}$$

with the volume specific forces f_{i} and the density of the material ρ.

The total time derivative of the rate of deformation describes the change in a volume element $\mathrm{d}V = \mathrm{d}x_1 \cdot \mathrm{d}x_2 \cdot \mathrm{d}x_3$ that is moving with the flow. This representation is called the *Lagrange description.* The partial time derivative of the rate of deformation with respect to time and the convective terms differentiated with respect to the space coordinates is called the *Euler representation* (see Section 3.1).

For the flow-structure coupled calculation, the boundary conditions at the edges of the fluid space of the ventricle are formulated using the Lagrange representation, while the flow is calculated using the Euler representation. This leads to the *Lagrange-Euler formulation* of the fundamental equations for the structure and the flow.

The rate of deformation v_{i}:

$$v_{\text{i}} = \begin{pmatrix} v_1 \\ v_2 \\ v_3 \end{pmatrix} \Longleftrightarrow \boldsymbol{v} = \begin{pmatrix} u \\ v \\ w \end{pmatrix}$$

corresponds to the flow vector $\boldsymbol{v}$. The stress tensor of structure σ_{ij} :

$$\sigma_{\text{ij}} \Longleftrightarrow \tau_{\text{ij}}$$

corresponds to the shear stress tensor of the flow τ_{ij}. Therefore the equation of motion of the structure mechanics (12.54) can be written as:

$$\rho \cdot \frac{\mathrm{d}v_i}{\mathrm{d}t} = \rho \cdot \left(\frac{\partial v_i}{\partial t} + v_j \cdot \frac{\partial v_i}{\partial x_j} \right) = \frac{\partial \sigma_{ij}}{\partial x_j} + f_i \quad , \tag{12.55}$$

and the Navier-Stokes equation of the fluid mechanics (12.53) can be written as:

$$\rho \cdot \frac{\mathrm{d}v_i}{\mathrm{d}t} = \rho \cdot \left(\frac{\partial v_i}{\partial t} + v_j \cdot \frac{\partial v_i}{\partial x_j} \right) = \frac{\partial \tau_{ij}}{\partial x_j} + f_i \quad . \tag{12.56}$$

For incompressible media, conservation of mass is identical for the structure mechanics and for the fluid mechanics:

$$\frac{\partial v_i}{\partial x_i} = 0 \quad . \tag{12.57}$$

If we bring equations (12.55) and (12.56) together to a single equation, we obtain the *Euler-Lagrange formulation* of conservation of momentum for both the structure mechanics and the fluid mechanics in vector notation:

$$\rho \cdot \left(\left. \frac{\partial \boldsymbol{v}}{\partial t} \right|_G + (\boldsymbol{v} \cdot \nabla)(\boldsymbol{v} - \boldsymbol{v}_G) \right) = \nabla \boldsymbol{\sigma} + \boldsymbol{f} \quad . \tag{12.58}$$

$\boldsymbol{v}_G$ is the reference velocity of the moving surface and G denotes the associated reference surface that is moving in the Lagrange formulation. The fundamental equations of structure mechanics and fluid mechanics are given in the Euler formulation relative to this surface. Concerning the coupling of the structure mechanical and fluid mechanical fundamental equations via the Lagrange representation of the moving surface, this so-called ALE (**A**rbitrary **L**agrange-**E**uler) mixed Lagrange-Euler formulation has the advantage that the various computational grids of each region on the surface G can be coupled. For the relative velocity $\boldsymbol{v} - \boldsymbol{v}_G$, the continuity equation $\nabla \cdot (\boldsymbol{v} - \boldsymbol{v}_G) = 0$ also holds.

In the ALE fundamental equation (12.58), ρ denotes the density of the structure and of the flowing medium. The tensor $\boldsymbol{\sigma}$ stands for

$$\boldsymbol{\sigma} = \sigma_{ij} \quad \text{for the structure} \quad ,$$

with the associated ansatz for the stress–extension law, and

$$\boldsymbol{\sigma} = \tau_{ij} \quad \text{for the flow} \quad ,$$

with the Stokes friction law for incompressible flow

$$\tau_{ij} = -p \cdot \delta_{ij} + \mu \cdot \left(\frac{\partial v_i}{\partial x_j} + \frac{\partial v_j}{\partial x_i} \right) \quad . \tag{12.59}$$

The coupling takes place via the boundary conditions at the interface G. The kinematic coupling condition states that the rate of deformation v_{i} must be equal to the flow velocity $\boldsymbol{v}$ at the interface:

$$v_{\mathrm{i}}|_G = \boldsymbol{v}|_G \quad . \tag{12.60}$$

The dynamic coupling condition relates the stress tensor $\boldsymbol{\sigma}$ with the shear stress vector $\boldsymbol{\tau}$ at the interface with the normal vector $\boldsymbol{n}$:

$$\boldsymbol{\sigma} \cdot \boldsymbol{n} = \boldsymbol{\tau} \cdot \boldsymbol{n} \quad . \tag{12.61}$$

The exchange of stresses with the hydrostatic pressure and with the shear stress components of the friction is a matter for the coupling models.

For the flow calculation, as shown in Figure 12.35, three regions are to be distinguished. In the first region the motion of the coupling interface leads to a substantial Lagrange description of the flow quantities. The second transition region requires a mixed Lagrange-Euler approach and at a sufficiently large distance from the interface in the third region the Euler formulation is used. Figure 12.35 shows the division of regions with a characteristic computational grid for the flow calculation of the human heart.

To calculate the interaction of the flow and the structure in the ventricle and atrium, a material law for the myocardium of the heart is necessary (see Section 12.3.2). The arrangement of muscle fibers and the lines of acceleration are shown in Figure 12.36. The muscle fibers are oriented in a spiral manner around the ventricle and cause radial and longitudinal contraction of the ventricle.

The qualitative shapes of the stress–strain curves of the human myocardium and epicardium are shown in Figure 12.37. The inner layer of the myocardium leads to different stress–strain behavior from that of the outer

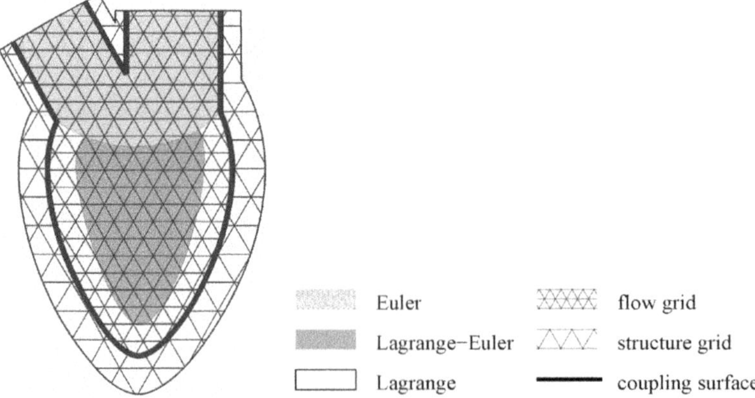

Fig. 12.35. Division of regions for the ALE Lagrange-Euler formulation of the flow-structure coupling for the human ventricle

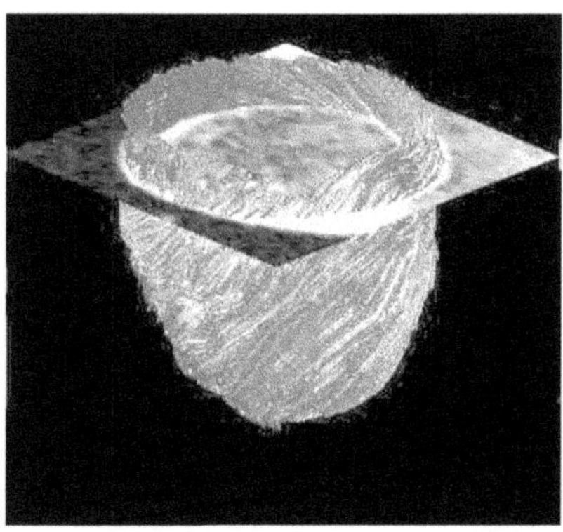
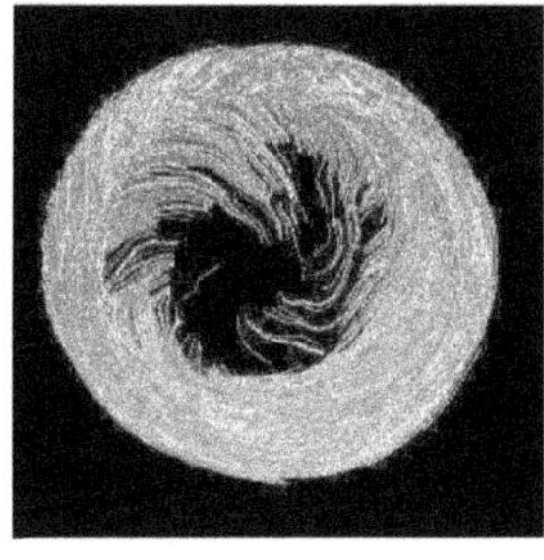

Fig. 12.36. Orientation of cardiac muscle fibers, *B. Jung et al.* (2006)

epicardial layer. In the myocardium there are different limiting values for the stress, depending on whether the load is along the muscle fibers or perpendicular to them. The nonlinear stress–strain relationship exhibited by the epicardial layer is more pronounced than that exhibited by the myocardial layer. In addition, there is hysteresis in the load curve of the external muscle layer of the cardiac ventricle.

Another ansatz for the calculation of the flow-structure coupling was introduced by *C. S. Peskin* and *D. M. McQueen* (1997). It approximates the muscle fibers of the heart as well as the cardiac valves in the Lagrangian description, with discrete elastic *fiber filaments* embedded in the flow. The discretization of the fiber filaments is chosen to be so fine that they have no volume or mass, but can still be used for a continuum-mechanical description of the biological material. The filaments are oriented along the flow and have local flow velocity $\boldsymbol{v}$. At each point of the filament–flow combination, a unique fiber direction is given, fixed by the unit vector $\boldsymbol{e}$.

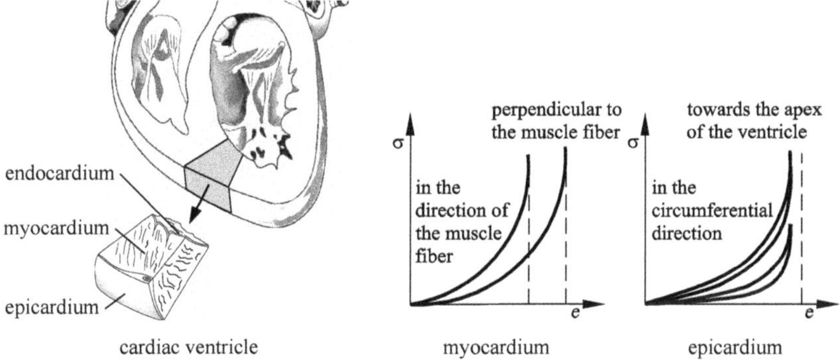

Fig. 12.37. Stress-strain curves for the human myocardium and epicardium

The force $\boldsymbol{F}$ that the fiber filaments exert on the flow is computed with the *interaction equations* of the filament–flow system:

$$\boldsymbol{F}(\boldsymbol{x},t) = \int\limits_V \boldsymbol{f}(q,r,s,t) \cdot \delta(\boldsymbol{x} - \boldsymbol{X}(q,r,s,t)) \cdot \mathrm{d}q \cdot \mathrm{d}r \cdot \mathrm{d}s, \tag{12.62}$$

with the filament coordinates q, r, s, the position of the filament in time $\boldsymbol{x} = \boldsymbol{X}(q,r,s,t)$, the unit vector $\boldsymbol{e} = (\partial \boldsymbol{X}/\partial \boldsymbol{s}/(|\partial \boldsymbol{x}/\partial s|)$, and the integration volume V.

Coupling with the velocity vector $\boldsymbol{v}$ takes place via

$$\begin{aligned} \frac{\partial \boldsymbol{X}}{\partial t}(q,r,s,t) &= \boldsymbol{v}(\boldsymbol{X}(q,r,s,t),t) \\ &= \int\limits_V \boldsymbol{v}(\boldsymbol{x},t) \cdot \delta(\boldsymbol{x} - \boldsymbol{X}(q,r,s,t)) \cdot \mathrm{d}\boldsymbol{x}. \end{aligned} \tag{12.63}$$

The fiber–filament equations are

$$\begin{aligned} \boldsymbol{F} &= \frac{\partial(\tau \cdot \boldsymbol{e})}{\partial s}, \\ \tau &= \sigma \cdot \left(\left| \frac{\partial \boldsymbol{X}}{\partial s} \right|, q, r, s, t \right). \end{aligned} \tag{12.64}$$

Note that the flow equations (12.63) and (12.64) are written down in the Eulerian description. Here $\boldsymbol{X} = (x_1, x_2, x_3)$ are Cartesian coordinates fixed in space. The variables to be computed are the velocity vector $\boldsymbol{v}(\boldsymbol{x},t)$, the pressure $p(\boldsymbol{x},t)$, and the filament force $\boldsymbol{F}(\boldsymbol{x},t)$. The constants ρ and μ are the density and the viscosity of the flow.

The fiber-filament equation (12.64) and its connection to the flow (12.62), (12.63) are given in the Lagrangian description, where q, r, s are time-dependent curved coordinates that determine the position of the material points of the fiber filaments. The unknowns of the system of equations, are

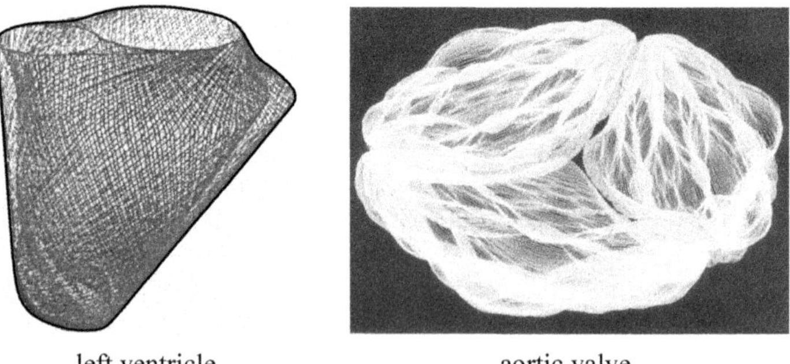

Fig. 12.38. Fiber-filament model of the inner wall of the left ventricle and the aortic valve, *C. S. Peskin, D. M. McQueen* (1994, 1997)

the fiber configuration $\boldsymbol{X}(q, r, s, t)$, the fiber stress $\tau(q, r, s, t)$, and the Lagrangian description of the fiber forces $\boldsymbol{f}(q, r, s, t)$. The interaction equations (12.62) and (12.63) connect the Lagrangian and Eulerian variables.

Figure 12.38 shows the simplified fiber-filament model of the heart that corresponds to the structure model of Figure 12.31. Fiber filaments of the inner layer of the left ventricle and the computed three sacks of the aortic valve are shown.

Figure 12.39 shows the computed flow. Streaklines from the particles spread through the flow are shown. The first figure shows the intake flow process in the left and right ventricles with open mitral and tricuspid valves. During the filling process, a ring vortex forms below the mitral valve. Particles that make the flow visible are added to the atria and ventricles of the heart. During contraction of the ventricles, the mitral and tricuspid valves are closed. A residual flow with low flow velocity remains. During the expulsion process, the aortic and pulmonary valves are open, and a jet flow with high flow velocity can be seen in the aortic and venal channels. But the details of flow bifurcations in the ventricles cannot be described with this model. During relaxation of the ventricles, the cardiac valves are closed, and the intake process begins again.

As there are only limited in vivo structural data of the human heart available to determine the constants of the strain-energy function (12.49), there is another possibility to calculate the flow in the heart, without modeling the structure of the myocardium. The volume force $\boldsymbol{f}$ in equation (12.53) is replaced by the knowledge of the time-dependent heart geometry that acts on the flow in the ventricles. Following this idea, a geometrical surface model of the heart for one cardiac cycle is derived from the human MRI geometry data. Using these prescribed time-dependent geometrical boundary conditions, the flow calculation in the ventricles is carried out.

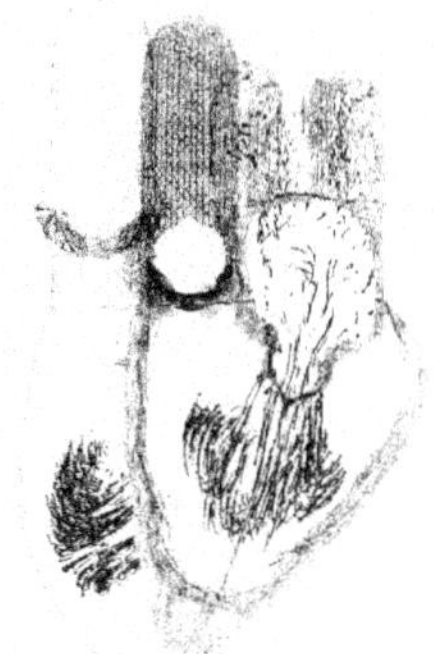

intake flow
mitral and tricuspid valve open

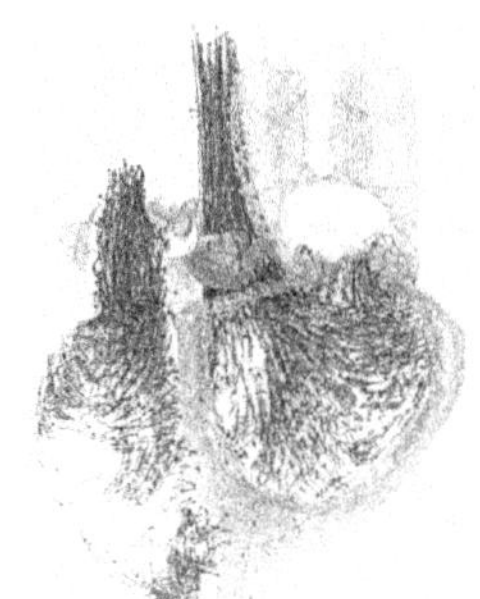

outward flow
aortic and pulmonary valve open

Fig. 12.39. Flow simulation in the heart, *C. S. Peskin, D. M. McQueen* (1997)

For the calculation of the pulsating flow, the continuity and Navier–Stokes equations (12.53) are made dimensionless with the characteristic diameter D of the aorta and the averaged velocity U in the left ventricle:

$$\boldsymbol{x}^* = \frac{\boldsymbol{x}}{D}, \quad \boldsymbol{v}^* = \frac{\boldsymbol{v}}{U}, \quad t^* = t \cdot \omega, \quad p^* = \frac{p}{\rho \cdot U^2}.$$

Using the dimensionless characteristic Reynolds number $\mathrm{Re}_D = U \cdot D/\nu_{\text{eff}}$ and Womersley number $\mathrm{Wo} = D \cdot \sqrt{\omega/\nu_{\text{eff}}}$ ($\omega = 2 \cdot \pi \cdot f$), we obtain the dimensionless fundamental equations

$$\nabla \cdot \boldsymbol{v} = 0,$$

$$\frac{\mathrm{Wo}^2}{\mathrm{Re}_D} \cdot \left(\frac{\partial \boldsymbol{v}}{\partial t} + (\boldsymbol{v} \cdot \nabla)\boldsymbol{v} \right) = -\nabla p + \frac{1}{\mathrm{Re}_D} \cdot \Delta \boldsymbol{v}. \tag{12.65}$$

Figure 12.40 shows the magnetic spin resonance (MRI) data of a horizontal slice through the human heart and the derived geometry model of the heart, representing the fluid space of the ventricles and the atria. At each point in time through one cardiac cycle, a total of 26 horizontal and vertical planes are analyzed. Several cardiac cycles are recorded at 20 points in time, and the data are transferred into the dynamic geometry model using image recognition software. An ECG of the test person is used as the trigger to record the images (*H. Oertel jr. et al* (2005), (2006)).

The dynamic geometry model of the heart consists of the ventricles, atria, aorta and vena cava. The motion of the contraction and relaxation of the ventricles and atria is given by the geometry model, while the motion of the aorta and vena cava caused by the flow pulse is calculated. The volume flux of the pressure controlled cardiac valves is modeled in each projection plane, as shown in Figure 12.44.

The calculated three-dimensional flow structure in the left and right human ventricles is shown for one cardiac cycle in Figure 12.41. When the mitral

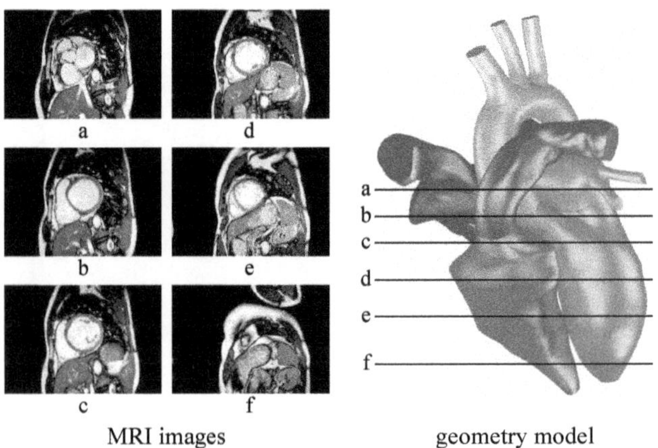

Fig. 12.40. Dynamic geometry model of the human heart

and tricuspid valves open at the time $t = 0.76$, first an intake jet forms in the *left* and *right ventricles* during the filling process; after a quarter of the cardiac cycle these intake jets are each accompanied by a ring vortex (three-dimensional focus F1, see Figure 3.7). These ring vortices arise to balance the intake jet braked in the motionless fluid. Further ring vortices, given by the Helmholtz vortex laws (see Section 4.4.3), arise in the atria. As the diastole progresses, the size of the vortices increases because of the motion of the myocardium. The expansion of the vortex in the axial direction is uniform, while in the radial direction the left side is strengthened in the left ventricle. As the vortices enter the ventricles their velocities decrease. There is no flow through the tops of the ventricles at this time. As the intake flow progresses, because of the large deformation in the left ventricle, the ring vortex tends towards the apex of the ventricle and the saddle surface S1 begins to form at the myocardial wall, which prepares for efficient expulsion of blood in the systole. The velocity of the three-dimensional flow decreases until finally the intake flow process is complete and the mitral valve closes. Further deformation of the vortex structure is determined by the inertia of the flow. In parallel, the upper part of the ring vortex induces a secondary vortex in the aortic canal F2 with the saddle point S2 on the wall of the aortic canal. At the beginning of the diastole, the intake process in the atrium causes the three-dimensional vortex F3.

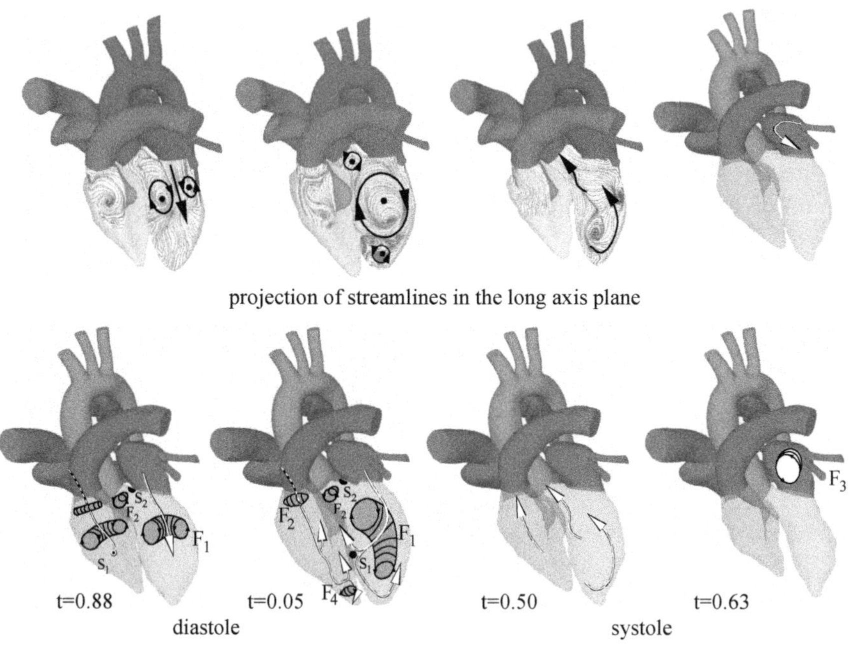

Fig. 12.41. Flow in the human heart $Re_D = 3470$, $W_0 = 25$, $T_0 = 1.0$ s

Because of the more complex geometry of the right ventricle (see Figure 12.29), the intake vortex is deformed along the ventricle contour. During the diastole, as the ring vortex rotates in the direction of the apex of the ventricle, this deformation causes the vortex axis to tend towards the outer wall of the myocardium and there produce a saddle surface S1. Calculation of the flow shows that because of this the ring vortex decays before the exit flow begins and causes a secondary flow in the apex of the ventricle F4, corresponding to the secondary flow in the pulmonary artery canal F2. Thus the interpretation of the three-dimensional flow structure in the right ventricle is not as clearly defined as that in the left ventricle.

At the time $t = 0.41$, the aortic valve opens and the flow begins to pass out into the aorta. The direction of motion of the vortices remains the same. First the vortex F2 and then the ring vortex F1 are washed out. The velocity maximum of the exit flow process is reached in the central region of the aortic valve and at time $t = 0.61$ the flow pulse in the aorta is fully developed. At the end of the systole the vortex structure in the left and right ventricles has completely disintegrated. In a healthy human heart about 62 % of the volume of the left ventricle is expelled.

In addition to the characteristic quantities of the dimensionless Navier-Stokes equation (12.65) Re_D and Wo, further dimensionless characteristic numbers are defined for medical evaluation of the ventricular flow. The ejection fraction

$$E = \frac{V_s}{V_d} \tag{12.66}$$

is the ratio of the stroke volume V_s to the end-diastolic volume V_d. It indicates the percentage of the ventricle volume expelled into the aortic channel.

With the ventricle pumping work A_p, which is calculated from the p-V diagram of Figure 12.27, the mixing time of the blood in the ventricle t_b in

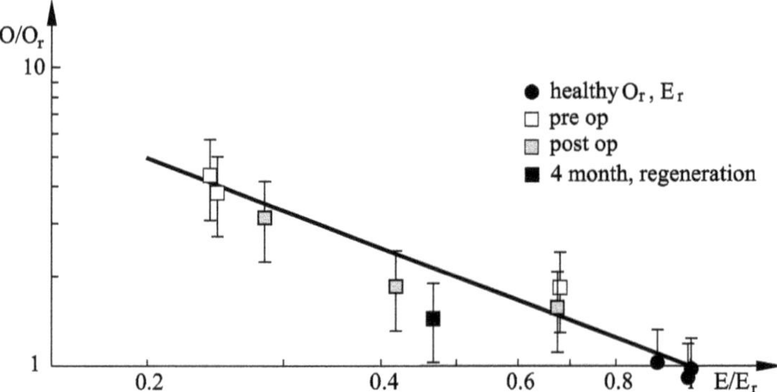

Fig. 12.42. Dependence of dimensionless pumping work O/O_r on the ejection fraction E/E_r; reference values $O_r = 3.4 \cdot 10^6$, $E_r = 62\%$

general over 2 – 3 pump cycles, the effective viscosity of the blood μ_{eff} (12.4) and the stroke volume V_s, a dimensionless pumping work can be defined:

$$O = \frac{A_p \cdot t_b}{\mu_{\text{eff}} \cdot V_s} \quad , \tag{12.67}$$

where $A_p \cdot t_b / \mu_{\text{eff}}$ has the dimension of a volume and O is a ratio of volumes. If the patient is suffering a heart attack, the pumping work of the ventricle decreases and the mixing time of the blood increases. The dimensionless pumping work assumes larger values than those of a healthy heart. Figure 12.42 plots the dimensionless pumping work O relative to the reference value of the healthy ventricle O_r for the patients with aneurysm, before and after ventricular reconstruction and after four months' regeneration time, over the ejection fraction E/E_r or the Reynolds number Re_D. On a plot with logarthimic scales on both axes a straight line is found, and therefore the linear power law:

$$\frac{O}{O_r} = \left(\frac{E}{E_r}\right)^{-1} \quad . \tag{12.68}$$

he heart of an athlete also has an increased value of the dimensionless pumping work and is approximately the value of the regenerated operated ventricle. As the diagram in Figure 12.42 shows, with the dimensionless pumping work and the power law (12.68), a quantitative fluid mechanical evaluation can be carried out for a patient before and after surgery. In this it is assumed that the Womersley number Wo is approximately constant.

12.3.5 Cardiac Valves

The operation of the four cardiac valves has already been described in Section 12.3.1. The modeling of flow relations in the cardiac valves of the left ventricle will now be discussed in this section.

Figure 12.43 shows the anatomy of the pressure-controlled mitral and aortic valves. The left mitral valve has two flaps. The mitral valve ensures that the left atrium can fill between heart beats, but prevents backflow of the blood during contraction of the ventricle. The tendons leading to the papillary muscles prevent the mitral valve turning inside out during the high pressure of the contraction phase.

The aortic valve consists of three semilunar tissue sacks. It prevents backflow of blood out of the aorta during the relaxation phase of the heart. Because of the higher pressure exerted on the aortic valve during the contraction phase, the valve sacks are considerably stronger than the flaps of the mitral valve.

In spite of the higher aortic pressure, the sacks of the aortic valve do not touch the aortic bulb while the aortic valve is open. The flow goes past the peaks of the sacks and forms a backflow region between the valve sack and the aortic bulb. The back pressure of this region prevents the sacks from flattening and attaching.

Because of the high shear rate of the intake jet into the aorta, the tips of the aortic valve flaps become unstable and begin to flutter while open.

To calculate the flow in the ventricles it is not necessary to reproduce every detail of the valve motion caused by the flow. It is sufficient to model the volume flux through the cardiac valves correctly, based on subsonic Doppler velocity measurements and MRI flux data of the human heart. In the models of the natural cardiac valves, we consider only their projections onto the valve plane. The opened shape of the two-flap mitral valve and the three-flap aortic valve are shown in Figure 12.44.

The model valves are realized by boundary conditions with which a variable resistance is associated. This resistance can be varied between 0 and ∞. By changing the resistances, the valves corresponding to their projections on the valve plane are opened. In the closed state the valve is given the resistance ∞ on the entire surface of the valve plane, while in the open state the resistance is 0.

Modeling the tricuspid and pulmonary valves in the right ventricle is performed in a similar manner.

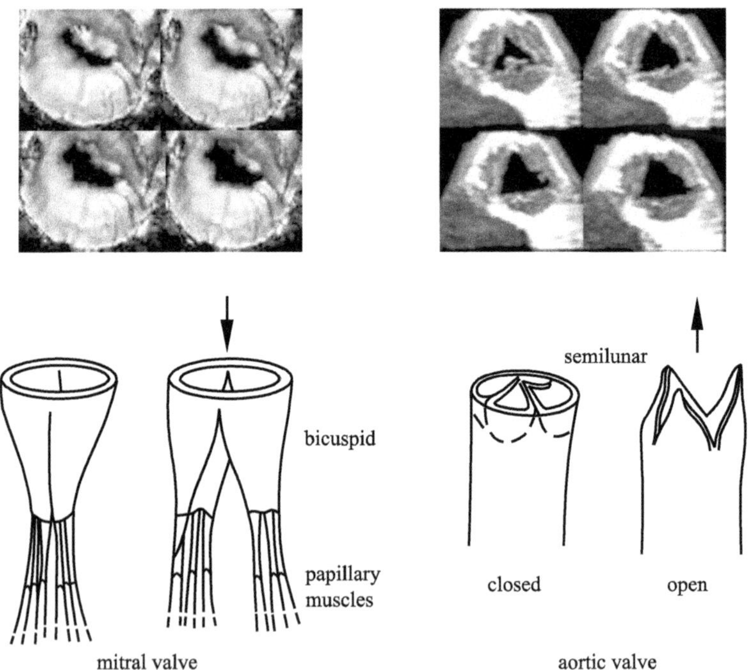

Fig. 12.43. Mitral and aortic valves in the heart

Diseases of the cardiac valves can lead to backflows into the ventricle or atria. An aortic valve stenosis leads to calcium deposits on the valve sacks, which then does not open fully. Whereas the flow out of the left ventricle is still laminar, a turbulent jet flow with increased flow losses forms downstream from the aortic valve.

Because of the stenosis, the left ventricle has to overcome higher pressure losses. The ventricle volume increases with time, and the cardiac muscle increases. The supply of oxygen to the increased cardiac muscle is possible only to a certain degree, since the number of coronary vessels remains the same.

If the aortic valve does not close completely, backflow occurs because of insufficence in the left ventricle, and again there is an increase in flow losses. These are compensated by an increase in the volume of the heart and a faster pulse.

With insufficiency of the mitral valve, the high pressure from the left ventricle is carried over into the atrium. This leads to an extension of the left atrium, and the volume strain of the right ventricle increases via the lung, and consequently to a higher pressure in the vascular system of the lung.

In cases of severe cardiac valve disease, or if the cardiac muscle increase has to be compensated after many years, operative correction is needed.

One common method of operation is an implant of an artificial cardiac valve. For many years, back pressure valves with spherical or disk-shaped flaps were used for the artificial aortic valve. These have high pressure peaks and considerable backflow regions, which lead to an aggregation of erythrocytes in the low shear rate flow regimes (Figure 12.11) and then to thrombosis. In the high shear rate regimes, deformation of the erythrocytes and their eventual destruction occurs.

The pendulum valve (Bjork–Shiley) was an improvement. However, because of sealing of the guiding clip and the noise from the valve, this was not used for long. Development led to the bipartite or tripartite pendulum valve, based on the natural aortic valve, whose pressure peaks and backflow

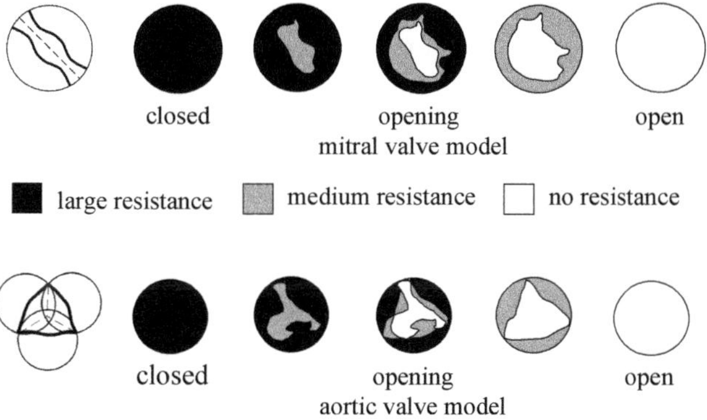

Fig. 12.44. Modeling the mitral and aortic valves of the heart

Fig. 12.45. Artificial aortic valve and experimental flow simulation, *F. Hirt* (1994)

regions were considerably reduced, but could still not be completely eliminated. The future will most likely bring genetically engineered cardiac valves with minimized flow losses.

Figure 12.45 shows a bipartite pendulum valve with experimental and numerical flow visualization in a heart-pressure chamber. In the laser light cut, the streaklines of the open aortic valve show regions of high flow velocity and large shear rates, as well as backflow regions. If the angle of inclination of the valves in their open state is too large, the flow separates at the leading edge of the valve and forms an extensive backflow region, which, because of shear instabilities, becomes turbulent and therefore has higher flow losses. At the optimized opening angle, flow separation at the leading edge is prevented, although the wake instabilities lead to a periodic backflow downstream.

12.4 Flow in Blood Vessels

The human blood circulation has already been introduced in Section 12.1.2 with Figure 12.1. The pressure exerted by the heart in the branching arteries is shown in Figure 12.9. The size ratios and wall strengths of the arteries and veins are also shown in Figure 12.46. As the blood passes through one pulse, the arteries expand, and the wall thickness decreases. The strain on the inner wall is greater than that on the outer wall. The stress–strain relation for the vessel walls can be approximately described with an exponential function. Because of the nonlinearity of the stress–strain curve, the stress at the inner wall is considerably greater than the extension.

According to *Y. C. Fung* (1993), the flow strain–energy function (12.49) for the blood vessels can be simplified to

$$\rho_0 W = q + c e^Q \quad . \tag{12.69}$$

where q and Q can be represented as polynomials in the strain components.

In this section we describe the flow relations in the arteries, arterial bends, and branchings as well as the microflows in capillaries in detail. In this section we describe the flow relations in the arteries, arterial bends, and branches as well as the microflows in capillaries in detail.

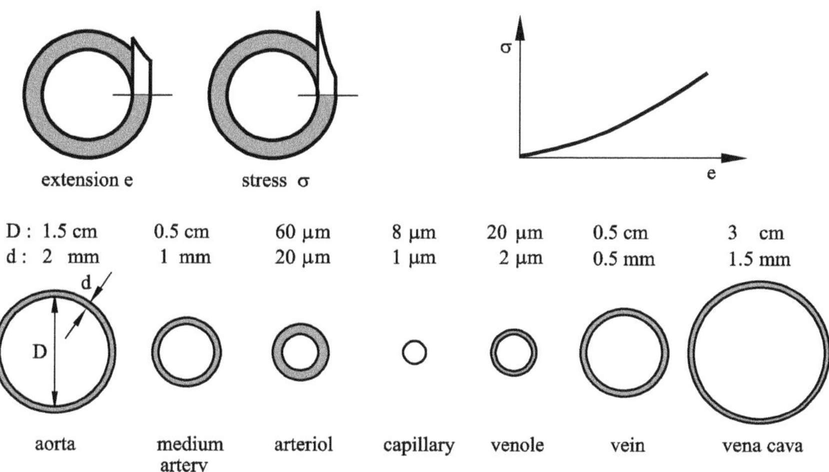

Fig. 12.46. Extension, stress, size ratios, and wall strengths of arteries and veins

Figure 12.47 shows the instantaneous *velocity profile* of a fully developed *arterial flow* as well as the time development of the velocity wave. The periodic flow pulse of the heart causes a laminar, unsteady pipe flow in the medium and small arteries, with Reynolds numbers of several hundred to a thousand. Fully developed flows without the effect of the intake flow or arterial branchings have as their time-averaged velocity profile the parabolic Poiseuille profile from Section 4.2.7. As the aorta ascends, the pipe flow exceeds the critical Reynolds number, and the laminar–turbulent transition begins close to the arterial walls in the turning point profiles while the heart relaxes. Before the turbulent flow can fully develop in the aorta, the secondary flow downstream in the curve of the aorta stabilizes the flow and so causes relamination of the flow.

Figure 12.48 shows the averaged velocity profile in the *aorta bend.* In the intake region, the boundary layers at the inner and outer walls of the aorta

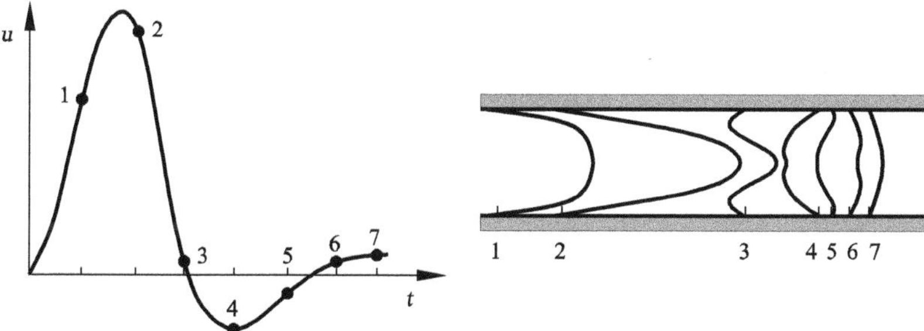

Fig. 12.47. Time development of the velocity wave and instantaneous velocity profiles in a medium artery

first develop. Since the inner curvature is greater than the outer curvature, the lower pressure causes the boundary-layer flow to be increasingly accelerated. Because of the centrifugal force, a *secondary flow* forms downstream. There is then a velocity component perpendicular to the streamlines that induces two secondary vortices superimposed on the main flow and rotating in opposite directions. Superimposing the pulsing velocity profile of Figure 12.47 onto the average profiles in the curved pipe, we find a complex three-dimensional secondary flow with temporary backflows close to the walls.

Figure 12.48 shows, in addition to the basic sketch of the secondary flow, the calculated velocity profiles in the model aorta during the systole shortly after opening and before closing the aortic valve. The streamline snapshot in the section of the descending aorta shows the structure of the secondary flow at each point in time of the cardiac cycle. Because of the branching of the aorta, the transverse flow velocity of the secondary flow may be neglected compared to the maximum velocity in the descending aorta.

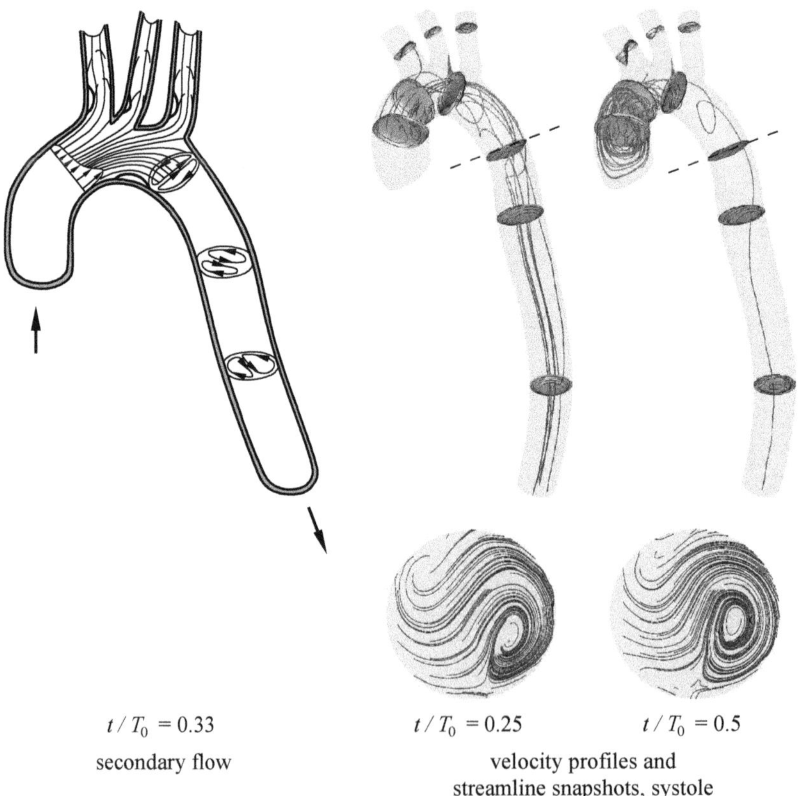

Fig. 12.48. Velocity profile and structure of the secondary flow in a model aorta, Wo = 27, $T_0 = 0.84$ s

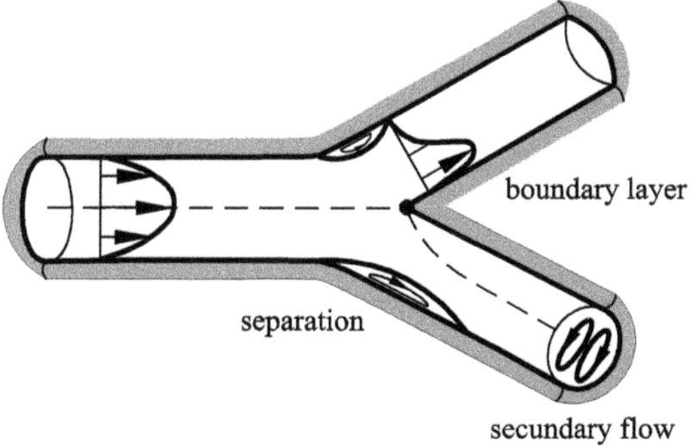

Fig. 12.49. Secondary flow separation downstream from artery branches

Similar secondary flows occur downstream from *arterial branchings* because of the curvature of the streamlines in the branchings (Figure 12.49). The resulting flow field depends on the ratio of the arterial diameter, the geometry of the branchings, and the volume flux. At arterial branchings at great angles, flow separation occurs. Figure 12.49 shows two examples of separation and reattachment lines, as well as the stagnation point. In the separation region, low shear rates occur at the wall, while the opposite wall has high shear rates. The flow separates at the inner wall of the branching. Because of the streamline curvature, a distinctive secondary flow again occurs downstream.

If there is arterial disease to the extent that a stenosis occurs in the artery, flow separation will also occur downstream from the narrowing. Figure 12.50 shows the averaged velocity profiles and the separation bubble downstream from the artery narrowing. In the region of the narrowing, acceleration of the flow takes place. The subsequent slowing down as the artery again widens and the associated pressure increase cause flow separation with corresponding low shear rates at the wall. In arteries with Reynolds numbers less than 100, the flow passes through the narrowing without separation.

In *veins* and *vein branchings*, a pulsing blood backflow towards the right ventricle occurs, corresponding to the arterial flows. However, because of the lower mean pressure and the smaller wall strength in the veins, the veins can collapse above the heart. This happens when, because of muscle contractions or when the arms are raised, the pressure difference in the vein wall $P = p - p_a$ between the inner and outer pressure becomes negative. Figure 12.51 shows

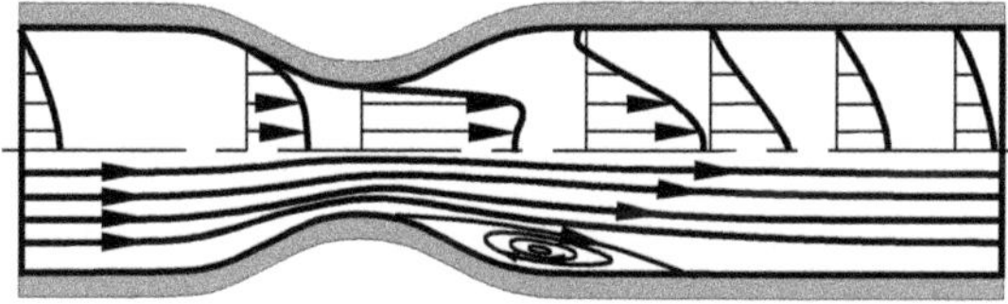

Fig. 12.50. Flow separation due to arterial stenosis

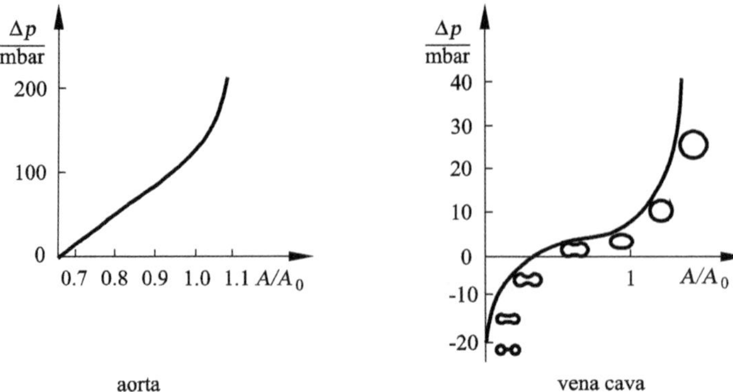

Fig. 12.51. Wall pressure of the aorta and cross-sectional shape of a collapsing vena cava

the pressure difference P across the cross-sectional ratio A/A_0 for the vena cava compared to that for the aorta. The starting point is an elliptical cross section A_0 for the vena cava and a circular cross section for the aorta. At higher pressure differences a circular cross-section occurs in the vena cava, while at negative pressure differences the vena cava collapses and only a small remaining volume flux of the blood remains. A partially collapsed vein occurs when the pressure difference during the inflow into the vein is still positive but, because of the friction downstream, a negative pressure difference occurs. New flow forms such as surge flows or self-induced oscillations occur, as described in Section 4.1.8.

The veins leading upward to the right ventricle have valves (see Figure 12.1) to prevent the blood flowing backwards at low mean pressures.

12.4.1 Unsteady Pipe Flow

There is an *exact solution* of the Navier–Stokes equation for the pulsing pipe flow of a Newtonian fluid. In cylindrical coordinates the Navier–Stokes equation (5.20) for axially symmetric fully developed flow is

$$\frac{\partial u}{\partial t} = -\frac{1}{\rho} \cdot \frac{\partial p}{\partial x} + \nu \cdot \left(\frac{\partial^2 u}{\partial r^2} + \frac{1}{R} \cdot \frac{\partial u}{\partial r} \right) , \qquad (12.70)$$

with radial coordinate r and pipe radius R. The no-slip condition holds at the pipe wall, $u(R,t) = 0$, and at the pipe axis $\partial u(0,t)/\partial r = 0$. As a further condition it is assumed that the flow is periodic in time. Let the volume flux $\dot{V}(t)$ be given. It can be expressed as a Fourier series:

$$\dot{V}(t) \sim -\frac{1}{\rho} \cdot \frac{\partial p}{\partial x} = a_0 \cdot \sum_{\omega=1}^{\infty} (a_\omega \cdot \cos(\omega \cdot t)) = F(t). \qquad (12.71)$$

With the separation trial solution

$$u(r,t) = \sum_i g_i(t) \cdot f_i(r) \tag{12.72}$$

we obtain two ordinary differential equations:

$$f'' + \frac{1}{r} \cdot f' + \lambda^2 \cdot f = 0, \tag{12.73}$$

with $f(R) = 0$, $f'(0) = 0$, and

$$\dot{g} + \nu \cdot \lambda^2 \cdot g = c. \tag{12.74}$$

Here $g(t)$ is a periodic function of time, and $F(t)$ expanded in $f_i(r)$ yields

$$F(t) = \sum_i c_i(t) \cdot f_i(r). \tag{12.75}$$

In the direction of the radial coordinate r we have a Sturm–Liouville eigenvalue problem with zeroth-order Bessel functions as the fundamental solution. The analytical solution of the eigenvalue problem is written as

$$u(r,t) = \sum_{i=1}^{\infty} q_i \cdot \left(\frac{a_0}{\sigma_i} + \sum_{\omega=1}^{\infty} \frac{a_\omega}{\sigma_i^2 + \omega^2} \cdot (\sigma_i \cos(\omega \cdot t) + \omega \cdot \sin(\omega \cdot t)) \right) \cdot I_0\left(k_i \cdot \frac{r}{R}\right), \tag{12.76}$$

with eigenvalues $\lambda_i = k_i/R - i$, the zeroth-order Bessel function I_0, and the abbreviations $q_i = 2/(k_i \cdot I_1(k_i))$ and $\sigma_i = r \cdot \lambda_i$.

For the periodic flow in the pipe, we assume in the simplest case the following time-dependent pressure gradient:

$$-\frac{1}{\rho} \cdot \frac{\partial p}{\partial x} = a_\omega \cdot \cos(\omega \cdot t). \tag{12.77}$$

The reference velocity used, $u_{\max}$, is the maximum velocity on the pipe axis of the steady Hagen–Poiseuille pipe flow (Section 4.2.9):

$$u_{\max} = \frac{R^2 \cdot a_\omega}{4 \cdot \nu} = \frac{R^2}{4 \cdot \nu} \cdot \left(-\frac{\partial p}{\partial x}\right). \tag{12.78}$$

The solution of the eigenvalue problem is a superposition of the steady Hagen–Poiseuille flow with a periodic oscillating flow. The characteristic number for the periodic part of the solution is the *Womersley number*

$$\mathrm{Wo} = D \cdot k = D \cdot \sqrt{\frac{\omega}{\nu}}, \tag{12.79}$$

and $\omega = 2 \cdot \pi f$ with the pulse frequency f and the pipe diameter D. Here $\sqrt{\omega/\nu}$ is the unsteady boundary-layer thickness. For very small Wo, that is, at very low frequencies, steady pipe flow occurs. The flow oscillates in the same phase as the exciting periodic pressure distribution. For Womersley numbers of order of magnitude 30, as in the case of pulsing blood flow, the flow portrait is qualitatively that shown in Figure 12.47. Figure 12.52 shows the deviation from the analytical solution of the time-averaged Hagen–Poiseuille flow for

Wo $= 27$ and $\mathrm{Re}_D = 3600$. Instantaneous backflow profiles occur during the relaxation phase of the heart, opposing the exciting pressure gradient. The reference velocity here is the maximum velocity (12.78).

For the pulsing *elastic pipe flow*, the structure mechanical equation of motion (12.43) also has to be solved. It is assumed that the elastic pipe wall is thin, i.e. $d/D \ll 1$. Radial compression effects and radial deflection gradients inside the wall are neglected. It is also assumed that the radial displacements u_r and the axial displacements u_x are small and that the material properties of the pipe wall are isotropic and homogeneous. Linear elasticity theory is valid, with modulus of elasticity E and the Poisson ratio $m \leq 0.5$ ($m = 0.5$ for rigid walls). The deflection of the pipe wall is due to a periodic pressure gradient.

The radial and axial stresses are:

$$\sigma_r = \frac{E \cdot d}{1 - m^2} \cdot \left(\frac{u_r}{R} + m \cdot \frac{\partial u_x}{\partial x} \right) \quad , \tag{12.80}$$

$$\sigma_x = \frac{E \cdot d}{1 - m^2} \cdot \left(\frac{\partial u_x}{\partial x} + m \cdot \frac{u_r}{R} \right) \quad . \tag{12.81}$$

The linearized equations of motion are:

$$\rho_\mathrm{w} \cdot d \cdot \frac{\partial^2 u_r}{\partial t^2} = p_\mathrm{i} - p_\mathrm{a} - \frac{\sigma_r}{R} \quad , \tag{12.82}$$

$$\rho_\mathrm{w} \cdot d \cdot \frac{\partial^2 u_x}{\partial t^2} = \frac{\partial \sigma_x}{\partial x} - \mu \cdot \left(\frac{\partial u}{\partial r} + \frac{\partial w}{\partial x} \right)_{r=R} \quad , \tag{12.83}$$

with the density of the wall ρ_w and the inner and outer pressures of the pipe p_i and p_a.

The coupling with the fluid mechanical fundamental equations (12.86) - (12.88) at the wall $r = R + u_r \approx R$ is effected through the boundary conditions

$$u = \frac{\partial u_x}{\partial t} \quad , \qquad w = \frac{\partial u_r}{\partial t} \tag{12.84}$$

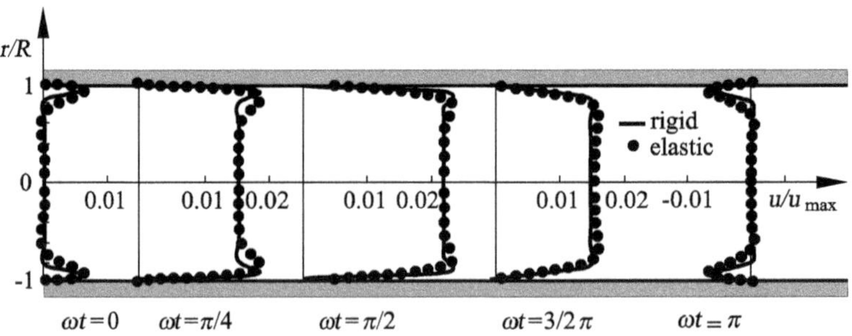

Fig. 12.52. Periodic part of velocity distribution of the pulsing rigid and elastic pipe flow at different times in a period of oscillation, $Wo = 27$, $Re_D = 3600$

and on the pipe axis $r = 0$:

$$w = 0 \quad , \qquad \frac{\partial u}{\partial r} = 0 \quad . \tag{12.85}$$

The solution of the coupled system of equations is due to *J.R. Womersley* (1955). A wave ansatz for the periodic flow pulse and the periodic displacement of the pipe wall leads, for rigid pipe walls, to an eigenvalue problem whose solution can be represented by Bessel functions.

As well as the solution for the rigid wall, Figure 12.52 shows, superimposed on the Poiseuille flow, the periodic velocity distribution of the elastic pipe for a half period $T_0 = 0.61\ s$ at Womersely number $W_0 = 27$. Comparison of the velocity profile with the corresponding change in time of the pressure gradient shows a phase shift close to the center of the pipe. This is true for both rigid and elastic walls. Differences in the solutions for rigid and elastic walls are only visible close to the wall.

For the analytical solution of the rigid and elastic pipe flow, an infinitely extending pipe was assumed, on which the periodic volume flux (12.71) is superimposed. In view of the calculation of the pulsing arterial flow, however, it is more realistic to select a finite elastic pipe section fixed at both ends. The pressure and velocity pulse of the heart is simulated by means of a deflection of the pipe wall at one end of the pipe section with a periodically fluctuating block profile for the velocity. For the numerical calcuation of the pulsing elastic pipe flow we use the flow-structure coupled formulation (12.58) of the nonlinear structure-mechanical and fluid-mechanical fundamental equations. The elastic pipe section has length $L = 10D$ and wall strength $d = 0.1D$. Because of the small wall strength $d \ll D$, the inertia of the wall may be neglected. For the calculation of the flow the density ρ and the viscosity μ_{eff} of blood are used, and for the calculation of the elastic wall the modulus of elasticity $E = 10^6$ and the transverse contraction number $m = 0.4$ are used. The result of the calculation is shown in Figure 12.53 for half a period of oscillation $T_0 = 0.61$ at each point in time in the middle of the pipe section. The core flow corresponds to the analytical solution of Figure 12.52.

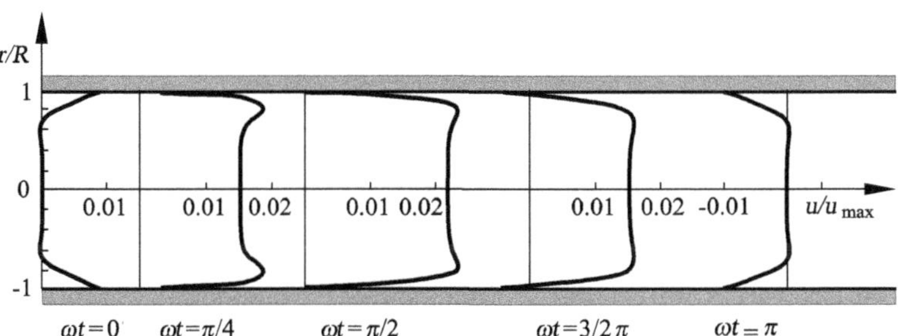

Fig. 12.53. Periodic part of velocity distribution of the finite pulsing elastic pipe flow $Wo = 27$, $Re_D = 3600$

As is to be expected, a phase shift of the velocity distribution at the wall appears at a Womersley number of $Wo = 27$. The specification of the pulsing velocity at the intake of the pipe section causes a transient oscillation in the numerical solution. While at small Womersley numbers the core velocity, wall shear stress and wall deflection are in phase, at a Womersley number of $Wo = 27$ there is a constant phase shift between the core velocity and the wall deflection that is due to the intake stretch in the finite pipe section.

12.4.2 Unsteady Arterial Flow

In calculating the flow in arteries, the elasticity of the vessels has to be taken into account. In contrast, in the heart, where the muscle contraction acts on the flow in the ventricles, the expansion of the arteries is caused by the pressure pulse generated within the heart.

Figure 12.54 shows the pressure and velocity waves in the aorta and in the descending arteries. The reflected waves from the arterial branchings of the aorta almost double the amplitude of the pressure wave. The amplitude increase carries on to the third arterial branching, and then decreases in further arterial branchings, as shown in Figure 12.9. Figure 12.55 shows the time development of the velocity profiles in a model aorta for a pulse duration of 0.8 s. The axial velocities u are made dimensionless with their maximum value $u_{\max} = 0.77$ m/s. Compared to the basic sketch in Figure 12.47, the non-Newtonian behavior of the blood with μ_{eff} of equation (12.4) has now been taken into account.

In order to calculate the expansion of pressure and velocity waves in large arteries, taking the viscosity of the arterial walls into account and assuming small perturbations, we use the linearized Navier–Stokes equations for Newtonian blood flow (blood plasma) or non-Newtonian behavior (12.4) and the linearized Navier equation for the wall, as well as the continuity equation. The axially symmetric wave expansion for incompressible flow in cylindrical coordinates yields

$$\frac{\partial u_r}{\partial t} = -\frac{1}{\rho} \cdot \frac{\partial p}{\partial r} + \nu \cdot \left(\frac{\partial^2 u_r}{\partial r^2} + \frac{1}{r} \cdot \frac{\partial u_r}{\partial r} - \frac{u_r}{r^2} + \frac{\partial^2 u_r}{\partial x^2} \right) , \tag{12.86}$$

$$\frac{\partial u_x}{\partial t} = -\frac{1}{\rho} \cdot \frac{\partial p}{\partial x} + \nu \cdot \left(\frac{\partial^2 u_x}{\partial r^2} + \frac{1}{r} \cdot \frac{\partial u_x}{\partial r} + \frac{\partial^2 u_x}{\partial x^2} \right) , \tag{12.87}$$

$$\frac{\partial u_x}{\partial x} + \frac{\partial u_r}{\partial r} + \frac{u_r}{r} = 0. \tag{12.88}$$

For the viscoelastic arterial wall, we have

$$\frac{\rho_w}{\mu_w} \cdot \frac{\partial^2 u_r}{\partial t^2} = \frac{\partial^2 u_x}{\partial r^2} + \frac{1}{r} \cdot \frac{\partial u_r}{\partial r} - \frac{u_r}{r^2} + \frac{\partial^2 u_r}{\partial x^2} - \frac{1}{\mu_w} \cdot \frac{\partial \Omega}{\partial r}, \tag{12.89}$$

$$\frac{\rho_w}{\mu_w} \cdot \frac{\partial^2 u_x}{\partial t^2} = \frac{\partial^2 u_x}{\partial r^2} + \frac{1}{r} \cdot \frac{\partial u_x}{\partial r} + \frac{\partial^2 u_x}{\partial x^2} - \frac{1}{\mu_w} \cdot \frac{\partial \Omega}{\partial x}, \tag{12.90}$$

$$\frac{\partial u_x}{\partial x} + \frac{\partial u_r}{\partial r} + \frac{u_r}{r} = 0. \tag{12.91}$$

In the case of the flow, u_r and u_x are the velocity components, and for the wall, u_r and u_x are the deviation components of the wall, μ_w is the stiffness coefficient, and ρ_w the density of the wall. The quantity Ω is a pressure that has to be introduced into equations (12.89) and (12.90), since the wall was assumed to be incompressible. The boundary conditions for the flow–wall coupling are the continuity of the shear and normal stresses, as well as the velocities at the interface between liquid and solid. At the outer wall of the artery, similar boundary conditions hold.

Until now, we have treated the linearized fundamental equations (12.86)–(12.91) for small amplitudes of the perturbation waves. Blood is a non-Newtonian medium with a nonlinear dependence on the blood viscosity. Its effect on the pulsing flow has particularly to be taken into account in the region of flow separation. In the equations for the vessel walls, there are significant nonlinear effects from the finite extension and the nonlinear viscoelasticity.

In computing the wave expansion in large arteries, the convective terms $u_i \cdot (\partial u_i / \partial x_j)$ may be neglected compared to the transient acceleration $\partial u_i / \partial t$.

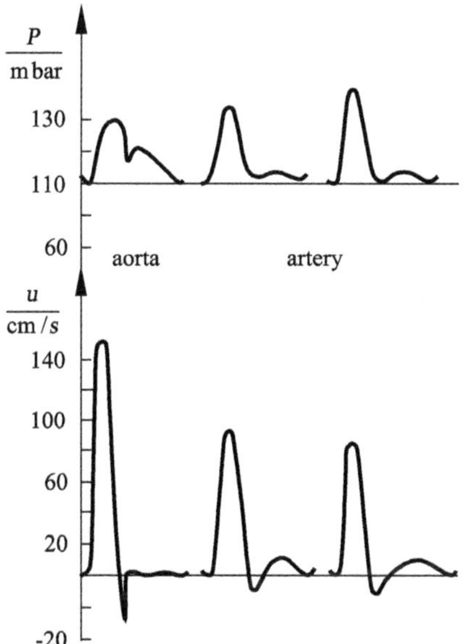

Fig. 12.54. Pressure and velocity waves in the aorta and descending arteries

Let u' be a characteristic velocity of the flow, ω the angular frequency, and c the phase velocity of the wave relative to the mean flow. The period of oscillation is $2\pi/\omega$, and the wavelength $2 \cdot \pi \cdot c/\omega$. Therefore, the transient acceleration $\partial u_i/\partial t$ is of order of magnitude $u'/(2 \cdot \pi \cdot c/\omega)$, and the convective acceleration $(u_j \cdot \partial u_i)/\partial x_j$ is of order of magnitude $u' \cdot u'/(2 \cdot \pi \cdot c/\omega)$. The condition that the convective acceleration may be neglected yields

$$\frac{u'}{c} \ll 1. \tag{12.92}$$

In the large arteries the maximum value of u'/c is 0.25, so that a smaller effect of the nonlinearity is expected. In the peripheral smaller arteries condition (12.92) is satisfied.

For inviscid flow, the wave expansion of a small pressure perturbation p' can be calculated approximately using the one-dimensional wave equation:

$$\frac{\partial^2 p'}{\partial t^2} = c^2 \cdot \frac{\partial^2 p'}{\partial x^2} \quad , \tag{12.93}$$

with the phase velocity

$$c^2 = \frac{A}{\rho} \cdot \frac{\mathrm{d}\Delta p}{\mathrm{d}A} \quad . \tag{12.94}$$

A denotes the cross-sectional area of the artery, ρ the blood density and $p = p_i - p_a$ the pressure in the artery wall, as shown in Figure 12.51. The solution of the wave equation (12.93) yields waves traveling to the left and to the right:

$$p' = \mathrm{f}_1(t - \frac{x}{c_0}) + \mathrm{f}_2(t + \frac{x}{c_0}) \quad . \tag{12.95}$$

Assuming that the wall is thin, $d/D \ll 1$, for homogenous, isotropic wall properties with modulus of elasticity E, we obtain the phase velocity in the wall:

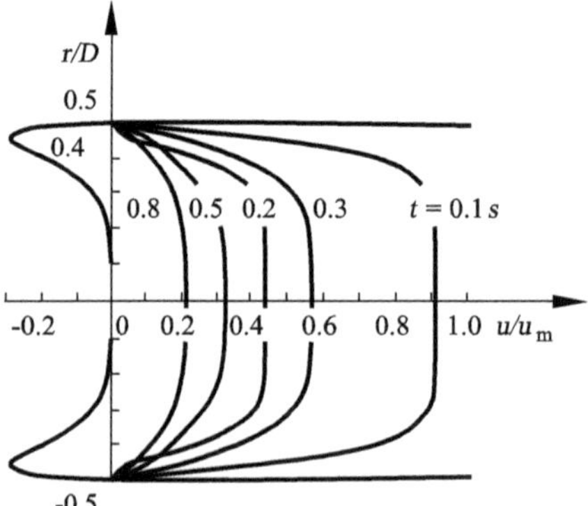

Fig. 12.55. Instantaneous profiles of the axial velocity in a model aorta, *S. C. Ling* and *H. B. Atabek* (1972)

$$c_0 = \sqrt{\frac{E \cdot d}{\rho \cdot D}} \quad , \tag{12.96}$$

known as the *Moens-Korteweg wave velocity.* The same phase velocity of the wave is also found in the solution of the viscous basic equations (12.82) and (12.83) for elastic pipe flow in the limit $\mu \rightarrow 0$ and $W_0 \rightarrow \infty$.

For the descending aorta we obtain the wave velocity $c_0 = 5m/s$, rising to $c_0 = 8m/s$ in the larger arteries.

The wave velocities in the wall of the aorta are 15 times larger than the pressure wave propagation (12.94) in the blood, and in the wall of the distal arteries 100 times larger. The wave resistance increases towards the peripheries of the circulation, and the pulse wave velocity increases more than the total cross-section of all branches of the arteries. This is due mainly to the increase in the ratio of wall thickness to diameter of the arteries.

The solutions of the wave equation (12.93) for small pressure perturbations have no nonlinear effects such as separation of the pressure waves or deformation of the pressure waves in the direction of propagation. If such effects are to be taken into account, the system of equations (12.86) - (12.92) has to be solved numerically.

12.4.3 Arterial Branchings

At branchings of vessels the pressure pulse of the heart is reflected. Figure 12.56 shows a sketch of such a vessel branching. The pressure wave 1 with volume flux $\dot{V}_1$ in the artery of cross-sectional area A_1 is divided in the arterial branching into the transmitted pressure waves 2 and 3 with volume fluxes $\dot{V}_2$ and $\dot{V}_3$. The longitudinal coordinate of each vessel is x and the arterial branching lies at $x = 0$. The incoming pressure wave in artery 1 is

$$p'_{\mathrm{I}} = \hat{p}_{\mathrm{I}} \cdot \mathrm{f}(t - \frac{x}{c_1}) \quad , \tag{12.97}$$

with the amplitude parameter $\hat{p}_{\mathrm{I}}$. The volume flux of the incoming pressure wave yields

$$\dot{V}_{\mathrm{I}} = A_1 \cdot u = Y_1 \cdot \hat{p}_{\mathrm{I}} \cdot \mathrm{f}(t - \frac{x}{c_1}) \quad , \tag{12.98}$$

with $Y_1 = A_1/(\rho \cdot c_1)$.

At the branching point $x = 0$ the reflected wave R and the transmitted waves 2 and 3 are superimposed on the incoming wave I:

$$\begin{aligned} p'_{\mathrm{R}} &= \hat{p}_{\mathrm{R}} \cdot \mathrm{g}(t + \frac{x}{c_1}) \quad , \\ p'_{\mathrm{j}} &= \hat{p}_{\mathrm{j}} \cdot \mathrm{h_j}(t - \frac{x}{c_{\mathrm{j}}}) \quad , \quad \mathrm{j} = 2, 3 \quad . \end{aligned} \tag{12.99}$$

The corresponding volume fluxes yield:

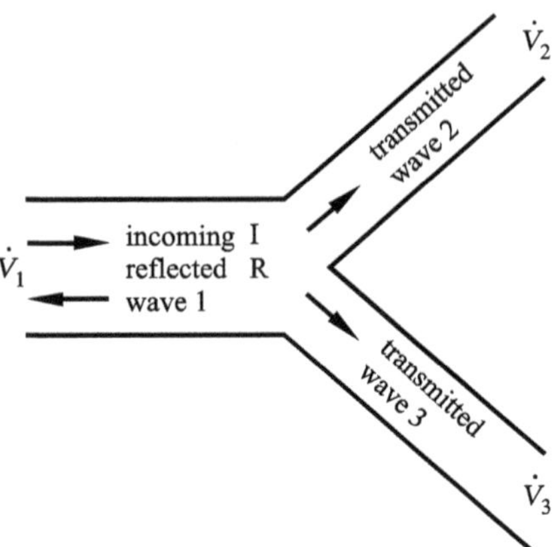

Fig. 12.56. Artery branching

$$
\begin{aligned}
\dot{V}_{\mathrm{R}} &= Y_1 \cdot \hat{p}_{\mathrm{R}} \cdot \mathrm{g}(t + \frac{x}{c_1}) \quad , \\
\dot{V}_{\mathrm{j}} &= Y_{\mathrm{j}} \cdot \hat{p}_{\mathrm{j}} \cdot \mathrm{h_j}(t - \frac{x}{c_{\mathrm{j}}}) \quad , \quad \mathrm{j} = 2, 3 \quad .
\end{aligned}
\tag{12.100}
$$

The boundary conditions at the branching point prescribe the continuity of pressure (avoidance of large accelerations) and of volume flux (mass conservation). With the assumption of large wavelengths the details of the flow at the branching point can be neglected. This results in the wave functions $\mathrm{g}(t)$ and $\mathrm{h_j}(t)$ being equal to the incoming wave function $\mathrm{f}(t)$. This yields the boundary conditions at the branching point:

$$
\begin{aligned}
\hat{p}_{\mathrm{I}} + \hat{p}_{\mathrm{R}} &= \hat{p}_1 + \hat{p}_2 \quad , \\
Y_1 \cdot (\hat{p}_{\mathrm{I}} - \hat{p}_{\mathrm{R}}) &= \sum_{\mathrm{j}=2}^{3} Y_{\mathrm{j}} \cdot \hat{p}_{\mathrm{j}} \quad .
\end{aligned}
\tag{12.101}
$$

Thus for the amplitude ratios

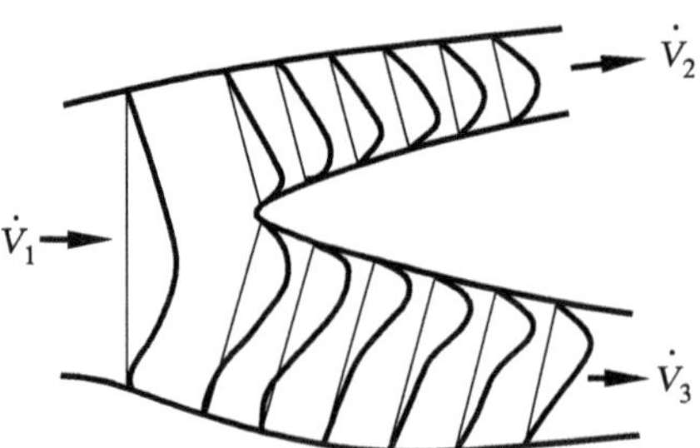

Fig. 12.57. Averaged velocity profiles in an artery branching, $\mathrm{Re}_D = 600$, $\dot{V}_2/\dot{V}_1 = 0.6$, *M. Motomiya* and *T. Karino* (1984)

$$\frac{\hat{p}_{\mathrm{R}}}{\hat{p}_{\mathrm{I}}} = \frac{Y_1 - \sum Y_{\mathrm{j}}}{Y_1 + \sum Y_{\mathrm{j}}} \quad ,$$
$$\frac{\hat{p}_{\mathrm{j}}}{\hat{p}_{\mathrm{I}}} = \frac{2 \cdot Y_1}{Y_1 + \sum Y_{\mathrm{j}}} \quad . \tag{12.102}$$

The pressure disturbance in the branching artery is calculated with (12.97) and (12.99):

$$\frac{p'}{\hat{p}_{\mathrm{I}}} = \mathrm{f}(t - \frac{x}{c_1}) + \frac{\hat{p}_{\mathrm{R}}}{\hat{p}_{\mathrm{I}}}(t + \frac{x}{c_1}) \tag{12.103}$$

and the volume flux with (12.97) and (12.100):

$$\dot{V} = Y_1 \cdot \hat{p}_{\mathrm{I}} \cdot \left(\mathrm{f}(t - \frac{x}{c_1}) - \frac{\hat{p}_{\mathrm{R}}}{\hat{p}_{\mathrm{I}}}(t + \frac{x}{c_1}) \right) \tag{12.104}$$

For the calculation of the velocity profile in an arterial branching, again the system of equations (12.86) - (12.92) needs to be solved numerically. Figure 12.57 shows an example of a time averaged velocity profile. On the opposing external artery wall of the branching, a time averaged backflow region can be seen with an associated smaller or negative wall shear stress. The curvature of the streamlines in turn causes a spiral-shaped secondary flow. In the time averaged backflow region, turning point profiles arise that are unstable and thus introduce the transition to turbulent flow. However this unstable transition process of the shearing flow is damped by the secondary flow, as in the case of a curved artery.

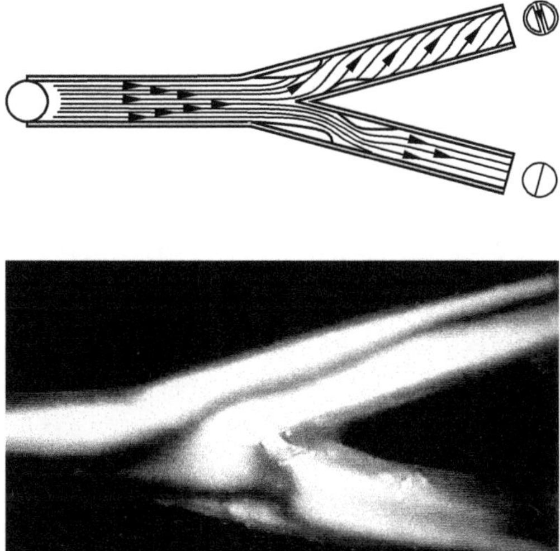

Fig. 12.58. Streamlines in an arterial branching with deposits and stress-optical visualization, *Liepsch* (1996)

If deposits on the arterial walls cause arteriosclerosis and thus lead to increased flow separation at the arterial branchings, the distinctive turning-point profiles will ensure that a turbulent flow with increased flow losses occurs, in spite of fully developed secondary flow. Figure 12.58 shows such a flow portrait in an elastic branching model. Stress-optical visualization allows the flow separation in the model experiment to be seen.

To calculate the flow in the cardiac ventricles in Section 12.3.4, the pressure boundary conditions at the exits from the arteries and the entrances to the veins are necessary. For this, the information from this section is summarized to a simplified *circulation model* that takes into account the blood flow in the human circulation from the left ventricle and the aorta into the connecting arterial system of the circulation, to the venous system, to the right cardiac ventricle, to the lung and back to the left ventricle.

This model calculates the flow resistance in the vessels as well as different parameters that influence the flow resistance. The subsystem of the arterial circulation is shown in Figure 12.59 and is represented as 128 segments. Each segment consists of a thin-walled elastic and cylindrical pipe section, where each pipe section is associated with a specific length, a wall strength, a specific diameter and a modulus of elasticity, according to the human anatomy. Peripheral branchings of the arterioles and capillaries with a diameter smaller than 2 mm are taken into account by means of a total peripheral resistance term.

The flow velocity u and the pressure p are represented by the electrical quantities current and voltage, in analogy to the Navier-Stokes equation. The solution of the Navier-Stokes equation for the elastic pipe flow for each segment of the circulation model is then associated with electrical resistance, the inductance and the capacitance, according to the physical properties of

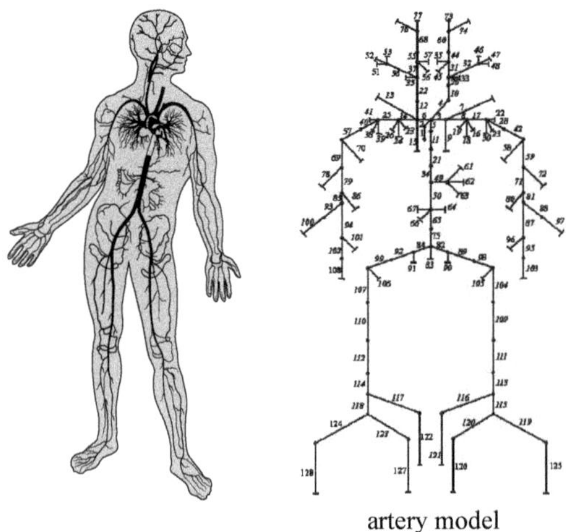

Fig. 12.59. Circulation model, *E. Nanjokat* and *U. Kreinke* (2000)

the arterial branching and the rheological properties (e.g. viscosity) of the blood. In analogy to the solution of the Navier-Stokes equations, for each pipe segment i we have the following ordinary differential equations for the blood pressure and the flow velocity:

$$p_{i-1} - p_i = \frac{9 \cdot \rho \cdot l}{4 \cdot \pi^2} \cdot \frac{du_i}{dt} + \frac{4 \cdot \mu_{eff} \cdot l}{\pi \cdot R^4} \cdot u_i = I \cdot \frac{du_i}{dt} + R_\Omega \cdot u_i \quad , \qquad (12.105)$$

$$u_i - u_{i+1} = \frac{3 \cdot \pi \cdot R^3 \cdot l}{2 \cdot E \cdot d} \cdot \frac{dp_i}{dA} = C \cdot \frac{dp_i}{dA} \quad , \qquad (12.106)$$

with the electrical resistance R_Ω, the inductance I and the capacitance C. l is the pipe length, R the pipe radius, d the wall strength, ρ the density of the blood and μ_{eff} the blood viscosity. E is the modulus of elasticity of the elastic pipe segment.

The modeling of the venous and pulmonary circulations is carried out in an analogous manner to the modeling of the arterial circulation, but with a lesser degree of detail.

The circulation model assumes a pulsing flow through the circulatory system, whereby the intake flow after each segment branching is not taken into account. The flow pulse of the heart is replaced by an average velocity in each segment.

12.4.4 Microcirculation

In the previous sections the blood flow in the large vessels was treated, in which there is an equilibrium between the pressure force, the inertial force and the forces of the elastic walls. The effect of friction at large Reynolds numbers is limited to wall boundary layers, which have an intake flow downstream after every branching of a vessel. As the degree of branching of the circulation system increases, the diameters and thus the Reynolds numbers and Womersley numbers become smaller, so that even for relatively short sections of vessels a fully developed flow forms. The inertial and centrifugal forces become insignificantly small and the flow is determined, as for the motion of single celled animals in Section 12.1.1, by the equilibrium of pressure gradient and friction. This flow regime is called *microcirculation* and it makes up 80 % of the pressure loss between the aorta and the vena cava.

Figure 12.60 shows the branchings of the arterioles and venules in muscle tissue with a diameter smaller than 50 μm as well as in the coronary vessels. The diameter of the connected capillaries lies between 10 μm and 4 μm, with Reynolds and Womersley numbers smaller than 0.01. In this region of the microflow the deformability in particular of the red blood cells (erythrocytes) and the exchange of the blood with the surrounding tissue have to be taken into account. The muscle cells regulate the flow in the capillaries locally.

The erythrocytes have a biconcave form with a diameter of 8 μm. The deformation of the visco-elastic cell membrane in the fully developed shearing flow in the capillaries depends on the pressure gradient and the geometry of the capillaries. Figure 12.61 shows the deformation of the erythrocytes in a capillary constriction from 12 μm to 6 μm.

To calculate the two-phase flow of solid particles and blood plasma flow in the capillaries, their interaction needs to be modeled. The *homogeneous flow model* introduced in Section 8.2.2 assumes that a mechanical equilibrium exists between the particle phase and the blood plasma phase. This means that all the particles have the same velocity as the homogeneous phase. To determine the change of particle concentrations in the flow, a transport equation is formulated that takes into account the effect of the shearing of the Stokes flow.

A more precise formulation of the two-phase flow of particles and blood plasma can be made by means of *separate modeling* of the two phases, which interact via exchange of momentum, that takes into account the deformation of the erythrocytes. The blood plasma is treated as an incompressible Newtonian medium that, neglecting the inertial force, leads to the Stokes equations of the fully developed capillary flow:

$$\nabla p + \mu \cdot \Delta \boldsymbol{u} = 0 \qquad (12.107)$$

and the continuity equation:

$$\nabla \cdot \boldsymbol{u} = 0 \quad . \qquad (12.108)$$

The stresses generated in the cell membrane are in equilibrium with the shear stresses of the Stokes flow. If we assume that the surface of the cell membrane does not change during the deformation (valid up to a capillary diameter of 3 μm), the components of the expansion stresses can be written as:

$$\begin{aligned} T_1 &= B \cdot \frac{(\lambda_1^2 - 1) \cdot \lambda_1}{2 \cdot \lambda_2} + T_0 \quad , \\ T_2 &= B \cdot \frac{(\lambda_2^2 - 1) \cdot \lambda_2}{2 \cdot \lambda_1} + T_0 \quad . \end{aligned} \qquad (12.109)$$

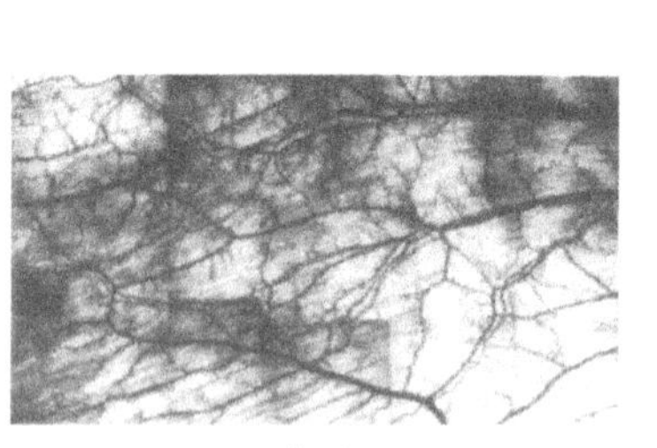

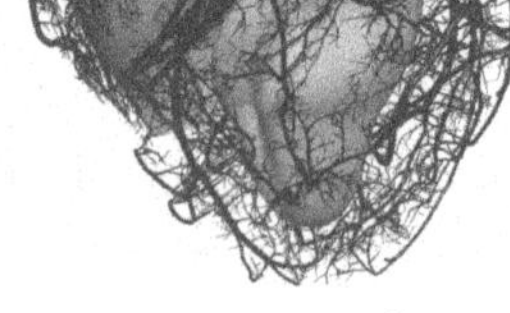

muscle tissue coronary vessels

Fig. 12.60. Arterioles and venules, *R. Skalak et al.* (1989), coronary vessels, *A. J. Pullan et al.* (2005)

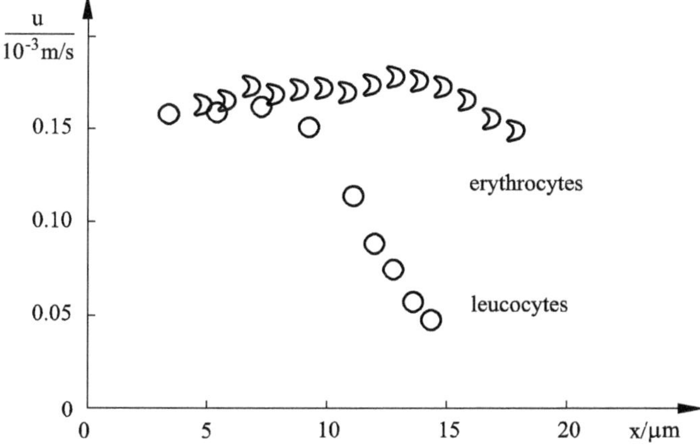

Fig. 12.61. Axial velocity of the red and white blood cells in a capillary constriction, *R. Skalak et al.* (1989)

Here λ_1 and λ_2 are the expansion ratios in the meridian and circumferential directions and T_0 is the isotropic stress, which takes into account the constancy of the surface of the membrane. The coefficient B is a modulus of elasticity.

The dependence of the bending moments on the curvature effects K_1 and K_2 yields:

$$M_1 = D \cdot \frac{K_1 + \alpha \cdot K_2}{\lambda_2} \quad , \qquad M_2 = D \cdot \frac{K_2 + \alpha \cdot K_1}{\lambda_1} \quad , \tag{12.110}$$

with the bending stiffness D and the Poisson ratio α.

Figure 12.62 shows the calculated axial velocities of the red and white blood cells in a capillary constriction from 9 μm to 5 μm. The red blood cells pass through the constriction without much reduction of their velocity, whereas the white blood cells come almost to a standstill. The time scales and the necessary shear stresses for the deformation of the white blood cells

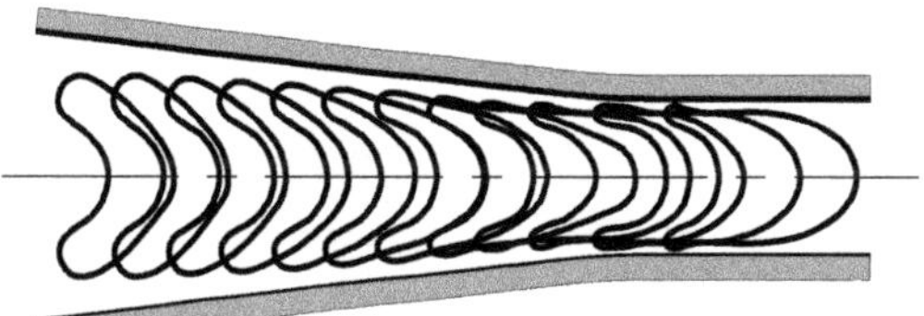

Fig. 12.62. Deformation of the red blood cells in a capillary constriction, *R. Skalak et al.* (1989)

are much larger than for the red blood cells. This means that the capillary resistance for white blood cells is two to three orders of magnitude larger.

Bibliography

L. Prandtl: Selected Bibliography

A. Sommerfeld. Zu L. Prandtls 60. Geburtstag am 4. Februar 1935. ZAMM, 15, 1–2, 1935.

W. Tollmien. Zu L. Prandtls 70. Geburtstag. ZAMM, 24, 185–188, 1944.

W. Tollmien. Seventy-Fifth Anniversary of Ludwig Prandtl. J. Aeronautical Sci., 17, 121–122, 1950.

I. Flügge-Lotz, W. Flügge. Ludwig Prandtl in the Nineteen-Thirties. Ann. Rev. Fluid Mech., 5, 1–8, 1973.

I. Flügge-Lotz, W. Flügge. Gedächtnisveranstaltung für Ludwig Prandtl aus Anlass seines 100. Geburtstags. Braunschweig, 1975.

H. Görtler. Ludwig Prandtl - Persönlichkeit und Wirken. ZFW, 23, 5, 153–162, 1975.

H. Schlichting. Ludwig Prandtl und die Aerodynamische Versuchsanstalt (AVA). ZFW, 23, 5, 162–167, 1975.

K. Oswatitsch, K. Wieghardt. Ludwig Prandtl and his Kaiser-Wilhelm-Institut. Ann. Rev. Fluid Mech., 19, 1–25, 1987.

J. Vogel-Prandtl. Ludwig Prandtl: Ein Lebensbild, Erinnerungen, Dokumente. Universitätsverlag Göttingen, 2005.

L. Prandtl. Über Flüssigkeitsbewegung bei sehr kleiner Reibung. Verhandlg. III. Intern. Math. Kongr. Heidelberg, 574–584. Teubner, Leipzig, 1905.

Neue Untersuchungen über die strömende Bewegung der Gase und Dämpfe. Physikalische Zeitschrift, 8, 23, 1907.

Der Luftwiderstand von Kugeln. Nachrichten von der Gesellschaft der Wissenschaften zu Göttingen, Mathematisch-Physikalische Klasse, 177–190, 1914.

Tragflügeltheorie. Nachrichten von der Gesellschaft der Wissenschaften zu Göttingen, Mathematisch-Physikalische Klasse, 451–477, 1918.

Experimentelle Prüfung der Umrechnungsformeln. Ergebnisse der AVA zu Göttingen, 1, 50–53, 1921.

Ergebnisse der Aerodynamischen Versuchsanstalt zu Göttingen. R. Oldenbourg, München, Berlin, 1923.

The Generation of Vortices in Fluids of Small Viscosity. Journal of the Royal Aeronautical Society, 31, 720–741, 1927.

Vier Abhandlungen zur Hydrodynamik und Aerodynamik. Ergebnisse der AVA zu Göttingen, 3, 1927.

Über Strömungen, deren Geschwindigkeiten mit der Schallgeschwindigkeit vergleichbar sind. Report of the Aeronautical Research Institute, Tokyo Imperial University, 5, 65, 1930.

Über Tragflügel kleinsten induzierten Widerstandes. ZFM, 24, 305, 1933.

Allgemeine Betrachtungen über die Strömung zusammendrückbarer Flüssigkeiten. ZAMM, 16, 129, 1936.

Theorie des Flugzeugtragflügels im zusammendrückbaren Medium. Luftfahrtforschung, 13, 313, 1936.

Über Schallausbreitung bei rasch bewegten Körpern. Schriften der Deutschen Akademie der Luftfahrtforschung, 7, 1938.

Über Reibungsschichten bei dreidimensionalen Strömungen. MoS RT 64, Betz Festschrift, Göttingen, 1945.

Mein Weg zu hydrodynamischen Theorien. Physikalische Blätter, 4, 89–92, 1948.

O. G. Tietjens, L. Prandtl. Applied Hydro- and Aeromechanics. Based on Lectures of L. Prandtl. McGraw-Hill, New York, 1934.

W. Tollmien, H. Schlichting, H. Görtler, L. Prandtl. Gesammelte Abhandlungen zur angewandten Mechanik, Hydro- und Aerodynamik, 1-3. Springer, Berlin, Göttingen, Heidelberg, 1961.

Selected Book Bibliography

J. Ackeret. Handbuch der Physik, 7. Springer, Berlin, Heidelberg, 1927.

W. Albring. Angewandte Strömungslehre. Steinkopff, Dresden, 1966.

J. D. Anderson jr. Introduction to Flight. McGraw-Hill, New York, 1989.

G. K. Batchelor. An Introduction to Fluid Dynamics. Cambridge University Press, Cambridge, 2005.

E. Becker. Technische Strömungslehre. Teubner, Stuttgart, 1982.

R. Betchov, W. O. Criminale Jr. Stability of Parallel Flows. Academic Press, New York, 1967.

E. C. Bingham. Fluidity and Plasticity. McGraw-Hill, New York, 1922.

G. Böhme. Strömungsmechanik nicht Newtonscher Fluide. Teubner, Stuttgart, 1981.

T. Cebeci, A. M. O. Smith. Analysis of Turbulent Boundary Layers. Academic Press, New York, 1975.

T. Cebeci, J. Cousteix. Computation of Boundary-Layer Flows. Springer, Berlin, Heidelberg, New York, 1999.

A. J. Chorin, J. E. Marsden. A Mathematical Introduction to Fluid Mechanics. Springer, New York, 2000.

R. Courant, K. O. Friedrichs. Supersonic Flow and Shock Waves. Springer, Heidelberg, Berlin, 1976.

F. Durst. Grundlagen der Strömungsmechanik. Springer, Berlin, Heidelberg, New York, 2006.

M. van Dyke. An Album on Fluid Motion. Parabolic Press, Stanford, 2005.

T. E. Faber. Fluid Dynamics for Physicists. Cambridge University Press, Cambridge, 1997.

K. Gersten. Einführung in die Strömungsmechanik. Shaker, Aachen, 2003.

H. Glauert. The Elements of Aerofoil and Airscrew Theory. Cambridge University Press, Cambridge, 1926.

H. Herwig. Strömungsmechanik. Springer, Berlin, Heidelberg, 2006.

J. O. Hinze. Turbulence. McGraw-Hill, New York, 1987.

T. von Kármán. Aerodynamik. Interavia, Geneva, 1956.

E. Krause. Strömungslehre. Vieweg+Teubner, Wiesbaden, 2003.

D. Küchemann. The Aerodynamic Design of Aircraft. Pergamon Press, Oxford, 1978.

P. K. Kundu, I. M. Cohen, H. H. Hu. Fluid Mechanics. Elsevier Academic Press, Amsterdam, Heidelberg, 2006.

H. Lamb. Hydrodynamics. Cambridge University Press, Cambridge, 1932.

L. D. Landau, E. M. Lifshitz. Fluid Mechanics. Pergamon Press, London, 1959.

L. D. Landau, E. M. Lifshitz. Hydrodynamik, Lehrbuch der theoretischen Physik. Akademie-Verlag, Berlin, 1974.

H. W. Liepmann, A. Roshko. Elements of Gasdynamics. John Wiley & Sons, New York, 1957.

J. Lighthill. Waves in Fluids. Cambridge University Press, Cambridge, 1987.

O. Lilienthal. Der Vogelflug als Grundlage der Fliegekunst. Gaertner, Berlin, 1889.

L. M. Milne-Thomson. Theoretical Hydrodynamics. Macmillan, New York, 1955.

R. von Mises. Mathematical Theory of Compressible Fluid Flow. Academic Press, New York, 1966.

H. Oertel jr., ed. Prandtl - Essentials of Fluid Mechanics. Springer, New York, 3 edition, 2009.

H. Oertel jr., ed. Prandtl - Führer durch die Strömungslehre, Chinese Translation. Science Press, Beijing, 2008.

H. Oertel jr., ed. Prandtl - Führer durch die Strömungslehre, Russian Translation. Russian Institute of Dynamics, Izhevsk, 2007.

H. Oertel jr., ed. Prandtl - Führer durch die Strömungslehre. Vieweg+Teubner, Wiesbaden, 12th edition, 2008.

H. Oertel jr. Aerothermodynamik. Springer, Berlin, Heidelberg 1994, Universitätsverlag, Karlsruhe, 2005.

H. Oertel jr., J. Delfs. Strömungsmechanische Instabilitäten. Springer, Berlin, Heidelberg 1996, Universitätsverlag, Karlsruhe, 1996.

H. Oertel jr. Introduction to Fluid Mechanics. Vieweg, Braunschweig, Wiesbaden 2001, Universitätsverlag, Karlsruhe, 2005.

H. Oertel jr., M. Böhle, U. Dohrmann. Strömungsmechanik. Vieweg+Teubner, Wiesbaden, 2008.

H. Oertel jr. Bioströmungsmechanik. Vieweg+Teubner, Wiesbaden, 2008.

K. Oswatitsch. Grundlagen der Gasdynamik. Springer, Berlin, Heidelberg, New York, Wien, 1967.

S. J. Pai. Viscous Flow Theory. Van Nostrand, Princeton, N. J., 1956, 1957.

J. Piquet. Turbulent Flows. Springer, Berlin, Heidelberg, New York, 1999.

R. Robinson, J. A. Laurmann. Wing Theory. Cambridge University Press, Cambridge, 1956.

J. C. Rotta. Turbulente Strömungen. Teubner, Stuttgart, 1972.

H. Schlichting, E. Truckenbrodt. Aerodynamics of the Airplane. McGraw-Hill, New York, 1979.

H. Schlichting, K. Gersten. Boundary Layer Theory. Springer, Berlin, Heidelberg, New York, 2003.

H. Schlichting, K. Gersten. Grenzschicht-Theorie. Springer, Berlin, Heidelberg, 2006.

A. H. Shapiro. The Dynamics and Thermodynamics of Compressible Fluid Flow. Ronald Press, New York, 1953.

S. L. Soo. Fluid Dynamics of Multiphase Systems. Blaisdell Publication, Waltham, Mass., 1967.

J. H. Spurk. Fluid Mechanics. Springer, Berlin, Heidelberg, New York, 2007.

J. H. Spurk. Strömungslehre. Springer, Berlin, Heidelberg, 2007.

M. B. Squire. Modern Development in Fluid Dynamics. Oxford University Press, New York, 1938.

O. Tietjens. Strömungslehre. Springer, Berlin, Heidelberg, 1960.

W. G. Vincenti, C. H. Kruger. Introduction to Physical Gas Dynamics. Huntington, New York, 1967.

F. M. White. Viscous Fluid Flow. McGraw-Hill, New York, 1974.

K. Wieghardt. Theoretische Strömungslehre. Teubner, Stuttgart, 1974.

C. Yih. Fluid Mechanics. McGraw-Hill, New York, 1969.

J. Zierep, K. Bühler. Grundzüge der Strömungslehre. Vieweg+Teubner, Wiesbaden, 2008.

Selected Bibiliography

Chapter 5 Fundamental Equations of Fluid Mechanics

D. A. Anderson, J. C. Tannehill, R. H. Pletcher. Computational Fluid Mechanics and Heat Transfer. McGraw-Hill, New York, 1984.

D. A. Anderson jr. Computational Fluid Dynamics. McGraw-Hill, New York, 1995.

B. S. Baldwin, H. Lomax. Thin Layer Approximation and Algebraic Model for Separated Turbulent Flows. AIAA Journal, 78–257, 1978.

R. B. Bird, W. E. Stewart, E. N. Lightfoot. Transport Phenomena. John Wiley & Sons, New York, 1960.

J. Boussinesq. Essai sur la théorie des eaux courantes. Mémoires présentés par divers savant à l'Académie des Sciences de l'Institut de France, 23, 1-680, 1872.

C. Crowe, M. Sommerfeld, Y. Tsuji. Multiphase Flows with Droplets and Particles. CRC Press, Boca Raton, Boston, New York, London, 1998.

B. J. Daly, F. H. Harlow. Transport Equations in Turbulence. Physics of Fluids, 13, 11, 2634–2649, 1970.

D. A. Drew, G. Wallis. Fundamentals of Two-Phase Flow Modeling. Multiphase Science and Technology, 8, 1–67, 1994.

D. A. Drew, S. L. Passman. Theory of Multicomponent Fluids. Springer, Berlin, Heidelberg, New York, 1999.

S. E. Elgobashi. On Predicting Particle-Laden Turbulent Flows. Appl. Scientific Research, 52, 309–329, 1994.

J. H. Ferziger, M. Peric. Computational Methods for Fluid Dynamics. Springer, Berlin, Heidelberg, New York, 1996.

S. Hosokawa, A. Tomiyama. Turbulence Modification in Gas-Liquid and Solid-Liquid Dispersed Two-Phase Pipe. Int. J. Heat and Fluid Flow, 25, 3, 489–498, 2004.

W. P. Jones, J. H. Whitelaw. Modelling and Measurement in Turbulent Combustion. Proceedings of the Combustion Institute, 20, 233, 1985.

B. E. Launder, D. B. Spalding. Mathematical Models of Turbulence. Academic Press, London, New York, 1972.

E. Laurien, H. Oertel jr.. Numerische Strömungsmechanik. Vieweg+Teubner, Wiesbaden, 2009.

M. Lesieur. Turbulence in Fluids. Kluwer, Dordrecht, 1997.

H. Lomax, T. H. Pulliam. Fundamentals of Computational Fluid Dynamics. Springer, New York, 2003.

M. Maxey, J. J. Riley. Equation of Motion for a Small Rigid Sphere in a Non-Uniform Flow. Phys. Fluids, 26, 883–889, 1983.

G. L. Mellor, H. J. Herring. A Survey of the Mean Turbulent Closure Models. AIAA Journal, 11, 590–599, 1973.

J. B. Moss. Simultaneous Measurements of Concentration and Velocity in an open Premixed Turbulent Flame. Combustion Science and Technology, 22, 115, 1979.

C. L. M. H. Navier. Mémoire sur les lois du mouvement des fluides. Mémoires de l'Académie des Sciences, 6, 389–416, 1823.

Sir I. Newton. Philosophiae Naturalis Principia Mathematica, II. Innys, London, 1726.

H. Oertel jr., M. Böhle, U. Dohrmann. Strömungsmechanik. Vieweg+Teubner, Wiesbaden, 2008.

M. Plesset, S. A. Zwick. The Growth of Vapour Bubbles in Superheated Liquids. J. of Applied Physics, 25, 4, 493–501, 1954.

J. W. S. Rayleigh. The Solution for the Motion of a Bubble under Constant Pressure Conditions. Phil. Mag., 34, 94, 1917.

A. Roshko. Structure of Turbulent Shear Flows: A New Look. AIAA Journal, 14, 1349–1357, 1976.

P. Sagaut. Large Eddy Simulations for Incompressible Flows. An Introduction. Springer, Berlin, Heidelberg, New York, 2001.

C. C. Shir. A Preliminary Numerical Study of Atmospheric Turbulent Flows in the Idealized Planetry Boundary Layer. J. Atmos. Sci., 30, 1327–1339, 1973.

W. A. Sirignano. Fluid Dynamics and Transport of Droplets and Sprays. Cambridge University Press, Cambridge, 1999.

G. Stokes. On the Theories of the Internal Friction of Fluids in Motion. Transactions of the Cambridge Philosophical Society, 8, 287–305, 1845.

J. Warnatz, U. Maas, R. W. Dibble. Combustion. Springer, Berlin, Heidelberg, New York, 2001.

Chapter 6 Instabilities and Turbulent Flows

P. H. Alfredsson, A. A. Bakchinov, V. V. Kozlov, M. Matsubara. Laminar-Turbulent Transition at a High Level of a Free Stream Turbulence. P. W. Duck, P. Hall, eds., Proceedings IUTAM Symposium on Nonlinear Instability and Transition in Three-Dimensional boundary layers, 35, 423–436, Dordrecht, 1996. Kluwer.

C. F. Barenghi, R. J. Donnelly, W. F. Vinen. Quantized Vortex Dynamics and Superfluid Turbulence. Springer, Berlin, Heidelberg, New York, 2001.

G. I. Barenblatt. Scaling Laws for Fully Developed Turbulent Shear Flows. Part 1. Basic Hypothesis and Analysis. J. Fluid Mech., 248, 513–520, 1993.

G. K. Batchelor. Recent Developments in Turbulence Research. H. Levy, eds., Proceedings of the 7th International Congress for Applied Mechanics, London, 1948.

G. K. Batchelor. Computation of the Energy Spectrum in Homogeneous Two-Dimensional Turbulence. Physics of Fluids, 12, II, II–233–II–239, 1969.

G. K. Batchelor, A. A. Townsend. The Nature of Turbulent Motion at Large Wave Numbers. Proceedings of the Royal Society of London, A 199, 238–255. Royal Society, London, 1949.

G. Boffetta. Energy and Enstrophy Fluxes in the Double Cascade of Two-Dimensional Turbulence. J. Fluid Mech., 589, 253–260, 2007.

A. V. Boiko, G. R. Grek, A. V. Dovgal, V. V. Kozlov. The Origin of Turbulence in Near-Wall Flows. Springer, Berlin, Heidelberg, 2002.

M. S. Borgas. A Comparison of Intermittent Models in Turbulence. Phys. Fluids A, 4, 2055–2061, 1992.

F. N. M. Brown. A Combined Visual and Hot-Wire Anemometer Investigation of Boundary-Layer Transition. AIAA Journal, 6, 1, 29–36, 1957.

H. H. Brunn. Hot Wire Anemometer: Principles and Signal Analysis. Oxford University Press, Oxford, New York, 1995.

G. F. Carnevale, J. C. McWilliams, Y. Pomeau, J. B. Weiss, W. R. Young. Evolution of Vortex Statistics in Two-Dimensional Turbulence. Physical Review Letters, 66, 2735–2737, 1991.

F. H. Champagne. The Fine-Scale Structure of the Turbulent Velocity Field. J. Fluid Mech., 86, 67–108, 1978.

S. Chandrasekhar. Hydrodynamics and Hydromagnetic Stability. Clarendon Press, Oxford, 1961.

S. Chen, G. D. Doolen. Lattice Boltzmann Method for Fluid Flow. Ann. Rev. Fluid Mech., 23, 539–600, 1991.

A. J. Chorin. Vorticity and Turbulence. Springer, Berlin, Heidelberg, New York, 1994.

P. G. Drazin, W. H. Reid. Hydrodynamic Stability. Cambridge University Press, 1981.

H. L. Dryden. Recent Advances in the Mechanics of Boundary Layer Flow. Advances in Applied Mechanics, 1, 1–40, 1948.

F. Durst. Principles and Practice of Laser-Doppler Anemometry. Academic Press, 1981.

H. W. Emmons. The Laminar-Turbulent Transition in a Boundary Layer - part i. Journal of the Aeronautical Sciences, 18, 490–498, 1951.

R. E. Falco. The Production of Turbulence Near a Wall. AIAA Journal, 80-1356, 1980.

G. Falkovich, K. Gawedzki, M. Vertgassola. Particles and Fields in Fluid Turbulence. Rev. Mod. Phys., 73, 913–975, 2001.

M. J. Feigenbaum. Quantitative Universality for a Class of Nonlinear Transformations. J. Stat. Phys, 19, 25, 1978.

T. M. Fischer, U. Dallmann. Theoretical Investigation of Secondary Instability of Three-Dimensional Boundary Layer Flows. DFVLR-FB84, DVFLR, 1987.

U. Frisch. Turbulence: The Legacy of A. N. Kolmogorov. Cambridge Univ. Press, Cambridge, 1995.

U. Frisch, M. Vergassola. A Prediction of the Multifractal Model: The Intermediate Dissipation Range. Europhysics Letters, 14, 439–444, 1991.

U. Frisch, P. L. Sulem. Numerical Simulation of the Inverse Cascade in Two-Dimensional Turbulence. Physics of Fluids, 27, 8, 1921–1923, 1984.

Y. Gagne, B. Castaing. Une Représentation Universelle sans Invariance Globale d'Échelle des Spectres d'Énergie en Turbulence Développée. Comptes rendus de l'Académie des Sciences Paris, 312, 441, 1991.

M. Gaster. A Note on the Relation Between Temporally-Increasing and Spatially-Increasing Disturbances in Hydrodynamic Stability. J. Fluid Mech., 14, 222–224, 1962.

M. Gaster. The Development of Three-Dimensional Wave Packets in a Boundary Layer. J. Fluid Mech., 32, 173–184, 1968.

H. Görtler. Instabilität laminarer Grenzchichten an konkaven Wänden gegenüber gewissen dreidimensionalen Störungen. ZAMM, 21, 250–252, 1941.

H. L. Grant, R. W. Stewart, A. Moilliet. Turbulence Spectra From a Tidal Channel. J. Fluid Mech., 12, 241–263, 1962.

S. Grossmann. The Onset of Shear Flow Turbulence. Rev. Mod. Phys., 72, 2, 603–618, 2000.

A. S. Gurvich, A. M. Yaglom. Breakdown of Eddies and Probability Distributions for Small-Scale Turbulence. Physics of Fluids, 10, 59–65, 1967.

T. C. Halsey, M. H. Jensen, L. P. Kadanoff, I. Procaccia, B. Shraiman. Fractal Measures and Their Singularities: The Characterization of Strange Sets. Physical Review A, 33, 1141–1151, 1986.

W. Heisenberg. Über Stabilität und Turbulenz von Flüssigkeitsströmen. Annalen der Physik, 74 of 4, 577–627. Barth, Leipzig, 1924.

H. G. E. Hentschel, I. Procaccia. The Infinite Number of Generalized Dimensions of Fractals and Strange Attractors. Physica D, 8, 435–444, 1983.

T. Herbert. Secondary Instability of Boundary Layers. Ann. Rev. Fluid Mech., 20, 487–526, 1988.

T. Herbert, F. P. Bertolotti. Stability Analysis of Nonparallel Boundary Layers. Bull. Amer. Phys. Soc.,, 32, 2079–2806, 1987.

J. O. Hinze. Turbulence. McGraw-Hill, New York, 1987.

T. von Kármán. Progress in Statistical Theory of Turbulence. Proceedings of the National Academy of Sciences of the United States of America, 34, 530–539. Academy, Washington, DC, 1948.

S. Kida, M. Takaoka. Vortex Reconnection. Ann. Rev. of Fluid Mech., 26, 169–189, 2004.

P. S. Klebanoff. Characteristics of Turbulence in a Boundary Layer with Zero Pressure Gradient. Report 1247, NACA, 1955.

P. S. Klebanoff, K. D. Tidstrom, L. M. Sargent. The Three-Dimensional Nature of Boundary Layer Instability. J. Fluid Mech., 12, 1–34, 1962.

B. G. B. Klingmann, A. V. Boiko, K. J. A. Westin, V. V. Kozlov, P. H. Alfredsson. Experiments on the Stability of Tollmien-Schlichting Waves. Eur. J. Mech. B, Fluids, 12, 4, 493–514, 1993.

A. N. Kolmogorov. Die lokale Struktur der Turbulenz in einer inkompressiblen zähen Flüssigkeit bei sehr großen Reynolds-Zahlen. Dokl. Akad. Wiss. USSR, 30, 301–305, 1941.

A. N. Kolmogorov. A Refinement of Previous Hypothesis Concerning the Local Structure of Turbulence in a Viscous Incompressible Fluid at High Reynolds Numbers. J. Fluid Mech., 13, 82–85, 1962.

L. S. G. Kovasznay, H. Komoda, B. R. Vasudeva. Detailed Flow Field in Transition. F. E. Ehlers, J. J. Kauzlarich, C. A. Sleicher Jr., R. E. Street, eds., Proc. of the Heat Transfer and Fluid Mechanics Institute, A 27, 1–26, Stanford, CF, 1962. Stanford University Press.

R. H. Kraichnan. Inertial Ranges in Two-Dimensional Turbulence. Physics of Fluids, 10, 1417–1423, 1967.

S. Kurien, K. R. Sreenivasan. Measures of Anisotropy and the Universal Properties of Turbulence. New Trends in Turbulence, 53–111, Les Houches, 2001.

L. D. Landau, E. M. Lifschitz. Lehrbuch der Theoretischen Physik: Hydrodynamik, 6. Akademie Verlag, Berlin, 1991.

E. Laurien, H. Oertel jr.. Numerische Strömungsmechanik. Vieweg+Teubner, Wiesbaden, 2009.

M. Lesieur. La Turbulence Développée. La Recherche, 139, 1412–1425, 1982.

M. Lesieur. Turbulence in Fluids. Kluwer, Dordrecht, 1997.

M. Lesieur, O. Metais. New Trends in Large Eddy Simulation of Turbulence. Ann. Rev. Fluid Mech., 28, 45–82, 1996.

C. C. Lin. On the Stability of Two-Dimensional Parallel Flows. Quarterly of Appl. Math., 3, 117–142, 1945.

C. C. Lin. The Theory of Hydrodynamic Stability, 5. Cambridge University Press, Cambridge, 1955.

H. J. Lugt. Vortex Flow in Nature and Technology. John Wiley & Sons, New York, 1983.

B. B. Mandelbrot. Intermittent Turbulence in Self Similar Cascades: Divergence of High Moments and Dimension of the Carrier. J. Fluid Mech., 62, 331–358, 1974.

J. C. McWilliams. The Vortices of Two-Dimensional Turbulence. J. Fluid Mech., 219, 361–385, 1990.

C. Meneveau, K. R. Sreenivasan. A Simple Multifractal Cascade Model for Fully Developed Turbulence. Physical Review Letters, 59, 1424–1427, 1987.

A. Michalke. On the Inviscid Instability of the Hyperbolic-Tangent Velocity Profile. J. Fluid Mech., 19, 543–556, 1964.

A. Michalke. The Instability of Free Shear Layers - A Survey on the State of the Art. DFVLR-Mitteilungen 70-04, DFVLR, Porz-Wahn, 1970.

P. Moin, K. Mahesh. Direct Numerical Simulation: A Tool in Turbulence Research. Ann. Rev. Fluid Mech., 30, 539–578, 1998.

H. K.Moffatt. Degrees of Knotedness of Tangles Vortex Lines. J. Fluid Mech., 36, 117–129, 1969.

A. S. Monin, A. M.Yaglom. Statistical Fluid Mechanics: Mechanics of Turbulence II. MIT Press, Cambridge, MA, 1975.

R. Narasimha. The Laminar-Turbulent Transition Zone in the Boundary Layer. Prog. Aero. Sci., 22, 29–80, 1985.

J. Nikuradse. Gesetzmäßigkeit der turbulenten Strömung in glatten Rohren. Forschungsheft 356, VDI, Berlin, 1932.

M. Nishioka, M. Asai, S. Iida. An Experimental Investigation of the Secondary Instability. R. Eppler, H. Fasel, eds., Laminar-Turbulent Transition, 37–46, Berlin, Heidelberg, 1990. Springer.

E. A. Novikov. Intermittency and Scale Similarity in the Structure of a Turbulent Flow. J. App. Mathematics and Mechanics, 35, 231–241, 1971.

E. A. Novikov, R. Stewart. Intermittency of Turbulence and Spectrum of Fluctuations in Energy-Dissipation. Izvestiya Akademii Nauk SSSR, Seria Geofiz., 3, 408, 1964.

A. Obukhov. Some Specific Features of Atmospheric Turbulence. J. Fluid Mech., 13, 77–81, 1962.

H. Oertel jr., ed. Prandtl - Führer durch die Strömungslehre. Vieweg Verlag, Braunschweig, Wiesbaden, 11th edition, 2002.

H. Oertel jr., J. Delfs Strömungsmechanische Instabilitäten. Springer, Berlin, Heidelberg 1996, Universitätsverlag, Karlsruhe, 1996.

H. Oertel jr., M. Böhle, U. Dohrmann. Strömungsmechanik. Vieweg+Teubner, Wiesbaden, 2008.

H. Oertel sen., H. Oertel jr. Optische Strömungsmesstechnik. Braun Verlag, Karlsruhe, 1989.

W. M. F. Orr. The Stability or Instability of the Steady Motions of a Perfect Liquid and a Viscous Liquid. Proc. R. Ir. Acad., A 27, 69–138, 1907.

S. A. Orszag. Accurate Solution of the Orr-Sommerfeld Stability Equation. J. Fluid Mech., 50, 684–703, 1971.

J. Paret, P. Tabeling. Experimental Observation of the Two-Dimensional Inverse Energy Cascade. Phys. Rev. Lett., 79, 4162–4165, 1997.

G. Parisi, U. Frisch. Fully Developed Turbulence and Intermittency. M. Ghil, R. Benzi, G. Parisi, eds., Turbulence and Predictability in Geophysical Fluid Dynamics and Climate Dynamics, Proceedings of the International School of Physics *Enrico Fermi*, Varenna, Italy, 1983, 84–87, North-Holland, Amsterdam, New York, 1985.

S. B. Pope. Turbulent Flows. Cambridge University Press, Cambridge, 2000.

M. Raffel, C. Willert, S. Wereley, J. Kompenhans. Partical Image Velocimetry: A Practical Guide. Springer, Berlin, Heiderlberg, New York, 2007.

J. W. S. Rayleigh. On Convection Currents in a Horizontal Layer of Fluid, when the Higher Temperature is on the Under Side. Philos. Mag. Ser, 32, 6, 529–546, 1916.

O. Reynolds. An Experimental Investigation of the Circumstances which Determine whether the Motion of Water shall be Direct or Sinuous, and of the Law of Resistance in Parallel Channels. Philosophical Transactions of the Royal Society of London, 174, 935–982. Royal Society, London, 1883.

O. Reynolds. On the Dynamic Theory of Incompressible Viscous Fluids and the Determination of the Criterion. Philosophical Transactions of the Royal Society of London, 186, 123–164. Royal Society, London, 1894.

L. F. Richardson. The Supply of Energy From and To Atmospheric Eddies. Proceedings of the Royal Society of London, A 97, 354–373. Royal Society, London, 1920.

D. Ruelle, F. Takens. On the Nature of Turbulence. Commun. Math. Phys., 20, 167–192, 1971.

W. Saric. Görtler Vortices. Ann. Rev. Fluid Mech., 26, 379–409, 1994.

G. B. Schubauer, H. K. Skramstad. Laminar Boundary-Layer Oscillations and Stability of Laminar Flow. Journal of the Aeronautical Sciences, 14, 2, 69–78, 1947.

A. Sommerfeld. Ein Beitrag zur hydrodynamischen Erklärung der turbulenten Flüssigkeitsbewegung. G. Castelnuovo, ed., Atti del IV Congresso internazionale dei matematici, 116–124, Roma, 1908.

J. Somméria. Experimental Study of the Two-Dimensional Inverse Energy Cascade in a Square Box. J. Fluid Mech., 170, 139–168, 1986.

C. G. Speziale. Analytical Methods for the Development of Reynolds Stress Closures in Turbulence. Ann. Rev. Fluid Mech., 23, 107–157, 1991.

K. R. Sreenivasan. Fractals and Multifractals in Fluid Mechanics. Ann. Rev. Fluid Mech., 23, 539–600, 1991.

K. R. Sreenivasan. On the Universality of the Kolmogorov Constant. Physics of Fluids, 7, 2778–2784, 1995.

K. R. Sreenivasan, R. A. Antonia. Phenomenology of Small Scale Turbulence. Ann. Rev. Fluid Mech., 29, 435–472, 1997.

K. R. Sreenivasan, R. J. Donnelly. Role of Cryogenic Helium in Classical Fluid Dynamics: Basic Research and Model Testing. Adv. Appl. Mech., 37, 239–275, 2000.

G. Stolovitzky, P. Kailasnath, and K. R. Sreenivasan. Kolmogorov's Refined Similarity Hypotheses. Phys. Rev. Lett., 69, 1178, 1992.

J. T. Stuart. Unsteady Boundary Layers. L. Rosenhead, ed., Laminar Boundary Layers, 349–408, Oxford, 1963. Clarendon Press.

P. Tabeling. Two-Dimensional Turbulence: A Physicist's Approach. Phys. Rep., 362, 1–62, 2002.

G. I. Taylor. Stability of a Viscous Liquid Contained Between Two Rotating Cylinders. Philosophical transactions of the Royal Society of London, A 223, 289–343. Royal Society, London, 1923.

G. I. Taylor. Statistical Theory of Turbulence, parts 1-4. Proceedings of the Royal Society of London, A 151, 421–478. Royal Society, London, 1935.

G. I. Taylor. Correlation Measurements in a Turbulent Flow Through a Pipe. Proceedings of the Royal Society of London, A 157, 537–546. Royal Society, London, 1936.

W. Tollmien. Über die Entstehung der Turbulenz. Nachrichten von der Gesellschaft der Wissenschaften zu Göttingen, Mathematisch-Physikalische Klasse, 21–44, 1929.

S. I. Vainshtein, K. R. Sreenivasan. Kolmogorov's Refined Similarity Hypotheses. Phys. Rev. Lett, 73, 3085–3088, 1994.

G. A. Voth, K. Satyanarayanan, E. Bodenschatz. Lagrangian Acceleration Measurements at Large Reynolds Numbers. Phys. Fluids, 10, 2268–2280, 1998.

F. Wendt. Turbulente Strömung zwischen zwei rotierenden koaxialen Zylindern. Ingenieur-Archiv, 4, 577–595, 1933.

P. K. Yeung. Lagrangian Investigations of Turbulence. Rev. Fluid Mech., 34, 115–142, 2002.

M. V. Zagarola, A. J. Smits. Scaling of Turbulent Pipe Flow. J. Fluid Mech., 373, 33–79, 1998.

Chapter 7 Convective of Heat and Mass Transfer

H. D. Baehr, K. Stephan. Heat and Mass Transfer. Springer, Berlin, Heidelberg, New York, 1998.

H. Bénard. Les tourbillons cellulaires dans une nappe liquide. Rev. Gén. Sci. Pures Appl., 11, 1261–1268, 1900.

C. O. Bennet, J. E. Myers. Momentum, Heat and Mass Transfer. McGraw-Hill, New York, 1966.

R. B. Bird, W. E. Stewart, E. N. Lightfoot. Transport Phenomena. John Wiley & Sons, New York, 1960.

E. Bodenschatz, W. Pesch, G. Ahlers. Developments in Rayleigh-Bénard Convection. Ann. Rev. Fluid Mech., 32, 709–777, 2000.

A. Busemann. Gasdynamik. L. Schiller, ed., Handbuch der Experimentalphysik, IV, 341–460. Akademische Verlagsgesellschaft, Leipzig, 1931.

F. H. Busse. Bénard Convection in Particular. Rep. Prog. Phys., 41, 1929, 1978.

F. H. Busse. Nonlinear Properties of Thermal Convection. Rep. Prog. Phys., 41, 1929–1967, 1978.

T. Cebeci, P. Bradshaw. Physical and Computational Aspects of Convective Heat Transfer. Springer, Berlin, Heidelberg, New York, 1988.

L. Crocco. Sulla trasmissione del calore da una lamina piana a un fluido scorrente ad alta velocit. L'Aerotechnica, 12, 2, 181–197, 1932.

E. Eckert. Einführung in den Wärme- und Stoffaustausch. Springer, Berlin, Heidelberg, 1959.

E. R. G. Eckert, R. U. Drake. Analysis of Heat and Mass Transfer. McGraw-Hill, New York, 1972.

B. Gebhart. Heat Transfer. McGraw-Hill, New York, 1971.

B. Gebhart, Y. Jaluwina, R. L. Mahajan, B. Sammakia. Boyancy Induced Flows and Transport. Hemisphere Publishers, Washington, 1988.

H. Gröber, S. Erk, U. Grigull. Die Grundgesetze der Wärmeübertragung. Springer, Berlin, Heidelberg, 1961.

R. Hilpert. Wärmeabgabe von geheizten drähten und rohren im luftstrom. Forsch. Geb. Ing.-Wes., 4, 215–224, 1933.

J. O. Hirschfelder, C. F. Curtis, R. B. Bird. Molecular Theory of Gases and Liquids. J. Wiley & Sons, New York, 1967.

J. P. Holman. Heat Transfer. McGraw-Hill, New York, 1976.

M. Jacob. Heat Transfer. J. Wiley & Sons, New York, 1959.

M. Jischa. Konvektiver Impuls-, Wärme und Stoffaustausch. Vieweg, Braunschweig, Wiesbaden, 1982.

R. Kirchartz, H. Oertel jr. Three-Dimensional Thermal Cellular Convection in Rectangular Boxes. J. Fluid Mech., 192, 249–286, 1988.

R. Kirchartz, U. Müller, H. Oertel jr., J. Zierep. Axisymmetric and Non-Axisymmetric Convection in a Cylindrical Container. Acta Mech., 40, 181–194, 1981.

J. G. Knudsen, D. L. Katz. Fluid Dynamics and Heat Transfer. McGraw-Hill, New York, 1958.

E. Koschmieder. Bénard Cells and Taylor Vortices. Cambridge monographs on mechanics and applied mathematics. Cambridge University Press, 1993.

G. P. Merker. Konvektive Wärmeübertragung. Springer, Berlin, Heidelberg, 1987.

A. Mersmann. Stoffübertragung. Springer, Berlin, Heidelberg, 1986.

U. Müller, P. Erhard. Freie Konvektion und Wärmeübertragung. Müller, Heidelberg, 1999.

H. Oertel jr., J. Delfs. Strömungsmechanische Instabilitäten. Springer, Berlin, Heidelberg 1996, Universitätsverlag, Karlsruhe, 2005.

J. A. Reynolds. Turbulent Flows in Engineering. J. Wiley & Sons, New York, 1974.

W. M. Rohsenow, J. P. Hartnett, E. N. Gamić. Handbook of Heat Transfer Fundamentals. McGraw-Hill, New York, 1985.

H. Schlichting, K. Gersten. Boundary Layer Theory. Springer, Berlin, Heidelberg, New York, 2003.

H. Schlichting, K. Gersten. Grenzschicht-Theorie. Springer, Berlin, Heidelberg, 2006.

E. K. Schlünder. Einführung in die Wärmeübertragung. Vieweg, Braunschweig, Wiesbaden, 1981.

E. Schmidt, W. Beckmann. Das Temperatur- und Geschwindigkeitsfeld von einer wärmeabgebenden senkrechten Platte bei natürlicher Konvektion. Technische Mechanik und Thermodynamik 1, 341–391. VDI, Berlin, 1930.

W. Schneider. Konvektive Wärme- und Stoffübertragung. L. Prandtl, K. Oswatitsch, K. Wieghardt, eds., Führer durch die Strömungslehre, 293–342. Vieweg, Braunschweig, Wiesbaden, 1990.

K. Stork, U. Müller. Convection in Boxes: Experiments. J. Fluid Mech., 54, 599–611, 1972.

J. S. Turner. Boyancy Effects in Fluids. Cambridge University Press, Cambridge, 1973.

J. S. Turner. Convection in Multicomponent Systems. Naturwissenschaften, 72, 70–75, 1985.

J. Unger. Konvektionsströmungen. Teubner, Stuttgart, 1988.

A. Walz. Strömungs- und Temperaturgrenzschichten. Braun, Karlsruhe, 1966.

J. R. Welty, R. E. Wilson, C. E. Wicks. Fundamentals of Momentum, Heat and Mass Transfer. J. Wiley & Sons, New York, 1976.

K. Wilde. Wärme- und Stoffübertragung in Strömungen. Steinkopff, Darmstadt, 1978.

Chapter 8 Multiphase Flows

A. J. Acosta, B. R. Parkin. Cavitation Inception - a Selective Review. J. Ship Res., 19, 193–205, 1975.

S. P. Antal, R. T. Lahey, J. L. Flaherty. Analysis of Phase Distribution in Fully Developed Laminar Bubbly Two-Phase Flow. Int. J. Multiphase Flow, 17, 635–652, 1991.

W. Archer. Experimental Determination of Loss of Heat due to Sudden Enlargement in Circular Pipes. Trans. Am. Soc. Civil Eng., 76, 999–1026, 1913.

R. E. A. Arndt, V. H. Arakeri. Some Observations of Tip-Vortex Cavitation. J. Fluid Mech., 229, 269–289, 1991.

B. J. Azzopardi, E. Hervieu. Phase Separation in T-Junctions. Multiphase Science and Technology, 8, 645–713, 1994.

L. Bertodano. Two-Fluid Model for Two-Phase Turbulent Sets. Nuclear Engineering and Design, 179, 65–74, 1998.

L. Bertodano, S. L. Lee, R. T. Lahey, D. A. Drew. The Prediction of Two-Phase Turbulence and Phase Distribution Phenomena Using a Reynolds Stress Model. Journal of Fluids Eng., 112, 107–113, 1990.

J. P. Best. The Formation of Toroidal Bubbles upon the Collapse of Transient Cavities. J. Fluid Mech., 251, 79–107, 1993.

J. R. Blake, B. B. Taib, G. Doherty. Transient Cavities Near Boundaries, Pt I. Rigid Boundaries. J. Fluid Mech., 170, 479–497, 1986.

L. Bolle, J. C. Moureau. Spray Colling of Hot Surfaces. Multiphase Science and Technology, 1, 1–97, 1982.

C. E. Brennen. Cavitation and Bubble Dynamics. Oxford University Press, Oxford, 1995.

P. von Böckh. Ausbreitungsgeschwindigkeit einer Druckstörung und kritischer Durchfluss in Flüssigkeits/Gas-Gemischen. Dissertation, Universität Karlsruhe (TH), Fak. f. Chemieingenieurwesen, 1975.

S. Chandrasekhar. Hydrodynamic and Hydromagnetic Stability. Oxford Univ. Press, 1968.

Y. Chen, S. D. Heister. A Numerical Treatment for Attached Cavitation. Journal of Fluids Eng., 116, 613–618, 1994.

B. Chexal, M. Merilo, J. Maulbetsch, J. Horowitz, J. Harrison, J. Westacott, C. Peterson, W. Kastner, H. Schmidt. Void Fraction Technology for Design and Analysis. TR 106326, EPRI, Palo Alto, Ca., 1997.

D. Chisholm. Pressure Losses in Bends and T's during Steam-Water Flow. Report 318, N. E. L., 1967.

D. Chisholm. A Theoretical Basis for the Lockhart-Martinelli Correlation for Two-Phase Flow. Int. J. Heat Mass Transfer, 10, 1767–1678, 1967.

R. Clift, J. R. Grace, M. E. Weber. Bubbles, Drops and Particles. Academic Press, San Diego, 1978.

J. G. Collier, J. R. Thome. Convective Boiling and Heat Transfer. Clarendon Press, Oxford, 1994.

R. Collins. The Effect of a Containing Cylindrical Boundary on the Velocity of a Large Gas Bubble in a Liquid. J. Fluid Mech., 28, 1, 97–112, 1967.

C. A. Coulaloglou, L. L. Tavlarides. Description of Interaction Processes in Activated Liquid-Liquid Dispersions. Chem. Eng. Sci., 32, 1289–1297, 1977.

C. Crowe, M. Sommerfeld, Y. Tsuji. Multiphase Flows with Droplets and Particles. CRC Press, Boca Raton, Boston, New York, London, 1998.

J. Delhaye, M. Giot, M. Riethmüller. Thermodynamics of Two-Phase Systems for Industrial Design and Nuclear Engineering. Hemisphere Publishing, New York, 1981.

D. A. Drew, R. T. Lahey. A Mixing Length Model for Fully Developed Turbulent Two-Phase Flow. ANS Trans., 35, 624–625, 1980.

D. A. Drew, S. L. Passman. Multi-Phase Flows with Droplets and Particles. Springer, New York, 1999.

R. A. Duckworth. Pressure Gradient and Velocity Correlation and their Application to Design. Pneumotransport, 1, 49–52, 1971.

A. E. Dukler, M. Wicks III, R. G. Cleveland. Frictional Pressure Drop in Two-Phase Flow: B. An Approach Through Similarity Analysis. AIChE J., 10, 1, 44–51, 1964.

E. Elias, G. S. Lellouche. Two-Phase Critical Flow. Int. J. Multiphase Flow, 20, 17–40, 1994.

S. Ergun. Fluid Flow through Packed Columns. Chem. Eng. Progress, 48, 89–94, 1952.

H. K. Fauske. Contribution to the Theory of Two-Phase, One-Component Critical Flow. Technical Report ANL-6633, Argonne National Laboratory, 1963.

J.-P. Franc, J.-M. Michel. Attached Cavitation and the Boundary Layer: Experimental Investigation and Numerical Treatment. J. Fluid Mech., 154, 63–90, 1985.

J.-P. Franc, J.-M. Michel. Fundamentals of Cavitation. Kluwer, Dordrecht, 2004.

L. Friedel. Druckabfall bei der Strömung von Gas-Dampf-Flüssigkeitsgemischen in Rohren. Chem. Ing. Tech., 50, 3, 167–180, 1978.

D. Geldert. Types of Gas Fluidization. Powder Technology, 7, 285–292, 1973.

G. Govier, K. Aziz. The Flow of Complex Mixtures in Pipes. Reinhold Van Nostrand Co., 1972.

J. R. Grace. Fluidization. G. Hetsroni, ed., Handbook of Multiphase Flow Systems, Hemisphere Publ., Washington, 1982.

J. R. Grace, F. Taghipour. Verification and Validation of CFD Models and Dynamic Similarity for Fluidized Beds. Powder Technology, 139, 2, 99–110, 2004.

H. Henry, H. Fauske. Two-Phase Critical Flow of One-Component Mixtures in Nozzles, Orifices and Short Tubes. Trans. ASME, J. Heat Transfer, 95, 179–187, 1971.

R. Henry, M. Grolmes, H. Fauske. Pressure-Pulse Propagation in Two-Phase One- and Two-Component Mixtures. Technical Report ANL-7792, Argonne National Laboratory, 1971.

J. O. Hinze. Turbulence. McGraw-Hill, New York, 2nd edition, 1987.

S. Hosokawa, A. Tomiyama. Turbulence Modification in Gas–Liquid and Solid–Liquid Dispersed Two–Phase Pipe Flows. Int. J. Heat and Fluid Flow, 25, 489–498, 2004.

W. Idsinga, N. Todreas, R. Bowring. An Assessment of Two-Phase Pressure Drop Correlations for Steam-Water Systems. Int. J. Multiphase Flow, 3, 401–413, 1977.

M. Ishii. Thermo-Fluid Dynamic Theory of Two-Phase Flow. Eyrolles, Paris, 1975.

M. Ishii. One-Dimensional Drift-Flux Model and Constitutive Equations for Relative Motion Between Phases in Various Two-Phase Flow Regimes. Technical Report ANL-7747, Argonne National Laboratory, Ill., USA, 1977.

M. Ishii, K. Mishima. Two-Fluid Model and Hydrodynamic Constitutive Relations. Nuclear Engineering and Design, 82, 2-3, 107–126, 1984.

M. Ishii, T. Hibiki. Thermo-Fluid Dynamics of Two-Phase Flow. Springer, New York, 2006.

H. A. Jakobsen, H. Lindborg, C. A. Dorao. Modeling of Bubble Column Reactors: Progress and Limitations. Ind. Eng. Chem. Res., 44, 5107–5151, 2005.

T. Knapp, J. W. Daily, F. G. Hammit. Cavitation. Mc Graw-Hill, New York, 1970.

H. A. Kocamustafaogullari, M. Ishii. Foundation of the Interfacial Area Transport Equation and its Closure Relations. Int. J. Heat Mass Transfer, 38, 3, 481–493, 1995.

J. D. Kulick, J. R. Fessler, J. R. Eaton. Particle Response and Turbulence Modification in Fully Developed Channel Flow. J. Fluid Mech., 277, 109–134, 1994.

D. Kunii, O. Levenspiel. Fluidization Engineering. Butterworth-Heinemann, Boston, 1991.

R. T. Lahey. The Analysis of Phase Separation and Phase Distribution Phenomena Using Two-Fluid Models. Nuc. Eng. and Design, 122, 17–40, 1990.

M. Lance, J. Bataille. Turbulence in the Liquid Phase of a Uniform Bubbly Flow. J. Fluid Mech., 222, 95–118, 1991.

B. E. Launder, G. E. Reece, W. Rodi. Progress in the Development of a Reynolds-Stress Turbulence Closure. J. Fluid Mech., 68, 537–566, 1975.

W. Lauterborn, H. Bolle. Experimental Investigation of Cavitation Bubble Collapse in the Neighbourhood of a Solid Boundary. J. Fluid Mech., 72, 391–399, 1975.

Y. Lecoffre. Cavitation: Bubble Trackers. Balkema, Rotterdam, Brookfield, 1999.

M. Ledinegg. Instabilität der Strömung bei natürlichem und Zwangsumlauf. Wärme, 61, 891–908, 1938.

H. Lemonnier, Z. Bilicki. Steady Two-Phase Choked Flow in Channels of Variable Cross Sectional Area. Multiphase Science and Technology, 8, 257–353, 1994.

S. P. Lin, R. D. Reitz. Drop and Spray Formation from a Liquid Jet. Ann. Rev. Fluid Mech., 30, 85–105, 1998.

T. J. Liu, S. G. Bankoff. Structure of Air-Water Bubbly Flow in a Vertical Pipe. Int. J. Heat Mass Transfer, 36, 1049–1072, 1993.

R. Lockhart, R. Martinelli. Proposed Correlation of Data for Isothermal Two-Phase Two-Component Flow in Pipes. Chem. Eng. Prog., 45, 39–48, 1949.

E. K. Longmire, J. K. Eaton. Structure of a Particle-Laden Round Jet. J. Fluid Mech., 236, 217–256, 1992.

P. A. Lottes. Expansion Losses in Two-Phase Flow. Nuclear Science and Engineering, 9, 26–31, 1961.

J. M. Mandhane, G. A. Gregory, K. Aziz. A Flow Pattern Map for Gas-Liquid Flow in Horizontal Pipes. Int. J. of Multiphase Flow, 1, 537–553, 1974.

J. L. Marié, E. Moursali, S. Tran-Cong. Similarity Law and Turbulence Intensity Profiles in Bubbly Boundary Layer at Low Void Fractions. Int. J. Multiphase Flow, 23, 227–247, 1997.

R. Martinelli, D. Nelson. Prediction of Pressure Drop During Forced Circulation Boiling of Water. Trans. ASME, 79, 695–702, 1948.

M. Maxey, J.J. Riley. Equation of Motion for Small Rigid Sphere in a Non-Uniform Flow. Phys. Fluids, 26, 883–889, 1983

O. Molerus. Fluid-Feststoff-Strömungen. Springer, Berlin, Heidelberg, New York, 1982.

F. J. Moody. Maximum Flow Rate of a Single-Component, Two-Phase Mixture. ASME J. Heat Transfer, 86, 134–142, 1965.

E. Moursali, J. L. Marié, J. Bataille. An Upward Turbulent Bubbly Boundary Layer Along a Vertical Flat Plate. Int. J. Multiphase Flow, 21, 107–117, 1995.

V. E. Nakoriakov, O. N. Kashinsky. Gas-Liquid Bubbly Flow in a Near-Wall Region Two-Phase Flow Modelling and Experimentation, Edition ETS, Pisa, 453–458, 1995.

W. von Ohnesorge. Die Bildung von Tropfen an Düsen und die Auflösung flüssiger Strahlen. ZAMM, 16, 6, 355–358, 1936.

M. Ozawa. Flow instability problems in steam-generating tubes. S. Ishigai, ed., Steam Power Engineering, Cambridge University Press, Cambridge, 1999.

F. N. Peebles, H. J. Garber. Studies on the Motion of Gas Bubbles in Liquids. Chem. Eng. Progress, 49, 88–97, 1953.

A. Phillipp, W. Lauterborn. Cavitation Erosion by Single Laser-Produced Bubbles. J. Fluid Mech., 361, 75–116, 1998.

M. Pilch, C. A. Erdman. Use of Breakup Time Data and Velocity History Data to Predict the Maximum Size of Stable Fragments for Acceleration-Induced Breakup of a Liquid Drop. Int. J. Multiphase Flow, 13, 714–757, 1987.

M. S. Plesset, R. B. Chapman. Collapse of an Initially Spherical Vapour Cavity in the Neighbourhood of a Solid Boundary. J. Fluid Mech., 47, 283–290, 1971.

A. Premoli, D. Francesco, A. Prino. An Empirical Correlation for Evaluating Two-Phase Mixture Density under Adiabatic Conditions. European Two-Phase Flow Group Meeting, Milan, Italy, 1970.

J. Qian, C. R. Law. Regimes of Coalescence and Separation in Droplet Collision. J. Fluid Mech., 331, 59–80, 1997.

B. Richardson. Some Problems in Horizontal Two-Phase Two-Component Flow. Technical Report ANL-5949, Argonne National Laboratory, 1958.

F. Risso, J. Fabre. Oscillations and Break up of a Bubble Immersed in a Turbulent Field. J. Fluid Mech., 372, 323–355, 1998.

Y. Sato, K. Sekoguchi. Liquid Velocity Distribution in Two-Phase Bubble Flow. Int. J. Multiphase Flow, 2, 79–95, 1975.

J. Sauer, G. H. Schnerr. Development of a New Cavitation Model Based on Bubble Dynamics. ZAMM, 80, 731–732, 2000.

W. Shyy, H. S. Udaykumar, M. M. Rao, R. W. Smith. Computational Fluid Dynamics with Moving Boundaries. Taylor & Francis, Washington, D.C., London, 1996.

W. A. Sirignano. Fluid Dynamics and Transport of Droplets and Sprays. Cambridge Univ. Press, Cambridge, 1999.

K. D. Squires, J. K. Eaton. Particle Response and Turbulence Modification in Isotropic Turbulence. Phys. Fluids A, 2, 1191–1203, 1990.

X. Sun, S. Paranjape, S. Kim, H. Goda, M. Ishii, J. M. Kelly. Local Liquid Velocity in Vertical Air-Water Downward Flow. J. Fluid Eng., 126, 539–545, 2004.

Y. Taitel. Flow Pattern Transition in Two-Phase Flow. Proc. of the 9th Int. Heat Transfer Conf., 237–254, Jerusalem, 1990.

Y. Taitel, A. Dukler. A Model for Prediction of Flow Regime Transitions in horizontal and Near Horizontal Gas-Liquid Flow. AIChE J., 22, 47–55, 1976.

Y. Taitel, D. Bornea, A. E. Dukler. Modelling Flow Pattern Transitions for Steady Upward Gas-Liquid Flow in Vertical Tubes. AIChE J., 26, 345–354, 1980.

H. Tennekes, J. L. Lumley. A First Course in Turbulence. MIT Press, Cambridge, 1972.

C. Tsouris, L. L. Tavlarides. Breakage and Coalescence Models for Drops in Turbulent Dispersions. AIChE J., 40, 395–406, 1994.

L. Velasco. L'écoulement Biphasique Travers un Elargissement Brusque. Ph.D. thesis, Université Catholique de Louvain, Dép. Thermodyn. et Turbomach., 1975.

G. B. Wallis. One-Dimensional Two-Phase Flow. McGraw-Hill, 1969.

G. B. Wallis. Critical Two-Phase Flow. Int. J. Multiphase Flow, 6, 97–112, 1980.

M. Weber. Strömungsfördertechnik. Krauskopf, Mainz, 1974.

J. Weismann, A. Husain, B. Harshe. Two-Phase Pressure Drop Across Area Changes and Restrictions. Two-Phase Transport and Reactor Safety, 4, 1281–1316, 1976.

P. B. Whalley. Data Bank for Pressure Loss Correlations, a Comparative Study. Unveröffentlicht, zitiert in Short Courses on Modelling and Computation of Multiphase Flows, ETH Zürich, 1981.

G. Yadigaroglou. Two-Phase Flow Instabilities and Propagation Phenomena. J. M. Delhaye, M. Giot, M. L. Riethmüller, eds., Thermohydraulics of Two-Phase Systems for Industrial Design and Nuclear Engineering, 353–403. Hemishpere Publishing, Washington, 1981.

G. Yadigaroglou. Short Course on Modelling and Computation of Multiphase Flows. Swiss Federal Institute of Technology, ETH, Zürich, Switzerland, 2006.

G. Yadigaroglou, J. R. T. Lahey. On the Various Forms of the Conservation Equations in Two-Phase Flow. Int. J. of Multiphase Flow, 2, 477–494, 1976.

C. H. Yih. Stratified Flows. Academic Press, New York, 1975.

H. Zuber, J. A. Findlay. Average Volumetric Concentration in Two-Phase Flow Systems. J. Heat Transfer, 87, 453–468, 1965.

Chapter 9 Reactive Flows

R. G. Abdel-Gayed, D. Bradley, N. M. Hamid, M. Lawes. Lewis Number Effects on Turbulent Burning Velocity. Proceedings of the Combustion Institute, 20, 505, 1984.

A. A. Amsden, P. J. O'Rourke, T. D. Butler. KIVA II: A Computer Program for Chemically Reactive Flows With Sprays. Technischer Bericht LA-11560-MS, Los Alamos National Laboratory, 1989.

S. A. Arrhenius. Über die Reaktionsgeschwindigkeit bei der Inversion von Rohrzucker in Säuren. Z. Phys. Chem., 4, 226–248, 1889.

W. T. Ashurst. Modelling Turbulent Flame Propagation. Proceedings of the Combustion Institute, 25, 1075, 1995.

W. P. Atkins. Physical Chemistry. Freeman, New York, 1990.

R. W. Bilger. Turbulent Flows with Nonpremixed Reactants. P. A. Libby, F. A. Williams, eds., Turbulent reactive flows. Springer, Berlin, Heidelberg, New York, 1980.

H. Bockhorn, C. Chevalier, J. Warnatz, V. Weyrauch. Bildung von promptem NO in Kohlenwasserstoff-Luft-Flammen. 6. TECFLAM-Seminar. DLR, Stuttgart, 1990.

R. Borghi, ed. Recent Advances in Aeronautical Science. Pergamon, London, 1984.

D. Bradley. How Fast Can We Burn. Proceedings of the Combustion Institute, 24, 247, 1993.

K. N. C. Bray. Turbulent Flows with Premixed Reactants. P. A. Libby, F. A. Williams, eds., Turbulent reacting flows. Springer, Berlin, Heidelberg, New York, 1980.

S. Candel, D. Veynante, F. Lacas, N. Darabiha. Current Progress and Future Trends in Turbulent Combustion. Combustion Science and Technology, 98, 245, 1994.

M. E. Coltrin, R. J. Kee, F. M. Rupley. SURFACE CHEMKIN (Version 4.0): A Fortune Package for Analysing Heterogeneous Chemical Kinetics at a Solid-Surface-Gas-Phase Interface. Report SAND 90-8003B, Sandia National Laboratories, 1990.

W. J. A. Dahm, E. S. Bish. High Resolution Measurements of Molecular Transport and Reaction Processes in Turbulent Combustion. T. Takeno, ed., Turbulence and molecular processes in combustion. Elsevier, New York, 1993.

W. J. A. Dahm, G. Tryggvason, M. M. Zhuang. Integral Method Solution of Time-Dependent Strained Diffusion-Reaction Equations with Multi-Step Kinetics. J. App. Math., 1995.

G. Damköhler. Der Einfluss der Turbulenz auf die Flammengeschwindigkeit in Gasgemischen. Z. Elektrochem, 46, 601–652, 1940.

O. Deutschmann, F. Behrendt, J Warnatz. Numerical Modelling of Catalytic Combustion. Proceedings of the Combustion Institute, 26, 1747–1754, 1996.

O. Deutschmann, U. Riedel, J. Warnatz. Modelling of Nitrogen and Oxygen Recombination on Partial Catalytic Surfaces. J. Heat Transfer (Transactions of the ASME), 117, 495–501, 1995.

R. W. Dibble, A. R. Masri, R. W. Bilger. The Spontaneous Raman Scattering Technique Applied to Non-Premixed Flames of Methane. Combustion and Flame, 67, 189, 1987.

C. Dopazo, E. E. O'Brian. An Approach to the Description of a Turbulent Mixture. Acta Astron., 1, 1239, 1974.

T. Dreier, B. Lange, J. Wolfrum, M. Zahn, F. Behrendt, J. Warnatz. CARS Measurements and Computations of the Structure of Laminar Stagnation-Point Methane-Air Counterflow Diffusion Flames. Proceedings of the Combustion Institute, 21, 1729, 1987.

E. Gutheil, H. Bockhorn. The Effect of Multi-Dimensional PDFs in Turbulent Reactive Flows at Moderate Damköhler Number. Physicochemical Hydrodynamics, 9, 525, 1987.

J. B. Heywood. Internal Combustion Engine Fundamentals. McGraw-Hill, New York, 1988.

K. H. Homann. Reaktionskinetik. Steinkopff, Darmstadt, 1975.

K. H. Homann, W. C. Solomon, J. Warnatz, H. G. Wagner, C. Zetzsch. Eine Methode zur Erzeugung von Fluoratomen in inerter Atmosphäre. Berichte der Bunsengesellschaft für Physikalische Chemie, 74, 585, 1970.

W. P. Jones, J. H. Whitelaw. Modelling and Measurement in Turbulent Combustion. Proceedings of the Combustion Institute, 20, 233, 1985.

J. H. Kent, R. W. Bilger. The Prediction of Turbulent Diffusion Flame Fields and Nitric Oxide Formation. Proceedings of the Combustion Institute, 16, 1643, 1976.

A. R. Kerstein. Linear-Eddy Modelling of Turbulent Transport - Part 7: Finite-Rate Chemistry and Multi-Stream Mixing. J. Fluid Mech., 240, 289–313, 1992.

B. E. Launder, D. B. Spalding. Mathematical Models of Turbulence. Academic Press, London, New York, 1972.

C. K. Law. Dynamics of Streched Flames. Proceedings of the Combustion Institute, 22, 1381, 1989.

P. A. Libby, F. A. Williams. Turbulent Flows Involving Chemical Reactions. Ann. Rev. Fluid Mech., 8, 351–376, 1976.

P. A. Libby, F. A. Williams. Fundamental Aspects of Turbulent Reacting Flows. P. A. Libby, F. A. Williams, eds., Turbulent reacting flows. Springer, Berlin, Heidelberg, New York, 1980.

P. A. Libby, F. A. Williams. Turbulent Reacting Flows. Academic Press, New York, 1994.

Y. Liu, B. Lenze. The Influence of Turbulence on the Burning Velocity of Premixed CH_4-H_2 Flames with Different Laminar Burning Velocities. Proceedings of the Combustion Institute, 22, 747, 1988.

P. Magre, R. W. Dibble. Finite Chemical Kinetic Effects in a Subsonic Turbulent Hydrogen Flame. Combustion and Flame, 73, 195, 1988.

P. A. McMurtry, S. Menon, A. R. Kerstein. A Linear Eddy Sub-Grid Model for Turbulent Reacting Flows: Application to Hydrogen-Air Combustion. Proceedings of the Combustion Institute, 24, 271, 1992.

U. Metka, M. G. Schweitzer, H.-R. Volpp, J. Wolfrum, J. Warnatz. In-Situ Detection of NO Chemisorbed on Platinum Using Infrared-Visible Sum-Frequency Generation SFG. Zeitschr. f. Phys. Chem., 214, 865–888, 2000.

J. B. Moss. Simultaneous Measurements of Concentration and Velocity in an open Premixed Turbulent Flame. Combustion Science and Technology, 22, 115, 1979.

U. Nowak, J. Warnatz. Sensitivity Analysis in Aliphatic Hydrocarbon Combustion. A. L. Kuhl, J. R. Bowen, J.-C. Leyer, A. Borisov, eds., Dynamics of reactive systems. American Institute of Aeronautics and Astronautics, New York, 1988.

H. Oertel jr. Aerothermodynamik. Springer, Berlin, Heidelberg 1994, Universitätsverlag, Karlsruhe, 2005.

I. Orlandini, U. Riedel. Chemical Kinetics of NO-Removal by Pulsed Corona Discharges. Journal of Physics D (Applied Physics), 33, 2467–2474, 2000.

N. Peters. Laminar Flamelet Concepts in Turbulent Combustion. Proceedings of the Combustion Institute, 21, 1231, 1987.

N. Peters, J. Warnatz. Numerical Methods in Laminar Flame Propagation. Vieweg, Braunschweig, Wiesbaden, 1982.

H. Pitsch. Large-Eddy Simulation of Turbulent Combustion. Ann. Rev. Fluid Mech., 38, 453–483, 2006.

T. Poinsot, D. Veynante, S. Candel. Diagrams of Premixed Turbulent Combustion Based on Direct Numerical Simulation. Proceedings of the Combustion Institute, 23, 613, 1991.

S. B. Pope. Computations of Turbulent Combustion: Progress and Challenges. Proceedings of the Combustion Institute, 23, 591, 1991.

W. C. Reynolds. The Potential and Limitations of Direct and Large Eddy Simulation. Whither Turbulence. Turbulence at Crossroads, 313. Springer, Berlin, Heidelberg, New York, 1989.

R. P. Rhodes, ed. Turbulent Mixing in Non-Reactive and Reactive Flows. Plenum Press, New York, 1979.

U. Riedel, U. Maas, J. Warnatz. Detailed Numerical Modelling of Chemical and Thermal Nonequilibrium in Hypersonic Flows. Impact of Computing in Science and Engineering, 5, 20–52, 1993.

U. Riedel, U. Maas, J. Warnatz. Simulation of Nonequilibrium Hypersonic Flows. Computers and Fluids, 22, 285–294, 1993.

P. J. Robinson, K. A. Holbrook. Unimolecular Reactions. Wiley-Interscience, New York, 1972.

B. Rogg, F. Behrendt, J. Warnatz. Turbulent Non-Premixed Cumbustion in Partially Premixed Diffusion Flamelets with Detailed Chemistry. Proceedings of the Combustion Institute, 21, 1533, 1987.

K. Rummel. Der Einfluß des Mischungsvorganges auf die Verbrennung von Gas und Luft in Feuerungen. Stahlcisen, Düsseldorf, 1937.

V. Sick, A. Arnold, E. Diessel, T. Dreier, W. Ketterle, B. Lange, J. Wolfrum, K. U. Thiele, F. Behrendt, J. Warnatz. Two-Dimensional Laser Diagnostics and Modelling of Counterflow Diffusion Flames. Proceedings of the Combustion Institute, 23, 495, 1991.

M. D. Smooke, R. E. Mitchell, D. E. Keyes. Numerical Solution of Two-Dimensional Axisymmetric Laminar Diffusion Flames. Combustion Science and Technology, 67, 85, 1989.

D. B. Spalding. Mixing and Chemical Reaction in Steady Confined Turbulent Flames. Proceedings of the Combustion Institute, 13, 649, 1970.

G. Stahl, J. Warnatz. Numerical Investigation of Strained Premixed CH_4-Air Flames up to High Pressures. Combustion and Flame, 85, 285, 1991.

H. Tsuji, I. Yamaoka. The Counterflow Diffusion Flame in the Forward Stagnation Region of a Porous Cylinder. Proceedings of the Combustion Institute, 11, 979, 1967.

J. Warnatz. The Structure of Laminar Alkane-, Alkene-, and Acetylene Flames. Proceedings of the Combustion Institute, 18, 369, 1981.

J. Warnatz. Critical Survey of Elementary Reaction Rate Coefficients in the C/H/O-System. W. C. Gardiner Jr., ed., Combustion chemistry. Springer, Berlin, Heidelberg, New York, 1984.

J. Warnatz. Resolution of Gas Phase and Surface Chemistry into Elementary Reactions. Proceedings of the Combustion Institute, 24, 553, 1993.

J. Warnatz, U. Maas, R. W. Dibble. Verbrennung. Springer, Berlin, Heidelberg, 2001.

C. K. Westbrook, F. L. Dryer. Chemical Kinetics and Modelling of Combustion Processes. Proceedings of the Combustion Institute, 18, 749, 1981.

F. A. Williams. Combustion Theory. The Benjamin Cummings Publishing Company, Menlo Park, Reading, Don Mills, 1985.

Y. B. Zeldovich, D. A. Frank-Kamenetskii. The Theory of Thermal Propagation of Flames. Zh. Fiz. Khim., 12, 100, 1938.

D. K. Zerkle, M. D. Allendorf, M. Wolf, O. Deutschmann. Understanding Homogeneous and Heterogeneous Contributions to the Platinum-Catalyzed Partial Oxidation of Ethane in a Short-Contact-Time Reactor. J. Catal, 196, 18–39, 2000.

Chapter 10 Flows in the Atmosphere and in the Ocean

B. W. Atkinson. Meso-scale Atmospheric Circulations. Academic Press, London, 1981.

B. W. Atkinson, J. W. Zhang. Mesoscale Shallow Convection in the Atmosphere. Reviews of Geophysics, 34, 403–432, 1996.

K. Balzer, W. Enke, W. Wehry. Wettervorhersage: Mensch und Computer - Daten und Modelle. Springer, Berlin, Heidelberg, 1998.

D. Brunt. The Period of Simple Vertical Oscillations in the Atmosphere. Quarterly Journ. of the Royal Meteorological Society, 53, 30–32, 1927.

C. Church, D. Burgess, C. Doswell, R. Davies-Jones. The Tornado: Its Structure, Dynamics, Prediction and Hazards. Geophysical Monograph 79, American Geophysical Union, Washington, 1993.

B. Cushman-Roisin. Introduction to Geophysical Fluid Dynamics. Prentice Hall, London, 1994.

Discovery Channel, ed. Naturkatastrophen: Tornados, Hurricanes, Flut. VCL, München, 1997.

V. W. Ekman. On the Influence of the Earth on Ocean Currents. Ark. Math. Astr. Fys., 2, 11, 1–52, 1905.

J. B. Elsner, A. B. Kara. Hurricanes of the North Atlantic. Oxford University Press, Oxford, 1999.

K. A. Emanuel. Atmospheric Convection. Oxford University Press, Oxford, 1994.

K. A. Emanuel. Divine Wind. The History and Science of Hurricanes. Oxford University Press, Oxford, 2005.

H. Ertel. Ein neuer hydrodynamischer Wirbelsatz. Meteorologische Zeitschrift, 59, 277–281, 1942.

D. Etling. Theoretische Meteorologie. Springer, Berlin, Heidelberg, New York, 2002.

D. Etling, R. A. Brown. Vortices in the Planetary Boundary Layer: A Review. Boundary Layer Meteorology, 65, 215–248, 1993.

P. Fabian. Atmosphäre und Umwelt. Springer, Berlin, Heidelberg, 1992.

J. R. Garratt. The Atmospheric Boundary Layer. Cambridge University Press, Cambridge, 1992.

A. E. Gill. Atmosphere-Ocean Dynamics. Academic Press, London, 1982.

E. E. Gossard, W. H. Hooke. Waves in the Atmosphere. Elsevier, Amsterdam, 1975.

T. E. Graedel, P.J. Crutzen. Chemie der Atmosphäre. Spektrum Akademischer Verlag, Heidelberg, 1994.

R. Grotjahn. Global Atmospheric Ciculations: Observations and Theory. Oxford University Press, Oxford, 1993.

H. Häckel. Meteorologie. Eugen Ulmer, Stuttgart, 1999.

J. Houghton. Globale Erwärmung. Springer, Berlin, Heidelberg, 1997.

IPCC (Intergovernmental Panel on Climate Change), ed. Climate Change 2007: The Scientific Basis. Cambridge University Press, Cambridge, 2007.

J. C. Kaimal, J. J. Finnigan. Atmospheric Boundary Layer Flows. Oxford University Press, Oxford, 1994.

E. Kalnay. Atmospheric Modeling, Data Assimilation and Predictability. Cambridge University Press, Cambridge, 2003.

H. Kraus. Die Atmosphäre der Erde. Vieweg, Braunschweig, Wiesbaden, 2000.

E. B. Kraus, J. Businger. Atmosphere-Ocean Interaction. Oxford University Press, Oxford, 1994.

K. Labitzke. Die Stratosphäre. Springer, Berlin, Heidelberg, 1999.

J. Lighthill. Waves in Fluids. Cambridge University Press, Cambridge, 1987.

G. H. Liljequist, K. Cehak. Allgemeine Meteorologie. Vieweg, Braunschweig, Wiesbaden, 1987.

A. S. Monin, A. M. Obukhov. Fundamental Laws of Turbulent Exchange in the Near-Surface Layer of the Atmosphere. Trudy Geophys. Inst. Acad. Nauk SSSR, 24, 163–187, 1954.

C. Nappo. An Introdution to Atmospheric Gravity Waves. Academic Press, London, 2002.

J. Pedlosky. Geophysical Fluid Dynamics. Springer, Berlin, Heidelberg, New York, 1994.

J. Pedlosky. Ocean Circulation Theory. Springer, Berlin, Heidelberg, New York, 1996.

J. Pedlosky. Waves in the Ocean and Atmosphere. Springer, Berlin, 2003.

J. P. Peixoto, A. H. Oort. Physics of Climate. American Institute of Physics, New York, 1992.

H. Pichler. Dynamik der Atmosphäre. Spektrum Akademischer Verlag, Heidelberg, 1997.

S. Pond, G. L. Pickard. Introductary Dynamical Oceanography. Pergamon Press, Oxford, 1991.

E. Roeckner, L. Bengtsson, J. Feichter, J. Lelieveld, H. Rodhe. Transient Climate Change Simulations with a Coupled Atmosphere-Ocean GCM Including the Tropospheric Sulfur Cycle. Journal of Climate, 12, 3004–3032, 1999.

R. S. Scorer. Cloud Investigation by Satellite. Ellis Horwood, Ltd., Chichester, 1986.

J. E. Simpson. Sea Breeze and Local Winds. Cambridge University Press, Cambridge, 1994.

J. E. Simpson. Gravity Currents in the Environment and the Laboratory. Cambridge University Press, Cambridge, 1997.

S. Solomon. Stratospheric Ozone Depletion: A Review of Concepts and History. Rev. Geophys., 37, 275–316, 1999.

C. Timmreck, H.-F. Graf, J. Feichter. Simulation of Mt. Pinatubo Volcanic Aerosol with the Hamburg Climate Model ECHAM 4. Theoretical and Applied Climatology, 62, 85–108, 1999.

K. E. Trenberth, ed. Climate System Modelling. Cambridge University Press, Cambridge, 1992.

V. Väisälä. Über die Wirkung der Windschwankungen auf die Pilotbeobachtungen. Soc. Sci. Fennica, Commentationes Phys.-Math. II, 19, 37, 1925.

C. D. Whiteman. Mountain Meteorology. Oxford University Press, Oxford, 2000.

M. G. Wurtele, R. D. Sharman, A. Datta. Atmospheric Lee Waves. Ann. Rev. Fluid Mech., 28, 129–176, 1996.

I. R. Young. Wind-Generated Ocean Waves. Elsevier, Amsterdam, 1999.

Chapter 11 Microflows

F. F. Abraham. The Interfacial Density Profile of a Lennard-Jones Fluid in Contact with a (100) Lennard-Jones Wall and its Relationship to Idealized Fluid/Wall Systems: A Monte Carlo Simulation. J. Chem. Phys., 68, 3713, 1978.

T. M. Adams, S. I. Abdel-Khalik, S. M. Jeter, Z. H. Qureshi. An Experimental Investigation of Single-Phase Forced Convection in Microchannels. Int. J. Heat Mass Transfer, 41, 851–857, 1998.

D. P. J. Barz. Ein Beitrag zur Modellierung und Simulation elektrokinetischer Transportprozesse in mikrofluidischen Einheiten. Dissertation, Universität Karlsruhe, 2005.

G. K. Batchelor. An Introduction to Fluid Dynamics. Cambridge University Press, Cambridge, 2005.

G. A. Bird. Molecular Gas Dynamics. Claredon Press, Oxford, 1976.

D. Brutin, F. Topin, L. Tadrist. Transient Method for the Liquid Laminar Flow Friction Factor in Microtubes. AIChE J., 49, 2759–2767, 2003.

D. Burgreen, F. Nakache. Electrokinetic Flow in Ultrafine Capillary Slits. J. Phys. Chem., 68, 1084–1091, 1964.

D. Y. C. Chan, R. G. Horn. The Drainage of Thin Liquid Films Between Solid Surfaces. J. Chem. Phys., 83, 5311–5324, 1985.

S. Chapman, T. G. Cowling. The Mathematical Theory of Non-Uniform Gases. Cambridge University Press, Cambridge, 1970.

S. B. Choi, R. F. Barron, R. O. Warrington. Fluid Flow and Heat Transfer in Microtubes. ASME AMD-DSC, 32, 123–134, 1991.

V. S. J. Craig, C. Neto, D. R. M. Williams. Shear-Dependent Boundary Slip in an Aqueous Newtonian Liquid. Phys. Rev. Lett., 87, 5, 054504, 2001.

P. G. de Gennes. Wetting: Statistics and Dynamics. Rev. Modern Phys., 57, 827, 1985.

P. Debye, E. Hückel. Zur Theorie der Elektrolyte. Gefrierpunktserniedrigung und verwandte Erscheinungen. Physikalische Z., 24, 185–206, 1923.

N. A. Derzko. Review of Monte Carlo Methods in Kinetic Theory. UTIAS Review 35, University of Toronto, 1972.

E. B. Dussan On the Spreading of Liquids on Solid Surfaces: Static and Dynamic Contact Lines. Ann. Rev. Fluid Mech., 11, 371, 1979.

E. B. Dussan, S. H. Davis. On the Motion of a Fluid-Fluid Interface Along a Solid Surface. J. Fluid Mech., 50, 977, 1974.

G. Fritz. Über den dynamischen Randwinkel im Fall der vollständigen Benetzung. Z. Angew. Physik, 19, 374, 1965.

M. Gad-el-Hak. The Fluid Mechanics of Microdevices – the Freeman Scholar Lecture. J. Fluids Engineering, 121, 5–33, 1999.

M. Gad-el-Hak. Flow Physics. M. Gad-el-Hak, ed., The MEMS Handbook, 4, 1–38. CRC, Boca Raton, 2002.

M. L. Gee, P. M. McGuiggan, J. N. Israelachvili, A. M. Homola. Liquid to Solidlike Transition of Molecularly Thin Films Under Shear. J. Chem. Phys., 93, 1895–1906, 1990.

H. Green. The Structure of Liquids. S. Flügge, ed., Handbuch der Physik, 10. Springer, Berlin, 2002.

H. Herwig. Flow and Heat Transfer in Micro Systems: Is Everything Different or Just Smaller? Z. Angew. Math. Mech., 82, 579–586, 2002.

G. Hetsroni, A. Mosyak, E. Pogrebnyak, L. P. Yarin. Fluid Flow in Micro-Channels. Int. J. Heat Mass Transfer, 48, 10, 1982–1998, 2005.

G. Hetsroni, A. Mosyak, E. Pogrebnyak, L. P. Yarin. Heat Transfer in Micro-Channels: Comparison of Experiments with Theory and Numerical Results. Int. J. Heat Mass Transfer, 48, 25-26, 5580–5601, 2005.

R. L. Hoffman. A Study of the Advancing Interface. I. Interface Shape in Liquid-Gas System. J. Colloid Interface Sci., 50, 228, 1975.

R. J. Hunter. Zeta Potential in Colloid Science: Principles and Applications. Accademic Press, London, 1981.

M. S. Ivanov, S. V. Rogasinsky. Theoretical Analysis of Traditional and Modern Schemes of the DSMC Method. Proc. 17th RGD Symposium, 1, Aachen, 1991. Verlag Chemie.

P. Joseph, P. Tabeling. Direct Measurement of the Apparent Slip Length. Phys. Rev. E, 71, 035303, 2005.

J. Judy, D. Maynes, B. W. Webb. Characterization of Frictional Pressure Drop for Liquid Flows Through Microchannels. Int. J. Heat Mass Transfer, 45, 3477–3489, 2002.

G. E. Karniadakis, A. Beskok. Micro Flows. Fundamentals and Simulation. Springer, New York, 2004.

P. J. Koplik, J. R. Banavar. Continuum Deductions From Molecular Hydrodynamics. Ann. Rev. Fluid Mech., 27, 257–292, 1995.

J. E. Lennard-Jones. Cohesion. Proc. Phys. Soc. London, 43, 461, 1931.

Z. X. Li, D. X. Du, Z. Y. Guo. Experimental Study on Flow Characteristics of Liquid in Circular Micro-Tubes. Microscale Thermophys. Eng., 7, 3, 253–265, 2003.

L. Löfdahl, M. Gad-el-Hak. Sensors and Actuators for Turbulent Flows. M. Gad-el-Hak, ed., The MEMS Handbook, 6, 1. CRC, Boca Raton, 2002.

W. Loose, S. Hess. Rheology of Dense Model Fluids Via Nonequilibrium Molecular Dynamics: Shear Thinning and Ordering Transition. Rheologica Acta, 28, 91–101, 1989.

C. Maier. Techniken der Hochgeschwindigkeitsmikrokinematographie zur Bewertung von Mikrodosiersystemen und Mikrotropfen. Fortschritts-Bericht 1037, VDI, Dissertation Universität Ulm, 2004.

A. Manz, H. Becker. Microsystem Technology in Chemistry and Life Sciences. Springer, Berlin, 1999.

J. Maxwell. On Stresses in Rarefied Gases Arising From Inequalities of Temperature. Phil. Trans. Royal Soc., 170, 1, 231–256, 1879.

I. Meisel, P. Ehrhard. Electrically-Excited (Electroosmotic) Flows in Microchannels for Mixing Applications. Eur. J. Mech. B: Fluids, 25, 491–504, 2006.

J. N. Moss, G. A: Bird. Direct Simulation of Transitional Flow for Hypersonic Reentry Conditions. 84-0223, AIAA, 1984.

K. Nanbu. Numerical Simulation of Boltzmann Flows of Real Gases – Accuracy of Models Used in the Monte Carlo Method. Rep. Inst. Fluid Science 4, Tohoku University, Sendai, 1992.

H. Oertel jr. Aerothermodynamik. Springer, Berlin, Heidelberg 1994, Universitätsverlag, Karlsruhe, 2005.

A. Oron. Physics of Thin Liquid Films. M. Gad-el-Hak, ed., The MEMS Handbook, 10, 1. CRC, Boca Raton, 2002.

A. Oron, S. H. Davis, S. G. Bankoff. Long-Scale Evolution of Thin Liquid Films. Rev. Modern Phys., 69, 931–980, 1997.

J. T. G. Overbeek. Electrokinetic Phenomena. H. R. Kruyt, ed., Colloid Science, 1. Elsevier, Amsterdam, 1952.

A. Ramos, H. Morgan, N.G. Green, A. Castellanos. AC Electrokinetics: A Review of Forces in Microelectrode Structures. J. Phys. D: Appl. Phys., 31, 2338–2353, 1998.

C. L. Rice, R. Whitehead. Electrokinetic Flow in a Narrow Cylindrical Capillary. J. Phys. Chem., 69, 4017–4024, 1965.

W. Rose, R. W. Heins. Moving Interfaces and Contact Angle Rate-Dependency. J. Colloid Sci., 17, 39, 1962.

S. A. Schaaf, P. L. Chambré. Flow of Rarefied Gases. Princeton University Press, Princeton, 1961.

K. Schubert, J. J. Brandner, M. Fichtner, G. Linder, U. Schygulla, A. Wenka. Microstructure Devices for Applications in Thermal and Chemical Process Engineering. J. Microscale Thermophys. Engineering, 5, 17–39, 2001.

A. M. Schwartz, S. B. Tajeda. Studies of Dynamic Contact Angles on Solids. J. Colloid Interface Sci., 38, 359, 1972.

K. V. Sharp, R. J. Adrian. Transition From Laminar to Turbulent Flow in Liquid Filled Microtubes. Experiments Fluids, 36, 741–747, 2004.

K. V. Sharp, R. J. Adrian, J. G. Santiago, J. I. Molho. Liquid Flows in Microchannels. M. Gad-el-Hak, ed., The MEMS Handbook, 6, 1. CRC, Boca Raton, 2002.

J. C. Shih, C.-M. Ho, J. Liu, Y.-C. Tai. Non-Linear Pressure Distribution in Uniform Microchannels. ASME AMD-MD, 238, 1995.

C. Sobhan, S. V. Garimella. A Comparative Analysis of Studies on Heat Transfer and Fluid Flow in Microchannels. J. Microscale Thermophys. Engineering, 5, 293–311, 2001.

L. H. Tanner. The Spreading of Silicone Oil Drops on Horizontal Surfaces. J. Phys. D: Appl. Phys., 12, 1473, 1979.

P. A. Thompson, S. M. Troian. A General Boundary Condition for Liquid Flow at Solid Surfaces. Nature, 389, 360–362, 1997.

D. C. Tretheway, C. D. Meinhart. Apparent Fluid Slip at Hydrophobic Microchannel Walls. Phys. Fluids, 14, L9–L12, 2002.

M. Vallet, B. Berge, L. Vovelle. Electrowetting of Water and Aqueous Solutions on Poly-ethylene-terephthalate Insulating Films. Poplymer, 37, 2465–2470, 1996.

T. Young. An Essay on the Cohesion of Fluids. Philosophical Transactions of the Royal Society of London, 95, 65–87. Royal Society, London, 1805.

D. Yu, R. Warrington, R. Barron, T. Anieel. An Experimental and Theoretical Investigation of Fluid Flow and Heat Transfer in Microtubes. Proc. ASME/JSME Thermal Engineering Conference, 1, 523–530, Hawaii, 1995.

S. Zheng, Y. C. Tai. Streamline Based Design of a MEMS Device for Continuous Blood Cell Separation. Twelth Hilton Head Workshop on the Science and Technology of Solid-state Sensors, Actuators, and Microsystems, 1, Hilton Head Islands, 2006.

Chapter 12 Biofluid Mechanics

D. W. Bechert, M. Bruse, W. Hage, R. Meyer. Fluid Mechanics of Biological Surfaces and their Technological Application. Naturwissenschaften, 87, 157–171, 2000.

R. H. J. Brown. The Flight of Birds: II. Wing Function in Relation to Flight Speed. J. Exp. Biol., 30, 90–103, 1953.

D. M. Bushnell. Drag Reduction in Nature. Ann. Rev. Fluid Mech., 23, 65–79, 1991.

C. G. Caro, T. J. Pedley, W. A. Seed. The Mechanism of the Circulation. Oxford University Press, Oxford, 1978.

R. T. W. L. Conroy, J. N. Mills. Human Circadian Rhythms. Churchill, London, 1970.

O. Dössel. Bildgebende Verfahren in der Medizin. Springer, Berlin, Heidelberg, 2000.

M. H. Friedman, C. B. Bargeron, D. D. Duncan, G. M. Hutchins, F. F. Mark. Effects of Arterial Compliance and Non-Newtonian Rheology on Correlations between Internal Thickness and Wall Shear. J. of Biomechanical Engineering, 114, 317–320, 1992.

Y. C. Fung. Biomechanics - Motion, Flow, Stress and Growth. Springer, New York, Berlin, Heidelberg, 1990.

Y. C. Fung. Biomechanics - Mechanical Properties of Living Tissues. Springer, Berlin, Heidelberg, New York, 1993.

Y. C. Fung. Biomechanics: Circulation. Springer, Berlin, Heidelberg, New York, 1997.

L. Glass, P. J. Hunter, A. D. McCulloch, eds. Theory of Heart: Biomechanics, Biophysics and Nonlinear Dynamics of Cardiac Function. Springer, Berlin, Heidelberg, New York, 1991.

J. Gray. Animal Locomotion. Weidenfeld & Nicolson, London, 1968.

M. Handke, D. M. Schäfer, G. Müller, A. Schöchlin, E. Magosaki, A. Geibel. Dynamic Changes of Atrial Septal Defect Area: New Insights by Three-Dimensional Volume-Rendered Echocardiography with High Temporal Resolution. Eur. J. Echocardiography, 2, 46–51, 2001.

K. Hayashi, Y. Yanai, T. Naiki. A 3D-LDA Study of the Relation between Wall Shear Stress and Intimal Thickness in a Human Aortic Bifurcation. J. of Biomechanical Engineering, 118, 273–279, 1996.

F. Hirt. Cardiac Valves in a Model Circulatory System. Sulzer Technical Review, 2, 36–40, 1994.

J. D. Humphrey, S. L. Delange. An Introduction to Biomechanics. Springer, New York, Berlin, Heidelberg, 2004.

P. J. Hunter, B. H. Smaill. Electromechanics of the Heart Based on Anatomical Models. D. P. Zipes, J. Jalife, eds., Cardiac Electrophysiology from Cell to Bedside, 277–283. Saunders, Philadelphia, 2000.

P. J. Hunter, M. P. Nash, G. B. Sands. Computational Electromechanics of the Heart. A. V. Panfilov, A. V. Holden, eds., Computational Biology of the Heart, 12, 345–407. John Wiley & Sons, 1997.

P. J. Hunter, P. M. F. Nielsen, B. H. Smaill, I. J. LeGrice, I. W. Hunter. An Anatomical Heart Model with Applications to Myocardial Activation and Ventricular Mechanics. T. C. Pilkington, B. Loftis, J. F. Thompson, S. L.-Y. Woo, T. C. Palmer, T. F. Budinger, eds., High-Performance Compution in Biomedical Research, 1, 3–26. CRC Press, 1993.

B. A. Jung, B. W. Kreher, M. Markl, J. Hennig. Visualization of Tissue Velocity Data from Cardiac Wall Motion Measurements with Myocardial Fiber Tracking: Principles and Implications for Cardiac Fiber Structures. European J. Cardio–Thoradic Surgery, 29, 158–164, 2006

J. P. Keener, A. V. Panfilov. The Effects of Geometry and Fibre Orientation on Propagation and Extracular Potentials in Myocardium. A. V. Panfilov, A. V. Holden, eds., Computational Biology of the Heart. John Wiley & Sons, Chichester, 1997.

U. Kertzscher, K. Affeld. Messung der Wandschubspannung in Modellen von Blutgefäßen. Biomedizinische Technik, 45, 75–126, 2000.

D. N. Ku. Blood Flow in Arteries. Ann. Rev. Fluid Mech., 29, 399–434, 1997.

D. Liepsch. Flow in Tubes and Arteries - A Comparison. Biorheology, 23, 395–433, 1986.

D. Liepsch, G. Thurston, M. Lee. Studies of Fluids Simulating Blood-like Rheological Properties and Applications in Models of Arterial Branches. Biorheology, 39–52, 1991.

D. Liepsch, C. Weigand. Comparison of Laser Doppler Anemometry and Pulsed Color Doppler Velocity Measurements in an Elastic Replica of a Carotid Artery Bifurcation. J. Vascular Investigation, 2, 3, 103–113, 1996.

M. J. Lighthill. Note on the Swimming of Slender Fish. J. Fluid Mech., 9, 305–317, 1960.

M. J. Lighthill. Mathematical Biofluidmechanics. Society for Industrial and Applied Mathematics, Philadelphia, 1975.

M. J. Lighthill. An informal introduction to theoretical fluid mechanics. Clarendon Press, Oxford, 1986.

S. C. Ling, H. B. Atabek. Nonlinear Analysis of Pulsatile Flow in Arteries. J. Fluid Mech., 55, 493–511, 1972.

J. Malmivuo, R. Plonsey. Bioelectromagnetism. Oxford University Press, New York, 1995.

J. N. Mazumdar. Biofluid Mechanics. World Scientific, Singapore, London, 1992.

D. A. McDonald. Blood Flow in Arteries. Edward Arnold, London, 1960.

C. J. Mills, I. T. Gabe, J. H. Gault, D. T. Mason, J. Ross jr., E. Braunwald, J. P. Shilingford. Pressure-Flow Relationships and Vascular Impedance in Man. Cardiovascular Research, 4, 405–417, 1970.

M. Motomiya, T. Karino. Flow Patterns in the Human Carotid Artery Bifurcation. Stroke, 15, 50–56, 1984.

W. Nachtigall. Technische Biologie von Umströmungsvorgängen und Aspekte ihrer bionischen übertragbarkeit. Abhandlungen der Mathematisch-Naturwissenschaftlichen Klasse / Akademie der Wissenschaften und der Literatur. Steiner, Stuttgart, 2001.

M P. Nash, P. J. Hunter. Computational Mechanics of the Heart. J. of Elasticity, 61, 113–141, 2000.

E. Naujokat, U. Kienke. Neuronal and Hormonal Cardiac Control Processes in a Model of the Human Circulatory System. J. Bioelectromagnetism, 2, 2, 1–7, 2000.

H. Oertel jr. Bioströmungsmechanik. Vieweg+Teubner, Wiesbaden, 2008.

H. Oertel jr. Modelling the Human Cardiac Fluid Mechanics. Universitäts-Verlag, Karlsruhe, 2005.

H. Oertel jr., K. Spiegel, S. Donisi. Modelling the Human Cardiac Fluid Mechanics. Universitäts-Verlag, Karlsruhe, 2006.

H. Oertel jr., S. B. S. Krittian, K. Spiegel. Modelling the Human Cardiac Fluid Mechanics. Universitäts-Verlag, Karlsruhe, 2009.

J. T. Ottesen, M. S. Olufsen, J. K. Larsen. Applied Mathematical Models in Human Physiology. Society for Industrial Mathematics, Philadelphia, 2004.

A. V. Panfilov, A. V. Holden. Computational Biology of the Heart. John Wiley & Sons, Chichester, 1997.

D. J. Patel, R. N. Vaishnav. Basic Hemodynamics and its Role in Disease Processes. University Park Press, Baltimore, 1980.

T. J. Pedley. The Fluid Mechanics of Large Blood Vessels. Cambridge University Press, Cambridge, 1980.

G. Pedrizzetti, K. Perktold. Cardiovascular Fluid Mechanics. Springer, Wien, Heidelberg, 2003.

K. Perktold, G. Rappitsch. Mathematical Modelling of Arterial Blood Flow and Correlation to Arteriosclerosis. Technology and Health Care, 3, 139–151, 1995.

K. Perktold, M. Resch, H. Florian. Pulsatile Non-Newtonian Flow Characteristics in a Three-Dimensional Human Catroid Bifurcation Model. J. of Biomechanical Engineering, 113, 464–475, 1991.

C. S. Peskin, D. M. McQueen. Mechanical Equilibrium Determines the Fractal Fiber Architecture of Aortic Heart Valve Leaflets. American Journal of Physiology, 363–6135, 1994.

C. S. Peskin, D. M. McQueen. Fluid Dynamics of the Heart and its Valves. Mathematical Modelling, Ecology, Physiology and Cell Biology. Prentice-Hall, New Jersey, 1997.

A. Poll, D. Liepsch, C. Weigand, J. McLean. Two and Three-Dimensional LDA Measurements and Shear Stress Calculations for a True Scale Elastic Model of a Dog Aorta with Stenosis. Automedica, 18, 163–210, 2000.

A. S. Popel, P. C. Johnson. Microcirculation and Hemorheology. Annu. Rev. Fluid Mech., 37, 43–69, 2005.

A. J. Pullan, M. L. Buist, L. K. Cheng. Mathematically Modelling the Electrical Activity of the Heart. World Scientific, New Jersey, London, 2005.

L. Schauf, D. F. Moffet, S. B. Moffet. Medizinische Physiologie. Walter de Gruyter, Berlin, New York, 1993.

E. Schubert, ed. Medizinische Physiologie. Walter de Gruyter, Berlin, New York, 1993.

R. de Simone. Three-Dimensional Color Doppler. Futura Publishing, Armonk, New York, 1999.

R. Skalak, N. Ozkaya, T. C. Skalak. Biofluid Mechanics. Ann. Rev. Fluid Mech., 21, 167–204, 1989.

M. Sugawara, F. Kajiya, A. Kitabatake, H. Matino. Blood Flow in the Heart and Large Vessels. Springer, Tokyo, Berlin, Heidelberg, 1989.

F. Torrent-Guasp, et al. Seminars in Thoracic and Cardiovascular Surgery, 13, 4, 301–319, 2001.

C. D. Werner, F. B. Sachse, C. Baltes, O. Dössel. The Visible Man Dataset in Medical Education: Electrophysiologie of the Human Heart. Proc. Third Users Conference of the National Library of Medcine's Visible Human Project, 1–81, 2000.

G. E. W. Wolstenholme. Circulatory and Respiratory Mass Transport. Churchill, London, 1969.

J. R. Womersley. Method for the Calculation of Velocity, Rate of Flow and Viscous Drag in Arteries when the Pressure Gradient is Known. J. Physiol., 127, 553–563, 1955.

M. Zacek, E. Krause. Numerical Simulation of the Blood Flow in the Human Cardiovascular System. J. of Biomechanics, 29, 13–20, 1996.

Index

GPSR Compliance
The European Union's (EU) General Product Safety Regulation (GPSR) is a set of rules that requires consumer products to be safe and our obligations to ensure this.

If you have any concerns about our products, you can contact us on

ProductSafety@springernature.com

In case Publisher is established outside the EU, the EU authorized representative is:

Springer Nature Customer Service Center GmbH
Europaplatz 3
69115 Heidelberg, Germany

www.ingramcontent.com/pod-product-compliance
Ingram Content Group UK Ltd.
Pitfield, Milton Keynes, MK11 3LW, UK
UKHW022316190726
13856UKWH00001B/32
9781461424871